CORROSION
Volume 2

Corrosion Control

CORROSION

Volume 2

Corrosion Control

Edited by

L. L. Shreir, PhD, FRIC, FIM, FICorrT, FIMF

Head of Department of Metallurgy and Materials,
City of London Polytechnic

NEWNES-BUTTERWORTHS

LONDON   **BOSTON**

Sydney – Wellington – Durban – Toronto

The Butterworth Group

United Kingdom   **Butterworth & Co (Publishers) Ltd**
London: 88 Kingsway, WC2B 6AB

Australia   **Butterworths Pty Ltd**
Sydney: 586 Pacific Highway, Chatswood, NSW 2067
Also at Melbourne, Brisbane, Adelaide and Perth

Canada   **Butterworth & Co. (Canada) Ltd**
Toronto: 2265 Midland Avenue, Scarborough,
Ontario M1P 4S1

New Zealand   **Butterworths of New Zealand Ltd**
Wellington: T & W Young Building,
77–85 Customhouse Quay, 1, CPO Box 472

South Africa   **Butterworth & Co (South Africa) (Pty) Ltd**
Durban: 152–154 Gale Street

USA   **Butterworth (Publishers) Inc**
Boston: 19 Cummings Park, Woburn, Mass. 01801

First published 1963 by George Newnes Ltd
Reprinted 1965
Second edition 1976 by Newnes-Butterworths
Reprinted 1977, 1978, 1979

© The several contributors listed on pages xiii–xvi of Volume 1

ISBN 0 408 00267 0 (for two-volume set)

Printed by Tien Wah Press (Pte) Ltd., Singapore.

CONTENTS

Volume 2. Corrosion Control

Introduction to Volume 2

10. Design and Economic Aspects of Corrosion

10.1   Economic Aspects of Corrosion

10.2   Corrosion Control in Chemical and Petrochemical Plant

10.3   Design for Prevention of Corrosion in Buildings and Structures

10.4   Design in Marine Engineering

10.5   Design in Relation to Welding and Joining

10.5A   Appendix—Terms Commonly Used in Joining

11. Cathodic and Anodic Protection

11.1   Principles of Cathodic Protection

11.2   Sacrificial Anodes

11.3   Power-impressed Anodes

11.4   Practical Applications of Cathodic Protection

11.5   Stray-current Corrosion

11.6   Cathodic-protection Interaction

11.7   Cathodic-protection Instruments

11.8   Anodic Protection

v

12. Pretreatment and Design for Metal Finishing

12.1   Pretreatment Prior to Applying Coatings

12.2   Pickling in Acid

12.3   Chemical and Electrochemical Polishing

12.4   Design for Corrosion Protection by Electroplated Coatings

12.5   Design for Corrosion Protection by Paint Coatings

13. Methods of Applying Metallic Coatings

13.1   Electroplating

13.2   Principles of Applying Coatings by Hot Dipping

13.3   Principles of Applying Coatings by Diffusion

13.4   Principles of Applying Coatings by Metal Spraying

13.5   Miscellaneous Methods of Applying Metallic Coatings

14. Protection by Metallic Coatings

14.1   The Protective Action of Metallic Coatings

14.2   Aluminium Coatings

14.3   Cadmium Coatings

14.4   Zinc Coatings

14.5   Tin and Tin Alloy Coatings

14.6   Copper and Copper Alloy Coatings

14.7   Nickel Coatings

14.8   Chromium Coatings

14.9   Noble Metal Coatings

15. Protection by Paint Coatings

15.1   Paint Application Methods

15.2   Paint Formulation

15.3   The Mechanism of the Protective Action of Paints

15.4   Paint Failure

15.5   Paint Finishes for Industrial Applications

15.6   Paint Finishes for Structural Steel for Atmospheric Exposure

15.7   Paint Finishes for Marine Application

15.8   Protective Coatings for Underground Use

15.9   Synthetic Resins

15.10   Glossary of Paint Terms

16. Chemical Conversion Coatings

16.1   Coatings Produced by Anodic Oxidation

16.2   Phosphate Coatings

16.3   Chromate Treatments

17. Miscellaneous Coatings

17.1   Vitreous Enamel Coatings

17.2   Thermoplastics

17.3   Temporary Protectives

18. Conditioning the Environment

18.1   Conditioning the Atmosphere to Reduce Corrosion

18.2   Corrosion Inhibition: Principles and Practice

18.3   The Mechanism of Corrosion Prevention by Inhibitors

18.4   Boiler Feed-Water Treatment

19. Non-Metallic Materials

19.1   Carbon

19.2   Glass and Glass-ceramics

19.3   Vitreous Silica

19.4   Glass Linings and Coatings

19.5   Stoneware

19.6   Plastics and Reinforced Plastics

19.7   Rubber and Synthetic Elastomers

19.8   Corrosion of Metals by Plastics

19.9   Wood

19.10   The Corrosion of Metals by Wood

20. Corrosion Testing, Monitoring and Inspection

20.1   Corrosion Testing

20.1A   Appendix—Removal of Corrosion Products

20.1B   Appendix—Standards for Testing

20.2   The Potentiostat and its Applications to Corrosion Studies

20.3   Corrosion Monitoring in Chemical Plant

20.4   Inspection of Paints and Painting Operations

21. Tables and Specifications

21.1   Tables

21.2   British and American Standards

21.3   Calculations Illustrating the Economics of Corrosion Protection

22. Terms and Abbreviations

22.1   Glossary of Terms

22.2   Symbols and Abbreviations

        Index

Introduction

Corrosion Control

In Section 1.1 *corrosion* was defined simply as the reaction of a metal with its environment, and it was emphasised that this term embraces a number of concepts of which the rate of attack/unit area of the metal surface, the extent of attack in relation to the thickness of the metal and its form (uniform, localised, intergranular, cracking, etc.) are the most significant. The rate of corrosion is obviously the most important parameter, and will determine the life of a given metal structure. Whether or not a given rate of corrosion can be tolerated will, of course, depend upon a variety of factors such as the thickness of the metal, the function and anticipated life of the metal structure and the effect of the corrosion products on the environment, etc.

With metals used as construction materials *corrosion control* may be regarded as the regulation of the reaction so that the physical and mechanical properties of the metal are preserved during the anticipated life of the structure or the component. In relation to the term 'anticipated life' it should be noted that this cannot be precise, and although the designer might be told on the basis of information available at that time that the plant should last, say, 10 years, it might be scrapped much earlier or be required to give more prolonged service. It is also evident that, providing there are no restrictions on costs, it is not difficult to design a plant to last at least 10 years, but quite impossible to design one that will last exactly 10 years. Thus although under-design could be catastrophic, over-design could be unnecessarily expensive, and it is the difficult task of the corrosion engineer to avoid either of these two extremes. A further factor that has to be considered is that in the processing of foodstuffs and certain chemicals, contamination of the environment by traces of corrosion products is far more significant than the effect of corrosion on the structural properties of the metal, and under these circumstances the materials selected must be highly resistant to corrosion.

Since corrosion involves a reaction of a metal with its environment, control may be effected through either or both of the two reactants. Thus control could be based entirely on the selection of a particular metal or alloy in preference to all others or the rejection of metals in favour of a non-metallic material, e.g. by a glass-reinforced polymer (g.r.p.). At the other extreme control may be effected by using a less corrosion-resistant material and reducing the aggressiveness of the environment by (*a*) changing composition, (*b*) removing deleterious impurities, (*c*) lowering temperature, (*d*) lowering velocity, (*e*) adding corrosion inhibitors, etc.

1

Although it has been found to be convenient to present this work in the form of two volumes entitled *Corrosion of Metals and Alloys* and *Corrosion Control*, it is evident that this separation is largely artificial, and that a knowledge of the various types of corrosion behaviour of different metals under different environmental conditions is just as important for corrosion control as the protective treatments that have been collated in this volume.

In many structures and components the choice of a metal and alloys is based largely on their engineering properties, but it is seldom that their resistance to corrosion can be ignored completely; at the other extreme corrosion resistance may be of predominant importance, but even so the engineering properties cannot be neglected. Availability is frequently of over-riding importance, and it is quite futile to specify a particular alloy and then to find that it cannot be manufactured and delivered to the fabricators for a year or more. Fabrication technology and fabrication costs will also have to be considered, and in certain cases a more expensive alloy will be preferable to a cheaper one with adequate corrosion resistance that is more difficult to fabricate, e.g. an 18% Cr–8% Ni austenitic stainless steel is frequently selected in preference to a cheaper ferritic 17% Cr stainless steel, since the latter is more difficult to weld than the former, although its corrosion resistance may be adequate.

Costs must always be considered, but it does not follow that an inexpensive metal or alloy will prove to be the cheapest in the long term; platinum and platinum alloys are used in certain applications and apart from their high corrosion resistance have been a wise investment for the purchaser. However, mild steel, which has good mechanical properties, is readily available in a variety of forms and easily fabricated, is frequently preferred to more corrosion-resistant alloys for large structures, and its poor resistance to corrosion is counteracted by means of protective coatings, cathodic protection, conditioning the environment, etc.

Classification of Practical Methods of Corrosion Control

In 1957 the late Dr. W. H. J. Vernon[1] presented an outline scheme of 'Methods of Preventing Corrosion' in which four categories were defined, i.e. (*a*) modification of procedure, (*b*) modification of environment, (*c*) modification of metal, and (*d*) protective coatings; the scheme also indicated the suitability of the method for protecting a metal in different natural environments.

Table 0.1 provides a more comprehensive scheme of methods of corrosion control, in which it has been considered appropriate to include 'Corrosion Testing and Monitoring' and 'Supervision and Inspection', since as will be discussed subsequently these can be of some importance in ensuring that the material, coating or procedure provides effective protection. No attempt has been made to include environments for the very good reason that a particular method may be equally suitable for a number of environments of diverse nature, e.g. a stainless steel may be used for a high-temperature oxidising environment or for an ambient-temperature aqueous environment; cathodic protection may be used for a variety of aqueous environments ranging from fresh waters to wet-clay soils.

Table 0.1 shows the enormous scope of corrosion control, and serves to emphasise the fact that it is just as important to avoid certain features in the design of a structure as to apply a particular protective scheme, and it is also apparent from Method 1 that many of the factors that determine the choice of a metal or a particular protective scheme are outside the realm of metallic corrosion.

It is almost axiomatic amongst corrosion engineers that corrosion control should be given due consideration at the design stage of the structure, and much has been said and written[2] about stimulating *corrosion consciousness* in design engineers who normally make the decisions concerning materials selection and methods of protection. There is no doubt that the incidences of corrosion failure could be substantially reduced if due consideration were given to corrosion hazards during the design stage, a point that was emphasised strongly in the Hoar Report[3]. However, the design engineer is frequently too involved in the stability and proper functioning of the structure to be overconcerned about corrosion protection, particularly when hazards such as stress-corrosion cracking are unlikely to arise. It follows that all too frequently corrosion control has to be effected when the design has been finalised and at a stage when the corrosion hazards that have been inadvertently built into the structure cannot be altered, and under these circumstances considerable ingenuity has to be exercised by the corrosion engineer. Fortunately, many methods of corrosion control such as cathodic protection, conditioning the environment, protective coatings, etc. can be applied after the structure is designed and constructed, and although this is by no means an ideal situation it is one that has to be lived with.

It is evident from Table 0.1, and bearing in mind the enormous variety of materials that are now available, that the choice of a particular method for controlling the corrosion of a given system is an extremely difficult task, and it is seldom that a particular method has so many advantages that it presents the obvious and only solution to the problem when all factors are taken into consideration; frequently, the final decision is based upon a compromise between effectiveness of protection and the cost of its implementation. For heat exchangers using sea-water as a coolant a variety of alloys are available ranging from the aluminium brasses to titanium; the latter might be the obvious choice for highly polluted sea-water, but at the present the cost could be prohibitive although for certain applications (de-salination of sea-water) titanium could be a serious competitor to the cupro-nickel alloys.

The continuous development of new materials has resulted in changing attitudes towards materials selection for corrosion control, and the range of materials now available can be gauged from the Materials Selector Review[4], which becomes considerably thicker each time it is updated. Plastics are replacing metals for a variety of applications and a recent application is the use of g.r.p. in place of metals for the construction of hulls of hovercrafts; the corrosive action of the high velocity spray of sea-water is such that very few metals are capable of withstanding it and the use of g.r.p. represents the best combination of strength, impact resistance, rigidity, lightness and corrosion resistance.

Table 0.1   Outline of methods of corrosion control

---

1. Selection of materials

Select metal or alloy (or non-metallic material) for the particular environmental conditions prevailing (composition, temperature, velocity, etc.) taking into account mechanical and physical properties, availability, method of fabrication and overall cost of structure. Decide whether or not an expensive corrosion-resistant alloy is more economical than a cheaper metal that requires protection and periodic maintenance.

2. Design

If the metal has to be protected make provision in the design for applying metallic or non-metallic coatings or applying anodic or cathodic protection.
Avoid geometrical configurations that facilitate corrosive conditions such as
   (a) Features that trap dust, moisture and water.
   (b) Crevices (or else fill them in) and situations where deposits can form on the metal surface.
   (c) Designs that lead to erosion–corrosion or to cavitation damage.
   (d) Designs that result in inaccessible areas that cannot be re-protected, e.g. by maintenance painting.
   (e) Designs that lead to heterogeneities in the metal (differencies in thermal treatment) or in the environment (differences in temperature, velocity).

3. Contact with other materials

Avoid metal–metal or metal–non-metallic contacting materials that facilitate corrosion such as
   (a) Bimetallic couples in which a large area of a more positive metal (e.g. Cu) is in contact with a small area of a less noble metal (e.g. Fe, Zn or Al).
   (b) Metals in contact with absorbent materials that maintain constantly wet conditions or in the case of passive metals that exclude oxygen.
   (c) Contact (or enclosure in a confined space) with substances that give off corrosive vapours, e.g. certain woods and plastics.

4. Mechanical factors

Avoid stresses (magnitude and type) and environmental conditions that lead to stress-corrosion cracking, corrosion fatigue or fretting corrosion.
   (a) For stress-corrosion cracking avoid the use of alloys that are susceptible in the environment under consideration, or if this is not possible ensure that the external and internal stresses are kept to a minimum.
   (b) For a metal subjected to fatigue conditions in a corrosive environment ensure that the metal is adequately protected by a corrosion-resistant coating.
   (c) Processes that induce compressive stresses into the surface of the metal such as peening, carburising and nitriding are frequently beneficial in preventing corrosion fatigue and fretting corrosion.

There is a great deal of information available on the corrosion resistance of metals and alloys in various environments and this aspect of corrosion control has been dealt with in Volume 1. Reference should also be made to Rabald's *Corrosion Guide*[5] which gives the corrosion resistance of metals and alloys in over 500 chemicals, to the N.A.C.E. *Corrosion Data Survey*[6] and to *Dechema Materials Tables*[7]. However, in spite of all this information environments and/or environmental conditions will be encountered for which corrosion data is not available, and under these circumstances it will be necessary to initiate a programme of corrosion testing (Table 0.1, Method 8), which must be regarded, therefore, as an aspect of corrosion control. Corrosion testing is, of course, vitally important in ensuring that an alloy

Table 0.1 (continued)

5. Coatings

If the metal has a poor resistance to corrosion in the environment under consideration make provision in the design for applying an appropriate protective coating such as

(a) Metal reaction products, e.g. anodic oxide films on Al, phosphate coatings on steel (for subsequent painting or impregnation with grease), chromate films on light metals and alloys (Zn, Al, Cd, Mg).

(b) Metallic coatings that form protective barriers (Ni, Cr) and also protect the substrate by sacrificial action (Zn, Al or Cd on steel).

(c) Inorganic coatings, e.g. enamels, glasses, ceramics.

(d) Organic coatings, e.g. paints, plastics, greases.

Note. Prior to applying coatings adequate pretreatment of the substrate is essential.

6. Environment

Make environment less aggressive by removing constituents that facilitate corrosion; decrease temperature, decrease velocity*; where possible prevent access of water and moisture.

(a) For atmospheric corrosion dehumidify the air, remove solid particles, add volatile corrosion inhibitors (for steel).

(b) For aqueous corrosion remove dissolved O_2*, increase the pH (for steels), add inhibitors.

7. Interfacial potential

(a) Protect metal cathodically by making the interfacial potential sufficiently negative by (i) sacrificial anodes or (ii) impressed current.

(b) Protect metal by making the interfacial potential sufficiently positive to cause passivation (confined to metals that passivate in the environment under consideration).

8. Corrosion testing and monitoring

(a) When there is no information on the behaviour of a metal or alloy or a fabrication under specific environmental conditions (a newly formulated alloy and/or a new environment) it is essential to carry out corrosion testing.

(b) Monitor composition of environment, corrosion rate of metal, interfacial potential, etc. to ensure that control is effective.

9. Supervision and inspection

Ensure that the application of a protective coating (applied *in situ* or in a factory) is adequately supervised and inspected in accordance with the specification or code of practice.

**Note.* For passive metals in solutions free from other oxidising species the presence of dissolved O_2 at all parts of the metal's surface is essential to maintain passivity and this can be achieved in certain systems by increasing the velocity of the solution.

---

conforms to specifications, particularly when maltreatment can result in the precipitation of phases that lead to intergranular attack or to a susceptibility to stress-corrosion cracking. It is also important when conditioning the environment (control of oxygen concentration and pH, addition of inhibitors) to ensure that this is being carried out effectively by monitoring the environment and/or the corrosion rate of the metal, monitoring the potential (as in cathodic and anodic protection), etc.

Paints are one of the most important methods of corrosion control, but it is well known that many cases of failure result from inadequate surface preparation of the metal and careless application of the paint system; procedures that are often carried out under adverse or unsuitable environmental conditions by labour that is relatively unskilled. A great deal of research and development followed by an extensive programme of corrosion testing is required before a paint system is incorporated in a specification or code of practice, but all this effort will be fruitless unless the work is carried out properly, and for this reason effective supervision and inspection is essential. Similar considerations apply, of course, to factory-applied coatings such as sprayed, hot-dipped and electroplated coatings.

Finally, it is necessary to point out that although a particular method of corrosion control may be quite effective for the structure under consideration it can introduce unforeseen corrosion hazards elsewhere. Perhaps the best example is provided by cathodic protection in which stray currents (inter-action) result in the corrosion of an adjacent unprotected structure or of steel-reinforcement bars embedded in concrete; a further hazard is when the cathodically protected steel is fastened with high-strength steel bolts, since cathodic protection of the latter could result in hydrogen absorption and hydrogen cracking.

More Fundamental Classification

Any fundamental classification of corrosion control must be based on the electrochemical mechanism of corrosion, and Evans diagrams may be constructed (Fig. 1.27, Section 1.4) illustrating

(a) Decreasing the thermodynamics of the corrosion reaction.
(b) Increasing the polarisation of the cathodic reaction (cathodic control).
(c) Increasing the polarisation of the anodic reaction (anodic control).
(d) Increasing the resistance between the cathodic and anodic sites (resistance control).

Tomashov[8] has produced a detailed scheme of control based on the electrochemical mechanism of corrosion, which has been set out in an abridged and modified form in Table 0.2. However, although more fundamental than Table 0.1, it has several limitations, since it is not always possible to define the precise controlling factor, and frequently more than one will be involved. Thus removal of dissolved oxygen (partial or complete) from an aqueous solution reduces the thermodynamics of the reaction and also increases the polarisation of the cathodic reaction, and both contribute to the decrease in the corrosion rate although the latter is usually the more significant.

The primary function of a coating is to act as a barrier which isolates the underlying metal from the environment, and in certain circumstances such as an impervious continuous vitreous enamel on steel, this could be regarded as thermodynamic control. However, whereas a thick bituminous coating will act in the same way as a vitreous enamel, paint coatings are normally permeable to oxygen and water and in the case of an inhibitive primer (red lead, zinc chromate) anodic control will be significant, whilst the converse applies to a zinc-rich primer that will provide cathodic control to the substrate.

Tomashov considers that greater effectiveness of control may be achieved by using more than one method of protection, providing that they all affect the same controlling factor. Thus chromium is alloyed with iron to produce an alloy that relies on passivity for its protection, and passivation can be enhanced by raising the redox potential of the solution, by alloying it with platinum or palladium, or by raising the potential by an external source of e.m.f. However, there is no reason why stainless steel should not be cathodically protected, and although this appears to be a contradiction it is sometimes necessary, particularly when the stainless steel is in contact with a mild steel.

Conclusions

1. The selection of a particular method of corrosion control is by no means simple and a variety of factors will have to be considered before a final decision is taken, particularly when there is no previous experience of the corrosiveness of the environment or the alloy under consideration.

2. It is just as important to avoid design features in the structure that facilitate corrosion as to apply positive protective schemes, an aspect of corrosion control that is frequently neglected.

3. Corrosion testing and monitoring, and supervision and inspection are essential aspects of corrosion control.

L. L. SHREIR

REFERENCES

1. Vernon, W. H. J., 'Metallic Corrosion and Conservation' in *The Conservation of Natural Resources*, Inst. of Civil Engineers, London, 105–133 (1957)
2. Shreir, L. L., *Brit. Corr. J.*, **5**, 11 (1970)
3. Hoar, T. P., *Report of the Committee on Corrosion and Protection*, D.T.I., Published by H.M.S.O., London (1971)
4. 'Materials Selector', *Mater. Eng.*, **74** (1972)
5. Rabald, E., *Corrosion Guide*, 2nd edn., Elsevier, Amsterdam (1968)
6. *Corrosion Data Survey*, N.A.C.E., Houston (1967)
7. Rabald, E., Bretschneider, H. and Behrens, D. (editors) *Dechema-Werkstofftabelle*, Frankfurt (1953–75) (in German)
8. Tomashov, N. D., *Corros. Sci.*, **1**, 77 (1961)

Table 0.2 More fundamental classification of corrosion control[8]

Principle of method	Part of system involved	Method of corrosion control	Examples
(a) Increase thermodynamic stability of the system	Metal	Alloy with a more thermodynamically stable metal	Additions of Au to Cu, or Cu to Ni
	Environment (aqueous)	Lower the redox potential of the solution, i.e. lower $E_{eq,c}$	Lower a_{H^+} by raising pH, remove dissolved O_2 or other oxidising species
		Increase the potential of the M^{z+}/M equilibrium, i.e. increase $E_{eq,a}$	Increase $a_{M^{z+}}$ by removing complexants (e.g. CN^- ions) from solution
	Environment (gaseous)	Remove O_2 or other oxidising gases in which the metal is unstable	Use of inert atmospheres (H_2, N_2, A) or of vacuo
	Metal surface	Coat with continuous film of a thermodynamically stable metal	Au coatings on Cu
(b) Increase cathodic control	Metal	Decrease kinetics of cathodic reaction	Change the nature of the cathode metal in a bimetallic couple; plate cathodic metal (Cd plating of steel in contact with Al); apply paint coatings. Reduce area of cathodic metal
		Remove cathodic impurities; ensure that cathodic phases do not precipitate	Remove heavy metal impurities from Zn, Al, Mg (for use as sacrificial anodes or in the case of Zn for dry cells); ensure that CuAl$_2$ phase in Duralumin and carbide phase in stainless steel are maintained in solid solution
		Increase cathodic overpotential	Amalgamation of zinc; alloying commercial Mg with Mn
	Environment	Reduce kinetics of cathodic reaction	Reduce a_{H^+}, reduce O_2 concentration or concentration of oxidising species; lower temperature, velocity, agitation
		Lower potential of metal	Cathodically protect by sacrificial anodes or impressed current; sacrificially protect by coatings, e.g. Zn, Al or Cd on steel
		Cathodic inhibition	Formation of calcareous scales in waters due to increase in pH; additions of poisons (As, Bi, Sb) and organic inhibitors to acids

Table 0.2 (continued)

Principle of method	Part of system involved	Method of corrosion control	Examples
(c) Increase anodic control	Metal	Alloy to increase tendency of metal to passivate	Alloying Fe with Cr and Ni
		Alloying to give more protective corrosion products	Additions of low concentrations of Cu, Cr and Ni to steel
		Introduction of electrochemically active cathodes that facilitate passivation	Additions of Pt, Pd and other noble metals to Ti, Cr and stainless steels
	Environment	Raise potential by external e.m.f.	Anodic protection of steel, stainless steel and Ti
		Increase redox potential of solution	Passivation of stainless steel by additions of O_2, HNO_3 or other oxidising species to a reducing acid
		Addition of anodic inhibitors	Additions of chromates, nitrates, benzoates, etc. to neutral solutions in contact with Fe; inhibitive primers for metals, e.g. red lead, zinc chromate, zinc phosphate
	Surface	Coatings of metals that passivate readily	Cr coatings on Fe
		Surface treatments to facilitate formation of passive film	Polishing stainless steel and removing Fe impurities by HNO_3; chromate treatment of Al
(d) Resistance control	Surface	Coatings	Organic coatings that increase IR drop between anodic and cathodic areas
	Environment	Removal of water or electrolytes that increase conductivity	Design to facilitate drainage of water; drainage of soils

10 DESIGN AND ECONOMIC ASPECTS OF CORROSION

10.1	Economic Aspects of Corrosion	**10**:3
10.2	Corrosion Control in Chemical and Petrochemical Plant	**10**:10
10.3	Design for Prevention of Corrosion in Buildings and Structures	**10**:33
10.4	Design in Marine Engineering	**10**:54
10.5	Design in Relation to Welding and Joining	**10**:68
10.5A	Terms Commonly Used in Joining	**10**:82

10.1 Economic Aspects of Corrosion

Deterioration as a result of corrosion has come to be accepted as an unavoidable fact of life whenever metals are used. This acceptance has resulted in a general lack of awareness of the importance of the economics of corrosion.

A survey of corrosion and protection in the United Kingdom revealed that only a limited number of firms are sufficiently corrosion conscious to be able to estimate the magnitude of the problem and the cost to their own enterprises. The majority of companies are unable to supply any economic information on corrosion problems and in many cases the cost is hidden under general maintenance[1].

Calculations on a national scale are extremely difficult, and any figure produced for the annual cost of corrosion to a country cannot be considered precise, but merely as an indication of the order of magnitude of expenditure. Such estimates have been made by Uhlig[2] in the USA, Worner[3] in Australia, and Vernon[4] and the Committee on Corrosion and Protection[1] in the UK. In these estimates the cost of protection and prevention have been added to the cost of deterioration due to corrosion. The sum of these costs is a measure of the total demands on the economy due to corrosion. It is obvious, however, that although this cost cannot be totally avoided there is considerable scope for reducing it by methods of corrosion control that are available at present. The total cost is of interest mainly for the estimate it allows of possible savings that could be made, (a) by better use of available knowledge and techniques and (b) by more fruitful direction of research and development.

The estimates of the national annual cost of corrosion vary from 1% to 3·5% of the Gross National Product (GNP). The lower values were estimated by individual scientists[2-4] from the cost of corrosion in a few major industries scaled up to a national level. The value of 3·5% of the GNP for the UK, which was obtained by the Committee on Corrosion and Protection in 1970, represents the only national attempt to estimate the cost of corrosion and the potential savings in a detailed manner.

The possible savings identified by the Committee were based entirely on better use of existing knowledge and were not dependent on the results of future research. The expenditure and potential savings in a variety of industries are shown in Table 10.1.

The costs referred to in Table 10.1 are mainly those arising in the industries concerned or, in certain cases, sustained by the users of the products because of the need for protection, maintenance and replacement of the materials of construction. In the oil and chemical industries the costs of using corrosion resistant materials, where the conditions of service make this essential, have been included. Indirect consequential losses, such as loss of goodwill, which in some industries can be very high, have not been included.

Table 10.1 Expenditure and estimated potential savings in various industries

Industry or agency	Estimated cost (£M)	Estimated potential saving	
		(£M)	Expressed as % of estimated cost
Building and construction	250	50	20
Food	40	4	10
General engineering	110	35	32
Government depts. and agencies	55	20	36
Marine	280	55	20
Metal refining and semi-fabrication	15	2	13
Oil and chemical	180	15	8
Power	60	25	42
Transport	350	100	29
Water supply	25	4	16
Total	1365	310	23

At first sight these costs appear to be far too high to be attributable to corrosion and it is worthwhile considering the indirect expenses which add so considerably to the total.

Loss of Production

The repair or replacement of a corroded piece of equipment may cost relatively little, but while the repair is being carried out a whole plant may shut down for a day or more. In a small plant it may sometimes be more profitable to use a cheap material and replace it regularly than to use a more expensive material with a longer life. In a large integrated factory, however, maintenance work on one plant may cause loss of production from several others. Under these circumstances it becomes essential to reduce periods when the plant has to be shut down for maintenance to a minimum. Thus the choice of materials may be dictated by requirements beyond the individual units, and the higher cost of corrosion resistant alloys justified in return for longer maintenance-free periods. The replacement of a corroded boiler or condenser tube in a modern power station capable of producing 500 MW could result in losses in excess of £3 000/h. Similar examples could be quoted for oil, chemical and other industries. These problems sometimes arise from lack of knowledge when the plant is designed, but all too often the cause is minimisation of capital outlay. The decisions on costing are usually taken by non-technical management who either do not obtain, or do not fully understand, the technical assessments of the factors involved.

Reduction of Efficiency

The accumulation of corrosion products can reduce the efficiency of operating plant. It has been estimated that the extra pumping costs due to clogging of the interior of water pipes amounts to £17 000 000/year in the United States[5]. Pipelines for conveying North Sea Gas are painted on the interior to reduce the costs of pumping. Other examples of loss of efficiency due to corrosion are the reduction of heat transfer through accumulated corrosion products and the loss of critical dimensions within internal combustion engines. The corrosion within internal combustion engines is caused by both the combustion gases and their products and has been claimed to be more detrimental than wear[2].

Product Contamination

Some industries, notably the fine chemicals and parts of the food processing industry, cannot tolerate the pick-up of even small quantities of metal ions in their products. To avoid corrosion, plants often have to incorporate lined pipework and reaction vessels, while in a slightly less demanding situation whole plants are made of an appropriate grade of stainless steel. The capital investment in these industries is thus considerably increased due to the necessity to avoid corrosion.

Very small quantities of corrosion can result in discoloration or staining of products and the less corrosion-conscious organisations can suffer heavy losses due to this form of product contamination. This loss may continue for some time before the cause is finally identified.

Overdesign

The principle of overdesign is the use of much thicker sections than would normally be rquired for mechanical strength to allow for the ravages of corrosion. Within the water treatment and oil industries, corrosion allowances of between 50% and 100% are made on susceptible areas of plant[1,4]. This approach to design results in extra consumption of material and is completely at variance with the concept of conservation. The United Kingdom, with limited resources, is very dependent in terms of living standard on the relationship between the import and export of real goods. Thus in national terms the concept of overdesign is less preferable than protective schemes for the prevention of corrosion, unless the latter are prohibitively expensive.

Maintenance of Standby Plant and Equipment

Regular shutdowns cannot be tolerated in large integrated factories and replacement sections of plant have to be maintained in readiness to operate when corrosion failure occurs. This method of dealing with corrosion can lead to a considerable increase in capital investment.

General Losses

The indirect consequential losses resulting from corrosion are less amenable to calculation but may well outweigh the direct costs. The unpredictable failure of critical parts of industrial equipment, aircraft or other means of transport can cause accidents costing both lives and money. The cost in human life and suffering cannot be assessed but the material damage alone probably amounts to many millions of pounds annually. The degree of corrosion involved may be very slight, such as pitting penetration of a washer or a tube, but the ramifications are large. Surface oxidation of an electrical contact has caused the failure of expensive and sophisticated equipment[1]. Thus the cost of corrosion can be, and often is, many orders of magnitude higher than the value of the material reacted.

Savings

The high national cost of corrosion has been described in terms of the industries in which the major expenditure occurs (Table 10.1) and the ways in which this expenditure arises. To those concerned with this expenditure the major interest is the value of possible savings and how these can be achieved.

From the data presented in Table 10.1 it might be assumed that a company with its major activities in, for example, the field of general engineering could achieve a saving of about 32% of its present expenditure on corrosion and protection by proper application of current knowledge and techniques. However, the estimated saving is very dependent on the present attitude of the company to corrosion and protection in particular, and to materials problems in general. A company with a lack of awareness of corrosion and protection could well make even bigger savings eventually, but only after a thorough investigation of processes and equipment by experts in the field.

In many ways the term *corrosion and protection* is too narrow to describe the activities of industrial scientists dealing with corrosion problems. In the larger chemical companies corrosion and protection are the responsibility of a 'materials of construction' organisation and some companies have claimed savings of up to 50% of corrosion costs as the result of forming such groups (c.f. estimated 8% in Table 10.1). However, in smaller companies corrosion problems are dealt with on a part-time basis by engineers, metallurgists or chemists whose primary job may be outside the materials field. These companies could well benefit from the engagement of corrosion consultants, in the same way that management consultants are used, to make a thorough survey of problems within the firm and to make recommendations.

The largest savings, in terms of percentage of estimated cost, may be expected in building and construction, general engineering, government agencies, marine, power and transport industries. These six groupings account for 81% of the estimated total expenditure on corrosion and protection and, with the exception of the oil and chemical industry, include all groupings with expenditure above £50 000 000/year. However, only the Government agencies and the power industry would benefit directly from increased awareness of corrosion and protection, since the other four groupings are able to include the cost of corrosion in the cost of their products.

The only driving force for improvement within manufacturing industries which are able to sell their products under relatively short warranty periods will be enlightened demands from purchasers. This illustrates clearly that the economics of corrosion are intimately bound-up with the general level of awareness of corrosion and protection throughout the country.

While almost any corrosion problem can be solved or avoided, it is vitally important from a commercial point of view that the economics of corrosion prevention are taken into account. This usually necessitates a full evaluation of initial protection costs together with maintenance charges throughout service life. This type of costing is very important when alternative protection schemes are available and will often reveal that an initial 'cheap' scheme can in reality prove to be very expensive.

The cost of protection of steelwork in a mild industrial atmosphere has been compared for a number of protection schemes[6] and the relevant data is shown in Table 10.2. The difference in aggregate costs illustrates the importance of considering protection on a whole life rather than on an initial cost basis; particularly when maintenance involves a labour intensive operation such as painting.

Table 10.2   Comparative schemes for protecting steelwork ($27 \cdot 4 \, m^2/t$)

Protective scheme	Initial cost (£/t)	Maintenance scheme	Cost of each maintenance (£/t)	Aggregate cost over 24 years (£/t)
1. Auto grit blast, hot galvanise (0·20 mm)	29·5	None	—	29·5
2. Pickle, galvanise (0·10 mm), 2 coats of paint (one on site)	35·0	Brushdown, spot prime, 1 top coat—after 12 years	13·7	48·7
3. Pickle, galvanise (0·10 mm)	22·0	Brushdown, prime, 1 top coat—after 12 years	17·1	39·1
4. Wire-brush, 3 coats of paint (two on site)	23·9	Brushdown, spot prime, 2 top coats—4 year intervals	20·2	124·9
5. Auto grit blast, 4 coats of paint (two on site)	35·2	Brushdown, spot prime, 1 top coat—8 year intervals	12·4	60·0

The management of a company should consider expenditure on corrosion prevention as an investment and will utilise appropriate accountancy techniques to assess the true cost of any scheme. The main methods used to appraise investment projects are payback, annual rate of return and discounted cash flow (DCF). The last mentioned is the most appropriate technique for considering protection schemes since it is based on the principle that money has a time value. This means that a given sum of money available now is worth more than an equivalent sum at some future date, the difference in value depending on the rate of interest earned (discount rate) and the time interval. A full description of DCF[7, 8] is beyond the scope of this section, but in essence this method of accounting can make a periodic maintenance scheme more attractive than if the time value of money were not considered. The concept is illustrated in general terms by considering a sum of money P invested at an interest rate of $r\%$ per annum that will have the value $P\left(1+\dfrac{r}{100}\right)^n$ after n years. Therefore the net present value (NPV) of a sum of money to be spent in the future is reduced by the interest the money can earn until the expenditure is required, hence

$$\text{NPV} = \frac{P}{\left(1+\dfrac{r}{100}\right)^n}$$

A consideration of the costs of the protective Schemes 1 and 3 given in Table 10.2 indicates that the latter is significantly more expensive than the former, and it is of interest, therefore, to apply the concept of the time value of money to these two schemes for 1 t of steel processed.

1. In Scheme 1 the company will spend £29.50 immediately, but will incur no further expense.

2. In Scheme 3 it will spend £22.00 immediately, and can invest the money saved—£29.50—£22.00 = £7.50 at, say, 8% for 12 years, which will yield approximately £18.90. It will then have to spend £17.10 on painting, but even so there will be a bonus at the end of 12 years of £18.90—£17.10 = £1.80 as compared to Scheme 1.

3. Alternatively, for Scheme 3, in order to have £17.10 in 12 years time the company need invest only £6.79 immediately so that the total outlay will be £22 + £6.79 = £28.79, which again is a lower outlay than that for Scheme 1.

Thus it can be seen that although the aggregate cost indicates Scheme 1 to be the cheaper, it is the more expensive when account is taken of the time value of money.

It should be noted that the calculation has not taken account of inflation or changes in labour costs and in a labour intensive project such as painting, wage increases could offset the saving revealed by the DCF method of accounting. Thus in certain circumstances the converse to the above would apply and it would be more economical in the long term to invest initially in relatively expensive corrosion resistant alloys that require no maintenance.

A recent NACE publication devoted entirely to the economic aspects of corrosion control contains several worked examples applicable to a number of industries[9]. These examples serve to illustrate that anti-corrosion procedure and materials should be selected on economic grounds, and not solely on performance grounds. In presenting a proposal to management, the corrosion technologist should show that a thorough investigation of possible solutions has been made in terms of both equipment and expense, and highlight the solution offering the greatest economic advantage to the company[10].

In modern industry decisions are based almost exclusively on economic considerations. The scientist or technologist who is unable to present a proposal based on sound economic projections will find that his recommendations are being rejected. Most scientists are ill-equipped in terms of accountancy language and techniques, because this subject is rarely introduced into school or university science courses. However, it will become increasingly necessary to acquire and use these techniques if scientists are to rise to the upper echelons of management.

Further information on methods of calculating costs of corrosion control is given at the end of Chapter 22.

S. ORMAN

REFERENCES

1. Report of the Committee on Corrosion and Protection, H.M.S.O., London (1971)
2. Uhlig, H. H., United Nations Scientific Conference on Conservation of Resources, Sectional Meeting, Lake Success (1949); *Chemical and Engineers News*, **27**, 2764 (1949) and *Corrosion*, **6**, 29 (1950)
3. Worner, H. K., Symposium on Corrosion, University of Melbourne (1955)
4. Vernon, W. H. J., *Metallic Corrosion and Conservation*, The Conservation of National Resources, Institution of Civil Engineers, London, 105 to 133 (1957)
5. Holme, A., *Anti-Corrosion, Methods and Materials*, 12, Feb. (1969)
6. Porter, F. C., *Corrosion Control Gives Cost Control*, Zinc Development Association, London, Dec. (1970); Brace, A. W. and Porter, F. C., '*The Economics of Protection by Zinc Coatings,*' Metals and Materials, **13**, 169 (1968)
7. Merrett, A. J. and Sykes, A., *The Finance and Analysis of Capital Projects*, Longman (1964)
8. Alford, A. M. and Evans, J. B., *Appraisal of Investment Projects by Discounted Cash Flow*, Chapman and Hall (1967)
9. 'NACE Standard RP-02-72', *Mat. Prot. and Perf.*, **11** No. 8 (1972)
10. 'Economics of Corrosion Control', *Autumn Review Course*, Series 3, No. 2, Institution of Metallurgists, Nov. (1974)

10.2 Corrosion Control in Chemical and Petrochemical Plant

Corrosion control in chemical and petrochemical plant is exercised in five distinct phases (Fig. 10.1) through the life of the plant, as follows:

1. Plant and process design, where the materials of construction, equipment design, process conditions and recommended operating practice can all be influenced to minimise the risk of corrosion. This is the most important phase.

   In large companies an internal project team may design the plant, otherwise contractors provide the design. In either case, the corrosion engineer must be involved from the inception of the project. Otherwise, the materials of construction will have to be chosen to satisfy process

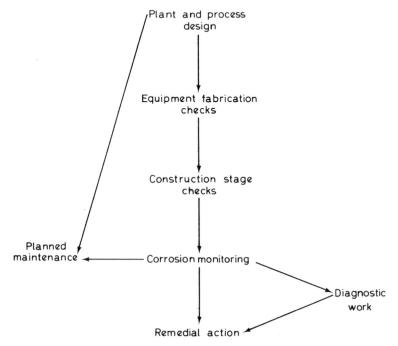

Fig. 10.1   Phases of corrosion control in chemical plant

conditions which may have been decided upon without consideration of the economic balance between process efficiency and capital cost of the plant. The importance of a continuing dialogue between the corrosion engineer and the other disciplines in the project team cannot be emphasised too strongly.

Contractor designs will be in the context of a competitive bidding situation and in-company checks of the design should cover not only design errors but also cases where calculated risks have been taken, which may not, however, be acceptable to the operating company.

The effort required to specify the materials schedule for a new plant or to check a design very much depends on how much experience there is of similar or identical units in operation. Factors such as process conditions and raw material sources are taken into consideration before extrapolating the experience of another unit.

2. Fabrication of the equipment for the plant and plant construction. An inspection system to ensure that fabricators are working to design codes and that their quality control systems are operating effectively, is of considerable value. At the construction stage, checks are made for materials correctly specified but wrongly supplied, on-site welding quality, heat treatments carried out as specified and for damage to equipment especially where vessels have been lined.

When specifying equipment to fabricators, it should be remembered that equipment may well lie exposed on-site before erection and temporary corrosion protective measures should be considered. Any equipment precommissioning treatments specified by the design, e.g. descaling, must be carried out.

Such points of detail can make for a smooth start-up and minimum trouble during the early operational period of a new plant.

3. Planned maintenance or regular replacement of plant equipment to avoid failure by corrosion, etc. is an essential adjunct to design, and constitutes the third phase of control. The design philosophy determines the emphasis placed on controlling corrosion by this means, as opposed to spending additional capital at the construction stage to prevent corrosion taking place at all. Where maintenance labour costs are high or spares may be difficult to procure, a policy of relying heavily on planned maintenance should be avoided.

4. Even with all these checks on design, fabrication and construction, errors are made which, with maloperation and changes in process conditions during the lifetime of the plant, can all lead to corrosion. The fourth phase of control therefore lies in monitoring the plant for corrosion in critical areas. Corrosion monitors should be regarded as part of plant instrumentation and located in areas of high corrosion risk or where corrosion damage could be particularly hazardous or costly. Monitoring should include a schedule of inspections once the plant is commissioned.

5. Corrosion monitors by themselves only warn of corrosion and must be coupled with the fifth phase of control, viz. remedial action, to be effective. In some cases of corrosion the remedial measure is known or easily deduced, but in others diagnostic work has to precede a decision on remedial action.

Phase One—Plant and Process Design

The factors influencing the final choice of design are summarised in Fig. 10.2.

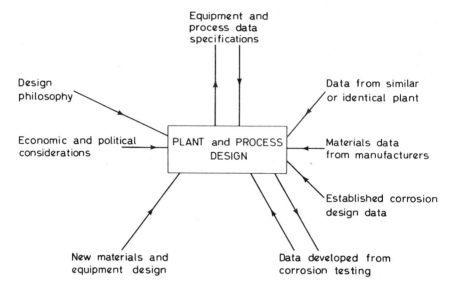

Fig. 10.2   Factors influencing plant and process design

Design Philosophy

The chemical and petrochemical industries are highly capital intensive and this has two important implications for the plant designer. Before the expenditure for any plant is approved, a discounted cash flow (DCF) return on capital invested is projected. The capital cost of the plant is a key factor in deciding whether the DCF return is above or below the cut-off value used by a company to judge the viability of projects. Thus, there is always strong pressure on the materials engineer not to 'overspecify' the materials of construction.

Conversely, however, the cost of downtime can be very high and this creates a 'minimum risk' philosophy which runs contrary to the capital cost factor. The balance between these two forces has to be clearly stated to allow the materials engineer to operate effectively.

The choice of material from the viewpoint of mechanical properties must be based on design conditions. However, from a corrosion standpoint it must be realised that the design conditions are limiting values and that for most of its life the equipment will operate under 'process conditions'. For the decision on the requisite corrosion resistance properties it is necessary to examine, by means of an operability study, how far process conditions may deviate from the normal and how often and for how long. The operability study is carried out using a line diagram for the projected plant.

The design life of the plant has to be stated so that corrosion allowances may be calculated:

Corrosion allowance = Design life × Expected annual corrosion rate

This calculation assumes, of course, that corrosion is uniform. Finally, implicit in the design will be boundary conditions on the way the plant can be run, outside of which the risk of corrosion is high. These should be clearly set out in the operating manual for the plant.

Influence of Process Variables

The rate of a chemical reaction is influenced by pressure, temperature, concentration of reactants, kinetic factors such as agitation, and the presence of a catalyst. Since the viability of a plant depends not only on reaction efficiencies but also on the capital cost factor and the cost of maintenance, it may be more economic to alter a process variable in order that a less expensive material of construction can be used. The flexibility which the process designer has in this respect depends on how sensitive the reaction efficiency is to a change in the variable of concern to the materials engineer.

Where, for example, chloride stress-corrosion cracking is a risk the process temperature becomes a critical variable. Thus it may be more economic to lower the process temperature to below 70°C, a practical threshold for chloride stress-corrosion cracking, than to incur the extra expense of using stress-corrosion cracking-resistant materials of construction.

Pressure has less influence on corrosion rates than temperature in most cases of aqueous corrosion, although it has a large effect on some forms of gaseous corrosion at high temperatures, e.g. hydrogen embrittlement[1]. However, impingement attack is influenced by pressure in specific instances. Thus, when a gas is dissolved under pressure, and the pressure is reduced (let-down) gas bubbles are released which can contribute significantly to impingement attack if released into a high velocity stream. Pressure let-down in such cases should take place where the velocity of the liquid stream is low.

An example where reactant concentration is solely governed by corrosion considerations is in the production of concentrated nitric acid by dehydration of weak nitric acid with concentrated sulphuric acid. The ratio of HNO_3:H_2SO_4 acid feeds is determined by the need to keep the waste sulphuric acid at $>70\%$ $^w/_w$ at which concentrations it can be transported in cast-iron pipes and stored after cooling in carbon-steel tanks.

Equipment Design

A recent survey by du Pont on all failures in their metallic piping and equipment taking place during a 4-year period showed corrosion accounting for 55% of total failures[2]. Table 10.3 lists the major causes of corrosion failure in this wide ranging survey.

For this one company stress-corrosion cracking alone cost £2M annually and the corresponding figure for the US chemical industry was £13M[3].

Some general points can be made about equipment design in connection with the more important types of corrosion.

Table 10.3   Corrosion failures analysed by type (after du Pont (1968 to 1971))

Type of corrosion	Failure rate (average %)	
General	15·2	
Stress-corrosion cracking	13·1	
Pitting	7·9	All other types <1%
Intergranular	5·6	
Erosion/corrosion	3·8	
Weld corrosion	2·5	

General corrosion   If the rate of general corrosion of a particular material in a duty is well enough characterised at the design stage, then the designer can, in some instances, use a corrodible material of construction with a suitable corrosion allowance. There are limitations to this approach. Thus, while a corrosion allowance of 6 mm on the shell of an exchanger is common practice, to make this allowance on the tube wall thickness is not practical and either a more resistant tubing material is used, or planned retubing of the unit is accepted as part of the design. This may mean carrying a spare installed exchanger or a spare tube bundle. Overall economics dictate the course to be taken.

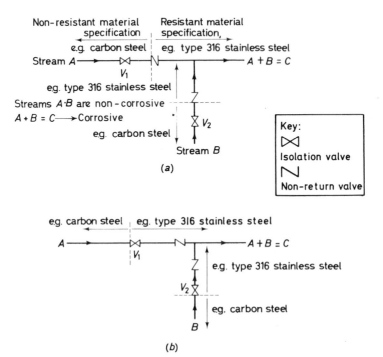

Fig. 10.3   Incorrect (*a*) and correct (*b*) designs where there is a sharp change in corrosivity.

Corrosion with the formation of insoluble corrosion products may be unacceptable where heat-transfer equipment is concerned. Fouling by corrosion products has to be allowed for when sizing the equipment and the extra cost of using resistant material may not be as great as the increased cost of a larger exchanger in the less resistant material plus the cost of downtime to clean fouled surfaces.

Product purity specifications determine how much soluble corrosion product can be tolerated.

In many plants, one section of a process will be relatively non-corrosive, allowing cheap materials of construction to be used, while the following section will be very corrosive, necessitating more corrosion-resistant materials. The interface between the two sections has to be carefully designed to avoid corrosion in the first section at a shutdown when there may be some backflow from the corrosive section. Figures 10.3*a* and 10.3*b* show an incorrect and a correct design. The basic principle here is that a non-return valve should not be treated as an isolation. Obviously, in this case the operating manual should include instructions to close the isolation valves V_1 and V_2 if flow is stopped for any length of time.

A common case where intense general corrosion is experienced in a very restricted section of plant is where an acidic vapour is condensing. As a vapour the acid is usually non-corrosive, but when condensed it can only be handled in expensive materials. Another variation on this theme is that only at the region of initial condensation is there a corrosion problem, either the condense/reboil condition being particularly corrosive or else corrosion only takes place at or near the boiling point. Several variations in design are possible to cope with these situations:

1. Where the acid condensate is corrosive, neutralisers, e.g. ammonia or neutralising amines, can be injected into the vapour stream to co-condense with the acid vapour. This is the practice with the overheads of a crude oil pipestill (Fig. 10.4).
2. Where the corrosion problem is limited to the condense/reboil situation, i.e. where, due to variations in vapour temperature (or temperatures of the surfaces with which the condensate can come into contact), the condensate reboils, the answer may be to use resistant material at the

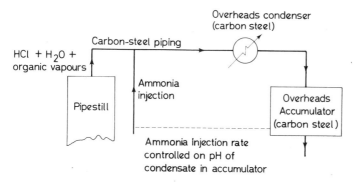

Fig. 10.4   Corrosion control in the overheads system of a crude oil distillation unit

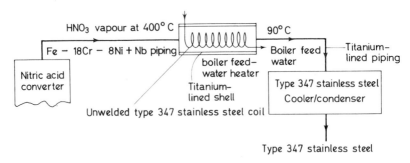

Fig. 10.5   Materials design to control corrosion in a nitric acid plant

critical region. This expedient is adopted in condensing nitric acid vapour (Fig. 10.5). Condensation first occurs in the boiler feed-water heater. The coil-carrying boiler feed water is weldless Fe–18Cr–8Ni+ Nb (type 347) to prevent attack on welds. Condensed acids drip off the coils onto the shell where they can reboil—a severe corrosion condition —the shell is therefore titanium lined.

3. Where the acid condensate is corrosive and neutralisers cannot be used, then a condenser of resistant materials has to be employed. However, by steam tracing the lines leading the vapour to the condenser, premature condensation can be avoided (Fig. 10.6) and in consequence a cheaper material can be used for them.

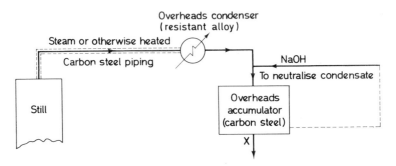

Fig. 10.6   Method of preventing corrosion by premature condensation of acidic vapours. pH measured continuously at point X automatically controls flow of sodium hydroxide

Localised corrosion   The various forms of localised corrosion are a greater source of concern to the plant designer (and operator) since it is usually difficult to predict an accurate rate of penetration, difficult to monitor, and (especially in the case of stress-corrosion cracking) consequently can be catastrophically rapid and dangerous.

Stress-corrosion cracking   New metal/environment combinations which produce stress-corrosion cracking are continually being found. Combinations discovered in service in recent years include titanium in red fuming nitric acid; carbon steel in liquid anhydrous ammonia[4] and in carbonate solutions containing arsenite as a general corrosion inhibitor[5]; and $CO-CO_2-H_2O$ systems. Also, polythionic acid has been identified as the specific environment in several cases of the stress-corrosion cracking of stainless steel in service. Chloride stress-corrosion cracking of stainless steels is a continuous source of trouble. Almost invariably it is caused by the concentration of chlorides from a bulk electrolyte which by itself would not cause cracking. The designer has, in theory, four degrees of freedom to avoid the problem:

1. Stress relief—although success is claimed for stress relief at 900°C in alleviating the problem of chloride-induced stress-corrosion cracking, there is considerable doubt that, except in cases of marginal risk, it is a reliable method of preventing stress-corrosion cracking of austenitic stainless steels. Working stresses would appear to be sufficient to cause cracking.
2. Avoid situations that increase the chloride concentration—some circumstances in which this can occur in practice are listed below, and it can be seen that control is difficult to ensure.
   (*a*) Under deposits on heat-transfer surfaces.
   (*b*) At a vapour/liquid interface in contact with a heat-transfer surface.
   (*c*) At a leakage point where evaporation of leaking liquid takes place, e.g. at a leaking joint.
   (*d*) Where liquid boils in a restricted space (e.g. a tube/tubeplate crevice), a thermowell, level control instruments, or in a condense/reboil situation at the point of initial condensation of vapour.
   (*e*) In a shutdown situation where vapour containing chloride condenses to be re-evaporated when the unit is restarted.
   (*f*) In a system where chloride can be continuously recycled, e.g. in a distillation process where there is recycle of a fraction of the distillate, chloride can concentrate in the still bottoms.
3. Maintain the chloride-containing liquid in contact with the stainless steel at <70°C. Process design considerations limit this approach.
4. Use an alloy of higher stress-corrosion cracking resistance or one which is immune. If their general corrosion resistance is adequate, ferritic steels may be used. Extra-low-interstitial-content ferritic stainless steels containing molybdenum are claimed to have corrosion resistance at least the equal of the 300 series steels and to be virtually immune from stress-corrosion cracking. Otherwise, duplex-structure alloys (typically Fe–18Cr–5Ni) are now available, having superior resistance to chloride stress-corrosion cracking, although cracking of them has been experienced in more acidic chloride-containing media. Stainless steels containing 18% Cr, 18% Ni and 2–3% Si are also reported to have been successfully used where type 304 has cracked.[6]

   Alloys of high nickel content also have improved chloride stress-corrosion cracking resistance and Incolloy 825 has replaced type 321 stainless steel for steam bellows on some plants. Occasionally cracking of the latter was experienced due to chloride-contaminated steam condensing in the convolutions on shut-down and being re-evaporated at start-up.

Titanium is immune to chloride induced stress-corrosion cracking but more expensive than type 300 series stainless steels.

Chloride stress-corrosion cracking under lagging on hot equipment is a classic problem. Rainwater leaches chlorides from the lagging, the solution percolating to the hot wall of the vessel or pipe where it concentrates by evaporation. Several remedies are available. Wrapping the pipe or vessel with aluminium foil before lagging has proved effective. The chloride solution thus concentrates on the aluminium surface instead of stainless steel. Pipe hangers and vessel flanges constitute points where this protection can be incomplete and attention to detail is essential to minimise failures. Alternatively, lagging materials containing soluble inhibitors of chloride stress-corrosion cracking have been used with success.

For carbon steels, however, a full stress-relief heat treatment (580–620°C) has proved effective against stress-corrosion cracking by nitrates, caustic solutions, anhydrous ammonia, cyanides and carbonate solutions containing arsenite. For nitrates, even a low-temperature anneal at 350°C is effective, while for carbonate solution containing arsenite the stress-relief conditions have to be closely controlled for it to be effective[5].

However, with large vessels, there are two areas where it is difficult to ensure adequate stress relief:

(a) Where welds are stress relieved on site—the large heat sink provided by the vessel and the difficulty of shielding the area being heat treated from draughts mean that very strict temperature monitoring is necessary.

(b) Vessels fitted with large branches—during pressure testing or even in normal operation, yield-point stresses can be reached at stress raisers provided by the configuration of such branches.

In addition, a surprisingly large number of stress-corrosion cracking failures have resulted from the welding of small attachments to vessels and piping after stress-relief heat treatment has been carried out.

Pitting   Pitting of carbon steel is seldom catastrophically rapid in service and can often be accommodated within the corrosion allowance for the equipment. It often takes place under scale or deposits so that regular descaling of equipment can be beneficial.

Pitting of carbon steel in cooling-water systems is a well-known problem which can be avoided by a correctly instituted and maintained inhibitor treatment. Correct institution includes descaling of the equipment before commissioning, since experience with chromate-inhibited systems has shown that a pre-existing rust layer prevents chromate reaching the metal surface, the equipment continuing to corrode as if no inhibitor was present in the cooling water.

'Oxygen pitting' of boiler tubes by boiler feed water due to inadequate de-aeration is also a problem, but soluble by proper maintenance of de-aerators, coupled with regular boiler feed water analysis, or, preferably, continuous dissolved-oxygen monitoring.

Pitting of stainless steels can usually be avoided by correct specification of steel type, and type 316 is the normal choice where pitting is at all likely.

Some duplex alloys have even better pitting resistance than type 316 and should be considered in severely pitting media. Titanium is virtually immune to chloride pitting and cupro-nickel alloys are used for condensers where sea-water is the coolant; high pitting resistance in this duty is claimed for Cu–25Ni–20Cr–4·5Mo.

Crevice corrosion without heat transfer   Since this is a phenomenon affecting alloys which depend on diffusion of an oxidising agent (usually oxygen) to the metal surface for maintenance of passivity, there are two degrees of freedom open to the designer to avoid this problem. The first is to choose an alloy that does not rely on an oxide film for its corrosion resistance. This, in the case of replacing conventional austenitic stainless steel, will be a more costly option. An alternative is to choose a passive alloy whose passivity is less critical in terms of oxidising agent replenishment at the metal surface. As an improvement on type 316, duplex-structure stainless steels, e.g. Ferralium or higher-alloy-content stainless steels based on a 25% Cr–20% Ni composition (e.g. *2RK65* and *904L*) are more crevice-corrosion resistant, in addition to improved general corrosion and stress-corrosion cracking resistance.

The second degree of freedom is 'to design-out' crevices where possible, although it must be remembered that crevice corrosion can go on underneath deposits. Crevice corrosion at a butt weld with incomplete root penetration is a common case (Fig. 10.7*a*). Where internal inspection is not possible and crevice corrosion is recognised as likely, *X*-radiography of each weld can be specified.

The correct flange design, in particular where crevice corrosion is known to be a problem, is important. Thus screwed flanges (Fig. 10.7*b*) and socket-welding flanges (Fig. 10.7*c*) present crevices to the fluid whereas slip-on-welding (Fig. 10.7*d*) and welding-neck flanges (Fig. 10.7*e*) are designs that avoid crevices. Welding-neck flanges have the advantage that the butt weld to the adjoining pipe can be radiographed whereas the fillet welds in the slip-on welding type cannot. Poor fusion resulting in a crevice cannot therefore be detected.

Crevice corrosion often occurs at gasketted joints. It can be alleviated as a problem by painting the flange faces with inhibited paints or coating the gasket and flange faces with impervious compounds, e.g. liquid rubbers, ensuring the gaskets are specified correctly from the design code and have the correct internal diameter. Figure 10.7*f* shows how a crevice is created when a gasket of a sub-standard specification (or the wrong size) is fitted. Figure 10.7*g* shows the correct configuration if crevice corrosion is thought likely. Branches for thermosheaths must be generously sized so that a crevice is not created between the sheath and the branch wall (Fig. 10.7*h*).

Crevice corrosion with heat transfer   This can give rise to catastrophically high rates of corrosion. A classic situation in which this occurs is at the crevice formed at the back of a tubeplate in a tube-and-shell heat exchanger (Fig. 10.7*i*).

A heat-exchanger tube is expanded into the tubeplate to effect a joint. However, to avoid bulging the tube outside the confines of the tubeplate, expansion never takes place through the whole tubeplate thickness and a

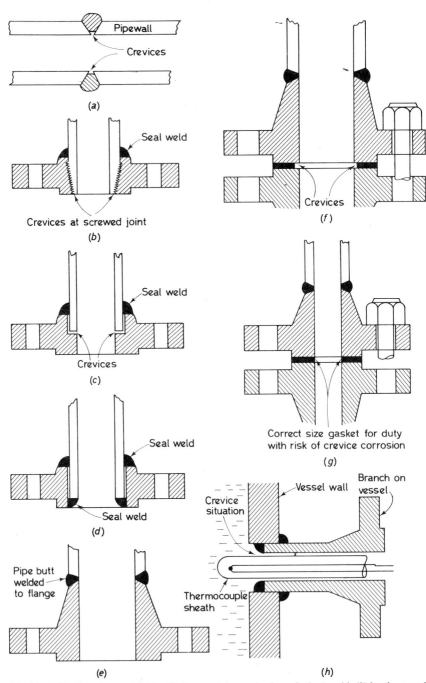

Fig. 10.7 Crevices. Formed (*a*) by the incomplete penetration of a butt weld, (*b*) by the use of a screwed flange and (*c*) by the use of a socket-welding flange. (*d*) Crevice-free slip-on-welding flange type, (*e*) crevice-free welding-neck flange type, (*f*) crevices created by choice of wrong size or wrong standard gasket for duty, (*g*) correct configuration if crevice corrosion is thought likely, (*h*) crevice situation created by too small a clearance between thermosheath and the containing branch, (*i*) crevice formed at the back of a tubeplate and (*j*) and (*k*) variants of sealing the tube/tubeplate crevice

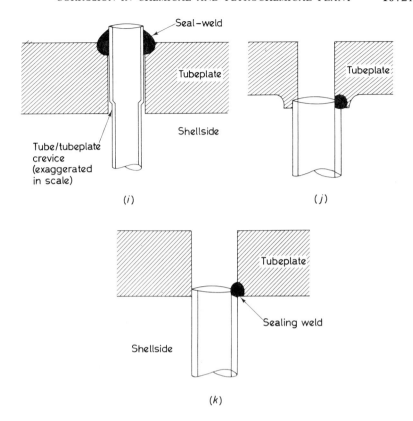

(i)

(j)

(k)

crevice is thus formed. Within this crevice local boiling can take place, concentrating corrodents (chloride stress-corrosion cracking of austenitic-stainless-steel exchangers can occur in this way). Even if local boiling does not take place, the high rate of heat transfer across the gas entry tubeplate of a gas/liquid exchanger with gas on the tubeside, may by itself stimulate corrosion in the crevice. Where the shellside liquid is inhibited, ingress of inhibitor into the tube/tubeplate crevice may be too slow to be effective in a heat transfer situation.

There are three degrees of freedom in designing-out these problems for tube and shell exchangers:

1. By reducing the heat-transfer rate at the front tubeplate using insulating ferrules in the tube ends.
2. By putting the corrodent on the tubeside and the hot vapour on the shellside provided the vapour is compatible with the shell material; this is the normal configuration for steam-heated exchangers.
3. By eliminating the crevice using the recently developed technique of seal welding the tubes to the back of the tubeplate (Fig. 10.7k). However, this would add markedly to the cost of the unit.

Weld corrosion   Crevice corrosion at butt welds due to poor penetration has already been discussed and was shown in Fig. 10.7(a). Conversely, if there is a large weld bead protruding in the pipe bore, erosion/corrosion can occur downstream due to the turbulence produced over the weld bead (Fig. 10.8). In either case, the fault probably lies in the incorrect spacing of the butts at welding.

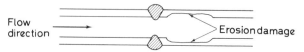

Fig. 10.8   Erosion/corrosion downstream of a butt weld with too much root penetration

Selective corrosion in the heat-affected zone of a weld occurs most commonly when unstabilised stainless steels are used in certain environments. The obvious answer is to use an extra-low-carbon grade of stainless steel, e.g. types 304L, 316L or a stabilised grade of steel, e.g. types 321 and 347. Knifeline attack at the edge of a weld is not commonly encountered and is seldom predictable, and it must be hoped that it is revealed during preliminary corrosion testing.

For both heat-affected zone corrosion (intergranular attack) and knifeline attack the heat flux during welding and the time at temperature can critically affect the severity of the attack. Both these factors may vary from one welder to another, and when preparing pieces for corrosion testing not only should fabrication welding conditions be accurately reproduced, but the work of more than one welder should be evaluated.

Erosion/corrosion   Erosion/corrosion by a single-phase liquid system is characterised by a maximum acceptable fluid velocity for a given material. Generally, velocities in straight pipes should not exceed 50% of this value because turbulence, i.e. high local velocities, is bound to be superimposed in some areas. However, where possible, sharp changes in section and of flow direction should be avoided. Bends should be swept rather than right-angled, 'T'-junctions should be avoided where possible, and section reducers should be gradually tapered. Where turbulence cannot be avoided, e.g. downstream of pumps, control valves and orifice plates used to measure flow, it is advisable to consider short sections (say 3 m) of a more erosion-resistant material. Any piece of equipment in which turbulence occurs, e.g. pumps and control valves, should similarly have a higher velocity-rating material of construction. Introduction of a sacrificial impingement plate should be considered when the velocity cannot be kept low enough to prevent erosion. For example, an impingement plate is often fitted opposite the liquid inlet on the shell of a tube-and-shell exchanger to protect the tubes on which the liquid would otherwise impinge.

The tube-side inlet to an exchanger, i.e. the tube ends, is a highly turbulent region and nylon ferrules in the tube ends of the inlet pass have been used in cupro-nickel-tubed condensers to prevent erosion. Where the flow is two phase the same rules will apply except that an erosion velocity limit is more difficult to specify.

Erosion/corrosion of pump impellers, casings and wear plates can be very troublesome. Positive-displacement pumps create much less turbulence than centrifugal or axial-flow pumps and should be used where possible in critical duties where erosion is particularly severe or where the tolerable deviation from the design delivery rate is low. However, positive-displacement pumps tend to have low delivery rates, and for large rates pumps creating more turbulence will have to be used. For very severe duties, planned maintenance may be the only way to live with the problem. Experience on severe duties is that closed-impeller centrifugal pumps are less prone to erosion than the open type. Replaceable liners may also be considered to accommodate casing erosion.

Corrosion Data

This is derived from four sources:

1. From similar or identical plants—such data must be treated with caution until it is verified that process conditions, mode of operation and raw materials are all such that direct comparisons can be made.
2. From data published by manufacturers to support the use of their materials, e.g. References 7 to 10. Again, case histories quoted must be treated with caution but such data are very useful for sorting out the possible materials for a particular duty.
3. From established corrosion design data. There are some very useful reference works containing corrosion data from a multitude of sources[11-13], and for simpler corrosive systems, well-established corrosion design charts[1, 13], e.g. limiting concentrations and temperatures above which carbon steel must be stress relieved to avoid stress-corrosion cracking in caustic solutions. These are invaluable to the corrosion engineer involved in design.
4. Corrosion testing data. The pitfalls in corrosion testing and the test methods are described in Chapter 20, but several points need underlining from experience in the design of chemical plants:
   (a) Liquids used for testing must reproduce all possible variations that are to be expected in the operating plant.
   (b) When testing for corrosion under heat-transfer conditions, the heat flux must be realistic. It is not good enough to merely reproduce the correct temperatures[14].
   (c) When testing for corrosion in a distillation process, very localised effects must be covered, e.g. the corrosion characteristics at the point of initial condensation of the overhead vapours.
   (d) Testing for erosion limits should include a reference condition, i.e. a fluid velocity/material combination whose erosive characteristics in a plant are known.

New Materials and Equipment Design

New alloys with improved corrosion-resistance characteristics are continually being marketed, and are aimed at solving a particular problem, e.g. improved stress-corrosion cracking resistance in the case of stainless steels, improved pitting resistance or less susceptibility to welding difficulties.

Composite materials are also becoming more freely available and explosive cladding offers possibilities of duplex-plate materials for such items as tube-plates[15]. Duplex-material heat-exchanger tubing is also now marketed and vessels can be satisfactorily clad or lined with an ever increasing list of metallic and non-metallic linings. Glass-reinforced plastics are in increasing use as confidence grows in their long-term performance for such items as low-pressure vessels. In short, the impasse of a design today may well be soluble tomorrow and the materials engineer must keep abreast of developments.

Economic and Political Considerations

This input to design refers to the long-term stability of the raw material sources for the plant. It is only of importance where the raw materials can or do contain impurities which can have profound effects on the corrosivity of the process. Just as the design should cater not only for the norm of operation but for the extremes, so it is pertinent to question the assumptions made about raw material purity. Crude oil (where H_2S, mercaptan sulphur and napthenic acid contents determine the corrosivity of the distillation process) and phosphate rock (chloride, silica and fluoride determine the corrosivity of phosphoric acid) are very pertinent examples. Thus, crude-oil units intended to process low-sulphur 'crudes', and therefore designed on a basis of carbon-steel equipment, experience serious corrosion problems when only higher sulphur 'crudes' are economically available and must be processed.

Phase Two—Construction Stage Checks

The question of safeguarding against wrong materials being installed in a plant has been a focus of much attention recently[16]. Mistakes can arise in two ways:

1. Items which cannot themselves be wrongly assembled are supplied in the wrong material by the fabricator due to a mix-up in his identification system.
2. Common items such as valves, piping and welding electrodes which may be supplied for a large plant in half a dozen material specifications can become mixed up due to poor identification marking.

This is a very serious problem in that, for example, in high-temperature hydrogen service, use of carbon steel when the duty demands a 1% Cr–Mo steel, can have disastrous consequences[16].

Corrosion failures in service are minimised by ensuring that all fabrication and erection work conforms to the codes of practice specified by the design. The corrosion engineer can only influence matters here by persuading the designers to specify more stringent codes if there are identifiable risks in using a less demanding code. For example, for duties in which crevice corrosion is a possible problem, it might pay to adopt a policy of radiographing all welds for defects rather than only 10% as specified normally.

Phase Three—Planned Maintenance

Strictly, 'planned maintenance' refers to a policy of shutting down a plant at regular intervals to replace or refurbish items of equipment which although not having failed by corrosion or any other mode, would have a high probability of doing so before the next shutdown.

The justifications for adopting a planned maintenance policy rather than spending extra capital to ensure that the component lasts the life of the plant are summarised in Fig. 10.9.

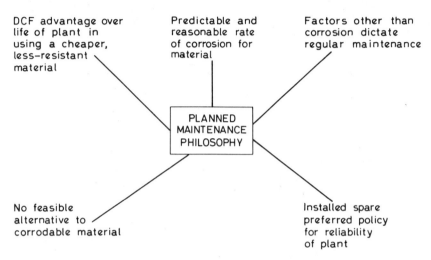

Fig. 10.9   Factors contributing to a policy of planned maintenance

A pre-requisite to a corrodable material being used is that it is known to have a useful and reasonably predictable life. Planned, or unplanned downtime costs money and the intervals between planned replacements must be of reasonable duration. In practice, the replacement interval is usually conservative at first and then as experience accumulates, the intervals between planned replacements will usually extend. The main reason for choosing a planned maintenance policy is that on a discounted cash flow calculation over the life of the plant, the cost of regular replacements including maintenance labour and downtime is less than the extra initial capital cost of a more durable material.

In some cases, the item of equipment may have to undergo maintenance at regular intervals for reasons other than corrosion damage, e.g. change of bearings and seals on pumps. This fact alters the basis of the DCF calculation, i.e. the cost of downtime and maintenance labour is no longer all set against the increased cost of a more durable material and the cheaper, corrodable material becomes a more attractive alternative. For some items of equipment, their importance to the plant may be such that the minimum risks of failure shutting the plant down are taken, and a spare, which can be rapidly brought on-line, is installed. In this case, maintenance can be carried out on the spare with the plant on-line. This is a variation on the strict definition of planned maintenance.

Lastly, but quite frequently, the sole justification for a planned maintenance policy is that there is no feasible alternative to the use of a corrodible material of construction, usually because the item of equipment in a non-corrodible material is not commercially available or delivery is slow such that the construction programme would be jeopardised.

Phase Four—Corrosion Monitoring

Figure 10.10 summarises the techniques available for monitoring corrosion in an operating plant. Visual inspection is a statutory obligation at regular intervals for some classes of chemical plant equipment, e.g. pressure vessels.

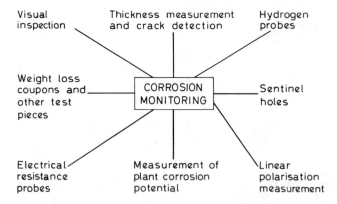

Fig. 10.10   Techniques for monitoring corrosion in process plant

However, much equipment is opened up mainly to gain reassurance that it is not suffering from serious corrosion damage. Due to the high cost of such downtime, there is a considerable financial incentive to develop on-line monitoring methods which will partially or wholly replace such visual inspections. (See also Section 20.3.)

Thickness Measurement and Crack Detection

The principal technique used is ultrasonics, but it has limitations as a monitor for the progress of corrosion. The sensitivity of the technique is commonly quoted as ± 0.005 in. (0.125 mm) although even this may be difficult to achieve where the surface is hot or where close coupling of the probe head and metal surface is difficult. This means that for a normally low corrosion rate, the useful interval between readings may be fairly long and an unexpected rapid increase in corrosion rate could be missed for some time.

γ-ray and β-ray backscatter and absorption techniques can give very accurate thickness definition, especially for thin gauge material. The γ-ray absorption method has been successfully developed to detect corrosion in the tube/tubeplate crevice of a heat exchanger and also of the shellside of heat-exchanger tubes[18]. Both situations are difficult to visually inspect. The

γ-source is placed in one tube while the detector is placed in an adjacent one.

For detecting stress-corrosion cracks and estimating their depth of penetration, the ultrasonic technique and, to a lesser extent, X-radiography, have proved successful.

Sentinel holes are used as a simple form of thickness testing. A small hole of about 1·6 mm diameter is drilled from the outer wall of the piece of equipment to within a distance from the inner wall (in contact with the corrodent) equal to the corrosion allowance on the equipment (Fig. 10.11). The technique has been used even in cases where the corrodent spontaneously ignites on contact with the atmosphere. The philosophy is that it is better to have a little fire than a big one which would follow a major leak from corrosion through the wall. When the sentinel hole begins to weep fluid a tapered plug is hammered into the hole and remedial maintenance planned. Siting the sentinel holes is somewhat speculative although erosion at the outside of a pipe bend is often monitored in this way.

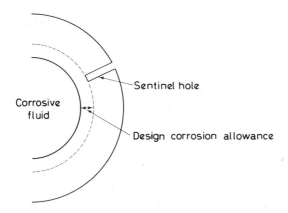

Fig. 10.11   Sentinel hole method of monitoring corrosion of a pipewall

Hydrogen probes are mainly used in refineries to detect the onset of conditions when H_2S cracking of carbon-steel equipment could become a real risk. As a qualitative monitoring technique, it has a long and proven service of worth.

Weight-loss coupons are the most used and most abused of corrosion-monitoring methods. The technique is abused by the often repeated mistake of coupons being placed in such a position that the fluid flow around them is totally unrepresentative of that experienced by the equipment they are intended to simulate. The flow around a specimen projecting into a flowing piped stream may result in totally different corrosion conditions from that experienced by the pipe walls. A less precise result from a spool piece inserted into a pipeline may be far more typical of true corrosion rates in the pipe than a highly precise result from a corrosion coupon. Sometimes, however, it is possible to get close to actual flow conditions. Thus, in agitated vessels, specimens bolted to the outer edge of the agitator blade, in the same orientation as the blade, will give very useful information on agitator corrosion rates. Corrosion coupons are probably most usefully used to rank materials of construction and to detect the permanent onset of a significant change in corrosivity. Coupons integrate corrosion damage over a period and are of only marginal use in a situation where rapid and large increases in corrosion rate can occur.

Electrical Resistance Monitors

Electrical resistance monitors use the fact that the resistance of a conductor varies inversely as its cross-sectional area. In principle, then, a wire or strip of the metal of interest is exposed to the corrodent and its resistance is measured at regular intervals. In practice, since the resistance also varies with temperature, the resistance of the exposed element is compared in a Wheatstone bridge circuit to that of a similar element which is protected from the corrodent but which experiences the same temperature.

In process streams where there are large changes in process temperature over a short time, the fact that the temperature of the protected element will lag behind that of the exposed element can give rise to considerable errors. The most recent development is the use of test and reference elements that are both exposed to the corrodent. The comparator element has a much larger area than the measuring element so that its resistance varies much less than that of the measuring element during their corrosion. Several drawbacks to this type of monitor, deduced from service experience, may be quoted:

1. If corrosion occurs with the formation of a conducting scale, e.g. FeS or Fe_3O_4, then a value of the measured element resistance may be obtained which bears little relation to the loss in metal thickness.
2. Pitting or local thinning of the measured element effectively puts a high resistance in series with the rest of the element and thus gives a highly inflated corrosion rate.
3. Wire form measured elements tends to suffer corrosion fatigue close to the points where it enters the support. This is particularly true in turbulent-flow conditions, and strip-type elements are preferred in such cases.
4. Where a solid corrosion product is formed, meaningful results are only obtained after a 'conditioning' period for a new measured element. Even so, the conditions under which the scale is laid down may not be the same as that for the original equipment. This objection applies equally to coupons or spools, and points to one of the basic objections of using anything other than the plant itself to monitor corrosion rates.
5. The more massive the measured element, the longer its useful life, but the less sensitive the monitor is to small changes in cross-sectional area. Thus, a compromise between long life and sensitivity has to be decided upon, depending on the application.

The advantages of this type of monitor are that it can be automated to produce print-outs of corrosion rate at regular intervals and that it can be used to monitor corrosion in any type of corrodent, e.g. gaseous, non-ionic liquid or ionic electrolyte. Such monitors are in wide use, especially in refinery applications.

Linear Polarisation Measurement

Linear polarisation measurement is based on the Stern–Geary equation:

$$\left(\frac{\Delta E}{\Delta i}\right)_{E_{\text{corr.}}} = -\frac{b_a b_c}{2\cdot 3 i_{\text{corr.}}(b_a + b_c)}$$

$$i_{\text{corr.}} = K\left(\frac{\Delta i}{\Delta E}\right)$$

There has been considerable talk recently in the literature about errors in this equation, but the modifications to it proposed are minor compared with the practical errors introduced by its use (see also Section 20.1):

1. The values of b_a and b_c, i.e. the Tafel constants of the anodic and cathodic polarisation curves, first have to be measured directly in the laboratory or deduced by correlating values of $\Delta E/\Delta i$ measured on the plant with $i_{\text{corr.}}$ values deduced from corrosion coupons. The criticism is that the K value is likely to be inaccurate and/or to change markedly as conditions in the process stream change, i.e. the introduction of an impurity into a process stream could not only alter $i_{\text{corr.}}$ but the K factor which is used to calculate it.
2. The equation assumes that for a given ΔE (usually 10 mV) shift, the corresponding in Δi is solely attributable to an increase in metal dissolution current. However, in solutions containing high redox systems, this may be very far from the case.

Practical experience with the technique has been that in some simple electrolyte solutions, 'reasonably good' correlation is achieved between corrosion rates deduced by linear polarisation and from corrosion coupons. 'Reasonably good' here seems to be considered anything better than a factor of two or three. However, the a.c. linear-polarisation technique has been used with considerable success to control inhibitor additions to overcome corrosion in ships' condensers while operating in estuarine waters[18] and the d.c. technique has been used in controlling the corrosivity of cooling waters. Although it can only be used in ionic electrolytes, results have indicated that the necessary conductivity is not as high as was once thought to be the case.

To summarise: the technique is very much in its infancy as a monitoring method and must be used with caution until proven in specific applications.

Corrosion Potential Measurement

The application of this method of corrosion monitoring demands some knowledge of the electrochemistry of the material of construction in the corrodent. Further, it is only applicable in ionised electrolytes.

The nature of the reference electrode used depends largely on the accuracy required of the potential measurement. In the case of breakdown of passivity of stainless steels the absolute value of potential is of little interest. The requirement is to detect a change of at least 200 mV as the steel changes from the passive to the active state. In this case a wire reference electrode, e.g.

silver if there are chloride ions in solution to give a crude silver/silver chloride electrode, may well be sufficient. Alternatively, the redox potential of the solution may be steady enough to be used as a reference potential by inserting a platinum wire in the solution as the wire electrode.

However, in the case of stress-corrosion cracking of mild steel in some solutions, the potential band within which cracking occurs can be very narrow and an accurately known reference potential is required. A reference half cell of the calomel or mercury/mercurous sulphate type is therefore used with a liquid/liquid junction to separate the half-cell support electrolyte from the process fluid. The connections from the plant equipment and reference electrode are made to an impedance converter which ensures that only tiny currents flow in the circuit, thus causing the minimum polarisation of the reference electrode. The signal is then amplified and displayed on a digital voltmeter or recorder.

Corrosion potential measurement is increasing in use as a plant monitoring device. It has the very big advantage that the plant itself is monitored rather than any introduced material. Some examples of its uses are:

1. To protect stainless-steel equipment from chloride stress-corrosion cracking by triggering an anodic protection system when the measured potential falls to a value close to that known to correspond to stress-corroding conditions.
2. To trigger off an anodic protection system for stainless-steel coolers cooling hot concentrated sulphuric acid when the potential moves towards that of active corrosion.
3. To prompt inhibitor addition to a gas scrubbing system solution prone to cause stress-corrosion cracking of carbon steel when the potential moves towards a value at which stress-corrosion cracking is known to occur.
4. To prompt remedial action when stainless-steel agitators in a phosphoric-acid-plant reactor show a potential shift towards a value associated with active corrosion due to an increase in corrosive impurities in the phosphate rock.

It can be seen that in each case considerable knowledge is required before the potential values associated with the equipment can be interpreted.

Monitor Retractability

Corrosion coupons require periodic weighting, resistance-probe elements require renewing and reference electrodes develop faults. Since the emphasis is on monitoring plants which remain on-line for long periods, careful consideration has to be given to how the monitor is going to be serviced. Systems are now marketed which enable such servicing to be carried out with the plant on-line and these do not rely on the monitor being installed in a by-pass line or in line with a duplicated piece of equipment such as a pump, which may not always be in use. Figure 10.12 shows a system based on a tool used for under-pressure break-in to operating plant.

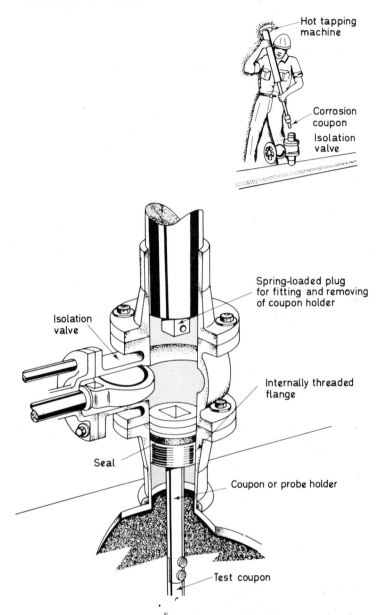

Fig. 10.12   System based on a tool used for under-pressure break-in to operating plant

Phase Five—Remedial Measures

Figure 10.13 summarises the tools which are at the corrosion engineer's disposal in the solution of a corrosion problem once it has appeared. The solution adopted will frequently be a combination of these, and economics and convenience will determine the course adopted if there is an option.

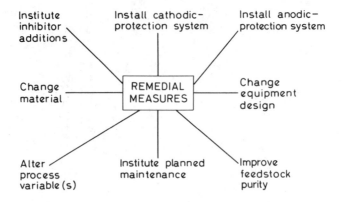

Fig. 10.13   Options for remedying corrosion problems in process plant

Summary

Corrosion control in chemical plant is a continuous effort from the inception of the design to the closure of the plant. Economics dictate the risks which are taken at the design stage with respect to corrosion and the extent of the precautions taken to prevent it.

Errors in design and changes in operation will occur which increase the risk of corrosion. Corrosion-monitoring systems give advance warning and enable remedial measures to be worked out and adopted.

D. FYFE

REFERENCES

1. American Petroleum Institute Publication No. 941, 1st edn, July (1970)
2. Collins, J. A. and Monack, M. L., *Mats. and Perf.*, **12**, 11, June (1973)
3. Bates, J. F., *Ind. and Eng. Chem.*, **55**, 18, Feb. (1966)
4. Loginow, A. W. and Phelps, E. H., *Corrosion*, **18**, 299 (1962)
5. Atkins, K., Fyfe, D. and Rankin, J. D., *Safety in Air and Ammonia*, Conf. Proc., Chem. Eng. Prog. Pub., Vancouver (1973)
6. Loginow, A. W., Bates, J. F. and Mathay, W. L., *Mats. Perf.*, **11**, 35, May (1972)
7. *Corrosion Resistance of Hastelloy Alloys*, Stellite Div., Cabot Corp.
8. *Design of Chemical Plant in Wiggin Nickel Alloys*, and *Wiggin Corrosion Resisting Alloys*, Henry Wiggin & Co. Ltd.
9. *Corrosion Resistance of Titanium*, New Metals Div., Imperial Metal Industries
10. *Uranus Stainless Steels for Severe Corrosion Conditions*, CAFL, France
11. Rabald, E., *Corrosion Guide*, Elsevier, Amsterdam, 2nd edn (1968)
12. Polar, J. P., *A Guide to Corrosion Resistance*, Climax Molybdenum Co.
13. *Corrosion Data Survey*, NACE Pub., Houston, USA, 2nd edn (1971)
14. Ross, T. K., *British Corrosion J.*, **2**, 131 (1967)
15. Hix, H. B., *Mats. Prot. and Perf.*, **11**, 28, Dec. (1972)
16. Clark, W. D. and Sutton, L. J., *Weld. and Met. Fab.*, **21**, Jan. (1974)
17. Charlton, J. S., Heslop, J. A. and Johnson, P., *Phys. in Tech.*, to be published
18. Rowlands, J. C. and Bentley, M. N., *British Corrosion J.*, 7 No. 1, 42 (1972)

BIBLIOGRAPHY

Henthorne, M., *Corrosion and the Process Plant*, collection of papers from *Chem. Eng.* (1971/ 1972)

10.3 Design for Prevention of Corrosion in Buildings and Structures

The prevention of corrosion must begin at the design stage, and full advantage should be taken of the range of protective coatings and corrosion-resistant materials available. Furthermore, at this stage attention should be paid to the avoidance of geometrical details that may promote or interfere with the application of protective coatings and their subsequent maintenance. Consideration should also be given to the materials to be used, the methods of protection, fabrication and assembly, and the conditions of service.

The corrosion of metal components in buildings may have important and far-reaching effects, since:

1. The structural soundness of the component may be affected.
2. Where the component is wholly or partly embedded in other building materials, the growth of corrosion products on the face of the metal may cause distortion or cracking of these materials; trouble may also arise when the metal is in contact with, although not embedded in, other building materials.
3. Failure of the component may lead to entry of water into the building.
4. Unsightly surfaces may be produced.
5. Stresses produced in the metal during manufacture or application may lead to stress-corrosion cracking.

The Corrosive Environment

The conditions to which a metal may be exposed can vary widely[1], but broadly the following types of exposure may arise.

Exposure to external atmospheres   The rate of corrosion will depend mainly on the type of metal or alloy, rainfall, temperature, degree of atmospheric pollution, and the angle and extent of exposure to the prevailing wind and rain.

Exposure to internal atmospheres   Internal atmospheres in buildings can vary; exposure in the occasionally hot, steamy atmosphere of a kitchen or bathroom is more severe than in other rooms. Condensation may occur in roof spaces or cavity walls. One particularly corrosive atmosphere created within a building and having its effect on flue terminals is that of the flue gases and smoke from the combustion of various types of fuel.

Embedment in, or contact with, various building materials Metal components may be embedded in various building mortars, plasters, concrete or floor compositions, or else may be in contact with these. Similarly, they may be in contact with materials such as other metals, wood, etc.

Contact with water or with water containing dissolved acids, alkalis or salts
Many details in building construction may permit rain water to enter and this may be retained in crevices in metal surfaces, or between a metallic and some other surface. Water may drip on to metal surfaces. These conditions, which can involve a greater risk of corrosion than exists where a metal is exposed to the normal action of the weather, are more severe when the water contains dissolved acids, alkalis or salts derived from the atmosphere or from materials with which the water comes into contact. Normal supply waters can also cause corrosion.

Contact between dissimilar metals Galvanic action can occur between two different bare metals in contact if moisture is present, causing preferential corrosion of one of them (see Section 1.7).

It is thus important to consider all types of exposure. If a building is to be durable and of good appearance, special attention must be paid to the design of details, especially those involving metals, and precautions must be taken against corrosion, since failure which is not due to general exposure to the external atmosphere often occurs in components within or structurally part of the building.

Ferrous Metals

Faulty geometrical design is a major factor in the corrosion of ferrous metals. A design may be sound from the structural and aesthetic points of view, but if it incorporates features that tend to promote corrosion, then unnecessary maintenance costs will have to be met throughout the life of the article, or early failure may occur.

Some of the more important points that should be observed are noted below[2]. Where these cannot be implemented, extra protection should be provided.

Air

1. Features should be arranged so that moisture and dirt are not trapped. Where this is not practicable consideration should be given to the provision of drainage holes of sufficient diameter, located so that all moisture is drained away (Fig. 10.14).
2. Crevices should be avoided. They allow moisture and dirt to collect with a resultant increase in corrosion. If crevices either cannot be avoided, or are present on an existing structure, they can often be filled by welding or by using a filler or mastic.

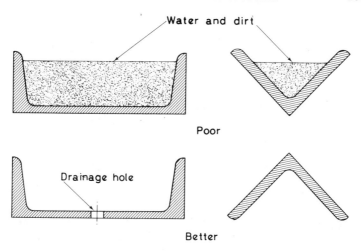

Fig. 10.14 Channels and angles

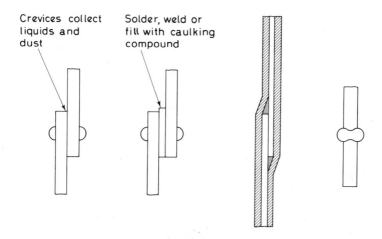

Fig. 10.15 Welded and riveted joints

3. Joints and fastening should be arranged to give clean uninterrupted lines. Welds are generally preferable to bolted joints, and butt welds to lap welds. If lap joints have to be used, then appropriate welding or filling may be necessary to avoid the entrapment of moisture and dirt (Fig. 10.15).

4. Condensation should be reduced by allowing free circulation of air, or by air-conditioning. Storage tanks should be raised from the ground to allow air circulation and access for maintenance and provision should be made for complete drainage (Fig. 10.16).

5. All members should either be placed so that access is provided for maintenance, or so thoroughly protected that no maintenance will be required for the life of the equipment or structure.

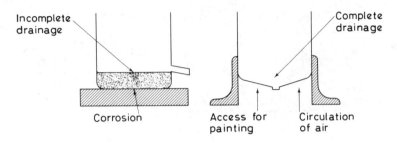

Fig. 10.16   Storage tanks

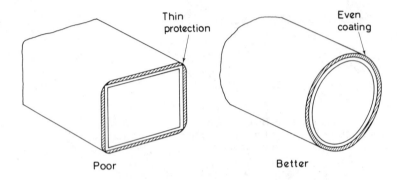

Fig. 10.17   Contour used in construction

6. Where practicable, rounded contours and corners are preferable to angles, which are subject to mechanical damage at edges and are difficult to coat evenly[3]. Tubular or rolled hollow sections could often advantageously replace '*I*' or '*H*'-sections (Fig. 10.17).

7. Corrosion is often particularly pronounced on sheltered surfaces where the evaporation of moisture is retarded. Design features of this kind should either be avoided or additional protection provided.

8. Steel should not be exposed to contact with water-absorbent materials and care must be exercised when using steel in contact with wood. Not only is wood absorbent, but the vapours from it may be corrosive in enclosed spaces.

9. Large box-section girders can be enclosed by welding-in bulkheads near the ends; the welds must not have gaps or condensation may occur within the box section.

10. Features that allow moisture to drip on to other parts of a structure should be avoided, and in this connection particular attention should be paid to the siting of drainage holes.

11. Where steel members protrude from concrete, or in similar situations, attention should be given to the position of abutment. This should be arranged so that water drains away from the steel.

12. When surfaces are being bolted, the holes should coincide and bolts

must not be forced into undersize holes, since the resulting stresses may result in stress-corrosion failure.

13. The corrosion of painted mild-steel window frames is often troublesome, especially on horizontal members where moisture tends to collect. This effect can be reduced by bevelling the edges with putty before painting, to help drainage. It is preferable, however, to use more resistant materials such as galvanised steel, stainless steel or aluminium for window frames.

Materials of Construction

There are three broad categories of steel[4]:

1. Mild steels to which only small amounts of alloying elements are deliberately added, e.g. manganese.
2. Low-alloy steels to which 1–2% of alloying elements are added.
3. Highly alloyed steels, such as stainless steels, which contain 12–20% Cr and sometimes up to 10% Ni and 3% Mo.

Mild Steels

Most steels fall into this category, ranging from large structural sections to thin sheet, and minor variations in composition do not markedly affect their corrosion resistance. This is not generally important since such steels are usually protected by some form of coating which is specific for the condition of service.

Account should be taken of this fact when planning to use this type of steel, so that the coating can be applied at the stage at which the maximum benefits will result. At the same time, thought should also be given to the type of coating and its method of application, since one geometrical design may be more suitable for one particular type of coating application than another.

The more automatic a coating application, the more economical and efficient it is, since automation lends itself more readily to even coatings than do manual methods, e.g. large surface areas lend themselves more readily to spraying techniques, whereas open work structures are more suitable for dipping methods. The coating should also be applied to a specified minimum thickness which is adequate for the service conditions and life envisaged.

Surface preparation is of prime importance, and optimum performance of modern protection coatings can be achieved only if the surface of the steel has been adequately treated. The method of surface preparation depends on the shape and size of the structure or component. Thus it is preferable to blast-clean an openwork steel structure by manual methods, since with this type of structure automatic blast cleaning would lead to excessive impingement of the abrasive on the machine itself.

Steel, whether in structural form or as a sheet, can be protected by many different coating systems, such as paint, plastic materials, concrete and other metals, either singly or in combination (such as a metal coating followed by a paint system, or a plastic coating). Examples of this compound type of protection are the new Forth road bridge and the Severn bridge, both of which are protected by sprayed metal plus a paint system. (See also Section 13.4.)

Low-alloy Steels

Low-alloy steels usually contain small percentages of alloying elements, such as copper, chromium and nickel, up to a total of 1–2%. Under favourable conditions they tend to corrode less rapidly than the ordinary carbon steels when exposed freely in air. Under sheltered conditions, or in crevices, they may well corrode at the same rate as mild steels[5].

When such steels are exposed bare, the initial appearance of the rust is similar to that on mild steel, although in time it tends to become darker, more compact, and of a more even texture than ordinary rusts.

When low-alloy steels are considered for use in the bare condition, the remarks made earlier with regard to design must be given even greater attention, particularly in relation to crevices and sheltered areas. Also, the appearance and performance must be acceptable for the specific application, and care should be taken to ensure that adjacent concrete and stonework is not stained brown in the early stages by moisture dripping from the rusted steel. This can be accomplished by the following: attention to design, the careful siting of the rainwater drainage system, the use of loose gravel that can be raked over, painting the concrete or in several other ways.

Low-alloy steels can be obtained as structural sections or in sheet form, and must be blast-cleaned to remove the millscale before exposure. Such material has been widely used in North America for highway bridges and for architectural purposes, and also to some extent in the UK and Europe.

After suitable surface preparation, e.g. blast cleaning, low-alloy steels can be coated by paints, sprayed metal coatings, etc. and there is some evidence that such coatings last longer than on mild steel under similar conditions of exposure[6].

Stainless Steels

There are many grades of stainless steel, and some are virtually non-corrodible under ordinary atmospheric conditions. Their resistance results from the protective and normally self-repairing oxide film formed on the surface. However, under reducing conditions, or under conditions that prevent the access of oxygen, this film is not repaired, with consequent corrosion.

Since stainless steels are generally unprotected, the design points discussed earlier are particularly applicable to them, and features such as crevices should be avoided.

It is recommended that advice be sought when choosing the type of stainless steel to be used. Under severe conditions it may be necessary to use an Fe–18Cr–10Ni–3Mo type, but under milder conditions a much lower grade such as an Fe–13Cr steel may be satisfactory. The following points should be considered:

1. The environment to which the steel will be exposed.

2. Types and concentration of solutions that may be in contact with the steel. This is particularly important where the failure may be due to local concentrations of dilute solutions. For example, the small chloride content of tap waters is unlikely to cause any trouble, but if it concentrates at the water level due to heating and evaporation of the water, then attack may occur.
3. Operating temperatures and pressures.
4. Mechanical properties required.
5. Work to be performed on the steel.
6. Fabrication and welding techniques to be used. In connection with welding it should be emphasised that the correct grade of steel and electrode or filler rod must be used.

Stainless steel is not generally made in the large sizes offered in the cheaper steels, but a range of sections, tubes, flats, rods and sheet is obtainable. Some savings in thickness and weight are possible, however, because of its superior corrosion resistance. If the strength requirements go beyond the point where the use of stainless steel becomes economic, it is possible to use clad material. Stainless steel is often used as cladding and for window frames, doors, etc. for prestige buildings.

Coated Steel Sheet

Probably the most familiar coated steel sheets are the ubiquitous galvanised corrugated roofing and cladding sheets which have been used for many years, particularly for farm buildings, either painted or unpainted.

Nowadays, however, zinc-coated steel sheets, either continuously galvanised or electroplated, are often used as a basis material for overcoating with plastic materials or paints. The coatings are usually applied continuously and have a range of uses both externally and internally. Many surface finishes are obtainable, e. g. plain or embossed, and in an extensive range of colours, to suit almost any requirement[7].

Some of the uses of such precoated materials are roofing, cladding, decking, partitions, domestic and industrial appliances, and furniture. The formability of these materials is excellent and joining presents no problems. The thickness of the coating varies according to the material used and the service conditions which the end product has to withstand.

Vast amounts of continuously galvanised steel sheets are produced, and unless they are painted or otherwise coated, their life depends on the thickness of the galvanising and the service environment in which they are used (see Section 14.4). A problem often associated with such material is corrosion at the cut edges. From work carried out by BISRA and others[8] it has been shown that providing the bare steel edge is less than 3 mm in width, the amount of corrosion is minimal and the life of the sheet is not adversely affected, although rust staining will occur. If staining is important because of the appearance, the cut edge should be orientated so that the stain does not run over the sheet. The edge could, alternatively, be beaded over or painted with a suitable painting scheme. If appearance is important it may be advantageous to paint overall.

Protective Coatings

Zinc, aluminium and other materials such as paints and plastic coatings are often used as protective coatings for steel. These metals act not only as a barrier, but where breaks occur in the coating, corrode preferentially under most conditions and thus sacrificially protect the underlying steel. Aluminium is normally less negative than zinc, but provides adequate sacrificial protection in industrial and marine environments. With both metals, the life expectancy depends on the coating weight, which is generally synonymous with thickness. The thickest coatings are produced by dipping or by spraying, thinner coatings by diffusion and in the case of zinc by electrodeposition (see Chapter 13).

The metal spraying operation using zinc or aluminium as a protective coating is usually followed by a painting scheme. The choice of sprayed metal and paint scheme depends on the service conditions[9], but normally this type of system is used on prestige buildings or structures, where longevity is of prime importance and maintenance requirements need to be kept to a minimum.

Paint is the most widely used protective coating for steelwork and normally acts as a barrier between the metal and environment. The choice of type of paint and the final thickness required depends on the conditions of service, and the more severe the conditions the thicker and more resistant the paint film needs to be. Also the more sophisticated the paint system the more demanding is the surface preparation required.

Often steelwork will initially be painted before final fabrication, and problems that may arise when maintenance painting becomes necessary may not be fully appreciated[10] (see Fig. 10.18). The 'I'-beam can be painted in the shop, but access for maintenance may be inadequate and either the distance t should be increased or the gap closed so that maintenance is not required.

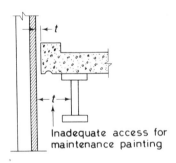

Fig. 10.18   Access for maintenance

An actual case of failure occurred where the rolled steel joists carrying the floor of a refrigeration chamber were placed so close together that they could not be reached for painting[11]. Heavy condensation led to dangerous rusting on the inner surfaces of the joints and in consequence the steelwork had to be replaced prematurely (Fig. 10.19).

Designers should always bear in mind the necessity to inspect and maintain all parts of a structure that may be corroded and should provide adequate access for these purposes.

The choice of protective system will be determined by many factors such as the importance of the structure, the environment and its proposed life.

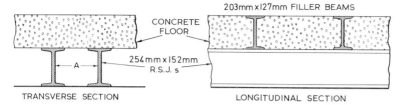

Fig. 10.19   Design of reinforced concrete floor. For the old joists *A* was 7 in (178 mm) leaving a 1 in (25 mm) gap between the toes; for the new joists *A* was increased to 18 in (457 mm)

Having chosen a suitable system or systems[12], it is essential that requirements including adequate inspection are specified exactly, and that there is the fullest possible collaboration between the paint suppliers, the contractors, the architects and all other parties concerned. It is not always appreciated that coatings do not have an inherent 'life' and that much depends on other factors such as surface preparation, and application of the protective system. Simplicity in design, however, will always assist in achieving optimum performance from protective coatings.

Bimetallic Corrosion

When other metals are used in conjunction with steel, careful consideration must be given to the possibilities of galvanic attack (Section 1.7).

The rate of corrosion and damage caused to the more negative metal will depend upon the relative sizes of the anodic (corroding metal) and cathodic areas. A small anode and a large cathode will result in intensive corrosion of the anodic area. On the other hand, if the anode is large compared with the cathode, the corrosion of the anodic area will be more general and less likely to result in rapid failure. For example, a steel rivet in a copper plate will be rapidly attacked in sea-water, whereas a copper rivet in a steel plate may lead only to slightly accelerated corrosion of the steel in the area adjacent to the rivet. Prediction of the rate of corrosion of the less noble metal in a galvanic cell is difficult, but there is always the possibility of serious trouble if two dissimilar metals are in contact, particularly under immersed conditions. The safest way of avoiding this is to ensure that dissimilar metals are not in contact. If this is impracticable, the following will help to reduce or stop attack on steel:

1. Use more noble metals for fastening.
2. Insulate the metals from each other by suitable gaskets, washers, etc.
3. Paint the surfaces of both metals. Avoid painting only the less noble metal because if the coating is damaged severe attack may result at the damaged area.
4. Prevent moisture dripping from the more noble metal on to the less noble metal.

From the reversible potential of zinc, accelerated corrosion would be expected to occur when zinc is coupled with many other metals commonly used in buildings. Aluminium, contrary to its reversible potential, is generally found to be slightly cathodic to zinc and is protected when the two metals are coupled together, as when aluminium sheet is fixed with galvanised nails. In practice, although some small acceleration in the corrosion rate of zinc will be expected in the immediate area of contact with another metal, the effect is usually severe only when it is in contact with copper. For example, where zinc and aluminium gutters or zinc and cast-iron gutters are fitted together, very little accelerated corrosion of the zinc is normally found. Brass[13], with its 30–40% zinc, is very much less active than copper, and brass screws and washers can be used for fixing zinc with little or no accelerated corrosion troubles; but with copper, rapid failure occurs.

Drainage water from copper affects zinc in a similar way. Zinc sheets must never be fixed with copper nails, nor should copper roofs drain into zinc or galvanised gutters. Copper lightning arrestors provide further potential hazards to zinc work; when a copper lightning strip has to pass over or near a zinc roof, it should be either well insulated or heavily tinned.

Non-ferrous Metals and Plastics

For some purposes where the strength and ductility of steel are not prerequisites, other metals or materials may be used to advantage, particularly when the component or article is not a load-bearing one. Some of the non-ferrous metals and plastics materials are extremely useful in this respect, especially the latter with their excellent corrosion-resistant properties and ease of formability.

Copper[14]   Although copper is weather-resistant under normal conditions of exposure, certain precautions are necessary to avoid the risk of premature failure. For instance, copper that is exposed to high concentrations of flue gases, as may happen within a metre or so of chimney exits, may become corroded within a relatively short time. To avoid this the chimney should be built to a reasonable height above the roof. For similar reasons ventilators should not be made of copper where highly sulphurous fumes may be encountered. The use of potentially corrosive materials as underlays for copper roofing may also result in failure.

Bare copper exposed indoors will slowly tarnish. Transparent lacquers may be used, however, to retain a bright surface without the need for frequent cleaning. Neither copper nor any copper alloy will remain bright and polished without maintenance or coating.

Lead[1]   Corrosion of lead gutters and weatherings is usually associated with slate roofs on which vegetable growth such as algae, moss or lichen is present. These produce organic acids and carbon dioxide which significantly increase the acidity of rain water running over the roof. New cedar-wood shingles also contain acids which are slowly washed out by rain, thus intensifying the attack that would in any case slowly occur owing to vegetable growth on the roof. Probably the simplest way of avoiding this type of failure is to protect the lead with a thick coating of bituminous preparation extending well underneath the edge of the roof.

Lead is relatively easily corroded where acetic acid fumes are present and under such conditions it either should not be used or should be efficiently protected. Generally, any contact between lead and organic material containing or developing acids will cause corrosion; for instance, unseasoned wood may be detrimental. Trouble from this cause may be prevented by using well-seasoned timber, by maintaining dry conditions, or by separating the lead from the timber by bitumen felt or paint. Lead is also subject to attack by lime and particularly by Portland cement, mortar and concrete, but can be protected by a heavy coat of bitumen. A lead damp-proof course laid without protection in the mortar joint of a brick wall may become severely corroded, especially where the brickwork is in an exposed condition and is excessively damp.

Aluminium   The resistance of aluminium and certain of its alloys to atmospheric corrosion is fairly high. Nevertheless, corrosion does occur, especially on under-surfaces, e.g. of bus shelters. Normally, in simple exposure the corrosion reaction stifles itself and the rate falls to a low value. With a few alloys, however, atmospheric corrosion may lead to severe attack, and layer corrosion may occur. It is important therefore to pay attention to materials, design and protection. Intermetallic contacts, crevice conditions, horizontal surfaces, etc. should be avoided. Materials for sections, plate and sheet, where strength is important, should be restricted to primary alloys. All neat-treated alloys should be painted, using first a chromate priming paint containing not less than 20% zinc chromate pigment, or an equivalent chromate paint. Crevices should be packed with a suitable composition such as chromate jointing compound or impregnated tape. By paying attention to these points, aluminium should behave satisfactorily.

Zinc   Zinc surfaces corrode more slowly in the country than in either marine atmospheres or in industrial areas where sulphur pollution constitutes the main danger both to them and to many other building materials. Sulphur and its compounds in the air can become oxidised to sulphuric acid; this forms soluble zinc sulphate, which is washed away by rain[13].

The purity of the zinc is unimportant, within wide limits, in determining its life, which is roughly proportional to thickness under any given set of exposure conditions. In the more heavily polluted industrial areas the best results are obtained if zinc is protected by painting, and nowadays there are many suitable primers and painting schemes which can be used to give an extremely useful and long service life under atmospheric corrosion conditions. Primers in common use are calcium plumbate, metallic lead, zinc phosphate and etch primers based on polyvinyl butyral. The latter have proved particularly useful in marine environments, especially under zinc chromate primers[15].

Zinc has been used extensively as a roofing material, but its life, especially in industrial areas, is somewhat dependent on the pitch or slope of the roof, those of steep pitch draining and drying more rapidly and therefore lasting longer. Irrespective of the locality, exterior zinc-work may fail prematurely if the design is unsuitable or the installation faulty. Many failures arise from a combination of purely mechanical reasons and secondary corrosion effects, white-rusting being the most important of these. This tendency can, however, be reduced by chromating. When used inside factories[16] zinc coatings have been found to be satisfactory in withstanding attack by many industrial gases and fumes. The protection of fabricated structural steelwork by hot-dip galvanising as is used in current constructions, has allowed lightweight concrete-clad steel sections to be used with complete safety.

Zinc in contact with wood   Zinc is not generally affected by contact with seasoned wood, but oak and, more particularly, western red cedar can prove corrosive, and waters from these timbers should not drain onto zinc surfaces. Exudations from knots in unseasoned soft woods can also affect zinc while the timber is drying out. When a timber preservative is specified the choice should preferably be made from among the preparations containing zinc and copper naphthenates which are not water-soluble and are inert to zinc. Zinc chloride treatment is unsuitable for preserving timber in contact with zinc or zinc coatings. Preservatives based on chromates and arsenates are unlikely to cause trouble since these salts quickly become insoluble and, in the case of chromate, help to protect the zinc from corrosion. The majority of proprietary preservatives of the tar or creosote type do not attack zinc.

Zinc-alloy diecastings used indoors   Zinc-alloy diecast fittings have good corrosion resistance. Generally, such castings may be used in buildings without further protection by painting, but it is of advantage, especially where conditions of permanent dampness may occur, that they should be chromate-treated or phosphated, and then enamelled, or coated with an etch primer and painted after installation. Where a chromium-plated finish is used, it is important that an adequate basis of copper and nickel plating is provided.

Soluble sulphates and chlorides in brickwork, plaster and other walling materials provide a more serious source of corrosion under damp conditions. Under such circumstances, or where zinc or zinc-alloy fittings are to be placed in contact with breeze, concrete or black mortar (made from ground ashes) the metal may be protected with two coats of hard-drying bitumen paint.

Zinc as a protective coating to building components   Perhaps the most important use of zinc in building is as a protective coating to steel. In spite of the initial cost, a substantial coating of zinc (or of aluminium) is of great value, and often saves the cost of remedying troubles caused by corrosion. For general purposes it can be accepted that the effectiveness of the coating depends on the weight of zinc coat applied and not on the method of application.

Metals in Contact with Concrete

Little information is available about the corrosion of metals in concrete, although it seems likely that all Portland cements, slag cement and high-alumina cement behave similarly[17]. Concrete provides an alkaline environment and, under damp conditions, the metals behave generally as would be expected; e.g. zinc, aluminium and lead will react, copper is unaffected, while iron is passivated by concrete.

Aluminium reacts vigorously with a wet, freshly prepared concrete mix and the reaction, in which hydrogen is evolved, has been used for preparing lightweight cellular concrete. When the concrete has set, however, its reactivity is much reduced, and aluminium placed in contact with it is not usually severely attacked. Aluminium embedded in concrete normally suffers only mild superficial corrosion during the setting period and the corrosion is generally so slight that even thin aluminium conduit does not require a protective coating, unless it is expected that the concrete might remain damp[18].

Zinc is not so severely attacked as is aluminium, and its use has frequently been suggested as an extra protection for steel from corrosion in concrete. Galvanised metals are attacked, but usually not severely, protection being regarded as advisable rather than essential. Complete destruction of zinc conduits in contact with wet concrete has sometimes occurred. Although galvanised steel reinforcement does not cause spalling of concrete if the zinc is attacked, solid zinc plates or rods may have this effect on account of the greater bulk of the corrosion products. For general protection, a coating of bitumen is probably desirable for zinc sheets or other metal thinly coated with zinc when embedded in concrete.

The reaction of lead with concrete differs from that of aluminium and of zinc in that it is not normally rapid during the early wet stage. It is, however, progressive in damp conditions, and this is said to be due to the fact that the concrete prevents the formation of a protective basic lead carbonate film on the surface of the lead. The packing of lead cables in plaster of Paris is reported to be of doubtful value in preventing corrosion from surrounding concrete.

Little information is available on the performance of copper and of copper alloys in contact with concrete, but concrete sometimes contains ammonia, even traces of which will induce stress-corrosion cracking of copper pipe. The ammonia may be derived from nitrogenous foaming agents used for producing lightweight insulating concrete.

The corrosion behaviour of iron and steel in contact with concrete is of great importance, not only because of the amount of metal involved, but also because the metal is frequently load-bearing, and the stability and durability of a structure may depend upon the control of corrosion. The corrosion process is essentially the same whether it affects embedded structurally important steelwork or unstressed steel attached to the surface of the concrete. Therefore the normal methods for protecting steel from rusting can be used successfully. The alkaline reaction of the adjacent concrete may, however, damage sensitive paints and protective finishes. The corrosion of steel reinforcements in concrete is discussed below.

Effects of Composition of Concrete

Steel can corrode badly within a few years when buried in damp concrete and the increase in volume on rusting can cause concrete to become fissured. The risk of attack, however, decreases as the thickness of the concrete is increased, and as the ratio of cement to aggregate is increased. The protection provided by Portland and by slag cements is similar, the environment provided by the slag-cement concrete being less alkaline but probably more stable.

There is no objection to the use of slag aggregates for reinforced concrete provided the slag meets the various sulphur-content specifications, and similar considerations apply to lightweight aggregates, although it has been claimed[19] that the sulphur content of blast-furnace slag is not dangerous.

Clinker aggregates, on the other hand, are not permitted in the UK because they cause corrosion of reinforcement. The corrosiveness of clinker and boiler slag is due probably to the high sulphur content.

Since concrete itself is a powerful inhibitor of corrosion, any additional procedure purporting to prevent corrosion must reinforce the protection already provided. Lime water has been used instead of plain water for mixing concrete, and it is claimed that in concrete so made the presence of 2–7% calcium chloride does not give rise to corrosion of steel, although in similar concrete made with plain water 2% calcium chloride can sometimes cause corrosion. It is similarly claimed that sodium silicate prevents corrosion when calcium chloride is present in reinforced concrete.

The Corrosion of Steel Reinforcements in Concrete

Normal Reinforcement

Large quantities of mild steel are used as an integral part of reinforced concrete structures. The condition of steel in many old structures suggests that corrosion is not normally progressive and that when steel is in good contact with concrete of reasonably low permeability, and remains in contact with it, corrosion is negligible. This is so even when calcium chloride has been used to accelerate the hardening of the concrete. Sometimes, however, corrosion of reinforcements has been comparatively rapid and progressive, necessitating repair or replacement of reinforced concrete members (Fig. 10.20). The customary criterion of concrete quality is compressive strength, but for the protection of reinforcement it is probably more important that the concrete should be sufficiently plastic to envelop the steel in a continuous film, even if this involves using more cement than would be necessary merely to provide a given strength. High strength is desirable, nevertheless, since it can prevent spalling of concrete by the corrosion products of steel, which, it has been estimated, can exert a pressure exceeding $15 \, MN/m^2$. The cohesion

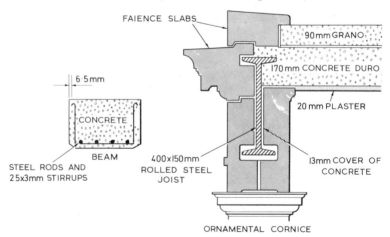

Fig. 10.20   Faulty designs in reinforced concrete. In both examples the concrete cover is too thin to exclude water, which permeates through it and causes corrosion of the steel

of steel and concrete is usually good but may be weakened by differential thermal expansion. If extra protection of steel is considered essential, paints have some value, but are rarely long-lived. Bitumen paints protect reinforcement from calcium chloride attack, but some paints of this type cause lack of bond between the steel and the concrete.

Nowadays, it is quite common to protect reinforcing bars, especially for use in concrete which is to be exposed in an aggressive atmosphere, by galvanising[20]. There are several advantages to be gained, not the least of which is that galvanised reinforcing bars do not need derusting. This is an expensive operation and is rarely carried out properly, and if the rust is heavily contaminated can lead to spalling of the concrete if the cover is insufficient and/or the concrete mix is of poor quality or the wrong materials have been used. The long-term performance of steel reinforcing bars in concrete immersed in sea-water and at the splash zone has been investigated by Browne and Domone[20(a)].

Prestressed Reinforcement

For prestressed concrete, either high-tensile steel wires or occasionally bars of steel alloy containing manganese and silicon, are normally used. Galvanised wires are also used for prestressed concrete, and it is recommended that they be chromated before use.

In a normal reinforced concrete structure, the tensile stress in the steel is comparatively low, but in prestressed concrete the steel is held permanently in tension with a stress equivalent to about 65% of its breaking load. It is necessary, therefore, in prestressed concrete, to consider the possibility of the occurrence of stress corrosion. Furthermore, surface rusting or corrosion of prestressed wires will affect the working cross-sectional area of the reinforcement, and pitting, which might be unimportant on a 12 mm bar, could cause failure on a 2·5 mm diameter wire. The number of reported failures of prestressed concrete due to fracture of reinforcement is very low and in general the behaviour of steel in prestressed concrete is no different from that of steel in ordinary reinforced concrete. Prestressed concrete is made from materials of slender section using higher working stresses than are customary for ordinary reinforced concrete.

Work carried out at the Building Research Station[21] suggests that the most significant influences on the corrosion of prestressed steel wire in concrete are: (1) the presence of chloride, (2) the composition of the concrete, (3) the degree of carbonation of the concrete, (4) the compaction of the concrete around the wire ensuring that voids are absent, and (5) chloride promoting pitting attack, leading to plastic fracture and not stress-corrosion cracking of prestressing steel. Gilchrist[21(a)] considers that prestressing steels may fail by either hydrogen cracking or active path corrosion, depending on conditions; most service failure is due to hydrogen cracking.

Prestressed steel in concrete should thus be durable if a dense, impervious and uniform concrete free of chloride surrounds the steel and adequate depth of concrete is given to the steel.

Materials in Water-supply Systems

The most important non-ferrous metals for handling water are lead, copper and zinc; the last, however, is used chiefly as a protective coating on steel or alloyed with copper to form brass.

The choice of materials for most applications in domestic water supply is governed by consideration of mechanical properties and resistance to corrosion, but the cost, appearance and ease of installation should also be considered when the final choice has to be made between otherwise equally suitable materials[22].

Many plastic materials are also now being used in domestic water systems, in the form of pipes and fitments.

Features that should be avoided for all materials (particularly ferrous metals) in liquid environments and points that should be followed are[23]:

1. Crevices, because they collect deposits and may promote corrosion by causing oxygen depletion in the crevice, thereby setting up a corrosion cell in which the areas receiving less oxygen corrode.

2. Sharp changes in direction, especially where liquids are moving at high velocities. Also to be avoided are re-entrant angles, dead spaces and other details where stagnant conditions may result. This is particularly important if inhibitors are to be used.

3. Baffles and stiffeners inside tanks should be arranged to allow free drainage to the bottom of the vessel. The bottom should slope downwards and have rounded corners. Any drain valves or plugs should fit flush with the bottom (Fig. 10.16).

4. Wherever possible different metals should not be connected in the same system. If they have to be used they should be insulated from each other, and the cathodic metals placed downstream of anodic ones.

Galvanised steel pipes Threaded mild-steel tube is the cheapest material for water pipes, but it is not normally used owing to the amount of rust introduced into the water as a result of corrosion. Galvanised mild-steel tube overcomes this problem and may be used for nearly all hard waters, but it is not satisfactory for soft waters or those having a high free-carbon-dioxide content. The ability of a water to form a 'scale' is, therefore, of prime importance when considering the suitability of galvanised steel for an installation. The Langelier index (Section 2.3) gives useful guidance to this, but it is only an approximation. The scale can be deposited either as nodules covering a relatively small area of metal, or as a thin scale covering a large area. Provided the deposit is not porous, the latter has the greater effect in reducing corrosion, and it has been found that waters derived from rivers tend to form more useful scales than those from wells.

Galvanised steel tubing is cheaper than lead or copper tubing, but is, however, more costly to install because it cannot be bent without damage to the galvanising. A full range of preformed bends, tees, etc. is available, but cutting the tubes to length, threading the ends and screwing up the joints are slow processes. Consequently, a galvanised steel installation is cheap for long straight runs of pipes, but for complicated systems it is liable to be more expensive than copper because of the high cost of installation.

Lead pipes These are supplied in long coils, which are flexible, and can be bent round corners. Soldered joints and tees are made by wiping, with no need for special fittings.

The flexibility of lead pipe facilitates the making of connections in awkward corners, but this flexibility implies low strength and the necessity of using rather heavy-walled pipes in order to confer sufficient strength to resist mechanical damage and bursting. The low strength of lead also renders it liable to failure by creep and fatigue.

The corrosion resistance of lead is generally excellent, but it is attacked by certain waters. This is usually of little significance so far as deterioration of the pipe is concerned, but is important because of danger to health, since lead is a cumulative poison; even very small doses taken over long periods can produce lead poisoning[24].

Copper pipes For plumbing above ground, copper is supplied in the half-hard condition. It has sufficient strength to require only few supports, and can be bent cold, in the small sizes, either by hand or with a portable bending machine. Copper is also supplied in the fully soft condition in coils for laying underground, for heating-panels, etc.

Light-gauge copper tube may be joined by autogenous welding or by bronze welding. These processes, which produce neat strong joints, are usually applied to the larger sizes of tube. For the tubes used for domestic water supply, capillary-soldered fittings, or compression fittings are normally employed.

Two types of corrosion may be experienced. The first is analogous to plumbo-solvency, with the copper being dissolved evenly from the surface of the tube. With some waters[25] it is potentially dangerous to use galvanised hot-water tanks and copper pipes. In domestic systems, premature failure of galvanised hot-water tanks connected to copper circulating pipes, due to pitting corrosion of galvanised steel, is encouraged by more than about 0·1 p.p.m. of copper in the water. Failures of galvanised cold tanks due to copper in the water are often the result of back circulation of hot, copper-bearing water in badly designed systems where the cold-water tank is installed too close to the hot-water cylinder. Hot water carried into the cold tank via the expansion pipes when the water is allowed to boil may also sometimes be responsible.

Secondly, under certain conditions copper may suffer intense localised pitting corrosion, leading sometimes to perforation of the tube, in quite a short time. This form of attack is not common and depends on a combination of unusual circumstances, one of which is the possession by the tube of a fairly, but not entirely, continuous film or scale that is cathodic to the copper pipe in the supply water; this can set up corrosion at the small anodes of bare copper exposed at faults or cracks in the film. Carbon films give rise to such corrosion, but since 1950, when the importance of carbon films was first discovered, manufacturers have taken precautions to avoid as far as possible producing tubes containing them. (See also Sections 1.6 and 4.2.)

Aluminium pipes Aluminium might become an important material for carrying water if its liability to pitting corrosion could be overcome. Very soft waters are difficult to accommodate when normal pipe materials are used, and it is for these that aluminium offers most promise[26]. The possibility of using it for domestic water pipes, however, appears at present to depend upon finding a cheap and effective inhibitor that could be added to the water, or upon the use of internally clad tube, e.g. Al–1·25 Mn alloy clad with a more anodic alloy, such as Al–1Zn. Such pipes are at present mainly used for irrigation purposes[27].

Stainless steels   Thin-walled stainless steel (Fe–18Cr–8Ni) tubes are now frequently used for domestic installations in place of copper pipe[27]. Care is required, however, in the design of stainless steel equipment for use in waters with a high chloride content, or where the concentration can increase, since pitting attack may occur. It may also be susceptible to failure by stress-corrosion cracking under certain conditions.

Plastic pipes   Pipes made from plastic materials such as unplasticised p.v.c., Polythene, ABS and GRP are now widely used for carrying domestic cold water, wastes and rain water. Joining varies according to pipe diameter and service condition, but is generally relatively simple (see Section 19.6).

Buried pipes   Pipes to be laid underground must resist corrosion not only internally but also externally. Light sandy soils, alluvium, or chalk are generally without appreciable action but made-up ground containing a high proportion of cinders is liable to be exceptionally corrosive as also is heavy clay containing sulphates. The latter provides an environment favourable to the growth of sulphate-reducing bacteria, which operate under anaerobic conditions, reducing sulphates in the soil to hydrogen sulphide, and causing severe corrosion especially of steel. Aluminium is believed not to be susceptible to this form of attack, but, like copper or galvanised steel, it is severely attacked by cinders. Lead is sometimes rather less attacked by cinders. Wet salt marsh, although it has little effect on copper and only slightly more on lead, causes severe corrosion of galvanised steel or aluminium. These materials are also severely corroded in London clay, in which lead or copper could probably be used unprotected.

When ferrous metal service pipes or piles, etc. are buried in the ground it is advantageous in almost all cases to coat it in some way even if the coating is just a simple dip into a bituminous solution. If, however, the soil is aggressive, or the component is vital or irreplaceable, a more resistant coating should be used, and consideration should be given to the application of cathodic protection (Section 11.4). The coating used can be an epoxide type or something similar, or a plastic coating which is wrapped or extruded onto the pipe wall.

Galvanised steelwork buried in the soil in the form of service pipes or structural steelwork withstands attack better than bare steel, except when the soil is more alkaline than pH 9·4 or more acid than pH 2·6. Poorly aerated soils are corrosive to zinc, although they do not necessarily cause pitting. However, soils with fair to good aeration containing high concentrations of chlorides and sulphates may do so. Bare iron may be attacked five times more rapidly than zinc in well-aerated soils low in soluble salts, or in poorly aerated soils; and if the soil is alkaline and contains a high proportion of soluble salts the rate may be even higher. Only in soils high in sulphide content does iron corrode less rapidly than zinc.

Plaster and concrete   Domestic water pipes are often used in contact with plaster, concrete or flooring materials[28]. Copper is unaffected by cement, mortars and concrete, which are alkaline in reaction, but it should be protected against contact with magnesium oxychloride flooring or quick-setting materials such as Keenes cement, which are acid in character. Materials containing ammonia may cause cracking at bends or other stressed parts of brass or copper tubes—some latex cements used for fixing rubber flooring come in this category and contact with these should be avoided. Lead is not affected by lime mortar but must be protected from fresh cement mortar and concrete, either by wrapping or by packing round with old mortar or other inert materials. Galvanised coatings are not usually attacked by lime or cement mortars once they have set, but aluminium is liable to be attacked by damp concrete or plaster, even after setting.

Materials for Tanks

Copper hot-water tanks These are usually made cylindrical with domed tops and bottoms, because this form of construction produces a strong tank from light-gauge sheet. They are normally trouble-free except for occasional cases of leakage at the seams, which are usually welted or overlapped and then brazed. Brazing brasses, containing 40–50% zinc, often give good service, but are susceptible to dezincification in some waters. Dezincification, which is most likely to occur in acid waters or waters of high chloride content, can be avoided if cylinders are brazed with an alloy such as Cu–14Ag–5P. It is more expensive but is fairly ductile, and if used in conjunction with capillary-gap seams, makes an economical as well as a sound job.

In water where copper tanks might be subject to pitting corrosion it is good practice to fit an aluminium rod[27] inside the tank. This corrodes sacrificially within the first few months of service, and during this period a protective film is built up on the copper surface.

Galvanised steel hot-water tanks These may be of cylindrical or rectangular form, the latter being popular where space is limited.

In hard or moderately hard waters, galvanised steel hot-water tanks, with galvanised circulating pipes and cast-iron boilers, usually give trouble-free service, but failure by pitting occurs occasionally. Sometimes this is due to extraneous causes, such as rubbish left inside the tank when it is installed. Iron filings left in the bottom of the tank or deposits of inert material are liable to interfere with the formation of the protective scale by the water, and can lead to failure.

Another cause of trouble is overheating, especially during the early life of the tank. Above 70°C a reversal of polarity may take place, the zinc becoming cathodic to the iron. Above this temperature protection of exposed iron is not to be expected. Persistent overheating is frequently the result of fitting a hot-water tank too small for the heating capacity of the boiler. A lowering of temperature by as little as 5–10°C can add years to the life of a tank. It has also been shown[29] that large-capacity immersion heaters operated intermittently are more beneficial to tank life than small-capacity heaters operated continuously.

Magnesium anodes[27] suspended inside a galvanised hot-water tank and in electrical connection with it afford cathodic protection to the zinc, the alloy layer and the steel, at high temperatures as well as in the cold. The magnesium is eventually consumed but it is probable that in the interim a good protective scale will have formed on the inside of the tank, so that the magnesium anode will then no longer be necessary. One of the difficulties of this method, however, is the maintenance of a sufficiently even current distribution over the inside of a tank to protect the whole surface, especially in waters of low conductivity. The method is therefore unlikely to be applicable to soft waters.

Cold-water tanks   Domestic cold-water tanks are usually made of galvanised steel. As with hot tanks, it is important to avoid leaving filings, etc. in the tank when it is installed and it should be covered to prevent rubbish falling in later. In most waters, galvanised cold-water tanks give good service, the zinc coating protecting the iron while a protective scale is formed. With very soft waters, however, or with waters of high free carbon dioxide content, which do not produce a scale, there may be trouble. Steel or galvanised steel tanks for use in such waters can be protected by coating with bituminous paint or, alternatively, reinforced plastics may be used. For larger cold-water storage tanks, sectional steel or cast-iron tanks protected by several coats of cold- or hot-applied bitumen or bituminous paints are often used. It is important, however, to ensure that all millscale, dirt, etc. is removed before applying the protective coating. Stainless steels can also be used for this purpose.

Cold-water tanks made from Polythene or GRP are generally available, especially in domestic sizes, and are now often used in domestic installations.

Water Fittings

Brass water fittings give no trouble except that dezincification may occur in acid waters or waters of high chloride content, especially when hot. This dezincification has three effects. Firstly, the replacement of brass by porous copper may extend right through the wall of the fitting and permit water to seep through. Secondly, the zinc which is dissolved out of the brass may form very voluminous hard corrosion products and eventually block the water-way—this is often the case in hot soft waters. Thirdly, and often the most important, the mechanical properties of the brass may deteriorate. For instance, a dezincified screwed union will break off when an attempt is made to unscrew it and a dezincified tap or ball-valve seat is readily eroded by the water.

In the past, brass water fittings were produced by casting, but they are now usually hot-pressed. This necessitates the use of two-phase brass which, unfortunately, is difficult to protect from dezincification by alloying. The older cast fittings were of single-phase brass which does not dezincify easily, especially if it contains 0·03% arsenic.

Plastic water fittings ranging from taps to lavatory cisterns are now available and are gradually replacing items previously made in metal in the domestic field, especially in situations where condensation is the cause of unsightly corrosion products.

E. E. WHITE
K. O. WATKINS

REFERENCES

1. Jones, F. E., *Chem. and Ind.* (*Rev.*), 1050 (1957)
2. Reinhart, F. M., *Prod. Engng.*, **22**, 158 (1951)
3. Rudolf, H. T., *Corrosion*, **11**, 347t (1955)
4. *B.i.s.r.a.*/BSC Booklet No. 1
5. Chandler, K. A. and Kilcullen, M. B., *Brit. Cor. J.*, **5** No. 1, 24–32 (1970)
6. Robinson, W. L. and Watkins, K. O., *Steel and Coal*, July 13, 65–68 (1962)
7. Watkins, K. O., B.I.S.F. Symposium, 'Developments in Methods of Prevention and Control of Corrosion in Building', Nov. 17th (1966)
8. Anderson, E. A. and Dunbar, S. R., American Zinc Institute Report N314, page 62
9. Watkins, K. O., *Brit. Cor. J.*, **9** No. 4, 204 (1974)
10. Chandler, K. A. and Stanners, J. F., I.S.I. Publication No. 122
11. Hudson, J. C. and Wormwell, F., *Chem. and Ind.* (*Rev.*), 1078 (1957)
12. *Protection of Iron and Steel Structures from Corrosion*, CP2008, British Standards Institution, London (1966)
13. Bailey, R. W. and Rudge, H. C., *Chem. and Ind.* (*Rev.*), 1222 (1957)
14. Baker, S. and Carr, E., *Chem. and Ind.* (*Rev.*), 1332 (1957)
15. Bonner, P. E. and Watkins, K. O., 8th Fatipec Congress, 385–394 (1966)
16. Stanners, J. F., *J. Appl. Chem.*, **10**, 461 (1960)
17. Halstead, P. E., *Chem. and Ind.* (*Rev.*), 1132 (1957)
18. Aluminium Development Association (London), Information Bulletin No. 2125
19. Larrabee, C. P. and Coburn, S. K., *Corrosion*, **17**, 155t (1961)
20. Building Research Station Digest, No. 32 (1968)
20(a). Browne, R. D. and Domone, P. L. J., *S.U.T. Conference on Offshore Structures*, Inst. Civil Engrs., London (1974); Internal Conference on 'Underwater Construction Technology', University College, Cardiff, Apr. (1975)
21. Treadaway, K. W. J., *Brit. Cor. J.*, **6** No. 2, 66–72, March (1971)
21(a). Gilchrist, J. D., C.I.R.I.A. Technical Note ISSN:0305–1781, May (1975)
22. Campbell, H. S., *Chem. and Ind.* (*Rev.*), 692 (1957)
23. Chandler, K. A. and Watkins, K. O., *Mach. Desgn. Engng.*, Aug. (1965)
24. Holden, W. S., *Water Treatment and Examination*, J. and A. Churchill, London, 55 (1970)
25. Kenworthy, L., *J. Inst. Met.*, **69**, 67 (1943)
26. Porter, F. C. and Hadden, S. E., *J. Appl. Chem.*, **3**, 385 (1853)
27. Campbell, H. S., B.N.F. Publication No. 544, Oct. (1968)
28. Non-Ferrous Metals Post War Building Studies, No. 13, 1944, H.M.S.O., London, 11 (1944)
29. Sereda, P. J., *Corrosion*, **17**, 30 (1961)

10.4 Design in Marine Engineering

The field of marine engineering involves an immense variety of structures, machinery and equipment, fulfilling widely diverse functions and operating under many different environmental conditions. The one factor common to nearly all these environments is the presence of chloride. These markedly increase the risk of causing corrosion or, depending on other factors, stress-corrosion cracking or corrosion fatigue.

Although most of the environments under consideration fall into two broad categories, i.e. either complete immersion in sea-water, or exposure to an atmosphere charged with salt particles, both categories are subject to many variables and pollutants. Thus, sea-water can be contaminated in in-shore waters by chemical and biological pollution; it can contain solid matter and floating debris that is liable to cause both abrasion or mechanical damage as well as deposit attack and blockages. It can be brackish in estuaries or highly concentrated in some equipment. Its speed in relation to the metal surface can be fast and turbulent, or stagnant. In some equipment it may be hot, or even boiling, and contaminated with various pollutants, and it may contain small organisms in an early stage of development which can result in troublesome fouling growths.

Similarly, so-called 'marine' atmospheres are far from being unvarying in nature. The force and direction of wind (which can influence the amount of airborne salt), the amount and duration of rainfall and the relative humidity can all affect the corrosivity of the atmosphere. In combination with these meteorological factors, other pollutants such as gaseous and solid-fuel combustion products can greatly increase the corrosive effects of chlorides. Hence conditions near port installations, etc. or in the wake of ships' funnel gases, can be particularly severe. The all-pervading salt-charged atmosphere in which ships and naval aircraft operate poses special problems. A notable example is the ingestion of salts in gas turbines, but many other items of equipment, particularly delicate electronic gear, must be protected against the ingress of salt. Salt-laden air also increases the tendency to corrosion in humid compartments of ships, where the condensation that is present will inevitably contain dissolved chlorides.

Yet a further group of environments frequently encountered consists of those which combine or alternate complete immersion with atmospheric exposure. Examples are the alternate wetting and drying of tidal zones, splash zones, boot-toppings (especially of large tankers) and, of course, the conditions of service of amphibious vehicles. In addition to these main categories, numerous other media of a heterogeneous character can be met with in marine engineering, especially in ships.

If the foregoing is considered in relation to Evans' statement[1] that 'It is difficult to produce, or even to maintain, passivity in presence of chlorides', it will be abundantly clear why corrosion is one of the major problems confronting the marine engineering industry. To combat it successfully demands the correct choice of materials, good design, good workmanship and attention to conditions of operation, and whilst these requirements are not unique to marine engineering, each is especially important in an industry in which many of the conditions of service are exceptionally severe.

In the case of ships, space and weight restrictions call for materials with properties that cannot always be met by alloys with the highest corrosion resistance, whilst the same restrictions in many instances accentuate the problem by increasing the severity of the conditions and by restricting accessibility for maintenance. Since the choice of materials is also subject to economic considerations and since the use of different metals in contact or close proximity in sea-water is frequently unavoidable, no method should be neglected that will contribute towards producing the conditions of service which the acceptable materials can withstand. By doing so it should be possible to reduce substantially the high proportion of the cost of ships' refits due to corrosion.

Design Principles

The broad principles of design that should be followed in order to minimise corrosion in marine engineering should also be subject to the overriding necessity to regard designing against corrosion as an integral part of the planning and costing procedures, which should be continuously followed at all stages from the initial plans to the finished construction. Failure to do so is likely to result in breakdown of plant (with consequential losses), costly maintenance or modifications in design (if these are practicable) and a possible lowering of safety standards. An attempt to design against corrosion as an afterthought is generally unsatisfactory and unrewarding.

Whilst careful design can often minimise or even prevent corrosion at little extra cost, where the environmental conditions or the conditions of service are severe (as in most forms of marine engineering) corrosion prevention cannot be achieved without expense, although even in these circumstances good design can help to reduce this. In general, the extra initial outlay involved in building structures and equipment with a level of corrosion resistance appropriate to the service concerned and the length of life required more than compensates for the cost and troubles that stem from employing cheaper materials of inadequate resistance.

The main principles to be observed can be summarised as follows:

1. Features should be avoided that entrap and retain corrosive agents. This can be done by (*a*) attention to the geometry of designs and methods of construction, (*b*) by the provision of adequate drainage, (*c*) by protection against contact with absorbent and/or corrosive materials, and

(*d*) by methods of preventing or reducing condensation.
2. Where appropriate, designs should facilitate the application of adequate coatings that can be readily maintained. This can be achieved by attention to the geometry of the design and methods of construction, and by making provision for good accessibility.
3. All methods of protection against corrosion, including paint systems, metal coatings and cathodic protection, should be regarded as an integral part of the design.
4. Care should be exercised in the use of dissimilar metals in contact or in close proximity.
5. Sea-water systems should be designed to avoid excessive water speeds, turbulence and aeration.
6. Undue static or cyclic stressing and features which give rise to stress concentrations should be avoided as these may lead to failures by stress-corrosion cracking or corrosion fatigue.
7. So far as is possible, components which operate in highly turbulent water should be designed with a view to eliminating cavitation.
8. Designs should have regard to the material being employed, e.g. designs and methods of construction suitable for steel will not be appropriate for aluminium alloys or for plastics.
9. Equipment for use in ships should, in general, be specifically designed for the purpose.

Some examples of how these principles should be applied, especially to ships, are described in the following pages.

Avoidance of Entrapment of Corrosive Agents

Many factors influence the corrosion of metals in the atmosphere, including the natural phenomena that make up the vagaries of climate and weather. Of these, the feature of greatest importance is moisture in its various forms, since, other factors apart, the amount of corrosion that takes place is largely a question of how long the surface of the metal remains wet.

Although the attack is not likely to be serious provided the moisture remains uncontaminated, this rarely happens in practice, and in marine environments sea salts are naturally present not only from direct spray but also as wind-borne particles. Moreover, many marine environments are also contaminated by industrial pollution owing to the proximity of factories, port installations, power stations and densely populated areas, and in the case of ships' upperworks, by the discharge from funnels. In these circumstances, any moisture will also contain sulphur compounds. In addition, solid pollutants such as soot and dust are likely to be deposited and these can cause increased attack either directly because of their corrosive nature or by forming a layer on the surface of the metal which can absorb and retain moisture. The hygroscopic nature of the various dissolved salts and solid pollutants can also prolong the time that the surface remains moist.

Designs should therefore avoid, as far as possible, all features that allow water (whether sea-water, rain-water or moisture) of any kind to be entrapped or retained. These conditions are not only corrosive towards bare metals;

they also adversely affect the life of protective coatings both directly and by the fact that it is often difficult at areas subject to these conditions to give adequate surface preparation for good paint adhesion.

Good designs in these respects do not differ for metal structures exposed in marine environments from those for similar structures exposed elsewhere, several examples of which are illustrated in Section 10.3 (Figs. 10.14 to 10.17) and in Reference 2, but the frequent presence of sea-water or salt-contaminated water makes observance of these designs especially necessary. The same principles also apply to ships' upperworks and internal fittings, although here there are many sites where it is impossible to ensure that water cannot collect and be retained. A few of many examples are ventilation trunking that is subject to the ingress of spray or rain, areas around wash-deck valves, junctions of mushroom ventilators with decks, junctions of horizontal stiffeners with vertical plates, and behind the linings of bathrooms and under deck coverings. In these areas either extra protection should be given, or designs should allow, so far as is practicable, ready accessibility for frequent maintenance.

Bilges are one of the most difficult areas of this nature to deal with, especially in machinery spaces, since not only are they almost impossible to keep dry or even to dry out while the ship is in operation, but effective maintenance of protective coatings at all areas is in any case quite impossible, except at major overhauls and refits, because of inaccessibility. Good initial protection of the bilges by metal spraying during building when all surfaces are accessible considerably reduces subsequent maintenance problems. Designs can help by arranging for water to collect in sumps where it can be pumped away. Alternatively, some authorities favour the use of cathodic protection with sacrificial anodes, but this is only effective if an adequate amount of water is present and if this has a low enough electrical resistivity. The method can also break down by the anodes becoming coated with oil or by being painted inadvertently.

Another common cause of the metal surface remaining wet is contact with absorbent materials, and this again can cause serious attack when sea-water or chloride-contaminated water is involved. In addition, the absorbent material itself can be corrosive, as in the case of wood. Examples of trouble that can occur from this cause in ships are wooden decks laid over steel, certain aluminium alloy frames in contact with wooden hulls, and zinc- or cadmium-coated fasteners in wooden hulls. The whole subject of the corrosion of metals by wood receives detailed treatment in Section 19.10.

Lagging of steam pipes can create similar trouble by absorbing moisture during shut-down periods or by becoming accidentally wet, especially in the case of asbestos which usually contains chloride. Calcium silicate is preferable, and moisture absorption should be prevented by the application of a waterproof coating.

The other main cause of metals remaining wet under atmospheric conditions is condensation. In ships this is particularly liable to occur in mess decks, laundries, galleys, etc. and in any enclosed space where humid conditions prevail. Such conditions also encourage the growth of fungi which are not only a source of deterioration of paint coatings but which can also result in attack of metals because of the corrosive nature of their products of metabolism. Cadmium plating, which is used extensively in electronic equipment, is particularly vulnerable to this form of attack[3]. The solution to condensation problems in ships, and in enclosed spaces in general, is air conditioning; failing this it is the use of de-humidifiers or the provision of as much ventilation as possible. Care should also be taken to site equipment in such a way that free circulation of air is not impeded. Condensation, usually contaminated with chlorides, is also prone to occur on the external surfaces of ships in dry dock because of the humid conditions that often prevail.

Designs to Facilitate the Application and Maintenance of Protective Coatings

Coatings employed in marine environments to protect steel are mainly sprayed aluminium and sprayed zinc, and galvanising (all of which may be exposed with or without an overcoat of paint), and paint systems applied directly. Both aluminium and zinc coatings are especially corrosion resistant in these environments and find extensive application both in ships and shore structures. The methods used for the necessary surface preparation of the steel prior to the application of these coatings, and the methods of application and properties of the coatings, are fully described in Chapters 13, 14 and 15. It is only necessary here to stress two main aspects of design concerning ships whereby coatings are used to the best effect.

The first of these is the avoidance of features that create severe local corrosive conditions on the underwater and boot-topping areas of the hull plating. Thus, surface irregularities or roughness such as undressed welds not only tend to receive thinner paint coatings than the surrounding surfaces, but they also suffer greater erosion during service and possess the further disadvantage of offering increased resistance to the ship's propulsion.

The second important aspect to remember is the problem of maintaining these coatings in service, and hence to design in such a way that when construction is completed all parts will still be accessible. This is not easy to achieve in the case of ships, where some areas such as bilges and outer bottom plating in contact with keel blocks during docking are unavoidably inaccessible. Nevertheless, close attention to this aspect when designing will yield long-term benefits, since even small areas that get neglected during service can lead to serious corrosion and to the need for carrying out repairs that can be time consuming and costly, especially if they involve dismantling and reassembling equipment or parts of the structure.

It should also be emphasised that although the time that ships are in dock is usually restricted, every effort, and sometimes more than is the case, should be made to plan the work to be done so that sufficient time can be given to carrying out any necessary corrosion preventive treatments properly. To this end, good design will help to make the best use of the available time.

Sea-water Systems

General Design and Layout

The conditions under which these systems operate can be extremely severe, and although the alloys at present available and in extensive use in sea-water systems offer good resistance to many forms of attack, even the most resistant can fail under the conditions that can arise from poor design and fabrication. In fact, these systems provide an outstanding example of the important part that design can play in minimising corrosion.

The ideal design is one in which all parts can be operated satisfactorily with water flowing with the least turbulence and aeration, and at a rate of flow within the limits that the materials involved can withstand. These limits, with regard to smooth flow, vary with the material, as described in Section 4.2, but turbulence and aeration can lower these limits considerably, and designs that eliminate these two factors go a long way towards preventing impingement attack, which can be the major cause of failures in sea-water systems. (See also Sections 1.6 and 2.1.)

Good designs should start at the inlets which should be shaped to produce smooth streamline flow. In the case of ships' inlets they should be located in the hull in positions, so far as is practicable with other requirements. where turbulence is least excessive and the amount of entrained air is as low as possible. Inlets located close under the bilge keel or immediately aft of a pump discharge are in particularly bad positions. The design should not impart a rotatory motion to the water stream, since the vortex formed (in which air bubbles will tend to be drawn) can travel along the piping until disrupted by some change in the conformation of the system, e.g. a bend or change of section in the piping or an irregularity in the bore; the energy will then be released resulting in an excessive local speed of highly aerated water and consequent rapid impingement attack. Some reduction in the amount of air in the water stream can be achieved by the maximum use of air-release pipes and fittings.

To ensure that the water flows through the whole of the system as smoothly as possible and with the minimum of turbulence, it is vital that the layout of pipework should be planned before fabrication starts. It should not be the result of haphazard improvisation to avoid more and more obstacles as construction proceeds. Pipes should run as directly as possible and every effort should be made to avoid features that might act as turbulence raisers. For this reason the number of flow controllers, bends, branches, valves, flanges and other fittings should be kept to a minimum.

In some systems it may be feasible to select the sizes of pipes to give the correct speed in all branches without the need for flow-regulating devices, or a bypass may be satisfactory thereby eliminating one possible source of turbulence. Where flow control is necessary, as is usually the case, this should be effected preferably either by valves of the glandless diaphragm type or by orifice plates, in either case set to pass the designed quantity of water under the conditions of maximum supply pressure normally encountered. Screwdown valves are not advised for throttling, nor are sluice and gate valves, as these in the partly open position cause severe turbulence with increased local speed on the downstream side. For auxiliary heat exchangers and sanitary services in ships fed by water from the firemain (which is at a high pressure), effective pressure-reducing valves must be fitted, otherwise rapid failures by impingement attack are almost inevitable. It is very important that all flow-regulating devices should be fitted only on the outlet side of equipment.

With regard to the various other features in a piping system that can set up turbulence, careful design of these can do much to reduce, if not completely eliminate, their harmful effects. Thus, pipe bends need cause little trouble if the radius of the bend is sufficiently large. A radius of four times the pipe diameter is a practice being followed in H.M. ships as well as by some authorities, with a relaxation to a radius of times three when space is restricted. Crimping of bends should be avoided by the use of a filler during bending, but care should be taken to remove all traces of filler residues before putting the pipe into service as these can initiate corrosion.

Branch pipes cause the minimum disturbance if they can be taken off the main piping by swept 'T'-pieces rather than by right-angled junctions. Where the latter are unavoidable the branch main should be as generous as possible. If connecting pieces are not used, branch pipes should be set at a shallow angle with the main piping and should not protrude into the latter.

Flanged joints are a very common cause of turbulence unless correctly made and fitted. Close tolerances should be placed on the machining of flanges to match the bore of the pipe; mating flanges should be parallel and correctly aligned, and gaskets should be fitted so that they are flush with the bore and do not protrude into the pipe. Alternatively, butt welding can be used provided pipe ends are accurately aligned and weld metal does not protrude into the bore. Screwed union fittings also give no trouble if pipes are correctly aligned. A further precaution to lessen the risk of impingement attack is to fit straight lengths of pipe down-stream of possible turbulence raisers.

A British Standards Specification, MA18:1973, covers salt-water piping in ships.

Piping Materials

The characteristics of the various metals commonly used for sea-water systems, chiefly copper alloys, nickel alloys, galvanised steel and to a lesser extent aluminium alloys and stainless steels, are fully described in their respective sections. Reference here will be confined to mentioning some of the advantages and limitations of clad piping and plastic piping.

As regards the former, a recent development has been the cladding of steel piping with a welded overlay of cupro-nickel[4-6] which shows promising application for components such as ships' inlet trunking. Non-metallic-clad steel piping, if correctly manufactured and fitted, has the advantage of being immune to deterioration by sea-water at the speeds normally encountered. However, if the coatings possess pin-holes or other discontinuities, or are too brittle to withstand a reasonable amount of shock, water may gain access to the steel causing rapid perforation or lifting of the coating with the possibility of a blockage ensuing.

The use of plastic piping for conveying sea-water in ships is limited to non-essential services because of its susceptibility to fire damage. Thus, it is not permitted to be used either in main or auxiliary circulating pipes, or in fire-mains, machinery spaces or other compartments where there might be a fire hazard. In other words, as stated in Lloyd's regulations it is not acceptable for services essential to safety. Where it can be used, however, it offers considerable advantages besides that of obviating corrosion, because of weight saving and ease of installation. It should, however, be borne in mind that the properties of the piping are very different from those of metal piping. Thus, plastic piping will not withstand the rough handling that metal piping is not infrequently subjected to on board ship. Moreover, although it needs supports at shorter distances apart than does metal piping, especially on horizontal lengths to prevent sagging, these must allow free movement since the thermal expansion of the piping may be at least eight times that of steel. Because of their relatively heavy weight, valves and other fittings must also be well supported to prevent the joints being strained.

Design of Components and Fittings

Even with the speed, turbulence and aeration of the water supplied to the equipment being within acceptable limits, if corrosion is to be avoided it is still necessary for the units themselves to be well designed. Some of the more important aspects involved are outlined below.

Condensers and heat exchangers The shape of the inlet water box and the shape and positioning of the entry should be designed to produce as smooth a flow of water as possible, evenly distributed over the tubeplate. Poor design can result in high-speed turbulent water developing at certain points, which may give rise to impingement conditions at some areas of the tubeplate and some tube ends; in other relatively stagnant pockets some tubes may receive an inadequate supply of water, leading to overheating or to the settlement of debris, with consequent deposit attack.

Incondensible gases can also accumulate in a stagnant area accentuating the tendency to local overheating, but this can be counteracted to some extent by the provision of adequate air-escape fittings. Local overheating may also be caused by impingement of steam due to poor baffling, and this can cause penetration of the tubes by erosion of wet steam. Local overheating of Cu–30Ni cupro-nickel tubes can lead to 'hot spot' corrosion[7-8]. More general overheating, which can occur with some auxiliary condensers, particularly drain coolers, may cause a rapid build up of scale on the inside of the tubes necessitating acid descaling.

Design faults in two-pass condensers and heat exchangers that can cause corrosion include poor division plate seals allowing the escape of water at high velocity between the passes, and flow patterns that produce stagnant zones.

Partial blockages of tubes by debris act as turbulence raisers and should be guarded against by fitting either weed grids or plastic inserts in the tube ends so that any object passing through the insert is unlikely to become jammed in the tube. These methods will not, however, prevent trouble from marine organisms that enter the system in their early stages of development. In shore installations this is usually dealt with by intermittent chlorination, and this is being increasingly used in ships. An alternative scheme is to fill the system with fresh (shore) water for a period, followed by removal of the dead organisms by water jetting or by an approved mechanical method.

An important point in the design of condensers and heat exchangers is the provision of facilities that allow ready access for cleaning and maintenance. Manhole doors should be well sited and should be clear of pipework and fittings. Covers of smaller heat exchangers should be easy to remove well clear of the units. Small doors should be fitted at the lowest point to enable debris to be flushed out, and to ensure complete drainage when this is required. Lack of these facilities understandably tends to result in maintenance procedures not being carried out as frequently as they should, and this can lead to serious corrosion.

The final point to be mentioned on the subject of marine condenser design is the danger of purchasing auxiliary units through 'package deals', which whilst adequate for the fresh water service for which they are designed, can have an extremely limited life when operated on sea-water. In addition, equipment designed for use on land even for sea-water service often proves unsatisfactory when installed in ships; it is generally preferable for ships' equipment to be specifically designed for the purpose.

Pumps and valves   In addition to designing these so as to cause least turbulence in the water stream, careful designing can also minimise corrosion of the pumps and valves themselves.

Pumps can be a major source of trouble, with rapid deterioration of parts by impingement attack and sometimes by cavitation. The latter may be the result of either poor design or the conditions under which the pump is operated. In either case the remedy is not to be found in material changes, since although some materials resist cavitation better than others, even the most resistant may have a short life if the cavitation is severe. Designs should avoid features that produce excessive turbulence which may induce cavitation, or that allow the passage of high-velocity water between the high- and low-pressure areas. Cavitation in pumps can develop if the water supply is not continuous, e.g. when pumping-out bilges, especially with choked strainers, or if the water supply is controlled by a badly designed throttling device. Pumps with high suction heads are particularly prone to suffer cavitation damage. Conditions in pumps can still be severe, however, in the absence of cavitation, and high duty materials may be necessary for a reasonable service life.

As regards valves, diaphragm types are the most satisfactory but most valves have to withstand extremely turbulent conditions, and as with pumps, even the best designs need to be constructed of resistant materials.

Designs to Prevent Bimetallic Corrosion

In general it is wise to avoid, as far as possible, the use of incompatible bimetallic couples in marine practice, since as these are often in contact with sea-water or water that contains chlorides which are good electrolytes it is necessary to take very considerable precautions to prevent corrosion. However, the widely diverse properties required of the materials used in ships make it impracticable to avoid all such couples.

In sea-water systems, for example, contact between different metals and alloys is very common, but trouble is mostly avoided by cathodic protection. In condensers, iron protector slabs are usually employed, although ferrous water boxes sometimes serve the purpose; impressed-current systems are also finding extended use. Different alloys are also to be found in contact in valves, pumps and other equipment, and between these components and copper-alloy piping; these seldom have equal electrode potentials, but many have slight differences that are of no consequence in practice. Even where components have a significant difference in potential this can often be accept-able if the less noble component is large in area compared with that of the more noble metal or is of thicker section and can be allowed to suffer a certain amount of attack without loss of efficiency. In fact, the galvanic pro-tection afforded by this means to a more noble and vital component can be most valuable, as for example if pump impellers are given some protection from the casing. The tendency for some otherwise highly corrosion-resistant alloys to suffer crevice attack can also often be overcome in this way.

Other combinations, however, do create difficulties. In the case of ships' outer bottoms, non-ferrous components and fittings such as copper-alloy valves and sea-tubes can cause serious corrosion of adjacent steel, the gal-vanic attack in many of these cases being accentuated by high water speeds and sometimes by the indifferent design of inlets which promote turbulence and contribute to the difficulty of maintaining paint films intact for any length of time.

Corrosion in these areas is sometimes countered with a zinc- or aluminium-alloy sacrificial anode in the form of a ring fixed in good electrical contact with the steel adjacent to the non-ferrous component. This often proves only partially successful, however, and it also presents a possible danger since the corrosion of the anode may allow pieces to become detached which can damage the main circulating pump impeller. An alternative method recently being tried in the Royal Navy and which shows considerable promise, is to coat the steel inlet with a welded overlay of Cu–30Ni[4-6].

Non-ferrous propellers can cause similar trouble, especially during the ship's fitting-out period. This may be countered by not fitting the propellers until the final dry docking prior to carrying out sea trials; by coating the propellers prior to launching and removing the coating at the final dry dock-ing; and by cathodic protection of the whole outer bottom. The protection of the outboard propeller shaft is most successfully achieved by a coating of epoxy resin reinforced with glass cloth. The earthing of shafts to the hull is advocated by some ship owners and by some navies, but the practical diffi-culties involved in maintaining good contact are substantial.

One of the most important and extensive applications of two metals in proximity in ship construction is the use of aluminium-alloy superstructures with steel hulls. This practice is now widely accepted in both naval and merchant services. The increased initial cost is more than offset by weight saving and fuel consumption, with the concomitant advantages of increased accommodation, a lowered centre of gravity giving greater stability (which is specially valuable for ships carrying deck cargo) and a lighter draft. These advantages are so substantial that the use of these two metals in juxtaposition even in a marine environment must be accepted. Serious trouble can, how-ever, usually be avoided if the following steps are taken:

1. The metal interface should be designed to be as close a fit as possible.
2. The faying surface of the steel should be given a metal coating of either zinc (by spraying or galvanising) or aluminium spray. In Royal Navy practice the superstructure is attached to a galvanised angle section.
3. Jointing compounds such as Neoprene tape or fabric strip impregnated

with chromate inhibitor should be inserted to ensure complete exclusion of water.

4. On the weather side of the joint the rivet should be of a similar material to the plate with which it is in contact.

5. The joint should be painted on both sides with the appropriate paint systems.

A joint corresponding to the above description is shown in Fig. 10.21[9] (see also Reference 10). An alternative method of jointing proposed both in the UK and the USA, but, at the time of writing untried, would be to bolt the superstructure and the steel to a narrow but thick strip of glass-reinforced plastic, leaving a small gap between the two metals[11].

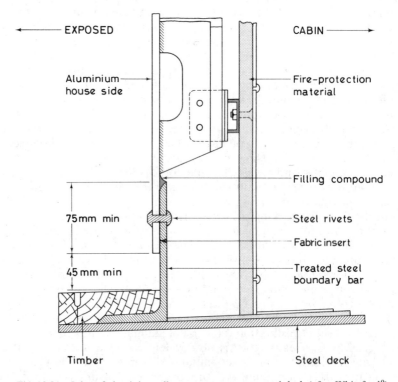

Fig. 10.21   Joint of aluminium alloy superstructure to steel deck (after Whiteford[9])

The riveted construction of aluminium-alloy superstructures was originally used, as in the SS *United States* built in 1952, but argon arc welding was introduced shortly afterwards and is now generally employed[11]. The alloy favoured in the UK nowadays, as used in the construction of the *Queen Elizabeth 2*, contains 4·0–4·9% magnesium (to BS 1477 NP8)[11–12].

Bimetallic corrosion arising from different metals not in electrical contact must also be guarded against in ships. In particular, water containing small amounts of copper resulting from condensation, leakage or actual discharge from copper and copper-alloy pipes and fittings, can cause accelerated attack of steel plating with which it comes into contact. Good designs can help in preventing this by making both the copper alloys and the plating accessible for applying and maintaining protective coatings.

Stress-corrosion Cracking, Corrosion Fatigue and Cavitation Damage

Some examples of how design can assist in counteracting these forms of attack in marine practice are given below. The three phenomena are dealt with comprehensively in Chapter 8.

Stress-corrosion Cracking

Mild steel, stainless steels, some copper alloys and some aluminium alloys are the materials most liable to be affected.

As regards mild steel, boilers suffer most, either from the so-called 'caustic cracking' of riveted boilers or from poor designs with features that interfere with smooth water circulation and set up thermal stresses.

All stainless steels are somewhat susceptible to stress-corrosion cracking in the presence of chlorides, although the austenitic Fe–18Cr–8Ni type are only likely to suffer if they have high residual stresses and are subjected to a high concentration of chloride and a high temperature. The cracking of boiler-feed de-aerators is a typical example.

Copper alloys used in marine engineering vary greatly in their susceptibility to stress-corrosion cracking and much trouble can be avoided by choosing the more resistant alloys where practicable. Similarly, the avoidance of stress-corrosion cracking in aluminium alloys in marine engineering is largely a matter of using the correct alloys rather than a question of design. The current practice in the UK of using for marine work the alloy complying with BS1477–NP8, which limits the magnesium content to a maximum of 4·9%, is thought will eliminate the risk of stress-corrosion cracking.

Corrosion Fatigue

Marine engineering is a sphere of activity in which corrosion fatigue problems can be a substantial source of trouble since the environments most frequently met with, such as sea-water, boiler water, condenser water and steam, are those in which corrosion fatigue is prone to occur. Sea-water in particular will stimulate corrosion fatigue in most metals and alloys. Good design, however, can prevent much trouble by reducing stresses.

Thus, cases of corrosion fatigue have occurred with lightly built high-speed ships in hull plating near the bows, and in rudder plating due to vibration caused by the plating being too thin in relation to the stiffener spacing. The trouble disappeared when thicker plating was substituted, or the spacing of the stiffeners was reduced[13]. Boiler components, especially fire rows, often develop corrosion fatigue as a result of pulsating thermal stresses arising from irregular circulation, sometimes augmented by residual stresses caused by uneven expansion and contraction. Fitting augmentors in steam drums, baffle plates in water drums and additional down-comers can improve the circulation and reduce the tendency to corrosion fatigue. Other examples are: condenser tubes that develop cracks as a result of vibration due to insufficient support; steam pipes (especially the corrugated type) and superheater tubes as a result of intermittent thermal stresses; steel propeller shafts especially when grooved by a loose rope guard (which could be prevented by better design); propeller and pump shafts that have suffered circumferential wastage as a result of bimetallic contact (especially alloy steels and irons in contact with non-ferrous fittings); boiler steam drums resulting from the rough machining of the arboring combined with operating stresses; blades of gas turbine compressors; and welded structures that have not been stress relieved.

Cavitation Damage

This type of damage is dealt with comprehensively in Section 8.8. It can be particularly severe in sea-water and hence can be a considerable problem in marine engineering. Among ships' components that may suffer in this way are the suction faces of propellers, the suction areas of pump impellers and casings, shaft brackets, rudders and diesel-engine cylinder liners. There is also evidence that cavitation conditions can develop in sea-water systems with turbulent high rates of flow.

Improvements in design that can assist in preventing cavitation damage may be concerned with the shape of the component itself, or with its surroundings. The design of propellers and impellers depends largely on the expertise of the manufacturers of these components. Where cavitation damage develops, this may occasionally be due to unsuitable design in relation to the conditions of service, but in most cases occurring in sea-water systems the trouble arises from other causes such as poor layout, air leaks in suction piping or faulty operation. Ships' underwater fittings should be designed to offer as little resistance as possible to movement through the water and to leave the minimum turbulence in their wake which might result in cavitation damage to other parts of the ship. Cathodic protection often proves of benefit in reducing damage to ships' propellers and underwater fittings.

L. KENWORTHY

REFERENCES

1. Evans, U. R., *The Corrosion and Oxidation of Metals*, Arnold, London, 238 (1960)
2. Rogers, T. H., *Marine Corrosion*, Newnes, London, 179–188 (Figs. 16 to 32) (1968)
3. Kenworthy, L., *Trans. Inst. Mar. E.*, **77** No. 6, 154 (1965)
4. Bradley, J. N. and Newcombe, G., Inst. Welding Symposium on Welding in Non-ferritic Materials, London, 77 (1967)
5. Bradley, J. N. and Newcombe, G., *The Engineer*, **229** No. 5920, 28, July 10th (1969)
6. Bradley, J. N. and Newcombe, G., Inst. Metals Conference on Copper and its Alloys, Amsterdam (1970)

7. Breckon, C. and Gilbert, P. T., *Proc. 1st Int. Congress on Metallic Corrosion*, London, 1961, Butterworth, London, 624 (1962)
8. Bem, R. S. and Campbell, H. S., *ibid.*, 630 (1962)
9. Whiteford, J. M., *Shipping World*, **129**, 192 (1953)
10. Rogers, T. H., *Marine Corrosion*, Newnes, London, 101 and 103 (Figs. 12 and 13) (1968)
11. Muckle, W., private communication.
12. Taylor, E. G., *Navy*, **71** No. 10, 328 (1966)
13. Forbes, W. A. D., Discussion of paper by A. J. Gould and U. R. Evans, *J.I.S.I.*, **165**, 294 (1950)

BIBLIOGRAPHY

Slater, I. G., Kenworthy, L. and May, R., *J. Inst. Metals*, **77**, 309 (1950)
Gilbert, P. T., *Trans. Inst. Mar. E.*, **66**, 1 (1954)
Evans, U. R. and Rance, V. E., *Corrosion and its Prevention at Bimetallic Contacts*, H.M.S.O., London (1958) (in the course of revision by the British Standards Institution)
Gilbert, P. T., *Chemistry and Industry*, **78**, 888 (1959)
LaQue, F. L. and Tuthill, A. H., *Trans. Soc. Nav. Architects and Mar. E.*, **69**, 619 (1961)
Kenworthy, L., *Trans. Inst. Mar. E.*, **77**, 149 (1965)
Recommended Practice for the Protection of Ships' Underwater and Boot-topping Plating from Corrosion and Fouling, British Ship Research Association (1966)
Taylor, E. G., contribution to 'Engineering Materials at Sea', *Navy*, **71** No. 10, 328 (1966)
West, E. G., *ibid.*, **71** No. 11, 353 (1966)
Wells, T., *ibid.*, **71** No. 12, 394 (1966)
Todd, B., *ibid.*, **72** No. 1, 8 (1967)
Craick, R. L., *ibid.*, **72** No. 2 (1967)
Kenworthy, L., *ibid.*, **72** No. 5, 141 (1967)
Rogers, T. H., *Marine Corrosion*, Newnes, London (1968)
Frederick, S. H. and Smedley, G. P., *Inst. Mar. E. Symposium*, 20th March, 1 (1968)
Gilbert, P. T., *ibid.*, 14 (1968)
Falconer, W. H. and Wong, L. K., *ibid.*, 26 (1968)
Recommended Practice for the Protection and Painting of Ships, British Ship Research Association (1971)

10.5 Design in Relation to Welding and Joining

A jointed fabrication* is one in which two or more components are held in position (*a*) by means of a mechanical fastener (screw, rivet or bolt), (*b*) by welding, brazing or soldering or (*c*) by an adhesive. The components of the joint may be metals of similar or dissimilar composition and structure, metals and non-metals or they may be wholly non-metallic. Since the majority of fabrications are joined at some stage of their manufacture, the corrosion behaviour of joints is of the utmost importance, and the nature of the metals involved in the joint and the geometry of the joint may lead to a situation in which one of the metals is subjected to accelerated and localised attack. Although corrosion at bimetallic contacts involving different metals has been dealt with in Section 1.7, it is necessary to emphasise the following in relation to corrosion at joints in which the metals involved may be either identical or similar:

1. A difference in potential may result from differences in structure or stress brought about during or subsequent to the joining process.
2. Large differences in area may exist in certain jointed structures, e.g. when fastening is used.

Furthermore, many joining processes lead to a crevice, with the consequent possibility of crevice corrosion.

Before considering the factors that lead to corrosion it is necessary to examine briefly the basic operations of joint manufacture.

Mechanical Fasteners

These require little description and take the form of boltings, screws, rivets, etc. Mechanical failure may occur as a result of the applied stress in shear or tension exceeding the ultimate strength of the fastener, and can normally be ascribed to poor design, although the possibility of the failure of steel fittings at ambient or sub-zero temperatures by brittle fracture, or at ambient temperatures by hydrogen embrittlement, cannot be ignored. If brittle failure is a problem then it can be overcome by changing the joint design or employing a fastener having a composition with better ductility transition properties.

*For definitions of terms used in this section see Section 10.5A.

The corrosion problems associated with mechanical fixtures are often one of two types, i.e. crevice corrosion or bimetallic corrosion[1-4], which have been dealt with in some detail in Sections 1.6 and 1.7, respectively.

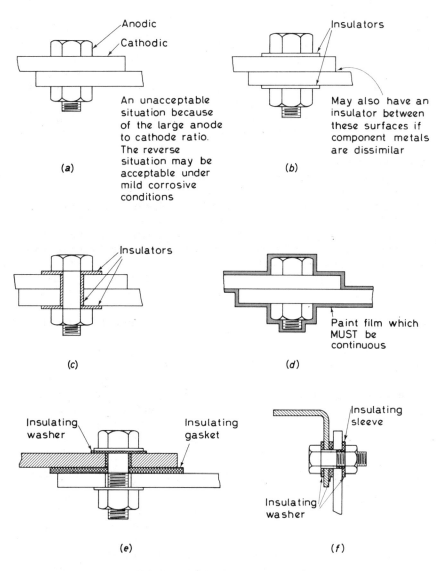

Fig. 10.22   Design of insulated joints[5]

The mechanical joining of aluminium alloys to steels using rivets and bolts, a combination which is difficult to avoid in the shipbuilding industry, represents a typical example of a situation where subsequent bimetallic corrosion could occur. Similarly, other examples of an ill-conceived choice of materials, which could normally be avoided, can be found in, for example, brass screws to attach aluminium plates or steel pins in the hinges of aluminium windows.

The relative areas of the metals being joined is of primary importance in bimetallic corrosion, and for example, stainless steel rivets can be used to joint aluminium sheet, whereas the reverse situation would lead to rapid deterioration of aluminium rivets. However, in the former case a dangerous situation could arise if a crevice was present, e.g. a loose rivet, since under these circumstances the effective anodic area of the aluminium sheet would be reduced, with consequent localised attack. In general, under severe environmental conditions it is always necessary to insulate the components from each other by use of insulating washers, sleeves, gaskets, etc. (Fig. 10.22)[5], and the greater the danger of bimetallic corrosion the greater the necessity to ensure complete insulation; washers may suffice under mild conditions but a sleeve must be used additionally when the conditions are severe.

The fasteners themselves may be protected from corrosion and made compatible with the metal to be fastened by the use of a suitable protective coating, e.g. metallic coating, paints, conversion coating, etc. The choice of fastener and protective coating, or the material from which it is manufactured, must be made in relation to the components of the joint and environmental conditions prevailing[6]. Thus high-strength steels used for fastening the fuselage of aircraft are cadmium plated to protect the steel and to provide a coating that is compatible with the aluminium. In the case of protection with paints it is dangerous to confine the paint to the more anodic component of the joint, since if the paint is scratched intense localised attack is likely to occur on the exposed metal.

In general, paint coatings should be applied to both the anodic and cathodic metal, but if this is not possible the more cathodic metal rather than the more anodic metal should be painted. The use of high-strength steels for bolts for fastening mild steel does not normally present problems, but a serious situation could arise if the structure is to be cathodically protected, particularly if a power-impressed system is used, since failure could then occur by hydrogen embrittlement; in general, the higher the strength of the steel and the higher the stress the greater the susceptibility to cracking.

A point that cannot be overemphasised is that, in the long term, stainless steel fasteners should be used for securing joints of stainless steel.

Soldered Joints

Soldering and brazing are methods of joining components together with a lower-melting-point alloy so that the parent metal (the metal or metals to be joined) is not melted (Table 10.4). In the case of soft soldering the maximum temperature employed is usually of the order of 250°C and the filler alloys (used for joining) are generally based on the tin–lead system. The components must present a clean surface to the solder to allow efficient wetting and flow of the molten filler and to provide a joint of adequate mechanical strength. To obtain the necessary cleanliness, degreasing and mechanical abrasion may be required followed by the use of a flux to remove any remaining oxide film and to ensure that no tarnish film develops on subsequent heating.

In the case of carbon steels and stainless steels, and many of the non-ferrous alloys, the fluxes are based on acidic inorganic salts, e.g. chlorides,

Table 10.4   Soldering and brazing

Process	Temp. range (°C)	Typical fillers	Fluxes
SOLDERING			
hot iron	60–300	70Pb–30Sn ⎫	Chloride based ⎫
oven		40Pb–60Sn ⎪	Fluoride based ⎬
induction		70Pb–27Sn–3Sb ⎬	Resin based ⎭
ultrasonic		40Pb–58Sn–2Sb ⎪	
dip		Sn–Zn–Pb ⎭	
resistance			
wave and cascade			
BRAZING			
torch	500–1 200	90Al–10Si ⎫	Borax based ⎫
dip		50Ag–15Cu–17Zn–18Cd ⎪	Fluoride based ⎪
salt bath		Ag–Cu–Ni–In ⎪	Hydrogen gas ⎬
furnace		60Ag–30Cu–10Zn ⎪	Town's gas ⎪
induction		50Cu–50Zn ⎬	Vacuum ⎭
resistance		97Cu–3P ⎪	
		70Ni–17Cr–3B–10Fe ⎪	
		82Ni–7Cr–5Si–3Fe ⎪	
		60Pd–40Ni ⎭	

which are highly corrosive to the metal unless they are removed subsequently by washing in hot water. For soldering tinplate, clean copper and brass, it is possible to formulate rosin-based fluxes having non-corrosive residues and these are essential for all electrical and electronic work. Activators are added to the rosin to increase the reaction rate, but these must be such that they are thermally decomposed at the soldering temperature if subsequent corrosion is to be avoided[7]. Corrosion is always a risk with soldered joints in aluminium owing to the difference in electrical potential between the filler alloy and the parent metal and the highly corrosive nature of the flux that is generally used for soldering. However, it is possible to employ ultrasonic soldering to eliminate use of flux. With aluminium soldering it is imperative that the joints be well cleaned both prior and subsequent to the soldering operation, and the design should avoid subsequent trapping of moisture.

Brazed Joints

When stronger joints are required, brazing may be used[8]. The filler alloys employed generally melt at much higher temperatures (600–1 200°C), but the effectiveness of the joining process still depends upon surface cleanliness of the components to ensure adequate wetting and spreading. Metallurgical and mechanical hazards may be encountered in that the filler may show poor spreading or joint filling capacity in a certain situation or may suffer from hot tearing, whilst during furnace brazing in hydrogen-containing atmospheres there is always the possibility that the parent metal may be susceptible to hydrogen embrittlement or steam cracking. Furthermore, brittle diffusion products may be produced at the filler base-metal interface as a result of the reaction of a component of the filler alloy with a base-metal component, e.g. phosphorus-bearing fillers used for steel in which the phosphorus diffuses into the steel.

Serious damage can be caused by (a) diffusion into the parent metal of the molten brazing alloy itself when either one or both of the parent metal(s) is in a stressed condition induced by previous heat treatment or cold working, and (b) by an externally applied load which need only be the weight of the workpiece. Nickel and nickel-rich alloys are particularly prone to liquid-braze-filler attack especially when using silver-based braze fillers at temperatures well below the annealing temperature of the base metal, since under these conditions there is then no adequate stress relief of the parent metal at the brazing temperature. The problem may be avoided by annealing prior to brazing and ensuring the maintenance of stress-free conditions throughout the brazing cycle. There is a whole range of silver-, nickel- and palladium-based braze fillers of high oxidation and corrosion resistance that have been developed for joining the nickel-rich alloys; however, the presence of sulphur, lead or phosphorus in the base-metal surface or in the filler can be harmful, since quite small amounts can lead to interface embrittlement (Section 7.5). In the case of the Monels, the corrosion resistance of the joint is generally less than that of the parent metal and the design must be such that as little as possible of the joint is exposed to the corrosive media.

When, in an engineering structure, the aluminium–bronzes are used for their corrosion resistance, the selection of braze filler becomes important and although the copper–zinc brazing alloys are widely used, the corrosion resistance of the joint will be that of the equivalent brass rather than that of the bronze. With the carbon and low-alloy steels, the braze fillers are invariably noble to the steel so that there is little likelihood of trouble (small cathode/large anode system), but for stainless steels a high-silver braze filler alloy is desirable for retaining the corrosion resistance of the joint, although stress-corrosion cracking of the filler is always a possibility if the latter contains any zinc, cadmium or tin.

An interesting example of judicious choice of braze filler is to be found in the selection of silver alloys for the brazing of stainless steels to be subsequently used in a tap-water environment[9]. Although the brazed joint may appear to be quite satisfactory, after a relatively short exposure period failure of the joint occurs by a mechanism which appears to be due to the breakdown of the bond between the filler and the braze metal. Dezincification is a prominent feature of the phenomenon[10] and zinc-free braze alloys based on the Ag–Cu system with the addition of nickel and tin have been found to inhibit this form of attack. A similar result is obtained by electroplating 0·007 mm of nickel over the joint area prior to brazing with a more conventional Ag–Cu–Zn–Cd alloy.

Brazing is generally considered unsuitable for equipment exposed to ammonia and various ammoniacal solutions because of the aggressiveness of ammonia to copper- and nickel-base alloys, but recently an alloy based on Fe–3·25B–4·40Si–50·25Ni has been shown to be suitable for such applications[11].

Upton[12] has recently studied the marine corrosion behaviour of a number of braze alloy–parent metal combinations and has shown that compatibility was a function of the compositions of the filler and parent metals, their micro-structures and chance factors such as overheating during the brazing operation.

Welded Joints

The welded joint differs from all others in that an attempt is made to produce a continuity of homogeneous material which may or may not involve the incorporation of a filler material. There are a large variety of processes by which this may be achieved, most of which depend upon the application of thermal energy to bring about a plastic or molten state of the metal surfaces to be joined. The more common processes used are classified in Table 10.5.

Table 10.5 Typical joining processes

Joining process	Types
Mechanical fasteners	Nuts, bolts, rivets, screws
Soldering and brazing	Hot iron, torch, furnace, vacuum
Fusion welding	Oxyacetylene, manual metal arc, tungsten inert gas, metal inert gas, carbon dioxide, pulsed arc, fused arc, submerged arc, electro slag and electron beam
Resistance welding	Spot, seam, stitch, projection, butt and flash butt
Solid-phase welding	Pressure, friction, ultrasonic and explosive

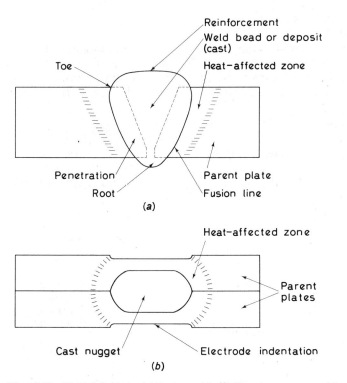

Fig. 10.23   Weld definitions. (a) Fusion weld and (b) resistance spot weld

The macrographic examination of a welded joint shows two distinct zones, namely the fusion zone with its immediate surroundings and the parent metal (Fig. 10.23). It is apparent therefore, that such processes produce differences in microstructure between the cast deposit, the heat-affected zone which has undergone a variety of thermal cycles, and the parent plate. Furthermore, differences in chemical composition can be introduced accidentally (burn-out of alloying elements) or deliberately (dissimilar metal joint). Other characteristics of welding include: (1) the production of a residual stress system which remains after welding is completed, and which, in the vicinity of the weld, is tensile and can attain a magnitude up to the yield point; (2) in the case of fusion welding the surface of the deposited metal is rough owing to the presence of a ripple which is both a stress raiser and a site for the condensation of moisture; (3) the joint area is covered with an oxide scale and possibly a slag deposit which may be chemically reactive, particularly if hygroscopic; and (4) protective coatings on the metals to be joined are burnt off so that the weld and the parent metal in its vicinity become unprotected compared with the bulk of the plate.

Therefore, the use of welding as a method of fabrication may modify the corrosion behaviour of an engineering structure, and this may be further aggravated by removal of protective systems applied before welding, whilst at the same time the use of such anti-corrosion coatings may lead to difficulties in obtaining satisfactorily welded joints[13-16].

Weld Defects

There is no guarantee that crack-free joints will automatically be obtained when fabricating 'weldable' metals. This is a result of the fact that weldability is not a specific material property but a combination of the properties of the parent metals, filler metal (if used) and various other factors (Table

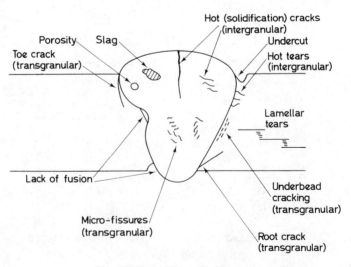

Fig. 10.24   Possible weld defects

Table 10.6 Factors affecting weldability*

Parent metal	Filler metal	Other factors
Composition	Composition	Degree of fusion
Thickness	Impact strength	(Joint formation)
State of heat treatment	Toughness	Degree of restraint
Toughness	Hydrogen content	Form factor
Temperature	Purity	(Transitions)
Purity	Homogeneity	Deposition technique
Homogeneity	Electrode diameter	Skill and reliability of the welder
	(Heat input during welding)	

*Data after Lundin[17].

Table 10.7 Weldability defects

Defect	Causes	Remedies
Hot cracks	Large solidification range	More crack-proof filler
	Segregation	Less fusion
	Stress	
Underbead cracks	Hardenable parent plate	Low hydrogen process
	Hydrogen	Planned bead sequence
	Stress	Preheating
Microfissures	Hardenable deposit	Low hydrogen process
	Hydrogen	Pre- and post-heating
	Stress	
Toe cracks	High stress	Planned bead sequence
	Notches	Preheating
	Hardenable parent plate	Avoidance of notches
Hot tears	Segregation	Less fusion
	Stress	Cleaner parent plate
Porosity	Gas absorption	Remove surface scale
		Remove surface moisture
		Cleaner gas shield

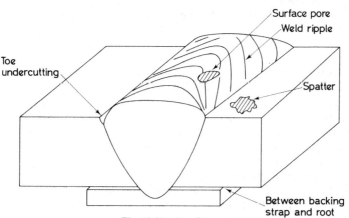

Fig. 10.25   Possible crevice sites

10.6)[17]. The consequence of the average structural material possessing imperfect weldability is to produce a situation where defects may arise in the weld deposit or heat-affected zone (Table 10.7 and Fig. 10.24).

It is obvious that these physical defects are dangerous in their own right but it is also possible for them to lead to subsequent corrosion problems, e.g. pitting corrosion at superficial non-metallic inclusions and crevice corrosion at pores or cracks. Other weld irregularities which may give rise to crevices include the joint angle, the presence of backing strips and spatter (Fig. 10.25). Butt welds are to be preferred since these produce a crevice-free profile and, furthermore, allow ready removal of corrosive fluxes.

Carbon and Low-alloy Steels

These usually present little problem since the parent and filler metals are generally of similar composition, although there is some evidence that the precise electrode type in manual metal-arc welding for marine conditions may be important; weld metal deposited from basic-coated rods appears to corrode more rapidly than that deposited from rutile-based coatings[18].

An environment containing H_2S, cyanides, nitrates or alkalis may produce stress-corrosion cracking in highly stressed structures and these should be first stress relieved by heating to 650°C.

An interesting development in weldable corrosion-resistant steels is the copper-bearing or weathering steels (Section 3.2) which exhibit enhanced corrosion resistance in industrial atmospheres in the unpainted condition. For optimum corrosion resistance after welding, the filler employed should be suitably alloyed to give a deposit of composition similar to that of the steel plate[19].

Stainless Irons and Steels

Since stainless irons and steels are widely used for resisting corrosive environments, it is relevant to consider the welding of these alloys in some detail. There are three groups of stainless steels, each possessing their own characteristic welding problem:

1. Ferritic type. Welding produces a brittle deposit and a brittle heat-affected zone caused by the very large grain size that is produced. The problem may be reduced in severity by the use of austenitic fillers and/or the application of pre- and post-weld heat treatments; the latter is a serious limitation when large welded structures are involved.
2. Martensitic type. Heat-affected zone cracking is likely and may be remedied by employing the normal measures required for the control of hydrogen-induced cracking.
3. Austenitic types. These are susceptible to hot cracking which may be overcome by balancing the weld metal composition to allow the formation of a small amount of δ-Fe (ferrite) in the deposit, optimum crack resistance being achieved with a δ-Fe content of 5–10%. More than this concentration increases the possibility of σ-phase formation if the weldment is used at elevated temperature with a concomitant reduction in both mechanical and corrosion properties.

The corrosion of stainless steel welds has probably been studied more fully than any other form of joint corrosion and the field has been well reviewed by Pinnow and Moskowitz[20], whilst extensive interest is currently being shown by workers at the British Welding Institute[21]. Satisfactory corrosion resistance for a well-defined application is not impossible when the austenitic and other types of stainless steels are fusion or resistance welded; in fact, tolerable properties are more regularly obtained than might be envisaged. The main problems that might be encountered are weld decay, knifeline attack and stress-corrosion cracking (Fig. 10.26).

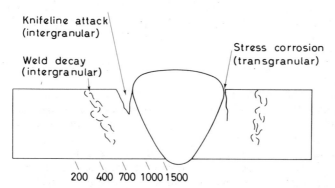

Fig. 10.26   Corrosion sites in stainless steel welds. The typical peak temperatures attained during welding (°C) are given at the foot of the diagram. Note that knifeline attack has the appearance of a sharply defined line adjacent to the fusion zone

Weld decay is the result of the intergranular precipitation of chromium carbide in the temperature range of 450–800°C and material in this condition is referred to as being 'sensitised'. Sensitisation depletes the matrix in the grain-boundary region of chromium and this region may eventually suffer intergranular corrosion. In corrosive environments (see Section 3.3) some zone in the vicinity of the weld area is inevitably raised within the sensitisation temperature range and the degree of severity of sensitisation will be dependent on a number of process factors that determine the time in this temperature range, e.g. heat input, thickness of plate. For most commercial grades of stainless steel in thin section (< 10 mm) the loss in corrosion resistance is slight and seldom warrants any special measures. For a high degree of corrosion resistance, or in welded thick plate, it becomes necessary to take one of the following courses of action:

1. Thermally treat the structures to effect a re-solution of the chromium carbide; this is often impractical in large structures unless local heat treatment is employed, but is not always satisfactory since a sensitised zone could be produced just outside the local thermally treated region.
2. Use extra-low-carbon steel.
3. Use stabilised steels, i.e. austenitic steels containing niobium, tantalum or titanium.

It is important to note that the filler metals should also be stabilised, particularly in a multi-run weld where previous deposits are obviously going to be thermally cycled as later runs are deposited. It may also be necessary to increase the nickel and chromium contents of the filler to offset losses incurred during welding.

It should be noted that sensitisation has very little effect on mechanical properties and that intergranular attack occurs only in environments that are aggressive. France and Greene[22] point out that the precautions taken to avoid sensitisation are frequently unnecessary, and they have carried out a potentiostatic study of a number of electrolyte solutions to evaluate the range of potential, composition and temperature in which intergranular attack occurs. They claim that by means of these studies it is possible to predict whether the environment will be aggressive or non-aggressive to the sensitised zone, and that in the case of the latter no precautions need be taken to avoid sensitisation. This work, which has been criticised by Streicher[23], is still controversial and generally the normal precautions concerning sensitisation are taken irrespective of the nature of the environment. (See also Section 20.1.)

Titanium stabilised fillers should not be used in argon-arc welding as titanium will be vaporised and its effectiveness as a stabiliser lost. Carburising the weld seam by pick-up from surface contamination, electrode coatings or the arc atmosphere leads to increased tendency to intercrystalline corrosion.

The effect of the welding process on the severity of weld decay varies according to the process and the plate thickness so that no single recommendation is possible for every thickness of plate if resistance to attack is essential. The severity of weld decay correlates quite well with sensitisation times as calculated from recorded weld heating cycles.

Under certain conditions it is possible for a weldment to suffer corrosive attack which has the form of a fusion line crack emanating from the toe of the weld; this is termed *knifeline attack*. It is occasionally experienced in welded stabilised steels after exposure to hot strong nitric acids. The niobium-stabilised steels are more resistant than the titanium-stabilised types by virtue of the higher solution temperature of NbC, but the risk may be minimised by limiting the carbon content of a steel to 0·06% maximum (ELC steel).

Stress-corrosion cracking is particularly dangerous because of the insidious nature of the phenomenon. The residual stresses arising from welding are often sufficiently high to provide the necessary stress condition whilst a chloride-containing environment in contact with the austenitic stainless steels induces the typically transgranular and branched cracking. An increased nickel content marginally improves the resistance of the steel to this type of attack, whilst at the opposite extreme, the ferritic chromium steels are not susceptible. The only sure means of eliminating this hazard is to employ either a stress-relief anneal or a molybdenum-bearing steel, but stabilised steels must be used since the required heat treatment is in the carbide-sensitisation temperature range.

Nickel Alloys

In the main, welding does not seriously affect the corrosion resistance of the high nickel alloys and stress relief is not generally required since the resistance to stress corrosion is particularly high; this property increases with increase in nickel content and further improvement may be obtained by the addition of silicon. The chromium-containing alloys can be susceptible to weld decay and should be thermally stabilised with titanium or niobium, and where conditions demand exposure to corrosive media at high temperatures a further post-weld heat treatment may be desirable. For the Ni–Cr–Mo–Fe–W type alloys, Samans[24] suggests that the material should be given a two-stage heat treatment prior to single-pass welding in order to produce a dependable microstructure with a thermally stabilised precipitate.

The Ni–28Mo alloy provides a special case of selective corrosion analogous to the weld-decay type of attack; it may be removed by solution treatment or using an alloy containing 2% V[25].

Of the weldability problems, nickel and nickel-based alloys are particularly prone to solidification porosity, especially if nitrogen is present in the arc atmosphere, but this may be controlled by ensuring the presence of titanium as a denitrider in the filler and maintaining a short arc length. The other problem that may be encountered is hot cracking, particularly in alloys containing Cr, Si, Ti, Al, B, Zr, S, Pb and P.

For optimum corrosion resistance it is recommended that similar composition fillers be used wherever possible, and obviously any flux residues that may be present must be removed.

Aluminium Alloys

These alloys are very susceptible to hot cracking and in order to overcome this problem most alloys have to be welded with a compensating filler of different composition from that of the parent alloy, and this difference in composition may lead to galvanic corrosion. A further problem in the welding of these materials is the high solubility of the molten weld metal for gaseous hydrogen which causes extensive porosity in the seam on solidification; the only effective remedy is to maintain the hydrogen potential of the arc atmosphere at a minimum by using a hydrogen-free gas shield with dry, clean consumables (e.g. welding rods, wire) and parent plate.

In general, the corrosion resistance of many of the alloys is not reduced by welding. Any adverse effects that may be encountered with the high-strength alloys can be largely corrected by post-weld heat treatment; this is particularly true of the copper-bearing alloys. Pure aluminium fillers impart the best corrosion resistance, although the stronger Al–Mg and Al–Mg–Si fillers are normally suitable; the copper-bearing fillers are not particularly suitable for use in a corrosive environment. Resistance welding does not usually affect the corrosion resistance of the aluminium alloys.

The heat-affected zone may become susceptible to stress-corrosion cracking, particularly the high-strength alloys, and expert advice is necessary concerning the suitability of a particular alloy for a certain environment after welding. In this context Al–Zn–Mg type alloys have been extensively studied[26] and it has been shown that maximum sensitivity appears to occur when there is a well-developed precipitation at the heat-affected zone grain boundaries adjacent to the fusion line, a fine precipitate within the grain and a precipitate-free zone immediately adjacent the grain boundaries. The action of stress-corrosion cracking then appears to be a result of local deformation in the precipitate-free zone combined with the anodic character of the precipitate particles.

Other Materials

Space does not permit a survey of all the various weldable metals and their associated problems, although some suggestions are made in Table 10.8. It is sufficient to state that with a knowledge of the general characteristics of the

Table 10.8 Possible problems in less commonly welded metals

Metal	Weldability	Corrosion
Copper alloys	Porosity Hot cracking Hot tearing Steam explosion	De-zincification De-aluminification Stress corrosion
Magnesium alloys	Porosity Hot cracking Lack of fusion	Stress corrosion Pitting
Titanium alloys	Porosity Embrittlement	Stress corrosion

welding process and its effects on a metal (e.g. type of thermal cycle imposed, residual stress production of crevices, likely weldability problems) and of the corrosion behaviour of a material in the environment under consideration, a reliable joint for a particular problem will normally be the rule and not the exception.

Corrosion Fatigue

A metal's resistance to fatigue is markedly reduced in a corrosive environment. Many welded structures are subjected to fluctuating stresses which, with the superimposed tensile residual stress of the joint, can be dangerous. In addition to this a welded joint is a discontinuity in an engineering structure containing many possible sites of stress concentration, e.g. toe or root of the joint, weld ripple.

Protection of Welded Joints

Structural steels are frequently protected from corrosion by means of a paint primer, but these materials can have an adverse effect on the subsequent welding behaviour and this is mainly observed as porosity[13]. Hot-dip galvanising for long-term protection can also lead to porosity and intergranular cracking after welding, in which case it may be necessary to remove the zinc coating from the faying edges prior to welding. The presence of zinc can also lead to operator problems due to the toxicity of the fume evolved unless adequate fume extraction is employed.

Prior to painting, all welding residues must be removed and the surface prepared by grinding, grit blasting, wire brushing or chemical treatment. This preparation is of fundamental importance, the method of applying the paint and the smoothness of the bead apparently having little effect on the final result[27].

Conclusions

Every type of corrosion and oxidation problem can be encountered in jointed structures and it is obvious that most engineering structures must be jointed. It would appear therefore, that all structures are on the verge of disintegration. Yet, for every jointed structure that fails by corrosion, there are many hundreds of thousands which have survived the test of time. With a reasonable knowledge of the mechanics of jointing, the possible design and process factors (e.g. crevices, dissimilar materials in contact, presence of fluxes), a basic understanding of corrosion science and, above all, commonsense, few problems in the established fabrication fields should be encountered. As aptly pointed out by Scully[28], as with all other scientific and technological problems, experience is often the final arbiter.

R. A. JARMAN

REFERENCES

1. Booth, F. F., *Br. Corros. J.*, **2** No. 2, 55 (1967)
2. Everett, L. H. and Tarleton, R. D. J., *Br. Corros. J.*, **2** No. 2, 61 (1967)
3. Layton, D. N. and White, P. E., *Br. Corros. J.*, **2** No. 2, 65 (1967)
4. Evans, U. R., *The Corrosion and Oxidation of Metals*, 1st Suppl. Edward Arnold, London (1968)
5. Layton, D. N. and White, P. E., *Br. Corros. J.*, **1** No. 6, 213 (1966)
6. Discussion, *Br. Corros. J.*, **2** No. 2, 71 (1967)
7. Allen, B. M., *Soldering Handbook*, Iliffe, London (1969)
8. Collard Churchill, S., *Brazing*, The Machinery Publ. Co., London (1963)
9. Sloboda, M. H., Czechoslovak Conf. on Brazing, 18 (1969)
10. Jarman, R. A., Myles, J. W. and Booker, C. J. L., *Br. Corros. J.*, **8** No. 1, 33 (1973) and Linekar, G. A. B., Jarman, R. A. and Booker, C. J. L., *Br. Corros. J.*, **10** (1975)
11. Stenerson, R. N., *Welding J.*, **48** No. 6, 480 (1969)
12. Upton, B., *Br. Corros. J.*, **1** No. 7, 134 (1966)
13. Gooch, T. G. and Gregory, E. N., *Br. Corros. J.*, Suppl. issue, 'Design of Protective Systems for Structural Steelwork', 48 (1968)
14. Baker, R. G. and Whitman, J. G., *Br. Corros. J.*, **2** No. 2, 34 (1967)
15. Hoar, T. P., *Br. Corros. J.*, **2** No. 2, 46 (1967)
16. Discussion, *Br. Corros. J.*, **2** No. 3, 49 (1967)
17. Lundin, S., *Weldability Questions and Cracking Problems*, ESAB, Göteborg, 2 (1963)
18. Bradley, J. N. and Rowland, J. C., *Br. Weld. J.*, **9** No. 8, 476 (1962)
19. Slimmon, P. R., *Welding J.*, **47** No. 12, 954 (1968)
20. Pinnow, K. E. and Moskowitz, A., *Welding J.*, **49** No. 6, 278 (1970)
21. Gooch, T. G. et al., *W.I. Res. Bull.*, **12** No. 2, 33 (1971) and *W.I. Res. Bull.*, **12** No. 5, 135 (1971)
22. France, W. D. and Greene, N. D., *Corrosion Science*, **8**, 9 (1968)
23. Streicher, M. A., *Corrosion Science*, **9**, 53 (1969)
24. Samans, C. H., Meyer, A. R. and Tisinai, G. F., *Corrosion*, **22** No. 12, 336 (1966)
25. Lancaster, J. F., *Metallurgy of Welding, Brazing and Soldering*, G. Allen & Unwin Ltd., London (1965)
26. Kent, K. G., *Met. Revs.*, **15** No. 147, 135 (1970)
27. Keane, J. D. and Bigos, J., *Corrosion*, **16** No. 12, 601 (1960)
28. Scully, J. C., *The Fundamentals of Corrosion*, Pergamon Press, London (1968)

10.5A Appendix—Terms Commonly Used in Joining*

Automatic Welding: welding in which the welding variables and the means of making the weld are controlled by machine.

Bead: a single run of weld metal on a surface.

Braze Welding: the joining of metals using a technique similar to fusion welding and a filler metal with a lower melting point than the parent metal, but neither using capillary action as in brazing nor intentionally melting the parent metal.

Brazing: a process of joining metals in which, during or after heating, molten filler metal is drawn by capillary action into the space between closely adjacent surfaces of the parts to be joined. In general, the melting point of the filler metal is above 500°C, but always below the melting temperature of the parent metal.

Brazing Alloy: filler metal used in brazing.

Butt Joint: a connection between the ends or edges of two parts making an angle to one another of 135° to 180° inclusive in the region of the joint.

Carbon Dioxide Welding: metal-arc welding in which a bare wire electrode is used, the arc and molten pool being shielded with carbon dioxide gas.

Covered Filler Rod: a filler rod having a covering of flux.

Deposited Metal: filler metal after it becomes part of a weld or joint.

Edge Preparation: squaring, grooving, chamfering or bevelling an edge in preparation for welding.

Electro-slag Welding: fusion welding utilising the combined effects of current and electrical resistance in a consumable electrode and conducting bath of molten slag, through which the electrode passes into a molten pool, both the pool and the slag being retained in the joint by cooled shoes which move progressively upwards.

Electron-beam Welding: fusion welding in which the joint is made by fusing the parent metal by the impact of a focused beam of electrons.

Filler Metal: metal added during welding, braze welding, brazing or surfacing.

Filler Rod: filler metal in the form of a rod. It may also take the form of filler wire.

*Data extracted from BS 499: Part 1 (1965). Complete copies of this standard can be obtained from The British Standards Institution, 101 Pentonville Road, London, N1 9ND.

Flux: material used during welding, brazing or braze welding to clean the surfaces of the joint, prevent atmospheric oxidation and to reduce impurities.

Fusion Penetration: depth to which the parent metal has been fused.

Fusion Welding: welding in which the weld is made between metals in a molten state without the application of pressure.

Fusion Zone: the part of the parent metal which is melted into the weld metal.

Heat-affected Zone: that part of the parent metal which is metallurgically affected by the heat of the joining process, but not melted.

Hydrogen Controlled Electrode: a covered electrode which, when used correctly, produces less than a specified amount of diffusible hydrogen in the weld deposit.

Manual Welding: welding in which the means of making the weld are held in the hand.

Metal-arc Welding: arc welding using a consumable electrode.

MIG-welding: metal-inert gas arc welding using a consumable electrode.

Oxyacetylene Welding: gas welding in which the fuel gas is acetylene and which is burnt in an oxygen atmosphere.

Parent Metal: metal to be joined.

Pressure Welding: a welding process in which a weld is made by a sufficient pressure to cause plastic flow of the surfaces, which may or may not be heated.

Resistance Welding: welding in which force is applied to surfaces in contact and in which the heat for welding is produced by the passage of electric current through the electrical resistance at, and adjacent to, these surfaces.

Run: the metal melted or deposited during one passage of an electrode, torch or blow-pipe.

Semi-automatic Welding: welding in which some of the variables are automatically controlled, but manual guidance is necessary.

Spatter: globules of metal expelled during welding onto the surface of parent metal or of a weld.

Spelter: a brazing alloy consisting nominally of 50% Cu and 50% Zn.

Submerged-arc Welding: metal-arc welding in which a bare wire electrode is used; the arc is enveloped in flux, some of which fuses to form a removable covering of slag on the weld.

TIG-welding: tungsten inert-gas arc welding using a non-consumable electrode of pure or activated tungsten.

Thermal Cutting: the parting or shaping of materials by the application of heat with or without a stream of cutting oxygen.

Weld: a union between pieces of metal at faces rendered plastic or liquid by heat or by pressure, or by both. A filler metal whose melting temperature is of the same order as that of the parent material may or may not be used.

Welding: the making of a weld.

Weld Metal: all metal melted during the making of a weld and retained in the weld.

Weld Zone: the zone containing the weld metal and the heat-affected zone.

11   CATHODIC AND ANODIC PROTECTION

11.1	Principles of Cathodic Protection	**11**:3
11.2	Sacrificial Anodes	**11**:18
11.3	Power-impressed Anodes	**11**:34
11.4	Practical Applications of Cathodic Protection	**11**:57
11.5	Stray-current Corrosion	**11**:85
11.6	Cathodic-protection Interaction	**11**:91
11.7	Cathodic-protection Instruments	**11**:97
11.8	Anodic Protection	**11**:112

11.1 Principles of Cathodic Protection

Sir Humphry Davy presented a paper to the Royal Society[1] in 1824 in which he described how zinc anodes could be used to prevent the corrosion of copper sheathing of the wooden hulls of British naval ships, and his paper shows a remarkable awareness of the principles of cathodic protection. Several practical tests were made on vessels in harbour and on sea-going ships, such as the effect of current density on protection of the copper. Davy also investigated the use of the impressed current system using a voltaic battery, but did not consider the method to be practicable.

The first 'full hull' installation on a vessel in service was applied to the frigate *Samarang* in 1824; four groups of cast-iron anodes were fitted, and virtually perfect protection was afforded to the copper sheathing. Unfortunately, the prevention of the corrosion of the copper resulted in the failure of the copper to act as a toxicide for marine growths, and consequently fouling increased. For this reason interest in cathodic protection waned, since the action of the corroding copper in preventing fouling was considered to be more important than its deterioration. The method was therefore neglected for almost 100 years, and was first used successfully by the oil companies in Texas to protect underground pipes.

Principles of Application

Cathodic protection is applied by one of two methods, viz. power-impressed current or sacrificial anodes. Figure 11.1 illustrates the use of an external power supply, usually a rectifier, which converts ordinary a.c. power to d.c. The protected structure is made electrically negative so that it acts as a cathode. A second electrode is made electrically positive and completes the circuit as an auxiliary anode. Current is carried in the external circuitry as electrons, and the applied current $I_{app.}$ is shown in Fig. 11.1 as electron current. Free electrons do not exist in an electrolyte solution; therefore, the current must be carried by positively- and negatively-charged ions. The current through the electrolyte solution, equal to that in the external circuit, is shown as positive current*, i.e. positively-charged ions carrying the current.

Both positive and negative ions carry current in the electrolyte. Whether or not one or the other carries more or less of the total current depends on the transport numbers of the ions. For simplicity only the direction of positive current flow, which results from conduction by positive ions, is shown in Figs. 11.1 and 11.2. (See also Section 9.1.)

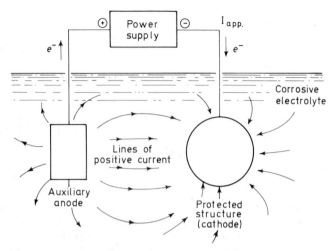

Fig. 11.1   Cathodic protection with an external power supply

Electrochemical reactions at the electrodes are responsible for the mechanism of cathodic protection and for the transfer of charges from electrons to ions at the electrode surfaces, and these will be discussed in more detail subsequently.

It should be emphasised that cathodic protection is possible only when the structure to be protected and the auxiliary anode are in both electronic and electrolytic contact; the former is achieved by a metallic conductor and the latter by the electrolyte solution in which the structure to be protected and the auxiliary anode are immersed.

Figure 11.2 shows the second method of applying cathodic protection. When two dissimilar metals are connected electrically in an electrolyte solution, a current flows between the two because of dissimilar electrochemical potentials. The metal with the more electropositive (noble) potential becomes a cathode and is protected from corrosion (protected structure in Fig. 11.2), and the more electronegative (active) metal becomes the anode. The current flowing between the two metals accelerates the dissolution (corrosion) of the

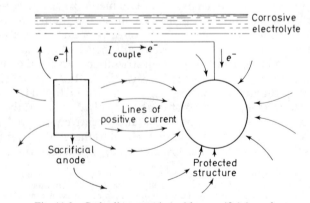

Fig. 11.2   Cathodic protection with a sacrificial anode

anode, which is thus sacrificed and must be replaced periodically. The flow of current in Fig. 11.2 is the same as that in Fig. 11.1. That is, electrons flow towards the cathode in the external circuitry and ions carry the current in the corrosive electrolyte.

Current distribution on the cathode becomes low at areas remote from the anode in either Fig. 11.1 or 11.2. Thus, in both cases multiple anodes are often used to even the current distribution.

Cathodic protection is monitored by measuring the electrode potential of the protected structure by determining the e.m.f. between the structure and a suitable reference electrode (Fig. 11.3). The reference electrode may be simply a corroding metal (e.g. pure zinc) which reaches a fairly steady potential during its corrosion. More often a special half cell (e.g. copper in saturated $CuSO_4$ solution) is utilised as a reference electrode. Details of reference electrode construction are given in other chapters. The voltmeter V in Fig. 11.3 must be of very high resistance to prevent drawing current which could polarise the reference electrode; as an alternative an electrometer may be used. The reference electrode must be specified in reporting the potential of the structure, e.g. $-0.85\,V$ *vs.* $Cu/CuSO_4$(sat.) electrode indicates that the potential of the structure is $0.85\,V$ more negative than the reference electrode, which corresponds with approx. $-0.55\,V$ *vs.* S.H.E. (standard hydrogen electrode).

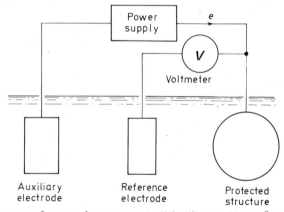

Fig. 11.3   Measurement of protected structure potential with respect to a reference electrode

The potential between reference electrode and protected structure measured by V in Fig. 11.3 includes an ohmic contribution, IR, when a cathodic current I is flowing through the effective resistance R between the reference electrode and protected structure. R and the ohmic contribution IR will become greater as the reference electrode becomes more remote from the protected structure*. Only the polarisation at the surface of the protected structure is effective in cathodic protection. Thus, to monitor the surface potential accurately and eliminate the ohmic contribution as much as possible, the reference electrode must be positioned quite near to the structure

*When the reference electrode is placed remotely from the protected structure, the effective resistance R increases. However, I, which is that portion of the total applied current $I_{app.}$ passing through R, also decreases so that IR may approach zero as I approaches zero. It is unclear, though, whether or not a remote reference electrode measures accurately the potential of a surface which is subject to high polarising current $I_{app.}$.

surface. This may be impossible for large complex structures which are deeply buried or immersed in sea-water.

Cathodic protection requires considerable current to protect a base metal surface of appreciable area. Consequently, cathodic protection is used most often in conjunction with a surface coating so that the area protected is confined to holidays and defects in the coating. A surface coating reduces the required current by several orders of magnitude, but as the coating deteriorates over a period of time, the cathodic current must increase to maintain the protection on exposed areas. However, in certain circumstances cathodic protection is applied to bare steel structures, e.g. North-Sea oil drilling rigs.

Electrochemical Principles[*]

Corrosion in aqueous solutions occurs by an electrochemical mechanism, which involves the exchange of electrons at the corroding surface. Electrons e^- are liberated by the anodic (oxidation) reaction and are absorbed by the cathodic (reduction) reaction:

$$M \rightarrow M^+ + e^- \quad \text{(anodic)} \qquad \qquad \text{...(11.1)}$$
$$Z^{2+} + e^- \rightarrow Z^+ \quad \text{(cathodic)} \qquad \text{...(11.2)}$$

where equation 11.1 represents the oxidation of metal atoms M to soluble positive ions M^+ and equation 11.2 is the reduction of some arbitrary dissolved ion Z^{2+} to Z^+. Both reactions occur simultaneously at the surface with the result that the metal corrodes by dissolution of surface metal atoms.

The mechanism of cathodic protection is simply understood by reference to equations 11.1 and 11.2. Making the surface more negative increases the access or concentration of electrons which accelerates the rate of the cathodic reaction and decreases the rate of the anodic reaction[2, 3]. Thus the application of a negative or cathodic potential consequently reduces the corrosion rate, and if the decrease in potential is sufficient corrosion can be completely arrested, i.e. the rate of the anodic reaction becomes zero and the whole surface of the metal becomes cathodic.

Electrons are reactants in both equations 11.1 and 11.2; therefore, the rate of either can be measured as a current (electrons/unit time, i.e. C/s or A). Reaction rates for equations 11.1 and 11.2 are known to obey a logarithmic (Tafel) relationship as a function of potential, i.e.

$$\eta_a = \beta_a \log i_a/i_{o,a} \qquad \qquad \text{...(11.3)}$$
$$\eta_c = -\beta_c \log i_c/i_{o,c} \qquad \qquad \text{...(11.4)}$$

where η_a and η_c are changes in potential (overvoltage or overpotential) caused by the current densities i_a and i_c, respectively[4]. The constants β_a, β_c, $i_{o,a}$, $i_{o,c}$, can be understood by inspecting a plot of logarithm of current vs. potential (Fig. 11.4). The straight lines in Fig. 11.4 each represent the rate of the indicated reaction as a function of potential, in accordance with equations 11.3 and 11.4. Where $i_a = i_c$, the rate of dissolution is equal to the rate of reduction, and corroding metal M reaches a steady state corrosion potential $E_{corr.}$. The rate i_a is simply the corrosion rate $i_{corr.}$ of the metal, i.e. $i_a = i_c = i_{corr.}$ at $E_{corr.}$.

---

[*]See also Section 9.1. Note that the Tafel slopes β_a and β_c for the anodic and cathodic reaction, respectively, are usually signified by b_a and b_c in books published in the UK.

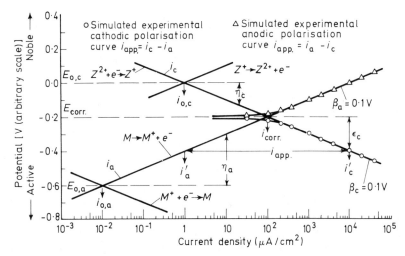

Fig. 11.4   Analysis of cathodic protection based on electrode kinetic theory (after Jones[2])

The meaning of the exchange current densities $i_{o,a}$ and $i_{o,c}$ is also evident. For example, at $i_{o,c}$ the rate of the forward reaction (equation 11.2), i.e. $Z^{2+} + e^- \to Z^+$ is exactly equal to the rate of the reverse reaction, i.e. $Z^+ \to Z^{2+} + e^-$. Thus, $i_{o,c}$ gives the rate at which these two reactions proceed in equal and opposite directions at equilibrium. Analogous explanations apply to $i_{o,a}$ and equation 11.1. $E_{o,a}$ and $E_{o,c}$ are the equilibrium or redox potentials for the anodic and cathodic reactions in equations 11.1 and 11.2 respectively. The slopes of the semi-logarithmic curves for equations 11.1 and 11.2 are designated β_a and β_c, respectively. These are commonly known as the Tafel constants for the reactions and are expressed in V/log i.

When a cathodic overvoltage $\varepsilon_c{}^*$ is impressed on the metal by passing a current $I_{app.}$, as in Fig. 11.1, the corrosion current is reduced from $i_{corr.}$ to i'_a. To maintain charge neutrality at the surface, $I_{app.} = i_c - i_a$ (at ε_c, $I_{app.} = i'_c - i'_a$). The simulated data points in Fig. 11.4 represent $I_{app.}$ plotted against potential. These points give an idealised polarisation curve which approximates to what is often observed for corroding metals. The polarisation curves for actual structures may not display the linear, semi-log behaviour of Fig. 11.4 because of limited access of the reducible species (e.g. Z^{2+}) or ohmic interferences, *IR* (discussed previously).

In practice the reduction reaction shown in equation 11.2 generally involves either hydrogen ions or dissolved oxygen:

$$2H^+ + 2e^- \to H_2 \qquad \text{...(11.5)}$$
$$O_2 + H_2O + 4e^- \to 4OH^- \qquad \text{...(11.6)}$$

Both cathodic reactions will result in an increase in pH, either by the removal of hydrogen ions (equation 11.5) or by the generation of OH⁻ ions (equation 11.6), and this increase will be facilitated if transport and diffusion are restricted, e.g. in soils or static waters. In the case of amphoteric metals

---

*Overvoltage is denoted here by both ε and η. The former is for a potential change from the corrosion potential, the latter for a change from a redox potential such as $E_{o,a}$ or $E_{o,c}$.

such as aluminium and lead, this increase in pH can lead to cathodic corrosion so that it is essential in the cathodic protection of these metals that the potential is not allowed to become too negative. Similarly, excessive alkalinity can result in the saponification of paint coatings with a consequent increase in the area to be protected and a lowering in the efficiency of cathodic protection.

In the case of ferrous metals a rise in pH can be beneficial, since the efficiency of cathodic protection increases with increase in pH. However, in the case of high-strength steels the possibility exists that hydrogen discharge may result in hydrogen entry into the steel, with the consequent danger of hydrogen embrittlement at stressed areas such as welds, and in these circumstances the potential should not be allowed to become too negative.

A further factor is that a rise in pH in water containing dissolved calcium bicarbonate and magnesium sulphate will result in the precipitation of calcium carbonate and magnesium hydroxide, respectively. If these salts deposit on the surface of the metal with the formation of an adherent and coherent scale, the surface area of the metal to be protected will be reduced, with a consequent saving in the cost of cathodic protection.

Protection Criteria

Figure 11.4 shows that the corrosion rate decreases continually as the cathodic polarisation ε_c increases. Even a moderate amount of polarisation results in considerable reduction in corrosion rate. As a result, any degree of cathodic protection is successful to some extent, and the optimum amount or 'criterion' of cathodic protection is rather subjective.

Ideally, polarisation to $E_{o,a}$ results in 'zero' corrosion rate[5], because the reverse discharge reaction, $M^+ + e^- \rightarrow M$ is exactly equal to the anodic dissolution reaction[2] $M \rightarrow M^+ + e^{-(2)}$. However, polarisation to $E_{o,a}$ is impractical as a cathodic protection criterion for two reasons. Firstly, $E_{o,a}$ varies depending on the environment and cannot be easily determined either by thermodynamic calculation or by experiment[2]. Secondly, referring to Fig. 11.4, the exchange current density $i_{o,a}$ for the anodic reaction is often very small (the value of $10^{-2} \mu A/cm^2$ shown in Fig. 11.4 is typical for iron and nickel[6]). Before polarisation reaches $E_{o,a}$, the corrosion rate may be sufficiently low for most practical purposes so that there is no difference between $1 \mu A/cm^2$ at i_c' (Fig. 11.4) and zero corrosion at $E_{o,a}$. In fact, a considerable economic penalty may result, because of the much higher current density $i_{app.}$ which is necessary to obtain $E_{o,a}$.

A potential of -0.85 V (Cu/CuSO$_4$(sat.) reference electrode) is customarily specified as the necessary potential which must be obtained for either optimum or absolute protection of ferrous structures in both soil[7] and seawater[8]. As already stated, $E_{o,a}$ cannot be determined with any accuracy for absolute protection. The 'optimum' value is usually that which gives negligible corrosion rate. Since -0.85 V is quite negative compared to most typical corrosion potentials, polarisation to -0.85 V is usually adequate to reduce corrosion below the 'negligible' level in most situations. However, in a survey of practical experience[9], no particular potential value emerged as superior for all applications; -0.95 V is normally used when the environment contains sulphate-reducing bacteria.

The polarisation curves in Fig. 11.4 (simulated data points) are generally curved until they reach the Tafel (linear) region on the plot of potential vs. log of current. This change of slope has been interpreted as a 'break' in the polarisation curve which, according to older interpretations[5], indicates that $E_{o,a}$ has been reached. Of course, Fig. 11.4 shows that the initial non-linear portion of the polarisation curve bears no relation to $E_{o,a}$. The non-linear portion is due to the ohmic and concentration interferences already discussed; thus, the 'break' is usually selected at a level of 100–200 mV cathodic polarisation, which is adequate to effect considerable cathodic protection.

A polarisation or change in potential of 100–300 mV has been called the *swing* criterion of cathodic protection[10]. Although rather inexact, this criterion is probably as accurate and useful as any of the others so far described.

The anodic Tafel constant β_a is the polarisation required to cause a factor of 10 reduction in the corrosion rate. Thus, β_a can be used to predict the polarisation or swing necessary to reduce corrosion by any desired amount. For example, in Fig. 11.4 with $\beta_a = 0.1$ V/log i, a polarisation of 200 mV reduces corrosion by 99%, or two orders of magnitude. It may not be always possible or realistic to determine β_a for a complex structure in service. However, a sample of the same alloy can be used to obtain at least an approximate value for β_a in the corrosive of interest[2]. The anodic Tafel constant β_a is a precise parameter for control of cathodic protection. Depending on how accurately it can be determined, β_a can make the swing criterion quantitative.

Current Requirements

The amount of current required to protect a metal depends on the corrosion rate and the area of the surface. Required current is directly proportional to corrosion rate, which is proportional to corrosion current density $i_{corr.}$. Secondary factors which increase the corrosion rate will also increase the current required for protection. Thus, most of the factors listed below have a direct effect on the corrosion rate.

Coating   The quality of the coating on a surface determines the amount of bare metal in contact with the electrolyte and also the current which will actually flow through the coating. The current required to protect a buried pipeline depends almost completely on the coating; all other factors are of lesser importance. For example, a well coated line of length 240 km has been brought to the protection potential by 2 A, whereas a similar bare line would require 1 000 A. As the coating deteriorates and more metal is exposed, the current requirements increase.

Resistance of aqueous environment   Cathodic protection is often applied in situations where the electrolyte resistance is high, e.g. in soils and domestic waters. Usually, as the resistance decreases the corrosion rate increases due to greater moisture content in soil and the presence of corrosive ions in both soil and water. Thus, an electrolyte of lower resistance generally requires a greater current than one with a higher resistance.

Access of reducible solutes   The cathodic current i_c (Fig. 11.4) is proportional to the amount of reducible species Z^{2+} which are present in solution.

Consequently, the required current to maintain a given potential is also proportional to the amount of reducible species ($i_{app.} = i_c - i_a$). In acid solutions high concentrations of H^+ ions are present, and very large (often prohibitive) currents are necessary for cathodic protection. In sea-water the reducible species is often dissolved oxygen (equation 11.6). Current is often limited by diffusion of dissolved oxygen to the surface, which can vary widely depending on turbulence, stirring and other hydrodynamic variables.

Controlled Potential Cathodic Protection

A constant applied direct voltage is used in power-impressed cathodic protection systems (Fig. 11.1), and provides a relatively constant protective current. However, as conditions change, the required protective current may vary widely, as discussed previously, and the structure may be under or over protected much of the time. A recent innovation is the use of automatic potential control (A.P.C.) rectifiers. This equipment employs control circuitry to maintain the structure potential constant with respect to a reference electrode, permitting the current to vary continuously to suit the needs of the system.

Some advantages of constant potential control are illustrated in Fig. 11.5[2], which shows iron (or steel) immersed in aerated salt water. The cathodic process is reduction of dissolved oxygen, i.e.

$$O_2 + H_2O + 4e^- \rightarrow 4OH^-$$

and oxygen diffusion to the iron surface limits the reduction current to a maximum value $i_{lim.}$ as shown in Fig. 11.5. Higher relative velocity (or turbulence) between the iron surface and salt water increases the limiting current, e.g. from $i_{lim., 1}$ to $i_{lim., 2}$. The current required to maintain cathodic protection correspondingly increases. Constant potential control at $E_{cath., 1}$ main-

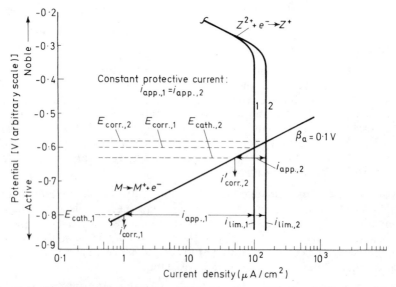

Fig. 11.5 Comparison of controlled potential and constant current cathodic protection in a system simulating steel in aerated sea-water (after Jones[2])

tains the corrosion rate at $i'_{corr., 1}$ and automatically increases (or decreases) the applied current from $i_{lim., 1}$ to $i_{lim., 2}$ as needed ($i_{lim.} \approx i_{app.}$). On the other hand, for constant current ($i_{app., 1} = i_{app., 2}$), an increase in surface velocity from 1 to 2 causes an increase in corrosion rate from $i'_{corr., 1}$ to $i'_{corr., 2}$, and the surface may not receive adequate protection at all times.

As a coating deteriorates, a larger area of the underlying metal is exposed. The applied current density $i_{app.}$ then decreases and the protection decreases in a conventional system. A constant, controlled-potential system compensates by maintaining the potential and current density constant. The total current is increased as the coating deteriorates.

If the electrolyte (soil or water) resistivity decreases, the applied current will increase in a conventional system because of less ohmic IR loss through the electrolyte. The increased current causes increased alkalinity at the protected surface which may be harmful to the coating. The controlled potential system maintains the potential constant and will reduce the applied current should electrolyte resistivity decrease.

A controlled potential system suffers from higher instrumentation and maintenance costs, but these are offset in some cases by the much better control provided. Potential control is important for the cathodic protection of the painted hulls of ships, which move at different velocities and in waters that vary in salinity and temperature.

Sacrificial Anodes

A galvanic couple is formed when the sacrificial anode is attached to the protected structure in Fig. 11.2. In order to utilise this method the anode must have a potential that is more electronegative than that of the protected

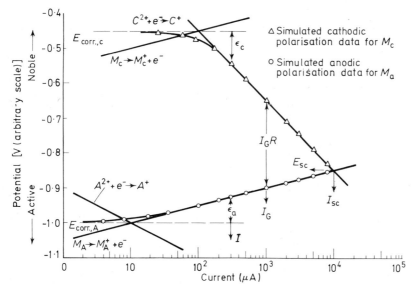

Fig. 11.6   Polarisation in a galvanic couple showing the behaviour of the sacrificial anode in cathodic protection (after Jones[2])

structure. When connected, the structure ($M_{c,\,cathode}$) is polarised cathodically, and the sacrificial anode ($M_{a,\,anode}$) is polarised anodically and the two will reach the same potential $E_{sc}^{(2)}$ (Fig. 11.6), providing the resistance of the electrolyte is sufficiently low. The resulting galvanic current is the short-circuit current, I_{sc}. The data points in Fig. 11.6 are simulated as in Fig. 11.4 by taking the difference $i_c - i_a$ for the cathode and $i_a - i_c$ for the anode.

The 'sacrificial' anode is consumed by dissolution during protection of the cathode and requires periodic replacement. For maximum efficiency of protection the self-corrosion of the anode should be a minimum, and the whole of the corrosion process should be utilised in decreasing the potential of the cathode by electron transfer.

If the electrolyte is of low conductivity or if the sacrificial anode is far from the protected structure, an ohmic $I_G R$ potential develops between the structure and the sacrificial anode, where R is the effective resistance between the two and I_G is the galvanic current in the structure–anode couple. I_G is the maximum current which can flow when the structure (cathode) and sacrificial anode are separated by a potential $I_G R$ as shown in Fig. 11.6[2].

Sacrificial anodes are characterised by four electrochemical properties:

1. The corrosion potential of the alloy must be sufficiently negative to drive the protective current through the electrolyte, i.e. the higher the resistance of the electrolyte and the greater the distance of the anode from the structure to be protected the more negative must be the potential of the anode. In this connection it should be emphasised that the corrosion potentials of aluminium and magnesium are far more positive than the corresponding reversible potentials for the $M^{z+} + ze \rightleftharpoons M$ equilibrium.
2. The degree of anode polarisation, which can limit the galvanic current to low values; it is essential therefore that the anodes should not become passive, which can occur in particular with aluminium.
3. The electrochemical equivalent (output) of the alloy, which is the charge theoretically available to provide galvanic current per unit weight of the alloy.
4. The efficiency of the alloy, which is the percentage of the theoretical output actually obtained in practice.

Sacrificial anodes are considered in more detail in the following section.

Potential Distribution

When current passes between the surface of the anode and the surface of the structure under protection, an ohmic potential gradient results, as discussed previously. The magnitude of the gradient at any point in the corrosive electrolyte depends upon the size and number of anodes, as well as the conductivity of the electrolyte.

Such potential gradients are an important consideration for buried pipelines and other structures in soils where conductivity is often low. The types of gradients encountered are shown in Fig. 11.7. Since the cathode (the structure) normally has a vastly greater area than the anode (except for extremely well-insulated structures) the potential gradients are much steeper around

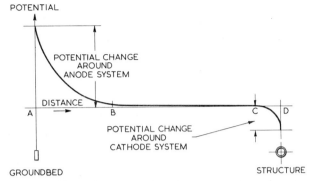

Fig. 11.7   Potential distribution between groundbed and protected structure, showing how the soil potential becomes more positive and more negative as the groundbed and structure are approached, respectively

the anode. The total path of current flow may be divided into three sections, viz. *AB*, which includes the large potential change around the groundbed (buried anodes); *BC*, where the current is flowing through an extremely large cross-section of soil so that the potential change is very small; and *CD*, the region where the cross-section of flow decreases and the potential drop increases appreciably as the current element approaches the structure. The potential between *A* and *B* is normally in the range (depending on anode size and soil resistivity) 10–50 V, and the potential *CD*, 1–2 V. The potential *CD* is not the same as the protective potential shift, which is measured by locating the half-cell in a particular position, i.e. vertically above the pipe. The 'peaking' of the potential curves when the groundbed is too close to the structure is due to the potential gradient of the groundbed cutting the pipe-line.

The number of current sources required to protect a surface, and their distribution, are determined by the requirement that the minimum potential for protection must be attained at all points of the surface at the minimum cost. The potential distribution depends on:

1. General geometry of protected surface, electrolyte and anodes.
2. Resistivity of the electrolyte.
3. Resistance of coating(s) at the metal/electrolyte interface.
4. Resistance through structure by metallic path.

No problem of practical importance is amenable to exact theoretical analysis but certain structures can be treated in a manner that yields useful results.

Pipelines

When current is drained from a single point on a pipeline and discharged to earth, both the current change ΔI and the potential change ΔE will be a maximum at the drainage point and will decrease in magnitude with distance x from that point. The relationship between ΔI, ΔE and Δi (the current density change) with distance is referred to as an *attenuation curve*. If the pipeline is infinitely long and the current is supplied from a remote ground-bed, then ΔI, Δi and ΔE will be exponential functions of distance, and $\log \Delta E$ vs. x and $\log \Delta I$ vs. x will be straight lines that have the same slope. If, however, the pipeline is of finite length the above exponential relationship will not apply and ΔI and ΔE will be hyperbolic functions of x.

Pipeline of infinite length   For a pipeline of infinite length, where (a) the coating conductance and soil resistivity are uniform thus giving uniform leakage conductance along the whole length, (b) the anode field has no influence (remote anode) and (c) the resistance of the pipeline is uniform, then

$$\Delta E_x = \Delta E_o e^{-\alpha x} \qquad \qquad \text{...(11.7a)}$$

$$\text{or} \quad x = \frac{2 \cdot 3 \, (-\log \Delta E_x + \log \Delta E_o)}{\alpha} \qquad \text{...(11.7b)}$$

$$\text{and} \quad \Delta I = \Delta I_o e^{-\alpha x} \qquad \qquad \text{...(11.8a)}$$

$$\text{or} \quad x = \frac{2 \cdot 3 \, (-\log \Delta I_x + \log \Delta I_o)}{\alpha} \qquad \text{...(11.8b)}$$

where ΔE_o and ΔI_o are the changes in potential and current, respectively, at the drainage point; ΔE_x and ΔI_x are the values at a distance x from the point; and α is the attenuation constant, which is a characteristic of the particular line under consideration. These equations show that as x increases, the magnitudes of ΔE_x and ΔI_x decay exponentially, and Fig. 11.8 shows the type of curve that may be obtained with a well coated pipeline (full lines) and a less well coated pipeline (dotted lines).

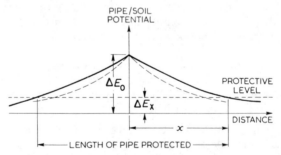

Fig. 11.8   Theoretical distribution of pipe/soil potential on infinite line

The attenuation constant is defined by

$$\alpha = \frac{R_S}{R_K} \qquad \qquad \text{...(11.9)}$$

where R_S is the longitudinal resistance of the pipe (Ω–unit length), which can be calculated from the specific resistivity of the steel or iron and by regarding the pipeline as an annular cylinder, and R_K is the characteristic resistance of the line, and is given by

$$R_K = (R_S R_L)^{\frac{1}{2}} \qquad \qquad \text{...(11.10)}$$

where R_L is the leakage resistance to remote earth of the line (Ω–unit length) at the drainage point and $R_L = \Delta E_o / \Delta I_o$.

Pipeline of finite length protected by two or more drainage points For a pipeline of finite length, the log ΔE_x vs. x and log ΔI_x vs. x curves are initially linear and have the same slope, but as the distance from the drainage point increases the curves diverge from linearity and no longer have the same slope. If the pipeline is terminated in an insulated joint, log ΔE_x attains a constant value and the curve becomes horizontal whilst log ΔI_x becomes zero. Figure 11.9 shows the individual attenuation curves between two drainage points A and B, and it can be seen that log ΔE_a vs. x and log ΔE_b vs. x are both linear and intersect at point O, at which the potential attains its least negative value and the line current becomes zero. If an insulated joint were inserted at point O it would have no effect on the attenuation curve. It can also be seen that the ΔE_x and ΔI_x vs. distance curves have different slopes, whereas when the pipe is infinitely long they are the same. The individual potential changes produced by the two drainage points at A and B can be combined to give the curve log ΔE_c vs. x, which provides an indication of the potential changes that occur in practice. If a section of line of length $2d$ lies between two vertical drainage points (Fig. 11.10) then the potential and current can be evaluated from the two simultaneous equations of the form shown in equations 11.7 and 11.8. In the case of ΔE_x, the potential at x from one of two drainage points (chosen as an arbitrary origin) is obtained by adding the

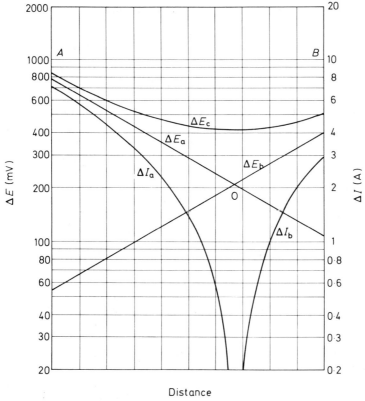

Distance

Fig. 11.9 Attenuation curves between two drain points (after Parker[11])

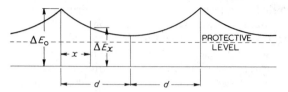

Fig. 11.10   Theoretical distribution of pipe/soil potential on line with stations at spacing of $2d$

equations for each individual drainage point, assuming the other is absent, whilst they must be substracted to give ΔI_x. It follows, bearing in mind that $2 \cosh u = e^u + e^{-u}$ and $2 \sinh u = e^u - e^{-u}$, that

$$\Delta E_x = \Delta E_o \cosh \alpha(d-x)/\cosh \alpha d \qquad \qquad \text{...(11.11}a\text{)}$$
$$\text{and} \quad \Delta I_x = \Delta I_o \sinh \alpha(d-x)/\sinh \alpha d \qquad \qquad \text{...(11.12)}$$

Equation 11.11a can be written as an approximation in a simple form

$$\Delta E_x \approx \Delta E_0 \cosh (d-x) \qquad \qquad \text{...(11.11}b\text{)}$$

Equations 11.11a and 11.12 have been used to draw the curves shown in Fig. 11.9, and a representation of the practical situation is given in Fig. 11.10.
   These equations are valid only when:

1. The coating resistance is high, uniform and ohmic, i.e. the resistance is independent of the voltage drop across it.
2. The groundbed is remote from the pipeline.

The effect of a coating of non-uniform resistance is to distort the exponential or hyperbolic shape of the curve, and in many practical instances the underlying mathematical shape of the curve is scarcely recognisable. If the groundbed is within a finite distance of the pipeline, the potential field around the groundbed cuts the pipeline, and the pipe/soil potentials are less negative than predicted by the theoretical equations, giving the type of curves shown dotted on Fig. 11.8. This is, of course, a disadvantage, since the potential distribution is less uniform than would otherwise be the case.
   The value of R depends upon whether the pipe is made in all-welded steel (in which case the value of resistance using the resistivity of mild steel is used); if the pipe is made conducting by the use of bonding wire between each pipe length then the average resistance per metre length including resistance of bonds must be employed. All units must be consistent, and this applies particularly to resistance per unit length, since the length may vary, e.g. Ωm, Ωft, etc.

Internal surface   Protection of the internal surface of pipelines has been carried out both by the use of a continuous axial conductor and by a series of discrete anodes spaced at intervals along the axis. In the first case the current to the pipe wall is uniform (for straight runs of pipe) and calculation is simple. For discrete anodes, at least two analyses have been made by Morgan[11, 12] and Plumpton and Wilson[13]. The 'spread' of potential along the pipe depends on the resistivity of the electrolyte and the polarisability/ coating resistance at the pipe wall. A distance between anodes of at least six pipe diameters has been found acceptable in practice.

Structures in Sea-water

Calculation of potential spread for complex structures such as jetties is not possible. When the electrolyte is sea-water (or similar low-resistance electrolyte) the location of anodes is of only secondary importance, although anodes are generally placed as near as possible to the protected areas.

Electrical Interference

The current flowing through an electrolyte from an anode to a structure under protection produces a potential field in the electrolyte. If other structures (which are not part of the protection scheme) exist in the vicinity of either the groundbed or the protected structure this potential field may cause current to enter and leave the secondary structures; at the points of current entry a measure of protection exists, and at points where current leaves increased corrosion will occur. The practical aspects of this subject are dealt with in Sections 11.5 and 11.6.

D. A. JONES

REFERENCES

1. Davy, H., *Phil. Trans. Royal Soc.*, **114**, 151, 242, 328 (1824)
2. Jones, D. A., *Corrosion Science*, **11**, 439 (1971)
3. Fontana, M. G. and Greene, N. D., *Corrosion Engineering*, McGraw-Hill, New York, 205 (1967)
4. Stern, M. and Geary, A. L., *J. Electrochem. Soc.*, **104**, 1, 56 (1957)
5. Mears, R. and Brown, R., *Trans. Electrochem. Soc.*, **74**, 519 (1938); Hoar, T. P., *J. Electrodepositors' Tech. Soc.*, **14**, 33 (1938)
6. Conway, B. E., *Electrochemical Data*, Elsevier, New York, 353 (1952)
7. Schwerdtfeger, W. J. and McDorman, O. N., *J. Res. Nat. Bur. Standards*, **47**, 2, 104 (1951)
8. Schwerdtfeger, W. J., *ibid.*, **60**, 3, 153 (1958)
9. Peterson, M. H., *Corrosion*, **15**, 483t (1959)
10. Morgan, J. H., *Cathodic Protection*, Macmillan, New York, 28 (1959)
11. Parker, M. E., *Pipe Line Corrosion and Cathodic Protection*, Gulf Publishing Co., Houston, Texas (1962)
12. Morgan, J. H., *Corrosion Technol.*, **3**, 194 (1956)
13. Plumpton, C. and Wilson, C., *Corros. Prevention and Control*, **6** No. 5, 53

11.2 Sacrificial Anodes

In view of the detailed exposition of the principles of cathodic protection given in Section 11.1, it is only necessary briefly to restate them with particular reference to sacrificial anodes.

When two different metals M_I and M_{II} are placed in an electrolyte and connected by a conductor there will be a transfer of charge (electrons) through the conductor. The direction of the transfer will depend upon the corrosion potentials of M_I and M_{II} in the electrolyte under consideration. If the corrosion potential of M_I is more positive than that of M_{II}, then the transfer of electrons will be

$$M_{II} - e \rightarrow M_I$$

and the potential of the former is raised whilst that of the latter is decreased. Since electrode potential and rate are related, it follows that the rate of corrosion of M_I is decreased, whereas the rate of corrosion of M_{II} is increased. Should the potential of M_I be decreased sufficiently, corrosion is completely arrested, and the metal is said to be cathodically protected, although the latter term may also be applied when the corrosion rate is decreased to an acceptable level.

Different metals have different electrical potentials, and their relative positions in most aqueous environments encountered in practice are as follows: magnesium (most negative) > zinc > aluminium > iron > steel (low alloy) > tin > lead > brass > copper > silver > gold (least negative). It is seen from this that magnesium, aluminium and zinc are negative to steel, and it is these metals that are used in practice as sacrificial anodes. It should also be noted that iron will sacrificially protect copper and copper alloys, a situation that may arise when the cast iron or steel of a water box sacrificially protects the copper-based tubeplates and tubes. The potentials will of course vary with composition of the metal and nature of the environment, and it is important to emphasise that they will not correspond with the reversible potentials given in the e.m.f. series of metals.

General Anode Requirement

The essential requirement is that the anode will polarise the steel to a point where it will either not corrode at all, or corrodes at an acceptable rate, for an acceptable period of time at an acceptable cost.

The characteristics of an anode can be classified as follows: (a) material composition, (b) manufacturing method, (c) physical shape, (d) electrical contact via the anode insert, (e) output, (f) capacity, (g) efficiency and (h) potential, and these will be defined more precisely in subsequent considerations.

The external factors that have to be taken into consideration in relation to anode performance are: (a) area of bare steel requiring protection, (b) system life or the period of time for which protection is required, (c) chemical composition of electrolyte, (d) temperature of electrolyte, (e) flow-rate of electrolyte, (f) aeration of electrolyte and (g) position in space of the anode relative to the steelwork.

Material Composition

Although aluminium, zinc and magnesium are sufficiently negative to act as anodes to protect steel in a number of environments[1,2], the chemical composition and distribution of alloying constituents are critical in anode performance. Because of the critical nature of the alloying elements, rigid quality control is required, and full spectrographic analysis of the materials introduced to each melt and coming from each melt is essential to control the level of impurities.

In order to meet most anode specifications, high grade primary metal must be used. Most primary metals are available to more than one specification, depending on the method of manufacture and the amount and method of purification or refining carried out. Because of this, it is essential to be sure to specify the right basic analysis for the main anode metal. Typical analyses of primary metals used for anodes are as follows:

99·99% zinc		99·80% aluminium	
Pb	0·003 max.	Fe	0·06 max.
Cu	0·001 max.	Si	0·08 max.
Cd	0·003 max.	Cu	0·004 max.
Fe	0·002 max.	Zn	0·02 max.
Sn	0·001 max.	Sn	0·01 max.
		Others	0·03 max.

99·80% magnesium			
Zn	0·002 max.	Mn	0·017 max.
Al	0·008 max.	Sn	0·002 max.
Si	0·072 max.	Ni	0·001 max.
Cu	0·003 max.	Ca	0·002 max.
		Pb	0·003 max.

The composition of an anode is clearly of fundamental importance in its performance, and there is a vast amount of patent literature on the subject in which claims are made for the beneficial effects of specific alloying additions. Ideally, the composition of the alloy should result in an anode having the following properties:

(a) A corrosion potential that is sufficiently negative for the specific application; in general, alloying additions are made to make the potential more negative than that of the unalloyed basis metal.

(*b*) A high anode efficiency, which means that impurities that result in self-corrosion must be absent or rendered innocuous.

(*c*) An ability to remain active and to corrode uniformly and not to become passive in the environment in which it is used for cathodically protecting a structure.

In general, heavy metal impurities that result in self-corrosion must be kept to a minimum and this is reflected in the compositions which specify maximum contents of copper, lead, iron, etc. or the deliberate additions of alloying elements to nullify their effects. Thus BP 813 657 (Dow Chemical Co. Ltd.) specifies the addition of 0·5–1·3% of manganese to magnesium and the claim is made that this counteracts the deleterious effect of the Fe–Al phase (iron must be kept to a low value) thus reducing self-corrosion, and also makes the potential of the alloy more negative. Alloying additions of mercury (0·005–0·05%) and zinc (0·03–0·5%) are claimed (BP 1 066 755) to prevent the passivation of aluminium anodes, increase the anode efficiency and to make the potential more negative. Similar claims are made for additions of indium and gallium (BP 1 066 722) to aluminium alloys, i.e. passivation is prevented, and the uniformity of corrosion is improved.

Typical analyses of commercially available zinc anodes are given in Table 11.1, and No. 2 in this table is the chemical composition given in US Military Specification 18001–H. The other analyses are of various proprietary alloys currently in use.

In Nos. 1 to 3 alloys aluminium is added to produce a more even corrosion pattern than would occur if it were not present. Cadmium is added to give a soft corrosion product which crumbles under its own weight and is self-removing. Silicon is added to remove iron as a type of ferro-silicon dross.

In the No. 4 alloy mercury is added to the zinc to allow the anode to maintain a constant potential which does not decrease with the passage of time.

In all anode alloys listed in Table 11.1 iron is controlled at a very low level in order to prevent anode self-corrosion.

Table 11.1   Typical zinc anode alloys[3–5]

	No. 1[6]	*No. 2*	*No. 3*	*No. 4*[7]
Cu	0·005 max.	0·005 max.	0·005 max.	0·002 max.
Al	0·30–0·50	0·10–0·50	0·4–0·6	
Si	0·003 max.	0·125 max.	0·125 max.	
Fe	0·002 max.	0·005 max.	0·0014 max.	0·0014 max.
Pb	0·005 max.	0·006 max.	0·3 max.	
Cd	0·025–0·100	0·025–0·15	0·075–0·125	
Hg	—	—	—	0·10–0·15
Zn	Remainder	Remainder	Remainder	Remainder
Efficiency*	95%	95%	95%	95%
Potential (V)†	−1·05 vs. Ag/AgCl	−1·05 vs. Ag/AgCl	−1·05 vs. Ag/AgCl	−1·05 vs. Ag/AgCl
Capacity* (Ah kg^{-1})	780	780	780	780

*The significance of anode efficiency and anode capacity is considered subsequently.

†The potentials given in Tables 11.1 to 11.3 are for the anodes and the Ag/AgCl, Cl⁻ in sea-water, in which the potential of the latter is approximately 0·25 V (*vs.* S.H.E.) compared to 0·30 V for Cu/CuSO₄, sat. (*vs.* S.H.E.).

Typical aluminium anode alloys are listed in Table 11.2. In these alloys zinc, tin, mercury and indium are added to make the anode potential more negative. In addition, mercury prevents the occurrence of the natural passive film on the anode surface and does this without reduction of anode efficiency. It should be noted that aluminium has a far greater capacity than magnesium or zinc, but tends to passivate although this disadvantage has now been overcome by alloying with mercury. Owing to the toxicity of mercury these anodes may pollute the environment, and under these circumstances the indium alloy is preferred.

Table 11.2   Typical aluminium anode alloys[8]

	No. 1[9-12]	No. 2[13-15]	No. 3[16]	No. 4	No. 5
Si	0·11–0·21	0·10 max.	0·10 max.	—	—
Fe	0·08 max.	0·13 max.	—	0·13 max.	—
Zn	0·35–0·50	0·50–5·0	5·00	4·00–5·00	7·00
Sn	—	—	—	0·1	0·1
Hg	0·035–0·50	—	—	—	—
In	—	0·005–0·05	0·02	—	—
Mg	—	—	0·80	—	—
Cu	0·006 max.	0·01 max.	0·01 max.	0·01 max.	0·01 max.
Al	Remainder	Remainder	Remainder	Remainder	Remainder
Efficiency	95% av.	90% max.	95% approx.	50%–80%	50%–80%
Potential	−1·05 vs. Ag/AgCl	−1·15 vs. Ag/AgCl	−1·15 vs. Ag/AgCl	−1·10 vs. Ag/AgCl	−1·10 vs. Ag/AgCl
Capacity (Ah kg^{-1})	2 830 av.	2 700 max.	2 700 approx.	Variable	Variable

Typical magnesium anode alloys are shown in Table 11.3. In Nos. 1 and 2 alloys aluminium and zinc are added to increase the efficiency, although at the same time the potential is slightly lowered. The other elements are kept to the minimum values in order to prevent the lowering of efficiency. In Nos. 3 and 4 alloys the manganese is added to give improved current efficiency and increased anode potential. The other elements are kept low in order to prevent the lowering of efficiency, which is only 50% at the best.

Table 11.3   Typical magnesium anode alloys

	No. 1	No. 2	No. 3[17]	No. 4
Cu	0·02 max.	0·01 max.	0·02	0·05 max.
Al	5·3–6·7	2·7–3·5	0·01 max.	—
Si	0·10 max.	0·03 max.	—	0·05 max.
Fe	0·003 max.	0·002 max.	0·03	—
Mn	0·15 min.	0·20 min.	0·5–1·3	1·20 min.
Ni	0·002 max.	0·001 max.	0·001	0·01 max.
Zn	2·5–3·5	0·7–1·3	—	—
Mg	Remainder	Remainder	Remainder	Remainder
Others	—	0·30 max.	—	0·30 max.
Efficiency	50%	50%	50%	50%
Potential (V)	−1·50 vs. Ag/AgCl	−1·50 vs. Ag/AgCl	−1·70 vs. Ag/AgCl	−1·70 vs. Ag/AgCl
Capacity (Ah kg^{-1})	1 230	1 230	1 230	1 230

Magnesium has the advantage of having a more negative electrode potential than aluminium or zinc, and finds particular application in waters and underground environments of high resistivity.

The No. 1 alloy listed in Table 11.3 is available in cast form, the No. 2 alloy in the extruded form, and the Nos. 3 and 4 alloys in either the cast or the extruded form[18]. The higher driving potentials of alloys 3 and 4 make them more suitable for use in higher resistivity environments, since fewer are required to give the required current output. In view of the higher driving potential of magnesium a resistance is sometimes placed in the circuit to limit the current and optimise the consumption of the anode.

Backfills for Anodes

When zinc or magnesium anodes are used for cathodic protection in soil conditions[19-21], they are usually surrounded by a backfill, which decreases the electrical resistance at the anode/soil interface. Small anodes are usually surrounded with backfill in bags and large anodes are usually surrounded with a loose backfill during installation in the soil. The backfill prevents the anode coming into contact with the soil and suffering local corrosion, and so reducing the efficiency. By surrounding the anodes with a material of known composition, the combination of the anode with soil salts is prevented and this prevents the formation of passive films on the anode surface.

As the backfill is of low resistivity relative to the soil, the effect is to increase the anode area, and this gives a lower resistance to a remote earth than if the bare anode was buried in the soil. Backfills tend to attract soil moisture and to increase the conductivity in the area immediately round the anode. Dry backfill expands on wetting, and the package expands to fill the hole in the soil and eliminate voids.

For use in high resistivity soils, the most common mixture is 75% gypsum, 20% bentonite and 5% sodium sulphate, and this has a resistivity of approximately 50Ω cm when saturated with moisture.

Manufacturing Method

Although most anodes are made by gravity casting, some anodes are made by continuous casting or extrusion. The method of casting effects the appearance and physical structure of the anode and hence its saleability and performance. From the point of view of performance, the anode should be cast so that the metal solidifies without segregation of alloying constituents. Also, there should be no inclusion of extraneous matter or blowholes. If the latter happen to occur, then the tendency of the anode to passivate or physically disintegrate in use will increase.

Physical Shape of the Anodes

Different shapes give different surface-area to weight relationships, and this results in different shapes giving different current outputs for the same weight. This means that different shapes have a different life for the same weight. In general, the shape is chosen in order to give a certain current output for a

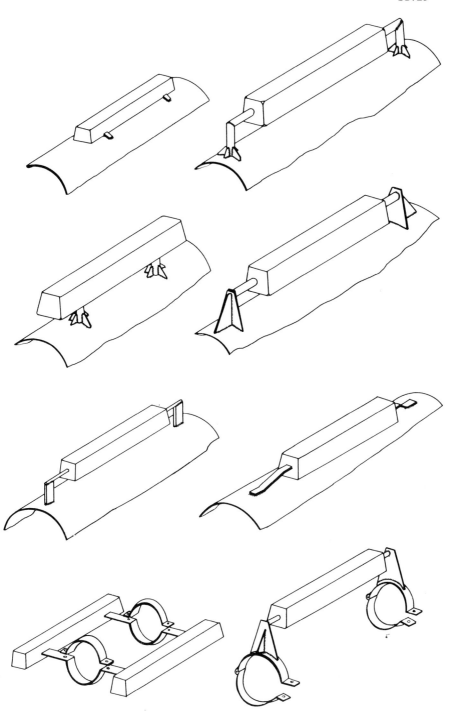

Fig. 11.11   Typical anode fixing methods for yard and underwater installations on curved steel surfaces

certain weight and thus have a certain life. In a number of other instances, the anode shape is designed to conform with the shape and room limitations of the steel structure that it is designed to protect.

The Anode Insert

All sacrificial anodes have a steel insert which is used for supporting it and for making electrical contact with the steelwork requiring protection. It is most important that the electrical contact of the steel insert to the steelwork being protected is perfectly sound. It is also essential that there is a sound mechanical bond between the anode material and the insert in order to ensure good conductivity. Anode inserts are of many shapes and sizes. The insert

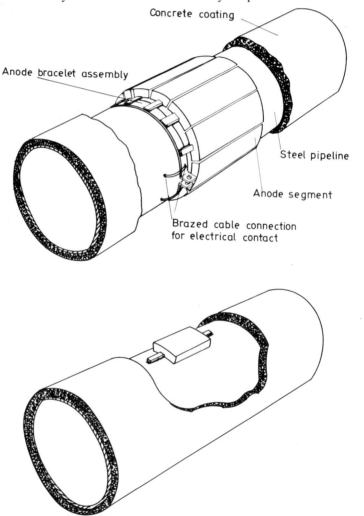

Fig. 11.12   Pipeline coated with concrete with the anodes attached to the steel pipeline and with the anode thickness the same as the coating thickness

must be strong enough to support the weight of the anodes, and must be capable of being welded, or mechanically fixed to the steel structure requiring protection.

Consideration must be given to the ease and speed of anode fixing, as this is a significant part of the total installation cost. In the case of protection of ships, the various classification societies have rules stating what type and method of fixing is permitted in different situations.

The methods of fixing anodes to flat, vertical or horizontal surfaces are relatively well known and simple. The methods of fixing anodes to curved surfaces of pipelines and immersed structures are more complex, and generally require more steel inserts. Figures 11.11 and 11.12 show some methods of attaching anodes to curved surfaces. Figure 11.12 shows a pipeline coated with concrete with the anodes attached to the steel pipeline and with the anode thickness the same as the concrete thickness. In practice, the coating would be brought up to the edge of the anode and cover the whole of the steel pipework.

Anode Output

Anode output $I(A)$ is the difference in potential (V) between the anode material E_1 and the steel at $-0.80\,V$ (vs. Ag/AgCl) E_2 divided by the resistance $R(\Omega)$ of the anode in the electrolyte, i.e.

$$I = \frac{(E_1 - E_2)}{R} \qquad \ldots(11.13)$$

The potential of steel is taken at $-0.80\,V$ (vs. Ag/AgCl), since it is considered that at this potential corrosion is arrested and the negative signs of the two potentials are disregarded.

Resistivity of Anodes in Waters

The resistance $R(\Omega)$ of slender rod anodes (where the ratio of length to mean effective radius is greater than 10) in an electrolyte can be obtained from the formula

$$R = \frac{\rho}{2\pi L}\left[\ln\left(\frac{2L}{r}\right) - 1\right] \qquad \ldots(11.14)$$

where ρ = resistivity of the water ($\Omega\,cm$),
L = length of anode (cm) and
r = mean effective radius of the anode (cm) which is taken as the radius that is left after 40% of the anode is consumed, i.e.

$$r = \left(\frac{60\% \text{ of original cross-sectional area}}{\pi}\right)^{\frac{1}{2}}$$

Water resistivity can be obtained from Fig. 11.13, provided that the temperature and water density are known.

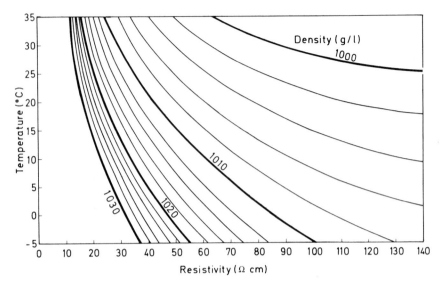

Fig. 11.13   Water resistivity

Example 11.1
Given density = 1 020 g/l and temperature = 20°C, then from Fig. 11.13 resistivity (ρ) = 25 Ω cm.

The formula given in equation 11.14 can be expressed in a graphical form for anodes with an original cross-sectional area of up to 105 cm^2.

Figure 11.14 gives the resistances of differently shaped rod anodes in sea-water, and it should be noted that different graphs are needed for different water resistivities.

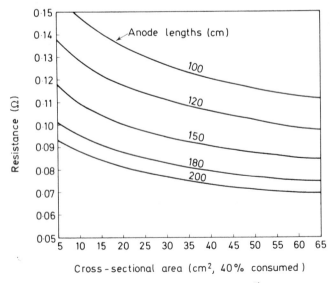

Fig. 11.14   Rod anode resistance in an electrolyte (sea-water); resistivity 25 Ω cm

Example 11.2   Given anode dimensions of $120 \times 6 \times 6$ cm and water resistivity of $25\,\Omega$ cm, then cross-sectional area reduced by $40\% = 6 \times 6 \times 0.6$ cm^2 $= 21.6$ cm^2. From the graph in Fig. 11.14 anode resistance $= 0.116\,\Omega$.

Example 11.3

Given: Aluminium anode

potential	$= -1.05$ V (Ag/AgCl)
Steel potential	$= -0.80$ V (Ag/AgCl)
Anode dimensions	$= 120 \times 6 \times 6$ cm
Water density	$= 1\,020$ g/l
Water temperature	$= 20°$C

Then: Water resistivity $= 25\,\Omega$ cm (see example 11.1)

Anode resistance $= 0.116\,\Omega$ (see example 11.2)

$$\text{Anode output} = \frac{\text{Aluminium potential} - \text{Steel potential}}{\text{Anode resistance}}$$

$$= \frac{1.05 - 0.80}{0.116}$$

$$= 2.15 \text{ A}$$

If the ratio of length to mean effective radius is small (less than $1:10$) then the resistance of the anode given in equation 11.14 may be simplified to apply to a rectangular bar anode

$$R = \frac{1.5}{(L + 0.8\,W + 0.8\,T)} \qquad \qquad ...(11.15)$$

where L = Length of anode (cm),
W = Width of anode (cm) and
T = Thickness of anode (cm).

The significance of anode output is that the higher the output (*a*) the smaller the number of anodes required to produce a specified current density relative to a fixed area of steel, and (*b*) the greater the anode consumption and shorter the anode life relative to an anode of the same weight, but of a smaller output.

Anode Capacity and Anode Efficiency

The *anode capacity* is the current which can be produced by a stated weight of anode material, and is usually expressed in ampere hours per kilogramme (Ah kg^{-1}). Owing to the fact that the efficiency is less than 100%, the capacity of an anode in practice is normally less than the theoretical faradaic capacity.

In theory, from Faraday's Law 1 gramme-equivalent of a metal $\equiv 96\,500$C. From this it can be calculated that 1 kg $Al \equiv 2\,981$ Ah. This means that if 1 kg of Al is dissolved electrochemically, and if all the charge is transferred to the metal to be protected, then 1 kg of Al will liberate $2\,981$ Ah of electrons for cathodic protection of the structure.

The significance of actual anode capacity (as opposed to the theoretical) is that this is a measure of the amount of cathodic protection an anode will give, and is the parameter against which the anode cost per unit anode weight should be evaluated.

The *anode efficiency* is the percentage of the theoretical anode capacity that is achieved in practice in the form of 'useful' current. This can be evaluated by means of a cell in which the anode to be studied is connected to steel in the electrolyte under consideration, and the weight loss of the anode is determined after a predetermined time. A copper coulometer is placed in series with the cell to give the total charge passed during the test. The significance of anode efficiency is that it provides a means of predicting the actual anode capacity of a given anode material under service conditions.

Anode and Cathode Potential

It must be emphasised that different reference electrodes are used as the standard against which the anode potential is related, and the reference electrode must always be specified. The most common reference cells used in practice are $Ag/AgCl$, $Cu/CuSO_4$, Hg/Hg_2Cl_2 and metallic zinc. In practice, the $Ag/AgCl$ electrode[22] is used for measuring anode or cathode potential in sea-water, and this is the reference electrode used in the potentials quoted in this section. The significance of anode potential is that the more negative the potential the greater the anode output with the associated benefits given previously for this, and the quicker the rate of polarisation of the steel to which the anode is attached. On the other hand, the more negative the potential the greater the current output and the greater the rate of consumption of the anode. Thus the potential of the anode should be sufficient to depress the potential of the structure to the protective value.

Area of Steel Requiring Protection

Sacrificial anodes are used to protect both bare steel and coated steel, provided it is submerged in an electrolyte or buried in soil. Anodes cannot protect steel exposed to the atmosphere.

The area of bare steel to be protected is usually calculated either from drawings or the actual structure itself. In practice the area is usually taken assuming the steel surfaces to be flat without corrugations, indentations or surface roughness. The area of exposed steelwork on a coated surface is calculated from (a) the amount of steelwork not covered by the original coating, and (b) the total area of steelwork that will become exposed during the life of the structure, due to coating breakdown and erosion. These areas of exposed steelwork are not calculated from any formula, but are based on experience gained from what has happened in practice in similar situations in the past. The significance of the area of steelwork is that the greater the area the greater the amount of anode material required for protection.

System Life

In different circumstances, sacrificial anodes are required to give cathodic protection for different periods of time. At the lower end of the scale protection for 1 year is required, and at the upper end of the scale 40 years' protection may be necessary. The significance of system life is that the greater the time of protection, the greater the quantity of anode material that is required, and the more sure one has to be about the quality of the product, and that it will perform in the long term in the same way as it performs in the short term.

Effect of Electrolyte Solution on Anode Performance

Sacrificial anodes will perform differently in electrolyte solutions of different chemical composition and of different concentrations. In practice, the concentration of the different constituents can vary over a period of time, and the only answer to this is to design for anticipated 'average' conditions. The significance of the chemical composition is that this governs the type and amount of anode material used. The most important parameter is the resistivity, but the specific nature of the electrolyte present in the solution must also be considered.

Anode performance is directly related to electrolyte temperature. The lower the temperature, the greater the resistivity of the electrolyte, and the lower the anode output. Some anode materials which operate at more elevated temperatures satisfactorily, passivate at lower temperatures.

Anode performance is directly related to flow rate of the electrolyte since this in turn affects the corrosion rate of the structure to be protected. In general the greater the flow rate, the greater the current density required for protection. It follows that the anode requirement is directly related to the aeration of the electrolyte. The greater the aeration of the electrolyte, the greater is the oxygen availability at the cathode and thus the greater the current density needed for protection.

Position in Space of the Anode Relative to the Steelwork[23]

If the sacrificial anodes are placed on or very close to steelwork, then the output from the face of the anodes next to the steelwork will be severely limited or restricted altogether. On the other hand, when anodes are located at an appreciable distance from the steelwork, part of the potential difference will be consumed in overcoming the water resistivity between the anode and cathode.

There is more art than science in distributing sacrificial anodes relative to one another to protect steelwork. In the simplest form, a standard geometrical form of distribution is used. Superimposed on this, and positioning relative to electrolyte flow, electrolyte aeration and areas of shielding have all to be taken into consideration.

Design Parameters[24]

Before a satisfactory cathodic protection design can be made using sacrificial anodes, the following information has to be available or decided upon:

1. Information on the area of the steelwork to be protected.
2. The type of coating, if any, that is to be used.
3. The length and frequency of time the steelwork is in contact with the electrolyte. In ships, certain ballast tanks are only full of ballast water for certain periods of time, and it is only in these periods of time that the sacrificial anodes operate.
4. The life required from the cathodic protection system.
5. The current density to be used to protect the structure has to be chosen, and different situations require different current densities. These current densities, in practice, are in the range $20 \, mA/m^2$–$4 \, 000 \, mA/m^2$.
6. The resistivity of the electrolyte must be determined in order that the right anode material is chosen. Tables 11.4 and 11.5 show the material to be used for various environments of different resistivities.

Table 11.4   Anodes for different resistivities of water

Material	Solution resistivity (Ω cm)
Aluminium	Up to 150
Zinc	Up to 500
Magnesium (-1.5 V)	Over 500

Table 11.5   Anodes for different resistivities of soil

Material	Soil resistivity (Ω cm)
Zinc with backfill	Up to 1 500
Magnesium (-1.5 V) with backfill	Up to 4 000
Magnesium (-1.7 V) with backfill	4 000–6 000

Method of Choosing the Weight and Number of Aluminium and Zinc Anodes

For any specific situation, the total anode weight, the total current required and the number of anodes which will meet these current and weight requirements are calculated as follows:

1. The wetted area of steel to be protected is calculated from drawings or direct measurements.

2. The total current $I(A)$ needed is:

$$I = \frac{\text{Area (m}^2) \times \text{Current density (mA/m}^2)}{1\,000}$$

3. The following formula gives the weight of *anode* material required (excluding insert):

$$\text{Anode weight (kg)} = \frac{\text{Area (m}^2) \times \text{Current density (mA/m}^2) \times \text{Life} \times 8\,760}{1\,000 \times \text{Capacity of material (Ah kg}^{-1})}$$

where 'life' is the design life in years $(1\,\text{y} = 8\,760\,\text{h})$.

4. The minimum number of anodes required per structure is assessed from the following formula:

$$\text{Minimum no. of anodes} = \frac{\text{Current density (mA/m}^2) \times \text{Area (m}^2)}{1\,000 \times \text{Anode output (A)}}$$

Note: the anode selected must satisfy both the total weight and total current output requirements as follows:

1. Weight requirement = No. of anodes × Individual net weight.
2. Current requirement = No. of anodes × Individual anode current output.

Current Densities for Protecting Steel

1. Current density requirements for ships can be summarised as follows:

Specific area	Current density (mA/m²)
External hull with high-duty paint	10
External hull with standard paint	15
Cargo/clean ballast tanks	86
Ballast only and ballast white-oil cargo tanks	108
Upper wing tanks	120
Fore and aft peak tanks	108
Tanks with painted surfaces	5
Lower wing tanks	86
Double bottom tanks, ballast only	86
Cargo/dirty ballast tanks	Dependent on trade

2. Current density requirements for off-shore structures can be summarised as follows:

Location	Current density (A/m²)
Gulf of Mexico	80–150
Nigeria	85
Cook Inlet, Alaska	250
Arabian Gulf	65–85
North Sea	90–150
Mud zone, all locations	20

Application of Cathodic Protection Using Sacrificial Anodes

In considering the most suitable anode for a given application a variety of factors have to be taken into consideration, amongst which the following are the most significant:

Magnesium: its major advantage is its negative potential, which makes it suitable for environments of high resistivity (up to $6000\,\Omega\,cm$); its major disadvantage is the high cost in relation to capacity; its incendivity is a serious disadvantage where there is a spark hazard.

Aluminium and zinc: these have more positive potentials than magnesium and their use is restricted therefore to low resistivity environments; aluminium is more economical than zinc (capacity 2830 and 780 A h per kg respectively), but has the tendency to passivate, which may be overcome by alloying with mercury (toxic) or indium; aluminium is unsatisfactory when immersed in mud or sand, and zinc is preferred under these circumstances.

Sacrificial anodes are used mainly to protect ships' externals[25-27], ships' internals[28-30], off-shore oil drilling platforms and rigs[31-33], underwater pipelines, underground pipelines[33], harbour piling and jetties[34], floating docks, dolphins, buoys, and lock gates. There are many other uses including a large range of industrial equipment where the surfaces are in contact with corrosive electrolytes.

W. B. MACKAY

REFERENCES

1. Hine, R. A. and Wei, M. W., 'How Effective Are Aluminium Anodes in Sea Water?' *Materials Protection*, Nov. (1964)
2. Carson, J. A. H., Phillips, W. L. M. and Wellington, J. R., 'A Laboratory Evaluation of Zinc Anodes in Sea Water', *Corrosion*, April (1960)
3. Wellington, J. R., 'The Low Potential Zinc Anode in Theory and Application', *Corrosion*, Nov. (1961)
4. Dempster, N. S. and Knuckey, P. J., *The Development of Zinc Alloy Anodes for Cathodic Protection of Marine Structures*, Fourth Annual Convention of the Australasian Corrosion Association, November (1963)
5. Bikkers, A. G., 'Cathodic Protection by Means of Zinc Anodes', *European Shipbuilding*, No. 5 (1962)
6. U.S. Military Specification MIL-A-18001H
7. Metallgesellschaft, Brit. Pat. No. 794 871 (14.5.58)
8. Groover, R. E., Lennox, T. J. Jr. and Peterson, M. H., *Electrochemical Characteristics of Six Aluminium Galvanic-Anode Alloys in the Sea*, National Association of Corrosion Engineers 26th National Conference, March (1970)
9. Newport, J. J. and Reding, J. T., The Influence of Alloying Elements on Aluminium Anodes in Sea Water, *Materials Protection*, Dec. (1966)
10. Dow Chemical Co., Brit. Pat. No. 1 006 755 (24.4.67)
11. Sumitomo Chemical Co. Ltd., Brit. Pat. No. 1 128 138 (25.9.68)
12. Reding, J. T. and Schrieber, C. F., 'Field Testing a New Aluminium Anode. Al-Hg-Zn Galvanic Anode for Sea Water Applications', *Materials Protection*, May (1967)
13. Mitsubishi Kinzoku Kogyo K.K., Brit. Pat. No. 994 401 (28.7.65)
14. Mitsubishi Kinzoku Kogyo K.K., Brit. Pat. No. 1 013 327 (15.12.65)
15. Hanada, M., Sakano, T. and Toda, K., 'Tests on the Effects of Indium for High Performance Aluminium Anodes', *Materials Protection*, Dec. (1966)

16. British Aluminium Co. Ltd., Brit. Pat. No. 1 118 302 (26.6.68)
17. Dow Chemical Co., Brit. Pat. No. 813 657 (21.5.59)
18. Robinson, H. A., 'Magnesium Ribbon Anode', *Gas*, April (1952)
19. Osborn, O. and Robinson, H. A., 'Performance of Magnesium Galvanic Anodes in Underground Service', *Corrosion*, April (1952)
20. Craven, D., *The Protected Gas Service*, Institution of Gas Engineers, Cardiff, March (1969)
21. Peabody, A. W., *Control of Pipeline Corrosion*, NACE, Houston (1971)
22. Groover, R. E. and Peterson, M. H., 'Tests Indicate the Ag/AgCl Electrode is Ideal Reference Cell in Sea Water', *Materials Protection and Performance*, May (1972)
23. Wilson, C. L., 'The Distribution of Current Densities at the Cathode Surfaces of Cathodic Protection Systems', *Anti-Corrosion*, Feb. (1970)
24. Davis, J. G., Doremus, G. L. and Graham, F. W., *The Influence of Environmental Conditions on the Design of Cathodic Protection Systems for Marine Structures*, Offshore Technology Conference, Dallas (1971)
25. Carson, J. A. H., 'Zinc as a Self-regulating Galvanic Anode for Ship Hulls', *Corrosion*, Oct. (1960)
26. Barnard, K. N., 'Prevention of Corrosion of Ships in Sea Water', *Research*, March (1952)
27. Preiser, H. S. and Tytell, B. H., 'Cathodic Protection of an Active Ship Using Zinc Anodes', *Journal of the American Society of Naval Engineers Inc.*, Nov. (1956)
28. Logan, A., 'Corrosion Control in Tankers', *Transactions of the Institute of Marine Engineers*, No. 5 (1958)
29. Cook, A. R. and Lattin, B. C., 'Zinc Anodes in a Crude Oil Tanker', *Materials Protection*, Nov. (1962)
30. Nelson, E. E., 'Galvanic Anodes and Fresh Water Rinse Reduce Tanker Corrosion', *Corrosion*, Oct. (1961)
31. Hanson, H. R., 'Current Practices of Cathodic Protection on Offshore Structures', NACE Conference, Shreveport, Louisiana (1966)
32. Compton, K. G., Reece, A. M., Rice, R. H. and Snodgrass, J. S., 'Cathodic Protection of Offshore Structures', Paper No. 71, *ibid.* (1971)
33. Tipps, C. W., 'Protection Specifications for Old Gas Main Replacements', Paper No. 27, *ibid.* (1970)
34. Nakagawa, M., 'Cathodic Protection of Berths, Platforms and Pipelines', *Europe and Oil*, July (1971)

11.3 Power-impressed Anodes

Impressed-current Anodes for the Application of Cathodic Protection

Numerous materials fall into the category of electronic conductors and hence may be utilised as power-impressed anode material. That only a small number of these materials have a practical application is a function of their cost per unit of energy emitted and their electrochemical and mechanical durability.

These major factors are interrelated and—as with any field of practical engineering—the choice of a particular material can only be related to total cost. Within this cost must be considered the initial cost of the cathodic protection system and the maintenance, operation and refurbishment costs during the required life of both the plant to be protected and the cathodic protection system.

There are obviously situations which demand considerable over-design of a cathodic protection system, in particular where regular and proficient maintenance of anodes is not practical, or where temporary failure of the system could cause costly damage to plant or product. Furthermore, contamination of potable waters by anodes such as iron-based or lead-based alloys may lead to the choice of the more expensive but more inert platinum-coated anodes. The choice of material is then not unusual in being one of economics coupled with practicability.

Although it is not, in all cases, possible to be specific regarding the choice of anode material, it is possible to make a choice based upon the comparative data which are at present available. Necessary factors of safety would be added to ensure suitability where lack of long-time experience or quantitative data necessitate extrapolation or even interpolation of an indefinite nature.

Bearing in mind that the choice of the most suitable material must usually be by comparison of characteristics, the data are presented in tabular or graphical form wherever applicable.

The manufacture, processing and application of a particular material as a power-impressed anode requires knowledge of several physical characteristics. Knowledge and attention to these characteristics is necessary to design for anode longevity with maximum freedom from electrical and mechanical defects.

The various types of materials used as anodes in power-impressed systems may be classified as follows:

1. *Rarer metals*
   - (*a*) Platinised tantalum   (see also Section 5.5)
   - (*b*) Platinised niobium   (see also Section 5.3)
   - (*c*) Platinised titanium   (see also Section 5.4)
   - (*d*) Platinised silver
   - (*e*) Platinum metals   (see also Chapter 6)
   - (*f*) DSA titanium   (see p. 11.41)
2. *Ferrous materials*
   - (*a*) High-silicon chromium iron ⎫
   - (*b*) High-silicon molybdenum iron ⎬ (see also Sections 3.8 and 3.9)
   - (*c*) High-silicon iron ⎭
   - (*d*) Cast iron
   - (*e*) Steel
   - (*f*) Iron
   - (*g*) Stainless steel
3. *Lead materials*
   - (*a*) Lead–antimony–silver ⎫
   - (*b*) Lead/platinum bi-electrodes ⎬ (see also Section 4.3)
   - (*c*) Lead dioxide/titanium
   - (*d*) Lead dioxide/graphite
4. *Carbonaceous materials*
   - (*a*) Graphite
   - (*b*) Carbon
   - (*c*) Graphite chips
   - (*d*) Coke breeze
5. *Reactive non-ferrous metals*
   - (*a*) Aluminium   (see also Section 4.1)
   - (*b*) Zinc   (see also Section 4.7)
6. *Combination anodes*

There are cases where a combination of materials is used, generally co-axially, to extend the life of the prime anode, to reduce resistance to earth, to facilitate installation and to improve mechanical properties, e.g.:

1. Canned anodes: steel casing/carbonaceous extender/graphite rod or silicon-iron rod current conductors.
2. Groundbeds: carbonaceous extender/graphite rod or silicon-iron rod anodes, scrap steel, or platinised titanium current conductors.
3. Co-axial anodes: copper-cored platinised titanium or platinised niobium.

For long lengths of anode it is sometimes necessary to extrude one material over another to improve a particular characteristic. Thus titanium may be extruded over a copper rod to improve longitudinal conductivity and current attenuation characteristics of the former; lead alloys may be treated similarly[1] to compensate for the poor mechanical properties of these alloys. It should be noted that these anodes have the disadvantage that should the core metal be exposed to the electrolyte by damage to the surrounding metal, rapid corrosion of the former will occur.

Where a tubular electrode is used for economy in cantilever in flowing water and is thus subjected to vibration, the mechanical characteristics can be modified by the insertion of other materials. The flexural strength and centres of mass may then be suitably altered to improve resistance to fatigue failure within the range of water velocities involved.

Platinum and Platinum-coated Anodes

The properties of platinum as an inert electrode in a variety of electrolytic processes is well known, and in cathodic protection we are concerned with utilising these whilst keeping material volume, and hence cost, to a minimum.

Tests carried out in the USA and initiated in 1953[2] indicate the following consumption rates of precious metals and their alloys: Pt, Pt–12Pd, Pt–5Ru, Pt–10Ru, Pt–5Rh, Pt–10Rh, Pt–5Ir and Pt–10Ir—6 to 7 mg $A^{-1}y^{-1}$; Pt–20Pd, Pt–50Pd, Pt–20Rh and Pt–25Ir—slight increase in rate; Pt–50Pd—greater increase in rate although the lower cost of palladium may offset this; Pd, Ag, Pd–40Ag, Pd–10Ru and Pd–10Rh—excessive rate. The tests were carried out for periods of some months at current densities of from 54 to 540 mA/cm^2, and the results appear to be independent of current density and duration of test.

Tests carried out in the UK[3] on electrodeposited platinum on a titanium substrate indicate a consumption of the order of 8·63 mg $A^{-1}y^{-1}$.

Platinised Titanium

Titanium, which was in commercial production in 1950[4], is intrinsically a very reactive metal (machining swarf can be ignited in a similar fashion to that of magnesium ribbon) that has a strong tendency to passivate (see Section 5.4). When made anodic in a chloride-containing solution it forms an anodic oxide film of TiO_2 (rutile form) that thickens with increase in voltage until 8–12 V when localised breakdown of the film occurs with subsequent pitting. The film has a high electrical resistivity, and this, coupled with the fact that breakdown occurs at voltages that can be achieved by rectifier-transformers used in cathodic protection, makes it unsuitable for use as an anode material. Nevertheless, it forms a most valuable substrate for platinum, and the composite anode is characterised by the fact that the titanium exposed at discontinuities in the platinum is protected by the anodically formed dielectric oxide. Platinised titanium provides an economical method of utilising the inertness and electron conductivity of platinum, which is applied to the titanium in the form of a very thin coating (0·0025 mm).

Titanium can be forged, bent, cut, stamped, rolled, extruded and successfully welded under argon, making possible the large variety of electrode shapes that have been manufactured. It is a very light but strong material. Both titanium and platinum have high resistance to abrasion; in particular, electrodeposited platinum is much harder than the cold-rolled form.

Cotton[5] was the first to publish results on the use of platinised titanium as an anode material and found that the breakdown voltage of the unplatinised titanium was in excess of 12 V in sea-water. Between 1956 and 1957 investigations were made into breakdown voltages and a patent was applied for in 1957. The first experimental installation was completed in 1957 at Thameshaven on a Thames-side jetty. Following this, at the same site, the first commercial installation was completed in 1958. This was still operating in 1966[4].

The availability of this material has allowed radical advancement in the cathodic protection of immersed surfaces[6, 7], and the material has been operated[8] as high as 540 mA/cm^2. Even though the applied voltage required for this is above the normal breakdown potential the potential at the titanium substrate exposed at pores in the platinum apparently does not exceed 2·5 V (S.H.E.)[9]. It has operated successfully at 270[10] and 215[11] mA/cm^2 in high velocity sea-water. At the necessarily high voltages there is, of course, danger of rapid failure should an area of titanium be bared by mechanical damage and the anode/cathode configuration be such as to produce a relatively high voltage at the exposed area[12]. Some figures for breakdown (as measured by a close electrode) that have been published are shown in Table 11.6.

Table 11.6 Breakdown potentials of titanium in various environments

Ref. no.	Electrolyte and conditions	Breakdown of commercial titanium (V)
3	Tests in pure sea-water at ambient temperatures	8·5–15
3	Tests in NaCl from saturated to 5 g/l, below 60°C	8·5–15
13	Sea-water	12–14
13	Sulphuric acid	80–100
14	Chlorides	8
14	Sulphates	60
14	Carbonates	60
14	Phosphates	60
14	Bromides	<2
14	Iodides	<2
15	Sulphates	>80
15	Phosphates	>80
16	Ratio of sulphate plus carbonate to chloride ions, >4:1	>35

Note: further published figures are available on behaviour in halide solutions[17].

Early failures of platinised titanium anodes have usually been found to occur for reasons other than for platinum corrosion rates as indicated in Table 11.7. The following are examples:

1. Attack of the substrate in over-acid conditions as mentioned earlier, e.g. when covered in mud. Marine growth prior to energising has been suggested as a further cause[20]; a commercial guarantee requires this period to be not longer than 8 weeks[14].
2. Attack on the substrate by contact with magnesium hydroxide (see Note 3 in Table 11.7). This can be removed without damage to the anode by means of hot nitric acid of suitable concentration.
3. Fatigue fracture from under-designing or changes in plant operation, particularly of cantilever anodes from Kármán vortex forces[22].
4. Attack of bare substrate used at voltages incompatible with a particular electrolyte.

5. Poor regulation of transformer allowing voltage to rise with fall of current during a period of high resistivity of the electrolyte (e.g. this would vary in estuarine regions with change of tide). It is not advisable to have any part of the substrate bare when working near the voltage limit.

6. Insufficient attention to field-end effects, particularly on cantilever anodes. If an anode of length some 10 times that of its diameter extends away from a cathode surface, the end of its surface close to the cathode may be operating at a current density of some 50 to 100 times that of the mean. The life of the platinising in this region would then be reduced in inverse proportion to the current density. The loss is progressive and produces a progressively higher mean current density if the voltage is raised to compensate for the resulting increase in resistance. This can, of course, lead to rapid reduction in output or complete failure should the breakdown voltage be exceeded.

Table 11.7 Corrosion rates for platinum and platinised titanium

Ref. no.	Conditions	Corrosion rate $(mg\,A^{-1}\,y^{-1})$
2	Pure Pt in sea-water 54–540 mA/cm²	6–7
3	Pure Pt in sea-water 500 mA/cm²	13·14
3	Pure Pt in sea-water 130 mA/cm²	8·76
3 and 18	Pt/Ti in sea-water 32–320 mA/cm² (0°–15°C—d.c. 100 Hz ripple and above)	8·76 average[1]
3	As above with high-frequency chopped supply	13·8[2]
19	Pt/Ti in highly acid conditions caused by deposits, etc.	up to 100 × normal rate[3]
20	Indicated by replacement sales in UK of Pt/Ti	8·76–17·52[4]

Notes: 1. Not in excess of wrought Pt, and frequently much lower.
2. More tests required to confirm.
3. Significant cause of shortened anode life is detachment of active layer by chemical attack of substrate via porosity in low pH stagnant environments[21]. Rapid loss of platinum has been noted in cathodic protection and brine electrolysis where magnesium and calcium hydroxide produced by the high current density on a close cathodic surface have bridged across to the anode. Recent techniques have made Pt/Ti available with low coating porosity and hence greater resistance to acid environments.
4. Reductions in life probably for reasons in 3 above or mechanical failure.

Platinised-titanium installations have now been in use for 15 years for jetties, ships and submarines[23] and for internal protection, particularly of cooling-water systems[24]. For the protection of heat exchangers an extruded anode of approximately 6 mm in diameter (copper-cored titanium–platinum) has shown a reduction in current requirement (together with improved longitudinal current spread) over cantilever anodes of some 30%[15]. This 'continuous' or coaxial anode is usually fitted around the water box periphery a few centimetres away from the tubeplate. Consideration has to be given to the maximum distance between feed points regarding current attenuation. This depends upon the relationship between anode leakance (dependent upon electrolyte resistivity and distance to cathode) and its longitudinal conductivity. Information on this has been produced in the form of computer plots[25].

A directional strip anode[24, 26], by means of which some determination of the leakance/conductance characteristics is possible, is now in use and is proving successful. Experimental and practical results show improved axial as well as longitudinal current spread with consequent further reduction in both polarisation period and total current.

Platinised titanium has been used in perhaps the largest variety of situations of any anode material. It is not usually recommended for underground applications, but has been utilised in this manner at low current densities; the examples below are typical[4, 15].

1. Expanded mesh used in a small coke breeze groundbed at 21.6 mA/cm^2 (calculated on the basis of the area of the platinised titanium); however, slight pitting of the unplatinised titanium lead wire was noticed after use for 21 months.
2. As d.c. earth returns for submarine repeater stations. Rods were buried in sand and operated at 2.16 mA/cm^2. Experiments at higher current densities showed rapid deterioration of the platinum.
3. A 3.2 mm diameter wire driven into soil with a steel driver was still operating at 6.5 mA/cm^2 after 3 years.

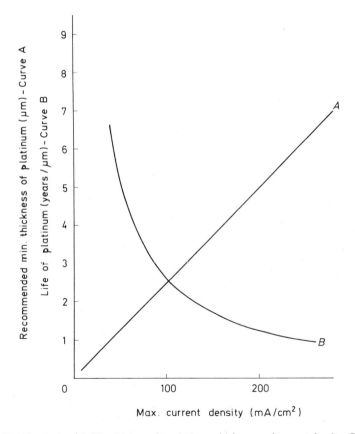

Fig. 11.15 Nominal anode life with regard to platinum thickness and current density. Curve *A* shows the relationship between thicknesses of platinum consumed and current density; curve *B* shows the relationship between years of platinum per micro-metre and current density

Based on information provided by Marston Excelsior Ltd.[3,14] a guide on nominal anode life with regard to platinum thickness and current density is given in Fig. 11.15 in which curve *A* shows the relationship between the thickness of platinum consumed and current density, and curve *B* the relationship between years/µm of Pt and current density.

Platinised Niobium and Platinised Tantalum

The principle of these anodes is similar to that of platinised titanium, since both niobium and tantalum are 'valve' metals that form dielectric oxide films. Platinised tantalum was suggested as an anode as early as 1922[27]. These are very much more costly than platinised titanium, but this may be offset where their superior electrical conductivities and higher breakdown voltages[28] are required. Table 11.8 gives the breakdown voltages in various solutions under laboratory conditions, which should be considered as comparisons only. Physical and mechanical properties of niobium, titanium and tantalum are given in Sections 5.3–5.5.

Table 11.8   Comparison of breakdown voltages (V)[29]

Solution	Ti	Nb	Ta
Sea-water	9	115	155
Sulphate/carbonate	60	255	280
Phosphate/borate	80	250	280
Drinking water (Essex)	37·5	250	280

Both niobium and tantalum are as versatile in fabrication as titanium, but have a tendency to gall and care is needed to produce a good finish, particularly in the case of tantalum. Both metals may be welded under argon. The costs of the special care required to manufacture anodes from these metals, including vacuum heat-treatment before plating, of course adds to the high initial costs. Niobium is approximately twice the density and one third of the resistivity of titanium, and tantalum four times and one quarter, respectively.

The application of these materials has been mainly restricted to marine work, e.g. shipping anodes which may pass into high resistivity water and thus require higher driving voltages, and on fixed structures for high current/electrode length anodes. Since niobium is significantly cheaper than tantalum the use of the latter as a substrate for platinum has become rare.

Metallurgically bonded co-axial composite rods 'Tibond' (Cu/Ti/Pt) and 'Niobond' (Cu/Nb/Pt) are now available (Marston Excelsior, England) and have a number of advantages over the electroplated analogues.

Platinised Silver

This material can be used only in sea-water or similar chloride-containing electrolytes. The reason for this is that passivation of the silver at discontinuities in the platinum is dependent on the formation of a film of silver chloride, which is only slightly soluble in sea-water so that corrosion is inhibited. This anode (Pt–10Pd on Ag) was tried as a substitute for rapidly consumed aluminium in the early method of trailing wire anodes for the protection of ships' hulls, and has been operated as high as $194\,mA/cm^2$ with success. However, the use of trailing anodes has been found inconvenient regarding ships' manoeuvrability and the consequent maintenance required.

With the advent of hull-mounted anodes this material has been supplanted by platinised titanium.

DSA-Titanium

A relatively new material is Dimensionally Stabilised Anode, a proprietary name sometimes shortened to DSA. This material was originally developed by H. Beer of the Netherlands (US Patent No. 3 632 498) and consists of a thin film of ruthenium oxides baked on to a titanium substrate[30].

To date, the major application has been in the electrolytic production of chlorine and hypochlorites, but tests are under way in the USA to determine the wear rate of the surface. Indications are that this may be of the same order as platinum, but at present there are few published results of practical trials that provide a means of comparing it with platinised titanium as a possible anode for cathodic protection.

Ferrous Materials

Steel

One of the earliest materials to be used in power-impressed cathodic protection was steel. Its economy lies in situations where scrap is available in suitable quantity and geometry.

The anode tends to give rise to a high resistance polarisation due to the formation of voluminous corrosion product, particularly when buried as opposed to immersed. This can be alleviated by closely surrounding the scrap with carbonaceous backfill; this of course reduces its economy if the backfill is not also a local by-product. It is necessary under conditions of burial to ensure compactness and similarity of backfill (earth or carbon) at all areas on the steel, otherwise particularly rapid loss of metal at the better packed areas could lead to decimation of the groundbed capacity.

The problem of the high resistance polarisation becomes less with increasing water content and salinity such as prevails during immersion in sea-water, where these anodes are particularly useful. Since no problems of burial arise in that environment an endless variety of disused iron-ware has been utilised for anodes: pipes, piling, machinery, rails and even obsolete shipping which it is not economic to salvage. Experimental installations have been established from time to time to demonstrate the possibility of using ferrous metals in anolytes[31-33] selected to minimise polarisation and to reduce metal ionisation by making the metal passive. In particular, caustic soda solutions contained in a coaxial porous pot have been used for iron anodes. With suitable solution concentration and current density, metal loss can be very low, since the behaviour of such anodes is similar to that of the positive plate of the alkaline storage battery[34].

The use of carbonaceous extender is of value if segregation of a steel anode in soil might be expected to result from high local corrosion rates. Continuity of the anode is facilitated by the bridging effect of the extender. One example of this is in deep-well groundbeds installed in stratifications of widely differing ground resistivities, where a well casing may be filled with coke breeze. Commercial examples of these are known to be working well after periods of 28 years[35].

One advantage of steel as an anode is the low gassing at the electrode during operation (the predominant reaction is the corrosion of the iron) so that the problem of resistive polarisation due to gas blocking, as may be the case with more inert materials, does not occur. Iron compounds do, of course, form, but these do not appreciably affect the anode/soil resistivity. Furthermore, the introduction of metallic ions by anode corrosion into the adjacent high resistivity soil is beneficial in lowering the resistivity.

It is necessary to ensure the integrity of anode cable joints and to give consideration to the number of connections related to longitudinal resistance and leakance if early failure is to be obviated.

Cast Iron

Cast iron may be used under similar circumstances, but has inferior mechanical properties. It has been used, although not in current practice, for internal cathodic protection, where it has been demonstrated that the presence of ferrous ions in the water is of benefit in passivating the copper alloy tube-plate and tubes[36]. Water treatment with ferrous sulphate (dosage$\simeq$1 p.p.m.) has now been found to be a more practical method.

During corrosion the anodes become graphitised with a consequent decrease in the corrosion rate of the cast iron (gas evolution occurs on the graphite layer). However, should the graphitised layer break up, the graphite particles may cause localised corrosion up-stream of the anode.

Iron

Swedish iron is still used extensively as galvanic wastage plates in heat exchangers, particularly for marine applications. This is possibly based on tradition, since it cannot be the most economical method in the light of current cathodic-protection practice. In addition, iron may be used as a power-impressed anode in a similar fashion to steel.

Stainless Steel

Stainless steel has been tried as an inert anode, mainly under laboratory conditions and with only partial success. Even at low current densities in fresh water the majority of alloys pit rapidly, although others show the ability to remain passive at a low current density[7, 37]. However, at practical current densities, the presence of chloride ions, deposits on the anode or differential aeration at support leads to rapid failure[38], but it may be possible that stainless steel could be of service under certain conditions and with particular alloys[39].

High-silicon Iron (HSI)

This is reported as being first developed in 1912[40]. It is a hard, brittle material unable to sustain thermal or mechanical shock (e.g. dropping on hard standing), although this is not a deterrent to experienced handlers. It has a similar performance to graphite when used in carbonaceous backfill, but is superior in the absence of the latter in low resistivity soils. It allows the use of higher current densities, but has a similar economic life. A typical analysis of HSI for anodes is: 14·4% Si, 0·7% Mn, 0·95% C, remainder Fe.

The only practical method of machining is by grinding, and to obviate machining it is cast into fairly standard sizes to suit the general requirements of industry. HSI has a long successful history as a corrosion-resistant material in the chemical industry (see Section 3.9) for such items as pumps for acids, and has been used in this capacity for more than 50 years. Used anodically it readily forms a protective film which is replenished if mechanically removed. This is grey-white in appearance and has a tendency to flake under the compressive stress produced at thickened areas. The film is 50% porous and comprises 75% silicon dioxide[40]; it is a fairly good electron conductor.

There is a tendency when buried, as opposed to immersed, for the surface resistance of such an anode to increase, but not to an extent that affects performance. The large resistance changes sometimes reported are usually due to gaseous polarisation (gas blocking) caused by poor venting or inadequate backfill.

HSI anodes are subject to severe pitting by halogens and this precludes their use in sea-water or other environments in which these ions may be present in quantity. They are ideal for fresh-water applications (below 200 p.p.m. Cl^-) although not for temperatures above 38°C. The addition of molybdenum or chromium to the alloy can improve performance under these conditions[40].

The wastage rate of HSI depends upon the current density, and from collated experience in fresh-water conditions the approximate rate is about $0.5\,kg\,A^{-1}\,y^{-1}$. This is, of course, dependent upon the water analysis and temperature. In a carbonaceous backfill a lower rate would be expected with the backfill correctly installed and an insignificant chloride content. This would be in the region of 0·25, and when buried directly in the soil in the region of $1\,kg\,A^{-1}\,y^{-1}$. Very much higher apparent rates of wastage are most likely due to high local current densities, caused when the anode is inadequately backfilled, partially submerged or where it has become partially silted up.

The anode effectiveness is only as good as the anode connection and loss of insulation at this point by deep pitting of the HSI or penetration of the anode cable seal will bring about rapid failure. Hydrostatic pressure should be borne in mind when considering the seal required for any depth of water. The useful life of HSI anodes is usually considered at an end after a 33% reduction in diameter has taken place, but this depends upon the original diameter, the amount of pitting sustained and the mechanical stresses to be withstood. Thus doubling the cross-sectional area may more than double the effective life of the anode.

High-silicon/Molybdenum Iron

The addition of 1–3% Mo to HSI results in an improvement in maintaining an oxide film in chlorides above 200 p.p.m. or temperatures of 38°C or above. However, the addition of chromium has shown even greater improvements. A wastage rate of $0.9 \, kg \, A^{-1} y^{-1}$ at $1.08 \, mA \, cm^{-2}$ has been reported for the molybdenum-containing irons in chloride-containing waters[41].

High-silicon/Chromium Iron (HSCI)

This alloy was put into commercial use about 1959. The chromium, together with the silicon, result in a film that is very resistant to pitting in waters containing halogen ions and these alloys can be used in sea-water with confidence. A typical analysis is: 14.4% Si, 0.7% Mo, 1.0% C, 4.25% Cr, remainder Fe.

Neglecting possible mechanical damage, and anode/cable joint failure, then it is possible in view of the very minor pitting in free suspension for the anode to continue operating until totally consumed.

Comparative tests between HSI and HSCI in sea-water at 93°C and 1.08 $mA \, cm^{-2}$ show consumptions of $8.4 \, kg \, A^{-1} y^{-1}$ and $0.43 \, kg \, A^{-1} y^{-1}$ respectively[40]. It can be seen from these figures that the wastage of HSI by pitting when used in sea-water without the addition of chromium can approach that of steel, but because of the very deep pitting and its fragility it is in most cases inferior to steel. However, in fresh-water conditions HSI has a far lower corrosion rate than steel.

Lead Materials

Investigations into the use of several lead alloys for cathodic-protection anodes were made in the early 1950s[42–44] and a practical material was developed by 1954. The general use of lead alloys in sea-water service had previously been established[45].

Lead has a relatively low melting point which facilitates casting with little problem. It is a malleable and ductile material and can be machined with ease. Sections can be joined by lead burning using a welding flame (e.g. hydrogen). It will readily bond to tinned copper connections.

In using the materials as an anode the formation and maintenance of a hard layer of lead dioxide (PbO_2) is essential, since this is the actual anode material, the lead acting as a source of PbO_2 and as an electrical conductor. Lead dioxide is insoluble in sea-water and its dissipation is more usually from mechanical wear and stress than electrolytic losses, Lead dioxide, a dark chocolate brown compound with an electronic conductivity about $\frac{1}{10}$th of that of lead, is a strong oxidising agent and unlike the lower oxides is unaffected by most acids. However, it is generally unsuitable for alkaline conditions approaching pH 10 and for this reason should be mounted to remain clear of calcareous deposits that may be formed on a cathodic area close to an anode, since the solution permeating the calcareous deposit will have a high pH and will thus attack and remove the lead dioxide film. Lead has a considerable history as an anode in a variety of electrometallurgical processes. Its resistance to wet or dry chlorine at temperatures of up to 100°C has been utilised in fabrications for brine electrolysis.

Pure lead has been tried as an anode, but even in sea-water it fails to passivate, and lead chloride forms beneath the lead dioxide thus insulating the latter from the lead. Several lead alloys have been tried as anodes in seawater with the following results: Pb—failure[46]; Pb+6 to 8Sb—required more than 20 mA/cm^2 to passivate[45]; Pb–6Sb–10Sn—high corrosion rate[45]; Pb–1Ag corrosion rate 10 times lower, lower passivating current density but soft PbO$_2$[46]; Pb–6Sb–1Ag—hard PbO$_2$, practical anode that has been put into commercial use[26, 47–50]; Pb–6Sb–2Ag might be slightly better but in the region of 50% more costly.

The first large scale industrial application of power-impressed cathodic protection was made possible by the Pb–6Sb–1Ag lead alloy[26, 50]. This work was backed up by extensive laboratory tests[47–49]. The following typical results from tests on Pb–6Sb–1Ag samples are of interest in recognising the scope of a practical lead alloy in waters of differing resistivities.

Table 11.9 Behaviour of Pb–6Sb–1Ag anodes

Resistivity of electrolyte at 35°C (Ω cm)	Average wastage at 10·8 mA/cm^2 (kg A^{-1} y^{-1})	Length of trial (days)	Note
16·3 (sea)	0·086	236	1
16·3 (NaCl)	1·99	1·75	2
50 (sea)	0·0145	236	1
50 (NaCl)	0·654	1·75	3
1 000 (sea)	23·80	5·75	4
1 000 (NaCl)	23·70	1·75	4
5 000 (sea)	0·10	236	5
5 000 (NaCl)	11·64	1·75	5

Notes: 1. Service indicates a practical consumption of between 0·057 and 0·114 kg A^{-1} y^{-1}. Under laboratory conditions PbO$_2$ has been formed at as low as 2·16 mA/cm^2. Typical operating current densities are 5·4 to 27 mA/cm^2 at wastage rates of 0·045[10] to 0·082[48].

2. Similar performance between 0·27 and 27 mA/cm^2; thin adherent film of PbO$_2$[49].

3. Similar performance between 0·27 and 16 mA/cm^2; thick nodules of PbO$_2$ in some areas, severe deterioration at >27 mA/cm^2[51].

4. Tests have indicated failure to form PbO$_2$, rapid deterioration, although at 10 mA/cm^2 it slows down after several weeks. Increasing silver content shows some improvement[49]. Anode passivated in 16·3 Ω cm water continues to operate whilst PbO$_2$ is undamaged[47].

5. Above 2·2 mA/cm^2 deterioration rate may be low, but PbO$_2$ coating is poor and interspersed with PbCl$_2$.

In electrolytes containing both sulphate and chloride ions the sulphate ion favours the formation of lead sulphate that is rapidly transformed to the dioxide. The continuing satisfactory operation of the anode depends upon the initial conditions of polarisation. The peroxide is of better quality and more adherent when formed at not exceeding 10·8 mA/cm^2, in solutions containing higher sulphate concentrations or when the water is agitated[47].

If lead alloy is used as a ship's hull anode, consideration should be given both to the make-up of the water in which the anode is initially passivated and that in which it will normally operate. The same consideration will apply for static structures in estuarine waters.

It should be noted that lead dioxide will discharge if electronically connected to a baser material, when in an unenergised state. The reverse current leakance of a rectifier will allow this to a small extent if the rectifier is faulty, with the consequent formation of lead chloride and corrosion of the anode.

Lead Dioxide on Other Substrates

Lead dioxide on graphite or titanium substrates has been utilised in perchlorate cells[52]. Lead dioxide on a titanium substrate has been tested for use in the cathodic protection of heat exchangers[10].

Lead/Platinum Bi-electrodes

The insertion of platinum micro-electrodes into the surface of lead and some lead alloys has been found to promote the formation of lead dioxide in chloride solutions[53–54]. Experiments with silver and titanium micro-electrodes have shown that these do not offer this improvement[47].

Similar results to those when using platinum have been found with graphite and iridium, and although only a very small total surface area of micro-electrodes is required to achieve benefit, the larger the ratio of platinum to lead surface, the faster the passivation[53]. Platinised titanium micro-electrodes have been utilised. Lead dioxide will readily form on lead with a platinum electrode as small as 0·076 mm in diameter[53–55].

It has been observed that the current density on the platinum is considerably less than on the lead peroxide once polarisation has been achieved; the proportion of current discharged from the platinum decreasing with increase in total current.

The additions of antimony, bismuth and tin to the lead appear to be detrimental. There is indication that 0·1% silver is almost as effective as 1%, and as low as 0·01% has been utilised. Dispersion-hardened lead alloys have been unsatisfactory, showing pronounced spalling in the direction of extrusion[56]. Pb–0·1Te–0·1Ag has been used with apparent success.

A typical anode for practical use would be in the order of 25 to 40 mm in diameter, with hard platinum alloy pins of 0·75 mm diameter by 10 mm long every 150 to 300 mm progressively positioned around the circumference. The pins are a press fit into holes in the lead alloy (approximately 0·1 mm diametric interference) and lie flush with the surface; the lead is then peened around the pins to improve mechanical and electrical contact.

The action of the platinum micro-electrode has been extensively studied, but the precise mechanism of its action is still not fully understood. Trials carried out by Peplow[11] have shown that the lead platinum bi-electrode can be used at high current densities, and that blister formation with corrosion under the blisters is decreased by the presence of the platinum micro-electrode. Work in this field[53, 56, 57, 59] can be summarised as follows:

1. It acts as a stable electrode for nucleation of PbO_2.
2. In the case of a lead anode (without a platinum micro-electrode) the PbO_2 thickens during prolonged polarisation with the consequent development of stresses in the film.
3. These stresses result in microcracks in the PbO_2 thus exposing the underlying lead, which corrodes with the formation of voluminous $PbCl_2$, resulting in blisters; the resistance of the anode increases and high voltages are required to maintain the current (if the voltage is maintained constant the current falls to a low value).

The platinum micro-electrode appears to act as a potentiostat and maintains the potential of the Pb/solution interface at a crack at a value that favours the re-formation of PbO_2 rather than the continuous formation of $PbCl_2$ with consequent excessive corrosion.

It is known that an increase in the resistance of the electrode indicates that corrosion is taking place, and this is confirmed in practice by observation of the anodes which reveal localised areas coated with white corrosion products although the PbO_2 remains intact at other areas. However, it is possible that an insulating film forms over the whole surface thus isolating the conducting PbO_2 from the lead, and Wheeler[58] has submitted the view that the sole function of the platinum is to provide a conducting bridge between the lead and the PbO_2. However, it has been demonstrated that although initially the PbO_2 nucleates at the surface of the platinum, the initially formed $PbCl_2$ is rapidly converted into PbO_2 that is in direct contact with the lead[53].

The formation of PbO_2 is favoured in solutions containing passivating anions such as SO_4^{2-} and in chloride solutions of intermediate concentrations; very high and very low concentrations of chloride inhibit formation of PbO_2. The platinum–lead bi-electrode performs best in sea-water, and is not recommended for use in waters of high resistivity.

Carbonaceous Materials

Carbon

The corrosion product is predominantly carbon dioxide, but considerable amounts of free oxygen are produced at the anode surface, particularly in fresh-water applications. This can attack both the carbon and any organic binders used to reduce its porosity. For this reason carbon anodes for underground service must be used in conjunction with a carbonaceous backfill.

If all the oxygen produced were to combine with the carbon the wastage rate would be of the order of $1 \text{ kg A}^{-1} \text{ y}^{-1}$[(60)]. However, in practice the rate is usually of the order of $0.2 \text{ kg A}^{-1} \text{ y}^{-1}$, and in coke breeze may be as low as $0.05 \text{ kg A}^{-1} \text{ y}^{-1}$. In sea-water where chlorine is the predominant gas produced, to which carbon is immune, any oxygen formed will be quickly removed and the corrosion rate may be very low.

Graphite

Graphite has now superseded carbon as a less porous and more reliable anode material, particularly in saline conditions; although their mechanical properties are similar, graphite has only 10% of the specific resistivity of carbon. The physical properties of graphite (or carbon) may vary by 50% transversely to longitudinally, according to the method of manufacture. It is principally manufactured from petroleum coke.

The performance of graphite in sea-water where chlorine is the principal gas evolved, is considerably better than in fresh water where oxygen is produced freely. Graphite is immune to chlorine and has a long history in the chemical industry in this and similar capacities[61].

It is current practice to impregnate the graphite, traditionally with linseed oil, although synthetic resins are also successful. The concept behind impregnation is to reduce porosity and hence inhibit sub-surface gas evolution or carbon oxidation which would initiate spalling and early anode failure. From this it can be seen that the ideal impregnant should corrode away at the same rate as the graphite; if slower an increase in surface resistance could occur giving rise to localised higher-than-average current densities and leading to a shortened life. Apparently electrode processes occur to a depth of 0·5 mm below the surface of the anode and the true current density can be shown to be only 1/400th of the geometrical value[62].

Acidity has been found to increase the wear rate[63] and so has the presence of sulphate ions[8]. These factors might be considered with regard to the operating environment and the chemical treatment of backfill.

The material can be easily machined being a natural lubricant, it has a negligible contact resistance and it is relatively simple to make a sound cable joint, although the comments regarding cable/anode joints raised under High-Silicon Iron still apply. The material can be d.c. welded under high-pressure argon. It is brittle but a little more shock resistant than silicon iron in that it can absorb energy by localised damage; it is of course a lighter material to handle.

The anode is not recommended for use in water at above 50°C, where the consumption increases rapidly and erratically. It is no longer the practice to use this material in cooling-water plant where secondary attack from contact with the relatively noble pieces of anode may occur, should damage take place.

Table 11.10   Performance of graphite

Environment	Wastage $(kg\,A^{-1}\,y^{-1})$	Current density (mA/cm^2)	Reference no.
Backfill	0·9	$\leq 1·08$	64
Hot water	0·9	—	46
Sea-water	0·045	0·43–10·7	41
Sea-water	Little	$< 1·08$	46
Fresh water	0·45	0·35	46
Fresh water	0·45	0·27	41
Mud	1·36	3·24	41

The wastage rate of graphite is lower in sea-water at the higher current densities because of the preferential evolution of chlorine. Table 11.10 gives some results obtained with graphite under different conditions. Results obtained with one particular installation using a 1 m × 100 mm diameter anode operating at 0·69 mA/cm^2 indicate a predicted life of 20 years[65].

Carbonaceous Backfills

Coke breeze is used as an anode extender thus producing an anode with an enormous surface area, its main component being carbon. By virtue of its porosity it gives a large volume to weight ratio of conducting media suitable for anodic conditions. This allows the economic extension of the ground anodes both linearly, for decrease in resistance to ground, and volumetrically for longevity. The grading of the coke is of some importance in that too large a grade offers large local contact resistance, leading to uneven consumption. Too much 'fines' leads to over-tight compaction and gas blocking (gaseous polarisation). Chemicals are sometimes added, e.g. slaked lime (10% by weight), to counteract the tendency to lose moisture by electro-osmosis, since it is essential that an aqueous electrolyte is present to conduct the current to the protected structure. Calcium sulphate is sometimes used in very dry conditions.

In using coke breeze the consumption of the primary anode is reduced as the majority of the conduction from the anode becomes electronic as opposed to electrolytic. The electrochemical and physical nature of the coke results in the dispersion of the anode reaction (formation of CO_2 and O_2) over a large surface area, thus reducing attack on the primary anode.

The coke is oxidised primarily to carbon dioxide which in a suitable groundbed will escape into the atmosphere together with any oxygen formed. If all the oxygen reacted, the coke consumption would be of the order of $1.0 \, kg \, A^{-1} y^{-1}$, however, in practice consumption can be of the order of $0.25 \, kg \, A^{-1} y^{-1}$[66], depending upon the environment.

Table 11.11 Densities of backfill

Backfill in bulk	Density (kg/m^3)
Coal coke breeze	650–800
Calcined petroleum coke granules	700–1100
Natural graphite granules	1100–1300
Man-made graphite, crushed	1100–1300

Table 11.12 Typical coal coke specifications

To pass 15 mm screen	100%
To pass 10 mm screen	85% max.
To pass 5 mm screen	15% max.
Volatile matter	3·25% max.
Fixed carbon	78% min.
Ash	19% max.
Sulphur	1·2% max.
Resistivity (bulk)	55 Ω cm max.

Some typical properties of coke breeze and similar materials are shown in Tables 11.11 to 11.13. The densities given in Table 11.11 are for bulk and are dependent upon grading. Flake graphite is not recommended as it tends to conglomerate and prevent gas emission.

When coke breeze is tamped down correctly a pressure of approximately $15 kN/m^2$ is to be aimed for. This will ensure integrity of the groundbed whilst in operation, remembering that it will be reducing in volume. A pressure of this magnitude will reduce the initial bulk resistivity of the coke.

Table 11.13   Resistivities of carbonaceous backfills (Ω cm)

Material	Dry	Tamped	Wet
Coal coke	55	45	15
Graphite granules	150	120	20

The usual main object in using coke breeze is economically to lower the resistance of the anode/earth resistance. In this case the cross-section of the coke is at the minimum (except in pre-canned assemblies) of 300 mm × 300 mm. In fresh-water soil conditions the higher than average current density at the ends of the primary anodes can be prevented by not exceeding an anode spacing of twice its length. This depends entirely on the care taken in preparation of the groundbed assembly and ratio of the anode/coke resistance to the anode/electrolyte resistance. As the electrolyte resistance decreases, with a consequent increase in current density at the ends of the primary anodes, either reduction in anode spacing or increase in backfill cross-section should be considered, particularly in fore-shore groundbeds where the electrolyte resistance will be similar in magnitude to that of the backfill.

Reactive Metals

Aluminium

This has been used extensively in fresh-water tank protection[67]. To reach the required circuit resistance it is necessary to use long extrusions of the order of 20 mm diameter. The alloys H14 and H15 have been used for this purpose, pure aluminium being preferred for sea-water service[46].

For tank protection the life of the anode is very much dependent on the extent of pitting. Necking can be a problem if the water level drops below part of the anode for long periods. The wastage rate in this area can be twice the normal.

In fresh water, voluminous corrosion products (mainly aluminium hydroxide) can cause quite large increases above the initial anode/electrolyte resistance; this could be two-fold or more. It may reduce at times as the product sloughs off, but contributing to local above-average current density and local high consumption of material. The product, whilst not toxic, in quantity could prove an embarrassment in potable waters. Resistive polarisation is negligible in sea-water use.

Claims of a reduction in scaling problems have been made when used in circulating water systems, but this apparently depends upon the nature of the water in consideration.

Theoretically the material would be expected to dissipate of the order of $6.4\,kg\,A^{-1}y^{-1}$, but in practice the rate has been found to be of the order of $4.5\,kg\,A^{-1}y^{-1}$ average, although a reasonably large factor of safety regarding anode integrity should be considered.

Aluminium has been used for 'trailing wire anodes' for the protection of shipping. Although the use of the anodes in this manner has proved successful, it is no longer considered a practical application. Further use has been made in the field of de-watering and clay stabilisation. An interesting example is in the use of aluminium-clad load-bearing piles. After installation, current is applied between the pile and a remote cathode. Water is driven away and the clay is permanently consolidated and cemented to the piles by the aluminium corrosion product.

Zinc

The normal potential of zinc can be raised by the use of an external e.m.f. source. This may be a convenient way of achieving a high initial current, particularly where descaling is involved, but it does, of course, require the anodes to be locally insulated from the cathode and for current carrying conductors to be brought to its surface so that an e.m.f. of large magnitude may be superimposed. It would seem more practical to use a temporary groundbed of, say, scrap steel.

Used as a power-impressed anode the energy wastage rate per unit energy tends towards the theoretical value of $10.8\,kg\,A^{-1}y^{-1}$ instead of the usual 90% efficiency when used as a galvanic anode.

Summary

The comparison of typical properties of cathodic protection materials given in Table 11.14 is by no means comprehensive. It is obvious that by the modification of an alloy, environment or other important factors will be reflected in the life and output characteristics. In some cases the maximum voltages and current densities recommended can be vastly exceeded. In others, particularly where abnormal departures from the norm of environmental dissolved solids may be met, factors of safety must be taken upon the figures proposed. Acceptance of a much reduced or uncertain life weighed against a possible economy, may also influence the chosen working limits. For example, it should be borne in mind that the life in particular of the ferrous alloys may in terms of a practical anode be only $\frac{2}{3}$ of that expected, by virtue of preferential attack eventually disconnecting the anode or part thereof from the source of e.m.f.

Table 11.14 must only be taken as a guide and interpreted in the best manner available, preferably that of experience in that particular environment or operational requirement. The table should be consulted in conjunction with the text. The relative costs of the materials can only be a guide as it does not take into consideration the costs of the supporting equipment compatible with its use. Consideration must be given to practicability, factor of safety required, environment, physical and electromechanical hazards, maintenance, operation, installation, availability and the economics of replacement[68].

J. W. L. F. BRAND

Table 11.14 Comparison of typical

	Platinised tantalum	Platinised niobium	Platinised titanium	Platinum	DSA-titanium	High-silicon/chromium iron
Approximate consumption $(kg\,A^{-1}\,y^{-1})$	See Pt	See Pt	See Pt	$8\cdot63\times10^{-6}$ B	$4\cdot75\times10^{-6}$	$0\cdot25-1\cdot0$ B
Suggested minimum factor of safety on cross-sectional area	See Pt	See Pt	See Pt	$1\cdot2$	M	$1\cdot8$
Max. recommended current density (sea, mA/cm²)	H	H	H	B	100	12
Max. recommended current density (fresh water, mA/cm²)	H	H	H	B	M	12
Max. recommended current density (soil, mA/cm²)	B	B	B	B	M	6
Max. recommended voltage (sea, V)	B	B	8 B	D, G	See Pt/Ti	D
Max. recommended voltage (fresh water, V)	B	B	B	D, G	See Pt/Ti	D
Specific resistivity at 20°C $(\Omega\,cm \times 10^{-6})$	$12\cdot5$	$15\cdot2$	$48\cdot2$	$9\cdot85$	See Pt/Ti	72
Density (g/cm³)	$16\cdot6$	$8\cdot57$	$4\cdot51$	$21\cdot45$	See Pt/Ti	7
Tensile strength (approx.) (kgf/cm²)	3 500–12 600	2 400–3 900	4 700–6 300	2 500 K		1100
Hardness (approx.)	80–100 HV	75–95 HV	—	200–400 HV J	See Pt/Ti	520 HB
Approx. cost of consumption per unit energy (scale of 100)	—	—	—	2–3	M	—
General uses						
Marine environment	Yes	Yes	Yes	No (N)	Yes	Yes
Potable waters	No (N)	No (N)	Yes	No (N)	M	No (N)
In carbonaceous backfill	No (N)	No (N)	No (N)	No (N)	M	Yes
Buried directly in soil	No (N)	No (N)	No (N)	No (N)	M	Yes
High purity liquids	Yes	Yes	Yes	No (N)	M	No

Notes: A Used with carbonaceous backfill, see text—depends on water resistivity/backfill resistivity.
    B See text.
    C Normal max. longitudinally.
    D Limited by local or environmental safety requirements regarding apparatus and/or earth voltage gradient regulations.
    E Min. current density to ensure passivation $5\cdot0\,mA/cm^2$.
    F Resistivity of PbO_2—40–$50\,\Omega\,cm$.
    G Voltage on Pt $1\cdot35$ V min.

properties of cathodic protection anode materials

High-silicon iron	Steel	Iron	Cast iron	Pb–6Sb–1Ag	Lead/platinum	Graphite	Aluminium	Zinc	Coke breeze
0·25–1·0 B	6·8–9·1	Approx. 9·5	4·5–6·8	0·09 B	0·09 B	0·1–1·0 B	4·5	10·8	0·5
1·8	1·8	1·8	1·8	1·5	1·5	1·5	2	1·2	1·5
—	L	L	L	20 E	50 E	3·0	2	2	—
12	L	L	L	—	—	0·25	2	2	—
6	0·5	0·5	0·5	—	—	1·0 A	0·25	0·25	0·25
—	D	D	D	D	D	D	D	D	—
D	D	D	D	—	—	D	D	D	—
72	17	12	55	25 F	22 F	700 C	3·3	6·2	B
7	7·7	7·82	7·1	10·9	11·3	1·56–1·67	2·7	7·1	B
1 300	5 000	3 000	1 500	300	250	280	850	—	—
450 HB	130–160 HB	120–170 HB	140–170 HB	13–10·7 HB	4 HB	80 Shore	—	—	—
40–100	—	—	—	5–15	—	20–100	—	—	—
No	Yes	Yes	Yes	Yes	Yes	Yes	No (N)	Yes	Yes
Yes	No	No	No	No	No	Yes	Yes	No	No
Yes	Yes	Yes	Yes	No	No	Yes	No	No	—
Yes	No (N)	No (N)	No (N)	No	No	No (N)	No (N)	No (N)	Yes
No	No	No	No	No	No	No	No	No	No

H 40 mA/cm² per micron of Pt.
J Electrodeposited.
K 50% cold rolled.
L If in free suspension in moving water, no limit, local effects under high current density may increase wastage rate.
M Information not currently available.
N May be used in this environment under special circumstances

REFERENCES

1. Cathodic Corrosion Control Ltd., British Pat. No. 880 519 (25.10.61)
2. Anderson, D. B. and Vines, R. F., *Extended Extracts of Second International Congress on Metallic Corrosion*, March, 146 (1963)
3. Warne, M. A. and Hayfield, P. C. S., *Durability Tests on Marine Impressed Current Anodes*, I.M.I./Marston Excelsior Ltd., March 15th (1971)
4. Lowe, R. A., *Materials Protection*, NACE, Houston, 23 and 24, April (1966)
5. Cotton, J. B., *Chem. and Ind. (Rev.)*, **68** (1958)
6. Walkiden, G. W., *Corrosion Technology*, **9** No. 14, 38 (1962)
7. Cotton, J. B., *Chem. and Ind.* (Lond.), **68**, 492 (1958)
8. Ksenzhek, O. S. and Solovei, Z. V., *Journal of Applied Chemistry*, USSR, **33**, 279 (1960)
9. Walkiden, G. W., 'Platinum in Anodes', *Corrosion Technology*, Jan. and Feb. (1962)
10. Dotemus, G. L. and Davis, J. G., *Materials Protection*, NACE, 30–39, Jan. (1967)
11. Peplow, D. B., *Brit. Power. Engng.*, **1**, 51–33, Oct. (1960)
12. Shreir, L. L., *Corrosion*, **17** No. 3, 118t–124t, March (1961)
13. Hames, W. T., *Aircraft Production*, London, **20**, 369 (1958)
14. From Commercial Guarantee—Marston Excelsior Ltd. (1972)
15. Proceedings of Symposium on Recent Advances in Cathodic Protection, Wilton, Marston Excelsior, May (1964)
16. Cerny, M., *Behaviour of Anodes of Platinised Titanium in River Waters*, State Research for Protection of Materials, Prague
17. Dugdale, I. and Cotton, J. B., 'Anodic Polarisation of Titanium in Halide Solutions', *Corrosion Science*, **4** No. 4, 397–412 Dec. (1964)
18. Juchniewicz, R. and Hayfield, P. C. S., 'Corrosion of Platinum and Platinised Titanium Under the Influence of Superimposed Alternating Current in Monopolar and Bipolar Operating Conditions', paper presented at Moscow Congress (1966)
19. Bittles, J. A. and Littauer, E. L., *Corrosion Science*, **10**, 29 (1970)
20. Warne, M. A. and Hayfield, P. C. S., *Br. Corros. J.*, **6**, 192–195, Sept. (1971)
21. Private correspondence with Marston Excelsior/CWE, Sept. (1970)
22. Love, T. J., *Power*, USA, 80–81, Feb. (1961)
23. Preiser, B. H. S. and Tutell, B. H., *Corrosion*, **15** No. 11, 596t–600t Nov. (1959)
24. Lowe, R. A. and Brand, J. W. L. F., *Materials Protection and Performance*, NACE, 45–47, Nov. (1970)
25. Sale, J. P., 'The Evaluation of Anode Configurations for the Internal Cathodic Protection of Pipes', paper presented at Conference on Corrosion and Protection of Pipes or Pipelines
26. CWE Ltd., British Pat. applied for
27. Baum, G., US Pat. No. 1 477 000 (22.8.22)
28. Shreir, L. L., *Platinum Metals Review*, **4**, 15 (1960)
29. Private correspondence with Marston Excelsior/CWE, May (1969)
30. Dreyman, E. W., *Materials Protection and Performance*, **11** No. 9, 19 and 20, Sept. (1970)
31. McAnenny, A. W., *P.I.E.A. News*, July (1941)
32. McAnenny, A. W., *P.I.E.A. News*, **10** No. 3, 11–29 (1940)
33. McAnenny, A. W., US Pat. No. 2 360 244 (1944)
34. 'Alkaline Iron Plate Storage Batteries', *Elec. Eng. Ref. Book*, Butterworths, London (1968)
35. Peabody, A. W., *Mat. Prot.*, **9** No. 5, 15 and 16, May (1970)
36. Bengough and May, *Journal of Inst. of Metals*, **32** No. 2, Part 5 (1924)
37. Brand, J. W. L. F. and Tullock, D. S., CWE International Report, London (1965)
38. Redden, J. C., *Mat. Prot.*, NACE, 51, Feb. (1966)
39. Applegate, L. M., *Cathodic Protection*, London, 119 (1960)
40. Bryan, W. T., *Mat. Prot. and Perf.*, **9** No. 9, 25–29 Sept. (1970)
41. Tudor, S. and Ticker, A., *Progress Report No. 7*, U.S. Naval Material Lab. Project 5589–1, 29 June (1961)
42. Moller, G. E. *et al.*, *Mat. Prot.*, **1** No. 2 (1962)
43. Littauer, E. and Shreir, L. L., *Proceedings of 1st International Congress on Metallic Corrosion*,
44. Shreir, L. L., *Corrosion*, **17**, 118t (1961)
45. Bernard, K. N., Christie, G. L. and Gage, D. G., *Corrosion*, **15** No. 11, 501t–586t (1959)
46. Morgan, J. H., *Cathodic Protection*, Leonard Hill, London (1959)
47. Private report on films on lead–silver–antimony electrodes, Fulmer Research Institute to CWE Ltd.
48. Tudor, S. and Ticker, A., *Mat. Prot.*, **3** No. 1, 52–59, Jan. (1964)

49. Private report on lead alloy produced by CWE Ltd., Materials Laboratory of New York Naval Shipyard to CWE Ltd. (1957)
50. CWE lead alloy classification
51. Bernard, K. N., Christie, G. L. and Gage, D. G., *Corrosion*, **15** No. 11, 581t–586t, Nov. (1959)
52. *Lead*, **29** No. 4, Lead Industries Association, New York, 7 (1965)
53. Shreir, L. L., *Corrosion*, **17** No. 3, 118t–124t, March (1961)
54. Metal and Pipeline Endurance Ltd., British Pat. No. 870 277 (14.6.61)
55. Fleischman, M. and Lilen, M., *Trans. Faraday Soc.*, **54**, 1370 (1958)
56. Shreir, L. L., *Platinum Metals Review*, **12** No. 2, 42–45, April (1968)
57. Shreir, L. L. and Weinraub, A., *Chemistry and Industry*, No. 41, 1326, Oct. (1958)
58. Wheeler, W. C. G., *Chem. and Ind.*, No. 75 (1959)
59. Shreir, L. L., *Platinum Metals Review*, **3** No. 2, 44–46 (1959)
60. Palmquist, W. W., *Petroleum Engineer*, **2**, D22–D24, Jan. (1950)
61. Hienks, H., *Ind. Eng. Chem.*, **47**, 684 (1955)
62. Bulygin, B. M., *ibid.*, **32**, 521 (1959)
63. Krishtalik, L. I. and Rotenberg, Z. A., *Russian Journal of Physical Chemistry*, **39**, 168 (1965)
64. Peabody, A. W., *Control of Pipeline Corrosion*, NACE, Houston, Texas (1967)
65. Oliver, J. P., *A.I.E.E.*, paper 52–506, Sept. (1952)
66. Applegate, L. M., *Cathodic Protection*, McGraw Hill, New York (1960)
67. Shepard, E. R. and Graeser, H. J., *Corrosion*, **6** No. 11, 360–375, Nov. (1950)
68. Brand, J. W. L. F., 'Cathodic Protection', *Elec. Rev.*, 781–783, Dec. (1972)

BIBLIOGRAPHY

Antler, M. and Butler, C. A., 'Degradation Mechanisms of Platinum and Rhodium Coated Titanium Anodes in the Electrolysis of Chloride and Chloride-Chlorate Solutions', *Corrosion*, **5** No. 3–4, 126–130, March–April (1967)
Applegate, L. M., *Cathodic Protection*, McGraw-Hill, New York (1960)
Brady, G. D., 'Graphite Anodes for Impressed Current Cathodic Protection. A Practical Approach', *Mat. Prot. and Performance*, **10** No. 10, NACE, Houston, Oct. (1971)
Berkeley, K. G. C., 'Electrochemical Corrosion Control in Effluent Treatment Plant', *Chemistry and Industry*, **13**, March (1971)
Bryan, W. T., 'Use of High Silicon Chromium Iron Anodes for Deep Groundbeds', *Mat. Prot. and Performance*, **9** No. 9, Sept. (1970)
Cathodic Protection. Its Range of Application and Modern Techniques, CWE, London (1968)
Cotton, J. B. and Bradley, 'Corrosion Resistance of Titanium', *Chem. and Ind.*, 1588 (1957)
Cotton, J. B., *Platinum Faced Titanium for Electrochemical Anodes*, I.M.I. Metals Division (1957)
Doremus, G. L. and Davis, J. G., 'Marine Anodes. The Old and New', *Mat. Prot.*, NACE, 30–39, Jan. (1967)
Dugdale, I. and Cotton, J. B., 'The Anodic Polarisation of Titanium in Halide Solutions', *Corrosion Science*, **4**, 397–412, Dec. (1964)
Dwight, H. B., 'Calculation of Resistance to Ground', *Electrical Engineering*, Dec. (1936)
Evans, U. R., *An Introduction to Metallic Corrosion*, Edward Arnold, London (1963)
Evans, U. R., *Metallic Corrosion Passivity and Protection*, Edward Arnold, London (1946)
Feige, N. G. and Murphy, J. J., 'Environmental Effects on Titanium Alloys', *Corrosion*, **22**, NACE, Houston, 320–324 (1966)
James, W. J., Straumanis, M. E. and Johnson, J. W., 'Anodic Disintegration of Metals Undergoing Electrolysis in Aqueous Salt Solutions', Paper No. 10 of 22nd conference, NACE, April 18th–22nd (1966)
Kimmel, A. L., 'A Study of Some Metals for Use as a Permanent Anode', *Corrosion*, **6** No. 11, 353, Nov. (1950)
Lowe, R. A., 'Platinised Titanium as an Anode Material', *Mat. Prot.*, **5** No. 4, NACE, April 23rd–24th (1966)

Luce, W. A., 'High Silicon Cast Iron Tested for Use with Impressed-Current', *Corrosion*, **10** No. 9, 325, Sept. (1954)

Luce, W. A., 'Use of High Silicon Cast Iron for Anodes', NACE Committee T-2B, **13** No. 2, NACE Publication 57-4, 103t, Feb. (1957)

Luce, W. A., 'Use of High Silicon Cast Iron for Anodes', Fourth Interim Report, NACE Committee T-2B-4, *Corrosion*, **16** No. 2, NACE Publication 60-3, 65t, Feb. (1960)

Morgan, J. H., *Cathodic Protection*, Leonard Hill, London (1959)

Morgan, J. H., 'Lead Alloy Anodes Tested up to 20 Amperes per Square Foot', *Corrosion*, **13** No. 7, 128, July (1957)

Mathewman, W., 'The Use of Platinised Titanium', *Corrosion Prevention and Control*, Oct. (1962)

Oliver, J. P., 'The Use of Graphite Anodes for Cathodic Protection of Bottoms of Inactivated Ships', AIEE paper No. 52-506, Sept. (1952)

Oliver, J. P. and Palmquist, W. W., 'The Value of Backfill with Carbon and Graphite Anodes', *Corrosion*, **6**, NACE, May (1950)

Oliver, J. P. and Palmquist, W. W., 'The Value of Backfill with Carbon and Graphite Anodes', *Corrosion*, **6** No. 5, 147, May (1960)

Parker, M. E., *Pipeline Corrosion and Cathodic Protection*, Gulf Houston (1954)

Pritula, V. A., *Cathodic Protection of Pipelines and Storage Tanks*, H.M.S.O., London (1953)

Peabody, A. W., *Control of Pipeline Corrosion*, National Association of Corrosion Engineers, Houston, Texas (1967)

Pourbaix, M., *Atlas d'Equilibres Electro-chimiques*, Pergamon Press (1966)

Palmquist, W. W., 'Graphite Anodes for Cathodic Protection', *Petroleum Engineer*, **22**, D22-D24, Jan. (1950)

Parker, M. E., 'Corrosion and Its Control', *The Oil and Gas Journal*, Tulsa, Oklahoma

Peabody, A. W., 'Use of Steel as an Anode Material In Deep Groundbeds', *Mat. Prot.*, **9** No. 5, May (1970)

Peplow, D. B. and Shreir, L. L., 'Lead/Platinum Anodes for Marine Applications', *Corrosion Technology*, April (1962)

Pope, R., 'Carbon Anodes', *Corrosion*, **10**, NACE, Houston, 443 (1954)

Potter, *Electro Chemistry*, Cleaver Hume (1946)

Preisen, H. S. and Cook, F. E., 'Cathodic Protection of an Active Ship Using a Trailing Platinum Clad Anode', *Corrosion*, **12** No. 3, 18, March (1956)

Rohrman, F. A., 'Bibliography of Cathodic Protection', *World Oil*, **131**, 179 (1950)

Shreir, L. L., 'Anodic Polarisation of Lead/Platinum Bi-electrodes in Chloride Solution', *Corrosion*, **17** No. 3, 118-124, March (1961)

Stewart, D. and Tulloch, D. S., *'Principles of Corrosion and Protection'*, MacMillan, London (1968)

Sunde, E. D., *Earth Conduction Effects in Transmission Systems*, Van Nostrand (1949)

Toncre, A. C., Impressed Current Anodes in Brackish Water, *Mat. Prot.*, NACE, April (1966)

Tudor, S., Miller, W. L., Ticker, A. and Preisen, H. S., 'Electrochemical Deterioration of Graphite and High Silicon Iron Anodes in Sodium Chloride Electrolytes', *Corrosion*, **14** No. 2, 93-99, Feb. (1958)

Tudor, S. and Ticker, A., 'Lead Alloy Anodes for Cathodic Protection in Various Electrolytes', *Mat. Prot.*, **3** No. 1, 52 and 59, Jan. (1964)

Uhlig, H. H., *Corrosion Handbook*, John Wiley (1963)

Vaaler, L. E., 'Graphite-Electrolytic Anodes', *Electrochemical Technology*, **5** Nos. 5-6, May-June (1967)

Walkiden, G. W., 'Platinum Anodes in Cathodic Protection Applications', *Corrosion Technology*, Jan. and Feb. (1962)

Anti-corrosion Manual, Corrosion Prevention and Control, London (1959)

Cathodic Protection, NACE, Houston (1949)

11.4 Practical Applications of Cathodic Protection

The complexity of the systems to be protected and the variety of techniques available for cathodic protection are in direct contrast to the simplicity of the principles involved, and, at present the application of this method of corrosion control remains more of an art than a science. However, as shown by the potential pH diagrams, the lowering of the potential of a metal into the region of immunity is one of the two fundamental methods of corrosion control.

In principle, cathodic protection can be used for a variety of applications where a metal is immersed in an aqueous solution of an electrolyte, which can range from relatively pure water to soils and to solutions of acids. Whether the method is applicable will depend on many factors and, in particular, economics—protection of steel immersed in a highly acid solution is feasible but too costly to be practicable. It should be emphasised that as the method is electrochemical both the structure to be protected and the anode used for protection must be in both metallic and electrolytic contact. Cathodic protection cannot therefore be applied for controlling atmospheric corrosion, since it is not feasible to immerse an anode in a thin condensed film of moisture or in droplets of rain water.

The forms of corrosion which can be controlled by cathodic protection include all forms of general corrosion, pitting corrosion, graphitic corrosion, crevice corrosion, stress-corrosion cracking, corrosion fatigue, cavitation corrosion, bacterial corrosion, etc. This section deals exclusively with the practical application of cathodic protection principally using the impressed-current method. The application of cathodic protection using sacrificial anodes is dealt with in Section 11.2.

Structures that are Cathodically Protected

The following structures are those which in given circumstances can benefit from the application of a cathodic-protection system:

Underground and underwater Underground fuel/oil tanks and pipelines; water, fire protection, gas and compressed underground air distribution schemes; underground metallic sewers and culverts; underground communication and power cables; deep wells; other buried tanks and tanks in

Table 11.15 Methods of application of cathodic protection

Method	Characteristics	Anode materials*	Current source	Installation	Possibilities of secondary interaction in foreign structures
Sacrificial anodes	Metal protected by sacrificial wastage of more electronegative metal	Magnesium, aluminium or zinc (iron for copper and copper alloys)	Faradaic equivalent of sacrificial metal—in practice the efficiency is seldom 100%	Extremely simple	Very improbable providing anodes properly located with respect to surface being protected
Impressed current ('power impressed')	Impressed currents using transformer-rectifiers, or any other d.c. source	Carbon, silicon–iron, lead–platinum, platinised titanium, platinised niobium, scrap iron	Source of low-voltage d.c. This may be generated or drawn from transformer-rectifier fed from main supplies	More complex	Very significant especially in built-up areas
Stray current ('drained current')	Buried structures bonded into traction system in such a way as to receive impressed-current protection	Bonded directly into stray d.c. supply	Drained from d.c. traction or stray-current supply	Simple	Stray-current effects are basically associated with primary power supply

* For protection of ferrous structures.

Table 11.15 (continued)

Method	Application for which scheme is economical	Major limitations	Potential distribution	Current limitation
Sacrificial anodes	Small schemes and in built-up areas where interaction problems are liable to be severe	Impracticable for large installation on account of number of anodes required, and in soils of high resistivity owing to the small driving e.m.f. Replacement required at frequent intervals when current output is high	Reasonably uniform	Cannot be applied in high-resistivity environments
Impressed current ('power impressed')	Especially suited to large schemes	Impracticable for small schemes on account of high installation costs	Varies—maximum at drainage point falling towards remote points, but not below the optimum potentials for protection, i.e. in most cases the potential -0.85 V	Can be used in high-resistivity environments
Stray current ('drained current')	Applicable only in proximity to stray d.c. areas			

contact with the ground; tower footings, sheet steel piling and 'H'-piling; piers, wharfs and other mooring facilities; submarine pipelines; intake screens (condenser/circulating water); gates, locks and screens in irrigation and navigation canals; domestic oil distribution lines or central heating systems.

Above ground (internal surfaces only)   Surface and elevated water storage tanks; condensers and heat exchangers; hot-water storage tanks, processing tanks and vessels; hot- and cold-water domestic storage tanks; breweries and dairies (pasturisers).

Floating structures   Ballast compartments of tankers; ships (active and in 'mothballs'); drilling rigs; floating dry-docks; barges (interior and exterior); dredgers; caisson gates; steel mooring pontoons; navigation aids, e.g. buoys.

Type of System

The use of an impressed-current system or sacrificial anodes will both provide satisfactory cathodic protection, but each has advantages and disadvantages with respect to the other (Table 11.15).

Sacrificial anodes and power-impressed anodes have been dealt with in detail in the previous sections, but some further comment is relevant here in relation to the choice of a particular system for a specific environment. In this connection it should be noted that the conductivity of the environment and the nature of the anode reactions are of fundamental importance.

The main anodic reactions may be summarised as follows:

Sacrificial anodes

primary reaction	$M \rightarrow M^{z+} + ze$	...(11.16a)
secondary reaction	$M^{z+} + zH_2O \rightarrow M(OH)_z + zH^+$	...(11.16b)

Impressed-current anodes

$$3H_2O \rightarrow 2H_3O^+ + 2e + \tfrac{1}{2}O_2 \qquad ...(11.17a)$$

and/or

$$2Cl^- \rightarrow Cl_2 + 2e \qquad ...(11.17b)$$

or, in the case of graphite anodes

$$C + O_2 \rightarrow CO_2 \qquad ...(11.17c)$$

It should be noted that when metals like zinc and aluminium are used as sacrificial anodes the anode reaction will be predominantly 11.16a and 11.16b, although self-corrosion may also occur to a greater or lesser extent. Whereas the e.m.f. between magnesium, the most negative sacrificial anode, and iron is ≈ 0.7 V, the e.m.f. of power-impressed systems can range from 20 V to 50 V or more, depending on the transformer–rectifier employed. Thus, whereas sacrificial anodes are restricted to environments having a resistivity of $<6\,000\,\Omega$ cm there is no similar limitation in the use of power-impressed systems.

In the case of sacrificial anodes the electrons that are required to depress the potential of the structure to be protected are supplied by reaction 11.16a, and providing the metal ions can diffuse away from the structure before they react with water to form insoluble hydroxides the reaction will be unimpeded and will take place at a low overpotential. If, however, the metal hydroxide precipitates on the surface of the metal as a non-conducting passive film the anode reaction will be stifled and this situation must be avoided if the anode is to operate satisfactorily. On the other hand, in the case of non-reactive impressed-current anodes, rapid transport of the reactants (H_2O and Cl^-) to, and the reaction products (O_2 and Cl_2) away from, the anode surface is essential if the anode reaction is to proceed at low overpotentials. This presents no problems in sea-water, and for this reason the surface areas of the anodes are comparatively small and the anode current densities correspondingly high. Thus, in sea-water inert anodes such as platinised titanium and lead–platinum can operate at ≈ 500–$1\,000\,Am^{-2}$, since the anode reaction 11.17b occurs with little overpotential, and there is rapid transport of Cl^- to and Cl_2 away from the anode surface. In this connection it should be noted that even in a water of high chlorinity such as sea-water, oxygen evolution should occur in preference to oxygen evolution on thermodynamic grounds. This follows from the fact that the equilibrium potential of reaction 11.17a in neutral solutions is 0·84 V, whereas the corresponding value for 11.17b is 1·34 V, i.e. 0·5 V higher. However, whereas the chlorine evolution reaction occurs with only a small overpotential, very appreciable overpotentials are required for oxygen evolution, and this latter reaction will occur therefore only at high current densities. Even in waters of low salinity chlorine evolution will occur in preference to oxygen evolution at low overpotentials.

In the protection of pipelines or other underground structures the anode reaction is dependent on diffusion of water to the anode surface and oxygen and CO_2 away from it, and since these processes do not occur with the same mobility as in water it is necessary to use a very large surface area of anode and a corresponding low current density. For this reason the actual anode is the carbonaceous backfill, and graphite or silicon–iron anodes are used primarily to make electrical contact between the cable and the backfill. It can also be seen from reaction 11.17a that the products of the oxidation of water are oxygen and the hydrated proton H_3O^+, which will migrate away from the anode surface under the influence of the field, thus removing two of the three water molecules that participate in the reaction, and this will tend to dehydrate the groundbed. This difficulty can be overcome, when feasible, by locating the groundbed below the water table.

Sacrificial Anode Systems

Advantages   No external source of power is required; installation is relatively simple; the danger of cathodic protection interaction is minimised; more economic for small schemes; the danger of over protection is alleviated; even current distribution can be easily achieved; maintenance is not required apart from routine potential checks and replacement of anodes at the end of their useful life; no running costs.

Disadvantages   Maximum anode output when first installed decreasing with time when additional current may be required to overcome coating deterioration; current output in high resistivity electrolytes might be too low and render anodes ineffective; large numbers of anodes may be required to protect large structures resulting in high anode installation and replacement costs; anodes may require replacement at frequent intervals when current output is high.

Impressed-current Systems

Advantages   One installation can protect a large area of metal; systems can be designed with a reserve voltage and amperage to cater for increasing current requirement due to coating deterioration; current output can be easily varied to suit requirements; schemes can be designed for a life in excess of 20 years; current requirements can be readily monitored on the transformer –rectifier or other d.c. source; a time switch can be incorporated into the d.c. source to enable interaction effects to be assessed.

Disadvantages   Possible interaction effects on other buried structures (Section 11.6); subject to the availability of a suitable a.c. supply source or other source of d.c.; regular electrical maintenance checks and inspection required; running costs for electrical supply (usually not very high except in the case of bare marine structures and in power stations where structures are often bare and include bimetallic couples); subject to power shutdowns and failures.

Stray Current or Forced Drainage

Stray current schemes are relatively rare in occurrence in the UK as few localities now have widespread d.c. transport systems. Such systems are extensively used in overseas countries where d.c. transport systems are in use, i.e. Australia and South Africa. Where stray current can be employed it is normally the most economical method of applying cathodic protection since the power required is supplied gratis by the transport system.

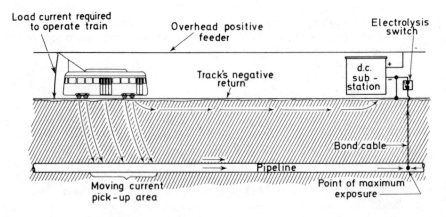

Fig. 11.16   Bond between pipeline and d.c. substation

In such systems it is necessary to provide a metallic bond between the pipe-line and the negative bus of the railway substation. By providing such a bond the equivalent of a cathodic-protection system is established whereby current discharged from the traction-system rails is picked up by all portions of the pipeline and drained off via the bond. The bond must have sufficient carrying capacity to handle the maximum current drained without damage. In order to ensure that the direction of current flow in the bond does not reverse, it is normal to employ a reverse-current prevention device or 'electrolysis switch'. This may take the form of a relay-actuated contactor which opens auto-matically when the current reverses. Diodes may also be used as blocking valves to accomplish the same purpose. They are wired into the circuit so as to ensure that current can flow to the negative busbar system only. Sufficient diodes must be used in parallel to handle the maximum amount of current anticipated. Also the inverse voltage rating of the diodes must be sufficient to resist the maximum reverse voltage between the negative busbar and pipe-line (Fig. 11.16).

Design of a Cathodic-protection System

To enable an engineer to design a cathodic-protection scheme, consideration should be given to the following points (see also Table 11.16).

Good practice in modern underground or underwater structures involves the use of good coatings in combination with cathodic protection. With a well-coated structure the cathodic-protection system need only protect the minute areas of steel exposed to the corrosive environment rather than the whole surface of an uncoated structure. The effect of coatings can be demon-strated by comparing the current density of a bare steel pipeline in average soil conditions, which could be up to $30\,\text{mA/m}^2$, with that achieved on a well-coated and inspected line where a current density of only $0.032\,\text{mA/m}^2$ or even lower may be required to obtain satisfactory cathodic protection. In all cases the current density for protection is based on the superficial area of the structure.

Surface area   In the case of underground pipelines, calculation of the super-ficial surface area can be obtained from the diameter and length of line in-volved. The superficial surface area should include any offtakes and other metal structures in electrical contact with the main line. For marine struc-tures the area should include all submerged steel work below full-tide level. In the case of power stations, details of the water boxes, number of passes on coolers and detailed drawings are required. In the case of ships, details of the full underwater submerged area at full load are needed.

Electrical continuity   It is essential for any structure to be fully electrically continuous. In the case of pipelines, welded joints are obviously no problem but mechanical joints require bonding. For marine structures individual piles and fendering must be electrically connected either by the reinforcing bars in the concrete deckhead or separately by cable. In power stations and ships, rotating shafts must be bonded into the structure by means of brush

Table 11.16   Steps in design of cathodic-protection installation

Sacrificial and impressed-current anodes

1. Establish soil or water resistivity.
2. Estimate total current requirements which will depend on aggressiveness of the environment, nature of protective coating, area of structure.
3. Establish electrical continuity of structure.
4. Determine requirements for insulating flanges (to limit the spread of cathodic-protection currents at the ends and branches) and for bonding to foreign structures, and assess extra current allowances.

Sacrificial	*Impressed current*
5. Select suitable anode metal; select size to give optimum life and output.	5. Select suitable groundbed locations; these should be: (i) In low resistivity soils. (ii) Reasonably near power supply. (iii) At points where interaction problems are unlikely to be significant. (iv) In a location where groundbeds and cables are reasonably immune from disturbance.
6. Determine total number of anodes required. (*Note:* in soils of below 1 500 Ω cm the use of Zn or Al should be considered.)	6. Decide on whether anodes should be installed vertically or horizontally.
7. Deduce anode spacing to give uniform current distribution.	7. Determine circuit resistances (i.e. depending on soil condition, etc.): (i) Cathode (usually insignificant owing to large area of cathode). (ii) Cable (depending on current, cable size, and length). (iii) Anode (depending on groundbed design). (iv) Consider cathode volt-drop effects at remote points.
8. Select test-point localities (usually mid-way between anodes and at line terminations).	8. Decide upon voltage to be used.
	9. Determine optimum anode material.
	10. Deduce optimum number and size of anodes.
	11. Decide upon anode spacing.

gear or a suitable alternative. In the case of modern offshore mooring installations, it may be necessary to install a bonding cable to bring the outer-lying dolphins, etc. into the system.

Estimate of current required   The surface area of the structure is calculated and the current density required for the particular environment is selected (Table 11.17). In the case of an existing structure the condition of the coating may be unknown and the application of a temporary cathodic-protection system may be necessary to determine the amount of current required for protection, as established by the potential. Such a test to determine the absolute amount of current required is known as a *current drain* test. Misleading information may, however, be obtained if the results from current drainage tests on bare or coated steel in sea-water are extrapolated, because long-term polarisation effects, together with the formation of a calcareous

Table 11.17 Typical values of current requirements for steel free from adverse galvanic influences in various environments

Environment	Current density required for adequate cathodic protection* based on superficial area (mA/m^2)
BARE STEEL	
Sterile, neutral soil	4·3–16·1
Well-aerated neutral soil	21·5–32·3
Dry, well-aerated soil	5·4–16·1
Wet soil, moderate/severe conditions	26·9–64·6
Highly acid soil	53·8–161·4
Soil supporting active sulphate-reducing bacteria	451·9
Heated in soil (e.g. hot-water discharge line)	53·8–269·0
Dry concrete	5·4–16·1
Moist concrete	53·8–269·0
Stationary fresh water	53·8
Moving fresh water	53·8–64·6
Fresh water highly turbulent and containing dissolved oxygen	53·8–161·4
Hot water	53·8–161·4
Polluted estuarine water	538·0–1614·0
Sea-water	53·8–269·0
Chemicals, acid or alkaline solution in process tanks	53·8–269·0
Heat-exchanger water boxes with non-ferrous tube plates and tubes	1345·0 overall
WELL-COATED STEELS	
Soils	0·01–0·2
HIGH-VOLTAGE HOLIDAY-DETECTED WELL-COATED STEELS	
Soils	0·01

*Higher current densities will be required if galvanic effects (i.e. dissimilar metals in contact) are present.

deposit on the structure, may considerably reduce eventual current requirements. On the other hand, in estuaries and polluted waters special care must be taken to allow for seasonal and other variable factors which may require higher current densities.

Establishing electrolyte resistivity To enable a satisfactory cathodic-protection scheme to be designed, it is necessary to determine the resistivity of the electrolyte (soil or water). This information is necessary to enable the current output of anodes to be determined together with their position and power source voltage, and it also provides an indication of the aggressiveness of the environment; in general the lower the resistivity the more aggressive the environment.

Economics After evaluating these variables, it must then be decided which type of system, i.e. sacrificial anode or impressed current, would be the most economical under the prevailing conditions. For instance, it would obviously be very expensive to install an impressed-current system on 100 m of fire main. Similarly, it would be equally uneconomic to install a sacrificial-anode system on hundreds of miles of high-pressure poorly coated gas main. Therefore, each system must be individually calculated taking note of all the factors involved.

Impressed-current Systems

Cathodic-protection schemes utilising the impressed-current method fall into two basic groups, dictated by the anode material:

1. Graphite, silicon-iron and scrap-steel anodes used for buried structures and landward faces of jetties, wharves, etc.
2. Platinised-titanium, platinised-niobium, lead and lead–platinum anodes used for submerged structures, ships and power stations.

These two groups will be discussed briefly since a more detailed account has been given in Section 11.3.

Group 1 Anodes

Scrap steel   In some fortunate instances a disused pipeline or other metal structure in close proximity to the project requiring cathodic protection may be used. However, it is essential in cases of scrap steel or iron groundbeds to ensure that the steelwork is completely electrically continuous, and multiple cable connections to various parts of the groundbed must be used to ensure a sufficient life. Preferential corrosion can take place in the vicinity of cable connections resulting in early electrical disconnection, hence the necessity for multiple connections.

Graphite   Graphite anodes are usually linseed oil or resin impregnated and supplied in standard lengths, e.g. 2·5 in (approx. 65 mm) dia. × 4 ft (1·2 m) long and 3 in (75 mm) dia. × 5 ft (1·5 m) long with a length of cable (called the anode *tail*) fixed in one end. Graphite anodes are still used and were particularly common in early cathodic-protection systems. However, they have tended to be replaced by silicon-iron, the main reasons being (*a*) graphite tends to spall in use, particularly in chloride environments and (*b*) its relatively low operating current density (10–20 Am^{-2}, Table 11.18).

Silicon-iron   Silicon-iron anodes are again generally supplied in standard lengths, e.g. 2·5 in (approx. 65 mm) dia. × 4 ft (1·2 m) long and 3 in (75 mm)

Table 11.18   Impressed-current anodes

Material	Max. working capacity (A/m^2)		Approx. consumption (kg/A year)	
	Soil	Water	Soil	Water
Scrap steel	5·4	5·4	8·0	10·0
Scrap cast iron	5·4	5·4	6·0	8·0
Silicon-iron	32	32–43	<0·1	0·1
Graphite	11	21·5	0·25	0·5
Lead	—	107–215	—	—
Lead–platinum	—	1 080	—	—
Platinum	—	<10 800	—	—
Platinised titanium	—	<10 800	—	—
Platinised tantalum	—	<10 800	—	—
Aluminium	—	21·5	—	4·0

× 5 ft (1·5 m) long and are complete with a cable tail. These anodes are made from cast iron with a high silicon content of 14–15%, together with small percentages of alloying elements such as chromium. The main disadvantage is their extreme brittleness, resulting in transport problems from the foundry to the cathodic protection site, especially if this is overseas.

Group 2 Anodes

This group includes platinised-titanium, platinised-niobium, lead alloys and lead–platinum anodes, which are used for immersed structures, e.g. jetties, sheet piling and power stations.

Platinised titanium   These anodes are usually in the form of a rod of titanium with a thin coating of platinum 2·5 μm thick, electrodeposited on to them. These rods vary in diameter from 6 to 18 mm and in length from 150 to 450 mm. In order to reduce the voltage drop in long length anodes, a copper cored variety is available. Platinised-titanium anodes may also be used in mesh form, particularly for internal protection of pasturisers and other vessels where space for installation is at a premium.

Platinised niobium and tantalum   Niobium and tantalum can be used as substrate materials where environmental conditions warrant a high current density. However, the occasions are rare due to high cost and manufacturing difficulties of these anodes.

Lead–platinum   The alloy is lead together with 0·1–2% silver usually in rod form with platinum micro-electrodes inserted every 150 mm. The purpose of these micro-electrodes, which take the form of pins, is to stabilise the formation of lead peroxide on the anode surface.

Cathodic Protection of Buried Structures

Soil or Water Resistivity

Resistivity measurements are carried out by the Wenner technique with an instrument such as a four pin megger. Soil resistivity not only enables the anode resistance and size to be calculated but also indicates the probable corrosivity of the soil or water. In the case of a pipeline route, variations in soil resistivity may stimulate localised corrosion and can cause long line currents; the area in the soil of lower resistivity is usually anodic (more negative) compared to that in the soil of higher resistivity. It is therefore impossible to give any precise ruling as to the rate at which corrosion will occur under any particular soil conditions, and the results of any survey must be taken as giving only a general indication of the probability of corrosion. Chemical and other forms of industrial contamination of the soil can significantly influence soil resistivities and corrosion characteristics and it is inadvisable to place too much faith in any results where such contamination may have occurred (e.g. in the proximity of gasworks and chemical plant, etc.).

A typical soil resistivity survey is shown in Fig. 11.17. Soil resistivities will normally indicate whether a cathodic-protection system is advisable in principle and whether impressed current or sacrificial anode schemes in particular are preferable. It may, as a result of the survey, be considered desirable to apply protection to the whole line or to limit protection to certain areas of low soil resistivity or 'hot spots'.

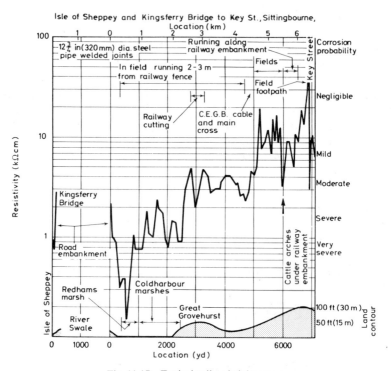

Fig. 11.17   Typical soil resistivity survey

The cost of protecting uncoated structures will, however, be high and expense may prohibit complete protection, in which case sacrificial anodes may be installed in the lowest resistivity areas only. The soil resistivity survey therefore indicates the most corrosive areas where cathodic protection should be applied. However, in the case of high-pressure gas and oil lines, it is becoming increasingly common practice to apply cathodic protection irrespective of the soil resistivity in view of the low cost involved with modern pipeline coatings and the potential hazards which could occur should even a minute pin hole perforate the pipe. In such cases a soil resistivity survey is only required in the probable areas of groundbeds in order to enable the size of the groundbed, rectifier voltage, etc. to be determined. The soil resistivity also determines the current output of magnesium or zinc anodes should this type of system be preferred (see Section 11.2).

The soil resistivity, measured using the Wenner technique, determines the average soil resistivity to a depth approximately equivalent to the distance between the pin spacings, and it must therefore be recognised that the design figure for groundbed design in a particular location may not be highly precise. At least 10% error in the calculated resistance may be expected.

Soil resistivity surveys are often impractical in built-up areas, but in such areas impressed-current cathodic protection is usually avoided on account of the danger of interaction. Under such conditions adequate protection can be achieved by installing magnesium anodes in the pipe trench should the soil resistivity measurements made when the trench is opened indicate that this is necessary.

Impressed-current Design

Designs can be prepared once soil resistivities and groundbed locations have been determined. Figure 11.18 illustrates typical installations. Design steps are as follows:

1. Select the current density to be applied from the results of cathodic-protection tests and from any available data. On pipeline structures attenuation is always a factor and the average current density is determined from the attenuation curves, or the total current to give protection at the points most distant from the drainage point may be computed.

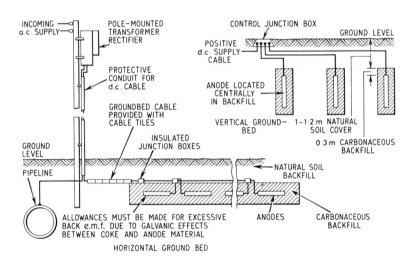

Fig. 11.18   Groundbeds for buried pipes using impressed current

2. Compute the total current requirement to achieve the required current density (total current = current density × superficial surface area).
3. Design the groundbed system in accordance with procedures outlined.
4. Design the d.c. wiring system for the most economical cable size in accordance with standard electrical practices and then calculate the total *IR* drop in the circuit.
5. Select rectifier voltage and current outputs.

6. Design the electrical circuits, fittings and switchgear in accordance with standard electrical practice.
7. Select the location of cathodic-protection test stations.
8. Prepare project drawings and specifications.

The total circuit resistance of a groundbed installation includes cable resistance, resistance of the anode to the carbonaceous backfill plus resistance to earth of the backfill column itself. In the case of sea-water installations, the anode resistance is between the sea-water and the anode surface only. The cable resistance can be calculated from Ohm's law.

Anode Backfill

The carbonaceous backfill surrounding an anode is essential and serves a number of functions, e.g.

1. Being of low resistivity it has the effect of increasing the anode size with resulting reduction in resistance to earth.
2. Most of the current is transmitted to the backfill from the anode by direct contact, so that the greater part of material consumption is on the outer edges of the backfill column, enabling the anodes themselves to have an increased life.

The backfill composition may be of several types varying from coke breeze to man-made graphite particles. A coke breeze backfill consists of high-temperature-fired coke breeze or calcined petroleum coke with not less than 95% dry weight of carbon with a maximum resistivity of 50 Ω cm when lightly tamped, dry and prior to the addition of lime. The physical properties of the backfill should be such that 100% will pass through a 16 mm aperture, 90% will pass through an 8 mm aperture and not more than 15% will pass through a 1 mm aperture. The coke breeze should be thoroughly mixed with 5% by weight of slaked lime.

Resistance of Groundbeds

The resistance of groundbeds for protection of pipelines or anodes for protection of jetties or other sea-water structures, is usually calculated in accordance with the formulae originally developed by Dwight. However, the following abridged formulae are normally used and are sufficient for all practical purposes:

Resistance of a single horizontal rod anode $R_H = \dfrac{P}{2\pi L}\left(\ell n\,\dfrac{4L}{d} - 1\right)$

and

Resistance of a single vertical rod anode $R_V = \dfrac{P}{2\pi L}\left(\ell n\,\dfrac{8L}{d} - 1\right)$

where R_H is the resistance of a single horizontal anode (Ω cm), R_V is the resistance of a single vertical anode (Ω cm), P is the earth resistivity (Ω cm), L is the length of rod (cm) and d is the diameter of the rod (cm).

Resistance of the anode to backfill is obtained by using the formula:

$$R_V = \frac{0\cdot0171\rho}{L}\left(2\cdot3 \log \frac{8L}{d} - 1\right)$$

where R_V is the resistance of the vertical anode to earth (Ω), ρ = resistivity of backfill material (or earth) (Ω cm), L = length of anode (m) and d = diameter of anode (m).

Resistances of several anodes in parallel can be calculated by the formula:

$$R = \frac{0\cdot0171\rho}{NL}\left(2\cdot3 \log \frac{8L}{d} - 1 + \frac{2L}{S} 2\cdot3 \log 0\cdot656N\right)$$

where R = resistance to earth (Ω) of the vertical anodes in parallel, ρ = soil resistivity (Ω cm), N = number of anodes in parallel, L = length of anode (m), d = diameter of anode (m) and S = anode spacing (m).

Deep Well Groundbeds

This type of groundbed is illustrated in Fig. 11.19 and is normally employed where the surface soil resistivities are very high, e.g. in desert areas. They

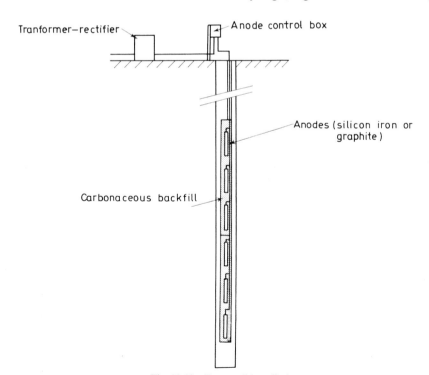

Fig. 11.19   Deep well installation

have the advantage of taking up little surface area and can be installed, in the case of a pipeline project, in the existing pipeline wayleave. They have the further advantage of minimising interaction on foreign structures. Abandoned oil or water wells can sometimes be used for this purpose. The type of groundbed can be of the design illustrated or alternatively can be assembled on the surface in a steel case and lowered down a suitably drilled hole.

Determining Rectifier Voltage

In determining the rectifier voltage, the following must be taken into consideration:

1. Voltage drop caused by groundbed resistance, as previously explained.
2. Back voltage polarisation between groundbed and pipeline. In the case of both graphite and silicon-iron anodes, an allowance of 2 V is normally used. This back voltage is that which exists between the anodes and the structure in opposition to the applied voltage.
3. Resistance to earth of the pipeline at the groundbed location. This resistance to earth depends on the quality of the pipeline coating. The better the coating the higher the resistance.
4. Resistance of the cable from the pipeline to the power source and from the power source to and along the anodes comprising the groundbed. This cable resistance must be determined from the standard tables supplied by the cable manufacturer.

Once the total circuit resistance is known, applying Ohm's law the rectifier or power-source voltage can be calculated.

Attenuation

When cathodic protection is applied to an underground metal structure the greatest effect on the pipe to soil potential is at the drainage point. This effect decreases, or attenuates, as the distance from the drainage point increases. Complex structures such as tank farms, tank bottoms and marine installations have complicated attenuation patterns and it is not feasible to use complex equations to determine the effect. This must be determined by a current drain test or by practical experience. The line current and pipe to soil potential are at a maximum at the drainage point and decrease with distance from this point. The attenuation depends on the lineal resistance of the pipe, coating resistance, and to some extent the resistivity of the soil. Connections to foreign structures and the method of termination of the line also affect the attenuation (see Section 11.1).

Field Measurements

In order to obtain the actual field attenuation characteristics, a pipeline can be placed under temporary cathodic protection and measurements made

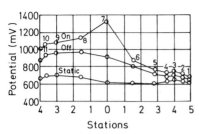

Fig. 11.20   Longitudinal distribution potential on pipeline. Note: 'Stations' refer to points at which the potential is measured

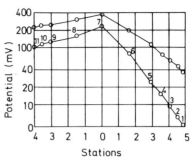

Fig. 11.21   Attenuation curve on pipeline shown in Fig. 11.20. Upper curve shows driving voltage; lower curve the pipe/soil potential

under different cathodic-protection current conditions. A 'natural' pipe to soil potential survey should be made over the length of the line to which cathodic protection is to be applied. These static measurements supply a reference level on which to base the subsequent readings. After the natural survey is completed, measurements are made at the same test points with the temporary cathodic-protection system installed. The pipe to soil potential should be read twice, once with the test current on and once with the current off. The data are then plotted as shown in Fig. 11.20.

The horizontal scale is the distance along the pipeline from the drainage point 0 and the vertical scale is the pipe to soil potential. The polarisation potential and the driving voltage are then plotted on semi-logarithmic paper using the same horizontal scale (Fig. 11.21). Attenuation is more rapid in low resistivity soils than in high resistivity soils. If non-uniform conditions prevail the curves will not plot as straight lines as shown in Fig. 11.21, in which the lines in the left of the figure are typically curved as a result of low resistance.

Measurement Between Drainage Points

The theoretical principles of attenuation have been considered in Section 11.1, but some illustrations are appropriate here. Figure 11.22 shows typical

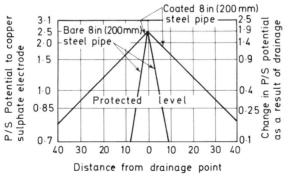

Fig. 11.22   Attenuation curves for bare pipe and poorly coated pipe in similar soils. Note: P/S (pipe/soil) potentials will be negative

curves for a bare pipe and a poorly coated pipe in similar soils, and it can be seen that the former has a much steeper attenuation curve than the latter.

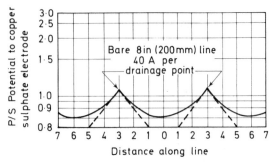

Fig. 11.23   Attenuation curves for multiple drainage points of moderate size

Figure 11.23 shows how the desired protection level of potential has been attained between widely spaced drainage points, in which the dotted curves show the attenuation curves for the individual drainage points.

Coating Resistance

The electrical resistance of a coating on a pipeline will depend upon the effectiveness of its initial application to the metal and on its condition at the time of testing; absorption of moisture may decrease the resistance by as little as 10% during a $2\frac{1}{2}$-year period. The leakage resistance of a given section of pipeline can be determined by the following procedure:

1. Applying temporary cathodic protection to the pipeline and measuring the pipe-to-soil potential for a given driving voltage of various test points spaced along the pipeline from the drainage point. The drainage current should also be measured.
2. Calculating the attenuation constant α from equations 11.9 and 11.10 given in Section 11.1.
3. Measuring or calculating the longitudinal resistance of the pipeline. The resistance can be calculated from

$$R = \frac{2\rho\ell}{dt}$$

   where ρ is the specific resistivity, ℓ is the length of the pipeline, d is the diameter and t is the thickness in appropriate and consistent units.
4. From equations 11.9 and 11.10 (Section 11.1) calculating R_L the leakage resistance. If it has been evaluated for 1 000 ft of pipeline the leakage resistance of the coating R_c is given by

$$R_c = 262dR_L \qquad\qquad ...(11.18)$$

   where R_c is in Ω ft and d is the external diameter of the pipe (in).

Types of Cable for Anode Installations

It is essential that cables used for anode groundbeds are chemically resistant to chlorine which may be generated from the anodes during operation. The cable in the vicinity of the anode itself is particularly susceptible and can be protected by an unplasticised p.v.c. tube (or something similar) installed at the time of manufacture. Alternatively, chlorine-resistant cable such as cross-linked Polythene can be used. This type of cable is essential for deep-well groundbeds where the cable is immediately above the anodes and is easily damaged by any chlorine generated from the anodes themselves.

Power Sources for Cathodic Protection

Where a.c. supplies exist, transformer–rectifiers are the most economical source of d.c. for cathodic protection systems. In the case of pipelines, standard transformer–rectifiers, either oil or air cooled, can be employed. They range in size from 5 A, 5 V for small systems to 100 A, 48 V for major pipeline schemes. A typical size for a well-coated cross-country pipeline in the UK would be 15 A, 48 V. Usually in the case of a pipeline they are pole mounted, but may equally be plinth mounted to suit site conditions. In the case of sea-water jetties where the voltage required is usually low because of the lower sea-water resistivity, a typical rectifier size for a major installation would be 500 A, 48 V. For offshore pipelines and loading platforms where a fire hazard exists, it is usual to employ certified flameproof or intrinsically safe rectifiers to overcome any possibility of fire hazard should faults develop in the unit.

Where a.c. supplies do not exist, other sources of power can be used such as d.c. generators, and either diesel or gas driven. Alternatively, thermo-electric generators may be considered if the power requirement is relatively low. Thermo-electric generators are available in relatively small outputs only as shown in Table 11.19. They have the advantage of being completely self-contained since they are powered by taking off some of the gas which passes through the pipeline. For this reason they are becoming widely used for protecting pipelines in desert areas.

Table 11.19 Thermo-electric generators

Electrical characteristics				Fuel consumption*		
Power† (W)	Load voltage† (V)	Load current† (A)	Power range‡ (W)	Propane (kg/h)	(litre/h)	Natural gas (m³/h)
9	1	9·0	8·5–10·0	0·033	0·077	0·035
15	?	7·5	14–18	0·059	0·105	0·054
28	3·8	7·3	26–32	0·082	0·191	0·1
50	6	8·3	46–60	0·136	0·322	0·2
100	12	8·3	92–120	0·272	0·636	0·4
200	6	33	188–240	0·544	1·273	0·8

*Fuel consumption may vary slightly depending upon air shutter adjustment.
†Values are typical at 24°C ambient temperatures, under a fixed matched load at constant gas pressure.
‡Typical power output range of generator for a variation in ambient temperature from 51·5 to −31·5°C.

Protection of Power Station or Refinery Sea-water-cooled Circulating Water Systems

Sacrificial Anodes

The use of sacrificial anodes in circulating water systems is limited to the application of cathodic protection to stop gates, coarse screens and other plant that are readily accessible so that the anodes can be replaced when they are consumed. Such anodes are not normally used in condensers, pumps and auxiliary coolers for the following reasons:

1. Frequent replacement required with consequent shut-down of the plant concerned.
2. Anodes required to be very large to provide the current density necessary, and with a reasonable life. Anodes can interfere with the water flow.
3. Corrosion products from anodes can cause tube blockage and subsequent failure.

Impressed-current Systems

Anodes for the internal protection of plant are normally platinised titanium. Continuous coaxial anodes are normally used for the protection of water boxes since they provide a more uniform level of protection and only one anode is required per water box. The terminations of the coaxial anodes pass through the water box at selected points by means of specialised unplasticised p.v.c. mounts to ensure that short circuiting does not occur. Anode retaining clamps keep the anode approximately 20 mm from the water-box surface. It is advisable to fit a plastic mesh over the anode to prevent shorting against the water box, should it become distorted during maintenance.

Rod Anodes

These consist of solid titanium rod, tube or they can be copper cored, portions of which can be platinised. They range from 6 to 25 mm in diameter and are normally supplied pre-assembled in a steel or unplasticised p.v.c. mount ready for screwing into prepared bosses on the plant under protection. Electrical connections are made via cables in the usual way. The anodes are spaced to give even current distribution throughout the water box space.

Tubular Anodes

Tubular anodes are supplied in diameters between 12·5 and 32 mm and have been designed for installations where water conditions on the plant under protection are known to be turbulent. The tubular anode has a number of holes drilled in the active portion of the anode and the non-active portion is filled with sand to act as a damping agent. As in the case of rod anodes they are supplied complete with mounts ready for installation in the prepared bosses on the plant under protection. They are particularly suitable for internal protection of pump casings and internal protection of pipelines, carrying salt or other low resistivity liquids.

Types of Installation

Impressed-current systems for power stations are somewhat more sophisticated than those required for pipelines or marine structures inasmuch that a large number of items of plant, with a wide range of current requirements, are protected by one transformer–rectifier. Each section of every water box in order to provide even current distribution requires one or more anodes. In the case of a large circulating water pump as many as 30 anodes may be required to provide the current distribution necessary. Three types of system should be considered as follows:

1. Manually controlled.
2. Automatically thyristor controlled.
3. Automatically controlled modular.

Manually Controlled System

A manually controlled system comprises one or more transformer–rectifiers each with its associated control panels which supply the d.c. to the various anodes installed in the water box spaces. Each transformer–rectifier is provided with its own control panel where each anode is provided with a fuse, shunt and variable resistor. These enable the current to each anode to be adjusted as required. Reference cells should be provided in order to monitor the cathodic protection system. In the case of a major power station, one transformer–rectifier and associated control panel should be provided for separate protection of screens, circulating water pumps and for each main condenser and associated equipment.

When coolers or condensers are shut-down but remain full of water, the amount of current required to maintain satisfactory cathodic protection is considerably reduced. If the current is not reduced over-protection occurs and excessive amounts of chlorine can be generated which would tend to accumulate in the upper section of the water boxes causing considerable corrosion, not only to the water boxes, but also possibly to the tubes. To ensure against this a stand-by condition should be included on the control panel which effectively reduces the current required under shut-down conditions. This control is effected by a limit switch fitted to the outlet valve of the condenser or cooler concerned. It is impossible to determine exact requirements for the protection of circulating water systems in advance and it is normal to adjust the current to provide protection during commissioning.

Automatically-thyristor-controlled System

This method is basically the same as the manual system. However, the current output of the transformer–rectifier is automatically maintained at a level to ensure satisfactory cathodic protection under all operating conditions. This is achieved by means of sensing devices located in the main item of plant, e.g. main condenser, which feeds back signals to an automatic control device within the transformer–rectifier. The control device is pre-set at the required potential and any incoming signals are compared with this pre-set potential and the level of current either raised or lowered until the incoming signal agrees with that of the pre-set potential in the automatic device.

Normally, a number of such sensing electrodes are installed in one condenser box and the controller will automatically select the potential indicating the highest amount of current required. Also, should one controlling sensing electrode fail, the controller will automatically select the next one showing the most demand. It is also usual to fit monitoring electrodes of the zinc type. In addition to automatic control, manual override control is normally provided.

Automatically Controlled Modular System

This method employs one large manually controlled transformer–rectifier used in conjunction with a number of modular cabinets located adjacent to each item of plant requiring protection. The main transformer–rectifier feeds d.c. to each of the module units and the modular unit provides the exact amount of current required by the item of plant in question.

Marine Structures

The method of applying cathodic protection to immersed structures will depend on several factors including:

1. Size of the project.
2. Availability of power supply.
3. Possibility of problems from interaction.
4. Necessity for safety from spark hazard.
5. Expected economic life of the system.

Where electric power is available, impressed-current schemes are adopted for most major installations and Fig. 11.24b shows basic designs for typical anode assemblies often adopted for protection of jetties and other major harbour and offshore installations.

Current density requirements depend on the environment, galvanic effects, volocities and other factors influencing polarisation. In the absence of galvanic influences or other secondary effects $30\,\mathrm{mA/m^2}$ is usually sufficient in sea-water to maintain adequate polarisation for protection once it has been achieved; it is however normally necessary to apply 100–$150\,\mathrm{mA/m^2}$ to achieve initial polarisation within a reasonable period and if rapid protection is required, current densities as high as $500\,\mathrm{mA/m^2}$ may be applied.

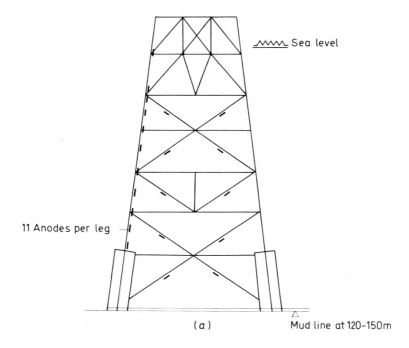

11 Anodes per leg

(a)   Mud line at 120-150m

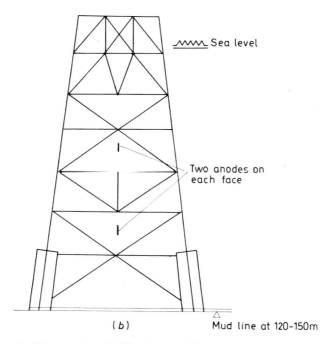

(b)   Mud line at 120-150m

Fig. 11.24 Protection of offshore marine oil-drilling rig. (*a*) With external zinc (or aluminium) anodes and (*b*) with impressed current using platinised titanium or platinised niobium. (Compare the large number of anodes used in (*a*) with the small number used in (*b*))

Groundbeds remote from the structure can be considered but usually with this type of installation problems arise due to damage to the connecting cable by ships' anchors, etc. The calculation of rectifier voltage / anode resistance is exactly as described for impressed current pipeline installations except that the voltage required is very small because of the low resistance of the electrolyte—normally 25–40 Ω cm for typical sea-water.

For protection of sheet-steel piling, the anodes are normally mounted within the re-entrants of the piles to prevent mechanical damage by berthing ships, dredging, etc. (Fig. 11.25).

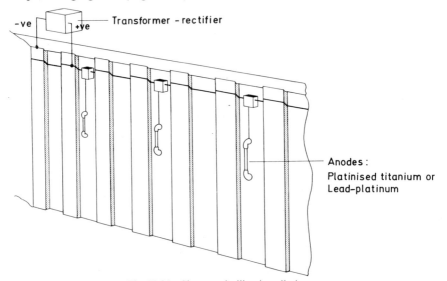

Fig. 11.25   Sheet steel piling installation

As in the case of power stations, where there is known to be considerable variation in operating conditions due to tidal changes, or in estuary waters variations in salinity, automatic control systems may be desirable. For such systems the current output of the transformer–rectifier is controlled by thyristor or transductors. Silver/silver chloride sensing electrodes are permanently installed on selected piles and transmit the electrode potential of the steel back to the controlling device. This type of system enables the most economic amount of current to be provided under all operating conditions.

Ships

Cathodic protection is effectively supplied to ships' hulls and also to the tanks of oil tankers and other vessels where sea-water is used for ballast or other purposes. For internal protection sacrificial anodes are normally employed of the zinc or aluminium type. Impressed current is not normally used because of the potential fire hazard and also because of the anodic generation of chlorine inside closed water spaces. External hulls can be protected either by sacrificial or impressed-current methods, although impressed current is now preferred (Fig. 11.26). Anodes of platinised titanium, lead–platinum or

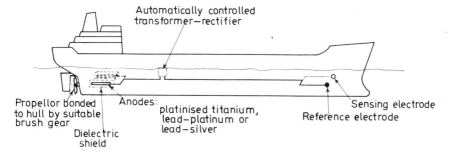

Fig. 11.26 Impressed-current system for ships

lead, are mounted on the hull below the water line and can be designed to have currents of up to 200 A. Due to access problems, the anodes are located either at the stern or the forward end of the ship. All cable or other entries through the hull plate must be carefully designed using a special backing plate and the complete system has to be submitted to the Ship Approval Authorities prior to installation. Because of the rapid variation in conditions, the condition of the paint surface, etc. all schemes for ships are automatically controlled using silver chloride sensing electrodes installed remotely from the anodes.

The transformer–rectifiers are either transistor or transductor controlled and operate in a similar manner to that described for power stations.

Water Storage Tanks

Cathodic protection can be satisfactorily employed for the internal protection of water storage tanks. Any of the conventional anodes may be employed for impressed-current systems, but normally silicon-iron or platinised titanium are preferred (Fig. 11.27). In the case of open-topped tanks, the

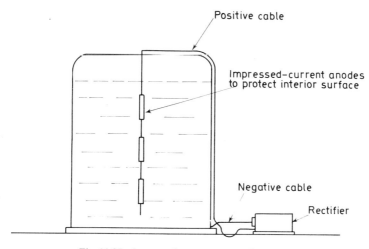

Fig. 11.27 Impressed-current system for water tanks

anodes are suspended from suitable supports. Alternatively, in the case of a fixed roof tank, it may be necessary to drill the tank surface and insert anodes and reference electrodes through the roof structure. Anodes should be distributed so as to ensure even current distribution throughout the tank surfaces. Current density is determined by the type of coating, if any, and is shown in Table 11.17. When the water resistivity is exceptionally high, copper-cored platinised-titanium continuous anodes may be used although the current spread will be relatively poor and considerable anode length will be required to achieve the necessary current distribution. Anode resistance calculation and cable resistance calculation is as previously described.

Internal Protection of Pipelines

The current spread from an internally immersed anode is directly related to pipe size and the resistivity and temperature of the water; with sea-water having a resistivity of about $30\,\Omega\,cm$, anodes are normally required on a 10 in (250 mm) pipe at about 1 m centres and in the case of a 30 in (760 mm) pipe, at 3 m (10 ft) centres (Fig. 11.28). Small bore or high resistivities and velocities further reduce the anode spacing.

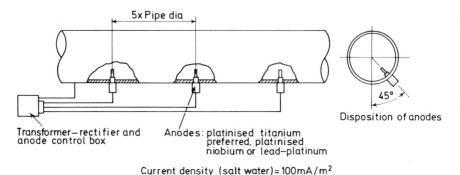

Fig. 11.28   Internal protection of pipeline

Cathodic protection is therefore normally practicable only in large bore pipes carrying salt water. Under special circumstances, however, it has been found necessary to use cathodic protection in fresh water and in these instances anodes have been run longitudinally down the entire length of the pipe, but the cost of such schemes is usually prohibitive.

Economics

Cathodic protection design involves achieving an economic balance between installation costs, maintenance costs, initial cost of power units and power consumption. Because both the cost of the rectifier and the cost of the electric power consumed are contingent on the operating voltage of the system, it is desirable to keep the operating voltage as low as possible; for this reason a

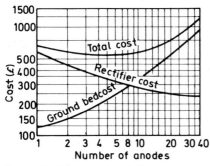
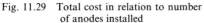
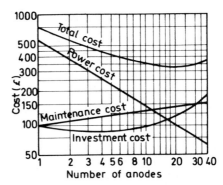

Fig. 11.29   Total cost in relation to number of anodes installed

Fig. 11.30   Variation of minimum annual cost with number of anodes installed

low resistance groundbed is desirable when it is economically feasible. Typical cost curves plotted on a cost versus the number of anodes in a groundbed for an assumed set of conditions are shown in Figs. 11.29 and 11.30. Figure 11.29 is based on installed costs only, whereas Fig. 11.30 is the minimum annual cost.

Conclusion

Although the principles of cathodic protection are essentially simple and were in fact first outlined by Sir Humphry Davy in 1824, the application of the method to practical problems remains more of an art than a science. A properly designed cathodic protection system will be both economical and effective. On the other hand an incorrectly designed scheme will be inefficient, uneconomical and under certain circumstances may accelerate corrosion instead of controlling it.

Practical considerations have so far prevented the application of impressed-current systems to oil and gas platforms in very deep water. This particular problem is receiving attention throughout the world. At the present time sacrificial anodes are used for such structures with a life of several years. For economic reasons suitable methods of applying impressed current will have to be found as it will be physically impossible to replace sacrificial anodes installed prior to the launching of these immense structures.

J. S. GERRARD*

*Mr. J. S. Gerrard died unexpectedly in April 1974, and it is appropriate here to acknowledge the important contributions that he made in the practice of cathodic protection.

BIBLIOGRAPHY

Applegate, L. M., *Cathodic Protection*, Leonard Hill, London, 22 (1959)
Brady, G. D., 'Graphite Anodes for Impressed Current Cathodic Protection. A Practical Approach', *Mat. Prot. and Perf.*, **10** No. 10, 21, Oct. (1971)

Burgbacher, J. A., 'Cathodic Protection of Offshore Structures', *Mat. Prot.*, 7 No. 4, 26, April (1968)

Costanzo, F. E., 'Using Graphite as Impressed Current Anode for Deep Groundbeds', *Mat. Prot.*, 9 No. 4, 26, April (1970)

Davy, H., *Phil. Trans.*, 114 (1824) and in a collection of papers on underground corrosion, St. Louis, 4, 11 (1960)

DeMille Campbell, E., 'High Silicon Iron Anodes for Cathodic Protection', *Corrosion*, 27 No. 4, 141 (1971)

Dietl, B. and Gleason, J. D., 'Compatibility of Impressed Current Cathodic Protection with Paint Systems', *Mat. Prot. and Perf.*, 10 No. 11, 44, Nov. (1971)

Doremus, E. P. and Canfield, T. L., 'The Surface Potential Survey Can Detect Pipeline Corrosion Damage', *Mat. Prot.*, 6 No. 9, 33, Sept. (1967)

Doremus, E. P. and Pass, R. B., 'Cathodic Protection of 516 Offshore Structures: Engineering Design and Anode Performance', *Mat. Prot. and Perf.*, 10 No. 5, 23, May (1971)

Geld, I. and Acampora, M. A., 'Current Density—A Factor in Hydrogen Embrittlement of Cathodically Protected Steels in Sea Water', *Mat. Prot.*, 7 No. 11, 31, Nov. (1968)

Groover, R. E., Lennox, T. J. (Jr.) and Peterson, M. H., 'Cathodic Protection of 19 Aluminium Alloys Exposed to Sea Water—Corrosion Behaviour', *Mat. Prot.*, 8 No. 11, 25, Nov. (1969)

Hammer, N. E., 'Applying Cathodic Protection to a Paper Making Cylinder', *Mat. Prot.*, 8 No. 2, 48, Feb. (1969)

Hausmann, D. A., 'Criteria for Cathodic Protection of Steel in Concrete Structures', *Mat. Prot.*, 8 No. 10, 23, Oct. (1969)

Husock, B., 'A Method for Selecting the Most Economical Magnesium Anode', *Mat. Prot.*, 7 No. 6, 33, June (1968)

Kurr, G. W., 'Sacrificial Anode Applications in Ships', *Mat. Prot.*, 8 No. 11, 19, Nov. (1969)

Lowe, R. A. and Brand, J. W. L. F., 'Internal Cathodic Protection of Water Cooled Plant', *Mat. Prot. and Perf.*, 9 No. 11, 45, Nov. (1970)

Morgan, J. H., 'Cathodic Protection—A Review', *Brit. Cor. J.*, 5 No. 11, 237, Nov. (1970)

McComb, G. B., (Ed.), *A Collection of Papers on Underground Pipeline Corrosion*, Vols. 1–5, St. Louis, Mo., USA (1958–1961). (Information on this publication may be obtained from Geo. B. McComb, 8038 Lafon Place, St. Louis 30, Mo., USA)

Parker, M. E., 'Innovations and the Future of Cathodic Protection', *Mat. Prot.*, 8 No. 2, 21, Feb. (1969)

Parker, N. E., *Pipe Line Corrosion and Cathodic Protection: A Field Manual*, Gulf, Houston (1954)

Peabody, A. W., *Control of Pipeline Corrosion*, National Association of Corrosion Engineers, Houston, Texas

Pritula, V. A., *Cathodic Protection of Pipelines and Storage Tanks* (translated from Russian), H.M.S.O., London (1953)

Reding, J. T., 'Sacrificial Anodes for Ocean Bottom Applications', *Mat. Prot. and Perf.*, 10 No. 10, 17, Oct. (1971)

Spencer, K. A., in *Anti-Corrosion Manual*, Scientific Surveys Ltd., London, 359 (1962)

Vrable, J. B., 'Protecting Underground Steel Tanks', *Mat. Prot.*, 6 No. 8, 31, August (1967)

West, L. H., 'Cathodic Protection—The Answer to Corrosion Prevention of Underground Structures', *Mat. Prot.*, 7 No. 7, 33, July (1968)

Cathodic Protection of Buried and Submerged Structures, CP 1021: 1973, British Standards Institution, London

Technical Committee Reports of the National Association of Corrosion Engineers, USA, on pipeline corrosion control, including: 'Statement on Minimum Requirements for Protection of Buried Pipelines', 'Some Observations on Cathodic Protection Criteria', 'Criteria for Adequate Cathodic Protection of Coated Buried Submerged Steel Pipelines and Similar Steel', 'Methods of Measuring Leakage Conductance of Coatings on Buried or Submerged Pipelines', 'Recommended Practice for Cathodic Protection of Aluminium Pipe Buried in Soil or Immersed in Water'

Symposium on Cathodic Protection, National Association of Corrosion Engineers, Houston, 93 (1949)

11.5 Stray-current Corrosion

If a continuous metallic structure is immersed in an electrolyte, e.g. placed in the sea or sea-bed or buried in the soil, stray direct currents from nearby electric installations of which parts are not insulated from the soil may flow to and from the structure. At points where the stray current enters the immersed structure the potential will be lowered and electrical protection (cathodic protection) or partial electrical protection will occur. At points where the stray current leaves the immersed structure the potential will become more positive and corrosion may occur with serious consequences.

In practice, corrosion of buried metallic pipes or cable sheaths is liable to occur when direct current flows from equipment associated with electric railway traction systems, trolley vehicle or tramway systems, electricity supply systems and electric welding or similar equipment in factories or shipyards. This type of attack is known as *stray-current electrolysis* and is particularly severe in the vicinity of traction systems running on rails in direct contact with the soil or when earth leakage faults develop on d.c. power supply systems. Small signalling currents may be detected flowing to or from buildings housing telecommunications or signalling apparatus and stray galvanic or earth currents may also flow from earth systems associated with protective or signalling equipment.

Stray-current electrolysis occurring as a result of the application of cathodic protection to a nearby immersed or buried structure is known as *cathodic-protection interaction* and is described in Section 11.6.

The corrosion of underground pipes and cables owing to the electrolytic action of stray currents from d.c. electric railway and tramway systems has long been a serious problem. Ministry of Transport regulations limiting the maximum potential between tramway rails and neighbouring buried structures and the maximum potential difference between points on the rail systems have been in operation in the UK since 1894.

In 1929, an international commission, the *Commission Mixte Internationale pour les expériences relatives à la protection des lignes de télécommunication et des canalisations souterraines* (C.M.I.), was enlarged and its scope extended for the purpose of making experimental studies concerning the corrosion of pipes and cables. The C.M.I. membership was composed of international technical experts from learned bodies, research laboratories and manufacturers of electrical equipment, and representatives of authorities responsible for electric power supply, electric railways, electric tramways, gas supply, and telecommunications.

Members of the C.M.I. co-operated with the *Comité Consultatif International Téléphonique et Télégraphique* (C.C.I.T.T.) to produce recommendations[1] for the protection of underground cables against the action of stray currents arising from electric traction systems. These recommendations are revised at regular intervals and, although they specifically refer to the protection of telephone cables, most of the preventive measures and testing methods are also applicable to pipe systems.

In the USA, interest has been particularly concentrated on leakage currents from inter-urban and street railway tracks, and the 1921 Report of the American Committee on Electrolysis[2] summarises the problem and the methods of controlling electrolytic corrosion applied both in America and in many European countries.

Mechanism of d.c. Stray-current Corrosion

An electric railway or tramway system with an adjacent buried pipeline or cable which may cross the running rails at intervals is illustrated in Fig. 11.31 in which the arrows indicate the general flow of stray currents when one vehicle is in service. Rapid variations of current and potentials will occur as the tram or train moves along the rails. Corrosion will occur at points near the sub-station or near negative feeders where the stray current leaves the buried structure to return to the negative busbar at the sub-station.

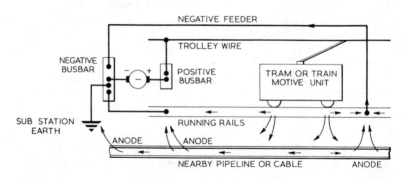

Fig. 11.31   Stray current from d.c. traction system

Damage is greatest with the 'Third Rail System' illustrated in the figure, since the negative pole of the generator is connected to the running rails which are not deliberately insulated from the surrounding soil, and in some instances cables and pipes have been severely pitted by stray-current corrosion in less than one year after installation. Some d.c. electric railways operate on the 'Fourth Rail System' in which the current is collected from specially insulated positive and negative conductor rails which are not in electrical contact with the running rails or the surrounding soil.

When the motive unit is taking current from a third rail system the potential of the rails immediately beneath the wheels of a tram or train may rise 10 or more volts above normal rail potential. Since the joint resistance to earth of the running rails will usually be less than 1 Ω there will be fluctuating potential changes in the surrounding soil, and tests in these circumstances have shown that the potential difference between two reference electrodes placed in the soil a hundred metres or so apart near the rails will vary rapidly in amplitude and polarity and may reach peak values of 3 or 4 V.

Most investigators agree that the electrolytic effects attributable to the operation of a d.c. traction system depend on the net amount of current per unit area passing from pipe or cable into the adjacent soil. The most common method of assessing the order of magnitude of this current density is measurement of the change in potential to nearby earth by means of a high-resistance recording voltmeter connected between the buried structure and a reference electrode [Cu/CuSO$_4$(sat.)] buried in the soil close to the structure. The natural structure/soil potential (the 'shut-down potential') is ascertained by recording the steady potential attained after the system has been de-energised for at least an hour.

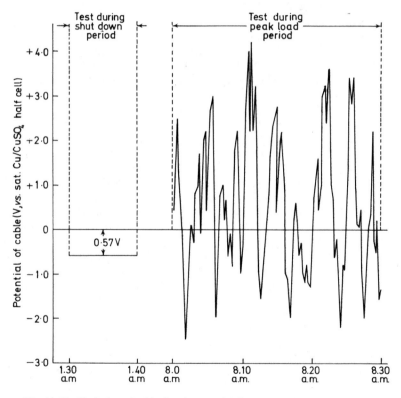

Fig. 11.32   Variation of cable sheath potential due to stray d.c. traction currents

The variation in potential between the sheath of a lead-covered cable lying parallel to a d.c. railway system and a reference electrode placed close to the cable is shown in Fig. 11.32. The shut-down potential of the cable in this test was − 570 mV [Cu/CuSO$_4$ (sat.)]. There is a marked positive change in mean cable sheath potential during the period 8.0 a.m. to 8.30 a.m. and similar but smaller potential changes were measured at off-peak periods.

a.c. Stray-current Corrosion

In nearly all known cases of stray-current corrosion the damage is caused by direct currents, but leakages of alternating currents at industrial frequencies have been suspected of causing corrosion of buried metallic structures. The mechanism of corrosion caused by a.c. is not clearly understood and fresh studies are being made[3]. However, the corrosion caused is much less severe than with stray d.c. and experiments[4] indicate that stray a.c. at 50 Hz will produce less than 1% of the corrosion caused on most buried metals by an equivalent d.c.

Methods of Control

The C.C.I.T.T. have recommended that the amount of stray current from the running rails should be controlled by limiting the average difference in potential between any point on the running rails and the nearest sub-station negative feeder bus-bar. The average difference in potential is defined as the average potential over all working days, each day counting as 24 h, whatever the effective period of service each day may be.

The amount of corrosion damage resulting from the operation of d.c. railways is generally less than the corrosion caused by street tramway systems since the railway track insulation is better, the frequency of service is less, and pipes and cables are usually buried further from the running rails.

A brief summary of the effects of stray currents from d.c. traction systems on buried structures is given by Evans[5] who describes methods by which the damage may be diminished. When corrosion occurs at points where the stray current from a d.c. electric railway or tramway leaves the buried structure to return to the traction system, attention should first be given to reducing the amount of stray current by improving the conductivity of the running rails and by regular rail-bond testing. Consideration should also be given to the provision of additional negative feeders and to increasing the resistance to earth of the running rails whenever possible, e.g. by improving drainage, embedding rails in bituminous material, etc.

In addition to these precautions, the most satisfactory way of preventing corrosion on nearby buried structures is by means of drainage bonds. The structure to be protected is made cathodic to the surrounding earth by being connected to the negative feeders or to suitable points on the running rails. The connection may be direct, through a resistance or similar current-controlling device, or through a suitable rectifying device (polarised electric drainage). The rectifier must be capable of passing a current of about 1 A from the buried structure to the negative feed of the traction system when the voltage of the structure rises by about a quarter of a volt above the potential of the adjacent rails. The rectifier must also be able to carry currents in excess of 50 A without breakdown.

Telephone cable sheaths have been successfully protected by this method, using germanium or silicon rectifier elements connected at intervals between the sheaths and the adjacent tramway rails. In a recent project a steam-operated railway system was electrified using 660 V d.c. Before electrification the potential of an adjacent lead-sheathed cable was steady and measured -470 mV [Cu/CuSO$_4$ (sat.)] as shown in Test 1 of Fig. 11.33. After electrification of the railway system positive and negative fluctuations of potential were measured on the cable sheath and there was a general shift of potential towards values more positive than the steady potential of -470 mV, as shown in Test 2 on the graph. Test 3 shows that a drainage bond containing a silicon rectifier element connected between the running rails and the cable sheath ensured that the sheath remained cathodic to the surrounding soil.

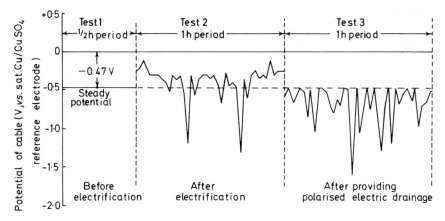

Fig. 11.33 Improvement effected by the provision of polarised electric drainage

A substantial reduction in the amount of stray current picked up by nearby buried pipes or cables may be achieved by interrupting the longitudinal conductivity of the structure by means of insulating gaps or joints. Care must be taken in siting the gaps, and they should preferably be placed in localities where the current tends to enter the structure and at points on each side of the track where the pipe or cable crosses under the rails.

Protection may also be obtained by insulating structures from the soil by means of protective coatings. Pipes may be dipped in bituminous compounds and wrapped with hessian, glass-fibre, or similar materials.

Cable sheaths may be covered with paper and hessian wrappings impregnated with bituminous compounds or with extruded or taped plastics outer sheaths. At pinholes or discontinuities in protective coatings the sheath will be particularly liable to electrolytic corrosion in stray-current areas, and it is desirable to supplement this form of protection by drainage bonds or direct cathodic protection.

Calculation of the amount of stray current entering or leaving a buried structure is a difficult matter, and a solution is usually possible only if the layout of the buried plant is simple. The very complicated cases normally met with in practice are not usually capable of solution, owing to heterogeneity of the soil, difference in pipe coatings, variable resistances between pipe joints, and variation in spacing between buried pipes or cables. Estimates of the potential changes on structures due to nearby traction systems are particularly difficult because of wide and rapid fluctuations in track potentials caused by movement of vehicles and by variable loads on motive units.

J. R. WALTERS

REFERENCES

1. C.C.I.T.T. *Recommendations Concerning the Construction, Installation and Protection of Telecommunication Cables in Public Networks*, I.T.U., Geneva (1974)
2. Report by the American Committee on Electrolysis (1921)
3. C.C.I.T.T. Study Group VI; working party on corrosion by alternating currents (C. Cabrillac). A bibliography with over 60 references giving particular emphasis on buried lead, iron, aluminium and zinc structures, I.T.U., Geneva
4. Morgan, P. D. and Double, E. W. W., *A Critical Resumé of a.c. Corrosion from the Standpoint of Special Methods of Sheath Bonding of Single Conductor Cables*, British Electrical and Allied Industries Research Association Technical Report F/T 73 (1934)
5. Evans, U. R., *Metallic Corrosion, Passivity and Protection*, Edward Arnold, London (1946)

11.6 Cathodic-protection Interaction

Stray currents are produced in the electrolyte during the operation of cathodic-protection systems and part of the protection current may traverse nearby immersed structures which are not being cathodically protected. The resultant corrosion produced on the unprotected structure is referred to as *corrosion interaction* or *corrosion interference*.

The cathodically protected primary structures may be the hulls of ships, jetties, pipes, etc. immersed in water, or pipes, cables, tanks, etc. buried in the soil. The nearby unprotected secondary structures subjected to inter-action may be the hulls of adjacent ships, unbonded parts of a ship's hull such as the propeller blades, or pipes and cables laid close to the primary structure or to the cathodic-protection anode system or groundbed.

The lines of flow of the protection current and the interaction current when cathodic protection is applied to a pipeline buried in the soil parallel to a buried secondary pipeline, are shown in Fig. 11.34. The distribution of

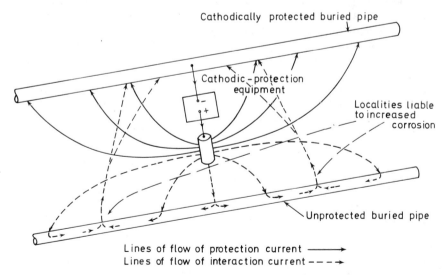

Fig. 11.34   Cathodic protection and interaction currents with parallel pipelines. Note that the pick-up area is the point on the unprotected pipe nearest to the groundbed; at this point the current flows in opposite directions and attains a maximum at the points of discharge

potentials along the two pipelines broadly follows the exponential distribution described in Section 11.1.

Methods of Estimating Interaction

Methods of estimating the amount of interaction between parallel buried pipeline systems have been described by Sunde[1] and Pope[2]. In practice the calculated changes in potential are distorted because of variations in spacing between the structures, changes in soil resistivity and lack of uniformity of the pipe coatings. Two sources of interaction are considered, (a) that caused by current flowing from the groundbed (the *anode* effect) and (b) that caused by the protection current flowing to and from unprotected buried structures in the vicinity (the *structure* effect). The net effect at any point on the unprotected structure will be the sum of the anode and structure effects.

Current due to the anode effect   The potential of the earth near the groundbed of a cathodic-protection system becomes more positive as the groundbed is approached (see Fig. 11.7, p. **11**.13). A structure buried near the groundbed will pick up current due to this variation in the soil potential and current will flow in the structure in each direction away from a point close to the groundbed (Fig. 11.34). The upper curve *AGA'* in Fig. 11.35 shows how the current in the unprotected structure changes owing to the anode effect.

Current due to the structure effect   Due to leakage of the protection current through the soil between the protected structure and nearby unprotected structures, part of the protection current may flow in an unprotected structure in a direction towards the point of connection to the cathodic-protection equipment. The change in current in the unprotected structure due to this effect is indicated by curve *SGS'* in Fig. 11.35.

The relative amplitude of the two curves will depend on the attenuation constant of the protected and unprotected structures, the spacing between them, and the distance between the two structures and the groundbed. The total effect at any point on the unprotected structure is obtained by adding together the two currents due to (a) and (b) as shown in curve *NGN'*.

The rate at which the net current is picked up or discharged will be proportional to the change in structure/soil potential along the unprotected structure and this is shown by the dotted curve *PCG'C'P'* in Fig. 11.35. At points where current is picked up the potential change is negative and the natural rate of corrosion is reduced as shown by the dotted curve below *XX'*. The structure/soil potential is made more positive where the interaction current is discharged and the rate of corrosion at such points is increased, as shown by the dotted curve above *XX'*. The maximum positive structure/soil potential occurs where the gradient of the net current curve *NGN'* is steepest, that is at points *C* and *C'* on the potential curve *PCG'C'P'*.

If, with impressed-current protection, the groundbed is installed near the unprotected structure, large negative changes may occur on the structure at points close to the anode. The maximum positive potential change is usually found on the unprotected installation at a distance of 270–450 m from the

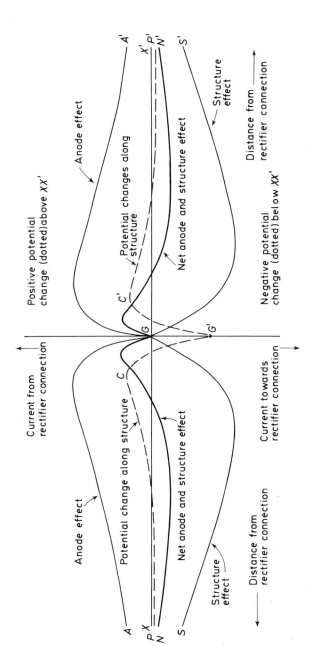

Fig. 11.35 Change of current and potential of an unconnected structure near a groundbed due to cathodic-protection interaction. Note that the dashed curve $PCG'C'P'$ gives the potential change along the structure and that the maxima at C and C' are the maximum positive changes in the structure/soil potentials and that they correspond with the points of maximum gradient on the net current curve NGN'. Note also that 'Current from rectifier connection' does not mean that structure is connected to the rectifier

point opposite the connection to the protection equipment. If the two structures are not parallel, positive or negative changes of potential will also occur at crossings and at other places where the structures are electrically close to each other.

The amount of interaction caused by a protection scheme using galvanic anodes will be much less than that involved in the case of impressed-current protection, because of the low current output obtained from each anode. Significant positive potential changes have, however, been measured on nearby structures in cases where galvanic anodes are closely spaced and the distance between structures is small.

The severity of corrosion interaction will depend on the density of the stray current discharged at any point on the secondary structure. This may be assessed by measuring the changes in structure/soil potential due to the application of the protection current. Potential tests should be concentrated on the portions of pipe or cable which are close to the structure to be cathodically protected, where the potential change is likely to be more positive.

Since the interaction tests may require the co-operation of owners of protected and unprotected structures, it is advisable to agree on the testing apparatus and methods to be employed. This is facilitated in the UK by a Joint Committee composed of representatives of the major underground services. The Committee receives active assistance from the D.T.I. and from a number of national research associations. Criteria have been agreed for determining the acceptability or otherwise of potential changes on installations buried near cathodically-protected structures. Recommendations[3] have been published by the Committee on testing methods and equipment, on the information that should be supplied to the interested authorities when a cathodic-protection scheme is being installed, and on methods of avoiding interaction.

The relationship of anode current density with electrode potential for mild steel in dilute aqueous soil electrolytes has been studied by Hoar and Farrer[4]. The study shows that in conditions simulating the corrosion of mild steel buried in soil the logarithm of the anode current density is related approximately rectilinearly to anode potential, and the increase of potential for a ten-fold increase of current density in the range 10^{-5} to 10^{-4} A/cm^2 is between 40 and 65 mV in most conditions. Thus a positive potential change of 20 mV produces a two- to three-fold increase in corrosion rate in the various electrolyte and soil solutions used for the experiments.

The results of these experiments have been considered by the Joint Committee for the Co-ordination of the Cathodic Protection of Buried Structures and, in view of the various types of buried structures concerned and the circumstances in which field tests are conducted, the Committee decided not to amend its provisional recommendation that when cathodic protection is applied to a buried structure the maximum permissible potential change in the positive direction on a nearby pipe or cable should be 20 mV. If there is a history of corrosion on the unprotected installation no detectable positive change in structure/soil potential should be permitted. These criteria of interaction have been adopted in the British Standard Code of Practice for Cathodic Protection[5].

The adoption of a maximum permissible positive potential change of 20 mV is based on theoretical and laboratory studies. However, there is evidence to show that in practice changes of $+20\,mV$ do not appear to cause noticeable corrosion damage even after the passage of many years, and in some countries larger changes with values of up to $+100\,mV$, are accepted. The rate of corrosion depends on the density of the current exchanged between structure and electrolyte and no practicable method has yet been found for measuring this. However, some account should be taken of the resistivity of the electrolyte and studies are being made to devise a better criterion based on potential change and also on the resistivity of pipe coatings, cable wrappings and soil.

Methods of Preventing or Reducing Interaction[6]

Interaction tests should be made on all unprotected structures in the vicinity of a proposed cathodic protection installation, and should be repeated annually or at some other suitable interval to ensure that alterations in the layout of plant or in the electrical conditions are taken into account. It is most convenient if the tests on all unprotected pipes or cables are made at the same time, the potential measurements being synchronised with the regular switching on and off of the protection current. It may then be convenient to continue with further tests to confirm that any remedial measures applied to one installation do not adversely affect other installations.

A number of methods may be used to reduce the interaction on neighbouring structures. In some circumstances it may be practicable to reduce the current output applied to the protected structure or to resite the groundbed so that the anode effect on an unprotected pipe or cable is altered as required*. The physical separation between the groundbed and nearby buried structures can be increased by installing anodes at the bottom of deep-driven shafts and substantial improvements can be made using this technique.

The electrical resistance between the structures at crossing points or elsewhere can be increased by applying insulating wrappings. If the wrappings are applied only to sections of the unprotected structure where a positive change in potential is measured, the wrapping must ensure uniformly high insulation in order to avoid localising the anodic attack at pinholes or similar discontinuities.

If the positive potential changes are very small and confined to a few points on a small unprotected structure, it may be practicable to reduce the potential at these points by installing reactive anodes. The anodes will probably be most effective if they can be buried between the two structures. In some circumstances a similar screen of zinc, aluminium or steel may be installed between the structures. The screen must be electrically connected to the unprotected structure since it is installed with the object of providing an electrolytic path to earth for the interaction current.

If both buried structures have been subject to corrosion damage the best solution may be to install a joint cathodic-protection scheme with sufficient current output to provide adequate protection for both installations. The application of separate cathodic-protection schemes to structures buried

---

*Although the geometry of buried structures cannot be altered the groundbed can be placed so that the anode effect and the structure effect on the unprotected structure tend to balance each other.

close to each other is liable to cause interaction damage because of variations in current output on the two systems. This may be avoided by connecting the two structures together at several points.

In the UK the most common method of reducing interaction is to connect the protected and unprotected structures together by means of metallic bonds. This method is more successful if care has been taken to ensure that the unprotected structure is electrically continuous. If possible, bonds should be connected to points on the unprotected structure where maximum positive changes in potential are observed, but it may be more practicable to install a shorter bond between points where access can conveniently be obtained for periodic inspection and tests.

If one of the structures to be bonded is the sheath or metallic armouring of an electric supply cable, special precautions will be necessary to ensure that the voltage rise at the bond in the event of an instantaneous earth fault on the power-supply system does not endanger personnel or equipment associated with other buried structures. The bond and any associated current-limiting device should be suitably insulated and of adequate current-carrying capacity.

Account must also be taken of small alternating currents which may be diverted from the sheath of a power supply cable by a bond connected to nearby buried structures. Such currents may be sustained for long periods and if they are diverted to the sheaths of telecommunication cables noise may be induced in the telephone circuits.

If the electrical continuity of a buried pipe or cable is broken at a point where a.c. is liable to flow owing to the presence of a bond, the gap should be bridged by means of a continuity bond. This will prevent the appearance of a dangerous voltage between the two sections of pipe or cable.

Interaction due to the use of reactive anodes can best be avoided by careful siting of each anode during installation. In particular, anodes should not be buried close to a point where the protected structure crosses an unprotected structure, nor should anodes be so placed that an unprotected pipe or cable passes between the anode and the protected installation.

J. R. WALTERS

REFERENCES

1. Sunde, E. D., *Earth Conduction Effects in Transmission Systems*, Van Nostrand, New York (1948)
2. Pope, R., *Corrosion*, **6**, 201 (1950)
3. Joint Committee for the Co-ordination of the Cathodic Protection of Buried Structures, London, Recommendations Nos. 1, 2, 3 and 4
4. Hoar, T. P. and Farrer, T. W., *Corros. Sci.*, **1**, 49 (1961)
5. *Code of Practice for Cathodic Protection*, CP 1021, B.S.I., Aug. (1973)
6. C.C.I.T.T. *Recommendations Concerning the Construction, Installation and Protection of Telecommunication Cables in Public Networks*, I.T.U., Geneva (1974)

11.7 Cathodic-protection Instruments

A number of measurements, principally electrical, are necessary in order to ensure that a cathodic-protection system is correctly designed and will provide full protection to the structure concerned, and to determine accurately the effect of such a system on other structures. This section deals with the instruments used for making these measurements and indicates in general terms the various types available to the corrosion engineer today. Brief details of instruments not directly connected with cathodic protection as such, but nevertheless associated with it, are also included.

Basic Requirements

All instruments used in cathodic-protection work must be accurate, reliable and easy to maintain.

For cathodic-protection work the best instruments are calibrated to an accuracy of 0·5% or less of full-scale deflection. In addition, scales are usually so selected that the majority of the readings taken fall within the upper half of the scale, where the least error will occur. Second-grade instruments are usually calibrated to an accuracy of approximately 2% of full-scale deflection.

Most instruments used in cathodic-protection work are essentially field instruments, and must therefore be portable and sufficiently robustly constructed to withstand rough handling and to be capable of operating reliably under a wide range of climatic conditions.

As far as possible instruments should be so constructed that repairs or replacements can be made quickly, and standard, readily obtainable components should be used.

Theory of d.c. Indicating Instruments

The application of cathodic protection generally involves the use of direct current. The movement in nearly all instruments used in cathodic protection is therefore of the moving-coil permanent-magnet type, which gives coil deflection (and thus pointer deflection) proportional to the current in the coil.

Damping of the movement to prevent overswing or oscillation of the needle is provided by the eddy currents induced in the metal former on which the coil is wound, and further damping (i.e. critical damping) can be obtained by placing a low resistance across the coil. The main advantages of this type of instrument are uniformity of scale, high torque : weight ratio, and low power consumption.

By the insertion of suitable shunts and/or resistances it is possible to use one instrument to measure both current and voltage over a wide range.

The coil (wound on a light metal former) can be suspended by a fine strip of phosphor bronze between the pole pieces. Attached to this suspension is a small mirror which reflects on to a scale a beam of light which is focused upon it. An instrument of this kind is known as a D'Arsonval galvanometer and is used in potentiometer circuits and various methods of measurement of resistance.

The use of a light beam as an indicator avoids the errors caused by friction with a normal pointer, while a hairline incorporated in the projection lens eliminates errors due to parallax, and with the scale at a distance of 1 m from the mirror, a sensitivity of up to $1\ 500\ mm/\mu A$ can be achieved.

This instrument, as described above, is quite unsuitable for field measurements.

Types of Instruments

Cathodic-protection instruments may be classified as potential-measuring, current-measuring, resistance/conductance-measuring, multicombination, recording and ancillary.

The various types are dealt with in some detail in the following pages. Except where noted, the instruments have a moving-coil movement.

Potential-measurement Devices

Reference electrodes   The generally accepted criterion for the effectiveness of a cathodic-protection system is the structure/electrolyte potential (Section 11.1). In order to determine this potential it is necessary to make a contact on the structure itself and a contact with the electrolyte (soil or water). The problem of connection to the structure normally presents no difficulties, but contact with the electrolyte must be made with a reference electrode. (If for example an ordinary steel probe were used as a reference electrode, then inaccuracies would result for two main reasons: first, electrochemical action between the probe and the soil, and second, polarisation of the probe owing to current flow through the measuring circuit.)

Figure 11.36 shows two patterns of the $Cu/CuSO_4$ half-cell and Fig. 11.37 an $Ag/AgCl$, Cl^- half-cell; both are commonly used as reference electrodes. The illustrations are intended only to show the general features of the many different patterns commercially available. Table 11.20 gives a comparison of the potential of a structure measured against $Cu/CuSO_4$ (sat.), $Ag/AgCl$, KCl(sat.) and corroding pure Zn, which is widely used as a reference electrode in chloride environments.

Table 11.20 Approximate comparison of potentials (V) using Zn/sea water, Cu/CuSO$_4$, sat., and Ag/AgCl, sat. KCl reference electrodes*

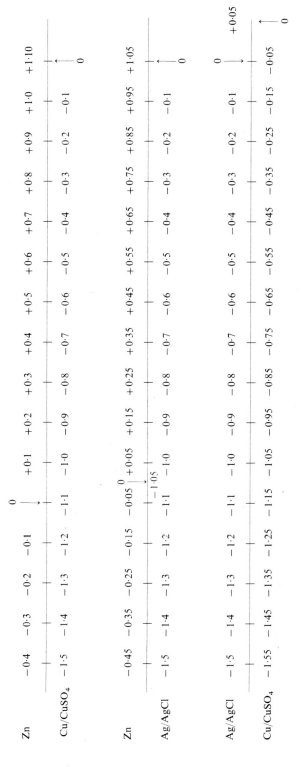

*If the potential of the structure is found to be +0·4 V $vs.$ Zn it will be −0·7 V $vs.$ Cu/CuSO$_4$, sat., and −0·65 V $vs.$ Ag/AgCl, sat. KCl. Approximate values of the above reference electrodes on the Standard Hydrogen Scale (S.H.E.) are as follows: Cu/CuSO$_4$, sat. = 0·30−0·32 V; Ag/AgCl, sat. KCl = 0·020 V (0·025 V in sea-water); Zn/sea-water = −0·75 V.

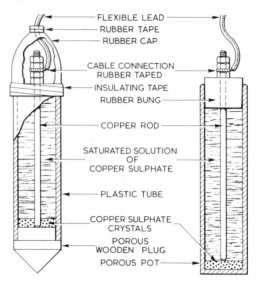

Fig. 11.36   Copper/copper sulphate half-cells

From Fig. 11.36 it will be seen that contact between the electrolyte (soil or water) and the copper-rod electrode is by porous plug. The crystals of $CuSO_4$ maintain the copper ion activity at a constant value should the half-cell become polarised during measurements. The temperature coefficient of such a cell is extremely low, being of the order of 1×10^{-5} V/°C and can thus be ignored for all practical purposes. To avoid errors due to polarisation effects, it is necessary to restrict the current density on the copper rod to a

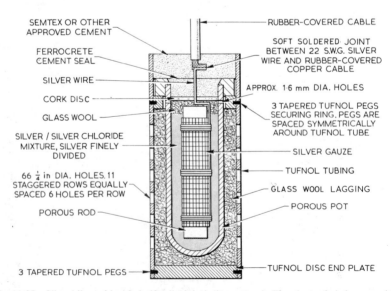

Fig. 11.37   Silver/silver chloride half-cell (Admiralty pattern). The electrode is immersed in a chloride-containing solution which diffuses through the porous pot and thus comes into contact with the Ag/AgCl mixture

value not exceeding $20\,\mu A/cm^2$. The $Cu/CuSO_4$ half-cell is in almost universal use for measurement of structure/electrolyte potentials on buried structures. Its main advantages are that it is cheap and simple to construct, that the materials are easily obtainable, and that it can be quickly made up at any time.

The $Ag/AgCl$, Cl^- half-cell has a chloridised silver electrode which is generally immersed in either a saturated potassium chloride solution or a mixture of finely divided silver and silver chloride. This type of half-cell is normally used for measurement of structure/electrolyte potentials of immersed structures, contact between the half-cell and electrolyte (fresh or sea-water) being made through the walls of a porous pot enclosing the cell. For measurements in sea-water a chloridised silver electrode is used and the electrode is immersed directly in the water thus dispensing with the saturated KCl solution; the electrode is prepared by fusing silver chloride onto a silver mesh and partially reducing the silver chloride by cathodic polarisation.

The $Ag/AgCl$, Cl^- electrode is probably the best reference electrode for measuring potentials of waters at elevated temperatures.

One further type of reference electrode is the zinc half-cell which consists merely of a high-purity zinc rod. Such an electrode will maintain a constant potential after a short 'settling down' period following installation (particularly in sea-water). The main advantages of a zinc reference electrode are that it requires no special preparation and no subsequent remaking during its life. It is ideal for use as a permanent reference electrode installed on a cathodic-protection scheme, as for example the water box of a large heat exchanger, where inspection is infrequent and access to the structure while it is operating is impossible. It should be noted that in the case of zinc the potential is a *corrosion* potential, whereas in the case of the other reference electrode the potential is a *reversible* potential (see Appendix 9.1B). In practice, it is therefore an advantage to maintain the surface of the zinc in a slightly active condition by deliberately reducing the impedance of the measuring circuit. Generally the zinc electrode is used where the electrolyte surrounding the protected structure is water, as this ensures good contact with the electrode. Its use in soils is generally restricted to damp ground and it cannot be commended as reliable in dry soils, even when placed in a special backfill.

Voltmeters and potentiometers The instruments described here are generally referred to as *corrosion voltmeters*. As mentioned previously, the current flowing through any potential-measurement circuit must be small to avoid errors due to polarisation. Moreover, if the current flow is too large, errors will be introduced owing to the voltage drop caused by the contact resistance between the reference electrode and the electrolyte. It is thus clear that the prime requirement of a potential measurement circuit is high resistance.

For a direct-reading instrument, sensitivity should be at least of the order of $50\,k\Omega/V$, and instruments are commercially available with sensitivities of up to and exceeding $1\,M\Omega/V$. Direct-reading meters are usually made to show several ranges, which are obtained by the use of suitable resistances placed in series with the indicating instrument (Fig. 11.38).

To measure structure/electrolyte potentials with electrolyte resistivities in excess of $2\,k\Omega\,cm$, a high-resistance potentiometer unit as shown in Fig. 11.39 or a potentiometric voltmeter as illustrated in Fig. 11.40 may be used.

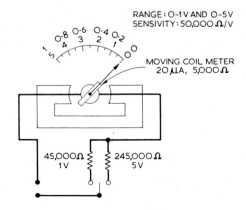

Fig. 11.38   Direct-reading voltmeter

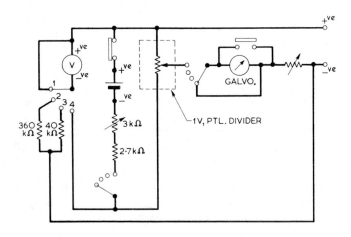

Fig. 11.39   Potentiometer voltmeter (courtesy The Post Office)

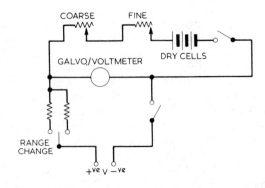

Fig. 11.40   Basic circuit potentiometer voltmeter

In both types of instrument the voltage to be measured is balanced against an external applied voltage (usually from batteries within the instrument). At balance, no current flows through the external circuit and thus errors due to contact resistance are eliminated.

The potentiometer must be calibrated against a standard cell, while with the potentiometric voltmeter the voltage to be measured is balanced against a battery e.m.f., using uncalibrated resistors and the voltmeter as a galvanometer. At balance the voltage is transferred to the voltmeter and read direct. The magnitude of voltage which can be measured by both instruments is limited by the maximum voltage of the (usually) dry cell which they contain. It is, however, possible to extend the range by using a potential divider or *volt box*.

For normal field work the potentiometric voltmeter is the more popular instrument, being usually of lighter construction and not requiring calibration against a standard cell. Where extremely small potentials (usually potential shifts) of the order of 1 mV are to be measured, however, the potentiometer is more suitable and accurate.

Valve voltmeters were widely used in the past, but have been replaced by transistor voltmeters. With instruments of this type it is possible to achieve an input resistance of 50 MΩ or more, the current required to operate the instrument being of the order of 10^{-14} A. The early instruments had a tendency to zero drift on the lower ranges, but this has been overcome in the modern transistor types. Such instruments are most often used to make potential readings in extremely high-resistance electrolytes. The accuracy of such instruments is of the order of 2% full-scale deflection. It is necessary to ensure that both types are so designed that they do not respond to alternating currents.

The technique adopted in measuring structure/electrolyte potential is illustrated in Fig. 11.41. While it is not truly within the scope of this chapter, it is

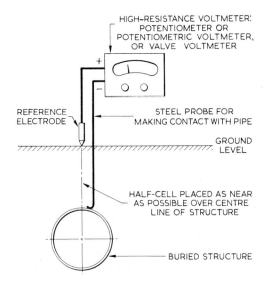

Fig. 11.41 Measurement of structure/electrolyte potential

as well to note that the position of the reference electrode in relation to the structure is important. Theoretically the half-cell should be placed as near to the structure as possible (i.e. within a few millimetres) to avoid IR drop in the electrolyte. This is often not possible in practice as, for example, with a buried pipeline. In such a case, the cell should be placed in the soil directly over the structure, and it is wise to allow a 'safety factor' of say 50 mV over and above the minimum protective potential to compensate for IR drop.

Current-measuring Instruments

These may be classified generally according to whether they are used to measure current delivered or drained by a structure under protection, or to measure current flowing within the structure itself.

By the use of suitable shunts, the basic moving-coil movement can be adapted to measure an almost unlimited range of currents. Figure 11.42 illustrates a direct-indicating instrument with shunt, to measure current up to 5 A d.c.

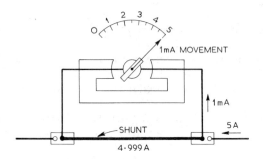

Fig. 11.42   Direct-indicating instrument

To ensure that the resistance of the circuit is not materially altered by the insertion of an ammeter, it is usual to install either a shunt or the meter itself (usually a moving-coil meter with internal shunt) permanently in the circuit.

Ammeter shunts are normally of the four-terminal type, to avoid contact resistance errors, i.e. two current terminals and two potential terminals, as shown in Fig. 11.42.

In making measurements of current flowing within a structure, it is extremely important that additional resistance, as for example a shunt, is not introduced into the circuit, as otherwise erroneous results will be obtained. One method is to use a tong test meter. Such instruments are, however, not particularly accurate, especially at low currents, and are obviously impracticable in the case of, say, a 750 mm diameter pipeline. A far more accurate method and one that can be applied to all structures, is the zero-resistance ammeter or, as it is sometimes called, the *zero-current ammeter* method. The basic circuit of such an instrument is shown in Fig. 11.43.

From Fig. 11.43 it will be seen that if I_b is adjusted until there is zero voltage on the voltmeter, then $I_s = I_b$. When this type of instrument is used to

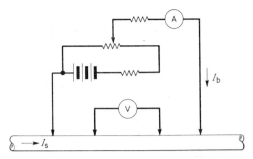

Fig. 11.43   Zero-resistance ammeter

measure current in, say, a bond between one structure and another, if simultaneous measurements are made at locations on the structure (i.e. interference testing) inaccurate results will be obtained. This is because at balance there is no current in the structure at the measurement point, and this could materially alter the current distribution to the structure.

Where measurements of current in massive structures having extremely low longitudinal resistances are required, and to avoid the errors referred to in the previous paragraph, a refinement of the zero-resistance ammeter is used. Instead of a voltmeter, a high-sensitivity D'Arsonval galvanometer with a centre-zero scale is used. The connections remain essentially the same as those in Fig. 11.43, except that a reversing switch is incorporated so that the current I_b can be applied first in one direction and then in the other.

Let I_s = current in structure, let I_b = meter current (calibrating current), let deflection of the galvanometer due to I_s be A_0 (note that the direction of deflection will indicate the direction of current flow), and let the deflection of the galvanometer with I_b applied in both directions be A_1 and A_2, then:

$$A_1 = I_b + I_s \text{ (say)}$$

and thus

$$A_2 = I_b - I_s$$

therefore

$$A_1 + A_2 = 2I_b$$

Thus calibration of galvanometer in amps or milliamps per division deflection is

$$2I_b/(A_1 + A_2)$$

therefore

$$I_s = A_0[2I_b/(A_1 + A_2)]$$

The zero-resistance ammeter is seldom employed for routine testing. This instrument requires careful handling to avoid damage, in particular to the galvanometer. Usually two permanent test leads are installed at a set distance apart, and by the initial use of a zero-resistance ammeter a calibration chart of potential between the two leads and current in the structure is drawn up. Thus when routine testing is made, it is only necessary to measure the potential difference between the leads. The current so measured will not be truly accurate, owing to the minute current in the measurement circuit, but for all practical purposes this can be ignored.

Resistivity/Conductivity-measuring Instruments

Instruments in this category are used for the measurement of electrolyte resistivity, resistance, and insulation (i.e. protective-wrap) conductivity.

Measurement of resistivity   The most usual method of measuring soil resistivity is by the four-electrode 'Wenner' method. Figure 11.44 indicates the basic circuit. The mean resistivity R_M is given by

$$R_M = 2a(E/I)$$

where I is the current applied between the current electrodes C_1 and C_2, and E is the potential developed between the potential electrodes P_1 and P_2 by the current I^*. The value given by this method is the average resistivity of the soil to a depth a equal to the spacing of the electrodes. It is most important to note that for accurate and consistent results, the electrodes must be equidistant from one another and placed in a straight line. This relatively simple set-up, illustrated in Fig. 11.44, suffers from inaccuracies arising from polarisation of the potential electrodes and effects of possible stray currents in the soil. To minimise these, the electrodes used for measuring potential P_1, P_2 should be $Cu/CuSO_4$ half-cells, and reverse polarity readings should be taken. One source of inaccuracy that cannot be overcome with this method is that due to the contact resistance of the potential electrodes P_1 and P_2 to earth.

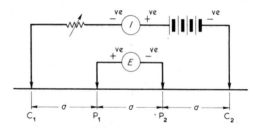

Fig. 11.44   'Wenner' method. Normally C_1, C_2 and P_1, P_2 are steel rods, but for higher accuracy $Cu/CuSO_4$ electrodes should be used for P_1, P_2

To avoid the errors of polarisation and stray currents, special resistivity meters are employed. One form of these uses an alternating current produced from batteries by a vibrator. The effective resistance is measured by a modified Wheatstone bridge with balance indicated by a galvanometer.

A further type of instrument employs a hand-generated current passed through the current coil of an ohmmeter and then through a current reverser so that an alternating current is delivered to the current electrodes. The alternating current due to alternating potential between the potential electrodes is then rectified by a commutator mounted on the generator shaft, and the

*For proof of this equation, *see* Wenner, F., *Nat. Bur. Stand. Publn.*, **12**, US Dept. of Commerce, 496 (1915).

current is passed through the potential coil of the ohmmeter. Thus the deflection of the instrument is the ratio of IR/I, or resistance R. To compensate for resistance to earth of the potential electrodes, the ohmmeter is connected directly across these potential electrodes. The resistance is adjusted (by means of a variable resistance) to a set value, which is usually denoted by a mark on the meter scale so that the resistance in the potential circuit is at the predetermined value used in calibrating the instrument. A circuit diagram is given in Fig. 11.45, but it should be noted that this type of device is now being replaced by solid-state battery-powered versions.

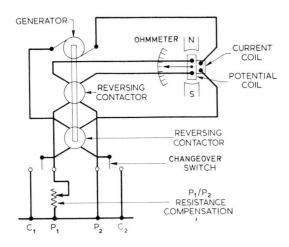

Fig. 11.45 'Megger' earth resistivity meter (courtesy Evershed and Vignoles)

Soil resistivity may be measured by an instrument known as the *Shepard Cane* resistivity meter, which uses only two electrodes. In principle, the instrument operates by measuring, with a low-resistance ammeter, the current flowing between two electrodes placed a set distance (usually 0·3 m) in the soil. To avoid polarisation effects the positive electrode has a very small area and the negative electrode a large area. This instrument is of light construction, and is easily portable and simple in use. Normally, however, it will indicate resistivity only near the surface, and it is of use only in giving a general indication. The basic circuit diagram of a Shepard Cane is given in Fig. 11.46.

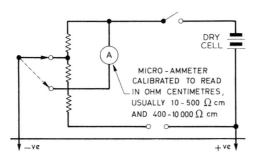

Fig. 11.46 Shepard Cane resistivity meter

To avoid errors due to polarisation, the instrument can be used with a vibrator to provide an alternating potential between the electrodes, in which case the areas of the electrodes can be the same.

A variation of the Shepard Cane incorporates both electrodes on a single rod. A circuit diagram of this type of instrument is shown in Fig. 11.47. The single rod has a calibration constant, determined by the area of the positive electrode. At balance the resistivity is given by the resistance multiplied by the rod constant. This instrument is portable and simple to operate, but it cannot be recommended when accurate results are required.

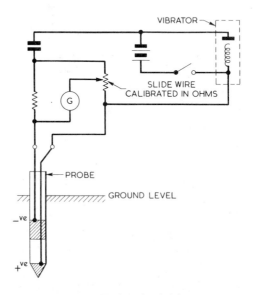

Fig. 11.47   Single-rod resistivity meter

Resistivity of an electrolyte can also be measured by passing a known current through a known length and cross-section of the electrolyte, and measuring accurately the voltage drop across the length. It is preferable to use alternating current to avoid polarisation effects. Alternatively, the resistance of a known length and cross-section of the electrolyte may be measured directly by using a four-electrode instrument with the potential and current electrodes P_1, C_1 and P_2, C_2 joined together in such a way that the instrument will measure resistance as a normal ohmmeter. Special cells also exist for use with resistivity meters, which are so constructed and calibrated that resistivity can either be read directly from the instrument or simply calculated from the product of the measured resistance and known cell constant. Measurements of resistivity on samples, as just described, will give accurate results only on liquid electrolytes and cannot be recommended for use on soil samples, where the value of resistivity measured will vary with the degree of compaction of the soil.

Measurement of resistance   As previously mentioned, the four-electrode resistivity meters can be used to measure resistances. For this purpose the most accurate instrument is the hand-generator type, as it usually has a large clear scale calibrated directly in ohms. Most vibrator-type instruments, however, rely on a calibrated dial and galvanometer and on low ranges the exact point of balance is often difficult to determine with any accuracy.

The accurate measurement of resistance can of course be carried out by means of a Wheatstone bridge, Carey Foster bridge, or a similar arrangement.

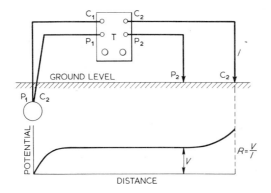

Fig. 11.48   Measurement of resistance of structure to remote earth

The measurement of resistance to remote earth of a metallic structure is normally carried out with a four-electrode instrument. The connections are shown in Fig. 11.48. A current I is passed between the structure and a remote electrode. The potential difference V is measured between the structure and a second remote electrode. In this way the ohmmeter records the resistance of the structure to earth, i.e. V/I. The spacing of the electrode from the structure is important and must be such that the remote potential electrode lies on the horizontal part of the resistance/distance curve, as shown in Fig. 11.48. Generally speaking, a minimum distance of 15 m from the structure is necessary for the potential electrode to lie on the flat part of the curve, with the current electrode usually at least twice the distance of the potential electrode.

Measurement of conductivity   The measurement of electrolyte conductivity —the reciprocal of the resistivity— is a fairly simple matter, being calculated from the resistivity as measured by some of the methods described above.

Often it is necessary in designing a cathodic-protection system to know the conductivity of a protective coating (e.g. bitumen enamel) on a structure. This measurement is usually carried out by finding the resistance between an electrode of known area placed in contact with the coating and the structure itself. The electrode placed on the structure can be either of thin metal foil or, preferably, of material such as flannel soaked in weak acidic solution. The resistance between the pad and the metal is measured by means of either a resistivity meter, as previously described, or a battery with a voltmeter and an ammeter or microammeter. Generally speaking, in field work where such measurements have to be made, a resistivity meter is preferable.

Multicombination Instruments

As their name suggests, these instruments are capable of carrying out a variety of measurements, e.g. structure/electrolyte potentials, current, resistivity and voltage. Most instruments of this type contain two meters in one case, one being a low-resistance millivolt/voltmeter and milliamp/ammeter, and the second a high-resistance voltmeter.

The advantages of this type of instrument are (*a*) that only one instrument is required for all necessary measurements, and (*b*) that it is portable and thus particularly suitable for field work.

The main disadvantage is that should a fault occur in one circuit of the instrument the complete instrument is often rendered inoperative.

Recording Instruments

Miniaturisation of electronic components has enabled the construction of a compact, portable, battery-operated recording voltmeter. The principal use of this instrument is to measure pipe/soil potential fluctuations over a period of time. The instrument can be modified to measure current variations.

Ancillary Instruments

Apart from the instruments described in previous paragraphs, there are others that, while not directly connected with cathodic protection as such, are extremely useful tools to a corrosion engineer. They include pH meters, Redox probes, protective-coating test instruments and buried-metal-location instruments.

High-voltage coating-testing equipment   When cathodic protection is applied to a structure which has a protective coating, the current required is proportional to the bare metal area on the structure. Thus whenever a protective coating is applied it should be of good quality, with very few failures or pin holes in it, so that the cathodic-protection system may be economic.

It is now fairly standard practice, particularly in the case of pipelines, to carry out an inspection of the protective coating (after application) with a high-voltage tester, known as a *holiday detector*. Basically, a high voltage is applied between an electrode placed on the coating and the structure. At any 'holidays' (flaws) in the coating, a 'spark over' occurs and is usually accompanied by a visual or audible alarm. For pipelines, the electrode often takes the form of a rolling spring clipped round the pipe. Other electrodes are in the form of wire brushes.

The high voltage is obtained either by a step-up transformer from the mains, or a motor-generator set, or alternatively from batteries and a vibrator with a high-tension coil. The battery-operated instruments are most popular as they are easily portable and most suited for field use. Such an instrument is illustrated in Fig. 11.49.

For the examination of paint films or special coatings, e.g. pipe linings, low-voltage holiday detectors with wet sponge electrodes are available.

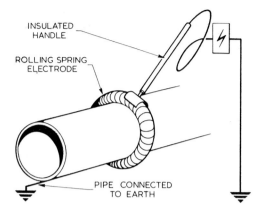

INSULATED HANDLE

ROLLING SPRING ELECTRODE

PIPE CONNECTED TO EARTH

Fig. 11.49   Holiday detector

The choice of voltage for use with a holiday detector depends on the thickness and type of coating applied to the structure. As a guide, in the case of pipelines, a voltage equivalent to approximately 5 kV/mm thickness is used in testing a coal-tar coating.

Buried metal locating instruments   Instruments in this category are used to locate buried metallic structures, to detect possible electrical discontinuities in buried structures, to locate possible junctions or points of contact between one structure and another, and to indicate the condition of a protective coating applied to a buried structure. A signal of a given frequency to a buried structure is either injected or induced into the structure and traced with a search coil and earphones. Most instruments use a high-frequency signal of 1 kHz or higher, although one or two use a simple interruptor. The transistor has made it possible to make very compact portable battery-operated instruments.

R. G. ROBSON
D. AMES

BIBLIOGRAPHY

Golding, E. W., *Electrical Measurements and Measuring Instruments*, Pitman, London, 4th edn (1955)
Reference Book On Instruments for Electrolysis Corrosion and Cathodic Protection Testing, American Gas Association, New York (1951)
Miller, M. C. 'Portable Testing Instruments for Corrosion Mitigation Testing', *Proceedings Appalachian Underground Short Course*, W. Virginia University, USA (1959)
Morgan, J. H., 'Instruments for Cathodic Protection', *Corros. Technol.*, **4**, 269 (1957)
Howell, P. P., 'Potential Measurements in Cathodic Protection Design', *Corrosion*, **8**, 300 (1952)
Parker, M. E., 'Corrosion Instruments', *Oil Gas J.*, **50**, 108, 136, 163 (1951)
Rasor, J. P., 'Selecting a Holiday Detector', *A Collection of Papers on Underground Corrosion* (Ed. George B. McComb, NACE), Vol. 4, 1st edn (1960)

11.8 Anodic Protection

Fontana and Greene[1] state that 'anodic protection can be classed as one of the most significant advances in the entire history of corrosion science', but point out that its adoption in corrosion engineering practice is likely to be slow. Anodic protection may be described as a method of reducing the corrosion rate of immersed metals and alloys by controlled anodic polarisation, which induces passivity. Therefore, it can be applied only to those metals and alloys that show passivity when in contact with an appropriate electrolyte. As recently as 1954 Edeleanu first demonstrated the feasibility of anodic protection and also tested it on small-scale stainless-steel boilers used for sulphuric acid solutions[2-3]. This was probably the first industrial application, although other experimental work had been carried out elsewhere[4-5].

Principles

Before considering the principles of this method, it is useful to distinguish between anodic protection and cathodic protection (when the latter is produced by an external e.m.f.). Both these techniques, which may be used to reduce the corrosion of metals in contact with electrolytes, depend upon the electrochemical mechanisms that result from changing the potential of a metal. The appropriate potential–pH diagram for the Fe–H_2O system (Section 1.4) indicates the magnitude and direction of the changes in the potential of iron immersed in water (pH about 7) necessary to make it either passive or immune; in the former case the stability of the metal depends on the formation of a protective film of metal oxide (passivation), whereas in the latter the metal itself is thermodynamically stable and egress of metal ions from the lattice into the solution is thus prevented.

A further difference is that in anodic protection the corrosion rate will always be finite, whereas a completely cathodically protected metal will not corrode at all. Raising the potential of an anodically protected metal may actually increase the corrosion rate if the metal exhibits transpassivity, whereas depressing the potential of a metal far below the protective potential will not affect the corrosion rate although it will be wasteful in terms of power consumption. Nevertheless it should be noted that a too negative potential can be detrimental in certain cases (see Section 11.1).

The significance of the Flade[6] potential E_F, passivation potential E_{pp}, critical current density $i_{crit.}$, passive current density, etc. have been considered in some detail in Sections 1.4 and 1.5 and will not therefore be considered in the present section. It is sufficient to note that in order to produce passivation (*a*) the critical current density must be exceeded and (*b*) the potential must then be maintained in the passive region and not allowed to fall into the active region or rise into the transpassive region. It follows that although a high current density may be required to cause passivation ($> i_{crit.}$) only a small current density is required to maintain it, and that in the passive region the corrosion rate corresponds to the passive current density ($i_{pass.}$).

Passivity of Metals

Since anodic protection is intimately related to passivity of metals it is relevant to review certain aspects of the latter, before considering the practical aspects of the former. The relative tendency for passivation depends upon both the metal and the electrolyte; thus in a given electrolyte, titanium passivates more readily than iron, and Fe–18Cr–10Ni–3Mo steel passivates more readily than Fe–17Cr steel. The ability to sustain passivity increases as the current density to maintain passivity ($i_{pass.}$) decreases, and as the total film resistance increases, as indicated for metals and alloys in 67 wt. % sulphuric acid (Table 11.21)[7]. The lower the potential at which a passive metal becomes active (i.e. the lower the Flade potential) the greater the stability of passivity, and the following are some typical values of E_F (V): titanium -0.24, chromium -0.22, steel $+0.10$, nickel $+0.36$ and iron $+0.58$[8]. These values are only approximate, since they depend upon the experimental conditions such as the pH of the solution[9]:

$$E_F = E_F^{\ominus} - n\,0.059\,\text{pH}$$

where $E_F^{\ominus}$ = the standard Flade potential at pH = 0, and
n = a number between 1 and 2 depending upon the metal and its condition.

Table 11.21 Current density to maintain passivity and film resistance of some metals and alloys in 67 wt.% sulphuric acid (after Shock, Riggs and Sudbury[7])

Metal or alloy	Current density to maintain passivity ($i_{pass.}$, Am^{-2})	Total film resistance ($\Omega\,cm$)
Mild steel	1.5×10^{-1}	2.6×10^4
Stainless steel (Fe–18Cr–8Ni)	2.2×10^{-2}	5.0×10^5
Stainless steel (Fe–24Cr–20Ni)	5×10^{-3}	2.1×10^6
Stainless steel (Fe–18Cr–10Ni–2Mo)	1×10^{-3}	1.75×10^7
Titanium	8×10^{-4}	1.75×10^7
Carpenter 20 (Fe–25Cr–20Ni–2.5Mo–3.5Cu)	3×10^{-4}	4.6×10^7

Only those metals which have a Flade potential below the reversible hydrogen potential ($0.00\,V$ at $a_{H^+} = 1$) can be passivated by non-oxidising acids, e.g. titanium can be passivated by hydrogen ions.

Table 11.22   Effect on critical current density and Flade potential
of chromium content for iron–chromium alloys in 10 wt. %
sulphuric acid (after West[8])

Chromium (%)	Critical current density ($i_{crit.}$, Am^{-2})	Flade potential (E_F, V)
0	$1·0 \times 10^4$	$+0·58$
2·8	$3·6 \times 10^3$	$+0·58$
6·7	$3·4 \times 10^3$	$+0·35$
9·5	$2·7 \times 10^2$	$+0·15$
14·0	$1·9 \times 10^2$	$-0·03$

Table 11.23   Effect on critical current density and passivation potential on alloying nickel with chromium in 1N and 10N H_2SO_4 both containing 0·5N K_2SO_4
(after Myers, Beck and Fontana[10])

Nickel (%)	Critical current density ($i_{crit.}$, Am^{-2})		Passivation potential (E_{pp}, V)	
	1N acid	10N acid	1N acid	10N acid
100	$1·0 \times 10^3$	$2·3 \times 10^2$	$+0·36$	$+0·47$
91	$9·5$	$3·9 \times 10$	$+0·06$	$+0·14$
77	$1·1$	$8·2$	$+0·07$	$+0·08$
49	2×10^{-1}	$2·0$	$+0·03$	$+0·06$
27	$1·2 \times 10^{-1}$	$4·1 \times 10^{-1}$	$+0·02$	$+0·05$
10	$1·3 \times 10^{-2}$	$1·1 \times 10^{-1}$	$+0·04$	$+0·08$
1	$1·0 \times 10$	$5·0 \times 10$	$-0·32$	$-0·20$
0	$1·5 \times 10$	$8·0 \times 10$	$-0·30$	$-0·20$

The addition of a more passive metal to a less passive metal normally increases the ease of passivation and lowers the Flade potential, as in the alloying of iron and chromium in 10 wt. % sulphuric acid (Table 11.22)[8]. Because
exceptions may exist, each system should be considered separately, as indicated by the fact that both the additions of nickel to chromium and also
chromium to nickel decrease the critical current density in a mixture of
sulphuric acid and 0·5 N K_2SO_4 (Table 11.23)[10].

These parameters depend upon the composition, concentration, purity,
temperature and agitation of the electrolyte. The current densities, required
to obtain passivity $i_{crit.}$, and to maintain passivity $i_{pass.}$ for a 304 stainless
steel (Fe–18 to 20Cr–8 to 12Ni) in different electrolytes, are given in Table
11.24[7]. From the data in this table, it can be seen that it is about 100 000
times easier to passivate instantaneously large areas of this steel in contact
with 115% phosphoric acid than in 20% sodium hydroxide. The concentration of the electrolyte is also important and for a 316 stainless steel (Fe–16 to
18Cr–10 to 14Ni–2 to 3Mo) in sulphuric acid, although there is a maximum
corrosion rate at about 55%, the critical current density decreases progressively as the concentration of acid increases (Table 11.25)[11]. The presence of
impurities, particularly halogen ions, that retard the formation of a passive
film, is often detrimental as illustrated by the fact that the addition of 3·2%
hydrochloric acid to 67% sulphuric acid raises the critical current density for

Table 11.24 Critical current density and current density to maintain passivity of stainless steel (Fe–18 to 20Cr–8 to 12Ni) in different electrolytes (after Shock, Riggs and Sudbury[7])

Electrolyte	Critical current density ($i_{crit.}$, Am^{-2})	Current density to maintain passivity ($i_{pass.}$, Am^{-2})
20% sodium hydroxide	4.65×10	9.9×10^{-2}
67% sulphuric acid (24°C)	5.1	9.3×10^{-4}
Lithium hydroxide (pH = 9.5)	8.0×10^{-1}	2.2×10^{-4}
80% nitric acid (24°C)	2.5×10^{-2}	3.1×10^{-4}
115% phosphoric acid (24°C)	1.5×10^{-4}	1.5×10^{-6}

Table 11.25 Effect of concentration of sulphuric acid at 24°C on corrosion rate and critical current density of stainless steel (after Sudbury, Riggs and Shock[11])

Sulphuric acid (%)	Corrosion rate (gm^{-2}d^{-1})	Critical current density ($i_{crit.}$, Am^{-2})
0	0	47
40	48	16
45	120	14
55	192	10
65	168	7
75	144	4
105	0	1

a 316 stainless steel[12] from 5 to 400 Am^{-2} and the current density to maintain passivity from 0.001 to 0.6 Am^{-2}. This is potentially dangerous, and the effect of the chloride ion on the passivation of iron has been studied by Pourbaix[13] who has produced a modified potential–pH diagram for the Fe–H$_2$O system. Therefore, the use of the calomel electrode in anodic-protection systems is not recommended because of the possible leakage of chloride ions into the electrolyte, and metal/metal oxide[14–15] and other electrodes[16–17] are often preferred. Because of this chloride effect the storage of hydrochloric acid requires a more passive metal than mild steel, and titanium anodically protected by an external source of current or galvanic coupling has been reported to be satisfactory[18–19] although even this oxide film has sometimes been found to be unstable[20]. Other additions, such as chromous chloride to chromic chloride, may result in the breakdown of passivity on titanium, but fortunately in this application, anodic protection gives repassivation and increases the corrosion resistance in the new solution by a factor of thirty[21].

An increase in the temperature of an electrolyte may have several effects: it may make passivation more difficult, reduce the potential range in which a metal is passive and increase the current density or corrosion rate during passivity as indicated in Fig. 11.50 for mild steel in 10% H$_2$SO$_4$. These changes are illustrated for several steels in different acids in Table 11.26[22] and it may

Table 11.26 Effect of temperature on different acids on the operating variables for anodic protection of different steels (after Walker and Ward[22])

Alloy	Acid concentration	Temp. (°C)	Critical current density ($i_{crit.}$, Am^{-2})	Current density to maintain passivity ($i_{pass.}$, Am^{-2})	Corrosion rate (mm y^{-1}) Unprotected	Corrosion rate (mm y^{-1}) Anodically protected	Passive potential range (V)
Stainless steel 304[7] (Fe–18 to 20Cr– 8 to 12Ni)	Phosphoric, 115%	24	1.5×10^{-4}	1.5×10^{-6}			
		82	3.1×10^{-4}	1.5×10^{-6}			
		177	6.5×10^{-1}	2.2×10^{-2}			
	Nitric, 80%	24	2.5×10^{-2}	3.1×10^{-4}			
		82	1.2×10^{-1}	1.1×10^{-3}			
	Sulphuric, 67%	24	5.1	9.3×10^{-4}			
		82	4.6×10^{-2}	2.9×10^{-3}			
Stainless steel 316 (Fe–16 to 18Cr– 10 to 14Ni– 2 to 3Mo)	Sulphuric, 67%	24	5.0	1×10^{-3}			0.26–1.09
		66	4.0×10^{1}	3×10^{-3}			0.27–1.04
		93	1.1×10^{2}	9×10^{-3}			0.32–0.72
	Phosphoric, 115%	93					0.26–1.14
		117					0.26–0.94
	Phosphoric, 75–80%	104		9×10^{-2}–1.4×10^{-1}	1.5	0.12	
		121		1.5×10^{-1}–3.5×10^{-1}	2.2	0.12	
		135		3.8×10^{-1}–4.4×10^{-1}	5.3	0.85	
Carbon steel[46]	Sulphuric, 96%	27		1.1×10^{-2}	0.15	0.01	
		49		1.16×10^{-1}	0.8	0.11	
		93		1.16	2.8	0.8	

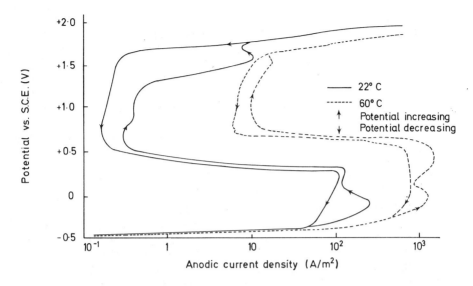

Fig. 11.50 Potentiostatic anodic polarisation curves for mild steel in 10% sulphuric acid. Note the magnitude of the critical current density which is 10^2–10^3 A/m^2; this creates a problem in practical anodic protection since very high currents are required to exceed $i_{crit.}$ and therefore to passivate the mild steel

be noted that, whereas the critical current density for the 316 steel increases with the temperature of the sulphuric acid, the opposite effect is observed with the 304 steel. During the storage of an acid, changes in the ambient temperature between day and night or summer and winter may double the current required for protection, and the increase may be even higher during manufacture or heat-transfer processes, so these should be considered at the design stage. Agitation or stirring of an electrolyte in certain conditions may increase the rate of corrosion of immersed metals and raise the passivation current density[23-24] (Table 11.27)[22]. Finley and Myers have recently found that both the temperature of the electrolyte[25] and cold working of the metal[26] have a marked effect on the anodic polarisation of iron in sulphuric acid.

Because these variables have a very pronounced effect on the current density required to produce and also maintain passivity, it is necessary to

Table 11.27 Effect of electrolyte agitation on corrosion rate and the current density to maintain passivity of mild steel in acid solutions at 27°C (after Walker and Ward[22])

| Acid | Condition | Corrosion rate (mm y^{-1}) | | Current density to maintain passivity ($i_{pass.}$, Am^{-2}) |
		Unprotected	Anodically protected	
Spent alkylation acid (sulphuric acid and organic matter)[23]	Stirred	3·0	0·15	2·48 × 10^{-1}
	quiescent	1·4	0·12	3·2 × 10^{-2}
Sulphuric acid, 93%[24]	Stirred	3·3	0·28	
	quiescent	0·9	0·07	

know the exact operating conditions of the electrolyte before designing a system of anodic protection. From an economic aspect, it is normal, in the first instance, to consider anodically protecting a cheap metal or alloy, such as mild steel. If this is not satisfactory, the alloying of mild steel with a small percentage of a more passive metal, such as chromium, molybdenum or nickel, may decrease both the critical and passivation current densities to a sufficiently low value. It is fortunate that the effect of these alloying additions can be determined by laboratory experiments before application on an industrial scale is undertaken.

Practical Aspects

It is essential that the throwing power of the system (the ability for the applied current to reach the required value over long distances) is good and that the potential of the whole of the protected surface is maintained in the passive region. This can normally be achieved with commercial potentiostats, providing the range of the potential over which the metal or alloy exhibits passivity is greater than 50 mV. In the case of stainless steels the corrosion rate will increase if the potential rises into the transpassive zone (or if it falls into the active zone). However, titanium does not show transpassivity and, therefore, has a large potential range over which it is passive.

In general, a uniform distribution of potential over a regular-shaped passivated surface can be readily obtained by anodic protection. It is much more difficult to protect surface irregularities, such as the recessions around sharp slots, grooves or crevices[27-33] since the required current density will not be obtained in these areas; therefore, a local cell is set up and corrosion occurs within the recess. This incomplete passivation can have catastrophic consequences, in the form of intergranular corrosion[34], stress-corrosion cracking[35-36], corrosion fatigue[28] or pitting[35,37-38]. This difficulty can be overcome by designing the surface to avoid these irregularities around bolt and rivet holes, threaded pipe sections and imperfect welds, or by using a metal or alloy which is very easily passivated having as low a critical current density as possible. In the rayon industry, crevice corrosion in titanium has been overcome by alloying with 0·1% palladium[39].

The throwing power of a system is particularly important in the anodic protection of pipelines and, therefore, has been widely studied[36,39-43]. The length of the pipe that can be protected by a single cathode placed at one end depends upon the metal, electrolyte and the pipe diameter; the larger the diameter the longer the length that can be protected. Thus, for mild steel in 93% sulphuric acid the length protected (or made passive) is 2·9 m for 0·025 m diameter, 4·8 m for 0·05 m diameter and possibly about 9 m for 0·15 m diameter, whereas for mild steel in a nitrogen fertiliser (a less aggressive medium than sulphuric acid) the protected length can be as much as 60 m with one cathode. As a result of recent field tests with an 0·30 m diameter carbon-steel pipe and 93% sulphuric acid at ambient temperatures, it is proposed to install anodic-protection systems for 650 m of pipeline[44].

The actual passivation of a surface is very rapid, if the applied current density is greater than the critical value. However, because of the high current requirements, it has been found to be neither technically nor economically practical to consider passivating the whole surface of a large vessel at the same time. This can be illustrated by the fact that for a storage vessel with an area of 1 000 m² a current of 5 000 A is necessary for some metal–environment systems, so it is therefore essential to use some other technique to avoid these very high currents. It may be possible to lower the temperature of the electrolyte to reduce the critical current density before passivating the metal. The feasibility of this is indicated from the values given for some acids in Table 11.26, but generally the reduction in the total current obtained by this method is insufficient. If a vessel has a very small floor area, it may be treated in a stepwise manner[29, 41] by passivating the base, then the lower walls and finally the upper walls, but this technique is not practical for very large storage tanks with a considerable floor area. Another method which has been successful is to passivate the metal by using a solution with a low critical current density (such as phosphoric acid), which is then replaced with the more aggressive acid (such as sulphuric acid) that has to be contained in the vessel (see Table 11.24). The critical current density can be minimised by pretreating the metal surface with a passivating inhibitor; for example, chromate solution has been applied to the floor and lower walls of a carbon-steel storage tank, which was then used to contain 37% nitrogen fertiliser solution[44].

Applications and Economic Considerations

Anodic protection can be applied to metals and alloys in mild electrolytes as well as in very corrosive environments, including strong acids and alkalis, in which cathodic protection is not normally suitable. The operating conditions, Flade potential, critical and passive current densities can be accurately determined by laboratory experiments and the current density during passivity is often a direct measure of the actual rate of corrosion in practice. The very low corrosion rate of a passive metal or alloy results in very little metal pick-up and solution contamination or discoloration. However, special care should be taken in the selection of the metal or alloy and in the design if there is a possibility of crevice or intergranular corrosion. A portable form of anodic protection is available that can be applied to rail and road tankers[45-47]. It can also be used for old vessels as well as new, so that a container designed for one liquid can be protected and used to hold a more corrosive solution. Because a system with a good throwing power can be designed, anodic-protection systems have been applied to pipelines[44] and spiral heat exchangers[36, 39]. It has been found possible to maintain protection of the vapour space above a liquid, once it has been completely immersed and passivated[36], and this is particularly important when the liquid level may rise and fall during storage and use.

One consequence of reducing the rate of corrosion of steel in an acid is to decrease the formation of hydrogen, which has been reported as the cause of explosions in phosphoric acid systems[48]. Anodic protection, which has been found to reduce the formation of hydrogen by 97%, can therefore prevent this form of accident[49].

The limitations of anodic protection arise from the inability to form a stable, continuous, protective, passive film on the metal to be protected. Thus

Table 11.28 Summary of anodic protection application (after N.A.C.E.[51])

Application	Vessel metal	Temperature range (°C)	Vessel type and purpose	Vessel size range‡ (m)	Number of systems	Type of controller	Power supply size range (kW)	Date of first installation
Oleum	Mild steel	≯50	Storage*	8D×6H–12D×6H	4	On-off and proportional	0·5–5·0	Oct. 1960
100% H_2SO_4	Mild steel	≯50	Storage*	4D×4H	1	On-off	0·5	March 1960
99% H_2SO_4	Mild steel	≯50	Storage*	2D×9L–9D×10H	5	On-off	2·5–5·0	Nov. 1962
98% H_2SO_4	Cast iron	≯93	Mix tank*†	3D×2D	1	On-off	1·0	Aug. 1964
98% H_2SO_4	Mild steel	≯50	Storage*	8D×6H–9D×9H	2	On-off	2·5	Sept. 1963
96% H_2SO_4	Mild steel	≯50	Storage*	4D×6H	2	On-off	1·0–2·5	Dec. 1961
93% H_2SO_4	Mild steel	≯50	Storage	2D×5L–15D×7H	14	On-off	0·5–5·0	Oct. 1962
93% H_2SO_4	304 stainless steel	≯66	Heat exchanger*†	61 m²		Proportional	1·0	Sept. 1966§
60°Bé H_2SO_4	Mild steel	≯66	Storage*	8D×6H–12D×6H	3	On-off	5·0	June 1965
60°Bé H_2SO_4	430 stainless steel	≯50	Storage*	2D×10L	1	On-off	0·5	June 1965

Table 11.28 (continued)

Application	Vessel metal	Temperature range (°C)	Vessel type and purpose	Vessel size range‡ (m)	Number of systems	Type of controller	Power supply size range (kW)	Date of first installation
Black and spent H₂SO₄	Mild steel	≯60	Storage*	4D×6L–12D×12H	6	On-off and proportional	2·5–10·0	Nov. 1962¶
Black H₂SO₄	304 stainless steel	≯163	Storage*	3D×3H	1	Proportional	5·0	Oct. 1966
75% H₃PO₄	304 stainless steel	≯50	Storage*	24D×10H	1	Proportional	5·0	Sept. 1963
N₂ fertiliser solutions	Mild steel	≯50	Storage*†	30D×5H–27D×12H	7	On-off	5·0	Dec. 1963
Kraft cooking liquor in digester	Mild steel	≯177	Reactor*	3D×14H	1	Time	30	1961‖
ClO₂ bleach	317 stainless steel	≯60	Washer wires*	Not applicable	1	Proportional	2·5	Nov. 1963

*Chemical purity of product.
†Corrosion control of vessel.
‡D = diameter, H = height.
§Exchangers failed after two weeks operation due to high chloride content of cooling water, causing stress-corrosion cracking of 304 stainless steel.
¶One application failed due to unanticipated composition variations. Relatively low unprotected rate did not provide incentive for further work.
‖There are both successful and unsuccessful pulp digester installations.

metals and alloys which neither form passive films nor non-conducting solutions cannot be used. For strongly aggressive acids, such as hydrochloric acid, a very stable anode film is required and, while steel is not satisfactory, titanium[18,50] may be suitable. A power failure may be a considerable danger, since it can result in a drop in potential from the passive region to the active region. This may be rectified by the use of a 100% effective 'fail-safe' back-up current source or by the selection of a basically more corrosion-resistant alloy, which would be marginally satisfactory in an unprotected state. Table 11.28[51], which was produced late in 1968, summarises the applications of anodic-protection systems which have been used in the USA. Most of these were for chemical purity and corrosion control in storage vessels and details are given of the size of the tank and the operating conditions.

Table 11.29 Comparison of the cost of protecting tanks by various methods
(after Reference 14)

	Cost $(£m^{-2}y^{-1})$				
	Mild steel protected	Mild steel lined	Stainless steel	Mild steel protected	Aluminium
Tank cost	1·70	1·70	4·57	1·70	3·5
Anodic-protection system	0·94			0·04	
P.V.C. lining		4·05			
Power	0·04			0·04	
Maintenance	0·54	0·81		0·54	
Total cost	3·22	6·56	4·57	2·32	3·5
		95 000 litre		3 800 000 litre	

Economically the installation of anodic protection is often very good. The use of a potentiostat and its associated equipment involves a high installation cost but low operating costs, because only very small current densities are required to maintain passivity. By the use of a relatively inexpensive switching mechanism Hays[52] has been able to use one control and power-supply system to anodically protect three separate tanks, and therefore reduced the high initial cost. It can be seen from Table 11.29 that it is more economical to anodically protect mild steel than to use mild steel with a p.v.c. lining or to use a more resistant and expensive metal or alloy such as aluminium or stainless steel. It is worth noting that because most of the expense of an anodic-protection system is due to the cost of the potentiostat, it is more economical per unit volume to use a larger instrument and a bigger tank. This is also illustrated by the relative costs of a 30 000 litre storage vessel which have been given[53] as £2 000 for an anodically protected mild steel, £2 500 for glass-lined mild steel, and £3 500 for unprotected stainless steel; for the same systems with a larger tank of 450 000 litre the figures were £6 500, £10 500 and £17 000, respectively. Not only is the anodic protection of a mild-steel tank cheaper than one with a glass or phenolic lining[54], but, because the steel conducts heat, it can be used for heat exchangers, and in addition it may be more stable at high temperatures for long time periods. In very corrosive conditions it may be necessary to use a very resistant alloy together with anodic protection, e.g. Corronel 230 was employed in the extraction of uranium using hot acidic solutions, which were too aggressive to be contained by Neoprene, Karbate and Teflon coatings[55].

Conclusion

Although the first industrial application of anodic protection was as recently as 1954, it is now widely used, particularly in the USA and USSR. This has been made possible by the recent development of equipment capable of the control of precise potentials at high current outputs. It has been applied to protect mild-steel vessels containing sulphuric acid as large as 49 m in diameter and 15 m high, and commercial equipment is available for use with tanks of capacities from 38 000 to 7 600 000 litre[44]. A properly designed anodic-protection system has been shown to be both effective and economically viable, but care must be taken to avoid power failure or the formation of local active–passive cells which lead to the breakdown of passivity and intense corrosion.

The extent of the interest in the application of anodic protection is indicated in the following list of recent publications. These include the application of anodic protection to titanium in the rayon industry[39], heating coils[56] and chromic chloride/chromous chloride solutions[21]; different steels with chrome ammonium sulphate[57] and in heat exchangers[34, 58]; the storage of ammonium carbonate/ammonium nitrate solutions[59], manufacture of acrylamide[60] and transport of nitrogen fertiliser[45]. Further details regarding the operating details can be obtained from other general articles and reviews[22, 61–69].

R. WALKER

REFERENCES

1. Fontana, M. G. and Greene, N. D., *Corrosion Engineering*, McGraw-Hill, New York, 214 (1967)
2. Edeleanu, C., *Nature*, **173**, 739 (1954)
3. Edeleanu, C., *Metallurgia*, Manchr., **50**, 113 (1954)
4. Cherrova, G. P., Dissertation, *Akad. Nauk.*, Moscow Institute of Physical Chemistry, SSSR (1953)
5. Novakovskii, V. M. and Levin, A. I., *Dokl. Akad. Nauk.*, SSSR, **99** No. 1, 129 (1954)
6. Flade, F., *Z. Phy. Chem.*, **76**, 513 (1911)
7. Shock, D. A., Riggs, O. L. and Sudbury, J. G., *Corrosion*, **16** No. 2, 99 (1960)
8. West, J. M., *Electrodeposition and Corrosion Processes*, Van Nostrand, New York, 81 (1965)
9. Uhlig, H. H., *Corrosion and Corrosion Control*, Wiley, New York, 61 (1967)
10. Myers, J. R., Beck, F. H. and Fontana, M. G., *Corrosion*, **21** No. 9, 277 (1965)
11. Sudbury, J. G., Riggs, O. L. and Shock, D. A., *Corrosion*, **16** No. 2, 91 (1960)
12. Shock, D. A., Sudbury, J. D. and Riggs, O. L., *Proceedings of the First International Congress on Metallic Corrosion*, London 1961, Butterworths, London, 363 (1962)
13. Pourbaix, M., *Corrosion*, **25** No. 6, 267 (1969)
14. Corrosion Control Systems Bulletin, No. 773, Magna Corporation, USA (1967)
15. Sudbury, J. D., Locke, C. E. and Coldiron, D., *Chemical Processing*, Feb. 11th (1963)

16. Togano, H., *J. Japan Inst. of Metals*, **33** No. 2, 265 (1969)
17. Kuzub, V. S., Tsinman, A. I., Sokolov, V. K. and Makarov, V. A., *Prot. Metals*, **5** No. 1, 45 (1969)
18. Cotton, J. B., *Chem. Ind.*, Lond., **18** No. 3, 68 (1958)
19. Stern, M. and Wissenberg, H., *J. Electrochem. Soc.*, **106**, 755 (1959)
20. Togano, H., Sasaki, H. and Kanda, Y., *J. Japan Inst. Metals*, **33** No. 11, 1280 (1969)
21. Letskikh, E. S., Komornokova, A. G., Kryazheva, V. M. and Kolotyrkin, Ya. M., *Prot. Metals*, **6** No. 6, 635 (1970)
22. Walker, R. and Ward, A., Metallurgical Review No. 137 in *Metals and Materials*, **3** No. 9, 143 (1969)
23. Locke, C. E., Banks, W. P. and French, E. C., *Mat. Prot.*, **3** No. 6, 50 (1964)
24. Sudbury, J. D. and Locke, C. E., *Oil Gas J.*, **61**, 63 (1963)
25. Finley, T. C. and Myers, J. R., *Corrosion*, **26** No. 12, 544 (1970)
26. Finley, T. C. and Myers, J. R., *Corrosion*, **26** No. 4, 150 (1970)
27. France, W. D. and Greene, N. D., *Corrosion*, **24** No. 8, 247 (1968) and also Report No. AD 665, 788 (1968)
28. Cowley, W. C., Robinson, F. P. A. and Kerrich, J. E., *Brit. Corrosion J.*, **3** No. 5, 223 (1968)
29. Makarov, V. A. and Kolotyrkin, Ya. M., *Media for Prevention of Corrosion*, Moscow, 5–15 (1966)
30. Anon., *Anticorrosion Methods and Mat.*, **15** No. 4, 5 (1968)
31. Karlberg, G. and Wranglen, G., *Corros. Sci.*, **11** No. 7, 499 (1971)
32. Ruskol, Y. S. and Klinov, I. Y., *J. Appl. Chem.*, USSR, **41** No. 10, 2084 (1968)
33. France, W. D. and Greene, N. D., *24th Annual NACE Conference*, Cleveland, Ohio, 1, March (1968)
34. France, W. D. and Greene, N. D., *Corros. Sci.*, **8**, 9 (1968)
35. Greene, J. A. S. and Haney, E. G., *Corrosion*, **23**, 5 (1967)
36. Stammen, J. M., *25th Annual NACE Conference*, Houston, 688, March (1969)
37. Schwenk, W., *Corrosion*, **20**, 129t (1964)
38. Banks, W. P. and Hutchison, M., *Mat. Prot.*, **7** No. 9, 37 (1968)
39. Evans, L. S., Hayfield, P. C. S. and Morris, M. C., *Werkst. u. Korrosion*, **21**, 499 (1970) and also *Proceedings of the Fourth International Congress on Metallic Corrosion*, Amsterdam, 625 (1969)
40. Edeleanu, C. and Gibson, J. G., *Chem. Ind.*, Lond., **21** No. 10, 301 (1961)
41. Mueller, W. A. J., *Electrochem. Soc.*, **110**, 699 (1963)
42. Timonin, V. A. and Fokin, M. N., *Prot. Metals*, **2** No. 3, 257 (1966)
43. Makarov, V. A., Kolotyrkin, Ya. M., Kryazheva, V. M. and Mamin, E. B., *Prot. Metals*, **1** No. 6, 592 (1965)
44. Stammen, J. M., private communication
45. Banks, W. P. and Hutchison, M., *Mat. Prot.*, **8** No. 2, 31 (1969)
46. Sudbury, J. D. and Locke, C. E., *Chem. Eng.*, **70** No. 11, 268 (1963)
47. Locke, C. E., *Mat. Prot.*, **4** No. 3, 59 (1965)
48. Lowe, J. B., *Corrosion*, **17** No. 3, 30 (1961)
49. Riggs, O. L., *Mat. Prot.*, **2** No. 8, 63 (1963)
50. Jaffee, R. I. and Promisel, N. E., *The Science, Technology and Application of Titanium*, Pergamon Press, London, 155 (1970)
51. Report compiled by NACE Task Group T–3L–2 (1968)
52. Hays, L. R., *Mat. Prot.*, **5** No. 9, 46 (1966)
53. Fisher, A. O. and Brady, J. F., *Corrosion*, **19**, 37t (1963)
54. Anon., *Mat. Prot.*, **2** No. 9, 69 (1963)
55. Robinson, F. P. A. and Golante, L., *Corrosion*, **20** No. 8, 239t (1964)
56. Schmidt, W., Hampel, H. and Grabinski, J., *Chem. Tech.*, Berlin, **22** No. 5, 296 (1970)
57. Letskikh, E. S. and Komornikova, A. G., *Prot. Metals*, **5** No. 3, 255 (1969)
58. Makarov, V. A., Egorova, K. A. and Kuzub, V. S., *Prot. Metals*, **6**, 5 (1970)
59. Kuzub, V. S., Moisa, V. G., Kuzub, L. G., Gnezdelova, V. I., Kozmenko, N. K. and Danielyan, L. A., *Zashchita Metallov*, **7** No. 3, 361 (1971)
60. Makarov, V. A. and Egorova, K. A., *Prot. Metals*, **6** No. 3, 302 (1970)
61. Stammen, J. M., *27th Annual NACE Conference*, Chicago, March (1971)
62. Franz, F., Bartel, V. and Koritta, J., *Proceedings of the Fourth International Congress on Metallic Corrosion*, 1969, Amsterdam, 705 (1969)
63. Togano, H., *Boshoku Gijitsu*, **17** No. 2, 51 (1968)
64. Reinoehl, J. E. and Beck, F. H., *Corrosion*, **25** No. 6, 233 (1969)

65. Grafen, H., Herbsleb, G., Paulekat, F. and Schwenk, W., *Werkst. u. Korrosion*, **22**, 16 (1971)
66. Dambal, R. P. and Rama Char, T. L., *Chemical Age of India*, **19**, 1175 (1968)
67. Ross, T. K., *Chem. Engineer*, No. 247, 95 (1971)
68. Henthorne, M., *Chem. Eng.*, **78** No. 29, 73 (1971)
69. Walker, R., *Metallurgia*, Manchr., **82** No. 490, 51 (1970)

12 PRETREATMENT AND DESIGN FOR METAL FINISHING

12.1	Pretreatment Prior to Applying Coatings	**12**:3
12.2	Pickling in Acid	**12**:16
12.3	Chemical and Electrochemical Polishing	**12**:24
12.4	Design for Corrosion Protection by Electroplated Coatings	**12**:28
12.5	Design for Corrosion Protection by Paint Coatings	**12**:35

12.1 Pretreatment Prior to Applying Coatings

Chemical and mechanical pretreatments are given to surfaces before painting or electroplating, either to remove existing natural oxide films, corrosion products, or tarnish, or to remove factitious coatings such as protective oils and greases or oxide films produced by heat treatment and manufacturing processes. Pretreatments to produce anodic, phosphate, and similar coatings are treated in Chapter 16.

The standard of cleanliness of a metal surface may be dictated by the metal-finishing process which is to follow. In general, electroplating processes and vacuum deposition demand the highest possible freedom from contaminants, whereas anodic oxidation processes and electrophoretic painting can tolerate the presence of a small amount of residuals. Phosphating processes may benefit from the presence, not of grease, but of a lightly adsorbed film of hydrocarbon which is left after the use of emulsion cleaners. The above statements should not be taken as a licence to skimp precleaning processes; it is true to say that the majority of metal finishing defects which are encountered in practice stem not from an injudicious choice of process or materials, but from a failure to clean the surface to be treated to the required standard.

The sections on individual metal treatment and anti-corrosive processes will indicate the appropriate cleaning treatments. A British Standard Code of Practice CP 3012 *Cleaning and Preparation of Metal Surfaces* brings together the requirements of Defence Standard DEF STAN 03–2/1 of the same title and therefore covers the details of the commonly-used degreasing processes for a wide range of industrially used metals.

Removal of Grease and Organic Contaminants

In general these contaminants will be oils or greases applied to assist deformation or to give protection in storage. Where processes involving the use of large forces such as extrusion have been used, this grease may sometimes be carbonised and burned on to the surface. The so-called 'carbon' or 'graphite' which is found as a fine, black smut on fully finished steel sheet is usually finely divided iron oxide which is difficult to wet with detergents.

Organic Solvents

Oil and grease solvents such as white spirit, paraffin and chlorinated hydro-carbons are widely used to prepare surfaces for further treatments. The methods employed range from wiping over with a solvent-soaked rag, through dipping into tanks of solvent, to degreasing by immersion in chlorinated hydrocarbon vapour and/or boiling liquid chlorinated hydro-carbons.

Cold solvent cleaners   The simple methods of wiping and dipping into liquids have obvious limitations. Hand methods expose operators to dangers arising from the defatting action of the solvents on the skin and, in the case of trichlorethylene and similar solvents used in this way, to the inhalation of toxic vapours. White spirit and paraffin give rise to possible fire hazards; many proprietary 'Safety Solvents' are available in which the fire risk is reduced by the addition of sufficient chlorinated hydrocarbon to raise the flash point and render the mixture incapable of propagating combustion. Some of these additives have the disadvantage of being toxic (see Table 12.1).

Table 12.1   Toxicity of various chlorinated hydrocarbons

Compound	Limiting threshold value (p.p.m.)
Carbon tetrachloride	10
Trichloroethylene	100
1,1,1-Trichloroethane	350
Perchloroethylene	100
Trichlorotrifluorethane	1 000

The efficacy of these mixtures in preventing the combustion of contents of the degreasing vat depends on the retention over the surface of a layer of heavy vapour in which the concentration of chlorinated hydrocarbon is sufficient to reduce the flammability to the point at which flame will not propagate. It follows therefore that the containing vessel should have a high freeboard in order to prevent the dissipation of this layer by draughts and for the same reason work should be lowered into and removed from the tank at a slow and regular rate, about 3 m/min is a reasonable figure to aim at; hand dipping should be avoided and electric or air hoists should be used to handle the racked or basketted work. The degreased work should be allowed to drain over the tank, and preferably within the free space below the edge of the tank.

A confusing situation existed with reference to the flash point of safety solvents. Under the conditions of the statutory Abel method of determining flash point some mixtures will show a distinct, though feeble, flash below 23°C. Although it is extremely difficult to ignite the same mixture in an open dish by playing a gas flame onto the surface, the degreasing solvents had to be handled under the conditions prescribed for inflammable liquids. The Highly Flammable Liquids and Liquefied Petroleum Gases Regulations, 1972, have gone far to rectify this anomalous position of safety solvents and water-borne paints, which have similar flash point characteristics, by the use of a special test for combustibility to supplement the flash point test.

The effectiveness of degreasing by these methods is at the best doubtful. The solvent reservoirs quickly become loaded with dissolved oil and grease which is in turn transferred back to the surfaces to be cleaned. Immersion degreasing is best carried out using at least two tanks on the counter-flow principle (Fig. 12.1).

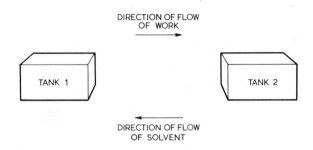

Fig. 12.1   Counter-flow principle of cold immersion solvent degreasing

The losses in solvent by drag-out and/or evaporation from the first tank are made up by transfer of liquid from the second tank, which is in turn made up from the following tank or with fresh solvent.

None of these methods will produce a surface sufficiently clean for electroplating.

Vapour and hot-immersion cleaning   These processes employ chlorinated hydrocarbons as the grease solvents. They have the advantage of chemical stability, good grease solvency, high vapour density and non-inflammability.

The principally used halogenated hydrocarbons and their limiting threshold values, which provide a measure of their toxicity, are shown in Table 12.1.

The value of the limiting threshold for carbon tetrachloride has been given as a basis for comparison, this compound should never be used either on its own or as a component of degreasing mixtures because of its high toxicity. Halogenated hydrocarbon solvents can be used for cleaning all industrially used metals in their massive form, but care should be taken to remove aluminium or titanium swarf before cold degreasing in halogenated hydrocarbons and it is additionally advisable to remove the swarf from zinc components before using hot trichloroethylene, since these metals in the finely divided form may react violently with the solvent.

Vapour degreasing   The equipment for industrial use (Fig. 12.2) consists essentially of a tall vessel, usually constructed of galvanised sheet or plate. The bottom acts as a sump in which the liquid can be boiled. Above the liquid level is placed a wire or perforated metal grille on which work can be rested so as to be in the middle zone between the liquid and the condenser coils. The sump can be heated by any convenient method such as gas, oil, steam, high-pressure hot water or electricity. Both the sump and the cooling coils are fitted with thermostats to limit the maximum temperature of the liquor and the cooling water.

The vapour from the boiling liquor rises into the centre portion of the vessel and there condenses on the cold work. The liquid dissolves any soluble oils or greases present and runs back into the sump for redistillation. Any vapour passing the work is caught by the condensing coils and also returned to the sump.

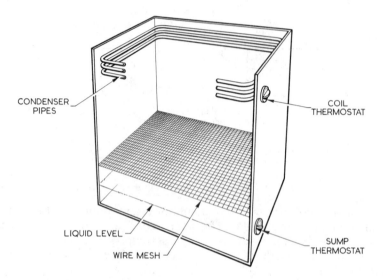

CONDENSER PIPES

COIL THERMOSTAT

LIQUID LEVEL

WIRE MESH

SUMP THERMOSTAT

Fig. 12.2   Cut-away drawing of vapour degreasing plant

Very light work, especially if the ratio of mass to surface is low, may heat up so quickly that condensation ceases before all the grease has been dissolved. In this case it may be necessary to remove the work from the vapour zone and retreat it after cooling to room temperature. This multiple treatment can be avoided by use of liquor/vapour processes. Some amelioration of this condition can also be obtained by the selection of higher boiling point solvents.

The scouring action of the liquid running off the surface is not sufficient to remove solid insoluble soils, and these therefore remain and are liable to interfere with subsequent processes. In order to overcome this defect, combined liquor/vapour processes have been introduced.

Liquor/vapour degreasing   In these plants, the work after passing through the vapour phase is passed, either manually or mechanically, through one or more compartments of boiling liquor. When more than one liquor compartment is used, the counter-flow principle is again employed, liquor from the last compartment which receives clean condensate flows into the next, and so on down the system. More complex plants, with arrangements for jets of solvent to impinge on work, have also been constructed.

Operating notes When the boiling point of the sump liquor reaches 105°C in the case of trichlorethylene, or about 140°C in the case of perchlorethylene, redistillation is necessary. The degreasing plant itself can be used for this purpose by fitting a pipe to lead the condensate from the coils to the outside of the tank, where it is collected; alternatively a separate distillation plant may be used. The grease left is raked out through cleaning doors and disposed of; care must be taken in view of the fact that the grease still contains some solvent. During the distillation it is convenient to add a little sodium carbonate solution to rectify any acidity which may have developed.

Vapour-degreasing plant should be sited in draught-free areas to avoid loss of vapour. Baskets of work and other pieces should be lowered into and removed from the plant at low speed for the same reason. About 3 m/min is a reasonable maximum hoist speed.

The precautions regarding plant operation, maintenance, smoking and naked flames in the vicinity of vapour-degreasing plant are laid down by the manufacturers and should be strictly observed.

Emulsion Cleaners (Water-rinsable Cold Solvent Cleaners)

Emulsion cleaners, or as they should more accurately be called, *emulsifiable* cleaners, present a means by which the accumulation of grease in a cold solvent cleaner can be prevented. The cleaning medium consists of a solvent, usually white spirit or kerosine, blended with cresylic or oleic acid soaps to form a grease solvent mixture which can be rinsed away with a stream of water. The work to be treated can be dipped into a tank of the cleaner, or the solvent mixture can be brushed or sprayed onto the surface. After the grease has been penetrated by the solvents the work is thoroughly washed by a good stream of water. Cleaning with this type of degreasing agent is particularly suitable for the removal of heavy grease coatings such as those applied as preservatives. The liquid cleaner is comparatively low in first cost and even very heavy coatings can be removed by racking in a basket and immersion in solvent without serious contamination of the bulk of the cleaner.

A variant of this type of degreaser is available in the form of a paste which is used for the removal of local grease deposits or for cleaning articles of composite construction which could be damaged by immersion or overall spraying with solvent. In practice, water-rinsable degreasing compounds are usually based on halogenated hydrocarbons and the health precautions usual with these materials must be observed. It is particularly important that operations should be carried out in well ventilated situations.

Emulsion cleaners containing cresylic acid soaps can give rise to effluent problems. The rinsings from the degreased work are not usually objectionable, particularly if diluted with a large volume of clean factory effluent. Difficulties may arise on the comparatively rare occasions when it is necessary to dispose of a tank of spent solution. Burning under conditions which will not create atmospheric pollution is often resorted to, but sending it to a specialist effluent treatment firm may prove to be the simplest method.

Emulsion cleaners may also be used in spray washing machines of the conventional type as a hot aqueous solution containing about $2\frac{1}{2}$ to 5% by volume of cleaner. This treatment is followed, as in degreasing by alkaline cleaners, by a water wash usually applied in a second stage of the cleaning machine.

An extension of the emulsion cleaner idea is the di-phase cleaner in which a solvent layer, paraffin or distillate, floats on an aqueous layer consisting of a solution of water-softening agents, emulsifiers and mild detergents. The di-phase cleaners are more efficiently used in spray-washing machines at a temperature in the region of 60–70°C; the mixed phases are pumped through spray nozzles over the work to be cleaned. The mode of action is generally assumed to be that the petroleum solvent dissolves the soil, which is then emulsified into the aqueous phase. Di-phase cleaners are also followed by water wash.

The use of emulsion cleaners does not result in the complete removal of all the contaminating material from the surface treated. Not only may traces of the original contaminant be left, but the products which are based predominantly on kerosine or light fuel oil, will leave an adsorbed layer of hydrocarbon solvent on the surface. This state is sometimes referred to as one of physical, compared with chemical, cleanliness. This layer is not always objectionable, it has a limited resistance to atmospheric corrosion and may therefore be helpful when it is necessary to store steel or cast iron components for a short period under reasonable conditions.

Emulsion and di-phase cleaners are particularly useful prior to phosphating as they leave the surface in a condition such that conversion commences at a large number of nuclei, and a coating of small uniform crystal size is thus produced.

The residue of contaminant and the adsorbed layer can readily be removed by a light cleaning with alkaline degreasants. Emulsion cleaners, therefore, provide a low cost method for the removal of heavy contamination as a preliminary to spray alkali cleaning. There use not only lengthens the life of the alkaline solution but also increases the periods for which a spray machine can be operated between shut down for cleaning.

Alkaline cleaners

Alkaline cleaners are probably the most widely used of all cleaning media and can, unaided, produce surfaces sufficiently clean for the electrodeposition or chemical deposition of metals. They are, in general, based on the hydroxides, carbonates, phosphates, both simple and complex, and the various silicates of sodium. Synthetic surfactants, complexing and chelating agents are used with these salts to assist the cleaning action. The causticity of the mixture to some extent governs its activity, but highly caustic compositions attack light alloys, and high-pH solutions used prior to phosphate treatments yield coatings which are extremely coarse in structure. In addition, removal of inorganics by rinsing may be difficult.

Used in spray-washing machines, alkaline cleaners can be effective at extremely low concentrations. Typical working conditions are: concentration, 6–12 g/l; temperature, 80°C (min.); spraying pressure, 140–170 kN/m^2; time of treatment, 1–1·5 min. Degreasing by immersion in solutions of alkalis is less efficient, concentrations of 50 g/l or more being usual, and the time may be as long as 30 min.

The economics of alkaline cleaning depend on a balance between the first cost of the solution, its life under operating conditions, the rate of make up required and the amount of plant maintenance called for. Make-up rate is not only a function of the consumption of alkali by the actual degreasing

Table 12.2 Typical alkaline cleaner composition (%)

	Heavy duty	Light duty	Electrolytic cleaner (steel)	Aluminium cleaner
Caustic soda	65	—	30	—
Soda ash	15	15	30	—
Sodium metasilicate	15	30	25	67
Trisodium phosphate	5	55	15	33

process, but is profoundly influenced by plant design and upkeep. Particularly in the case of spray installations, the design of plant, by causing unnecessary losses through the ventilating system or heavy loss of solution by overspray between zones, can be the governing cost factor as well as giving rise to inefficient cleaning.

Modern solutions containing complexing agents prevent the build up of precipitated hardness salts on heat transfer surfaces and in spray nozzles, which is the cause of much poor cleaning due to failure to maintain operating temperatures and pressures. The cost of plant down-time to rectify these faults is a considerable operating overhead, which may easily offset the attraction of a low priced chemical.

One way of accelerating immersion degreasing and producing a higher degree of cleanness than is otherwise possible is to pass a direct or alternating electric current through the solution, using the work to be degreased as one of the electrodes. Usually the work is made cathodic for most of the direct current cycle in order to utilise the greater volume of gas evolved at the cathode; a few seconds' final anodic treatment is then given in order to dissolve off any metal deposited during the cathodic cycle. Lead, tin and zinc may be badly etched if the anodic part of the cleaning cycle is too prolonged.

Table 12.3 Cleaning high-tensile steels with electrolytic cleaners

Specified tensile stress (kN/mm²)	Cleaning methods
1–1·4	Cathodic cleaning must not be used. Instead, use either anodic cleaning or preferably non-electrolytic alkaline cleaning
1·4–1·8	Neither cathodic nor alternating current cleaning may be used. Instead, use anodic direct current cleaning
>1·8	No electrolytic treatment

The degreasing of high tensile steels using electrolytic cleaners is subject to special conditions due to the danger of hydrogen embrittlement. These conditions are rigidly defined in CP 3012 and may be summarised in tabular form.

It is not good practice to use the tank as an electrode, and sheets of steel or nickel are preferable. The solutions are generally made more caustic than non-electrolytic alkaline cleaners so as to obtain the maximum conductivity, since current densities of the order of 500–1 000 A/m^2 are usual.

Electrolytic cleaning gives rise to large volumes of oxygen and hydrogen which are evolved at the electrodes. In the presence of surfactants which are used in the formulation of the cleaner or of soaps produced during the cleaning process, a blanket of foam can be produced over the liquid surface. This foam may be explosive, and while it does not normally represent a serious hazard, care should be taken to prevent ignition.

Ultrasonic cleaning

This method of cleaning produces rapid renewal of the cleaning liquid at the soiled surface coupled with a violent scouring action. The effect is obtained by the generation of high-frequency vibration in the liquid by means of a transducer fed with electrical oscillations of the required frequency, which may be between 100 kHz and 2 MHz. The efficiency of conversion is very low, and the capital cost of any but the smallest installation, say up to 200 litres capacity, is prohibitive except for very special purposes.

The high-frequency waves cause the rapid formation and collapse of minute bubbles on the surface, and the effect can penetrate into blind holes. The cleaning is extremely thorough, and solid adherent soils are completely removed. A feature of ultrasonic cleaning is the ability to remove soil from crevices, small bores or even blind holes. Any type of detergent solution or grease solvent can be employed, provided that it has a high coefficient of compressibility and is not too volatile, as the energy dissipated in the solution can cause a marked rise in temperature. Cooling coils in the degreasing tank will help to dissipate this heat.

Mechanical Methods of Removing Rust and Scale

Weathering   This commonly employed method of removing millscale from heavy structural members and the like consists of exposing castings and rolled sections to the weather for a period of some months in order to allow millscale to crack off under the stress of expansion and contraction.

Flame descaling   This method uses a fishtail or bar-type oxyacetylene burner to heat the scale layer and crack it off by differential expansion. It is an excellent method of descaling substantial structures and leaves a warm, dry surface on which to paint. The process employs a mixture of equal volumes of acetylene and oxygen, consuming about 10 litre/s to descale 1 m^2 at a traverse rate of about 0·75 m/min.

Shot and grit blasting   The basis of shot and grit blasting is the mechanical projection of hardened steel shot, cut wire grit, alumina or other non-metallic abrasive on to the surface to be cleaned. Shot or grit blasting is essential for the preparation of work for metal spraying and care must be taken to ensure that the grit is coarse enough and sufficiently sharp to provide a surface texture onto which the sprayed metal particles will key. The grit size used will depend on the surface finish required, but will normally fall within the range G17 of BS 2451: *Chilled Iron Shot and Grit*, or larger. Coarse blast cleaning of this nature is not suitable for use on thin material where the impact forces would distort the article being treated. Fine surface finishes cannot be obtained and if these are required, fine abrasive blasting should be used.

Non-metallic grit should be used for all metals other than cast iron or non-corrosion-resisting steels. Aluminium surfaces should be blasted only with alumina grit which has not been previously used for other metals, in order to minimise the risk of surface contamination and consequent galvanic corrosion due to metal impurities.

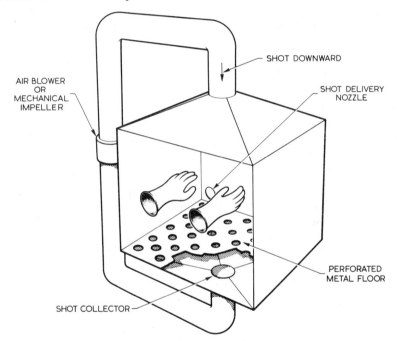

Fig. 12.3   Shot-blast cabinet

The process can be completely mechanised or operated as a manual process. The grit is thrown at the work either by allowing it to drop on to revolving paddles or by feeding it into a stream of compressed air. In either case the shot and scale drop on to riddles which screen off the scale and broken shot, allowing the whole shot to return to a hopper for re-use. From time to time the shot or grit should be picked over and worn pieces removed.

Various pieces of equipment have been designed for carrying out the process. Individual items may be grit blasted by introducing them into a cabinet through one wall of which projects a pair of heavy rubber gloves by means of which the operator holds the work-piece, turning it from side to side so as to expose all faces to the blast (Fig. 12.3); alternatively, the shot is delivered through a flexible pipe which can be held in the gloved hand and directed as required.

Fine abrasive blasting   This method is essentially similar to coarse abrasive blasting but uses only non-metallic abrasive particles such as glass beads, walnut shells, corn husks or other materials capable of producing the fine surfaces required. The abrasive can be projected dry, in the form of a water slurry or entrained in steam. In the latter case the process may be referred to as *vapour blasting*. The use of dry steam rather than wet blasting produces a similar surface quality but the heat of the steam dries the work rapidly and so prevents re-rusting. Corrosion inhibitors can be added to slurries for the same purpose.

The surface produced by vapour blasting is very fine and uniform and there is no danger, such as exists with more robust forms of blasting, of driving scale into the surface. Fine abrasive blasting is extremely useful in cleaning machined surfaces of refractory metals and finds application in removing the complex oxides from Nimonic turbine blades. It is essentially a small-scale process carried out in cabinets similar to those used for shot blasting.

Chemical Descaling and Derusting

Acid derusting   The descaling of iron and steel by acid pickling in tanks is treated in Section 12.2 and in this section only the removal of rust by application processes is dealt with. Generally speaking, only superficial rust can be removed by application de-rusters, but the ability to work on the surface with brushes facilitates the removal of rust from the bottom of pits into which immersion de-rusting solutions might not penetrate. Solutions of phosphoric acid, usually containing alcohols, wetting agents and sometimes thickeners, are used to clean large steel articles. Butyl or isopropyl alcohols are usually used and the thickening agents are cellulose ethers. The storage life of such cellulose-ether thickened solutions is limited to a few weeks as the thickening media are decomposed by hydrolysis in the acid medium. If the thickening is carried to the point of making a paste, bentonite or similar clays are often used, and these pastes are quite stable.

The method of use is simple; the material is applied by brushing to the surface to be cleaned, allowed to remain until the scale or rust is loosened, and then washed away. Pastes may be allowed to dry and then removed by brushing. The action of the acid may be facilitated by wire brushing. Washing is not always used if non-thickened solutions have been employed and paint is to be applied as soon as the surface has dried (Section 12.2). While this is an excellent method for removing rust from structures which could not otherwise be treated, it is too slow to be a practical method of removing thick scale.

Sodium hydride descaling   This is a most versatile method of scale removal, although it is expensive to operate. It will work well on metals ranging from mild steel to high-nickel-chromium alloys of the Nimonic type which cannot readily be descaled by other methods. The process employs the intense reducing activity of a bath of molten caustic soda containing 1·5–2·5% of sodium hydride at a temperature of 350–370°C. The hydride is generated in the bath by the interaction of hydrogen and sodium; the hydrogen is derived from cracked ammonia, and the sodium is contained in a perforated steel box welded on to the inner side of the tank. The cracked ammonia is passed over the sodium by means of pipes which enter at the top of the generating box and then dip below the level of the molten caustic soda.

The normal sequence of operation is:

Preheat → Sodium hydride bath → Drain → Water quench

Small work can be treated in baskets and larger pieces may be hung from hooks.

The process is potentially dangerous unless the proper precautions are taken, when it can be operated with perfect safety. Of these, the most important is a thorough preheating in order to remove every trace of water, which can otherwise project molten caustic soda from the bath with explosive violence. The simplest safeguard is to enclose the whole of the operation in a wired glass enclosure and arrange for push-button operation of the hoists from outside the enclosure. The use of a monorail over the tanks and a hoist to handle the work obviates the need for workmen to go inside the enclosure and allows the sequence to be so interlocked that it is impossible for work to be put into the descaling bath without having spent a preset time in the preheating chamber.

Descaling Non-ferrous Metals

In industrial practice the only non-ferrous metals that require chemical descaling are brasses, bronzes, aluminium alloys prior to spot welding and high-nickel-chromium alloys.

Copper alloys The scale on these consists mainly of a mixture of cupric and cuprous oxides together with the oxides of any alloying elements present. The simplest method of removing this is pickling in hot 10–15% sulphuric acid, and the techniques described in Section 12.2 are precisely applicable to pickling copper alloys.

If a bright surface is required, a subsequent dip in a sulphuric/nitric acid mixture of approximate composition sulphuric acid 60%, nitric acid 10%, and water 30% is used. A thorough washing must follow this treatment. This bath evolves oxides of nitrogen during use and should therefore be operated under a hood. The solution must be cooled as the evolution of heat is considerable.

Various additions have been suggested from time to time to improve the basic composition, chromic acid, sodium chloride and hydrochloric acid being among the commonest.

Aluminium alloys Aluminium and its alloys are covered by a thin coating of oxide. This must be removed before spot welding, but before painting the film must be thickened either by anodising or by oxidising solutions. These latter processes will be dealt with in Section 16.1. Another method of preparing aluminium and its alloys for painting is a deep chemical etch. Similar solutions can be used for preparation prior to either spot welding or painting.

Two such solutions are described in the Ministry of Aviation Specification DTD 915; they are based on mixtures of chromic acid or sodium bichromate with sulphuric acid. The compositions given are:

	Pickling solution (i)	Pickling solution (ii)
Concentrated sulphuric acid, s.g. 1·82	15% v/v	15% v/v
Chromic acid	5% w/w	—
Sodium bichromate ($Na_2Cr_2O_7 \cdot 2H_2O$)	—	$7\frac{1}{2}$% w/w
Water	remainder	remainder

This solution is worked at a temperature of 43–70°C, the work being first degreased and then immersed for 20 min in the pickling solution. The method of control and other details may be found in the specification.

If painting is to follow this treatment, a supplementary cleaning treatment with a butanol/acetone/white-spirit mixture made into a paste with Paris white is called for.

Other proprietary etching media based in the main on phosphoric or hydrofluoric acid or mixtures of the two are approved for aircraft work and listed in CP 3012.

Caustic soda solutions can be used to give an etch suitable for painting, but the action is somewhat uncontrollable unless the solution is so dilute that it has a very short life. Another defect of caustic soda etches is that a sludge of sodium aluminate quickly accumulates in the bath. This can be avoided by the addition of chelating agents such as ethylene diamine tetra acetic acid (EDTA), gluconates or heptonates.

Rinsing after Pretreatment

All the processes mentioned, with the exception of coarse abrasive blasting and dry fine abrasive blasting, call for some form of rinsing after the treatment stage. The removal of chemical residues from the surface must be complete if subsequent finishing difficulties are to be avoided. Hosing with water which is allowed to run to waste is probably the most efficient method of rinsing. The same effect can be produced by copious spraying in a washing machine or lowering the work into a tank, the upper part of which is ringed with spraying nozzles. Immersion or swilling cannot be considered a sound method of washing unless at least two tanks are used and a rate of water change allowed for which will maintain the concentration of contaminant at an acceptable level. Economy in the use of water may be achieved by operating the tanks on the counter-flow principle or by the use of conductivity sensing devices to control the water flow. In order to achieve the maximum assurance against subsequent finishing troubles, either in the course of processing or from service complaints, a final rinse with demineralised water should be used.

A. A. B. HARVEY

BIBLIOGRAPHY

The following is a selective list of publications in this field over the past few years:

General
Harvey, A. A. B., *Paint Finishing in Industry*, Robert Draper, London (1958)
Spencer, G. F., 'Surface Preparation of Metals', *Metal Finish.*, **58** No. 6, 53 (1960)
Radcliffe, J. Q., 'Pre-treatment of Metals', *Prod. Finish., Lond.*, **14** No. 8, 72 (1961)
Bishop, E. E., 'Cleaning of Surface', *Metal Ind.*, **98**, 127 (1961)
Cairns, G. A., 'What's New in Metal Cleaning', *Industr. Finish. (Indianapolis)*, **37** No. 7, 39 (1961)

Degreasing
Saxby, G. F., 'Non-flammable Solvents in Finishing', *Prod. Finish., Lond.*, **13** No. 4, 69 (1960)
Hanley, W. B., 'Efficient Degreasing Essential for Quality Finishing', *Industr. Finish. (Lond.)*, **13** No. 156, 46 (1961)

Emulsion Cleaners
Radcliffe, J. Q., 'Cold Emulsion Cleaning and its Application to Mass Production Lines', *Corrosion Technol.*, **6**, 272 (1959)
Radcliffe, J. Q., 'Single and Diphase Emulsion Cleaners', *Industr. Finish. (Lond.)*, **13** No. 152, 34 (1961)

Electro-cleaning
'Electro-cleaning: A Report on Current Practice', *Prod. Finish., Lond.*, **14** No. 4, 50 (1961)

Spray Cleaning
Holman, C., 'Spray Cleaning in Automation', *Industr. Finish. (Lond.)*, **13** No. 153, 46 (1961)

Ultrasonic Cleaning
Stein, P. W., 'Ultrasonic Cleaning', *Metal Ind.*, **96**, 355 (1960); 'Ultrasonic Cleaning', *Prod. Finish., Lond.*, **13** No. 5, 88 (1960)

Flame Cleaning
Widdecombe, J. H., 'Economics of Flame Cleaning in Shipyards', *Paint J.*, **10** No. 76, 388 (1959)

12.2 Pickling in Acid

Mechanism of Scale Removal from Steel with Acid

When mild steel is heated in air at between 575 and 1 370°C an oxide or scale forms on the steel surface. This scale consists of three well-defined layers, whose thickness and composition depend on the duration and temperature of heating. In general, the layers, from the steel base outwards, comprise a thick layer of wüstite, the composition of which approximates to the formula FeO, a layer of magnetite (Fe_3O_4), and a thin layer of haematite (Fe_2O_3).

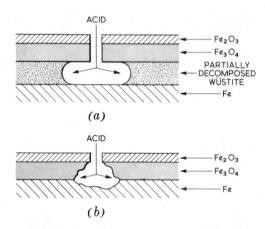

Fig. 12.4 Mechanism of scale removal with acid. (a) High-temperature scale and (b) low-temperature scale

When the steel is rapidly cooled, the thickness and composition of these layers remain more or less unchanged, but when it is slowly cooled through 575°C the scale becomes enriched in oxygen and the remaining wüstite layer breaks down to some extent into an intimate mixture of finely divided iron and magnetite[1].

Holding of the temperature between 400 and 575°C causes the iron particles to coagulate and the scale becomes further enriched in oxygen. Since wüstite is unstable below 575°C, scales produced at temperatures lower than this contain magnetite and haematite only[1]. In addition, the scales are often cracked and porous. This is due to the difference in contraction between

12:16

scale and metal on cooling and to the change in volume when the metal is oxidising.

When a steel which has been slowly cooled through 575°C is immersed in mineral acid, the acid penetrates through the cracks and pores in the upper layers of scale and rapidly attacks the decomposed wüstite layer, thus releasing the relatively insoluble magnetite and haematite layers (Fig. 12.4). This rapid dissolution of the wüstite layer is due to the setting up of many minute electrolytic cells between the finely divided iron particles, magnetite and acid. The iron, being anodic, dissolves to form ferrous ions, and the magnetite, being cathodic, is reduced, forming more ferrous ions. Since the three constituents of these cells are good electrical conductors, the resistance of the cells is so small that the rate of dissolution of the decomposed wüstite layer is largely governed by the rate at which acid diffuses through the cracks to it, and the rate at which spent acid diffuses from it. A similar but slower action occurs between the exposed metal and the magnetite and haematite layers which have not been detached[2].

The pickling rate of steels which have been rapidly cooled or held between 400 and 575°C is slower. This is due in the former case to the absence of the iron/magnetite cell action, and in the latter to the increased cell resistance resulting from coagulation of the iron. Similarly, the pickling rate of steels scaled at temperatures below 575°C is slow, because the resistance of the few larger cells formed between the magnetite and the base metal is high.

Apart from this cell mechanism in the scale, and between metal and scale, another cell action occurs on the exposed steel surface. In this ferrous ions are produced at the anodic areas and hydrogen at the cathodic areas.

Hydrogen Embrittlement

Although the majority of the hydrogen produced on the cathodic areas is evolved as gas and assists the removal of scale, some of it diffuses into the steel in the atomic form and can render it brittle. With hardened or high-carbon steels the brittleness may be so pronounced that cracks appear during pickling. Austenitic steels, however, are not so subject to embrittlement.

If the acid contains certain impurities such as arsenic, the arsenic raises the overvoltage for the hydrogen evolution reaction. Consequently, the amount of atomic hydrogen diffusing into the steel, and the brittleness, increase.

As well as causing brittleness, the absorbed gas combines to form molecular hydrogen on the surface of inclusions and voids within the steel. Thus a gas pressure is set up in the voids and this may be sufficient to cause blisters to appear either during pickling or during subsequent processing such as hot-dip galvanising.

The embrittlement effect can be largely removed by ageing the steel at about 150°C, but even then the original ductility is not entirely restored.

In the estimation of the degree of embrittlement, the temperature and rate of testing have an important effect. Thus the embrittlement tends to disappear at very low and very high temperatures, and it is reduced at high strain rates.

Several theories of the mechanism of embrittlement have been put forward[3-5] and further details are given in Section 8.4.

Acids Used for Pickling

Before steel strip or rod can be cold rolled, tinned galvanised, or enamelled, etc. any scale formed on it by previous heat treatment must be removed. This can be done by mechanical and other special methods, but if a perfectly clean surface is to be produced, acid pickling is preferred, either alone or in conjunction with other pretreatment processes.

Sulphuric Acid

Sulphuric acid is used to a very large extent for pickling low-alloy steels. The rate at which it removes the scale depends on (a) the porosity and number of cracks in the scale, (b) the relative amounts of wüstite, decomposed wüstite, magnetite and haematite in the scale, and (c) factors affecting the activity of the pickle.

Temperature is the most important of the factors affecting pickle activity. In general, an increase of 10°C causes an increase in pickling speed of about 70%. Agitation of the pickle increases the speed since it assists the removal of the insoluble scale and rapidly renews the acid at the scale surface. Increase in acid concentration up to about 40% w/w in ferrous-sulphate-free solutions, and up to lower concentrations in solutions containing ferrous sulphate, increases the activity. Increase in the ferrous sulphate content at low acid concentrations reduces the activity, but at 90–95°C and at acid concentrations of about 30% w/w it has no effect.

For economic reasons, continuous wide mild steel strip must be pickled within 0·5–1·0 min. To achieve this, the strip is flexed to increase the number of cracks in the scale and then passed through four or five long tanks. Acid of 25% w/w strength enters the last tank and flows countercurrent to the strip, and finally emerges from the first tank as waste pickle liquor containing about 5% w/w acid and almost saturated with ferrous sulphate. To increase the activity of the pickle to a maximum, live steam is injected to agitate the pickle and to raise the temperature to about 95°C. The scale on the edges of the strip and on the leading and trailing ends is usually more difficult to pickle than that in the centre. Consequently, whereas the centre scale is removed in the first or second tank, the remainder is removed only in the last tank. After pickling, the strip is thoroughly rinsed and dried.

For the batch pickling of rod, sheet, tube or strip in coil form, short pickling times are not so important, and pickling times of several minutes at 60–80°C in 5–10% w/w acid are common. The acidity is maintained by the addition of fresh strong acid, until the pickle is nearly saturated with ferrous sulphate, and then the acidity is worked down to 1 or 2%.

Electrolytic pickling

Anodic pickling   Sulphuric acid is also used in electrolytic pickling. Anodic pickling, which is suitable only for lightly scaled steel, has the advantage that no hydrogen embrittlement is produced, but the base metal is attacked as the scale is being removed. Under 'active' conditions the steel is rapidly attacked, and the surface is left rough and covered with smut. Under 'passive' conditions the attack on the steel is reduced, and the evolved oxygen mechanically removes the smut and other surface contaminants, leaving the steel with a clean satin finish which provides good adhesion for electrodeposits.

Cathodic pickling Cathodic pickling protects the base metal from acid attack while the scale is being reduced to spongy iron, but there is a danger of hydrogen embrittlement, and particularly if the acid contains arsenic. In the Bullard–Dunn[6] process the steel is made cathodic at $0.065\,A/cm^2$ in hot 10% w/w acid containing a trace of tin or lead. The tin or lead plates out on the descaled areas as a very thin coating, and owing to the high hydrogen overvoltage of these metals the formation of hydrogen ceases and so the current is diverted to the remaining scaled areas. The tin or lead can be removed by means of anodic alkali treatment, or may be left on and used as a base for subsequent painting.

A.C. pickling Alternating current pickling can be used where it is difficult to feed current into the steel by direct electrical connection, e.g. in the case of strip moving at high speed. In this process the electrodes are placed above and below the strip and so while one face of the strip is anodic the other is cathodic, the polarity being reversed during each cycle of alternation. The application of 0.11–$0.16\,A/cm^2$ to strip in 10% v/v H_2SO_4 at $88°C$ has been claimed to increase the pickling speed by 35%[7]. Alternatively, cathodic/anodic pickling may be employed on moving strip without the use of contacts. The moving strip is made cathodic with anodes in one tank and anodic with cathodes in the following tank, the strip itself carrying the current from tank to tank.

Hydrochloric Acid

Although hydrochloric acid is more expensive than sulphuric acid, it is gradually replacing the latter for pickling mild-steel strip, because the waste liquor can be recovered more economically. It is more active than sulphuric acid at an equivalent concentration and temperature, probably because the rates of diffusion of acid to, and ferrous ions from, the steel surface are greater. Consequently it is used cold for pickling in open tanks and for high-speed pickling of mild steel strip it is used hot in covered tanks to prevent loss of acid by volatilisation. It is more suitable than sulphuric acid for pickling articles which have to be tinned or galvanised since it gives less smut on the steel. In addition, any residual iron chloride left on the steel can be rinsed off more readily than residual iron sulphate deposits. Hydrochloric acid, however, readily dissolves the detached magnetite and haematite and, consequently, the ferric ion produced increases the rate of attack on the steel and thus increases the acid consumption.

Phosphoric Acid

Although phosphoric acid can be used for pickling steel, it is seldom used simply for scale removal since it is so expensive and slow in action. Steel plates are often initially descaled in sulphuric acid and then, after rinsing, immersed in 2% phosphoric acid containing 0.3–0.5% iron at $85°C$ for 3–$5\,min$[8]. The plates are then allowed to drain and dry without further rinsing. This treatment produces a grey film of iron phosphates on the steel surface, which provides a good base for subsequent painting.

Nitric Acid

Nitric acid does not dissolve scale so readily as mineral acid. A cold 5% w/v nitric acid solution is used to etch bright mild-steel strip when the smut resulting from the acid attack is easily and completely removed with a light brushing. It is also used in conjunction with sulphuric acid for cleaning bright annealed strip, which is difficult to pickle in mineral acid. This difficulty arises when certain types of rolling lubricant have not been thoroughly removed before annealing. During annealing, these lubricants polymerise to gum-like materials which are unattacked by mineral acid but are oxidised and removed with nitric acid. For this type of steel, pickling in a bath containing 20% w/v H_2SO_4 and 4% w/v HNO_3, with a trace of HCl, at 30°C for 4–6 min has been recommended[9].

Pickling of Alloy Steels

The furnace scales which form on alloy steels are thin, adherent, complex in composition, and more difficult to remove than scale from non-alloy steels. Several mixed acid pickles have been recommended for stainless steel, the type of pickle depending on the composition and thickness of the scale[10]. For lightly-scaled stainless steel, a nitric/hydrofluoric acid mixture is suitable, the ratio of the acids being varied to suit the type of scale. An increase in the ratio of hydrofluoric acid to nitric acid increases the whitening effect, but also increases the metal loss. Strict chemical control of this mixture is necessary, since it tends to pit the steel when the acid is nearing exhaustion. For heavy scale, two separate pickles are often used. The first conditions the scale and the second removes it. For example, a sulphuric/hydrochloric mixture is recommended as a scale conditioner on heavily scaled chromium steels, and a nitric/hydrochloric mixture for scale removal. A ferric sulphate/hydrofluoric acid mixture has advantages over a nitric/hydrofluoric acid mixture in that the loss of metal is reduced and the pickling time is shorter, but strict chemical control of the bath is necessary.

Electrolytic pickling of stainless steel in 5–10% w/v sulphuric acid at 50°C can be used for removing the majority of the scale. The strip is first made anodic, when a little metal dissolves, and then cathodic, when the evolved hydrogen removes the loosened scale. To complete the pickling, a nitric plus hydrofluoric acid dip is given for austenitic steels and a nitric acid dip for ferritic steels. Austenitic and ferritic stainless steels are not subject to hydrogen embrittlement with reducing acids, but steels of relatively high carbon content in the hardened state may be.

Organic Inhibitors

During the pickling of scaled steel the thinner and more soluble scale is removed before the thicker and less soluble scale. Consequently, some exposed base metal is attacked before the pickling operation is complete. In order to reduce this acid attack to a minimum, organic inhibitors are used. Their use also leads to less acid being consumed and less smut and carbonaceous matter is left on the steel. Because of the reduced hydrogen evolution, the amount of acid spray and steam consumption are also reduced.

Although a good inhibitor reduces the acid attack, it does not prevent the attack of oxidising agents on the exposed base metal. Thus the ferric ions resulting from the gradual dissolution of the detached magnetite and haematite attack the exposed steel even in the presence of an inhibitor, and are reduced to ferrous ions.

The inhibitor should not decompose during the life of the pickle nor decrease the rate of scale removal appreciably. Some highly efficient inhibitors, however, do reduce pickling speed a little. It would be expected that since the hydrogen evolution is reduced the amount of hydrogen absorption and embrittlement would also be reduced. This is not always the case; thiocyanate inhibitors, for example, actually increase the absorption of hydrogen.

Since inhibitors form insulating films on the steel, they interfere with any subsequent electroplating. In many cases, however, the films can be removed prior to plating by anodic cleaning or by a nitric acid dip.

Surface-active agents are often added to the pickle if the inhibitor has no surface-active properties. They assist the penetration of the acid into the scale, reduce drag-out losses, and form a foam blanket on the pickle. This blanket reduces heat losses and cuts down the acid spray caused by the hydrogen evolution.

Many organic substances soluble in acid or colloidally dispersible have been shown to have inhibiting properties. The most effective types contain a non-polar group such as a hydrocarbon chain and a polar group such as an amine. They contain oxygen, nitrogen, sulphur, or other elements of the fifth and sixth groups of the Periodic Table. They include alcohols, aldehydes, ketones, amines, proteins, amino acids, heterocyclic nitrogen compounds, mercaptans, sulphoxides, sulphides, substituted ureas, thioureas and thioazoles.

The efficiency of an inhibitor under a given set of conditions is expressed by the formula

$$ I = \frac{A - B}{A} \times 100 $$

where I is the per cent inhibition efficiency, A the corrosion rate in un-inhibited acid, and B the corrosion rate in acid containing a certain concentration of inhibitor.

In general the efficiency increases with an increase in inhibitor concentration—a typical good inhibitor gives 95% inhibition at a concentration of 0·008% and 90% at 0·004%. Provided the inhibitor is stable, increase in temperature usually increases the efficiency although the actual acid attack may be greater. A change in acid concentration, or in type of steel, may also alter the efficiency. Thus, the conditions of a laboratory determination of efficiency should closely simulate the conditions expected in commercial practice.

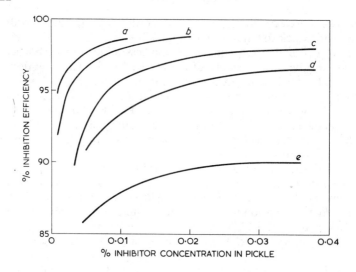

Fig. 12.5 Relationship between % inhibitor efficiency and inhibitor concentration in 6% w/w H₂SO₄. Curve (a) di-o-tolyl thiourea; (b) mono-o-tolyl thiourea; (c) commercial inhibitor containing 20% di-o-tolyl thiourea; (d) commercial inhibitor containing 20% di-phenyl thiourea; (e) gelatin

Table 12.4 Pickling solutions for non-ferrous metals

Metal	Acid	Temperature	Time
Copper and brass (60–90% Cu, 10–40% Zn)	7–25% w/w H₂SO₄ or 15–25% w/w HCl	15–60°C 15°C	1–10 min 1–3 min
Aluminium bronze (82–95% Cu, 5–10% Al, 0–5% Fe, 0–5% Ni)	Scale conditioned with 10% w/w NaOH followed by H₂SO₄ or HCl as above	75°C	2–5 min
Copper–silicon alloys (96–97% Cu, 1–3% Si)	7–25% w/w H₂SO₄ + 1–3% w/w HF	15°C	1–5 min
Nickel–copper alloys (55–90% Cu, 10–30% Ni, 0–27% Zn)	10% w/w HCl + 1·5% w/w CuCl₂	80°C	30 min
Nickel–chromium alloys (35–80% Ni, 16–20% Cr, 0–45% Fe, 0–2% Si)	Scale conditioned with 20% NaOH + 5% w/w KMnO₄ followed by 20% w/w HNO₃ + 4% w/w HF	100°C 50°C	1–2 h 5–30 min
Aluminium alloys (0–10% Cu, 0–10% Mg, 0–6% Zn, 0–12% Si)	25% w/w H₂SO₄ + 5% w/w CrO₃ 40% w/w HNO₃ + 1–5% w/w HF	65°C 15°C	20 min 1–5 min
Magnesium alloys (0–10% Al, 0–3% Zn, 0–0·2% Mn)	10–20% w/w CrO₃ + 3% w/w H₂SO₄	100°C 25°C	1–30 min 15 s

Figure 12.5 shows the relationship between efficiency and concentration of some thiourea derivatives and gelatin in the pickling of cold-reduced and annealed strip in 6% w/w sulphuric acid at 85°C. The thiourea derivatives, diluted with sodium chloride, gelatin and a wetting agent, are used commercially. Mono- and di-o-tolyl thioureas are stable in this pickle for at least 50 h, but diphenyl thiourea and gelatin decompose after four or five hours.

Inorganic Inhibitors

Inorganic inhibitors are salts of metals having a high hydrogen overvoltage, e.g. antimony and arsenic. The inhibiting action is associated with the formation of a coating of the metal, which, being cathodic to the steel and having a high hydrogen overvoltage, prevents the discharge of hydrogen ions and so stops the dissolution of the steel. These inhibitors are seldom used in commercial practice, but antimony chloride dissolved in concentrated hydrochloric acid is used in the laboratory for stripping deposits of zinc, cadmium, tin and chromium from steel, and with the addition of stannous chloride for removing scale and rust[11].

Further details of inhibitors for acid solution are given in Section 18.2.

Acid Pickling of Non-ferrous Metals

Table 12.4 summarises the pickling conditions for removing oxide and scale from some of the more important non-ferrous metals and alloys.

W. BULLOUGH

REFERENCES

1. Pfeil, L. B., *J. Iron St. Inst.*, **123**, 237 (1931)
2. Winterbottom, A. B. and Reed, J. P., *J. Iron St. Inst.*, **126**, 159 (1932)
3. Zapffe, C., *Trans. Amer. Soc. Metals*, **39**, 191 (1947)
4. Petch, N. J. and Stables, P., *Nature, Lond.*, **169**, 842 (1952)
5. Morlet, J. G., Johnson, H. H. and Troiano, A. R., *J. Iron St. Inst.*, **189**, 37 (1958)
6. Fink, C. G. and Wilbei, T. H., *Trans. Electrochem. Soc.*, **66**, 381 (1934)
7. Neblett, H. W., *Iron St. Engr.*, **16** No. 4, 12 (1939)
8. Footner, H. B., Iron and Steel Institute, 5th Report of the Corrosion Committee, London (1938)
9. Liddiard, P. D., *Sheet Metal Ind.*, **22**, 1 731 (1945)
10. Spencer, L. F., *Metal Finish.*, **52** No. 2, 54 (1954)
11. Clarke, S. G., *Trans. Electrochem. Soc.*, **69**, 131 (1936)

12.3 Chemical and Electrochemical Polishing

Electrolytic Polishing

Unlike mechanical polishing which works the surface and causes incorporation of grease, oxide and polishing compound into the surface, electrolytic polishing removes the surface layers without mechanical or thermal damage and without contamination. Mass-produced small articles or components of complex shape that are difficult to polish mechanically are readily polished electrolytically. Mechanical barrel polishing removes exposed burrs but electrolytic polishing has the advantage of deburring and smoothing the internal surfaces of holes, slots and recesses.

The smooth, clean surface produced by electrolytic polishing has inherently advantageous properties which are exploited in many industrial applications. These include non-stick, easy to clean, free-rinsing surfaces for the chemical, photographic and food processing industries, surfaces of high vacuum equipment where the smooth surface reduces adsorption and therefore outgassing time, non-arcing and high a.c. conducting surfaces of h.t. and h.f. electrical equipment, surfaces of cryogenic equipment where the reduction in absolute surface area markedly reduces evaporation of coolant, pharmaceutically clean surfaces of surgical instruments, etc. In certain cases electrolytically polished surfaces have the further advantage of somewhat better corrosion resistance compared with mechanically polished surfaces.

Electrolytic polishing processes cover bright dipping, chemical polishing and electropolishing, and the term 'brightening' is often used instead of polishing to describe these processes. More strictly, polishing consists of brightening together with levelling of the surface, although under certain conditions it is possible to have either separately. The degree of brightening or measured specular reflectivity depends on the composition and metallurgical condition of the material and the polishing conditions. All grades of metals and alloys can be electrolytically polished to a certain degree but special grades have been developed for optimum polishing. Design is important as welds and joins tend to be emphasised as do defects on large flat areas. For maximum reflectivity some prior mechanical polishing is advantageous to cut down electrolytic polishing time and reduce metal loss.

Equipment for electrolytic polishing is similar to that used for plating or anodising and requires similar solution control, fume extraction and effluent treatment. Degreasing and cleaning are required before electrolytic polishing and in some cases passivation afterwards can further improve corrosion resistance.

Bright Dipping

Bright dipping is simpler and cheaper than chemical polishing or electropolishing, and simple immersion in a suitable acid mixture will result in a dissolution process that gives an evenly bright but not fully specularly reflecting surface. Bright-dip solutions are usually mixtures of cold nitric and sulphuric acid. Commercial and patented solutions may also contain inhibitors to suppress pitting and etching tendencies, wetting agents and sequestering agents. Similar additions are made to sodium hydroxide solutions used to pretreat aluminium prior to chemical polishing. Suppression of pitting and slight smoothing action reduce subsequent chemical polishing time. Sometimes orthophosphoric acid is added to acid bright-dip solutions to confer a degree of levelling action or chemical polishing.

Transfer and immersion times are kept to a minimum to avoid etching and pitting. The rate of metal dissolution is high and, in general, is increased by a decrease in water or oxidising acid content or by an increase in temperature.

Chemical Polishing

Chemical polishing provides a fully specularly reflecting surface and has the advantage over electropolishing that immersion only is required, without the need for an external cathode, rectifier and applied current. However, it is more difficult to formulate suitable solutions and to control them, although chemical polishing is widely used prior to anodising for brightening aluminium for decorative purposes. Chemical polishing solutions for aluminium are usually based on mixtures of concentrated phosphoric and sulphuric acid operated at 95–100°C. The degree of brightening can be improved by controlled additions of copper either to the aluminium or more usually to the solution (about 0·5 g/l). Maximum reflectivity is reached in 2–5 min, and more prolonged treatment leads to non-uniform attack and a loss of brightness. Copper deposited on the surface is removed by dipping in nitric acid. The degree and rate of brightening also depends on the amount of aluminium dissolved in the solution.

Electropolishing

Although aluminium, copper, nickel and their alloys are electropolished, most industrial electropolishing is of stainless steel. Several commercial and patented solutions are available. Typical compositions include:

(*A*)

H_3PO_4 (s.g. 1·70)—500 ml/l
H_2SO_4 (s.g. 1·84)—450 ml/l
60–90°C
20–30 A/dm^2

(*B*)

H_3PO_4 (s.g. 1·70)—600 ml/l
CrO_3 —150 g/l
35–75°C
1–10 A/dm^2

The operating conditions required for electropolishing depend on the composition of the stainless steel. The solutions must be carefully controlled and the iron content limited to 4%.

Mechanism of Electrolytic Polishing

Electrolytic polishing is an anodic process, in which brightening results from the suppression of crystallographic etching. The mechanism involves the formation on the anode surface of a compact solid film of a composition and structure that allows metal cations to leave it at the film/solution interface at the same rate as they enter it. Levelling results from the preferential removal of surface asperities, assisted by the influence of a relatively thick liquid film that controls mass transfer. Electrolytic polishing solutions are formulated and operated to give the requisite films and balanced rate of cation transport.

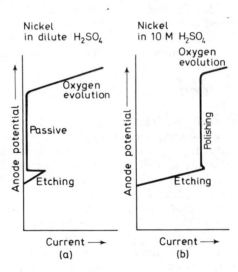

Fig. 12.6 Diagrammatic anode potential/current curves showing relationships between passivity, etching and polishing for (*a*) nickel in dilute H_2SO_4 and (*b*) nickel in 10 M H_2SO_4

Too low a rate of solid film formation may lead to pitting; too low a rate of film dissolution leads to film growth and passivation. Thus alloy composition, the presence of impurities, the electrochemical potential, the applied current density, temperature and all other operating conditions, have to be determined and controlled. While it is easier to polish pure metals electrolytically it is possible, with skill, to adjust the conditions to polish impure metals and alloys to a certain degree. Potentiostatic control has proved helpful in selected cases, particularly in the laboratory for metallography.

The mechanism of chemical polishing is similar to electropolishing: generally the solutions are strongly oxidising to produce the requisite films at the polishing surface. Anodic dissolution takes place through the film; the cathodic reaction, usually hydrogen evolution, occurs on its surface.

Hoar, *et al.*[1], on the basis of *E* vs. *I* curves, have drawn attention to the similarity between anodic passivity and brightening (Fig. 12.6).

D. J. ARROWSMITH

REFERENCE

1. Hoar, T. P., Mears, D. C. and Rothwell, G. P., 'The Relationships Between Anodic Passivity, Brightening and Pitting', *Corros. Sci.*, **5**, 279 (1965)

BIBLIOGRAPHY

Hoar, T. P., 'Anodic Polishing', from *Modern Aspects of Electrochemistry*, Edited by J. O'M Bockris, Butterworths, London, 313 *et seq.* (1959)
Surface Finishing Stainless Steel, Information Book No. 1, Stainless Steel Development Association, London
Tegart, W. J. McG., *The Electrolytic and Chemical Polishing of Metals in Research and Industry*, Pergamon, London, 2nd edn (1959)
Wernick, S. and Pinner, R., *The Surface Treatment and Finishing of Aluminium and its Alloys*, Robert Draper, London, 3rd edn (1964)

12.4 Design for Corrosion Protection by Electroplated Coatings

Introduction

The use of electrodeposited metals to protect corrodible basis metals from their service environments has been well established for many years and accounts for by far the larger part of the activities of the plating industry. There are many reasons for using an electroplated metal finish in preference to an organic finish or to making the articles concerned from inherently corrosion-resistant materials.

Some of the reasons are aesthetic, and on many larger articles, such as motorcars, an attractive appearance is achieved by a careful visual balance between parts which are finished and protected by electroplating, and parts which are protected by organic finishes. In many other instances the manufacture of parts from corrosion-resistant materials is ruled out because of the relatively high cost of such materials, and in some cases the physical properties of the appropriate corrosion-resistant materials may render them either unsuitable for economic production or unable to perform the function for which the article is required.

Satisfactory service of an electroplated article is not achieved, however, unless adequate care is given to the choice of deposited metal, the thickness of the deposit, the technique of application, and the design of the article. The choice of metal deposit is primarily determined by the basis metal, i.e. the metal from which the article is made, and the actual conditions to which the plated article will be subjected during service. In addition, however, attractive appearance and reasonable cost are also important considerations.

Deposit Thickness

Deposit thicknesses appropriate to various conditions of service are laid down in a number of specifications. Those of most general application are the standards issued by the British Standards Institution recorded in Table 12.5.

The protection of screw threads by electroplated coatings presents peculiar problems arising from the geometry of threads and the components of which the threaded portion is a part. The first six parts of BS 3382 specify thicknesses which may be electroplated on to screw threads of standard form, dimensions, tolerances and plating allowances, without risk of interference on assembly and which will provide a degree of protection adequate for many purposes. Part 7 of BS 3382 is devoted to special methods of providing protection by electroplated coatings against more severe corrosive environments.

All the other standards stipulate a minimum local thickness, i.e. a value below which the deposit thickness should never fall at any part of the significant surface. The significant surface is defined as the surface over which satisfactory corrosion resistance must be achieved. The standards also stipulate that the minimum thickness requirement shall apply only to those parts of the surface which can be touched by a ball of 1 in (or 20 mm) diameter.

Table 12.5 British Standards for electrodeposited coatings

BS 1224	Electroplated Coatings of Nickel and Chromium
BS 1706	Electroplated Coatings of Cadmium and Zinc on Iron and Steel
BS 1872	Electroplated Coatings of Tin
BS 2816	Electroplated Coatings of Silver for Engineering Purposes
BS 3382	Electroplated Coatings on Threaded Components:
	Part 1 Cadmium on Steel Components
	2 Zinc on Steel Components
	3 Nickel or Nickel Plus Chromium on Steel Components
	4 Nickel or Nickel Plus Chromium on Copper and Copper Alloy (Including Brass) Components
	5 Tin on Copper and Copper Alloy (Including Brass) Components
	6 Silver on Copper and Copper Alloy (Including Brass) Components
	7 Thicker Platings for Threaded Components
BS 3597	Electroplated Coatings of 65/35 Tin–nickel Alloy
BS 4292	Electroplated Coatings of Gold and Gold Alloy
BS 4641	Electroplated Coatings of Chromium for Engineering Purposes
BS 4758	Electroplated Coatings of Nickel for Engineering Purposes

There are two reasons for making this reservation in the minimum thickness requirement—and both of them are practical. In the first place, none of the methods of measuring deposit thickness quickly, e.g. by recording the time required to penetrate the deposit by the impingement of a jet of a suitable solution (known as the BNF jet test), can normally be applied to recesses such that a 1 in (or 20 mm) ball cannot touch all the surface. (The thickness could, of course, be measured by sectioning the article and examining it under the microscope, but such a procedure would destroy the article, and also take so long to perform that it could not be used for satisfactory control of the plating process.) Secondly, a recess which cannot be penetrated by a 1 in (or 20 mm) ball cannot normally receive an adequate thickness of deposit unless a grossly excessive thickness of metal is deposited on other parts of the article.

Deposit Distribution

An electrodeposited coating is never, in practice, completely uniform in thickness. The actual thickness of metal deposited at a particular point in a given time is dependent on the current density at that point, and the current density is not uniform over the whole surface of an article being plated—it tends to be highest at sharp points, edges and corners (see Fig. 12.7) and lowest in re-entrant corners and recesses [see Fig. 12.7(b) and (d)]. Even a flat plate will not have a uniform deposit, since the current density will be higher towards the edges than in the centre, and highest of all at the corners [see Figs. 12.7(a) and 12.7(c)]. Figure 12.7(d) shows diagrammatically the sort of deposit distribution which is experienced at a slot in a plane surface which has been plated.

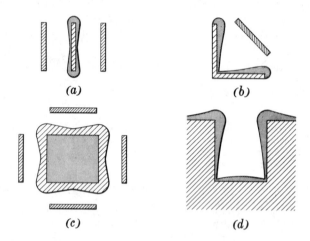

Fig. 12.7   Relation between current density and plating thickness

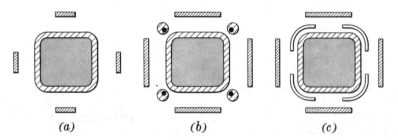

Fig. 12.8   Methods of securing more uniform current distribution

The distribution of current, and hence of the electrodeposited metal, can be made more uniform in several different ways. Sharp corners should always be avoided and curved corners used instead (see Figs. 12.8 and 12.10). In many instances plating anodes which conform approximately to the shape of the article being plated or which are suitable in size and carefully placed in relation to the workpiece will greatly improve the deposit distribution [see Figs. 12.8(a) and 12.9]. Another method of effecting an improvement is to suspend 'burners' or 'robbers' near projections on the work to be plated [see Fig. 12.8(b)]. These burners are electrically connected to the workpiece and take a high proportion of the metal deposited in regions of high current

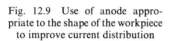

Fig. 12.9 Use of anode appropriate to the shape of the workpiece to improve current distribution

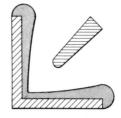

density. Better than the use of burners is the use of plastics or other non-conducting shields [see Fig. 12.8(c)]. These shields placed near projections which would otherwise be regions of high current density impede the current and solution flow to them and hence result in a more uniform distribution of current and deposited metal.

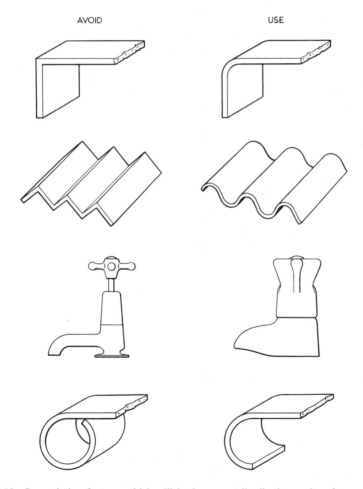

Fig. 12.10 Some design features which will lead to poor distribution and performance of electroplated coatings, and alternatives which will yield satisfactory results

If conforming anodes or plastics shields are to be effective they must be fairly accurately located in relation to the workpiece. The practice, common in many plating shops, of wiring the work directly to the electrical bus bars over the plating tanks is, therefore, not satisfactory. Instead, the work should be mounted on rigid jigs or racks which are then suspended from the bars over the tanks.

Some electroplating solutions produce more uniform deposits than others, and the ability of a solution to deposit uniform coatings is measured by its 'throwing power' (see Section 13.1).

Design Considerations

Uniformity of Deposit

It is clear that for satisfactory service and resistance to corrosion, it is necessary not only to choose the right metal to be electrodeposited, and the right thickness of deposit, but also to ensure that the design of the basic article to be plated is such that a reasonably uniform deposit can be achieved without too much difficulty. The more complex the shape to be plated, the higher will be the cost, either because the total weight of metal over the whole article will have to be greater in order to provide the minimum thickness in the most difficult places, or because special plating jigs and anodes will be required to achieve a uniform deposit. The designer should therefore strive to avoid deep or narrow recesses, sharp edges and corners, sharp points, and generally complex shapes; he should aim to provide relatively simple shapes without sudden changes of contour or cross-section, and generous radii on all corners and edges, both external and re-entrant. Ideally, the worse the throwing power of the solution to be used for electroplating, the more simple should be the shape to be plated. Figure 12.10 illustrates some of the points mentioned.

Surfaces other than significant ones   The ease of achievement of a reasonably uniform deposit over the significant surface is, however, not the only way in which the design can influence the corrosion resistance of plated articles. Although the part of the surface which must be protected from corrosion is described as the significant surface, the remainder cannot be considered of no importance. Some protection must be given to such surfaces to prevent severe corrosion which might ultimately result in penetration of the article from the back. It can usually be arranged that a thinner deposit is plated on the less important surfaces while the full specified thickness is plated on the significant surface. An alternative method is to protect the less important surface after plating in some other way, e.g. by painting, if it is not possible to ensure an adequate thickness of plating without materially increasing the cost.

Physical properties   The corrosion resistance of an electrodeposit depends not only on suitability for the service environment, thickness, and uniformity, but also on its physical and mechanical qualities such as adhesion to the basis metal, ductility, internal stress and porosity. All these properties depend primarily on the plating shop—the skill and care of the operators and the exercise of proper control over the actual plating processes. The designer can, however, influence these factors to some extent, and he can certainly help to make the plater's task easier. (The easier the plater's job, the more likely a satisfactory result is.) In order to ensure good adhesion of the electro-deposit to the basis metal, it is necessary to remove all grease, oxides and any other contaminants from the article to be plated. This can be difficult with articles of very complex shape. Furthermore, the processes and solutions used for cleaning are usually different for different basis metals, and hence it is often not easy (and sometimes not possible) to clean satisfactorily an article assembled from parts made of different materials. It is preferable to assemble such parts after plating.

High stress and low ductility of electrodeposited coatings are often the result of impurities in the electroplating solution. Such contamination of plating solutions is often the result of solutions being carried over from one process to the next in recesses and blind holes in work being plated. Con-tamination can also be caused in some cases by attack of the plating solution on parts of the article being plated which are not receiving a sufficient share of the total current to achieve an adequate deposited coating. This trouble can arise in deep recesses and holes, and particularly in the bores of tubes which are being plated externally (an extra anode running through the bore of the tube would be required to ensure a deposit inside). Tubes to be plated externally should always be rendered solution-tight if possible; alternatively provision should be made by suitably placed holes for easy rinsing and draining (see Fig. 12.11). Good platers, of course, purify their plating solu-tions at intervals to remove contamination, but any help designers can render in reducing the rate of contamination is obviously beneficial.

DRAIN AND AIR
HOLES

Fig. 12.11 Drainage holes in a tube to be plated externally

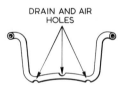

Some porosity is inherent in thin deposits of certain metals, e.g. chromium, nickel and tin, but this usually disappears with increasing thickness. Porosity can also be caused by inadequate cleaning and by the condition of the surface of the basis metal, i.e. its degree of roughness, its porosity and the extent to which it is free from inclusions. Hence the designer can help by ensuring that the materials used for manufacture are of good quality and that machining is to a reasonable surface finish.

Retention of corrosive media   Apart from their effect on deposit thickness, the construction and fabrication of plated articles can also influence corro-sion resistance by affecting the retention of dirt and liquids in certain parts during service. Thus if two straps crossing each other have to be joined it is better that the joint should be welded along each edge [Fig. 12.12(*a*)] rather than made by riveting or spot welding [Fig. 12.12(*b*)]. The latter arrange-ment allows process solutions to be trapped during plating, and rain, con-densed moisture and dirt to be trapped during use. The closed welded joint of Fig. 12.12(*a*) avoids all these troubles. (Spot-welded and riveted joints are, of course, extensively used and often with satisfactory results. The point to be made is that for *maximum* resistance to corrosion they should be avoided.) Similarly, any form of ledge, cup or recess which could trap processing solutions and thus impair the quality of the plate, or which could trap dirt or corrosive liquids and hence lead to premature failure, should be avoided.

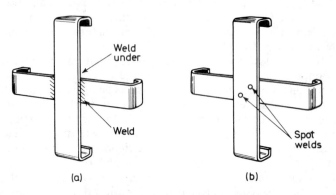

Fig. 12.12   Methods of joining straps. (*a*) Continuous welds prevent entrapment of process solution during plating and corrosive liquid during service, and (*b*) spot-welded joint leaves space in which process solution and corrosive liquids can be trapped

Reference may be made to BS 4479: *Recommendations for the Design of Metal Articles that are to be Coated*, Section 2 of which deals with electroplated coatings.

Conclusion

It will be seen that the design of articles to be electroplated can have a considerable effect on the corrosion resistance of the electrodeposited coating. The chief effects are the result of variations in deposit thickness, but also important are features which can influence the adhesion, porosity and physical properties of the deposit. Good design will also avoid features of the plated article capable of trapping liquids or solid contaminants which might cause more rapid corrosion.

D. N. LAYTON

12.5 Design for Corrosion Protection by Paint Coatings

Introduction

The fact that an article is to be finished by a painting process should be borne in mind at the design stage. Bad design, or perhaps even more often, bad styling, can make the painting process more difficult and costly or so reduce the efficacy of the applied coatings in their function of corrosion protection that premature failure can result. That this failure is usually only local to the area of bad design increases the feeling of waste and frustration engendered by the failure. Despite the economic penalties and sales resistance created by these design faults it is only necessary to examine a representative sample of sheet metal or cast consumer durables to find evidence that the needs of the finisher have been overlooked in the drawing office. The fact that the period since the first edition of this work has seen enormous advances in both finishing materials and their application makes this neglect of the influence of design on good finishing practice all the more lamentable.

In the ensuing pages, the elementary principles of design which must be observed if the resulting article is to be finished easily, economically and efficiently by large-scale production processes, will be considered. Although it is convenient to treat the subject under the headings of the various production processes the rules to be obeyed are of very general application and are relevant to a greater or lesser extent whatever the finishing process to be employed; thus, although the scope clause of BS 4479:1969 *Recommendations for the Design of Metal Articles that are to be Coated*[1] does not include painting as one of the processes which have been considered, the sections on conversion coatings and hot dipping may be studied with profit by the painter.

Pretreatment Processes and Design

These processes have been dealt with fully in Chapters 12 and 16. Here it is only necessary to state that, whatever the method of treatment, any liquids applied must be allowed to drain freely from the surfaces, and must not be allowed to become entrapped in crevices. The active solutions must be completely washed away so as to leave a clean uncontaminated surface for application of paint, and just as the treatment solutions must not be allowed to lodge in crevices or hollows, so neither must the wash waters. Special attention should be paid to articles which are built up from simpler structures and painted as assemblies. These may be constructed with tubes or channel sections used either as strengthening ribs, or as a skeleton to support the structure. Such constructional elements form ready traps for processing solutions. If these remain during the subsequent painting, they may leak out and cause damage before or after the article is put into service, or they may remain *in situ* to give rise subsequently to corrosion. Therefore such spaces should be sealed off before pretreatment. If a watertight seal—produced for example by welding a cap on to the end of a tube—cannot be achieved, the interior must be protected during assembly by a suitable plug. If it is necessary to treat the inner surfaces, drain holes must be provided so as to allow free drainage for the treatment solutions.

Accumulation of pretreatment liquid in the form of pools or runs leads to concentration of salts as the solution evaporates at the points of final drying. These act as foci for blistering or filiform corrosion under paint films. Design features which can give rise to such accumulations include too-rapid changes in contour, surface protuberances such as rivet or screw heads, lapped joints, undrained swages or recesses, or joints so formed that drainage is much slower from them than from the remainder of the article (see the section on dip painting, p. **12**.40). Where these undesirable features cannot be altogether avoided, the harmful effects may be counteracted to a considerable extent by careful hanging during processing and hand-wiping or compressed-air blow-off operations after the final pretreatment processes and before drying.

Attention to design should not stop at the design of the article to be treated. The design of jigs and holders for the work can play an important part in the production of efficient and attractive organic coatings. In the pretreatment stages jigs should hold the work at such an angle that the processing solutions can drain freely from all surfaces over as broad a front as possible. The formation of drainage lines should be avoided as these lead to an unwelcome concentration of liquid in the final stages of draining. If the jigs cannot be completely shielded from the processing solutions (the usual condition) deflector plates should be incorporated in their design so as to throw the drainings clear of the work. Drips from jigs are a frequent cause of small rings of tiny blisters which develop under paint coatings.

Jigs for immersion treatments must be sufficiently strong to keep work pieces liable to float completely immersed and the work should be held at such an angle that air traps are not formed and solutions can readily reach all the significant surfaces.

Designing for Painting[2]

Properties of Liquid Films

The designer of articles which are to be painted must take into account not only the actual method of application of the paint film, but also the fundamental fact that all paints are applied in the liquid form and later converted to a dried film.

Surface tension is the property of liquids which has the most significance in design for painting. When a liquid paint film is applied to a sharp edge, e.g. Fig. 12.13(a), the contractile forces of surface tension tend to drag the paint back from the edge in order to diminish the area of free surface, and even if the edge is not completely denuded of paint, as may often happen, the film will become dangerously thinned in this region.

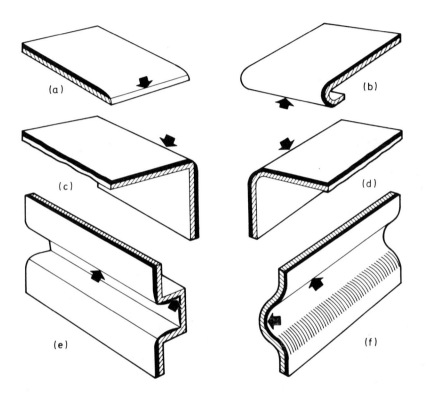

Fig. 12.13   Examples of good and bad design for paint finishing. (a), (c) and (e) bad design, (b), (d) and (f) good design (after *Design for Metal Finishing*)

The remedy for this is simple. Edges should be chamfered and not square-sheared, and where possible the sheet edge should be rolled or beaded; if these measures are too expensive, a simple fold is better than a square-sheared raw edge. By rolling, beading or folding over, the weakest portion of the paint film is shielded from the worst exposure to the environment [Fig. 12.13(b)]. Generous radii should be allowed or the rolls in their turn may act as solution traps.

A similar state of affairs is also found where a sharp angle or corner has been specified; here the conditions are worse than along an edge [Fig. 12.13(c) and (e)]. To avoid the dangerous film thinning in these regions generous radii should be provided at bends, whether internal or external, and edges and corners should be rounded and not left sharp. The preferred forms are shown in Fig. 12.13(d) and (f).

Effect of Surface Condition on Paint Films

As a paint film dries, whether by the evaporation of solvents at room temperature or by the driving off of the volatile constituents by heat, it shrinks. The shrinkage may be a considerable fraction of the wet film thickness, and at any given point is roughly proportional to the wet film thickness. When a paint film is applied to rough metal, the film, being mobile, will level out to form an unbroken, uniform surface, but the film thickness will vary at different points, being greatest over the valleys in the underlying surface and least over the peaks. When the film dries and shrinks, these differences in thickness will give rise to differing amounts of shrinkage and so reproduce on a diminished scale the contour of the original surface (Fig. 12.14).

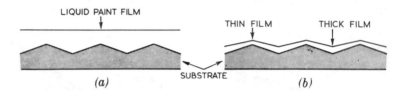

Fig. 12.14   Reproduction of original surface contour on reduced scale owing to film shrinkage. (*a*) Wet paint and (*b*) dried film showing differences in shrinkage over peaks and valleys (after *Design for Metal Finishing*)

The only remedy for this state of affairs is sanding the primer surface in order to produce a level ground on which to apply the subsequent paint coatings. This is expensive in labour and often results in dangerous weakening of the primer coating due to excessive reduction in thickness by the removal of too much stock. This may extend to cutting through the primer film at edges or raised portions of the surface being sanded. A semi-quantitative treatment of this phenomenon has been given by Lloyd and Stein[3], who suggest that a normally applied stoving primer or primer surfacer coating should adequately fill a surface of 750 nm roughness. There will always be some sanding to do, since deep scratches made during manufacture, and surface waviness, as distinct from roughness, may exceed this figure.

Designers should therefore study carefully the finishing processes and the standard of appearance required on the finished article. The apparent savings achieved by the specification of a cheaper grade of steel with a poorer surface finish than the 750–2 500 nm characteristic of deep drawing steel sheet may easily be swallowed up in the paint shop by excessive sanding and priming costs.

Castings The consumer-goods industries are making increasing use of castings which have to be finished to a high standard; both aesthetic and hygienic considerations combine to demand a smooth surface.

The surface quality of pressure diecastings in zinc or aluminium alloys is usually sufficiently good to cause little trouble in the paint shop. The same cannot be said of ferrous castings, whose method of manufacture must of necessity give poorer surfaces than those of the more tractable lower-melting-point metals. Nevertheless a stricter control of the foundry and the specification by the designer of a maximum permissible degree of surface roughness, could materially reduce the cost of the multiple filling and rubbing which are so characteristic of the preparation of ferrous castings.

In cases where the surface roughness cannot be economically removed by mechanical means or filled by successive coats of stopper or surfacer, the use of plastisols which contain no solvent or organosols containing only small amounts of solvent, should be considered. These compositions which dry with little or no shrinkage will not reproduce the contours of any but a very rough surface. Their use is limited by the fact that they must be cured at a sufficiently high temperature to allow the resin system to flow out to form a smooth surface.

Designing for Particular Methods of Finishing

Design for finishing must be considered in relation to two possible situations: (a) design for the use of a given production system with the maximum efficiency, and (b) choice by designer and production engineer of the most appropriate method of finishing for a particular article. However, design problems raised by the requirement to coat inaccessible surfaces may be eliminated by the use of pre-finished sheet or strip, which is obtainable in a wide variety of finishes, both metallic and organic, and the finish may be specified for application to one or both sides.

The available methods of paint application are fully treated in Section 15.1; in the ensuing paragraphs, the various methods will be discussed strictly in relation to design.

Spray Finishing

The basic rules of design for articles to be painted by spraying are little affected by the variations in equipment used, although the newer methods may allow certain relaxations.

The cloud of paint coming from a conventional spray gun, however used, may be imagined as spreading out in the form of a right circular cone with its apex at the gun tip. The particles are borne outward on an expanding air blast. The volume of air passing through an industrial-size gun is quite considerable, amounting to as much as 1 m^3/min of free air. If this expanding air blast is directed into a crevice or re-entrant angle it will be baffled by the proximity of the walls and the full depth of the crevice or re-entrant will not be coated. Electrostatic spraying offers some advantages in this respect, as, to a lesser extent, do hot spraying and airless spraying.

Thus in designing for spray painting sharp re-entrants should be avoided and replaced by shallow recesses with generous radii at the angles. In particular, joints of the form shown in Fig. 12.15(a) are very poor design features. Sprayed paint is most unlikely to penetrate the joint. At the best a weak bridge of paint without support will be produced, which will collapse under the natural movements occasioned by use or even under the stress of varying temperature. If such open joints are unavoidable the joint area including the overlapping surfaces should be primed before assembly and the gap filled with a plastisol which is cured before the colour coat is applied [Fig. 12.15(b)].

If this cannot be done it is often a good plan to specify that the joint shall be inspected and any paint bridge broken. The resulting damage can then be hand repaired.

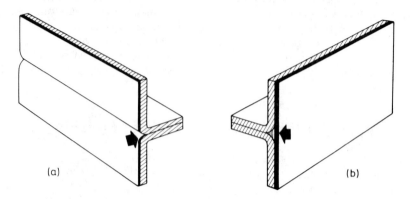

Fig. 12.15   Joint design. (*a*) Bad and (*b*) good (after *Design for Metal Finishing*)

Some articles built up from pressings or stampings contain pockets and recesses which have been created by the methods of jointing. During the course of manufacture it is inevitable that they will be subjected to many dust-producing operations such as disking or grinding the metal surfaces, levelling off soldered areas by filing, and, in the paint shop itself the sanding of primer and undercoats. The dust and dirt from these operations will tend to accumulate in the pockets or recesses, from which it is extremely difficult to remove. When the final glossy finishing coat is spray-applied this dust will be stirred up by the air blast from the spray gun. Some at least will be deposited on the wet paint film giving rise to a dirty or 'bitty' finish.

When designing for spraying, care should be taken that all such constructional pockets occur where the disturbed dust cannot fall on a significant surface, or are sufficiently open in structure to permit easy or complete cleaning. It is preferable for such spaces to be completely sealed so as to prevent the ingress of dust and manufacturing dirt. Electrostatic and airless spraying, which do not employ an air blast to atomise the paint into globules, will naturally cause less disturbance of trapped dirt than compressed-air spraying, but even if the use of these processes is envisaged, the presence of dirt in the paint shop should not be countenanced.

Very small articles are, in general, not suited for spraying as they would have to be handled and jigged individually and this is costly in labour. Such pieces are better coated by dipping or flow coating, or they can be processed in a perforated basket and coated by barrel painting or centrifugal painting.

Dipping or Flow Coating

These processes differ from one another only in the method of paint application. In dipping, the pieces to be coated are immersed bodily in a tank of paint, while in flow coating the articles are passed in front of jets which flood the surface with paint (in the related technique of curtain coating, they pass through a falling curtain of paint). In every case the surplus is allowed to drain away. The need for total immersion means that hollow ware or articles buoyant enough to float in a paint bath should not be dipped, unless the jigs and fastenings provided are strong enough to anchor them firmly and plunge them beneath the surface.

Similarly, designs which cannot be jigged so as to prevent the occurrence of airlocks when immersed in the paint bath, should be flow-coated rather than dipped, unless auxiliary devices such as directed jets of paint can be provided in the dip tank.

In general, the difficulties encountered in producing a uniform, unblemished surface by dipping, curtain coating or flow coating, arise during the draining period. Runs, sags or other film irregularities are caused by draining being interrupted by a sudden alteration in the contour of the surface over which draining is taking place. Typical examples of these interruptions are screw or rivet heads, stepped joints, holes and especially blind holes, which may hold up paint for a period and then release it when draining has been completed in the surrounding areas.

Riveted or screwed joints should be replaced by overlapped joints finished off to a smooth contour by welding or soldering (Fig. 12.16).

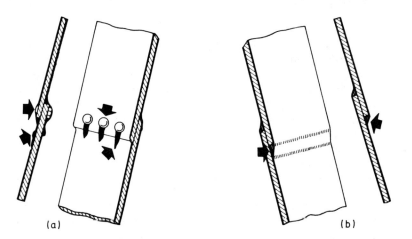

Fig. 12.16   Joint design. (*a*) Riveted or screwed joints should be avoided, (*b*) overlapped joints given smooth contour by dipping or soldering (after *Design for Metal Finishing*)

Spot-welded joints, despite their apparently inconspicuous appearance, should be avoided however, as they can be unpleasantly visible through a glossy paint film. It is possible by the use of large flat electrodes and other special devices to produce spot welds between two metal sheets without marking the face of one of them. Any breakdown in the conditions of welding may still lead to a state of affairs where spot-weld marks show through a dipped paint coat, and the method should be avoided unless one is prepared to grind the weld marks out of the surface.

The observations on spray painting of open joints apply with equal or even greater force to painting by dipping or flow coating. Bridging of the paint across the gap can occur with even greater ease in these cases than when application is by means of a spray gun.

Hollows and recesses are points demanding particular attention; they may fill with paint and be unable to empty before the paint film sets, or the paint trapped in them may remain liquid and drain out later. Recesses, hollows and swages should be provided with drain holes of such size that the paint can drain freely from them in step with the paint on adjacent surfaces. One large hole is more effective than a number of smaller holes of the same aggregate area.

A similar state of affairs may be created by the presence of tubes in the design. If these are open, draining may be prolonged and produce runs over the paint surface. It is good practice to specify that all tubes shall be sealed if dipping or, to a lesser extent, flow coating is contemplated as the method of finishing. If the tubes are for any reason required to be open, then temporary removable caps can be fitted during the painting operation. In this case, in order to provide adequate corrosion protection, the tubes should be painted before assembly. Dipping followed by draining in a vertical position is probably the simplest method of coating the interior of a tube.

Design for Miscellaneous Methods of Coating Small Objects

Coating a large number of comparatively small articles with a good and uniform layer of paint at a reasonable cost is one of the more difficult problems of metal finishing. In general, any method which would involve individual handling or jigging is too expensive in labour. This rules out dipping, flow coating or spraying. The simplest methods of handling large quantities of small articles are barrel painting and centrifugal painting. The design points requiring consideration if either method is to be successfully used are the same. Since the pieces are emptied into a container in random fashion and then tumbled over each other (in the case of barrelling) or spun (in the case of centrifuging) they must be sufficiently robust to withstand this treatment. When the paint barrel or centrifuge is emptied the pieces must readily separate or they will be cemented together by the paint as it hardens. Similarly, if the shapes are merely blanked out of flat sheet or are of such a form that they will nest readily, it will be impossible to produce an unblemished surface by these methods. Interlocking of the articles short of complete nesting may also be objectionable for the same reasons, but a small amount of interlocking may be permissible, depending on the degree of surface finish required and the position of the blemish which it causes.

Deep recesses, re-entrant areas and deep holes of small diameter may be missed by the barrelling or centrifugal methods, and these design features should be avoided.

Coating the Interior of Hollow Ware

The application of coatings to the interior of hollow ware which presents the condition of a large area to be coated and a small orifice through which solvent vapour can escape, should be avoided by, where possible, coating the parts of the article separately and then assembling by seaming or some similar non-destructive operation. A typical example of this type of product is a 20 litre can which if interior painted as a complete unit will have bare patches on the upper parts of the interior due to solvent washing. If the body and upper part are coated separately and then seamed together a perfect coating can be obtained.

A. A. B. HARVEY

REFERENCES

1. *Recommendations for the Design of Metal Articles that are to be Coated*, BS 4479 (1969)
2. *Design for Metal Finishing*, Institute of Metal Finishing, London (1960)
3. Lloyd, D. H. and Stein, W., 46th *Ann. Tech. Proc. Amer. Electropl. Soc.*, 104 (1959)

13   METHODS OF APPLYING METALLIC COATINGS

13.1	Electroplating	**13**:3
13.2	Principles of Applying Coatings by Hot Dipping	**13**:48
13.3	Principles of Applying Coatings by Diffusion	**13**:58
13.4	Principles of Applying Coatings by Metal Spraying	**13**:78
13.5	Miscellaneous Methods of Applying Metallic Coatings	**13**:92

13.1 Electroplating

Approach[1-7]

There are books and reviews dealing with electroplating which approach it as a commercial technology, a useful art or an interesting process for scientific enquiry. In this case the emphasis is on the production of coatings for the control of corrosion and the consequences for service corrosion of events at various stages of electroplating. Corrosion control is not the sole function of electroplated coatings; indeed most authors play down corrosion in favour of 'more positive' virtues such as decoration (sales appeal), reflectivity, wear resistance, solderability or low contact resistance. However, to preserve these properties it is essential to prevent corrosion; in essence, electroplated metal coatings are required to confer corrosion resistance *together* with one or more other property, the nature of which provides the criteria on which selection of a coating is made. Corrosion protection is a common factor.

Origin[8-16]

The production of electricity by metallic corrosion in Volta's pile (1786) was followed by the reduction of metal ions to metal by electrons at a metal/solution interface. Electroplating is an application of electrodeposition to produce thin, coherent, adherent coatings of solid metal, and is one of the important methods for producing coatings. Brugnatelli[9], a Professor of Chemistry at the University of Pavia, published an account of silver plating in 1800, and is said to have gold plated silver coins using Volta's pile as early as 1805. Wollaston deposited copper (presumably a very small quantity) on silver wire using current from an electrostatic generator. Electrodeposition was developed energetically in the succeeding three decades, and by 1840 commercial electroplating had advanced to the stage where patent protection was being sought for various processes. The main source of electricity for the first six or seven decades was the primary galvanic cell, and progress was aided by the invention of cells such as those of Daniell, Meidinger, Grove and Bunsen, which were capable of delivering large currents for long periods without polarisation. Faraday discovered electromagnetic induction in 1831 and Pixii used the commutator in 1832 to produce a direct current generator, but it was not until about 1870 that several independent inventors used electromagnetic field coils to generate the heavy currents needed for electrolysis. Large scale electroplating enterprises were established by the latter part of the 19th century. In 1883 the Postal Telegraph Company of New York used 200 copper plating cells to deposit 250 kg/d onto steel wire. Aqueous solutions have always been by far the most important electrolytes. There have been many investigations of non-aqueous solutions or ionic melts, and Davy deposited the alkali metals from the latter in 1807. Aqueous solutions, and their associated pre- and post-plating treatments are so well established that they are likely to remain pre-eminent for the immediate future.

Substrates

Metallic substrates which are good electron conductors and non-metallic ones which are not, can both be electroplated with adherent coatings, but the preliminary procedures or 'pretreatment' differ markedly. The commonly plated metal substrates are mild and low-alloy steels, zinc alloys for die-casting, and copper or high-copper alloys like brasses, bronzes and beryllium coppers. A large number of other alloys can be electroplated, but their use is restricted to particular industries, and they often require special pretreatments. Aluminium alloys, stainless steels, titanium alloys and refractory metals are examples. The three major substrates are plated with corrosion control as one of the important objects, but with less common substrates this may not be the case. There has been a rapid growth in electroplating parts made from plastics material. The main object is to make the part look metallic for reasons of appearance rather than function. A complicated acrylonitrile–butadiene–styrene copolymer dispersion (ABS) was the first plastics material to be plated on a large scale. It has been followed by polypropylene, and the development of plating grades of other plastics (e.g. polysulphone, polycarbonate, epoxide resins, glass-reinforced polyester) is in hand. Metallic corrosion problems arise as soon as plastics have been plated, creating a new field for the corrosion enthusiast to investigate.

By exercising ingenuity, most non-conducting objects may be electroplated, and this is a long-established small-scale operation. The usual aim is artistic or a search for novelty, a good example being the preparation of a heavily gold-plated haggis for the 1973 conference of the Institute of Metal Finishing which took place in Edinburgh.

Pretreatments

Metallic Substrates[17-21]

This is considered in detail in Chapter 12, but a brief review is relevant here in relation to electroplating.

Metallic articles arrive at the electroplating stage with a surface film of *soils* derived from earlier production processes. These are divided into three classes:

(*a*) Organic films, oils, greases and perhaps polymeric films.
(*b*) Fine particles suspended in (*a*), often of silicaceous material, i.e. *dirt*.
(*c*) Films of the substrate oxide or other corrosion product.

With few exceptions it is more efficient and economical to use at least two different pretreatments, i.e. *degreasing* processes which deal with (*a*) and (*b*), and *pickling* processes which remove oxide and corrosion films. Degreasing comes first, as pickling processes fail on hydrophobic surfaces.

Degreasing Hydrocarbon or mineral oils are removed by solvents; *vapour* degreasing with chlorinated solvents, or emulsification, are common alternatives. Greases of animal or vegetable origin, which are fatty acids, alcohols or esters, are removed with hot aqueous solutions of high pH (*alkaline* degreasing). They react with alkali to form water-soluble soaps. Electrolytic alkaline degreasing is considerably faster than soak cleaning. The work, immersed in hot alkali, is usually the cathode. A mass of hydrogen bubbles formed at the solution–metal interface undermines and removes the grease very effectively; saponification takes place more slowly later. Ferrous metals may be degreased anodically, the metal surface becoming passive, but non-ferrous metal anodes corrode in hot alkaline solutions. Anodic cleaning produces only half the volume of gas (oxygen) and is less effective on that count. There is however an advantage. During use small amounts of metal enter the cleaning solution; other metals enter as impurities in the salts and make-up water. A thin metal film may be electrodeposited during cathodic cleaning, which if it remains can decrease the corrosion resistance of the main deposit. Anodic cleaning avoids this danger even when the cleaning bath is contaminated. For steel, and occasionally for copper alloys, a cleaning cycle is employed, with the work cathodic for degreasing but reversing the current for the final few seconds, during which any thin film of impurity metal is removed by anodic dissolution.

Alkaline cleaning solutions are compounded from sodium hydroxide, trisodium phosphate (TSP), sodium silicate, sodium carbonate, soaps and detergents, and other materials. The higher the pH the more effective is saponification, but with non-ferrous metals the greater is the danger of corrosion. Zinc diecastings and aluminium need much less aggressive alkalies than are safe with steel. For aluminium and its alloys a high concentration (10%) of sodium silicate acts as a corrosion inhibitor and a cleaner.

To check the efficacy of grease removal, the alkali solution is rinsed away or neutralised by dipping in dilute acid. If, after removal from the acid, the draining metal surface remains wetted evenly all over for 30–60 s (or until it dries by evaporation), hydrophobic soils have been removed. Traces of grease cause the surface to de-wet, and surface tension draws the water into separate droplets. This is the *water-break* test. Traces of grease which remain when the work is plated do not prevent electrodeposition, but are detrimental to adhesion and corrosion resistance.

Particulate dirt is usually removed together with the grease which binds it, although there are a few exceptions. Sheet steel may carry a 'smut' of finely divided carbon (or perhaps iron carbide) left from pickling processes in steelmaking. This is not removed with grease, and if evenly distributed is not apparent until the work is rubbed locally. Left in place it leads to porous and poorly adherent coatings. Whilst it can be removed by mechanical means, e.g. vigorous brushing, this is impracticable when automatic plating machines are used. Steel which shows this defect may be unsuitable for plating.

Removal of oxides and corrosion products Oxide and corrosion-product films are removed by dissolution in aqueous solutions. Hydrochloric and sulphuric acid are the most common. Concentrations and temperatures are varied according to the substrate. For mild steel, for example, cold 15 wt% HCl is suitable, but for zinc diecastings the concentration must be reduced to 0·25 wt%, and the pickling time must be kept very short to avoid excessive attack. Rust is more quickly removed by acid pickling when in the prior degreasing stage the work is made the cathode, than when it is the anode. Cathodic cleaning partially reduces rust to magnetite and iron, which undergoes rapid reductive dissolution in the acid. Mixed acids containing wetting agents are supplied as proprietary mixtures; hydrochloric, sulphuric and phosphoric acids are common. Inhibitors—generally amino compounds—may be added, but strongly absorbed films of inhibitors or their breakdown products may cause trouble later. Electrolytic pickling is used for special purposes, but hydrochloric acid is unsuitable for this purpose because of its volatility and the possibility of chlorine evolution. Cathodic pickling of steel in 10–20 wt% sulphuric acid enables thick rust or scale to be dissolved without losing metal, which is cathodically protected. Anodic pickling in 42 vol% (about 55 wt%) sulphuric acid is used to remove a thin surface layer from steel. A high current density is used and dissolution, which is under diffusion control, is uniform. After 10–20 s the metal becomes passive and dissolution ceases. Disordered and fragmented metal produced by abrasion or machining is removed to leave a surface which favours good adhesion of an electrodeposit. Oxide and corrosion products on copper and its alloys may be removed in hydrochloric or sulphuric acids, less concentrated than used for steel. Sulphuric acid allows the dissolved copper to be recovered and the acid to be regenerated by electrolysis. However, when copper alloys are cathodically degreased, cuprous oxide is reduced to loose copper particles which do not dissolve in acid. An electroplated coating over loose metal is likely to be defective. A much more aggressive mixture of sulphuric and nitric acids known as a *bright dip* or sometimes (wrongly) *aqua fortis* is used. This is a rudimentary chemical-polishing system and produces a bright surface from which the loose particles have been removed.

Non-conductors[22-32]

Plating plastic articles has become a widespread commercial process fairly recently. The main plastics in use at the time of writing are ABS and polypropylene. Both replace diecast and pressed metal in various fields. Moulded parts have smooth surfaces unsuited to producing adherent plate, and the first step is to 'etch' the plastic surface by using strongly oxidising acids, usually chromic plus sulphuric acid mixtures. The plastic must have dispersed in its surface small areas more susceptible to oxidation than the surrounding matrix, and these are introduced in a variety of ways during manufacture. Etching produces a pattern of small pits. The second step is to produce within the pits and over the surface a fine metallic precipitate. One method is to dip in solutions of stannous and palladium salts in succession, which produces fine palladium particles, but there is a variety of proprietary processes available. The metal particles become nuclei for the deposition of a metal coating from an electroless plating bath (see Section 13.5). This is an aqueous solution of a metal salt containing a reducing agent that is able to reduce the metal ions to metal. The solution is unstable but is compounded and used so that homogeneous electron transfer does not occur, and there is neither homogeneous precipitation of metal, nor heterogeneous precipitation on non-conducting surfaces but only on the metallic nuclei on the plastic surface. Several electroless plating processes have been devised but only those for copper or nickel are widely used for plastics. Once a continuous coating of metal has been produced, the substrate can be transferred to an electroplating bath for the application of any desired coating. Most plastic articles are finished with nickel and chromium.

Mechanical Pretreatments[33-40]

Mechanical processes which cold-work a substrate have important effects on electrodeposits. Examples are grinding and abrasive polishing, grit and shot blasting, cold rolling and severe cold deformation. They alter the metallurgical structure of the substrate, reducing the surface grain size and in some cases produce small crevices filled with non-metallic debris. Abrasive processes which act parallel to the surface (e.g. grinding, polishing) may leave splinters and leaves of metal attached at one end but otherwise separated from the surface on which they lie. In addition, non-metallic abrasive material is embedded in the surface. Surfaces which are neither annealed nor otherwise treated to remove mechanically disturbed surface layers affect the structure and properties of metal electroplated over them, as mentioned below. In many cases one result of the modification of the electrodeposit is to reduce the corrosion protection it affords. Where the same topographical alteration of a substrate can be achieved by non-mechanical means, e.g. electrochemical polishing, electrochemical machining, chemical milling, the surface left is not cold-worked and does not disturb an electrodeposit to the same extent.

Plating Processes

Electroplating[41-43]

The metallic substrate, clean and rinsed, is immersed wet in the plating cell. The base metals which are usually plated present an essentially metallic surface to the electrolyte, and the slight corrosive action of the rinse water in preventing the formation of any substantial oxide film is important. A critical balance of corrosion processes in the initial stages is vital to successful electroplating, and for this reason there is a severe restriction on the composition of the electroplating bath which may be used for a particular substrate. This will be discussed later. The substrate is made the cathode of the cell; it may be immersed without applied potential ('dead' entry) or may be already part of a circuit which is completed as soon as the substrate touches the electrolyte ('live' entry). Live entry reduces the tendency for the plating electrolyte to corrode the substrate in the period before the surface is covered by the coating. The main cathodic process is usually the reduction of dissolved ions to metal in the form of an adherent, coherent coating.

The ions reduced may be aquo cations, e.g.

$$[\text{Ni}(\text{H}_2\text{O})_x]^{2+} + 2e^- = \text{Ni} + x(\text{H}_2\text{O})$$

or oxyanions, e.g.

$$\text{Sn}(\text{OH})_6^{2-} + 4e^- = \text{Sn} + 6\text{OH}^-$$

or complex ions, usually cyanides, e.g.

$$\text{Au}(\text{CN})_2^- + e^- = \text{Au} + 2\text{CN}^-$$

Cations are assisted by the electric field to migrate to the cathode. On the other hand, the field impedes the migration of anions, and diffusion has to overcome this. It is rare for metal reduction to be the sole cathode process, since water and other dissolved substances are reduced simultaneously. Many compounds are added intentionally to take part in the cathode process, with the object of modifying the nature of the coating. Such materials are called *addition* agents and are subdivided into classes based on their main effect (e.g. brighteners, levellers, grain refiners, stress reducing agents). In all aqueous solutions, water may be reduced:

either
$$2\text{H}_2\text{O} + 2e^- = \text{H}_2 + 2\text{OH}^-$$
or
$$2\text{H}_3\text{O}^+ + 2e^- = \text{H}_2 + 2\text{H}_2\text{O}$$

This reaction becomes thermodynamically possible whenever the cathode potential falls below

$$E = +\frac{RT}{2F} \ln \frac{a_{\text{H}_3\text{O}^+}^2}{\bar{P}_{\text{H}_2} a_{\text{H}_2\text{O}}^2} \qquad \ldots(13.1)$$

Where R is the gas constant, T the temperature (K), F the Faraday constant and $\bar{P}_{\text{H}_2}$ is the relative partial pressure (strictly, the fugacity) of hydrogen in solution, which for continued evolution becomes the total external pressure against which hydrogen bubbles must prevail to escape (usually 1 atm). The activity of water $a_{\text{H}_2\text{O}}$ is not usually taken into account in elementary treatments, since it is assumed that $a_{\text{H}_2\text{O}} = 1$, and for dilute solutions this causes little error. In some concentrated plating baths $a_{\text{H}_2\text{O}} \neq 1{\cdot}0$ and neither is it in baths which use mixtures of water and miscible organic liquids (e.g. dimethyl formamide). However, by far the most important term is the hydrogen ion activity; this may be separated so that equation 13.1 becomes

$$E = -\frac{RT}{2F} \ln \bar{P}_{\text{H}_2} a_{\text{H}_2\text{O}}^2 - \frac{2{\cdot}303\, RT}{F} \text{pH} \qquad \ldots(13.2)$$

As $\bar{P}_2 \approx a_{\text{H}_2\text{O}} \approx 1$, we have at 298K, $E \approx -0{\cdot}059\,\text{pH}$ ...(13.3)

These considerations have been based entirely on thermodynamics and take no account of the overpotential, which is dependent on the rate of the process and the nature of the surface at which the reaction occurs. For this reason, the rate of reduction of H_3O^+ or H_2O is usually low, and remains so to potentials from 0·5 to 1·0 V below that given in equation 13.1. Even so, the instability of water is an insuperable obstacle to electrodepositing metals whose ions are so stable in aqueous solutions (e.g. Al^{3+} aq.) that water reduction becomes the sole cathode process (see Section 9.1 for the kinetics of the hydrogen evolution reaction).

Much time has been given in recent years to studying the mechanism of electrodeposition. Most investigators have assumed that electrodeposition should follow a mechanism akin to that for the deposition of a crystalline coating by condensation of a vapour. The solvated metal ion approaches and adsorbs on the cathode, losing some of its solvation sheath as the cathode gains the requisite electrons. The *adion* which is mobile, diffuses over the cathode surface until it reaches an atomic step. It adsorbs on the step, losing more water of solvation, and having its freedom reduced to diffusion along the step. Further desolvation and co-ordination follows when it reaches a kink in the step, at which stage it is immobilised. When other adions following this path eventually join and submerge the first, co-ordination with water in the electrolyte is exchanged fully for co-ordination with metal ions in the metallic lattice. This view of the mechanism of electrodeposition is plausible and compatible with both current views on metallic crystals and their defects, and certain properties of electrodeposits. It is, however, a preconceived notion, and considerable experimental difficulties in the way of producing evidence are responsible both for the numerous different investigations in the field, and for the fact that the evidence produced is impressive more for its volume than its conclusive force.

Current enters through the metal–electrolyte interface of the anode, which is usually made from the same metal as is plated on the cathode. The anode dissolves replacing the metal lost at the cathode:

$$M = M^{n+} + ne^-$$

The overall process is metal transfer from anode to cathode via the solution. The form of anode corrosion is important, and materials may be added both to the anode metal and to the electrolyte, to influence it. There are important instances where an insoluble anode is used, and the anode reaction becomes the oxidation of water or hydroxyl ions:

either
$$6H_2O = O_2 + 4H_3O^+ + 4e^-$$

or
$$4OH^- = O_2 + 2H_2O + 4e^-$$

and also the oxidation of any other susceptible materials. Oxidation of water may occur at an anode which was intended to be soluble, if the metal becomes passive. The minimum potential above which the anode must rise before oxidation of water occurs, is:

$$E = E^0 + \frac{RT}{4F} \ln \frac{\bar{P}_{O_2} a^4_{H_3O^+}}{a^6_{H_2O}} \qquad \qquad \text{...(13.4)}$$

In this equation E^0 is the free energy of formation of water expressed in V (equiv.)$^{-1}$, i.e. $-\Delta G^0_{H_2O}/nF$. Simplifying, equation 13.4 at 298K becomes

$$E = 1.23 - 0.059\,pH \qquad \qquad \text{...(13.5)}$$

There is often an overpotential of about 0·5 V before the rate of oxidation of water becomes rapid.

Aqueous Electrolytes[44]

Aqueous solutions have a complex structure. Liquid water is anomalous; properties estimated by interpolation from those of neighbouring hydrides in the periodic table fall wide of the observed properties. For example, estimated melting and boiling points are $-43°C$ and $-11°C$, respectively. Molecular interaction (hydrogen bonding) imposes short range order in the liquid, which the anomalous properties reflect. Some of the crystal structure of ice is retained in the liquid, though the structure is less open, and water shares this peculiarity with diamond, silicon and germanium, since in each case the liquid is denser than the solid at the melting point. A simple view of water is as a fluid with two species, small local regions with an ice structure and others with a strongly associated but irregular structure $(H_2O)_n$. The 'icebergs' and random groups are in dynamic equilibrium, exchanging individual molecules throughout the lifetime of the larger groups, which are not themselves permanent.

Water has a permanent dipole moment (strictly it is a quadrupole moment) caused by the asymmetry of the molecule and the greater electron affinity of oxygen. When soluble strong electrolytes, e.g. nickel sulphate, dissolve they dissociate completely, and the interactions between the charged ions and the dipole water molecules considerably modify the water structure. The small nickel ions with its high charge density will cause more disruption than the larger sulphate ion, but both become strongly associated with a sheath of water molecules (Fig. 13.1). The bonding between ion and solvation sheath confers a high degree of stability on the ions in aqueous solutions. The dipolar nature of the solvent, able to stabilise ions of either

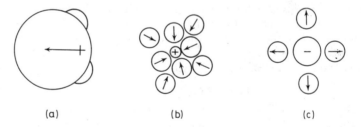

(a)   (b)   (c)

Fig. 13.1   (a) Water molecule, dipole structure, (b) solvation of a small cation, e.g. Ni^{2+} and (c) solvation of a larger anion, e.g. SO_4^{2-}

charge sign equally well, makes water an excellent ionic solvent. Ionic solvation destroys the ice-like component of the liquid with a result similar to that observed on melting, i.e. a decrease in volume as the 'space-wasting' ice structure is removed, and many strong electrolytes have a negative volume of mixing.

Aqueous solutions of non-electrolytes, especially of non-polar solutes, may show the reverse effect and increase the proportions of ice-like component. The non-polar part of organic electrolytes such as soaps and wetting agents may predominate in increasing the ice component. Thus solutes can be divided into two classes: *structure making* and *structure breaking*, and in some metal-finishing process solutions both types of solute may be added.

In certain cases it is impossible to satisfy both and there is a decrease in solubility of the least successful component. For example, a soap such as sodium palmitate, which in water is mainly

$$Na^+ \ ^-OOC \cdot (CH_2)_{14}CH_3$$

is very soluble, and the proportion of ice-like water increases to minimise interaction with the hydrophobic hydrocarbon chain. If sodium chloride is added to a concentrated soap solution, the ice-like structure is broken, and the increased concentration of 'free water' causes the palmitate to separate. Polar non-electrolytes like the lower alcohols have an even greater effect on the water structure than soaps. Being themselves capable of forming hydrogen bonds, they enter the 'iceberg' structure, increasing its stability. Methyl or ethyl alcohol added to solutions saturated with structure-breaking electrolytes, e.g. $NiSO_4$ in a plating bath, can withdraw water from the ionic solvation sheaths to form ice-like water, and now it is the electrolyte whose solubility falls. Soap, on the other hand, is more soluble in alcohol–water mixtures.

The ability of water to be oxidised or reduced at the plating-bath electrodes is in practice an important advantage, though it has a few drawbacks. The drawbacks are overemphasised in textbooks and the advantages ignored. An ion which is much more stable towards reduction than water is unaffected in the region of the cathode during electrolysis. With the margin of polarisation of 0.5–1.0 V, the manganous aquocation is the most stable which can be reduced in aqueous solutions, but because of the shortcomings of the metallic coating, manganese is not used, and the zinc aquocation is the most stable ion that can be reduced in large scale electroplating. A drawback is that the aquocations of aluminium, titanium, zirconium, niobium and tantalum are too stable. Their known corrosion resistance would make them important coatings but they cannot be electrodeposited. On the other hand there are numerous electrolytes whose presence is desirable in a plating bath but whose cations must not be reduced during electrolysis. The complete stability of cations like Na^+, K^+, NH_4^+, Li^+, Mg^{2+}, Ca^{2+}, Al^{3+} is then an advantage. Should the cathode be depressed (accidentally) below the normal working value, they are safe from reduction as hydrogen evolution acts as a 'safety valve' for excessive currents. The products of water reduction, hydrogen gas and OH^- ions are less likely to contaminate a large volume of valuable solution than are the reduction products of other materials.

Similar considerations apply to oxidation. An anion which is considerably more stable than water will be unaffected in the neighbourhood of the anode. With a soluble anode, in principle, an anion only needs be more stable than the dissolution potential of the anode metal, but with an insoluble anode it must be stable at the potential for water oxidation (equation 13.4 or 13.5) plus any margin of polarisation. The metal salts, other than those of the metal being deposited, used for electroplating are chosen to combine solubility, cheapness and stability to anode oxidation *and* cathode reduction. The anions most widely used are SO_4^{2-}, Cl^-, F^- and complex fluorides BF_4^-, SiF_6^{2-}, Br^-, CN^- and complex cyanides. The nitrate ion is usually avoided because it is too easily reduced at the cathode. Sulphite, which is used for gold plating, is nevertheless too readily oxidised at the anode, and provides an example of an ion used in one instance despite its drawbacks. The oxidation of water at the anode is also a reaction which does not contaminate the solution.

Much laboratory work has been performed to develop non-aqueous plating baths, using either organic solvents or ionic melts. In so far as the reduction of water is concerned, there are two rather different aims in view. Firstly, there is a search for solvents or melts sufficiently stable to allow electrodeposition of coatings of aluminium, titanium, etc. which are impossible with water. Secondly, there is a need for a non-aqueous solvent for metals such as cadmium, whose electrodeposition from water, while practicable and satisfactory, is always accompanied by hydrogen which can embrittle cathodes of certain high strength steels and other alloys (see Sections 8.4 and 14.3). Here the aim is to avoid hydrogen discharge. A practical drawback to the use of non-aqueous solvents is the accumulation of the by-products of solvent–electrode reactions leading eventually to the bath being poisoned.

Simple and Complex Ions[45]

Amongst the common metals of the electroplating industry, only nickel is invariably reduced from its aquocation. Copper, silver, gold, cadmium and zinc are normally deposited from solutions of complex cyanides; tin, and chromium from oxyanions, and tin, in other cases, from a complex fluoride. Platinum-metal plating baths contain ions, all of which are complex; it is doubtful if any platinum-metal aquocation can exist in aqueous solution, such is the high tendency of these metals to form complexes. If a ligand, such as cyanide, can displace water from an aquocation to form a complex ion, the complex must be more stable, and the deposition potential is always more negative for a complex ion than for the equivalent simple ion. Let the equilibrium between aquocation M^{n+}, ligand X^{x-} and complex ion $\{M(X)_p\}^{(px-n)-}$ be

$$\{M(X)_p\}^{(px-n)-} \rightleftharpoons M^{n+} + pX^{x-}$$

and
$$K = \frac{a_{M^{n+}} \times a_{X^{x-}}^p}{a_{M(X)_p^{(px-n)-}}}$$

where K is the *instability* constant for the complex ion, p is the *co-ordination* number. The potential below which deposition becomes possible is

$$E = E^0 + \frac{RT}{nF} \ln \left\{ \frac{K \times a_{M(X)_p^{(px-n)-}}}{a_{X^{x-}}^p} \right\}$$

or

$$E = \left(E^0 + \frac{RT}{nF} \ln K \right) + \frac{RT}{nF} \ln \left\{ \frac{a_{M(X)_p^{(px-n)-}}}{a_{X^{x-}}^p} \right\} \qquad \ldots(13.6)$$

where E^0 is the standard electrode potential for the simple ion/metal equilibrium. The two bracketed terms on the right of equation 13.6 constitute a sort of 'E^0' for the complex ion, and as K is usually very small the second term is negative.

Complex ions used for electroplating are anions. The cathode tends to repel them, and their transport is entirely by diffusion. Conversely, the field near the cathode assists cation transport. Complex cyanides deserve some elaboration in view of their commercial importance. It is improbable that those used are covalent co-ordination compounds, and the covalent bond breaks too slowly to accommodate the speed of electrode reactions. The electronic structure of the cyanide ion is:

$$\left(\begin{array}{c} x \\ o \\ x \ x \\ \bullet C o \quad N^o_o \\ x \\ o \end{array}\right)^{-}$$

where electrons are contributed by the carbon atom (x), the nitrogen atom (o) and the cation (●), e.g. Na^+ in the case of a sodium cyanide solution.

A soluble cyanide added to silver nitrate solution precipitates silver cyanide as an ionic compound:

$$(Ag)^+ \ (\overset{x}{\bullet}C \equiv N^o_o)^-$$

The precipitate redissolves in excess soluble cyanide, and the complex ion is probably an ion–dipole co-ordination compound, i.e.

$$\overset{\longrightarrow}{^-(C \equiv N)} \ (Ag^+) \ \overset{\longleftarrow}{(N \equiv C)^-}$$

The solubility of $Ag(CN)_2^-$ in water stems from the overall negative charge encouraging solvation with water dipoles, which uncharged AgCN does not. It is likely that the other cyanide complex ions of low co-ordination number have a similar structure.

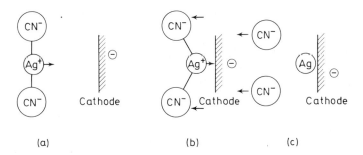

Fig. 13.2   Ion distortion by the field in the vicinity of a negative cathode. (*a*) Diffusion of anion to cathode, (*b*) diffusion and migration of distorted complex and (*c*) release of CN^- ions and incorporation of Ag into the lattice

When ions with this structure diffuse to the vicinity of a negative cathode, the ion is distorted by the field to become polarised, with the positive silver core nearest the cathode (see Fig. 13.2). Once within a critical distance the field can assist transport of the distorted complex, discharge the silver core, and eventually repel the released cyanide anions. There are several practical advantages in plating from complex cyanides. The reduction in deposition potential is most important in the application of relatively noble metals to base substrates, while avoiding severe cathode corrosion. The important case of the copper cyanide bath is discussed later. The impeded diffusion of the complex anion, the energy needed to polarise and reduce the anion, and the diffusion 'barrier' produced by the high concentration of cyanide near the cathode, all contribute to a high overpotential for electro-deposition, which aids the production of even coatings on cathodes of uneven shape. The cyanide ions released after discharge of the metal from the complex modify the structure of the coating in a manner similar to that of addition agents, and it is probable that some cyanide is adsorbed. The result is that cyanide bath coatings are fine grained, hard, and may contain small amounts of carbon.

Some disadvantages stem from the same phenomena: impeded diffusion reduces the maximum practical rate of plating to well below that possible with aquocation baths. The cyanide ion is not entirely stable; both oxidation and reduction products accumulate, including carbonate. Carbonate is also formed in the alkaline cyanide baths (all cyanide baths are alkaline except some based on aurocyanides) by absorption of CO_2 from the air, and it is necessary either to replace or purify baths periodically. Much has been made of the toxicity of cyanides, but the other process solutions used in plating are generally extremely toxic and corrosive or caustic, and it is necessary to treat them all with respect.

Oxyanions are important in chromium and tin plating. Although chromium plating baths have a simple formulation (chromium trioxide, sulphuric acid and water), hydrolysis and side reactions form a variety of polychromate ions. The cathode reactions are numerous and imperfectly elucidated; only about 5% of the charge passed produces chromium. As in cyanide baths the by-products behave as addition agents. The grain size of chromium electroplated under the normal conditions is the smallest encountered amongst metals. Tin coatings plated from stannate solutions are reduced from the $Sn(OH)_6^{2-}$ ions, and are fine-grained.

The platinum-group metals are necessarily plated from complex solutions, since all platinum-metal salts form complex ions in water. The development of aqueous baths has hinged upon finding complex ions sufficiently *unstable* to be reduced; all the platinum group cyanides are far too stable. Because all the platinum metals are good catalysts for hydrogen evolution (high exchange current density) there is no useful margin of hydrogen overpotential over that predicted by equation 13.2. Nevertheless, aqueous plating baths have been devised for all the platinum group, although only palladium, rhodium and platinum find present commercial use.

The complex cyanides of transition metals, especially the iron group, are very stable in aqueous solution. Their high co-ordination numbers mean the metal core of the complex is effectively shielded, and the metal–cyanide bonds, which share electrons with unfilled inner orbitals of the metal, may have a much more covalent character. Single electron transfer to the ferricyanide ion as a whole is easy (reducing it to ferrocyanide, with no alteration of co-ordination), but further reduction does not occur.

Conducting Salts, Addition Agents and Other Ingredients of Plating Baths [46-49]

Plating baths contain a number of salts and compounds in addition to those of the ion to be reduced to metal. Much commercial electroplating is from 'proprietary' solutions whose use is covered by patents, and which are supplied completely or partly compounded. The precise composition is generally a trade secret, and the patents may sometimes be very widely drawn to include a larger range of compositions and ingredients than is successful. A broad classification of additional ingredients includes:

(a) Those which maintain certain conditions but do not take part in electrode reactions; examples are electrolytes added to improve conductivity, pH buffer systems and ions which maintain complex ion equilibria.
(b) Those which take part in the anode reaction, added to promote efficient dissolution of a soluble anode.
(c) Those which take part in the cathode reaction by adsorption, co-deposition, or both, or by modifying the reaction; these ingredients alter the structure and properties of the coating and have an important bearing on its corrosion properties. The term *addition agents* is generally used for this class.

A particular ingredient often acts in more than one category.

Conducting salts are usually sodium, potassium or ammonium salts, or the acid of a stable ion already present from other ingredients. Apart from energy conservation, the higher the conductivity of a bath the better the distribution of thickness on complex shapes.

Addition agents are subdivided according to the main effect they have on the coating, i.e. grain refiners, brighteners (primary), brighteners (secondary), levellers, stress reducers and anti-pitting agents. Apart from the last, all these addition agents modify the growth process profoundly. They may be the ions of foreign metals which can co-deposit, or polar or ionic organic materials. They introduce irregular atoms or molecular fragments into the metal lattice or grain structure of the main metal of the coating, and alter the crystal structure. The macroscopic results are suggested by the names: smaller grain size, mirror-like surface, or a relatively smooth (level) surface on an initially rough substrate. These characteristics are not achieved without corresponding disadvantages. Foreign material from the addition agent is incorporated, up to 5% by weight for metallic co-deposition, less for organic agents. First- and second-order tensile stresses usually increase, hardness rises, ductility falls. First order tensile stress can exceed the tensile strength of the coating, but for some metals, particularly nickel, addition agents (stress reducers) are known which decrease first-order stress, though second-order stress is not reduced. The mechanical results are detrimental to corrosion protection, but the topographical results—fine grain, level and bright coatings, are favourable. The chemical results, on the evidence available, vary. The most extensively investigated case is nickel. Here it is almost universal practice to employ several addition agents together in a bright plating bath, one of which causes the incorporation of a small amount (about 0·02%) of sulphur, and reduces the corrosion resistance of bright nickel (see

Section 14.7). There are other systems (benzotriazole brighteners in copper, aldehyde–amine brighteners in tin) which enhance corrosion resistance.

Electroplating Anodes[50–51]

The anode is usually soluble, and is made from a high purity form of the metal being deposited, or occasionally from an alloy. A soluble anode is often the cheapest and most convenient means of replacing the metal reduced at the cathode. Effective anode corrosion is important, and different examples present a variety of types of dissolution.

Copper anodes in the acid sulphate bath are an example of active anodic dissolution. They etch uniformly with low polarisation at 100% efficiency, forming little anode debris or sludge. Idle anode corrosion is very slow with dissolved oxygen reduction as the cathodic reaction. Copper anodes in the cyanide bath corrode easily providing there is sufficient excess or 'free' cyanide present. Polarisation is higher than in the acid bath, with a much lower maximum current density. At high current densities the rate of dissolution of copper exceeds the rate of supply of cyanide ions needed to form the cuprocyanide complex, and blue cupric aquocations form. Eventually the anode becomes passive and evolves oxygen. For special purposes where anode area is limited (e.g. in a Hull cell) an insoluble mild-steel anode avoids these troubles. A proportion of insoluble anodes or alternate use of soluble and insoluble ones is necessary, because anode efficiency exceeds cathode efficiency, and the metal content of the solution would rise continually if soluble anodes alone were used.

Nickel is normally plated from mixed solutions of nickel sulphate and chloride using soluble anodes. The standing potential of pure nickel anodes indicates they are passive in the idle bath, while in operation dissolution is by pitting corrosion brought about by chloride (Fig. 13.3). In all sulphate solutions, nickel anodes are passive and insoluble. Dissolution is aided by adding small amounts of sulphur or carbon to the nickel anode, which aid the breakdown of passivity ('depolarised anodes'). Nickel anodes produce a fine particulate 'anode sludge'; anode bags of finely woven cloth are used to retain much of this, and continuous filtration is needed to remove the rest, otherwise the corrosion resistance of the coating is severely degraded. Nickel can be used in the form of small 'chips' in baskets of titanium mesh. The titanium is passive and the surface is effectively insulated from electrolytic current exchange, but electrons released by the dissolving chips of nickel are able to pass via the metal–semiconductor contact with the basket to the outside circuit. Anode efficiency slightly exceeds cathode efficiency, but not sufficiently to increase the metal concentration, unless rigorous precautions are taken to return all the solution lost on surfaces removed from the bath.

Tin anodes dissolve by etching corrosion in acid baths based on stannous salts, but in the alkaline stannate bath they undergo transpassive dissolution via an oxide film. In the latter the OH^- ion is responsible for both film dissolution and for complexing the tin. Anodes must not be left idle because the film dissolves and thereafter corrosion produces the detrimental divalent stannite oxyanion. Anodes are introduced 'live' at the start of deposition, and transpassive corrosion is established by observing the colour of the film

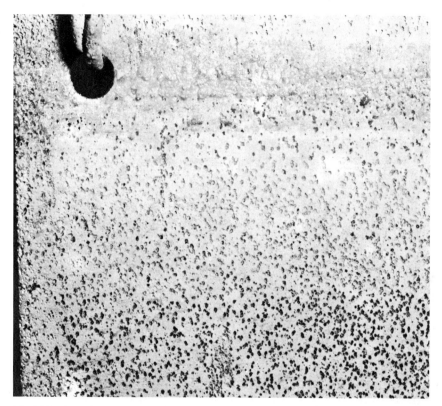

Fig. 13.3   Part of an anode from a nickel plating bath, showing dissolution by pitting corrosion

(pale yellow for correct operation) and the anode polarisation and by adjusting the immersed area. The anodes retain a smooth, quasi-polished surface, and no anode bags are needed. An alternative where anode filming is inconvenient is to use insoluble (passive) mild-steel anodes, and replace tin by adding a colloidal dispersion of stannic oxide.

Chromium plating from hexavalent baths is carried out with insoluble lead–lead peroxide anodes, since chromium anodes would be insoluble (passive). There are three main anode reactions: oxidation of water, reoxidation of Cr^{3+} ions (or more probably complex polychromate compounds) produced at the cathode and gradual thickening of the PbO_2 film. The anode current density must balance the reduction and reoxidation of trivalent chromium so that the concentration reaches a steady state. From time to time the PbO_2 film is removed as it increases electrical resistance.

Gold plating is another process where insoluble anodes are the rule. Soluble gold anodes would be technically satisfactory in some baths, but too tempting to thieves. A factor in their decline is the disappearance of gold coins, whose temporary sojourn in the anode circuit is alleged to have provided a form of corrosion beneficial to gold platers of earlier times. Insoluble stainless-steel anodes are now widely used, with graphite, platinised titanium and platinum mesh as alternatives. All suffer gradual deterioration. Stainless-steel anodes are pitted, especially in areas of high current density, and the corrosion products may eventually give trouble. Platinised titanium sheds its platinum coating, becoming steadily more polarised. Platinum anodes used with cyanide solutions slowly corrode.

The potentials and corrosion data for some anodes in typical plating baths are listed in Table 13.1.

Table 13.1 Behaviour of anodes in plating baths

Plating bath	Anodes	Corrosion process		Current density (A/m²)	Potential (V) (vs.S.H.E.)
		Idle	Working		
1. Watts' nickel	Nickel	Passive	Pitting	0	+0·08
$NiSO_4 \cdot 7H_2O$, 330 g/l	slab			65	+0·23
$NiCl_2 \cdot 6H_2O$, 45 g/l				130	+0·27
H_3BO_3, 30 g/l				195	+0·30
pH 2·5					
55°C					
2. Bright nickel	Nickel	Passive	Pitting	0	+0·03
as (1) plus saccharin	slab			390	+0·36
and 2–3 butyne 1–4 diol					
pH 2·2					
50°C					
3. Acid copper	Pure	Slow	Uniform	0	+0·28
$CuSO_4 \cdot 5H_2O$, 150 g/l	copper	etching	etching	130	+0·32
H_2SO_4, 50 g/l				260	+0·35
pH < 0				390	+0·38
18°C					
4. Rochelle copper	Pure	Slow	Uniform	0	−0·38
CuCN, 27 g/l	copper	etching	etching		
NaCN, 38 g/l					
Na_2CO_3, 35 g/l					
$KNaC_4H_4O_6 \cdot 4H_2O$, 45 g/l					
pH 12·0	Mild	Passive	Passive,	0	+0·21
55°C	steel		evolving O_2	130	+1·09
5. Silver	Fine	Slow	Uniform	0	−0·24
AgCn, 50 g/l	silver	etching	etching	32	−0·01
KCN, 30 g/l			↓	↓	
K_2CO_3, 50 g/l			passive,	65	+6·0
pH 11·5			evolving O_2		
25°C					
6. Acid zinc	Zinc	Etching,	Uniform	0	−0·63
$ZnSO_4 \cdot 7H_2O$, 250 g/l		rather	etching	163	+0·30
NH_4Cl, 15 g/l		rapid, H_2		325	+1·86
$Al_2(SO_4)_3 \cdot 18H_2O$, 30 g/l		evolved			
Dextrose, 75 g/l					
pH 4·0					
18°C					
7. Chromium	Pb/PbO_2	Passive	Passive,	0	+1·5
CrO_3, 400 g/l			evolving O_2,	1 075	+2·65
H_2SO_4, 4 g/l			reoxidising		
pH < 0			Cr^{3+}, etc.		
37°C					

Corrosion of the Cathode[52-57]

At the start the cathode is invariably a metal different from that to be deposited. Frequently, the aim is to coat a base metal with a more noble one, but it may not be possible to do this in one step. When a metal is immersed in a plating bath it will corrode unless its potential is sufficiently low to suppress its ionisation. Fortunately, a low rate of corrosion is tolerable for a brief initial period. There are cases where even when a cathode is being plated at a high cathodic (nett) current density, the substrate continues to corrode rapidly because the *potential* (determined by the metal deposited) is too high. No satisfactory coating forms if the substrate dissolves at a high rate concurrently with electrodeposition. This problem can be overcome by one or more of the following procedures:

    (*a*) The use of a complex anion bath.
    (*b*) The use of a 'strike' bath.
    (*c*) The use of intermediate electrodeposits (undercoats or underplates).

The principles are illustrated by the following important commercial examples.

Zinc diecastings: complex baths and undercoats   Diecastings are made from zinc with up to 8% aluminium, but from the viewpoint of corrosion they behave like zinc. Diecast parts are often plated with nickel and chromium for protection and decoration, but when zinc is immersed in a typical nickel plating solution it corrodes rapidly. There are two cathodic reactions: hydrogen is evolved and spongy nickel precipitates. If a diecasting is immersed 'live', corrosion and electrodeposition occur together as the potential for nickel deposition still leaves the cathode too positive to suppress zinc corrosion (Fig. 13.4). Standard practice is first to plate the zinc with an undercoat of copper, which is too noble to evolve hydrogen or reduce nickel ions. In the bath to which the data of Fig. 13.4 applied, the potential of copper was $+0.042\,\mathrm{V}$. Prolonged immersion led only to slow copper corrosion, with the reduction of dissolved oxygen as the cathodic reaction. This does not interfere with satisfactory nickel plating, and is entirely suppressed by live entry. The problem of applying a copper undercoat is solved by using the cuprocyanide complex bath. The acid cupric bath would present a worse problem than the nickel bath, with its lower pH and the greater oxidising power of the cupric ion. The stabilities of zinc and copper are reversed in alkaline cyanide, and zinc will not displace copper from cuprocyanide. Zinc immersed in the cuprocyanide bath can corrode to form either zincate or a complex zinc cyanide ion $Zn(CN)_4^{2-}$, with two possible cathodic corrosion reactions, i.e. dissolved oxygen reduction or hydrogen evolution. However, high polarisation prevents either supporting rapid corrosion, and the situation is like that for copper in the nickel bath; a sound coating is possible and live entry suppresses corrosion. Once a coherent copper coating envelopes the zinc, the part is rinsed and transferred to the nickel bath.

    The cuprocyanide plating bath is invaluable in numerous similar cases where a base substrate cannot be plated directly with the chosen coating. Steels, brass, bronze, beryllium copper and other substrates are copper

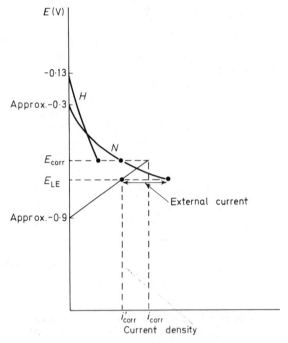

Fig. 13.4   Corrosion diagram for a zinc diecasting in a nickel plating bath, pH 2·2. There are two possible cathodic reactions, hydrogen evolution (H) and nickel ion reduction (N). The corrosion current i_{corr} is the sum of the partial cathode currents. Even with 'live entry' the potential E_{LE} is still too high to suppress corrosion, though the rate is reduced to i'_{corr}

underplated. Aluminium articles are ennobled, usually in two stages, i.e. first coated with zinc, which is in turn copper plated as described. Such is the utility of the cuprocyanide bath in circumventing cathode corrosion that although copper finds no significant use as a coating in its own right, cuprocyanide baths are found in almost every electroplating establishment.

Silver plating: 'strike' baths   A 'strike' is usually a solution of special composition in which electroplating is initiated. After a short period of 10–150 s, the cathode is transferred to a normal bath. The term *strike* is also used when plating is initiated in the normal bath, but for 10–150 s under very different conditions (often much higher current density), continuing normally afterwards. Silver is electroplated from argentocyanide anions, i.e. $Ag(CN)_2^-$. The bath is alkaline and contains 'free' CN^-. The argentocyanide ion is the least stable of the soluble complex cyanides, and therefore $Ag(CN)_2^-$ is a strong oxidising agent towards other metals which form soluble cyanide complexes. Generally, the reaction

$$Ag(CN)_2^- + (p-2)CN^- + M = Ag + M(CN^-)_p^{(p-n)-} \quad \dots(13.7)$$

proceeds as written where M is a more base metal of valency n. Copper, gold, zinc, iron, etc. are all base metals by comparison, so a copper undercoat does not solve the problem. An immersion deposit of silver leads to blistering and loss of adhesion if electrodeposition is continued over it. The 'nobility gap' between copper and silver can be bridged by combining a shift in the equilibrium of equation 13.7 with live entry. As a first step the cathode may be given a copper undercoat. The deposition potential for silver (special case of equation 13.7) is

$$E = \left(E^0 + \frac{RT}{F} \ln K_{instab} \right) + \frac{RT}{F} \ln \frac{a_{Ag(CN)_2^-}}{a_{CN^-}^2} \qquad ...(13.8)$$

where $\quad K_{instab} = \dfrac{a_{Ag^+} a_{CN^-}^2}{a_{Ag(CN)_2^-}} = 1 \cdot 8 \times 10^{-19}$

At 298K, equation 13.8 is

$$E = -0 \cdot 310 + 0 \cdot 059 \log \frac{a_{Ag(CN)_2^-}}{a_{CN^-}^2} \qquad ...(13.9)$$

and may be depressed by reducing the ratio in the last term, so a silver strike bath has a small concentration of metal and a large concentration of free cyanide. Equation 13.7 shows that a high concentration of free cyanide further decreases the nobility of the substrate, so some of the advantage is lost. One way of redressing the balance would be to have the substrate complex $M(CN)_p^{(p-n)-}$ present in solution. Silver strike baths are formulated with cuprocyanide in solution, but curiously these are usually recommended for striking steel, while those recommended for copper alloys omit the cuprocyanide. The author uses the cuprocyanide silver strike for ferrous and copper alloys alike. The combination of low silver and high cyanide concentrations means the cathode potential at a relatively high current density is very low, and both charge transfer and diffusion polarisation are high. With live entry, the low potential suppresses substrate ionisation, and a detrimental immersion deposit of silver is avoided. A thin underplate— called a *flash*—is deposited from the strike, after which the normal bath is used. Dilute strike baths working at high current density cannot deposit thick coatings; continued electrolysis produces incoherent, powdery material. They have a low cathode efficiency and evolve considerable hydrogen.

Electroplating passive alloys Another application of strike baths reverses the case illustrated in the previous example. The strike is used to *promote* a small amount of cathode corrosion. When the passivation potential of a substrate lies below the cathode potential of a plating bath, deposition occurs onto the passive oxide film, and the coating is non-adherent. Stainless steel plated with nickel in normal baths retains its passive film and the coating is easily peeled off. A special strike bath is used with a low concentration of nickel and a high current density, so that diffusion polarisation (transport overpotential) depresses the potential into the active region. The bath has a much lower pH than normal. The low pH *raises* the substrate passivation potential E_{pass}, which theoretically follows a relation

$$E_{pass} = E_{pass}^0 - 2 \cdot 303 \frac{RT}{F} pH \qquad ...(13.10)$$

When stainless steel is 'struck' the passive film is reduced and an adherent flash of nickel forms on the active metal surface. Deposition is continued in a normal bath.

Table 13.2 Corrosion potentials of substrates of copper and steel, plated and unplated in same plating solutions. Deposition potential is accompanied by current density (A/m^2) in parentheses; the plated substrate's coating thickness was 2·5 µm. The final column gives the potential below which hydrogen evolution is *possible;* only in the cuprocyanide is it observed

Plating bath	Copper	Steel	During plating	Plated copper	Plated steel	Hydrogen evolved below:
Watts nickel, pH 2·5, 55°C (not agitated)	+0·112	−0·238	−0·65 (130)	+0·132	−0·173	−0·148
Bright nickel, pH 2·2, 50°C (air agitated)	+0·042	−0·293	−1·22 (388)	−0·003	−0·138	−0·130
Cupric sulphate, pH < 0, 20°C	+0·275	approx. −0·35	−0·15 (388)	+0·265	—	> 0
Cuprocyanide, pH 12, 55°C	−0·378	−0·353	−1·21 (130)	−0·378	−0·388	−0·71
Argentocyanide, pH 11·5, 20°C	—	— (immersion deposit impedes measurement)	−0·58 (65)	−0·24	—	−0·68
Zinc sulphate, pH 4·0, 20°C	+0·122	−0·578	−1·213 (162)	−0·64	−0·65	−0·24
18% w/v HCl pickle for ferrous metal, pH < 0	—	−0·17				> 0
3% w/v HCl pickle for copper alloys, pH 0	+0·17					0

Electroplating aluminium and its alloys requires a similar technique. In aqueous solutions it is impossible to lower the potential sufficiently to reduce an alumina film, so the substrate is immersed in a strongly alkaline solution capable of dissolving it:

$$Al_2O_3 + 2OH^- = 2AlO_2^- + H_2O$$

The solution also contains a high concentration of zinc (as zincate), which is noble relative to aluminium. As metallic aluminium is exposed, it corrodes, reducing zincate ions and forming a coating of zinc:

$$Al + 4OH^- = AlO_2^- + 2H_2O + 3e^-$$
$$ZnO_2^{2-} + H_2O + 2e^- = Zn + 4OH^-$$

The immersion deposit is necessarily somewhat defective, for the reasons already mentioned, though immersion deposits from complex ions are finer grained and more satisfactory than those reduced from aquocations. The zinc coating is, under the best conditions, an acceptable basis for a copper undercoat from the cuprocyanide bath, on which other coatings can be plated, but there is usually a fair proportion of rejects in commercial operation. Other processes similar in principle use tin or bronze immersion coatings.

Service corrosion effects Undercoats, 'flash' deposits produced by strike baths, and immersion deposits are potential sources of weakness. If their structure is faulty it affects the subsequent layers built on the faulty foundation. The greater the number of stages, the higher the probability of faults.

Additional metal layers can create bimetallic corrosion cells if discontinuities appear in service. The layer of copper beneath cadmium plate on aluminium (using a zincate plus cuprocyanide deposit technique) can cause corrosion troubles. When aluminium is plated with nickel and chromium, rapid service corrosion in the zinc layer causes exfoliation.

Corrosion potentials in plating baths   The standing potentials of steel and copper (before application of current) are shown in Table 13.2, together with the standing potential of the plated metal and the potential below which hydrogen should, in theory, be evolved. The potential of the cathode during deposition at a typical current density is also given.

Factors influencing Structure[58–64]

Substrate effects: epitaxy and pseudomorphism   Both the words *epitaxy* and *pseudomorphism* are derived from classical Greek, the former meaning literally *close to* or *close upon* an *arrangement, row* or *series* (technically an arrangement imposed upon a skin or layer, e.g. an electrodeposit, which is close upon a substrate) and the latter *false form* (technically a mineral or crystal displaying a form more characteristic of another material than its usual one). For many years the two terms were held to be synonyms for one phenomenon in electrodeposits. Since 1936 it has become clear that there are two related phenomena, on each of which one of the names is bestowed. Not all authors recognise this, nor is the usage employed here adopted uniformly. Both phenomena are of great practical importance.

Pseudomorphism received methodical study from about 1905. A microsection taken across the interface between a substrate and an electrodeposit shows the grain boundaries of the former continue across the interface into the deposit (Fig. 13.5). As grain boundaries are internal faces of metal crystals, when they continue into the deposit the latter is displaying the form of

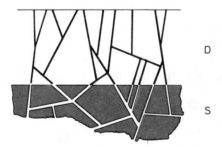

D

S

Fig. 13.5   Pseudomorphism; grain boundaries in the substrate (*S*) are continued in the electrodeposit (*D*)

the substrate. Hothersall's 1935 paper contains numerous excellent illustrations with substrates and deposits chosen from six different metals, crystallising in different lattice systems and with different equilibrium spacing. Grain boundary continuation and hence pseudomorphism is evident despite the differences.

Epitaxy is a relation on the atomic scale between substrate and electro-deposit. Imagine that the interface of the micro-section were magnified about 10^7 times so that the rows of atoms in the metal lattice become visible. If the deposit shows epitaxy, there will be an ordered and regular relation between substrate and deposit atom positions (Fig. 13.6(*a*)). A non-epitaxial deposit shows no such relation (Fig. 13.6(*b*)). Direct experimental demonstration of epitaxy was first made in 1936 by Finch and Sun. Earlier, metallographers argued that pseudomorphism (which they could see) meant there must be epitaxy (which they could not), as grain boundaries are surfaces where the direction of lattice rows of atoms changes; if epitaxy were assumed to exist, pseudomorphism should result. Reversing the argument, pseudomorphism was taken as evidence for epitaxy (Fig. 13.6(*c*)).

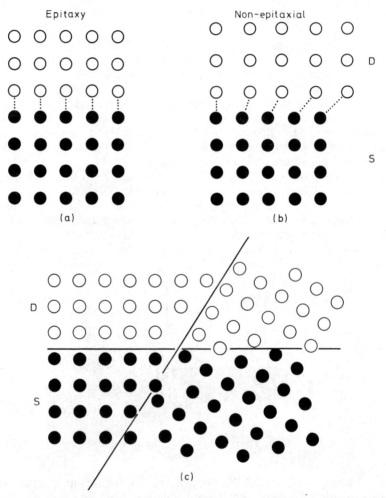

Fig. 13.6 (*a*) Co-ordination across a substrate *S*–electrodeposit *D* interface on the atomic scale produces epitaxy, (*b*) a non-epitaxial deposit has no co-ordination and (*c*) epitaxy would be expected to produce grain boundary continuation at the interface, though in fact grain boundaries often continue to thicknesses far greater than those at which epitaxy disappears

Electron diffraction investigations showed that epitaxy did indeed exist when one metal was electrodeposited on another, but that it persisted for only tens or hundreds of atomic layers beyond the interface. Thereafter the atomic structure (or lattice) of the deposit altered to one characteristic of the plating conditions. Epitaxy ceased before an electrodeposit is thick enough to see with an optical microscope, and at thicknesses well below those at which pseudomorphism is observed.

Epitaxy reflects the formation of metallic bonds between the dissimilar atoms at the interface. When the two metals crystallise in different systems, their relative orientation is that which promotes the maximum co-ordination and the maximum metallic bonding. The stability achieved by epitaxy overrides any lost due to the lattice strains imposed. These strains may be considerable; 'stresses' calculated from the bulk elastic moduli are correspondingly high, and sometimes puzzle the uninitiated if they exceed the bulk tensile strength. It is an oversimplification to regard the interface as being highly stressed; were the 'stress' which seems to be parallel to the interface reduced by some means to zero, the energy that would have to be put into the bonds normal to the interface would be much greater than that released. The simple concept of stress in a homogeneous alloy is not applicable to the peculiar case of a substrate–electrodeposit interface. The latter is unique in having metallic bonds carried across a very sharp boundary.

The practical result of epitaxy is a very high degree of adhesion between coating and substrate. The force needed to separate the interface is similar to that needed to break the metals on either side. Where a true metallic bond forms at an epitaxial interface it is only possible to measure adhesion if the bond is the weakest of the three near the interface. An adhesion test based on breaking the joint indicates only which of the three is weakest. For practical purposes any epitaxial joint will have a strength more than adequate for service conditions.

Non-epitaxial electrodeposition occurs when the substrate is a semiconductor. The metallic deposit cannot form strong bonds with the substrate lattice, and the stability conferred by co-ordination across the interface would be much less than that lost by straining the lattices. The case is the converse of the metal–metal interface; the stable arrangement is that in which each lattice maintains its equilibrium spacing, and there is consequently no epitaxy. The bonding between the metallic lattice of the electrodeposit and the ionic or covalent lattice of the substrate arises only from secondary or van der Waals' forces. The force of adhesion is not more than a tenth of that to a metal substrate, and may be much less.

Epitaxial growth is prevented if semiconducting films of grease, oxide, sulphide, etc. cover the cathode surface. These occur when pretreatment is inadequate, when plating baths are contaminated, or when, as with stainless steel, aluminium, titanium, etc. an oxide film reforms immediately after rinsing. Low adhesion resulting from non-epitaxial electrodeposition is used in electroforming to promote easy separation of deposit and substrate. When semiconductors or non-conductors are to be electroplated, a form of dovetail mechanical joint (achieved as outlined on p. **13**.6) is essential. Means similar to those for stainless steel and aluminium have been devised to deal with other alloys which passivate readily. Sometimes, even with special methods, some oxide remains so that the electroplated coating is anchored only by small epitaxial areas. There is risk of failure. Thermal stress or relatively mild abrasion may part the interface and cause the unanchored areas to blister. Adhesion is improved by post-plating annealing. The oxide at the interface is dissolved in one or other metal, or diffuses to grain boundaries, etc. and alloying at the interface produces the desired metallic bond.

Pseudomorphism has less desirable consequences, and usually means are sought to suppress it. If the substrate has been scratched, ground or abrasively polished, or if it has been cold rolled or cold formed, the surface is left in a peculiar state. Cold working reduces the surface grain size, and produces deformed, shattered and partly reoriented metal. It may produce micro-crevices between the deformed grains, and, with some processes, non-metallic impurities and oxides are embedded in the surface. The disturbed state of the substrate is copied by a pseudomorphic electrodeposit with several consequences (Fig. 13.7). One is aesthetic; it has often been noted that almost invisible abrasion of the substrate develops as more prominent

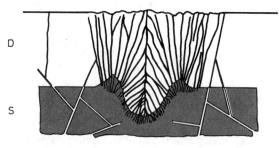

Fig. 13.7   The disturbed structure of a scratch, with fragmented and distorted grains, is perpetuated by a strongly pseudomorphic electrodeposit

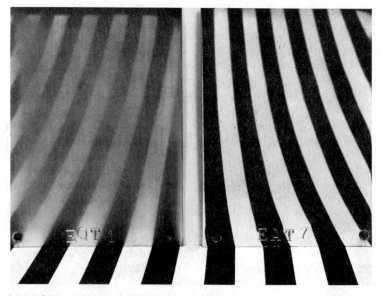

Fig. 13.8   A fairly strongly pseudomorphic bright tin deposit (left) has its brightness impaired by the shattered surface layer produced on steel by cold rolling. When this layer is removed, the deposit is mirror bright (right). Coating 5 μm thick

Fig. 13.9 Corrosion resistance of tin–nickel electrodeposit impaired by pseudomorphic porosity originating on cold-rolled steel surface (left). Panel on right has had the shattered grain surface removed by chemical polishing (0·125 μm removed). Coating thickness 15 μm; panels exposed 6 months to marine atmospheric corrosion (Hayling Island)

markings in the deposit. A chalk mark on steel produces local abrasion, hardly noticeable when the chalk is wiped away. If a strongly pseudomorphic electrodeposit is applied the chalk mark reappears indelibly on its surface. A bright deposit may have its lustre greatly reduced by pseudomorphic growth on a deformed surface (Fig. 13.8). The corrosion protection is reduced if pseudomorphism with a deformed substrate leads to discontinuities at ill-fitting deposit grains (Fig. 13.9). A pseudomorphic coating usually presents a dull or rough crystalline appearance. When the crystal form of the substrate is copied in the deposit, growth generates faces of simple index. An artificial face of high index soon grows out when plated. Tradition demands a featureless mirror surface on metal coatings, and a way of producing this which has attracted much commercial effort is by using brightening addition agents. Micro-sections of electrodeposits from the more effective bright plating baths do not exhibit pseudomorphism. The deposit usually shows no grain structure, but instead a series of light and dark bands parallel to the substrate (Fig. 13.10). Pseudomorphism is suppressed by the addition agent adsorbing on and blocking areas taking part in pseudomorphic growth. In the initial stages of bright plating the addition agents adsorb at similar points on the substrate. Growth commences from fewer substrate nuclei when annealed nickel is plated in a bright nickel bath than in a dull (Watts') bath without additions. In the earliest stages of deposition, replicas of the surface show evidence of pseudomorphism even in bright baths (the substrate grain boundaries are carried into the deposit) but this is suppressed rapidly as the thickness increases. The aim with bright plating baths is to inhibit growth sufficiently to suppress pseudomorphism, but not so much as to suppress epitaxy and adhesion. An excessive concentration of addition agent will also suppress epitaxy, so that deposition occurs on to an adsorbed layer of brightener. Brightener adsorption is often potential dependent and trouble may occur first at high current density (low potential) areas.

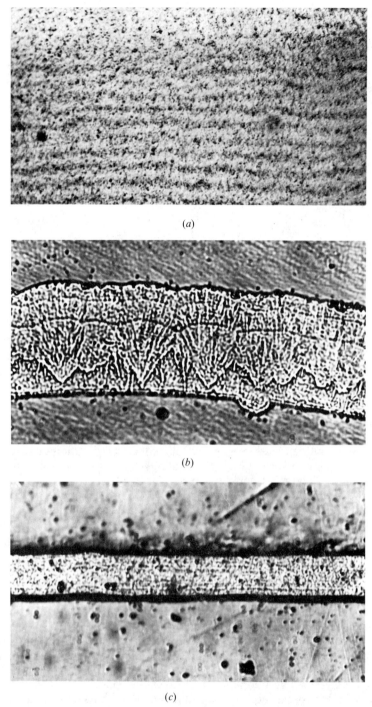

(a)

(b)

(c)

Fig. 13.10 Banding often observed in micro-sections of bright electrodeposits. (a) Bright tin (courtesy of the Tin Research Institute), and (b) and (c) Bright gold

Electrolyte effects[65-69] As a deposit becomes thicker, the influence of the substrate diminishes, and eventually the structure is characteristic only of the electrolyte composition, the temperature, current density and mode of agitation. A great variety of structure is observed; some are analogous to those seen in cast metals, but others are obtained only by electrodeposition. Crystalline deposits from baths containing little or no addition agent often develop a preferred orientation texture. Some bright deposits show a texture, but in general as growth processes are progressively inhibited by increasing addition agent concentration or by using more active materials, the deposit becomes progressively finer grained and loses preferred orientation textures.

The compositions of baths chosen for practical use result in initial rates of lateral growth much greater than the rate of outward growth. This is a desirable feature; it causes the coating to become continuous at low thicknesses. The opposite condition of a faster rate of outward growth is undesirable, and results in a non-coherent deposit. Predominantly outward growth occurs when the transport of metal ions becomes slow compared with their rate of discharge, i.e. it is favoured by high current density, low temperature and lack of agitation. Lateral growth processes are then starved of material to support them, but outward growth moves the deposit towards the supply, and the prominences formed benefit from greater diffusive flux. There are strong pressures in industrial production to increase electroplating rates, which carry a danger of using high current density and causing a shift to outward growth. In baths where the coating is electroplated from aquocations at high cathode efficiency, the onset of lateral growth is fairly sharp. Cathodes have a range of local current density, and the coating on the high current density areas becomes friable, dark coloured and rough as the transition is reached. Such coatings are termed *burnt* and the corrosion protection is degraded. With baths working in the acid pH range there is the complication that once an appreciable part of the current is used to reduce water, the pH at the cathode rises and insoluble hydroxides are precipitated and incorporated in the coating. With complex cyanide baths the onset of 'burning' is less sharp. There is normally considerable simultaneous hydrogen discharge, and as the current density rises there is no sharp limiting current density for metal discharge. Addition agents raise the lateral–outward transition to higher current densities, by inhibiting outward growth. Nevertheless all electroplated coatings show signs of deteriorating properties if the baths in which they are produced are worked at sufficiently high current density.

Form of current passed through cell[70-79] Commercial electroplating began with pure d.c. from galvanic cells. Later, for many years d.c. generators were used. Their current output is unidirectional but with a superimposed ripple. Part of the ripple stems from the angular motion of the armature coils during the period they supply current to a commutator segment, and part from variations of contact resistance at the commutator. Generators have been superseded by transformers and rectifiers. Copper-oxide, mercury-arc, selenium, germanium and silicon rectifiers have been used, and examples of each are to be found in service. These devices supply varying unidirectional current whose form depends on the number of phases in the input and the circuit used. A half-wave single-phase rectifier provides a pulsating current; a full-wave three-phase set has a much smoother output.

Alternating currents with asymmetric forms have been used, mainly for electroforming and thick engineering deposits. Where the cycles are slow, e.g. several seconds, the term *periodic reverse current* (p-r-c) is used. The benefit claimed for p-r-c plating is that smoother, thick deposits result from selective dissolution of peaks in the reverse part of the cycle. This assumes the electrode process reverses during the anodic period, which is not always the case. In chromium plating the coating becomes passive in anodic periods, while in acid gold baths based on aurocyanide, the process is also irreversible. More recently, asymmetric a.c. with a much higher frequency of 500 Hz was found to alter beneficially the properties of nickel from chloride baths.

Pulses of unidirectional current have been used to modify coating properties. When plating starts it is possible, for a time, to use a current much higher than the steady state limit, drawing on the stock of ions near the cathode. Provided sufficient time is allowed between pulses, a coating can be built of layers plated at much higher current density than normal. Improved gold coatings were produced by relatively rapid pulses. The technique of *barrel plating* results in pulse plating of an irregular sort, with pulse durations of the order of a second and inactive periods rather longer.

Chromium plating from chromic acid baths is more sensitive to the source of current than most other processes, sufficiently so for commercial operators to use at least three-phase rectifiers as a rule, and to take precautions against any temporary break of current during voltage regulation. A recent investigation showed that the ripple introduced by thyristor control of rectifiers was detrimental to chromium electrodeposits.

Industrial Electroplating Techniques[80–82]

Electroplating is usually a finishing technique applied after an article has been completely fabricated. Fairly large articles, from cutlery to motorcar bumpers, are dealt with by vat plating. They are suspended by a conducting connection in a rectangular tank or vat of electrolyte. The anodes are arranged about the periphery of the tank. For small runs the cathodes may be suspended by copper wire wrapped round a suitable part, but for longer runs a plating jig is used. This is a copper frame with phosphor bronze spring contacts to hold the work, and insulated, usually with a p.v.c. coating, on all but the contact points. The point of contact between wire or jig and the article becomes a weak part in the coating, and some thought should be given to providing or selecting contact points in insignificant areas.

Vat plating is used sometimes with articles too large for complete immersion. Printing, calendering, drying and similar rolls are part-immersed and revolved continuously during plating. However, it is much more difficult to plate half an object, reverse it, and complete the other half later; the 'join' between the two deposits is rarely satisfactory.

Small objects, nuts, bolts, screws and small electrical parts are plated in a revolving barrel. Electrical connection is made by a conductor immersed in the tumbling mass, and electrodeposition, which is confined to the outer layer of the mass at any instant, takes place in intermittent stages for any individual object. The coating is abraded during the process. The peculiarities of chromium deposition set it apart, and the normal barrel-plating processes are not used. In so-called *chromium* barrels the small parts travel and tumble along a helix inside a rotating cylinder during deposition, and are electroplated for a much greater proportion of the time than are parts in normal barrels.

Brush plating is a special technique which dispenses with a container and uses a swab soaked in electrolyte applied to the work. In *jet* plating a stream of electrolyte is applied to the cathode. Both are methods of *selective* plating, applying an electrodeposit to only a part of an article. Little has been published about the techniques or the properties of coatings they produce.

Continuous plating of wire and strip is, unlike the preceding techniques, a prefabrication process. The production of tinplate is the largest scale continuous operation, but any electrodeposit may be applied this way. Subsequent fabrication processes are likely to damage the coating, so that pre-coating is best reserved for ductile coatings which are anodic to the substrate in service, as is the case for tin.

Rinsing[83–92]

Between all stages of immersion (cleaning, pickling, plating, post-plating treatment) work has to be rinsed. Once the hydrophobic solid has been removed, metal surfaces withdrawn from solutions carry a film of liquid. The solution lost this way is known as *drag-out*. A film 10 μm thick is the minimum retained by smooth, well-drained, vertical surfaces. On rough or horizontal surfaces and in recesses it is much thicker, as it is also with viscous solutions. During rinsing the film is diluted, and the ratio of the final concentration to that present initially is the *dilution* ratio. The dilute material is carried forward to the next process, and clearly the highest concentration of impurity permissible before the subsequent process is affected adversely determines the maximum dilution ratio which can be allowed. Sometimes there is a minimum dilution ratio; between nickel plating and chromium plating it is essential that the rinsed metal surface does not become passive, and prolonged rinsing carries a danger of eliminating the slight but important amount of rinse water corrosion which keeps the surface active between stages.

Usually rinsing troubles are caused by a dilution ratio that is too high. If incoming work passes through a process stage, and the drag-out from that stage is in turn discarded in a subsequent rinse, the maximum concentration of material carried into the bath is equal to that in the film carried over. However, there is an increasing tendency to conserve materials and steps are taken to return drag-out losses. In so doing the impurities are also returned, so conservation measures require a reduction in the dilution ratio of the preceding rinse. Inadequate intermediate rinses are detrimental to the corrosion resistance of the coating because carried-over impurities impair the functioning of plating baths. Inadequate final rinsing leads to increased corrosion of the coating, and to staining. Staining, which is a serious aesthetic problem with decorative coatings, may itself arise from corrosion. Some stains are caused by the precipitation of dissolved solids when rinse water evaporates, but in other cases they are caused by corrosion supported by the presence of an electrolyte in the rinse water.

Post-Plating Treatments[93-96]

Where the corrosion resistance of a coating depends upon its passivity, it is common to follow plating with a conversion coating process to strengthen the passive film. Zinc, cadmium and tin in particular are treated with chromate solutions which thicken their protective oxides and also incorporate in it complex chromates (see Section 16.3). There are many proprietary processes, especially for zinc and cadmium. Simple immersion processes are used for all three coatings, while electrolytic passivation is used on tinplate lines. Chromate immersion processes are known to benefit copper, brass and silver electrodeposits, and electrolytic chromate treatments improve the performance of nickel and chromium coatings, but they are not used to the extent common for the three first named.

The tin coatings as deposited in tinplate manufacture are not bright. Until comparatively recently bright tin electrodeposition was not practised commercially, there being no reliable addition agents. To produce bright tin on tinplate and other products, the process of *flow melting* or *flow brightening* is used; tinplate is heated by induction or resistance, and plated articles by immersion in hot oil to melt the tin, which flows under surface tension to develop a bright surface. While the tin is molten it reacts to form an alloy layer with the substrate. The alloy layer alters the corrosion behaviour.

Other electroplated articles are heated after plating to expel hydrogen which has entered the substrate during cleaning, pickling and plating, and which embrittles some metals, mainly high-strength steels. Generally speaking alteration of the deposit structure and properties is not desired. Another use of post-plating heat treatment is to improve adhesion, as already mentioned (p. **13**:26).

Mechanical polishing, formerly the principal means of producing bright coatings, has become less important with the extension of the use of brightening addition agents. Mechanical polishing reduces the thickness of a coating, and may cut through to the substrate. As corrosion resistance is related to thickness, mechanical polishing can be detrimental. It may also increase porosity (p. **13**:41).

Properties of Electrodeposits

Thickness[97]

Coating thickness is one of the most important quantities connected with corrosion resistance, and its measurement and control is a feature common to all electroplating operations and in all quality specifications. In some cases coating thickness has functional importance, e.g. where there are fitting tolerances, as with screw threads. In most cases however it is the connection with corrosion resistance that makes thickness important. Where the coating is anodic to an area of substrate exposed at a discontinuity the coating is slowly consumed by corrosion, but the criterion of failure is the appearance of substrate corrosion product. This does not form until almost all the coating is consumed. Coatings which are cathodic to the substrate must have no discontinuities if substrate corrosion is to be suppressed.

The criterion of failure is usually the same. Freedom from discontinuity is also related to thickness. Discontinuities have three origins: spontaneous cracking to relieve internal stress, pores formed during the growth of the coating (see p. **13**:41), and abrasion and wear. The last two causes, i.e. porosity and wear, both exhibit diminishing incidences as thickness rises. Apart from the peculiar case of electrodeposited chromium, internal stress cracking is a sign of incorrect plating conditions. Broadly speaking, thickness and corrosion resistance increase together. The thickness of an electroplated coating is never uniform. On the *significant* area (i.e. that on which corrosion resistance and other special properties are important) of a plated surface there are two important thicknesses, i.e. (*a*) *average thickness*, which determines the production rate and plating costs; and (*b*) *minimum local thickness*, which, as the weakest link in the chain, determines the corrosion resistance. The ideal is to make these equal; the larger the difference the greater the waste of metal. The difference can be reduced by special procedures, but at a cost.

When the cathode is being plated, the electrical field is not uniform. Both electrodes are equipotential surfaces, so that prominent parts of the cathode, e.g. corners, edges, protuberances, etc. which are relatively nearer the anode are plated at a higher average current density, resulting in a thicker coating. Recesses and more distant parts are more thinly plated. The distribution of thickness tends to be the reverse of that found with paints, hot-dipped and other coatings which are applied as liquids. Liquid-applied coatings are thin on sharp edges, and thick in recesses because of the effects of surface tension and radii of curvature.

The numerous factors which contribute to the thickness distribution can be divided into two groups, i.e. (*a*) those connected with the nature of the plating bath (p. **13**:33) and (*b*) those to do with the geometry of current paths in the bath, including the shapes of the electrodes (p. **13**:35).

Throwing Power[98-105]

In a given plating cell, thickness distribution is found to vary with bath composition, current density, temperature and agitation. It is common to speak of the *throwing power* of a plating bath. The throwing power of chromic acid baths is *poor*, i.e. there is a relatively large difference between maximum and minimum local thickness; conversely, the throwing power of alkaline stannate baths is *good*, i.e. there is much less difference in the local thicknesses. Strictly speaking the bath composition should be qualified by the conditions of use, as they affect throwing power. Otherwise, the usual conditions are implied. A numerical *throwing index* can be calculated from the performance of a plating bath in a cell of standard geometry. Two widely used cells are (*a*) the Haring–Blum cell and (*b*) the Hull cell (Fig. 13.11). The Haring–Blum cell was devised for throwing index measurement; the Hull cell is used mainly to study the effects of varying bath composition.

The Haring–Blum cathode is divided into two equal plane areas, distant ℓ_1 and ℓ_2 from a common anode, and a quantity called the *primary current density ratio P* is defined as

$$P = \ell_2/\ell_1$$

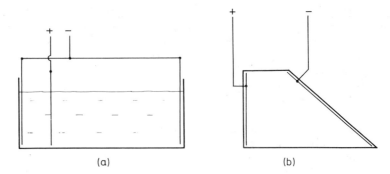

Fig. 13.11   (*a*) Haring-Blum cell for throwing index measurement, in elevation and (*b*) Hull cell (plan view) which can also be used for measuring throwing indices

This is the ratio in which the current would divide, if electrolytic resistance were to control its flow entirely. The metal distribution ratio M is the ratio of the thicknesses of the coating actually deposited during a measurement. There are several numerical scales of throwing index T, but Field's is widely adopted:

$$T = 100 \, \frac{P-M}{P+M-2} \% \qquad \qquad ...(13.11)$$

On this scale, zero represents the case when $M = P$, and electrolyte resistance is the main factor. Throwing power can be worse, down to a limit $T = -100\%$ when $M = \infty$, i.e. no deposit at all on the far cathode. Conversely, when $M < P$, T is positive. Were M to reach $1 \cdot 0$ despite the difference in position, $T = +100\%$. At one time $+100\%$ was regarded as an unrealisable limit, but conditions have been found for which $T = +150\%$ in a Haring–Blum cell.

The Hull cell cathode has a continuous variation of current density along its length, and there are equations which give the primary current density at any point not too near the end. If the local thickness is measured at two points for which P is known, T can be calculated. The real current distribution is a function of cathode and anode polarisation as well as of the resistance of the electrolyte. The metal distribution ratio will be

$$M = \frac{(V - \Delta E_1) \, \ell_2 \, \varepsilon_1}{(V - \Delta E_2) \, \ell_1 \, \varepsilon_2} \qquad \qquad ...(13.12)$$

where V = cell potential difference between anode and cathode,
         ΔE = total potential difference caused by polarisation (anode and cathode) on the cathode area indicated by the subscript and
         ε = cathode efficiency as indicated by the subscript.

As ΔE will be a function of current density, T will be a function of electrode area, and comparisons should therefore be made with cells of standard size. Equation 13.12 shows that high throwing indices will result when polarisation rises steeply with current ($\Delta E_1 \gg \Delta E_2$) and cathode efficiency falls steeply ($\varepsilon_2 \gg \varepsilon_1$). The primary current ratio, $P = \ell_2/\ell_1$, affects the result because by altering the currents the polarisation terms are altered. For example, with an acid copper bath in a Haring–Blum cell, $194 \, \text{A/m}^2$ average c.d.:

$$P = \quad 2 \quad\quad 5 \quad\quad 11 \quad\quad 23$$
$$T\% = +7 \quad +11 \quad +22 \quad +41$$

An increase in conductivity usually increases T because it increases the proportion of polarisation in the total cell potential difference and lowers the ratio $(V - \Delta E_1)/(V - \Delta E_2)$. Changing the conductivity of an acid copper bath with sulphuric acid produced the following result ($291 \, \text{A/m}^2$ average c.d., $P = 5$):

Conductivity (S/cm)	0·08	0·15	0·26	0·30
$T(\%)$		+5	+11	+13	+27

where S is the SI unit of conductance.

Many baths in which metal is reduced from complex anions (e.g. cyanide baths, stannate baths) give high throwing indices because both polarisation and cathode efficiency variation favour a low value of M. The cathode efficiency for a typical copper cyanide bath (40°C) was:

Current density (A/m^2)	32	65	129	258	388
Cathode efficiency	76	68	56	34	21

The throwing index for the cyanide bath is usually about $+40\%$ and rises as the cell current is increased to as high as $+85\%$. Aquocation baths give values near $T = 0$, though conditions may be selected which give much higher figures if there is a steeply rising section of the polarisation curve. Chromium plating baths invariably have large negative throwing indices, despite deposition from a complex ion. The cause is the anomalous *rising* trend of cathode efficiency with current density and the existence of a minimum current density *below* which the efficiency is zero. A typical bath ($400 \, \text{g/l} \, CrO_3$, $4 \, \text{g/l} \, H_2SO_4$, 38°C) gave:

Cathode current density (A/m^2)	199	253	384	763	1 785	5 130	30 800
Cathode efficiency (%)	0	5·9	11·9	13·9	18·8	22·7	24·4

If the current density on the far cathode in a Haring–Blum cell was $199 \, \text{A/m}^2$ or less, $T = -100\%$.

Throwing indices measured in a Hull cell differ from those in a Haring–Blum cell because of the differences in geometry. In a Hull cell several pairs of points can be found which have the same primary current ratio, but for which M and hence T are found to vary because of polarisation changes.

Current Path Geometry[106-110]

The polarisation and cathode efficiency terms in equation 13.12 cannot be altered in practice to improve thickness distribution, as they tend to be decided by overriding considerations. It is usual to accept the distribution obtained without special precautions as being the best commercial solution, although the average thickness needed to achieve the necessary minimum local thickness may be high. Where this approach does not serve there are a number of methods of altering the term ℓ_2/ℓ_1 in equation 13.12:

(a) By using shaped (conforming) anodes, additional (auxiliary) anodes or 'bipolar' anodes to bring anode areas nearer to cathode recesses. Insoluble anodes are better where they are applicable as, once made, they do not alter shape during use.

(b) By using non-conducting shields of plastic or glass to equalise the current path lengths.

(c) By placing auxiliary cathodes ('robbers' or 'thieves') near high-current-density points to divert deposition. This does not save metal, but has the merit that auxiliary cathodes can be incorporated into jigs for long runs in automatic plating machines. Auxiliary cathodes are used in heavy chromium deposition where metal waste is secondary to the cost of removing excess chromium when grinding to precise dimensions. Where a number of small parts are plated together on a jig, it is usually possible to dispose them so that they serve as 'robbers' for each other.

(d) By attention to certain 'rules' when designing articles which will be finished by electroplating. Many external contours are chosen for reasons of style. It helps to avoid features like sharp recesses, which are bound to cause trouble. A simple rule is the '1 in ball test' or perhaps the '25 mm ball test': if there is any part of a surface which a ball of this diameter cannot touch when rolled over it, there will be difficulties. There are other design aspects, covered in specialist publications, attention to which improves the corrosion resistance which can be imparted by plating (see also Section 12.4).

Structure-dependent Properties[111–116]

Composition of the electrodeposit Attention has been drawn to the dependence of structure on both substrate and plating conditions, and the transition in properties which occurs across the section of a deposit. Most commercial electrodeposits have a high purity, yet in a sense impurities are vital to their successful application. Alloy electrodeposition possesses a literature whose bulk attests the subject's fascination for research (which the author shares), but is out of proportion to the extremely limited commercial applications. Alloys in general metallurgical practice provide a variety of mechanical properties; in electroplating the range of properties desired is narrower, and it can generally be achieved by altering the structure of a single metal deposit through changes in the plating bath composition or plating conditions. The microstructure of an electrodeposit can be altered much more than that of a cast and worked metal. This is because the deposit forms well below its melting point, where crystallisation processes are hindered by the virtual absence of solid-state diffusion. Consequently, very small amounts of 'impurity' absorbed at important growth sites on the surface cause large changes in the structure of what is, chemically, almost pure metal. The structure is metastable, but permanent as long as the electrodeposit is not heated. A variety of mechanical and physical properties are a reflection of the structure: hardness, ductility, tensile strength, internal stress, electrical and thermal conductivity, etc. As the structure of an electro-deposited metal is altered by changing the plating conditions, the mechanical and physical properties also alter. A plot of structure-dependent properties against the plating variable usually shows the various properties moving in parallel or inverse motion, and over ranges not accessible in cast and worked metal of the same composition. However, if electrodeposits are heated to temperatures where moderate mobility of the atoms is possible, their proper-ties rapidly revert to 'normal'. The corrosion resistance of electrodeposits depends much more on chemical composition rather than on structure, so that the corrosion resistance of a particular metal is retained for a wide range of mechanical and physical properties.

The 'impurities' responsible for modifying the structure may originate from: water (dispersed oxides); adsorbing ions, especially cyanides; organic addition agents parts of which are incorporated; or ions of a second metal which are co-deposited. Some regard deposits in which the impurity is a small amount of a second metal as an alloy, but generally they have the same sort of metastable structures as are obtained with non-metallic impurities, rather than those of stable alloys of the same composition. The 'alloying' metal serves to cause and perpetuate a non-equilibrium structure whose real basis is the low temperature of the electrocrystallisation process. Generally, the corrosion properties of the various different structures of a given metal are much the same, with the notable exception of nickel containing sulphur from addition agents, which has already been mentioned.

Internal Stress[117-120]

Electrodeposits are usually in a state of internal stress. Two types of stress are recognised. First order, or macro-stress, is manifest when the deposit as a whole would, when released from the substrate, either contract (tensile stress) or expand (compressive stress) (Fig. 13.12). Second order or micro-stress, occurs when individual grains or localities in the metal are stressed, but the signs and directions of the micro-stresses cancel on the larger scale. The effects of first order stress are easily observed by a variety of techniques.

Second-order stress is difficult to observe and much less extensively studied. The causes of internal stress are still a matter for investigation. There are broad generalisations, e.g. 'frozen-in excess surface energy' and 'a com-bination of edge dislocations of similar orientation', and more detailed mechanisms advanced to explain specific examples.

Tensile first-order stress is a corrosion hazard in coatings cathodic to the substrate. Compressive stress is not usually troublesome, nor is stress of either sign in anodic coatings. Less can be said about high second-order stress, though it may well cause brittleness. If tensile stress is large enough, the coating cracks and a cathodic coating will fail to protect, as illustrated in Fig. 13.13. Tensile stress below the level needed for spontaneous cracking lowers the fatigue limit of a substrate. Tensile stress can in several cases be reduced to safe values by fairly minor changes in microstructure and plating conditions, insufficient to upset other desirable properties. Saccharin is an addition agent for reducing stress in nickel; additions of ammonium chloride

Fig. 13.12   An electrodeposit showing unusually high compressive stress. A 150 × 150 mm copper sheet was insulated with lacquer on one side and electroplated with Sn–35 Ni alloy. The high compressive stress has caused the sheet, originally flat, to coil in the manner shown, with the electrodeposit *outside*

reduce stress in tin–nickel alloy, and small changes in bath temperature and $CrO_3 : SO_4$ ratio reduce stress in chromium.

The effects of tensile stress in the various layers of nickel plus chromium coatings are complex, and internal stress in both chromium and nickel (post-nickel strike or PNS) layers can be harnessed to produce beneficial cracking ('microcracking').

Ductility, Hardness, Wear, Strength[121–124]

The mechanical properties reflect very closely the structures of electro-deposits. The softest, most ductile, weakest form of a particular metal is that with a large crystal size, deposited with minimum polarisation from baths which have no addition agents. This is the type of deposit in which pseudo-morphism is strongest. In terms of the accepted deposition mechanism, there is the least inhibition of adion mobility as the deposit grows, and least inhibition of those sites at which equilibrium growth would occur. This electrodeposit has properties the nearest to those for the annealed metal, but even so tends to be somewhat harder. Because of pseudomorphism the properties near the substrate interface may be greatly modified if the latter has a metastable structure, especially one with very small grains produced by mechanical working. The deposit in turn becomes 'work hardened' by pseudomorphic growth.

When electrodeposition is inhibited the metal becomes harder, less ductile and increases in tensile strength. Metals deposited from acidic solutions of

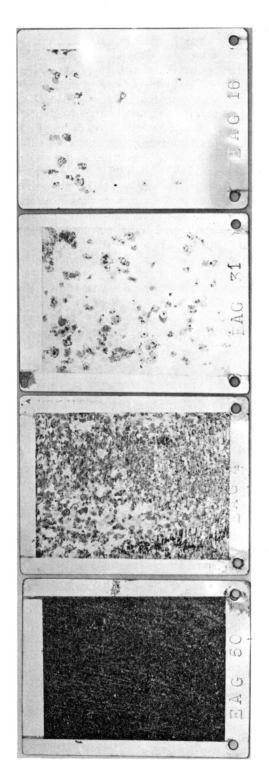

Fig. 13.13 Cracking in a cathodic coating due to tensile stress which exceeds the coating's strength, leading to loss of protection. All 5 μm coatings of tin–nickel on steel, with internal stresses. From left to right: EAG 50, 134 N/mm² tensile; EAG 4, 86 N/mm² tensile; EAG 31, 57 N/mm² tensile; EAG 16, 127 N/mm² compressive

aquocations become harder when the pH is raised to near the value at which the hydroxide precipitates. Co-deposited oxide acts as an addition agent, giving small grained, hard deposits. Hard nickel is produced for engineering surfacing from high pH baths. Many metals can be electrodeposited in extremely hard forms from inhibited baths, but they tend to become brittle, with high internal stress, so that the true tensile strength is hard to establish. Ductility necessarily falls as hardness rises, and coatings become more susceptible to damage by impact, reducing their protective value if they are cathodic to the substrate. Some applications of electroplating depend on the production of unusually hard and wear-resistant forms of corrosion-resistant metals. Thick coatings of chromium and nickel are applied to numerous steel parts to combine wear resistance with corrosion resistance. Thick or engineering chromium electrodeposits crack repeatedly during deposition, but the cracks are subsequently sealed and none should traverse the entire coating. Thick chromium coatings have practically no ductility, and because of their defective structure they have a low effective strength. They serve best on stiff substrates.

Gold coatings on separable electric contacts and slip rings make use of the high hardness possible with electrodeposition to resist wear. Rhodium is another metal which can be exceptionally hard. Thick coatings have a cracked-sealed structure similar to that of chromium.

Interdiffusion with the Substrate[125–126]

A thin metal coating on a metal substrate is not a stable entity; greater stability would be attained if the coating were to diffuse evenly throughout the substrate. Fortunately, at ambient temperatures most of the usual combinations interdiffuse so slowly as to present no practical problem. At high temperatures however, many coatings diffuse quickly. Diffusion in a few systems at moderate temperatures causes corrosion problems. Difficulties can occur with tin, which, with its low melting point of 231°C, is relatively 'hot' at room temperature. On copper and copper-alloy substrates diffusion transforms the tin into the intermetallic phases Cu_6Sn_5 and Cu_3Sn. At 100°C the transformation is accelerated, and 5 μm of tin may become wholly alloyed within a year. The alloy coating may pass as tin having a silvery colour, but it is much harder and has a very stable passivity. One use of tin on copper is to facilitate easy joining by soldering, but the alloy has a high melting point and is not easily wet by solder. Thin 'tin' coatings on copper which have become wholly alloyed in storage are difficult to solder. Sometimes extremely thin coatings (0·25 μm) used purely for solderability become wholly alloyed in a few weeks. Parts should not be stored too long, and very thin coatings are false economy.

Tin will protect copper from corrosion by neutral water. Pure tin is anodic to copper, and protects discontinuities by sacrificial corrosion. Both intermetallic phases are strongly cathodic to copper, and corrosion is stimulated at gaps in wholly alloyed coatings. An adequate thickness of tin is needed for long service, e.g. 25–50 μm. Another diffusion problem occurs with tin-plated brass. Zinc passes very quickly to the tin surface, where under conditions of damp storage zinc corrosion products produce a film which greatly impairs solderability. An underplate of copper, or better still nickel, usually cures this trouble.

A similar problem, i.e. diffusion of the substrate through the coating to corrode at the surface, arises with gold-plated copper. Many gold coatings are used to ensure a low electric contact in electrical connectors. Gold is pre-eminent because of the absence of stable-corrosion-product films under most service conditions, but it is expensive, so coatings are kept as thin as possible. Electronic devices may operate at fairly high temperatures (100–150°C), and significant amounts of copper may diffuse through the coating to produce a film of oxide on the surface, nullifying the contact value of the gold. Nickel underplate mitigates this trouble (though increasing plating difficulties). To reduce costs, attempts have been made to dilute gold with cheaper metals, while retaining gold-like corrosion properties. Cadmium has been used as a diluent, but while quite high cadmium–golds are gold-like at 25°C, at higher temperatures cadmium oxidises at the surface. Pure gold is preferred for high-temperature contacts.

Porosity[127–139]

In the very earliest stages of electroplating the substrate carries discontinuous areas of deposit growing around nuclei. Lateral growth causes the great majority of growing edges to coalesce with sufficient perfection to be impervious to corrosive gases and liquids. On normal metallic substrates a few edges do not grow together, and a gap remains in the coating. As the coating thickens the gap is propagated as a channel through the coating, to form a *pore*. Under the conditions chosen for practical electroplating, pores diminish in cross section as deposition continues, and pore density (pores per unit area) falls as thickness increases. The corrosion which occurs when pores allow liquid and gaseous corrosive agents to reach the substrate varies in importance according to the relation between the corrosion potentials of deposit and substrate, the corrosive environment and the function of the coating.

If the environment favours wet corrosion processes, relative polarity is the main consideration. If the coating is anodic, porosity is seldom of any serious consequence. The cathode is the very small area of substrate exposed at the base of the pore, and the restricted channel limits the diffusion of reactants and products. The large anode area provided by the coating reduces the bimetallic corrosion current density thereon. Two important examples of this type are zinc coatings on steel in cold waters or the atmosphere and tin coatings on steel on the inside (but not the outside) of a sealed, air-free can of wet food. In the first case oxygen is the cathodic reactant; in the second it is hydrogen ions (or water). Where the coating is cathodic, porosity enables the exposed substrate to corrode. In most cases this is detrimental; the exception is found in some multi-layer nickel plus chromium coatings where certain forms of porosity in the chromium layer are harnessed to divert the direction of corrosion to the overall benefit of coating life. In other cases corrosion at pores causes trouble. In wet atmospheric corrosion, substrate corrosion product, if coloured and insoluble, spoils decorative appearance. In immersed conditions or condensing atmospheres, if the corrosion product is soluble intense pore corrosion will perforate sheet metals. Here a porous coating may accelerate corrosion when compared with the uncoated substrate.

Porosity causes little trouble when corrosion is restricted to dry processes (oxidation). Corrosion products block the pores and stifle the reaction.

There was much research into the causes of porosity in nickel deposits when it was thought to be the main cause of failure in nickel and chromium plate. Much was discounted as it became clear that nickel pitting at discontinuities in the chromium was the factor determining service life. Porosity remains relevant to the corrosion resistance of simpler cathodic coatings, and especially for gold. The use of gold for contact surfacing since about 1950 has revived the importance of studies of porosity. Pores in gold coatings allow films of substrate corrosion product to contaminate the surface and to destroy the low contact resistance of the gold. Sulphides, which are one of the products of corrosion by service atmospheres, have a particularly high rate of spreading over gold in the solid state (Fig. 13.14).

Pores originate on substrate areas known as *precursors*, which are of at least three types. Firstly, an obvious cause is an inclusion of foreign material which is a semiconductor or insulator—particles of oxide, sulphide, slag, polishing abrasive, etc. When electrodeposition starts, inclusions will not be nucleation sites, and they will impede the lateral growth and coalescence of crystals from neighbouring nuclei. Secondly, substrates whose surface grain structure has been severely disturbed by cold working (abrasion, cold rolling, drawing, etc.) have precursors whose physical state (rather than chemical difference as in the first type) precludes coalescence of the electrodeposit. This is probably an effect of pseudomorphic growth. Relatively low-temperature annealing (as low as 210°C for steel) greatly reduces the effect, and further cold work increases it again (Fig. 13.15). The third type of precursor

Fig. 13.14 Spread of silver sulphide from discontinuities in gold electrodeposits on silver substrates. The gold was deliberately scratched and the specimen exposed for 24 h to an atmosphere containing 10% SO_2. Immediately after this the sulphide stain extended 0·2 mm. Five years later, the stain extends to about 13 mm, after storage in a normal indoors atmosphere

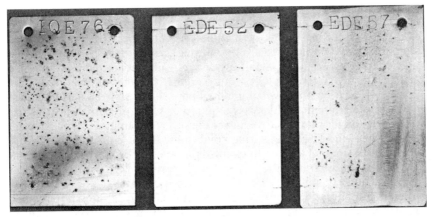

Fig. 13.15 Porosity caused by a cold-worked substrate. Left (EQE 76) cold-rolled steel as received; centre (EDE 52) steel bright-annealed in vacuum before plating, 2·5 h at 700°C; right, annealed steel, further cold-rolled (0·914 mm to 0·864 mm) produces porosity again. No steel was removed from the surface; 5 μm tin–nickel electrodeposit

is a crevice in the substrate. If the depth is great relative to the width, the electric field is excluded and deposition does not occur within the crevice. Lateral growth is impeded once the edges from neighbouring nuclei reach it in much the same way as with a non-conducting inclusion. A pore caused by any type of precursor in one electrodeposit becomes in turn a precursor for a second deposit plated over it. There may be other forms of precursor.

In a particular area of substrate there will be a number of precursors, distributed over a range of sizes, and reflecting the nature, composition and

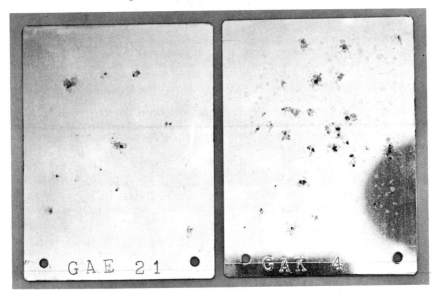

Fig. 13.16 Increase in porosity of an electrodeposit caused by mechanical polishing. Left, 7·5 μm unpolished coating; right, polished with lime finishing compound. The average thickness removed by abrasion was 0·1 μm

history of the metal. In principle anything affecting the substrate surface will affect porosity in an electroplated coating.

As deposition continues growth gradually diminishes the surface opening of a pore, and if continued to a sufficient thickness, closes it, leaving a sealed cavity filled with solution. Small precursors will generate pores which seal relatively early, large ones will require greater thicknesses. The total pore density revealed by a test which renders pore sites visible falls as thickness increases. The minimum thickness required to seal a precursor of fixed size will depend on the rate of narrowing of the surface opening, and, as a growth process this will reflect the plating conditions. Because of this, the density of pores still open at a fixed thickness is a function of all the plating conditions, i.e. of the composition of the plating bath, of temperature, current density, agitation and anything affecting deposit growth.

Post-plating treatments affect pore density, either by closing pores which are still open or opening sealed ones. It has been asserted that mechanical polishing in general, or flow-melting for tin, are both processes which could seal pores and reduce porosity. It is also conceivable that polishing might cut-open sealed pores, and likewise under flow-melting conditions the vaporisation of solution trapped in sealed pores could disrupt the coating and recreate discontinuities. The author has come across no convincing demonstration of porosity reduction by either treatment, but has found experimental evidence for porosity increases (Fig. 13.16).

M. CLARKE

REFERENCES

1. Blum, W. and Hogaboom, G. B., *Principles of Electroplating and Electroforming*, McGraw-Hill, New York (1949)
2. Brenner, A., *The Electrodeposition of Alloys*, Academic Press, New York (1963)
3. Dennis, J. K. and Such, T. E., *Nickel and Chromium Plating*, Newnes-Butterworths, London (1972)
4. Graham, A. K. and Pinkerton, H. L., (Eds.), *Electroplating Engineering Handbook*, Reinhold, New York (1956)
5. Lowenheim, F. A. (Ed.), *Modern Electroplating*, John Wiley, New York (1974)
6. Raub, E. and Müller, K., *Fundamentals of Metal Deposition*, Elsevier, Amsterdam (1967)
7. Vagramyan, A. T. and Soloveva, Z. A., *Technology of Electrodeposition*, Robert Draper, Teddington (1961)
8. Barclay, W. R., *J. Electrodepositors' Tech. Soc.*, **12**, 169 (1937)
9. Brugnatelli, L., *Ann. Chim. (Pavia)*, **18**, 152 (1800) (silver)
10. Dubpernell, G., *Plating*, **47**, 35 (1960) (history of chromium plating)
11. Dubpernell, G., *Plating*, **46**, 599 (1959) (history of nickel plating)
12. Durose, A. H., *Plating*, **57**, 793 (1970) (USA developments)
13. Elkington, G. R. and Elkington, H., UK Pat. 8 447, March (1840)
14. Passal, F., *Plating*, **46**, 629 (1959) (copper)
15. Smee, A., *Elements of Electrometallurgy*, Longmans Green, London, 2nd edn (1843)
16. Wormwell, R., *Electricity in the Service of Man*, Cassell, London (1897)
17. Edwards, J. and Swanson, C. J., *J. Electrodepositors' Tech. Soc.*, **29**, 290 (1953)
18. Graham, A. K., *Trans. Inst. Met. Fin.*, **31**, 259 (1954)
19. Meyer, W. R., *ibid.*, 290 (1954)
20. Moreley, E., *Trans. Inst. Met. Fin.*, **33**, 102 (1956)
21. Westman, A. E. R. and Mohrnheim, F. A., *Plating*, **42**, 154, 281 and 417 (1955)

22. Beacom, S. E., *Plating*, **56**, 129 (1969)
23. Hepfer, I. C., Hampel, K. R., Vollmer, T. L. and Boehm, D. R., *Plating* **58**, 333 (1971)
24. Jostan, J. L. and Bogenschutz, A. F., *Plating*, **56**, 399 (1969)
25. Logi, G. R. and Rantell, A., *Trans. Inst. Met. Fin.*, **46**, 91 (1968)
26. Murphy, N. F. and Swansey, E. F., *Plating*, **58**, 583 (1971)
27. Perrins, L. E., *Trans. Inst. Met. Fin.*, **50**, 38 (1972)
28. Prince, D. E., *Plating*, **58**, 588 (1971)
29. Rantell, A., *Trans. Inst. Met. Fin.*, **47**, 197 (1969)
30. Saubestre, E. B., *Trans. Inst., Met. Fin.*, **47**, 228 (1969)
31. Saubestre, E. B. and Khera, R. P. *Plating*, **58**, 464 (1971)
32. Woldt, G., *Trans. Inst. Met. Fin.*, **47**, 236 (1969)
33. Moore, A. J. W., *J. Electrodepositors' Tech. Soc.*, **28**, 117 (1952)
34. Pinner, W. L., *Proc. Amer. Electroplater's Soc.*, **40**, 83 (1953)
35. Raub, E., Mietka., M. and Beeskow, H., *Metalloberfläche*, **13**, B153 (1959)
36. Samuels, L. E., *Electroplating and Metal Finishing*, **11**, 130 (1959)
37. Samuels, L. E., *J. Inst. Metals*, **85**, 177 (1956)
38. Samuels, L. E., *Plating*, **48**, 46 (1961)
39. Research Report No. 38, American Electroplaters' Soc. (1958)
40. Research Report No. 39, American Electroplaters' Soc. (1959)
41. Fleischmann, M. and Thirsk, H. R., *Advances in Electrochemistry and Electrochemical Engineering*, Ed. Delahay, P. and Tobias, C. W., Vol. 3, Chapt. 3, Interscience, New York (1963)
42. Vetter, K. J., *Electrochemical Kinetics*, English edn., Academic Press, New York (1967)
43. Bockris, J. O'M. and Razumney, G. A., *Fundamental Aspects of Electrocrystallisation*, Plenum Press (1967)
44. Eisenberg, D. and Kauzmann, W., *The Structure and Properties of Water*, Oxford University Press (1969)
45. Lyons, E. H., in *The Chemistry of Co-ordination Compounds*, Ed. Bailar, J. C., Reinhold, New York (1956)
46. Edwards, J., *Trans. Inst. Met. Fin.*, **41**, 169 (1964)
47. Kendrick, R. J., *Plating*, **48**, 1 099 (1961)
48. Saubestre, E. B., *Plating*, **45**, 927 (1958)
49. Saubestre, E. B., *Plating*, **45**, 1 219 (1958)
50. Hardisty, D. W., *Plating* **56**, 705 (1969)
51. Knapp, B. B., *Plating* **58**, 1 187 (1971)
52. Bailey, G. L. J., *J. Electrodepositors' Tech. Soc.*, **27**, 233 (1951)
53. Beyer, S. J., *Plating* **56**, 257 (1969)
54. Longkind, J. C., *Trans. Inst. Met. Fin.*, **45**, 155 (1967)
55. Vandenberg, R. V., *Trans. Inst. Met. Fin.*, **45**, 161 (1967)
56. West, E. G., *J. Electrodepositors' Tech. Soc.*, **21**, 211 (1946)
57. Wyszynski, A. E., *Trans. Inst. Met. Fin.*, **45**, 147 (1967)
58. Epelboin, I., Froment, M. and Maurin, G., *Plating*, **56**, 1 356 (1969)
59. Finch, G. I. and Williams, A. L., *J. Electrodepositors' Tech. Soc.*, **12**, 105 (1937)
60. Finch, G. I. and Sun, C. II., *Trans. Faraday Soc.*, **32**, 852 (1936)
61. Finch, G. I., Wilman, H. and Yang, L., *Discussions Faraday Soc.*, **1**, 144 (1947)
62. Hothersall, A. W., *Trans. Faraday Soc.*, **31**, 1 242 (1935)
63. Nystrom, W. A., *Plating*, **56**, 285 (1969)
64. Thomson, G. P., *Proc. Roy. Soc.*, A 1931, **113**, 1 (1931)
65. Brenner, A., Zentner, V. and Jennings, C. W., *Plating*, **39**, 865 (1952)
66. Clarke, M. and Chakrabarty, A. M., *Trans. Inst. Met. Fin.*, **48**, 124 (1970)
67. Fischer, H., *Elektrolytische Abscheidung und Elektrokristallisation von Metallen*, Springer-Verlag, Berlin (1954)
68. Weil, R., *Metal Finishing*, **53** No. 11, 60 and No. 12, 60 (1955)
69. Wilman, H., *Trans. Inst. Met. Fin.*, **32**, 281 (1955)
70. Avila, A. J. and Brown, M. J., *Plating*, **57**, 1 105 (1970)
71. Baeyens, P. A., *Trans. Inst. Met. Fin.*, **31**, 429 (1954)
72. Beckmann, M. E. and Maass-Craefe, F., *Metalloberfläche*, **5**, A161 (1951)
73. Crossley, J. A., Kendrick, R. J. and Mitchell, W. I., *Trans. Inst. Met. Fin.*, **45**, 59 (1967)
74. Davies, G. R., *Trans. Inst. Met. Fin.*, **51**, 47 (1973)
75. Hicklin, A. and Rothbaum, H. P., *ibid.*, **34**, 53 and 199 (1957)

76. Jernstedt, G. W., US Pats. 2 451 340 and 2 451 341 (1948)
77. Kendrick, R. J., *Trans. Inst. Met. Fin.*, **44**, 78 (1966)
78. See Chapter 3 of Reference 7
79. Wernlund, C. J., US Pat. 2 701 234 (1955)
80. Geissmann, W. and Carlson, R., *Proc. Am. Electroplaters' Soc.*, **39**, 153 (1952)
81. Nanis, L., *Plating*, **58**, 805 (1971)
82. Rasmerova, N. J., *J. Electrodepositors' Tech. Soc.*, **13**, paper 11 (1937)
83. Clarke, M., *Chemistry and Industry* 329 (1970)
84. Clarke, M., *Trans. Inst. Met. Fin.*, **46**, 201 (1968)
85. Clarke, M. and Ashburn, R. J., *ibid.*, **47**, 18 (1969)
86. Clarke, M. and Kieszkowski, M., *ibid.*, to be published
87. Clarke, M. and Zaya, P., *ibid.*, **50**, 54 (1972)
88. Soderberg, K. G., *Proc. Amer. Electroplaters' Soc.*, **24**, 233 (1936)
89. Tallmadge, J. A. and Buffam, B. A., *J. Water Pollution Control Fed.*, **33**, 817 (1961)
90. Tallmadge, J. A. and Mattson, R. A., *ibid.*, **34**, 723 (1962)
91. Tallmadge, J. A. and Barbolini, R. R., *ibid.*, **38**, 1 461 (1966)
92. Walker, C. A. and Tallmadge, J. A., *Chem. Eng. Progress*, **55** No. 3, 73 (1959)
93. Hoare, W. E., *Tinplate Handbook*, Tin Research Institute, Greenford
94. Jenkins, H. A. H. and Freeman, D. B., *Trans. Inst. Met. Fin.*, **42**, 163 (1964)
95. Raub, E. and Müller, K., *Fundamentals of Metal Deposition*, Elsevier, Amsterdam, Section 5.6 (1967)
96. Thwaites, C. J., *The Flow Melting of Electrodeposited Tin Coatings*, Tin Research Institute, Greenford
97. Reference should be made to publications on electrodeposit specifications of the British Standards Institution, American Standards Institution and other standardising bodies.
98. Agar, J. N. and Hoar, T. P., *Disc. Faraday Soc.*, **1**, 162 (1947)
99. Clarke, M. and Bernie, J. A., *Electrochemica Acta*, **12**, 205 (1967)
100. Field, S., *Metal. Ind. (London)* **33**, 564 (1934)
101. Garmon, L. B. and Leidheiser, H., *Plating*, **48**, 1 003 (1961)
102. Leffler, E. B. and Leidheiser, H., *Plating*, **44**, 388 (1957)
103. Wagner, C., *J. Electrochem. Soc.*, **98**, 116 (1951)
104. Wagner, C., *Plating* **48**, 997 (1961)
105. Watson, S. A., *Trans. Inst. Met. Fin.*, **37**, 28 (1960)
106. Bruijn, W. de, *J. Electrodepositors' Tech. Soc.*, **27**, 1 (1951)
107. Layton, D. N., *Design for Electroplating*, International Nickel Co. Ltd., London
108. Leadbeater, C. J., *J. Electrodepositors' Tech. Soc.*, **19**, 35 (1944)
109. Tope, N. A., *ibid.*, **22**, 29 (1947)
110. *Design for Metal Finishing*, Institute of Metal Finishing, London
111. MacNaughtan, D. J. and Hothersall, A. W., *Trans. Faraday Soc.*, **31**, 1 168 (1935)
112. MacNaughtan, D. J., Gardam, G. E. and Hammond, R. A. F., *ibid.*, 729 (1935)
113. Mohrnheim, A. F., *J. Electrochem. Soc.*, **117**, 833 (1970)
114. O'Sullivan, J. B., *Trans. Faraday Soc.*, **26**, 89 (1930)
115. Schlotter, M., *ibid.*, **31**, 1 177 (1935)
116. *Nickel Plating for Engineers*, Mond Nickel Co. Ltd, London
117. Kendrick, R. J., *Trans. Inst. Met. Fin.*, **40**, 19 (1963)
118. Walker, R., *Internal Stress in Electrodeposited Metal Coatings*, Industrial Newspapers Ltd., London (1968)
119. Watson, S. A., *Trans. Inst. Met. Fin.*, **40**, 41 (1963)
120. Weil, R., *Plating*, **57**, 1 231 (1970); **58**, 50 (1971); **58**, 137 (1971), Review and Bibliography
121. Angus, H. C., *Trans. Inst. Met. Fin.*, **39**, 20 (1962); **43**, 135 (1965); **44**, 41 (1966)
122. Antler, M., *Proc. 6th Int. Conf. Electr. Contact Phenom.*, Chicago (1972)
123. Fairweather, A., *Proc. Inst. Elec. Eng.*, **100A**, 174 (1953)
124. *Nickel Plating*, International Nickel Co. Ltd., London (1963)
125. Britton, S. C. and Clarke, M., *Trans. Inst. Met. Fin.*, **36**, 230 (1959)
126. Clarke, M. and Britton, S. C., *ibid.*, **40**, 205 (1963)
127. Berdick, M. and Lux, G. A., *Proc. Am. Electroplaters' Soc.*, **30**, 19 (1942)
128. Britton, S. C. and Clarke, M., *Proc. 3rd European Corrosion Congress* (1963)
129. Clarke, M., *Trans. Inst. Met. Fin.*, **51**, 150 (1973)
130. Clarke, M. and Britton, S. C., *ibid.*, **36**, 58 (1959); **37**, 110 and 230 (1960)
131. Clarke, M. and Chakrabarty, A. M., *ibid.*, **48**, 99 (1970); **50**, 11 (1972)

132. Clarke, M. and Leeds, J. M., *ibid.*, **43**, 50 (1965); **46**, 1 and 81 (1968); **47**, 163 (1969)
133. Clarke, M. and Sansum, A. J., *ibid.*, **50**, 211 (1972)
134. Evans, U. R. and Shome, S. C., *J. Electrodepositors' Tech. Soc.*, **26**, 137 (1950)
135. Frant, M. S., *J. Electrochem. Soc.*, **107**, 1 009 (1960)
136. Hothersall, A. W. and Hammond, R. A. F., *Trans. Electrochem. Soc.*, **73**, 449 (1938)
137. Jones, M. H. *et al.*, *Proc. Am. Electroplaters' Soc.*, **44**, 53 (1957); **45**, 45 (1958); **46**, 113 (1959)
138. Shome, S. C. and Evans, U. R., *J. Electrodepositors' Tech. Soc.*, **27**, 129 (1951)
139. Clarke, M. and Subramanian, R., *Trans. Inst. Met. Fin.*, **52**, 48 (1974)

13.2 Principles of Applying Coatings by Hot Dipping

General

The application of a metallic coating to a metallic base by 'hot dipping' is the oldest, simplest and generally the cheapest method. It has certain obvious limitations: the coating metal or alloy must melt at a reasonably low temperature and the basis metal must withstand the temperature without undesirable changes in properties. The process is thus limited to coatings of tin, zinc, lead and aluminium.

Steel is by far the most important basis metal. Cast iron and malleable cast iron are also coated by hot dipping, and copper is coated with tin on a fairly large scale. For special purposes, metals such as titanium and molybdenum may be coated, e.g. with aluminium, to avoid oxidation or combination with nitrogen from the atmosphere.

Outline of the Hot-dipping Process

The hot-dipping process may be divided into four stages which are common to the three processes of tinning, galvanising and hot-dip aluminising.

1. *Pretreatment*. This includes degreasing, pickling and rinsing. Shot-blasting is advisable for cast iron and malleable iron. The 'Sendzimir'[1] pretreatment of steel strip by oxidation followed by reduction at high temperature eliminates the necessity for these operations and also the fluxing stage which follows.

2. *Fluxing* may be by immersion in an aqueous prefluxing bath followed by drying, or by passage through molten flux floating on the bath of molten metal, or both of these. In hot-dip galvanising the choice of flux procedure is affected by the composition of the bath.

3. *Dipping*. Strip or wire is coated continuously, often at high speeds. Sheets are dipped singly in rapid sequence, mechanically or by hand. Fabricated articles are dipped in batches mounted on suitable jigs if small, or singly if large, and the operation may be mechanised. Very small articles such as nuts and bolts are dipped in perforated baskets.

4. *After-treatments*. These include processes intended to produce one or more of the following results:

(*a*) Reduction of coating thickness by reducing the amount of molten metal adhering to the article as it leaves the bath. This may be done by rolling, wiping or centrifuging, or by air-blast, while the coating is still molten.

(b) Improvement of the properties or appearance of the coating. Such treatments include chromating, phosphating, light rolling and roller levelling.

(c) Changing the character of the coating—as distinct from the more superficial treatments listed under (b). Hot-dipped zinc coatings are sometimes 'annealed' to convert the whole of the coating into alloy. Aluminium coatings intended for heat resistance may be converted into alloy in the same way. Aluminium coatings can be anodised and if necessary dyed in attractive colours.

All these after-treatments, except anodising, follow immediately after the hot-dipping process and are regarded as part of the coating process.

Principles of the Hot-dipping Operation

Some of the principles of hot dipping are common to tinning, galvanising and aluminising. Assuming that the article to be coated has been properly cleaned by the processes previously mentioned, and that it is properly fluxed either before dipping or by passing through a molten flux blanket floating on the molten bath, then the following sequence of events occurs.

1. Molten flux attacks the surface, cleaning it and promoting the wetting of the basis metal by the molten metal into which it is dipped.

2. The comparatively cold article freezes out a 'skin' of metal upon itself, which rapidly melts again. For strip or wire the time involved here is exceedingly short—a fraction of a second.

3. The basis metal and the molten coating metal interact, producing an alloy layer, at a rate which increases in the order tin, zinc, aluminium.

For any given pair of metals, the total amount of alloy increases with the duration of immersion and the temperature of the bath. The interaction of iron and zinc is abnormally rapid within a narrow temperature range, slightly above the normal dipping temperature.

The amount and nature of the alloy layer may be profoundly affected by additions to the molten bath, such as silicon to an aluminium bath, or aluminium to a galvanising bath. A change in the composition of the basis metal may also have a marked effect, e.g. mild steel containing a little silicon alloys with zinc much more rapidly than does steel without silicon.

Although the alloy layer is always referred to as if it were a single layer, it is usually made up of two or more layers corresponding to known metallurgical phases.

4. As the article moves upward out of the molten bath, it carries with it a layer of liquid metal. This will vary in amount according to the roughness of the surface being coated and the viscosity of the molten metal. Two other factors are (a) the speed of withdrawal of the article and (b) the temperature of the bath, which affects its viscosity and also decides the time interval during which the outer layer is liquid and free to drain.

It is necessary to control the thickness of this liquid layer, reducing it in some cases to a fraction of its original thickness, but always ending with the desired amount and with an acceptable appearance. Zinc coatings on wire are reduced by wiping, on sheet by the use of 'exit' or skimming rolls, on strip by jets of superheated steam, on tubes and a few other simple shapes by an air blast, and on small articles by rapid centrifuging in baskets.

5. After the article leaves the coating bath, the coating cools and finally solidifies. Tin coatings are sometimes liable to 'retract' or 'de-wet' owing to surface tension, i.e. the thin liquid film develops gross discontinuities and in extreme cases forms a collection of minute droplets. De-wetting of tin coatings is a consequence of imperfect preparation of the metal, and when it is experienced, improved cleaning measures are necessary. Zinc and aluminium coatings may sometimes be completely converted to alloy, the original alloy layer growing and consuming the outer layer of zinc or aluminium.

Zinc coatings usually contain an appreciable amount of lead (up to about 1%, the solubility of lead in zinc at galvanising temperature), often some tin or antimony, and small amounts of other elements. These constituents cause the coating to solidify with characteristic spangles.

When it is known that the conditions of working are likely to cause excessive alloy growth in zinc or aluminium coatings, the article is quenched as soon as possible after leaving the bath; a bath of oil or of oil floating on water is used for 'boshing' in this way.

Hot Tinning[2]

The Hot Tinning of Fabricated Steel Articles

After degreasing by either trichloroethylene vapour or electrolytically in a hot alkaline solution, e.g. 25 g/l trisodium phosphate + 12·5 g/l sodium carbonate at 80–90°C, the articles are pickled in 50% (by vol.) hydrochloric acid at ambient temperature and rinsed. They are then dipped in an aqueous flux solution containing, for example, 240 g/l zinc chloride, 60 g/l sodium chloride, 30 g/l ammonium chloride and 6–12 ml/litre hydrochloric acid (sp. gr. 1·14). The articles, wet with flux solution, are lowered carefully into a bath of molten tin (at about 280°C) having a layer of fused flux floating on it and covering about two-thirds of its surface. The flux cover is composed of a mixture of dry chlorides in the same proportions by weight as in the flux solution being carried over with the work. Thus, in the example given, finely ground zinc, sodium and ammonium chlorides are mixed in the proportions 8:2:1 to form the flux blanket.

When the article has attained the temperature of the tinning bath, the flux cover is drawn to one side of the bath with a paddle and the article is immediately withdrawn through the flux-free area of the tin surface. The quality of the coating may be improved by dipping the tinned article into a second tinning pot in which the tin is covered with a layer of molten grease, such as tallow, and is at a temperature of 240–270°C.

Small articles may be tinned in batches, on jigs or in baskets, and then spun in a heated centrifuge to remove excess tin. The brightness of the tin coating may be preserved by quenching the articles in a layer of paraffin or white spirit, 50–100 mm thick, floating on water. Small parts can be separated from one another, before the tin coating solidifies, by allowing them to fall on to a paddle wheel rotating at 500–1 000 rev/min.

The Hot Tinning of Steel Sheet and Strip

It is not easy consistently to control coating weight in hot tinning of steel sheet or strip, below about 22 g m^{-2} of tinplate (11 g m^{-2} of surface $\equiv 1\cdot 5$ µm coating thickness), and line speeds are only about 10 m min^{-1}. Therefore, electrolytic tinning lines, in which strip can be coated uniformly with tin as thin as 0·4 µm at a speed of 7·5 m s^{-1}, have now largely replaced hot-tinning units. Hot tinning is continued on a limited scale to supply customers who still specify hot-tinned sheet.

In preparation for hot tinning, steel sheets are cathodically pickled in 3–5% (by vol.) hydrochloric acid and washed with a spray of water. They are then fed, automatically, into a Poole–Davis* single-sweep tinning unit. In this they pass downwards through a flux layer, which is mainly a concentrated solution of $ZnCl_2$, into molten tin at 320°C. As they emerge from the tin, the sheets pass up through three pairs of pressure rolls, arranged one above the other, in a box of palm oil at 240°C. Tin removed by the rolls is wiped from their under-surfaces by asbestos brushes. On leaving the palm oil the tinned sheets are cooled by air jets to solidify the coating. Excess palm oil is removed by passing the sheets first through a hot, dilute alkaline solution, where they are scrubbed with wool-fibre-covered rollers, and then through a series of wool-fibre-covered scrubbing rolls fed with sawdust.

A small amount of narrow (up to 300 mm wide) steel strip is hot tinned as a continuous strand at a speed of about 5 m min^{-1}. The strip is first electrolytically degreased in a hot alkaline solution, washed and then electrolytically pickled in dilute sulphuric acid at ambient temperature. Problems arising from the need to make electrical contact with the strip can be avoided by using a bi-polar system in which adjoining lengths of strip are alternately anodically and cathodically polarised. After being rinsed the wet strip is tinned in a manner similar to that described for sheets.

The structure of tin coatings on steel   A hot-dipped tin coating on steel has a layer of $FeSn_2$ at the interface between the tin and the steel. This alloy grows very slowly in comparison with the more complex alloy layers formed in galvanising and in aluminising. It usually comprises from 10 to 20% of the total coating thickness.

The Tinning of Cast Iron

The surface of cast iron is often contaminated with sand from the casting process and normal pickling processes cannot deal with this. The addition of some HF to a sulphuric acid bath helps to remove a siliceous skin. Making the pickling process more drastic or more prolonged often makes matters worse because it increases the sludge and smut left on the surface by the pickling action. The simplest cleaning method is by fine grit-blasting after which the iron may be fluxed and tinned directly[3].

Tinning trouble due to graphite may be overcome by rough mechanical cleaning followed by electrodepositing a thin coating of Fe or Cu. Alternatively, it may be removed, either by surface decarburisation (as in malleable cast iron) or by treatment in molten salt baths, e.g. equal parts of sodium and potassium nitrates. The nitrate salts are then washed off and the article is lightly pickled, rinsed and fluxed in the usual way.

*Details in *The Technology of Tinplate* (see Bibliography).

The Tinning of Copper

Degreasing by cathodic treatment in a hot alkaline solution is desirable when the copper contains inclusions of cuprous oxide. Pickling may be carried out in either cold hydrochloric (50% by vol.) or nitric (20% by vol.) acids, or in hot (70°C) sulphuric acid (6% by vol.) containing 1·5% by wt. of sodium dichromate.

An aqueous solution of zinc chloride, acidified with hydrochloric acid, may be used as a flux. A light flux cover is used on the tinning bath and this is normally produced and maintained by the carry-in of flux solution on the work. The bath temperature should be 250–260°C.

Alloy formation is faster than with steel and more irregular. The thickness of the alloy layer may amount to half that of the complete coating. It consists of a thin layer of Cu_3Sn adjacent to the copper and a thicker layer of Cu_6Sn_5 adjacent to the tin.

Bright-annealed copper wire is tinned by passing through a flux solution of stannous chloride, into a tinning bath at 300°C, and wiping off excess tin to give a smooth coating by pinching the wire between tightly clamped rubber blocks as it emerges from the bath. Twelve strands, in two sets of six, may be run side by side at a speed of $3 \cdot 5 \, \text{m s}^{-1}$. Instead of the multistrand system, a single wire may be drawn down to the required diameter, annealed in-line, and hot tinned at $25 \, \text{m s}^{-1}$, excess tin coating being removed by a diamond die.

An alternative flux, which is non-corrosive when cold and is thus not deleterious if traces are dragged out on the finished wire, is a mixture of fatty acids which melts to a low-viscosity liquid at about 40°C and becomes a highly active flux when the wire enters the molten tin.

Terne Coating

Coatings of lead alloyed with up to 20–25% tin are known as *terne* coatings. The tin content may be only 2% and such coatings are often described as *lead* coatings.

The coating process is essentially similar to tinning. It is, however, necessary to raise the bath temperature to suit the freezing range of the particular alloy in use.

Coatings much richer in tin are sometimes applied, e.g. to articles intended for subsequent soldering, but these are not true 'terne' coatings.

Hot-dip Galvanising of Sheet Steel and Fabricated Articles

The basic principles of hot-dip tinning apply equally to hot-dip galvanising. The chief differences which arise in the galvanising of various products will be discussed in the order in which they occur.

1. *Pickling*. Pickling may be more vigorous and inhibitors may be omitted. Although H_2SO_4 is used in the sheet industry, HCl is used in the galvanising of fabricated articles and wire. Cast iron should be grit-blasted and then given a *light* pickle.

2. *Fluxing*. The fluxing reagent is a double chloride of zinc and ammonium. A prefluxing operation using an aqueous solution of flux containing a wetting agent is often employed. This is followed by drying before immersion in the molten zinc. This practice is usually essential when the galvanising bath contains aluminium.

3. *Dipping*. *Alloy formation*. When the cleaned and fluxed steel surface meets the molten zinc, at a temperature in the range 430–470°C, alloying rapidly commences. This is a much more vigorous and complex reaction than that which occurs at the corresponding stage of tinning. Even when the total time of immersion in the zinc is only a few seconds, as in strip and wire galvanising, an appreciable thickness of the ζ (FeZn$_{13}$) phase which contains 6·25% Fe is formed (this phase has a narrow range of composition). The actual thickness and nature of the alloy formed depends upon (*a*) time, (*b*) temperature, (*c*) the composition of the steel base, and (*d*) the composition of the bath[4]. With increasing time a second alloy layer δ (FeZn$_7$), forms beneath the first, and later a third, γ (FeZn$_3$), layer is formed. Increase in temperature leads to an increase in alloy formation. When the zinc temperature exceeds 480–490°C the rate of alloying suddenly increases substantially.

The composition of the steel, more especially its silicon content, markedly affects alloy formation. Increasing the silicon content from 0·02% to, say, 0·20% has a pronounced accelerating effect. This effect may be undesirable or otherwise according to circumstances.

By adding comparatively small amounts (e.g. 0·2%) of aluminium to the galvanising bath, the normal alloying action can be almost completely prevented, provided that neither the duration of immersion nor the temperature is excessive.

Alloy formation is thus a much more important factor in galvanising than in tinning.

4. *Control of coating thickness and uniformity*. The grease-pot mechanism used in tinning is replaced, in sheet galvanising, by a single pair of rolls partially immersed in the bath at the exit end. By this means the amount of liquid zinc drawn out of the bath by the sheet or strip can be greatly reduced. It is obvious that it is not possible to reduce the total coating thickness to below that of the solid alloy layer which has formed. In orthodox sheet galvanising this is about 13–18 μm and the total thickness is in most cases controlled at 25–31 μm. This can be expressed as 380–460 g/m^2 of sheet (both sides).

The result obtained is largely decided by the level of the zinc relative to the 'nip' of the exit rolls. The condition of these rolls and their grooving is important.

Wire may have an appreciably thinner zinc coating and the immersion period is usually very short, which minimises alloy formation. This is partially cancelled by the fact that steel containing some silicon, which speeds up alloying, is favoured. For some purposes the outer layer of zinc is drastically reduced by wiping, but in other cases it is given only a slight smoothing action in a 'charcoal' wiper, which leaves a comparatively thick coating. The zinc coating on wire may also be reduced in thickness as a result of heavy cold reduction of the original dipped wire.

The liquid zinc layer on the outside of a tube is reduced by passing it through a 'die', i.e. an annular air-jet situated immediately above the bath.

When the coating thickness cannot be controlled by any of these methods, the total thickness is kept as low as possible by removing the article very slowly from the bath so that maximum drainage occurs. This also improves coating uniformity. The actual coating weight will then depend upon the surface roughness and the duration of immersion, which in turn depends upon the mass of the article and the difficulty of handling it. Prolonged immersion obviously permits the growth of more alloy.

Threaded articles such as bolts are galvanised in baskets and then centrifuged to free the threads from excessive liquid zinc. Allowance is made for the change in dimensions produced by the coating.

5. *The influence of aluminium additions to the bath*[5]. When the bath contains 0·15–0·25% Al the normal alloy layer is absent and only a minute amount of alloy can be found in the coating. This means that (*a*) thinner coatings can be produced, if desired, and (*b*) the physical properties of the coating are improved by the absence of the brittle alloy layer. This type of coating has superior fabricating properties and is the one applied in modern strip galvanising.

When aluminium is present in the bath in this quantity, it is not possible to maintain a molten flux blanket of the usual composition, owing to its reaction with the aluminium, in which aluminium is rapidly removed from the bath and the flux also becomes ineffective. Prefluxing and drying is the usual remedy, although special fluxes have been suggested for use on baths containing aluminium.

6. *Cooling*. In general, galvanised articles are cooled in air. Massive articles are quenched momentarily to prevent growth of alloy to the surface. The opposite result is achieved by 'Galvannealing', which is used to produce a coating which is harder, more resistant to heat, and, despite its brittleness, less liable to 'flake'. It fails by forming microcracks, and this is less injurious than flaking. In this process, sheet (and sometimes wire) enters a furnace immediately after galvanising in order to ensure the complete conversion of the coating into alloy.

The customary air-cooling should not be delayed in any way, e.g. by nesting of newly galvanised buckets, within the temperature range 300–400°C. This practice may cause a peculiar flaking of the outer zinc layer, which should not be confused with the flaking of the whole coating which occurs during severe bending.

7. *Chemical treatment*. It is a distinct advantage to treat the galvanised surface with an acidulated chromate solution. This operation confers resistance to humidity under those conditions which usually lead to the unsightly and damaging *white rust* or *wet storage stain*. The protection is not permanent but it is sufficient for the period of storage and transit, when white rusting generally occurs.

Phosphating of the galvanised surface permits early painting, improves paint adhesion, and leads to better results in general (see also Sections 14.4 and 16.2).

Hot-dip Galvanising of Continuous Strip[6, 7]

Modern lines are capable of speeds of 2 m s^{-1} or more. The steel strip is first heated to about 400°C in an oxidising atmosphere to burn off rolling oil. It is then reduced in an atmosphere of cracked ammonia at about 730°C and is annealed in line. After cooling to about 460°C the strip enters a pot of molten zinc, without coming into contact with air, via a chute which dips below the surface of the zinc. The zinc is contained in a ceramic-lined induction-heated pot. After passing under a sinker roll in the pot the strip moves vertically upwards through the zinc, and as it leaves the bath, it is subjected to jets of superheated steam to control the coating weight. The coated strip is cooled by passing through numerous sections containing groups of air jets. It then passes through a chemical treatment section where it is either chromated or phosphated. Finally the strip is roller-levelled.

Coating weights may be as low as $214 \, g \, m^{-2}$ (including both sides), which is equivalent to a thickness of about $17 \, \mu m$. Differentially coated strip can also be produced. By heat treatment in the line, lightweight coatings can be completely alloyed to matt grey coatings which have improved welding and painting properties. Normal zinc coatings can be produced without spangle by spraying the strip with steam at a suitable temperature as it emerges from the bath.

Hot-dip Aluminising

Although hot-dip aluminising might appear to resemble hot-dip galvanising very closely, there are several important differences which combine to make the process more difficult to operate[8]. The main difficulties are:

1. The higher melting point of aluminium (660°C). The high operating temperature, which usually exceeds 700°C, causes a loss in tensile properties of cold-drawn wire. On the other hand, if cold-worked material which is to be subsequently annealed is used in this process the annealing and coating operations may be combined, with obvious economic advantage.

2. Very rapid reaction occurs between molten aluminium and iron. This leads to rapid alloying and increased dross formation. An iron pot, as used in galvanising, is rapidly attacked and a ceramic-lined container is desirable.

3. Fluxing is much more difficult with aluminium than with tin and zinc.

4. The oxide layer on molten aluminium, though thin, is most tenacious. Any article leaving the bath is liable to be contaminated with streaks of this oxide or with globules of metal entangled in the oxide film. Hot-dip aluminising of fabricated articles does not appear to be carried out any more either in this country or on the continent.

Hot-dip Aluminising of Continuous Strip[9, 10]

The line speed is only about one tenth that of a galvanising line. The pre-treatment of the strip is similar, except for the inclusion of spray pickling with dilute hydrochloric acid between the oxidation and reduction stages. This removes the bulk of the oxide and thus diminishes the amount of finely-divided iron left on the surface of the strip after its passage through the reduction furnace. Less alloy is then formed at the coating/substrate interface. After passing through molten aluminium at about 720°C the coated strip is cooled rapidly by air jets to minimise alloy formation. The coating weight is $153\,\mathrm{g\,m^{-2}}$ (including both sides) which is equivalent to a thickness of about 25 µm.

Alloy Formation in Hot-dip Aluminising

Mild steel rapidly alloys with molten aluminium to produce a comparatively thick layer of a very hard alloy (Fe_2Al_5) which penetrates into the steel on an irregular front. The alloy layer is much thinner with steel containing more carbon and, in addition, it is usually more uniform in thickness.

The presence of a hard and brittle alloy layer seriously affects the forming qualities of aluminised sheet and wire. Any failure of the coating on deformation is accompanied by a crumbling of the alloy layer. This differs from the flaking behaviour of the δ plus ζ alloy in galvanised coatings, when there is almost always a clean separation of the coating, usually at the steel/δ interface.

The thickness and irregularity of the Fe/Al alloy layer are reduced by the addition of silicon to the molten aluminium bath. An addition of about 3% Si reduces the thickness of the alloy and makes it much more uniform, and also greatly reduces its hardness. Two alloy layers are then formed, Fe_2Al_5 next to the steel and $FeAl_3$ next to the aluminium[11]. The total coating thickness is also reduced and formability is increased. Some further advantage is obtained by increasing the silicon to 5–7%. A disadvantage of silicon-containing aluminium coatings is that they darken when subjected to atmospheric corrosion. Beryllium produces the same results and is much more effective, under 0.1% being sufficient to halve the alloy thickness, but its toxicity and greater cost have hindered its commercial use. Although the influence of numerous other elements has been investigated[12] nothing has been found to equal the results obtained with beryllium and silicon.

Whereas in hot-dip galvanising the alloy layer can be almost completely eliminated (by the addition of aluminium to the bath), this cannot be achieved in hot-dip aluminising. It is therefore necessary, if the coating is to undergo deformation, to keep the alloy to a minimum by reducing the time of immersion and the temperature of operation. A coating which has a total thickness of 25 µm, in which the alloy layer occupies about 25% of the total, has satisfactory bending properties; this coating would, however, not be suitable if the longest possible life were required. The useful life of a coating is generally proportional to coating thickness and it is often necessary to sacrifice some degree of formability to coating thickness in order to obtain maximum resistance to atmospheric corrosion. Thicker coatings are also desirable for some heat-resisting purposes (see below).

Heat-resistant Coatings

Hot-dipped aluminium coatings have excellent resistance to heat, a property not shared by tin, terne or zinc coatings. When exposed to heat, the coating is converted to alloy which provides the heat resistance[13,14]. The life of coatings, especially at the higher temperatures, depends upon the maintenance of a minimum aluminium content in the alloyed surface. Owing to diffusion into the body of the steel, the amount of aluminium at the surface diminishes and it is obvious that for the maximum life the initial aluminium-rich coating should be as thick as possible. The thickness is, however, limited by the increasing tendency to flaking with increased coating thickness. The alloy layer, and the outer aluminium in the initial stages, have coefficients of expansion higher than that of the basis steel, and on heating severe stress may be set up, owing to differential expansion.

The Structure of Hot-dip Aluminised Coatings

When the aluminising bath does not contain any added silicon the coating consists of an alloy layer of appreciable thickness, depending upon the duration and temperature of immersion. It is very hard and brittle and the interface between it and the steel is usually irregular. The outer layer is basically similar to the bath metal and contains a certain amount of iron. The ratio of alloy to outer layer increases with the duration of immersion.

As previously stated, the addition of silicon to the bath modifies the alloy-layer structure. The outer layer now includes some Al/Si eutectic, and some free silicon.

The extent to which the coating can be deformed depends, as with gal-vanised coatings, upon its structure and total thickness. When the coating is thin and the proportion of alloy is low the coating behaves well under tension but is inclined to crumble under pressure. Its behaviour on the inside of a bend is thus inferior to that of a comparable zinc coating. Deep drawing of sheet and cold drawing of coated wire require extra care.

M. L. HUGHES
F. W. SALT

REFERENCES

1. UK Pat. 453 803 (18.9.36)
2. *Hot Tinning*, Tin Research Inst. Publication No. 102 (1966)
3. Thwaites, C. J. and Day, J. J., *Metallurgia, Manchr.*, **56**, 263 (1957)
4. Horstmann, D., *Fourth International Conference on Hot Dip Galvanising, Milan, 1956*, Zinc Development Association, London (1957)
5. Hughes, M. L., *First International Conference on Hot Dip Galvanising, Copenhagen, 1950*, Zinc Development Association, Oxford (1950)
6. *Sheet Metal Industries*, **46**, 387 (1969)
7. Baughman, M. D., Jr., *Iron and Steel Engineer*, **47** No. 8, 134 (1970)
8. Hughes, M. L. and Moses, D. P., *Metallurgia, Manchr.*, **48**, 105 (1953)
9. Silman, H., *Trans. Inst. Met. Fin.*, **40**, 85 (1963)
10. *Sheet Metal Industries*, **45**, 485 (1968)
11. Lamb, H. J. and Wheeler, M. J., *J. Inst. Met.*, **92**, 150 (1964)
12. Gittings, D. O., Rowland, D. H. and Mack, J. O., *Trans. Amer. Soc. Metals*, **3**, 587 (1957)
13. Sykes, C. and Bampfyylde, J. W., *J. Iron St. Inst.*, **130**, 389 (1934)
14. Hughes, M. L. and Thomas, D. F. G., *Metallurgia, Manchr.*, **52**, 241 (1955)

BIBLIOGRAPHY

Hoare, W. E., Hedges, E. S. and Barry, B. T. K., *The Technology of Tinplate*, Arnold, London (1965)
Bablick, H., *Galvanizing (Hot-dip)*, Spon, London (1950)

13.3 Principles of Applying Coatings by Diffusion

General

Since corrosion is a surface reaction, all types of protective coating must involve a change in the surface composition of the metallic component. This change can be brought about by addition of a different material, metallic or otherwise, applied in the form of an outer 'skin', which provides a barrier between the body of the component and the surrounding corrosive medium. This form of coating is by far the most common; it includes paints, varnishes, enamels, plastics, and metals applied by electroplating, hot dipping, flame spraying etc.

It is also possible to modify the chemical composition of the surface to be protected by diffusing into it a suitable metal or element which in combination with the parent metal or alloy will provide the required resistance to the corrosive medium. Such surface alloys are called *diffusion* coatings. 'Surface alloying' would probably be a more correct designation since the modified surface layers are in fact an integral part of the component and cannot be stripped off the parent metal as could a paint film or an electroplated metal deposit. The dimensional change in the protected article is less than the thickness of the effective surface alloy, and may be negligible. The term 'coating thickness' is usually applied to the surface alloy as revealed by suitable etching (Fig. 13.17). Often there is a convenient metallographic feature which is taken as the limit of the coating thickness although the coating element has diffused deeper at low concentration, as can be shown by chemical analysis and more recently by electron micro-probe studies.

These properties of integration with the underlying metal and negligible dimensional change are the most important features which distinguish diffusion coatings from other types of protection.

Metals Usually Applied by Diffusion

So far, few of the commercially operated diffusion processes have been applied to the lower-melting-point metals. While they are being used to an increasing extent for protection of nickel, cobalt and refractory alloys, the bulk of present-day applications is still concerned with the treatment of ferrous materials.

Table 13.3 has been compiled in order to give a general idea of the field covered by diffusion processes. Chromium diffusion (*chromising*), which has been applied for the purposes of resistance to corrosion, thermal oxidation and abrasion, is probably one of the most versatile forms of diffusion coating, and has been much more extensively studied than the other processes. Reference may be made to several general accounts of diffusion coatings[1,2] and chromising[3–7]; papers on specific aspects are indicated where appropriate. A recent review[8] has a good bibliography covering aluminium, chromium and zinc diffusion coatings.

Zinc diffusion (*sherardising*)[8–11] is mainly used for protection of ferrous metals against atmospheric corrosion. It has, in some respects, properties related to other types of zinc coating such as galvanising, but owing to the small dimensional change involved, it is of particular value for the treatment of machined parts, bolts, nuts, etc.

Aluminium diffusion (*calorising, aluminising*)[8,12,13] protects steels against oxidation at elevated temperatures, and the more recently developed processes for aluminising and chromaluminising 'superalloys' are widely used to increase the life and operating temperature of aircraft gas turbine vanes, etc.

Table 13.3 Diffusion processes

Coating element(s)	Substrates	Properties obtained or improved	Stage of development	References
Al	Cu and alloys	Th.ox.	Com.	
	Fe, steels, cast iron	Th.ox.	Com.	8, 12, 13
	Nb, Ti and alloys	Th.ox., frict.	Exp.	
Al, Al+Cr	Stainless steels	Th.ox.	Com.	
	Superalloys*	Th.ox., th.corr.	Com.	2, 35–43
B	Fe, steels, Ni- and Co-alloys, Mo	Hard., th.corr.	Exp. (Com. Russia)	48
Cr	Fe, steels, cast iron	Corr., th.ox., hard.	Com.	1–8
	Stainless steels	Corr., th.ox., frict.	Com.	4, 33
	Ni and alloys	Th.ox., th.corr.	Com.	1, 7, 34
	Mo, W and alloys	Th.ox.	Com.	1
Cr+Al, Si, Zr, etc.	Steels	Th.ox. (higher temp.)	Com.	26
	Superalloys*	Th.ox., th.corr.	Com.	40
	Mo	Th.ox.	Com.	45
Mo	Steels, stainless steels	Special corr.	Exp.	
Mo, Mo+Cr	Ni and alloys	Corr. (acids)	Exp.	34
Si	Fe, steels	Corr. (acids), hard.	Com. (USA)	14, 15
Si (+others)	Mo, W, Nb, Ta	Th.ox.	Com.	2, 16, 42, 46
Ta	Ni alloys, superalloys*	Pretreatment for chromising, etc.	Exp.	40
Ti	Steels, Ni alloys	Hard., pretreatment for chromising	Exp.	
Zn	Fe, steels	Corr.	Com.	8–11
	Nb	Th.ox.	Exp.	47

Abbreviations: Corr. = resistance to corrosion
Frict. = low friction, anti-galling
Hard. = hardness, resistance to abrasion
Th.corr. = resistance to thermal corrosion (e.g. sulphidation)
Th.ox. = resistance to thermal (high temp.) oxidation
Com. = commercial; Exp. = experimental
*Superalloys: high temperature creep-resisting alloys based on Ni, Co, and including (here) the Nimonic series.

Silicon diffusion (*Ihrigising*)[14,15] is not commonly applied to steels, but is increasingly used to protect the refractory metals[2,16] (*disilicide coatings*).

Mechanism of Coating Formation

Diffusion coatings are formed as a result of interaction of two distinct processes; the solute metal is brought into contact with the surface of the solvent, and this is followed by diffusion proper which consists in the gradual absorption of the solute into the lattice of the solvent.

Whatever method is used to provide an adequate supply of diffusing metal, the diffusion mechanism remains identical for any given solute/solvent system, and it is appropriate to discuss the diffusion aspect in the first place.

Theory of Diffusion [17-19]

Diffusion is a process whereby the distribution of each component in a phase tends to uniformity.

For such a condition of equilibrium to be reached, the atoms must acquire sufficient energy to permit their displacement at an appreciable rate. In the case of metal lattices, this energy can be provided by a suitable rise in temperature. In the application of coatings the diffusion process is arrested at a suitable stage when there is a considerable solute concentration gradient between the surface and the required depth of penetration.

In metals, the distance between the individual atoms in the lattice is of the order of 0·4 nm and only atoms of very small size are able to penetrate interstitially. This takes place, for instance, in the diffusion of hydrogen into iron, and of carbon into austenite, etc. This type of interstitial diffusion is usually rapid, since the inward movement of the solute atoms is relatively unhampered.

Interstitial diffusion is rarely possible when two metals interdiffuse, since their atomic radii are usually of the same order. Several mechanisms have been proposed, but it is now generally accepted that interdiffusion is due to the motion of vacant sites within the lattice, solvent and solute atoms moving as the vacant sites migrate. The diffusion process is thus dependent upon the state of imperfection of the solvent metal and the alloy being formed.

The kinetics of the diffusion process (whether interstitial or substitutional) can be expressed by Fick's equations:

$$P = -D\frac{dc}{dx} \qquad ...(13.13)$$

where P is the rate of permeation of solute (e.g. coating metal) through a unit of the solvent metal in the direction x, D is the diffusion coefficient and c is the concentration, and

$$\frac{\partial c}{\partial t} = D\frac{d^2c}{dx^2} \qquad ...(13.14)$$

which expresses the rate of increase of concentration at any point (when D is independent of concentration).

For diffusion into semi-infinite solids, the depth of diffusion is related to the time t by the equation

$$x^2 = 4kDt \qquad \qquad ...(13.15)$$

where k is a constant which is determined by the concentration at the surface and at a depth x.

The diffusion coefficient varies with temperature according to the following Arrhenius-type equation:

$$D = D_0 \exp\left[-E^{\ddagger}/RT\right] \qquad \qquad ...(13.16)$$

where T is the absolute temperature, R is the gas constant, $E^{\ddagger}$ the energy of activation and D_0 the diffusion or frequency factor.

It must be noted that the values of D_0 and $E^{\ddagger}$ are influenced by the concentration of the solute metal and also by the presence of alloying elements in the solvent. It has also been shown that the diffusion coefficient for a given solute is in inverse proportion to the melting point of the solvent. D is least for metals forming continuous series of solid solutions and for self-diffusion.

Strictly the diffusion coefficient D measured for any type of binary system A/B is in fact the resultant effect of two partial diffusivities D_A and D_B, representing respectively the diffusivity of A into B and of B into A. For most practical purposes, however, a single diffusion coefficient is sufficient to define a given diffusion system.

Methods of Deposition

As mentioned earlier, the diffusion process depends on an adequate supply of solute metal at the surface of the solvent. So long as such a supply is available, the diffusion will proceed at a rate which, as shown above, is largely determined by the temperature selected for the process. Various methods of providing this supply may be employed. A metal may be electrodeposited on another metal, and the coated material may be subsequently heated at a temperature sufficient to effect diffusion. This method has the disadvantage of requiring two operations. Another method consists in heating the solvent metal in the presence of the solute metal in powder form (sherardising when zinc is the coating metal, and calorising when aluminium forms the coating). This technique, designated by the general name of cementation, is mainly applicable to the diffusion of low-melting-point metals.

Solid–solid diffusion Sherardising is a good example of a solid–solid diffusion process. The iron or steel articles, after thorough degreasing and pickling, are packed in a powder containing zinc dust and a diluent such as silica or sand. The powder and articles are then placed in a steel drum which is slowly rotated while the temperature is raised to between 350 and 400°C, i.e. below the melting point of zinc. The temperature is maintained for 3–10 h depending on the size of the components and on the coating thickness required.

Usually, rapid cooling is obtained by removing the drum from the furnace and spraying it under a water jet while rotating slowly.

Substantially all the zinc is absorbed by the surface of the components, and the resulting coating has a matt-grey appearance and consists essentially of a zinc–iron alloy averaging 90–95% Zn. If excess zinc is made available and the treatment is prolonged, pure zinc is deposited at the surface.

Temperatures well in excess of 400°C can be used for processing; in this case much deeper coatings are obtained, but the iron content of the surface alloy is higher and the diffusion layer is very brittle and less corrosion-resistant. This effect is easily explained when it is remembered that the rate of interdiffusion is far more rapid when the temperature is above the melting point of zinc (420°C).

Although zinc has an appreciable vapour pressure at the temperatures of treatment, it is unlikely that zinc vapour plays any significant part in the diffusion process and it is generally accepted that the mechanism relies almost exclusively on intimate contact of finely divided zinc dust with the steel surface. In spite of this requirement, coatings of even thickness and composition are obtained on the most intricate shapes, on fine threads, inside blind holes, and in the bore of small-diameter tubes. Large articles of uniform section, e.g. rods, tubes, etc. can be coated by this process.

Zinc diffusion produces slight dimensional changes, but the increase of thickness due to the addition of zinc is controllable and independent of the surface contours. This property which is common to most diffusion processes is of special value for the treatment of threaded parts and articles with intricate contours.

Gas-phase deposition   In this process, a halide of the solute metal is passed in vapour form over the surface of the metal to be coated, which is heated to a temperature at which diffusion can take place. Temperatures of 500–1 300°C or more can be used, depending on the particular system considered. Generally, 'filler' atmospheres are provided to carry the halide vapour; these atmospheres are usually reducing gases such as hydrogen, cracked ammonia, etc. or inert gases (helium, argon).

Three main types of reaction can take place in which the metal halide, BX_2, is reduced to metal B, which then diffuses into the solvent metal A. The following considerations, which have been confined to bivalent metals, are of general application.

Interchange	$A + BX_2 \text{ (gas)} \rightleftharpoons AX_2 + B$	...(13.17)
Reduction	$BX_2 + H_2 \rightleftharpoons 2HX + B$	...(13.18)
Thermal dissociation	$BX_2 \rightleftharpoons X_2 + B$	...(13.19)

The interchange reaction* implies the removal of one atom of A at the surface for each atom of B deposited. It therefore takes place with a minimum change in weight or dimensions of the article (A). If A and B have similar atomic weights, as in the case of iron and chromium, interchange reaction will produce little change in weight and no measurable increase in dimension, whatever the thickness of the diffusion layer.

*Reaction 13.17 might be considered as the result of two reactions, i.e. reaction 13.18 or 13.19 followed by reaction of HX or X_2 with A to form AX_2.

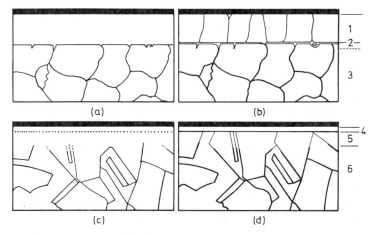

(a) (b) (c) (d)

LEGEND

1. Columnar ferrite formed from austenite during chromising
2. Boundary between coating and core, with intergranular diffusion into austenite of core
3. Core; ferrite formed from austenite on cooling
4. β-phase layer; apparently the 'coating' in (d)
5. α-phase coating, not etched by nitric acid [(c)]
6. α-phase core

Fig. 13.17 Illustration of grain patterns for chromised low-carbon steel and nickel. (a) Chromised steel, etched with nital; (b) chromised steel etched with Marble's reagent; (c) chromised nickel etched with nitric and acetic acids; (d) chromised nickel etched with Marble's reagent

On the other hand, both reduction and thermal-dissociation reactions will result in an increase in weight (equivalent to the solute deposited) and a slight increase in dimensions which will depend on the average composition of the diffused layer.

The thermodynamics of interchange and reduction reactions are of particular interest, since knowledge of the feasibility 'areas' of these reactions is of great assistance in determining the best conditions of processing for any given system.

Some interesting conclusions can be drawn by plotting log K_p against temperature for thermal dissociation reactions of the type $M + Cl_2 \rightleftharpoons MCl_2$ (M is any given metal). (See Fig. 13.18)[20].

The equilibrium constant K_p can be derived from the free energy equation

$$-\Delta G^\circ = 2 \cdot 303 \, RT \log K_p \qquad \text{...(13.20)}$$

Considering B as a coating metal and A as the solvent, we have for

$$B + Cl_2 \rightleftharpoons BCl_2, \qquad K_{p,B} = \frac{p_{BCl_2}}{a_B p_{Cl_2}} \qquad \text{...(13.21)}$$

$$A + Cl_2 \rightleftharpoons ACl_2, \qquad K_{p,A} = \frac{p_{ACl_2}}{a_A p_{Cl_2}} \qquad \text{...(13.22)}$$

and for the interchange reaction,

$$BCl_2 + A \rightleftharpoons ACl_2 + B, \qquad K_{p,1} = \frac{a_B p_{ACl_2}}{p_{BCl_2} a_A} = \frac{K_{p,A}}{K_{p,B}} \quad ...(13.23)$$

or

$$\log K_{p,1} = \log K_{p,A} - \log K_{p,B} \qquad ...(13.24)$$

For coatings where $a_A \approx a_B$ at the surface, a difference of -1 unit in equation 13.24 means that approximately 10% of the chloride vapour has been converted into coating metal. A difference of -2 units represents a conversion of approximately 1%, which we can regard as a minimum required for practical interchange reaction, in a continuous flow reaction. On this basis, it can be seen from Fig. 13.18 that chromising falls within the limits of feasibility, and that siliconising and stannising of iron should proceed even more efficiently, as is in fact borne out experimentally. Chromising of nickel or of molybdenum should not be possible by interchange reaction since the difference is well over -2 units, and in fact interchange does not occur.

The feasibility of the reduction reaction can be considered by a similar method.

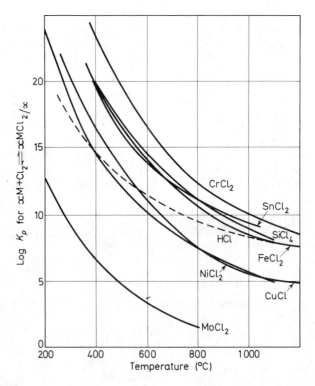

Fig. 13.18   Log K_p for $xM + Cl_2 \rightleftharpoons xMCl_2/x$ [after *Aust. Eng.*, Dec., 63 (1950)]

For $H_2 + Cl_2 \rightleftharpoons 2HCl$,

$$K_{p,H_2} = \frac{p^2_{HCl}}{p_{H_2} p_{Cl_2}} \qquad \qquad ...(13.25)$$

For the reduction $BCl_2 + H_2 \rightleftharpoons 2HCl + B$ we have

$$K_{p,R} = \frac{a_B p^2_{HCl}}{p_{BCl_2} p_{H_2}} = \frac{K_{p,H_2}}{K_{p,B}} \qquad \qquad ...(13.26)$$

or

$$\log K_{p,R} = \log K_{p,H_2} - \log K_{p,B} \qquad \qquad ...(13.27)$$

p_{H_2} is equal to 1 if hydrogen is at atmospheric pressure. If no interchange occurs, $a_B \approx 1$. The reduction reaction will proceed effectively if $\log K_{p,B}$ is not more than 1 or 2 units greater than $\log K_{p,H_2}$. Thermal dissociation is not feasible, except for iodides.

In many instances, more than one type of reaction can take place at any given temperature. When the reduction reaction plays an important part, it is usually possible to alter the composition and characteristics of the coating by controlling the hydrogen supply to the reaction chamber.

When the coating metal halide is formed *in situ*, the overall reaction represents the transfer of coating metal from a source where it is at high activity (e.g. the pure metal powder, $a_A = 1$) to the surface of the substrate where a_A is kept less than 1 by diffusion. The formation of carbides or intermetallic compounds such as aluminides or silicides as part of the coating reaction may provide an additional driving force for the process.

The reactions work both ways, and constituents of the alloy being coated may be removed by the halide atmosphere even when 'interchange' does not occur. For example, a nickel–chromium alloy may be superficially de-chromised by nickel powder in a chloride atmosphere. Thus loss of important alloying constituents may have to be controlled during diffusion coating processes.

Molten bath deposition The interchange reaction also takes place when the coating metal halide is dissolved in a fused salt[21]. Alternatively, deposition may be by electrolysis[22]. Another technique uses the coating metal dissolved in molten calcium[23].

Chromising offers a good example of the types of reaction involved and of the general characteristics of diffusion coating.

Chromising as an Example of Diffusion Coating

Most modern methods of chromising rely on the reaction of a chromium halide at the surface of the metal to be coated. Three main methods are commercially operated.

Purely gaseous method[24-26] Articles are suitably jigged inside a retort chamber and vapours of chromium halide, usually diluted with a reducing gas, are circulated. The temperature is maintained within the range 950–1 200°C for chromising of iron and ferrous alloys (recent Russian work on vacuum chromising utilises chromium metal vapour derived from ferro-chromium in a vacuum furnace together with the articles being coated, at about 1 350°C[27]).

Semi-gaseous method   The articles, surrounded by a solid 'compound' containing chromium metal, are packed in a retort chamber, and vapours of chromium halide or halogen compound are circulated in the chamber, which is maintained at 950–1 200°C.

Pack method   The articles are packed in a 'powder compound' containing chromium metal or ferrochrome and a source of halogen in solid form, e.g. NH_4Cl, NH_4Br, NH_4I. Gas-tight boxes fitted with suitable sealing fixtures are used. The range of temperatures is the same as in the other two methods.

It is usual to treat low-carbon steel between 980 and 1 050°C, although the process is operable within the temperature limits stated above. Techniques based on all three methods, and on the use of fused salts, have been proposed for the treatment of steel in coil form[28].

The iron–chromium constitution diagram is shown in Fig. 13.19. At around 1 000°C, it can be seen that when chromium is deposited and diffused

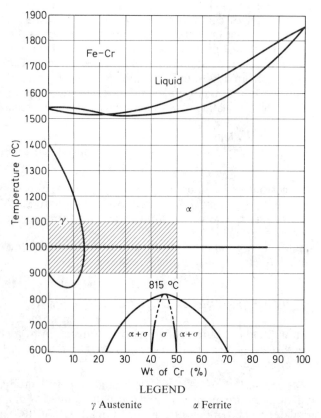

LEGEND

γ Austenite   α Ferrite

Fig. 13.19   Iron/chromium diagram

inwards, the austenite structure of the steel remains unchanged until a concentration of approximately 12% chromium is reached. At higher chromium concentrations, the structure becomes ferritic. Since the rate of diffusion of chromium is greater in ferrite than in austenite, there is a rapid rise in the chromium concentration of the coating which is only partly offset by the slow diffusion beyond the 12% Cr boundary. As a result, there is a sharp concentration drop at the ferrite/austenite boundary. Grain boundary diffusion occurs[30], but has little effect on the coating thickness.

The chromium concentration as a function of distance from surface is shown in Fig. 13.20. The 12% chromium boundary (dotted line) represents

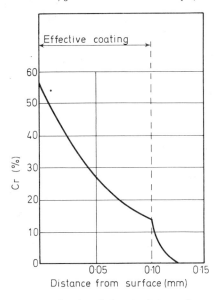

Fig. 13.20 Chromium concentration in relation to distance from surface for a chromised low-carbon steel

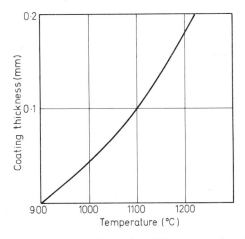

Fig. 13.21 Chromised low-carbon steel; coating thickness as a function of temperature; treatment for 4 h

the effective coating thickness, the shallow transition zone to the right being ignored for practical purposes. If a chromised specimen is etched in nital, only the effective coating will remain unattacked. The core of such a specimen can be dissolved in boiling nitric acid, leaving the effective coating as a complete 'shell' which can be analysed. The average chromium content of a chromised coating on low-carbon steel is of the order of 25%. By comparing the overall chromium uptake with the weight increase of a complete specimen, it is possible to determine the relative contributions of interchange and reduction (or thermal dissociation) reactions during processing.

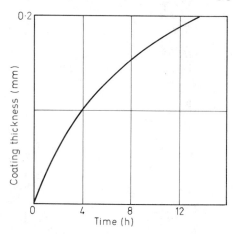

Fig. 13.22   Chromised low-carbon steel; coating thickness as a function of time; temperature 1 100°C

The influence of temperature on coating thickness is shown in Fig. 13.21 which relates to a 4 h treatment at temperature. Figure 13.22 shows the variations of thickness as a function of time at a constant temperature of 1 100°C. This curve is in good agreement with the third of Fick's equations (13.15):

$$x^2 = 4kDt$$

From such curves, taking $k = 1$, the values of D for different temperatures can easily be derived and an energy of activation for the process calculated. This has been found to be of the order of 239–247 kJ/mol[4, 21, 24]. Increasing the carbon content of the steel progressively reduces both the rate of diffusion and the energy of activation[29].

Characteristics of Diffusion Coatings

The structure and composition of diffusion coatings depends of necessity on the metal or alloy from which the article is made. Thus, for example, it is not possible to speak of chromised coatings generally; the material into which chromium is diffused must be specified. Some data on methods of application and properties of commercially chromised irons and steels are given in Table 13.4.

Table 13.4 Diffusion coatings obtainable on ferrous materials

Base material	Coating element	Processing temperature range (°C)	Method	Coating characteristics
Iron Low-carbon steels Low-carbon alloy steels Decarburised malleable iron	Chromium	980–1 050	(Halide) 1. Gaseous 2. Semi-gaseous 3. Pack	25–75 μm Solid-solution (ferritic) 20–25% Cr Ductile (200–300 HV) Weldable Heat treatment acceptable*
	Zinc	350–400	Cementation (rotary furnace)	12–50 μm Single-phase intermetallic 90–95% Zn Fairly brittle Can be soldered and spot welded
	Aluminium	850–950	1. Cementation 2. (Halide) pack	125–750 μm Single-phase intermetallic Not exceeding 25% Al Brittle Not weldable
	Silicon	900–1 050	(Halide) 1. Gaseous 2. Semi-gaseous 3. Pack	125–250 μm Solid-solution (ferritic) Average 15% Si Brittle (160–180 HV) Not weldable
	Boron	800–1 050	(Halide) 1. Gaseous 2. Semi-gaseous 3. Pack 4. Salt-bath electrolysis	Up to 500 μm Matrix plus borides Brittle Hardness up to 1 500 HV Heat treatment acceptable*
Medium and high-carbon steels Cast irons	Chromium	850–980	(Halide) 1. Gaseous 2. Semi-gaseous 3. Pack	4–40 μm Matrix plus chromium carbides 60–80% Cr 1 500–1 800 HV Heat treatment acceptable*
	Zinc	350–400	Cementation (rotary furnace)	12–50 μm Single-phase intermetallic 80–90% Zn Brittle Spot-welding and soldering possible

Table 13.4   (*continued*)

Base material	Coating element	Processing temperature range (°C)	Method	Coating characteristics
	Aluminium	850–950	1. Cementation 2. (Halide) pack	50–350 μm Single-phase intermetallic Average 25% Al Brittle Not weldable
	Boron	800–1 050	(Halide) 1. Gaseous 2. Semi-gaseous 3. Pack 4. Salt-bath electrolysis	Up to 500 μm Matrix plus borides plus boron carbide Brittle Hardness up to 2 000 HV Heat treatment acceptable*

*The mechanical properties of the core material can be subsequently restored by suitable heat treatment.

Chromising of Iron and Low-carbon Steels

The coating is a chromium–iron alloy, showing a columnar ferrite grain pattern (Fig. 13.17*b*) and has most of the properties of a stainless chromium–iron. Carbides may be present in the coating depending on the carbon content of the steel and on the conditions of processing, and also nitrides[30,31]. The carbides reduce coating ductility and corrosion resistance and so titanium-stabilised steels in which the carbon is 'fixed' as titanium carbide, have been developed; large quantities of such steel are now chromised as flat sheets and fabricated by pressing, e.g. to make heat exchanger shells.

Being of stable ferritic structure, the coating is not affected by heat treatment, and chromised articles can be normalised, air-hardened or quenched without detrimental effect to the coating. Spot-welding, arc or gas welding (preferably with 18 Cr–8 Ni-stabilised rods), and most methods of brazing and of silver soldering can be used on chromised low-carbon steel surfaces.

Chromising of Carbon Steels and Cast Irons

Carbon has a great affinity for chromium. When chromium diffuses into a steel containing appreciable amounts of carbon, the latter diffuses outwards and combines with the inward stream of chromium atoms to form a continuous carbide 'barrier' which effectively blocks any further diffusion in depth. Accordingly, when medium- or high-carbon steels are chromised the coating contains a high proportion of chromium carbides, and the average chromium content may be as high as 70–80%. There may be a layer of chromium-containing pearlite just beneath the coating and beneath this again a decarburised zone (carbon having diffused into the coating). Given sufficient time after the carbide barrier has formed, this zone may be recarburised from the core. Thus the final structure depends on the kinetics of diffusion and carbide formation[31,32]. The high hardness and low co-efficient of friction of the chromised surface give it useful abrasion-resisting properties. The hardness of the coating is not affected by subsequent heat treatments which may be required in order to recover the mechanical characteristics of the core material.

Most tool and die alloy steels can be classified within this group. Care should always be exercised during processing to avoid a decarburised layer beneath the coating.

Most cast irons, except those fully decarburised during malleabilising, give coatings of the chromium carbide type. In view of the great variations in composition of cast irons, reproducibility of results can be achieved only by careful control of specification. High phosphorus and sulphur contents are detrimental to the formation of non-porous coatings.

Chromising of Alloy Steels

Apart from carbon, other elements present in commercial steels exert a great influence on the nature of the chromised coating. Generally speaking, austenite formers (manganese, nickel, cobalt, copper, etc.) have a retarding effect on the rate of diffusion. Chromised coatings on austenitic stainless steels may vary from relatively thin coatings of high chromium content to thicker coatings averaging 35% Cr[33], depending on such factors as carbon content and stabilisation and whether interchange (Cr with Fe, but not with Ni) or reduction reaction predominates. Maraging steels, because of their extremely low carbon content, chromise well and can be age-hardened afterwards. Ferrite formers (chromium, vanadium, aluminium) and strong carbide formers (molybdenum, tungsten, titanium, zirconium) tend to promote faster diffusion with smoother concentration gradients.

Chromising of Nickel, Cobalt and Alloys

Nickel and cobalt and their alloys lend themselves readily to chromising. The reaction is primarily one of reduction. The effective coating on pure nickel contains an average of 35–45% chromium and has good ductility. There is no sharp diffusion boundary as might be expected from consideration of the nickel–chromium constitution diagram, but an outer chromium-rich β-phase layer may be present, depending on the conditions of treatment, a sharp boundary and concentration gap separating this from the α-phase layer (Fig. 13.17c and d). A two-phase layer may develop at the boundary at certain rates of cooling[7]. The nickel–chromium surface alloy has useful oxidation- and thermal-corrosion-resisting properties.

This brief summary illustrates the fact that a wide range of coating properties can be obtained by applying the same type of processing treatment to different materials. Table 13.4 shows the types of coating obtainable on ferrous materials.

Aluminising of 'Superalloys'

Nickel- and cobalt-based high-temperature alloys are commonly protected against high temperature corrosion by 'aluminide' coatings, generally applied by pack methods[2,35]. For the stringent operating conditions of components such as gas turbine vanes, the aluminising process must be controlled to suit the particular alloy[36,37]. The coating should consist of *high* melting aluminides of nickel or cobalt, modified by the presence of chromium and other constituents of the alloy; the aluminium rich aluminides of lower melting point are to be avoided. Thus the rate of aluminium uptake must be limited. The structure of the coatings is complex; carbides often form beneath the aluminide layer[38,39].

Chromium powder is often included in the aluminising mixture. Chromium and aluminium may diffuse together (*chromaluminising*[40]) or by balancing the activities of chromium in the alloy and the pack, de-chromising may be prevented. Alloying of the chromium and aluminium powders[40] may be used to control aluminising.

Aluminising is also applied to stainless steels to increase oxidation resistance.

Properties of Diffusion Coatings

Some of the more important properties of diffusion coatings have already been mentioned. Generally speaking, the properties of a diffusion coating are those to be expected from a wrought or cast alloy of the same composition. Thus the corrosion-resisting properties of chromised low-carbon steels are very similar to those of a high-chromium stainless iron[4], and sherardised materials behave much in the same way as galvanised steel or iron–zinc alloys. This generalisation of course assumes that the coatings are substantially non-porous. For example, the presence of carbon in steel may result in the formation of slightly imperfect chromised layers which are prone to develop pin-point corrosion in strong electrolyte corrodents such as sodium chloride, whereas the use of steels in which carbon is stabilised by a strong carbide-former such as titanium gives coatings of perfect continuity and structure, which provide adequate protection for all types of wet-corrosion conditions. Much of the value of modern techniques lies in the controlled production of homogeneous non-porous coatings.

It must always be remembered that diffusion coatings are produced by a form of heat treatment and that, with the exception of low-temperature zinc diffusion (sherardising), the treated ferrous materials are usually in the annealed condition. Whenever the mechanical properties of the parts must be restored to their original level, a subsequent heat treatment is necessary[7]. This does not as a rule present any difficulty with chromised or boronised steels. In order to prevent undue distortion and internal stresses during treatment and subsequent hardening, it is recommended that high-carbon and alloy steels should be processed in the 'normalised' condition.

Fig. 13.23 shows the variation of hardness with distance from surface for a chromised high-carbon steel (1% C). The full line represents the material 'as treated', the dotted line corresponds to conditions after full-hardening and tempering. The apparent greater depth of the hard zone in the heat-treated material is due to the effect of small concentrations of chromium

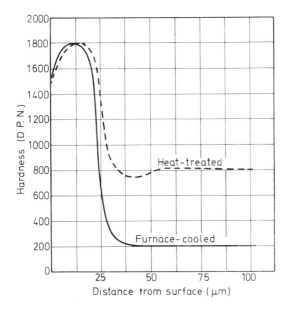

Fig. 13.23   Hardness as a function of distance from surface-chromised high-carbon steel (1% C)

immediately underneath the chromium carbide layer. With suitable processing conditions, the hardness may approach 2 000 HV and the drop at
the surface is minimal.

Among the most important purposes for which diffusion coatings
are applied are high-temperature-oxidation and thermal-corrosion resistance[35,41–43]. Their resistance behaviour in this context differs in some
respects from that of plain alloy materials. The diffusion mechanism which
is responsible for the formation of the protective coating will also take place
whenever a treated part is reheated within the same range of temperature.
This creates a gradual diffusion of the solute atoms in the parent metal
without compensation for their reduced concentration at the surface, so that
the nett result is an impoverishment of the alloy composition throughout
the coating.

The practical implications of this 'rediffusion' effect are twofold. In the
first place, the nominal composition of a diffusion coating may be such as
to suggest by analogy with an alloy that it will have good oxidation resistance
at a temperature of, say, 900°C. This is, however, true only if the rate of
diffusion of the coating metal at that temperature is negligible or small
enough to be compatible with the anticipated life of the component. The
deep coatings (which generally presuppose a high rate of diffusion) are not
necessarily the most suitable for high-temperature stability, and it is often
preferable to utilise coatings where the diffusion process has been arrested
by the formation of a diffusion 'barrier' (see p. **13**.70).

These features are well illustrated in the case of the chromising of ferrous
materials. The weight increase of chromised low-carbon steel as a function
of time at different temperatures is shown schematically in Fig. 13.24. At a
temperature of 700°C, the weight increase in air is practically negligible, and

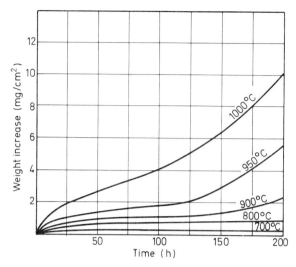

Fig. 13.24   Rate of oxidation of chromium-diffused mild steel (0·15% C) when heated in air at various temperatures

it is very small at 800°C, the curve showing the parabolic weight-increase/time relationship which is typical of good heat-resisting characteristics. When the temperature is raised to 900°C, the first part of the curve follows the normal parabolic pattern, but after some 100–150 h the rate of oxidation increases. This effect becomes more pronounced at higher temperatures, and at 1 000°C the oxidation rate after 25–100 h is appreciably higher than that of a 30% Cr ferritic steel. This gradual deterioration in oxidation resistance can also be followed by analysing the oxide scale at the surface. A thin, strongly adherent chromium oxide film is predominant at temperatures up to 900°C. At higher temperatures there is a gradual increase in the iron content of the superficial oxides which is due to continuous diffusion of chromium inwards and diffusion of iron towards the surface. In time, the composition will reach a level where the chromium concentration is too low to provide adequate oxidation resistance. It is of interest that for chromised *pure* iron, continued diffusion is much faster below 900°C (core ferritic) than at just above 900°C (core austenitic)[44]. However, for chromised steel this effect is evidently off-set by the very low solubility of the thin carbide diffusion barrier in ferrite, as indicated by service life at elevated temperatures. The

Table 13.5   Temperature range at which 'rediffusion' becomes appreciable

Coating metal	Basis metal	Temperature (°C)
Aluminium	Ferrous materials	700
	Nickel-base materials	950
Chromium	Low-carbon steel	870
	High-carbon steel	950
	Nickel-base materials	950

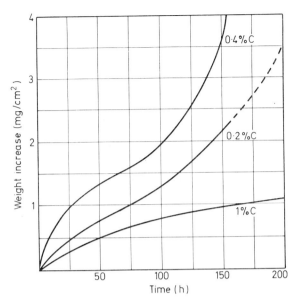

Fig. 13.25   Influence of carbon content on the rate of oxidation of chromium-diffused steel at 950°C

temperature above which 'rediffusion' becomes important in service is shown in Table 13.5 for several systems.

The effect of coating composition and of substrate structure on the oxidation rate is shown in Fig. 13.25 which gives weight-increase/time curves for oxidation in air at 950°C of three steels of different carbon contents.

The low-carbon steel (0·20% C) has a ferritic-type coating, and the core structure is primarily ferrite at temperatures below 880°C. After some 100 h exposure to oxidation the 'rediffusion' effect becomes appreciable, and the oxidation rate increases accordingly. In marked contrast, the high-carbon (1% C) steel has a primarily carbide-type diffusion coating which has good oxidation resistance. The low solubility of chromium carbide in austenite at 950°C stabilises the coating composition, which remains substantially unchanged for several hundred hours. At an intermediate carbon content (0·4% C), the coating is primarily ferritic but contains a certain percentage of carbides, mainly at the coating/core interface. Although the coating possesses good intrinsic resistance to oxidation and the carbide layer provides a diffusion barrier, the rate of oxidation is higher than for the low-carbon steel. In this case, failure is due to poor matching between coating and core structure, the ferrite–pearlite structure of the parent metal which develops on cooling introducing stresses in the coating. On reheating, these stresses cause cracks in the protective coating with consequent failure in a comparatively short time. The inherent stresses in chromised coatings[30] are not of great importance in practice.

In addition to providing coatings resistant to air oxidation, diffusion methods are of considerable value in reducing the detrimental effects of sulphur corrosion at high temperatures. Both aluminised and chromised coatings are also used for protection against lead-corrosion and in preventing intergranular oxidation of steels and nickel-base alloys in certain atmospheres.

The present treatment has been limited to a general survey of diffusion coatings; for a more complete account on specific aspects, the reader should consult the references.

N. A. LOCKINGTON

REFERENCES

1. Samuel, R. L., *Murex Rev.*, **1** No. 18 (1958)
2. Krier, C. A., Chapter 15 in *Vapour Deposition* by Powell, C. F., Oxley, J. H. and Blocher, Jr., J. M., Wiley, New York.
3. Becker, G., Daeves, K. and Steinberg, F., *Stahl u. Eisen*, **61**, 289 (1941)
4. Samuel, R. L. and Lockington, N. A., *Met. Treat.*, **18**, 354, 407, 440, 495 (1951) and **19**, 27, 81 (1952)
5. Samuel, R. L., Lockington, N. A. and Dorner, H., *Met. Treat.*, **22**, 233, 288, 336 (1955)
6. Galmiche, P., *Rev. Mét.*, **47**, 192 (1950)
7. Sully, A. H. and Brandes, E. A., *Chromium*, 2nd ed., Chapt. 7, Butterworths, London (1967)
8. Drewett, R., *Anti-Corros. Meth. Mat.*, **16**, April, 11–16; June, 10–14; Aug., 11–14, 16 (all 1959)
9. Cowper-Coles, S., *Electrochem. and Metall.*, **3**, 828 (1904)
10. *Metals Handbook*, Amer. Soc. Metals, Cleveland, Ohio, 714 (1948)
11. Price, G. C., *Corros. Techn.*, **10**, 171 (1963)
12. *Metals Handbook*, Amer. Soc. Metals, Cleveland, Ohio, 703 (1948)
13. Burns, R. M. and Schuh, A. E., *Protective Coatings for Metals*, Reinhold, New York (1939)
14. Ihrig, H. K., *Metal Prog.*, **33**, 367 (1938)
15. Ihrig, H. K., *Iron Age*, **14**, April, 4 (1946)
16. Wood, R. J., *Mat. Prot.*, **3** No. 7, 25 (1964)
17. Shewman, P. G., *Diffusion in Solids*, McGraw-Hill, New York (1963)
18. Le Clair, A. D., 'Diffusion of Metals in Metals', *Progress in Metal Physics*, Butterworths, London, 306 (1949)
19. Birchenall, C. E., *Met. Rev.*, **3**, 235 (1958)
20. Hoar, T. P. and Croom, E. A. G., *Aust. Engr.*, Dec., 63–66 (1950)
21. Campbell, I. E., Barth, V. D., Hoekelman, R. F. and Gonser, B. W., *Trans. Electrochem. Soc.*, **96**, 262 (1949)
22. *Iron Age Metalw. Internat.*, **7**, 32, March (1968) and *Metals and Materials*, **2**, 187 (1968)
23. Carter, G. F., *Met. Prog.*, **93**, 117 June (1968)
24. Hoar, T. P. and Croom, E. A. G., *J. Iron Steel Inst.*, **169**, 101 (1951)
25. Galmiche, P., *Met. Fin.*, **49**, 61 (1951)
26. Fabian, D., Ullrich, G. and Ebersbach, G., *Technik*, **24**, 392 (1969)
27. Belov, Yu. K., Plyshevskii, A. I., Ponomarenko, E. P., Panfilova, S. Ya. and Filina, L. I., *Stal (in English)*, No. 3, 227 (1968)
28. Williams, E. W., *Trans. Inst. Met. Fin.*, **48**, 199 (1970)
29. Menzies, I. A. and Mortimer, D., *Corros. Sci.*, **5**, 539 (1965)
30. Weiss, B.-Z. and Meyerson, M. R., *J. Iron Steel Inst.*, **208**, 1 069 (1970)
31. Komem, Y., Weiss, B.-Z. and Niedzwiedz, S., *J. Iron Steel Inst.*, **206**, 487 (1968)
32. Davies, G. A., Ponter, A. B. and Menzies, I. A., *Acta Met.*, **15**, 1 799 (1967)
33. Prenosil, B., *Anti-Corros. Meth. Mat.*, **15** No. 10, 4 (1968)
34. Samuel, R. L. and Lockington, N. A., *Trans. Inst. Met. Fin.*, **31**, 153 (1954)
35. Samuel, R. L. and Lockington, N. A., *Chem. and Process Engng.*, **45**, 249 (1964)
36. Fleetwood, M. J., *J. Inst. Metals*, **98**, 1 (1970)
37. Brill-Edwards, H. and Epner, M., *Electrochem. Techn.*, **6**, 299 (1968)
38. Llewelyn, G., *Hot Corrosion Problems associated with Gas Turbines*, A.S.T.M.S.T.P. 421, 3–20 (1967)
39. Margalit, J., Kimmel, G. and Niedzwiedz, S., *J. Inst. Metals*, **98**, 126 (1970)

40. Galmiche, P., *Metals and Materials*, **2**, 241 (1968)
41. Sully, A. H., Brandes, E. A. and Brown, R. H., *Metallurgia*, **49**, 165 (1954)
42. Samuel, R. L. and Hoar, T. P., *Metallurgia*, **60**, 75, 80 (1959)
43. Derry, L. W. and Samuel, R. L., *Chem. and Process Engng.*, **43** (1962)
44. Menzies, I. A. and Mortimer, D., *J. Iron Steel Inst.*, **204**, 599 (1966)
45. Spring, A. I. and Wachtell, R. L., *Metal Finishing J.*, **8**, 357 (1962)
46. Cox, A. R. and Brown, R., *J. Less-common Metals*, **6**, 51 (1964)
47. Northcott, L., *J. Less-common Metals*, **3**, 125 (1961)
48. Minkevitch, A. N., *Rev. Mét.*, **60**, 807 (1963)

13.4 Principles of Applying Coatings by Metal Spraying

The metal spraying process originated in the work of the late Dr. M. U. Schoop of Zurich, who took out the original world patents for it, between the years 1910 and 1913. The idea was then so revolutionary that many exaggerated claims were made for the coatings produced, with the result that it was a long time before the method really made commercial headway. The first experimental practical use of the process was made in France and Germany during World War I, but commercial development on a considerable scale only took place in England in the early 1920s. Great use was made of the process during World War II and expansion has continued in most countries ever since. At reasonable intervals the method forms the topic of international conferences, the most recent being held in Paris in September 1970.

Methods of Spraying

Four main methods are used in metal spraying, the variation being in the initial form in which the metal is used. The earliest process employs molten metal. The metal is poured into a container and is allowed to flow through a small nozzle surrounded by an annular orifice, through which compressed air or other compressed gas is fed. The stream of molten metal is divided into small particles, as in a scent spray, and as the gas pressure used is high, each particle travels forward at a high velocity. If the droplet strikes a suitable surface while in the molten condition, it will adhere and form one member of a deposit. The process, originally employed in the UK for zinc spraying steel window frames, finds a limited outlet in the repair of dents in motorcar panels and for the backing of metallic rectifiers with fusible alloys. The disadvantages of the process are: (a) the tools are rather awkward to handle as they contain molten metal, (b) the metal must be easily fusible and (c) there is a good deal of erosion on the nozzle orifice. Among the advantages of the method are cheapness, as the metal used has not to be processed in preparation for its use in spraying in any way and the heat is usually obtained from towns gas at normal pressure and oxygen is not required.

The second process (the Schori process)* of metal spraying is that in which

*F. W. Berk Ltd. (Engineering Division), London, W5.

metallic powders are used as the raw material and are mixed with a stream of gas, which may be air or a combustible gas. This stream is fed into a central nozzle, again surrounded by an annular orifice. The annular orifice is fed with a fuel-gas and oxygen mixture which would normally give a blow-pipe flame on ignition, and the passage of the powder through the flame results in the melting of most of the particles. Compressed air or other gas is fed through a second annular orifice surrounding the gas ports, with the result that the molten powder particles are projected forward as in the molten metal process.

With modern methods of powder production it has become possible to increase the output and efficiency of powder pistols by introducing the powder, not through a central duct in the nozzle, but through four or more small ports interspaced with a similar number of gas ducts in a ring form-ation. The nozzle bearing this multiplicity of ducts is surrounded by an outer case forming an annular space through which the propellant gas (usually air) flows. The tools are fed from a specially designed powder con-tainer fed by positive pressure, and outputs of 50 kg/h are not unusual.

The advantages of the modern form of this method are that many alloys that cannot be conveniently drawn into wire form can be used in the process, that the hand tool contains no moving parts, and that high outputs can be obtained. The disadvantage of the powder system is that it is not very suitable for high-melting-point metals, and the losses are higher than with wire, because not all the particles are melted.

The third and most highly developed method uses a metal wire as the raw material. The wire is fed into the central orifice of a nozzle resembling that of the powder pistol; it passes through an intense oxy-gas flame and is sprayed by an air blast provided by the feed of air into an orifice surrounding the gas ports. In the case of hand pistols, the wire is fed by a mechanism in the tool itself, powered by compressed air which is fed through the same supply that provides air to the nozzle. The feed of the wire can be syn-chronised exactly with the rate of melting. This process will deal with all metals that can be drawn into wire, and which can be melted in any oxy-gas flame. As all the metal sprayed must first be melted, there is much less loss with this process than with the others. The tools weigh from 1·4 to 1·8 kg and can be handled without difficulty. The disadvantage of this process is that it is applicable only to metals ductile enough to be produced in wire form. It is essential that the feeding mechanism in the pistol is kept in good order. Some modern pistols also have automatic governors incorporated so that the wire speed remains constant.

The demand for higher output and greater efficiency has been met by a steady improvement in the design of wire pistols*. The small high-speed turbine and its attendant reduction gear is now often replaced by a special air-driven displacement motor giving a more positive drive to the wire. The group of valves controlling the gaseous feeds have been modified to give positive action and freedom from leaks. While the nozzle design is basically similar, improvements in wire quality and accuracy of nozzle part manu-facture allow outputs from the hand tools similar to those from the powder gun and comparable to those from early mechanised tools. Wire pistols need

*As used by Metallisation Ltd., Dudley, and Wall Colmonoy Ltd., Pontardawe.

three gas feeds:

1. A fuel gas under pressure which may be hydrogen, towns gas, acetylene or propane. With special nozzles, low-pressure acetylene or butane can be used, but it is not common practice. The use of single cylinders of fuel gas is now giving way (except under site conditions) to the gases drawn from bulk liquid storage. All modern installations use flow-meters in the gaseous feeds.
2. Oxygen from cylinders or bulk storage for the tools.
3. Compressed air at pressures of from 4 to $6 \times 10^2 \, kN/m^2$. The air supply must be free from suspended moisture and also be reasonably dry.

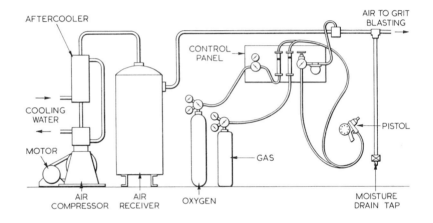

Fig. 13.26   Diagrammatic arrangement of compressed air and gases for wire pistols

The general arrangement of the apparatus is shown in Fig. 13.26. In recent years both the powder and wire processes have been automated for large projects. Mechanisation allows the use of large nozzles of special contours and the replacement of pneumatic drives by electric motors. Controls are usually from consoles by pneumatic or electronic signal systems.

Improvements of the manual pistols have made it possible to use a much thicker wire which can be up to 8 mm dia for the softer metals. Recently, the range of the wire pistol has been extended by the spraying of plastic cords containing the material to be sprayed as a fine powder. This development, as yet, is not much used in the anticorrosives field.

Although at the time of his early inventions Dr. Schoop envisaged that an electric arc could be used to produce the molten metal for spraying, forty years passed before the method became commercially important. Then, in Germany, Russia and Japan tools were made based on the arc. Although in Japan alternating current is used, the noise is nearly intolerable and elsewhere direct current from motor generators is employed. The fundamental idea is simple; two wires, carefully insulated from each other, are advanced to meet at a point where an arc is formed, immediately in advance of a jet of compressed gas, usually air. The metal melted by the arc is immediately broken down into a spray of fine droplets which are projected onto the work surface. In the UK the process had a limited use for spraying metals of high melting point for reclamation, but when metallic rectifiers and step-down transformers became common the future of arc spraying was assured. A transformer taking 3-phase current and delivering single phase to a rectifier giving out a direct current of up to 600 A at around 27 V is ideal for a spraying tool, and together with the controls can be a convenient size for work-shop use. In general, arc deposits tend to be a little coarser than those from the best gas pistol, but because of the high temperature there is some incipient welding to the work face and adhesion is high. Unfortunately, as yet, spraying losses are higher than with the gas gun and the arc does not seem to have economic advantages over flame guns for the production of zinc coatings. There is a rapidly expanding demand for the system for aluminium, and by using wires of two metals, for coatings of mixed metals. The diameter of the wire used in arc pistols is normally about 1·6 mm and outputs of 8 kg/h are possible with aluminium.

Metals which are subject to oxidation or attack by nitrogen can be sprayed in a closed system so that air is exluded. The heat necessary to melt the wire is produced by current generated in the wire itself by high-frequency currents flowing in small water-cooled coils. By this means, titanium, niobium and even uranium, can be sprayed without gaseous contamination.

Before leaving this brief consideration of spraying methods it should be noted that metal can be deposited by plasma or constricted arc and by detonation. In the plasma process* an arc is struck between a central electrode of a refractory metal such as tungsten and the rim of a water cooled nozzle, often of copper. The nozzle forms a small chamber in which the arc is constricted by the pressure of the spraying gas. This is one of the inert gases such as argon, helium or mixtures of these with nitrogen. The temperature of the arc plasma can greatly exceed 5 500°C and the material to be sprayed is introduced into the nozzle chamber as a fine powder of diameter in the range of 5–40 μm.

While plasma deposits are widely used, especially in the American aero industry to provide wear resistance, there is not at the moment any great demand for the exotic materials deposited to be used as a protection against corrosion. However, M. A. Levinstein of General Electric (USA) reports the successful use of sprayed chromium carbide as a protection for ventilator blades operating in corrosive conditions.

Finally, it should be noted that there is now a method of coating by detonation† which in some ways is akin to spraying. Briefly, mixtures of explosive gases charged with the coating material as a fine powder are fired in a special strong tubular chamber with a small exit. At each explosion the powder leaving the exit strikes the work face at high velocity. After each detonation (many times per minute) the chamber is swept clean with nitrogen. As yet the demand for these deposits is for wear resistance but they could be important as a protection against high temperatures. Unlike most spraying processes the apparatus is not mobile and the operation is carried out in

*Plasma process utilised by Metco Ltd., Chobham and Ministry of Defence, Royal Armament Research and Development Establishment, Fort Halstead.
†Detonation process utilised by Union Carbide (UK) Ltd., Stratton St. Margaret and Glossop.

sound-insulated chambers. In some cases when plasma deposits fail detonation may provide an answer.

Surface Preparation

As with all metal coating processes, the success of the finished deposit depends considerably on the surface to which it is applied. Most metal coating processes depend upon interatomic forces or alloy formation for adhesion; in metal spraying, on the other hand, the bond is entirely mechanical. Although each small particle arriving at the surface is molten, there is not sufficient heat present to produce incipient welding to the surface and the process is virtually cold. Adhesion is thus normally the result of a mechanical interlocking of metal particles with a clean and rough surface.

A suitable surface for the reception of ordinary metal-sprayed coatings as used for protection is obtained by blasting with angular grit, but not by shot blasting. When spraying is carried out in a works the normal abrasive used is chilled iron grit (BS 2451) Grade 2. On site, or where recirculation of abrasive is not easy, an aluminous grit (24 mesh) or some type of hard copper slag is used. Blasting with iron grit can be performed by means of compressed air or by airless blasting, in which case the abrasive is thrown by centrifugal force from the periphery of fast rotating wheels. This process is not suitable for the lighter abrasives. It goes without saying, that the function of the blasting is not only to give the required angular roughness but to ensure that the work surface is clean and free from all scale.

In cases where very high adherence is necessary an undercoating of sprayed molybdenum is applied. Owing to the volatile nature of its oxide this metal presents a clean surface to the workface and with most metals (except copper) very high adhesion is obtained. To a limited extent arc-sprayed aluminium bronze will also form a strongly adherent base coating.

Deposition by metal spraying can also be used for the reclamation of worn parts; in this case, surface preparation is often accomplished by machining, i.e. by cutting a rough thread on the surface or by increasing the surface area of the part by grooving. Such methods are not, however, normally used in corrosion prevention, except in the case of pump rods, which can be built up with nickel or stainless steel.

Sprayed Metal Coatings

Aluminium- and Zinc-sprayed Coatings

The metals most used for corrosion protection by metal spraying are aluminium and zinc, both of which are anodic to steel in most environments. Physical properties of these coatings are shown in Table 13.6.

Metal spraying is, however, capable of producing deposits of nearly all the metals and alloys of commerce. The great advantage which the process possesses over almost any other is that zinc and aluminium coatings can be

Table 13.6 Physical properties of sprayed zinc and sprayed aluminium

Material	Property				
	Density (g/cm^3)	Coating weight (kg/m^2) 0·025 mm thickness	Brinell hardness	Ratio of contraction stresses in sprayed deposits 0·51 mm thick	Compressive strength (stress to collapse) (MN/m^2)
Sprayed zinc	6·35	0·159	20–28	1	159
Sprayed aluminium	2·35	0·059	25–35	4	185

applied to very large structures, such as bridge members. Usually these are treated in unit form, and after erection the coating is touched up on site. Examples of well-known structures that are zinc sprayed are the bridges over the Menai Straits and the Volta river in Africa, and the new Forth and Severn suspension bridges. The structure of the British Steel Corporation's mills at Margam, Trostre and Velindre in South Wales are sprayed with aluminium. Sprayed deposits of zinc and aluminium have protective properties proportional to their thickness. They have the advantage that, being of a matt finish, they absorb paint well.

The coatings produced by metal spraying have an unusual structure which is characteristic of the method of formation. They are composed of small particles usually not more than 0·01 mm in diameter which, having reached the surface in the molten condition, have splashed outwards and then solidified. Figure 13.27 (left) shows in section the irregular form of the flattened particles. In transverse section the surface profile is undulating (Fig. 13.27 (right)).

In all metal spraying processes the particles emerge from the nozzle in a conical stream, and although the particles near the centre are molten, those at the periphery have solidified. In the powder process there are in addition

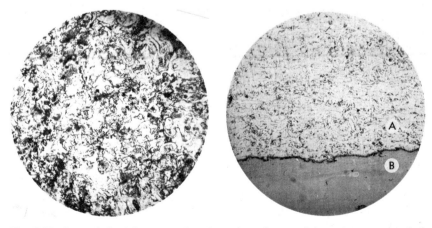

Fig. 13.27   Sprayed aluminium on steel. *Left:* section taken parallel to substrate and etched, × 75. *Right:* section taken transverse to substrate and etched in HF, × 75; *A* aluminium and *B* steel

solid particles which have not melted. The solid particles tend to become entrapped in the coating, making it porous. The effect is more pronounced in the powder process owing to the larger number of solid particles present.

In the case of the reactive metals like aluminium and zinc, the particles themselves become coated with a thin oxide skin, so that the finished deposit contains very thin layers of oxide. In the early days it was thought that the presence of pores and oxide layers (Fig. 13.28) would cause trouble, but this has been found to be untrue, at least for anodic metals. In a zinc coating, in fact, pores should have little effect. Zinc protects by sacrificial action, and

Fig. 13.28   Transverse section of sprayed 0·4% C steel, unetched and × 75

immediately an electrolyte permeates a zinc coating, corrosion will commence, giving rise to relatively insoluble products which will seal the pores automatically. Therefore if the zinc coating has sufficient weight per unit area, it will give complete protection. The life of a zinc coating applied by metal spraying, weight for weight, is equal to the life of a zinc coating applied by any other method.

Pores play an important part in the protective value of sprayed aluminium. Its behaviour when sprayed is entirely different from what would be expected from the massive metal. Sprayed aluminium is in fact used to protect wrought aluminium from corrosion. Aluminium coatings are slightly more porous than zinc coatings, and the interconnected porosity may be as high as 10%, although it is usually nearer 5%. Each particle of aluminium on the coating is surrounded by a very thin oxide layer. This layer is, however, discontinuous in places, so that the electrical conductivity of sprayed aluminium coating is quite high.

From its position in the electrochemical series, aluminium would be expected to protect steel at discontinuities more effectively and over a wider area than zinc. Aluminium with an oxide film is, however, more noble than zinc, and thus although sprayed aluminium will protect steel by sacrificial action, its action in this respect is not so marked as that of zinc. Thus an electrolyte passing through a sprayed aluminium coating in the first few hours of its life will cause corrosion, with the formation of very insoluble substances, which completely seal the pores of the aluminium, so that after a little time the aluminium coating becomes absolutely impervious to moisture. In the event of mechanical damage this self-healing mechanism is supplemented by sacrificial action, insoluble products form and a scratch is healed almost immediately. Aluminium does not give heavy corrosion products and therefore coatings of paint on the sprayed layer do not tend to lift. Aluminium-sprayed coatings on steel have been exposed for over 20 years in very severe atmospheric conditions (such as those at Congella near Durban) and have given perfect protection; the only result of long exposure has been the appearance of a few small nodules of aluminium oxide which appear to have little or no significance as sites of future corrosion.

Aluminium coatings are extremely attractive from the point of view of protection in both immersed and atmospheric conditions, but they are most advantageous when the electrolyte of the corrosive medium is of high conductivity. While aluminium-sprayed coatings give good results in sea water, and excellent results in sulphurous atmospheres, a combination of sulphur and chlorine seems to reduce the insolubility of the corrosion product, and for resistance to such combined attack zinc is preferable to aluminium. If a freshly sprayed aluminium coating on steel is exposed to pure water for a few hours it sometimes becomes covered with a brown stain. This is due to the aluminium acting cathodically to the steel during the first few hours. The action appears to be due to the oxide layers. A very small quantity of iron is corroded during the initial period, but after a little time the aluminium acts normally as anode. The insoluble aluminium oxides formed are coloured by a small amount of iron, and it has been proved that these brown stains, although unsightly, do not have a significant effect on the expectation of life of the aluminium coating.

The thicknesses of the deposits are normally those recommended in BS 2569, Part 1*, i.e. 0·004 in (0·1 mm) in the case of aluminium and 0·003 in (0·075 mm) for zinc. If the surface is to be subsequently painted, the thickness is sometimes reduced to 0·002 in (0·05 mm). It should be realised that these thicknesses are the average, the high and low spots varying 0·001 in (0·025 mm) up and down. For immersed conditions some specifications call for 0·008 in (0·2 mm) coatings of aluminium. Recently, it has been shown that layer corrosion in high-tensile aluminium members is completely prevented by spraying with aluminium, and it is worthy of note that the best results for this purpose and for the protection of steel are obtained with an aluminium of commercial purity, containing 99·5% of aluminium (the main impurity should not be of copper). Commercial aluminium of this purity is much more effective than the 'super-pure' metal. It is now suggested that aluminium–zinc alloys containing from 25 to 50% Zn give better protection in certain conditions than either of the components alone. There appears to be some evidence that this is the case in stagnant sea water, but much more practical experience will be necessary before these alloys can be generally recommended. The raw materials would be available in powder or wire form, but it is probable that, although there is no difficulty in the actual spraying process, the cost would be greater.

An interesting application of sprayed aluminium is for resistance to high-temperature oxidation up to 900°C. The article is grit-blasted and aluminium sprayed, usually to a thickness of 0·008 in (0·2 mm). It is then treated with a sealing composition which may be bitumen or water-glass, and is diffusion

---

*At the time of writing this standard was still in Imperial units, and the SI units given here in parentheses are the nearest direct equivalents.

Fig. 13.29   Aluminised steel. A Al_2O_3 + aluminium, B Fe–Al alloys and C steel; × 75

annealed in a furnace at 850°C for approximately 30 min. The final coating consists of a gradation of aluminium–iron alloys with a skin of aluminium oxide (Fig. 13.29). Such deposits will withstand oxidation for very long periods at temperatures up to 900°C. Above this temperature the diffusion of iron into the aluminium becomes so rapid that the alloy layer becomes impoverished and the upper layer contains insufficient aluminium to provide further protection. A refinement of this process which produces better results is the use of aluminium containing 0·75% of cadmium; with this alloy no sealer is necessary and the sprayed article is put straight into the furnace. Deeper penetration is obtained using this method. The process can be used on some cast irons, but if the free graphite content is too high the aluminium layer will not prevent growth.

Lead Sprayed Coatings

Lead is a toxic metal and rigid precautions against lead poisoning are essential. Sprayed lead is, like other sprayed metals, porous, and the sprayed layers will not as a rule withstand attack by strong acids. On the other hand, lead from 0·005 to 0·01 in (0·13 to 0·25 mm) in thickness has proved extremely useful in atmospheres containing sulphuric acid. In this case the pores in the lead become blocked with lead sulphate, with the result that complete protection is assured. In the event of mechanical breakdown, lead does not exhibit any sacrificial action and therefore corrosion may lift the lead layer.

Tin Sprayed Coatings

Tin can be sprayed by all the processes mentioned above, and is very often used for food vessels. The advantage of the spraying process is that heavy layers of pure tin can be deposited. As the layer is porous, it is usual to apply 0·015 in (0·38 mm) and polish the deposit. The polishing of tin, which is a soft metal, completely seals the pores. When tin coatings have to be used in deep-freeze conditions, there is a likelihood that pure tin will be transformed to a grey powder, owing to allotropic change. This is completely prevented by using tin containing 0·75% of bismuth, and this alloy can readily be used in the metal spraying process.

Copper Sprayed Coatings

Copper and its alloys are all cathodic to steel, therefore sprayed coatings of these materials are not used for protection, except for ornamentation work in the interior of buildings, or in conditions such as where there is a minimum humidity.

Stainless Steel, etc.

Most stainless steels can be sprayed, but it is necessary to use material which has been stabilised against weld decay, because during spraying the stainless steel particles pass through the critical temperature range. Stainless steel coatings are porous, hence, it is not economical to spray tanks with these alloys, as it would be necessary to polish the surface to ensure that no discontinuities were present. Pump rods, shafts and rolls are now reclaimed and resurfaced with really heavy deposits of several millimetres in thickness which are afterwards ground and polished. Under these conditions stainless steel gives good service. The above remarks apply, in addition, to coatings of nickel and Monel. These coatings give good results on rods where a ground and final high finish is possible.

For resistance to high-temperature oxidation in air, at temperatures above 900°C, good protection may be obtained by spraying nickel–chromium alloys and then heat treating at 1 100°C so that diffusion takes place. Such coatings are extremely useful under high-temperature conditions where it is not economical to use solid nickel–chromium alloys. Under some furnace conditions, solid nickel–chromium parts sometimes give unsatisfactory results owing to attack by sulphurous gases. This can be largely overcome by spraying them with aluminium.

Mechanical Properties of Sprayed Metal Coatings

The strength and adhesion of sprayed metal coatings are extremely difficult to measure with precision, and the properties of sprayed metals vary greatly with the spraying conditions and with the conditions of test. It is difficult, therefore, to correlate the values taken from the literature on the subject. For instance, American workers produce tensile test pieces by depositing on to $\frac{3}{8}$-in (9·5 mm) steel tube and then machining out the tube. By this method the results shown in Table 13.7 were obtained.

On the other hand, the ultimate strength obtained by testing solid sprayed bodies in the UK gives values about one-third of those in Table 13.7. The density of sprayed metal is between 10 and 15% lower than that of the wrought metal.

Spraying conditions make hardness values so variable that unless they are accurately known no comparisons are possible. Brinell hardness figures for sprayed molybdenum vary from 350 when produced with a reducing flame to 725 with an oxidising flame, and while a thick sprayed deposit of 0·8% carbon steel can give a figure of 330, the hardness of a particle obtained

Table 13.7   Mechanical properties of sprayed metals

Metal or alloy	Ultimate tensile strength (MN/m²)	Elongation (%)
Fe–18Cr–8Ni	206	0·27
Fe–13Cr	274	0·50
0·1% carbon steel	205	0·30
0·25% carbon steel	239	0·46
0·80% carbon steel	188	0·42
Aluminium	134	0·23
Bronze	139	0·45
Zinc	90	1·43
Molybdenum	51	0·30

Table 13.8   Hardness values (Brinell) for solid blocks of sprayed metal

Metal	Hardness value (HB)
Zinc	21
Aluminium	18
Copper	93
Monel	192
Mild steel	320
Fe–18Cr–8Ni	301
0·75% carbon steel	336 (kept cold during spraying)
0·75% carbon steel	280 (allowed to get hot during spraying)

by micro hardness methods will be about 550.

Hardness values ascertained for solid blocks of sprayed metal are given in Table 13.8. Values obtained in America on deposits 0·76 mm in thickness were about two-thirds of those given in this table. Figures for shrinkage of various deposits are given in Table 13.9.

Thermal conductivty can be as low as one-eighth that of solid metal; in the case of steel 7 W/m°C. The electrical resistance (specific) of copper, zinc and silver is about twice that of the cast metal, and of aluminium as much as five times, depending on spraying conditions. Adhesion in tension should be

Table 13.9   Shrinkage of sprayed metal deposits

Metal	Shrinkage (mm/mm)
Fe–18Cr–8Ni	0·012
Fe–13Cr	0·001 8
Ni + Cr + Mo steel	0·002
0·10% carbon steel	0·008
0·25% carbon steel	0·006
0·80% carbon steel	0·001 4
Aluminium	0·006 8
Bronze	0·006
Zinc	0·010
Molybdenum	0·003

from 7·7 to 35 MN/m^2, but shear can be five times this according to the method of preparation.

For 30 years, large constructional engineering projects have been sprayed with zinc and with aluminium. Spraying has been done in metal spraying works, and no preferential treatment after transport to the site has been given. Experience has shown that the amount of touching up and repair of coatings on site necessary is extremely small. One can assume that if the grit-blasting process has been carried out satisfactorily the adhesion will be sufficient for all normal anti-corrosive work. It should be realised that the spraying process is essentially the casting of a great number of small particles under pressure, and the coatings have a strength comparable to that of a cast metal of similar thickness. It is clear that thin sprayed metal coatings cannot be expected to withstand rubbing friction or heavy mechanical damage. If coatings are applied to sharp edges some breakdown is to be expected. In constructional engineering most edges are radiused, and the problems connected with the corners do not arise. The spraying process can be applied wherever a pistol can reach the surface, and in order to extend its adaptability special nozzles which deflect the spray in angular directions are used. Straight tubes up to 2·5 m in length can often be sprayed internally by the wire process, by use of special nozzles, provided their internal diameter exceeds 32 mm.

Painting of Sprayed Coatings

While zinc sprayed coatings give excellent protection in themselves to steel, the matt surface is very reactive and hence becomes unsightly due to the formation of corrosion products. Galvanised steel gives a smooth surface and is therefore not an ideal base for painting and it is normal to use an etch primer to give the necessary adhesion. The etch primer is usually based on a butyral resin and free phosphoric acid, and it leaves the zinc surface with a thin inhibitive film. It is obvious that there is no need to etch a sprayed surface, and the late Dr. Jordan of the Paint Research Association suggested reduction of the free phosphoric acid by a quarter to 3·5–4%. This primer is now normally applied to zinc sprayed material before it leaves the metallising site. There is still no definite ruling as to the ideal paint system to follow the priming wash, but research on this is still being pursued at the Paint Research Station by Bullett, Wallington and others. Most paint systems give good results over sprayed zinc but certain oily media which react with the metal should be avoided. The use of inhibitors such as chromates has been found to be good in practice, but the less soluble zinc tetroxychromate is preferable to zinc potassium chromate. After much thought and in order to make the system used on the Forth and Severn bridges near-perfect, an undercoat of zinc tetroxychromate in a phenolic vehicle, followed by two coats of micaceous iron ore in a similar vehicle, were used. Very promising results have been obtained with the inert paints based on polyurethanes and vinyl copolymers.

The texture and drying properties of the paint are important. If it is too thin it will leave the peaks of the matt coating uncovered and if too thick, gas will be trapped in the valleys giving a tendency to blister. In view of the success of the silicate-bonded zinc-dust paints, tests are now being made on paints with a silicate vehicle, usually ethyl silicate. All over the world, painted sprayed zinc has given good results in many types of exposure conditions and with many types of paint, and it is because of the need to produce the best paint that speculation and research continues.

Aluminium coatings exposed as sprayed show remarkable protective properties and in many cases the only evidence of age is the appearance of small hard nodules of aluminium oxide, and these do not appear to be deleterious. However, the dead white of the sprayed surface soon becomes discoloured by dirt and so a paint covering is usual. A chromate inhibitor is not essential and two coats of a simple vinyl-based sealing paint have given remarkable results in the tests of the American Welding Society over periods of 12 and 15 years.

Thin zinc sprayed deposits of 0·05 mm in thickness are used as a base for stove enamelled finishes on such structures as radar cabinets.

It is necessary to choose the type of paint with care. If it skins-over too quickly in the oven the gases cannot escape and blistering results, but if setting is deferred for too long the paint will flow to the base of the article. Some experimentation is advisable and the conditions once established should be retained. Some users treat the sprayed coating with a passivator of 200 g sodium dichromate in 1 litre of 6% sulphuric acid and dry it before enamelling.

Recent Developments

In recent years, the spraying process has been adapted for hard facing, using the chromium–nickel–boron alloys which have become known as Colmonoy. More recently still, the cobalt-base Stellite alloys have also been used. These materials in powder form are sprayed on to the surface in the usual way. The deposit is afterwards heat treated by a torch, so that fusion takes place. The process is often known as *spray-welding*. Such coatings are primarily used for hard facing under wear conditions, but as the final surface is nickel–chromium or cobalt–chromium they exhibit very high anticorrosive properties.

The spraying process provides a method of treating steel with coatings of zinc or aluminium, which can afterwards be painted. A combination of such a metallic primer and a good paint system is the most effective means of combating corrosion of constructional steelwork at a reasonable cost that is yet known.

W. E. BALLARD

BIBLIOGRAPHY

Textbooks
Ballard, W. E., *Metal Spraying and the Flame Deposition of Ceramics and Plastics*, Griffin, London (1963)
Brennek, J., Brodzki, Z., Gebalski, S., Drazkiewicz, T. and Kowalski, Z., *Poradnik Metalizacji*

Natryskowej, Panstwowe Wydawnictwa Techniczne, Warsaw (1959)
Kretzschmar, E., *Metall, Keramik und Plastspritzen*, V.E.B. Verlag Technik, Berlin (1970)

Papers
(*a*) *Formation of Coatings*
Ballard, W. E., *Proc. Phys. Soc.*, **157**, 67 (1945)
Matting, A. and Raale, W., *Schweissen u. Schneiden*, **8**, 569 (1956)
(*b*) *Applications, etc.*
Ballard, W. E., *Chem. and Ind. (Rev.)*, 1 606 (1955)
Ballard, W. E., *Trans. Inst. Metal Finish.*, **29**, 174 (1953)
Champion, F. A., *Electroplating*, **8**, 180 (1955)
Himsworth, F. R., *Chem. and Ind. (Rev.)*, 1 618 (1955)
Mayer, W. B., *Corrosion*, **5**, 282 (1949)
Rivet, F. A., *Chem. and Ind. (Rev.)*, 1 612 (1955)
Shepard, A. P. and McWaters, R. J., *Corrosion*, **11**, 115 (1955)
Stanton, W. E., *Corrosion Tech.*, **3**, 311 (1956)
(*c*) *Behaviour of Sprayed Coatings*
Evans, U. R., *Chem. and Ind. (Rev.)*, 706 (1951)
Gilbert, P. T., Lecture. Assoc. Belge pour L'Étude et L'Emploi des Matériaux, June (1954)
Moss, A. R., *Arc Plasma Technology*, Welding Institute Autumn Meeting, London (1963)
Sprayed Aluminium Coatings, brochure published by Aluminium Wire and Cable Co. Ltd., Swansea

13.5 Miscellaneous Methods of Applying Metallic Coatings

The more important metallic coating methods such as electroplating, spraying, hot-dipping and diffusion have already been considered. There are, however, several other methods which have certain advantageous features and therefore find exclusive application in specific fields.

The miscellaneous coating methods to be considered here are:

1. Immersion plating.
2. Chemical reduction.
3. Peen (or impact) plating.
4. Vacuum evaporation or metallising.
5. Gas (or vapour) plating.
6. Cathode sputtering.
7. Brush plating.
8. Plasma spraying.

Their characteristics are summarised in Table 13.10.

Immersion Plating

Principles Immersion plating resulting from a displacement reaction involving the metal to be coated can continue only as long as the less noble substrate remains accessible to the plating solution, and therefore as plating proceeds, the quantity of M_1 deposited, and of M_2 dissolved, falls. Dissolution of M_2 can be avoided by coupling it with a less noble metal M_3, so that only M_3 dissolves, i.e. by internal electrolysis.

The mechanisms of these displacement reactions have been described elsewhere[1]. Radio-tracers have been used to study silver deposition[2], and it has been shown that when certain inhibitors or complexing agents are present the displacement reaction may be markedly affected[3].

If a displacement reaction is to take place uniformly, the surface of M_2 should be chemically clean initially; in addition, a wetting agent may be employed in the plating solution.

Features of the method The solutions and prodecures are uncomplicated, and can often be adapted for barrel plating. Addition agents[4] are usually necessary in order to restrict the high initial rate of plating, which might otherwise cause poor adhesion and spongy deposits containing trapped electrolyte, although it has been reported that adherent deposits of copper on aluminium can be plated from a solution of $CuCl_2$ in alcohol[5]. Some difficulty in achieving a uniform coating may arise with articles composed of two or more different metals. Coating thickness is limited to about 0·001 2 mm (average), because plating cannot proceed after the last exposed M_2 atom has been replaced by an M_1 atom.

Table 13.10   Miscellaneous coating methods*

Method	Significant plating mechanism involved	Common examples
Immersion plating	Displacement of metal ions in solution by metal to be coated: $$xM_1^{n+} + M_2 \rightleftharpoons M_1 \downarrow + yM_2^{m+},$$ where $xn = ym$	M_1: Cu from aqueous $CuSO_4$ M_2: Fe
Chemical reduction	Chemical reduction, often based on simple redox reactions: $$M_1^{n+} + R = M_1 \downarrow + \text{oxidised } R$$ where R = a suitable reducing reagent	M_1: Cu from aqueous $CuSO_4$ R HCHO
Peen plating	Adhesion brought about by the intimate contact of two chemically clean solid surfaces: $$M_{1\,powder} + M_2 \text{ (or } M_{1\,solid}) \xrightarrow{\text{impacted}} M_{1\,solid\ coating}$$	M_1: Zn M_2: Fe
Vacuum evaporation	Evaporation at reduced pressure in accordance with the Clausius equation, followed by condensation on a cold surface: $$M_{1\,solid} \rightarrow M_{1\,vapour} \xrightarrow{} M_{1\,solid} \text{ on } M_2$$	M_1: Al M_2: Fe
Gas plating	Chemical reaction causing gaseous M_1 compound to split up and release M_1 to deposit on M_2: $$XM_{1\,gas} \xrightarrow{\text{Heated } M_2} M_{1\,solid} \text{ on } M_2$$	M_1: Cr M_2: Fe
Cathode sputtering	M_1 is volatilised by d.c. arcing, followed by condensation on cool M_2: $$M_{1\,cathode} \rightarrow M_{1\,vapour} \rightarrow M_{1\,solid} \text{ on } M_2$$	M_1: Au M_2: Se
Brush plating	M_1 is electroplated on local areas by a brush or tampon	M_1: Ni M_2: Fe or Cu
Plasma spraying	Metals, or refractory materials or composites are applied by melting in an ionised inert gas	M_1: tantalum, molybdenum, alumina, zirconia M_2: a variety of metallic substrates

*M_1: coating metal and M_2: metal to be coated.

Properties of deposits These deposits invariably possess an appreciable number of discontinuities, which diminish their protective value. At such discontinuities there is an inherent susceptibility to corrosion because of the galvanic relationship that necessarily exists between M_1 and M_2. Corrosion resistance may, however, sometimes be improved by sealing[6] (cf. anodising) or by chromate passivation.

Owing to the thinness of the coatings, properties such as reflectivity and smoothness are chiefly governed by the substrate, although addition agents find useful application in improving these properties.

Applications The method is clearly restricted to those systems in which a displacement reaction can occur[7], but it is possible to extend the range by pre-coating M_2 with a thin coating of another baser metal by an alternative coating method. Some alloys can be deposited[8]. The displacement coating is often followed by a further metal-finishing step. Applications include 'nickel dipping' of ferrous metals prior to vitreous enamelling to assist bonding; 'copper lacquering' of ferrous rod for wire drawing in order to assist lubricant carrying and improve the final appearance and corrosion resistance of the drawn wire; deposition of certain precious metals and alloys on copper and its alloys in engineering applications, and of copper alloys to facilitate their soldering[8]; plating of aluminium with zinc using sodium zincate solution to facilitate the subsequent electrodeposition of a Cu–Ni–Cr system; 'quicking' of copper alloys with mercury to promote adhesion and covering power in subsequent electrodeposition; deposition of gold on to steel pen nibs; deposition of copper on to plastics substrates containing a suitable base-metal powder for electrical conduction[9].

Chemical Reduction

Principles The reduction reaction is controlled essentially by the usual kinetic factors such as concentration of reactants, temperature, agitation, catalysts, etc. Where the reaction is vigorous, as, for example, when a powerful reducing agent like hydrazine is used, wasteful precipitation of M_1 may occur throughout the whole plating solution followed by deposition on all exposed metallic and non-metallic surfaces which can provide favourable nucleation sites. In order to restrict deposition and aid adhesion, the selected areas are pre-sensitised after cleaning; the sensitisers used are often based on noble metal salts.

Features of the method With proper plating control there is usually no limit to the coating thickness that can be built up, although the simple plating solutions seldom give acceptable thick deposits. Certain addition agents can be used; in the 'electroless' nickel process, for example, such additions include sequestering agents to prevent pre-precipitation of M_1; exaltants to enhance the reducing power of the reducing agent; stabilisers to inhibit certain active nuclei, thus preventing overall decomposition of the plating solution; buffers to control the pH of the plating solution; and wetting agents[10]. While high plating rates, e.g. 0·025 mm/min, can be achieved, moderate rates avoid poor adhesion and entrapment of solution in the deposit. Although the actual plating operation may be uncomplicated, the plating solution requires regular replenishment of M_1 salts, and filtration.

The most extensively used reducing agent for the electroless deposition of nickel is hypophosphite[11], and the reaction is as follows:

$$3H_3PO_2^- + 3H_2O + Ni^{2+} \rightarrow 3H_2PO_3^- + 2H^+ + 2H_2 + Ni$$

The most important parameters of deposition are the temperature and the pH of the solution. Generally speaking, raising the temperature increases the rate of deposition. The solution can be alkaline or acid, the latter being preferred. The composition and operating conditions of a typical bath are:

$NiSO_4 \cdot 6H_2O$	20–25 g/l
$NaH_2PO_2 \cdot H_2O$	25–30 g/l
Complexing and buffering agents	30–50 g/l
Accelerators	3–5 g/l
Stabiliser	3–5 g/l
Surfactant	30–50 mg/l
pH value	4·3–4·7
Temperature	93–95°C
Deposition rate	25–28 μm/h

Nickel can also be deposited by reduction with the aid of boranates[12] such as sodium boranate ($NaBH_4$) or N-diethyl borazane, i.e. $(C_2H_5)_2NH \cdot BH_3$, the basic reaction proceeding as follows:

$$NaBH_4 + 4NiCl_2 + 8NaOH \rightarrow 4Ni + NaH_2BO_3 + NaCl + 5H_2O$$

The sodium boranate bath is operated at a pH of 14 and a temperature of 90–95°C. The rate of deposition is 10–30 μm/h. A metallic catalyst is required for the reaction.

Chemical reduction is used extensively nowadays for the deposition of nickel or copper as the first stage in the electroplating of plastics, especially ABS resin mouldings and, to a lesser extent, polypropylene. The plastic is first etched in a strong chromic–sulphuric acid mixture, sensitised in a stannous chloride solution and then activated in a palladium or silver bath. It is then ready to proceed to the electroless copper or nickel plating stage, further coatings then being applied by electrodeposition in the conventional manner.

Properties of deposits Deposits can be produced that are adherent, coherent and finely crystalline. Addition agents, e.g. organic sulphonamides[13], can improve the deposit structure so that thick coatings can be produced free of nodules and blisters. The production of very smooth thick deposits of copper has been reported[14]. Thin deposits tend to reproduce the substrate topography, but some cases of levelling have been reported. The brightness tends to fall with increasing thickness.

Ductile and easily buffed chromium deposits having satisfactory corrosion resistance have been produced; thus 0·005 mm-thick chromium deposits applied to steel by chemical deposition or by electrodeposition gave similar results when subjected to a salt-spray test[15].

Applications The method can be adapted to barrel plating and to the mirror-spray technique[7]. The development of printed circuitry has stimulated demand for means of rendering parts of insulating surfaces conducting and solderable. Developments seem to be confined to deposition of single metal species, e.g. Sn, Ni, Co, Ag, Cu, Pd and Au. The hypophosphite method[10] gives deposits which contain 8–10% of phosphorus. Such deposits are hard (e.g. 500 HV) and brittle; they are relatively pore-free and have good corrosion resistance, particularly if heat-treated. In general, the corrosion resistance is proportional to the phosphorus content, and such coatings can comply with Fed. Spec. QQ–N–290. Exposure tests have shown that 0·012 mm thick Ni–P coatings give better protection to steel than 0·025 mm of electrodeposited nickel. Their high resistance to wear has been proved by their application in, for example, surfacing moulding dies. These coatings have been used to protect a low-alloy pearlitic steel from corrosion in air at 650°C, and in superheated steam for up to 1 000 h[16] (see also Section 14.7).

Peen Plating

Principles   The process consists of 'tumbling' the metal to be coated with a powder of the coating metal[17]. It is considered that a form of welding is involved, but the type of conditions conducive to successful deposition indicates that the deposit adheres by mechanical keying (M_1 must therefore be relatively soft) and adhesive forces[18]. Thus pretreatments such as abrading or pickling enhance the keying effect, and the use of a soft metal 'strike' coating (e.g. Cu electrodeposited on steel) can also aid bonding. The presence of grease or oxide films on M_2 prevents adhesion of M_1, and the use of 'promoters' and wetting agents ensures that such films do not interfere.

The optimum quantities and grades of M_1 powder, impact media, water and promoter, and plating conditions such as barrel-rotation speed, are best decided by trial runs.

Features of the method   This method has the advantage that sintered, carburised, nitrided and non-metallic articles can be plated, and that no involved surface preparation is required. Futhermore, the possibility of hydrogen embrittlement is avoided[19]. Plating speeds are of the order of 0·007 5 mm/h, and thick deposits (0·05 mm) can be produced, although subsequent heat treatment may be required to enhance adhesion. Alloy coatings can be produced by using a coating powder prepared from its appropriate alloy.

Properties of deposits   These are usually satisfactorily adherent, coherent and continuous on accessible surfaces. Micro-structures are often similar to those of metal-sprayed coatings. The coatings sometimes contain traces of included promoter solution which may partly account for the good tarnish and corrosion resistance. Exposure tests in a severe industrial atmosphere containing sulphurous gases showed that, for zinc on steel, similar corrosion resistance was provided by electrodeposition, hot-dipping and impact plating. In other tests, e.g. in marine atmospheres, impact plating provided superior corrosion resistance, and this method has met with approval for certain applications. Post-plating chromate passivation has been observed to confer more benefit on impact deposits than on electrodeposits of zinc on steel.

Owing to the slight abrasion during tumbling, very thin coatings are rarely mirror-bright, and the coating smoothness and reflectivity usually fall slightly as the coating thickness increases.

Applications   Since M_1 must be relatively soft the method is restricted to metal powders such as Zn, Cd, Sn, Al, Pb and certain alloys. The geometry and strength of M_2 govern whether this method can be usefully applied. Generally impact deposits are applied for their corrosion resistance rather than for their decorative value. Thus the method has found particular application for coating ferrous nails, washers, chain links, and high-tensile steel springs and hose-clips, particularly where hydrogen embrittlement must be avoided (see Section 14.3).

Vacuum Evaporation

Principles   Details of the process and plant, which consists essentially of a heated M_1 vapour source[20] contained within a closed coating chamber capable of evacuation to $13–1.3 \, kN/mm^2$, for example, have been given elsewhere[21,22]. The subject has also been reviewed by Fabian[23].

Special preparation of M_2 is necessary to obtain maximum adhesion and to avoid outgassing of foreign matter during evacuation. Outgassing can be prevented by pre-coating the substrate with a lacquer which can further provide a smooth base for M_1.

Condensation of M_1 vapour can occur on all cool surfaces and the plating process is therefore not selective with regard to M_2. Some fundamental aspects concerning the formation and properties of these coatings have recently been considered[24].

Features   Good control of average coating thickness can be obtained and there is no excessive build-up at sharp discontinuities, but irregularly shaped articles should be so located in the coating chamber as to avoid 'shadowing' and to minimise loss of M_1 on the coating chamber's inner surfaces. The coating rates depend on the nature of M_1; for example, $0.025–0.075 \, mm$ thickness of Cd, Zn or Se can be deposited in 10 min, but considerably lower rates are observed for Cu, Al or Pt. The cost of depositing thick, protective coatings of, for example, Al is estimated to be similar to that for electroplating. The process has the advantage of being a dry one, and subsequent steps such as rinsing, de-watering and drying can be eliminated. Adhesion can be still further improved by, for example, prior shot blasting (where hydrogen embrittlement must be avoided) or by post-plating heat treatment. The plating can be carried out in stages to allow inspection, because lamination is not usually encountered.

Properties of deposits   Deposits are normally smooth (often highly reflecting, depending on the topography of M_2), adherent, coherent, non-porous (although porosity may be affected by plating conditions), and free from inclusions. Significant alloy formation at the M_1/M_2 interface can be avoided. The electrical resistance is often greater than that of the same metal as cast or rolled.

Compared with other methods, vacuum evaporation produces coatings that have a most satisfactory corrosion resistance, e.g. 0·005 mm of evaporated Cd gives a degree of protection to steel similar to that afforded by 0·01 mm of electrodeposited Cd. Cadmium coatings on ferrous and other substrates can meet authoritative specifications concerning corrosion resistance, adhesion and other factors, and the National Research Corporation has found that 0·012–0·025 mm of Al on steel withstood the standard 20% salt-spray test for 1 600 h.

Applications   While any metal can be evaporated irrespective of its physical form, the method is restricted chiefly by the capacity of the vacuum plant and by the economic time available for the required thickness. Alloys such as Al–Cu, Sn–Cu and Ni–Cr, which do not fractionate unduly can also be deposited, while other alloys showing a greater tendency to fractionate may be dealt with by rapid evaporation, which minimises the fractionating effect.

The relatively low temperature rise of M_2 during plating allows the coating of temperature-sensitive materials. Furthermore, composite articles, even those having porous and otherwise reactive surfaces, can be successfully plated.

Aluminium is widely applied for decorative and protective requirements, while cadmium[25], zinc and titanium[26] have been applied to ferrous materials chiefly for their protective value. The method finds particular application in the plating of high-tensile steels used in aviation and rocketry, car fittings and lamp reflectors, and gramophone record master discs, as well as in the preparation of specimens for electron microscopy and in rendering insulated surfaces electrically conducting, e.g. 'metallising' of capacitors and resistors.

Gas Plating

Principles   Plant and procedure have been described in the literature[27] and in Section 13.3. In a given process one of the following reactions steps may predominate:

1. Reduction, e.g. $CrCl_3(g) + H_2(g) \rightarrow Cr \downarrow + HCl(g)$
2. Thermal decomposition, e.g. $CrI_2(g) \rightarrow Cr \downarrow + I_2(g)$
3. Displacement, e.g. $CrCl_2(g) + Fe_{M_2} \rightarrow FeCl_2(g) + Cr \downarrow$

Similar steps can sometimes be distinguished in cementation procedures such as calorising and sherardising.

Plating is carried out in a closed system whose atmosphere is adjusted to contain the metal-gas and a second gas which may be an inert diluent or a reactive gas (as in 1 above). M_2 is heated, for example by high frequency, and this then initiates deposition of M_1 by one of the above steps. Spent reaction products are exhausted and where possible reclaimed and recycled.

The usual kinetic factors govern reaction and therefore plating rates.

Features   High plating rates (1·75 mm of tungsten/h) can be sustained[28], but efficiency can vary widely (between 5 and 90%) unless optimum control over reactant concentrations is exercised. The average coating thickness can be controlled accurately, but where the M_2 article is irregularly shaped, agitation and proper positioning are essential. Reaction step 3 does not require strict cleanness of M_2 for satisfactory adhesion. When M_1 is required for its corrosion-protective value then strict plating control is essential to ensure that M_1 is continuous and pure. Hydrogen embrittlement can be avoided.

Properties of deposits Deposits are often more adherent, coherent and temperature-stable than those produced by alternative coating methods. Adhesion can be adversely affected by spurious reactions between the metal-gas and impurities in M_2 (e.g. as observed during the deposition of molybdenum on steel[29]) and also where the thermal coefficients of expansion of M_1 and M_2 differ widely. The purity of reactants can affect that of M_1. M_1 crystal size is reduced by raising the reactant concentrations, or by lowering the plating temperature.

Very few corrosion performance figures have been published, but one established process for depositing Cr on steel is stated to give a product possessing similar properties to those of 18/8 stainless steel.

Applications Important applications include chromium deposition for protecting ferrous substrates against oxidation, wear and abrasion, production of a thick ductile coating of Ni [e.g. from $Ni(CO)_4$] for surfacing moulds, dies and tools[30], and deposition of metals which are very difficult to deposit by alternative methods, for example, tungsten, which can be used to confer protection on certain jet and rocket engine parts which are exposed to exhaust gases.

In general, many metals and alloys (e.g. of Al, Ta and Mo) can be deposited on metallic and some non-metallic substrates. M_1 may also be a metal compound having special useful properties (e.g. borides, nitrides, oxides, silicides and carbides), or even a non-metal such as Si (as in 'Ihrigising').

Cathode Sputtering

Principles Procedure and plant involved are similar to those used for vacuum evaporation. The pressure of the coating-chamber atmosphere, which may be air or an inert gas, is reduced and an arc is struck. The M_1 vapour formed subsequently deposits on surrounding cool surfaces, including those of M_2.

Probably mechanisms which have been discussed[21] may be either (a) electrochemical, with active gas particles forming unstable transient volatile M_1 compounds which finally decompose to deposit M_1 on M_2, or (b) physical, volatilisation of M_1 (e.g. by thermal evaporation) being induced by arcing; but in this case high arc voltages are probably essential. The fundamental theory was developed by the experimental work of Wehner[31] on measurement of sputtering yields and its dependence on ion energy.

Features The rate of deposition is low compared with that of vacuum evaporation, and is affected by variables such as pressure and temperature of coating-chamber atmosphere, arc voltage, cathode current density and geometry of cathode and M_2. A high vacuum is not essential, particularly where an inert gas can be used. The M_1 disintegration rate is affected by the atomic weight of coating-chamber gas.

Properties of deposits These are usually adherent and coherent. M_1 is pure provided that all adverse chemically reactive gases are removed prior to sputtering; nitrogen, for example, can form a nitride with copper, and oxygen can form oxides with most metals.

Applications   Although a wide range of metals can be sputtered, the method is often commercially restricted by the low rate of deposition. Applications include the coating of insulating surfaces, e.g. of crystal vibrators, to render them electrically conducting, and the manufacture of some selenium rectifiers. The micro-electronics industry now makes considerable use of sputtering in the production of thin-film resistors and capacitors.

Sputtering has also been employed for the deposition of dry lubricants, and of hard and wear-resistant coatings. The use of multi-component alloys for the protection of gas turbine blades has been investigated[32].

Brush Plating

Principles   In brush or tampon plating, a pad or stylus soaked in the plating solution is attached to the anode (usually inert), the article to be plated being the cathode. A voltage of about 6 V is applied and the pad moved over the localised area to be plated. The pad can be of cotton wool or plastic foam, and any part which does not require to be plated can be masked off with a suitable lacquer. After plating for half a minute or so, the solution in the pad is replenished.

Features of the method   Most metals can be deposited by this technique, usually on to steel or copper alloys as substrates. The most commonly deposited metals are nickel, chromium, tin, copper, zinc, cadmium, gold and silver.

The solutions used are generally highly concentrated, except in the case of the precious metals (for reasons of expense) and when zinc and cadmium are deposited from cyanide solutions, since soluble anodes are employed in the cases of these two metals. Proprietary solutions are available, particularly for such difficult metals as chromium, rhodium and indium, and it is advisable to use them.

The metal to be plated is first cleaned carefully and then activated with a weak acid. Steel can be treated with 3–5% HCl, whilst a 10% fluoboric acid solution is suitable for copper alloys. It is then ready for the electrodeposition process.

Properties of the deposits   The deposits are generally similar to those obtained from conventional electrolytes, and are for the most part matt or semi-bright.

Applications   Brush plating is mainly used on expensive assemblies where dismantling is either too expensive or impracticable. It can also be employed for building-up worn parts, and for the repair of local defects in printed circuits.

Plasma Spraying

Principles   Plasma is a gas which has been raised to such a high temperature that it becomes electrically conductive as a result of ionisation. In plasma spraying a gas is passed through an electric arc, the resulting plasma being formed into a jet as it emerges from a nozzle. The materials to be deposited are introduced into the jet in the form of a powder, which melts and strikes the surface to be coated at high speed. The result is a tough and adherent deposit.

Features of the method An inert gas such as argon is generally employed to produce the plasma to avoid contamination or oxidation of the substance to be applied. The coating material reaches the surface at a comparatively low temperature, which can be below 100°C, although the plasma itself is at a temperature of many thousands of degrees; hence sensitive or low-melting-basis materials can be plasma coated. The coatings are also usually denser and more adherent than those obtained with chemical flame spraying. Two or more powders can be applied simultaneously, so that metal composites can be produced in this way which are either in the form of coatings or free standing.

Properties of the deposits Almost any material which can be melted is suitable for plasma spraying, giving a vast range of possible coatings of single or mixed metallic or non-metallic substances. It is often possible to produce types of coatings which are not obtainable in any other way. Typical of the materials which are plasma sprayed are copper, nickel, tantalum, molybdenum, Stellites, alumina, zirconia, tungsten and boron carbides, and stainless steels.

Applications Plasma spraying is used to apply coatings for protection against wear and corrosion, to prevent erosion or cavitations, and to provide electrical insulation or conductivity. It can also be employed to produce bearing surfaces, abrasive properties or resistance to wetting by molten metals. The coatings can also be applied to facilitate the joining of different materials.

J. K. PRALL
H. SILMAN

REFERENCES

1. Potter, E. C., *Electrochemistry*, Cleaver-Hume, London, 87 (1956)
2. Simnad, M., Spillers, A. and Ling Yang, *Trans. Inst. Met. Fin.*, **31**, 82 (1954)
3. Antropov, L. L., *Kinetics of Electrode Processes and Null Points of Metals*, C.S.I.R., New Delhi, India, 77 (1960)
4. Kozawa, A. and Takahashi, T., *J. Metal Finish. Soc. Japan*, **11**, 301 (1960)
5. Tsvetkov, N. S. and Zarechnyuk, O. S., *Zh. Prikl. Khim. Leningr.*, **33**, 636 (1960)
6. Johnson, R. W., *J. Electrochem. Soc.*, **108**, 632 (1961)
7. *Metal Finishing Guidebook and Directory*, Metals and Plastics Publications, New Jersey, N.Y. (1961)
8. Gray, A. G., *Modern Electroplating*, Chapman and Hall, London, 432 (1953)
9. Judd, N. C. W., O.S.T.R. (1960)
10. *Electroless Nickel*, Amer. Soc. Test. Mater., Spec. Tech. Publ. No. 265 (1959)
11. Gutzeit, G., *Plating*, **46** No. 10, 1 158 (1959); **47** No. 1, 63 (1960)
12. Hoke, R. M., US Pat. 3 150 994 (1964)

13. Machu, W. and El-Geudi, S., *Werkst. u. Korrosion*, Weinheim, **12**, 223 (1961)
14. Saubestre, E. B., *Proc. 46th. Ann. Conv. Emer. Electropl. Soc.*, **46**, 264 (1959)
15. West, H. J., *Metal Finish. N.Y.*, **53** No. 7, 62 (1955)
16. Ryabchenkov, A. V. and Velemitsina, V. I., *Metalloved. i Obrabotka Metal.*, No. 11, 39 (1960)
17. Jenner, G. H. and Hoar, T. P., *Trans. Inst. Met. Fin.*, **34**, 253 (1956-1957)
18. Gregg, S. J., *The Surface Chemistry of Solids*, Chapman and Hall, London, Chapt. 3 (1951)
19. *Iron Age*, **187** No. 13, 107 (1961)
20. Holland, L., *Vacuum*, **6**, 161 (1959)
21. Holland, L., *Vacuum Deposition of Thin Films*, Chapman and Hall, London (1956)
22. Bunshah, R. F., *Vacuum Metallurgy*, Reinhold, New York (1958)
23. Fabian, D. J., *Met. Rev., Mets. and Mats.*, **1** No. 3, 27 (1967)
24. Neugebauer, C. A., Newark, J. B. and Vermilyea, D. A., *Structure and Properties of Thin Films*, Wiley, New York (1960); Finch, G. I., Wilman, H. and Ling Yang, *Disc. Faraday Soc.*, **1**, 144 (1947); Wilman, H., *Trans. Inst. Met. Fin.*, Adv. copy. No. 9 (1955)
25. *Machinery, Lond.*, **97**, 1 411 (1960)
26. Wilburn, D. K. and Horn, C. W., *Prod. Finish., Lond.*, **24**, 90 (1960); *Corrosion*, **16**, 46 (1960)
27. Powell, C. F., Campbell, I. E. and Gonser, B. W., *Vapour Plating*, Chapman and Hall, London (1955)
28. *Industr. Finish., London*, **12**, 54 (1960)
29. Wlodek, S. T. and Wulff, J., *J. Electrochem. Soc.*, **107**, 565 (1960)
30. *Steel*, **146** No. 18, 78 (1960); *Machinery, Lond.*, **97**, 911 (1960)
31. Wehner, G. K. and Rosenberg, D. J., *J. Appl. Phys.*, **31**, 177 (1960)
32. Krutenat, R. C., Wielonski, R. F. and Lawrence, F. M., Pratt and Whitney Aircraft, Research Report (1968)

14 PROTECTION BY METALLIC COATINGS

14.1	The Protective Action of Metallic Coatings	**14**:3
14.2	Aluminium Coatings	**14**:17
14.3	Cadmium Coatings	**14**:31
14.4	Zinc Coatings	**14**:36
14.5	Tin and Tin Alloy Coatings	**14**:47
14.6	Copper and Copper Alloy Coatings	**14**:62
14.7	Nickel Coatings	**14**:69
14.8	Chromium Coatings	**14**:87
14.9	Noble Metal Coatings	**14**:97

14.1 The Protective Action of Metallic Coatings

The application of metallic coatings for the protection of metals may be required for one or more of the following reasons:

(*a*) To prevent or reduce corrosion of the substrate metal.
(*b*) To modify the physical or mechanical properties of the substrate metal.
(*c*) To achieve and maintain some desired decorative effect.

Although the initial choice of coating material applied for reasons (*b*) or (*c*) may be dictated by the particular properties required, the corrosion behaviour of the composite metal coating/metal substrate system must also be taken into consideration in so far as it may affect the maintenance of the desired properties. Consequently, in all cases where protective metal coatings are used the corrosion performance of *both* coating and substrate require careful consideration.

Choice of substrate metal is usually governed by cost, weight and general physical, mechanical or fabricational properties, and these factors will normally dictate a very limited number of possible materials none of which may be ideal in resisting the corrosive environment which will be encountered in service. Ideally, a protective metal coating should exclude completely the corrosive environment from the substrate metal and, if this can be achieved, only the resistance of the coating metal itself to that corrosive environment needs to be considered. However, in practice, discontinuities in the coating may occur during application (Chapter 13) or be produced subsequently by mechanical damage or by the corrosion of the coating itself. In these cases the corrosion performance of the bimetallic system so produced becomes of major importance.

Anodic and Cathodic Coatings

A detailed discussion of galvanic corrosion between dissimilar metals in contact in a corrosive environment has been given in Section 1.7, but in the case of coating discontinuities the effect of the anode/cathode area relationship and the nature of any corrosion products formed at small discontinuities may modify any choice made on strict considerations of general galvanic corrosion theory based on the potentials of the coating and substrate in the environment under consideration.

Thus, coatings which are anodic or cathodic to the substrate are both used in practice, but whereas the former will provide sacrificial protection at a coating discontinuity the latter may stimulate attack on any exposed substrate. The sacrificial consumption of an anodic coating at a pore or discontinuity results in a gradual increase in the area of substrate exposed and a corresponding decrease in the corrosion current density which, in time, may become insufficient to maintain protection in the centre of the exposed area. The rate of sacrificial consumption of an anodic coating is reduced by several factors such as the resistance of the electrolyte solution within the discontinuity, blocking of the discontinuity by corrosion products or the formation of protective films on the exposed substrate which may be encouraged by increase of electrolyte pH by the cathodic reaction at the discontinuity. Under these circumstances the life of the anodic coating will be prolonged.

Zinc and cadmium are both anodic to steel and provide sacrificial protection to the substrate when used as coating metals. In exposure to industrial atmospheres, zinc will protect steel for a longer period than cadmium, whereas in many marine or rural environments cadmium provides a longer period of protection. Layton[1] attributes this difference in behaviour to the nature of the corrosion products formed in the different environments. In industrial atmospheres soluble sulphates of both zinc and cadmium are produced and are removed by rain so that corrosion can continue freely. Under these conditions zinc, which has a more negative potential than cadmium in most environments, is a more efficient anode than cadmium and gives a longer period of protection to the substrate. In rural and marine exposure, however, the cadmium carbonates and basic chlorides which are formed are insoluble and corrosion is stifled to a greater extent than is the case with the more soluble zinc carbonates and basic chlorides, so tending toward a longer life with cadmium coatings.

When cathodic coatings are used the sacrificial action is reversed, the substrate being attacked and the coating protected. This attack on the substrate can be highly localised and can lead to rapid penetration through the thickness of the metal. The effect is appreciably reduced by electrolyte resistance and by the stifling action of corrosion products, as mentioned above for anodic coatings, and the production of substrate corrosion products within the discontinuity can significantly decrease the corrosion rate and delay penetration.

Decorative coatings of nickel plus chromium are cathodic to steel or zinc alloy substrates and with these protective systems deliberate use may be made of discontinuities in the chromium topcoat where corrosion of the underlying nickel will occur. If the number of these discontinuities in the chromium layer is greatly increased the current density at each individual corrosion site is reduced, penetration of corrosion through the thickness of the nickel layer is thus slowed down and the period of protection of the substrate metal is prolonged.

In general, the choice between cathodic or anodic coatings will be governed by the service application. Where cathodic coatings are used any attack on the substrate will be highly localised, leading to rapid perforation of thin sections with a consequent loss of functional integrity. Anodic coatings, on the other hand, will protect exposed areas of substrate metal by sacrificial action until the area exposed exceeds that over which cathodic polarisation of the corrosion reaction can be maintained, after which time the substrate itself will corrode freely.

With both anodic and cathodic coating systems, account must be taken of the extent to which the presence and nature of corrosion products on the surface may impair either the decorative aspects or the functional use of the article. For example, the presence of small amounts of corrosion products on the surface of gold-plated electrical contacts can markedly increase the electrical contact resistance and cause malfunctioning; the problem may be overcome by increasing the thickness of the gold deposit—thus reducing its porosity—or by interposing an undercoat of a more resistant metal to act as a barrier layer between the gold and the substrate.

Because of the many variables which can influence the corrosion reaction, the use of the e.m.f. series of metals to predict the behaviour of galvanic couples in a given service environment can be hazardous and misleading. Numerous examples of coatings expected to act cathodically which have, in fact, been anodic have been reported in the literature[2-5] and specialised lists of galvanic couples in different environments have been compiled[6-8].

Factors Affecting Choice of Coating

Many factors are involved in the choice of coating material to be used for any particular application, and these will now be described.

Resistance to the Corrosive Environment

The principal difficulty in assessing the resistance of a coating material to a corrosive environment lies in an adequate accurate definition of that environment. Metals exposed to natural atmospheres will corrode at markedly different rates dependent upon the degree of pollution present and a number of interrelated meteorological factors. Atmospheric corrosion rates for most of the metals in common use have been published by many authors and by official organisations such as the A.S.T.M.[9] These tables may be consulted for general guidance but must always be used with caution, choosing data for environments which most nearly approach the service conditions concerned. However, purely local conditions can markedly affect the rate of attack. An example of this in the author's knowledge involved the corrosion of galvanised-steel air-intake louvres on the roof of a building in a severe industrial environment; most of the louvres gave satisfactory service but premature rusting occurred on those which faced in a westerly direction where the prevailing winds exposed them to additional chloride contamination picked up from a nearby river estuary. A detailed knowledge of both the macro- and micro-environmental conditions to be encountered is thus seen to be essential if the best choice of coating is to be made and it is often desirable to make site inspections before recommending a coating system for a particular application.

When the corrosive environment consists of waters or other liquids, the effect of the presence of minor constituents in the liquid as well as the degree of aeration and rate of flow must be taken into account.

Practicability of Application

It is essential to choose a material and a method of application which will provide a coating of adequate thickness with good coverage and distribution over the surface of the article.

Hot-dipped tin coatings are difficult to apply outside the thickness range of 8–38 µm[10] and hot-dipped zinc coatings do not normally greatly exceed 50 µm in thickness. Hollow sections and excessively large articles may be impractical to handle by the hot-dipping process and very thin sections may be subject to much distortion.

Electrodeposited coatings may range in thickness from about 0·1 µm to about 25 µm for decorative and protective purposes, though considerably greater thicknesses may be applied in the case of coatings for wear resistance purposes. The shape of the article to be plated greatly influences the thickness and coverage of the electrodeposit. Copper and nickel deposits cover well, throw well into recesses and levelling can be obtained during electrodeposition. Zinc gives good coverage but poor levelling and chromium has a very poor throwing power leading to bare areas in deep recesses of the plated article. Precious metals, because of cost, tend to be electrodeposited in thicknesses of less than 1 µm and at these thicknesses porosity of the electrodeposit is a significant factor; acid gold electrodeposits tend to be less porous than deposits from alkaline baths.

Sprayed metal coatings may be applied without limitations of size of article to be processed such as may apply for hot-dipping processes and the thickness and coverage of the coating can be readily controlled during application. Problems can arise, however, with applications involving complex shapes or hollow sections. Coating thicknesses are normally in the range 50–250 µm but considerably greater thicknesses may be built up for applications involving wear resistance, and diffusion heat-treatments may be subsequently applied in order to improve wear resistance further. Sprayed metal coatings are of a porous nature and contain a high proportion of oxides produced by the method of application. In service, corrosion products may build up in the pores of the coating which can contribute materially to a stifling of the corrosion reaction, but even so, sacrificial protection of the substrate exposed at any coating discontinuities may still be maintained.

Cladding by pressing, rolling or extrusion can produce a coating in which the thickness and distribution can be readily controlled over wide ranges and the coatings so produced will be completely free from porosity. Although there is very little practical limit to the thickness of coatings which can be produced in this way, the application of the process is limited to comparatively simple shaped articles which do not require much subsequent mechanical deformation. Among the principal uses are lead and aluminium sheathing for cables, lead-sheathed sheets for architectural applications and composite extruded tubes for heat-exchangers.

Compatibility of Galvanic Coupling

Most of the published data on galvanic corrosion concern solid metal couples rather than bimetallic coating systems, and it is important to bear in mind that the same galvanic relationships do not necessarily apply in both cases; nevertheless, useful guidance can be obtained from the data for the solid-couple systems exposed to suitable environments. Data have been reported[11] for combinations of metals commonly used in the electrical industry; couples involving plated brass, copper and aluminium were exposed to a 1% salt spray and the corrosion currents (mA) were measured over a one-week test period. Relative ratings in this test are shown in Table 14.1 and items of particular interest in this table are that coatings of aluminium on either brass or copper are unsatisfactory because of the active galvanic corrosion which occurs at discontinuities. If tin plating is used as an undercoat to the aluminium, corrosion of the substrate is prevented, but if tin is applied as a topcoat over the aluminium (particularly by means of the conventional zincate process) it is sufficiently porous to allow the aluminium to continue to function as an active anode and corrosion can

Table 14.1 Performance of various crimped metallic couples in 1% salt spray

A. Completely Satisfactory Combinations

Copper/nickel-plated copper
Copper/gold-plated copper
Tin-plated copper/aluminium
Tin-plated copper/nickel-plated copper
Tin-plated copper/solder-dipped copper
Tin-plated brass/aluminium
Solder-dipped copper/nickel-plated copper
Nickel-plated copper/gold-plated copper
Nickel-plated copper/silver-plated copper
Gold-plated copper/silver-plated copper
Aluminium/tin-plated aluminium (no copper undercoat)

B. Satisfactory Combinations, Slight Galvanic Corrosion

Copper/silver-plated copper
Solder-dipped copper/tin-plated aluminium
Copper/tin-plated copper
Copper/solder-dipped copper
Copper/reflowed tinned copper
Silver-plated copper/tin-plated copper
Silver-plated copper/solder-dipped copper
Gold-plated copper/tin-plated copper
Aluminium/tin-plated aluminium (zincate process)

C. Borderline, Moderate Galvanic Corrosion

Gold-plated copper/solder-dipped copper
Tin-plated aluminium/nickel-plated copper
Aluminium/solder-dipped aluminium

D. Unsatisfactory, Severe Galvanic Corrosion

Aluminium/brass
Aluminium/copper
Tin-plated aluminium/copper
Aluminium/nickel-plated copper
Aluminium/nickel-plated brass
Aluminium/silver-plated copper
Tin-plated aluminium/silver-plated copper
Aluminium/gold-plated copper
Tin-plated aluminium/gold-plated copper

continue. The relationships given in Table 14.1 apply to the specific environment quoted and it must always be remembered that if the conditions are varied, even to only a small extent, different galvanic effects may be produced.

Effect of Coating Process on Substrate Properties

The application of any coating process may affect the physical or mechanical properties of the substrate material and any such effects should be considered when choosing the type of coating to be used and its method of application.

With hot-dipping processes, apart from the risk of distortion previously mentioned, the high temperatures involved can produce annealing, e.g. softening of brass and copper during hot-tinning. Furthermore, hard and brittle intermetallic-alloy zones are produced during hot-dipping as a result of diffusion of the liquid coating metal into the solid basis metal, e.g. $FeSn_2$ in the tinning of steel, Cu_6Sn_5 and Cu_2Sn in the tinning of copper—the extent and depth of the alloy formation depending on temperature and time of dipping. Thus, too long a dipping time in hot-dip galvanising can lead to flaking during subsequent mechanical deformation[12, 13].

Evolution of hydrogen during some electrodeposition processes can cause embrittlement if it diffuses into the substrate; the effect has been reported for chromium and cadmium plating of high-strength steels[14] and provision is made in relevant standards[15, 16] for diffusion heat-treatments after plating to reduce the hazard (see Sections 8.4 and 14.3). Alternatively, zinc or cadmium coatings may be applied by vacuum deposition, thus avoiding any embrittlement of the steel—a process of this nature having been developed by the Royal Aircraft Establishment, Farnborough. Cases have also occurred where cracks in a highly stressed electrodeposit have acted as stress-raisers which initiate stress-corrosion of susceptible substrate metals.

Although the annealing effects of overheating are avoided when coatings are applied by metal spraying processes—provided that those processes are properly applied—it has been reported[17] that compressive stresses imparted to the substrate by the grit-blasting pretreatment can alter the fatigue properties of the material. Cladding involves extensive cold-working which may necessitate annealing of the composite material before use.

Coating Properties

A number of physical and mechanical properties of coating metals need to be considered when making a choice of metal to be used in a particular application.

Appearance, colour and brightness are important in decorative applications. Copper, zinc, cadmium, nickel, silver and gold can be readily plated in a bright condition whilst tin normally plates as a dull deposit but may be brightened by flash melting after electroplating (flow brightening process). Aluminium and lead deposits are always dull, but reflective aluminium coatings can be produced by roll cladding using highly polished rolls. Colours may range from the blue-white of chromium through yellows for gold or brass to the reds of bronzes. Reflectivity after polishing also varies with the coating metal, being very high for silver and rhodium and progressively decreasing in the order aluminium and palladium, tin, zinc, gold, iron and lead.

Hardness, strength and wear resistance are prime properties, not necessarily interrelated. For example, rubbing contact between two hard surfaces may produce more wear than with two soft surfaces though, in general, rubbing contact between one hard and one soft surface causes wear in the softer material. However, mechanical design factors can alter this wear relationship so that the harder material wears to a greater extent, e.g. the case of the rapid wear imposed on a steel record needle rubbing against a vinyl record surface. In general, the hardest deposits are those of chromium, nickel and rhodium; with iron, copper, zinc, cadmium and silver in an intermediate hardness group; with tin, lead, gold and indium being relatively soft.

Temperature resistance, i.e. a combination of melting point and oxidation resistance, may be of prime importance. A general correlation exists between melting point and hardness since both reflect the bond strength of the atoms in the crystal lattice, and the preferred order of coating metals for use in high temperature applications as temperature is increased is: silver, aluminium, nickel, rhenium, chromium, palladium, platinum and rhodium.

The electrical conductivity of coatings is often of secondary importance since they are of thin section and are in parallel with a metallic substrate of larger cross-section which is generally a good electrical conductor. A more important property for coatings used as electrical contacts is surface hardness and the ability to remain free from oxide and tarnish films. Thus, although aluminium has almost four times the conductivity of tin it is often tin-plated to improve its electrical contact properties. Other coating metals commonly used for low-voltage applications are gold, tin/lead, silver, palladium, copper, rhodium and nickel.

Economics

Economic factors are obviously of prime importance when choosing both the coating material and its method of application. Individual items in the economic balance sheet will vary not only with the material and the process but also with availability, local labour costs and factors unique to the design and use of the articles concerned. A further factor which frequently does not receive adequate importance in costing is the ease or otherwise of maintaining the finish so as to ensure an adequate and efficient service life for the component. In general, though with many exceptions, processing costs may be ranged in ascending order from hot-dipping to plating, spraying and cladding. The lowest cost group of metals includes zinc, copper, iron and lead, the intermediate group contains nickel, tin, tin/lead, cadmium and aluminium, and in the highest cost group are silver, palladium, gold and rhodium, though cost relationships may vary from time to time as a result of price fluctuations in response to supply and demand.

Experience in the application of metal-coating processes can materially affect economics. Thus, although it is possible to apply aluminium by electrodeposition, the process is difficult to operate and few metal finishers apply the process; the application of aluminium coatings by hot-dipping or by metal spraying is much more readily accomplished and more of these types of installations are becoming available on the metal finishing market.

Coatings in Practical Use

In modern coating technology the range of materials used is ever increasing and specific coatings may be chosen and applied, often by specially designed techniques, for particular applications. Details of the behaviour of various specific metal coatings are given elsewhere in this book but some general information on a number of the more commonly applied coating metals is as follows.

Zinc

This is an anodic coating material that may be applied by hot-dipping, metal spraying or electrodeposition, with a good corrosion resistance to most neutral environments, particularly when used in combination with chromate or phosphate passivation treatments. In most cases of atmospheric exposure, zinc will provide good protection to steel, particularly where any sulphur pollution is present, but in rural and pure marine environments the conditions of humidity and chloride pollution level can reduce the effectiveness of zinc coatings and make the use of cadmium more suitable, though the same degree of protection may often be achieved at lower cost by increasing the thickness of the zinc coating. Zinc is the preferred coating for steel used under immersed conditions in scale-forming waters or sea-water and, since it is less toxic than cadmium it should be used in applications involving welding.

The life of zinc coatings is generally proportional to thickness and independent of the method of application, though it has been reported[18, 19] that zinc electrodeposited from the sulphate bath gives a better performance than when deposited from the cyanide bath. Hudson[20] has reported lives for 42 µm thick zinc coatings on steel ranging from $3\frac{1}{2}$ years in a severe industrial environment (Sheffield) to more than 10 years in a rural environment (Llanwrtyd Wells) and Gilbert[21] quotes lives of 4–5 years in London, 9 years in Cambridge and 18 years in Brixham.

Cadmium

Cadmium also provides a sacrificial coating to steel which gives better protection than zinc in applications where strong acids and alkalis may be encountered and those involving immersion in stagnant or soft neutral waters. It should be used in applications involving bimetallic contact with aluminium and in electrical applications where ease of solderability is important. Cadmium has a low torque resistance and should be used as a coating material in cases where bolted assemblies have to be frequently dismantled. It also provides better protection than zinc in enclosed spaces where condensation can occur, particularly when there is contamination by organic vapours.

Cadmium is more expensive than zinc. It is usually applied by electro-deposition in thicknesses up to about 25 μm and has a superior tarnish and stain resistance to that of zinc. As with zinc, the life of cadmium coatings is proportional to thickness; Hudson[20] quotes a life of only 9 months at Sheffield for a 25 μm thick coating and approximately 8 years for a 42 μm thick coating exposed to a marine environment at Calshot.

Tin

Tin is applied by hot-dipping or electrodeposition and has a similar corrosion behaviour to that of zinc. Coating thicknesses are usually in the range 12–50 μm, and in the lower portion of this range coating porosity can be a factor to be taken into account (see discussions by Kochergin[22], and Gonser and Strader[23]).

Tin coatings are widely used in the electrical industry because of their good contact properties and in the food industry because of low toxicity. In addition to pure tin coatings a number of alloy coatings have been de-veloped for special applications, e.g. tin–lead (terne plate), tin–zinc, tin–cadmium, tin–bronze and tin–nickel. Reference should be made to Section 14.5 and to the publication by Britton[24] for data on the corrosion of tin and its alloys.

Aluminium

Aluminium may be applied as coatings by metal spraying, cladding, hot-dipping and electrodeposition, though the last-named process is difficult to apply and by far the largest proportion of aluminium-coated metals are produced by the first two methods.

In atmospheric exposure to industrial environments its corrosion rate is only about one-third that of zinc and the corrosion reaction is stifled by the tenacious oxide which is produced; nevertheless it can frequently function as an anodic coating both for steel and for the less corrosion-resistant aluminium alloys.

Hudson[20] reported lives of about $4\frac{1}{2}$ years for 38 μm thick metal-sprayed aluminium coatings on steel exposed at Sheffield, and more than $11\frac{1}{2}$ years for coatings 75 μm thick. Sprayed aluminium coatings (approximately 125 μm thick) have also provided complete protection against exfoliation and stress corrosion to aluminium–copper–magnesium (HE15) and aluminium–zinc–magnesium (DTD 683) alloys in tests lasting up to 10 years in industrial and marine environments[25,26].

Nickel

Nickel has an inherently high corrosion resistance, particularly in chloride-free atmospheres and is widely used as a coating material in the chemical industry. When exposed to the atmosphere, rapid tarnishing and slow super-ficial corrosion occur; for this reason, nickel coatings are seldom used alone, but they are widely used as undercoats beneath bright chromium to give decorative and protective schemes for steel, zinc-alloy and copper-alloy consumer goods notably in the automobile and domestic hardware indus-tries. Used in this way, corrosion of the nickel undercoat is confined to localised pitting which develops at discontinuities in the chromium layer and which will eventually penetrate to the substrate. Many special processing variations have been developed to improve the corrosion resistance of these composite coatings and recommended systems are detailed in standards documents such as BS 1224 (1970)[16].

Lead

Lead coatings are mainly applied by cladding and find principal use in the chemical industry for resistance to sulphuric acid, for cable sheathing resistant to attack by soils and in architectural applications where resistance to industrial atmospheres is particularly good. They rely for their protective action on the formation of insoluble corrosion products which stifle the corrosion reaction and lead to very long service lives, but the corrosion resistance is impaired when chlorides are present.

Copper

Except in the case of certain decorative and electrical applications, copper is seldom used as a coating material in its own right owing to the rapidity with which it tarnishes, particularly in sulphur-polluted environments. Nevertheless its atmospheric corrosion resistance is good owing to the development of the well-known green patina of basic copper salts which gives protection against further corrosion of the metal. When copper coatings are used for their decorative effect the high lustre and distinctive colour are retained by applying a protective coating of transparent lacquer which may contain an inhibitor, e.g. benzotriazole.

By far the largest use of copper as a coating metal is in the form of under-coats to other protective schemes, such as the nickel plus chromium systems, where they offer great benefit by levelling the surface so as to improve the brightness of the finished article. Their rôle in the corrosion protection of the substrates is complex; they are themselves often preferentially attacked when overlay coatings are penetrated by corrosion and can stimulate en-hanced corrosion of the substrate when penetration through their thickness occurs. On the other hand, however, in the case of coatings of bright nickel plus micro-discontinuous chromium, the use of a copper undercoat is known to improve corrosion resistance and to extend the period of protection of the substrate[27].

Chromium

Chromium is highly resistant to atmospheric corrosion, being almost inert in most atmospheres, and is therefore used as a thin, bright overlay to other

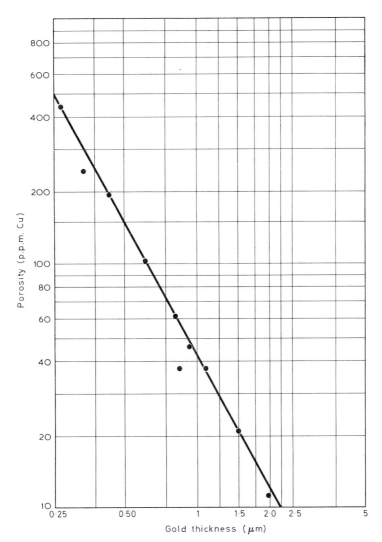

Fig. 14.1 Log/log plot of gold-plate porosity against thickness for the conventional hot-cyanide gold-plating bath on copper substrate. The porosity was determined by the amount of copper (p.p.m.) dissolved under standard conditions by an ammonia–ammonium persulphate test solution

coatings to retain decorative appeal for long periods. The thickness of these coatings, applied by electrodeposition, is normally in the range 0·3–1·3 μm. In the lower thickness range the coating contains minute discontinuities which cannot be eliminated by increasing the thickness, since spontaneous cracking of the deposit occurs as the thickness builds up. The tendency to cracking of chromium electrodeposits is encouraged and put to good use by inducing cracking on a micro scale by processing modifications. When this is done the micro-cracked deposits so produced provide greater protection to nickel-plated steel and zinc-alloy substrates exposed to the

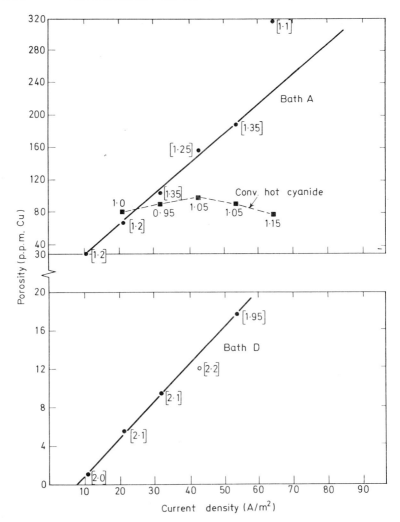

Fig. 14.2   Linear plots of the effect of current density on the porosity, expressed as p.p.m. Cu, for three different gold-plating baths. The numbers next to each point show the actual average thickness (μm) for each test. Bath *A* was a proprietary alkaline cyanide bath using silver as a brightener. Bath *D* was an acid gold bath containing cobalt and an ethylenediamine tetra-acetic acid gold complex

atmosphere by increasing the area of nickel exposed at the micro-discontinuities and so reducing the corrosion current density at individual corrosion sites with a consequent reduction in the rate of penetration through the nickel layer.

Chromium is also a very hard metal with excellent wear resistance, and so is widely used as a coating material for engineering applications. For these purposes, coatings are applied by electrodeposition which may be several millimetres thick. These hard engineering chromium coatings invariably contain fine cracks and fissures which can allow corrodents to attack the substrate, but this is not often a hazard in service and in many cases they are advantageous in providing a means of retaining lubricant on the working surface during use.

Precious Metals

Gold and platinum, being highly noble metals, can provide highly corrosion-resistant coatings, but are rarely used for this application alone because of cost. Silver, though cheaper than gold or platinum, has a somewhat lower corrosion resistance since it is very prone to attack by sulphides which cause dark tarnishing.

The principal use of gold is as a very thin coating about 0·05 μm thick for electrical and electronic applications. Because of the thinness of gold electrodeposits, porosity must be very carefully controlled since seepage of corrosion products from substrate or undercoat exposed at these pores can have serious adverse effects on both appearance and electrical properties of the composite. The porosity can vary with the thickness of the deposit (Fig. 14.1), and with the type of plating bath and with its method of operation (Fig. 14.2), and the phenomenon has been extensively studied by Clarke and many other workers.

V. E. CARTER

REFERENCES

1. Layton, D. N., *Trans. I.M.F.*, **43** No. 14, 153 (1965)
2. Hoar, T. P., *Trans. Faraday Soc.*, **30**, 472 (1934)
3. Koehler, E. L., *J. Electrochem. Soc.*, **103**, 486 (1956)
4. La Que, F. L., *Corrosion*, News Section, **8**, 1, April (1952)
5. Compton, K. G., *et al.*, *Corrosion*, **11**, 35 (1955)
6. La Que, F. L. and Cox, G. L., *Proc. Amer. Soc. Test. Mat.*, **40**, 670 (1940)
7. Evans, U. R. and Rance, V. E., *Corrosion and its Prevention at Bimetallic Contacts*, H.M.S.O., London (1958)
8. McKay, R. J. and Worthington, R., *Corrosion Resistance of Metals and Alloys*, A.C.S. Monograph Series (1936)
9. *Metal Corrosion in the Atmosphere*, A.S.T.M. Special Technical Publication No. 435 (1968)
10. Hoare, W. E., *Hot Tinning*, Tin Research Institute, Perivale, Middlesex, 7 (1947)
11. Frant, M. S., *Electron. Design*, **20**, 112 (1961)
12. Mathewson, C. H., *Zinc*, Reinhold, New York, 462–465 (1959)
13. Spencer, L. F., *Metal Prod. Manuf.*, **17**, 27 (1960)
14. Cotton, W. L., *Plating*, **47**, 169 (1960)
15. *Electroplated Coatings of Cadmium and Zinc on Iron and Steel*, BS 1706 (1960)
16. *Electroplated Coatings of Nickel and Chromium*, BS 1224 (1970)
17. Whittaker, J. A., *et al.*, *Proc. 2nd. Internat. Congress on Metallic Corrosion*, 229–235 (1966)
18. Hippensteel, C. L. and Borgmann, C. W., *Trans. Amer. Electrochem. Soc.*, **58**, 23 (1930)
19. Biestek, T., *Met. Finishing*, **68** No. 4, 48 (1970)
20. Hudson, J. C. and Stanners, J. E., *J. Iron and Steel Inst.*, **175**, 381 (1953)
21. Gilbert, P. T., *J. Appl. Chem.*, **3**, 174 (1953)

22. Kochergin, V. P. K., Prostkov, M. F. and Nimvitskaya, T. A., *Konserv. Ovoshchesushil. Prom.*, **14**, 22 (1959)
23. Gonser, B. W. and Strader, J. E., Chapt. on 'Tin Coatings' in *The Corrosion Handbook*, Ed. H. H. Uhlig, Wiley, New York, and Chapman and Hall, London (1948)
24. Britton, S. C., *The Corrosion Resistance of Tin and Tin Alloys*, Tin Research Institute, Perivale, Middlesex (1951)
25. Carter, V. E. and Campbell, H. S., *J. Inst. Met.*, **89**, 472 (1960)
26. Carter, V. E. and Campbell, H. S., *Br. Corros. J.*, **4** No. 1, 15 (1969)
27. Clauss, R. J. and Klein, R. W., *Proc. 7th. International Metal Finishing Conference*, Hanover, (Interfinish 68), 124 (1968)

14.2 Aluminium Coatings

Aluminium is used as a protective coating for iron and steel and also for some high- and medium-strength aluminium alloys; in some situations the optimum protection is obtained by using an aluminium alloy as the protective coating. Aluminium is applied also as a decorative coating on both metallic and non-metallic surfaces. Several methods of applying the aluminium coating are available, and the method selected depends to a large extent on whether the protective or decorative aspect is more important. Some methods of coating will be excluded by the geometry of the article to be coated, while others will be excluded by the chemical or physical nature of the article.

Coating Methods

Methods available for coating other metals with aluminium include spraying; spray aluminising (heat-treated sprayed coatings); hot dipping; Calorising (diffusion or cementation); vacuum deposition; electroplating; electrophoretic deposition; chemical deposition (gas or vapour plating); cladding or mechanical bonding; casting.

Painting with aluminium-pigmented paint is considered in Chapter 15 because its properties are essentially those of a paint and not of a metallic coating; in this respect aluminium paint differs from zinc-rich paint which can be formulated to provide galvanic properties similar to those of metallic zinc.

Sprayed, vacuum-deposited and plated coatings can be applied to most metals and to many non-metals, e.g. vacuum deposition is applied to many substrates including plastics; spray application can be used for coating fabric, plastic and paper. Hot dipping and other diffusion processes are dependent on the nature of the substrate for the properties of the coating. Most commercial applications of aluminium coatings are on iron and steel with smaller quantities applied to aluminium alloys and plastics.

Sprayed coatings These are of the greatest importance, particularly for the protection of structural steel or certain aluminium alloys. The metal to be coated must be grit blasted shortly before spraying to provide a clean rough surface. Chilled iron grit is used for most steelwork, while alumina or silicon carbide may be used for metals having a hardness exceeding 360 HV. Calcined alumina should be used for cleaning aluminium in order to avoid contamination of the surface by residual iron grit (see Sections 12.1 and 13.4). Grades of pure aluminium or aluminium alloys are applied by means of pistols fed by wire or powder; the two application methods give similar results. Coatings of 0·075–0·1 mm give good protection and it is preferable that not more than 0·05 mm should be applied in any one pass of the gun. In most atmospheric conditions no advantage is obtained by the use of aluminium coatings thicker than 0·25 mm and there is danger of flaking under some circumstances if the local film thickness exceeds 0·3 mm. For general purposes, the optimum performance is obtained using 99·5% purity aluminium applied within the limits of 0·1–0·2 mm.

Recent advances in metal spraying include arc spraying, which claims higher application speeds, better adhesion and reduced oxide content in the coating.

Spray-aluminised coatings For service temperatures in the range 550°C to about 900°C it is recommended that the coated steel item should be heated first to 800–900°C either in a mildly oxidising atmosphere or after coating with a solution of coal tar pitch in order to cause diffusion at the steel–aluminium interface. Oxidation of the aluminium coating during this heat treatment can also be reduced by the use of Al–0·75 Cd for the coating or by a protective coating of calcium hydroxide plus sodium silicate. The heat-treated sprayed coatings are sometimes known as *aluminised*, but the term *spray aluminised* is desirable in order to distinguish them from *hot-dip aluminised* coatings which are obtained (after pretreatment of the metal surface) by immersion in a bath of molten aluminium.

Hot-dipped coatings These coatings are applied after cleaning the work, e.g. pickling in hydrochloric acid in the case of steel, and then preheating. The work is then immersed in a molten salt bath, a flux or a reducing atmosphere, prior to immersion in the bath of molten aluminium. The bath temperature is usually in the range 620–710°C, depending on whether the coating material is an aluminium–silicon alloy (for use in high-temperature conditions) or pure aluminium (for corrosion prevention). Alloying occurs between iron and aluminium at the interface. The aluminium coating thickness is generally in the range 0·025–0·075 mm.

This type of coating is applied to both individual items and on continuous-coil plant. The process is in use in North America, Great Britain and other European countries.

Calorised coatings (diffusion, cementation) These coatings are produced by packing the articles to be coated in a mixture of powdered aluminium, aluminium oxide and ammonium chloride for periods of 4–6 h at 800–950°C, followed by 12–48 h at 815–1 000°C after removal from the mixture. Coatings of 0·025–0·15 mm are reported with diffusion to a depth of 1 mm. The resistance of such coatings to scaling at temperatures up to 950°C depends on the maintenance at the surface of an aluminium content sufficient to give self-healing properties associated with the formation of a protective aluminium oxide layer.

Vacuum deposited coatings This type of coating can be deposited from

high purity aluminium vaporised by passing a heavy current through tungsten filaments around which the aluminium is wound. The application chamber is operated at very low pressure and the vaporised aluminium condenses on the cold surface of the work. Coatings of this type can be used to give an extremely thin bright finish to metallic or non-metallic surfaces. It is usual to apply a coat of stoving lacquer before and after depositing the aluminium coating. Coatings of the order of 0·025 mm are normally produced, but deposits up to 0·025–0·075 mm can be applied and are more appropriate where corrosion protection as well as decorative properties is required. Silicon monoxide is sometimes used as a final protective coating at thicknesses around 100 μm. Vacuum deposited coatings of aluminium can be applied to irregular static shapes or to plastic strip passing through the application chamber at speeds of 7·6 m/s or greater.

Sputter application is somewhat analogous to vacuum deposition. The deposition is carried out in a partial vacuum of about 13 N/m^2 and at a potential of around 1 000 V.

Electroplated coatings   Aluminium can be electroplated from molten salts or organic solvents. It can be plated on to other metals from fused aluminium chloride melts, e.g. 75% AlCl$_3$, 20% NaCl, 5% LiCl, operated at 170–180°C and 210 A/m^2.

Aluminium can be deposited from complex organic solutions if sufficient precautions are taken, and such coatings are now being produced commercially in North America. Two systems on record are (1) aluminium chloride, benzoyl chloride, nitrobenzene and formamide used at 50°C and 3·2–3·5 kA/m^2, and (2) aluminium chloride, n-butylamine and diethyl ether used at 20°C and 970 A/m^2. Deposits of 0·010 mm can be obtained on mild steel or copper at 20°C and 970 A/m^2 using aluminium-wire anodes and nitrogen or argon atmospheres.

Electrophoretic coatings   Coatings of this type can be produced by electrophoretic deposition of aluminium powder from a carrier system based on methyl alcohol and water. The material to be coated, e.g. steel, can be passed through the system using aluminium anodes and a potential of about 50 V. The deposited aluminium and strip then pass through compacting rolls, and are coiled prior to heat treatment at temperatures in the region of 400–600°C; the lower temperatures give maximum ductility to the coating while the higher temperatures improve adhesion at the expense of the formation of a brittle intermetallic layer between the coating and the steel substrate. Further improvement in adhesion can be obtained by heating the coated coil at 750–950°C for 30 min to 5 h with a corresponding increase in the intermetallic layer between the substrate and coating. A similar deposition of powder has been applied by electrostatic-spray techniques in air or in an inert atmosphere. Deposits have also been made as a slurry in an organic binder, e.g. polyethylene oxide, followed by drying and compacting.

Chemical deposition, gas or vapour plating   This is a chemical process whereby the aluminium is deposited when an aluminium compound is decomposed. An alkyl gas such as aluminium diethyl hydride (b.p. 55–56°C) is introduced into the work chamber after purging with an inert gas such as argon or nitrogen. The hydride is then decomposed by heating at around

180°C to deposit the aluminium: deposits of from 0·075 to 2·5 mm are claimed on a variety of substrates. The coating is said to be ductile, adherent and lustrous.

In a similar process the work is dipped in a catalytic titanium compound and then transferred to a bath of aluminium hydride solution. The solution dries leaving the hydride on the surface and mild heating then turns the hydride into metallic aluminium. The conversion stage occurs slowly at room temperature or in seconds at 100°C.

The technique lends itself to the coating of surfaces which are relatively inaccessible, and it is stated that the process has been used to coat parts for turbine engines and could be used to coat parts for steel mills and coil coaters.

Cladding (mechanical bonding) The mechanical cladding of a surface with aluminium can be achieved by rolling, extrusion or drawing. Aluminium cladding can be bonded to other aluminium products and to a variety of substrates including copper and steel; the products can vary from sheet or strip to extruded shapes and wire. The cladding process is carried out by the metal supplier and this type of material is discussed in Section 4.1. There is an increasing interest in explosive bonding as a means of bonding aluminium-steel and other composite cladding systems.

Casting Casting around steel parts, which are first hot dipped in aluminium or in aluminium-silicon alloy (the Al-Fin process), gives good bonding but requires careful design because of the different thermal-expansion coefficients of the two metals.

Nature of Coatings and Physical Properties

Sprayed aluminium coatings consist of small flattened globules of metal surrounded by aluminium oxide which forms about 1 to 3% of the coating. The coating density is typically 2·3-2·4 g/cm³. The outer surface is uniformly slightly rough with a whitish appearance. The coating is a reasonable conductor of heat and electricity, although inferior to bulk aluminium of similar purity. Coatings are cohesive and normally 0·10-0·20 mm thick. They have good adhesion on a properly prepared surface, but above 0·3 mm there may be danger of adhesion being weakened. Only slight deformation of the coating is possible. The mechanical properties of the base metal remain substantially unimpaired except for a slight loss in fatigue strength. Softening of strong, heat-treated aluminium alloys can occur with overheating through poor spraying technique. The structure of sprayed aluminium coatings is illustrated in Figs. 14.3 and 14.4.

Hot-dip aluminised coatings are obtained preferably by continuous methods. They are normally about 0·05 mm when unalloyed metal is used, but in the case of aluminium-silicon alloys only 0·025 mm is applied. As dipped, the surface may be rough; a bright appearance can be obtained by mechanical treatment such as rolling or drawing. Silicon appears to alter the viscosity of the bath and also inhibits the growth of the iron-aluminium alloy layer. The structure of the aluminium-silicon alloy coatings is distinctive. They show dark grey acicular needles of a silicon-rich constituent, and dispersed islands of an iron-aluminium constituent are also present in

the aluminium solid solution outer layer. The two layers close to the steel base are of different iron–aluminium–silicon alloys with an average iron content of 33·5%. In the absence of silicon the plates and stringers in the aluminium solid solution outer layer are iron–aluminium constituents containing about 45% iron. X-ray examination of both types of coating shows the presence of FeAl and lesser amounts of Fe₃Al, although this is not in accord with the equilibrium diagram. The coating is more dense than aluminium but no precise figures can be given. The coatings will withstand moderate forming (up to bends of twice the metal thickness, depending on the thickness of the brittle alloy layer); failure occurs more readily in compression than in tension. In the former case, shearing stresses develop at the alloy/base metal interface, but in tension, cracks which do not immediately affect adhesion develop at right angles to the surface in the iron–aluminium layer. The mechanical properties of the aluminised product are those of the base metal except for a slight loss of fatigue strength and a loss of tensile strength due to the annealing effect of the coating operation. At temperatures up to 500°C, coated steel will reflect about 80% of incident radiant heat. The coating is a good electrical conductor.

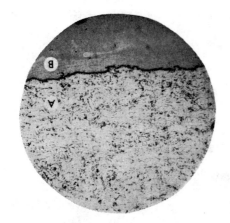

Fig. 14.3 Sprayed aluminium; section taken parallel to substrate and etched; × 75

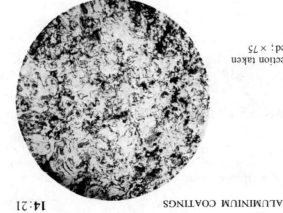

Fig. 14.4 Transverse section of sprayed aluminium on steel; etched in HF. *A* aluminium and *B* steel; × 75

Calorised and heat-treated mechanically-clad products have coating structures similar to hot-dip aluminised coatings, but the degree of alloying with iron is variable (Fig. 14.5). With Calorised products the surface layers usually contain 25–50% aluminium.

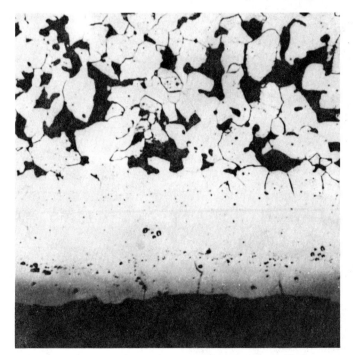

Fig. 14.5 Section through Calorised coating on mild steel; approximately × 500 (courtesy Babcock and Wilcox Ltd.)

Vacuum-deposited and electroplated coatings are pure metal with no chemical bond to the underlying surface. The properties will be those of pure aluminium. The presence of lacquer, in the case of vacuum-deposited coatings will, however, afford resistance to the passage of electricity and limit the maximum temperature of use.

Steel with up to 0·075 mm thick aluminium coatings can be gas-cut by both hand and machine methods, and welded, without removal of the aluminium. No fume problems arise in welding aluminium-coated material; in practice, however, it is usual to leave the edges bare and to spray the joint with aluminium after welding.

Corrosion Resistance

The corrosion resistance of aluminium coatings is generally related to that of solid aluminium of similar thickness. Additional factors arise with sprayed coatings associated with texture, and with aluminised and other coatings when diffusion from the substrate can occur.

Ordinary temperatures   The protection of steel by aluminium coatings

depends partly on sacrificial action and partly on a physical barrier associated with the aluminium and oxidation products. Initially, sprayed aluminium coatings in atmospheric or immersed conditions may show slight superficial rust staining through pores in the coating before penetration through the oxide skin around each aluminium particle enables aluminium to protect the steel cathodically. Subsequently, insoluble aluminium corrosion products block the pores and reduce corrosion currents to a negligible value, thus conserving the aluminium coating. This result is readily observed in practice when slight initial rust staining of the coating may occur and then either diminish or remain unaltered for many years. Rust staining can be avoided by the use of sealing lacquers. It is generally agreed that the cathodic polarisation of steel by aluminium is less than, for example, that effected by zinc although the standard electrode potentials indicate the reverse. This results in a long life for the aluminium coatings although they may allow some rust to form at breaks in the coating if the nature of the electrolyte in the environment is such that there is no galvanic protection of the steel, e.g. at cut edges of sheets in a mild atmosphere. The initial rusting is not progressive and in some cases the rust is reduced to magnetite leaving only a grey colour.

In selecting sprayed-metal coatings for steel, aluminium is preferred for use in aggressive environments such as marine environments, acidic environments and sulphur-containing industrial environments. Aluminium coatings are less suitable than zinc in most strongly alkaline environments. Coatings of aluminium–zinc mixtures or compounds of composition of approximately 65Zn–35Al have been used commercially, and it is claimed that the initial rust staining sometimes associated with aluminium coatings is avoided. Coatings of aluminium–zinc mixtures can also provide galvanic protection to some aluminium base alloys where aluminium alone does not provide the required sacrificial protection. Duplex sprayed-metal coatings of aluminium and zinc, or of two grades of aluminium, have been tested to see if they could prevent rusting or give improved protection; such duplex systems have not given worthwhile improvement over that of a single coating material.

Scott shows that unpainted sprayed coatings, 0·075 mm thick, of aluminium and zinc are giving good service in marine and rural atmospheres after 15 years, with the aluminium coating being slightly superior. In an industrial atmosphere BISRA tests show that the aluminium coating after 15 years is superior to zinc. When immersed in the sea, 0·075 mm zinc lasted less than four years, whereas the same thickness of aluminium was still affording protection to steel after 14 years.

Other investigations indicate that half-tide conditions give results similar to fully immersed conditions, and that in sea-water the 99·5% aluminium coating is preferred to aluminium–zinc or aluminium–magnesium alloy coatings.

Hudson, in tests at Sheffield (severe industrial atmosphere), observed that an 0·075 mm coating of aluminium had not failed after 18 years service (Fig. 14.6).

Sprayed coatings can be used as a base for painting and it is recommended that a minimum of 0·075 mm of aluminium be used. In practice a combination of 0·1–0·2 mm of aluminium plus paint coatings will give maximum economic durability on long-term structures.

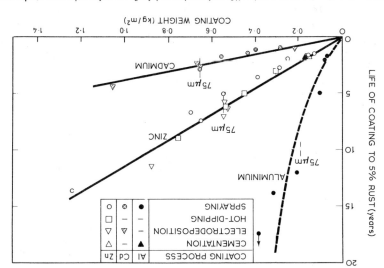

Fig. 14.6 Relation between coating life and coating weight for metal coatings on steel exposed at Sheffield. The line for aluminium has been amended and drawn as a curve (after Fig. 7 of Special Report No. 66 of the Iron and Steel Institute)

Tests by Clark for the Corrosion Sub-committee of the American Welding Society were carried out at severe marine and industrial sites. After four years, the greatest protection to steel was given by sprayed aluminium coatings combined with aluminium vinyl paint in the following environments: (a) sea air, (b) sea-water immersion, (c) alternate sea-water immersion and exposure to air (tidal conditions) and (d) industrial atmospheres contaminated with sulphur compounds.

Exposure tests of 5-years duration in the atmosphere and during immersion in fresh water or sea-water by the Institute of Welding and by Stanners and Watkins have indicated the best paint systems for aluminium-coated steel. For immersed conditions two systems giving good performance are:

1. Modified pretreatment primer plus zinc chromate primer plus two coats of micaceous iron oxide/polyurethane.
2. Modified pretreatment primer plus three coats of aluminium-pigmented vinyl copolymer.

Under atmospheric conditions the following systems gave good performances:

1. Pretreatment primer plus micaceous iron oxide/long oil tung phenolic.
2. Zinc chromate primer plus micaceous iron oxide/alkyd or phenolic.
3. Aluminium-pigmented vinyl.

Hot-dip aluminised coatings have been tested by the American Society for Testing and Materials over a 23-year period in rural and marine atmospheres. In most cases, coatings of 0.025–0.050 mm thickness show no deterioration, but at the more severe marine site some rusting has occurred. Calorised coatings have a substantial iron content at the surface. Iron–aluminium alloys generally have superior corrosion resistance to that of the

steel substrate, and atmospheric exposure tests have shown Calorised coatings to afford useful protection.

Successful anodising of aluminium coatings produced by hot dipping and by electrophoretic deposition has been claimed, but this is subject to having an aluminium coating free from pores and a coating of sufficient thickness to enable the coating to be anodised; an anodic film 0·025 mm thick is formed at the expense of approximately 0·05 mm from a uniform aluminium surface. The steel must be fully stopped off in the anodising bath. Similar observations can be made with respect to aluminium coatings applied by other techniques.

High temperatures Diffusion coatings producing iron–aluminium alloys are used to provide protection to steel at high temperatures. These coatings can be produced by hot dipping, Calorising or by metal spraying followed by heat treatment.

Alloying of the coating and steel substrate commences at temperatures in the range 300–480°C, and the resistance of the substrate to oxidation increases with increasing aluminium content at the surface. An aluminium content of 8–10% is said to markedly reduce subsequent oxidation at temperatures as high as 1 000–1 100°C. The diffusion and alloying processes enable the aluminium coating to give good protection to steel at temperatures well beyond the melting point of aluminium and such coatings find service in atmospheres where sulphur-containing fuels are being used.

Up to 550°C, aluminium coatings may be used in the 'as applied' condition, and the hot-dipped aluminium–silicon alloy may be used up to at least 680°C, but for service at higher temperatures additional diffusion treatment is recommended for all except Calorised coatings.

Up to 750°C, the performance of all aluminium diffusion coatings is considered to be very good, but above this temperature the results appear to be dependent on the coating thickness, diffusion treatment and the specific service environment. Sprayed aluminium coatings can be used up to 900°C after diffusion treatment. Hot-dipped coatings also benefit from additional diffusion treatment, and omission of silicon from the coating alloy improves performance at the elevated temperatures.

Above 900°C, the life of aluminium-coated steel is more limited although the coating can still provide significant protection to steel, e.g. at 900–980°C the life is said to be increased 20 times, and at 980–1 000°C the life is said to be increased 5 times.

Applications

The main application of sprayed aluminium is for the protection of structural steel, and the process can also be utilised to protect high-strength aluminium alloys. The process has the important advantage that it can be carried out on site.

Sprayed aluminium coatings were applied over twenty years ago for a number of buildings in steelworks, notably at Margam, Trostre and Velindre in South Wales. At Margam all steelwork above crane girder level has been aluminium sprayed to 0·10 mm thickness and painted with two coats of

commercial-quality aluminium paint. Two sections in a severely aggressive environment were repainted about ten years after installation, although there was no rust present on the sprayed steelwork; the steelwork below crane level had, meanwhile, been repainted twice. It has been estimated that with an average interval of 8½ years between repainting operations, the total painting cost over 50 years will be reduced by nearly 50%. The more recent Spencer Steelworks has been similarly protected following the experience gained in the earlier buildings.

Aluminium-sprayed steel windows in Sheffield remained in good condition some 17 years after erection. Aluminium-sprayed coating on a steel structure at a steelworks in Sheffield is intact after service for 34 years with paint maintenance at approximately 10-year intervals.

The Spanish National Steelworks has been given similar protection using sprayed aluminium coatings. Railway bridges at Carlisle and Hook showed no corrosion when examined after five years with an 0·075 mm aluminium coating on main surfaces and an 0·15 mm coating where exposed to direct locomotive blast. Aluminium-sprayed coatings have been used on two large road bridges on the southern end of the M1 motorway. Parts of the Severn Bridge have been sprayed with 65Zn–35Al alloy. Harbour bridges in Australia and the Near East sprayed with aluminium are in good condition 10–12 years after application. It is estimated that most structures protected by aluminium-sprayed coatings could have an extremely long service life if a simple paint coating was maintained at intervals of the order of 10 years.

The main roof girders of the BEA engineering base and workshop at London's Heathrow airport are aluminium sprayed.

Soft peaty water, e.g. as in Scottish hydroelectric schemes, has been satis-factorily carried in painted duplex-coated steel pipe (0·075 mm of zinc followed by 0·075 mm of aluminium), while a similar duplex coating was reported successful in a steam–ammonia atmosphere. A French railway bridge, however, showed peeling of the aluminium in a few months, owing to corrosion of the underlying zinc, suggesting that the duplex coatings should not be used indiscriminately; the powder-sprayed technique is recommended for duplex coatings.

Hydroelectric-plant pipelines of 0·4 km in length and 1·8 m in diameter were in good condition after service for eight years. At another hydroelectric plant, steel and cast-iron pipes carrying high-pressure soft water from a mountain reservoir had been in service for about 50 years with considerable internal corrosion. For safety reasons, it was decided that the pipes should be replaced or the corrosion stopped. In 1959 the pipeline was gritblasted internally and sprayed with 0·25 mm of 99·5% Al followed by two coats of petroleum bitumen. In 1972 the 0·25 mm aluminium coating was still giving satisfactory service on these surfaces which were prepared and coated under the most arduous conditions.

Aluminium-sprayed coatings are used successfully on gas cylinders and on ammunition boxes under a wide variety of atmospheric conditions. Car-ferry vessels in Australia have 0·15 mm of aluminium-sprayed coatings on the inside and outside of the hull followed by a compatible anti-fouling paint. Steel vertical legs on the S.S. President Cleveland and S.S. President Wilson are aluminium sprayed followed by two coats of aluminium-pigmented vinyl paint. Aluminium-sprayed steel is also giving good service

in tropical marine atmospheres on sheds erected in 1960 at Tema Harbour, Ghana.

Aluminium is particularly resistant to sulphur-polluted atmospheres, and sprayed coatings are used in sulphuric-acid plants for the main convectors; for hot, intermediate and cold heat exchangers; and for the internal surfaces of interconnecting ducting. Coatings of 0·15 mm thickness have given good service.

There are many instances of the use of aluminium spraying for the protection of steel in gasworks. These include the Dudley Gasworks' Retort House, which is in good condition after nine years or more; the largest welded gas holder in Europe located at Bristol was in excellent condition 10 years after treatment despite the severely corrosive environment, and a steel structure at the Coxside Gasworks at Plymouth needed no maintenance after more than 13 years service on a site only a few hundred metres from the sea. Other aluminium-sprayed applications include the gas holders, gas mains and blast furnace components at the Spanish National Steelworks at Aviles, and the components for the converter and heat exchangers of sulphuric-acid plants in Austnalia.

Chimney stacks have been sprayed externally with aluminium at many gasworks, oil refineries, laundries and petrochemical plants, to resist the combined effects of heat, condensation and atmospheric attack.

Steel structures and components subject to high temperature and corrosive attack on which paint would fail rapidly are given excellent protection by sprayed aluminium applied in accordance with BS 2569:Part 2:1965. Such coatings are used in blast-furnace downcomers, conveyor cooling hoods, and offtake ducting in reheating furnaces.

Converter shells for production of sulphur trioxide from sulphur dioxide have been in operation for over 12 years after spraying with 0·20 mm of aluminium. Operating conditions are sulphur trioxide at 600–650°C inter-nally and 7% dry sulphur dioxide at 200–450°C externally, conditions which formerly resulted in lives of only two years without the metal spraying. A similar coating thickness has been used on coal-wagon tipper gear to provide protection against wet sulphurous conditions in the steel industry, and a coating life of 20 years is expected.

Aluminium coatings are not favoured in atmospheres containing explosive mixtures because contact with rusty steel can cause incendiary sparking, and for this reason aluminium coatings are not used for protection of structures in coalmines (cf. CP 2008:1966).

Aluminium spraying of steel street-lighting columns has been used since the 1950s and it is estimated that one producer alone has supplied up to 200 000 such columns; repainting is simpler, even on neglected columns, than on columns not metal sprayed. Aluminium spraying has been used on reflector towers used in the television link between Manchester and Edin-burgh, and on similar structures.

Spray-aluminised coatings are used for exhaust valves in automobile engines, exhaust and silencer systems (double and triple life), tyre moulds, gas ducting, heat-treatment pots, furnace ladles, carburising boxes and fans handling hot gases. Similar applications utilise Calorised and hot-dip aluminised coatings. Hot-dip aluminised steel wire has been used in steel-cored aluminium conductors for overhead transmission lines. For some

applications the greater heat-transfer efficiency of the aluminium coating compared with that of solid steel is an asset.

Electrophoretic and similar compacted coatings are in early stages of development but will no doubt take their place alongside other coating methods.

Vacuum deposition of high-purity aluminium has been used as a bright finish of a decorative nature on domestic items and some car accessories, as well as special items for space missions where opacity to solar radiation was required. Continuous deposition on plastic strip at speeds up to 450 m/min has been achieved.

Plating of aluminium has been developed for electroforming wave-guides with wall thicknesses up to 0·10 mm, and for the aluminium coating of reflectors.

Aluminium cast onto steel or cast iron is used to produce integral aluminium/steel drums and bimetallic pistons. Aluminium clad onto other metals by mechanical bonding is used in heat-exchanger systems subject to multiple atmospheres or environments. The clad products are also used for cooking utensils and functional presswork.

The non-toxicity of aluminium coatings and freedom from taste or taint means freedom from health hazards during application and provides hygienic finishes for contact with foodstuffs, e.g. baking tins, oven trays, containers. Aluminium coatings also provide a suitable key or pretreatment for subsequent coatings, e.g. aluminised steel provides a good base for vitreous enamel.

E. W. SKERREY

BIBLIOGRAPHY

General

Burns, R. M. and Bradley, W. W., *Protective Coatings for Metals*, Reinhold, New York (1955)

Lowenheim, F. A., *Modern Electroplating*, Wiley/American Electrochemical Society, 2nd edn. (1963)

Wernick, S. and Pinner, R., *Surface Treatment of Aluminium*, Robert Draper, London, 4th edn. (1972)

Spraying

Andrews, D. R., 'The Protection of Iron and Steel by Aluminium Coatings', *Metallurgia*, **62**, 153 (1960)

Ballard, W. E. *Metal Spraying and the Flame Deposition of Ceramics and Plastics*, Griffin, London, 4th edn. (1963)

Carter, V. E. and Campbell, H. S., 'Protecting Strong Aluminium Alloys Against Stress Corrosion with Sprayed Metal Coatings', *British Corrosion Journal*, **4** No. 1, 15–20, Jan. (1969) and **4** No. 4, 196–198, July (1969)

Franklin, J. R., 'Metallized Coatings for Heat Corrosion Protection', *Corrosion Technol.*, **2**, 326 (1956)

Hoar, T. P. and Radovici, O., 'Zinc–Aluminium Sprayed Coatings', *Trans. Inst. Met. Fin.*, **42**, 211–222 (1964)

Hudson, J. C., Sixth Report of the Corrosion Committee of BISRA, Special Report No. 66. I.S.I. (1959)

Mansford, R. E., 'Sprayed and Diffused Metal Coatings', *Metal Ind.*, London, **93**, 413 (1958)
Mansford, R. E., 'Sprayed Metal Coatings in the Gas Industry', *Chem. and Ind.* (Rev.), 150 (1961)
Porter, F. C., 'Aluminium Coatings on Iron and Steel', *Metal Finishing Journal*, **9** No. 104, 303–312, Aug. (1963)
Porter, F. C., 'Aluminium Sprayed Steel in Marine Conditions', *Engineer*, Lond., **211**, 906 (1961)
Reininger, H., 'Further Developments in Metal Spraying Technique', *Metalloberfläche*, **15**, 52, 88, 118, 148 (1961). (Translation available as TM460, Aluminium Federation, Birmingham)
Scott, D. J., 'Aluminium Sprayed Coatings: Their Use for the Protection of Aluminium Alloys and Steel', *Trans. Inst. Met. Fin.*, **49** No. 3, 111–122 and **49** No. 4, 173–175 (1971)
Sprowl, J. D., *Aluminized Steel—A Description*, Report of the Department of Metallurgical Research, Kaiser Aluminium and Chemical Corporation, April (1958)
Stanners, J. F. and Watkins, K. O., 'Painting of Metal Sprayed Structural Steelwork', *British Corrosion Journal*, **4** No. 1, 7–14, Jan. (1969)
Aluminizing, DTD 907B
Defence Guide DG-8, Part 2, H.M.S.O., April (1971)
How to Prevent Rusting, BISRA (1963)
Metallizing: Aluminium and Zinc Spraying, DTD 906B
Methods of Protection Against Corrosion for Light Gauge Steel Used in Building, PD 420 (1953)
Painting of Metal Sprayed Structural Steel, BISRA Corrosion Advice Bureau (1966)
Proceedings of the Second International Metal Sprayers' Conference, Birmingham, Association of Metal Sprayers (1958). (Specific references to aluminium notes in TM 420, Aluminium Federation, London)
Protection by Sprayed Metal Coatings, The Institute of Welding, London (1968)
'Protection of Iron and Steel against Corrosion and Oxidation at Elevated Temperatures', *Sprayed Metal Coatings*, BS 2569: Part 2 (1965)
'Protection of Iron and Steel by Aluminium and Zinc against Atmospheric Corrosion', *Sprayed Metal Coatings*, BS 2569: Part 1 (1964)
The Protection of Steel by Metal Coatings, No. 5 of Series by Corrosion Advice Bureau of BISRA (1967)
The Protection of Iron and Steel Structures from Corrosion, CP 2008 (1966)

Hot-Dip Aluminising
Coburn, K. G., 'Aluminized Steel—Its Properties and Uses', *Metallurgia*, Manchr., **60**, 17 (1959)
Drewett, R., 'Diffusion Coatings for the Protection of Iron and Steel', Part I, *Anti-Corrosion*, **16** No. 4, 11–16, April (1969)
Edwards, J. A., 'Coated Engine Valves', *Auto. Engr.*, **45**, 441 (1955)
Gittings, D. O., Rowland, D. H. and Mack, J. O., 'Effect of Bath Composition on Aluminium Coatings on Steel', *Trans. Amer. Soc. Metals*, **43**, 587 (1951)
Hughes, M. L., 'Hot Dipped Aluminized Steel: Its Preparation, Properties and Uses', *Sheet Metal Industry*, **33**, 87 (1956)
Schmitt, R. J. and Rigo, J. H., 'Corrosion and Heat Resistance of Aluminium-coated Steel', *Materials Protection*, **5** No. 4, 46–52, April (1966)
Serra, M., 'Considerations Arising from Experiments on a New Process of Metallic Protection', *Rev. Cienc. Appl.*, **12**, 222 (1958). (Translation available as TM398, Aluminium Federation, Birmingham)
Whitfield, M. G., 'Rolling of Hot Dipped Aluminized Steels to Make Them More Durable', *Anti-Corrosive Materials and Processes*, **2**, 31, Oct. (1963) and US Pat. 2 170 361
Anon., 'Largest Aluminium Line in UK', *Product Finishing*, **21** No. 7, 61–65, July (1968)
Anon., 'Welded Aluminized Steel Sheet', *Anti-Corrosive Materials and Processes*, No. 2, 4–7, Oct. (1963)
'A.S.T.M. Field Tests and Inspection of Hardware', *Proc. Amer. Soc. Test. Mater.*, **52**, 118 (1952)
Hot Dip Aluminizing, Preprint, A.S.T.M. Committee A5 (1963)

Calorising (*cementation*)
Drewett, R., 'Diffusion Coatings for the Protection of Iron and Steel., Part I, *Anti-Corrosion*, **16** No. 4, 11–16, April (1969)
Porter, F. C., 'Aluminium Coatings on Iron and Steel', *Metal Finishing Journal*, **9** No. 104, 303–312, Aug. (1963)

Vacuum Deposition

Holland, L., *Vacuum Deposition of Thin Films*, Chapman and Hall, London (1956)

Porter, F. C., 'Aluminium Coatings on Iron and Steel', *Metal Finishing Journal*, **9** No. 104, 303–312, Aug. (1963)

Remond, G. B. and Johnson, A. R., 'Vacuum Deposition of Metals', *Metal Finishing Journal*, **4**, 393 (1958)

Weil, F. C., 'Recent Developments in Vacuum Metal Coatings', *Electroplating Metal Finish*, **9**, 6, 25 (1956)

Anon., *Vacuum Coating of Formed Metal Parts*, National Research Corporation, Cambridge, Massachusetts (1959)

Electrodeposition (electroplating)

Couch, D. E. and Brenner, A., 'Hydride Bath for the Electrodeposition of Aluminium', *J. Electrochem. Soc.*, **99**, 234 (1952)

Couch, D. E. and Connor, J. H., 'Nickel-Aluminum Alloy Coatings Produced by Electro-deposition and Diffusion', *J. Electrochem. Soc.*, **107**, 272 (1960)

Holland, B. O., 'Aluminium Coating of Steel with Special Reference to Electrodeposition', *Broken Hill Pty. Technical Bulletin*, **3** No. 1, 29 (1959)

Lowenheim, F. A., *Modern Electroplating*, Wiley/American Electrochemical Society, 2nd edn. (1961) .

Menzies, I. A. and Salt, D. B., 'The Electrodeposition of Aluminium', *Trans. Inst. Met. Fin.*, **43**, 186–191 (1965)

Safranek, E. H., Schiekner, W. C. and Faust, C. L., 'Electroplating Aluminium Wave Guides Using Organo-Aluminium Plating Baths', *J. Electrochem. Soc.*, **99**, 53 (1952)

Utz, J. J. and Kritzer, J., 'New High Purity Aluminium Coatings', *Mater. Design Engn.*, **49**, 88 (1959)

Electrophoretic Coatings and Other Compacted Coatings

Bright, A. W. and Coffee, R. A., 'Electrostatic Powder Coatings', *Trans. Inst. Met. Fin.*, **41**, 69–73 (1964)

Brown, D. R. and Jackson, A. E., 'The Elphal Strip-Aluminizing Process', *Sheet Metal Ind.*, **39**, 249 (1962)

Sugano, G., Mari, K. and Inoue, K., 'A New Aluminium Coating Process for Steel', *Electro-chemical Technology*, **6** Nos. 9 and 10, 326–329, Sept.-Oct. (1968)

Process for Forming Sintered Metal Coatings, Texas Instruments Inc., UK Pat. 1 163 766 (10.9.69)

Chemical Deposition, Gas or Vapour Plating

Dow Chemical Co., 'Catalytic Plating of Aluminium', *Financial Times*, Oct. 21 (1969)

Hiler, M. J. and Jenkins, W. C., *Development of a Method to Accomplish Aluminium Deposition by Gas Plating*, US Air Force WADC Tech. Report, 59–88, June (1959) and US Pat. 2 929 739 (1955)

Powell, C. F., Campbell, I. E. and Gonser, B. W., *Vapour-Plating*, Chapman and Hall, London (1955)

Anon., 'Aluminium Plating Via Alkyd Gas', *The Iron Age*, 52–53, Dec. 23 (1965)

Method of Aluminium Plating with Diethylaluminium Hydride, Continental Oil Co., UK Pat. 1 178 954 (28.1.70)

Mechanical Bonding

BIOS Final Reports, Nos. 1467 and 1567, H.M.S.O. London (1947)

Bonded Aluminium/Steel Composites and Methods of Making Same, Du Pont de Nemours and Co., UK Pat. 1 248 794 (6.10.71)

Casting

Little, M. V., 'Bonding Aluminium To Ferrous Alloys', *Machinery*, N.Y., **56**, 173 (1950)

Drewett, R., 'Diffusion Coatings for the Protection of Iron and Steel', Part 1, *Anti-Corrosion*, **16** No. 4, 11–16, April (1969)

14.3 Cadmium Coatings

In some environments cadmium gives better protection than zinc; it is, however, some ten times more costly. It does not compete with zinc on articles on which a high degree of protection can be achieved by the use of a thick film deposited by hot dipping or metal spraying. Where only thin coatings of 25 μm or less are tolerable, the greater protection of cadmium in some environments is worth while, and as uniform thin coatings must be deposited by relatively expensive processes such as electroplating, the greater cost of cadmium then has little effect on the cost of the finished article.

Nature and Degree of Protection

Coatings of both cadmium and zinc protect steel mainly by simple physical exclusion. At gaps in the coating, however, whether these are in the form of natural pores or areas of weakness, scratches or cut edges, protection is by sacrificial action of the coating followed probably by the plugging of gaps with sparingly soluble corrosion product.

It is not at once clear why cadmium should be sacrificially protective to steel. Standard electrode potentials of iron and cadmium in contact with solutions of their salts of normal activity, given in Table 14.2, suggest that iron should be sacrificial to cadmium, but Hoar[1] has shown by means of E/I curves that the mixed potential of corroding cadmium will be more electronegative than the mixed potential of corroding iron. This follows from the higher exchange current for Cd $\rightarrow$ Cd^{2+} + 2e. Under these circumstances iron will be sacrificially protected by cadmium (see also Section 1.4). What-ever the explanation, the fact of sacrificial protection is reflected in the potentials, also given in Table 14.2, found for the two metals in sea-water.

Table 14.2 Potentials of iron, cadmium and zinc

Metal	Standard electrode potential, hydrogen scale (V)	Corrosion potential* in flowing sea-water, hydrogen scale (V)
Iron	− 0.44	− 0.36
Cadmium	− 0.40	− 0.45
Zinc	− 0.76	− 0.76

*The corrosion potential will vary with aeration and velocity of the sea-water.

Table 14.3 Corrosion rates of zinc and cadmium coatings in various atmospheres

Location	Rate of corrosion of electrodeposited coating (μm/y)	
	Zinc	Cadmium
Industrial	5·1	10·2
Suburban	1·8	2·3
Marine	2·5	1·3

The degree of protection given in practice by zinc and cadmium, whether by physical exclusion or by sacrificial action at gaps, depends on the durability of the coatings themselves against corrosive attack. It is now well established that, thickness for thickness, cadmium is more resistant to marine and tropical atmospheres and zinc more resistant to industrial atmospheres. This is well demonstrated by comparative tests made by Biestek[2] in various laboratory conditions, and by Clarke and Longhurst[3] in practical tropical exposure tests. Table 14.3 gives the order of corrosion rates, based on the results of these and other[4] tests. It must be emphasised that these figures give only a broad comparison; actual corrosion rates will be much affected by the exact environment.

If the corrosion mechanism in an industrial atmosphere is mainly a straight chemical dissolution in sulphur acids, then the relative chemical equivalents present in a given thickness of the two metals account for a large part of the difference in corrosion rate. In an unpolluted humid atmosphere the slightly greater corrosion resistance of cadmium compared with zinc at unit thickness (and therefore much greater resistance per unit chemical equivalent) must be ascribed to a greater insolubility and protective power of the first corrosion product. The solubility data in Table 14.4 (quoted from the *Handbook of Chemistry and Physics*, 40th edition) show that cadmium hydroxide is more soluble in water than zinc hydroxide, but that the cadmium carbonate is the less soluble, so it is concluded that the protective films formed are carbonates or possibly basic carbonates. The considerably greater comparative resistance of cadmium in a marine atmosphere must be postulated as being due to a greater insolubility of the basic chloride of cadmium compared with that of zinc.

In conclusion, relative cost and relative behaviour towards different conditions of exposure lead to the use of zinc on parts on which thick films can be tolerated and for general industrial use, and of cadmium on fine-tolerance

Table 14.4 Solubility in water of cadmium and zinc carbonates and hydroxides

Metal	Solubility (g/100 ml)	
	Carbonate	*Hydroxide*
Cadmium	insoluble	0·000 26
Zinc	0·001	0·000 000 26

parts, such as aircraft and instrument parts, required to withstand unpolluted humid and marine atmospheres.

Other Factors Governing the Choice between Zinc and Cadmium

As well as the reasons already given, other considerations influence the choice between zinc and cadmium.

In the electronics industry cadmium is preferred to zinc because it is easier to solder and has a lower contact resistance.

On very strong steels cadmium is also preferred because it appears that cadmium electroplating from a given type of electroplating solution, e.g. a cyanide solution, causes less embrittlement than zinc plating from the same type of solution[5]. On the other hand, on steels subject to elevated temperatures in use, the possibility of intergranular penetration of stressed steel which occurs above (and even, if the steel is highly stressed, somewhat below) the melting point of the coating, may lead to the choice of zinc (m.p. 419·5°C) in preference to cadmium (m.p. 321°C).

Acid vapours emitted by wood, oleoresinous paints and some plastics (cf. Section 19.10) attack both zinc and cadmium. The relative behaviour varies, and appears to depend on the nature and concentration of the acid vapours and on the relative humidity. For these conditions of exposure, therefore, no advice can be given as to which metal should be used. It should not be assumed that because one metal has failed therefore the other would be better.

Both metals are applied to copper-base alloys, stainless steels and titanium to stop bimetallic corrosion at contacts between these metals and aluminium and magnesium alloys, and their application to non-stainless steel can serve this purpose as well as protecting the steel. In spite of their different potentials, zinc and cadmium appear to be equally effective for this purpose[6], even for contacts with magnesium alloys[7]. Choice between the two metals will therefore be made on the other grounds previously discussed.

Protection of Cadmium Coatings

Full chromate passivation (Section 16.3) improves the corrosion resistance of both zinc and cadmium towards all environments and is normally applied to cadmium-plated parts for aircraft (see DTD 904). Passivated zinc has a resistance to humid and marine atmospheres roughly equal to that of unpassivated cadmium. Passivation improves the adhesion of normal types of priming paints, but for best adhesion and protection an etch primer should be used (Section 16.3).

Methods of Deposition

Electroplating Like zinc, cadmium is usually electroplated from a cyanide solution. Typical solutions for both metals, taken from specifications DTD 903 and 904, are given in Table 14.5.

Table 14.5 Typical plating solutions for zinc and cadmium

Solution	Constituents				
	Zn or Cd (as metal) (g/l)	Total cyanide (as NaCN) (g/l)	Caustic soda (g/l)	Temperature of operation (°C)	Current density (A/dm²)
Zinc plating	25	56	40	32–48	100–200
Cadmium plating	14–17	56–63	11–14	15–35	100

Other solutions, some based on cyanide, some on sulphate, fluoborate, etc. will be found in textbooks and handbooks of electroplating, and a comprehensive review of methods and of their relative advantages has been published by Such[8]. Much work has been devoted to the development of cadmium plating processes which cause little or no hydrogen embrittlement to very strong steels; references are given in Section 8.4 and in a paper by Yaniv et al.[9]

Vapour deposition Hydrogen embrittlement can be avoided by depositing cadmium by vacuum evaporation. Vapour plating is carried out in a chamber to below 2.7×10^{-3} N/m². Cadmium metal is placed in mild-steel boats arranged along the chamber and heated to about 200°C. The evaporating metal moves in straight lines, so the parts to be coated are held in jigs that rotate on their own axes and revolve round the chamber, thus presenting all surfaces to the moving vapour. Before evaporation is begun, the parts must be cleaned by ion bombardment in a high-tension ($\simeq$ 1 kV) glow discharge at a pressure of approx. 4 N/m². Formerly, the glow discharge was stopped before the chamber was pumped down to evaporation pressure, but adhesion of the coating was poor unless the parts had first been roughened by fine abrasive blasting. In an improved process[10] the glow discharge is maintained concurrently with pumping down and the start of evaporation; under these conditions there is no interval during which oxide reforms on the steel by reaction with residual oxygen in the chamber, cadmium atoms arrive on a surface still under bombardment, and adhesion of the coating is good even on smooth machined surfaces.

Specifications for Cadmium Plating

Cadmium plating for general engineering use is covered by BS 1706:1960, and for aircraft parts by Ministry of Aviation Supply Specification DTD 904. Special requirements for very strong steels are given in Defence Standard 03–4 (Directorate of Standardisation, Ministry of Defence).

H. G. COLE

REFERENCES

1. Hoar, T. P., *J. Electrodep. Tech. Soc.*, **14**, 33 (1938)
2. Biestek, T., *Proceedings of the First International Congress on Metallic Corrosion*, London, 1961, Butterworths, London, 269 (1962)

3. Clarke, S. G. and Longhurst, E. E., *Proceedings of the First International Congress on Metallic Corrosion*, London, 1961, Butterworths, London, 254 (1962)
4. Uhlig, H. H. (Ed.), *The Corrosion Handbook*, Wiley, New York; Chapman and Hall, London, 803 and 837 (1948)
5. Maroz, I. I. *et al.*, see Domnikov, L., *Metal Finish.*, **59** No. 9, 52 (1961)
6. Evans, U. R. and Rance, V. E., *Corrosion and its Prevention at Bimetallic Contacts*, H.M.S.O., London (1958)
7. Higgins, W. F., *Corrosion Technol.*, **6**, 313 (1959)
8. Such, T. E., *Electroplating Met. Finish.*, **14**, 115 (1961)
9. Yaniv, A. E. *et al.*, *Trans. Inst. Met. Finish.*, **45**, 1 (1967)
10. UK Pat. 1 109 316 and specification DTD 940

14.4 Zinc Coatings

It is estimated that approximately 40% of the world production of zinc is consumed in hot-dip galvanising of iron and steel, and this adequately demonstrates the world-wide use of zinc as a protective coating. The success of zinc can be largely attributed to ease of application, low cost and high corrosion resistance.

Methods of Application

The principal method for applying zinc coatings to iron and steel is hot-dip galvanising. There are four other important methods, each of which has its own particular applications; these methods are spraying, plating, sherardising and painting with zinc-rich paints. The choice of method in any given case is determined by the application envisaged, and the five processes may be said to be complementary rather than competitive, for there is usually little doubt as to which is the best for any particular purpose. Processes of applying coatings by various methods are discussed in detail elsewhere[1] and are considered in Chapter 13, and will not, therefore be considered here. The reactions inherent in galvanising tend to ensure a thick and even coating but guides to the inspection of both galvanising and zinc spraying are available[2].

Characteristics of Zinc Coatings

In practice the thickest zinc coatings can be obtained by hot-dip galvanising or spraying, while sherardising can be normally expected to produce somewhat lighter coatings. Table 14.6 compares the essential aspects of each coating.

The thickness of hot-dip galvanised coatings depends on the nature of the steel and the dipping conditions. It can be controlled to a certain extent in practice. Heavier coatings are obtained on grit-blasted steel or on steels of high silicon content, and at higher operating temperatures and longer dipping times. In strip galvanising, aluminium is deliberately added to the bath to suppress the action between molten zinc and steel, with the result that lighter coatings are produced compared with those on fabricated assemblies galvanised after manufacture. Mechanical wiping of the surface on withdrawal from the bath, as employed in wire or sheet galvanising, also causes a reduction of coating thickness.

Table 14.6 Comparison of zinc coatings

Characteristics of coating	Hot-dip galvanising	Metal spraying	Plating	Sherardising	Zinc dust painting
1. Process considerations	Parts up to 29 m long and fabrications of 18 m × 2 m × 5 m can be treated. Care required at design stage for best results. Continuous galvanised wire and strip up to 1·4 m wide) in UK	No size or shape limitations. Access difficulties may limit its application, e.g. inside of tubes. Best method for apply very thick coatings. Little heating of the steel	Size of bath available. Process normally used for simple, fairly small components suitable for barrel plating or for continuous sheet and wire. No heating involved	Batch processing is mainly suitable for fairly small complex components. Semi-continuous process for rods, etc.	Can be brush, spray or dip applied on site when necessary. No heating involved. Performance varies with media used
2. Economics	Generally the most economic method of applying metallic zinc coatings 20–200 μm thick	Most economic for work with high weight to area ratio. Uneconomic on open mesh	Used where a very thin zinc coating is sufficient. Thick coatings are expensive	More expensive than galvanising for equivalent thicknesses. Generally used when control of tolerances is more important than thickness of coating	Low overheads but high labour element in total cost as with all paints. Thixotropic coatings reduce number of coats and hence labour costs
3. Adhesion	Process produces iron–zinc alloy layers, over-coated with zinc; thus coating integral with steel	Good mechanical interlocking provided the abrasive grit-blasting pretreatment is done correctly	Good, comparable with other electroplated coatings	Good—the diffused iron–zinc alloy coating provides a chemical bond	Good—abrasive grit blasting preparation of the steel gives best results
4. Thickness and uniformity	Normally about 75–125 μm on products, 25 μm on sheet. Coatings up to 250 μm on products by prior grit-blasting. Very uniform—any discontinuities due to poor preparation of the steel are readily visible as 'black spots'	Thickness variable at will, generally 100–200 μm but coatings up to 250 μm or more can be applied. Uniformity depends on operator skill. Coatings are porous but pores soon fill with zinc corrosion products; thereafter impermeable	Thickness variable at will generally 2–25 μm. Thicker layers are possible but generally uneconomic. Uniform coating within limitations of 'throwing power' of bath. Pores not a problem as exposed steel protected by adjacent zinc	Usually about 12–40 μm closely controlled. Thicker coatings also possible. Continuous and very uniform even on threaded and irregular parts	Up to 40 μm of paint (and more with special formulations) can be applied in one coat. Good uniformity—any pores fill with reaction products

Table 14.6—*(continued)*

Characteristics of coating	Hot-dip galvanising	Metal spraying	Plating	Sherardising	Zinc dust painting
5. Formability and joinability	Applied to finished articles, not formable: alloy layer is abrasion resistant but brittle. Sheet has little or no alloy layer and is readily formed. All coatings can be arc or resistance welded	When applied to finished articles, forming not required. Can weld through thin coating if necessary but preferable to mask edges to be welded and spray these afterwards	Electrogalvanised steel has excellent formability and can be resistance spot welded. Small components are usually finished before plating	Applied to finished articles: forming not required. Excellent abrasion resistance	Abrasion resistance better than conventional paints. Painted sheet can be formed and resistance welded with little damage
6. British standards	General work: BS 729 Continuous galvanised: plain sheet: BS 2989 Corrugated sheet: BS 3083 Wire: BS 443 Tubes: BS 1387	BS 2569: Part 1	BS 1706 Threaded components: BS 3382	BS 4921	BS 4652 for the paint
7. Extra treatments	Conversion coatings; chromates used to prevent storage stains on new sheet; phosphates used as base for paints, but some paints (e.g. calcium plumbate) applied directly to new coating. Weathered coatings painted for long service	Frequently used when new base for paints on long-life structures. Alternatively a sealer can be used	Conversion coatings, e.g. chromates used to prevent 'wet storage' stain. Frequently used as a base for paints	Can be painted if required	Can be used alone, or as primer under other paints

Properties and Nature of the Coatings

The actual structure and composition of zinc coatings depends upon the method of deposition. Zinc coatings produced by hot-dip galvanising and sherardising consist partly or wholly of iron–zinc alloys. Sprayed and plated zinc coatings contain no alloys, plated coatings consisting essentially of pure zinc. The characteristic properties of each type of coating are discussed below.

Hot-dip galvanised coatings (see Section 13.2) Here the coating is not uniform in composition but is made up of layers of zinc–iron alloys becoming progressively richer in zinc towards the surface, so that the actual surface layer is composed of more or less pure zinc. Because of this alloy formation there is a strong bond between the coating and the steel. The alloy layers are harder than mild steel.

The nature and thickness of the alloy layers are influenced greatly by the composition of the underlying steel, and also by the galvanising conditions. Notably the presence of silicon in the steel encourages the formation of iron–zinc alloys and thereby leads to the formation of heavier coatings, and indeed a steel with a high silicon content is often used intentionally when very heavy zinc coatings are required. In such cases the coating may consist entirely of iron–zinc alloys, and this is seen in the uniform grey spangle-free appearance of the galvanised surface obtained under these conditions.

The addition of up to 0.2% of aluminium to the galvanising bath, on the other hand, depresses the formation of alloys and produces lighter and more ductile coatings, which are more suitable for galvanised sheet since they render it more amenable to bending.

Sprayed zinc coatings Details of the method and the nature of the coatings are given in Section 13.4. In this method there is no alloy formation and the bond is primarily mechanical. Although porous, the coating is protective partly due to its sacrificial action and partly due to the zinc corrosion products which soon block up the pores, stifling further attack.

Zinc plating Plating with zinc normally gives a dull-grey matt finish, but lustrous deposits can be obtained by adding brighteners to the electrolyte. The coating is of uniform composition throughout, containing no alloy layer, and is united with the underlying surface by a metal/metal bond (Section 13.1). Plated coatings are very ductile and zinc-plated sheet can therefore be readily fabricated.

Sherardising This is another alloy-forming process, a typical coating containing alloys with 8 or 9% of iron. Like galvanised coatings, the deposits are very hard (Section 13.3).

Relative Advantages of the Coating Methods

Each coating method has its own particular advantages, which are really the decisive factors in determining which one is used for a given purpose. Consideration must be given to the complexity and size of the work, the corrosion resistance, and hence the coating thickness needed, and the quality of finish required.

Hot-dip galvanising produces a thick coating which thoroughly covers the work, sealing all edges, rivets, seams and welds when fabricated articles are treated. The size of the article which can be treated is limited to a certain extent by the size of the galvanising tank, but the technique of double dipping, i.e. dipping first one end and then the other, makes it possible to treat surprisingly large items, up to 30 m long, successfully. Hot-dip galvanising is the most widely used method for coating with zinc.

Zinc spraying possesses the important advantage that, since the equipment is essentially portable, it can be applied on site to either small or large structures. Thick coatings can be applied where desired, and it is possible to ensure that welds, edges, seams and rivets on finished articles receive sufficient coverage. It is not normally a suitable process for coating the inside of cavities or for coating open structures, such as wire meshes, since a large amount of zinc would be wasted. In hand spraying the uniformity of the coating depends largely on the skill of the operator.

Zinc coatings produced by plating have the advantages that the thickness can be accurately controlled according to the protection desired.

The acid zinc sulphate bath is used for applying coatings to uniform sections, e.g. box strapping, wire, strip, etc. The plating rates in these solutions can be very high. The throwing power of this bath is poor, and for more intricate shapes and where appearance is important the cyanide bath is preferred. Bright deposits can be produced from the latter by the use of either addition agents or bright dips.

Sherardising is distinguished particularly by the uniformity of the coatings which it produces. This makes it an ideal process for work, such as screw threads, where close tolerances are required and where complex or recessed parts are involved, since the inside surfaces of pipes and other hollow articles receive coatings comparable with those on the outside. The coating is very hard and offers a high resistance to abrasion. The maximum size of the articles which can be sherardised is limited by the size of the drums. In general, sherardising is best suited to the treatment of small castings, forgings and pressings, and fixings, such as wood screws, nuts, bolts, washers, chains, springs, etc.

The outstanding virtue of zinc-rich paints is simplicity in application. No special equipment is required and the operation can, of course, be carried out on site, large or small structures being equally suitable for treatment. While there is some evidence that the zinc-rich paints will reduce iron oxides remaining on the steel surface, proper surface preparation is as important here as with traditional paints if the best results are to be achieved. The main use of zinc-rich paints is to protect structural steel-work, ships' hulls, and vulnerable parts of car bodies, and to repair damage to other zinc coatings.

Corrosion Resistance of Zinc Coatings

There are two main reasons why zinc is chosen as a protective coating for iron and steel. The first is the natural resistance of zinc itself against corrosion in most atmospheric conditions, and the second is the fact that zinc is electronegative to iron and can protect it sacrificially*.

*The reversal of polarity of Zn and Fe which can occur in certain circumstances is discussed in Section 1.3.

The corrosion resistance of zinc is discussed in Section 4.7, and it is only necessary here to say that zinc is protected against further attack by a film of corrosion products. It is remarkably resistant to atmospheric corrosion except perhaps in the most heavily contaminated industrial areas, and even there its use as a protective coating is still a sound practical and economic proposition. The value of zinc coatings as a basis for painting under very aggressive conditions has been clearly demonstrated.

The natural corrosion resistance of zinc is, therefore, its most important property in relation to zinc coatings. The electrochemical property becomes important when the zinc coating is damaged in any way to expose the steel, when sacrificial corrosion of the zinc occurs and the steel is thereby protected. Moreover, the corrosion product of the zinc normally fills the break in the coating and prevents or retards further corrosion of the exposed steel.

Life of Zinc Coatings in the Atmosphere

As the protective value of the zinc coating depends largely on the corrosion resistance of zinc, the life of a coating is governed almost entirely by its thickness and by the severity of the corrosive conditions to which it is exposed. Extensive tests and field trials which have been carried out have shown that the life of a zinc coating is roughly proportional to its thickness in any particular environment[3] and is independent of the method of application.

The corrosion rates and lives of zinc coatings in UK atmospheres are given in Table 14.7. These are based on practical experience as well as exposure trials. The figures should be taken only as a guide because of the difficulty of defining atmospheres in a word or two (indeed there is now a tendency for research workers to define the corrosivity of an atmosphere in terms of the corrosion rate of zinc) because of unpredictable local variations from place to place and time to time. For example, moorland which is frequently covered with acid-laden mist can be very corrosive.

Corrosion tests have also shown that there is a difference in the rates of corrosion throughout the year. This is partly because the sulphur content of the air is greater in winter than summer and partly because more of the

Table 14.7 Typical corrosion rates and lives of zinc coatings in the UK

Atmosphere	Corrosion rate		Life of coating with average thickness (years)‖			
	$gm^{-2}y^{-1}$	$\mu m\,y^{-1}$	200 μm*	100 μm†	25 μm‡	5 μm§
Rural	14	2	50–150	25–75	6–20	1–3
Urban	40	5	30–50	15–25	4–6	≃1
Marine	40	5	30–50	15–25	4–6	≃1
Industrial	80	10	10–30	5–15	1–4	0·25–1

*Can be produced either by grit-blasting before galvanising or by zinc spraying.
†Typical thickness of coating on galvanised or zinc-sprayed structural steel.
‡Typical thickness of coating on galvanised sheet or sherardised components.
§Typical thickness of zinc plating.
‖The lives given are *additional* to the life of the unprotected steel.

Table 14.8 Average loss of zinc in two years and steel/zinc ratio for 45 test sites[4]

Location	Described by authors as:	2-year test: zinc lost per year (μm)	Steel/zinc loss ratio (by weight)	Life of 100 μm zinc coating (calc.) without maintenance (years)
Norman Wells, N.W.T., Canada	Rural	0·2	10·3	500
Phoenix, Ariz., USA	Rural	0·3	17·0	300
Saskatoon, Sask., Canada	Rural	0·3	21·0	300
Esquimalt, Vancouver Is., Canada	Rural/ marine	0·5	31·0	200
Fort Amidor Pier, Panama C.Z.	Marine	0·7	25·2	150
Melbourne, Aust.	Industrial	0·8	37·4	125
Ottawa, Canada	Urban	1·1	19·5	90
Miraflores, Panama C.Z.	Marine	1·2	41·8	80
Cape Kennedy, 0·8 km from ocean, USA	Marine	1·2	84·0	80
State College, Pa., USA	Rural	1·2	22·0	80
Morenci, Mich., USA	Rural	1·2	18·0	80
Middletown, Ohio, USA	Semi-industrial	1·3	26·0	75
Potter County, Pa., USA	Rural	1·3	18·3	75
Bethlehem, Pa., USA	Industrial	1·3	32·4	75
Detroit, Mich., USA	Industrial	1·4	12·2	70
Manila, Philippine Is.	Marine	1·5	39·8	65
Point Reyes, Calif., USA	Marine	1·6	364·0	60
Halifax (York Redoubt) N.S., Canada	Urban	1·6	18·5	60
Durham, N.H., USA	Rural	1·6	19·0	60
Trail, B.C., Canada	Industrial	1·6	24·2	60
South Bend, Pa., USA	Semi-rural	1·8	20·8	55
East Chicago, Ind., USA	Industrial	1·8	52·1	55
Brazos River, Texas, USA	Industrial/ marine	1·9	56·0	50
Monroeville, Pa., USA	Semi-industrial	2·0	28·4	50
Daytona Beach, Fla., USA	Marine	2·1	164·0	45
Kure Beach, N.C. 240-m lot, USA	Marine	2·1	80·0	45
Columbus, Ohio	Urban	2·2	16·8	45
Montreal, P.Q., Canada	Urban	2·5	10·9	40
Pilsea Island, Hants., UK	Industrial/ marine	2·5	21·6	40
Waterbury, Conn., USA	Industrial	2·6	9·8	40
Pittsburgh, Pa., USA	Industrial	2·7	13·1	35
Limon Bay, Panama C.Z.	Marine	2·7	25·9	35
Cleveland, Ohio, USA	Industrial	2·8	15·7	35
Dungeness, UK	Marine	3·7	148·0	27
Newark, N.J., USA	Industrial	3·8	15·1	25
Cape Kennedy, 55 m from ocean 9 m up, USA	Marine	4·1	45·5	24
ditto, ground level	Marine	4·3	117·0	23
ditto, 18 m up	Marine	4·5	33·0	22
Bayonne, N.J., USA	Industrial	4·9	17·9	20
Battersea, UK	Industrial	5·8	20·0	17
Kure Beach, N.C., 24 m lot, USA	Marine	6·5	93·0	15
London (Stratford), UK	Industrial	7·1	17·8	14
Halifax (Federal Bldg.), N. S., Canada	Industrial	7·6	17·0	13
Widnes, UK	Industrial	10·5	39·0	9
Galeta Point Beach, Panama C.Z.	Marine	15·9	49·4	6

zinc corrosion products are dissolved under the wetter winter conditions. Thus if unpainted zinc coatings are first exposed to the atmosphere in spring or early summer a more protective film will be formed.

Detailed test results for a 2-year exposure period[4] are given in Table 14.8. It should be remembered that test sites are sometimes chosen because they are believed to represent particularly corrosive examples of the type of atmosphere being studied. The ratios of steel:zinc losses are particularly interesting. It shows that zinc is far less affected than steel by many chloride-containing atmospheres. On a global view the single word description of the site is often misleading; particularly, it gives no indication of the times each year that objects remain wet, which varies considerably from country to country and also within countries.

An extensive compilation of atmospheric exposure test data on zinc is now available[5] and complements the slightly earlier critical study by Schikorr[6].

Water Zinc-coated steel, like zinc, behaves less favourably in distilled and soft waters than in hard waters, where the scale-forming ability of the hardness salts provides considerable protection. Hot-dip galvanised tanks, cisterns and pipes are very widely used for storing and carrying domestic water supplies throughout the world, and as a rule such equipment gives long and trouble-free service (Section 10.3).

Sea-water The protective properties of zinc coatings in sea-water have been shown to be very good, and zinc is widely used as a coating metal in the shipbuilding industry and for protecting structural steel work on docks and piers, etc. In BISRA tests at Gosport[7], specimens of steel coated with aluminium, cadmium, lead, tin and zinc were immersed for two years. In this time all but the zinc-coated specimens had failed. The zinc-coated specimens were then transferred to Emsworth and immersed for a further four years—a total of six years—before the coatings ceased to give complete protection. The coating on these specimens was about 900 g/m^2, indicating a rate of attack of about 20 µm/y in this sea-water. Other tests[5, 8] show corrosion rates of $10–25 \text{ µm}$.

Conditions within a few hundred metres of the surf line on beaches are intermediate between total immersion in sea-water and normal exposure to a marine atmosphere. High corrosion rates can occur on some tropical surf beaches where the metal remains wet and where inhibiting magnesium salts are not present in the sea-water.

Soil Galvanised pipe is frequently used for underground water services. Table 14.9 gives results of tests[9] carried out with galvanised pipes and plates buried at different sites.

The specimens were removed after five years, when the only ones that had failed were some plates buried in made-up ground, consisting of ashes, at Corby and one pipe at Benfleet. At Corby no galvanised pipes were exposed and most of the coatings on the plates had corroded away. For this reason no figures are recorded for Corby in Table 14.9. The high rate of corrosion at Benfleet was attributed to the fact that the specimens were below the soil-water level for about half their life as the tide rose and fell.

Similar tests have been carried out in the United States[10]; in these the maximum depth of pitting was also measured. Except in the most corrosive

Table 14.9 Loss of coating thickness of galvanised specimens after five years in various soils

Location	Soil conditions	Galvanised pipes		Galvanised plates		Uncoated steel loss in thickness (μm)
		Initial thickness of coating (μm)	Loss of coating thickness (μm)	Initial thickness of coating (μm)	Loss of coating thickness (μm)	
Benfleet	Alluvium or reclaimed salt marsh	82	47	90	52	200
Gotham	Keuper Marl (gypsum)	77	17	90	17	50
Pitsea	London clay	77	17	90	17	160
Rothamstead	Clay with glacial flints	82	13	95	13	120
Corby	Made up ground (ashes)	—	—	—	*	300

*Most of coating on plates corroded away.

soils the maximum depth of pitting in steel specimens exposed for about $12\frac{1}{2}$ years was more than 11 times that in zinc-coated specimens, even though the ratio of the rates of corrosion was only about half that figure. This resistance to pitting, combined with the fact that rusting appears to start only when nearly all the zinc and zinc–iron alloy layers have corroded away, reduces the risk of premature failure in galvanised piping. The coatings on galvanised specimens remained virtually intact during exposure for $2\frac{1}{2}$ years in about half the 15 soils in which they were buried. Their corrosion resistance was most marked in alkaline soils. In clays and loams, where little or no organic material was present, a $600\,g/m^2$ coating could be expected to provide protection for 10 years or more.

Protective Systems Applied Subsequently to Zinc Coatings

Chromating Chromating is considered in Section 16.3. The chromate film on zinc is adherent and can be drab, yellow-green or colourless in appearance; the colour varies considerably with the method of application. It retards 'white rust', the white deposit which sometimes forms on fresh zinc surfaces which are kept under humid conditions (see Section 4.7). A chromate film is damaged by heat and if used as a basis for paint adhesion, should not be stoved above 150°C, nor for longer than 1 h.

Painting In mildly corrosive conditions zinc coatings will probably have a life longer than that expected of the coated article, and no further treatment of the coating will be necessary. When, however, the coating is subjected to a more strongly corrosive environment one or more coats of paint can to advantage be applied over the zinc.

Paint films used in conjunction with zinc coatings give systems whose lives are longer than the sum of the lives of the coatings used independently.

Paint applied to a suitably prepared zinc coating will last longer than would be the case if it were applied direct to iron or steel, and the need for repainting thus becomes less frequent. With hot-dip galvanised or zinc-plated coatings, however, it is necessary either to use special primers or to prepare the surface before painting. This is primarily because most oil-based paints react with the unprepared zinc surface to form zinc soaps resulting in poor adhesion.

Weathered galvanised steel surfaces suffice to give good adhesion for many paint systems, but where new galvanised or zinc-plated steel is to be painted it is necessary either to convert the surface into an adherent phosphate or chromate coating or to use specially developed primers[11]. Many commercial phosphating processes are available, but all consist essentially of an etch in a phosphoric acid solution containing zinc salts and certain accelerators. These treatments produce uniform, fine-grained and strongly adherent phosphate films on the surface of the work. Many chromate finishes also give a satisfactory base for painting (see Section 16.3).

Etch primers are widely used. They are mostly based on polyvinyl butyral and contain chromates and phosphoric acid. They are said to act both as primers and as etching solutions because it is believed that the chromates and phosphoric acid form an inorganic film, which provides adhesion, while oxidised polyvinyl butyral provides an organic film. Of the primers developed for direct application to new galvanised steel, the best known are those based on calcium-orthoplumbate pigment. Metallic lead paints and some zinc-dust paints have also been successfully applied and encouraging results are now reported with zinc-phosphate pigmented paints.

Applications of Zinc Coatings

Zinc coatings are successfully used in a very wide field to protect iron and steel goods from corrosive attack.

The building trade is one of the largest users of zinc-coated steel. The frameworks of large modern buildings can be either galvanised or zinc-sprayed before erection, or sprayed on site. Where a framework is accessible, zinc-rich paints provide an excellent way of renovating old buildings. Apart from the structural aspect, galvanised sheeting is used for roofing, ventilation ducts and gutters, as well as for water tanks and cisterns, and galvanised pipe is used for public and domestic services.

In the Forth, Severn and most other suspension bridges, zinc coatings have an important function. The whole main structure is of steel and has been zinc-sprayed on the external surfaces, while the main cable and hanger ropes have been coated by continuous hot-dip galvanising. Case histories of galvanised multi-truss bridges cover more than 30 years.

Zinc-coated structures are used in pylons carrying electrical transmission lines, in masts for radio, television and radar aerials, and for supports for overhead wires on electrified railways.

An interesting new use of galvanised steel is for reinforcement of concrete. This reduces the risk of spalling and staining or can enable the depth of cover to be reduced leading to slimmer structures of lower cost.

Economics

All coatings cost money and a true appraisal of both the initial and main-tenance costs is essential when specifying the protective scheme for a new structure. An example showing the comparative costs of galvanising and painting is given in Section 10.1.

A. R. L. CHIVERS
F. C. PORTER

REFERENCES

1. Technical Notes on Zinc. Separate leaflets entitled *Zinc Coatings*, *Galvanising*, *Zinc Spraying*, *Sherardising*, *Zinc Plating* and *Zinc Dust* are available from Zinc Development Association, 34, Berkeley Square, London, W1X 6AJ
2. *General Galvanizing: User's Guide*, Galvanizers Association (1967) and *Inspection of Zinc Sprayed Coatings*, Z.D.A. (1968)
3. Hudson, J. C., *Sixth Report of the Corrosion Committee*, I.S.I. Special Publication No. 66 (1959)
4. *Metal Corrosion in the Atmosphere*, A.S.T.M., STP435 (1967)
5. Slunder, C. J. and Boyd, W. K., *Zinc: Its Corrosion Resistance*, Z.D.A. (1971)
6. *Atmospheric Corrosion Resistance of Zinc*, Z.D.A.
7. Hudson, J. C. and Banfield, T. A., *J. Iron St. Inst.*, **154**, 229 (1946)
8. Wiederholt, W., *Korrosionverhalten von Zink*, Metall-Verlag GmbH, Berlin (translation available on loan from Z.D.A. Library)
9. Hudson, J. C., *Corrosion of Buried Metals*, I.S.I. Special Publication No. 45 (1952)
10. Denison, I. A. and Romanoff, M., *J. Res. Nat. Bur. Stand.*, **49**, 299 (1952)
11. Porter, F. C., *Br. Corr. J.*, **4**, 179–186 (1969). (Reprint available from Z.D.A.)

14.5 Tin and Tin Alloy Coatings

Tin Coatings

Methods of application Tin coatings are applied by hot-dipping, electro-deposition, spraying and chemical replacement. A variant of hot-dipping called *wiping*, in which the tin is applied either solid or melted to the fluxed and heated surface and is wiped over it, is also used for local application, e.g. to one face of a sheet or vessel. The hot-dipped and wiped coatings are bright; electrodeposited coatings may be matt or bright.

The electrolytes used for electrodeposition of matt coatings[1] for general purposes are: (*a*) an alkaline bath containing tin as stannate; (*b*) an acid bath containing stannous sulphate, free sulphuric acid, cresol-sulphonic acid, gelatin and β naphthol; and (*c*) an acid fluoroborate-bath containing organic addition agents. For high-speed plating on rapidly moving strip in the production of tinplate, the baths used are the 'Ferrostan Bath' based on stannous sulphate and phenolsulphonic acid, the 'Halogen Bath' based on chloride and fluoride (both with appropriate addition agents) and to a lesser extent, the alkaline stannate bath and the acid fluoroborate bath.

Bright coatings are deposited from acid stannous sulphate baths containing combinations of organic addition agents. The electrodeposited matt coatings may be brightened by momentary fusion. This process of flow-brightening or flow-melting is achieved with most of the electrolytic tinplate production by conductive or inductive heating; for manufactured articles, it is usually carried out by immersion in a suitable hot oil[2].

The hot-dipped coatings[3] are distinct from the others in having practical thickness limits and in possessing an inner layer of intermetallic compound, usually described as the *alloy layer*. The flow-melted electrodeposited coatings also have an alloy layer, which is somewhat thinner than that obtained in hot dipping.

Coatings of tin produced from tin-containing aqueous solutions by chemical replacement may be used to provide special surface properties such as appearance or low friction, but protect from corrosion only in non-aggressive environments. Copper and brass may be tinned in alkaline cyanide solutions or in acid solutions containing organic addition agents such as thiourea. Steel may be first coated with copper and then treated as copper,

14:47

or it may be tinned in acid tin salt solutions with or without contact with zinc. Aluminium alloys may be tinned by immersion in alkaline stannate solutions.

Articles of steel, copper or brass which require a thicker coating than is possible by chemical replacement, and which are difficult to tin by normal electrodeposition, may be coated by immersion in alkaline sodium stannate solution in contact with aluminium suitably placed to act as anode.

Thickness of tin coatings　The thicknesses of the various types of tin coating are shown in Table 14.10.

Table 14.10　Thicknesses of tin coatings

Hot dipped	1·5–25 μm
Electrolytic tinplate	0·4–2 μm
Wiped coatings	1–12 μm
Electrodeposits other than tinplate	2·5–75 μm
Flow-melted electrodeposits	2·5–7·5 μm
Sprayed coatings	75–350 μm
Chemical replacement coatings	trace–2.5 μm

Properties of tin coatings　When the choice of coating is not governed by the size and geometry of the article to be coated, it depends upon appearance, the thickness required, and the degree of porosity which can be tolerated. Bright coatings, as produced by hot dipping, flow-melting or bright electrodeposition, have the advantage of smoothness, good appearance and resistance to finger-marking. The presence of an alloy layer in hot-dipped and flow-melted coatings also confers some advantage in the making of soldered joints. On the other hand, with hot-dipped coatings it is rarely possible to ensure absence of coating porosity, whereas electrodeposition can build the coating up to the thickness, above 25 μm, at which pores are unlikely to penetrate the coating.

Sprayed coatings have structures in which fine pores thread tortuous paths through the deposit, and it is necessary to apply a coating thickness of about 350 μm if all these paths are to be closed. Scratch-brushing of the deposit, however, makes it possible to consolidate the surface and to achieve adequate continuity in thinner deposits, e.g. 200 μm.

Tin coatings are ductile and are able to contribute a lubricating effect in the deep drawing of steel. The presence of the thin alloy layer in flow-melted tinplate coatings does not impair this property appreciably but bright electrodeposited coatings may be less ductile than others.

Sometimes a spontaneous outgrowth of metal filaments about 1 μm diameter, commonly called *whiskers*, occurs on tin coatings in a time after application which may vary from days to several years. This growth does not affect the protective quality of coatings but the whiskers are able to short-circuit compact electronic equipment. The character of the substrate is influential and tin coatings on brass should always be undercoated, e.g. with nickel or copper. The introduction of some impurities, e.g. 1% lead, into the tin coating is some safeguard. Hot-dipped or flow-melted coatings are rarely affected[49].

Corrosion Resistance of Tin Coatings

General considerations Influential factors in the behaviour of tin coatings are the variation according to environment in the relative polarity of coating and substrate, the nature of any intermetallic compound layers formed between coating and substrate and the extremely low rate of corrosion of tin in alkaline and mildly acid media in the absence of cathode depolarisers. The depression of the corroding potential of tin, when the tin ion concentration in solution is reduced by the formation of complex ions, has been referred to previously (Section 4.6). Iron may also be complexed and the potential of iron is affected by the presence of tin ions in solution. The extent of the potential shifts[4] depends on the complexing agents present, the solution concentration and pH. The electrochemical relationship of tin and iron is therefore a complicated one, but for practical purposes, tin can be regarded as being anodic to iron in contact with such products as fruit juices, meat and meat derivatives and milk, in solutions of citric, tartaric, oxalic and malic acids and their salts, and in alkaline solutions. In solutions of inorganic salts, natural waters or atmospheric water, tin is cathodic to iron.

Supplementary protection can be given to tin coatings either by passivation treatments or by organic finishes. Passivation in chromate solutions gives some protection to the steel exposed at the base of pores as well as to the tin coating. Electrolytic tinplate is passivated on the production line by rapid passage through acid solutions, usually dichromate, with applied cathodic current. Similar treatments may be employed on other forms of tinned cathodic metal, and a process of immersion in hot alkaline chromate[5], which combines cleaning and passivation, is useful for treating metal coated by oil or other contaminants in manufacturing operations.

Tinplate for containers and closures is often decorated by colour printing and protected by clear lacquers. No surface preparation is carried out and difficulties with wetting and adhesion, sometimes associated with the character of the oxide layers on the surface, are rare.

The corrosion of tinned steel

Atmospheric corrosion During full exposure to the weather, some rust at pores in the coating soon appears. In coatings of thicknesses less than about 5 μm the rust spreads out from the pores and in due course the whole surface becomes covered by rust. With thicker coatings, this spread of rust does not occur. In industrial atmospheres, penetration of the steel may cease after a few weeks, the surface becoming covered by a growing grey layer of tin corrosion product with faint rust stains; tin coatings of upwards of 12 μm will outlast zinc coatings of comparable thickness[6, 7]. In marine atmospheres, however, attack at pores persists even with the thicker coatings and pits are formed.

In most of its uses, e.g. the external surfaces of tinplate cans, tinned steel has only to resist condensed moisture. In the absence of pollution of the atmosphere by unusually large amounts of sulphur dioxide or chlorides, or of several days of continuous wetting, tinned steel remains unrusted; even the thin porous coatings on the common grades of tinplate remain bright and unmarked over the periods involved in the commercial handling and domestic storage of cans, and the domestic use of kitchenware. When wetting persists for long periods, especially if pools of water collect, rusting at pores begins. This situation can easily arise in the holds of ships in transit through the tropics unless proper precautions are taken; shipment in large sealed containers seems likely to avoid most of the trouble[8]. The conditions needed to ensure complete absence of pore rust are similar to those needed to preserve uncoated steel, although with the tinned steel, rust-promoting conditions can be tolerated for a much longer period without the general appearance of the metal being spoilt.

Condensed moisture rarely produces serious pitting of the steel at pore sites, but for many purposes maintenance of appearance is important. The change in aspect which takes place on rusting is much influenced by the degree of porosity of the coating, which is usually dependent on coating thickness. The thinnest coatings of electrolytic tinplate of 5g/m^2 of sheet (equivalent to a coating thickness of 0.4 μm) will develop a continuous rust coating in conditions where a hot-dipped coating of 30 g/m^2 will show only inconspicuous rust spots. A coating heavier still may show no visible change.

The oil film present on both types of tinplate and on newly hot-dipped tinware has a slight protective value. The passivation processes have much more effect but this is unlikely to compensate for a substantial reduction in coating weight. The effects of oil and passivation on the outside of tinplate cans may be reduced during can manufacture, filling and sterilisation.

The resistance of tin to organic acid vapours emanating from wooden cases and from some insulating materials and paints gives tin an advantage over zinc and cadmium as a coating for equipment likely to be exposed to these vapours. There is, however, some risk of rust-spotting at pores in tin coatings; one method of trying to secure immunity of the coating from organic vapour corrosion and of the pores from rusting, consists in plating a layer of tin over a layer of cadmium or of zinc.

Immersion in aqueous media open to air   Solutions in which tin is cathodic to steel cause corrosion at pores, with the possibility of serious pitting in electrolytes of high conductivity. Porous coatings may give satisfactory service when the corrosive medium deposits protective scale, as in hard waters, or when use is intermittent and is followed by cleaning, as for kitchen equipment, but otherwise coatings electrodeposited or sprayed to a sufficient thickness to be pore-free are usually required.

Sometimes it is possible to add corrosion inhibitors to an aqueous product that is to remain in contact with tinned steel. The normal inhibitors used for protecting steel, e.g. benzoate, nitrite, chromate, etc. are suitable, provided that they are compatible with the product and that the pH is not raised above 10. In a closed container with an air-space, such inhibitors will not protect the zone above the water-line, and possibly not the water-line zone itself, against condensate. Volatile inhibitors have been used to give protection to these areas.

Fruit juices, meat products, milk and milk products, fish and most vegetables, in which tin is likely to be anodic to steel, can be handled open to the air in tinned steel vessels. Some corrosion of the tin occurs at rates similar to those found for pure tin and in due course retinning may be necessary. The alloy layer in hot-dipped tin coatings is cathodic to both tin and steel and, under aerated conditions may stimulate the corrosion of both metals, but this effect appears to be unimportant in practice.

Tin is anodic to steel in alkaline solutions, the corrosion rate for a continuous coating being similar to that of pure tin, and tinned articles that are washed in aerated alkaline detergents slowly lose their coating.

Tinplate containers The behaviour of tinplate is basically similar to that of other types of tinned steel, but performance requirements of tinplate containers are special. Containers are used in several forms:

1. Cans with replaceable closures for such products as dry foodstuffs, pharmaceuticals, tobacco, solvents, liquid fuels and paint. These usually contain an appreciable amount of oxygen. Tinplate closures for bottles and jars made of non-metallic materials may also be considered in this category.
2. Sterilised sealed cans of foodstuffs, including fruit, vegetables, meat, fish and milk, which should contain only residual traces of oxygen.
3. Cans for beer and soft drinks.
4. Aerosol cans which may contain a propellant together with products such as paints, cleaners, cosmetic preparations and foodstuffs. These may also contain some oxygen.

With all of these containers, both the can and its contents must reach the user in a visibly good condition. Cans must therefore resist external rusting, and methods of achieving this (e.g. adequate coating thickness, passivation treatment, attention to packaging and storage conditions and, if need be, lacquering) have already been mentioned. In other respects, both the requirements and the methods of achieving them may differ for the several classes.

For categories 1 and 4, the relative polarity of tin and steel may be in either direction, depending on the product contained, but, more commonly, steel is anodic to tin and sufficient oxygen is present to make perforation by corrosion a possibility with water-containing products. Small quantities of water in nominally non-aqueous products can be seriously damaging because they are able to use all the oxygen present in the contents. Change of formulation, including addition of corrosion inhibitors, is possible for many non-food products and protection by lacquering is a generally available means of protection.

Containers of foodstuffs should not be unduly stained or etched and must not be perforated or allowed to become distended by pressure due to evolution of hydrogen, and the contents must not suffer unacceptable changes of colour or flavour. Long storage periods, e.g. two years, may be required.

Yellow-purple staining of can interiors may be produced by adherent films of tin sulphides formed by S^{2-} and HS^- compounds derived from proteins in meat and vegetable products. It may be prevented by suitable passivation treatment of the tinplate[9] or by the use of appropriate lacquers. Loose iron sulphide, occasionally formed by sulphur-containing products at pHs above about 5·5 in the headspace of a can where there is some residual oxygen, is more objectionable and is not prevented by passivation or by normal lacquer since it occurs at breaks in the coating. Careful control of can-making and canning procedures is the best safeguard.

Discolouration of products inside cans may follow the reduction of colouring matters or the formation of new coloured compounds with tin or iron. This is a problem with strongly coloured fruits and the remedy is to use fully lacquered cans. Other than this effect, dissolved tin has no objectionable action on the quality of canned products, but very small amounts of dissolved iron have adverse effects on flavour.

Except in special circumstances, the anodic relation of tin to steel and the inhibition of steel corrosion by dissolved tin[10-12] protects the unlacquered tinplate can of food from risk of perforation or from taking up appreciable quantities of iron. The main hazards are excessive dissolution of tin, which may impair the appearance of the can and breach food regulations, and evolution of hydrogen which may distend the can and make it an unsalable 'hydrogen swell'. The amount of hydrogen collecting in a can to produce a 'swell' is usually roughly proportional to the amount of iron dissolved, and the high rate of hydrogen evolution responsible for swells seems to arise from self-corrosion of the steel when protection by tin has been lost, and not from the combination of tin anode with steel cathode[13].

In general, tin dissolution inside an unlacquered can has a high but diminishing initial rate followed by a steady slow rate[14]. The initial phase is associated with the reduction of cathodic depolarisers, including residual oxygen, and its duration and the corrosion rate reached depend on the nature of the product and on canning technology. In the second phase the cathode reaction is hydrogen ion reduction and the slow rate of tin dissolution, often equivalent to corrosion currents of the order of 10^{-9} A/cm^2, is due to the scarcity of effective cathodes. The area of steel exposed at pores and scratches may be expected to have some influence on the corrosion rate, and small grain size of the tin coating has been considered to be associated with high rates.

Many compounds capable of acting as cathodic depolarisers are naturally present in foodstuffs; they vary in character from product to product and, even in the same product, may vary in amount under such influences as season of growth, harvesting conditions[13] and sterilising procedures[15, 16]. The reduction of colouring matters in fruit has already been mentioned; other organic compounds in fruit and vegetables may be reduced and, in fish, trimethylamine oxide is a known large stimulator of corrosion. Inorganic nitrate, which is reduced to ammonia, is a most damaging promoter of corrosion in many vegetables and in fruit at pH values below 5·5[17]. If cathodic depolarisers are present in amounts sufficient to promote dissolution of a substantial amount of tin coating then the best means of obtaining satisfactory can appearance and shelf-life is to use lacquered tinplate. Passivation films are not a reliable means of preventing etching of the tin coating. In the more acid media they are removed wholly and, in some slightly corrosive products such as milk, the films break down locally where the surface has been slightly damaged in can manufacture and unsightly local corrosion then occurs.

Although lacquering is used increasingly for can interiors, there are advantages in cost and in preserving the flavour and colour of some products for the use of plain tinplate. With plain cans, deferment of serious hydrogen evolution can be obtained by increasing the thickness of the tin coating but the preferred method is to control characteristics of the coating and steel base in manufacture, checking achievement by suitable tests. Control measures in use are:

1. To limit the content in the steel of phosphorus, sulphur and 'tramp' elements such as copper, nickel and chromium[18].
2. To avoid the slight oxidation of the steel surface during the bright-annealing process that precedes tinning[19, 20]. The harmful, so called 'pickle-lag layer' so produced is detected by its resistance to 10N HCl.
4. To limit the total porosity of the coating, checking by the Iron Solution Value (ISV) test in which samples are immersed under standard conditions in a solution of sulphuric acid, hydrogen peroxide and ammonium thiocyanate, and the amount of iron dissolved is measured[19].
5. To ensure maximum continuity of the tin–iron compound layer between tin and steel. This layer is itself corrosion resistant and appears to act as a nearly inert screen limiting the area of steel exposed as tin is removed by corrosion. Its effectiveness is measured by the Alloy–Tin Couple (A.T.C.) test, in which the current flowing is measured between a sample of tinplate from which the unalloyed tin layer has been removed, and a relatively large tin electrode immersed in an anaerobic fruit juice[21–23].

Tinplate that meets the rigid specifications imposed by these controls is sometimes supplied as special quality material and undoubtedly can give improved shelf-life, particularly with citrus fruits. The A.T.C. value has probably more effect on shelf-life determined by hydrogen swell than any other factor.

A limited degree of control over the corrosivity of the product packed is possible. Minor pH adjustments may be helpful, especially in ensuring an anodic relation of tin to steel; corrosion promoters, like nitrate, sulphur and copper may be excluded from necessary additives, such as water and sugar, and from sprays applied to crops approaching harvest. The effect of sulphur compounds which may remain from spray residues is complex[24] but often includes reversal of the tin–iron polarity.

The use of lacquered tinplate does not automatically guarantee freedom from serious corrosion. The covering of the tin surface largely denies both corrosion inhibition by dissolved tin and cathodic protection to any steel exposed at coating discontinuities. Consequently, if discontinuities exist, perforation of cans and hydrogen swells are possible. Lacquer is applied to the tinplate before it is made into cans so that there is a risk, especially at seam areas, of scratching through or cracking of the coating. The dangers are minimised by suitable choice of lacquer and, for critical packs, by double coating and by applying a stripe of lacquer to the seam after can manufacture. The tin coating is not entirely without influence and coating thicknesses may still influence performance[25, 26]. In general, the coating properties found desirable for the plain can are not likely to be so important for the lacquered can, although steel quality remains an important factor.

Requirements for cans for beer and soft drinks differ from those for food cans in that (a) only low tin and iron contents can be tolerated in the product and (b) the anticipated shelf-lives are much shorter. Specialised lacquering techniques including striping the seams are used to give complete cover to the metal. For soft drinks it is sometimes possible to select colouring matters and acids least likely to give rise to corrosion troubles, and rapid methods of testing formulations have been devised[27]. Steel quality is also controlled by special tests.

Tinned copper and copper alloys   Copper itself has a fair corrosion resistance but traces of copper salts are often troublesome and a tin coating offers a convenient means of preventing their formation. Thus copper wire to receive rubber insulation is tinned to preserve the copper from sulphide tarnish and the rubber from copper-catalysed oxidation, and also to keep the wire easily solderable. Vessels to contain water or foodstuffs, including cooking vessels, water-heaters and heat exchangers, may all be tinned to avoid copper contamination accompanied by possible catalysis of the oxidation of such products as milk, and discolouration in the form of, for example, green stains in water and food.

Tin is anodic to copper in water supplies and in all solutions except those in which copper is dissolved as a complex, e.g. strong ammonia solutions. In water supplies the corrosion of the tin coating is, like that of tin, localised, but once the copper is reached it may spread slowly. This simple behaviour can, however, be considerably altered by the action of tin–copper compound layers in the coating. A hot-dipped or wiped coating will have from the outset a layer of Cu_6Sn_5 and perhaps also another, nearer the copper, of Cu_3Sn. Even an electrolytic coating will in time develop a compound layer by diffusion, at a rate depending on temperature; in boiling water the formation of the compound consumes about 2·5 μm of the coating per year.

The compounds are always cathodic to tin; in a wiped coating, which usually has streaks of compound in the surface, this has the effect of increasing the extent of local corrosion of tin with the production of unsightly black streaks. In addition, the compound Cu_6Sn_5 can be cathodic to copper; this behaviour is favoured by mild oxidising conditions, which ennoble the compound, and water movement, which anodically depolarises the copper. So long as some tin coating remains it will protect the copper, and a complete coating of compound is protective, but if all the coating is converted to compound and if there is a break in it which exposes copper, then pitting can occur. Adequately thick tin coatings and re-tinning of equipment when necessary are the proper safeguards.

The unfortunate action of the compound layer is observed only rarely, usually in hot water. In cooking vessels (domestic or industrial) the copper is protected satisfactorily at some sacrifice of tin, and occasional re-tinning ensures long service. In atmospheric corrosion the arrival of compounds at the surface of the coating results in some darkening and in loss of solderability.

With tin coatings on brass, the interdiffusion of coating and substrate brings zinc to the surface of the tin: the action can be rapid even with electro-deposited coatings. The effect of zinc in the surface layers is to reduce the resistance of the coating to dulling in humid atmospheres, and the layer of zinc corrosion product formed makes soldering more difficult. An intermediate layer of copper or nickel between brass and tin restrains this interdiffusion[28].

Since galvanic action between tin and aluminium alloys is slight, tin coatings are often applied to copper and copper alloys which are to be used in contact with these metals. Both direct galvanic action and corrosion resulting from copper dissolving and re-depositing on the light alloy are prevented by this means.

Applications of tin coatings The properties of tin coatings which are advantageous in most of their applications are: fair general resistance to corrosion except in strongly alkaline or acid environments, lack of colour, toxicity or catalytic activity of any corrosion product formed, and ease of soldering. The ready availability of coated steel sheets in the form of tinplate, which has a bright appearance and is easy to form and receptive to decoration and protective finishes, is also an advantage.

The main application is in the form of tinplate. Apart from its use for containers mentioned earlier, tinplate is made into domestic and industrial kitchen equipment, light engineering products and toys. For most of these purposes, coatings in the thickness range 0·4–2·5 μm, with or without organic finishes, are used. For returnable containers and more permanent articles such as fuel tanks and gas-meter cases, heavier coatings of up to 15 μm may be necessary.

Hot-dipped and electrolytic coatings are applied to vessels and equipment made of steel, cast iron, copper or copper alloys for use in the food industry, and to wire and components for the electrical and electronics industries, where ease of soldering is an essential property. Although tin coatings are not immune from damage by fretting corrosion, and fretting between tin-plate sheets in transit sometimes produces patterns of black spots, tin coatings may be used to reduce the risk of fretting damage in press fits and splined joints of steel components[29]. The coating packs the joint and any movement takes place within the coating. An allied application is the tinning of aluminium alloy or iron pistons to provide a suitable working surface during the running-in period[30].

Sprayed coatings find a use in large vessels and some equipment used in the food industry. The necessity for these coatings to be thick enough to be pore-free has already been mentioned.

As a general guide to the thickness of coating desirable for various applications, the requirements of BS 1872:1964 for electrodeposited tin coatings are shown in Tables 14.11 and 14.12.

For many purposes involving contact with food and water, coatings

Table 14.11 Thickness suggested for electrodeposited tin coatings on ferrous components

Purpose	Minimum local thickness (μm)
Contact with food or water where a complete cover of tin has to be maintained against corrosion and abrasion	30
Protection in atmosphere	20
Protection in moderate atmospheric conditions with only occasional condensation of moisture	10
To provide solderability and protection in mild atmospheres	5
Coatings flow-brightened by fusion (solderability and protection in mild atmospheres)	2.5 (maximum 8)

Table 14.12 Thickness suggested for electrodeposited
tin coatings on copper and copper alloys with at least
50% copper

Purpose	Minimum local thickness (μm)
Contact with food or water where a complete cover of tin has to be maintained against corrosion and abrasion	30
Protection in atmosphere and in less aggressive immersion conditions	15
To provide solderability and protection in mild atmospheric conditions	5*
Coatings flow-brightened by fusion (solderability and protection in mild atmospheres)	2·5* (maximum of 8)

*On brass an undercoat of copper, nickel or bronze of thickness 2·5 μm is required.

thinner than those specified in the first category are sometimes sufficient; much depends on the expected amount of abrasion, or loss during cleaning processes. Hot-dipped coatings in the usual thickness range of 10–25 μm give good protection to water-heaters, dairy equipment and much industrial plant, and the thinner coatings of the tinplates in common use are usually sufficient with proper care to preserve appearance in storage and transport. On the other hand, on copper for hot-water service it may sometimes be desirable to use coatings thicker than those recommended in view of the risk in interdiffusion between tin and copper.

Tin Alloy Coatings

Tin–lead

Tin–lead coatings with upwards of 5% lead may be applied by hot dipping to steel, copper and copper alloys. Steel sheets are commonly coated with alloys containing 7%, 10% or 25% tin; these are called *terne-plate*, with the name *tin-terne* sometimes applied to the higher tin-content coating. Tin–lead alloys may also be electrodeposited from a fluoroborate solution containing organic addition agents and bright deposits are possible.

These alloy coatings have advantages over tin in atmospheric exposure where there is heavy pollution by oxides of sulphur. They are cathodic to steel and anodic to copper. In industrial atmospheres, however, formation of a layer of lead sulphate seals pores and produces a generally stable surface[5] and terne-plate has been used extensively as roofing sheet, especially in the USA. It is easily and effectively painted when additional protection is required. Copper heat exchangers in gas-fired water-heaters may be coated by hot dipping in 20% tin alloy[31].

Tin–lead alloy coatings have some of the susceptibility of lead to vapours of organic acids such as acetic acid, and may be attacked by vapours from wood and insulating materials when enclosed in wooden cases or in electrical apparatus. They are, however, widely and successfully used as protective and easily solderable coatings on wire, electronic components and printed circuit boards.

Tin–lead can be substituted for tin for other purposes, although the toxicity of lead limits the field of application. The corrosion resistance is usually no better than that of unalloyed tin, but there may be some saving of cost in applications such as wash-boilers and other vessels for non-potable liquids and light engineering components formed from sheet metal. Heavily coated terne-plates may be used for the fuel tanks of stoves and vehicles.

Tin–zinc

Tin–zinc alloys of a wide range of composition can be electrodeposited from sodium stannate/zinc cyanide baths; only the coatings with 20–25% zinc have commercial importance[32, 33].

There is no intermetallic compound formation and the electrodeposit behaves as a simple mixture of the two metals. It can be considered as basically a stable wick of tin through which zinc is fed to be consumed at a rate lower than its consumption from a wholly zinc surface. If the conditions are such that zinc is rapidly consumed, and no protective layer of corrosion products is formed, the coating may break down, but in mildly corrosive conditions some of the benefits of a zinc coating, without some of its disadvantages, are obtained.

In condensed moisture, there is sufficient corrosion of zinc to give protection at pores in a coating on steel without the formation of as much zinc corrosion product as would develop on a wholly zinc surface. In solderability the coating is tin-like when new or stored dry, but the selective corrosion of zinc in humid conditions may produce a layer obstructive to easy soldering.

In full weathering in industrial areas, the zinc is taken from the coating too quickly and the alloy coatings do not endure as long as either zinc or tin coatings of comparable thickness; they may, however, outlast cadmium[34]. By the sea, the alloy coatings are somewhat better, and in more continuously wet conditions, such as at half-tide positions, they may outlast zinc coatings; possibly here the corrosion product is protective. It is, however, in sheltered conditions and special environments that tin–zinc is most useful. Its easy solderability combined with protection at pores makes it applicable in electrical and radio equipment and in components of tools and mechanisms. It is also used on the bodies of water-containing fire extinguishers, and on components exposed to hydraulic fluids.

The coating is, in addition, useful in preventing galvanic corrosion[35]. Plated on steel which is to be used in contact with aluminium alloys, it protects the steel and does not stimulate the corrosion of the light alloy and is itself not consumed as rapidly as a 100% zinc coating.

Tin–cadmium

Tin–cadmium alloys of a range of compositions can be deposited from stannate/cyanide solutions or fluoride/fluorosilicate solutions[36]. The behaviour of the coatings is rather similar to that of tin–zinc, but as cadmium is less effective than zinc in giving cathodic protection to steel, a 25% cadmium coating is barely able to protect pores and a 50% content is better for this purpose. The coatings in some conditions form an extremely dense layer of corrosion product, and give an outstanding performance in laboratory salt-spray tests[37], but there has been no substantial practical application. Coatings of tin over cadmium, which combine an inert outer surface with protection from rusting at pores, have been used on containers of solvents and to protect electrical components against organic vapour corrosion.

Tin–copper

Tin–copper alloys may be electrodeposited from copper cyanide/sodium stannate baths[38] or from cyanide/pyrophosphate baths[39] to give a range of compositions. Alloys with 10–20% tin have a pleasant golden colour but are not tarnish-resistant unless coated with lacquer. The alloy with 42% tin known as *speculum* is silver-like in colour and is resistant to some forms of corrosion. At this composition the deposit is formed as the intermetallic compound Cu_3Sn. It has a useful hardness (about 520 HV). The deposit becomes dull on exposure to atmospheres containing appreciable amounts of sulphur dioxide, but resists hydrogen sulphide, and remains bright in the more usual indoor atmospheres. Although out of doors it becomes dull and grey if not cleaned frequently, the coating is very suitable for metalwork used indoors; it resists the action of most foodstuffs and is suitable for tableware. Like many intermetallic compounds, the deposit shows a corroding potential which becomes increasingly noble with duration of immersion in electrolyte. It is strongly cathodic to steel, and pore-free deposits are desirable. Recommended minimum thicknesses are 12 μm on brass, copper, nickel silver, etc. and 25 μm on steel.

The fact that the composition of the speculum deposit must be closely controlled to obtain the best results has been a serious drawback to development. The coating finds uses on decorative hollow-ware, oil lamps and tableware. The bronze deposits with 10 or 20% tin are used lacquered in decorative metal-ware for domestic and personal ornament and, in thick layers to protect hydraulic pit props against corrosion and abrasion. They have also been used with success as undercoatings for nickel–chromium[39,40] or tin–nickel alloy deposits.

Tin–nickel

Tin–nickel alloy coatings are deposited from a bath containing stannous chloride, nickel chloride, ammonium bifluoride and ammonia[41,42]. The useful deposit contains 65% tin and the conditions are maintained to obtain this composition only; control is, fortunately, easy. A special feature of the process is the good throw of deposit into recesses.

The deposit plates out as the intermetallic compound NiSn, which is white with a faintly pink tinge, and has a hardness of about 710 HV. Deposits from new baths are usually in tensile stress but those from baths used for some time are in compressive stress; the stress can be controlled if desired by adjustment of solution composition[43]. The properties of the intermetallic compound differ from those of both the constituent metals. It is easily passivated, resisting concentrated nitric acid and becoming considerably ennobled during immersion in solutions of neutral salts, including sea-water. In a wide range of solutions, the potential of NiSn with reference to the normal hydrogen electrode was, on immediate immersion, $+0.33 - 0.055$ pH and, after some hours, $+0.59 - 0.056$ pH[44]. Higher potentials are reached after long immersion or in oxidising conditions, but ennoblement occurs in solutions with extremely low oxygen concentrations; evidence for five oxidation states for the surface film has been obtained, one at the low potential of $-0.42 - 0.06$ pH[45]. The film thickening that accompanies this change to a more noble potential may become visible, and in hot water or steam a purple film may be produced.

The deposit resists atmospheric tarnish even in the presence of high pollution by sulphur dioxide (in contrast to nickel) and hydrogen sulphide, and coatings exposed to the outdoor atmosphere remain bright indefinitely, sometimes taking on a slightly more pink colour as the oxide film thickens.

The passivity at pH values above about 1·5 is maintained in a great variety of solutions, including fruit juices, vinegar, sea-water, alkalis, and even ferric chloride[46]. Hot caustic alkali solutions above about 10% attack the coating slowly, and the halogens etch it.

The nobility of the coating brings with it the handicap that corrosion of base metal exposed at pores is stimulated. In an electrolyte of good conductivity, steel, brass or copper are attacked freely at pore sites; steel plates 1 mm thick were perforated after 12 months in the sea. In the outdoor atmosphere, the rate of penetration of the basis metal is slow, but disfigurement by the appearance of corrosion products at pore sites may occur[46,47]. Since the coating itself is not attacked, new pores do not develop during atmospheric exposure, so that the risks of corrosion at pores can be mitigated by attention to the original condition of the coating. Deposits more than 30 µm thick will usually be pore-free, and for deposits on steel for outdoor exposure an undercoating of copper is decidedly advantageous[47,48]. The copper undercoat, preferably about 12 µm thick, reduces the number of pores penetrating from the surface to the steel, and in industrial atmospheres tends to reduce corrosion at such pores as remain. Tin or tin–copper alloy undercoats may also be used and in marine environments are somewhat better than copper.

Indoors, pore corrosion is troublesome only if there are prolonged periods of wetting by condensed moisture, and coating thicknesses may safely be much less than those desirable out of doors. The coating will not, however, withstand much deformation, and even with the thinner coatings plating should, if possible, be carried out after all forming operations are complete.

The application of the tin–nickel coating for out of doors service has been restricted by fear of pore corrosion and of physical damage, and by the poor colour match with chromium. For indoor use, the coating has many applications, e.g. laboratory instruments, balance weights, the valves of wind instruments, internal mechanism of watches, electrical instruments, lighting fittings, interiors of cooking vessels and decorative hollow-ware. Many of these are special applications in which, in addition to corrosion resistance, the hard, smooth surface, non-magnetic quality and the good covering power in deposition of the coatings, may have been required.

These qualities have also lead to its use on printed circuit boards and on electrical connectors, although the persistent oxide film obstructs easy solderability and produces too high a contact resistance for satisfactory switching at low voltages.

S. C. BRITTON

REFERENCES

1. *Instructions for Electrodepositing Tin*, Tin Research Institute, Greenford (1971)
2. Thwaites, C. J., *The Flow-melting of Electrodeposited Tin Coatings*, Tin Research Institute, Greenford (1959)
3. Thwaites, C. J. *Hot-tinning*, Tin Research Institute, Greenford (1965)
4. Willey, A. R., *Br. Corros. J.*, **7**, 29 (1972)
5. Britton, S. C. and Angles, R. M., *J. Appl. Chem.*, **4**, 351 (1954)
6. Hudson, J. C. and Banfield, T. A., *J. Iron St. Inst.*, **154**, 229P (1946)
7. Britton, S. C. and Angles, R. M., *Metallurgia, Manchr.*, **44**, 185 (1951)
8. Middlehurst, J. and Kefford, J. F., *Condensation in Cargoes of Canned Foods*, CSIRO, Tech. Paper No. 34 (1968) and *Proc. Conference on the Protection of Metal in Storage and in Transit*, London, Brintex Exhibitions Ltd., 83 (1970)
9. Rocquet, P. and Aubrun, P., *Br. Corros. J.*, **5**, 193 (1970)
10. Hoar, T. P. and Havenhand, D., *J. Iron St. Inst.*, **133**, 253 (1936)
11. Buck, W. R. and Leidheiser, H. J., *J. Electrochem. Soc.*, **108**, 203 (1961)
12. Koehler, E. L., *J. Electrochem. Soc.*, **103**, 486 (1956) and *Corrosion*, **17**, 93t (1961)
13. Liebmann, H., *Proceedings 3ème Congrès International de la Conserve*, Rome–Parma, 1956, Comité International Permanent de la Conserve, Paris, 133 (1956)
14. Frankenthal, R. P., Carter, P. R. and Laubscher, A. N., *J. Agr. Food Chem.*, **7**, 441 (1959)
15. Cheftel, H., Monvoisin, J. and Swirski, M., *J. Sci. Fd. Agric.*, **6**, 652 (1955)
16. Dickinson, D., *J. Sci. Fd. Agric.*, **8**, 721 (1957)
17. Strodtz, N. H. and Henry, R. E., *Food Techn.*, **8**, 93 (1954)
18. Hartwell, R. R., *Surface Treatment of Metals*, Amer. Soc. Metals, Cleveland, Ohio, 69 (1941)
19. Willey, A. R., Krickl, J. L. and Hartwell, R. R., *Corrosion*, **12**, 433t (1956)
20. Koehler, E. L., *Trans. Amer. Soc. Metals*, **44**, 1 076 (1952)
21. Kamm, G. G. and Willey, A. R., *Corrosion*, **17**, 77t (1961)
22. Kamm, G. G., Willey, A. R., Beese, R. E. and Krickl, J. L., *Corrosion*, **17**, 84t (1961)
23. Carter, P. R. and Butler, T. J., *Corrosion*, **17**, 72t (1961)
24. Board, P. W., Holland, R. V. and Elbourne, R. G. P., *J. Sci. Fd. Agric.*, **18**, 232 (1967)
25. Salt, F. W. and Thomas, J. G. N., *J. Iron St. Inst.*, **188**, 36 (1958)
26. Koehler, E. L., *Werkst. u. Korrosion*, **21**, 354 (1970)
27. Koehler, E. L., Daly, J. J., Francis, H. T. and Johnson, H. T., *Corrosion*, **15**, 477t (1959)
28. Britton, S. C. and Clarke, M., *Trans. Inst. Metal Finishing*, **40**, 205 (1963)
29. Wright, O., *J. Electrodep. Tech. Soc.*, **25**, 51 (1949)
30. Smith, D. M., *Trans. Soc. Automot. Engrs.*, *N.Y.*, **53**, 521 (1945)
31. Kerr, R. and Withers, S. M., *J. Inst. Fuel.*, **22**, 204 (1949)
32. Angles, R. M., *J. Electrodep. Tech. Soc.*, **21**, 45 (1946)
33. Cuthbertson, J. W. and Angles, R. M., *J. Electrochem. Soc.*, **94**, 73 (1948)

34. Britton, S. C. and Angles, R. M., *Metallurgia, Manchr.*, **44**, 185 (1951)
35. Britton, S. C. and de Vere Stacpoole, R. W., *Metallurgia, Manchr.*, **52**, 64 (1955)
36. Davies, A. E., *Trans. Inst. Met. Finishing*, **33**, 75, 85 (1956)
37. Britton, S. C. and de Vere Stacpoole, R. W., *Trans. Inst. Met. Finishing*, **32**, 211 (1955)
38. Angles, R. M., Jones, F. V., Price, J. W. and Cuthbertson, J. W., *J. Electrodep. Tech. Soc.*, **21**, 19 (1946)
39. Safranek, W. H. and Faust, C. L., *Plating*, **41**, 1159 (1954)
40. Chadwick, J., *Electroplating*, **6**, 451 (1953)
41. Parkinson, N., *J. Electrodep. Tech. Soc.*, **27**, 129 (1951)
42. Davies, A. E., *Trans. Inst. Met. Finishing*, **31**, 401 (1954)
43. Clarke, M., *Trans. Inst. Met. Finishing*, **38**, 186 (1961)
44. Clarke, M. and Britton, S. C., *Corrosion Science*, **3**, 207 (1963)
45. Clarke, M. and Elbourne, R. G. P., *Corrosion Science*, **8**, 29 (1968)
46. Britton, S. C. and Angles, R. M., *J. Electrodep. Tech. Soc.*, **27**, 293 (1951)
47. Britton, S. C. and Angles, R. M., *Trans. Inst. Met. Finishing*, **29**, 26 (1953)
48. Lowenheim, F. A., Sellers, W. W. and Carlin, F. X., *J. Electrochem. Soc.*, **105**, 339 (1958)
49. Britton, S. C., *Trans. Inst. Met. Finishing*, **52**, 95 (1974)

14.6 Copper and Copper Alloy Coatings

Copper coatings are usually applied by electrodeposition, although for more limited purposes 'electroless' or immersion deposits are used. Less frequently, copper may also be applied by flame spraying[1].

Applications

Copper deposits are applied predominantly for the following purposes:

1. As an undercoat for other metal coatings. The main use of copper plating is as an undercoating prior to nickel–chromium plating steel and zinc-base die castings. On steel, the primary purpose is to reduce polishing costs. Other advantages are that with a copper-plated undercoating, cleaning is less critical for achieving a well-adherent nickel deposit and metal distribution is frequently improved. Nickel–chromium plating standards of most countries permit some part of the nickel thickness to be replaced by copper[2,3]. On zinc-base die castings a copper undercoat is almost universally used, as an adherent nickel deposit cannot be deposited directly from conventional baths. For a similar reason copper is deposited on aluminium which has been given an immersion zinc deposit[4] before nickel plating is applied.

Under micro-discontinuous chromium coatings, copper undercoats improve corrosion resistance. On non-conductors, especially on plastic substrates, copper is often applied before nickel–chromium plating over the initial 'electroless' copper or nickel deposit in order to improve ductility and adhesion, e.g. as tested by the standard thermal-cycling test methods[5].

2. As a decorative finish on steel and zinc-base alloys for a variety of domestic and ornamental articles. The finish may be protected by clear lacquers or may be coloured by metal colouring techniques for use on, for example, door handles, luggage trim, etc.

3. As a 'stop-off' for nitriding or carburising of steel. The 10–40 μm deposits, which are electroplated on selected areas, are removed after the heat treatment.

4. For protection of engineering parts against fretting corrosion, on electrical cables and on printing cylinders. Temporary protection allied with lubrication is provided by immersion deposits of copper on steel wire.

5. Chemical deposits of copper are applied to provide conducting surfaces on non-metallic materials.

6. Copper is plated on printed circuit boards to provide electrical conductors and for a variety of other electrical and electronic applications[6].

14:62

Plating Solutions

Copper is electrodeposited commercially mainly from cyanide, sulphate and pyrophosphate baths. For rapid deposition in electro-forming, a fluoborate bath may also be used.

The sulphate bath The sulphate bath, the earliest of electroplating solutions and the simplest in composition, contains typically 150–250 g/l of copper sulphate and 40–120 g/l of sulphuric acid. The composition is not critical and the higher concentrations are used for plating at higher current densities, normally up to 6 A/dm².

Addition agents used to produce smooth and fine-grained (though dull) deposits include gelatin, glue, phenol sulphonic acid, hydroxylamine and triethanolamine. These are believed to inhibit crystal growth by forming colloids in the cathode layer, and, in some cases, to change the crystallographic orientation.

Modern bright acid copper plating baths contain both organic and inorganic addition agents which act as brighteners and levellers. The two functions are largely distinct, the latter being the more important when copper is plated as an undercoat for decorative nickel–chromium coatings. Additives of this type include organic sulphur compounds, e.g. thiourea derivatives. Such solutions are sensitive to the chloride ion concentration which must be maintained at a low level.

On ferrous metals immersion deposition in the copper sulphate bath produces non-adherent deposits, and a cyanide copper undercoat is therefore normally used. Where the use of a cyanide strike cannot be tolerated, an electroplated or immersion nickel deposit has been used[7,8]. Additions of surface-active agents, often preceded by a sulphuric acid pickle containing the same compound, form the basis of recent methods for plating from a copper sulphate bath directly on to steel[9–11].

While the sulphate bath has a high plating speed, its throwing power is poor, and this limits its application to articles of simple shapes.

Cyanide baths Most general copper plating, other than that applied, for example, to wire and strip or for electroforming, is carried out in a cyanide bath. Its main advantages are (a) that it can be used to plate directly on to steel and zinc-base alloys, and (b) that it has good throwing power, which renders it suitable for plating a large variety of shapes.

Modern solutions fall mainly into three types: (a) the plain cyanide bath which contains typically 20–25 g/l of copper cyanide, 25–30 g/l total sodium cyanide (6·2 g/l 'free' sodium cyanide), and is operated at 21–38°C and 110–160 A/m²; (b) the 'Rochelle' copper bath to which is added 35–50 g/l of Rochelle salt and which is used at 66°C at up to 645 A/m²; and (c) the high-efficiency cyanide baths which may contain up to 125 g/l of copper cyanide, 6–11 g/l of 'free' sodium or potassium cyanide, 15–30 g/l of sodium or potassium hydroxide, and are operated at up to 6–9 A/dm² and 65–90°C. Most bright cyanide copper baths[12] are of the high-efficiency type and, in addition, contain one or more of the many patented brightening and levelling agents available. Periodic reverse (p.r.) current is also sometimes used to produce smoother deposits.

Plating speeds for the high-efficiency baths are high, partly because higher current densities can be used without 'burning', but mainly because the cathode efficiency of the more concentrated solution is higher at higher current densities (e.g. 90–98% compared to 30–60% for the 'plain' and 'Rochelle' type solutions). However, a more dilute solution must generally be used as a 'strike' bath on steel and zinc-base alloys to avoid immersion deposition.

Pyrophosphate bath The pyrophosphate bath is intermediate in throwing power between the sulphate and cyanide baths. A typical bath contains 80–105 g/l of copper pyrophosphate, 310–375 g/l of potassium pyrophosphate, and 25 g/l of potassium citrate, pH 8·7–9·4. Similar baths containing nitrate, ammonia and oxalate are also employed. The solutions are used at 50–60°C with vigorous air agitation when current densities of up to 10 A/dm^2 are permissible. A proprietary bath is available with excellent brightening and good levelling characteristics.

A more dilute strike bath is employed for obtaining the initial deposit on steel, while for strongly recessed parts, e.g. tubular work, an immersion nickel deposit has been used[8]. A short cyanide copper strike is used before plating on zinc-base die castings.

Other electroplating solutions Other solutions[12], which are more rarely used for plating copper, include the fluoborate bath, the amine bath, the sulphamate bath and the alkane sulphonate bath.

Chemical deposition Simple immersion deposits of copper may be obtained on iron and steel in a solution containing, for example, 15 g/l of copper sulphate and 8 g/l sulphuric acid, and on zinc-base alloy in a solution containing copper sulphate 300 g/l, tartaric acid 50 g/l and ammonium hydroxide 30 ml/l[13]. Such deposits are thin and porous and are mainly plated for their colour, e.g. for identification, or for their lubricating properties, e.g. in wire drawing.

Solutions containing tetrasodium E.D.T.A. have also been used for this purpose and give slightly superior coatings.

On non-conductors, copper may be deposited by chemical reduction from a modified Fehling's solution. Such solutions have gained wide application in the plating of ABS and other plastics which are 'electrolessly' copper plated before nickel–chromium plating. Pretreatment of the plastic is important in order to gain adequate adhesion and includes steps for etching the surface as well as for providing a conducting substrate by treatment in stannous chloride and palladium chloride solutions.

Properties of Copper Deposits

Deposit uniformity The uniformity of a deposit is an important factor in its overall corrosion resistance and is a function of geometrical factors and the 'throwing power' of the plating solution. A distinction is made here between macro-throwing power, which refers to distribution over relatively large-scale profiles, and micro-throwing power, which relates to smaller irregularities[14].

The copper cyanide bath has excellent macro-throwing power and is chosen whenever irregular-shaped parts are to be plated. The sulphate bath is not inferior when parts with very narrow recesses, i.e. with width of opening less than 6 mm, are to be plated, although its macro-throwing power is poor. Pyrophosphate baths are intermediate between the two in macro-throwing power.

Porosity As is the case with all cathodic deposits, the corrosion resistance of a copper deposit is reduced in the presence of continuous porosity. Experience has shown that porosity is least when attention is paid to adequate cleaning, and the solution is kept free from solid or dissolved impurities (see Section 13.1). Porosity of copper deposits is also related to polarisation[15].

Corrosion resistance The corrosion resistance of a copper deposit varies with the conditions under which it is deposited and may be influenced by co-deposited addition agents (see, for example, Raub[16]). Copper is, however, plated as a protective coating only in specialised applications, and the chief interest lies in its behaviour as an undercoating for nickel–chromium on steel and on zinc-base alloy. Its value for this purpose has long been a controversial issue.

A thin copper deposit, e.g. 2·5 μm, plated between steel and nickel, improves corrosion resistance during outdoor exposure[17], and many platers also believe that a copper undercoating improves the covering power of nickel, particularly on rough steel.

Where heavier copper coatings are plated as a partial replacement for nickel, as is permitted under most nickel plating specifications, the effects are not clearly established. According to Blum and Hogaboom[18] the protective value of nickel on steel is reduced by the presence of a copper undercoating, but this does not apply when the nickel is chromium plated. This is largely corroborated by more recent corrosion tests[17,19] and the detrimental effect in the absence of chromium is probably due to attack on the nickel by the copper corrosion products. In the presence of conventional chromium plate, on the other hand, the fact that statistical evidence on many thousands of chromium-plated motor components has not established any difference in the behaviour of parts in which nickel formed respectively 95–100% and 50% of the copper–nickel coating[20], bears out the view that after chromium plating the differences in protective value tend to disappear. Moreover, when, as frequently happens in practice, the copper coating is polished, the protective value of the copper–nickel coating is higher than that of nickel alone, owing to the pore-sealing effect of the polishing operation.

The case is different again under micro-discontinuous (i.e. micro-cracked or micro-porous) chromium, on which a definite improvement in corrosion resistance can be achieved when copper is present under the nickel coating[21,22].

As an undercoating for chromium, i.e. in place of nickel, copper is not to be recommended. On the other hand, both accelerated and outdoor corrosion tests have shown that a tin–bronze deposit, containing 80–90% copper, is considerably better for this purpose and it has been claimed to be approximately equal to nickel in this respect.

Mechanical properties The hardness and strength of copper deposits may vary widely according to the type of bath used (see Table 14.13). In the presence of addition agents which decompose in use, the hardness may, moreover, vary appreciably with the age of the bath[23].

In copper sulphate solutions, hardness and tensile strength are increased

Table 14.13   Mechanical properties of electrodeposited copper[12]

Plating bath	Hardness (HV)	Elongation [% on 50·8 mm (2 in)]	Tensile strength (MN/m^2)
Sulphate bath	40–65	20–40	230–310
Sulphate bath with addition agent	80–180	1–20	480–620
Fluoborate bath	40–75	7–20	240–275
Cyanide bath	100–160	9–15	415–550
Cyanide bath with p.r. current	150–220	6–9	690–760
Pyrophosphate bath	125–165	—	—

by raising the current density and reducing the temperature. As will be seen from Table 14.13, particularly high hardness values can be obtained in the cyanide bath by using periodic reverse current.

Annealing of electrodeposited copper reduces the mechanical properties. As an example, the tensile strength has been reported to decrease from 275–330 MN/m^2 to 180–255 MN/m^2 on heating at above 300°C[24] while the hardness of deposits obtained in the presence of addition agents may drop from as high a value as 300 HV to 80 HV after annealing at 200°C.

Internal stress of copper deposits may vary between $-3\cdot4$ MN/m^2 (compressive) and $+100$ MN/m^2 (tensile). In general, tensile stress is considerably lower in deposits from the sulphate bath than in those from cyanide solutions[25–27], while pyrophosphate copper deposits give intermediate values. In cyanide solutions, tensile stress increases with metal concentration and temperature decreases if the free cyanide concentration is raised. P.R. current significantly lowers tensile stress[28]. With some exceptions, inorganic impurities tend to increase tensile stress[29]. Thiocyanate may produce compressive stress in cyanide baths[25].

In the sulphate bath the tensile stress increases if the temperature is reduced or the current density is increased, and gradually diminishes with increase in deposit thickness[25]. Addition of thiourea (1 g/l) or gelatin to the acid bath results in compressively stressed deposits, though at higher concentrations of addition agent this effect may be reversed[30]. Dextrose and gum arabic increase tensile stress[31]. The effect of other organic compounds may similarly depend on the operating conditions[32, 33]. The relationship between ductility and stress is complex, e.g. thiourea additions increase ductility over a wide range[34].

Despite the large differences in respect of other mechanical properties, it has been established[35] that the wear resistance of copper deposits, which is markedly inferior to, for example, that of electrodeposited nickel, is not significantly affected by either type of bath or addition agents.

Embrittlement by hydrogen absorbed by the substrate during pretreatment, e.g. in acid pickling baths or during plating, is generally important only on copper-plated wire or where copper is plated for lubrication before drawing operations on high-strength steels. For these purposes the acid copper bath is slightly preferable to the cyanide bath. Hydrogen may be removed and ductility restored by heat treatment in air (140–200°C for 0·5–1 h), in water (80–100°C for 0·5–2 h), or in oil (175–230°C for 1·5–2 h)[12]. Other properties have been comprehensively summarised in the literature[12, 36].

Copper Alloy Deposits

Copper–zinc Copper–zinc alloys are deposited for two main purposes: (*a*) as a decorative finish, e.g. on steel and (*b*) as a means of obtaining an adhesive bond of rubber to other metals.

Cyanide solutions are used almost exclusively. One typical solution contains copper cyanide 26 g/l, zinc cyanide 11 g/l, sodium cyanide (total) 45 g/l and sodium cyanide ('free') 7 g/l[12]. This bath is operated at pH 10·3–11·0, 110 A/m^2 and 27–35°C, with 75 Cu–25 Zn alloy anodes. Many other solutions are used[12], including a special rubber-bonding bath[37] and a high-speed bath which is capable of being used at up to 16 A/dm^2[2, 38]

Brass deposits normally contain 70–80% copper and 30–20% zinc; the colour does not normally match solid brass of the same composition and may, moreover, vary with the operating conditions and solution composition.

White brass deposits containing 85% zinc and 15% copper have also been plated to a limited extent[39], mainly as an undercoating for chromium during the nickel shortage, but they did not prove fully satisfactory.

While brass deposits have a somewhat higher protective value on steel than the equivalent thickness of copper, the deposits tend to tarnish, and when used for decorative purposes bright deposits are normally protected by a clear lacquer.

Copper–tin Although a wide range of copper–zinc alloy deposits can be plated[40], most experience has been gained with two compositions, i.e. the red copper-rich tin–bronze which contains 90–93% copper and 10–7% tin and the white speculum which contains 50–60% copper and 50–40% tin.

While tin–bronze has been successfully plated as an undercoating for chromium during the nickel shortage[41, 42] its main use now is as a decorative finish in its own right, because of its pleasing red-gold colour. As in the case of brass, however, the deposits must be protected against tarnishing by a clear lacquer.

Speculum deposits are similar in appearance to silver, but are harder and have good tarnish resistance. Alloys containing only 2% copper and 98% tin are plated on bearing surfaces.

Copper–tin deposits can be plated from cyanide or pyrophosphate[43, 44] baths and deposits are of good corrosion resistance (approximately equivalent to the same thickness of nickel). Hardness values of up to 314 HV are obtainable for the copper-rich alloys[45], and up to 530 HV for the tin-rich alloys can be obtained. (See also Section 14.5.)

Other alloys Other copper alloys can be plated, including copper–tin–zinc (Alballoy)[46], copper–nickel[47], copper–cadmium[48, 49], copper–gold and copper–lead[50].

R. PINNER

REFERENCES

1. Ballard, W. E., *Metal Spraying and Sprayed Metal*, Griffin, London (1948)
2. *Electroplated Coatings of Nickel and Chromium*, BS 1224 (1970)
3. A.S.T.M. 166

4. Wernick, S. and Pinner, R., *Surface Treatment and Finishing of Aluminium and its Alloys*, Draper, Teddington, 2nd edn (1959)
5. Crouch, P. C., *Trans. Inst. Metal Finishing*, 49 No. 4, 141 (1971)
6. Saubestre, E. B. and Khera, K. P., 'Plating in the Electronics Industry', *Symposium of the Am. Electroplaters' Soc.*, 230 (1971)
7. Clauss, R. J. and Adamowicz, N. C., *Plating*, 57 No. 3, 236 (1970)
8. O'Dell, C. G., *Electroplating and Metal Finishing*, 24 No. 7, 14 (1970)
9. Dehydag Deutsche Hydrierwerke, UK Pats. 784 091 (1957) and 811 773 (1959); US Pat. 2 903 403 (1959)
10. Antropov, L. I. and Popopov, S. Ya., *Zh. Prikl. Khim.*, *Leningr.*, 27, 55, 527 (1954)
11. Pantshev, B. and Kosarev, C., *Metalloberfläche*, 24 No. 10, 383 (1970)
12. Pinner, R., *Copper and Copper Alloy Plating*, Copper Development Assoc., London (1962)
13. Saubestre, E. B., *Proc. Amer. Electropl. Soc.*, 46, 264 (1959)
14. Raub, E., *Metalloberfläche*, 13 No. 10 (1959)
15. Kovaskii, N. Ya. and Golubev, V. N., *Zh. Priklad. Khim.*, 43 No. 2, 348 (1970)
16. Raub, E., *Z. Metallk.*, 39, 33, 195 (1948)
17. Knapp, B. B. and Wesley, W. A., *Plating*, 38, 36 (1951)
18. Blum, W. and Hogaboom, G. B., *Principles of Electroplating and Electroforming*, McGraw-Hill, New York, 3rd edn, 136 (1949)
19. Pray, H. A., Rept. of Subcomm, II of Comm, B-8, *Proc. Amer. Soc. Test. Mat.*, 49 (1949)
20. Phillips, W. M., *Plating*, 38, 56 (1951)
21. Turner, P. F. and Miller, A. G. B., *Trans. Inst. Metal Finishing*, 47 No. 2, 50 (1969)
22. Preprints, Discussion Session, Annual Conf. of the Inst. of Met. Fin. (1972)
23. Spähn, H. and Tippmann, H., *Metalloberfläche*, 13 No. 2, 32 (1959)
24. Prater, T. A. and Read, H. J., *Plating*, 36, 1 221 (1949)
25. Kushner, J. B., *Metal Finish.*, 56 No. 4, 46; No. 5, 82; No. 6, 56 (1958)
26. Nishiharaud, K. and Tsuda, S., *Suyokuro-shi*, 12 No. 12, 25 (1952)
27. Phillips, W. M. and Clifton, F. L., *Proc. Amer. Electropl. Soc.*, 34, 97 (1947)
28. Bachvalov, G. T., All Union Sci./Tech. Conference on Corrosion and Protection of Metals, Moscow, 1958 (cf. *Plating*, 46, 157 (1959))
29. Fujino, T. and Yamomoto, J., *J. Metal Finishing Soc. Japan*, 20 No. 1, 18 (1969)
30. Lizlov, Yo. V. and Samartsev, A. G., All Union Sci./Tech. Conference on Corrosion and Protection of Metals, Moscow, 1958 (cf. *Plating*, 46, 266 (1959))
31. Graham, A. K. and Lloyd, R., *Plating*, 35, 449, 506 (1948)
32. Walker, R. and Ward, A., *Electrochimica Acta.*, 15 No. 5, 673 (1970)
33. Walker, R., *Plating*, 57 No. 6, 610 (1970)
34. Sard, R. and Weil, R., *Plating*, 56 No. 2, 157 (1969)
35. Ledford, R. F. and Dominik, E. A., *Plating*, 39, 360 (1952)
36. Lamb, V. A., Johnson, C. E. and Valentine, D. R., *J. Electrochem. Soc.*, 117 No. 9, 291C; No. 10, 341C; No. 11, 381C (1970)
37. Compton, K. G., Ehrhardt, R. A. and Bittrich, G., *Proc. Amer. Electropl. Soc.*, 41, 267 (1954)
38. Roehl, E. J. and Westbrook, L. R., *Proc. Amer. Electropl. Soc.*, 42, 3 (1955) and Roehl, E. J., *Electroplating and Metal Finish.*, 11, 299 (1958)
39. Saltonstall, R. B., *Proc. Amer. Electropl. Soc.*, 39, 67 (1952)
40. Batten, H. M. and Welcome, C. J., US Pats. 1 970 548 and 1 970 549 (1934)
41. Schmerling, G., *Electroplating and Metal Finish.*, 5, 115 (1952)
42. Lee, W. T., *Trans. Inst. Metal Finish.*, 36, 2, 51 (1958–1959)
43. Faust, C. H. and Hespenheide, W. G., US Pat 2 658 032 (1953)
44. Safranek, W. H. and Faust, C. L., *Proc. Amer. Electropl. Soc.*, 41, 201 (1954); *Plating*, 41, 1 159 (1954)
45. Rama Char, T., *Electroplating and Metal Finish.*, 10, 347–9 (1957)
46. Diggin, M. B. and Jernstedt, G. W., *Proc. Amer. Electropl. Soc.*, 31, 247 (1944)
47. Priscott, B. H., *Trans. Inst. Metal Finish.*, 36, 93 (1959)
48. Hogaboom, G. B., Jr. and Hall, N., *Metal Finishing Guidebook and Directory*, Metal Finishing, Westwood, N. J., USA, 295–298 (1959)
49. German Pat. 876 630 (1953)
50. Krasikov, B. S. and Grin, Yu. D., *Zh. Prikl. Khim.*, *Leningr.*, 32, 387 (1959)

14.7 Nickel Coatings

Nickel coatings have long been applied to substrates of steel, zinc and other metals in order to provide a surface that is resistant to corrosion, erosion and abrasion. Most of the nickel is used as decorative coatings 5–40 μm thick, usually under a top coat of chromium about 0·5 μm thick so as to give a non-tarnishing finish. Such coatings are applied to metal parts on cars, cycles, perambulators and a wide range of consumer items; they have also been applied increasingly to plastic components during recent years in order to give an attractive metallic appearance[1,2]. Decorative nickel coatings are also applied without chromium top coats to products such as spanners, screwdriver blades, keys and can-openers.

About 3% of all nickel used in the form of coatings is employed in engineering applications where brightness is rarely needed and the deposits are relatively thick; these coatings are used for new parts and for reclamation.

Most nickel electroplating is carried out in solutions based on the mixture of nickel sulphate, nickel chloride and boric acid proposed by O. P. Watts[3]. Typical composition and operating conditions are:

Composition
Nickel sulphate ($NiSO_4 \cdot 7H_2O$): 240–300 g/l
Nickel chloride ($NiCl_2 \cdot 6H_2O$): 40–60 g/l
Boric acid (H_3BO_3): 25–40 g/l

Operating conditions
Temperature: 25–50°C
Air agitation
pH: 4·0–5·0
Cathodic current density: 3–7 A/dm^2
Mean deposition rate: 40–90 μm/h

The Watts solution is a relatively cheap, simple solution which is easy to control and keep pure. The nickel sulphate acts as the main source of nickel ions, though nickel chloride is an additional source. Higher deposition rates can be used when the ratio of nickel chloride to nickel sulphate is raised and some proprietary bright nickel solutions are available in a 'high-speed' version which contains an increased concentration of nickel chloride.

Chloride ions are also needed to ensure satisfactory dissolution of some nickel anodes at usual values of pH and solution temperature. Where sulphur is deliberately incorporated in the anode during manufacture however, anodic dissolution of the nickel is activated and the chloride in the solution may be reduced or entirely eliminated, depending upon the degree of anodic activation achieved and the maximum anodic current density required. Nickel anodes are usually either (a) bars or sheets fabricated by casting, rolling or extrusion, or (b) strips of electrolytic nickel, pieces of carbonyl-nickel pellets or electrolytic-nickel contained in a basket of titanium mesh. The anodes are held in bags of cotton twill, polypropylene or Terylene in order to prevent metallic particles from entering the solution and causing deposit roughness. Accounts of the anodic dissolution of nickel are given by Raub and Disam[4], and Sellers and Carlin[5]. (See also Section 13.1.)

At normal current densities, about 96–98% of the cathodic current in a Watts solution is consumed in depositing nickel; the remainder gives rise to discharge of hydrogen ions. The boric acid in the solution buffers the loss of acidity arising in this way, and improves the appearance and quality of the deposit. Although phosphates, acetates, citrates and tartrates have been used, boric acid is the usual buffer for nickel solutions.

A detailed discussion of the function of the constituents of the Watts bath is given by Saubestre[6]. In addition to inorganic constituents, organic wetting agents are often added to prevent pitting of the deposit that might otherwise arise from adhesion to the cathode of small bubbles of air[7] or hydrogen evolved cathodically. Elimination of pitting and other defects is discussed by Bouckley and Watson[8].

Decorative Plating

The majority of decorative nickel plating is carried out in solutions containing addition agents which modify growth of the nickel deposit so that a fully bright finish is obtained that is suitable for immediate chromium plating without mechanical finishing. At one time, wide use was made of deposits with brightness achieved through additions of cobalt salts plus formates and formaldehyde[9,10], but the use of a mixture of organic addition agents enables deposits to be obtained which are smoother, more lustrous, give bright deposits over a wider range of current densities, and have lower internal stress. In consequence, the bulk of bright nickel plating is carried out in organic bright nickel solutions.

Organic bright nickel solutions   Several organic substances are used at appropriate concentrations in these solutions in order to give brightness, levelling and control of deposit stress. Portions of the addition agent molecules are incorporated in the deposit, resulting in a hard, fine-grained coating which has a finely striated structure when etched in section and which usually contains incorporated sulphur. The sulphur causes the deposit to be electrochemically less noble than pure nickel deposits. Decomposition products of the additives form in the solution with use, and at one time they accumulated and impaired the mechanical properties of the plate, eventually necessitating batch purification. In modern solutions however, continuous carbon filtration can be used to remove deleterious organic substances without significant removal of the addition agents themselves.

Brighteners   Modern solutions contain a brightener system comprising several additives which together enable bright deposits to be obtained over a wide range of current densities such as that occurring over a component having a complicated shape with deeply recessed areas. Brighteners are broadly divided into primary brighteners and secondary brighteners, but the division is not sharp.

Primary brighteners have a powerful effect on the deposit and are normally used at low concentrations which are carefully controlled. Metals such as cadmium and zinc act as primary brighteners, as do organic substances such as amino polyarylmethanes, quinoline and pyridine derivatives, and sulphonated aryl aldehydes. Primary brighteners often, especially at higher concentrations, affect adversely the mechanical properties of the deposit.

Secondary or carrier brighteners have a milder effect on the deposit when used alone, and modify the effect of primary brighteners. Judicious combination of primary and secondary brighteners gives fully bright but relatively ductile deposits having low internal stress. Aryl sulphonic acids and sulphonates, sulphonamides and sulphimides frequently act as secondary brighteners.

Combinations of brighteners often behave synergistically, so that the final brightening effect is greater than might have been expected from the individual effects.

Stress reducers Many organic substances used as secondary brighteners also reduce the tendency for the internal stress in the deposit to become tensile. In the absence of primary brighteners, they are able to give zero or even compressive stress in nickel deposits, and thereby find wide application in electroforming where accurate control of deposit stress is vital. Saccharin, p-toluene sulphonamide, and mono-, di- and tri-sulphonates of benzene and naphthalene are common stress-reducing agents. Stress is usually measured in nickel deposits by observing the bending induced by plating one side only of a metal strip. Convenient and sensitive developments of this technique are available[11-14].

Levelling agents A nickel plating solution is said to have levelling action if deposits from it, when applied to an uneven cathode surface, become increasingly smooth as plating proceeds. Levelling agents are therefore widely used to eliminate expensive final polishing of the nickel surface and to reduce the fineness of the surface finish needed on the substrate surface. Both features reduce the cost of producing a bright and smooth finish on a plated article.

Levelling agents increase cathode polarisation and are consumed at the cathode by decomposition or incorporation in the deposit. They are used at a concentration sufficiently low that a diffusion layer is established at the cathode surface, and then the levelling agent is able to diffuse at a greater rate to peaks than to recesses on the surface. In order that the layer of solution adjacent to the cathode surface shall remain an equipotential surface, the current density at recesses rises above that at peaks, giving progressive smoothing of the deposit as deposition proceeds.

The levelling action of an addition agent[15] depends upon its concentration C in the solution, the rate of change of cathode potential with change of concentration (dE/dC) and the rate of change of cathode potential with current density (dE/dI). Levelling power (L.P.) at a given current density, defined as deposit thickness in recesses minus thickness at peaks divided by average thickness, may be expressed as

$$\text{L.P.} = KC(dE/dC)(dI/dE)$$

where K is a constant. Typical levelling agents include coumarin, quinoline ethiodide, butyne 1:4 diol and its derivatives, and thiourea and its derivatives at certain concentrations. Extensive studies of the mechanism of levelling have been carried out in the United States[16], Britain[15] and the Soviet Union[17].

Semi-bright solutions　Maximum levelling action is often found in solutions which do not give a fully bright deposit, but the deposit is smooth and can easily be lightly buffed to give a lustrous finish; moreover, many levelling agents used are sulphur-free, so that the deposits are also free from sulphur and as noble as a Watts deposit when subjected to corrosive attack. This feature is exploited in double-layer nickel coatings (see below).

Wetting agents　As mentioned earlier, wetting agents are added to nickel solutions to prevent pitting. These wetting agents can be cationic, non-ionic or anionic in nature. In general, the best anti-pit agents tend to produce the most foaming, and a compromise must be struck. Where mechanical agitation of the solution is used, by stirrers or by cathode movement, a greater tendency to foaming can be tolerated than when air-agitation is employed.

Interaction of addition agents　The success of modern proprietary bright nickel solutions has resulted in large measure from the skill of the research departments of plating supply houses in balancing the effects of various additives to give optimum results. The detailed findings are usually kept confidential, but the broad principles of addition agent action and interaction are discussed in published work[18-20].

A commercial-scale operation with bright and semi-bright solutions based not on Watts but on a solution having nickel sulphamate as the main constituent (430–450 g/l), is described by Siegrist[21].

Decorative Coating Systems that give Improved Resistance to Corrosion

Double-layer nickel coatings　These coatings have an undercoat of highly-levelled sulphur-free nickel covered with sufficient bright nickel to give a fully bright finish with minimum requirement for expensive mechanical finishing of the part. They were initially produced simply to reduce costs, but it was soon noticed that, because the undercoat of sulphur-free semi-bright nickel is electrochemically more noble than the final bright nickel above it, corrosive attack when it does occur is preferentially directed towards the bright nickel, and penetration to the basis metal is markedly delayed.

Figure 14.7 shows how pits in a single-layer nickel deposit start at small pores or other imperfections in the chromium top coat[22]. The pits are initially hemispherical; those shown here were produced by 6 months in an industrial atmosphere on a copper plus nickel plus chromium plated car bumper.

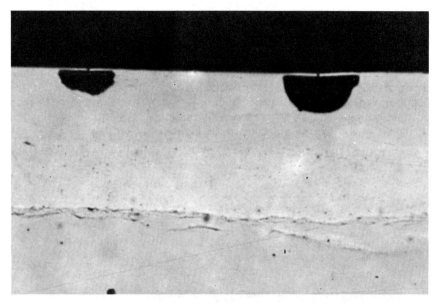

Fig. 14.7 Commencement of corrosion at discontinuities in chromium topcoat over nickel; × 1 000 (after Reference 22)

In double-layer nickel coatings however, a flat-based pit is formed in the nickel coating, giving marked resistance to penetration to the basis metal. Figure 14.8 shows a pit in a double-layer nickel plus chromium coating after 58 months service.

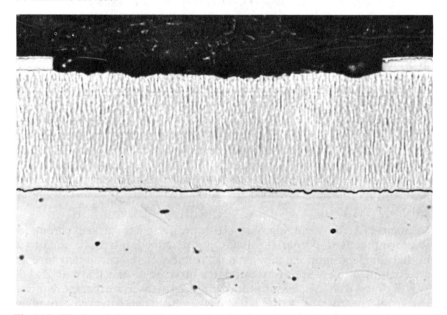

Fig. 14.8 Flat-based pit in double-layer nickel plus chromium coating after 58 months service; × 300 (after Reference 22)

Triple-layer nickel coatings In order to minimise the effect of corrosive attack on the appearance of the deposit while still retaining the resistance to penetration to the substrate afforded by double-layer nickel, triple-layer nickel coatings have been developed in which the semi-bright and bright layers are separated by a thin nickel layer electrochemically less noble than both of them. This thin layer of nickel, highly activated by incorporated sulphur, is described by Brown[23]. Figure 14.9 shows a section through such a triple-layer coating. In service, corrosive attack is substantially confined to that part of the coating adjacent to the highly-activated layer.

Fig. 14.9   Triple-layer nickel deposit consisting of semi-bright and bright nickel layers with a thin, highly activated layer of nickel between them (after Reference 23)

Nickel coatings that induce microporosity in chromium topcoats In addition to the methods invoked in double- and triple-layer nickel coatings to ensure that the inevitable corrosion currents developed in a corrosive environment are directed away from the basis metal, another method of protecting the basis metal is to ensure that the conventional chromium top-coat ($0 \cdot 3 \mu m$) is made sufficiently porous for the corrosion current to be dissipated over a large number of exposed nickel sites. This is achieved conveniently by applying, between the nickel coating and the chromium, a further thin nickel layer containing incorporated solid particles which are inert and which induce in the chromium a large number of pores. The rate of attack at any one pore is then small. Such coatings are increasingly used on decorative trim on cars and are described by Odekerken[24] and Williams[25], among others.

Microcracked chromium topcoats Historically, microcracked chromium preceded the micro-porous chromium just described, but it is related to it in that the deposition conditions and thickness of the chromium topcoat are controlled to give porosity through a network of very fine cracks.

A thickness of at least $0 \cdot 8 \mu m$ is normally needed to ensure that the required crack pattern is formed all over a shaped part. Such microcracked chromium coatings have a slightly lower lustre than the thinner conventional chromium deposits and take longer to deposit. The improved resistance to corrosion that they impart to nickel coatings[26-29] has been chiefly of interest to the automotive industry. In an attempt to avoid the slightly diminished lustre of thick microcracked coatings, an alternative process has been developed whereby a thin, highly stressed nickel layer is deposited upon the normal bright nickel layer. A conventional chromium topcoat is then applied, causing the thin nickel layer to crack, thereby cracking the chromium layer itself so as to give a microcrack pattern[30].

Supplemental films The Batelle Memorial Institute[31] has developed a post-treatment for nickel plus chromium coatings in which the plated part is made cathodic in a solution containing dichromate. A film thereby formed on the surface seals pores in the coating through which corrosion of the nickel might otherwise occur. Later work[32] suggests, however, that micro-cracked chromium gives superior results.

Control of quality of decorative nickel coatings Increasing international effort has been spent during the past few years in drawing up agreed recommendations aimed at ensuring that incorrect plating procedures do not diminish the high performance of nickel, or nickel plus chromium, coatings. During 1970, the International Standards Organisation issued Recommendation 1456 *Electroplated Coatings of Nickel plus Chromium* and Recommendation 1457 *Electroplated Coatings of Copper plus Nickel plus Chromium on Steel (or Iron)* which were used as guidelines by the British Standards Institution in drawing up BS 1224:1970 *Electroplated Coatings of Nickel and Chromium* and BS 4601:1970 *Electroplated Coatings of Nickel Plus Chromium on Plastic Substrates*. These British standards specify the type and thickness of deposits required for various service conditions, appropriate accelerated corrosion test procedures, and methods of measuring other important properties. The quality of nickel salts and anodes for plating is specified in BS 558 and 564:1970 *Nickel Anodes, Anode Nickel and Salts for Electroplating*.

Engineering Electroplating

Engineering nickel coatings are used to improve load bearing properties and provide resistance to corrosion, erosion, scaling and fretting. The coatings are applied to new parts such as rolls for glass making, laundry plates, wire and tube. They are also used for reclaiming worn gears, shafts and other parts of buses and ships, and as undercoats for engineering coatings of chromium.

Deposits from Watts-type solutions Most coatings of nickel for engineering applications are electrodeposited from a Watts-type bath[3]. Typical mechanical properties of deposits from Watts and sulphamate solutions are compared with those of wrought nickel in Table 14.14.

The uncertain effects of impurities are avoided by periodic or continuous electrolysis of the solution at low current densities to remove metallic contaminants and by filtration through active carbon to remove organic substances. A concise review of the effects of impurities and their removal is given by Greenall and Whittington[33].

Table 14.14   Typical mechanical properties of nickel deposits and wrought nickel

	Appearance as plated	Hardness (HV)	Ductility (% elongation)	Tensile strength (MN/m²)	Tensile stress (MN/m²)
Hot rolled and annealed Nickel 200	—	90–140	47	460	—
Watts nickel	Dull, matt	130–200	25	420	150
Conventional sulphamate nickel	Dull, matt	160–200	18	420	14

The mechanical properties of Watts deposits from normal, purified solutions depend upon the solution formulation, pH, current density and solution temperature. These parameters are deliberately varied in industrial practice in order to select at will particular values of deposit hardness, strength, ductility and internal stress. Solution pH has little effect on deposit properties over the range pH 1·0–5·0, but with further increase to pH 5·5, hardness, strength and internal stress increase sharply and ductility falls. With the pH held at 3·0, the production of soft, ductile deposits with minimum internal stress is favoured by solution temperatures of 50–60°C and a current density of 3–8 A/dm² in a solution with 25% of the nickel ions provided by nickel chloride. Such deposits have a coarse-grained structure, whereas the harder and stronger deposits produced under other conditions have a finer grain size. A comprehensive study of the relationships between plating variables and deposit properties was made by the American Electroplaters' Society and the results for Watts and other solutions were reported in 1952[34].

Hard nickel deposits   When the plating variables are adjusted to give deposits with a hardness much above 200 HV with a Watts solution, internal stress is usually too high and ductility too low for the deposits to be fully satisfactory. Higher hardness coupled with reasonable ductility can be achieved by addition of ammonium salts and operation at higher solution pH. A solution used for this purpose[35] and some deposit properties are as follows:

Composition
Nickel sulphate ($NiSO_4 \cdot 7H_2O$): 180 g/l
Nickel chloride ($NiCl_2 \cdot 6H_2O$): 30 g/l
Ammonium chloride (NH_4Cl): 25 g/l
Boric acid (H_3BO_3): 30 g/l

Operating conditions
Temperature: 60°C
pH: 5·6
Cathodic current density: 5 A/dm²

Deposit properties
Hardness: 400 HV
Tensile strength: 1·1 GN/m²
Elongation: 6%

Values of hardness higher than 400 HV (up to 600 HV) can be obtained by addition of organic substances to a conventional Watts solution. Similarly, internal stress can be made less tensile, zero or compressive, by the use of organic addition agents of the type used in organic bright nickel solutions. In practice, such hard nickel deposits are seldom used in engineering applications unless the required coating is so thin that no machining will be required.

Increased hardness and wear resistance may also be achieved by incorporating approximately 25–50% by volume of small non-metallic particles. These may be carbides, oxides, borides or nitrides, and hardness values up to 560 HV have been reported[36].

Deposits from sulphamate solutions The concentration of nickel ions in a conventional sulphamate plating solution is similar to that in a Watts solution, but nickel coatings deposited from the sulphamate bath have lower internal stress. Consequently higher plating rates than those employed in the Watts solution may often be used and this compensates for the higher initial cost of nickel sulphamate compared with nickel sulphate. Typical solution compositions and suitable operating conditions are given in Table 14.15 for the conventional solution and for a concentrated solution used for deposition at high rates:

Table 14.15 Typical solution compositions and suitable operating conditions for the conventional and the concentrated sulphamate baths

	Conventional solution (g/l)	Concentrated solution (g/l)
Compositions:		
Nickel sulphamate [Ni(NH$_2$SO$_3$)$_2$4H$_2$O]	300	600
Nickel chloride (NiCl$_2 \cdot 6$H$_2$O)	30	10
Boric acid (H$_3$BO$_3$)	30	30
Operating conditions:		
Temperature (°C)	25–50	60–70
Agitation	air	air
pH	3·5–4·5	3·5–4·5
Cathodic current density (A/dm^2)	2–15	2–80
Mean deposition rate (μm/h)	25–180	25–1 000

Replacement of nickel chloride by nickel bromide has been claimed[37] in the USA to reduce deposit stress, but subsequent German work[38] was unable to substantiate this finding.

Deposition of nickel at rates up to 1 mm/h in the concentrated solution is described by Kendrick[39]. If pure nickel anodes are operated at a current density between 0·5 and 1·0 A/dm^2 in sulphamate solutions, a substance which behaves as a stress reducer is produced continuously in sufficient quantity that the stress in deposits can be varied at will from compressive to tensile by adjusting cathode current density and solution temperature. This finding is exploited with the concentrated sulphamate solution in the Ni-Speed process[40], and in a recent development[41] cobalt is added to give deposits of hardness up to 500 HV. The nature of the stress reducer conveniently produced at the nickel anode is unknown but it differs[42] from the azo-disulphonate produced[43] at an insoluble anode such as platinum.

A comprehensive and authoritative study of the sulphamate bath has been made by Hammond[44].

Deposits from all-chloride solution   Nickel deposits from a solution of nickel chloride and boric acid are harder, stronger and have a finer grain size than deposits from Watts solution. Lower tank voltage is required for a given current density and the deposit is more uniformly distributed over a cathode of complex shape than in Watts solution, but the deposits are dark coloured and have such high, tensile, internal stress that spontaneous cracking may occur in thick deposits. There is therefore little industrial use of all-chloride solutions.

Deposits from other solutions   Nickel can be deposited from solutions based on salts other than the sulphate, chloride and sulphamate. Solutions based on nickel fluoborate, pyrophosphate, citrate, etc. have been extensively studied but none of them is used to any significant extent in Europe for engineering deposits.

Resistance to corrosion   Oswald[45] has surveyed the resistance of engineering coatings of nickel to corrosion by various chemical environments. Environments in which nickel has proved satisfactory include: (a) dry gases including ammonia, the atmosphere, carbon dioxide, coal gas, fluorine, hydrogen, nitrous oxide; (b) carbon tetrachloride, cider, creosote, hydrogen peroxide, mercury, oil, petrol, soaps, trichlorethylene, varnish; (c) alkalis (incl. fused), nitrates (incl. fused at 500°C), cheese, cream of tartar, eggs, fish, gelatin, fused magnesium fluoride, synthetic resins.

On heating in air, nickel forms a protective oxide and gives good service up to 700°C. Nickel is not recommended for exposure to chlorine, sulphur dioxide, nitric acid, sodium hypochlorite, mercuric or silver salts.

Where nickel is provided as a corrosion-resistant finish, a thickness of 120–130 μm is usually applied, but for well-finished basis metals and in mild environments, a lower thickness may be adequate. For parts machined after plating however, up to 0·5 mm may be required.

Effect of nickel coatings on fatigue strength   In general, a coating of high fatigue strength raises the fatigue resistance of a basis metal having low fatigue strength, and vice versa. Thus nickel coatings applied to steels of tensile strength greater than about 420 MN/mm^2 can lead to reduced fatigue strength. In practice, this reduction in fatigue resistance is often taken to be negligible for industrial components because the safety factor used in design is high enough to accommodate the degree of loss[45]. The loss can also be minimised either by using high-strength nickel deposits with compressive internal stress obtained by using appropriate addition agents, or by shot peening the surface of the steel before plating.

Effect on corrosion fatigue   The combination of corrosion and fatigue can cause rapid failure, and a coating of nickel, by preventing corrosion, can increase the life of the parts. Figure 14.10 shows results obtained by the National Physical Laboratory on mild steel Wöhler specimens sprayed with

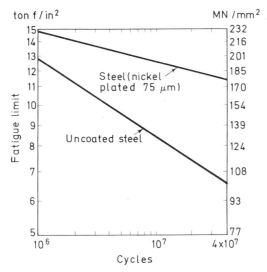

Fig. 14.10 Corrosion-fatigue tests; specimen sprayed with 3% sodium chloride (after Fescol Ltd.)

3% sodium chloride solution during testing at 2 200 cycles/min; the benefit given by the 75 µm nickel coating is clearly shown.

Effect on galling and fretting corrosion   Even when well lubricated, nickel tends to gall, i.e. stick, when rubbed against some metals, including other nickel surfaces. Nickel also tends to give galling in contact with steel and it is necessary to chromium plate the nickel. Nickel does not form a good combination rubbing against chromium or against phosphor-bronze, owing to the action on the nickel of the hard particles contained in the phosphor-bronze. Good performance is given by well-lubricated nickel against normal white-metal bearings, brasses or bronzes.

When two metals in intimate contact are subjected to vibration, a dark powder forms at the areas of contact. The effect is referred to as fretting corrosion though it is due to wear rather than true corrosive attack. The galling effect between nickel and steel ensures good resistance to fretting corrosion and lubricated nickel against steel is a very satisfactory combination used widely in industry for components assembled by press-fitting.

Heat treatment after plating   Heat treatment may be necessary after plating to improve the adhesion of coatings on aluminium and its alloys when certain processes, e.g. the Vogt process, are used, or to minimise hydrogen embrittlement of steel parts. Care is needed since heating may distort the part and impair the mechanical properties of the substrate.

Heat treatment to improve adhesion on aluminium and its alloys is normally carried out at 120–140°C for 1 h.

Heat treatment to minimise hydrogen embrittlement should be carried out immediately after plating and before any mechanical finishing operation. Delay is especially undesirable with steels having a tensile strength exceeding $1.4 \, GN/m^2$. Steels with tensile strengths below $1 \, GN/m^2$ are usually not heat treated. For the stronger steels, heat treatment is carried out at 190–230°C for not less than 6 h with steels of tensile strengths in the range $1-1\cdot85\,GN/m^2$, and for not less than 18 h in the case of even stronger steels.

Other aspects of engineering electrodeposited coatings   A great deal of information has been published on important, but specialised, aspects of engineering nickel coatings.

General guidance is provided by BS 4758:1971 *Electroplated Coatings of Nickel for Engineering Purposes*[46]. Cleaning, stopping off, etching, plating and subsequent machining of the coatings are discussed by Oswald[45], and the special pretreatments for maraging steels are described by Di Bari[47]. Detailed recommendations for turning, grinding, milling and boring nickel coatings are given by Greenwood[48]. Treatments that promote strong adhesion of subsequent nickel deposits after intermediate machining operations are discussed by Carlin[49]. The physical and mechanical properties of nickel at elevated and sub-zero temperatures, determined with electroformed test pieces, have been described by Sample and Knapp[50]. Other details of the properties of electrodeposited nickel coatings are given in Reference 51.

Electroless Nickel

In contrast to electrodeposited nickel, electroless nickel is deposited without application of electric current from an external supply. The metal is formed by the action of chemical reducing agents upon nickel ions in solution and, although several substances including hydrazine[52-54] and its derivatives will give metallic nickel, commercial processes use either sodium hypophosphite which gives a nickel–phosphorus alloy, or, sodium borohydride or various alkyl aminoboranes which give a nickel–boron alloy. These reducing agents can be used in either batch or continuous deposition processes. The amount of boron (typically 3–7%) or phosphorus (usually 5–12%) incorporated in the deposits depends upon solution composition and deposition conditions, and it determines to a large extent the properties of the deposit.

A major advantage of the electroless nickel process is that deposition takes place at an almost uniform rate over surfaces of complex shape. Thus, electroless nickel can readily be applied to internal plating of tubes, valves, containers and other parts having deeply undercut surfaces where nickel coating by electrodeposition would be very difficult and costly. The resistance to corrosion of the coatings and their special mechanical properties also offer advantages in many instances where electrodeposited nickel could be applied without difficulty.

Commercial processes   Commercial electroless nickel plating stems from an accidental discovery by Brenner and Riddell made in 1944 during the electroplating of a tube, with sodium hypophosphite added to the solution to reduce anodic oxidation of other bath constituents. This led to a process available under licence from the National Bureau of Standards in the USA. Their solutions contain a nickel salt, sodium hypophosphite, a buffer and sometimes accelerators, inhibitors to limit random deposition and brighteners. The solutions are used as acid baths (pH 4–6) or, less commonly, as alkaline baths (pH 8-10). Some compositions and operating conditions are given in Table 14.16[55].

Table 14.16   Brenner and Riddell electroless nickel solutions[55]

	Alkaline solution	Acid solutions		
		1	2	3
Composition (g/l)				
Nickel chloride, $NiCl_2 \cdot 6H_2O$	30	—	30	30
Nickel sulphate, $NiSO_4 \cdot 7H_2O$	—	30	—	—
Sodium hypophosphite, $NaH_2PO_2 \cdot H_2O$	10	10	10	10
Sodium acetate, $NaC_2H_3O_2 \cdot 3H_2O$	—	10	—	—
Sodium hydroxyacetate, $NaC_2H_3O_3$	—	—	10	—
Sodium citrate, $Na_3C_6H_5O_7 \cdot 5\frac{1}{2}H_2O$	—	—	—	10
Ammonium chloride, NH_4Cl	50	—	—	—
Operating conditions				
pH	8–10	4–6	4–6	4–6
Temperature (°C)	90	90	90	90
Plating rate (μm/h)	7·5	25	12·5	5
Appearance	Bright	Rough, dull	Semi-bright	—

Further development was made by the General American Transportation Corporation, and their Kanigen process[56,57] has been available since 1952. Other commercial processes based on the use of hypophosphite have since been developed. Work with reducing agents containing boron has given rise to the Nibodur process[58,59] which has been available since 1965.

Plating on plastics   Electroless nickel is used in thin deposits in order to provide an initial electrically-conducting surface layer in the preparation of plastics parts for electroplating. A typical procedure has as its first step an etching treatment of the plastic moulding in a solution of chromic and sulphuric acids in order to give a surface into which subsequent metallic deposits can key. The surface is then made catalytically active for electroless nickel deposition, usually by successive treatments in solutions containing tin compounds and compounds of a platinum group metal. Electroless nickel deposition is then followed by electrodeposition of the required coating which is usually copper plus nickel plus chromium.

Thorough rinsing between the pretreatment steps is essential to prevent carry-over of solutions. The commonest plastic plated is ABS (acrylonitrile butadiene styrene copolymer) but procedures are also available for polypropylene[2] and other plastics. In some proprietary processes, electroless copper solutions are used to give the initial thin conducting layer.

Engineering coatings   In the field of engineering coatings of electroless nickel, use of boron compounds as reducing agents has up until now been confined largely to Germany. A comprehensive account of electroless nickel–phosphorus and nickel–boron plating in Germany was published by International Nickel[60]. Electroless nickel–boron deposits have broadly similar mechanical, physical and chemical properties to those of electroless nickel–phosphorus deposits, and in the following discussion of deposit properties, data refer to nickel–phosphorus coatings unless otherwise stated.

Preparation of basis metals for plating   Preliminary cleaning of various basis metals follows the broad principles used for electrodeposited nickel.

Electroless nickel deposition may then be carried out directly onto steel, aluminium, nickel or cobalt surfaces. Surfaces of copper, brass, bronze, chromium or titanium are not catalytic for deposition of nickel–phosphorus and the reaction must be initiated by one of the following operations:

1. Apply an external current briefly so as to electrodeposit nickel.
2. Touch the surface with a metal such as steel or aluminium while immersed.
3. Dip in palladium chloride solution (this gives only modest adhesion and carries the danger of contamination of the bath by solution carry-over).

Antimony, arsenic, bismuth, cadmium, lead, tin and zinc cannot be directly plated by these techniques and should be copper plated.

Resistance to corrosion   Most authors who compare resistance to corrosion of electroless nickel with that of electrodeposited nickel conclude that the electroless deposit is the superior material when assessed by salt spray testing, seaside exposure or subjection to nitric acid. Also, resistance to corrosion of electroless nickel is said to increase with increasing phosphorus level. However, unpublished results from International Nickel's Birmingham research laboratory showed that electroless nickel–phosphorus and electrolytic nickel deposits were not significantly different on roof exposure or when compared by polarisation data.

Resistance to corrosion of electroless nickel, both as-deposited and, in most cases, after heating to 750°C, is listed by Metzger[61] for about 80 chemicals and other products. Resistance was generally satisfactory, with attack at a rate below 13 µm/year. The only substances causing faster attack were acetic acid, ammonium hydroxide or phosphate, aerated ammonium sulphate, benzyl chloride, boric acid, fluorophosphoric acid, hydrochloric acid, aerated lactic acid, aerated lemon juice, sodium cyanide and sulphuric acid.

Electroless nickel–phosphorus should not be used with either fused or hot, strong, aqueous caustic solutions because the coating offers lower resistance to attack than does electrodeposited nickel. As-deposited electroless nickel–boron, however, offers good resistance to hot aqueous caustic solutions[60]. It is also resistant to solutions of oxidising salts such as potassium dichromate, permanganate, chlorate and nitrate.

Heat treatment, e.g. 2 h at 600°C, improves the resistance to corrosion of nickel–boron and nickel–phosphorus electroless nickel deposits, especially to acid media. This presumably results from formation of a nickel–iron alloy layer[61].

Mechanical properties

Ductility   The ductility of electroless nickel deposits is low, but the brittleness of deposits containing less than 2% phosphorus can be reduced by heating to approx. 750°C for some hours followed by slow cooling.

Hardness   The hardness of electroless deposits is higher after heating to intermediate temperatures, the final value depending upon temperature and time of heating. Values of maximum hardness of nickel–phosphorus after heating to various temperatures[61] are plotted in Fig. 14.11; the variation of

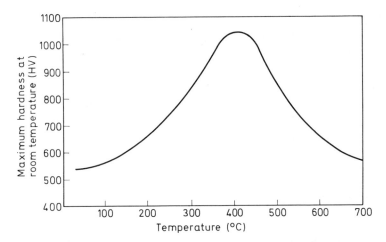

Fig. 14.11 Heat-treatment curve for electroless nickel (after Reference 61)

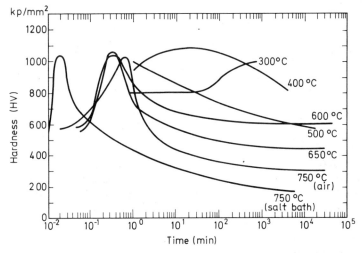

Fig. 14.12 Relationship between hardness and heat-treatment time for electroless nickel
(after Reference 62)

hardness with heating time[62] is shown in Fig. 14.12 for various heat-treatment temperatures. These curves show that hardness can be made to exceed 1 000 HV by appropriate heat treatment. Nickel–boron deposits can similarly be heat treated to values up to 1 200 HV.

Resistance to abrasion The resistance to abrasion of electroless nickel–phosphorus hardened to 600 HV, assessed by Taber abrasion tests, has been found to be double that of electroplated nickel[63]. However, electroless nickel coatings are not suitable for applications where two electroless nickel surfaces rub together without lubrication unless the values of hardness are made to differ by over 200 HV units. Galling of aluminium, titanium or stainless steel may be overcome by applying electroless nickel to one of the two mating surfaces.

Applications

In 1970, according to best estimates, 60 000 t of nickel coatings were deposited in the western world. This figure corresponds to 13% of the nickel consumed for all purposes.

Decorative coatings　It is impossible to give a comprehensive list of the uses of nickel coatings but applications of decorative nickel coatings, usually with a chromium top-coat, are given below:

1. Automotive: bumpers, grills, handles, over-riders, hubcaps, exhaust trim, locks, aerials, ash-trays, knobs.
2. Bicycles: rims, handlebars, spokes, cranks, hubs, bells, brake levers.
3. Perambulators: wheels, handles, springs, wing-nuts, body trim.
4. Door furniture: numbers, letter boxes, handles, bells, locks, keys.
5. Bathrooms: shower attachments, taps, chains, handles, locks, holders for soap and toothbrushes, mirror surrounds.
6. Kitchens: window fasteners, toasters, can-openers, trim for cooker, washing machine and dishwasher, clips.
7. General household: irons, needles, pins, press-studs, birdcages.
8. Tools: spanners, nuts, bolts, screws, screw-drivers, hacksaw bodies.

Toys, office equipment, sports equipment and shop furniture also provide large markets for decorative coatings of nickel or nickel plus chromium.

Engineering electrodeposits　Engineering electrodeposits are used to give improved properties on new components, or to replace metal lost by wear, corrosion or mis-machining, or as an undercoat for thick chromium deposits.

For new components, the nickel coating is usually 25–250 μm thick. Normally, the deposits are not machined. Applications include pump bodies, laundry plates, heat exchanger plates, evaporator tubes, alkaline battery cases and food-handling equipment of various sorts.

Machined deposits on new equipment, including undercoats for chromium, are usually 125–500 μm thick. Applications include cylinder liners (on the water side), cylinders used in the rubber, pulp and paper handling industries, compressor rods and armatures for electric motors.

Machined deposits for salvaging worn parts, with or without a chromium topcoat, are limited in thickness only by the economic limitation when it becomes cheaper to manufacture a new part; thicknesses up to 5–6 mm are used. Applications include axles, swivel pins, hydraulic rams, shafts, bearings and gears. Among larger installations, the repair shop of the London Transport Executive at Chiswick houses many nickel plating tanks devoted to reclaiming worn engine parts from 8 000 buses, and repairs are carried out on about 15 buses each week.

Electroless nickel engineering deposits　Electroless nickel is not usually deposited to thicknesses greater than about 125 μm. Where a greater total thickness is required, an electrolytic nickel undercoat should be used.

The number of applications has been growing at a considerable rate in recent years[60] and amongst the most common are hydraulic cylinders, tools for handling plastics, machine parts, printing cylinders, internal plating of valves and tubes, cooling coils, compressor housings, parts for pumps, storage vessels for chemicals, braking equipment, industrial needles, reaction vessels, filters, moulds for glass, and precision gears. Electroless nickel is also used as a pretreatment stage in the preparation of some printed circuits[64].

Electroforms Electroforming is electrodeposition onto a suitable mandrel which is subsequently removed so that the detached coating becomes the desired product. The process has the advantages that an object of intricate form can be produced in a single stage, a variety of desired surface textures can be reproduced simultaneously, a high order of accuracy is obtained in reproducing mandrel shape, and tools can be replicated exactly for mass-production work. Nickel has the particular advantages that its internal stress, hardness and ductility can be varied at will between wide limits and the final electroforms are strong, tough and highly resistant to abrasion, erosion and corrosive attack. The many applications of electroforming with nickel in Europe have recently been reviewed by Bailey, Watson and Winkler[42].

S. A. WATSON

REFERENCES

1. Saubestre, E. B., *Trans. Inst. Metal Finishing*, **47** No. 5, 228–235 (1969)
2. Innes, W. P., Grunwald, J. J., D'Ottavio, E. D., Toller, W. H. and Carmichael, L., *Plating*, **56**, 51–56, January (1969)
3. Watts, O. P., *Trans. Electrochem. Soc.*, **29**, 395 (1916)
4. Raub, E. and Disam, A., *Metalloberfläche*, **13**, 308–314, Oct. (1959)
5. Sellers, W. W. and Carlin, F. X., *Plating*, **52** No. 3, 215–224 (1965)
6. Saubestre, E. B., *Plating*, **45** No. 9, 927 (1958)
7. Tucker, W. M. and Beuckman, F. O., *Proc. 43rd Ann. Meeting Amer. Electroplater's Society*, 118–122 (1956)
8. Bouckley, D. and Watson, S. A., *Electroplating and Metal Finishing*, **20**, 303–310 and 348–353 (1967)
9. Weisberg, L., *Trans. Electrochem. Soc.*, **73**, 435–444 (1938)
10. Hinrichsen, O., UK Pat. 461 126 (1937)
11. Brenner, A. and Senderoff, S., *J. Res. Bur. Standards*, **42**, 89–104 (1949)
12. Hoar, T. P. and Arrowsmith, D. J., *Trans. Inst. Metal Finishing*, **36**, 1–6 (1959)
13. Fry, H. and Morris, F. G., *Electroplating*, **12**, 207–214 (1959)
14. Sykes, J. M., Ives, A. G. and Rothwell, G. P., *Journal of Physics E. Scientific Instruments*, **3**, 941–942 (1970)
15. Watson, S. A. and Edwards, J., *Trans. Inst. Metal Finishing*, **34**, 167 198 (1957)
16. Foulke, D. G. and Kardos, O., *Proc. Amer. Electroplater's Soc.*, **43**, 172–180 (1956)
17. Kruglikov, S. S., Kudryavtsev, N. T. and Semina, E. V., *Proc. 'Interfinish '68'*, 66–71, May (1968)
18. Brown, H., *Electroplating and Metal Finishing*, **15** No. 1, 14–17 (1962)
19. Edwards, J., *Trans. Inst. Metal Finishing*, **41**, 169–181 (1964)
20. Brown, H., *Trans. Inst. Metal Finishing*, **47**, 63–70 (1969)
21. Siegrist, F. L., *Metal Progress*, **85**, 101–104, March (1964)
22. Flint, G. N. and Melbourne, S. H., *Trans. Inst. Metal Finishing*, **38**, 35–44 (1961)
23. Brown, H., *Electroplating and Metal Finishing*, **15** No. 11, 398 (1962)
24. Oderkerken, J. M., *Electroplating and Metal Finishing*, **17** No. 1, 2 (1964)
25. Williams, R. V., *ibid.*, **19** No. 3, 92–96 (1966)
26. Seyb, E. J., *Proc. Amer. Electroplater's Soc.*, **50**, 175–180 (1963)
27. Millage, D., Romanowski, E. and Klein, R., 49th Ann. Tech. Proc. A.E.S., 43–52 (1962)
28. Hairsine, C., Longland, J. E. and Postins, C., *Electroplating and Metal Finishing*, **21**, 41–43, Feb. (1968)
29. Carter, V. E., *Trans. Inst. Metal Finishing*, **48** No. 1, 19–25 (1970)
30. UK Pats. 1 122 795 and 1 187 843

31. Safranek, W. H. and Miller, H. R., *Plating*, **52**, 873–878, Sept. (1965)
32. Davies, G. R., *Electroplating and Metal Finishing*, **21**, 393–398, Dec. (1968)
33. Greenall, G. J. and Whittington, C. M., *Plating*, **53**, 217–224, Feb. (1966)
34. Brenner, A., Zentner, V. and Jennings, C. W., *Plating*, **39**, 865–899, 933 (1952)
35. Wesley, W. A. and Roehl, E. J., *Trans. Electrochem. Soc.*, **82**, 37 (1942)
36. Kedward, E. C. and Kiernan, B., *Metal Finishing Journal*, **13**, 116–120, April (1967)
37. Searles, H., *Plating*, **53**, 204–208, Feb. (1966)
38. Brugger, R., *Nickel Plating*, Robert Draper (1970)
39. Kendrick, R. J., *Trans. Inst. Metal Finishing International Conf.*, **42**, 235–245 (1964)
40. Kendrick, R. J. and Watson, S. A., *Electrochimica Metallorum*, **1**, 320–334, July–Sept. (1966)
41. Belt, K., Crossley, J. A. and Watson, S. A., *Trans. Inst. Metal Finishing*, **48** No. 4, 132–138 (1970)
42. Bailey, G. L. J., Watson, S. A. and Winkler, L., *Electroplating and Metal Finishing*, **22**, 21–34 and 38, Nov. (1969)
43. Greene, A. F., *Plating*, **55**, 594–599, June (1968)
44. Hammond, R. A. F., *Metal Finishing Journal*, **16**, Part 1, June, 169–176; Part 2, July, 205–211; Part 3, August, 234–243 and Part 4, September, 276–285 (1970)
45. Oswald, J. W., *Heavy Electrodeposition of Nickel*, International Nickel Ltd., Publication No. 2 471 (1962)
46. BS 4758, *Electroplated Coatings of Nickel for Engineering Purposes* (1971)
47. Di Bari, G. A., *Plating*, **52** No. 11, 1157–1161 (1965)
48. Greenwood, A., *Metal Finishing Journal*, **11**, 484–490, Dec. (1965)
49. Carlin, F. X., *Plating*, **55**, 148–151, Feb. (1968)
50. Sample, C. H. and Knapp, B. B., *A.S.T.M. Special Technical Publication No. 318*, 32–42 (1962)
51. Watson, S. A., 'Engineering Uses of Nickel Deposits', *Electroplating and Metal Finishing*, May (1972)
52. Dini, J. W. and Coronado, P. R., *Plating*, **54**, 385–390 (1967)
53. Levy, D. J., *Electrochem. Technology*, **1**, 38–42 (1963)
54. Kozlova, N. I. and Korovin, N. V., *Zh. Prikl. Khim.*, **40**, 902–904 (1967)
55. Kreig, A., *A.S.T.M. Special Technical Publication No. 265*, 21–37 (1959)
56. Colin, R., *Galvanotechnik*, **57** No. 3, 158–167 (1966)
57. Heinke, G., *Metalloberfläche*, **21** No. 9, 273–275 (1967)
58. Lang, K., *Galvanotechnik*, **55**, 728–729 (1964)
59. Lang, K., *Metalloberfläche*, **19**, 257–262 (1965)
60. *Stromloses Dickvernickeln*, International Nickel Deutschland GmbH, Publication No. 63, (1971)
61. Metzger, W. H., *A.S.T.M. Special Technical Publication No. 265*, 13–20 (1959)
62. Wiegand, H., Heinke, G. and Schwitzgebel, K., *Metalloberfläche*, **22** No. 10, 304–311 (1968)
63. Chinn, J. L., *Materials and Methods*, **41**, 104–106, May (1965)
64. Lönhoff, N., *Trans. Inst. Metal Finishing*, **46**, 194–198 (1968)

BIBLIOGRAPHY

Dennis, J. K. and Such, T. E., *Nickel and Chromium Plating*, Butterworths, London (1972)

14.8 Chromium Coatings

It is not economically or technically feasible to use chromium in a fabricated form, but the high resistance of the metal to corrosion can be utilised by applying a thin coating of chromium to less resistant metals. Although the metal is base ($E^\circ_{Cr^{3+}/Cr} = -0.74$ V) it is protected by a thin, stable, tenacious, refractory, self-sealing film of Cr_2O_3. This is preserved by oxidising con-. ditions, and the metal is very resistant to high-temperature oxidation and to atmospheric exposure in most natural environments. Unlike silver and copper, it is not tarnished by hydrogen sulphide, nor is it 'fogged' like nickel by atmospheres containing sulphur dioxide.

The high reflectivity, pleasing blue-white colour, and the oxidation- and tarnish-resistance of the metal are the main reasons for its application in the form of thin coatings to cheaper and less resistant metals, for decorative purposes.

In addition, the extreme hardness of the metal, its low coefficient of friction and its non-galling property, combined with its corrosion resistance, make it particularly valuable as a coating where resistance to wear and abrasion are important. Thick deposits applied for this purpose are referred to as *hard* chromium to distinguish them from the thin decorative deposits.

Methods of Applying Chromium Coatings

The only methods of significance for producing chromium coatings are electrodeposition, chromising and vapour deposition. The last-mentioned is used only to a negligible extent for special high-temperature applications, as the coatings are less porous than electrodeposited chromium and are less liable to spall (see Section 13.5). The metal is deposited *in vacuo* from chromous or chromic iodide. Chromising produces coatings which are essentially alloys, and which are considered in Section 13.3. Electrodeposited chromium is one of the most widely used metallic coatings.

Electrodeposition

Electrodeposited chromium, both decorative and 'hard', is produced with the use of a solution of chromic acid containing a small amount of catalyst which is usually sulphuric acid, although fluosilicic or fluoboric acid may be used. A typical electrolyte contains 250–400 g/l of chromic acid and

2·5–4·0 g/l of sulphuric acid; the $CrO_3:SO_4{}^{2-}$ ratio is important and for satisfactory plating it should be maintained at about 100:1. If the catalyst content is too low no metal will be deposited, and if it is too high throwing power will be considerably reduced. The cathode efficiency is usually only 10–12% although up to 20% can be achieved with a silicofluoride catalyst. The evolution of hydrogen at the cathode and oxygen at the anode (6% antimonial–lead, which becomes coated with lead peroxide) necessitates provision for removal of the toxic spray by extraction or for the suppression of bubble formation by the addition to the baths of a perfluoro–carbon type of surface-active agent, the only type known to be stable under the prevailing conditions[1]. The chromium content of the bath is replenished by addition of chromic acid, as a chromium anode is not technically feasible.

The voltage used is 4–8 V, current density 9–22 A/dm^2, and temperature 38–43°C. Higher current densities, up to 55 A/dm^2, are used for thick deposits. A considerable amount of heat is generated during electrodeposition and provision must be made for cooling of the electrolyte during operation.

The covering and throwing power of the electrolytes is low, and bright plating which is required for decorative purposes can be obtained for a given composition and temperature only within a relatively narrow range of current densities. Outside this range, the deposits are not bright and the hardness of chromium is such that polishing is very difficult and uneconomical. Hence special care must be taken in the racking of irregularly shaped articles to avoid unplated areas, or dull and burnt deposits.

Self-regulating chromium The self-regulating chromium solutions were introduced to eliminate the need for maintaining the correct catalyst concentration by periodic analysis; they depend on the addition of a sparingly soluble sulphate to the bath which supplies the correct amount of $SO_4{}^{2-}$ automatically. Initially strontium sulphate (solubility approx. 1·75 g/l at 30°C and 21 g/l at 40°C) was employed for this purpose[2]. The strontium sulphate forms a layer on the bottom of the bath, which must be stirred from time to time. A bath with a CrO_3 concentration of 250 g/l would have a catalyst content of 1·52 g/l $SrSO_4$ and 4·35 g/l of K_2SiF_6. Potassium dichromate and strontium chromate have also found application as additives for the control of the saturation solubility of the catalyst.

Succinic acid has also been proposed[3] for the stabilisation of a self-regulating bath, a recommended bath composition consisting of 375 g/l of CrO_3, 8 g/l of $SrSO_4$, and 40 g/l of succinic anhydride; the bath is operated at 35°C. Self-regulating solutions generally have a higher current efficiency (18–25%) than the conventional bright solutions[4].

Tetrachromate electrolytes The alkaline tetrachromate baths are used to a small extent chiefly for the direct chromium plating of zinc die-castings, brass or aluminium, since the solutions do not attack these metals[5]. The original bath was developed by Bornhauser (German Pat. 608 757) and contained 300 g/l of chromic acid, 60 g/l of sodium hydroxide, 0·6–0·8 g/l of sulphuric acid and 1 ml/l of alcohol.

The essential constituent of the bath is sodium tetrachromate, $Na_2Cr_4O_{13}$, which is, however, only stable at temperatures below about 25°C. This temperature should therefore not be exceeded in the operation of the bath. Current densities of 75–150 A/dm^2 are used. The current efficiency of the bath is high (30–35%) so that the metal is deposited at the rate of about 1 μm/min. The deposits are normally matt in appearance, but are comparatively soft and readily polished.

A proprietary tetrachromate bath has been used in Germany under the name of the *D* process[6]. By the use of additions of magnesium oxide and sodium tungstate it is claimed that the current efficiency of the bath can be raised to as high as 35–40%. Other additives such as indium sulphate, sodium selenate or sodium hexavanadate enable bright deposits to be obtained.

Trivalent chromium baths in non-aqueous solvents Considerable attention has been given recently to the possibility of depositing chromium from trivalent chromium solutions containing non-aqueous solvents. One promising bath uses a dimethyl formamide–water solvent system having chromic chloride as an active salt with additions of ammonium chloride, sodium chloride and boric acid to improve current efficiency and conductivity.

Plating efficiencies are of the order of 30–40% based on Cr(III) and bright deposit can be obtained over the normal plating range of $25-1\cdot25\,A/dm^2$ at a plating speed of at least $0\cdot3\,\mu m/min$. The deposits are micro-discontinuous[7–10]. More recently, wholly aqueous trivalent chromium baths have been developed.

Properties of electrodeposited chromium

Structure Although massive chromium has a body-centred cubic structure, electrodeposited chromium can exist as two primary modifications, i.e. α-(b.c.c.) and β- (c.p.h.). The precise conditions under which these forms of chromium can be deposited are not known with certainty. Muro[11] showed that at 40°C and $2\cdot0-22\,A/dm^2$ the deposit was essentially α-chromium but small amounts of β- and γ- were present, while Koch and Hein[12] observed the β- form at 50°C and $40\,A/m^2$. This form is unstable, however, and is converted rapidly by heating or more slowly by storage at room temperature to the α- form.

The crystal structure is exceedingly fine and cannot be revealed by the microscope; Wood[13] has shown by X-ray diffraction that the grain size is $1\cdot4\times10^{-9}\,m$.

Porosity and discontinuities Chromium plate of $0\cdot5\,\mu m$ or less in thickness is invariably porous. An increase in thickness above this value, however, when plating is carried out under conventional conditions (i.e. 38–43°C, $11-16\,A/dm^2$, and a $CrO_3:SO_4^{2-}$ ratio of 100:1 to 120:1) results in a cracked deposit which can be revealed by microscopical examination at about × 350 magnification. Cohen[14] considers that the cracks are filled with a transparent film, probably of hydrated chromic oxide, which dehydrates on heating to form Cr_2O_3. According to Snavely[15], cracks and included material in the cracks are caused by the formation of unstable chromium hydrides during plating. A hexagonal form of the hydride (CrH to CrH_2) is formed initially, but decomposes spontaneously to α-chromium and free hydrogen. This involves a decrease in volume of over 15%, and since the plate is restrained by the basis metal, surface cracks form normal to the surface. The chemical constituents found in the electrodeposit are due to the drawing of electrolyte into the cracks, which are then covered over by subsequent layers of electrodeposit.

Black chromium plating   Black chromium deposits are frequently required for the optical and instrument industries. The deposits contain large amounts of chromium oxides and are not strictly speaking chromium deposits. Graham[16] recommends a solution consisting of 250 g/l of chromic acid, 0·25 g/l of hydrofluosilicic acid and a $CrO_3:H_2SiF_6$ ratio of 1 000:1. The bath is operated at about 32°C with a current density of about 30 A/dm^2 and a bath voltage of 6 V. The electrolyte solution must be free from sulphuric acid, excess sulphate ions being removed by treatment with barium sulphate. Silvery deposits of chromium containing some nickel are obtained at 70–100 A/dm^2 from a bath consisting of 200 g/l of chromic acid, 20 g/l of nickel chloride and 5 ml/l of glacial acetic acid. By a short immersion (5–30 s) in concentrated hydrochloric acid, the deposit becomes greyish black. Good black deposits are produced from a bath containing 200 g/l of chromic acid, 20 g/l of ammonium vanadate and 6·5 ml/l of glacial acetic acid at a current density of 95 A/dm^2 and a temperature of 35–50°C[17]. Some types of black chromium deposits are claimed to have very good corrosion resistance.

Hard chromium plating   The so-called 'hard' (or thick) chromium deposits are applied on carbon and alloy steels, cast iron and light alloys, to improve resistance to wear, abrasion and corrosion. The solutions employed generally contain 150–500 g/l of chromic acid and employ a $CrO_3:H_2SO_4$ ratio of 80–120. Deposit thicknesses of 12–150 μm are applied, but the use of thicker deposits is limited to parts which are not subject to bending or stress. Plastic moulds are generally plated with coatings of 10–15 μm, which are considered adequate. Hard chromium deposits are normally ground or lapped before being put into service, and allowances must be made for this operation to be carried out. Applications for hard chromium deposits include cylinder liners, crankshafts, pump shafts, plastic moulds, dies, cams, rockers, journals and bearings.

    Plating is carried out by suspension in the bath in the usual manner, areas which are not to be plated being protected by 'stopping off' materials, such as lacquers, waxes or plastics. Chromium can also be deposited locally without the use of a tank by the tampon method. The Dalic[18] process makes use of an insoluble anode and an absorbent pad containing the electrolyte solution, which is a slightly alkaline organic complex amino-oxalate compound of chromium dissolved in an alcohol, with a wetting agent added. High current densities are used, and rates of deposition of up to 2·5 μm/min are practicable. The deposit is slightly softer than the conventional hard chromium deposits.

    Hard chromium plating provides excellent resistance to atmospheric oxidation both at normal temperatures and at temperatures of up to 650°C. It is unattacked by many chemicals, owing to its passivity. When attack takes place, this usually commences at cracks in the chromium network; hence the most corrosion-resistant deposits must have a very fine structure, such as is obtained from relatively high solution temperatures using low current densities.

Corrosion

Electrodeposited chromium, if it is required to protect an underlying metal against corrosion, has to be applied in considerable thicknesses, owing to its high porosity and tendency to crack. Such deposits are expensive to produce and are not fully bright, and, as the polishing of chromium is difficult, it is the general practice to use a protective undercoat, usually nickel, when ferrous or non-ferrous metals have to be protected. For wear resistance or engineering applications, however, 'hard' chromium coatings are usually plated directly on to steel and other metals at thicknesses of up to approximately 0·50 mm as against 0·000 25–0·002 0 mm for decorative chromium deposits on a nickel undercoat.

When corrosion of a chromium-coated metal takes place, the corroding current concentrates its action on fissures in the deposit. There appears to be an incubation period, after which rapid attack occurs in the form of pits, and sometimes a network of corrosion can be observed. The chromium becomes cathodic, the underlying metal (usually nickel) which is exposed at the pores or stress cracks of the chromium plate being anodic (see Fig. 14.13).

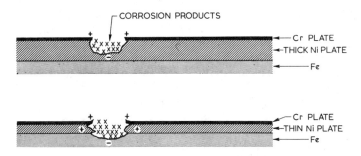

Fig. 14.13 Two stages in the corrosion commencing at a discontinuity in the chromium plating (after *Electroplating and Metal Finishing*, **12** No. 1, 3 (1959))

Dettner[19] claims that the degree of polish of the base metal has a definite influence on the corrosion resistance of chromium deposited directly on steel, a high degree of polish leading to improved protection; electrolytic polishing is said to lead to particularly good durability[20]. Polishing of the chromium layer usually has little effect, but excessive heat generation can lead to reduced corrosion resistance. It is desirable, from a practical point of view, that the chromium be deposited within the bright plating range, and although this does not always coincide with the conditions necessary for maximum protection, a reasonable compromise can be reached.

Crack-free Chromium

As has already been stated, attempts to reduce the porosity of chromium plating by increasing its thickness much above 0·000 5 mm result in cracked coatings when normal solutions and conditions are employed. It is, however, possible to obtain crack-free deposits in thicknesses up to 0·002 5 mm, with a consequent improvement in corrosion resistance as shown by accelerated tests, by the operation of the bright chromium plating bath at 49–54°C, and higher $CrO_3:SO_4^{2-}$ ratios of 150:1 to 200:1[21-23]. Crack-free deposits can be obtained under conditions outside these ranges, but for practical operation these are the ones which it is most convenient to employ.

The main drawback to plating chromium under these conditions is that the current requirements are greater owing to the need to work at twice or three times the conventional current density. There is also some tendency for the deposit to be rather more blue in colour, while frostiness can develop at high-current-density areas. Better results can be obtained by allowing the work to enter the plating tank at a lower voltage (2–3 V) before applying the full plating voltage, or by working at a temperature of around 49°C.

Some confirmation of these findings has been reported by Safranek et al.[24] in their work on the corrosion resistance of plated die-castings. They found that 0·000 64 mm (minimum) of bright crack-free chromium deposited in a high-ratio bath at 54°C will extend the 'corrosion-free' life of plated die-castings in highly corrosive environments to at least one year, as compared with less than six months for normal deposits. Bright, crack-free chromium deposited on 0·007 6 mm of copper and 0·020 mm minimum of bright nickel furnished good protection against accelerated corrosion.

Deposits of more than around 0·002 0 mm in thickness cannot be applied in this way without the initiation of cracks visible to the naked eye particularly at high-current-density areas. This is a disadvantage, since owing to the poor 'throw' of chromium it is sometimes necessary to exceed this thickness at such areas on articles of complex shape in order to secure an adequate deposit in recesses. A method of overcoming this problem known as *duplex* chromium plating (see below) has recently been developed.

Microcracked Chromium

Duplex Chromium

It has been claimed that better corrosion protection than that afforded by the high-temperature chromium-plating method can be obtained by the use of a double chromium plate. In this system, bright crack-free chromium is deposited as described above, followed by an equal thickness of a bright, finely cracked chromium plate. The total chromium thickness should be not less than 0·000 75 mm, and should preferably be greater. The initial crack-free deposit may be obtained by plating from a solution of chromic acid and sulphuric acid (250–400 g/l of chromic acid, chromic acid to sulphate ratio 125 to 175:1), operating temperature 49–54°C. Baths containing fluoborates, which are self-regulating so far as the ratio of chromic acid to catalyst is concerned, can also be used successfully for producing crack-free deposits[25].

Immediately after this deposit has been plated, a cracked chromium layer is applied from a dilute bath containing about 200 g/l of chromic acid at 46–52°C. The articles to be plated should enter the bath at a low voltage to prevent streakiness. Fluoboric or fluosilicic acid in the bath, in addition to sulphuric acid, helps to produce the required fine stress cracking.

Single-layer Chromium

A number of proprietary solutions are now available for producing the same result from a single bath. The plating time tends to be rather longer, but this can be reduced either by increasing the current density (which may upset the crack pattern), by decreasing the chromium thickness at which cracking occurs, or by increasing the cathode efficiency.

The effect of the finely cracked chromium layer is to equalise the anode and cathode areas more nearly, so that corrosion of the nickel under the chromium takes place more slowly than it would at larger, isolated cracks. Moreover, the corrosion proceeds laterally along the nickel surface and not in depth as is the case with conventional chromium; hence failure of the coating under adverse conditions is less likely to occur.

A crack count of 30–80 cracks/mm is desirable to maintain good corrosion resistance. Crack counts of less than 30 cracks/mm should be avoided, since they can penetrate into the nickel layer as a result of mechanical stress, whilst large cracks may also have a notch effect[26]. Measurements made on chromium deposits from baths which produce microcracked coatings indicate that the stress decreases with time from the appearance of the first cracks[27]. It is more difficult to produce the required microcracked pattern on matt or semi-bright nickel than on fully bright deposits[28]. The crack network does not form very well in low-current-density areas, so that the auxiliary anodes may be necessary.

Corrosion tests have shown that a system based on copper, double nickel and microcracked chromium gives good corrosion resistance, although automobile parts plated with microcracked chromium are not as easy to clean as those plated with crack-free chromium deposit.

Microporous Chromium

One of the best methods of improving the corrosion resistance of nickel–chromium deposits is to apply a uniformly porous layer, rather than a microcracked chromium layer, this having the advantage that the microporosity is not greatly dependent on the current density at which the chromium plating is carried out. Hence the chromium can be deposited in microporous form on quite complex-shaped articles from a single, conventional chromium solution. The method of achieving this is to suspend inert particles in the underlying nickel coating; the presence of these, being non-conducting, results in the formation of a highly microporous chromium deposit. Severe electrochemical attack of the underlying nickel at large cracks or pores in the chromium is thus prevented, and a substantial improvement in the corrosion resistance of the combined coating is obtained. A relatively thick copper deposit (75 µm) underneath the nickel layer has been found to add considerably to the protective value of the coating. Thereafter very little improvement occurs. The large number of microscopic anode nickel sites which develop when about 0·25 µm of chromium is applied results in very weak corrosion currents with extremely low corrosion penetration. The number of pores in the chromium can be varied from about 3 000 per square centimetre to several million per square centimetre[29]. The variation in the porosity of microporous chromium with thickness is shown in Fig. 14.14.

In practice a special nickel solution containing the suspended particles is applied over the normal bright nickel deposit. The plating time in this

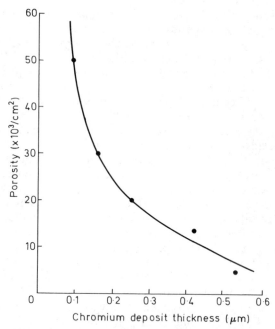

Fig. 14.14   Variation of porosity of microporous chromium with thickness of deposit

solution is from 20 s to 5 min; the most suitable ratio of the two deposits has to be determined in each particular case.

The use of a chromium deposit with a fine porosity pattern of 15 000 to 45 000 pores per square centimetre in the usual thickness results in a sharp slowing down of the corrosion rate. Such corrosion as does occur develops laterally, thus very greatly delaying the downward penetration into the vulnerable base metal (Fig. 14.15).

Fig. 14.15   Lateral corrosion in nickel deposit layer containing inert particles beneath microporous chromium

Since the theory of the mechanism of the microporous chromium system depends on the fact that the occlusions in the underlying nickel provide a large number of sites where nickel can be corroded at discontinuities in the nickel deposit, it was at one time believed that increasing the chromium thickness excessively would be disadvantageous, as it would seal some of the active nickel sites. Carter[30] has shown that this is not the case provided that the porosity in the chromium is not reduced below about 15 000 pores per square centimetre. When copper was present under the nickel plus chromium coating, full protection of steel was obtained in industrial atmospheres for two years in all environments. This effect on copper is not found under conventional chromium coatings. The reason for this is ascribed to the fact that the copper remains cathodic to the nickel because of the large area of nickel involved in the corrosion reaction when the chromium layer is discontinuous[31]. With conventional deposits, however, the smaller area of the corroding nickel allows high current densities to occur in the pits so that the copper becomes anodic and is readily penetrated. The effect is reduced when the nickel layer adjacent to the copper is of a less corrodible type (i.e. semi-bright, dull nickel) and hence the advantages of the copper undercoat are less in the systems employing double nickel deposits. There is some deterioration or dulling in the appearance of articles plated with microporous chromium (as is also the case with the microcracked deposits), but this is only significant on exposure in the severest environments. Dulling was progressively reduced by increasing thickness of the chromium deposit within the range studied, without adverse effect on the protection of the basis metal.

Chromium–nickel–chromium

A further development (1959) is the use of a combined chromium–nickel–chromium or nickel–chromium–nickel–chromium deposit on steel- or zinc-base alloy articles[32]. An advantage of this system is that the first chromium layer need not be plated within the bright range of the chromium bath, so that plating can be carried out under conditions giving deposits of maximum corrosion resistance; such conditions do not coincide with those under which fully bright chromium plate is obtained.

Knapp[33] reports that a chromium deposit of 0·000 25 mm from the usual type of chromium bath, followed by 0·013 mm nickel and a further 0·000 25 mm of chromium gave protection equal to that of a nickel coating of double the thickness applied in the form of normal nickel and chromium plate.

Porous Chromium

Porous chromium is largely used on cylinder liners for automobile engines, its advantage being that it retains lubricants better than normal chromium[34]. The porosity is usually created by etching the metal. Appropriate etching methods include reversal of current to make the work anodic in the plating solution, and cathodic or chemical treatment in a separate bath. Hydrochloric, sulphuric or oxalic acid may be used as the etching electrolyte in a separate bath, the work being made cathodic; alternatively, chemical etching without current in a hot dilute sulphuric acid or hydrochloric acid bath, to which an inhibitor such as antimony oxide is added, can be employed. Whether pin-point or channel porosity is produced depends primarily on the conditions of deposition, solution temperature and composition being the principal factors. Generally, higher temperatures and higher sulphate ratios in the bath favour channel-type porosity. The degree of porosity must be carefully controlled in order to ensure that excessive roughness is not produced. The ideal condition is one where the chromium becomes adequately receptive for oil but remains smooth. It is usual to hone or lap the porous chrome; careful cleaning is then essential to remove the débris produced by honing.

H. SILMAN

REFERENCES

1. UK Pat. 758 025 (1953)
2. Stareck, J. E., Passal, F. and Mahlstedt, H., *Proc. Am. Electroplaters' Soc.*, **37**, 31–49 (1950)
3. Dutch Pat. 6 513 035 (1966)
4. Griffin, J. L., *Plating*, **53**, 196–203 (1966)
5. Twist, R. D., *Prod. Fin.*, **25** No. 20, 37 (1972)
6. Kutzelnigg, A., *Metalloberfläche*, **5** No. 10, 156–160 (1953)
7. UK Pat. 1 144 913, March (1969)
8. Bharncho, N. R. and Ward, J. J. B., *Prod. Fin.*, **33** No. 4, 64–72, Jan. (1969)
9. Ward, J. J. B., Christie, I. R. A. and Carter, V. E., *Trans. Inst. Met. Fin.*, **49**, 97 (1971)
10. Ward, J. J. B. and Christie, I. R. A., *Trans. Inst. Met. Fin.*, **49**, 148 (1971)
11. Munro, Z. and Batsuri, O., *J. Appl. Phys.*, **21**, 321 (1952)
12. Koch, L. and Hein, G., *Metalloberfläche*, **7**, A145 (1953)
13. Wood, W. A., *Trans. Farad. Soc.*, **31**, 1 248 (1935)
14. Cohen, J. B., *Trans. Electrochem. Soc.*, **86**, 441 (1944)
15. Snavely, C. A., *Trans. Electrochem. Soc.*, **92**, 537 (1947)
16. Graham, A. K., *Proc. Am. Electroplaters' Soc.*, **46**, 61–63 (1959)
17. Queseley, M. F., *Plating*, **40**, 982 (1953)
18. *Electroplating*, **6** No. 4, 131 (1953)
19. Dettner, H. W., *Metalloberfläche*, **4**, A33 (1950)
20. Eilender, W., Arend, H. and Schmidtmann, E., *Metalloberfläche*, **2**, 141 (1948)
21. Dow, R. and Stareck, J. E., *Proc. Amer. Electropl. Soc.*, **40**, 53 (1953)
22. Brown, H., Weinberg, M. and Clauss, R. J., *Plating*, **2**, 144 (1958)
23. Brown, H. and Millage, D. R., *Trans. Inst. Met. Fin.*, **37**, 21 (1960)
24. Safranak, W. H., Miller, H. R. and Faust, C. L., *46th Ann. Proc. Amer. Electropl. Soc.*, 133 (1959)
25. Morriset, P., Oswald, J. W., Draper, C. R. and Pinner, R., *Chromium Plating*, Robert Draper, London, 123–129, 325–326 (1954)
26. Gabe, D. R. and West, J. M., *Trans. Inst. Met. Fin.*, **40**, 197–202 (1963)
27. Such, T. E. and Partington, M., *Trans. Inst. Met. Fin.*, **42**, 68–75 (1964)
28. Dennis, J. K., *Trans. Inst. Met. Fin.*, **43**, 84–96 (1965)
29. Brown, H. and Silman, H., *Proc. 6th International Conference on Electrodeposition and Metal Finishing*, 50–56 (1964)
30. Carter, V. E., *Trans. Inst. Met. Fin.*, **48**, 19 (1970)
31. Claus, R. J. and Klein, R. W., *Proc. 7th International Conference on Metal Finishing*, 124 (1968)
32. Weinberg, M. and Brown, H., US Pat. 2 871 550 (1959)
33. Knapp, B. B., *Trans. Inst. Met. Fin.*, **35**, 139 (1958)
34. Gray, A. G., *Modern Electroplating*, Wiley, New York, 135–188 (1953)

BIBLIOGRAPHY

Dennis, J. K. and Such, T. E., *Nickel and Chromium Plating*, Butterworths, London (1972)

14.9 Noble Metal Coatings

The most widely used methods for the application of coatings of gold, silver and the platinum group metals (platinum, palladium, rhodium, iridium, ruthenium, osmium) to base metals are mechanical cladding and electroplating.

In cladding, the coating is applied in the form of sheet, which may be bonded to the underlying metal by brazing or by elevated-temperature working processes such as swaging and drawing for the production, for example, of platinum-clad molybdenum or tungsten wire, or hot-rolling followed by spinning, cupping or drawing, for the production of coated dishes, tubes, etc. Silver coatings are extensively applied in this way in the lining of chemical reaction vessels, distillation and evaporation equipment, etc. particularly in fine chemical manufacture and food processing where product purity is vital and the protective coating must therefore be completely impervious. The main advantage of silver in this type of application, apart from its relative cheapness as compared with other metals of the group, is its good resistance to organic acids and other compounds and to chloride-containing media. Its high heat-transfer capacity is also a useful asset. Platinum and gold find application in a similar sense where particular conditions warrant their cost. Coating thicknesses may vary from less than 0·025 mm to 0·640 mm, depending on service requirements.

Palladium can be applied in the same way, but is not employed to a significant extent in this form since its corrosion resistance is inferior to that of platinum. Application of other metals of the platinum group, i.e. rhodium, ruthenium and iridium, as protective claddings is hindered by limitations in working technology. In experimental work on the protection of soldering iron bits by ruthenium, the expedient has been adopted of fabricating small hollow cones by compacting and sintering ruthenium powder, and fixing these to the tips by brazing[1].

In the case of silver and gold, thick coatings of equivalent protective value to those produced by cladding can be obtained by electrodeposition; both metals have, in fact, been successfully employed for electroforming. In the more general case, however, electrodeposited coatings, particularly those of the platinum group metals and, to a lesser extent, gold plated in the bright condition, are subject to some degree of porosity and, with increasing thickness, to the possibility of spontaneous cracking due to internal stress in the as-deposited condition. Nevertheless, the bulk of precious metal coatings used for decorative and industrial purposes, including tonnage use in the electronics field, are applied by electroplating, since protective requirements, though arduous, are in most cases less critical than those demanded by long term exposure to liquid or high-temperature corrosive environments, and some degree of porosity can often be tolerated.

Processes for electrodepositing silver, gold, platinum, palladium and rhodium have been long established. In this group the most striking development of recent times has been the emergence of bright gold plating solutions which utilise the stability of gold cyanide at relatively low pH values to plate under acid conditions[2,3], and more recently still, of non-cyanide electrolytes based on sulphite complexes[4]. Relatively new electrolytes have also been formulated for the deposition of ruthenium, iridium and even osmium, though these are subject to limitations with regard to the thickness of sound coatings. Platinum metal coatings may also be produced from fused cyanide electrolytes[5], a technique which is useful in those cases (e.g. ruthenium and iridium) where coatings of sufficient thickness cannot be produced from aqueous solutions. Iridium coatings of this type have been studied in connection with the high-temperature protection of molybdenum[6], and thick coatings of rhodium have been produced in a similar way[7]. Since they are deposited at a temperature of the order of 600°C from a non-aqueous medium, such coatings tend to be less stressed, softer and less porous than coatings from aqueous solutions. For example, rhodium from a bath of this type shows a hardness of approximately 300 HV compared with about 900 HV from a conventional sulphate electrolyte.

Limitations imposed upon the thickness of coating obtainable from aqueous solutions due to internal stress have been overcome in several directions. Atkinson[8] has reported the production of ductile, crack-free platinum coatings from a chloro-platinic acid plus hydrochloric acid electrolyte. Tripler, Beach and Faust[9] have achieved improvement in the protective value of platinum coatings from a di-ammino di-nitrito platinum electrolyte by the use of the periodic reverse current technique, which is also widely applied in gold plating. Patent claims have been made for the production of crack-free rhodium deposits from sulphate electrolytes modified by the addition of magnesium salts[10] or of selenic acid[11]. Highly ductile palladium coatings have been produced by Stevens in thicknesses up to 5 mm from a tetrammino-palladous bromide electrolyte[12].

Electroless plating processes of the chemical displacement type and of the auto-catalytic type have also been described. In the former, a thin film of the noble metal is formed on a base metal substrate by chemical replacement. The reaction may cease when the substrate is completely covered, or, as in the processes described by Johnson[13] for the platinum group metals, attack of the substrate may continue through an essentially porous top-coat, which may exfoliate on prolonged treatment. Such processes are of main utility for short-term protection purposes, e.g. retention of solderability of electronic components during storage. In auto-catalytic processes, further deposition of metal is catalysed by the initial layer of the coating itself. Processes of this type have been described for both gold[14] and palladium[15,16].

Electrodeposited Coatings

Silver and gold Apart from their traditional decorative applications, both silver and gold find important industrial use in various types of chemical processing equipment. In the electrical and electronics industries they are employed as plated coatings on contacts and as finishes on wave guides, hollow conductors for high-frequency currents, etc. Silver plating is particularly used in the latter application where its high electrical and thermal conductivity are required in addition to its protective value. The thickness of coating necessary for adequate protection depends on the conditions of service and the nature and condition of the basis metal to which the coating is applied. For electrical purposes for the protection of aluminium, steel and copper, DTD 919A specifies a minimum thickness of 0·000 3 in (0·007 5 mm) of silver, with a total thickness of undercoating deposit (copper or nickel) and silver of 0·001 5 in (0·038 mm) in the case of aluminium and 0·000 8 in (0·020 mm) for steel. BS 2816:1957 *Electroplated Coatings of Silver for Engineering Purposes* is also relevant in this context.

Laister and Benham[17] have shown that under more arduous conditions (immersion for 6 months in sea-water) a minimum thickness of 0·025 mm of silver is required to protect steel, even when the silver is itself further protected by a thin rhodium coating. In similar circumstances brass was completely protected by 0·012 5 mm of silver. The use of an undercoating deposit of intermediate electrode potential is generally desirable when precious metal coatings are applied to more reactive base metals, e.g. steel, zinc alloys and aluminium, since otherwise corrosion at discontinuities in the coating will be accelerated by the high e.m.f. of the couple formed between coating and substrate and by the electrochemical activity of the previous metal coating. The thickness of undercoat may have to be increased substantially above the values indicated if the basis metal shows special defects such as porosity.

In view of its susceptibility to sulphide tarnishing, silver may itself require some measure of protection in many decorative and industrial applications. Chromate passivation processes are commonly employed, but as an alternative, thin coatings of gold, rhodium or palladium may be used.

Although usage of gold plating in industrial applications has long outstripped that in traditional decorative fields, it is only recently that an appropriate British standard has been issued to cover both spheres of application[18]. The high reflectivity of gold in the infra-red region accounts for its use on reflectors in infra-red drying equipment, for which purpose a coating of 0·005 mm gives excellent service on beryllium–copper. This order of thickness is general in the protection of electrical contacts in the electronics field, where the main area of industrial gold plating is to be found. The basis metals involved are most commonly copper or copper-base alloys, e.g. brass, nickel–silver, beryllium–copper and phosphor bronze, and coating thickness is dictated not only by environmental conditions but by the need for mechanical wear resistance in sliding and wiping contacts, in which context the softness of pure gold deposits from cyanide electrolytes may be a disadvantage.

Numerous proprietary electrolytes have been developed for the production of harder and brighter deposits. These include acid, neutral and alkaline solutions and cyanide-free formulations and the coatings produced may be essentially pure, where maximum electrical conductivity is required, or alloyed with various amounts of other precious or base metals, e.g. silver, copper, nickel, cobalt, indium, to develop special physical characteristics.

The hardness of such coatings may reach a maximum of about 400 HV as compared with approximately 50 HV for a soft gold deposit. A series of corrosion studies in industrial and marine atmospheres by Baker[19] has indicated that the protective value of hard gold coatings is comparable with that of the pure metal, and that a thickness of only 0·002 5 mm gives good protection to copper base alloys during exposure for six months.

In view of the high cost of gold there is a continuing urge to reduce coating thickness in industrial applications to the bare minimum consistent with adequate service life. It is claimed for example that the thickness of gold on a wiping contact can be reduced by using an undercoating of silver, e.g. 0·007 5 mm of silver plus 0·000 25 mm of gold. In this case a special problem arises, particularly at elevated temperature, due to diffusion of silver outwards through the gold layer, with formation of a tarnish film at the surface. This effect can be obviated by interposing a thin deposit of palladium or rhodium between the gold and silver layers[20].

At gold thicknesses below 0·005 mm significant porosity is likely to be present, and a great deal of recent work has been directed to the study of factors affecting the degree of porosity of gold coatings[21-24], and to possible means of reducing this, or at least minimising its practical effect. Reduction of porosity can be achieved by the use of copper or nickel undercoats and recent patents claim that a coating of platinum only 0·38 μm thick will substantially reduce porosity and improve the high temperature stability of 0·002 5 mm gold coatings on copper[25]. The effect of corrosion through pores in thin gold coatings on copper- or silver-base substrates can be minimised by applying a thin coating of palladium or rhodium, since sulphide tarnish products do not spread on these metals[26], whereas they readily spread over gold to form large areas of high contact resistance. Gold coatings on sliding contacts are often lubricated, and it is claimed that pores in the coatings may be effectively sealed, with marked increase in service life, by incorporating a suitable corrosion inhibitor in the lubricating system[27].

The Platinum Metals

Rhodium   Rhodium is the most important of the platinum group of metals as an electrodeposited coating for protective purposes as shown by the fact that it is the only metal of the group for which a DTD Process Specification exists (No. 931). Major fields of application are the protection of silver from tarnishing in both decorative and industrial spheres, and the finishing of metallic reflectors and electrical contacts (particularly sliding or wiping contacts subject to mechanical wear and concerned with the transmission of very small electrical signals, e.g. in radar, telecommunication, and allied equipment, where freedom of the contact surface from films is a critical requirement). The special properties of the electrodeposited coating on which these applications depend are its high reflectivity, virtual immunity from attack by corrosive environments, its consequently low and stable contact resistance, and its extremely high hardness (approximately 900 HV). A disadvantage of the deposit, as produced from conventional acid sulphate or phosphate plus sulphate electrolytes, is a high internal tensile stress, which may give rise to cracking in deposits thicker than 0·002 5 mm and which, as indicated earlier, places strict limitations on the usefulness of the coating for protection against severely corrosive liquid environments. The value of rhodium against atmospheric corrosion in environments ranging from domestic to marine and tropical exposure has, however, been amply demonstrated by the experience of the last two decades, and it appears probable that the recent developments in technology referred to in the opening paragraphs may permit wider application.

In view of the high cost, when tarnish resistance of the surface is the only requirement it is customary to use the thinnest possible coatings of rhodium (0·000 25–0·000 5 mm). Since rhodium deposits in this thickness range, like thin electrodeposits of almost any metal, show significant porosity, readily corrodible metals, e.g. steel, zinc-base alloys, etc. must be provided with an undercoating deposit, usually of silver or nickel, which is sufficiently thick to provide a fairly high level of protection to the basis metal even before the final precious metal deposit is applied, and, in this way, to prevent accelerated electrochemical corrosion at pores in the rhodium deposit.

It is not possible to plate rhodium directly on to reactive metals of the type mentioned above, in view of the acid nature of the electrolyte, but copper and its alloys, e.g. nickel–silver, brass, phosphor-bronze, beryllium–copper, which are of special importance in the electrical contact field, may be plated directly. Even in this case, however, an undercoat is generally desirable.

Whether nickel or silver is selected for use as an undercoating is determined by a number of factors, relative resistance to particular corrosive environments being clearly of primary importance. Laister and Benham[17] have discussed the respective merits of the two metals on the basis of corrosion tests in a number of environments. Generally speaking, silver is preferred when the composite coating is required to resist exposure to marine or other chloride-containing atmospheres, the potential difference between silver and rhodium in sea-water at 25°C being only 0·05 V[28]. Rhodium–nickel coatings are superior against sulphide atmospheres and as coatings required to operate at elevated temperatures (up to 500°C). In this connection, it should be noted that rhodium itself will begin to oxidise at temperatures in the range 550–600°C.

Silver is often preferred as an undercoat for rhodium by reason of its high electrical conductivity. A further advantage of silver in the case of the thicker rhodium deposits (0·002 5 mm) applied to electrical contacts for wear resistance is that the use of a relatively soft undercoat permits some stress relief of the rhodium deposit by plastic deformation of the under-layer, and hence reduces the tendency to cracking[29], with a corresponding improvement in protective value. Nickel, on the other hand, may be employed to provide a measure of mechanical support, and hence enhanced wear resistance, for a thin rhodium deposit. A nickel undercoating is so used on copper printed connectors, where the thickness of rhodium that may be applied from conventional electrolytes is limited by the attack by the plating solution on the copper/laminate adhesive, and by the lifting effect of internal stress in the rhodium deposit.

A thickness of 0·000 38 mm may be regarded as a good quality finish for general decorative and industrial use for tarnish protection at normal temperatures. For optimum tarnish resistance at temperatures up to 500°C,

0·001 25 mm of rhodium on a nickel undercoat is recommended. In sliding contact applications, where the ability of the coating to withstand some degree of mechanical wear is almost as important as tarnish resistance, the order of thickness employed is 0·002 5–0·005 mm and, in a few special circumstances, this may be increased to 0·012 5 mm or more.

Palladium　Although satisfactory palladium plating processes have existed for many years, the metal has only recently attained industrial significance as an electrodeposited coating, and at the present time it is of considerable interest as an alternative to rhodium or gold in the finishing of electrical contacts, especially in copper end connectors of printed circuits[30]. Apart from its relatively low cost, palladium has special technical advantages in this type of application. It may be deposited from neutral or slightly alkaline non-cyanide electrolytes which virtually do not attack the copper-laminate adhesives, the deposit shows only a low tensile stress, and it may readily be soldered, whereas rhodium presents some difficulty in this respect. Palladium has good contact properties and, in the electrodeposited condition, shows a hardness of 200–300 HV which, while considerably lower than that of rhodium, is higher than that of most gold deposits, and affords a useful degree of wear resistance. Thicknesses of 0·002 5–0·005 mm are usual, and the comments made previously regarding the porosity of thin coatings and the importance of undercoatings are applicable here too.

In sliding electrical contact applications, palladium plating has been criticised on the basis of a tendency due to its catalytic activity to cause polymerisation of organic vapours from adjacent equipment with the formation of insulating films on the surface[31]. This effect may be of importance in very specialised circumstances, but is not serious in most practical applications[32].

Platinum　Since the ready workability of platinum permits cladding of base metal with sound coatings which may be as thin as 0·002 5 mm uses of the metal in the electrodeposited condition for corrosion protection are relatively few. As in the case of palladium, electrolytes for platinum plating have been available for many years and interest in the process has recently been stimulated, chiefly in connection with the plating of titanium for the preparation of inert anodes for electrolytic processes[33]. Attempts to use bare titanium as an anode in aqueous solutions result in the formation of a resistive oxide coating on the metal which prevents the passage of useful currents below about 15 V applied potential. At this figure, complete breakdown of the film occurs, with the onset of catastrophic corrosion. The presence of a thin layer of platinum on the titanium surface permits the passage of high currents at voltages well below the critical value, and in this application the presence of discontinuities in the electrodeposited platinum coating does not affect performance, since the exposed basis metal is sealed by the protective anodic film. This composite material, with a coating of platinum up to 0·002 5 mm thick, is proposed for many electrode applications in which platinum-clad base metals or graphite are at present used, e.g. in brine electrolysis, peroxide and per-salt production, electrodialysis, cathodic protection, etc.

Recent studies have suggested that under certain conditions platinum may become mechanically detached from titanium anodes due to attack of the substrate through pores in the coating. Anodes are now available with a mechanically-clad platinum coating, and alternative coatings, e.g. of platinum–iridium alloy or ruthenium oxide, are under evaluation.

Ruthenium, iridium and osmium   The use of a fused cyanide electrolyte is the most effective means for the production of sound relatively thick coatings of ruthenium and iridium, but this type of process is unattractive and inconvenient for general purposes and does not therefore appear to have developed yet to a significant extent for industrial application. This is unfortunate, since these metals are the most refractory of the platinum group and in principle their properties might best be utilised in the form of coatings. However, in recent years several interesting improvements have been made in the development of aqueous electrolytes.

For ruthenium, electrolytes based on ruthenium sulphamate[34] or nitrosyl-sulphamate[35] have been described, but the most useful solutions currently available are based on the anionic complex[36, 37] $(H_2O \cdot Cl_4 \cdot Ru \cdot N \cdot Ru \cdot Cl_4 \cdot OH_2)^{3-}$. The latter solutions operate with relatively high cathode efficiency to furnish bright deposits up to a thickness of about 0·005 0 mm, which are similar in physical characteristics to electrodeposited rhodium and have shown promise in applications for which the latter more costly metal is commonly employed. Particularly interesting is the potential application of ruthenium as an alternative to gold or rhodium plating on the contact members of sealed-reed relay switches.

Iridium has been deposited from chloride–sulphamate[38] and from bromide electrolytes[39], but coating characteristics have not been fully evaluated. The bromide electrolytes have been further developed by Tyrell[40] for the deposition of a range of binary and some ternary alloys of the platinum metals, but, other than the platinum–iridium system, no commercial exploitation of these processes has yet been made.

Electrodeposition of osmium[41], has been reported from a strongly alkaline electrolyte based on an anionic complex formed by reaction between osmium tetroxide and sulphamic acid. No information is yet available concerning the general soundness of such coatings, but they appear to show excellent mechanical wear-resistance, since in comparative abrasion tests an osmium coating lost only one-quarter the thickness of a hard chromium deposit. Both iridium and osmium have very high melting points and high work functions, which suggest application in the coating of tungsten valve grids to suppress secondary electron emission, but in both cases application is likely to be restricted by the high cost and limited availability of the metals.

Other Coating Techniques

'Brush' plating[42] is a variant of electrodeposition in which the electrolyte is held in a pad of cotton wool or other absorbent material and applied by wiping over the article to be plated. Though very old in principle, modern developments in equipment and applicational techniques render the method extremely useful in the case of precious metals in view of the possibility of localising the coating to selected areas. It is also useful in repair and salvage operations in the plating of electronic components.

Another method entails application of the coating by spraying, brushing or silk-screen printing onto the surface a liquid composition containing organic salts of the metal in a suitable vehicle, which, on firing, decomposes to produce a metal film. This process has been used for many years to apply very thin coatings of gold and other precious metals to non-conductors for decorative purposes, and has served as a basis for technological improvement designed to make it possible to apply thick coatings of platinum in a single application[43]. Though developed primarily for the coating of refractories for critical applications in the glass industry, the process could also be of use in the coating of metals carrying refractory oxide films, e.g. titanium, zirconium. It has the merit that the properties of the coating may approximate more closely to those of the pure metal than is generally the case for electrodeposits.

Protection at High Temperatures

Although the platinum metals have high melting points, covering a range from 1 552°C (palladium) to approximately 2 500°C (ruthenium and iridium) only platinum retains its freedom from oxide films at temperatures up to the melting point. Palladium and rhodium form stable protective oxide films over a temperature range of approximately 500–1 000°C, above which the oxides dissociate. The oxides formed by ruthenium and osmium are readily volatile, hence these two metals are quite unsuitable for high temperature application. The behaviour of iridium in this respect is intermediate between that of rhodium and ruthenium.

At temperatures above the melting point of gold, which represents the chief range of interest, the life of a platinum coating on a base metal is limited by the extent to which inter-diffusion with the substrate metal, and gaseous diffusion through the outer coating (leading to formation of base metal oxide initially along grain boundaries of the coating and ultimately at the surface) is possible. The problems involved are exemplified in the application of platinum coatings for protecting molybdenum against oxidation at temperatures in the region of 1 200°C in gas turbines, and in the preparation of clad-molybdenum stirrers for molten glass. Useful life of the composite material is obtained only with claddings 0·25–0·5 mm thick, and in this connection Rhys[44] has demonstrated the importance of an intermediate layer of gold or an inert refractory oxide as a barrier to outward diffusion of molybdenum[45].

Although electrodeposition permits the coating of relatively complex shapes, the permeability of coatings so applied to gases at temperatures of the order of 1 200°C is, in the present stage of development, too great for them to have a protective value comparable to that of wrought metal coatings. For example, an electrodeposit of platinum 0·10 mm thick protected molybdenum for only 16 h in air at 1 200°C, whereas a mechanical cladding of this thickness had a life of some 300 h under similar conditions. It is possible, however, that the modified coatings referred to previously might be more akin to the pure metal in this respect. Coatings produced by vapour-phase deposition would also appear to be promising in this type of application.

An interesting approach to the inter-diffusion problem was made by

Rhys[46] who protected ruthenium-rich ruthenium–gold alloys by palladium–gold coatings of composition corresponding to the opposite ends of the 'tie-lines' in the palladium–gold–ruthenium ternary system. Since substrate and coating compositions are in thermal equilibrium, diffusion between the two does not occur to an appreciable extent over long periods at high temperatures. Unfortunately, coating life is again limited by diffusion of oxygen through the coating.

F. H. REID

REFERENCES

1. Angus, H. C., Berry, R. D. and Jones, B., *Engineering Materials and Design*, **11**, 1965, Dec. (1968)
2. Rinker, E. C., and Duva, R., US Pat. 2 905 601 (1959)
3. Erhardt, R. A., *Proc. Amer. Electropl. Soc.*, **47**, 78 (1960)
4. US Pat. 3 057 789 (1962)
5. Rhoda, R. N., *Plating*, **49** No. 1, 69 (1962)
6. Withers, J. C. and Ritt, P. E., *Proc. Amer. Electropl. Soc.*, **44**, 124 (1957)
7. Smith, G. R. *et al.*, *Plating*, 56 No. 7, 805 (1969)
8. Atkinson, R. H., *Trans. Inst. Met. Finishing*, **36**, 7 (1958)
9. Tripler, A. B., Beach, J. G. and Faust, C. L., *J. Electrochem. Soc.*, **195**, 1 610 (1958)
10. US Pat. 2 895 889 and 2 895 890 (1959)
11. UK Pat. 808 958 (1959)
12. Stevens, J. M., *Trans. Inst. Met. Finishing*, **46** No. 1, 26 (1968)
13. Johnson, R. W., *J. Electrochem. Soc.*, **108**, 632 (1961)
14. Okinaka, Y., *Plating*, **57** No. 9, 914 (1970)
15. Rhoda, R. N., *Trans. Inst. Met. Finishing*, **36**, 82 (1959)
16. Pearlstein, F. and Weightman, R. F., *Plating*, **56** No. 10, 1 158 (1969)
17. Laister, E. H. and Benham, R. R., *Trans. Inst. Met. Finishing*, **29**, 181 (1953)
18. *Electroplated Coatings of Gold and Gold Alloy*, BS 4292 (1968)
19. Baker, R. G., *Proceedings of 3rd E.I.A. Conference on Reliable Electrical Connections*, Dallas, Texas (1958)
20. US Pat. 2 897 584 (1959)
21. Garte, S., *Plating*, **53** No. 11, 1 335 (1966)
22. Garte, S., *Plating*, **55** No. 9, 946 (1968)
22a. Antler, M., *Plating.*, **56** No. 10, 1 139 (1969)
23. Clark, M. and Leeds, J. M., *Trans. Inst. Met. Finishing*, **43**, 50 (1963) and **47**, 163 (1969)
24. Leeds, J. M., *Trans. Inst. Met. Finishing*, **47**, 222 (1969)
25. UK Pats.1 003 848 and 1 003 849 (1965)
26. Egan, T. F. and Mendizza, J., *J. Electrochem. Soc.*, **107**, 253 (1960)
27. Krumbein, S. J. and Antler, M., *Proceedings of NEP/CON '67 West Meeting*, Long Beach, Calif., Feb. (1967)
28. *Corrosion and its Prevention at Bi-Metallic Contacts*, Admiralty and Ministry of Supply Inter-Service Metallurgical Research Council, HMSO, London (1956)
29. Reid, F. H., *Trans. Inst. Met. Finishing*, **33**, 195 (1956)
30. Philpott, J. E., *Platinum Met. Review*, **4**, 12 (1960)
31. Hermance, H. W. and Egan, T. F., *Bell System Tech. Journal*, **37**, 739 (1958)
32. Reid, F. H., *Plating*, **52** No. 6, 531 (1965)
33. Cotton, J. B., *Chem. and Ind. (Rev.)*, **17**, 492 (1958)
34. US Pat. 2 600 175 (1952)
35. Reid, F. H. and Blake, J. C., *Trans, Inst. Met. Finishing*, **38**, 45 (1961)
36. Reddy, G. S. and Taimsalu, P., *Trans. Inst. Met. Finishing*, **47**, 187 (1969)
37. Bradford, C. W., Cleare, M. J. and Middleton, H., *Plat. Met. Rev.*. **13**, 80 (1969)
38. Conn, G. A., *Plating*, **52** No. 12, 1 258 (1965)

39. Tyrrell, C. J., *Trans. Inst. Met. Finishing*, **43**, 161 (1965)
40. Tyrrell, C. J., Paper presented at International Metal Finishing Conference, Hanover, May (1968)
41. Greenspan, K. L., *Engelhard Industries Tech. Bull.*, **10** No. 2, 48–49, Sept. (1969)
42. Hughes, H. D., *Trans. Inst. Met. Finishing*, **33**, 424 (1956)
43. UK Pat. 878 821 (1957)
44. Rhys, D. W., *Proceedings of Symposium sur la Fusion du Verre*, Brussels, 6–10 Oct. (1958), Union Scientifique du Verre, 677
45. Safranek, W. H. and Schaer, C. R., *Proc. Amer. Electropl. Soc.*, **43**, 165 (1956)
46. UK Pat. 1 150 356 (1965)

15 PROTECTION BY PAINT COATINGS

15.1	Paint Application Methods	**15**:3
15.2	Paint Formulation	**15**:10
15.3	The Mechanism of the Protective Action of Paints	**15**:24
15.4	Paint Failure	**15**:38
15.5	Paint Finishes for Industrial Applications	**15**:50
15.6	Paint Finishes for Structural Steel for Atmospheric Exposure	**15**:62
15.7	Paint Finishes for Marine Application	**15**:72
15.8	Protective Coatings for Underground Use	**15**:81
15.9	Synthetic Resins	**15**:93
15.10	Glossary of Paint Terms	**15**:100

15.1 Paint Application Methods

The various methods of applying paint today are so numerous that it would be impossible to list and describe them in detail in a section of this size. A great number of these methods have either been developed or adapted to meet special industrial requirements. Corrosion resistance of the finished article must always be a major consideration, but, with one or two exceptions, it is not the deciding factor on which the final choice of application system is made. With all methods of paint application, it is possible to study the peculiarities of each individual system and apply controls to ensure that maximum corrosion protection is obtained from the selected process.

Application Methods

These can be classified in various ways, but for simplicity they can be grouped under two main headings, cycling and non-cycling systems, i.e. systems in which all the paint is continuously returned to the plant as opposed to those in which the paint is applied to the article and the excess paint is not returned to the plant.

Cycling Processes

For these processes, the paint must not only meet the specification and end requirements of the finished product, but it must also be exceptionally stable over long periods under operating conditions. The simplest form of the cycling paint process is the standard dip tank which can vary from a simple hand dip in a container of paint, to a sophisticated mechanised system. In the conveyorised process, articles pass into the dip tank and are withdrawn at a controlled rate, and after draining are allowed to air dry or are cured in a stoving oven. In such a system a large volume of paint is involved with a large surface area exposed to atmosphere. The bulk paint is continuously contaminated by the excess paint draining from the articles and any extraneous substances introduced with the articles. These contaminants are limited to some extent by diverting the final draining away from the main dip tank and by filtering and circulating the paint which is essential to maintain uniform paint composition. Constant control of viscosity, paint composition and composition of the solvent mixture are necessary if uniform results are to be obtained and some of these factors are continuously controlled by automatic equipment in modern sophisticated plants.

Other Cycling Processes

Roto-dip   In this process, motorcar bodies are cleaned, phosphated, primed and stoved on a line production. The bodies are mounted on central spits on a conveyor system and are rotated while they are immersed in the various tanks. With this system, the volume of paint required is considerably

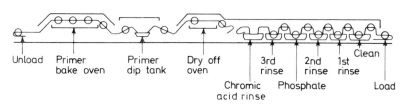

Fig. 15.1   Automatic phosphating and painting of car bodies ('Rotodip' process) showing pretreatment, drying, paint dip and baking oven (courtesy Carrier Engineering Co. Ltd.)

reduced and the tendency for sags and runs greatly reduced resulting in a more uniform film thickness. Improved coverage is obtained in areas which are not readily accessible with simple immersion (Fig. 15.1).

Slow dipping   Various plants of this type have been developed in which either the workpieces are lowered into the paint tank or the paint tank raised to the articles at a controlled very slow speed of the order of 2–25 mm/min. Paints of much higher viscosity and higher solids are used in this process resulting in thicker and more uniform paint films.

Flow coating   In this process, paint is directed on to the workpiece from a series of strategically placed jets in an enclosed area and the excess paint drains back to the main supply tank. The workpieces are then allowed to drain in a solvent-saturated zone before passing on to a flash-off zone and final stoving. A much smaller paint supply tank is required than with normal dipping.

Curtain coating   With this process, paint falls in a continuous curtain from a closely machined gap in a header tank on to the flat article passing below on a horizontal conveyor; the excess paint is collected in the main tank and then passed up to the header tank.

Electrodeposition   This more recent method of paint application is basically a dipping process. The paint is water based and is either an emulsion or a stabilised dispersion. The solids of the paint are usually very low and the viscosity lower than that used in conventional dipping. The workpiece is made one electrode, usually the anode, in a d.c. circuit and the cathode can be either the tank itself or suitably sized electrodes sited to give optimum coating conditions[1]. The current is applied for a few minutes and after withdrawal and draining the article is rinsed with de-ionised water to remove the thin layer of dipped paint. The deposited film is firmly adherent and contains a minimum of water and can be stoved without any flash-off period. This process is used for components, car bodies, etc. and with the latest paint formulations which are usually based on epoxies or acrylics, produces

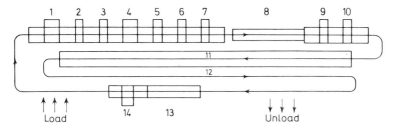

Fig. 15.2   Typical plant layout for electropainting (courtesy Stein, Atkinson Stordy Ltd.)

LEGEND

1. Alkali degrease	8. Paint dip
2. Cold water rinse	9. First rinse (town water)
3. Hot water rinse	10. Second rinse (demineralised water)
4. Zinc phosphate	11. Stoving oven
5 and	12. Cooling leg
6. Cold water rinses	13. Jig strip
7. Demineralised water rinse	14. Jig rinse

superior anti-corrosive properties giving uniform films on sharp edges with complete coverage of otherwise inaccessible parts of the car body (Fig. 15.2)[2].

Fluidised bed   This process is used for powder coating utilising thermoplastic powders such as polyvinyl chloride, nylon, etc. and thermosetting powders such as some of the epoxies. Basically, the equipment consists of a dip tank with a perforated shelf near the bottom. The powder is placed on this shelf and low pressure air is fed under the perforated shelf, resulting in a cloud of fine powder in the body of the dip tank. With the thermoplastic coatings, the article is heated to a little above the melting point of the powder and is then dipped into the fluidised bed for a short period. The article is then withdrawn and the coating fused in an oven operating at a temperature of 10–20°C above the melting point of the polymer. Thick films are formed, but it is difficult to obtain uniform film thickness with varying gauge metals. This process has been further modified by inducing electrostatic charges on to the powder particles and this modified process gives improved deposition with the thermosetting polymers[3].

Non-cycling Processes

Processes under this heading basically include those in which the paint is used once only and the excess material is not returned to the main bulk. A typical example is the normal spray system in which the paint is fed to the spray gun and atomised by air jets and applied to the article as a stream of small droplets. The excess paint and overspray are deposited on the walls of the booth and are collected by various methods as a dried cake or wet sludge depending on the type of spray booth used[4]. There are many modifications of the conventional spray system which include the following.

Hot spray   The paint is heated to 60–85°C enabling higher solid films to be applied, resulting in a heavier coating and less over-spray.

Airless spray   In this process, a high pressure of the order of 35 MN/m^2 is applied to the paint which is forced through a fine orifice in the spray gun. This process gives thicker films and considerably less bounce back and over-spray.

Electrostatic spraying   This process takes advantage of electrostatic attraction. In the Ransburg No. 2 process, rotating flat discs or bells are charged to a high negative potential of approximately 100 000 V. Paint is pumped to the centre of the disc or bell, which is sprayed by centrifugal force to the periphery which is in the form of a sharp edge. The paint is atomised as a result of the concentrated electrostatic field at the sharp edge. The article is earthed on the conveyor and the finely divided charged particles are attracted along the lines of force to the surface of the article. The path of the charged particles is influenced by the centrifugal force. Over-spray and paint wastage are virtually eliminated by this process (Fig. 15.3).

Electrostatic hand guns   During recent years the same principle has been adopted to portable hand guns and these are now produced incorporating air assistance, the airless spray principle and even the hot spray principle. These modifications facilitate the coating of electrically shaded areas.

E.A.G. electrostatic plant   This plant is of fairly simple design and is economical with paint, spraying only taking place when the workpiece is in position and the only moving part being a fixed-speed paint pump. It consists essentially of a hollow wedge-shaped atomising head precision machined and is available in lengths of up to 3 m. A precision-formed 0·5 mm slot with a sharpened lip runs along the length of the head. The paint is pumped at a uniform pressure to the top of the head and excess paint flows out by gravity from an outlet at the base of the head and is then recirculated. Due to the small width of the slot at the base, paint does not run out of the slot unless the plant is operating. The wedge is negatively charged to a potential of approximately 150 000 V and when an earthed workpiece passes within range, paint is electrostatically attracted from the wedge, atomised and deposited on to it.

Brush   The original brush process comes within this category and the normal roller application can be considered in the same class. More recently both these forms have been produced with fluid feed arrangements.

Paint Processes

In addition to the individual processes mentioned earlier, there are a number where one or more of these have been combined for specific end requirements; a typical example being the painting of ships' plates, where sheets are shot blasted and this operation is immediately followed by the application of an epoxy based primer by means of an automatic, traversing, airless hot-spray plant.

With a few exceptions, a single coat of paint is not sufficient to give the desired finish and protection from corrosion, but in all cases, the original cleaning of the substrate, pretreatment and initial priming coat are of prime importance in corrosion resistance. Where possible, separate components

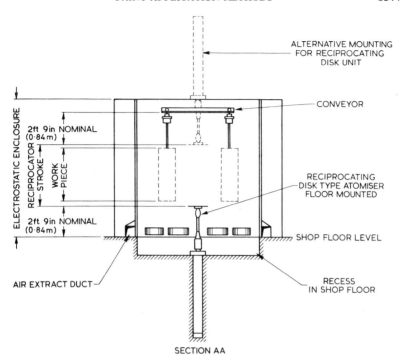

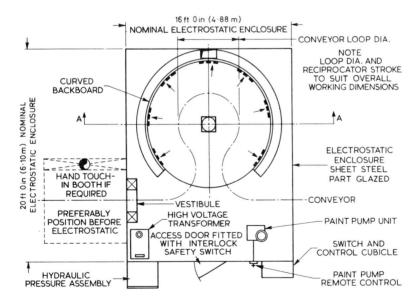

Fig. 15.3 Ransburg No. 2 Reciprocating Disc Plant (courtesy Peabody-Modernair Ltd.)

should be primed before assembly[5], and where joints occur in the finished article, suitable jointing compound or mastics should be used in these areas.

It is essential to avoid bimetallic corrosion by using a suitable jointing compound when joints are made between dissimilar metals. In the case of welds, special welding primers should be used and a commonly used primer of this type is highly pigmented, the pigment containing over 90% of finely divided metallic zinc.

With certain products such as car bodies, it is essential to obtain some measure of image clarity in the gloss component. It is necessary to obtain a level surface which involves the application of one or more additional coats of paint such as a filler or surfacer which is levelled by a mechanical process which may range from a dry denib to a full scale wet flatting involving washing, rinsing and drying processes, during each stage of which the undercoats are exposed to the danger of contamination. The most deleterious contaminants are water soluble salts, but the danger can be virtually eliminated by final rinsing with de-ionised water and removing any liquid water before final drying. In some cases, a coat of sealer is applied over the flatted surface prior to the application of the final colour coat to prevent sinkage.

Although the results obtained from a particular process depend almost entirely on the nature and design of the article, the plant layout and other local conditions play an important part. Table 15.1 gives some indications of the limitations of the processes mentioned.

Table 15.1   Summary of process limitations

Process	Uses	Limitations and defects
Dipping	All types of articles of suitable shape and size	1. Requires large throughput 2. Gives uneven film thickness on large flat sheets (from top to bottom) 3. Does not cover sharp edges or interior of channel sections, etc. 4. Possibility of solvent wash, i.e. solvent vapour from the hotter area condensing on cooler areas during flash-off and stoving
Slow dipping	Generally only used for small articles	Gives more uniform film than normal dipping, but does not give complete protection on interiors of channels, etc.
Flow coating	Suitable for use on most articles. Gives good penetration into pores of castings	Similar to dipping, but the defects not so marked and tendency for greater solvent loss
Curtain coating	Only suitable for flat sheets of uniform dimensions	On suitable articles this method will produce a uniform paint film providing the surface of the article is perfectly clean. This is the only cycling process where two-component systems can be used economically
Roller coating	Suitable only for flat surfaces	Gives uniform thin film, not adequate for protection against corrosion on rough surfaces
Electrodeposition	Suitable for most articles	Gives uniform film even on rough surfaces. With suitable plant will completely coat interior surfaces, sharp edges, etc. but generally only economic on mass production

Table 15.1 (continued)

Process	Uses	Limitations and defects
Fluidised bed	Only suitable for small articles	Produces thick films. Varying metal gauge could produce uneven films and weak spots
Conventional spray	Suitable for most components	Very difficult to obtain adequate cover on inside corners, etc. and results depend on the skill of the operator
Airless spray	Suitable for most components	Superior penetration in awkward areas to normal spraying. Less dependent on the skill of the operator
Electrostatic spraying (automatic)	With suitably designed plant, most articles can be painted with this process	Only economical in long runs. Very difficult to obtain adequate cover in electrically shaded areas and interiors of hollow articles
Air assisted electrostatic automatic hand gun	Similar to that for normal electrostatic processes	With suitably designed plant, it is possible to obtain more efficient cover in most shaded areas
Electrostatic airless spray	Similar to other electrostatic processes	Combines the advantages of electrostatic spraying with some of the advantages of airless application
Brushing	Suitable for use on most articles, but use limited for economic reasons. Results rely almost entirely on skill of the operator	Not possible to coat awkward or interior areas and labour cost is high

Dipping at one time was thought to be the best method of applying the priming paints, but today there are a number of processes which, if correctly controlled, give superior results. There are many examples today in the car industry, consumer durables, civil engineering and ship building, where superior corrosion resistance is obtained by using these more modern application techniques. The improved results are the combined results of new and improved methods of application, improved metal cleaning and pretreatment and improved paint technology. The decorative characteristics of different paint systems have not been discussed. The painting process and application method can influence the final finish. Imperfections in the final finish do not necessarily cause corrosion of the base metal[6].

W. H. TATTON

REFERENCES

1. Yeates, R. L., *Electropainting*, Robert Draper (1966)
2. Stordy, J. M., *Surface Coatings*, Stein Atkinson Stordy, March (1967)
3. Tatton, W. H. and Drew, E. W., *Industrial Paint Application*, 2nd edn, London, Butterworths (1971)
4. Lloyd, D. H., *Journal of the Oil and Colour Chemists' Association*, No. 40, 433 (1957)
5. Coleman, L. J., *Journal of the Oil and Colour Chemists' Association*, No. 42, 10–59 (1957)
6. Hess, M., *Paint Film Defects—Their Causes and Cure*, Chapman and Hall (1965)

15.2 Paint Formulation

Constituents of Paint

Paint consists essentially of a pigment dispersed in a solution of a binding medium. The binding medium or binder, which in most instances is organic, will decide the basic physical and chemical properties of the paint, but these will be modified by the nature and proportion of pigments present. In a finish, the primary function of the pigment is to provide colour, but in a primer it should contribute to the durability of the whole system in a variety of ways depending on the substrate to which it is applied. The sole function of the volatile component is to control the viscosity of the paint for ease of manufacture and for subsequent application. Thereafter the solvent evaporates and is lost. A further class of paint is based on a binder emulsified in water. This class of paint binders has increased in importance recently and there is considerable evidence that good anticorrosive properties can be built into paints which themselves are thinned with water. The development of these paints is attracting considerable attention because of the potential absence of fire hazards and their low level of toxic vapours.

Since the possible variations in binder alone are limitless, it is possible to produce an infinite number of paints. As the range of raw materials available to the formulator becomes wider, their chemical purity is continually being improved. Mathematical models of binders can be constructed using computers and, in some cases, it is possible to predict accurately the properties of a particular formulation before it is made. Nevertheless, the formulation of paints for specific purposes is still considered to be very much a technological art.

Although formulation is an art, science finds its place in the characterisation of the raw materials, in the design and testing of the series of experimental formulae and in the interpretation of the results. In addition to possessing an intimate knowledge of pigments, binders and solvents, the paint formulator must also be well acquainted with raw material costs and availability, paint making machinery, and the market's performance requirements.

Basic Principles of Formulation

Before any attempt is made to formulate a paint it is necessary to know a great deal about the circumstances under which it will be used and the conditions to which it will subsequently be exposed. The more comprehensive the information relating to requirements, the greater the probability of achieving complete success with the first practical trial.

The conditions under which the paint will be dried, i.e. air drying or stoving, and the properties subsequently demanded in service will dictate the choice of binder. This, in turn, will limit the choice of solvents, and a further restriction may be imposed by a consideration of fire hazards at the user's works, or the problem of toxic fumes in enclosed working spaces. The quantity of solvent in the paint will depend on the intrinsic viscosity of the binder and the paint viscosity appropriate to the method of application, e.g. brushing, spraying, dipping, electrostatic spray, electrocoating, flow coating, etc.

A single paint will rarely possess a majority of the required properties and it therefore becomes necessary to formulate a system comprising a primer, a finish, and possibly one or more intermediate coats.

A primer, as its name implies, is the first coat of a system. Its principal functions are to obtain adequate adhesion and afford good protection to the substrate. The manner in which these properties are obtained will vary with the substrate, but frequently involves the use of a large proportion of a specific pigment. This imposes a severe restriction on the colour and gloss, and probably on other desirable properties such as durability.

The finish or final coat must make up the deficiencies of the primer by affording it protection and providing the required colour and degree of gloss. These last two requirements will dictate the quality and quantity of the pigments to be used.

In the majority of cases maximum durability will be afforded by a multi-coat system comprising priming and finishing paints only. Other considerations, however, such as uniformity of colour and smoothness of surface, may make it desirable to introduce intermediate coats, e.g. putty, filler and undercoat. The appearance of the final film or of the final painted structure is of some importance, and a final colour coat, which may afford very little resistance to corrosion, may be necessary.

Putties are heavy-bodied pastes of high pigment content that are applied by knife for rough filling of deep indentations, more especially in rough castings.

Fillers are used to level-out shallower imperfections. Ease of rubbing is an important consideration and to a large extent dictates the composition and proportion of the pigment mixture.

Undercoats are invariably of high pigment content and low gloss. Their function is to provide a foundation that is uniform in both colour and texture for the finishing coats, thereby adding to the final appearance of the completed system. Frequently, it is possible to achieve the same result by substituting an additional coat of finish for the undercoat, which may improve the durability of the system.

When considering the number of coats of paint to achieve adequate protection it is worth noting that the cost of applying the paint often far outweighs the cost of the paint. This is leading to a class of relatively more expensive paints which can be applied in a minimal number of very thick coats. The increasing mechanisation of painting methods, such as the airless spraying of structural areas, influences the paint formulator in the selection of the most suitable formulations.

However comprehensive the information relating to requirements, the paint technologist cannot proceed with the problem of formulating a suitable paint unless he is in possession of considerable data on the properties of the raw materials at his disposal, but within the scope of the present work it is impossible to do more than indicate the important properties of the more commonly used ingredients*.

Binding Media

The most important component of the majority of paints is the binding medium, which determines the physical and chemical properties of the paint. Blends of binding media are often used to impart specific properties to the dry paint film or to suit a particular application method. The compatibility of chemically different types of binders is an important factor affecting the paint formulator's task. These properties will be modified, however, to a greater or lesser extent by the nature and quantity of the other components, more especially the pigment. The general characteristics of various binding media are given in Table 15.2.

Drying Oils (treated and untreated)

Apart from being basic ingredients of oil varnishes and alkyd resins, drying oils are occasionally used as such in paint. Linseed oil is by far the most important of the drying oils and is the only one used to any extent in its natural state. Its main use is in corrosion-inhibiting primers to BS 2523:1966. Disadvantages of paints based on raw linseed oil are their very slow drying, lack of gloss, and inability to flow sufficiently for brushmarks to level out. The mechanism of the protective action of these primers is considered in Section 15.3.

Heat-treated oils fall into three categories: boiled oils, stand oils and blown oils. Boiled oils are prepared by heating linseed oil in the presence of salts of the 'drier' metals, notably cobalt and lead. They have somewhat higher viscosities and better drying properties due to their higher molecular weights and more complex molecular structure than raw linseed oil. They are commonly used in oil-based primers and, in conjunction with oil varnishes, in undercoats. Stand oils range in viscosity up to about $20\,Ns/m^2$ and are prepared by heat-polymerising linseed oil either alone or in admixture with tung oil. They are used mostly in combination with oil varnishes and alkyd resins to improve application properties and, when desirable, to increase the total oil:resin ratio.

Blown oils differ from stand oils in that they are partially oxidised in addition to being polymerised. The oxidation is achieved by blowing air through the heated oil. This treatment results in a product having poor drying properties, and blown oils are therefore effective plasticisers and are used as such in nitrocellulose finishes.

---

*A list of relevant standard works is given at the end of this paper, but the principal sources of detailed information are in the form of technical data sheets issued by raw-material suppliers.

Table 15.2 General characteristics of binding media

Type of binder	Mode of drying	Solvents	Acid resistance	Alkali resistance	Water resistance	Solvent resistance	Exterior weathering resistance	Special features
Raw linseed oil Boiled linseed oil Stand oils	Air drying Oxidative polymerisation	Aliphatic hydrocarbons	Fair	Bad	Fair	Poor	Poor/fair	Vehicle for anti-corrosive primers for wire-brushed steel Slow drying
Oleoresinous varnishes	Air drying and/or stoving Condensation and/or oxidative polymerisation	Aliphatic and/or aromatic hydrocarbons	Fair	Bad	Fair/good	Poor	Fair/good	Pale-coloured finishes that yellow on exposure
Long oil length alkyds	Air drying Oxidative polymerisation	Aliphatic hydrocarbons	Fair	Bad	Fair	Poor	Very good	
Medium oil length alkyds	Air drying and/or stoving Oxidative and/or condensation polymerisation	Aliphatic and aromatic hydrocarbons	Fair	Poor	Fairly good	Fair	Very good	
Short oil length alkyds	Stoving Condensation polymerisation	Aromatic hydrocarbons	Fair	Fair	Good	Fairly good	Very good	
Modified alkyds	Air drying and/or stoving Oxidative and/or condensation polymerisation	A wide range of solvents depending on modification	Fair	Fair	Usually good	Fair	Usually good /very good	

Table 15.2 (continued)

Type of binder	Mode of drying	Solvents	Acid resistance	Alkali resistance	Water resistance	Solvent resistance	Exterior weathering resistance	Special features
Urea formaldehyde /alkyd blends	Stoving Condensation polymerisation	Aromatic hydrocarbons and alcohols	Fairly good	Fairly good	Very good	Good	Fair	Water white Gives white finishes of excellent colour
Melamine formaldehyde/ alkyd blends	Stoving Condensation polymerisation	Aromatic hydrocarbons	Fairly good	Fairly good	Very good	Good	Very good	Water white Gives white finishes of excellent colour
Epoxide/aliphatic amine or polyamide blends	Air drying Addition polymerisation	Blends rich in higher ketones	Fairly good	Very good	Poor	Very good	Fairly good/ good	Finishes need to be supplied in two separate containers and mixed just prior to use
Epoxide/amino or phenolic resin blends	Stoving Addition and condensation polymerisation	Blends rich in higher ketones and alcohols	Good	Good	Very good	Very good	Good	
Epoxide/fatty acid esters	Air drying and/or stoving Oxidative polymerisation	Aliphatic and/or aromatic hydrocarbons	Fair	Fair	Fairly good	Poor	Poor/fairly good	
Polyester/ polyisocyanate blends	Air drying or stoving Addition polymerisation	Blends rich in ketones and esters Alcohols excluded	Fairly good	Good	Fairly good	Very good	Very good	Finishes need to be supplied in two separate containers and mixed just prior to use

Table 15.2 (continued)

Type of binder	Mode of drying	Solvents	Acid resistance	Alkali resistance	Water resistance	Solvent resistance	Exterior weathering resistance	Special features
Vinyl resins	Air drying Solvent evaporation	Blends usually rich in ketones	Very good	Very good	Very good	Poor	Good	Fire hazard Flash point usually below 23°C
Chlorinated rubber	Air drying Solvent evaporation	Aromatic hydrocarbons	Good	Good	Very good	Poor	Good	Very poor heat resistance
Cellulose nitrate	Air drying Solvent evaporation	Blends of esters, alcohols and aromatic hydrocarbons	Fairly good	Bad	Good	Poor	Very good	Fire hazard Statutory regulations governing use

Oil Varnishes

The current practice is to classify as 'oil varnishes' all varnishes and paint media prepared from drying oils and natural or preformed oil-free synthetic resins. Examples of such resins are rosin, gum congo, rosin-modified phenolics and oil-soluble 100% phenolics. The introduction of the resin results in improved drying and film properties.

Oil varnishes are capable of producing primers for ferrous metals which perform excellently on clean or pretreated surfaces, but they have not the same tolerance as their oil-based counterparts for wirebrushed rusted surfaces. The undercoats that follow are frequently also based on oil varnishes.

Since the individual members of this group of media differ considerably in properties, so also do the finishes that can be made from them. As a class, however, they are generally inferior to the better alkyds for durability under normal conditions. A particular exception is the tung-oil 100% phenolic type of medium, which produces finishes with very good resistance to water and mildly acidic or alkaline conditions; pale colours, however, discolour by 'yellowing' on exposure.

Alkyd Resins*

Introduced some 40 to 50 years ago, alkyd resins quickly established themselves and are still widely used. They are essentially polyesters of moderate molecular weight prepared by the reaction of polyhydric alcohols with the mixtures of monobasic fatty acids and dibasic acids. Ethylene glycol (dihydric), glycerol (trihydric) and pentaerythritol (tetrahydric) are the more commonly used alcohols. Phthalic anhydride is the most commonly used dibasic acid. Isophthalic acid and adipic acid are also used for special purposes. An unsaturated dibasic acid called maleic anhydride is widely used and can give polymers of high molecular weight. There is a very wide range of fatty acids available, the ultimate choice being dependent upon the properties required. The fatty acid is frequently added in the form of a vegetable oil which is a tri-ester of fatty acid and glycerol.

Individual alkyds are usually described in terms of the proportion and type of fatty acid and of the alcohol that they contain. Thus a 70% linseed-oil pentaerythritol alkyd would be expected to comprise linseed oil fatty acids, pentaerythritol and phthalic anhydride, with an equivalent of 70% linseed oil calculated on the weight of the non-volatile resin.

The members of this family are so diverse that only the fundamental properties can be considered here. For convenience they will be subdivided according to their use, i.e. (a) air-drying, (b) stoving, (c) plasticising and (d) modified alkyds.

Air-drying alkyd resins Alkyds capable of air drying do so through the oxidation of the drying oils that they contain. Such alkyds are consequently

*The synthesis of various types of resins is given in Section 15.9.

usually of long oil length*, i.e. 65 to 75% and based on the tetrahydric alcohol pentaerythritol. The oil most commonly used is linseed, but soya bean imparts more freedom from yellowing to white- and pale-coloured finishes, especially where there is little natural light. Tung oil is less frequently used because it promotes yellowing. Sunflower oil, cottonseed oil, safflower oil and tall-oil fatty acids are being used more and more frequently for high quality white gloss paints. In some cases, the fatty acid is partly replaced by a synthetic organic monobasic acid which modifies the polymer solubility and film properties. Among the outstanding properties of air-drying alkyds are (a) their convenience in use and (b) their ability to give finishes of un-rivalled durability in all but heavily polluted atmospheres. Where premature failure does occur, it probably results from poor surface preparation or an inadequate priming system.

Air-drying alkyds may also be used for the production of primers and undercoats. In the case of primers, the shorter the oil length of the binder the faster the drying, but the lower the tolerance for wire-brushed rusted-steel surfaces. Alkyd-based undercoats are not significantly different in perform-ance from those based on oil varnishes; the choice is frequently dictated by economic considerations.

Stoving alkyd resins   When drying is to be effected by stoving, the oxidative properties of drying oils are of less importance, and advantage can be taken of the tougher properties of the phthalic ester component of the resin. Hence stoving alkyds may be based on drying or semi-drying oils, and the oil length is invariably shorter than for air-drying finishes, usually in the range 50 to 65%.

For high-quality stoving finishes the alkyd is frequently blended with a lesser quantity of an amino resin. This reduces the stoving schedule and enhances most of the physical properties of the finish.

The inclusion of a small proportion of rosin during the manufacture of the alkyd will also improve application and initial film properties. Such vehicles are commonly used for stoving primers and for cheap stoving finishes that will not be subjected to exterior exposure.

Plasticising alkyd resins   The term *plasticising* alkyd is a loosely used one, embracing those alkyds that are employed in conjunction with a larger pro-portion of another, and usually harder, stoving resin, e.g. an amino resin. In certain compositions the shorter-oil-length stoving alkyds referred to before may function as plasticisers, but in general plasticising alkyds are of even shorter oil length, usually 40 to 50%, and consist of fatty acids of non-drying oils, e.g. coconut oil.

Oil length is the relative proportion of oil to resin in a binding medium. It is expressed in a variety of ways, including simple ratios and, as in the present text, the percentage oil calculated on the weight of the non-volatile binder. In the case of traditional varnishes it is a precise value calculated directly from the relative quantities of oil and resin used. With more complex binders, including alkyds, such simple calculations are not possible; various assumptions must be made and the values then obtained are essentially theoretical. The terms *long, medium*, and *short* oil length are used loosely to indicate respectively, high, medium and low proportions of oil. There are no generally agreed limits but in the present context *long* oil length is applied to binders containing more than 65% oil, *medium* oil length to binders having between 65 and 50% oil and *short* oil length to those containing less than 50% oil.

Modified alkyd resins In this group one finds styrenated alkyds, V.T. alkyds, oil-modified vinyl resins, acrylic alkyds, silicone alkyds and polyurethane alkyds. The modifying component usually has a number of effects. It always increases the molecular weight of the alkyd polymer, and may impart hardness, durability, or chemical resistance. It also affects the solubility of the polymer in solvents.

Amino Resins

The two amino resins in common use are urea formaldehyde and melamine formaldehyde, and most stoving finishes contain one or the other. They have many properties in common; urea formaldehyde, however, while substantially cheaper, has poor exterior durability, whereas melamine formaldehyde imparts excellent exterior durability. As they are both water white they give white finishes of excellent colour, with the additional advantage of retaining their colour on over-stoving. Urea formaldehyde is commonly used in conjunction with a lesser quantity of an alkyd to give finishes with excellent resistance to water and mild chemicals, which are therefore well suited to use on domestic equipment, e.g. washing machines.

Alkyds can easily be made water soluble for use in electrocoating processes. Water soluble urea formaldehyde resins are commonly used in conjunction with water soluble alkyds in the electrocoat stoving paint system. This process enables areas which are normally inaccessible to conventional painting methods to be painted, and also ensures that an adequate thickness of paint is applied to sharp edges where corrosion would otherwise normally occur.

Melamine formaldehyde is also used in conjunction with an alkyd, but the ratio varies considerably according to the ultimate use of the finish. Stoving car finishes, which have an onerous task to perform, are among the finishes having the highest melamine content and are correspondingly expensive.

Epoxide Resins

Epoxide resins are essentially long-chain polyhydric alcohols with epoxide groups at either end. They make useful building blocks because both the hydroxyl and the epoxide groups are available for reaction with other compounds.

Aliphatic polyamines, amine adducts and polyamides react with epoxide resins at normal temperatures to give complexes with outstanding chemical resistance. Paints based on this type of reaction must be supplied in two separate containers, one containing the epoxide resin and the other the 'curing agent', the two being mixed in prescribed proportions immediately before use.

Amino resins and certain phenolics react with epoxide resins at elevated temperatures to give somewhat similar results. As the combination is non-reactive at normal temperatures this type can be supplied in the form of ready-for-use stoving finishes.

Epoxide resins can be esterified with fatty acids to give media ranging from air-drying to stoving types. The presence of fatty acid reduces the chemical resistance to the same order as that of the alkyds. It is nevertheless sometimes found advantageous to use an epoxy ester for certain specialised purposes.

Polyurethanes

Polyurethanes are essentially the reaction products of polyisocyanates and polyesters containing free hydroxyl groups. They are comparable with the epoxide types in that they possess excellent chemical resistance; it is also necessary to supply the air-curing types in two-pack containers. One-pack stoving types are formulated by using less reactive 'masked' isocyanates.

Another group of products which are rapidly gaining in importance are the polyurethane oils and polyurethane alkyds. In these binders, the chemical linkages are a mixture of the highly chemically resistant urethane links and the less resistant ester links. It is very misleading and difficult to classify their properties because the ratio of urethane to ester linkages varies widely from one product to another. In many properties, they are very similar to alkyds, but usually possess more rapid drying, even at low temperatures, and give a slightly harder film initially. They are, however, less flexible than comparable alkyds and often slightly worse for exterior durability. The name *polyurethane* on a product cannot be taken as an indication of chemical resistance unless it is a two-pack polyurethane or a moisture-curing polyurethane.

Vinyl Resins

A wide range of resins prepared by polymerisation of compounds containing vinyl groups is now available. Those most commonly used in paint manufacture are of the following types:

(a) Essentially copolymers of vinyl acetate and vinyl chloride.
(b) Emulsified vinyl acetate copolymers.
(c) Acrylic modified alkyds, etc.

A characteristic of the group (a) of resins is that they air-dry solely by solvent evaporation and remain permanently solvent soluble. This fact, combined with the need to use strong solvents, makes brush application very difficult, but sprayed coats can be applied at intervals of one hour. A full vinyl system such as (a) possesses excellent chemical and water resistance. Many members of group (a) have very poor adhesion to metal, and have therefore been exploited as strip lacquers for temporary protection. Excellent adhesion is, however, obtained by initial application of an etching primer; the best known of such primers comprises polyvinyl butyral, zinc tetraoxychromate and phosphoric acid.

The chemical resistance of group (b) is often upset by the presence of water-soluble emulsion stabilisers and thickeners, which remain water soluble in the dried paint film.

Group (c) has already been discussed under the heading *modified alkyds.*

Chlorinated Rubber

Chlorinated rubber is soluble in aromatic solvents, and paints made from it dry by solvent evaporation alone. In contrast to the vinyls, there is less difficulty in formulating systems that are suitable for brush application. It has excellent resistance to a wide range of chemicals and to water, but as it is extremely brittle it needs to be plasticised. To preserve chemical resistance it is necessary to use inert plasticisers such as chlorinated paraffin wax or chlorinated diphenyls. Admixture with suitable air-drying alkyds gives cheaper finishes possessing excellent durability and water resistance, but 'lifting' on recoating is often a problem.

Nitrocellulose

Paints containing nitrocellulose are of importance in relation to the protection of metals because of their excellent durability combined with very fast drying. They may, on this account, be used for mass-production work where stoving facilities are not available, and it is interesting to recall that had such paints not been available the mass production of motorcars would inevitably have been delayed. Their rapid drying makes them unsuitable for brush application to large areas, but a more serious disadvantage is the fire hazard associated with nitrocellulose, and users of such paints must comply with stringent statutory regulations.

Nitrocellulose alone will not give a continuous coating. It must, therefore, be used in admixture with other components comprising a plasticiser and a hardening resin. An extensive range of such products is available, the ultimate choice depending on the properties required.

Miscellaneous Binders

These consist of the following:

(a) Silicone polymers having high heat stability and excellent chemical resistance are available. They are very expensive and hence are not commonly found in paint coatings.

(b) Silicate binders are used in conjunction with zinc powder to give paints of excellent corrosion resistance. The alkali metal silicates (particularly sodium silicate) are more commonly used, but organo-silicates, e.g. ethyl orthosilicate, are gaining in importance. The full potential of this type of binder has not yet been exploited.

(c) Thixotropic binders.

(d) Fluorinated polymers such as polytetrafluorethylene are available for specialised applications. Titanium polymers with excellent heat stability are available. New polymers are being developed all the time, especially by the plastics industry, and the aforementioned groups of binding media are merely those commonly used, and do not constitute a complete list. It may, however, illustrate the range of products and properties available to the paint formulator for the selection of the most appropriate binders in the paint.

Pigments

In a finish, the function of the pigment is to provide colour, but in a primer it should contribute to the preservation of the metal substrate and enhance the adhesion of the finishing system. Pigments are essentially dry powders which are insoluble in the paint medium and which consequently need to be dispersed in it by a 'grinding' technique. They range from naturally occurring minerals to man-made organic compounds and may be subdivided broadly into priming pigments, colour pigments, extenders and metal powders.

Red lead, zinc chromate, calcium plumbate and zinc dust are of special importance as pigments for metal primers. When dispersed in raw or lightly-treated linseed oil, the first three possess the ability to inhibit the corrosion of mild steel and will function very well on wire-brushed rusted surfaces. In other media the tolerance towards rusted surfaces decreases with decreasing quantities of available oil, but performance on clean steel will usually be maintained and often improved.

Zinc chromate is the recognised pigment for primers of non-ferrous metals and is also an important constituent of etching primers, which are sometimes used as an alternative to pretreatment for obtaining adhesion to such metals. The proportion of the potassium chromate impurity in zinc chromate can have a significant effect on the corrosion inhibiting ability of zinc chromate paints on steel. In fact the addition of potassium chromate to anticorrosive paints is not uncommon. Lead silicochromate is used in stoving primers in the motorcar industry to improve corrosion resistance.

Zinc phosphate and zinc chromate phosphate are rapidly gaining importance in anticorrosive paints. The selection of the correct binder for use with these pigments is very important and can dramatically affect their performance. It is claimed that good metal primers can be formulated on calcium phosphate. Red lead is likely to accelerate the corrosion of non-ferrous metals, but calcium plumbate is unique in providing adhesion to newly galvanised surfaces in the absence of pretreatment, and is claimed to behave similarly on other metals in this group.

Primers containing 93–95% zinc dust by weight in non-saponifiable media afford sacrificial protection to clean steel (see Section 15.3).

Pigments for finishes are selected on the basis of their colour, but special attention must be paid to inertness in the chosen vehicle and stability and light fastness under the conditions of application and exposure. Flake pigments such as aluminium and micaceous iron oxide give finishes of lower moisture-vapour permeability than do conventional pigments, and consequently contribute to better protection.

Paint Additives

A paint rarely consists solely of pigment dispersed in a solution of a binder. For one reason or another, small quantities of ancillary materials called *additives* are included. The oldest and still the most important are the 'driers' which are used in all air-drying and many stoving paints containing drying oils. They are organic salts of certain metals, notably cobalt, lead and manganese.

Anti-oxidants are of value in preventing skinning in containers, but care must be taken to ensure that they do not adversely affect the drying properties of the paint. They are also used to reduce the oxidation of the excess paint that drains from dip-coated articles back into the dip tank.

Surface-active agents are used to facilitate the dispersion of pigments, to keep the pigment in suspension during storage of the paint, and to preserve the homogeneity of pigment mixtures while a paint is drying.

Another group of additives are used as thickeners and antisettle agents. They affect flow and reduce sagging of thick films.

Solvents

The term *solvent* is loosely applied to the volatile component of a paint, though this component may in fact consist of a true solvent for the medium plus a non-solvent or diluent. When such a mixture is used, usually with the aim of reducing cost or obtaining a higher solids content at a given viscosity, care must be taken to ensure that the diluent is more volatile than the true solvent in order that the medium shall remain in solution during the drying process.

A small amount of a particular solvent may be needed to aid application, to enable the release of small air bubbles in sprayed films, or to activate thickeners.

Classification of solvents is normally by chemical composition, e.g. aliphatic or aromatic hydrocarbons, alcohols, esters, ketones, etc. In addition to knowing which are appropriate for use with particular media, the paint formulator must also be acquainted with the fire hazards associated with the individual solvents and mixtures thereof, and the toxicity of various mixtures. Regulations against pollution will undoubtedly play an important rôle in the choice of solvents in the future. There are both statutory and transport regulations relating to the use and carriage of paints, according to their 'flash point' and the composition of their solvent.

Paint-making Machinery

For the purpose of paint formulation the most important units of equipment are the laboratory ball mill, bead mills and high speed dispersers. The most common, the ball mill, consists of a cylindrical porcelain vessel a little more than half filled with steel, porcelain balls or pebbles. Pigment, together with sufficient vehicle and solvent to make a free-flowing mix, is loaded into the mill until it is approximately two-thirds full. The mill is then closed and fixed into a device whereby it is made to rotate about its major axis. Normally, a period of about 16 h is required for thorough dispersal of the pigment, whereupon the mill-base is emptied out and blended with the remainder of the ingredients.

The selection of the appropriate type of machinery and the determination of the optimum conditions for bulk manufacture may be the responsibility of the paint formulator, but it is more probable that this task will be vested in a member of the production department.

Formulating a Paint

The paint technologist entrusted with the task of formulating a paint to meet a specified set of conditions must first decide what type of binders he should use and the type of solvent blend that this will entail. In the particular case of a finish, he must then select the pigments most likely to give the required colour, bearing in mind any limitations imposed by his choice of vehicle or by the conditions to which the paint will be subjected.

With the aid of a palette knife, weighed quantities of the several pigments are ground by hand into a binder such as linseed oil until an approximate match to the colour pattern is obtained. The consistency of this paste can be adjusted instrumentally to obtain the maximum work from the particular dispersion unit to be used. On the basis of this rough estimate, a ball mill is loaded with the appropriate quantities of pigment, binder and solvent, and a high-pigment-content mill-base is produced. From this and subsequent mill-bases, ordered series of paint samples are prepared and tested to establish the following data:

(a) The most appropriate pigment:binder ratio.
(b) In the case of a composite vehicle, the optimum proportion of each.
(c) The optimum addition of additives, e.g. driers, that may be necessary.
(d) The appropriate viscosity and solvent composition.

If, as is possible, the first mill-base gives a poor colour match, the relative proportions of the several pigments are suitably adjusted in subsequent batches. The ultimate aim should be to obtain a colour that is slightly deficient in the stronger staining pigments, it being more convenient to produce an exact match to the pattern by making small additions of strong staining mill-bases than by making larger additions of weaker bases.

Assuming that a paint with satisfactory properties has now been produced, there remains the possibility that it may deteriorate on storage. This must be investigated, and any faults that develop must be corrected.

The ability to apply knowledge gained by practical experience is the hallmark of a good paint formulator, for it frequently enables him to proceed to an acceptable basic formulation without delay. The greater part of the limited time that he has been allowed can then be devoted to perfecting his product. It is worthy of note, however, that the development of new products for exterior exposure is inevitably a slow process because there is no accelerated weathering cycle that can be relied upon to reproduce faithfully the effects of natural weathering.

M. W. O'REILLY

15.3 The Mechanism of the Protective Action of Paints

From time to time astronomical estimates are made of the annual destruction of metals, particularly iron and steel, by corrosion. Paint is one of the oldest methods used for delaying this process and consequently it is somewhat surprising that its protective action has only recently been systematically examined.

Since iron is the commonest structural material, the following discussion will be limited to the behaviour of this metal. The general principles can readily be extended to non-ferrous metals.

The Corrosion of Iron and Steel

Corrosion is essentially the conversion of iron into a hydrated form of iron oxide, i.e. rust. The driving force of the reaction is the tendency of iron to combine with oxygen.

It has long been known that iron is not visibly corroded in the absence of either water or oxygen. The overall reaction in their presence may be written:

$$4Fe + 3O_2 + 2H_2O \rightarrow 2Fe_2O_3.H_2O$$

When the supply of oxygen is restricted the corrosion product may contain ferrous ions.

The overall reaction can be broken down into two reactions, one producing electrons and the other consuming them:

$$4Fe \rightarrow 4Fe^{2+} + 8e \quad \text{(anodic reaction)}$$
$$2O_2 + 4H_2O + 8e \rightarrow 8OH^- \quad \text{(cathodic reaction)}$$

or

$$4Fe + 2O_2 + 4H_2O \rightarrow 4Fe(OH)_2$$

In the presence of oxygen the ferrous hydroxide will be converted into rust, $Fe_2O_3.H_2O$.

Ferrous hydroxide is soluble (9%) in pure water, but slight oxidation renders it appreciably less soluble. Thus in the presence of water and oxygen alone the corrosion product may be formed in close contact with the metal and attack will consequently be stifled. In the presence of an electrolyte such

15:24

as sodium chloride, however, the anodic and cathodic reactions are modified, ferrous chloride being formed at the anode and sodium hydroxide at the cathode. These two compounds are very soluble and not easily oxidised, so that they diffuse away from the sites of formation and react at a distance from the metal surface to form ferrous hydroxide, or a basic salt, which then combines with oxygen to form rust, with the regeneration of sodium chloride:

$$FeCl_2 + 2NaOH \rightarrow Fe(OH)_2 + 2NaCl$$
$$4Fe(OH)_2 + O_2 \rightarrow 2Fe_2O_3.H_2O + 2H_2O$$

Consequently rust is formed at a distance from the metal and stifling cannot occur.

It follows that when iron rusts, the conversion is accompanied by a flow of electrons in the metal from the anodic to the cathodic regions, and by the movement of ions in solution. This conclusion has been firmly established by Evans[1] and his co-workers, who have shown that, in the case of a number of metals under laboratory conditions, the spatial separation of the anodic and cathodic zones on the surface of the metal was so complete that the current flowing was equivalent to the corrosion rate (see Section 1.6).

In order to inhibit corrosion, it is necessary to stop the flow of current. This can be achieved by suppressing either the cathodic or the anodic reaction, or by inserting a high resistance in the electrolytic path of the corrosion current. These three methods of suppression are called *cathodic, anodic* and *resistance* inhibition respectively (Section 1.4).

The effect of paint films on the cathodic and anodic reactions will now be considered and the factors which influence the electrolytic resistance of paint films will be discussed.

The Cathodic Reaction

The cathodic reaction in neutral solutions usually involves oxygen, water and electrons:

$$O_2 + 2H_2O + 4e \rightarrow 4OH^-$$

If a paint film is to prevent this reaction, it must be impervious to electrons, otherwise the cathodic reaction is merely transferred from the surface of the metal to the surface of the film. Organic polymer films do not contain free electrons, except in the special case of pigmentation with metallic pigments; consequently it will be assumed that the conductivity of paint films is entirely ionic. In addition, the films must be impervious to either water or oxygen, so that they prevent either from reaching the surface of the metal.

The rate of corrosion of unpainted mild steel immersed in sea-water was found by Hudson and Banfield[2] to be 0·089 mm/y. Hudson[3] obtained a similar average value for steel exposed in the open air under industrial conditions (0·051 mm/y at Motherwell and 0·109 mm/y at Sheffield). This rate of corrosion corresponds to the destruction of 0·07 g/cm² per year of iron. Assuming that the corrosion product was $Fe_2O_3.H_2O$, this rate of attack represents the consumption of 0·011 g/cm² per year of water and 0·03 g/cm² per year of oxygen.

Diffusion of Water

The diffusion of water through paint films has been measured by various workers. The weight of water which could diffuse through three clear vehicles and eight paint films, each 0·1 mm thick, at 85–100% r.h. has been calculated on the assumption that the water would be consumed as soon as it reached the metal surface, i.e. that the rate-controlling step was the rate of diffusion of water through the film, and is shown in Table 15.3[4, 5].

Table 15.3   Diffusion of water through paint films of thickness 0·1 mm

Vehicle	Pigment	Rate of water consumed $(g\,cm^{-2}\,y^{-1})$	Reference
Glycerol phthalate varnish	None	0·825	4
Phenolformaldehyde varnish	None	0·718	4
Epoxy coal tar	None	0·391	5
Glycerol phthalate varnish	Flake aluminium	0·200	4
Phenolformaldehyde varnish	Flake aluminium	0·191	4
Linseed oil	Lithopone	1·125	4
Ester gum varnish	White lead/ zinc oxide	1·122	4
Linseed penta-alkyd	Iron oxide 15% p.v.c.	0·840	5
Linseed penta-alkyd	Iron oxide 35% p.v.c.	0·752	5
Epoxypolyamide	Iron oxide 35% p.v.c.	1·810	5
Chlorinated rubber	Iron oxide 35% p.v.c.	1·272	5

Note. Unpainted steel consumes water at a rate of $0·008–0·023\,g\,cm^{-2}\,y^{-1}$

By means of an ingenious instrument which measured the 'wetness' of a painted surface, Gay[6] found that although the relative humidity of the atmosphere varies appreciably, this is not reflected in the behaviour of paint films. He found that under normal conditions paint films are saturated with water for about half their life, and for the remainder the water content corresponded with an atmosphere of high humidity; furthermore, the relative humidity of sea-water is about 98%. It follows from Table 15.3 that the rate at which water passes through paint and varnish films is many times greater than the water consumed by an unpainted specimen exposed under industrial conditions or immersed in the sea.

Diffusion of Oxygen

The diffusion of oxygen through polymer films has been examined by a number of workers, who have been chiefly concerned with the permeability of plastic foils for packaging purposes. Recently Guruviah[5] has carried out measurements on films produced from several vehicles pigmented with iron oxide.

However, from the data available the weight of oxygen which could diffuse through unit area of film, 0·1 mm thick, under a pressure gradient of $2\,kN/m^2$ of oxygen, has been calculated and is shown in Table 15.4[7, 8], from which it can be seen that the rate at which oxygen passes through the film varies from about one tenth to a value greater than that consumed by an unpainted specimen. Since painted films upon exposure do not corrode at that rate (unless the film has been degraded) it is concluded that the rate of oxygen diffusion is not the controlling factor.

Table 15.4   Diffusion of oxygen through paint films of thickness 0·1 mm

Vehicle	Pigment	Rate of oxygen consumption $(\text{g cm}^{-2}\text{y}^{-1})$	Reference
Asphalt	None	0·053	7
Epoxy coal tar	None	0·002	5
Polystyrene	None	0·013	8
Polyvinyl butyral	None	0·027	8
Asphalt	Talc	0·039	7
Linseed penta-alkyd	Iron oxide 15% p.v.c.	0·003	5
Linseed penta-alkyd	Iron oxide 35% p.v.c.	0·003	5
Epoxypolyamide	Iron oxide 35% p.v.c.	0·002	5
Chlorinated rubber	iron oxide 35% p.v.c.	0·006	5

Note. Unpainted steel consumes oxygen at a rate of $0.020-0.030 \text{ g cm}^{-2}\text{y}^{-1}$.

The general conclusion drawn from these considerations is that paint films are so permeable to water and oxygen that they cannot inhibit corrosion by preventing water and oxygen from reaching the surface of the metal, that is to say they cannot inhibit the cathodic reaction.

The Anodic Reaction

The anodic reaction consists of the passage of iron ions from the metallic lattice into solution, with the liberation of electrons, which are consumed at the cathode by reaction with water and oxygen.

There are two ways in which the anodic reaction can be suppressed:

(a) If the electrode potential of iron is made sufficiently negative, positively charged iron ions will not be able to leave the metallic lattice, i.e. cathodic protection.

(b) If the surface of the iron becomes covered with a film impervious to iron ions, then the passage of iron ions into solution will be prevented, i.e. anodic passivation.

Cathodic Protection

In order to make the potential of iron more negative, the iron must receive a continuous supply of electrons. As has already been pointed out, polymer films do not contain free electrons; there remains the possibility of obtaining these from a pigment. The only pigments which contain free electrons are metallic ones, and such pigments will protect iron cathodically if the following conditions are fulfilled:

(a) The metallic pigment must be of a metal less noble than iron, otherwise the iron will supply electrons to the pigment, which will be protected at the expense of the iron.

(b) The pigment particles must be in metallic, i.e. electronic, contact with each other and with the coated iron; if they are not the movement of electrons cannot occur.

It has been shown[9] that zinc dust is the only commercially available pigment which fulfils both conditions. Paints capable of protecting steel cathodically can be prepared with zinc dust, provided that the pigment content of the dried film is of the order of 95% by weight; both organic and inorganic binders have been used, the latter being very useful when resistance to oil or organic solvents is required.

These paints are quite porous and function satisfactorily only in the presence of an electrolyte—e.g. water containing a trace of salt, or acid—which completes the circuit formed by the two metals. It might be thought that the useful life of these paints is limited to the life of the electronic contact between the zinc particles, but this is not correct. Under normal conditions of exposure the electrons supplied by the zinc to the steel are consumed at the surface of the steel by reaction with water and oxygen (cathodic reaction), with the formation of hydroxyl ions. Consequently the surface becomes coated with a deposit of the hydroxides, or carbonates, of zinc, calcium, or magnesium, which blocks the pores in the film and renders it very compact, adherent and impervious. Thus, although metallic contact between the steel and the zinc dust particles is essential in the *early* stages of exposure, the paints provide good protection after that contact has been lost. Paints containing less zinc dust have been known for a long time, but as the zinc dust concentration is decreased, protection at scratch lines or at gaps in the coating, decreases; however, such paints frequently afford good general protection owing to the formation of deposits (consisting of oxides and carbonates) on the metal at the base of the coating.

Recently it has been pointed out that manganese satisfies both conditions, since the oxide film around the particles contains ions in two states of oxidation, and it has been claimed that cathodically protective paints can be prepared with this pigment[10]. Exposure trials in this country have indicated that at an inland site their behaviour is comparable with the zinc dust controls, but that they were inferior to zinc-rich paints under severe marine conditions. It has been suggested that they might be of interest where zinc was unsuitable owing to toxicity[11].

Anodic Passivation

When a piece of iron is exposed to the air, it becomes covered with an oxide film. Upon immersion in water or solutions of certain electrolytes, the air-formed film breaks down and corrosion ensues. In order to prevent corrosion the air-formed film must be reinforced with similar material, or a ferric compound, and there are two ways in which this may be achieved:

(*a*) The pigment may be sufficiently basic to form soaps when ground in linseed oil; in the presence of water and oxygen these soaps may autoxidise to form soluble inhibitive degradation products.

(*b*) The pigment itself may be an inhibitor of limited solubility.

Basic pigments Typical pigments in this class are basic lead carbonate, basic lead sulphate, red lead and zinc oxide.

It has been established that water becomes non-corrosive after contact with paints prepared by grinding basic pigments in linseed oil[12]; it was also shown that lead and zinc linoleates, prepared by heating the oxide with linseed oil fatty acids in xylene, behave in a similar way. Later this observation was extended to the linoleates of calcium, barium and strontium[13].

Determinations have been made of the solubility of lead linoleate prepared in the absence of oxygen and extracted with air-free water[14]. Under these conditions, lead linoleate had a solubility of 0·002% at 25°C and the extract was corrosive when exposed to the air. When, however, the extraction was carried out in the presence of air, the resulting extract contained 0·07% solid material and was non-corrosive. It was concluded that in the presence of water and oxygen lead linoleate yielded soluble inhibitive degradation products.

In order to obtain information regarding the composition of these degradation products, aqueous extracts of the lead soaps of the linseed oil fatty acids were analysed, mainly by chromatography. The extracts contained formic acid 46%, azelaic acid 9% and pelargonic acid and its derivatives 27%, the remaining 18% consisting of a mixture of acetic, propionic, butyric, suberic, pimelic and adipic acids. It was shown that whereas the salts of formic acid were corrosive, those of azelaic and pelargonic acid were very efficient inhibitors.

More recently Ramshaw[15] has obtained information regarding the origin of these various acids by examining the degradation products of the lead soaps of the individual acids present in linseed oil. He found that it was only the unsaturated acids which degraded to give inhibitive materials, and that the lead soaps of linoleic and linolenic acid yielded in addition short-chain acids which were corrosive. He also examined the relative inhibiting powers of the lead, calcium and sodium salts of a range of mono- and di-basic acids in the pH range 4–6 at concentrations of 10^{-3} to 10^{-5} N[16]. Under these conditions the lead salts were always more efficient than the sodium and calcium salts, and the optimum efficiency occurred when both the mono- and di-basic acids had a chain length of 8–9 carbon atoms.

The mechanism of inhibition by the salts of the long chain fatty acids has been examined[17]. It was concluded that, in the case of the lead salts, metallic lead was first deposited at certain points and that at these points oxygen reduction proceeded more easily, consequently the current density was kept sufficiently high to maintain ferric film formation; in addition, any hydrogen peroxide present may assist in keeping the iron ions in the oxide film in the ferric condition, consequently the air-formed film is thickened until it becomes impervious to iron ions. The zinc, calcium and sodium salts are not as efficient inhibitors as the lead salts and recent work has indicated that inhibition is due to the formation of ferric azelate, which repairs weak spots in the air-formed film. This conclusion has been confirmed by the use of ^{14}C labelled azelaic acid, which was found to be distributed over the surface of the mild steel in a very heterogeneous manner[18].

Zinc phosphate is a relatively new inhibitive pigment. It has been found that this pigment is insufficiently soluble in water to act as an inhibitor, but that the aqueous extract from a zinc phosphate/linseed oil paint was inhibitive[19]. It was concluded that this pigment inhibits through soap formation, which then degrades to yield soluble inhibitive products; in addition, during soap formation phosphoric acid is liberated and this may enhance the protective properties of paints prepared with this pigment.

The function of calcium plumbate has also been examined[19]. This pigment decomposes in the presence of water to form lead peroxide, which is insoluble and not inhibitive, and calcium hydroxide. The aqueous extract is therefore alkaline and inhibitive. When ground in linseed oil, soaps are formed and the aqueous extract has a pH of 8·8. It seems that protection arises from the formation of calcium soaps and their alkaline degradation products; this would account for the success of this pigment on galvanised iron.

Soluble pigments   The most important pigments in this class are the metallic chromates, which range in solubilities from 17·0 to 0·00005 g/l CrO_3[(20)]. An examination has recently been carried out of the mechanism of inhibition by chromate ions and it has been shown by chemical analysis of the stripped film, Mössbauer spectroscopy and electron microprobe analysis that the air-formed film is reinforced with a more protective material in the form of a chromium-containing spinel[21]. The situation is, however, complicated by the possibility that some chromates, particularly the basic ones, may inhibit through the formation of soaps. There is evidence that lead chromate can function in this way.

It has been found that red lead, litharge and certain grades of metallic lead powder render water alkaline and inhibitive[12]; this observation has been confirmed by Pryor[22]. The effect is probably due to a lead compound, e.g. lead hydroxide, in solution. Since, however, atmospheric carbon dioxide converts these lead compounds into insoluble basic lead carbonate, thereby removing the inhibitive materials from solution, these pigments may have only limited inhibitive properties in the absence of soap formation.

Work by Beckmann[23] indicated that lead hydroxide was only very slightly better as an inhibitor than sodium hydroxide, and the mechanism of inhibition is probably similar to that suggested for alkaline solutions[24].

Owing to the low dielectric constant of organic vehicles, these pigments can ionise only after water has permeated the film, consequently their efficiency is associated with the nature of the vehicle in which they are dispersed, a point which is sometimes overlooked when comparing the relative merits of chromate pigments.

Resistance Inhibition

It has been shown that paint films are so permeable to water and oxygen that they cannot affect the cathodic reaction, and that the anodic reaction may be modified by certain pigments. There are, however, many types of protective paint which do not contain inhibitive pigments. It is concluded that this class of paint prevents corrosion by virtue of its high ionic resistance, which impedes the movement of ions and thereby reduces the corrosion current to a very small value.

It is assumed that conduction in polymer films is ionic—it is difficult to see how it could be otherwise—and the factors which break down this resistance, or render it ineffective, will now be considered.

The effective resistance of paint films may be influenced by ions derived from three sources:

(a) Electrolytes underneath the film.
(b) Ionogenic groups in the film substance.
(c) Water and electrolytes outside the film, i.e. arising from the conditions of exposure.

Electrolytes Underneath the Film

Atmospheric exposure trials, carried out in Cambridge, established the fact that when rusty specimens were painted in the summer, their condition, after some years' exposure, was very much better than that of similar specimens painted in the winter[25]. It was found that steel weathered in Cambridge carried spots of ferrous sulphate, deeply imbedded in the rust, and that the quantity of ferrous sulphate/unit area was very much greater in the winter than in the summer[26]; this seasonal variation was attributed to the increased sulphur dioxide pollution of the atmosphere in the winter, caused by the combustion of coal in open grates. It was concluded that there was a causal relationship between the quantity of ferrous sulphate and the effective life of the paint. It was suggested that these soluble deposits of ferrous sulphate short-circuit the resistance of the paint film and, since paint films are very permeable to water and oxygen, the ferrous sulphate will become oxidised and hydrolysed with the production of voluminous rust, which will rupture the film at numerous points, thus giving rise to the characteristic type of failure seen on painted rusty surfaces.

It can be claimed that the problem of painting rusty surfaces is now understood. A method for estimating the ferrous sulphate content of any rusty surface has been put forward[26], but the amount of ferrous sulphate which can be tolerated by various paints has not yet been established. Thus it is bad practice to apply paints to surfaces carrying electrolytes.

Ionogenic Groups in the Film Substance

Ionogenic (ion-producing) materials may be present, in the form of electrolytes, in both the pigments and the vehicle. Their presence in the pigments may be eliminated by the selection of suitable raw materials by the paint manufacturer, consequently it does not concern us here, but it is of importance to consider the possibility of the existence of ionogenic groups, such as carboxyl groups, in the polymer itself.

When paint films are immersed in water or solutions of electrolytes they acquire a charge. The existence of this charge is based on the following evidence. In a junction between two solutions of potassium chloride, 0·1N and 0·01N, there will be no diffusion potential, because the transport numbers of both the K$^+$ and the Cl$^-$ ions are almost 0·5. If the solutions are separated by a membrane equally permeable to both ions, there will still be no diffusion potential, but if the membrane is more permeable to one ion than to the other a diffusion potential will arise; it can be calculated from the Nernst equation that when the membrane is permeable to only one ion, the potential will have the value of 56 mV.

It is easy to measure the potential of this system and it has been found[27] that membranes of polystyrene, linseed oil and a tung oil varnish yielded diffusion potentials of 43–53 mV, the dilute solution being always positive to the concentrated. Similar results have been obtained with films of nitro-cellulose[28], cellulose acetate[29], alkyd resin and polyvinyl chloride[30].

This selective permeability is ascribed to the presence on the membrane of a negative charge, which is attributed to carboxyl groups attached to the polymer chains. Paint films can, therefore, be regarded as very large anions.

It has been shown[31] that the charge influences the distribution of the primary corrosion products, and recent work has indicated that the existence of carboxyl groups in the polymer film has an important influence on its behaviour when immersed in potassium chloride solutions.

Water and Electrolytes outside the Film

Here we are concerned with the effect of ions in the environment on the resistance of polymer films.

Kittelberger and Elm[32] measured the rate of diffusion of sodium chloride through a number of paint films. Calculations based on their results[27] showed clearly that the rate of diffusion of ions was very much smaller than the rate of diffusion of either water or oxygen. Furthermore, they found that there was a linear relationship between the rate of diffusion and the reciprocal of the resistance of the film. This relationship suggests that the sodium chloride diffused through the membrane as ions and not as ion pairs, since the diffusion through the film of un-ionised material would not affect the resistance, because if a current is to flow, either ions of similar charge must move in one direction, or ions of opposite charge must move in opposite directions.

An examination has, therefore, been made of the effect of solutions of potassium chloride on the electrolytic resistance of films cast from a penta-erythritol alkyd, a phenolformaldehyde tung oil and an epoxypolyamide varnish[33, 34]. Potassium chloride was chosen because its conductivity is well known and unpigmented films were first examined in order to eliminate the complexities of polymer/pigment interaction. However, the work is now being extended to pigmented systems.

The experimental procedure consisted of casting the varnish on glass plates by means of a spreader bar having an 0·004 in (0·051 mm) gap; this produced a wet film 0·002 in (0·102 mm) thick that yielded a dried film of 0·001 in (0·025 mm). This standard thickness was used throughout and resistances are quoted in $\Omega\,cm^2$. The cast films were dried for 48 h in a glove box followed by a further 48 h in an oven at 65°C.

The films were then soaked in water and removed from the plates. Portions were mounted in glass cells which were filled with potassium chloride solution; two Ag/AgCl electrodes were inserted into the limbs of the cells and the unit was placed in a thermostat. The resistance of the films was determined, from time to time, by connecting the cells in series with a known resistance and applying a potential of 1 V to the combination; the potential drop across the standard resistance was measured by means of a valve potentiometer.

When samples of about 1 cm^2 were taken from a single cast film of 100 × 200 mm of a pentaerythritol alkyd varnish, their resistances varied with the concentration of potassium chloride solution in one of two ways (Fig. 15.4).

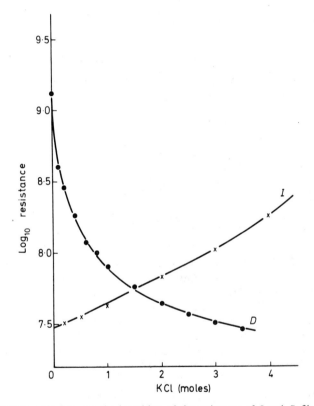

Fig. 15.4 Relationship between the logarithm of the resistance of *I* and *D* films and the concentration of the potassium chloride solution in which they were immersed (courtesy *Brit. Pol. J.*, **196** No. 1, 173 (1968))

Either the resistance increased with increasing concentration of the electrolyte (inverse or *I* conduction) or the resistance of the film followed that of the solution in which it was immersed (direct or *D* conduction). The percentage of *I* and *D* samples taken from different castings varied, but average values for a number of castings were 50% *D* for the pentaerythritol alkyd and the tung oil phenol formaldehyde varnishes, and 76% *D* for the epoxy polyamide varnish[35].

A careful examination has been made of the properties of *I* films when immersed in solutions of electrolytes. It was found that when a film of a penta-erythritol alkyd varnish was transferred from 0·001N KCl to 3·5N KCl its resistance rose, fell upon returning it to the 0·001N KCl, rose again to the same high value when immersed in a sucrose solution isotonic with 3·5N KCl and fell to the original value when returned to the dilute KCl solution (Fig. 15.5). It was concluded that the changes in resistance were dependent only upon the available water in the solution and were associated, therefore, with the entry of only water into the varnish film[33].

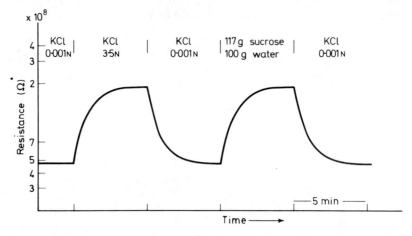

Fig. 15.5   Relationship between the resistance of an *I* film (log scale) and the concentration of the solute (courtesy *Official Digest*, **34** No. 452, 972–990, Sept. (1970))

In contrast, *D* films followed the resistance of the solution in which they were immersed, and this behaviour was originally explained by assuming that *D* films contained holes, or pores, filled with solution that controlled the resistance of the film. Thus a typical value for the resistance of a *D* film in 3·5N potassium chloride is $10^8 \, \Omega \, cm^2$ and if this resistance was due to a pore, then it would have a radius of about 500 Å. In order to test this explanation the distribution of *I* and *D* areas in a given piece of film has been determined by means of a series of gaskets fitted into a dismountable cell[36]. It was found that *I* films were free from *D* areas, but that in the case of the three vehicles examined, samples of *D* films always contained a mixture of *I* and *D* areas in an interlocking mosaic structure. It was concluded that those portions of the film having *D* properties were distributed over an appreciable area of the sample and not confined to a single area, as would have been the case had the sample contained a single pore. It was concluded that *D* conduction cannot be attributed to the presence of pores, unless they were of molecular dimensions.

In general, the water uptake of *D* films tended to be higher than that of *I* films, but a more significant difference was shown by microhardness measurements. The results obtained with all three vehicles showed that the *D* areas were significantly softer than the *I* areas and that the distribution of the hardness values corresponded to that of the resistances. It was concluded that these films have a very heterogeneous structure and that *I* and *D* areas are brought about by differences in crosslinking density within the film.

An investigation has been made of the factors which control I and D conduction and it has been found that the difference is only one of degree and not of kind[35]. Thus, if the varnish films are exposed to solutions of decreasing water activity, then the resistance falls with increasing concentration of electrolyte, but a point is eventually reached when the type of conduction changes and the films exhibit I-type behaviour. It appears that D films can be converted into I films, the controlling factor being the uptake of water.

The discussion so far has been limited to the behaviour of polymer films after immersion in potassium chloride solutions for only a short time. When varnish films were immersed in potassium chloride solutions for a month or more a steady fall in resistance took place. Further experiments indicated that the effect was reversible and dependent on both the pH of the solution and the concentration of potassium chloride. It was concluded that an ion exchange process was operative[33].

In view of this, the properties of I films were examined after they had been subjected to increasing amounts of ion exchange[34]. In order to do this, detached films were exposed at 65°C for 7 h to a universal buffer adjusted to a suitable pH and the resistance of the film measured at 25°C in 3N and 0·001N potassium chloride. The results obtained with a pentaerythritol alkyd are shown in Fig. 15.6 from which it can be seen that as the pH of the conditioning

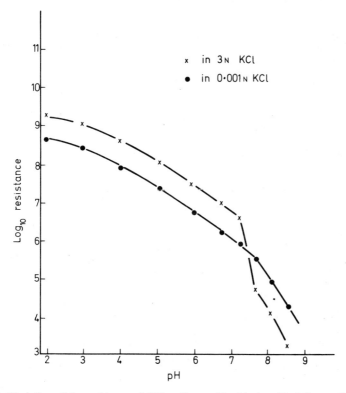

Fig. 15.6 Variation of the resistance of I films (log scale) with the pH of the conditioning solutions

solution increased, the resistance of the film fell, until at a pH of about 7·5 it suddenly dropped, the resistance of the film then followed that of the solution in which it was immersed, i.e. it became a D-type film. Similar results were obtained with films from the phenolformaldehyde tung oil and epoxypolyamide varnishes, although in both cases the change-over point occurred at higher pHs.

The phenomenon of ion exchange has been confirmed by chemical analysis[37] and by measurements[38] of the uptake of radioactive potassium in the form of [42]KCl; furthermore, the ion exchange properties of free films of a soya alkyd have been examined[39].

The general conclusions are that when varnish films are immersed in solutions of electrolytes, both water and electrolyte penetrate the film, but the electrolyte can only enter those areas where the crosslinking density is low. The penetration of electrolyte opens up the film and after some time an ion exchange process becomes operative in which hydrogen ions, probably derived from carboxyl groups attached to the polymer network, exchange with potassium ions present in the solution. This brings about a fall in resistance which slowly destroys the protective value of the film.

It appears that films of greatly improved protective value would be obtained if the crosslinking could be rendered more homogeneous.

J. E. O. MAYNE

REFERENCES

1. Evans, U. R., *The Corrosion and Oxidation of Metals*, Chapt. 21, Arnold, London (1960)
2. Hudson, J. C. and Banfield, T. A., *J. Iron St. Inst.*, **154**, 229 (1946)
3. Hudson, J. C., *The Corrosion of Iron and Steel*, Chapman and Hall, London, 66 (1940)
4. Edwards, J. D. and Wray, R. I., *Industr. Engng. Chem.*, **28**, 549 (1936)
5. Guruviah, S., *J. Oil Col. Chem. Ass.*, **53**, 669 (1970)
6. Gay, P. J., *J. Oil Col. Chem. Ass.*, **31**, 481 (1948)
7. Anderson, A. P. and Wright, K. A., *Industr. Engng. Chem.*, **33**, 991 (1941)
8. Davis, D. W., *Mod. Packag.*, 145, May (1946)
9. Mayne, J. E. O., *J. Soc. Chem. Ind.*, **66**, 93 (1947)
10. Kelkar, V. M. and Putambekar, S. V., *Chemy. Ind.*, 1 315 (1964)
11. Wild, G. L. E., *Paint Technology*, **30**, 9 (1966)
12. Mayne, J. E. O., *J. Soc. Chem. Ind.*, **65**, 196 (1946); **68**, 272 (1949)
13. Mayne, J. E. O., *J. Oil Col. Chem. Ass.*, **34**, 473 (1951)
14. Mayne, J. E. O. and van Rooyen, D., *J. Appl. Chem.*, **4**, 384 (1954)
15. Mayne, J. E. O. and Ramshaw, E. H., *J. Appl. Chem.*, **13**, 553 (1963)
16. Mayne, J. E. O. and Ramshaw, E. H., *J. Appl. Chem.*, **10**, 419 (1960)
17. Appleby, A. J. and Mayne, J. E. O., *J. Oil Col. Chem. Ass.*, **50**, 897 (1967)
18. Mayne, J. E. O. and Page, C. L., *Br. Corros. J.*, **5**, 94 (1970) and **7**, 111, 115 (1972)
19. Mayne, J. E. O., *Br. Corros. J.*, **5**, 106 (1970)
20. Sherman, L. R., *Official Digest*, **28**, 645 (1956)
21. Bancroft, G. M., Mayne, J. E. O. and Ridgway, P., *Br. Corros. J.*, **6**, 119 (1971)
22. Pryor, M. J., *J. Electrochem. Soc.*, **101**, 141 (1954)
23. Beckmann, P. and Mayne, J. E. O., *J. Appl. Chem.*, **10**, 417 (1960)
24. Gilroy, D. and Mayne, J. E. O., *Br. Corros. J.*, **1**, 161 (1966)
25. Mayne, J. E. O., *J. Iron St. Inst.*, **176**, 143 (1954)
26. Mayne, J. E. O., *J. Appl. Chem.*, **9**, 673 (1959)
27. Mayne, J. E. O., *Research*, Lond., **6**, 278 (1952)
28. Sollner, K., *J. Phys. Chem.*, **49**, 47, 147 and 265 (1945)
29. Meyer, K. H. and Sievers, J. F., *Helv. Chim. Acta.*, **19**, 665 (1936)
30. Nasini, A. G., Poli, G. and Rava, V., *Premier Congrès Technique International de l'Industrie des Peintures*, Paris, 299 (1947)
31. Mayne, J. E. O., *J. Oil Col. Chem. Ass.*, **40**, 183 (1957)

32. Kittleberger, W. W. and Elm, A. C., *Industr. Engng. Chem.*, **44**, 326 (1952)
33. Maitland, C. C. and Mayne, J. E. O., *Official Digest*, **34**, 972 (1962)
34. Cherry, B. W. and Mayne, J. E. O., *First International Congress on Metallic Corrosion*, Butterworths, London, 539 (1962)
35. Kinsella, E. M. and Mayne, J. E. O., *Br. Polym. J.*, **1**, 173 (1969)
36. Mayne, J. E. O. and Scantlebury, J. D., *Br. Polym. J.*, **2**, 240 (1970)
37. Cherry, B. W. and Mayne, J. E. O., *Second International Congress on Metallic Corrosion*, National Association of Corrosion Engineers, Houston 2, Texas, 680 (1966)
38. Rothwell, G. W., *J. Oil Col. Chem. Ass.*, **52**, 219 (1969)
39. Ulvarsson, U., Khullar, M. L. and Wåhlin, E., *J. Oil Col. Chem. Ass.*, **50**, 254 (1967)

15.4 Paint Failure*

In view of the wide scope of the subject, paint failure can be treated here only in general terms; detailed accounts will be found in the literature[1-6].

Forms of Paint Failure

A frequent defect of paintwork is cracking in all its forms, including checking, crazing and alligatoring, which produces such effects as flaking, scaling and peeling. These defects expose the underlying metal surface to the environment so that corrosion is not prevented. It should be observed that checking and crazing begin in the upper coat and extend gradually down towards the substrate, the fissures being wider on top and narrower towards the base. If such signs of breakdown become noticeable after a coating system has had a reasonable length of life in relation to the given conditions of exposure, it is not proper to consider them as film defects. Paint films start their gradual decomposition due to oxidation, erosion, weathering, etc. from the moment of exposure onwards at a rate dependent on their constituents, the environmental conditions and circumstances of application. With increasing age the elasticity of the film usually decreases (in the case of an oil-based or modified paint this results, in the main, from continued oxidation). Expansion and contraction of the metal base caused by severe temperature changes will result in the formation of discontinuities in a relatively inelastic paint film unless the paint has been formulated to withstand these conditions. Excessively high temperatures cause unsuitable paint films to become brittle, crack and loose adhesion. Loss of adhesion can also be caused by swelling.

The initial penetration of rust through the protective paint film (rust staining) may be a signal that indicates the necessity for checking the condition underneath the paint film. Rust staining is said to be characteristic of weathered and hand-cleaned steel[7].

Causes of Paint Failure

A consideration of the most important causes of paint failure must include the following: inadequate surface preparation, application of the paint

*A glossary of the terms frequently used in this field will be found in Section 15.10.

under unfavourable conditions or by inappropriate methods, use of unsuitable paints, adhesion difficulties, the nature of the corrosive environment, etc[8]†.

Premature failure can also occur as a result of lack of attention to design. Facilities should, therefore, be provided for ventilatory drainage of water (rain, condensation, etc.), and all structures should be designed so as to permit ready access for repainting. Due consideration by architects and structural engineers at the design stage can indeed help to obviate certain of the causes of paint failure mentioned in this section (see also Sections 10.3 and 12.5).

Pretreatment and Paint Failure

The majority of failures of paint applied to metal surfaces are undoubtedly due to insufficient or unsatisfactory preparation of the metal surface, and it is essential that residues of dirt, grease, oil, silicone compounds, etc. be removed from the metal and that all loose paint be removed from surfaces which have been painted previously. The most common cause of premature failure is omitting to remove (as far as is practicable) corrosion products, e.g. rust and millscale, before painting. If a metal has been adequately pretreated, it is then desirable to apply the primer immediately at the factory, if this is possible, in order to ensure that the metal surface remains free from contamination and corrosion products. This is particularly important after grit blasting or any mechanical operations where a protective air-formed film may have been disrupted so that the metal is sensitive to corrosion. The condition of shop-applied primers should be examined before further painting, e.g. on site, and defective areas made good. Edges, welds and rivets need special attention. Prepainted structural steel should not be left exposed unduly long to the inclemency of the weather, particularly under damp conditions or in a marine or industrial atmosphere.

Primers

Red lead, zinc chromate, zinc phosphate[10], basic lead silicochromate, metallic lead, calcium plumbate, metallic zinc, etc. are used in primers, and general considerations of their action have been dealt with in Section 15.3. Zinc is considered to reduce rust to metallic iron or magnetite, so that it can be applied, e.g. as zinc-rich paint, to a lightly rusted surface, and to protect the steel base sacrificially[11]. Zinc chromate and etch primers containing zinc chromate (amongst others) are used for steel and are especially suitable for aluminium and its alloys and for aluminium coupled to steel. They are not so effective as primers for zinc, particularly when zinc is phosphated, owing to the poorer adhesion[12]. Calcium plumbate can be applied to zinc or zinc coatings without prior treatment with mordant solutions or weathering, which is usually necessary to ensure good adhesion. Any inhibitive primer used must be applied directly to the metal surface.

†An exemplary colour photographic standard for the identification of the state of rusting of iron and steel surfaces, and of the degree of care with which they have been prepared for painting, is in use. It distinguishes between manual scraping and wire brushing and shot blasting[9].

Effects of Climatic Conditions on Paint Films

Paint failure is related to climatic conditions, and the weather prevailing during application of the paint and during subsequent exposure will determine the life of the paint system. This applies, of course, particularly to outdoor work. In unfavourable weather conditions, cracking and blistering can be promoted as a consequence of the expansion of the products of corrosion, and in the case of iron and steel this can lead to under-rusting.

Low Temperatures and Wet Weather

When severe drops in temperature occur, outdoor work should if possible be halted, as hail, hoar frost and freezing conditions at the time of painting or shortly afterwards will greatly reduce the life of any paint film, as well as being detrimental to its appearance. Wet (rain, fog, dew deposits) and cold weather (7–10°C and below) during or immediately after application will prolong the drying time and will leave the film tacky for a long time. During this period dirt will adhere to the tacky film and rain or condensed moisture will tend to reduce the gloss by displacement of some of the paint.

Moisture-curing polyurethane paints and bituminous paints, specially formulated for the purpose, are also suited for application to damp substrates; other polyurethane paints should not even be applied to dry surfaces if the relative humidity is high.

Suitable paints for use underwater include vinyl resin systems, coal tar paints over inorganic zinc-rich primers, and some coal-tar epoxy primers have also proved themselves[13]. Special paints are available for application under water, e.g. epoxy modifications with polyamides.

Loss of matter by weathering induces hazing and loss of gloss, which are followed by chalking (black chalking occurs with bituminous paints and with paints containing a fair proportion of black pigments, e.g. lamp black). The chalking caused by the presence of titanium dioxide (especially anatase) in the top coat initiates rapid erosion. As the pigments in such defective films become more exposed to rain and wind and wash away, the films become increasingly permeable to moisture, with consequent corrosion of the underlying metal although they have the advantage of looking clean. In special cases (see below) controlled chalking may be desirable.

Painted metal exposed at coastal areas, ports and docks often suffers most from such hazards, which are aggravated by the corrosive action of the saline atmosphere and the erosive action of blowing sand. Such conditions can prevail up to approximately 3 km inland.

Stripping of top coats soaked by rain or sea-water has occurred with alkyd-resin-based paint systems, mainly on ships. It is often found that this loosening of intercoat adhesion may be prevented by the use of a different undercoat, but the reason for this is obscure.

A good protective paint system should dry reasonably well, should not swell owing to moisture uptake, and should, for all practical purposes, be continuous.

Factors which can contribute to unsatisfactory drying, in particular of paints containing drying oils, include application of too-heavy paint films (especially during the cold seasons when viscosities and relative humidities are higher), overdoses of driers, gaseous pollution, use of unsuitable heavy thinners and residues of tar oil, wax or grease present on metal surfaces. It must be borne in mind that black and brown enamels, e.g. alkyd-based ones, dry more slowly than the white and most other colours. This difference in rates of drying is, of course, much more marked during colder periods of the year. Some polyurethane resin paints can cure at temperatures as low as 50°C.

Direct Sunlight

Drying of linseed-oil-based paints in direct sunlight, especially under tropical conditions, can produce resoftening and tackiness within a fortnight of application, owing to syneresis (separation of the gelled film into solid and liquid). This may be followed by cracking if the top layer hardens while the lower film strata remain soft. Tropical conditions in general contribute to faster paint breakdown, owing to high temperatures, moisture-laden atmosphere, high ultra-violet content in the solar radiation, or to a combination of some or all of these effects. Thus in the tropical regions, fading and discoloration, matting, chalking and cracking, followed by peeling and general embrittlement, can take place rapidly.

Effect of Industrial Atmospheres

In towns which are not heavily industrialised, the life of a paint film may be about equal to that in rural areas, but, except in the new smokeless zones, dirt collection from soot and dust will be noticeable earlier. The use of self-cleaning (chalking) paints can overcome the premature loss of some of the decorative effect without noticeably reducing protection.

Industrial towns, especially those having heavy or chemical industries, have an acid atmosphere and the pH of rain water is frequently about 3, owing to the presence of sulphuric acid. This can cause gradual attack on certain pigments and extenders, resulting in discoloration (e.g. red lead can be transformed into white lead sulphate, and atmospheric hydrogen sulphide results in the blackening of lead pigments) and decomposition of the paint film, and can lead to premature failure. Aluminium finishing paints applied over, for example, a red lead primer, are liable to be attacked in industrial atmospheres, owing to the formation of water-soluble aluminium salts, and the aluminium colour may disappear quickly. Similarly, top coats of zinc-rich paints may lose their metallic colour by formation of zinc salts (e.g. on iron chimney stacks), and contamination in the atmosphere may also endanger intercoat adhesion.

Effects of Moisture

Hot steam and severe condensation acting on a film surface exert a very destructive effect, comparable with that of a paint remover; they are particularly liable to cause swelling. Dry steam, in contrast to condensed steam, does not cause corrosion[14]. Less severe attack by water vapour can cause blistering, which can be of two types: intercoat blisters between paint films, and blisters through the complete film system. Only the latter leads to corrosion of the underlying material.

Paint films exposed to condensation often fail unexpectedly by very early blistering between primer and finishing coat, which is possibly due to impermeability of the primer. This is one of the few conditions for which a degree of permeability in a film system on metal appears to be desirable.

Dampness often accounts for the promotion of mould growth on painted surfaces, e.g. in breweries, laundries and dairies; fungi develop faster under tropical conditions. There are special media which are resistant to mould growth, in particular those which are based on the chlorinated compounds, such as chlorinated rubber, polyvinyl chloride, its various copolymers and other halogen-containing polymers. By addition of suitable fungicides and careful selection of the pigments, traditional hard-drying paints and varnishes can also be made to resist mould growth. Infected surfaces and films may be washed with antiseptics to control mould growth. Some fungi may even be dangerous to health if they occur indoors.

Factors Which Cause Paint Failure in Industrial Applications

Anti-oxidising Environments

Where fumes or deposits which act as anti-oxidants are present, no orthodox paint which drys by oxidation can give satisfactory service. Instead, a coating which dries either by evaporation (e.g. a selected chlorinated rubber paint), or by a cross-linking reaction (e.g. a catalysed epoxy or two-component polyurethane paint) must be used.

Oxidising and Acid Environments

Atmospheres polluted by oxidising agents, e.g. ozone, chlorine, peroxide, etc. whose great destructive power is in direct proportion to the temperature, are also encountered. Sulphuric acid, formed by sulphur dioxide pollution, will accelerate the breakdown of paint, particularly oil-based films. Paint media resistant to both acids, depending on concentration and temperature, and oxidation include those containing bitumen, methacrylate resins, chlorinated or cyclised rubber, epoxy and polyurethane coal tar combinations, phenolic resins and p.v.c.

Acid conditions occur in the vicinity of, for example, coke ovens, gas works, oil-fired plant, galvanising plant and paper pulp mills, and in these conditions, cracking is a frequent form of failure; the cracking and peeling in acid environments is usually much more severe and occurs much earlier in the life of the paint film than is the case in other environments. Failure in acid environments results from the specific properties of pigment, medium, or drier used in the paint, e.g. in sulphuric acid environment zinc pigments form zinc sulphate, which appears on the paint surface.

Non-oxidising and weak acids, in contrast to oxidising acids, can penetrate paint films without destroying them; they then react with the metal base to form salts with resultant stresses which cause cracks. Magnesium-rich alloys are particularly prone to attack by acids; their salts, having considerable volume, in severe cases effloresce through the broken paint films.

For resistance to acid conditions alone, traditional filled and unfilled bituminous solutions (which have economic advantages), gum damar and shellac may be employed. Cross-linking surface coatings, e.g. those of the amine-curing epoxy type, which develop insolubility after curing, can successfully protect storage tanks against damage by the spilling of such substances as tar acids.

Alkaline Environments

Oil-base (including oil-modified alkyd resin) paint films should not be used in alkaline environments as the paint will deteriorate owing to saponification; alkali-resistant coatings are provided by some cellulose ethers, e.g. ethyl cellulose, certain polyurethane, chlorinated rubber, epoxy, p.v.c./p.v.a. copolymer, or acrylic-resin-based paints. In particular, aluminium and its alloys should be protected by alkali-resistant coatings owing to the detrimental effects of alkali on these metals.

Salt Solutions

Corrosive solutions, e.g. salt solutions as present in salterns, refrigeration plant and sea-water, are particularly active at the water-line (cathodic zone), where alkali may accumulate and creep up between paint and metal[11] and cause softening and loosening of the paint. This process may also occur where the metal is completely immersed, particularly below paint films pigmented with zinc or aluminium[15]. Caustic soda is formed at the steel surface (which is made cathodic by the zinc) resulting in the softening of oil-base paints and consequent loss of adhesion. In sea-water, at the local cathodes the total concentration of ions will exceed that in the surrounding sea-water, and water may be drawn in by osmosis, with resultant alkaline blistering[15]. This is usually the first sign of electrochemical corrosion; alkaline peeling and corrosion of the metal become apparent only later. Good results in the salt-rich Mediterranean have been reported[16,17] with anticorrosive primers containing a proportion of chromium fluoride, including those for ships' bottoms.

Marine Atmosphere

Iron girders, etc. are frequently supplied to a site in the grit-blasted and primed condition, but occasionally this work is carried out on site. If the structures lie about afterwards for some time in a salt-laden environment, e.g. a marine atmosphere, and are not thoroughly washed with fresh water and dried before further painting in order to remove all traces of sodium chloride, the latter will soon play havoc with the steel and anticorrosive film system. This will occur after erection and possibly even inside buildings owing to under-rusting accompanied by severe blistering and followed by flaking with rustbacking. The rust can be in various states. Analogies can be drawn in connection with the repainting of ships in dry docks. High relative humidity has an aggravating effect.

Corrosion-promoting Pigments

Some pigments promote corrosion owing to their content of soluble salts, their tendency to swell, or their electrochemical action, and thus should be avoided. Rust of the spotted type can be the consequence of their presence in a paint, especially the first coat, e.g. of graphite (noble to steel), some red oxides of iron, gypsum, ochre or lamp black.

Heavy metals, such as copper or mercury which are used in antifouling compositions, should not be present in paints which are in direct contact with steel exposed to sea-water. Antifoulings are always separated from the steel by an effective anticorrosive primer, but interaction between the two must be avoided by suitable formulation to avoid corrosive and excessive leaching, i.e. making the antifouling ineffective[17]. Mercury compounds, which are sometimes used as fungicides, also tend to corrode aluminium and its alloys, particularly in the presence of sea-water, or in moist tropical conditions.

Effects of Stoving and Storage Conditions

During stoving in convection-type box ovens, drying can be delayed (as it can on air drying when the ventilation is insufficient, e.g. in a ship's hold) if the vents are closed too far, or if the coated articles are too closely packed. In the latter case there may even be trouble caused by solvent wash, i.e. redissolution of the uncured film by stagnant solvent vapours, which occurs mostly on surfaces near the top of the oven. This can lead to the establishment of practically unprotected areas.

Damp conditions contribute to 'gas-checking'* of some synthetic stoving lacquers, quite apart from the effects of foul oven gases, or the presence of detrimental solvent vapours, e.g. from a trichlorethylene degreasing plant.

Overstoving, too, can result in embrittling due to overpolymerising or oxidising, followed by cracking or crazing. Stoving enamels, etc. which are based mainly on cross-linking epoxy resin combinations, behave for all practical purposes in exactly the contrary manner to this, forming almost the only exception to the general rule. They are brittle when undercured and become elastic after complete cure, and even remain so when they have been somewhat overcured. Certain members of this class of coatings, therefore, do not perform too satisfactorily on air drying. It has been observed, for instance, that air-dried amine-cured anticorrosive epoxide paints over new steel were not able to hold down millscale, which appeared still adherent at the time of painting, for any practicable length of time in contrast to the

*A fine or coarse wrinkling due to irreversible swelling of a surface-dried film.

performance of traditional anticorrosive oil paints. Very premature flaking occurred, the brittle paint flakes being backed with millscale. Epoxide resin esters, however, perform quite well, apart from a tendency to chalking.

If infra-red heating or any other radiation curing method is employed, areas which are shaded from the rays or are outside the area of greatest flux density, cannot dry as hard as the fully radiated surfaces, and may form weak spots susceptible to corrosion.

After long storage in their packages, certain oxidising, i.e. drying-oil or drying-oil-modified alkyd-resin-based paints containing certain pigments, of which iron oxides, iron blues, toluidine red and carbon blacks are the most important, lose some of their drying properties, probably owing to inactivation of driers by adsorption on the pigment surface, followed by slow deactivation of the adsorbed catalysts[18]. Such paints, often used as primers, dry and harden satisfactorily when freshly made, but storage may make them increasingly sensitive to the application of a second coating. Discoloration due to mixing of the films, drag of the brush, and in severe cases even lifting, may result.

Lifting may also occur if a paint containing strong solvents (xylol or solvent naphtha, not to mention such active solvents as esters and ketones) is applied (not necessarily by brushing) over a paint which is not resistant to them. The older an oxidising paint film becomes, the more solvent-resistant it will be. Short-oil media and pigment-rich paints are not so prone to lifting. This type of failure is not restricted to oil-base materials; it can, for example, also occur with chlorinated rubber paints.

Effects of Application Methods

Excessive thinning of a paint of good quality is often the cause of the application of films which are too thin. The temptation for operators to do this is great as it often increases the ease of application and their bonuses, especially in the case of paints for brushing. Overthinning is particularly common when surface coatings based on a great number of synthetic resins, e.g. medium to short oil-modified alkyd resins, or coatings which dry by evaporation are being used, especially if the thinners contain a high proportion of fast-evaporating solvents.

Overthinning is also frequently responsible for running and sagging, which in turn promotes excessive pigment flotation. If application is by spraying, this can be countered by the use of thinners which evaporate quickly. In brushing, however, such thinners would cause dragging of the brush. If the evaporation rate of thinners is too fast, they may promote cobwebbing when highly polymerised resins such as the vinyl or acrylic families are being sprayed. Again, if heavy thinners containing strong solvents are used on the second coating, lifting trouble may be experienced in addition to sagging.

Some paints, particularly alkyd resin types, can be applied wet on wet by spraying, without the aforementioned disadvantages.

Certain thinners, such as those containing lower aromatic or alcoholic solvents, esters or ketones, reduce viscosity rapidly, and should therefore be used sparingly.

If paint is insufficiently stirred before use, over-pigmented paint from the bottom of the container will, when it comes to be used, act as a short-oil non-elastic coating of poor binding power, while under-pigmented mixture from the upper strata will perform more as a longer-oil, more elastic coating, and will possibly run. Similarly, excessively long dipping times for a top coating can cause resoftening of an insufficiently hardened undercoat, in which case the result might be one of the forms of cracking or lifting referred to in relation to thinners.

If an elastic or insufficiently hard primer or paint has been applied under a less elastic top coat, or if the first coat (or set of coats) of oil-base paint has been second-coated before it is completely dry, not only will the paint-work remain soft for an unduly long period, but cracking will also follow, as the upper layer cannot follow the movement. Red lead primers are rather liable to cause this defect. If the last coat is very thick this fault will frequently manifest itself in the form of alligatoring, i.e. the formation of cracks which do not penetrate all the films down to the substrate, and which may be present in the top layer only. It is advisable to allow such primers as long a period of hardening as possible.

Repainting

For painted structures it is essential that an additional paint coating be applied as soon as there is evidence of paint breakdown. The Protective Coating Sub-committee of BISRA[4] recommend painting of steel surfaces when 0·2–0·5% of the surface area shows evidence of rust. Delay in repainting may be a false economy, as if rusting is extensive it may be necessary to clean down to bare metal before paint can be applied.

Should an old bituminous paint layer have to be recoated, this should be done only with another bituminous paint, unless the surface is first insulated with one of the special primers which are available for the purpose. Bleeding and premature checking may otherwise occur.

Damage to prefabrication primers or even the whole film system can be caused on transport or on erection, leaving for example, bare edges. Good supervision is necessary to ensure that defective areas are conscientiously touched-up before applying further paint films. It is even recommended to disregard the coat of prefabrication primer when deciding the number of coats to be specified[19].

Adhesion Difficulties

It is important to realise that various factors contribute to good adhesion of paint films. These include:

1. Cleanness of the base, i.e. freedom from grease, which improves the wettability of the metal surface, and the removal of oxides, dust or loose paint, etc. already described. The closer the surfaces of paint film and metal, the more secondary valencies originating in the polar constituents of the medium are brought into play.

2. Mechanical pretreatment of the metal by weathering, sanding, shot-blasting[20], etc. for the removal of corrosion products and loose mill-scale, and chemical pretreatment by phosphating, pickling, mordant solutions, etc. to create a mechanical, or, in the case of etch primers, a chemical key. Mechanical pretreatments must always be followed by thorough washes with de-ionised water. Degrees of cleanliness of steel surfaces can be compared with BS 4232, etc.[21]
3. Selection of suitable coatings possessing good wetting properties, which are elastic enough to expand and contract with the metal base over a reasonably long period and which, as far as priming is concerned, have an affinity with the metal to be painted. It is often not appreciated that the adhesion properties of a given coating material may vary according to the type of metal to which it is applied, although it is suggested that the degree of retention of non-polar decontaminants is the real cause[22].

So far as iron and steel are concerned, the adhesion problem is simple, and the oleoresinous coatings which are generally applied to them form a good bond with them. Mechanical pretreatments are always extremely useful.

Cracking, flaking, scaling or blistering due to under-rusting (the latter often being accompanied by brown discoloration of the film) is, as has already been explained, due to mechanical action by the products of corrosion. This may at times pose the problem of weather the paint or the painting system was responsible for the corrosion, or whether, on the other hand, it was the corrosion (possibly residual) which was responsible for the unsatisfactory performance of the paintwork. The better the adhesion of the paint to the metal, the less damage there will be to the paint film, and the less premature corrosion will ensue. This is similarly the case with non-ferrous metals. Rough (especially blasted) steel surfaces which have received too thin a paint coverage will be indicated by the presence of pinpoint rust spots in the film surface, wherever the metal peaks have not been sufficiently protected.

A patchy form of rust that attacks paint films from underneath, can be caused by sweaty hands, residues from fluxes, etc. Examples of the latter include residues from phosphating and soluble salts (including those from unsuitable rinsing water) and they can manifest themselves on steel in the form of a creeping filiform corrosion, i.e. as progressing threads of rust which loosen the coating. This can be followed visually through transparent films. It occurs, however, only when the relative humidity of the surroundings is above 82%, and if oxygen diffuses the film. Diffusion of carbon dioxide seems, however, to suppress filiform corrosion[23, 24]. A somewhat similar type of corrosion that causes the destruction of paint films as a result of the presence of salts beneath, is termed *filigran corrosion*, and has been observed on painted ships[25]. Filiform corrosion is considered in Section 1.6.

Aluminium and magnesium alloys, copper and its alloys, and zinc and zinc-base diecastings, including galvanised iron, to name the most important groups of non-ferrous metals, can offer serious adhesion problems. These are aggravated if the surfaces are very smooth, as, for example, on diecastings or hard rolled sheets. For light metals, p.v.b.-based* etch primers are ideal;

---

*p.v.b. is an abbreviation for *polyvinyl butyral*.

long-oil alkyd-resin-based zinc chromate primers are also very satisfactory. Etch primers and alkyd-resin-based coatings are very suitable for zinc and its alloys and the latter for copper and its alloys. If the first coat has been selected for good adhesion, the subsequent ones may be chosen from a wider range of products to satisfy other requirements involved in the particular application.

A number of cold-rolled alloys based on aluminium, copper and zinc are susceptible in varying degrees to recrystallisation on exposure to heat. This can have a detrimental effect on the adhesion of paint films. While there may, at first, be no sign of trouble, the defect will become obvious by brittleness of the film after some storage time has elapsed.

To avoid peeling of oleoresinous top coats from zinc-rich primers, a sufficient interval should be allowed between coats to permit the zinc-rich primer to weather first.

Lacquers drying by evaporation to rather rigid films, e.g. some nitrocellulose lacquers, may not be able to follow the movements of metals caused by changes in temperature, and rapid cracking, followed by flaking of the paint film, can result. In all such cases the smoother the metal surface, and the less affinity the coating has for the grease- and oxide-free metal surface, the more likely breakdown is. The presence of various proportions of minor constituents in alloys, including those of iron, can have a profound effect on the behaviour of the main metal in this respect.

Reference has already been made to the detrimental consequences which bad weather conditions occurring during or shortly after application usually have on the life and protective value of a film[26].

To protect buried metals from premature breakdown it must suffice to say that protective coatings and other methods must be applied against factors such as the effects of galvanic currents, composition of the moisture in the ground[27], humus acids, bacteria, etc. (See Section 15.8.)

In conclusion, it should be emphasised that surface coatings which are of the highest quality, and which where necessary have the special protective properties required, should always be used.

Good supervision, careful working, and common sense can contribute a great deal to reduce paint failures and the wasteful work which is necessary to put a job right.

Acknowledgment   A number of suggestions by W. A. Edwards have been incorporated in this Chapter, and these are gratefully acknowledged.

M. HESS

REFERENCES

1. Hess, M., *et al.*, *Paint Film Defects, Their Causes and Cure*, 2nd edn, Chapman and Hall, London (1965)
2. Hudson, J. C., *The Corrosion of Iron and Steel*, Chapman and Hall, London (1940)
3. Third Report of the Corrosion Committee of the Iron and Steel Institute, London (1959)
4. Fancut, F. and Hudson, J. C., for the Protective Coatings (Corrosion) Sub-Committee of B.I.S.R.A., *Protective Painting of Structural Steel*, Chapman and Hall, London (1957)
5. Evans, U. R., *The Corrosion and Oxidation of Metals*, Arnold, London (1960)
6. Mayne, J. E. O., *J. Oil Col. Chem. Ass.*, **3**, 183 (1957); Mayne, J. E. O. *et al.*, ibid., **7**, 649 (1967)

7. *Protection of Iron and Steel Structures from Corrosion*, 130, B.S. Code of Practice CP2008:1966
8. *Breakdown of Paint Films on Steel, the B.I.S.R.A. Scale, Degrees of Rusting*, British Iron and Steel Research Association (1949)
9. Comité Européan des Associations de Fabricants de Peintures et d'Encres d'Imprimerie, Echelle Européenne de Degrés d'Enrouillement pour Peintures Antirouille, Paris, 1963, formerly I.V.As, Korrosionsnämnd, Stockholm, 2nd edn (1959)
10. Goodlas Wall & Co. Ltd., Brit. Pats. 904861 (29.8.62) and 915512 (16.1.63) and other foreign patents
11. Mayne, J. E. O., *Chem. & Ind. (Rev.)*, 109 (1944)
12. Costelloe, P. and Pace, E., *Bull. Metal Finish.*, **8**, 107 (1958)
13. Suggitt, J. W. and Graft, C. M., *J. Paint Techn.*, **38**, 150 (1966)
14. Waeser, B., *Rostschäden und Rostschutz*, Wilh. Pansegrau Verlag, 115, W. Berlin (1956)
15. Evans, U. R., *Metallic Corrosion, Passivity and Protection*, Arnold, London (1946)
16. Communicated by Paint Research Association, Haifa
17. Munk, F. and Rothschild, W., 'Interaction between Anticorrosive and Antifouling Coatings in Shipsbottom Painting', *J. of Paint Techn.*, **43**, 557 (1970)
18. Bell, S. H., *J. Oil Col. Chem. Ass.*, **38**, 599 (1955)
19. B.S. Code of Practice CP 2008:1966, p. 72
20. *Comparisons of Pretreatments: A Background to the Corrosion of Steel and its Prevention*, No. 3, 'Effect of Surface Preparation and Paint Performance', 20, The Corrosion Advice Bureau of B.I.S.R.A.
21. In BS 4232, *Surface Finish of Blast Cleaned Steel for Painting*, first quality corresponds to SA3, 2nd to SA2·5 and 3rd to SA2 of the much more extensive Swedish Standards Commission's SIS 055 900–1962, *Rust Grades for Steel Surfaces and Preparation Grades Prior to Protective Coating*, Stockholm (1962)
22. Bullett, T. R. and Prosser, J. I., *Problems in Adhesion to Metals, Adhesion Fundamentals and Practice*, 1966, 37 (1969)
23. Slabaugh, W. H. and Chan, E. J., *Off. Dig.*, **38** No. 499, 417 (1966)
24. Slabaugh, W. H. and Kennedy, G. H., *Amer. Chem. Soc., Org. Coatings Div.*, **26** No. 1, 1–9 (1966)
25. Rathsack, H. A., *Schiffsanstriche, Korrosions und Bewuchsschäden am Schiffsboden*, Berlin, 18 (1967)
26. *Comparison of Pretreatments: A Background to the Corrosion Control of Steel and its Prevention*, No. 3, 'Rates of Rusting in Different Environments', 16, Corrosion Advice Bureau of B.I.S.R.A.
27. *Comparison of Pretreatments: A Background to the Corrosion Control of Steel and its Prevention*, No. 3, 'Water Composition', 11, Corrosion Advice Bureau of B.I.S.R.A.

15.5 Paint Finishes for Industrial Applications

Introduction

Industrial finishes may be divided into three main functional classes: (*a*) protective, (*b*) protective and decorative and (*c*) decorative. The majority of industrial finishes today fall into category (*b*), i.e. protective and decorative. The bulk of such finishes are used in the automobile, domestic-appliance, furniture and packaging industries. With the exception of the finishes for furniture, stoved finishes are now almost invariably used. Stoving of the enamel film is preferred, in order to develop maximum film properties and to reduce drying time to a minimum. In addition the hazard of dust pick-up by the wet film is reduced.

In view of the importance of synthetic resins in the formulation of modern industrial finishes, special emphasis has been placed on these resins in this section.

Finishing Systems in Relation to Application

The appearance required of a finish will depend, of course, on the design of the product to which it is to be applied. From the point of view of protection, in many cases very careful consideration must be given not only to conditions of use but also to conditions of storage and transit. The nature of the base material must in all cases be taken into account. The accent may be on physical properties such as hardness and abrasion resistance, or on chemical factors such as resistance to moisture, alkalis, acids, organic solvents, fats, staining, perspiration, etc. In nearly all cases the finish formulated must represent a compromise in relation to a number of such properties, with due regard to cost (see Section 15.2).

Finishes for articles such as domestic vacuum cleaners or office furniture, for example, must have maximum resistance to abrasion and impact, combined with a reasonable resistance to moisture. The moisture resistance is required mainly because of the high humidities, often with condensation, which are likely to be encountered during storage and transit.

The demand for good resistance to both abrasion and impact in itself calls for a precise balance of film properties, since in general good impact resistance can only be achieved by some sacrifice of film hardness, and vice versa.

Enamel finishes now available, however, combine these two properties to a high degree. In addition to good impact and abrasion resistance, finishes for washing machines must have excellent detergent and alkali resistance, which is not always easy to achieve in combination with maximum moisture resistance. Finishes for this purpose invariably consist of a pretreatment followed by a stoved primer and stoved finishing enamel.

For the automobile industry, the emphasis must be on maximum resistance to weathering, moisture and ultra-violet light, as well as good impact, abrasion, petrol and oil resistance. For refrigeration and other equipment which comes into contact with food, good mechanical properties must be supplemented by maximum resistance to fats, staining, organic acids and alkali detergents.

In addition to the qualities necessary for satisfactory performance, all these more or less specialised finishing systems must have adequate moisture and humidity resistance, in order to withstand transit and storage conditions. While it is possible to design a pack to eliminate such hazards, such a procedure is rarely economical, and the cost of a given finishing system must be equated against the cost of packaging in relation to the severest transit conditions likely to be encountered.

Finishing of Plastics

Until relatively recently, industrial finishes and their application and curing or stoving methods were largely for metal or wood substrates. The finishing of plastics substrates is now of special importance and a technology specifically related to this is now being built up. It is reliably estimated that by 1984 the volume of plastics processed will equal that of steel. In the last decade, the inroads of plastics as replacements for steel pressings, and pressure die castings in zinc and aluminium alloys, has already become significant. In addition, many components are now used as minor or even major parts of assemblies which were not possible in previously available materials. As an example, the average automobile now contains between 25 to 35 kg of visible components in plastics, mainly as semi-structural interior trim items. While plastics are normally self-coloured, for exterior applications many polymers require painting to give them adequate ultra-violet resistance. Thus the injection moulded ABS body panels used by Honda in Japan are given a coating of a pigmented acrylic lacquer and force dried, while the FRP body panels being introduced in the USA and Europe are finished by a stoved primer and finishing enamel. In addition in many cases it is now found more economical to enamel injection mouldings for interior automobile applications. This enables cheaper natural (i.e. unpigmented) polymers to be used for the mouldings, and at the same time gives a more positive colour match and a greater range of colours with reduced inventory, loss of material and machine time from colour changeover at the injection moulding stage.

There are, in addition, many other requirements for special industrial finishes in relation to plastics, including lacquers for vacuum metallising, paints for polyurethane self-skin foam mouldings, paints for electroplated plastics and paints for selective plated plastics. All such finishes have to be specially formulated, bearing in mind the special problems of adhesion and curing encountered with the finishing of plastics.

Types of Industrial Finishes

Modern industrial finishes are formulated on synthetic resins which are based mainly on combinations or modifications of phenol, urea and melamine formaldehyde, alkyd and epoxy compounds. Polyesters, isocyanates and acrylics are also being used to some extent.

The reactions involved in the formation of the various synthetic resins mentioned will be found in Section 15.9, pp. **15**:93 to **15**:99.

Phenol Formaldehyde Resins

The use of synthetic resins began with the introduction of the phenol formaldehyde resins, and materials of this type are still finding useful applications. Phenol formaldehyde resins, suitable for surface-coating materials, are manufactured by condensing 1 mol of phenol with 1·5 mol of formaldehyde under alkaline-catalysed conditions (Stages 1 and 2, p. **15**:93).

Resins manufactured in this way may be dissolved or dispersed in only a limited number of solvents. To overcome this disadvantage, modified phenol formaldehyde resins are produced by, for example, condensing phenol and formaldehyde in the presence of abietic acid (rosin) or other acidic resins or fatty acids such as stearic and linoleic.

Instead of phenol, certain alkyl and aryl derivatives of phenol may be condensed with formaldehyde to give resins from which durable and fast-curing finishes may be prepared. In order to ensure good solubility, however, it is essential for the substitution to be in the ortho or para position, as meta substitution yields insoluble materials.

Urea Formaldehyde Resins

Resins of this type are suitable for use in industrial finishes and are prepared by condensing urea and formaldehyde in the presence of monohydric or complex polyhydric alcohols (Stages 1 and 2, p. **15**:94). These in turn are obtained by the condensation of phthalic anhydride with ethylene glycol or glycerol. Such resins may be used for stoved finishes, or air dried, using acid catalysis. The cured resin alone is very brittle and the film flexibility has therefore to be improved, usually by the admixture of suitable alkyd resin. The enamels based on urea formaldehyde show good colour stability, hardness and wear resistance.

Melamine Formaldehyde Resins

Although melamine has been produced commercially since 1940 by heating dicyandiamide under pressure, the mechanism of its formation is still obscure. When it is reacted with formaldehyde, hexamethylol derivatives are formed, and it is assumed that such compounds are the starting point for the production of the melamine formaldehyde resin polymer (see Stage 1, p. **15**:94 and Stage 2, p. **15**:95). Resins suitable for the formulation of industrial finishes are produced by adding polyhydric alcohols to the condensation reaction, as in the case of the urea formaldehyde resins.

Alkyd Resins

In recent years the trend has been towards the increasing use of the alkyd enamels for stoved finishes. In addition to the excellent adhesion to metal and flexibility of such materials, the composition of the resin can be easily manipulated to obtain a wide variety of specific characteristics. Modification is usually by copolymerisation with either urea or melamine formaldehyde, when increased hardness is obtained without loss of flexibility or adhesion. The light colour and colour stability of these resins is of particular importance with the advent of pastel shades.

The largest single production of resins for enamel is, in fact, now made up of the urea and melamine-modified alkyds. The alkyd resins are formed by the condensation of polybasic acids or anhydrides with polyhydric alcohols, and the most common member of the group is a glycerol/phthalic acid polymer. Varying the acid and alcohol allows production of a wide range of resins, thermoplastic and thermosetting.

If both the acid and the alcohol have only two functional groups, long-chain molecules that are thermoplastic only are produced (Stages 1 and 2, p. **15**:95). For thermosetting materials suitable for stoving enamel resins, it would appear essential that one of the reactants should have more than two functional groups and the other at least two, e.g. the esterification of glycerol with phthalic, succinic or maleic acid (see Stage 1, p. **15**:95 and Stages 2 and 3, p. **15**:96).

Acrylic Resins

Where exceptional hardness and mar resistance are required, acrylic resins are now well established (Stages 1 and 2, p. **15**:96). The β-substituted acrylic esters, such as crotonic acid esters, show little tendency to polymerise, but α-substituted derivatives polymerise readily. The methyl homologues or methylacrylate esters are the only ones which can be manufactured sufficiently economically for industrial use.

Epoxy Resins

The epoxy enamels are of particular interest where the highest standards of adhesion and of alkali and moisture resistance are required. Such enamels are, however, expensive, and are more difficult to apply. Consequently the trend has been towards the development of finishes based on modified epoxy enamels, in which just sufficient of the expensive epoxy resin is incorporated to ensure the desired improvement in film properties. The structure of the epoxy resins is essentially that of a polyhydric alcohol, the carbon chain being terminated by epoxide groups. They are produced by reacting epichlorhydrin with a polyhydric phenol, usually dihydroxydiphenylpropane, under alkaline conditions. The reaction appears to proceed by the formation of a polyether which reacts with excess of the polyhydric phenol present, linear polycondensation occurring to form a long-chain structure (see Stages 1 and 2, p. **15**:97).

Various methods of cross-linking the resin are employed in industrial finishes based on the epoxy resins. According to the methods used, specific outstanding properties are exhibited by the cured film. For example, polyfunctional amines can cause direct cross-linking by addition reaction with the epoxy groups, giving amine alcohols (see p. **15**:98). If tertiary or quaternary amines are used instead, catalytic polymerisation will occur at elevated temperatures. The addition of carboxylic acids will result in esterification (see p. **15**:98). Curing of the enamel film may then be accomplished by subsequent heat polymerisation or oxidation. Finishes having excellent solvent resistance, flexibility and adhesion may be produced in this way.

The addition of epoxy resins to phenol, urea or melamine formaldehyde resins, or vice versa, results in cross-linking on stoving of the film (p. **15**:98). The highest chemical resistance is obtained when phenol formaldehyde resins are used, and materials for drums and cans are formulated on such resins. Melamine gives a lower chemical resistance but better flexibility. Such materials find application for the finishing of collapsible tubes.

When ultimate flexibility of the enamel film is required, this may be obtained by the appropriate cross-linking of an epoxy resin with a linear polyamide at low reaction temperature. It is presumed that the length of the polyamide chain is largely responsible for the improved flexibility. Suitable polyamides are prepared from ethylene diamine and dimeric fatty acids and are of the type

$$H_2NCH_2CH_2NH(CO.R.CONHCH_2CH_2NH)_nCO.R.CONHCH_2CH_2NH_2$$

Finally, isocyanates may be used to react additively with the hydroxyl groups of the epoxides to give the cross-linked polyurethane films (see p. **15**:99). The flexibility and mar resistance of these materials are excellent and they are particularly suitable for a number of special applications, e.g. for the enamelling of phenolic mouldings when pastel colours are required.

Pretreatment (see also Sections 12.1, 16.2 and 16.3)

The term *pretreatment* covers all the operations which make the surface of the component ready to receive the enamel, i.e. scurfing and grinding, degreasing, phosphating, chromate treatments, etc. Etch primers can be regarded as a chemical pretreatment combined with the application of a primer. The performance of the enamel system is largely dependent on efficient pretreatment.

Many components, particularly steel pressings and aluminium diecastings, are first subjected to a scurfing or grinding operation in order to remove score and drag marks resulting from previous operations. Carborundum grit of about 190 mesh is usually used, either on emery bobs or by means of back-stand idler belts. Such abrasion of the metal surface is most useful, particularly when single-coat finishes are being used on aluminium without chemical pretreatment, as the adhesion of the enamel is appreciably improved.

Efficient degreasing must be carried out immediately before enamelling or chemical pretreatment. If the components are heavily soiled, such degreasing is most economically carried out in two stages, a preliminary stage to remove the bulk of the heavy contamination, followed by a final stage to give a completely clean surface.

If some form of chemical pretreatment is to be subsequently used, the most economical method of degreasing is generally by means of an alkali cleaning solution. Care must be exercised, however, in the selection of such cleaners for aluminium components, as the stronger alkalis cause excessive etching of the base metal.

If the components are not to receive any chemical pretreatment prior to enamelling, solvent degreasing will be found superior to alkali cleaning as the resulting surface will be completely free from any water-soluble salts. Traces of such salts, which are invariably left behind after alkali cleaning, cause deterioration of the enamel film when it is subsequently exposed to moisture. It is relevant to mention at this stage the relative permeability to water of all enamel films. The reaction between water diffusing through the enamel film and the metal substrate results, in the case of aluminium and magnesium, in an alkaline underfilm environment, which is no doubt in part responsible for the difficulty of maintaining good adhesion when enamelled aluminium or magnesium is exposed to moisture. This type of reaction does not take place with a steel substrate.

The chemical pretreatment used will depend on the metal to be treated. For steel, phosphating is used, while in the case of aluminium, treatment in an alkali chromate solution, or in some cases anodising, is employed. For magnesium-base alloys, chromate pretreatments are used.

Etch Primers

There are three main types of application for etch primers in industrial finishing. Pressings manufactured from zinc-coated steel, by the time they are ready for enamelling, invariably have areas where the steel is exposed as a result of scurfing operations. Phosphate solutions which are sufficiently reactive to phosphate the bare steel, strip off the remainder of the zinc, while those whose degree of activity is appropriate for phosphating the zinc coating properly do not phosphate the bare steel. Etch primers are ideal for the preparation of such components prior to finishing with an enamel system. They are also of value for pretreating such components as instrument cabinets, which, because of the complexity of assembly work involved, remain in the shop for a considerable time before final enamelling. Finally these materials find use in the pretreatment of phenolic mouldings for enamelling.

The formulation of etch primers is considered in Section 16.3.

Methods of Application

While two distinct methods for the application of industrial finishes, viz. dipping and spraying, have been in general use for a considerable time, in recent years several variants of these original techniques have been developed, e.g. hot spray, 'air-less' spray and electrostatic spray, automatic dipping and flow coating.

Spraying

With conventional air-atomisation spraying methods it is necessary, for efficient atomisation, to reduce the viscosity of the enamel with thinners, and this limits the film thickness which can be built up at each coat without the occurrence of sags and runs. The limit is usually about 0·013 to 0·018 mm, but by means of a hot-spray technique a considerably thicker film may be obtained, 0·025 to 0·038 mm being possible. The method involves using a much higher viscosity enamel, which is passed through a heat exchanger prior to atomisation in the spray gun. Immediately after contact with the article being sprayed, the enamel is chilled and its viscosity rises, thus preventing runs and sags even at the higher film thickness. Besides permitting a satisfactory finish to be obtained from a single enamel application instead of two, the method shows a saving owing to the lower volume of thinners used.

Maximum uniformity of enamel coverage with minimum enamel loss as overspray can be achieved by these means. The most efficient approach to these two objectives, however, has been the introduction of electrostatic spraying, in which a high potential is used to develop an electrostatic field between the article to be sprayed and the spraying heads.

A further spraying technique now finding some application is the so-called 'air-less' spraying method. The heated enamel is forced under pressure through a swirl-type nozzle and is atomised by ebullition of the solvent as a result of the drop in pressure which occurs as the material issues from the nozzle, assisted by the centrifugal force developed in the swirl chamber. The technique has the advantage of eliminating the effects of the atomising air-stream on the enamel already deposited on the article being sprayed, and is particularly useful when enamelling the insides of cans and drums.

Dipping

The application of enamels by dipping is used whenever possible because of its cheapness, but there are serious limitations to its application, since on many shapes dipping would result in cupping, windowing, runs and beading. Such defects can, however, often be tolerated to some extent in primer coats which are to be rubbed down prior to the application of a finishing enamel, and a wide range of articles, even of complex shape, may be dipped for the application of the finishing enamel if adequate attention is paid to the enamel formulation, and in particular the solvent balance. In some cases it is possible and worth while to redesign a component to permit it to be dipped successfully. In some cases electrostatic detearing is used to remove windows and beads during the drain after dipping. This process is in essence the reverse of electrostatic spraying.

One of the disadvantages of dipping large articles is the large volume of enamel required. This problem has been overcome by the introduction of flow coating.

Further details of methods of application are given in Section 15.1.

Industrial Finishes Specific to Plastics

Vacuum Metallising

The vacuum metallising of plastics as films or as mouldings has been a commercial process for quite some time, and the subject has been reviewed by Pinner and Simpson[1]. Recently, however, there has been a substantial improvement in the quality levels achievable and a resulting penetration into new markets. These improvements are related largely to the preparatory sequences used to form the surface on which the metal film is actually deposited, and the covering lacquers used to protect the deposited metal. The standards available are now so high in respect of outdoor weather resistance and abrasion resistance of 'super vacuum metallised' parts, that they are finding application for exterior automotive components.

Self-skinning Polyurethane Foams

Recent developments in polyurethane foam technology has allowed the replacement of car interior trim components such as arm rests, crash pads, door liners, etc. previously made by vacuum forming a surface layer from p.v.c.- or ABS-embossed sheet and backing up with polyurethane foam, by single shot components. These are made in special polyurethane foam formulations, in such a way that a solid skin is formed integrally with the foam backing. Although a limited range of self-colours are now available, the usual practice is to paint such components to ensure a wider colour range, and improve ultra-violet resistance. Special paint formulations are used compatible with the mould release agents required for the foaming operation.

Plated Thermoplastics

Substantial numbers of thermoplastic components in both ABS and polypropylene are now finished by electroplating, using chromium as the final coating. Many such components are also finished by the painting of certain areas and special stoving finishes are now available giving adequate adhesion to the chromium substrate. Such finishes are either single colours applied by spraying, or a sequence of colours applied by offset printing techniques to give a wood grain finish.

Techniques for selective plating have now been developed in order to reduce the cost of electroplating large components such as automobile grills, where large areas are subsequently painted. These techniques involve the use of special masking paints. In some cases these are adequate to withstand the hot plating solutions and are stripped subsequent to plating and replaced by a finish having adequate outdoor durability. The latest developments are finishes capable of withstanding both the plating conditions and subsequent outdoor use so that only a single coating operation is necessary.

Finishes for the painting of plastics such as styrene and ABS present little problem because of solvent compatibility. Polyolefins such as polypropylene, however, require special treatment because of the low reactivity of the surface. Surface oxidation treatment by flame or oxidising dips are used, followed by special primers.

Stoving finishes are now normally used for thermosets and present no problem. In the case of thermoplastics, however, there are problems of heat distortion that severely limit peak stoving temperatures. This has led to the development of polyester base finishes which are cross-linked by electron beam radiation. This is known as the *cold cure* or *Electrocure* process[2]. A number of industrial plants are now in operation. Probably the largest plant has recently been installed at the Ford Motor Company's plastic facility at Saline, Michigan[3,4].

Economic Factors

In considering the economic factors associated with paint finishes for industrial applications, it is important to realise that the cost of applying the finish represents an appreciable proportion of the total cost. It is evident that the cost, besides varying with the quality of the enamel applied, will also depend on the size and complexity of the component being enamelled. This is particularly so when the method of application is by normal-temperature spray or hot spray owing to the variation in spraying time and to loss of enamel due to overspray.

In comparing the economics of the various methods in greater detail, it is instructive to consider conveyorised plants using normal air spraying, hot spraying and electrostatic spraying. In all three cases, for similar work the same labour will be required to load and unload the conveyor. In the case of electrostatic spraying, no further labour will be required for spraying. So far as the other two methods are concerned, it is usually found that the ratio of spraying labour to loading/unloading labour is of the order of 2:1. The enamel build achieved in each coat when the hot-spray method is used is approximately double that attained with the other two. For the same build of enamel, therefore, the direct labour cost will be approximately in the ratio of 2:3:6 for electrostatic, hot spray and normal spray respectively. Efficiency in utilisation of enamel will be greatest with the electrostatic method because of the actual elimination of overspray losses. The hot-spray method has an advantage, however, in its lower consumption of thinners.

With automatic conveyorised dipping, no labour is required for application and the build obtainable is similar to that given by hot spraying, i.e. approaching 0·038 mm per coat. In comparison with the other methods of application therefore, for the same film thickness the direct labour costs on comparable items will be in the approximate ratio of 1:2:3:6 for dipping, electrostatic spraying, hot spraying and normal spraying, respectively.

Types of Failure

The mode of failure of an enamel system is usually quite obvious on superficial inspection, and may normally be easily classified into an appropriate group such as impact, abrasion, moisture attack, staining, etc. It is not always so easy to decide, however, whether the failure results from exposure of the component to an environment for which it was not intended, from the fact that one or more of the components of the finishing system are of inferior quality, or from incorrect application and stoving conditions (see Section 15.4).

All enamel films are more or less permeable to moisture, and if traces of water-soluble salts are left on the surface of the metal prior to enamelling, they will draw moisture through the film, which will be locally lifted off the metal substrate by the resulting salt solution. This phenomenon is known as *osmotic blistering*, the film acting as a semi-permeable membrane. The resulting water-filled blisters often show up as a pattern corresponding to the drain marks left on the component during alkali cleaning or phosphate pretreatment. This particular form of defect is referred to as *snail tracking* and is one of the drawbacks of chemical pretreatment prior to enamelling. If this defect is to be avoided it is essential to provide facilities for adequate rinsing and uniform draining when chemical pretreatments are used. It is perhaps significant that it is usual practice to employ a multi-coat system which gives a film thickness of at least 0·025 mm in conjuction with chemical pretreatments. The ease with which water permeates films thinner than this is such that it is considered most unwise to use either chemical pretreatments or alkali cleaners with single-coat finishes. It is the author's experience that even with the most effective rinsing, enough solids are left behind from the usual water supplies to lead to failure of the enamel finish on exposure to high humidities. Final rinsing in distilled water, or a steam rinse, appears to solve the problem, but this is usually impracticable from a production point of view. A further common cause of failure is related to the application of the enamel (for details, see Section 12.5).

Poor enamel adhesion shows itself not only in the form of low resistance to chipping and impact, but also by lowered resistance to moisture and humidity. If such a condition is experienced with an enamel known to have an inherently good adhesion, then the fault is nearly always the result of imperfect degreasing. While, as already mentioned, for most applications solvent degreasing is recommended, it should be borne in mind that not all soils are freely removed by solvent degreasing. For example, some of the extra deep drawing lubricants are not sufficiently soluble in trichlorethylene to permit the use of this solvent for degreasing. The usual approach with deep drawing and pressing is to use a drawing lubricant that contains sufficient polar-type materials, such as fatty oils, waxes and concentrated soaps, to ensure its adherence to the surface of the pressing. When the drawing pressures used are very high, however, the frictional heat developed between the drawing blank and tool becomes sufficient to reduce the adherence of such materials appreciably, and resort has to be made to 'extreme pressure' lubricants. Some materials rely on combined sulphur or chloride to react with the tool and drawing blank and so hold the lubricant film in place. Solvent degreasing is unsuitable for such materials, and alkali cleaning has to be used.

Even when steps have been taken to ensure that the colour of a finishing enamel is sufficiently stable to allow a reasonable tolerance on stoving temperatures, colour changes may still occur as a result of exposure to ultraviolet light, which may take place out of doors, in shop windows or indoors when fluorescent light sources are used. An assessment of the tendency of a given formulation to such colour changes may be readily made by exposing panels to ultra-violet light in a 'fadeometer'. With unsuitable formulations, not only is fading or colour change of the pigments likely to occur, but a certain amount of yellowing due to degradation of the resin medium may also result. In the case of white enamels, this yellowing of the medium must of course be guarded against.

Failure of an enamel system by breakdown of the enamel film itself, as distinct from the bond between the film and the substrate, is nowadays an unusual occurrence in the field of industrial finishing. It is occasionally met with, however, and is then usually the result of chemical attack by strong alkalis or synthetic detergents, accentuated by understoving of the enamel film.

New Developments

Curtain Coating

One of the latest developments now in production use is the so-called *curtain coating* technique. This process is suitable for the coating of flat panels, and is at the moment limited to wood panels for the furniture trade. The panels are passed, nose to tail, on a high-speed conveyor through a curtain of enamel. The width of the curtain is arranged to correspond to the width of the panels, while the speed of flow of the curtain is adjusted in conjunction with the conveyor speed so that a film of enamel of suitable thickness is, as it were, 'laid on' to the panel. Polyester-catalysed materials are being applied in this way.

Curing

The waxed-type polyesters have now been applied to wood for some time by electrostatic spraying, the resin and catalyst being mixed during atomisation. The cure of the formulations used up to now is air inhibited and the addition of wax is made to prevent the access of air. The wax has to be removed from the cured film before full gloss is developed. The latest development is in the use of wax-free formulations which are suitable for electrostatic application to either wood or metal, and which cure to give an immediate full gloss. A further development of particular interest in this field is the addition of photo 'activators' instead of a catalyst, curing being effected by exposure to ultraviolet irradiation, which results in cross-linking.

The cold curing process has already been mentioned in relation to plastic mouldings. It should be noted that for the moment it is limited to resins of the polyester type.

Organosols

In the field of 'organosols', much interest is being shown in dispersions of p.v.c. in the form of 'near solutions' in ketone solvents. Although it is essential to use a primer with such materials, the p.v.c. used in the top-coat dispersion need be only lightly plasticised, and this results in a very hard and tough film.

Hot Dipping

Finally, a certain amount of hot dipping, using trichlorethylene as a solvent, is being done. The procedure involves conveyorised dipping of the components in the enamel, which is maintained at the boiling point of the trichlorethylene solvent. On withdrawal, the heat retained in the components flashes off the trichlorethylene, giving a solvent-free film which is subsequently stoved. Withdrawal is made through a series of condensing coils which condense the trichlorethylene vapour and allow it to drain back into the dip tank. The process is limited to enamels based on resins which are not only soluble in trichlorethylene but are also stable when held for prolonged periods in the boiling solvent. 'Under the bonnet' finishes are already being produced in this way.

H. J. SHARP

REFERENCES

1. Pinner, S. H. and Simpson, W. G., *Plastics: Surface and Finish*, Butterworths (1971)
2. Ford Motor Co., US Pat. 3 247 012 (19.4.66)
3. 'Electron Radiation: Fast New Way to Finish Plastics', *Modern Plastics*, Dec. (1969)
4. 'Ford Moves Ahead with "Electrocure" Gaining High Speed, Large Volume, Low Cost', *Plastics*, June (1970)

BIBLIOGRAPHY

Fieser, L. F. and Fieser, M., *Organic Chemistry*, D. C. Heath, Boston, USA (1954). (A number of the structural formulae and reaction equations in Section 15.9 are derived from this work, for which due acknowledgement is made.)
Heaton, N., *Outlines of Paint Technology*, Charles Griffin, London (1956)
Ellis, C., *The Chemistry of Synthetic Resins*, Chapman and Hall, London (1954)
Krumbhaar, W., *The Chemistry of Synthetic Surface Coatings*, Chapman and Hall, London (1957)
Gailer, J. W. and Vaughan, E. J., *Protective Coatings for Metals*, Charles Griffin, London (1950)
Harvey, A. A. B., *Paint Finishes for Industry*, Robert Draper, London (1958)
Tatton, W. H. and Drew, E. W., *Industrial Paint Application*, 2nd edn, Butterworths, London (1971)

15.6 Paint Finishes for Structural Steel for Atmospheric Exposure

Paint for structural steelwork is required mainly to prevent corrosion in the presence of moisture which, in an industrial or urban atmosphere, carries acids or salts resulting from combustion and in a marine atmosphere carries chlorides in addition. Paint is therefore required to prevent contact between steel and corrosive electrolytes over an indefinite period, and to stifle corrosion in the event of its onset because of incomplete covering, mechanical damage or breakdown of the coating through age and exposure. Paint coatings outdoors may at times remain continuously wet for many days[1], and in some factories they may be permanently wet, which thus gives time for water to penetrate deeply.

For an adequate barrier against moisture, sufficient thickness of paint is necessary. A minimum of about 125 μm of oil-based paint, representing four normal coats has been found necessary[2] in free atmospheric exposure, but in wet conditions, particularly where condensed water hangs for a long time, this is not sufficient. The modern trend is to demand much greater coating thicknesses in such situations, and the development of high-build coatings based on media having high intrinsic water resistance has made this possible without undue increase in initial costs. Structural steel is not smooth and therefore paint-film thicknesses will vary widely over quite small areas; thus for adequate cover well over the mean thickness is necessary. Surface preparation of the steel before painting is a matter of utmost importance. Rust is hygroscopic and usually contains measurable quantities of sulphates[3] and other soluble corrosion stimulants including traces of ammonium salts[4] which will stimulate corrosion if left under a paint film. Broken millscale is also a corrosion stimulant. The metal is pretreated to remove rust and millscale and to clean the surface, but no known cleaning method entirely eliminates the effect of previous corrosion[5] and residues of aggressive salts can still prove harmful unless controlled by the use of a carefully chosen primer.

Types of Pretreatment

Pretreatments fall into four groups which will now be considered.

Mechanical Processes

These consist of shot- and grit-blasting and flame descaling. In shot- and grit-blasting, hard abrasive particles such as selected steel shot or specially hardened steel-wire pieces are projected at high speed against the surface to remove rust, scale, old paint coatings and the like, leaving a clean, dull, slightly roughened surface, which is, however, sensitive to re-rusting and must be protected immediately. In flame cleaning, differential expansion shears off scale and rust without the steel itself becoming sufficiently hot to distort or re-oxidise and form another harmful layer. The process is best applied to heavy structural steelwork with a large heat capacity. The slightly elevated surface temperature after flame descaling is excellent for immediate painting (Section 12.1).

Chemical Processes

These involve de-rusting by dipping the steel in mineral acid containing an inhibitor which prevents attack upon the unoxidised steel. De-rusting is usually combined with or followed by the formation of a chemical surface layer by interaction of the steel with phosphates or chromates or mixtures of the two. Facilities for dipping large units are, however, not commonly available, and the practice is less used than its effectiveness warrants (see also Section 12.2).

Metal Coating

This process entails coating the steel with a layer of non-ferrous metal after it has been thoroughly cleaned. The cleaning is usually by shot or grit-blasting, but in certain factory processes carried out before erection, the steel may be cleaned by acid pickling. The non-ferrous metal may be applied by dipping the steel in molten metal (Section 13.2) as in galvanising, by dry rolling hot steel in powdered metal, or by spraying (Section 13.4). The last two processes give a somewhat rough granular surface. Galvanised surfaces need rather careful treatment if failure of adhesion is to be avoided.

Although sprayed metal coatings may, when freshly applied, have good paint holding qualities, there is a danger, because of their porous or discontinuous structure, of rapid corrosion of the coating metal, particularly zinc. The coatings need to be sealed before exposure to prevent the formation of corrosion products harmful to the adhesion of coatings. Vinyl-resin-based etch primers are most suitable for this purpose. Subsequent protective coats need to be of adequate thickness to cover the profile of the surface and should preferably be based on highly water-resistant materials. The practice of using oil-varnish-based zinc chrome primers followed by oil varnish or alkyd-based finishing systems has not been outstandingly successful particularly in wet situations[6] where there is frequent prolonged contact with condensed water. This is probably because of the water solubility of zinc chromate and insufficient protection by the finishing paints. There is evidence[7] that there is no advantage in using an additional inhibitive priming paint over the etch primer provided the finishing system is of sufficient thickness.

Weathering

The weathering of steel, followed by mechanical cleaning such as wire brushing, chipping, etc. is still the commonest method of surface preparation, but it is somewhat risky because frequently millscale and contaminants are incompletely removed and the surface is more corrodible in its partially weathered state than it is in the initial condition. Weathering must be sufficient for all millscale to be removable by mechanical means and subsequent cleaning must be thorough. Many paint failures on steelwork can be attributed to incomplete removal of millscale by weathering. It is desirable to follow weathering and cleaning by chemical pretreatment with phosphates or chromates, or, preferably, with an etch primer containing phosphoric acid and a chromate. Properly carried out, this combination of treatments gives a satisfactory surface for receiving traditional oil-based paints.

The merits and limitations of the various methods of surface preparation are discussed in detail in British Standard Code of Practice CP 2008–1966 which also gives guidance on subsequent treatments[8].

In practice, shot and grit-blasting are rapidly gaining in importance for treating new construction and are established as the principal methods of cleaning steel for large structures. Site blasting for maintenance repainting is also being increasingly used. New steelwork is usually blasted in automatic equipment in the construction shops where there is the advantage of a clean dry atmosphere and the opportunity of applying at least the priming coat while the steel is still clean and dry and before re-rusting can commence. There is a growing tendency to apply all except the last coat in these conditions.

Blast-cleaned steel has a textured surface whose profile depends upon the type and size of blasting medium. The degree of cleanliness and the texture of the surface should always be carefully specified. Standards are given in BS 4232[9]. Many authorities specify cleanliness to Second Quality, corresponding to the USA Standard 'Near White' and Swedish Standard SA 2·5, with a surface roughness defined by a maximum amplitude of 100 μm. This control is important since paint equivalent to a coating thickness of 35 μm is required to fill the hollows in a surface of the amplitude stated above produced by blasting with normal commercial grades of grit[10].

Types of Paint

Many factors go to determine the type of paint required, but in most cases air-drying paints applicable by brush or spray are needed.

A complete paint system involves:

 (i) A priming coat in contact with the metal. This usually contains a cor-
     rosion-inhibiting pigment, capable of stifling either the anodic or the
     cathodic reactions in electrolytic corrosion.
(ii) A finishing paint system capable of adhering to the priming coat, re-
     sisting the ambient exposure conditions and providing the necessary
     decoration, light reflection, etc. where necessary.

It is usual to define priming paints in terms of the principal inhibiting pigment only, e.g. red lead, calcium plumbate, zinc dust or zinc chromate, and the finishing paint system in terms of the medium, e.g. oil, alkyd, chlorinated rubber, etc. but this practice is misleading and leads to such serious errors as the selection of incompatible coatings, with consequent early failure. The whole system, comprising the priming and subsequent coats needs to be chosen with a view to conditions of use. Marine and industrial conditions call for most discrimination. Paint may be exposed to alkali, acid or both, or subjected to prolonged contact with steam or water, often highly contaminated, or to abrasion by industrial grits, etc. and the appropriate systems must be selected in relation to the conditions which will be encountered.

In general, interior steelwork is exposed to less severe conditions than exterior, but in some chemical factories the reverse is true and here special types of paint are needed. Much structural steel is encased in concrete; it is therefore hidden from view and is given some protection while the concrete remains alkaline. Many cases are, however, known in which encased steel has corroded and ruptured the concrete (see Section 10.3). Where the concrete is thick, corrosion may be delayed, but as the concrete becomes carbonated and particularly if it is penetrated by acidic rain water, the metal will corrode[11]. In general it is advisable that steel which is to be encased in concrete, especially for industrial plants, should be surface treated according to one of the procedures outlined above and coated with an anticorrosive alkali-resisting composition. Alternatively, it is recommended that the concrete be coated with an alkali and water-resisting enamel, which will also lengthen the life of the concrete by preventing attack by acid-bearing rain water.

Air-drying Paints

The selection of paint is a matter for the expert, but some knowledge of composition is of help to the user. Paints based on the drying oils, usually linseed and tung oil, are still the most widely used for decoration and protection, though the traditional oil paints have now been largely superseded by those based on varnishes and synthetic resins. Of these the alkyd resin and phenolic resin paints are the most widely used because they have excellent durability. In recent years, alkyd-resin paints have become the principal decorative and protective paints, though frequently a combination of alkyd and phenolic or other hard resin is used. The normal decorators' paints, however, frequently do not have the necessary resistance to chemical attack required for protecting steelwork in industrial conditions. Alkyd paints, for instance, are sensitive to alkali attack and are frequently softened and degraded by prolonged exposure to hot steamy conditions where there may be serious condensation. Alkali formed locally at the cathodic area of a steel surface may destroy the adhesion between such paints and the metal.

A high degree of resistance to water and chemical attack is provided by some oil-based paints, notably those based on tung oil and pure phenolic resin or some hydrocarbon resins such as isomerised rubber and coumarone resin, but for the greatest resistance to these forms of attack, oil-free paints based on ester- and acid-free resins are recommended. Of these, bitumen is the most widely used, but while it is cheap, it has the disadvantage that all paints based on it are black or of dark colour with little decorative value, and most have poor resistance to weather unless used in thick coatings. Moreover, it is difficult to apply any other type of coating over bitumen. Bituminous coatings nevertheless fulfil an important rôle in protecting hidden steelwork, where appearance is of little account. In recent years there have been considerable advances in the technology of bituminous compositions, and heavy-duty compositions now available give hard, tough coatings which can withstand rough handling without damage and virtually exclude all water from the steel, and which can be decorated with aluminium or light-coloured paints. They are eminently suitable for protecting heavy industrial steelwork in gas works, engineering works, oil installations, power stations, docks and other public utilities, and for protecting the interiors of hollow sections, box girders, etc. from corrosion by condensed water.

Chlorinated-rubber-based paints, when properly compounded with non-ester-type plasticisers and associated resins, have the advantage of combining acid and alkali resistance with weather resistance and decorative qualities. Highly impermeable anticorrosive systems can be built up with these materials, and it is not surprising that they are finding increasing use in industrial plants where lightness of surface combined with low maintenance costs is needed. The alkali resistance of chlorinated rubber paints makes them suitable for protecting concrete where it is desirable to safeguard embedded steel from corrosion. Chlorinated rubber finishes are now also available as high-build coatings capable of application in thick films up to 125 μm thickness in single coats, and the combination of high intrinsic resistance with thickness provides excellent protection using only a few coats.

Other types of polymer resin paints having many of the characteristics of chlorinated rubber are also available, but they are of less significance as paint coatings, although there have recently been interesting developments in their use as so-called plastic coatings.

Chemically Cured Paints

These are supplied as dual or even multi-pack materials, the separate components of which have to be mixed just before use. The paints dry or cure by interaction of the components—a process which also occurs in the can and so limits the time available for application after mixing. Coating with these materials is really cladding with a plastic resin. The films are tough and can be built up to great thicknesses. There are three main types of these coatings:

   (i) The epoxide-resin-based materials, which cure by the interaction of amino compounds or their derivatives with epoxide resins.
  (ii) Polyurethane finishes which cure by the interaction of polyisocyanates with hydroxylated resins.
 (iii) Polyester resin finishes which cure by peroxide-stimulated polymerisation.

All these materials are capable of giving hard durable coatings, which are finding increasing use in industrial situations. Cured films of polyester resins and often of polyurethane resins have an ester structure and are therefore in varying degrees subject to saponification by alkali. The epoxide resin finishes are highly resistant to alkali and acid and, like the other chemically cured finishes, are resistant to a wide range of oils, greases and solvents. On this account they are increasingly used for protecting steelwork, where they compete strongly with chlorinated rubber finishes, since they have the advantage of greater resistance to oils and solvents. Time may well show that the ester-type materials adhere better than epoxide-resin-based materials to normally-cleaned industrial steelwork, and that a suitable protective system could be built up using them as priming coats for epoxide resin paints. The adhesion of epoxide resin paints to steel is, however, good if proper attention is paid to preparation of the surface and if due attention is given during formulation to the ultimate structure of the cured film. In this respect both curing agents and solvents play a significant part.

Thick Coatings

Chemically cured coatings differ from air-oxidised coatings in that they dry throughout the film regardless of thickness, whereas oil-based paints dry from the surface downwards by the action of the air. In thick coatings, oil paints may not cure satisfactorily. The chemically cured materials lend themselves to the development of protective coatings of considerable thickness with the consequent advantages of good performance and long life, and they have recently contributed significantly to the protection of steel in corrosive conditions. It is possible to apply single coatings which equal in thickness and performance many coats of orthodox paints, with consequent savings in labour costs. The extra cost of materials is more than compensated for by savings in time and application costs, and where scaffolding and shut-down time are involved this may be a matter of great importance. Particularly interesting are the solvent-free finishes based on low-viscosity resins, and the combination of epoxide resins and pitch which gives high-duty coatings at reasonable cost. Quite apart from the economic advantage of thick films, the lower the solvent content the lower the intrinsic permeability to moisture and aggressive ions. Solvents, particularly polar solvents as used in many polymer resin-based paints, influence the structure of films over the early critical weeks of their life. Small quantities of many solvents are retained in the cured films for a long time, and water and aqueous solutions are able to penetrate the solvated films more easily. Films free from such solvents therefore have the advantage over similar films laid down with the aid of solvent. The newly developed solventless coatings resemble the traditional types of white lead in linseed oil paints in that these too were virtually free from solvent; part at least of the good performance of the traditional paints could be attributed to this fact.

The Paint System

The priming coat provides the bond between the metal and the coatings later applied, and gives electrochemical control of corrosion. Adhesion is dependent largely on the nature of the organic binder and the metal on the surface. Zinc-sprayed steel demands a different type of paint from that which can be applied over shot-blasted steel. Drying oil-based primers are unsuitable for use under chlorinated rubber paints or under many types of bituminous or chemically cured paints.

The pigment is the principal agent in the electrochemical control of corrosion by primers (see Section 15.3). The lead-based pigments, of which red lead is still the most widely used, are effective for ferrous surfaces particularly where blast cleaning or chemical processes of preparation cannot be used. Calcium plumbate and metallic lead are increasingly used, and are also useful pigments in primers for galvanised steel[12,13]. However, the dominance of the lead pigments is being seriously challenged by a number of lead-free pigments such as metallic zinc and, more recently, zinc phosphate which is also proving a valuable pigment in primers for galvanised steel. The case for zinc chromate on structural steel is not established, although it is an effective ingredient of primers for non-ferrous surfaces such as galvanised, zinc or aluminium-sprayed steelwork. Zinc-rich priming coats initially act anodically to steel, but for effective use as part of a paint system the selection of medium and the quantity of pigment are critical. Best results are obtained with epoxy-resin-based zinc primers on blast-cleaned steel, but care must be taken not to over-pigment the primer as this leads to adhesion failure of subsequent coats on exposure to wet conditions. For this reason there has been a mixed reception of these materials by engineers, some of whom have gone so far as to prefer epoxy-resin-based primers free of zinc metal. It has to be recognised, however, that when lead pigments cannot be used, the choice of anti-corrosive pigments is limited, and from the purely anti-corrosive aspect, zinc dust is a highly satisfactory material. Thus where exposure to an external atmosphere is not expected, zinc-dust primers may prove to be the best choice, but it should be remembered that zinc is affected by both acids and alkalis and may therefore not be suitable in chemical works or in many types of textile factory.

BISRA[2] has published results of several years' exposure of painted steel panels at Brixham and Derby. Mixtures of pigments were used in the priming paints, and a finishing paint based on white lead was applied. Interim results indicated that several of the primers with mixed pigments gave better results than red lead in linseed oil. To what extent the superiority is due to differences in film structure resulting from the incorporation of a fibrous extender and polymerised oil in the primer is not known, but these variables have to be taken into account in assessing the value of the pigments.

Protective coatings for steel are built upon a clean surface by the successive application of priming coat and subsequent coatings thick enough to resist the exposure environment, usually the weather. Structural steel usually has a rough surface with many edges, bolts and crevices, and there must be a sufficient thickness of paint even on edges and pinnacles to give full protection. For this reason, unless thick, heavy-duty coatings such as the high-build materials mentioned earlier are used, it is necessary to apply several coats. The system may consist of undercoats and finishing coat or several coats of the same finishing material. Properly prepared undercoats frequently have the advantage that, being highly pigmented, they do not flow away from sharp edges and so help to build up the system in these difficult places, though this property can now be provided by other means in the formulation.

Welds on steelwork need special attention because of the different composition of weld metal and adjacent steelwork, the rough surface and spatter caused by welding and the presence of welding flux. The latter is often alkaline and destructive to many paints and adversely affects the adhesion even of alkali-resistant paints. It is necessary to clean thoroughly, preferably by re-blasting for 25–50 mm each side of the weld, fare the rough metal and wash off residual flux. The cleaned surface must then be stripe primed with the primer used on the remainder of the surface.

Methods of Application

Paint is still mainly applied by brush to structural steelwork but the development in recent years of airless spraying has led to its wide adoption particularly on broad surfaces. This method of application is most successful with high build coatings where the combination of rapid working and great film thickness allows work to be completed in short time with reduced cost. Airless spraying does not suffer from the disadvantage of the older compressed air spray process in that there is no tendency for dew formation due to cooling at the nozzle of the gun and there is a marked reduction in overspray. The enormous velocity of impact between paint and the surface[14] promotes intimate contact and good adhesion to the surface and cohesion of the film. Thus airless spray is equally suitable for priming as for finishing paints and is often to be preferred to brushing. Roller application is not suitable for protecting steelwork as it produces attenuated coatings of irregular thickness.

Priming and painting up to the penultimate coat in the construction shop has much to commend it, but provision must be made for careful transportation to the site and for care in erection. Provision must also be made for cleaning and patch painting.

The weather has an important effect on the drying of paint and on subsequent performance. Paint applied in bad weather may be slow in drying and remain susceptible to damage by rain and fog for a long time. Heavy steelwork has a large heat capacity and follows temperature changes of the ambient air only slowly. Thus warm air coming after cold weather does not necessarily imply warm conditions for the paint film, and such a warm, humid air stream causes condensation on cold metal surfaces during painting and while the film is drying, with damaging results. Careful consideration of weather conditions and planning of work is frequently repaid by improved results.

Economic Considerations

The costing of painting structural steelwork is a complex subject. The main items of costing are:

1. Scaffolding.
2. Labour, which may be further subdivided into surface preparation and application labour charges.
3. Materials.
4. Supervision and transport.

The proportion of the whole contributed by each of these items clearly varies with each job, but it will be immediately apparent that the cost of scaffolding and labour far outweighs the cost of materials and supervision. Therefore a relatively great increase in the cost of the last two items will produce only a fractional increase in total cost. On the other hand, first-quality materials and rigid supervision will give greatly increased protection and the best value from the expensive items of scaffolding and labour. The matter has been sufficiently explained in *Post War Building Studies No. 5*[15], and what was true in 1946 is increasingly true as labour costs rise. It is economically sound to consider not the initial cost of protection, but the annual cost over the life of the structure, taking into account initial work, maintenance charges and the cost of shutdown. There is now overwhelming evidence that high-quality initial preparation and protection lead to reduction of total costs on an annual basis.

Maintenance Repainting

All the foregoing has been concerned with the initial protection of steelwork, but there is far more maintenance repainting than new work. The same principles apply to maintenance repainting, with the exception that it is often only in isolated patches and in complicated situations, such as around flanges, etc. that the steelwork is bare of paint, and then it is frequently heavily contaminated with corrosion product. The first necessity, therefore, is to clean down these areas to bare steel, but often it is not possible to use chemical cleaning or blasting methods though blasting is being increasingly used in situations accessible to power and air lines. Flame cleaning can be used either locally or for removing all paint and then proceeding as for new steel. Often, however, hand cleaning is all that can be done. Careful supervision is needed, and the cleaned areas must be quickly patched with a chemical primer such as an etch primer and brought forward with a suitable anti-corrosive system before the whole surface is repainted. All remaining paint must be firmly adherent to the steel and its surface must be thoroughly cleaned before new coats of paint are applied. Tests should be made to ensure that the new paint will not disturb the old or fail to adhere to it, as otherwise rapid failure may result. It is generally good practice to build up successive coats of paint, as the greater thickness of coating gives added likelihood of good protection.

P. J. GAY

REFERENCES

1. Gay, P. J., Paint Research Association, Technical Paper No. 156 and *J. Oil Col. Chem. Ass.*, **31**, 481 (1948)
2. *Paints for Structural Steelwork*, British Iron and Steel Research Assoc., Third Interim Report of Joint Technical Panel J/P1, Feb. (1955) and others
3. Mayne, J. E. O., *J. Appl. Chem.*, **9**, 673 (1959)
4. Harrison, J. B. and Tickle, T. C. K., *Chem. and Ind.* No. 45, 1383 (1960)
5. De Vlieger, J. H., Congress Book, International Congress on Marine Corrosion and Fouling, Athens, 181 (1968)
6. Claxton, A. E. and Carter, R. V., *Brit. Corr. J.*, **4** No. 4, 194 and 195 (1969)

7. Rudram, A. T. S., *Some Painting Requirements of Metal Sprayed Steel* (pamphlet)
8. *The Protection of Iron and Steel Structures from Corrosion*, B.S. Code of Practice CP. 2008, Sect. 505 (1966)
9. *Surface Finish of Blast Cleaned Steel for Painting*, BS 4232 (1967)
10. Bullett, T. R., *Brit. Corr. J.*, Suppl. Issue, 5 (1968)
11. Stratfull, R. F., *Corrosion*, **13**, 173t (1957)
12. Read, N. J., *J. Oil Col. Chem. Ass.*, **33**, 295 (1950)
13. Newnham, H. A., *Anticorrosion Methods and Materials*, **18**, 3, 13 (1971)
14. Gay, P. J., *J. Oil Col. Chem. Ass.*, **51**, 287 (1968)
15. Post War Building Studies No. 5, *The Painting of Buildings*, H.M.S.O., London, 49 (1944). (Revised 1946)

15.7 Paint Finishes for Marine Application

In considering the requirements of paints for marine use it is necessary to distinguish between the parts of ships that are subject to different conditions of service. The exterior area of ships may be divided broadly into three parts: (a) the bottom, which is continuously immersed in the sea; (b) the boot-topping or waterline area, which is immersed when the ship is loaded and exposed to the atmosphere when cargo has been discharged; and (c) the topsides and superstructure areas, which are exposed to the atmosphere but subject to spray. In addition to these weather factors, the outsides of ships are also subjected to attack arising from the conditions of use, e.g. the boot-topping is subject to abrasion by rubbing from quays, wharves and barges, while the topsides, superstructures and decks may receive mechanical damage during cargo handling. The interior surfaces, too, present varying requirements according to the conditions of use; cabins and accommodation spaces for crew and passengers call for treatment other than that demanded by cargo holds. A particular problem of ship interiors, to which special attention has been devoted in recent years, is the protection of the cargo tanks of oil and chemical tankers.

Although light alloys and non-metallic materials such as reinforced plastics are finding increasing applications in shipbuilding, the principal construction material is generally mild steel. Hence the protective painting of ships is basically a special aspect of the painting of steel. In relation to atmospheric exposure, the main principles of the subject are:

    (i) Proper surface preparation.
    (ii) Appropriate composition of the paint, in particular the use of an inhibitive priming paint.
    (iii) Adequate film thickness.
    (iv) Good conditions of application.

These apply also to marine painting, but here additional factors must be taken into account. The present section refers specially to differences between ships' painting and structural steel painting.

Surface Preparation and Pretreatment

This is the most important factor determining the life of a protective paint system on steel. The best surface is free from rust, scale, grease, dirt and moisture, i.e. it is completely clean and dry. The removal of millscale is particularly important under marine conditions[1], especially for ships' bottoms, because the environment has a high conductivity which enables corrosion currents to pass easily between cathodic scale-covered and anodic scale-free areas. This results in pitting when the ratio of scale-covered to scale-free areas is high. A small scale-covered area with a large scale-free area is not so serious because the corrosion is spread over the larger area.

Millscale and rust can be completely removed from steel by acid pickling or by blast cleaning. Pickling was formerly used in some shipyards, but during the years 1960–65 nearly all shipbuilders installed automatic airless blast-cleaning machines for the treatment of steel plates and sections prior to fabrication. In these machines the abrasive[2], generally steel shot, is thrown against the steel by impeller wheels. A series of wheels directs the shot against each side of the plates as they pass through the machine at about 2 m/min, this speed being adjusted in relation to the quantity, size and velocity of the shot so that the millscale and rust are properly removed. The finish produced by these machines is normally Second Quality of BS 4232:1967 or SA2·5 of Swedish Standard S.I.S. 05 59 00–1967, and with a surface profile not exceeding 100 µm. The process is rapid and dry, and the machines are totally enclosed to prevent particles of abrasive and millscale getting into the atmosphere—accordingly they can be installed in the steel fabrication shops of modern shipyards. (Acid pickling, on the other hand, is a wet process requiring the steel to be immersed for some hours in a bath of acid and then rinsed thoroughly in water—it tended to be messy and was often banished to a corner of the shipyard.) Automatic blast cleaning of plates in these machines is much cheaper than blast cleaning after erection because labour charges are low and the abrasive is recovered, graded and re-used, fresh abrasive being added to make up for the fine particles rejected with the millscale. The cleanliness of the surface may be checked (a) visually using a hand lens, with which residual millscale and rust can be seen, (b) by the copper sulphate test[3], or (c) by a reflectance method[4]. The surface profile may be checked (a) by examining the surface, or a replica, using a stylus type of surface profile instrument[5], (b) by a simple probe type instrument[6], or (c) by using a roughness gauge[4] depending on the rate of leakage of gas from a cup held against the surface.

The clean, dry, slightly rough steel surface produced by blast cleaning is ideal for the application of paint, but will not remain in this state for more than a few hours under average shipyard conditions. General practice[7] is to apply a thin coat of prefabrication primer (also known as a *blast* or *shop* primer) to the steel as it emerges from the blast-cleaning machine. The primary function of this primer is to protect the surface of the steel for the six to nine months during the fabrication and erection of the ship, but it must also meet other requirements to permit its use under practical conditions in shipyards, e.g. it must dry rapidly to permit the steel to be handled in 2–3 min, must withstand abrasion, must not affect the speed of flame cutting or welding, must not affect weld quality, must not cause any health hazards from fumes when coated steel is welded or flame-cut, and must be compatible with any type of paint system likely to be used on the different parts of ships.

The principal types of prefabrication primer in commercial use are (a) cold-cured epoxies pigmented with zinc dust, (b) zinc silicates, (c) phenolic-reinforced wash primers pigmented with red iron oxide and (d) cold-cured epoxies pigmented with red iron oxide and inhibitive pigment. In many shipyards there are objections to the zinc types because zinc oxide fumes are evolved during welding and flame cutting, and for this reason the red oxide types are more widely accepted. The wash primer types are not universally compatible with marine paint systems, and the epoxy types are therefore recommended.

Selection of Paint Systems for Use on Ships

Exterior Surfaces above the Waterline

As indicated earlier in this section, the choice of paints for marine use depends upon the conditions of service to which the part in question will be subjected. Thus the paints used on the exteriors above the waterline and on most of the interiors do not differ fundamentally from those used on structures ashore. Inhibitive priming paints are used on steel, including those based on red lead, calcium plumbate, zinc phosphate or zinc chromate. The best known structural steel primer, i.e. red lead in linseed oil, is still used on ships, although it requires a long drying time. Slow drying is a disadvantage for marine paints, particularly on ships in service which have to be painted between voyages, since when out of commission, ships are not earning any revenue. Zinc chromate primers, usually based on alkyd or phenolic media, dry more quickly than red lead in linseed oil; they are frequently used on the interiors of ships because they may be sprayed without any risk of lead poisoning and may be applied either to steel or to aluminium alloys. Lead-based priming paints should not be used on aluminium. Finishing paints are also similar to those used ashore. Good-quality alkyds are used in accommodation spaces, and the standard of workmanship is high. Colour and decorative schemes receive careful attention, and the finish is kept up to standard by frequent cleaning and regular repainting. For exterior use on topsides and superstructures, finishing paints based on alkyd media are generally used; good water resistance is essential here. White is used extensively on the superstructures of ships; owing to the pollution of many estuaries and docks with sewage and the consequent evolution of hydrogen sulphide in warm weather, it is necessary to make marine white paints 'lead-free' in order to avoid discoloration by sulphide staining. Another feature of modern marine white paints is that they are usually made from alkyds based on a 'non-yellowing' oil such as soya-bean oil in order to prevent the yellowing which occurs on exposure of linseed-oil-based white paints. The British Navy's topsides grey paint consists of rutile-type titanium dioxide in an alkyd medium based on non-yellowing oil. Black topsides paint which is used on many merchant ships may be based on phenolic media or alkyds reinforced with phenolics.

Newer types of high-performance paints[8] used on ship exteriors include those based on epoxy resins, polyurethane resins, vinyl resins (also vinyl/alkyd or vinyl/acrylic blends) or chlorinated rubber. Epoxies and polyurethanes are chemically-curing types and present curing problems at low temperatures, whilst the overcoating intervals are critical for best adhesion between coats. Chlorinated rubber[9] does not suffer from these practical difficulties and is becoming widely used. A complete system based on one of these special coatings must normally be applied, and first class surface preparation is essential if the optimum performance is to be obtained from them. Simpler types of oil-based paints are generally less sensitive to the standard of surface preparation and may give better results than these special paints when imperfect surface preparation must be tolerated.

Interior Surfaces

Aluminium finishing paints are frequently used for the interior of dry-cargo holds because they help to improve lighting. Aluminium paint is also used in engine rooms; the general requirement here is for hard-drying paints resistant to oils and to heat.

Cargo and Ballast Tanks

Severe corrosion may occur in unprotected cargo and ballast tanks of oil tankers[10] as a result of the combined corrosive effects of the cargoes, fresh or salt-water ballast, and tank washing by cold or hot sea-water. Ships which carry cargoes of refined oil products ('white oils') suffer general corrosion, since these cargoes do not leave any oily film on the interior surfaces of the tanks. Corrosion rates vary widely according to the conditions of service, rates of up to about 0·4 mm/y being reported. Cargoes of crude oil ('black oil') leave an oily or waxy film on tank interiors, and this has some protective action. As this film is not continuous over the whole surface, severe local corrosion may occur at areas of bare steel exposed to the action of sea-water ballast. The mechanism of the attack at these bare areas may be likened to that on small bare areas on steel which is almost completely covered with millscale; the oil or wax-covered areas function as cathodes in the same way as millscale, and corrosion is concentrated on the anodic bare areas. Some crude oils contain appreciable quantities of sulphur compounds, and residues may react with water and oxygen to produce sulphuric acid. The attack in black-oil tanks therefore takes the form of pitting; rates vary widely, up to as much as 5 mm/y being known, depending upon the conditions of service. Corrosion in oil tankers is therefore a serious problem entailing costly steel renewals in unprotected tanks. Protective measures include (a) the use of cathodic protection, (b) oxygen elimination by the injection of inert gases, (c) dehumidification of the air above oil cargoes or in tanks when empty, (d) the addition of inhibitors to the oil cargoes or to the ballast water, or the spraying of inhibitors on to the interiors of tanks, or (e) protective coatings. Methods (a)–(d) reduce the corrosion, but only (e) offers the prospect of complete protection. The coatings must have good resistance to many types of petroleum or other liquid-chemical cargoes, to ballast water and to normal tank cleaning, must not contaminate cargoes, and must be capable of being applied under shipyard conditions. Two main types of paint coating have been developed for this service, viz. epoxies and zinc silicates.

Epoxy resin paints are supplied as two components, a base and hardener, to be mixed at the time of application. Curing of the film to a tough, oil-, chemical- and water-resistant state occurs by chemical reaction between the epoxy resin of the base component and a curing agent (amine or polyamide) forming the hardener. This reaction does not require the access of oxygen, so that the film cures right through, irrespective of thickness. It is, however, dependent on temperature, 10°C being the usual minimum practical recommendation. To ensure good intercoat adhesion, successive coats must be applied before the previous coat has fully cured, so that in practice there are maximum as well as minimum over-coating intervals, both varying with temperature. The early epoxy tank systems required application of four or even five coats to give a total dry film thickness of 200–250 μm, but common practice now is to apply two high-build coats to achieve the same film thickness. Solventless types are also available which may be applied as single coats of 200–300 μm. Coatings based on epoxy resins modified with coal tar pitch may be used in tanks for the carriage of crude oils, but are not suitable for refined oils because the pitch would contaminate the cargoes.

Zinc silicate tank coatings show good resistance to petroleum cargoes and many organic solvents, although their resistance to acids and alkalis is inferior to that of epoxies. The paints are supplied as two components, zinc dust being stirred into a silicate solution at the time of use; reactions take place during drying, the dry film consisting essentially of metallic zinc and silicic acid, together with zincates. Single coats with a thickness of 80–100 μm are normally applied.

The choice of tank coating[11] depends upon the cargoes to be carried, and must be determined by the ship operator with the advice of paint manufacturers. The application of epoxy or zinc silicate tank coatings demands special techniques to ensure control of surface preparation, ventilation, overcoating intervals, curing times and temperatures if satisfactory service is to be obtained, and much of the work is undertaken by contractors with the necessary knowledge and equipment. When properly applied, tank coatings not only prevent corrosion of the tanks for up to 8–10 years, but also render tank cleaning easier and quicker since cargo residues are not retained by corrosion products on the interior steel surfaces.

Ships' Bottoms

Paints used for protecting the bottoms of ships encounter conditions not met by structural steelwork. The corrosion of steel immersed in sea-water with an ample supply of dissolved oxygen proceeds by an electrochemical mechanism whereby excess hydroxyl ions are formed at the cathodic areas. Consequently, paints for use on steel immersed in sea-water (pH $\approx 8\cdot0$–$8\cdot2$) must resist alkaline conditions, i.e. media such as linseed oil which are readily saponified must not be used. In addition, the paint films should have a high electrical resistance[12] to impede the flow of corrosion currents between the metal and the water. Paints used on structural steelwork ashore do not meet these requirements. *It should be particularly noted that the well-known structural steel priming paint, i.e. red lead in linseed oil, is not suitable for use on ships' bottoms*[13]. Conventional protective paints are based on phenolic media, pitches and bitumens, but in recent years high performance paints based on the newer types of non-saponifiable resins such as epoxies, coal tar epoxies, chlorinated rubber and vinyls have become widely used. With conventional paint systems the usual interval between drydockings is about 9 to 12 months, but with a high performance system used in conjunction with impressed-current cathodic protection, Lloyds Register and other Classification Societies permit this interval to be extended to $2\frac{1}{2}$ years.

Antifouling compositions The finishing paints on ships' bottoms are required to prevent attachment of marine growths. These paints, known as *antifouling compositions*[14, 15], contain chemicals poisonous to the settling stages of marine plants and animals. The poisons are slowly released into the sea-water, maintaining a thin layer of water next to the surface of the paint in which the spores and larvae cannot survive; settlement and further growth are thereby prevented. The most widely used poison is cuprous oxide but its action, particularly against some types of plant growths, may be reinforced by other poisons, e.g. compounds of mercury, arsenic, tin, lead or zinc. The arsenic, tin and lead poisons are organometallic compounds. In addition, many hundreds of purely organic compounds have been examined as possible antifouling poisons, but none has yet proved so non-selectively effective against a wide range of organisms as the metallic poisons mentioned. It will be realised that antifouling compositions must have a limited effective life, because when the bulk of the poison in the film has been released, the poison release rate falls below that necessary to prevent attachment of marine organisms. On merchant ships the compositions are generally effective for about 9 to 15 months, but special long life types are effective for $2\frac{1}{2}$–3 years.

Details of typical marine painting systems are set out in Table 15.5.

Table 15.5 Typical marine painting systems

Type of paint	Method of application	Coats	Dry film thickness (μm)
1. SHIPS' BOTTOM SYSTEMS			
(a) Conventional bituminous system			
Bitumen or pitch solution pigmented with aluminium flake	Brush, roller or airless spray	3–5	150–200
Antifouling composition	Brush, roller or airless spray	1	50–80
(b) Conventional non-bituminous system			
Tung oil/phenolic medium pigmented with basic lead sulphate, aluminium flake and extenders	Brush, roller or airless spray	2–4	150–200
Antifouling composition	Brush, roller or airless spray	1	50–80
(c) High performance epoxy system			
Coal tar epoxy (2-pack)	Airless spray	2	200–300
Antifouling composition	Brush, roller or airless spray	1	80–100
(d) High performance chlorinated rubber system			
Chlorinated rubber primer	Brush, roller or airless spray	1	50
High build chlorinated rubber	Airless spray	2	175–225
Antifouling composition, chlorinated rubber based	Airless spray or brush	1	80–100

Table 15.5 (continued)

Type of paint	Method of application	Coats	Dry film thickness (μm)
2. TOPSIDES AND SUPERSTRUCTURE SYSTEMS			
(a) *Conventional system*			
Red lead primer in quick-drying alkyd or phenolic medium	Brush, roller or airless spray	2–3	100–125
Gloss finish, alkyd medium pigmented with rutile titanium dioxide (white) and tinting pigments as required	Brush, roller, conventional or airless spray	2	50–80
(b) *High performance epoxy system*			
High build epoxy (2-pack)	Airless spray	2	200–250
Gloss finish, epoxy or polyurethane (2-pack)	Airless spray or brush	1	40–60
(c) *High performance chlorinated rubber system*			
Chlorinated rubber primer	Brush, roller or airless spray	1	50
High build chlorinated rubber	Airless spray	1	80–120
Gloss finish, chlorinated rubber	Airless spray, brush or roller	1	50
3. INTERIOR ACCOMMODATION SYSTEMS			
(a) *Conventional system*			
Zinc chromate primer in quick-drying alkyd or phenolic medium	Brush, roller or airless spray	2	80–100
Semi-gloss undercoat, alkyd medium pigmented with titanium dioxide and tinting pigments	Brush, roller or airless spray	1	40–60
Gloss finish, alkyd medium pigmented with titanium dioxide and tinting pigments	Brush, roller, conventional or airless spray	1	40–60
(b) *High performance system*			
Epoxy primer (2-pack)	Brush, roller or airless spray	2	100–120
Gloss finish, epoxy or polyurethane (2-pack)	Airless spray or brush	1	40–60
4. DRY CARGO HOLD SYSTEM			
Zinc chromate primer in quick-drying alkyd or phenolic medium	Brush, roller or airless spray	2	80–100
Bright aluminium finish, leafing aluminium flake in oleoresinous medium	Brush, roller or preferably airless spray	2	50–80
5. SYSTEMS FOR CARGO/BALLAST TANKS			
(a) *Crude oil carriers*			
Coal tar epoxy (2-pack)	Airless spray	2	250–300
(b) *Refined oil and chemical carriers*			
High build epoxy (2-pack)	Airless spray	2	250–300

Note. The above systems are for application to steel blast-cleaned to a 'near-white' finish (Second Quality of BS 4232:1967) and immediately shop-primed before fabrication. The shop primer must be thoroughly cleaned and degreased at the time of painting.

Methods of Application

The paints used on ships may be applied by brush, roller or spray—airless spraying in particular being widely used when large areas are to be coated. High performance coatings are formulated to permit application of the full system in only a few coats, i.e. the paints must be capable of airless spray application at wet film thicknesses of 200–500 μm, without sagging or running on vertical surfaces, to give dry film thicknesses of 100–300 μm per coat. Time in drydock is generally restricted owing to high costs—figures of £20 000–£30 000 per day being quoted for a 200 000 t tanker—so ships' paints must dry rapidly and must tolerate application under non-ideal weather conditions since owners are unwilling to incur extra costs from delays in painting. Possible health hazards, particularly when spraying some types of antifouling compositions, must be guarded against by wearing protective masks and equipment.

Economics

In the painting of the general interior spaces and the exterior surfaces of ships above the waterline, protective and decorative aspects cannot be separated. Thus, on passenger liners the frequency of repainting the accommodation, superstructure and topsides is determined primarily by the decorative appearance, while on cargo ships this is usually less important than protection. For ships' bottoms the maintenance of a smooth surface free from marine fouling growths is important because a rough or fouled bottom leads to reduced speed and/or increased fuel consumption. Fouling may easily cause a 50% increase in fuel consumption, involving an appreciable increase in running costs. For this reason the intervals at which ships' bottoms are repainted depend on the efficiency of the antifouling compositions and on the degree of fouling encountered in service, marine growth being more vigorous in warm tropical seas than in temperate or polar waters.

The cargo tanks of oil tankers present a special case, because of the high cost of steel renewals in unprotected tanks. For an 18 000 t tanker, costs in the region of £100 000 for the initial painting of the tanks have been quoted; if the life of the paint system is 6–8 years, the total cost over the normal 20-year life of a tanker is expected to be appreciably less than the sum usually spent on steel renewals, which may amount to several hundred thousand pounds.

Types of Failure

Paints correctly applied to well-prepared surfaces on the above-water part of ships will normally fail first by chalking, with checking and crazing of the finishing paint following. Of the high performance systems, polyurethanes have better gloss retention than epoxies or chlorinated rubbers. In spite of a general improvement in conditions of application during recent years, however, ships' paints are still liable to be applied to damp or otherwise imperfectly prepared surfaces, and this leads to failure by adhesion breakdown and rust formation beneath the paint film. Intercoat adhesion failure is also likely with epoxy systems if recommended intervals between coats are exceeded.

On ships' bottoms the antifouling coat fails when its poison release rate (or leaching rate) falls below the value needed to prevent attachment and growth of marine fouling organisms. At this stage it becomes necessary to drydock the ship, clean the bottom and re-apply antifouling composition; the underlying protective paint system should normally only need renewal after about four or more years, depending on whether a conventional or a high performance system is used. For economic reasons (docking charges, interest, insurance, loss of earnings, etc.) no delay can be accepted in the repainting of ships' bottoms, so painting sometimes proceeds under adverse weather conditions to a poorly prepared surface—in consequence failure may occur from loss of adhesion. Paints capable of application to damp surfaces are being developed to overcome this difficulty. It may also be mentioned that promising results have been obtained by cleaning and recoating ships' bottoms under water, and this could eventually eliminate drydocking of ships for repainting[16,17].

T. A. BANFIELD

REFERENCES

1. Ffield, P., *Trans. Soc. Nav. Archit.*, N.Y., **50**, 608 (1950)
2. Singleton, D. W. and Wilson, R. W., *Br. Corros. J.*, Supplementary Issue, 12 (1968)
3. Singleton, D. W., *Iron and Steel*, **41**, 17 (1968)
4. Bullett, T. R., *Br. Corros. J.*, Supplementary Issue, 5 (1968)
5. Wilson, R. W. and Zonsveld, J. J., *Trans. N.E. Cst. Instn. Engrs. Shipb.*, **78**, 277 (1962)
6. Chandler, K. A. and Shak, B. J., *Br. Corros. J.*, **1**, 307 (1966)
7. Banfield, T. A., *Proc. Conf. Protn. Met.*, London, 95 (1970)
8. Banfield, T. A., *Fairplay International Shipping J.*, Anti-corrosion Survey, **233**, 37 (1969)
9. Banfield, T. A., *Shipping*, **59** No. 1, 29 (1970)
10. Logan, A., *Trans. Inst. Mar. Engrs.*, **60**, 153 (1958)
11. Rogers, J., *Trans. Inst. Mar. Engrs.*, **83**, 139 (1971)
12. Mayne, J. E. O., *J. Oil Col. Chem. Ass.*, **40**, 183 (1957)
13. Dechaux, G., *Peint. Pigm. Vern.*, **17**, 758 (1942)
14. Banfield, T. A., *Ind. Fin. and Surface Coatings*, **22** No. 266, 4 (1970)
15. Banfield, T. A., *Oceanology International 72*, Conference, Brighton (1972)
16. Rudman, J. A., ibid.
17. Jones, D. F., ibid.

15.8 Protective Coatings for Underground Use

Introduction

The general conception of a paint is of a cold-applied material containing thinners which evaporate to leave a higher molecular-weight base protective, of 25–50 µm thickness per coat. For buried or submerged structures, where maintenance is difficult or even impossible and a degree of physical protection is also necessary, such thin protective paint barriers between metal and the corrosive electrolyte environments of soil or water are usually quite inadequate. In relatively non-corrosive soil, thin bituminous coatings on thick cast iron may be satisfactory, but this is the exception rather than the rule. In dealing with underground structures, therefore, the thicker protectives needed are regarded as *coatings* rather than as paint finishes.

The most usual forms of buried metal structures are pipelines, piles, tanks and power and telephone cables. Power cables must usually have some metal protection, covered by expensive continuous factory-applied sheathings of considerable thickness. Since water, gas and petroleum pipelines provide the greatest area of metal surfaces to be protected below ground, a detailed discussion of the protection given to them would appear to be the best means of dealing with coatings for underground use.

Improvements are continually being made in the quality of coating materials and their application, but it is still difficult to produce at economic cost a permanent coating for a buried pipeline. The disruptive effects of handling, construction, penetration by rocks, soil stress, material ageing, etc. inevitably result in areas of bare metal being exposed to corrosive soil electrolyte at isolated locations, with ultimate pitting or holing of the metal. The aim is to supply the best possible coating at economic cost and to provide for any initial or later failures by application of cathodic protection. The combination of coating with cathodic protection shows the greatest economic advantage.

In pipelining, the trend is towards all-welded steel for long lines, and since the wall thickness is less than that of cast iron, protection is the more important. Many types of coating are used, from thick concrete to thin paint films, and each has its own particular suitability, but the majority of pipelines throughout the world today are coated with hot-applied coal tar or petroleum asphalt-base-filled pipeline enamels, into which reinforcing wraps, such as glass fibre are applied.

The use of coatings applied in the form of tape is also increasing. Poly-ethylene and polyvinyl chloride films, either self adhesive or else supporting films of butyl adhesive, petrolatum or butyl mastic are in use as materials applied 'cold' at ambient temperatures. Woven glass fibre or nylon bandage is also used to support films of filled asphalt or coal tar and these are softened by propane gas torches and applied to the steel surface hot, cooling to form a thick conforming adherent layer.

Recently, sheets of high density polyethylene extruded on to the pipe surface over an adhesive have become available and the use of polyethylene or epoxy powders sintered on to the steel surface is becoming more frequent.

Some use has been made in the water industry of loose envelopes of poly-ethylene sheeting and with the increasing lengths of submarine pipeline requiring heavy concrete coatings for reducing buoyancy, the use of a heavily filled bituminous coating is projected.

In the special case of pipelines operating at relatively high temperatures such as for the transmission of heavy fuel oil at up to 85°C, heat insulation and electrical insulation are provided by up to 50 mm of foam-expanded polyurethane. As a further insurance against penetration of water, and to prevent mechanical damage, outer coatings of polyethylene (5 mm), butyl laminate tape (0·8 mm) or coal-tar enamel reinforced with glass fibre (2·5 mm) have been used.

Properties Required of Buried Coatings

The aim in applying a coating to a buried metal such as a pipeline is to prevent electrical contact with an electrolyte such as soil and/or water. The characteristics required are as follows:

1. *Ease of application.* It must be possible to apply the coating in the factory or in the field at a reasonable rate and to handle the pipe reasonably quickly after the coating has been applied without damaging the coating.
2. *Good adhesion to the metal.* The coating must have an excellent bond to steel. Priming systems are frequently used to assist adhesion.
3. *Resistance to impact.* The coating must be able to resist impacts without cracking.
4. *Flexibility.* The coating must be flexible enough to withstand such deformation as occurs in bending, testing or laying, as well as any expansion or contraction due to changes in temperature. It must not develop cracks during cooling after application or curing.
5. *Resistance to soil stress.* The coatings are often subject to very high stresses, due, for instance, to the contraction of clay soil in dry weather, and they must be able to resist such stresses without damage.
6. *Resistance to flow.* The coating should show no tendency to flow from the pipe under prevailing climatic conditions. It must not melt or sag in the sun and it must have sufficient resistance not to be displaced from the underside of large-diameter pipes.
7. *Water resistance.* Coatings must show a negligible absorption of water and must be highly impermeable to water or water-vapour trans-mission.

8. *High electrical resistance.* The coating must be an electrical insulator and must not contain any conducting material.
9. *Chemical and physical stability.* The coating must not develop ageing effects, e.g. denaturing due to absorption of the lower-molecular-weight constituents, or hardening with resultant cracking from any cause including oxidation. It should be stable at operating temperatures.
10. *Resistance to bacteria.* The coating must be resistant to the action of soil bacteria.
11. *Resistance to marine organisms.* In the case of submarine lines, the coating should not be easily penetrated by marine life, e.g. mussels, borers, barnacles, etc.

These characteristics cover the general ideal for a pipeline coating, but obviously modified conditions may impose requirements which are more, or less stringent; this of course also applies to other types of buried structures.

Preparation of Metal Surface

Before applying a protective coating it is essential to ensure that the surface is free from rust, millscale, moisture, loose dust, or any other incompatible material which might prevent the electrically non-conducting coating from bonding properly with the metal surface or which might produce defects in the continuous film.

The following cleaning methods are available and each may have a particular advantage in given circumstances:

(a) *Mechanical cleaning.* Hand or mechanical wire brushing, impacting or abrading are methods suitable for hot applied coatings, for repairs to damaged areas or for relatively small or inaccessible areas. Visual standards to assess the degree of cleanliness are available but are not commonly used.
(b) *Blast cleaning.* Air-blast or centrifugally-impacted sand, shot or grit are appropriate for thin-film multicoat systems or for continuous factory production. Several visual standards are available. The cost of attaining a very high standard of cleanliness is considerable, and careful consideration should always be given to specifying the correct level of blasting for the particular application.
(c) *Pickling.* Dipping in inhibited hydrochloric or sulphuric acid is commonly used in factory production, particularly in conjunction with hot phosphoric acid dipping (Footner process). The considerable facilities necessary for this method limit its use to the larger steel producers. Published standards are available for the phosphate surface conversion coating process.
(d) *Flame cleaning.* This is appropriate only for field repair work where a dry or warm surface can be obtained only by flame application and must be preceded usually by mechanical cleaning.
(e) *Pipeline travelling machine.* For long runs on continuously-welded pipelines, a machine with rotary wire brushes and/or impact tools and cutting knives may be used to prepare the surface. These machines,

which are self-propelled along the pipe itself, are commonly combined with drip or spray apparatus to apply the primer which is spread over the surface by rugs or brushes so that the prepared surface is immediately primed.

No matter which method of cleaning is adopted, it is desirable to apply the primer or coating immediately after the cleaning operation.

The preparation of the metal surface to receive the protective coating is of prime importance since a coating which is not bonded to the metal surface can allow electrolytes to contact the metal, with resultant corrosion. If water films develop between the metal and the electrically non-conductive coating, cathodic protection becomes ineffective.

Coating Techniques

Dipping, Spraying and Brushing

These methods are generally appropriate for either thin-film solvent-based paints or for coatings up to about 150 μm thickness. The techniques are more usually used for the priming layer of the coating systems.

Factory or Yard Application

Protective coatings applied at a factory have the advantage that the work can be carried out under strictly controlled conditions but suffer from the disadvantage that they may be damaged during transport to the site.

Pipes are frequently shot-blasted or descaled by acid pickling, then phosphated, either sprayed with primer or dipped into a bath of hot asphalt to provide a thin prime coat. The dry primed pipes are then slowly rotated by a lathe head, while hot enamel, mastic or asphalt/micro-asbestos paste is applied from a hopper travelling alongside the pipe. A pipe coating approximately 5 mm thick is produced by use of a heated pallet attached to the hopper feed. Reinforced-glass wrapping materials may also be spirally wound on to the coating according to requirements.

'Rolling Rig', 'Fixed-head' and 'Rotating-head' Coating Machines

The coating equipment under this heading may be used in permanent factories, but is often set up at temporary coating yards close to the location where the pipes are to be laid. The coating produced is usually 2–3 mm per pass.

Rolling rig machines   The rolling rig machine rotates the cleaned and primed pipe on mechanically driven 'dollies', while a tank travelling alongside the pipe floods it with hot asphalt or coal-tar-base enamel.

At the same time internal and external reinforcing wraps may be spirally wound into or on to the hot enamel.

Fixed-head machines   Fixed-head machines are fed with the cleaned and primed pipe, which mechanically rotates as it passes through the fixed coating head which floods the hot enamel on to the pipe. At the same time reinforcing wraps are pulled on to the rotating pipe.

Rotating-head machines   In rotating-head machines the coating head and wrapping spindles rotate as the pipe is fed through the machine.

Pipeline Travelling Machines

In the case of long continuously-welded steel pipelines the above pipe-coating methods present the disadvantage that the joints have to be coated in the field after welding. To overcome this difficulty equipment which travels along the welded pipeline has been developed.

A mechanically propelled cleaning machine travels along welded lengths of the pipeline. The machine has counter-rotating cutting knives or brushes, and also applies by rotating swabs, a thin coating (cold application) of primer to the clean metal surface. When the primer is dry, a coating and wrapping machine travels along the pipeline.

The wrapping materials usually consist of staple glass tissue, pulled half-way into the hot enamel, and an outer wrap of glass impregnated with coal-tar or asphalt enamel to produce a coating of approximately 2·5 mm as shown in Fig. 15.7.

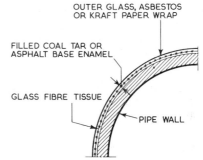

OUTER GLASS, ASBESTOS OR KRAFT PAPER WRAP

FILLED COAL TAR OR ASPHALT BASE ENAMEL

GLASS FIBRE TISSUE

PIPE WALL

Fig. 15.7   Type of coating produced by mechanical flood coat and wrap machine

These machines can coat and wrap up to 5 km of pipeline per day. After the coating has been checked for pin holes by a high-voltage rolling-spring electrode, the pipe may be lowered directly into the trench, so that undue handling is avoided. The line travelling machine is usually used with coal tar or asphalt-base pipeline enamels. Similar line travelling machines are in use for the cold application of tape coatings.

Types of Coating Materials

Plasticised Coal Tar and Petroleum Asphalt Enamels

The majority of pipelines today are coated with hot-applied plasticised coal tar or petroleum asphalt enamels. Both coal-tar pitch and petroleum asphalt have been used as protectives with and without filling materials. When filled they are termed *enamels* or *mastics*. The term *bitumen* or *bituminous* has always been loosely applied and it is preferable to specify petroleum asphalt base or coal-tar pitch base.

Straight and filled enamels   Fillers are normally added up to a maximum of about 30% weight (calculated on the mixture) which is equivalent to about 15 to 20% by volume. A filled coal-tar pitch has a higher softening temperature (as shown by the 'ring and ball' test) than the unfilled material, which results in a reduced tendency to flow. This fact is important in tropical countries or if a pipe is to operate at a somewhat elevated temperature. Resistance to impact and abrasion of a coating is improved by the filler. The viscosity of the pipe coating is also increased; this entails a higher application temperature (193–249°C).

A satisfactory filler must have the following characteristics:

1. Low water absorption. In this respect certain fine clays are unsuitable.
2. Ability to be readily wetted by the enamel.
3. Finely-ground composition, particles preferably of laminar shape to prevent settling when the enamel is molten.
4. Relatively low specific gravity, so that there is the minimum tendency for the filler to settle-out in the melting kettle.

In present-day practice the materials which are commonly used and which satisfy most closely these requirements are talc, pumice powder, micro-asbestos and slate powders.

It must be appreciated that there is an optimum percentage of filler which imparts to a coating the required melting point and toughness; beyond this point application becomes more difficult and watertightness may be impaired.

Petroleum asphalt or coal-tar pitch as coatings   The question of whether coal-tar pitch or petroleum asphalt is the more suitable for the coating of underground pipelines has raised a good deal of controversy. Asphalt and pitch are both waterproof materials, and they resemble one another in physical type. In the right circumstances both can be very effective in preventing the access of water to buried or submerged steel surfaces.

Petroleum asphalts are manufactured in two general types: (*a*) a straight residue from distillation, which can be of the hard, high-melting type, and (*b*) so-called 'blown' grades which are prepared by partially oxidising the asphalt base by blowing in air. The general difference between the two grades is that 'blown' asphalt has a higher softening point than straight asphalt of the same penetration (i.e. hardness). In assessing a pipeline coating the softening point is of considerable importance, since it determines the tendency to flow, and a certain minimum softening point is therefore necessary. A 'blown' asphalt has the advantage over straight material of the same softening point in that it has a better resistance to impact, since it is of a more rubbery nature. For this reason most petroleum asphalt coatings are based on the 'blown' variety.

So far as coal tar is concerned, it was formerly the custom to use the straight residual pitch, but nowadays shock resistance is improved by a so-called plasticising process.

The differences between asphalt and coal tar in relation to their application as pipeline coatings require comment.

1. It is often claimed that a coal-tar-base coating absorbs less water than an asphalt coating and there is evidence in practice to support this claim, but some asphalt enamels in practice have been as good as the best coal-tar enamels.
2. Coal-tar enamels are claimed to have better adherence than the asphaltic enamels to clean metal, probably because of the presence of polar compounds, but little difference can be noted in practice under proper pipelining conditions.
3. The asphaltic enamels are easier to apply since they do not produce so much obnoxious fume and are usually applied at slightly lower temperatures.

The field performance of the asphalt-base pipeline enamels was, at one time, erratic, probably because the material had been drawn from varying sources, without a close specification being used. The plasticised coal-tar-base enamel to the American Water Works Association Specification C–203 thus gained some favour in major pipelining organisations.

The AWWA C–203 Standard remains a widely used specification suitable not only for the materials, but also for their associated reinforcing wraps and application procedures. The standard has been regularly updated. Hot-applied asphaltic and coal-tar coatings with their priming systems are now well classified, described and specified in BS 4164:1967 (coal tar) and BS 4147:1967 (asphalt), but no guidance is given in these specifications to application procedures.

Reinforcing materials

Internal   At one time open-weave hessian cloth was very largely used as an internal reinforcement material, but experience showed that this is subject to rotting in the soil. Even when the material appears to be covered with enamel, some of the fibres must protrude, and thus moisture is absorbed so that after a period of years the hessian is generally found to be in a water-logged condition and forming food for bacteria.

The type of material to be used depends very largely on whether coating is carried out mechanically or by hand. For hand application it is not possible to use comparatively fragile staple tissues made of glass or asbestos and it is necessary to use a strong open-mesh fabric, such as woven asbestos or woven glass. The woven wraps are a great deal more expensive than the staple tissues, which are mechanically applied.

It is not economical to use expensive woven material for long lines, which can be, and normally are, coated by mechanical means. For such lines the most commonly used material nowadays is a glass-fibre tissue of a nominal 0·5 mm thickness, consisting of glass fibres bonded together with a phenolic resin or starch.

The improvement in coating quality achieved by using the internal glass wrap is illustrated by the following results. The tensile strength of a 3·2 mm thickness of 104°C softening-point enamel, 300 mm × 300 mm is virtually nil. A piece of 300 mm × 300 mm glass tissue 0·5 mm in thickness will break at about 50 kg under steadily increasing tensile load, but if it is embedded in

3·2 mm of the enamel a tensile strength of the order of 150 kg is obtainable. These wraps are now longitudinally reinforced to prevent tearing on line-travelling or other coating machines.

Where the pipeline is expected to have to withstand unusual dimensional variation due, perhaps, to temperature changes or near yield point pressure testing, the use of a woven glass or nylon reinforcement in place of the glass tissue is said to increase the flexibility of the coating system considerably.

External wrap   The purpose of an external overlapping wrap is to provide a shield against the penetration of the enamel by stones and to prevent the pulling of the enamel away from the pipe by soil stress. It also reduces flow of the enamel owing to the weight of the pipe, and damage to the coating caused by handling can be more easily observed.

The properties required of an external wrap are as follows:

(*a*) Compatibility of impregnant to bond with the enamel used.
(*b*) Tensile strength to prevent breaking while wrapping.
(*c*) Hardness to resist penetration.
(*d*) Flexibility to allow wrapping without cracking.
(*e*) Free rolling from the reel while wrapping.
(*f*) Resistance to soil conditions and bacterial attack.
(*g*) Non-absorption or low absorption of water.

These properties apply to a reinforcing outer wrap such as coal tar or asphalt-impregnated glass or asbestos bonded lightly to the outside of the hot-applied enamel. For some conditions kraft paper is adequate to facilitate handling and reduce soil stress. Were it not for its screening effect on cathodic protection with a consequent decrease in the effectiveness of the latter, the external wrap could be loose around the coating. It has become conventional to have the external wrapping lightly bonded to the coating to prevent lamination and water entry.

Armour wrapping   In rocky ground it has always been considered good practice to pad the trench for a buried pipeline with clean sand. This procedure can be very expensive if the sand has to be hauled long distances, and an armour wrap has been developed to supplement the normal outer wrap to meet such conditions.

A typical wrap is supplied in sheets about 6 mm thick, consisting of a sandwich of mastic enamel between sheets of asbestos about 1·5 mm thick. It may be longitudinally indented to allow the material to be wrapped around the pipe and secured by steel ribbon straps.

An objection to this form of wrap is that its mode of application renders it extremely difficult to obtain a good uniform bond between the wrap and the enamel. In view of this, water could become trapped under the armour wrap, and because of the non-conducting nature of the wrap itself the effective application of cathodic protection would be difficult.

Cold-applied Tapes

Hot-applied coatings require special melting and handling equipment to be available at the construction site. Clearly, considerable economies are possible if this equipment can be dispensed with, particularly in remote areas with difficult access. Thus, the availability of cold-applied tapes for use either at the joints between factory-coated pipes or continuously over the pipeline has led to the increased usage of this type of wrapping.

The tapes are usually relatively thin (0·5 mm) and easily damaged. It is, therefore, essential to take elaborate precautions to provide physical protection to the tape once it has been applied both during construction and after burial. Good results have been obtained when the tape is applied by line travelling machine and without further handling, immediately lowered into a sand padded trench and covered over with fine sand before the trench is back filled.

Initial effective electrical resistance of tapes, as evidenced by the cathodic protection current demand, has been outstanding. There have been reports of increasing current demand with time which indicate a need for investigation. The current demand increase has been found, on occasion, to be due to poor construction practice, but not all tapes are affected in this way.

On large diameter pipes having a raised seam weld, difficulty is encountered in covering the weld 'shadow' effectively.

Petrolatum-type tapes Petrolatum has, like lanolin, long been recognised as a means of preventing corrosion. It is easily cold-applied and has a definite place in corrosion engineering, but it is not suitable for buried structures, unless it is screened from soil and water by a woven glass or nylon cloth or an impervious membrane such as p.v.c. The polythenes normally tend to swell in contact with it. Earlier petrolatum coatings were frequently applied with cellulosic backing material; there were several objections to this type of protection, e.g. attack by sulphate-reducing bacteria on the cellulose, absorption of the grease by dry bentonite-type clays, lack of physical strength against stones, and water absorption.

Petrolatum-type tape coatings now incorporate inhibitors against bacteria and with their backing film have high electrical and water resistance and therefore find extensive applications in the UK. A great advantage of the petrolatum-type coatings is ease of application and conformability to irregular surfaces.

Pressure-sensitive tapes Unlike the more recently developed petrolatum tapes which rely on both the petroleum and backing films, the pressure-sensitive tapes offer protection which depends almost entirely upon the prevention of ingress of moisture to the metal surface by the tape itself.

The tapes are cold-applied, either by hand or by mechanically-operated equipment moving along the cleaned pipeline.

The tapes are usually produced from polythene or polyvinylchloride films of 25 μm to 0·5 mm in thickness and the inner surface is coated with an adhesive, frequently rubber-based. The adhesive is usually between 25 and 100 μm thick.

Earlier tapes frequently suffered from the migration of plasticiser from the tape to the adhesive with the result that the tape became detached from the metal, to which the adhesive remained attached. This has now been overcome by using a barrier between the tape and adhesive which itself may contain inhibitors against soil bacteria.

Spiral corrosion due to inadequate overlap has been detected with self-adhesive tapes, and a 25 mm (or preferably half-tape-width) overlap is to be advocated. Within normal limits, the thicker the adhesive the better.

The self-adhesive tape coatings are thin and the adhesive itself does not necessarily come into contact with the valleys in the cleaned metal surface. Under these circumstances, the transmission of water vapour through the film to the metal may be possible. Moisture-transmission characteristics and other properties of p.v.c. and polyethylene tapes, as given by major manufacturers, are provided in Table 15.6.

Table 15.6 P.V.C. and polyethylene tapes

Material	Thickness of film plus adhesive (mm)	Tensile strength (kg/cm width)	Elonga-tion at break (%)	Moisture absorp-tion (%)	Moisture-vapour transmission rate (g/m² per 24 h)	Dielectric strength (V)
			Physical property			
P.V.C.	0·229 + 0·025	10	175	0·19	24·0	10 000
Polyethylene	0·203 + 0·100	10	70	0·02	3·1	14 000

Table 15.6 is only indicative of general properties, and the latest developments of specific manufacturers of self-adhesive tapes may show advances on these.

P.V.C. tends to be more conformable to irregularities than polyethylene. Both types have their right and proper application for buried structures.

Laminated tapes In more general use now than pressure sensitive tapes are tapes consisting of polyvinyl chloride or polyethylene films in conjuction with butyl rubber. These tapes are applied with an adhesive butyl rubber primer. Thicknesses of up to 0·75 mm are in use and loose protective outer wraps of p.v.c. or polyethylene sheet are commonly applied. Tape quality control is exercised with reference to ASTM standard test methods and may include water vapour transmission rate and elongation.

Conventional holiday-detection is of little value in the field but great attention should be given to preventing damage to the applied tapes.

Coal-tar Epoxy Coatings

The epoxy resins when mixed with the correct amine produce tough films which adhere closely to metal. The chemistry of these resins is considered in Sections 15.5 and 15.9.

The thickness and water resistance of the normal air-cured film can now be much improved by the incorporation of suitable coal-tar pitch material. A typical coal tar/epoxy coating material would be constituted as follows:

Epoxy resin	30
Coal-tar pitch	25
Filler	25
Solvent	20

and to the above would be added the amine curing mix.

The coating is of the two-pack type, consisting of resin plus curing hardener. In practice the resin and amine may be mixed together and used for application by brush or spray, or by mechanical means at ambient temperature. Sometimes the clean metal is heated, as are the coating components, which are then sprayed separately on to the metal to reduce curing time. Little reaction occurs below 4°C. For pipeline coating the pipes can usually only be handled after a few hours, depending on the mix and temperature, but it takes anything from two to seven days before the best characteristics of the coating develop.

Information to date indicates that the total thickness of the coating should not be less than 0·3 mm and this requires several applications. These coatings are very tough and closely adherent (one pipeline company states that they handle coal tar/epoxy-coated pipe like bare pipe, including bending in the field). The first coal tar/epoxy coatings came into use only in 1953, and although they seemed most promising they have been little used to date compared to other materials. This is undoubtedly due to their relatively slow setting and curing time.

Polyethylene Sheet

The practice has been developed amongst some water undertakings to envelop uncoated spun iron pipes in 0·5 mm thick polyethylene sheet, the ends of which are tied down to the pipe with a substantial overlap by means of adhesive tape. This method has great advantages in cost and simplicity. No long term performance figures have been published but many have grave doubts about the effectiveness of this method since the possibility of aggressive soil water entering at perforations or through overlaps, appears to be very high.

Foam Polyurethane

These materials have been finding extensive use on transmission pipelines supplying heated heavy fuel oils to power stations. To prevent damage to the 50 mm thick coating, a mechanically stronger outer wrap which can also prevent water ingress is usually necessary. In one method of production, the foam is manufactured inside a polythene tube over the steel tube. In other methods where the foam is produced by spraying on to the steel surface, conventional tape or enamel coatings have been used.

Weight Coatings

For pipelines to be placed under water, it is necessary to provide negative buoyancy. This is commonly achieved by placing lightly reinforced concrete up to 150 mm thick over the 3–5 mm hot enamel coating on the steel. Joints at the welded tube ends have to be coated with a minimum of delay due to the high production rate required on the laying barge, and tapes have therefore found application at this point. Where submarine pipelines are 'pulled' into position off the land, joint repair is more commonly carried out by means of the same hot enamel used as the pipeline coating. For the final joint between towed 'strings' of up to 300 m, fast setting epoxies have been used.

A composite asphaltic mastic filled with high-density aggregate is now available as a combined insulation and weight coating, and this could be the development area in this field.

Internal Pipeline Coatings

In some instances it is necessary to coat pipelines internally, and materials widely used are red lead, hot-applied enamels, concrete and epoxy resins.

Internal coatings are usually applied at the factory and no difficulty exists in field construction if flanges, screwed, or spigot and socket joints are used, nor is there any difficulty with welded pipes above, say, 750 mm diameter, where patching can be carried out on the joints from the inside. Repair of internal coating on smaller-bore welded pipes presents many problems, which have not yet been satisfactorily overcome for all conditions.

Pipelines in the ground can be mortar lined *in situ* by the use of travelling devices. Epoxy resin paints for long welded pipelines already laid have been applied *in situ* by placing two plugs in the pipeline with the paint between them, and then forcing them to travel through the pipeline by the use of compressed air.

D. A. LEWIS

BIBLIOGRAPHY

Bigos, J., *Steel Structures Painting Manual*, Steel Structures Painting Council (1954)
Coal Tar Based Hot Applied Coatings, BS 4164:1967
Hall, R. E., Scott, F. S. and Weir, C. J., *Materials Protection*, **6** No. 8, 35 (1967)
Hot Applied Bitumen Based Coatings, BS 4147:1967
Peabody, A. W. and Woody, C. L., *Corrosion*, **5**, 369 (1949)
Romanoff, M., NBS Circular 579 (1957)
Spencer, K. A. and Footner, H. B., *Chem. Ind., Lond.*, **19** (1953)
Sparrow, L. R., *Petroleum, Lond.*, **21**, 357 (1958)
Shideler, N. T. and Whittier, F. C., *Pipeline Ind.*, **6**, May (1958)
Shideler, N. T., *Corrosion Technol.*, **17**, 52 (1960)

15.9 Synthetic Resins

Since the synthetic resins occupy a position of such importance in modern paint formulation, and are relevant to all the topics discussed in this Section, stages in their formation are diagrammatically set out below.

The Phenol Formaldehyde Resins

Stage 1. PRODUCTION OF THE MONOMER BY CONDENSATION OF THE PHENOL AND FORMALDEHYDE UNDER ALKALINE CONDITIONS.

Stage 2. PRODUCTION OF THE POLYMER BY POLYMERISATION DURING MANUFACTURE OF THE RESIN.

The Urea Formaldehyde Resins

UREA   FORMALDEHYDE   POLYHYDRIC
                                       ALCOHOL

Stage 1. PRODUCTION OF THE MONOMER BY CONDENSATION OF UREA AND FORMALDEHYDE IN THE PRESENCE OF A POLYHYDRIC ALCOHOL.

Stage 2. POSSIBLE METHOD OF POLYMERISATION TO A THREE-DIMENSIONAL STRUCTURE.

The Melamine Formaldehyde Resins

DICYANDIAMIDE   MELAMINE
                        $(2,4,6-$TRIAMINO$-1,3,5-$TRIAZINE$)$

Stage 1. MELAMINE IS FIRST PREPARED BY HEATING DICYANDIAMIDE UNDER PRESSURE.

MELAMINE HEXAMETHYLOLMELAMINE

Stage 2. SUBSEQUENT CONDENSATION WITH FORMALDEHYDE IN THE PRESENCE OF POLYHYDRIC ALCOHOLS PRODUCES RESINS SUITABLE FOR FILM FORMING, PROBABLY THROUGH HEXAMETHYLOL-MELAMINE.

The Alkyd Resins

Stage 1. ESTERIFICATION. PRODUCTION OF A THERMOPLASTIC ALKYD FROM ETHYLENE GLYCOL AND PHTHALIC ANHYDRIDE.

Stage 2. POLYMERISATION TO A CHAIN STRUCTURE:

$$HOCH_2CH_2O \left(COC_6H_4COOCH_2CH_2O\right)_n COC_6H_4COOH$$

Possible mechanism of formation of a thermosetting alkyd

PHTHALIC ANHYDRIDE   GLYCEROL

Stage 1. INITIAL ESTERIFICATION.

Stage 2.   FORMATION OF A LINEAR POLYMER.

Stage 3.   CROSS-LINKING WITH PHTHALIC ANHYDRIDE.

The Acrylic Resins

Stage 1.   PRODUCTION FROM ACETONE CYANOHYDRIN.

AND A SMALL AMOUNT OF

Stage 2.   POLYMERISATION.

Production of the Basic Epoxy Resins

Stage 1. PRODUCTION OF THE MONOMER BY REACTION OF EPICHLORHYDRIN WITH DIHYDROXY-DIPHENYLPROPANE UNDER ALKALINE CONDITIONS.

Stage 2. POLYMERISATION TO THE BASIC RESIN.

Stage 3. FURTHER POLYMERISATION TO A THREE-DIMENSIONAL NETWORK DURING CURING OF THE ENAMEL FILM.

Curing an Epoxy Resin by Direct Cross-linking with a Polyfunctional Amine

Curing an Epoxy Resin by Esterification

Curing an Epoxy Resin by Cross-linking with Phenol, Urea or Melamine Formaldehyde

Curing an Epoxy Resin by Cross-linking with an Isocyanate

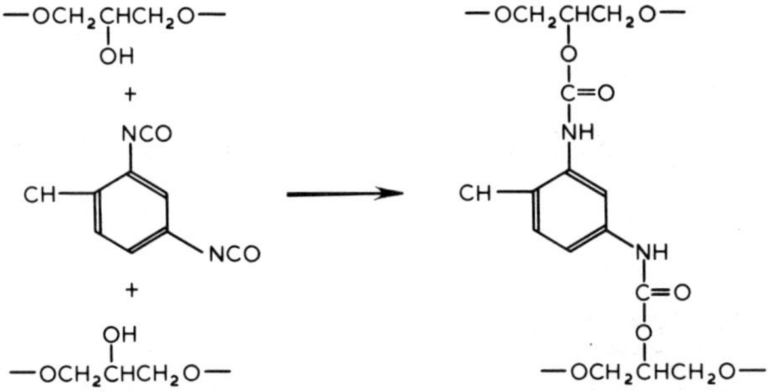

—OCH$_2$CHCH$_2$O—
$\quad\quad$ OH

+

NCO

CH—

NCO

+

OH
—OCH$_2$CHCH$_2$O—

$\longrightarrow$

—OCH$_2$CHCH$_2$O—
$\quad\quad$ O
$\quad\quad$ C=O
$\quad\quad$ NH

CH—

NH
C=O
O
—OCH$_2$CHCH$_2$O—

H. J. SHARP

15.10 Glossary of Paint Terms*

Alligatoring: see *cracking*.

Bitty: the description applied to a paint or varnish containing bits of skin, gel, flocculated material or foreign particles, which project above the surface when the paint or varnish is applied in a manner appropriate to its type and purpose. The term *peppery* is sometimes used when the bits are small and uniformly distributed. The term *seedy* specifically denotes bits which have developed in a paint or varnish during storage.

Bleaching:

    (a) Loss of colour of a paint or varnish. This may be due to internal chemical or physical action in the paint itself, to influences from the surface on which it is applied or to weathering or contamination from the atmosphere. (See also *fading* and *whitening in the grain*.)

    (b) The intentional lightening of the colour of a material, usually wood in the decorating trade. It can be done on new or old untreated timber or on previously painted or stained wood after stripping. Oxalic acid is most often used by painters although excellent proprietary brands of bleaching solutions are now available.

Bleeding: the process of diffusion of a 'soluble' coloured substance from, into and through a paint or varnish coating from beneath, thus producing an undesirable staining or discolouration. Examples of 'soluble' materials which may give rise to this defect are certain types of the following classes of materials: bituminous paints, wood preservatives, pigment dyestuffs and stains.

Blistering: the formation of dome-shaped projections or blisters in paints or varnish films by local loss of adhesion and lifting of the film from the underlying surface. Such blisters may contain liquid, vapour, gas or crystals.

Bloom: a deposit like the bloom on a grape which sometimes forms on glossy enamel, paint or varnish films, causing loss of gloss and dulling of the colour. Sometimes bloom may be removed by wiping with a damp cloth.

Blushing: a milky opalescence which sometimes develops as a film of lacquer dries and is due to the deposition of moisture from the air and/or precipitation of one or more of the solid constituents of the lacquer; usually confined to lacquers which dry solely by evaporation of solvent.

Bridging: the covering over of an unfilled gap such as a crack or corner with a film of paint. This introduces a weakness in the coating which may

*For a full range of definitions see BS 2015:1965.

lead to an eventual cracking of the dried paint.

Bubbling: a film defect, temporary or permanent, in which bubbles of air or solvent vapour, or both, are present in the applied film.

Chalking: the formation of a friable, powdery coating on the surface of a paint film caused by disintegration of the binding medium due to disruptive factors during weathering. The chalking of a paint film can be considerably affected by the choice and concentration of the pigment.

Cheesy: the character of a paint or varnish film which although dry is mechanically weak and rather soft.

Cissing: a defect in which a wet paint or varnish film recedes from small areas of the surface leaving either no coating or an attenuated one.

Clouding: the development in a clear varnish or lacquer film or liquid of an opalescence or cloudiness caused by the precipitation of insoluble matter.

Cold Checking: the development of hair cracks in a lacquer film when it is subjected to a 'cold check' test, e.g. when a furniture lacquer is subjected to defined cycles of alternating cold and normal temperatures.

Cold Cracking: the cracking or checking of paint, varnish or lacquer films caused by sudden or repeated reduction in the temperature of the film.

Cracking: generally, the splitting of a dry paint or varnish film, usually as a result of ageing. The following terms are used to denote the nature and extent of this defect:

Hair-cracking. Fine cracks which do not penetrate the top coat; they occur erratically and at random.

Checking. Fine cracks which do not penetrate the top coat and are distributed over the surface giving the semblance of a small pattern.

Cracking. Specifically, a breakdown in which the cracks penetrate at least one coat and which may be expected to result ultimately in complete failure.

Crazing. Resembles checking but the cracks are deeper and broader.

Crocodiling or alligatoring. A drastic type of crazing producing a pattern resembling the hide of a crocodile.

Cratering: the formation of small bowl-shaped depressions in a paint or varnish film.

Crawling: a pronounced form of *cissing (q.v.)*.

Crinkling: the development in a film of paint or varnish during its formation, of ridges and furrows. These vary in size and frequency with the composition of the film and conditions during film formation including temperature and contamination of the atmosphere. This term is synonymous with *wrinkling*.

Crowsfooting: a type of film defect where small wrinkles occur in a pattern resembling that of a crow's foot (see *wrinkling*).

Efflorescence: the development of a crystalline deposit on the surface of brick, cement, etc. due to water, containing soluble salts, coming to the surface, and evaporating so that the salts are deposited. In some cases, the deposit may be formed on the top of any paint film present, but usually the paint film is pushed up and broken by the efflorescence under the coat.

Filiform Corrosion: a form of corrosion under paint coatings on metals characterised by a thread-like form advancing by means of a growing head or point.

Flaking: lifting of the paint from the underlying surface in the form of flakes or scales.

Frosting: the formation of a translucent finely-wrinkled surface on a film of oil or paint during drying, particularly when exposed to gas fumes, etc. This defect is especially characteristic of paints and varnishes containing certain oils which have not received adequate heat treatment.

Gas Checking: see *webbing*.

Grinning Through: the showing through of the underlying surface due to the inadequate opacity of a paint film which has been applied to it.

Holidays: skipped or missed areas, left uncoated with paint.

Orange Peel: the pock-marked appearance, in particular of a sprayed film, resembling the skin of an orange due to the failure of the film to flow out to a level surface. (See also *spray mottle*.)

Pinholing: the formation of minute holes in a film during application and drying. Sometimes due to air or gas bubbles in the wet film which burst, forming small craters that fail to flow out before the film has set.

Pitting: the formation of holes or pits in a metal surface, by corrosion.

Pock-marking:
  (*a*) *Orange peel (q.v.).*
  (*b*) A film defect in the shape of irregular and unsightly depressions formed during the drying of a paint or varnish film.

Popping:
  (*a*) Eruptions in a film of paint or varnish after it has become partially set so that craters remain in the film.
  (*b*) Of plaster. A mild form of blowing *(q.v.).*

Rain Spotting: the particular case of *water spotting (q.v.)* caused by rain.

Rivelling: see *wrinkling*.

Run: a narrow downward movement of a paint or varnish film; may be caused by the collection of excess quantities of paint at irregularities in the surface, e.g. cracks, holes, etc. the excess material continuing to flow after the surrounding surface has set.
  Tear. A small characteristically shaped run.

Sagging: a downward movement of a paint film between the times of application and setting, resulting in an uneven coating having a thick lower edge. The resulting sag is usually restricted to a local area of a vertical surface and may have the characteristic appearance of a draped curtain, hence the synonymous term *curtaining*.

Seedines: a defect in a clear varnish or lacquer caused by small particles sometimes visible when examined by transmitted light. On application, varnished or lacquered surfaces may present a bitty, specky or sandy appearance due to this defect.

Sleepy: the description of a recently applied glossy coating which has lost its initial gloss other than by bloom and has become dulled or lacking in lustre.

Spotting: the development of small areas on a painted surface which differ in colour or gloss from the major portion of the work.

Spray Mottle: the irregular surface of a sprayed film resembling the skin of an orange. The defect is due to the failure of the film to flow out to a level surface. (See also *orange peel*.)

Sweating:
  (*a*) Exudation of oily matter from a film of paint, varnish or lacquer after the film has apparently dried.
  (*b*) Development of gloss in a dry film of paint or varnish after it has been flattened down.
  (*c*) Often incorrectly used to describe condensation of moisture from humid atmospheres on relatively cold surfaces, e.g. walls.

Tack: slight stickiness of the surface of a film of paint, varnish or lacquer, apparent when the film is pressed with the finger.

Water Spotting: the spotty appearance of a paint film which is caused by drops of water on the surface and which remains after the water has evaporated; the effect may or may not be permanent. Water spots usually appear lighter in colour than the surrounding paint.

Webbing: the development of wrinkles, often in a well-defined pattern, in the surface of a paint or varnish during drying. This condition results from the irreversible swelling of a partially dried surface skin (see also *frosting*) and may be aggravated by impure gas fumes during stoving in a gas oven, in which case it is termed *gas checking*. Webbing is generally regarded as a paint defect but is made use of in some paint finishes to give a textured coating which obscures minor faults and indentations in the surface to be coated.

Wrinkling: the development of wrinkles in a film during drying, usually due to the initial formation of a surface skin. (See *crinkling*.)

E. F. REDKNAP

16 CHEMICAL CONVERSION COATINGS

16.1	Coatings Produced by Anodic Oxidation	**16**:3
16.2	Phosphate Coatings	**16**:19
16.3	Chromate Treatments	**16**:33

16.1 Coatings Produced by Anodic Oxidation

Practice of Anodising

Anodic oxidation or anodising, as applied to metallic surfaces, is the production of a coating, generally of oxide, on the surface by electrolytic treatment in a suitable solution, the metal being the anode. Although a number of metals[1], including aluminium, magnesium, tantalum, titanium, vanadium and zirconium, can form such anodic films, only aluminium and its alloys, and to a lesser extent magnesium, are anodised on a commercial scale for corrosion protection.

The anodic oxidation of magnesium does not normally produce a film that has sufficient corrosion resistance to withstand exposure without further protection by painting, and the solutions used are complex mixtures containing phosphates, fluorides and chromates. In the case of aluminium, a relatively simple treatment produces a hard, compact, strongly adherent film of oxide, which affords considerably increased protection against corrosive attack[2,3].

A further advantage of this process lies in the decorative possibilities of the oxide film, which may be almost completely transparent on very high purity aluminium (99·99% Al) and certain alloys based on this purity, and thus protects the surface without obscuring its polish or texture. On metal of lower purity, and other alloys, the oxide layer may become slightly milky, or coloured grey or yellowish, although the deterioration is hardly apparent with purities down to 99·7–99·8% Al. The appearance and character of the film may also be influenced by the type of anodising treatment, and the oxide film may be dyed to produce a wide range of coloured finishes. Anodising characteristics of a number of aluminium alloys are listed by Wernick and Pinner[3] (see also Section 4.1, Tables 4.1 to 4.4).

The anodising procedures in general use are shown in Table 16.1, sulphuric acid being the most commonly used electrolyte. Treatment time is 15 min to 1 h.

The articles to be anodised should be free from crevices where the acid electrolyte can be trapped*. They may be given a variety of mechanical and chemical pretreatments, including polishing, satin-finishing, etching, etc. but before anodising, the surface must be clean and free from grease and polishing compound.

---

*The chromic acid process is preferred where the electrolyte is likely to be entrapped in crevices as it is an inhibitor for aluminium whereas sulphuric acid is corrosive.

After the anodic treatment, the work is removed from the tank and carefully swilled with cold water to remove all traces of acid. At this stage, the anodic film is absorptive, and care should be taken to avoid contamination with oil or grease, particularly if the work is to be dyed. Dyeing may be carried out by immersion for about 20 min in an aqueous solution of the dyestuff at a temperature of 50–60°C. Inorganic pigments may also be incorporated in the oxide layer by a process involving double decomposition.

Table 16.1   Traditional anodising processes

Electrolyte	Temp. (°C)	E.M.F. (V)	Current density (Am^{-2})	Film thick-ness (μm)	Appearance
5–10% (v/v) sulphuric acid	17–22	12–24	110–160	3–25	Transparent, colourless to milky
3–10% chromic acid	30–45	30–45	32	2–8	Opaque, light to dark grey
2–5% oxalic acid	20–35	30–60	110–215	10–60	Transparent, light yellow to brown

Finally, both dyed and undyed work are sealed by treatment in boiling water (distilled or deionised) or steam, which enhances the corrosion resistance and prevents further staining or leaching of dye. Solutions of metal salts, usually nickel or cobalt acetates, are often used to seal work after dyeing, and sealing in 5–10% dichromate solution, which gives the coating a yellow colour, is sometimes employed where the highest degree of corrosion resistance is desired[4].

In the architectural field, increasing use is being made of integral colour anodising which is capable of producing self-coloured films in a number of fade-resistant tints ranging from grey, through bronze and brown, to a warm black. The electrolytes are developments of the oxalic acid solution and consist of various dibasic organic acids, such as oxalic, malonic or maleic, or sulphonated organic acids such as sulphosalicylic acid, together with a small proportion of sulphuric acid. For constant and reproducible results, a close analytical control of the electrolyte must be maintained, particularly with respect to aluminium which dissolves as treatment proceeds, and ion-exchange resins are frequently used to regenerate the relatively expensive electrolyte and keep the aluminium in solution between controlled limits. Some typical colour anodising treatments are summarised in Table 16.2*.

Alloys are generally of the Al–Mg–Si type with additions of copper and chromium or manganese. Colour varies with the particular alloy and the film thickness. For optimum control of colour, the alloy must be carefully produced with strict attention to composition, homogenisation and heat-treatment, where appropriate, and the anodising conditions must be maintained within narrow limits. It is usual to arrange matters, preferably with automatic control, such that current density is held constant with rising

*Coloured metal compounds may also be introduced into the film by a.c. treatment in a suitable electrolyte [Fuji process, UK Pat. 1 022 927 (26.2.63)].

voltage up to a selected maximum, after which voltage is held steady; the whole cycle being for a fixed time. Refrigeration of the electrolyte may be necessary to maintain the temperature at the working level, owing to the relatively high wattage dissipation.

Hard anodic films, 50–100 μm thick, for resistance to abrasion and wear under conditions of slow-speed sliding, can be produced in sulphuric acid electrolytes at high current density and low temperature[8]. Current densities

Table 16.2 Integral colour anodising processes

Process	Electrolyte	Temp. (°C)	Current density (Am^{-2})	Voltage (V)	Time (min)
Kalcolor[5]	Sulphosalicylic acid, 100 g/l Sulphuric acid, 50 g/l	22–25	215–320	25–70	20–45
Duranodic[6]	4- or 5-sulphophthalic acid, 75–100 g/l Sulphuric acid, 8–10 g/l	15–30	130–370	≯70	30
Alcanadox[7]	Oxalic acid, 80 g/l to saturation	15–25	130–160	34–67	50–90

range from 250 to 1000 Am^{-2}, with or without superposed alternating current in 20–100 g/l sulphuric acid at $-4-+10°C$. Under these conditions, special attention must be paid to the contact points to the article under treatment, in order to avoid local overheating.

The films are generally dark in colour and often show a fine network of cracks due to differential expansion of oxide and metal on warming to ambient temperature. They are generally left unsealed, since sealing markedly reduces abrasion resistance, but may be impregnated with silicone oils[9] to improve the frictional properties. Applications include movable instrument parts, pump bodies and plungers, and textile bobbins.

Decorative self-coloured films[10] can also be produced in sulphuric acid under conditions intermediate between normal and hard anodising.

Continuously anodised strip and wire, which may be given a dyed finish, are produced by special methods, and are now available commercially with a film thickness up to about 6 μm. Uses include electrical windings for transformers and motors, where the light weight of aluminium and the insulating and heat-resistant properties of the film are of value, and production of small or light-section articles by stamping or roll-forming.

Mechanism of Formation of Porous Oxide Coatings

The irreversible behaviour of an aluminium electrode, which readily passes a current when cathodically polarised, but almost ceases to conduct when made the anode in certain aqueous solutions, has been known for over a century.

It has been established that in the case of electrolytes, such as boric acid or ammonium phosphate solutions, in which aluminium oxide is insoluble, this anodic passivity is due to the formation of a thin compact layer of aluminium oxide whose thickness is proportional to the applied voltage. In neutral phosphate solutions, for example, film growth practically ceases when the thickness corresponds to about 1.4 nm/V*, and a similar value has been found for many other electrolytes of this type. These thin films have a high electrical resistance, and can withstand several hundred volts under favourable conditions.

In electrolytes in which the film has a moderate solubility, film growth is possible at lower voltages, e.g. in the range 12–60 V, since the rate of formation of the oxide exceeds its rate of solution and current flow continues owing to the different structure of the oxide layer. Electron microscopy has revealed the characteristic porous structure of these films[11]. The pore diameter appears to be a function of the nature and concentration of the electrolyte and of its temperature, being greatest in a solution of high solvent activity, while the number of pores per unit area varies inversely with the formation voltage. In any given electrolyte, the lower the temperature and concentration, and the higher the voltage, the more dense will be the coating, as both the pore diameter and the number of pores per unit area are reduced under these conditions. Table 16.3, taken from a paper by Keller, Hunter and Robinson[12], illustrates these points.

Table 16.3   Number of pores in anodic oxide coatings

Electrolyte	Temp. (°C)	E.M.F. (V)	Pores/cm$^2 \times 10^{-9}$
15% sulphuric acid	10	15	77
		20	51
		30	28
3% chromic acid	49	20	22
		40	8
		60	4
2% oxalic acid	24	20	36
		40	12
		60	6

Note. Data reproduced courtesy *J. Electrochem. Soc.*, **100**, 411 (1953)

In order to account for the relatively high potential required to maintain the current it was suggested by Setoh and Miyata[13] that a thin *barrier-layer*, similar to that formed in non-solvent electrolytes, is present below the porous layer. This view has been supported by later work involving capacity and voltage-current measurements, which have allowed the thickness of the barrier-layer to be computed[14]. As in the case of electrolytes which produce barrier films, the thickness has been found to be proportional to the anodising voltage, but is lower than the limiting growth rate of 1.4 nm/V, and varies with the anodising conditions (Table 16.4).

The structure of the anodic film, according to present views, is shown diagrammatically in Fig. 16.1.

---

*The limiting thickness expressed in nm/V is of some practical value, but has little theoretical significance—at constant potential the rate of growth, although extremely small, is still finite.

The more or less regular pattern of pores imposes a cellular structure on the film, with the cells approximating in plan to hexagons, each with a central pore, while the bases which form the barrier-layer, are rounded. The metal surface underlying the film, therefore, consists of a close-packed regular array of nearly hemispherical depressions which increase in size with the anodising voltage. The thickness of the individual cell walls is approximately equal to that of the barrier-layer[12].

Table 16.4 Barrier-layer thickness in various electrolytes

Electrolyte	Temp. (°C)	Unit barrier-layer thickness (nm/V)
15% sulphuric acid	10	1·00
3% chromic acid	38	1·25
2% oxalic acid	24	1·18

Note. Data reproduced courtesy *J. Electrochem. Soc.*, **101**, 481 (1954)[14]

In view of its position in the e.m.f. series ($E°_{Al^{3+}/Al} = -1·66V$), aluminium would be expected to be rapidly attacked even by dilute solutions of relatively weak acids. In fact, the rate of chemical attack is slow, owing to the presence on the aluminium of a thin compact film of air-formed oxide. When a voltage is applied to an aluminium anode there is a sudden initial surge of current, as this film is ruptured, followed by a rapid fall to a lower, fairly steady value. It appears that this is due to the formation of a barrier-layer. Before the limiting thickness is reached, however, the solvent action of the electrolyte initiates a system of pores at weak points or discontinuities in the oxide barrier-layer.

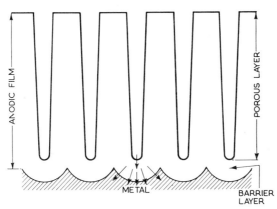

Fig. 16.1   Diagrammatic cross-section of porous anodic oxide film

The formation of pores appears to start along the sub-grain boundaries of the metal, followed by the development of additional pores within the sub-grains. Growth of oxide continues on a series of hemispherical fronts centred on the pore bases, provided that the effective barrier-layer thickness between the metal surface and the electrolyte within the pores, represented by the hemisphere radius, is less than $1 \cdot 4 \, nm/V$. As anodic oxidation proceeds at a uniform rate, a close-packed hexagonal cell-pattern is produced, the downward extension of the pore due to solution of oxide keeping pace with the downward movement of the oxide/metal interface, as shown by the arrows in Fig. 16.1. It is fairly clear that the thickness of the individual cell walls cannot exceed the thickness of the barrier-layer if columns of unchanged metal are not to be left behind in the anodic film. The inverse relationship between number of pores and anodising voltage also implies that cells with much thinner walls cannot be formed. Growth of pores in excess of the limiting number appears to be inhibited at an early stage of development, but the actual mechanism is still in doubt.

Radiochemical studies[15] indicate that the pore base is the actual site of formation of aluminium oxide, presumably by transport of aluminium ions across the barrier-layer, although transport of oxygen ions in the opposite direction has been postulated by some authorities[1]. The downward extension of the pore takes place by chemical solution, which may be enhanced by the heating effect of the current and the greater solution rate of the freshly formed oxide, but will also be limited by diffusion. It has been shown that the freshly formed oxide, γ'-Al_2O_3, is amorphous and becomes slowly converted into a more nearly crystalline modification of γ-Al_2O_3[16].

Prolonged action of the acid electrolyte on thick films may cause the pores to become conical in section, widening towards the upper surface of the film. This will impose an upper limit on film thickness in solvent electrolytes, as found in practice.

Although it might seem at first sight that dyestuffs are merely held mechanically within the pores, and this view is probably correct in the case of inorganic pigments, there is some support for the opinion that only those dyestuffs which form aluminium/metal complexes produce really light-fast colorations.

The effect of hot water sealing is to convert anhydrous γ-Al_2O_3 into the crystalline monohydrate, $Al_2O_3 . H_2O$, which occupies a greater volume and blocks up the pores, thus preventing further absorption of dyes or contaminants. The monohydrate is also less reactive.

Properties of Coatings

Composition   The main constituent of the film is aluminium oxide, in a form which varies in constitution between amorphous Al_2O_3 and γ-Al_2O_3, together with some monohydrate, $Al_2O_3 . H_2O$. In the presence of moisture, both the anhydrous forms are gradually transformed into the monohydrate, and the water content of as-formed films is, therefore, somewhat variable.

After sealing in boiling water, the composition of the completely hydrated film obtained when using sulphuric acid approximates to:

Al_2O_3	70
H_2O	17
SO_3	13

It is probable that the SO_3 is combined with the aluminium as a basic sulphate.

Films produced in oxalic acid contain smaller amounts (about 3%) of the electrolyte and only traces of chromium are found in chromic acid films. Sealed films show the electron diffraction pattern of the monohydrate, böhmite.

Density   Owing to the variable degree of porosity of the anodic film, it is only possible to determine the apparent density, which varies with the anodising conditions and also with the film thickness.

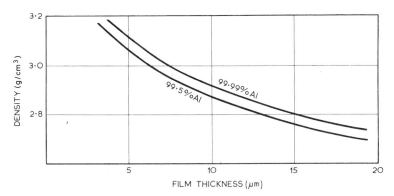

Fig. 16.2   Apparent density of anodic film as a function of film thickness (courtesy *Aluminium*, Berl., **32**, 126 (1938))

Fig. 16.2, taken from a paper by Lenz[17], shows the variation in density with thickness for steam-sealed anodic films produced in sulphuric acid on aluminium of 99·99% and 99·5% purity. A mean figure of 2·7 g/cm^3 for sealed, and 2·5 g/cm^3 for unsealed films is accepted by the British Standard for anodised aluminium[18].

Hardness   It is not possible to obtain a reliable figure for the hardness of anodic coatings with either the indentation or scratch methods, because of the influence of the relatively soft metal beneath the anodic film, and the presence of a soft outer layer on thick films. On Moh's Scale, the hardness of normal anodic films lies between 7 and 8, i.e. between quartz and topaz.

Methods are available for the determination of relative abrasion resistance using either a mixed jet of air and abrasive, as recommended in the appropriate British Standard[18] or an abrasive wheel or disc. Owing to variations in the quality of the abrasive, and the performance of individual jets, a standard comparison sample is included in each batch.

The hardness of the film is markedly affected by the conditions of anodising. By means of special methods involving dilute electrolytes at low temperatures and relatively high voltages[8], with or without superimposed alternating current, it is possible to produce compact abrasion-resistant films with thicknesses of 50–7ʋ μm and hardnesses of 200–500 VPN, for special applications.

Flexibility   The normal anodic film begins to crack if subjected to an extension exceeding about 0·5%. Thinner films up to 5 μm in thickness appear to withstand a greater degree of deformation without obvious failure, and are often used for dyed coatings on continuously anodised strip from which small items may be punched or stamped. Continuously anodised wire can be bent round a radius of 10–15 times its diameter without visible crazing. A greater degree of flexibility is also shown by the more porous coatings produced in 20–25% v.v. sulphuric acid at 35–40°C, while hard films are much less flexible. Unsealed films are only slightly more flexible than films sealed in water or dichromate solution.

Breakdown voltage The breakdown voltage of an anodic film varies with the method of measurement and conditions of anodising, and shows fluctuations over the surface. In the case of unsealed films, breakdown voltage also depends on the relative humidity at the time of measurement. It is normally measured by applying a slowly increasing alternating voltage between a loaded hemispherical probe on the upper surface of the film, and the underlying metal, contact to which may be established by removing a portion of the film[18]. The breakdown voltage/thickness relationship for sealed films up to about 20 μm is approximately linear, and the slope of the curve for sulphuric acid films varies from 30 to 40 V/μm. These results were obtained with a relatively high loading on the probe*; with reduced load (approx. 60 g and below on a hemispherical probe of 1·6 mm radius) values of 60–100 V/μm can be reached. The higher figures probably represent limiting values which will apply to the conditions between adjacent laps or turns on coils wound from anodised strip or wire.

Resistance The specific resistance of the dry anodic film is

$$4 \times 10^{15} \, \Omega \, \text{cm at } 20°\text{C}[19]$$

Dielectric constant The dielectric constant of anodic oxide films has been found to be 5·0–5·9 for sulphuric films, and 7–8 for oxalic films. A mean value of 7·45 has been quoted for barrier-layer films[1], but more recent work favours a value of 8·7[20].

Thermal expansion The thermal expansion of the film is only about one-fifth that of aluminium[1], and cracking or crazing is observed when anodised aluminium is heated above 80°C. The fine hair-cracks produced do not seem to impair the protective properties of the coating if anodising conditions have been correct.

Heat conduction The heat conductivity of the film is approximately one-tenth that of aluminium[2].

Heat resistance Apart from hair-cracks, little change is observable in the anodic film on heating up to 300–350°C, although some dyed finishes may change colour at 200–250°C, but at higher temperatures up to the melting point of the metal, films may become opaque or change colour, owing to loss of combined water, without losing their adhesion.

Emissivity Table 16.5 shows the total heat emissivity of various aluminium surfaces, as a percentage of that of a black body. The figures have been recalculated from the data of Hase[21]. The emissivity of anodised aluminium rises rapidly with film thickness up to 3 μm after which the rate of increase diminishes.

---

*Several hundred grams, BS 1615 suggests 50–75 g.

Table 16.5 Relative heat emissivity of various
aluminium surfaces

Surface	Heat emissivity (%)
Highly polished	4·3–6·4
Etched	6·4–8·5
Bright roll finish	5·3–7·4
Matt roll finish	8·5–16
Aluminium paint	17–32
Diecast	16–26
Sandcast	26–36
Anodised, according to film thickness	38–92
Black body	100

Heat reflectivity   The heat reflectivity of as-rolled aluminium is about 95%, but this high value may not be maintained for long in a corrosive atmosphere, although it is less affected by surface finish than is optical reflectivity. Anodising reduces the heat reflectivity, owing to absorption by the oxide layer; this effect increases with film thickness. There is a deep absorption trough in the region corresponding to a wavelength of 3 µm; this is probably due to the —OH grouping in the hydrate, the effects of which may be minimised by sealing the heated film in oil instead of water[22]. This treatment is particularly valuable for heat reflectors in apparatus using sources running at 900–1 000°C, which show a peak emission in the 2–3 µm region. Fig. 16.3

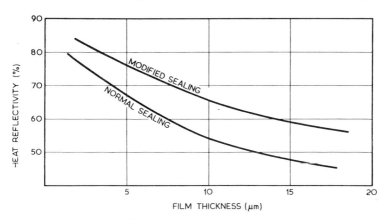

Fig. 16.3   Heat reflectivity of anodised aluminium

shows the heat reflectivity of anodised super-purity aluminium for a source of this type[23], plotted against film thickness. The benefits of the modified sealing treatment are obvious.

Refractive index   The refractive index of the clear anodic film produced on aluminium of the highest purity in sulphuric acid is 1·59 in the as-formed condition, rising to 1·62 after sealing[24].

Reflectivity   The total and specular reflectivities of an anodised aluminium surface are controlled by both the condition of the metal surface, polished or matt, and the absorption or light-scattering properties of the oxide layer. Total reflectivity may be defined as the percentage of the incident light reflected at all angles, while specular reflectivity is that percentage reflected within a relatively narrow cone with its axis along the angle of reflection. For many years the standard instrument for measuring specular reflectivity has been that designed by Guild[25], but more recently a modified gloss head giving rather greater discrimination has been described by Scott[26]. Other instruments, while placing a number of surfaces of varying specularity in the same relative order, may give different values for the specular reflectivity.

The general brightness of a surface is chiefly dependent upon the total reflectivity T, while specular reflectivity S controls the character of the reflected image. In assessing the subjective brightness of a surface the eye tends to be influenced more by the S/T ratio or image clarity than by the total reflectivity.

For a high degree of specularity, the metal surface must be given a high polish by mechanical means; this may be followed (or replaced) by electrochemical or chemical brightening. When such a brightened surface is protected by anodising, however, insoluble impurities (mainly iron and silicon) present in the aluminium will be incorporated in the anodic film and will increase its tendency to absorb or scatter light. Only metal of the highest purity, 99·99% Al, produces a fully transparent oxide film, while lower purities show decreased total reflectivities and S/T ratios after anodising because of the increased opacity of the anodic film.

Table 16.6   Effect of metal purity and anodic film thickness on reflectivity

Metal purity (%)	Film thickness (µm)					
	2		5		10	
	T	S/T	T	S/T	T	S/T
99·5	80	0·84	79	0·83	77	0·78
99·8	82	0·95	83	0·95	—	—
99·99 (super purity)	84	0·99	84	0·99	84	0·99
Super purity + 0·5 Mg	84	0·98	84	0·98	83	0·97
Super purity + 1·25 Mg	83	0·99	83	0·99	82	0·99
Super purity + 0·7 Mg, 0·3 Si, 0·25 Cu	82	0·99	79	0·98	—	—

Notes. 1. T = total, S = specular reflectivity.
2. Data reproduced courtesy *Met. Rev.*, **2** No. 8 (1957)[23].

Table 16.6, taken from a monograph by Pearson and Phillips[23], demonstrates these effects. The figures were obtained using the Guild meter on electrobrightened and anodised metal.

Effect of anodising on mechanical properties   The tensile strength of thin sections may be somewhat reduced by anodising, owing to the brittleness of the coating, but this effect is normally very slight. Thin sheet, less than about 0·6 mm with a relatively thick anodic coating, also has a tendency to break more easily on bending. The incompressibility of the anodic film on the inside of the bend probably enhances this effect, which is also seen on anodised wire.

Anodising should be used with caution on components likely to encounter high stresses, owing to the deterioration in fatigue properties liable to result under these conditions, but under light loading and with the thinner coatings, the reduction is negligible. In some cases[27], an actual improvement has been reported.

Friction The coefficient of friction of the sealed anodic film is 0·76, falling to 0·19 after impregnation with silicone oil[28]. These results were obtained with anodised wire.

Measurement of film thickness The thickness of an anodic film may be determined by a variety of non-destructive methods. Some of these are capable of a high degree of precision, while simpler methods are available for rough sorting. A number of instruments employing the eddy-current principle, with which, after prior calibration, a rapid estimate of film thickness may be made, are now available. With the best instruments, an accuracy of ± 1 μm can be obtained. For approximate determinations of thickness, the breakdown voltage of the film may be measured. Breakdown voltage shows wide variations with anodising conditions and metal or alloy composition. A separate calibration curve is, therefore, needed for each treatment. Accuracy is comparatively low, rarely being greater than $\pm 2\%$ of the total film thickness.

For control or calibration purposes, film thickness can be determined by mounting a sectioned specimen and measuring the oxide film thickness directly on the screen of a projection microscope at a known magnification. Alternatively, the loss in weight[18] of an anodised sample of known area may be found after the film has been stripped in a boiling solution made up as follows:

> Phosphoric acid (s.g. 1·75)   3·5% v/v
> Chromic acid in distilled water   2·0% w/v

Immersion for 10 min is usually sufficient to remove the film without the metal being attacked.

Corrosion Resistance

Since the natural passivity of aluminium is due to the thin film of oxide formed by the action of the atmosphere, it is not unexpected that the thicker films formed by anodic oxidation afford considerable protection against corrosive influences, provided the oxide layer is continuous, and free from macropores. The protective action of the film is considerably enhanced by effective sealing, which plugs the mouths of the micropores formed in the normal course of anodising with hydrated oxide, and still further improvement may be afforded by the incorporation of corrosion inhibitors, such as dichromates, in the sealing solution. Chromic acid films, in spite of their thinness, show good corrosion resistance.

The protective action of sulphuric films is mainly controlled by the anodising conditions, compact films formed at temperatures below 20°C in 7% v/v sulphuric acid being more resistant than the films formed at higher temperatures in more concentrated acid. The wider pores of the latter result in less protection but these films are more readily dyed. Greater protection is also given by thicker films, and a thickness of about 25 μm is generally considered adequate for architectural work in a normal urban environment. In a heavily polluted industrial area, even thicker films may be desirable, while in rural areas some reduction would be permissible. Bright anodised motorcar trim is generally given a film thickness of about 7 μm.

Alumina monohydrate in the mass is very unreactive, being rapidly attacked only by hot sulphuric acid or caustic soda solutions, and the anodic coating shows similar characteristics to some degree. The presence in the film of macropores due to localised impurities or imperfections in the metal and overlying oxide can bring about rapid penetration, owing to the concentration of attack at the few vulnerable points. Metal of good quality specially produced for anodising should therefore be used in order to ensure that such weak points are absent. For vessels and tanks for holding liquids, it may be preferable to use unanodised aluminium, and to accept generalised corrosive attack rather than run the risk of perforation, which may occur with anodised metal.

For ordinary atmospheric exposure, it is usually possible to arrange that thin spots of the film, such as the contact points of the anodising jigs, are located in relatively unimportant positions on the article and are hidden from view.

Since the corrosion resistance of anodic films on aluminium is markedly dependent on the efficacy of sealing (provided the film thickness is adequate for the service conditions), tests for sealing quality are frequently employed as an index of potential resistance to corrosion. While it is admitted that an unequivocal evaluation of corrosion behaviour can only be obtained by protracted field tests in service, accelerated corrosion tests under closely controlled conditions can also provide useful information in a shorter time within the limitations of the particular test environment employed.

Tests for sealing include dye staining tests such as that specified in BS 1615: 1972, Method *F*, involving preliminary attack with acid, followed by treatment with dye solution. Nitric acid[29] or a sulphuric acid/fluoride mixture may be used for the initial attack, and a rapid spot test[30] has been developed using the acid/fluoride mixture, followed by a solution of 10 g/l Aluminium Fast Red *B3LW*. Poor sealing is revealed by a deep pink to red spot, while good sealing gives nearly colourless to pale pink colorations. The test can be applied to architectural or other material on site.

Physical tests of film impedance[31] using an a.c. bridge have also been recommended, although the correlation with corrosion resistance is necessarily empirical. Film impedance increases at an approximately linear rate with sealing time and film thickness.

Exposure of the samples to a controlled moist atmosphere containing sulphur dioxide, as recommended in BS 1615:1972, Method *H*, is an example of a test bridging the gap between sealing tests and accelerated corrosion tests. After exposure for 24 h at $25 \pm 2°C$, poorly sealed films show a persistent heavy white bloom, while good sealing produces at the most a slight superficial bloom.

A rapid immersion test in a hot aqueous solution containing sulphur dioxide has also been developed by Kape[32] and is specified in BS 1615, Method *E*. Results are similar to those obtained in the preceding test,

Method *H*. The method can also be made quantitative by measuring the weight loss.

The accelerated corrosion test in most general use is the CASS test[33] in which the articles are sprayed intermittently with a solution made up as follows:

NaCl	50 g/l
$CuCl_2,2H_2O$	0·26 g/l
Acetic acid to pH 2·8–3·0	

The specimens are clamped at an angle of 15° to the vertical in a baffled enclosure maintained at 50°C, and the exposure time is 24–96 h. Corrosive attack of inadequately sealed or thin films is shown by pitting.

An interesting derivative of the CASS test, known as the *Ford Anodised Aluminium Corrosion Test* (FACT)[34] has been developed in the U.S.A. This makes use of a controlled electrolytic attack using the CASS solution. The electrolyte is contained in a glass test cell and clamped against the anodised surface with a Neoprene sealing gasket. A d.c. voltage of 200 V in series with a high resistance is maintained between an anode of platinum wire and the aluminium test piece as cathode. The integrated fall in potential across the cell over a fixed period of 3 min as corrosion proceeds and an increasing current flows, is taken as a measure of the corrosion resistance. A British version of this test using simplified circuitry for the integration is available commercially as the *Anodisation Comparator**. Remarkably good correlation has been obtained between the readings of this instrument and the amount of pitting after exposure at a number of outdoor sites[35]. Comprehensive reviews of sealing techniques including test methods and corrosion behaviour have been published by Thomas[36] and Wood[37].

The behaviour of samples under the actual conditions of service is the final criterion, but unfortunately such observations take a long time to collect and assess, and the cautious extrapolation of data from accelerated tests must be relied on for forecasting the behaviour of anodised aluminium in any new environment.

Atmospheric Exposure

Table 16.7 shows the effects of thin anodic oxide films on the resistance to industrial and synthetic marine atmospheres (intermittent salt spray) of three grades of pure aluminium. The results are taken from a paper by Champion and Spillett[38] and show how relatively thin films produce a marked improvement in both environments.

In an industrial atmosphere, an anodic film only 6·5 μm thick provides a two-fold increase in life over unprotected metal, and the effect under salt-spray conditions is even greater. It is interesting to note that both the industrial atmosphere and salt-spray results show parallel trends.

A similar improvement in expectation of life for thin anodic coatings has been reported by Phillips[39] for 99·5% Al, and for alloys of the following

*SIBA Ltd., Camberley, Surrey.

compositions: Al–1·25 Mn; Al–2 Mg–1 Mn; Al–1 Mg–1 Si. The results for a high-copper alloy were less good.

An interesting paper by Lattey and Neunzig[40] shows that the better the surface finish of the aluminium the thinner the coating required for protection. Neunzig[41] has also studied the effect of the hair-cracks produced by heating or bending on corrosion resistance. Although pitting was initiated by such cracks in thin films (5 μm), serious pitting in thicker films (15 μm) was observed only if anodising had been carried out at 25°C; films produced at 16–17°C were more resistant to corrosive attack. This re-emphasises the importance of maintaining correct anodising conditions for maximum corrosion resistance.

More recently, results of exposure tests for 10 years in a severe industrial environment at Stratford, London, have been reported by the Fulmer Research Institute[42]. A range of pure and alloy specimens, anodised to a maximum film thickness of about 25 μm, was exposed at an angle of 45°.

Table 16.7 Corrosion tests on unprotected and anodised pure aluminium

	Corrosive effect	Grade 1B (99·5%) Film thickness (μm)		Grade 1A (99·8%) Film thickness (μm)			Super purity (99·99%) Film thickness (μm)	
		0	6·5	0	4	6·5	0	4
Industrial atmosphere (7 years exposure)	Appearance* (life in years)	2·5	5	2·5	5	5	3·5	6
	Mechanical properties† (life in years)	2·75	5·5	3	4·5	6	3	5
	Pitting‡ (depth in mm)	0·18	0·20	0·18	0·25	0·25	0·20	0·13
Marine atmosphere (11 years exposure)	Appearance* (life in years)	<1	4	1	4	5	3	4
	Mechanical properties† (life in years)	5	>11	8	7	>11	>11	>11
	Pitting‡ (depth in mm)	0·30	0·18	0·15	0·33	0·15	0·15	0·08

*No. of years to deterioration of surface appearance to a fixed arbitrary level.
†No. of years to deterioration of mechanical properties to a fixed arbitrary level.
‡Mean depth of pitting obtained statistically.

Corrosion was assessed visually, by determination of weight loss after cleaning, and by reflectivity measurements. All specimens showed signs of pitting, and there was a considerable loss of reflectivity, the under surface being more affected than the upper. A striking feature of the results was the accelerating rate of deterioration in the last five years of exposure. Although none of the samples was completely protected, results were better for the purer specimens and the thicker films.

Maintenance

In architectural work, particular care must be taken to avoid destructive attack of the anodic film by alkaline mortar or cement during erection, and temporary coatings of spirit-soluble waxes, or acetate-butyrate lacquers are frequently applied to window frames and the like to protect against mortar splashes, which in any event should be removed at the earliest possible moment.

The resistance of properly anodised aluminium exposed to the weather can be considerably enhanced by correct and regular cleaning. Deposits of soot and dirt should be removed by washing with warm water containing a non-aggressive detergent; abrasives should not be used. For window frames this washing may conveniently be carried out when the glass is cleaned in the normal way. In such circumstances the life of the coating may be prolonged almost indefinitely, as exemplified by the good condition of the chromic-anodised window frames of Cambridge University Library which were installed in 1933, and of the sulphuric-anodised window frames of the New Bodleian Library, Oxford University, installed in 1938.

B. A. SCOTT

REFERENCES

1. Young, L., *Anodic Oxide Films*, Academic Press, New York (1961)
2. Schenk, M., *Werkstoff Aluminium und seine Anodische Oxydation*, Francke, Berne (1948)
3. Wernick, S. and Pinner, R., *The Surface Treatment and Finishing of Aluminium and its Alloys*, Robert Draper, Teddington, 3rd edn (1964)
4. *Processes for the Anodic Oxidation of Aluminium and Aluminium Alloy Parts*, DTD. 910C, H.M.S.O., London (1951)
5. Kaiser Aluminum Co., US Pat. 3 031 387 (7.12.59)
6. Alcoa, US Pat. 3 227 639 (24.10.61)
7. Aluminium Laboratories Ltd., UK Pat. 970 500 (29.3.62)
8. Campbell, W. J., Conference on Anodising Aluminium, A.D.A., Nottingham, Paper 11, Sept. (1961); Csokan, P., *Metalloberfläche*, **19** No. 8, 252 (1965) and *Trans. Inst. Met. Fin.*, **41**, 51 (1964)
9. Tsuji, Y., *Trans. Inst. Met. Fin.*, **40**, 225 (1963)
10. Scott, B. A., *Trans. Inst. Met. Fin.*, **43**, 1 (1965)
11. Edwards, J. D. and Keller, F., *Trans. Amer. Inst. Min. (Metall.) Engrs.*, **156**, 288 (1944)
12. Keller, F., Hunter, M. S. and Robinson, D. L., *J. Electrochem. Soc.*, **100**, 411 (1953)
13. Setoh, S. and Miyata, A., *Sci. Pap. Inst. Phys. Chem. Res. Tokyo*, **17**, 189 (1932)
14. Hunter, M. S. and Fowle, P., *J. Electrochem. Soc.*, **101**, 481 (1954)
15. Lewis, J. E. and Plumb, R. C., *J. Electrochem. Soc.*, **105**, 496 (1958)
16. Verwey, E. J. W., *Z. Kristallogr.*, **91**, 65 (1935)
17. Lenz, D., *Aluminium, Berl.*, **32**, 126 (1956)
18. *Anodic Oxidation Coatings on Aluminium*, British Standard 1615:1972
19. Franckenstein, G., *Ann. Phys.*, **26**, 17 (1936)
20. van Geel, W. Ch. and Schelen, B. J. J., *Philips Res. Rep.*, **12**, 240 (1957)
21. Hase, R., *Aluminium, Berl.*, **24**, 140 (1942)
22. Gwyer, A. G. C. and Pullen, N. D., *Metallurgia, Manchr.*, **21**, 57 (1939)
23. Pearson, T. G. and Phillips, H. W. L., *Metallurg. Rev.*, **2** No. 8, 348 (1957)
24. Edwards, J. D., *Mon. Rev. Amer. Electropl. Soc.*, **26**, 513 (1939)
25. Guild, J., *J. Sci. Inst.*, **17**, 178 (1940)
26. Scott, B. A., *J. Sci. Inst.*, **37**, 435 (1960)

27. Stickley, G. W. and Howell, F. M., *Proc. Amer. Soc. Test. Mat.*, **50**, 735 (1950)
28. Vevers, H. H., Conference on Anodising Aluminium, A.D.A., Nottingham, Discussion on Section 4, Sept. (1961)
29. Neunzig, H. and Rohrig, V., *Aluminium*, **38** No. 3, 150 (1962); Sacchi, F. and Paolini, G., *Aluminio*, **6**, 9 (1961)
30. Scott, B. A., *Electroplating and Metal Finishing*, Feb. (1965)
31. Wood, G. C., *Trans. Inst. Met. Fin.*, **41**, 99 (1964)
32. Kape, J. M., *Metal Industry*, **95** No. 6, 115 (1959)
33. ASTM Method B368
34. *Quality Laboratory and Chem. Eng. Physical Methods*, MA–P, BQ7–1, Ford (USA), Feb. (1961)
35. Carter, V. E. and Edwards, J., *Trans. Inst. Met. Fin.*, **43**, 97 (1965) and Carter, V. E., *Ibid.*, **45**, 64 (1967)
36. Thomas, R. W., Symposium on Protecting Aluminium, Aluminium Federation, London (1970)
37. Wood, G. C., *Trans. Inst. Met. Fin.*, **36**, 220 (1959)
38. Champion, F. A. and Spillett, E. E., *Sheet Metal Ind.*, **33**, 25 (1956)
39. Phillips, H. W. L., *Institute of Metals Monograph*, No. 13 (1952)
40. Lattey, R. and Neunzig, H., *Aluminium, Berl.*, **32**, 252 (1956)
41. Neunzig, H., *Aluminium, Berl.*, **34**, 390 (1958)
42. Liddiard, E. A. G., Sandersen, G. and Penn, J. E., Annual Technical Conference, Institute of Metal Finishing, Brighton, 28th May (1971)

16.2 Phosphate Coatings

Introduction

The use of phosphate coatings for protecting steel surfaces has been known for over 50 years, and during this period commercial utilisation has steadily increased until today the greater part of the world production of motorcars, bicycles, refrigerators, washing machines, office furniture, etc. is treated in this way. By far the greatest use of phosphate coatings is as a base for paint, although other important applications are in conjunction with oil, grease, wax and spirit stains to provide a corrosion-resistant finish, with soaps to assist the drawing and pressing of steel, and with lubricating oil to decrease the wear and fretting of sliding parts such as piston rings, tappets and gears.

Applications

Phosphate treatments are readily adaptable to production requirements for articles of all sizes, and for large or small numbers. Economical processing can be achieved, for example, by treating thirty car bodies per hour in a conveyorised spray or immersion plant, or by immersion treatment of small clips and brackets. Mild steel sheet is the material most frequently subjected to phosphate treatment, but a great variety of other ferrous surfaces is also processed. Examples include cast-iron plates and piston rings, alloy steel gears, high-carbon steel cutting tools, case-hardened components, steel springs and wire, powdered iron bushes and gears, etc. Phosphate treatments designed for steel can also be used for the simultaneous treatment of zinc dic-castings, hot-dipped zinc, zinc-plated and cadmium-plated articles, but if there is a large quantity of these non-ferrous articles it is more economical to phosphate them without the steel.

Phosphate solutions containing fluorides are used for processing steel, zinc and aluminium when assembled together, but chromate solutions are generally preferred when aluminium is treated alone. The increasing use of electrophoretic painting has not led to the adoption of fundamentally changed processes, but rather to closer control to obtain thinner and smoother phosphate layers.

Methods

The usual method of applying phosphate coatings is by immersion, using a sequence of tanks which includes degreasing and phosphating stages, with their respective rinses. The treatment time ranges from 3 to 5 min for thin zinc phosphate coatings up to 30 to 60 min for thick zinc, iron, or manganese phosphate coatings. The accelerated zinc phosphate processes lend themselves to application by power spray, and the processing time may then be reduced to 1 min or less. Power spray application is particularly advantageous for mass production articles such as motorcars and refrigerators, as the conveyor can run straight through the spray tunnel, which incorporates degreasing, rinsing, phosphating, rinsing and drying stages.

Flow-coating and hand spray-gun application is sometimes employed where a relatively small number of large articles has to be phosphated.

Mechanism of Phosphate Coating Formation

All conventional phosphate coating processes are based on dilute phosphoric acid solutions of iron, manganese and zinc primary phosphates either separately or in combination. The free phosphoric acid in these solutions reacts with the iron surface undergoing treatment in the following manner[1]:

$$Fe + 2H_3PO_4 \rightarrow Fe(H_2PO_4)_2 + H_2 \qquad \qquad ...(16.1)$$

thus producing soluble primary ferrous phosphate and liberating hydrogen. Local depletion of phosphoric acid occurs at the metal/solution interface. As the primary phosphates of iron, manganese and zinc dissociate readily in aqueous solution, the following reactions take place:

$$Me(H_2PO_4)_2 \rightleftharpoons MeHPO_4 + H_3PO_4 \qquad \qquad ...(16.2)$$
$$3MeHPO_4 \rightleftharpoons Me_3(PO_4)_2 + H_3PO_4 \qquad \qquad ...(16.3)$$
$$3Me(H_2PO_4)_2 \rightleftharpoons Me_3(PO_4)_2 + 4H_3PO_4 \qquad \qquad ...(16.4)$$

The neutralisation of free phosphoric acid by reaction 16.1 alters the position of equilibrium of equations 16.2, 16.3 and 16.4 towards the right and thereby leads to the deposition of the sparingly soluble secondary phosphates and insoluble tertiary phosphates on the metal surface.

As reaction 16.1 takes place even when the phosphating solution contains zinc or manganese phosphate with little or no dissolved iron, it will be seen that the simple or 'unaccelerated' phosphate treatment gives coatings which always contain ferrous phosphate derived from the steel parts being processed. After prolonged use, a manganese phosphate bath often contains more iron in solution than manganese and produces coatings with an iron content two or three times that of manganese.

The relation between free phosphoric acid content and total phosphate content in a processing bath, whether based on iron, manganese or zinc, is very important; this relation is generally referred to as the *acid ratio*. An excess of free acid will retard the dissociation of the primary and secondary phosphates and hinder the deposition of the tertiary phosphate coating; sometimes excessive loss of metal takes place and the coating is loose and powdery. When the free acid content is too low, dissociation of phosphates (equations 16.2, 16.3 and 16.4) takes place in the solution as well as at the metal/solution interface and leads to precipitation of insoluble phosphates as sludge. The free acid content is usually determined by titrating with sodium hydroxide to methyl orange end point, and the total phosphate by titration with sodium hydroxide to phenolphthalein end point. Using this test, non-accelerated processes operated near boiling generally work best with a free-acid titration between 12·5 and 15% of the total acid titration.

A zinc phosphate solution tends to produce coatings more quickly than iron or manganese phosphate solutions, and dissociation of primary zinc phosphate proceeds rapidly through reaction 16.2 to 16.3 or directly to tertiary zinc phosphate via reaction 16.4. Even so, a processing time of 30 min is usual with the solution near boiling.

Another factor in the initiation of phosphate coating reaction is the presence in the processing solution of tertiary phosphate, either as a colloidal suspension or as fine particles[2]. This effect is most apparent in zinc phosphate solutions, which produce good coatings only when turbid. The tertiary zinc phosphate particles can be present to a greater extent in cold processing solutions and act as nuclei for the growth of many small crystals on the metal surface, thereby promoting the formation of smoother coatings.

Similarly, the ferric phosphate sludge formed during the processing of steel in a zinc phosphate solution can play a useful part in coating formation[3]. The solubility of ferric phosphate is greater at room temperature than at elevated temperatures, and is increased by the presence of nitrate accelerators. To allow for saturation at all temperatures it is desirable always to retain some sludge in the processing bath. Coatings with optimum corrosion resistance are produced when the temperature of the bath is rising and causing super-saturation of ferric phosphate.

With zinc/iron/phosphate/nitrate baths the iron content of the coating comes predominantly from the processing solution and very little from the surface being treated[4]. This greatly diminished attack on the metal surface by accelerated baths has a slight disadvantage in practice in that rust is not removed, whereas the vigorous reaction of the non-accelerated processes does remove light rust deposits.

The solution of iron represented in equation 16.1 takes place at local anodes of the steel being processed, while discharge of hydrogen ions with simultaneous dissociation and deposition of the metal phosphate takes place at the local cathodes[1]. Thus factors which favour the cathode process will accelerate coating formation and conversely factors favouring the dissolution of iron will hinder the process.

Cathodic treatment in a phosphating solution exerts an accelerating action as the reaction at all cathodic areas is assisted and the formation of a phosphate layer is speeded accordingly. Conversely, anodic treatment favours only the solution of iron at local anodes and hinders phosphate coating formation. An oxidising agent acts as an accelerator by depolarisation of the cathodes, raising the density of local currents so that rapid anodic passivation of active iron in the pores takes place. This inactivation of local anodes favours the progression of the cathodic process. The accelerating effect of alternating current is explained by the practical observation that the cathodic impulse acting protectively greatly exceeds in its effect the anodic impulse which dissolves iron. In a similar manner the electrolytic pickling of iron with alternating current can dissolve iron at a slower rate than when no current is used.

Reducing agents have the same ultimate effect as cathodic depolarisation in that they convert anodic regions to cathodic and increase the ratio of cathodic to anodic areas.

Nitrogenous organic components such as toluidine, quinoline, aniline, etc. all act as inhibitors to the anodic reaction between metal and acid and thereby favour the cathodic reaction and accelerate the process.

Accelerators

The majority of phosphate processes in use today are 'accelerated' to obtain shorter treatment times and lower processing temperatures. The most common mode of acceleration is by the addition of oxidising agents such as nitrate, nitrite, chlorate and hydrogen peroxide. By this means, a processing time of 1 to 5 min can be obtained at temperatures of 43–71°C. The resultant coatings are much smoother and thinner than those from unaccelerated processes, and, while the corrosion resistance is lower, they cause less reduction of paint gloss and are more suited to mass-production requirements.

Table 16.8   Amount and composition of the gases evolved on phosphating of 1 m² of sheet metal for deep drawing

Phosphate solution	Pointage*	Time (min)	Amount of gas (cm³/m²)	H₂ (%)	N₂ (%)	O₂ (%)	N₂O+NO (%)
Manganese phosphate	30	60	7 000	87·5	11·4†	1·1	—
Zinc phosphate	40	30	2 540	92·7	6·4†	0·9	—
Manganese phosphate (accelerated with nitrate)	30	15	3 500	84·6	9·1	1·3	5·0
Zinc phosphate (accelerated with nitrate)	70	5	78	16·7	75·3	8·0	—
Zinc phosphate containing 1·5–2 g/l iron (accelerated with nitrate)	70	5	85	32·1	57·0	1·6	9·3

*A measure of the total acidity of a phosphating solution, as indicated by the number of ml of 0·1N sodium hydroxide (4·0 g/l) needed to neutralise 10 ml of the phosphating solution to phenolphthalein. †Presumably from nitrides present in the steel.

The presence of nitrate as accelerator has a pronounced effect on the amount and composition of gas evolved from the work being treated[1] (Table 16.8). It will be observed that hydrogen evolution drops to a very low figure with the zinc/nitrate baths. The formation of nitrite arises from decomposition of nitrate by reaction with primary ferrous phosphate to form ferric phosphate:

$$2Fe^{2+} + NO_3^- + 3H^+ \rightarrow 2Fe^{3+} + HNO_2 + H_2O$$

In an acid solution sodium nitrite acts as a strong oxidising agent by the following reaction:

$$2NaNO_2 + 2H_3PO_4 \rightarrow 2NaH_2PO_4 + H_2O + N_2O + 2(O)$$

A slight degree of acceleration can be obtained by introducing traces of metals which are more noble than iron, for example nickel, copper, cobalt, silver and mercury. These metals are deposited electrochemically over the iron surface undergoing treatment, thereby providing more active cathodic centres and promoting phosphate deposition. This method of acceleration has the disadvantage of leaving minute particles of the noble metal in the coating, and, in the case of copper, this can seriously inhibit the drying of some types of paint coatings. Copper also forms local cells with the iron and so reduces corrosion resistance.

Acceleration by addition of reducing agents, organic compounds, or by application of a cathodic or alternating current, is not nowadays used to any great extent.

Nature of Coatings

Effect of Metal Surface

The state of the metal surface has a pronounced effect on the texture and nature of phosphate coating produced by orthodox processes. Heavily worked surfaces tend to be less reactive and lead to patchy coatings. Grit blasting greatly simplifies treatment and gives uniform phosphate coatings. Accidental contamination of sheet steel with lead has been shown to have an adverse effect on the corrosion resistance and durability of phosphate coatings and paint[5].

Cleaning operations which make use of strong acids or strong alkalis tend to lead to the formation of excessively large phosphate crystals which do not completely cover the metal surface and therefore show inferior corrosion resistance; this is particularly serious if rinsing is inadequate between the preparatory treatment and the phosphating. Adherent dust particles can also lead to the formation of relatively large phosphate crystals, and surfaces which have been wiped beforehand show much smoother and more uniform phosphate coatings.

On the other hand, the provision of vast numbers of minute nuclei assists the phosphate coating reaction to start at a multitude of centres, resulting in a finely crystalline coating. This effect can be obtained chemically by a pre-dip in a solution of sodium phosphate containing minutely dispersed traces of titanium or zirconium salts[6] or in weak solution of oxalic acid. This type of pre-dip entirely eliminates any coarsening effect due to previous treatment in strong alkalis or acids.

Effect of Phosphate Solution

Improved nucleation within the phosphate solution itself can produce smoother coatings without the necessity of recourse to preliminary chemical treatment. This may be accomplished by introducing into the phosphating bath the sparingly soluble phosphates of the alkaline earth metals or condensed phosphates such as sodium hexametaphosphate or sodium tripolyphosphate. Such modified phosphating baths produce smoother coatings than orthodox baths and are very much less sensitive to cleaning procedures.

Very thin coatings of 'iron phosphate' can be produced by treatment with solutions of alkali metal phosphate. These serve a useful purpose for the treatment of office furniture, toys, etc. where a high degree of protection is not required, and also as a base for phenolic varnishes, or resin varnishes requiring stoving at over 204°C. The coating is of heterogeneous nature and contains less than 35% iron phosphate ($FePO_4 \cdot 2H_2O$) with the remainder probably γFe_2O_3[7].

Thin phosphate coatings can be formed by application of phosphoric acid solution alone, i.e. not containing metallic phosphates, to a steel surface, sufficient time being allowed after application to enable complete reaction to take place. In this way a thin film of iron phosphate can be formed. In practice it is difficult to obtain complete conversion and the remaining traces of phosphoric acid can cause blistering of paint coatings. This effect may be insignificant on rough, absorbent steel surfaces, e.g. ship's plating, where heavy coats of absorbent paint are applied, and under these circumstances the treatment can enhance the corrosion resistance of the finishing system.

Chemical Nature of Coatings

The simplest phosphate coating, that formed from solution containing only ferrous phosphate and phosphoric acid, consists of dark grey to black crystals of tertiary ferrous phosphate, $Fe_3(PO_4)_2$, and secondary ferrous phosphate, $FeHPO_4$, with a small proportion of tertiary ferric phosphate, $FePO_4$. Coatings formed from manganese phosphate solutions consist of tertiary manganese phosphate, and those from zinc phosphating solutions consist of tertiary zinc phosphate. With both the manganese and zinc type of coating, insoluble secondary and tertiary iron phosphates, derived from iron present in the bath, may be present in solid solution. Iron from the surface being treated can also be present in the coating, particularly at the metal/phosphate interface. The PO_4^{3-} content of coatings may vary from 33 to 50%, whereas the theoretical PO_4^{3-} content is lowest, at 41%, in $Zn_3(PO_4)_2 \cdot 4H_2O$ and highest, at 63%, in $FePO_4$.

Crystal Structure

It has been suggested that the zinc phosphate coating has the composition $Zn_3(PO_4)_2 \cdot Zn(OH)_2$, but X-ray diffraction studies have given very good correlation between $Zn_3(PO_4)_2 \cdot 4H_2O$ and the zinc phosphate coatings on steel[8].

$Zn_3(PO_4)_2 \cdot 4H_2O$ appears in three crystal forms, α-hopeite (rhombic plates), β-hopeite (rhombic crystals), and p-hopeite (triclinic crystals). Their transition points are at 105, 140 and 163°C respectively. It has been observed[9] that zinc phosphate coatings heated in the absence of air lose their corrosion resistance at between 150 and 163°C.

Manganese phosphate coatings heated in the absence of air lose their corrosion resistance at between 200 and 218°C. At these temperatures, between 75 and 80% of the water of hydration is lost and it is assumed that this results in a volume decrease of the coating which causes voids and thereby lowers the corrosion resistance. Fig. 16.4 shows the loss of water of hydration from zinc, iron and iron-manganese phosphate coatings.

Table 16.9 Analytical tests on industrial phosphate coatings

Process*	P	S	T	Q	V	R
Main cation in phosphate bath	Fe	Mn	Zn	Zn	Zn	Zn
Method of application	Immersion	Immersion	Spraying	Immersion	Immersion	Immersion
Duration of treatment (min)	15	30	1·5	4	5	12
Change in weight on phosphating (g/m²)	−26·1	−26·4	2·61	3·37	1·63	5·87
Coating weight (g/m²)	14·2	21·2	4·46	5·43	3·48	12·28
PO_4^{3-} (g/m²)	7·0	8·9	1·96	2·07	1·20	4·46
Moisture (mg/m²)	81·5	76·1	152·2	369·6	173·9	771·7
PO_4^{3-} content of coating (%)	49·0	42·0	44·0	38·0	34·0	36·0
Moisture content of coating (%)	0·6	0·4	3·4	6·9	5·0	6·4
Hygroscopicity of coating (%)	0·3	0·2	1·2	1·3	1·0	1·5
Absorption value (diacetone alcohol) (g/m²)	11·4	10·9	10·9	13·04	10·87	11·96

*The letters used for designation indicate proprietary processes.
Data reproduced courtesy *J.I.S.I.*, **170**, 11 (1952).

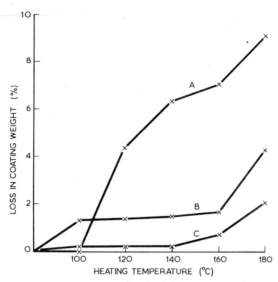

Fig. 16.4   Effect of heating on phosphate coatings for 16 h at various temperatures, showing loss of water of hydration. Curve *A* zinc phosphate, *B* iron phosphate and *C* iron manganese phosphate (courtesy *J.I.S.I.*, **170**, 11 (1952))

The heating of phosphate coatings in the absence of air provides conditions similar to those prevailing during the stoving of paint on phosphated articles, but in general the paint stoving temperatures and times are well below those at which damage to zinc phosphate coatings takes place.

The loss of water from conventional zinc and manganese phosphate coatings heated in air is from 10 to 20% higher than the loss on heating in the absence of air. It is thought that this greater loss may be due to oxidation of the iron phosphate present in the coatings.

The most important uses for phosphate coatings entail sealing with oil or paint and it is therefore of interest to study absorption values. Table 16.9 compares the absorption of diacetone alcohol into coatings of widely differing thicknesses and composition; despite these differences, values of 10·8–12·9 g/m^2 are obtained throughout. It is therefore evident that absorption is predominantly a surface effect and not appreciably influenced by coating thickness.

Rinsing

After phosphating, thorough rinsing with water is necessary in order to remove soluble salts which would otherwise tend to promote blistering under a paint film. Care should also be taken to ensure that the water supply itself is sufficiently free from harmful salts. Experience has shown that a water supply is potentially injurious if it exceeds any one of the three following limits:

1. 70 p.p.m. total chlorides and sulphates (calculated as $Cl^- + SO_4^{2-}$).
2. 200 p.p.m. total alkalinity (calculated as $CaCO_3$).
3. Maximum of 225 p.p.m. of (1) and (2) together.

Improved corrosion resistance and reduced tendency to blistering can be obtained by treating the final rinse with chromic acid, or preferably with phosphoric and chromic acids combined. Normally a total acid content of 0·05% is used. Higher concentrations of chromic acid in the rinse will increase corrosion resistance, partly by passivation of any bare metal or pores in the phosphate coating, but mainly by absorption into the coating[10, 11]. The corrosion resistance rises steadily with increase of chromic acid strength, but above 0·2% chromic acid the phosphate coating tends to dissolve. Absorbed chromic acid is removed only with difficulty by hot or cold water rinsing and is not affected by trichlorethylene vapour treatment. Advantage may be taken of the higher corrosion resistance given by chromic acid, whether or not the metal is to be painted, but care must be taken with white finishing paints, as chromic acid residues may cause local yellowing of the paint in the form of streaks. British Standard requirements for chromic rinsing are shown in Table 16.10.

Table 16.10 Concentrations of chromate solution (BS 3189:1959)

Nature of phosphate coating and of sealing coat	Concentration in terms of CrO_3* (%)	
	Min.	Max.
1. Phosphate coatings of all classes to be sealed with paint, varnish or lacquer	0·0125	0·05
2. Zinc phosphate coatings to be sealed with oil or grease	0·0125	0·25
3. Manganese and/or iron phosphate coatings to be sealed with oil or grease	0·0125	0·5

*The substitution of an equal weight of phosphoric acid for up to one half of the chromic acid is permissible.

In recent years there has been a great increase in the use of demineralised water for rinsing, especially before electrophoretic painting. The demineralised water is generally applied by misting jets at the end of all other pretreatment stages and allowed to flow back into the last rinse tank.

In certain cases rinsing may be dispensed with after non-accelerated phosphate treatment, but blistering of paint due to local concentration of solution in seams and crevices may occur. Rinsing is generally applied, regardless of the type of phosphate process employed[12].

Corrosion Protection

The corrosion protection provided by phosphate coatings without a sealing treatment is of a low order; their value when sealed is considerably greater. Unsealed corrosion tests are therefore of little value except perhaps for studying porosity or efficiency of coatings destined to be sealed only with oil.

Mention has been made of the necessity for controlling the acid ratio of phosphating baths, particularly those of iron, manganese and zinc operating near boiling point to produce heavy coatings. At a 'pointage' (see Table 16.8) of 30 in these solutions the free acidity is usually maintained between 12·5

Table 16.11   Typical phosphate coating processes

Phosphate coating solution	Accelerator	Immersion time (min)	Type of coating	Coating weight (g/m^2)
Iron	None	30	Heavy	10·87–32·61
Iron/manganese	None	30	Heavy	10·87–32·61
Manganese	Nitrate	15	Heavy	8·70–32·61
Zinc	Nitrate	15	Medium	3·26–32·61
Zinc	Nitrate/nitrite or chlorate	3	Light	1·09–6·52
Sodium/ ammonium	None	1–2 (spray)	Very light	0·22–0·65

and 15%; above this figure coatings with progressively lower corrosion resistance are obtained.

Heavy phosphate coatings do not necessarily have better corrosion resistance than lighter coatings. Even with a single process, e.g. zinc/iron/ phosphate/nitrate, no consistent relationship has been found between corrosion resistance and either coating weight or weight of metal dissolved.

Phosphate processes containing little or no oxidising agent and based on manganese or zinc tend to accumulate iron in solution from the work being processed. With a manganese content of from 0·2 to 0·5% it is best to control the iron at from 0·2 to 0·4%; a higher iron content reduces the corrosion resistance and may lead to the formation of thin powdery coatings, while a lower iron content gives soft coatings. Similarly, a zinc process operates best with 0·15–0·5% zinc and 0·4–0·5% iron. Again, with a higher iron content corrosion resistance falls off and powdery coatings may be formed, and soft coatings result from a lower iron content.

Jaudon[13] tested phosphate coatings with and without paint and found the salt-spray resistance, as judged by the first appearance of rust, to be as follows:

Bare steel	Few minutes
Phosphated steel	12 h average
Painted steel	150 h
Phosphated and painted steel	300 h

Table 16.12   Typical uses of phosphate coatings on steel

Coating weight (g/m^2)	For corrosion resistance	For wear prevention and metal forming
21·74–32·61	—	Critical cold extrusion
10·87–21·74	Military equipment, etc. requiring oil or grease finish	Normal cold extrusion 'Running in' treatment for piston rings, gears and tappets
5·43–10·87	Nuts, bolts, clips, brackets	Wire and tube drawing Sheet steel pressing
2·17–2·72	Cars, refrigerators, washing machines	Light metal pressing
1·63–2·17	Steel drums, bicycles, office machinery	
0·22–0·65	Toys, office furniture Strip steel, for painting and forming	

Within broad limits, phosphate processes can be classified according to the main metallic radical of the processing solution and the type of accelerator used; typical processes are given in Table 16.11. The selection of process and of coating weight is mainly dependent on the end-use of the article being processed; the general requirements for corrosion resistance and wear prevention are given in Table 16.12.

Testing

Heavy phosphate coatings are generally used as protection against corrosion in conjunction with a sealing film of oil or grease. The porosity or free pore area of these coatings should be kept to a minimum. Machu[14] devised a method of examination based on the quantity of electricity necessary to effect passivation of the bare steel and used this to determine the 'free pore area' which, in the phosphate coatings tested, varied from 0·27 to 63%. Attempts to use this method for the evaluation of the more widely used thin zinc phosphate coatings have not been successful, as these coatings show a porosity of less than 1·5% and the technique of measurement was not adequate for this range[15]. A method for making rapid measurements of the electrical resistance of phosphate coatings has been described by Scott and Shreir[16].

Akimov and Ulyanov[17] proposed an acidified copper sulphate spot test for assessing the corrosion resistance of phosphated articles by timing the colour change from blue to light green, yellow or red owing to the precipitation of copper. The assumption was that the longer this change took to occur, the higher the corrosion resistance. The test has been thoroughly examined in this country and rejected because of variation in results and poor correlation with corrosion resistance. Sherlock and Shreir[17(a)] consider that the hydrogen permeation technique (see Section 9.1) could provide a useful means of studying and evaluating the porosity of phosphate coatings.

The most widely used accelerated tests are based on salt spray, and are covered by several Government Specifications. BS 1391:1952[18] gives details of a hand-atomiser salt-spray test which employs synthetic sea-water and also of a sulphur-dioxide corrosion test. A continuous salt-spray test is described in ASTM B 117-61 and BS AU 148: Part 2(1969). Phosphate coatings are occasionally tested by continuous salt spray without a sealing oil film and are expected to withstand one or two hours spray without showing signs of rust; the value of such a test in cases where sealing is normally undertaken is extremely doubtful.

The main value of salt-spray tests is in the evaluation of the effectiveness of phosphate coatings in restricting the spread of rust from scratches or other points of damage in a paint film. This feature is of particular interest to the motorcar industry, as vehicles are often exposed to marine atmospheres and to moisture and salt when the latter is used to disperse ice and frost from road surfaces. Great care is needed in the interpretation of a salt-spray test, as it has been found to favour thin iron phosphate coatings more than is justified by experience with natural weathering. In the motorcar industry the present custom is to use zinc phosphate coatings on the car bodies and all other parts exposed to the outside atmosphere.

Humidity tests are generally of more practical use than salt-spray tests, particularly where painting is employed, as the thoroughness of rinsing may be checked by this means. The use of contaminated water can leave water-soluble salts in the phosphate coating and lead to blistering of the paint film under humid conditions, as paint films are permeable to water vapour. Immersion in water, or subjection to high humidity in a closed cabinet, will generally show any defects of this kind within a few days. The British Automobile Standard specifies freedom from blistering after 200 h in distilled water at 100°F (38°C).

Table 16.13   Weights of phosphate coatings
(Defence Specification DEF-29)

Class	Type	Minimum coating weight (g/m^2)
I	Mn or Fe	7·6
II	Zn, etc.	4·3
III	—	1·6
		0·5*

*A lower range of 0·5 to 1·6 g/m² may be permitted where thin sections are to be fabricated or formed after the application of paint, varnish or lacquer.

The texture or crystal size of phosphate coatings can conveniently be recorded by making an impression on clear cellulose tape moistened with acetone. Uniformity of crystal size is of importance for coatings which are to resist wear and assist metal working. Surface roughness may also be studied by means of a 'Talysurf' meter.

Phosphate coating weight determinations are generally performed by dissolving the coating from weighed panels by immersion in a solution of 20 g/l of antimony trioxide in concentrated hydrochloric acid at a temperature of 13–21°C[19]. The solution is used once only. Thin iron or zinc phosphate coatings can be removed for weight determination by immersion in 5% chromic acid solution at 70°C, but this solution should also be used once

Table 16.14   Salt-spray resistance
of phosphate coatings under various
finishes
(Defence Specification DEF-29)

Finish	Period of test (days)
Oil	1
Shellac	1
Lanolin	1
Air-drying paint	3
Stoving lacquer	6
Stoving paint	6

only, as the presence of more than a trace of phosphate leads to pitting of the steel and false results. Zinc phosphate coatings can be removed by immersion in 10% sodium hydroxide at boiling temperature, aided by rubbing during rinsing.

The Ministry of Defence requirements for phosphating are covered by Defence Specification DEF-29 and are divided into three classes as shown in

Table 16.13. This specification follows good industrial practice, with additional safeguards in rinsing to remove residues of treatment solutions. Non-accelerated treatments must be followed by a single rinse which may contain chromate; accelerated treatments must be followed by three rinses—cold water, hot water and a final chromate rinse. Table 16.14 shows the salt-spray test requirements for phosphate coatings with various finishes without formation of rust; the paints and lacquer have the additional requirement that no rust shall be visible beyond 0·2 in (5 mm) from the deliberate scratches and no blistering, lifting or flaking beyond 0·05 in (1·27 mm) from the original boundaries of the scratches.

The American Aeronautical Material Specification AMS 2480 A calls for 150 h salt-spray test without rusting extending more than 0·125 in (3·175 mm) on either side of scratch marks, using a black enamel finish for the phosphate coating.

Table 16.15 Weights of phosphate coatings
(BS 3189:1959)

Class of phosphate process	Coating weight (g/m^2)	
	Min.	Max.
A 1. Heavyweight (Mn or Fe)	7·61	—
A 2. Heavyweight (Zn)	7·61	—
B Medium weight (Zn, etc.)	4·34	—
C Lightweight (Zn, etc.)	1·09	4·34
D Extra lightweight (Fe)	0·33	1·09

British Standard 3189:1959[19] contains valuable information on the operation of phosphate processes to obtain optimum results, and on the testing of phosphate coatings. The classification of coatings according to composition and weight is shown in Table 16.15. Recommendations for chromate rinsing are given in Table 16.10. The inspection and testing includes determination of coating weight, freedom from corrosive residues as shown by a humidity test, and resistance to corrosion by salt spray.

R. E. SHAW

REFERENCES

1. Machu, W., *Die Phosphatierung—Wissenschaftliche Grundlagen und Technik*, Verlag Chemie, Weinheim (1950)
2. Wusterfeld, H., *Arch. Metallk.*, **3**, 223 (1949)
3. 'Determination of the Solubility of Ferric Phosphate in Phosphating Solutions Using Radioiron', US Department of Commerce, Office of Technical Services, Rep. No. PB 111, 399 (1953)
4. 'Radiometric Study of Phosphating Problems', US Department of Commerce, Office of Technical Services, Rep. No. PB 111, 396 (1951)
5. Wirshing, R. J. and McMaster, W. D., *Canad. Paint Varn. Mag.*, **30** No. 5, 42, 55 (1956)
6. Jernstedt, G., *Chem. Engng. News*, **21**, 710 (1943); *Trans. Electrochem. Soc.*, **83**, 361 (1943)
7. 'A Radiometric Study of the Iron Phosphating Process', US Dept. of Commerce, Office of Technical Services, Rep. No. PB 111, 400 (1953)

8. 'X-ray Diffraction Study of Zinc Phosphate Coatings on Steel', US Department of Commerce, Office of Technical Services, Rep. No. PB 111, 486 (1954)
9. Doss, J., *Org. Finish.*, **17** No. 8, 6 (1956)
10. B.S.I. Phosphate Coatings (Drafting) Panel, 'Phosphate Coatings as a Basis for Painting Steel', *J. Iron St. Inst.*, **170**, 10 (1952)
11. 'Radiometric Evaluation of the Effectiveness of the Chromic Acid Rinse Treatment for Phosphated Work', US Department of Commerce, Office of Technical Services, Rep. No. PB 111, 397 (1952); 'A Study of the Effect of Chromic Acid and Chromic-phosphoric Acid Rinse Solutions upon the Subsequently Applied Paint Coatings', US Dept. of Commerce, Office of Technical Services, Rep. No. PB 111, 578 (1954)
12. B.S.I. Phosphate Coatings (Drafting) Panel, 'Phosphate Coatings as a Basis for Painting Steel', *J. Iron St. Inst.*, **170**, 13 (1952)
13. Jaudon, E., *Peint.-Pigm.-Vernis.*, **25**, 224 (1949)
14. Machu, W., *Korros. Metallsch.*, **20**, 1 (1944)
15. 'Development of Accelerated Performance Tests for Paint–Phosphate–Metal Systems. Surface Preparation of Metals', Aberdeen Proving Ground, USA (1955)
16. Scott, J. W. and Shreir, L. L., *Chem. and Ind. (Rev.)*, 807 (1957)
17. Akimov, G. V. and Ulyanov, A. A., *C. R. Acad. Sci. U.R.S.S.*, **50**, 271 (1945)
17(a). Sherlock, J. C. and Shreir, L. L., *Corros. Sci.*, **11**, 543 (1971)
18. *Performance Tests for Protective Schemes used in the Protection of Light Gauge Steel*, BS 1391:1952, British Standards Institution, London
19. *Phosphate Treatment of Iron and Steel for Protection against Corrosion*, BS 3189:1959, British Standards Institution, London

16.3 Chromate Treatments

Introduction

The addition of chromates to many corrosive liquids reduces or prevents attack on metals, and chromates are often added to waters in contact with metals as corrosion inhibitors. Under atmospheric exposure an alternative method is used; this consists of depositing on the metal a chromate film which acts as a reservoir of soluble chromate. Although the quantity of chromate which can be held in this way at the metal surface is small, the film nevertheless improves the performance of metals with a high intrinsic corrosion resistance, e.g. cadmium, copper and some aluminium-base materials. With metals which are more liable to corrode, however, such as magnesium alloys and high-strength aluminium alloys, chromate films are used primarily for improving the adhesion of paint, their own inhibiting action making a useful contribution to the total protection.

Chromate treatments can be applied to a wide range of industrial metals. They are of two broad types: (a) those which are complete in themselves and deposit substantial chromate films on the bare metal; and (b) those which are used to seal or supplement protective coatings of other types, e.g. oxide and phosphate coatings. Types of treatment for various metals are summarised in Table 16.16.

Principles of Chromate Treatment

Chromate ions, when used as inhibitors in aqueous solutions, passivate by maintaining a coherent oxide film on the metal surface. Passivation is maintained even in a boiling concentrated chromic acid solution*, in which many of the oxides in bulk form are soluble. The passivity breaks down rapidly, however, once the chromate is removed.

In order that a chromate film may be deposited, the passivity which develops in a solution of chromate anions alone must be broken down in solution in a controlled way. This is achieved by adding other anions, e.g. sulphate, nitrate, chloride, fluoride, as activators which attack the metal, or by electrolysis. When attack occurs, some metal is dissolved, the resulting hydrogen reduces some of the chromate ion, and a slightly soluble golden-brown or black chromium chromate ($Cr_2O_3 \cdot CrO_3 \cdot xH_2O$) is formed.

*Vigorous attack can occur with industrial-grade chromic acid, which can contain sulphuric acid as an impurity.

16:33

This compound is deposited on the metal surface unless the solution is sufficiently acid to dissolve it as soon as it is formed. The film also usually contains the oxide of the metal being treated, together with alkali metal (when this is present in the treatment solution) perhaps in the form of a complex basic double chromate analogous to zinc yellow.

Table 16.16   Summary of types of chromate treatment

Metal	Type of treatment	Solution, radicals	Type of deposit
Aluminium and its alloys	(a) Alkaline dip	Alkaline chromate	Oxide/hydroxide with perhaps some chromate
	(b) Acid dip, 1	Acid chromate/ fluoride/phosphate	Phosphate with perhaps some chromate
	Acid dip, 2	Acid chromate/ fluoride/nitrate	Not known, but contains substantial chromate
	(c) Acid pickle	Acid chromate/ sulphate and chromate/phosphate	Very thin, may contain chromate
	(d) Sealing of anodic films	Chromate/ dichromate in pH range 5 to 7	Blockage of pores with hydroxide/ chromate
Cadmium and zinc	Acid dip	Acid chromate/ sulphate, sometimes with additions	Thin hydrated chromium chromate
Copper	Acid pickle	Acid chromate/ sulphate	Very thin, may contain chromate
Iron and steel	Rinse after phosphate treatment	Dilute acid chromate with or without phosphate	Probably some basic chromate left in the phosphate coating
Magnesium alloys	(a) Strongly acid dip	Acid chromate/ nitrate	Thin chromium chromate
	(b) Moderately acid dip	Chromate/sulphate with buffer, pH 4 to 5, also Dow No. 7	Thick chromium chromate
	(c) Slightly acid dip	Chromate/sulphate, pH 6, at boiling or with anodic current	Thick chromium chromate
	(d) Sealing of anodic films	Neutral chromate	Chromate retained by oxide, etc. coating
Silver	(a) Dip	Chromate and complexing salt	Very thin, may contain chromate
	(b) Anodic	Alkaline chromate	Very thin, may contain chromate
Tin	Dip	Alkaline chromate	Very thin, may contain chromate

The stability of the natural oxide film reinforced by the chromate ion determines the conditions of pH, ratio of activating anion to chromate, and temperature at which the oxide is broken down and a chromate film deposited. Thus magnesium alloys can be chromate-treated in nearly neutral solutions, whereas aluminium alloys can be treated only in solutions of appreciable acidity or alkalinity.

The same principle tends to apply to the protective efficiency of the chromate film, i.e. the greater the intrinsic corrosion resistance of the metal, the greater the protection conferred by the soluble chromate in the chromate film.

Aluminium

Chromate Treatment of Aluminium

Several immersion treatments using solutions containing chromates[1] have been developed for aluminium. It is not always clear to what extent the films formed can properly be called chromate films, i.e. films containing a substantial amount of a slightly soluble chromium chromate, but even if the film consists largely of aluminium oxide or hydroxide or other salt with chromate physically absorbed, it will still provide a reservoir of soluble chromate at the metal surface. Treatments fall into two classes: alkaline and acid. The latter are of more recent development.

Alkaline treatments These are all based on the original Bauer–Vogel process in which a boiling solution of alkali carbonate and chromate is used. The best known is the Modified Bauer–Vogel process (DTD 913); others contain silicate (E.W. process), fluoride, chromium carbonate, and/or disodium phosphate (Pylumin processes). The films formed are light to dark grey in colour, depending on the process and the composition of the alloy being treated, and consist substantially of aluminium oxide or hydroxide and probably some soluble chromate, either combined or adsorbed. The protection against mild atmospheres is fair, and is improved by sealing in hot sodium silicate solution. The films produced provide a good basis for paint.

Acid treatments The principal acid processes were developed in the USA under the name Alodine, and are marketed in the UK as Alocrom and under other names. The original solutions were based on acid solutions containing phosphate, chromate and fluoride ions. Immersion for up to 5 min in the cold or warm solution leads to the deposition of a greenish film containing the phosphates of chromium and aluminium, and possibly some hexavalent chromate. The more recent Alocrom 1 200 process uses an acid solution containing chromate, fluoride and nitrate. Room-temperature immersion for 15 s to 3 min deposits golden-brown coatings which contain chromate as a major constituent.

The success of the Alocrom 1 200 process has prompted the introduction of several other commercial processes which deposit similar substantial chromate-bearing films.

Acid pickles   Some of the acid pickles used to clean and etch aluminium alloy surfaces and remove oxide and anodic films, such as the chromic/ sulphuric acid pickle (method *O* of DEF STAN 03–2) and other chromic-acid bearing pickles (App. *F* of DEF–151) probably leave on the surface traces of absorbed or combined chromate which will give at least some protection against mild atmospheres.

Sealing of Anodic Films

In view of the porous nature of anodic films, especially those produced by the sulphuric acid process (Section 16.1), sealing treatments have been developed in an attempt to improve their protective value. Although not very effective on the relatively dense films produced by the chromic acid process, the sealing treatments enhance the protection afforded by films produced by the sulphuric acid process. For conferring protection against corrosion the most effective treatment is immersion for 5–15 min in a boiling chromate/ dichromate solution just on the acid side of pH 8, i.e. at a pH value at which aluminium oxide and hydroxide just begin to be slightly soluble. Defence Specification DEF–151 quotes two solutions, one a 7–10% dichromate/ chromate solution at pH 6–7, and the other a 5% dichromate solution containing a small amount of chromate to bring the pH from 4 (dichromate only) to between 5·6 and 6.

The chromate sealing treatment imparts to the anodic film a distinct yellow to brown colour, which is probably due to a basic aluminium chromate or alkali chromate adsorbed on to aluminium hydroxide. The film gives appreciable protection against marine exposure.

Chromate Passivation of Cadmium and Zinc

Cadmium and zinc coatings are widely used to protect steel from rusting, and for preventing accelerated corrosion when two dissimilar metals, e.g. copper and aluminium are in contact. It is important that zinc and cadmium should themselves be preserved from corroding, so that they may give protection by physical exclusion and sacrificial action. The durability of cadmium and zinc coatings depends on their thickness and their intrinsic corrosion resistance under any given exposure. On close-tolerance parts, the thickness is of necessity limited to 25 µm or often appreciably less. Zinc corrodes quite rapidly in humid and marine conditions, and cadmium, though more resistant, is not immune. Both metals are attacked by the organic vapours emitted by some plastics and paints, and by wood[2]. It is therefore often highly desirable to apply a protective coating.

The best protection is given by paint. An etch-primed paint scheme can be applied directly to the metal; for other paints an inorganic treatment must be given to ensure good adhesion. Of the two classes of inorganic treatment, phosphate treatment has little protective value in itself, but chromate passivation gives appreciable protection and in mildly corrosive surroundings may be sufficient in itself.

The most commonly used chromate passivation process is the Cronak process developed by the New Jersey Zinc Co. in 1936, in which the parts are immersed for 5–10 s in a solution containing 182 g/l sodium dichromate and 6 ml/l sulphuric acid. A golden irridescent film is formed on the zinc or cadmium surface. Many variants (all fairly acidic) have been developed subsequently; all are based on dichromate (or chromic acid) with one or more of the following: sulphuric acid, hydrochloric acid (or sodium chloride), nitric acid (or nitrate), phosphoric acid, formic acid and acetic acid. A survey by Biestek[3] shows that several of these variants are as good as the Cronak process, although none is superior.

Practical details of the Cronak process are given in Specification DEF-130, and a comprehensive account of the process as applied to zinc plate has been published by Clarke and Andrew[4]. Fig. 16.5 shows the loss of zinc and the

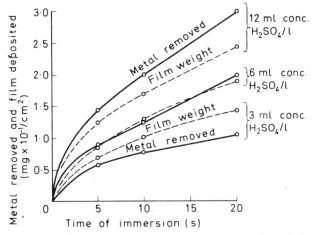

Fig. 16.5   Effect of sulphuric acid concentration on chromate passivation of zinc. Solution: 182 g/l of $Na_2Cr_2O_7 \cdot 2H_2O + H_2SO_4$ as indicated; temp. 18°C; $1 \cdot 0 \times 10^{-1}$ mg $Zn/cm^2 \equiv$ 0·145 μm thickness

weight of film deposited as a function of immersion period and of variation of sulphuric acid content above and below the normal 6 ml/l. The curves show that in the normal bath the weight of film deposited is equal to the weight of zinc dissolved, and that as the acid is consumed, the solution becomes more efficient in converting metal to film. The curves also show that the dissolution of zinc during film formation is small, less than 0·25 μm, which is an important consideration when small parts such as nuts and bolts are being treated. (On such parts, for reasons of tolerance, relevant specifications are forced to allow minimum thicknesses of down to 4 μm of cadmium and zinc plate.)

Claims for other passivation solutions should always be considered in relation to the quantity of metal consumed, unless, of course, the solutions are intended solely for use on zinc-base die-castings, where tolerance on thickness is unimportant.

The chromate film deposited by the Cronak process on zinc consists largely of a hydrated chromium chromate and contains some 10% by weight of hexavalent chromium, equivalent to 20% of CrO_4^{2-}. At least part of this chromate is soluble in water and available for protecting the underlying zinc or cadmium; on account of this solubility, passivated parts should not be washed in very hot water. Heating at 100°C or higher tends to dehydrate the film and render the chromate in it insoluble, with consequent reduction in protective value; any heat treatment after plating, e.g. for de-embrittlement, should therefore be completed before chromate passivation. If the yellow colour of the chromate film is considered undesirable, treated parts can be subjected to an aqueous extraction 'bleaching' treatment, but much of the protective value will be lost thereby.

The quantitative results quoted above all refer to zinc surface. It is likely that the behaviour of cadmium would be similar; in view of the fact that the equivalent weight of cadmium is double that of zinc, it is even more important that the passivation solution shall not attack and dissolve the metal to any appreciable extent.

Cleaning Etch for Copper and its Alloys

Copper and its alloys can be cleaned and brightened by immersion in solutions of substantial quantities of dichromate with a little acid (see, for instance method Q of DEF STD 03-2/1). Such solutions impart some resistance to tarnishing, ascribed to the formation of very thin chromate films.

Clarke and Andrew[5] have developed a similar solution further activated by addition of chloride ions which deposits more substantial films shown to contain hexavalent chromium. The films give appreciable protection against salt spray and tarnishing by sulphur dioxide.

Iron

Chromate Treatment

In spite of the effectiveness of chromates in stopping the rusting of steel in aqueous solutions, no successful chromate filming process has been developed for this purpose.

Chromate Rinsing of Phosphated Steel

The protective value of a phosphate coating is enhanced by a dip or rinse in an acid chromate solution. Joint Service Specification DEF-29 makes such a rinse mandatory for steel parts treated by an accelerated process, and optional after treatment by a non-accelerated process. Details of rinses are given in Section 16.2 (Table 16.10, p. **16**.27).

Magnesium Alloys

(See also Section 4.4.)

Chromate Treatments

Chromates are very effective inhibitors of the corrosion of magnesium alloys by saline and other waters, and many treatments have been developed by means of which substantial films containing slightly soluble chromate are formed in the metal surface. Except on parts which are to be exposed only to a rural atmosphere, chromate treatment must be supplemented by paint, for which it provides a good base.

Magnesium is a relatively reactive metal, and can be chromated in nearly neutral solutions as well as in acid solutions. The range of treatments possible illustrates well the rôle of pH, activating anion, temperature and duration of treatment in promoting the breakdown of passivity in the chromate solution and the consequent formation of a chromate film.

Strongly acid bath This class is represented by a treatment developed in Germany over 40 years ago and widely used since. The solution contains 15% sodium or potassium dichromate and 20–24% v.v. concentrated nitric acid. Parts to be treated are immersed for 30 s to 2 min at room temperature and then allowed to drain for 5 s or more before being washed. Most or all of the film formation occurs during the draining period and the chief function of the immersion period is to clean the surface by etching. The film is thin and of a golden or irridescent grey colour. The process is not suitable for close tolerance parts and is mainly used for protection during storage prior to matching. The process is used in the UK as bath (iv) of DTD 911C, in the USA as the Dow No. 1 treatment and in the USSR as treatment MOKH-1.

Medium acid baths, pH 4–5 At this acidity a dichromate solution plus sulphate ion as activator is sufficient to deposit chromate films in 30 min or so at room temperature or in a few minutes at boiling point. Unfortunately, a solution of alkali dichromate and alkali sulphate is quite unbuffered, and other substances must be added to give the bath a useful life over the working pH range. Acetates have been used successfully, but salts of aluminium, chromium, manganese and zinc have been more commonly employed. The pH of the solution rises slowly during use until basic chromates or sulphates begin to precipitate. The solution can then be rejuvenated by the addition of chromic or sulphuric acid or acid salts.

A successful bath of this class is the Magnesium Elektron Chrome Manganese Bath, bath (v) of DTD 911C, which contains 10% sodium dichromate, 5% magnesium sulphate (as a source of sulphate) and 5% manganese sulphate (as a source of sulphate and as a buffering agent). Treatment is by immersion for up to 2 h at room temperature or up to 10 min at boiling point, the treatment being continued until the appearance of the deposited film has passed the thin golden stage and reached the dark brown to black stage. A second bath of this class is the Dow No. 4 which contains sodium dichromate and potassium chrome alum; this solution is used at boiling point.

The Dow No. 7 treatment, popular in the USA, also falls within this class. The process differs from other chromate treatments in that the activator, magnesium fluoride, is formed on the metal surface by immersion in 20% hydrofluoric acid solution, the parts then being immersed in a 10–15% alkali dichromate solution with or without sufficient alkaline earth fluoride to saturate it. A slow action occurs on the surface and the fluoride film is replaced by a chromate or mixed chromate/fluoride film.

The dichromate solution is quite unbuffered over the working pH range of 4·0–5·5, but the degree of attack on the metal is so slight that in practice appreciable surface areas can be treated before readjustment of the pH by addition of chromic acid becomes necessary. The process is used in the USSR under the code name MFKH-1.

Slightly acid baths, pH 6 At this pH, a boiling temperature must be used to aid the activation of the sulphate present; alternatively, activation can be accomplished by use of an anodic current.

The R.A.E. 'hot half-hour bath', bath (iii) of DTD 911C falls into this class. The solution contains 1·5% each of ammonium and alkali dichromates, 3% ammonium sulphate, and enough ammonia to raise the pH from 4 (dichromate stage) to 6. Parts to be treated are immersed in the boiling solution for 30 min. The solution is well buffered against rise of pH due to magnesium dissolving in the solution, partly by the chemical reaction dichromate → chromate, and partly by loss during boiling of ammonia. A closely similar process is used in the USSR under the code name MOKH-6. In the USA the process is applied after a hydrofluoric acid pretreatment, either as above (Dow No. 8 treatment and USSR MFKH-3) or at 50–60°C with the aid of the galvanic current generated when the parts under treatment are connected electrically to the steel tank or to steel cathodes in the solution (Dow No. 9 treatment).

A cold treatment relying on electric current for activation has been developed at R.A.E.; the solution consisted of 15% sodium dichromate and 5% potassium permanganate with added caustic soda to bring the pH to the lower end of the treatment range of 6·0–7·1. Parts were made anode at a current density of 30–80 A/dm^2 for a treatment time of 20 min.

Chromate Sealing

A large number of electrolytic treatments of magnesium, anodic or a.c., have been developed, in which adherent white or grey films consisting of fluoride, oxide, hydroxide, aluminate or basic carbonate are deposited from alkaline solutions containing caustic alkali, alkali carbonates, phosphates, pyrophosphates, cyanides, aluminates, oxalates, silicates, borates, etc. Some films are thin, and some are relatively thick. All are more or less absorbent and act as good bases for paint, though none contributes appreciable inhibition. All can, however, absorb chromates with consequent improvement of protective efficiency.

The simplest method of chromate sealing involves immersion in a dilute alkali chromate or dichromate solution followed by washing; retained chromate imparts a yellow colour to the film. More substantial amounts of slightly soluble chromate can be deposited in the thicker type of absorbent anodic film by a method developed by Dr. L. Whitby at High Duty Alloys Ltd[6]. In this, anodised parts are immersed first in a boiling 30% solution of sodium chromate and then in a boiling 2% solution of zinc nitrate. Residues of the first solution in the film react with the second solution to give a substantial yellow deposit of a basic zinc chromate, probably similar in composition to zinc yellow.

Silver

Chromate treatments have been developed for protecting silver against sulphide tarnishing by the deposition of very thin films which are assumed to contain chromate. A Dutch–American immersion treatment[7] uses a chromate solution and a complexing agent, e.g. cyanide, ammonia or E.D.T.A. Working pH values depend on the nature of the agent and lie within the range pH 1–12.

Another treatment consists of making the silver parts cathode in an alkaline chromate solution.

Tin

Alkaline chromate treatments for tin, e.g. the Protecta-Tin processes[8], have been developed by the Tin Research Institute. The solutions resemble the M.B.V. compositions for treating aluminium, but are more alkaline. Thin invisible films which resist staining by heat and sulphur-bearing compounds and give protection against humid atmospheres at pores are deposited.

Etch Primers

While etch primers, also known as pretreatment primers and wash primers, can be regarded as priming paints which promote their own adhesion by etching the metal surface, they may also be regarded as phosphate/chromate etching treatments which leave an organic residue on the surface to form the basis of the subsequent paint scheme. A detailed account of the etch primers has been given by Coleman[9].

The standard etch primer (WP-1, DEF-1408) consists of two solutions, one containing polyvinyl butyral resin and zinc tetroxychromate in ethyl alcohol with n-butanol, and the second containing phosphoric acid and ethyl alcohol. It is essential that a small critical amount of water be present in the latter. The two solutions are mixed in appropriate ratio for use; the mixture deteriorates and should be discarded when more than 8 h old. Single-pack etch primers of reasonable shelf life are available but contain less phosphoric acid than the above and are not considered to be so effective.

The reactions which take place when the mixed etch primer is applied to a metal are complex. Part of the phosphoric acid reacts with the zinc tetroxychromate pigment to form chromic acid, zinc phosphates and zinc chromates of lower basicity. The phosphoric acid also attacks the metal surface and forms on it a thin chromate-sealed phosphate film. Chromic acid is reduced by the alcohols in the presence of phosphoric acid to form chromium phosphate and aldehydes. It is believed that part of the chromium phosphate then reacts with the resin to form an insoluble complex. Excess zinc tetroxy chromate, and perhaps some more soluble less basic zinc chromes, remain to function as normal chromate pigments, i.e. to impart chromate to water penetrating the film during exposure. Although the primer film is hard enough for over-coating after drying for 1 h, the above reactions continue in the nominally dry film for two to three days, during which time the film remains rather sensitive to water.

Etch priming is widely used on aluminium alloy, and is particularly effective on cadmium and zinc. The adhesion to stainless steel and titanium is good. It has also been used quite widely on bare steel and on magnesium alloy, but on these metals its performance is not, in the opinion of some investigators, always quite reliable. For best protection the etch primer coating is followed with a full paint scheme.

H. G. COLE

REFERENCES

1. Wernick, S. and Pinner, R., *Surface Treatment and Finishing of Aluminium Alloys*, Robert Draper, Teddington (1956)
2. Rance, Vera E. and Cole, H. G., *Corrosion of Metals by Vapours from Organic Materials*, London, H.M.S.O. (1958)
3. Biestek, T., *Prace Inst. Mech.*, **6** (3/1956), 39 (1957)
4. Clarke, S. G. and Andrew, J. F., *J. Electrodep. Tech. Soc.*, **20**, 119 (1945)
5. Clarke, S. G. and Andrew, J. F., *Proceedings of the First International Congress on Metallic Corrosion, London, 1961.* Butterworths, London, 173 (1962)
6. High Duty Alloys Ltd., UK Pat. 570 054 (1945)
7. N. America Phillips Co. Inc., US Pat. 2 850 419 (1958)
8. Britton, S. C. and Angles, R. M., *J. Appl. Chem.*, **4**, 351 (1954)
9. Coleman, L. J., *J. Oil Col. Chem. Ass.*, **42**, 1, 10 (1959)

17 MISCELLANEOUS COATINGS

17.1   Vitreous Enamel Coatings   **17**:3
17.2   Thermoplastics   **17**:13
17.3   Temporary Protectives   **17**:21

17.1 Vitreous Enamel Coatings

Nature of Vitreous Enamels

A vitreous enamel coating is, as the name implies, a coating of a glassy substance which has been fused onto the basis metal to give a tightly adherent hard finish resistant to many abrasive and corrosive materials. The purpose of modern vitreous enamels is twofold, i.e. to confer corrosion protection to the metal substrate and at the same time to provide permanent colour, gloss and other aesthetic values.

Most of the corrosion resistance, and indeed other properties of the finish, are determined by the composition of the vitreous enameller's raw material *frit*, although other factors can influence them to a minor degree. Frit, for application to sheet and cast iron, is essentially a complex alkali-metal alumino borosilicate and is prepared by smelting together at temperatures between 1 100 and 1 450°C an intimate mixture of refractory materials such as silica, titania, felspar, china clay, etc. with fluxes exemplified by borax, sodium silicofluoride and the nitrates and carbonates of lithium, sodium and potassium. The smelting continues until all the solid matter has interreacted to form a molten mass, but unlike true glass this liquid does contain a degree of bubbles. At this stage the melt is quenched rapidly by either pouring into water or between water-cooled steel rollers to form 'frit' or 'flake'.

In the case of dry process enamelling which is usually restricted to cast iron baths, this frit, after drying, is milled alone or with inorganic colouring oxides in cylinders using pebbles of flint, porcelain, steatite or more dense alumina, to a fine powder of predetermined size.

In the more common wet process the frit is milled with water, colloidal clay, opacifier, colouring oxide, refractory and various electrolytes in a ball mill to a closely controlled fineness or coarseness.

Typical frit and mill formulae are given in Table 17.1. Frits are tailor-made for each application so that the most desired properties are at their maximum in each case and thus the formulae presented must be regarded as examples of general composition.

Metal and Metal Preparation

To obtain a defect-free finish it is essential that the basis metal is of the correct composition and suitably cleaned.

17:3

Table 17.1 Typical enamel frit compositions (%) and a mill addition

	Chemical plant	Sheet-iron (groundcoats)			Sheet iron (white)	Sheet iron (acid resistant black)	Cast iron (semi-opaque)
Na_2O	15·8	17·5	21·8	16·0	7·0	16·0	17·5
K_2O	—	—	—	—	5·0	—	3·0
Li_2O	—	—	—	—	1·0	1·0	—
CaO	1·2	—	—	—	—	3·0	—
BaO	—	—	—	5·5	—	6·0	—
CaF_2	3·4	6·0	5·5	4·0	—	2·0	2·0
Na_2SiF_6	—	—	—	—	5·5	2·0	—
Al_2O_3	2·9	5·0	9·0	1·0	2·5	1·0	4·5
B_2O_3	0·9	20·0	18·2	25·0	15·0	7·0	7·5
SiO_2	60·0	50·0	44·0	47·0	46·0	53·0	43·5
TiO_2	15·8	—	—	—	18·0	8·0	13·5
CoO	—	0·4	0·3	0·5	—	0·4	—
NiO	—	0·5	0·6	0·5	—	0·6	—
MnO	—	0·6	0·6	0·5	—	—	—
Sb_2O_5	—	—	—	—	—	—	8·5
	100·0	100·0	100·0	100·0	100·0	100·0	100·0

Sheet iron mill addition

Frit	100
Water	35
Titania	1
Clay	2·5
Bentonite	0·3
Sodium nitrite	0·05
Potassium carbonate	0·1

} Grind to fineness of 1 g residue on 200 mesh sieve (50 ml sample)

Cast Iron

For cast iron enamelling the so-called grey iron is preferred. Its composition varies somewhat depending upon type and thickness of casting, but falls within the following limits: 3·25–3·60% total C, 2·80–3·20% graphitic C, 2·25–3·00% Si, 0·45–0·65% Mn, 0·60–0·95% P and 0·05–0·10% S.

The standard method of cleaning cast iron for enamelling is by grit or shot blasting which may be preceeded by an annealing operation.

Steel

Two general types of sheet steel are in current use, viz. cold rolled mild steel and decarburised steel. The former is exemplified by the following composition: 0·02% Si, 0·05% S, 0·04% P, 0·40% Mn and 0·20% C. It can be obtained in regular, deep drawing or extra-deep drawing grades. The newer decarburised steel is, as the name suggests, mild steel which has undergone heat treatment in a controlled atmosphere to reduce the carbon content to a very low level.

Sheet steel is cleaned by thorough degreasing followed by acid pickling, and is frequently given a nickel dip which results in the deposition of a thin layer of porous nickel (≈ 0.01 g/m^2). (See Section 14.7.)

Enamel Bonding

For effective performance the enamel must be firmly bonded to the underlying metal and this bond must persist during usage. The bond is formed by the molten enamel flowing into the 'pits' in the metal, i.e. mechanical adhesion, and by solution of the metal in the glass, i.e. chemical adhesion. The coefficient of thermal expansion of the enamel in relation to the cast iron or sheet steel and enamel setting temperature determines the stress set up in the coating. As enamel, like glass, is strongest under compression, its thermal expansion should be slightly less than the metal.

Enamel Application and Fusion

Dry process enamelling consists of sieving finely powdered frit onto the preheated casting then establishing the required smooth finish and adhesion by insertion of the whole into a furnace at say 900°C.

In the more usual wet process enamelling, the slurry is applied to the cleaned metal by spraying or dipping. After drying, the enamel is fused onto the steel at 750–850°C.

Properties of Enamel Coatings Affecting Corrosion

Mechanical Properties

This group includes such items as surface hardness, i.e. scratch and abrasion resistance, adhesion and resistance to chipping, crazing and impact. All of these and other properties depend upon the adhesion between the vitreous enamel layer and the metal being good and remaining so.

There is no single test that will give a quantitative assessment of adhesion, and those which have been proposed all cause destruction of the test piece. It has already been stated that this property is dependent upon mechanical and chemical bonds between the enamel and the metal. One must, however, also consider the stresses set up at the interface and within the glass itself during cooling after fusion or after a delayed length of time.

The coefficient of thermal expansion is primarily determined by the frit composition, although mill additions can have a minor influence. As a general rule, superior acid and thermal shock resistance obtain with low expansion enamel, and the skill of the frit manufacturer is to obtain good resistance and also to maintain a sufficiently high expansion to prevent distortion of the component (pressing or casting). Several workers have produced a set of factors for expansion in relation to the enamel oxides that constitute the frit, which provides a guide to the frit producer. However, as these factors are derived from a study of relatively simple glasses smelted to homogeneity it must be emphasised that they are only a guide. The effect of substituting certain oxides for others in a standard titanium superopaque enamel is given in Table 17.2. The use of a nickel dip improves adhesion by minimising iron oxide formation, but it should be noted that some iron oxide formation is necessary to produce enamel/metal adhesion. In the commonest method of testing for adherence to sheet iron, the coated metal is distorted by bending, twisting or impact under a falling weight. In the worst cases the enamel is removed leaving the metal bright and shiny, but in all others a dark coloured coating remains with slivers of fractured enamel adhering to a greater or lesser degree. With cast iron enamelling it is not possible to distort the metal and in this case an assessment of adhesion is obtained by dropping a weight on to the enamel surface and examining for fractures. Erroneous results can obtain in that often thicker enamel coatings appear to be better bonded and resistant to impact, whereas in fact the converse is true. Providing the bond is adequate this test really gives an indication of the strength of the enamel itself.

Table 17.2
Effect of frit ingredients on enamel expansion

Constituent varied	Expansion change
Increase alkali metal	Increase
Replace Na_2O by Li_2O	Increase
Replace Na_2O by K_2O	Decrease
Increase fluorine	Decrease
Increase B_2O_3	Decrease
Replace SiO_2 by TiO_2	Increase
Increase TiO_2	Slight increase
Replace SiO_2 by Al_2O_3	Slight increase
Introduce P_2O_5	Slight increase
Introduce BaO	Increase
Increase SiO_2	Decrease

According to Andrews[1] a typical sheet iron groundcoat has a tensile strength of about $10 \, kg/mm^2$. In small cross section, however, the tensile strength of glass is improved and fine threads, e.g. as in glass fibre, are quite strong. Enamels under compression are 15–20 times stronger than an equal thickness under tension.

The hardness of an enamel surface is an important property for such items as enamelled sink units, domestic appliances, washing machine tubs which have to withstand the abrasive action of buttons, etc. On Moh's scale most enamels have a hardness of up to 6 (orthoclase). There are two types of hardness of importance to users of enamel, viz. surface and subsurface. The former is more important for domestic uses when one considers the scratching action of cutlery, pans, etc. whereas subsurface hardness is the prime factor in prolonging the life of enamelled scoops, buckets, etc. in such applications as elevators or conveyors of coal and other minerals.

Of the several methods of measuring this property those specified by the Porcelain Enamel Institute and the Institute of Vitreous Enamellers are the best known and most reliable. They both consist of abrading a weighed enamel panel with a standard silica or other abrasive suspended in water and kept moving on an oscillating table with stainless steel balls. The loss in weight is measured periodically and a graph of time versus weight loss indicates both the surface and subsurface abrasion resistance. Pedder[2] has quoted relative weight loss figures for different types of enamel and they are shown in Table 17.3.

Table 17.3

Comparison of abrasion resistance of different enamels*

Types of enamel	Average loss in weight (g)†
Acid resisting titania based	56×10^{-4}
Acid resisting non-titania	342×10^{-4}
Antimony white cover coat	582×10^{-4}
High refractory enamel	129×10^{-4}
Plate glass	70×10^{-4}

*Table after Pedder[2].
†Overall figure for tests under standardised conditions for each grade of enamel.

Fine bubbles uniformly distributed throughout the coat improve elasticity and thus mill additions and under and over firing influence this property. The greatest effect on elasticity is enamel thickness and most developments are aimed at obtaining a satisfactory finish with minimum thickness.

Appen et al. have produced factors for calculating the elastic properties of enamel.

Thermal Properties

These properties are made use of in many applications ranging from domestic cookers to linings which must withstand the heat from jet engines. There is simple heat resistance, i.e. the ability of the enamel to protect the underlying metal from prolonged heat and also thermal shock resistance, which is the ability to resist sudden changes in temperature without failure occurring in the coating. These thermal properties depend upon the relative coefficient of thermal expansion of enamel and metal, enamel setting point, adhesion, enamel thickness and geometry of the shape to which the finish is applied.

It is obvious that the adhesion must be good in order to prevent rupture at the enamel/metal interface during heating and cooling. Thick coatings are liable to spall when subjected to thermal change due to differential strain set up within the enamel layer itself, caused by the poor heat conductivity of the glass. Thus again thin coatings are desirable.

Compressive forces on enamel applied to a convex surface are less than when a concave surface is coated, and it is therefore apparent that the sharper the radius of the metal the weaker the enamel applied to it will be. This fact is also relevant to mechanical damage.

Thermal shock resistance is important for gas cooker pan supports and hotplates where spillage is liable to occur, but in oven interiors heat resistance is more relevant.

The softening point of cast and sheet iron enamels is $\approx 400°C$, depending upon the load applied, but special compositions are obtainable which operate successfully at 600°C. Other more specialised enamels withstand service conditions ranging from being in excess of dull red heat, e.g. as obtains in fire backs, to those capable of enduring short exposure to temperatures of around 1 000°C, e.g. in jet tubes, after burners, etc.

Chemical Resistance

That examples of glass and glazes manufactured many centuries ago still exist is an indication of the good resistance of such ceramics to abrasion, acids, alkalis, atmosphere, etc.

In this section, chemical resistance will be divided into three parts, viz. acid, alkali (including detergents) and water (including atmosphere). Normally an enamel is formulated to withstand one of the corrosive agents more specifically than another, although vitreous enamel as a general finish has good 'all round' resistance, with a few exceptions such as hydrofluoric acid and fused or hot concentrated solutions of caustic soda or potash.

Acid resistance This property is best appreciated when the glass structure is understood. Most enamel frits are complex alkali metal borosilicates and can be visualised as a network of SiO_4 tetrahedra and BO_3 triangular configurations containing alkali metals such as lithium, sodium and potassium or alkaline earth metals, especially calcium and barium, in the network interstices.

Fused silica may be regarded as the ultimate from the acid resistance aspect but because of its high softening point and low thermal expansion it cannot be applied to a metal in the usual manner. Rupturing or distorting this almost regular SiO_4 lattice makes the structure more fluid. Thus to reduce its softening point B_2O_3 is introduced whereby some of the Si—O bonds are broken and an irregular network of

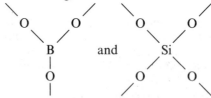

is formed. Further distortion of the network is obtained by introducing alkali and alkaline earth metals into the lattice. If fluorine is included in the frit, more bonds are broken; in this case an oxygen atom (—O—) linking two silicon or boron atoms is replaced by a fluorine atom (F—) which being monovalent cannot joint two Si or B atoms, hence causing bond rupture. A study of the relevant phase diagrams and eutectics proves useful in formulating low firing enamels.

Thus all frit ingredients act as either network formers or modifiers and with the principal exception of silica, titania and zirconia, all cause a diminution in acid resistance. The reacting acid causes an exchange between metal ions in the network modifier of the glass and hydrogen ions from the acid. This naturally occurs at the enamel surface, but as the etching or leaching reaction proceeds, a resulting thin layer of silica-rich material inhibits further reaction. Thus acid attack is dependent upon enamel composition and pH, with time and temperature playing a part. Boric oxide is also leached out, and it has been found that the Na_2O/B_2O_3 ratio is constant for any one enamel and is dependent upon enamel composition.

An increase in titania content of the frit acts in a similar way to increasing silica in enhancing acid resistance with the added advantage that the coefficient of expansion is also raised slightly and the glass viscosity not increased as much as by the equivalent SiO_2 increment. This only applies to the titania remaining in solution in the glass and does not necessarily hold when the frit is supersaturated with TiO_2, which occurs with the modern opaque sheet iron covercoats when some of the pigment recrystallises and causes opacification on cooling from the firing process.

In formulating holloware enamels the degree of acid resistance required is less than for chemical plant, e.g. reaction vessels, and consequently the RO_2 (SiO_2 and TiO_2) is lower thus permitting increased quantities of fluxes to be incorporated which confer improved 'workability'. Furthermore, they can be fired at lower temperatures and have superior chip resistance. Conversely, chemical plant enamels are higher in silica and dissolved titania and require harder firing. An example of such an enamel is shown in Table 17.1.

The acid resistance called for on domestic appliances varies with the particular component, e.g. the oven interior of a gas cooker necessitates a higher resistance than the outside sides—the former being at least Class A using 2% sulphuric acid while the latter can have a lower grading based on the less aggressive citric acid test, both of which are detailed in the relevant section of BS 1344.

The enamel mill addition, degree of firing and furnace atmosphere all affect acid resistance. An increase in clay and alkaline electrolyte detracts from this property and underfiring also has an adverse effect. The use of organic suspending agents is thus preferable to clays, from this aspect, but other factors must also be considered. Similarly the replacement of 1% milling clay by $\frac{1}{4}$% of the more colloidal bentonite is beneficial. Large additions of quartz at the mill improve heat resistance and, provided the firing temperature is increased to dissolve a sufficient quantity of this silica in the glass, the acid resistance is also enhanced.

In the glass-bottle industry the bottles can be cooled in a dilute SO_2/SO_3 atmosphere to increase chemical resistance. A similar effect has been noted with vitreous enamel. It has been postulated that a thin layer of $-OH$ groups or $-OH-H_2O$ (hydronium) ions is adsorbed on the surface of a fired enamel. These ions are transformed into $-OSO_2$ or $-OSO_3$ in the presence of oxides of sulphur which are more resistant to further acid attack. It is known that the acid resistance of a recently fired enamel improves on ageing, probably due to the enamel reaction with SO_2/SO_3 in the atmosphere and it is quite common for the grading to improve from Class A to Class AA (BS 1344).

In enamels for chemical plant such as autoclaves it is not only the degree of acid resistance which is important but also the freedom of the finish from minute flaws detectable by high frequency spark testing or chemical methods. The chemical methods depend upon a colour change when the reagent such as ammonium thiocyanate reacts with the iron exposed at the bottom of the pinhole or flaw in the finish. Alternatively, an electric cell can be formed via the exposed iron in the flaw and detected chemically.

In general, strong mineral acids are more severe in their attack on enamel than weak organic acids. Vargin[3] has stated that the severity of action of organic acids on enamel increases with the increase in the dissociation constant of the acid. Temperature plays a major part in acid resistance, the nearer the boiling point the greater the rate of attack. It is more significant than acid concentration.

It is recognised that vitreous enamel possesses good acid resistance, but an exception occurs with hydrofluoric acid. This is due to the relative ease of reaction between this acid and the silica (which is the largest constituent in the frit) to form silicon tetrafluoride. This reaction is made use of in some 'de-enamelling' plants.

Alkali and detergent resistance   The usual method of de-enamelling sheet iron is by immersion in fused or hot strong aqueous solutions of caustic soda when the silica network is broken down to form sodium silicate. However, in spite of this fact, enamels are capable of withstanding detergents and mild alkalis and this finish is often used very successfully in washing machines, baths, sink units, etc. where alkaline conditions prevail. Such enamels are usually higher in alumina than acid-resisting enamels and often contain zirconia in the frit. Other elements which aid alkali resistance are barium, calcium, lead and zinc[4] and their function in this context is to increase the bond with the essentially silica network and form insoluble silicates which act as a protective coating slowing down the formation of soluble sodium silicate. The necessity for alkali resistance is relatively limited when compared with detergent resistance and it has been shown that whilst these two properties are similar, a finish resistant to one is not necessarily as resistant to the other.

The Institute of Vitreous Enamellers produced a report on Detergent Resistance in 1959[5] and the following facts are taken from it:

1. Semi-opaque acid-resistant titania enamels and alkali-resistant frit generally have good detergent resistance whereas non-acid-resistant sign enamels and $Al_2O_3/B_2O_3/P_2O_5$-based finishes have poor resistance.
2. Initially, detergent attack is accompanied by a deposit on the enamel surface which can be abraded off resulting in an apparently unaffected glossy appearance. This contrasts with acid attack when a progressive weight loss occurs and original gloss cannot be restored once it has been lost or diminished. After more prolonged detergent attack it is not possible to restore the original high gloss.
3. The rate of attack is very dependent upon temperature, that at boiling being several times greater than that at room temperature.
4. An increase in milling clay has a marked effect on improving this property.
5. Increased detergent concentration, coarser grinding of the frit and non-standard firing all cause minor deterioration in resistance.

In the design of an enamel for a washing machine tub, detergent resistance alone is not sufficient and the enamel must also be capable of withstanding the possible abrasive action of buttons, zip fasteners, etc.

Resistance to water and atmosphere   These properties are of particular importance in enamelled signs, architectural panels, cooking utensils and hospital ware subjected to repeated sterilisation. That such enamelled signs as 'Stephen's Inks', etc. are still in existence and in good condition after many years outside exposure coupled with the fact that the use of vitreous enamel as a finish for architectural panels is growing are ready pointers to the good water and atmospheric resistance of enamel. Enamelled hospital utensils such as kidney bowls score over organic finishes because of their ease of sterilisation and also because they are less accommodating to germs, bacteria, etc. on account of their lower electrostatic type attraction for such microbes.

The action of water on enamel is in many ways similar to that of acids in that the network modifier is the weak link and through hydrolysis can be removed from the glass system resulting in loss of gloss and a porous surface. As with acids and alkalis, the attack on the glass by water can be continued in extreme cases, by an attack on the inorganic colouring matter initially liberated or made more active. In an enclosed system the soluble salts first leached out from the enamel by water become in turn the corrosive element and further attack is dependent upon the pH of such a salt, or, for example, on the Na_2O/B_2O_3 ratio.

The introduction of divalent metals into frits in preference to monovalent metals generally increases water resistance. Furthermore, oxides of tetravalent and pentavalent metals have a favourable effect on the resistance of glasses and enamels to water. The influence of B_2O_3 and fluorine in the frit upon chemical resistance is variable and is dependent upon the content of them and the balance of the frit constituents, but they usually cause a diminution in resistance. In general, mill-added clay, silica and opacifier increase water resistance provided the firing or fusing of the enamel is at the optimum.

As is expected, atmospheric resistance is related to water and the acid formed from CO_2, SO_2, SO_3, etc. The action of ultraviolet light has no apparent effect on vitreous enamel unlike the case with organic finishes.

There is good correlation between atmospheric resistance and acid resistance, and this fact is helpful to manufacturers of architectural panels who can easily and quickly determine the latter property and not have to carry out lengthy exposures to the relatively unpolluted air. An exception, however, occurs with reds and yellows where a strict correlation is not always true, and in these cases a test based upon exposure to a saturated copper sulphate solution under illumination by a white fluorescent light has been advocated.

In the main the comments recorded in this section apply to enamels fused onto sheet and cast iron. Enamel is, however, applied to aluminium, stainless steel, copper and noble metals on account of its aesthetic value and also to confer durability to the base metal. With low melting point metals such as aluminium it is obvious that superb resistance to chemicals is not so feasible as if iron was the base. Nevertheless, such metals are vitreous enamelled in growing quantities and sold, indicating that the range of colour and durability obtained is superior to that possible with alternative finishes.

It can justly be claimed that a vitreous enamel coating applied to sheet or cast iron (or indeed any other metal) will confer to the basic shape colour, gloss, texture and a high degree of resistance to corrosive influences.

N. S. C. MILLAR

REFERENCES

1. Andrews, A. I., *Porcelain Enamels*, Gerrard Press, Champaign, Ill., USA
2. Pedder, J. W. G., 'Wear and Tear of Enamelled Surfaces', *Inst. Vit. Enam.*, **9** No. 9, May (1959)
3. Vargin, V. V. (Ed.), *Technology of Enamels*, Maclaren & Sons Ltd., 31 and 78 (1967)

4. Kreuter, J. C. and Kraaijveld, Th. B., 'The Corrosion Resistance of Enamelled Articles', *Inst. Vit. Enam.*, **21** No. 2, Summer (1970)
5. I.V.E. Technical Sub-committee Report, 'An Investigation into the Effect of Detergents on Vitreous Enamel', *Bull. Inst. Vit. Enam.*, **10**, 285 (1960)

BIBLIOGRAPHY

Hughes, W., 'A Report on the Status of Electrodeposition for Porcelain Enamels', *Inst. Vit. Enam.*, **20** No. 2, Summer (1969)
Maskell, K. A., 'Practical Experiences with Electrocoating of Vitreous Enamel', *Inst. Vit. Enam.*, **20** No. 3, Autumn (1969)
Vitreous Enamels, Borax Consolidated Ltd, London SW1 (1965)

17.2 Thermoplastics

The use of plastics for corrosion-resistant coatings is a new and expanding field, consumption for this purpose having risen from ≈ 250 t/year in the early 1950s to 10 000 at the present time. While the maximum economic thickness of paint films rarely exceeds 0·25 mm, coatings of many thermoplastic materials can be applied in much greater thicknesses, and application, by dipping or spraying, is very rapid.

These coatings combine good anti-corrosive properties with good adhesion, high abrasion resistance, and attractive appearance. Their use permits simultaneous advantage to be taken of the mechanical strength of metal and the corrosion resistance of plastics.

Application Methods

The coatings can be applied in several ways which will now be described.

Dipping into a Liquid Plastisol

The plastisol method of coating consists of dipping a heated and primed article into a tank of cold liquid p.v.c. The action of the heat causes the plasticiser and polymer to cross link forming a gelatinous deposit. A subsequent curing operation completes the reaction, giving a tough resilient coating. The only material that is applied by the plastisol method is p.v.c. and it should be remembered that while this technique is ideally suited to anticorrosive or protective coatings, it does not give such an attractive appearance as can be achieved with a powder technique. Tanks, pipes or fabrications can be coated up to 12 mm thick in one operation.

Dipping into a Fine Powder

The powder technique is used principally for polythene, nylon, c.a.b., Penton and some grades of p.v.c.* It consists of dipping the heated article after the necessary metal preparation into a gas-agitated bed of fine powder. The article is subsequently reheated to fuse all the powder into a smooth

*See p. 17.15 for full names of abbreviations.

coating. Automatic machines have been developed to handle up to 10 000 articles per day by this method.

Spraying

The high labour costs of spraying tend to make this method of application uneconomic when compared to dipping, but most thermoplastic materials can be applied by spraying if size or other features make it desirable. Some fluorocarbons, e.g. p.t.f.e. and p.t.f.c.e., can be applied only by this method. Other typical examples of sprayed coatings include the lining of tanks which are too large to dip or require a one-side coating only of nylon, c.a.b., Penton or h.-d. polythene.

Electrostatic Spraying

A high voltage is applied to the nozzle of a powder spray gun and the powder spray, produced at a low velocity, acquires a strong charge. The object to be coated is earthed and so attracts the charged powder. As all parts of the object are earthed the powder will envelop it completely and not just coat the visible surfaces. The coated object is then heated to fuse the powder into·a continuous film.

The process has achieved commercial success with epoxy powders which are thermosetting and outside the scope of this review, but special p.v.c. and nylon powders are becoming available.

Vacuum Coating

This process relies upon a high vacuum to draw powder from a fluidised bed into the preheated article. When the powder has entered the article and has been dispersed the residual vacuum is maintained whilst the powder fuses onto the hot metal. The absence of air during this process yields a coating devoid of air entrapments. The process, which is patented, is best used for pipes and chemical vessels and is normally confined to Penton powder and to a lesser degree to nylon powder.

Flame Spraying

The major advantage of flame spraying is that work can be coated on site. A powder of approximately 60 mesh size is agitated and forced through a flame in front of the pistol, which is surrounded by an inert gas shield. The powder is therefore heated by convection rather than by passage through the flame itself, which would denature it. The powder in the molten state falls on to the metal in the form of globules which are subsequently fused together by gentle passes of the naked flame. Great skill is required if the coating material is not to be carbonised or embrittled by excessive heat and this is the principal limitation to the wide adoption of this process.

The most suitable method of application to be used in any instance is dependent upon many factors but, in general, it has been found that where both spraying and dipping can be employed, the cost of raw material saved by spraying only part of a component is more than offset by the extra hand work involved. Wherever possible, dipping is preferable as masking techniques are now available which enable almost any part of the component to be left entirely free of the coating medium.

Whichever method of application is chosen, pretreatment is of the utmost importance. This is carried out by degreasing and/or shot blasting, the latter being usually essential where adhesion of the coating is important. After such treatment, the coating operation must be carried out with the minimum of delay in order that rusting shall not reoccur.

Coating Plastics

At the present time there are six thermoplastics which provide satisfactory commercial coatings.
1. Plasticised polyvinylchloride (p.v.c.)
2. Penton
3. Nylon
4. Cellulose acetate butyrate (c.a.b.)
5. Polythene (high density and low density) and
6. Polytetrafluoroethylene (p.t.f.e.) and polytrifluoromonochlorethylene (p.t.f.c.e.)—a closely related material.

General characteristics of thermoplastic coatings are given in Table 17.4.

Polyvinylchloride (p.v.c.)

This is probably the most versatile of plastics coating materials. Its resistance to corrosion is slightly impaired by its plasticiser content, but it is none the less markedly superior to natural rubber and is not subject to the same deterioration with age. Its three outstanding characteristics of resilience, corrosion resistance and electrical insulation enable it to compete with rubber in many situations.

One of the advantages of p.v.c. for coating is that its properties can be varied widely, e.g. Shore hardnesses of between 35 and 95 can be obtained by varying the plasticiser and solvent percentages and by use of an appropriate polymer.

The metal to be treated is usually well cleaned by pickling, degreasing, or shot blasting, and then primed with adhesive primers based on a mixture of nitrile rubbers and phenolic resins. In addition to the spraying technique both the liquid-plastisol and the fluidised-bed method can be used for applying p.v.c. The former method is used for industrial finishes and those where the application calls for a substantial protective, rather than decorative, coating.

Coatings up to 12·5 mm thick can be applied in one dip on substrates having suitable thermal capacity. The ease with which thick, uniform coatings can be achieved with p.v.c. makes it in many cases very much cheaper than conventional rubber lining, as it eliminates the hand-work associated with the latter. The dip coating process necessitates simultaneous coating of external and internal surfaces of the metal, which can be advantageous when the article is intended for service in corrosive conditions, as the outside coating obviates the necessity for further painting or mainten-ance. If the external coating is not required, adhesive primers are not applied to the external surfaces before coating.

If a thinner attractive drip-free coating is required, then the powder coating method is the more suitable and the choice of whether p.v.c., nylon, c.a.b. or polythene is to be used will depend upon what other conditions of temperature, corrosion and abrasion are to be encountered. The technique of coating is similar to that for the other powder coatings but as the powdered raw material is more expensive, for industrial applications it is generally applied as a liquid plastisol. Air-curing cold-applied p.v.c. solutions can be applied by spraying. Its properties are not quite as good as those of the other grades available and it is rather more costly, but it can be used for on-site application. Coatings of up to 0·25 mm per pass can be obtained by this method but there is a danger of thin spots on sharp angles and corners. Complete cure takes approximately 24 h.

P.V.C. coatings have been applied as replacements for rubber linings in situations where ambient temperatures do not exceed 65°C, but limited exposure at higher temperatures of up to 100°C is harmless under certain circumstances. Typical uses are on fencing, partitioning and guards; on pumps, pipelines, fans and fume ducting; on stirrers, paddles, vessels and plating jigs; as insulation on electrical equipment and cathodic-protection plant; and as anti-abrasion coating on dredger slurry lines, fullers' earth plant, cement works and coke plants.

P.V.C. powder coatings applied to road poles and lighting columns demon-strate the excellent weatherability and toughness of this material. Powder coatings on automatic dishwashers also reveal the heat and detergent re-sistance of p.v.c.

Penton

Penton is a trade name for a chlorinated polyether—a thermoplastic which is a semitranslucent, hard, reasonably non-brittle material normally applied to a thickness of approximately 0·65 mm. Its primary interest is that it has an unusual combination of chemical and mechanical properties which opens new fields for the development engineer. By coating a relatively cheap sub-strate metal in Penton, resistance to a great many corrosive liquids at temperatures up to 120°C can be achieved at a fraction of the cost of stainless steel or glass lining. Penton is entirely non-toxic and can, of course, be steam-sterilised. It can be sprayed as a straight powder, or as a powder in a liquid carrier and subsequently sintered, or, preferably, by the simpler pro-cess of fluidised dipping as a powder and subsequently sintering.

Fluidised-bed dipping is the most suitable for quantity production of relatively small articles. Coatings up to 1·12 mm can be achieved by one dip with this process on metals which have sufficient thermal capacity, and as the Penton is in powder form there is no danger of drips.

Table 17.4 General data on thermoplastic coatings

Coating	Safe working temperature range (°C) Min.	Max. Continuous	Intermittent	Normal thickness range (mm)	Process temperature (°C)	Volume resistivity at 20°C (Ω cm)	General chemical resistance	Abrasion resistance	Toxicity	Inflammability	Shore hardness	Surface finish	Application method
P.V.C. 40% plasticised	−35	65	80	0·38–12·5	170	3×10^9	Good; poor for solvents	Very good	Very slight	Slow to self-extinguishing	60–70	Gloss	Dipping or spraying
Penton	−60	120	150	0·20–1·15	350	10^{15}	Very good	Good	Non-toxic	Slow	98	Gloss	Dipping or spraying
Nylon	−50	100	120	0·25–1	300	6×10^{13}	Good for solvents; unsuitable for strong mineral acids	Excellent	Non-toxic	Fair	95	Gloss or matt	Dipping or spraying
C.A.B.	−40	85	100	0·38–1·25	325	10^{13}	Fair	Good	Non-toxic	Slow	100	Gloss	Dipping
Polythene l.d.	−70	70	105	0·38–2·5	200	3×10^{17}	Very good; unsuitable for trichlorethylene	Fair	Non-toxic	Medium	70	Gloss	Dipping or spraying
Polythene h.d.	−70	70	115	0·30–2	250	10^{16}	Very good	Good	Non-toxic	Medium	85	Matt	Dipping or spraying
P.T.F.E.	−80	250	300	0·0125–0·125	400	10^{17}	Very good	Poor	Non-toxic	Non-flammable	75	Gloss	Spraying
P.T.F.C.E.	−70	200	225	0·125–0·25	270	$1·2 \times 10^{18}$	Very good	Fair	Non-toxic	Non-flammable	97	Gloss	Spraying

The normal surface preparation as outlined is the most satisfactory method in order to obtain maximum adhesion of these coatings.

Sprayed coatings are applied directly to the shot-blasted surface and subsequently sintered by stoving. Spray coating is normally used where the article is too large or has insufficient thermal capacity to enable the dip coating process to be carried out.

Penton can be used as an anti-corrosive coating on pumps, pipelines, valve bodies and plugs, and on stirrers, paddles and other items which require many of the outstanding corrosion-resistant properties to acids, alkalis and solvents of the fluorocarbons, but at rather lower temperatures and at a fraction of the cost.

Nylon

Although nylon is not as corrosion resistant as the other finishes available, excepting possibly c.a.b., it has other properties which make it a useful coating material. Most nylon coatings are based on nylon Type 11 or nylon Type 12 owing to their low water absorption. It can be applied to a thickness of 0·9 mm, it is easily coloured and will not chip. Its primary interest as a coating material lies in its excellent oil and solvent resistance, and in the fact that it can be used at temperatures up to 120°C, which makes steam sterilisation possible. At the same time the hardness of the coating and its excellent low frictional properties make it extremely useful as a light bearing or guide material. Nylon is applied, like other thermoplastic powders, by fluidised dipping or by spraying. It is a particularly useful material where coatings of attractive appearance and only mild corrosion resistance are required. Self lubrication can be enhanced when necessary by the impregnation of the nylon with molybdenum disulphide.

Advanced masking techniques, which have been developed over several years, now make it possible to leave any parts of an article entirely free of coating. This not only results in raw-material savings, but also leaves the coated edges smooth and bevelled. Natural adhesion to steels, aluminium and mazak is outstandingly good, but it is less satisfactory on copper and some brass alloys owing to oxidation, though the use of special primers can often overcome this.

Nylon has been applied in the form of coatings to resist trichlorethylene and methanol, erosion-resistant coatings for heat-exchanger heads, corrosion-resistant linings for pumps and pipework, decorative and semi-corrosion-resistant coatings for industrial metering equipment, electrical switchgear, etc., non-toxic coatings for the foodstuffs industry, non-contaminatory coatings for de-ionised-water equipment and non-chip coatings for metal furniture.

Cellulose acetate butyrate (c.a.b.)

Cellulose acetate butyrate is used principally as a decorative coating in similar applications to nylon. The same techniques as for nylon are used in applying this coating, and the resultant finish is self-polishing and has a higher gloss surface finish, but the adhesion is generally poor and uses are therefore restricted to those in which a completely enveloping coating is required. In general it can be said that the corrosion resistance is somewhat inferior to that of nylon, but c.a.b. has its uses on, for example, hand-wheels and levers, where the finish provides a degree of thermal insulation which protects the operators' hands from extremes of temperatures. The use of c.a.b. in powder coating has declined over the years and has now been almost entirely superseded by nylon and p.v.c.

Polythene

Two grades of polythene are available—high-density, which has a higher melting point and is tougher, and, while not giving such a decorative coating, has better adhesion to substrate metals and can withstand sterilising temperatures; and low-density which is simpler to apply and gives a more attractive coating, but is less tough. Like Penton, nylon and c.a.b., the coatings are applied as powders either by spraying or dipping, and by this means there are no drips and runs to detract from the attractive appearance of the material. Generally, high-density polythene is best for industrial applications where corrosion resistance is required. The thickness of deposit is dependent upon the thermal capacity of the substrate metal but it can be varied from 0·0004 to 0·0032 mm.

Polythene is prone to environmental stress cracking under certain adverse chemical conditions. This is aggravated by the difference between the coefficients of thermal expansion of the coating and the substrate and this defect is particularly apparent when coating sheet metal. The problem is generally overcome by annealing to produce a stress-free coating and such process control is vital when lining tanks and pipework.

High-density polythene is used as an anti-corrosive coating on fume ducting, pipework, chemical vessels and tanks, machine guards, cable trays, expanded metal, etc.

Low-density polythene is widely used in domestic products such as refrigerator shelves, self-service display wirework, etc. where a comparatively cheap decorative finish is required.

Polytetrafluoroethylene (p.t.f.e.) and Polytrifluoromono-chlorethylene (p.t.f.c.e.)

P.T.F.E. is largely used as an anti-stick or low-friction surface-coating material, and the application of p.t.f.c.e. is limited to specialised highly corrosion-resistant coatings where cost is of secondary importance. Both materials are sprayed as dispersions on to a suitably cleaned substrate, and subsequently heat sintered to fuse the particles together. In the case of p.t.f.e. these dispersions can be applied to metal and other surfaces which will withstand a processing temperature of around 400°C. Layers approximately 0·01 mm thick are applied, with sintering between each application. Dispersions applied to suitable substrates provide surfaces of outstanding anti-stick and low-friction properties, high heat stability (up to 250°C) and good low-temperature characteristics. The resistance of bulk p.t.f.e. to attack by acids, alkalis and solvents is well known, but dispersions cannot be guaranteed to give protection against chemical attack on a substrate because they cannot be built up to a thickness sufficient to eliminate micro-porosity. It is, however, possible to obtain a limited measure of protection which may prove worthwhile for some applications.

It is difficult to assess the abrasion resistance of p.t.f.e. coatings. Although they are relatively soft and might be expected to have low abrasion resistance, they have a very low coefficient of friction and may be considered self-lubricating. Therefore, unless a surface is damaged by hard, sharp or gritty substances, it may be expected to have a relatively long life.

It has been claimed that nothing sticks to p.t.f.e., but experience shows that some substances do tend to stick, especially when hot. Among these are glues and resins, but when these materials return to room temperature they can generally be very easily removed by a light tap. This aids cleaning operations so that expensive stripping down of machines is often unnecessary.

P.T.F.E. coatings applied to moulds give easy release to rubbers and parts potted in epoxy resins. The use of p.t.f.e. on rubber moulds is particularly valuable as in addition to increased speeds of production because of easy release, mouldings made are not stained by release agents and the moulds require far less maintenance.

Most of the desirable properties mentioned may be achieved with a coating 0·05 mm thick. Typical applications where full use is made of these properties are on paper-mill rollers, powder hoppers, baking tins, tyre moulds and chocolate moulds.

P.T.F.C.E. is without doubt the best of the anti-corrosive coatings. A completely non-porous coating can be applied at between 0·15 and 0·3 mm depending on the substrate metal. P.T.F.C.E. is suitable for use at temperatures up to and including 180°C, and can be applied to all metals capable of withstanding the stoving temperatures of approximately 270°C.

The stoving process is extremely long and, for this reason, the coating is by far the most expensive of all those listed. It has slightly less resistance to corrosion than p.t.f.e. and it is widely used for its electrical insulating properties on complex components of electronic equipment. With the advent of Penton, which is far less costly, the chemical uses of p.t.f.c.e. are now limited to those comparatively few instances in which Penton does not give the required degree of protection. Some typical applications are on such articles as bursting discs, diaphragm valves, high-temperature pipework, immersion heaters and mixing vessels.

The use of plastics as heavy protective coatings continues to grow at a rapid rate, as their remarkable properties when combined with substrate metals become increasingly recognised.

G. E. BARRETT

BIBLIOGRAPHY

Barrett, G. E., *Engineering Materials and Design*, **10** No. 11, 1 750 (1967)
Gemmer, E., *Kunstoffe*, **59**, 655 (1969)
Palmer, M. L., *Corrosion Technology*, 24, Aug. (1965)

17.3 Temporary Protectives

Definition

Many metal articles have to be transported and stored, sometimes for long periods, and are then used with their working surfaces in the bare state. Unless these surfaces are protected between manufacture and use, most of them will rust and corrode due to the effect of humidity or atmospheric pollution. The materials used for such protection are called temporary protectives as they provide protection primarily for the transportation and storage period. The significance of the term *temporary* lies not in the duration of the efficacy of the protective, but in the fact that it can easily be removed, so that the protected surfaces, can if necessary, be restored to their original state. They provide a water and oxygen-resistant barrier by reason of their blanketing effect and/or because of the presence of naturally occurring or added inhibitors which form an adsorbed layer on the metal surface.

Types of Temporary Protectives

There are many temporary protectives on the market and it would be impracticable to describe them individually. However, they may be classified according to the type of film formed, i.e. soft film, hard film and oil film; the soft film may be further sub-divided into solvent-deposited thin film, hot-dip thick film, smearing and slushing types. All these types are removable with common petroleum solvents. There are also strippable types based on plastics (deposited by hot dipping or from solvents) or rubber latex (deposited from emulsions); these do not adhere to the metal surfaces and are removed by peeling. In addition there are volatile corrosion inhibitors (V.C.I.) consisting of substances, the vapour from which inhibits corrosion of ferrous metals.

Soft-film Materials

Those deposited in the cold from a solvent usually consist of lanolin or petrolatum mixtures in such solvents as white spirit or coal tar naphtha. The film is thinner than other soft films deposited by different methods.

17:21

Materials applied by dipping the article to be protected in the hot molten material are usually based on petrolatum. Corrosion prevention depends largely on the barrier provided by the film, but for improved protection, corrosion inhibitors are added. The film may be relatively hard and waxy or quite soft like pharmaceutical petroleum jelly.

The smearing types of material are usually lubricating grease compositions, i.e. blends of soaps and lubricating oil, but may be mixtures containing petrolatum, oil, lanolin or fatty material. They are softer than the hot-dip materials to permit cold application by smearing.

The slushing compounds are a variant of the smearing types, and possess some flow properties at room temperature so that brush marks produced during application are reduced. Some materials contain solvent, so that they are free-flowing as applied, but stiffen when the solvent evaporates.

Hard-film Materials

These were developed to facilitate handling after treatment and to avoid contamination of adjacent components. The films are deposited in the cold and should be tough and neither sticky nor brittle. The deposited films may be plasticised resins, bitumens, etc. which are varied according to the subsidiary properties required, such as transparency and colour. The solvents used vary according to the solubility of the ingredients, drying time requirements, flammability and permissible toxicity in given circumstances. As with the soft-film solvent-deposited materials, the surface coverage is large, and for this reason, and because they can be applied at room temperature, hard and soft-film solvent-deposited protectives are widely used.

Oil-type Materials

These are usually mineral oils of medium or low viscosity, which contain specific corrosion inhibitors and anti-oxidants. In spite of the relatively low protective properties of the fluid films, which are not nearly so great as those of the previously described solid films, these materials have an established field of use on the internal surfaces of tanks and assembled mechanisms, and where solid material or solvent cannot be tolerated.

Strippable Coatings

The most important of these to date are those applied by hot dipping. Many are based on ethyl cellulose and the dipping temperature is comparatively high (about 190°C). They rely mainly on the thickness (≈ 2 mm) and toughness of the coatings for their extremely good protective properties, and they have the added advantage of giving protection against mechanical damage so that little added packaging is required for transport. Re-use of the material is frequently possible. The disadvantages are the necessity for special dipping tanks and cost; this latter may, however, be offset by saving in packaging materials.

The strippable films deposited from solvents in the cold are much thinner (≈ 0.05–0.25 mm) than those from the hot-dip materials, and their protective properties are not nearly so good. A possible difficulty which must be watched for is the development of brittleness on ageing and consequent difficulty of stripping. Latex films containing inhibitors such as sodium benzoate have been found to deteriorate under tropical conditions, but may have a use in more temperate climates.

Special Modifications of the Aforementioned Types

These have been developed for special uses. For example, since petroleum-based materials harm natural rubber, a grease based on castor oil and lead stearate is available for use on the steel parts of rubber bushes, engine mountings, hydraulic equipment components, etc. (but not on copper or cadmium alloys). Some soft-film solvent-deposited materials have water-displacing properties and are designed for use on surfaces which cannot be dried properly, e.g. water-spaces of internal combustion engines and the cylinders or valve chests of steam engines.

A recent application of this type of fluid is assistance in the removal of ingested salt spray from jet aircraft compressors and the neutralisation of corrosive effects. Other types of water-displacing fluids are claimed to have fingerprint neutralising properties or to be suitable for use on electrical equipment. Some oil-type materials serve temporarily as engine lubricants and contain suitable inhibitors to combat the corrosive products of combustion encountered in gasoline engines.

Volatile corrosion inhibitors (see also Section 18.1) are a special type of protective, which when present as a vapour inhibit the rusting of ferrous metals. They are generally used as an impregnant or coating on paper or synthetic film; as a powder, either loose or in a porous container; or in the form of a 5% w/v solution in non-aqueous solution (e.g. methylated spirits) with application by either swab or spray. Their effectiveness in preventing corrosion depends not only upon the inherent activity of the material but also upon their volatility and rate of release from the supporting medium. Being volatile, some form of enclosure is necessary for continued effectiveness whether it is the closing of orifices with bungs or overwraps when protecting internal surfaces, or by sealing the outer container for other packed stores. Volatile corrosion inhibitors should be used with caution in the presence of non-ferrous metals which may be attacked, particularly in the presence of free water. Care should also be taken with painted surfaces and with some plastics and other organic materials which may become discoloured or damaged.

The types of temporary protectives in general use are given in Table 17.5.

General Scope of the Materials

Temporary protectives against corrosion should be used only where removal is subsequently necessary for the fitting or the working of surfaces to which they are applied.

Table 17.5   Types of temporary protectives in general use

Type of protective	Typical ingredients*	Method of application	Properties of film
Solvent-deposited hard film		Dipping, spraying, brushing	Solid, thin, tough, non-sticky, removable by wiping with solvent
(a) ordinary grade	(a) Plasticised bitumens, plasticised resins, white spirit, coal tar naphtha, chlorinated solvents		
(b) water-displacing grade	(b) As (a) above together with water-displacing agents		
Solvent-deposited soft film		Dipping, spraying, brushing	Solid, thin, greasy, removable by wiping with solvent
(a) ordinary grade	(a) Lanolin, petrolatum, with and without specific corrosion inhibitors and anti-oxidants, white spirit, coal tar naphtha, chlorinated solvents		
(b) water-displacing grade	(b) As (a) above together with water-displacing agents		
Hot-dipping soft film	Petrolatum, lanolin, with and without specific corrosion inhibitors	Dipping in molten material	Solid, thick, waxy or greasy, removable by wiping with solvent or immersing in hot oil
Smearing	Metallic soap and mineral oil, soft petrolatum, lanolin (castor oil/lead stearate for rubber-containing components)	Smearing, brushing	Solid, thick, greasy, removable by wiping with solvent

These coatings are designed to protect packaged engineering materials against corrosion due to a humid atmosphere, in both rural and general industrial conditions, during transit and storage in temperate and tropical climates. Where conditions are severe, extra packaging may be required or, in the case of thick soft-film materials, extra thicknesses may be applied. The coatings are also often used to protect unpackaged spares during shelf storage.

In normal thicknesses, temporary protectives are unsuitable for outdoor exposure and they should be protected against gross liquid water by coverings or wrappings. The petrolatum-based thick-film material and some greases, however, will give adequate protection outdoors if they are applied extra thickly. Protection cannot be expected if the surfaces remain in contact with waterlogged packing material.

Corrosion preventives should be applied to surfaces which are clean and dry or corrosion may well continue beneath the coating. Materials with

Table 17.5   (continued)

Type of protective	Typical ingredients*	Method of application	Properties of film
Slushing	Metallic soap and mineral oil, oil-softened petrolatum, lanolin, small amounts of solvent	Smearing, brushing	As for smearing protective
Oil	Mineral oil, specific corrosion inhibitors and anti-oxidants	Dipping, rinsing, spraying	Liquid, thin, oily
Strippable (a) hot-dipping grade	(a) Ethyl cellulose, cellulose acetate butyrate, mineral oil, plasticiser, resins, stabilisers	(a) Dipping in molten material	(a) Solid, tough, non-adherent, often leaves oily film with lubricating properties; film removed by stripping
(b) cold applied grade	(b) Vinyl copolymer resins, plasticisers, stabilisers— flammable or non-flammable solvents	(b) Spraying, dipping	(b) Solid, tough, non-adherent film, removed by stripping
Volatile corrosion inhibitor (V.C.I.)	Organic amino salts (e.g. dicyclohexylamine nitrite, cyclohexylamine carbonate)	From solution by spraying, as a powder by sprinkling, by wrapping with V.C.I-impregnated paper	Adsorbed, non-visible film

*Some details of typical compositions, where these are available, are given in *Petroleum, Oils and Lubricants (POL) and Allied Products*, Defence Guide DG-12, Section IV, Ministry of Defence, H.M.S.O., London (1968).

special properties such as water displacement or the ability to neutralise fingerprints should not be used in place of drying and clean handling, but only where the application demands it.

Causes of Failure

In practice it is usually difficult to establish the reasons for failure as a number of factors may be simultaneously responsible, such as (a) application of the protective to dirty surfaces, (b) carelessness in application, (c) inherent inadequacy of the material, (d) exposure to unreasonably severe conditions, (e) inevitable difficulties in application. Point (c) includes inadequacy not only in protective properties but, in the case of the hard-film materials, in certain physical properties, e.g. the film may become brittle and flake when handled, may remain too sticky and become contaminated with dirt or adhere to the wrapping paper more strongly than to the surface to be protected, may age to form an insoluble material and become difficult to remove, or may not remain flexible and adherent at low temperatures. Point (*d*) includes, for example, the use of soft-film materials in hot conditions at temperatures too near to their melting point. As regards (*e*), it may be difficult to avoid thin places in the film arising from contact with other surfaces during the process of application, drying-off of the solvent, or cooling; when such thinning occurs, good surface-active properties are advantageous. In this connection, it may be pointed out that scraping in transit and stacking, and local thinning due to grit, dirt, etc. are common; it follows therefore that shelf storage of unpacked items should be avoided if possible.

General Comments on Application

Application by dipping gives the most complete film, is the most economical in material, and is usually the quickest for large quantities of articles. This method should be chosen whenever possible. Spraying is the next best. Brushing and hand-smearing should be adopted only when dipping or spraying is not feasible.

During the dipping process, articles with recesses should be rotated in the bath so that air can escape. Dipping baths should be kept covered when not in use to prevent contamination, and, in the case of solvent-containing materials, to prevent concentration by evaporation of the solvent, as this would lead to excessive film thicknesses and long drying times. The composition of a bath of solvent-containing material should be checked periodically. Unaided evaporation of the solvent from solvent-deposited films is usual, but the process can be speeded up by blowing air over the articles or by gentle warming; the heating, however, should not be excessive.

During hot-dipping in petrolatum-based materials, film thickness can be varied by altering the temperature of dipping and the duration of immersion. The petrolatum will first chill on to a cold article put in the bath, the solid coating bridging small crevices. This may give sufficient protection, but it may be desirable for the article to attain the temperature of the bath so that the molten petrolatum will penetrate into all the crevices, e.g. between the ball and race of a rolling bearing. The article may then be withdrawn, allowed to cool and given a quick dip to build up the film thickness.

Choice of Temporary Protective

Hard-film protectives can be applied to most types of single articles and are especially suitable in mass-production systems. They should not be applied to assemblies because the hard film is liable to cement mating surfaces together and considerable difficulty may arise in the removal of the protective film. This type of protective should be removed before the article is put into use.

The soft-film solvent-deposited type can be used broadly for the same purposes as the hard-film type. A grease-resistant wrapping is required as an inner wrapping (as for all soft-film types) in packaging. Grades of this material, consisting essentially of lanolin in a solvent, have been found to give better protection to packaged articles than some of the best available hard-film materials, and are to be preferred for articles with very high precision surfaces. The film is usually dispersable in lubricating oil and it is therefore not so important to remove it from surfaces when an article comes into use except when it has become contaminated with grit and dirt.

The thick soft films produced by hot dipping are suitable for highly finished as well as normal machined surfaces. Grades with drop-points substantially higher than 50°C are preferable for tropical storage as otherwise marked softening and possible thinning of the protective film is likely to occur. These films can be applied to many types of assemblies, the chief exceptions being assemblies with inaccessible interiors that cannot readily be blanked-off and fine mechanisms where any residue might interfere with the free movement of parts or their subsequent lubrication with low viscosity oil. These films can also be used on parts which might be affected by the solvent from the thin or soft film protectives, but they should not be applied to items having plastics or leather components.

Greases are usually applied by brush or smearing; the brush must be sufficiently stiff to give intimate contact with the surface yet not so stiff as to leave deep brush marks. Greases should not be melted and therefore cannot be applied by dipping or spraying; also, no attempt should be made to dissolve them in a solvent for application. They are particularly useful where only part of the surface of the item requires protection, because of the ease of application by cold smearing. They can be used in this way also in conjunction with solvent-deposited protectives for assemblies of a low degree of complexity, by coating screw threads and filling clearance spaces before dipping the article in the solvent-containing protective. Grease films can be made thick enough to give the desired level of protection. Wrapping is desirable to protect the very soft film. Removal before use is chiefly for the purpose of removing grit and dirt.

The slushing material finds its most useful application on big machinery requiring protection of large areas during storage or during intervals of idleness in machine shops. The effect of dust and dirt contamination should therefore be considered an important factor in assessing the quality of these materials.

The lower protective quality of oil-type materials largely restricts their use on internal surfaces of, for example, internal combustion engine cylinders, and gear-box and back-axle assemblies of motor vehicles. Such materials are widely used to fulfil the simultaneous function of a protective and a lubricating oil; e.g. in sewing machines the protective can also serve as a lubricant during its initial period of use. The functions of corrosion inhibitor and hydraulic oil are also often combined. Oil-type materials are also used on small nuts, screws and washers which cannot easily be protected by solid-film materials; in this case protection must be reinforced by good packaging.

The hot-dip strippable coating is applicable when a high standard of protection from corrosion and mechanical damage is required, as on gauges and tools which so often have their working surfaces facing outwards. Assemblies must have orifices plugged so that molten material cannot penetrate during the dipping.

Volatile corrosion inhibitors are particularly useful when oil, grease or other adherent films are unsuitable. They should be used in conjunction with a primary wrap which should form as close an approach to a hermetically-sealed pack as possible. They are widely used to provide protection to precision tools, moulds and dies, and also on a larger scale to car body components.

General Remarks

The listing of so many types of protective might indicate some complication in use. It should, however, be realised that the materials are to some extent interchangeable, and in most works it is seldom necessary to have more than two or three materials. It is emphasised that protection should be given by the manufacturer of the article as soon as possible after its fabrication; if stocks have to be held in a part-finished state, protection should also be given during this period. This is important for cast iron because corrosion once started is difficult to stop. If the conditions at the receiver's works or depot are particularly severe, the maker's protective processes should be appropriately supplemented.

The bibliography given below is classified according to the aspect of the subject mainly dealt with, but some references, of course, deal with several aspects. In addition there is a considerable body of patent literature concerning specific inhibitors.

E. W. BEALE

BIBLIOGRAPHY

General Description of Types and Mode of Use

Albin, J., *Iron Age.*, **155** No. 23, 52 (1945)
Anon., *Mod. Packag.*, 1764 (1944)*
Boyer, J. R. C., *Steel*, **116** No. 24, 129 and 176 (1945)
Carpenter, H. B., *Iron Steel Engng.*, **24** No. 9, 73 (1947)
Gould, B., *Iron Age*, **155** No. 24, 66 (1945)*
Houghton, E. F. *et al.*, *Steel*, **116** No. 14, 106 and 149 (1945)
Larson, C. M., *Nat. Petrol. News*, **37**, R609 (1945)
Lurchek, J. G., *Iron Steel Engng.*, **26** No. 5, 82 (1949)
Malm, C. J., Nelson, H. B. and Hiatt, G. D., *Industr. Engng. Chem.*, **41**, 1 065 (1949)*
Petroleum, Oils and Lubricants (POL) and Allied Products, Defence Guide DG–12, Ministry of Defence, H.M.S.O. London (1968)
Pohl, W., *Erdol u. Kohle*, **8**, 552 (1955)
Prince, W. H., *Mod. Plast.*, **22**, 116 (1944)*
Rhodes, C. M. and Chase, G. F., *Mod. Packag.*, **18**, 117 (1945)
Sellei, H. and Lieber, E., *Corros. Mat. Prot.*, **5**, 10–12 and 22 (1948)
Shearon, W. H. and Horberg, A. J., *Industr. Engng. Chem.*, **41**, 2 672 (1949)
Stroud, E. G. and Vernon, W. H. J., *J. Appl. Chem.*, **2**, 173 (1952)†
Temporary Protection of Metal Surfaces Against Corrosion (During Transport and Storage), BS 1 133: Section 6: 1965 (also deals extensively with testing)
Waring, C. E., *Mod. Packag.*, **19**, 143 and 204 (1946)*

*Hot-dip strippable coatings
†Rubber-latex-based strippable coatings

Clarke, S. G. and Longhurst, E. E., Selected Government Research Reports (London), 3: *Protection and Electrodeposition of Metals*, 135, H.M.S.O., London (1951)†
Hickel, A. E., *Petrol Refin.*, **27**, 424 (1948)
Inst. Petrol. Protectives Panel, *J. Inst. Petrol.*, **40**, 32 (1954)
McConville, H. A., *Gen. Elect. Rev.*, **49** No. 10, 30 (1946)*
Schwiegler, E. J. and Berman, L. U., *Lubric. Engng.*, **11**, 381 (1955)
Stroud, E. G. and Rhoades-Brown, J. E., *J. Appl. Chem.*, **3**, 287 (1953)†
Symposium on the Testing of Temporary Corrosion Preventives (15 authors), *J. Inst. Petrol.*, **36**, 423 (1950)
Walters, E. L. and Larsen, R. G., *Corrosion*, **6**, 92 (1950)
Wright, W. A. S., Amer. Soc. Test. Mater., Spec. Tech. Pub. No. 84, 18 (1948)*

*Lubricating greases
†Lanolin solutions

Investigations of Mode of Action

Baker, H. R., Jones, D. T. and Zisman, W. A., *Industr. Engng. Chem.*, **41**, 137 (1949)
Baker, H. R., Singleterry, C. R. and Solomon, E. M., *Industr. Engng. Chem.*, **46**, 1 035 (1954)
Baker, H. R. and Zisman, W. A., *Industr. Engng. Chem.*, **40**, 2 338 (1948)
Barnum, E. R., Larsen, R. G. and Wachter, A., *Corrosion*, **4**, 423 (1948)
Bigelow, W. C., Pickett, D. L. and Zisman, W. A., *J. Colloid Sci.*, **1**, 513 (1946)
Cessina, J. C., *Industr. Engng. Chem.*, **51**, 891 (1959)
Hackerman, N. and Schmidt, H. R., *Corrosion*, **5**, 237 (1949)
Hackerman, N. and Schmidt, H. R., *J. Phys. Chem.*, **53**, 629 (1949)
Kaufman, S. and Singleterry, C. R., *J. Colloid Sci.*, **7**, 453 (1952)
Kaufman, S. and Singleterry, C. R., *J. Colloid Sci.*, **10**, 139 (1955)
Pilz, G. P. and Farley, F. F., *Industr. Engng. Chem.*, **38**, 601 (1946)
Van Hong, Eisler, L., Bootzin, D. and Harrison, A., *Corrosion*, **10**, 343 (1954)

18 CONDITIONING THE ENVIRONMENT

18.1   Conditioning the Atmosphere to Reduce Corrosion   **18**:3
18.2   Corrosion Inhibition: Principles and Practice   **18**:9
18.3   Mechanism of Corrosion Prevention by Inhibitors   **18**:34
18.4   Boiler Feed-Water Treatment   **18**:57

18.1 Conditioning the Atmosphere to Reduce Corrosion

The impurities normally present in uncontrolled atmospheres are capable of producing serious corrosion on many metals and alloys which do not corrode significantly in clean, dry air (Section 2.2). It is therefore in principle possible to prevent corrosion by purifying the atmosphere, or by using a volatile corrosion inhibitor. In extreme cases, pure, dry nitrogen under positive pressure can be used. These methods will seldom be practicable with working equipment, but they may offer the most attractive solution in transport or storage, especially since they are often very effective against the particular hazards of these conditions. Temporary protectives (Section 17.3) may also be used.

The most important corrosive agents to be considered are water vapour, acid fumes (particularly sulphur dioxide) salts and hydrogen sulphide. Water plays an essential part in stimulating attack by all the other agents, except hydrogen sulphide, so that drying the atmosphere is the most important single means of preventing corrosion. Control of other contaminants will, however, be important where satisfactory drying is not practicable.

Control of Relative Humidity

At high relative humidity the common corrosive agents produce a film of aqueous electrolyte on exposed metal surfaces. No significant corrosion results on iron, zinc, aluminium, copper or their alloys (apart from tarnishing by hydrogen sulphide), unless the relative humidity is above 60% (Section 2.2). In packaging and storage, the relative humidity is usually kept below 50%. Packages are most conveniently protected with desiccants, but for larger volumes, drying by cooled surfaces may be used, and in store-rooms, the relative humidity can be kept down by heating.

Desiccants and Desiccated Packages

Desiccating agents used in corrosion prevention must be cheap, easy to handle and non-corrosive. These requirements rule out many of the familiar laboratory desiccants, and in practice the most common packaging desiccants are silica gel, activated alumina and quicklime (calcium oxide). Activated clays are sometimes also used, and for very low relative humidities, molecular sieves.

Silica gel and activated alumina present few practical problems. They are easily reactivated after use by heating in a ventilated oven, to 130–300°C for silica gel, and 150–700°C for activated alumina. British standard specifications have been published for desiccants for packaging[1, 2], which regulate the contents of soluble chloride and sulphate, dust content and absorptive capacity.

Quicklime is less easy to handle, and swells considerably on hydration. It is cheap, however, and is often used on open trays to protect process equipment, machinery, furnaces, etc. during shut-down periods. If it is accidentally flooded with water, the slurry of hydrated lime provides an alkaline medium in which uncoated steel surfaces will remain without rusting.

Packages intended for use with desiccants must have low permeability to water vapour. It is therefore necessary to consider the design of the package in relation to the storage life required. This subject is beyond the scope of the present work, and guidance should be sought from standard textbooks on packaging[3, 4]. The B.S.I. Packaging Code[5] includes sections on desiccants, temporary protectives, and the use of various types of packaging materials.

The following formulae are used for calculating the weight of desiccant required for a given package:

1. For tropical storage with average water-vapour pressure $3·2 \, \text{kN/m}^2$
   $$W = 40ARM + \text{Dunnage Factor}$$
2. For temperate storage with average water-vapour pressure $1·0 \, \text{kN/m}^2$
   $$W = 11ARM + \text{Dunnage Factor}$$
3. For completely impervious packages:

$$W = \frac{V}{6} + \text{Dunnage Factor}$$

Where W = weight (g) of 'basic desiccant' (i.e. one which absorbs 27% of its dry weight of moisture in an atmosphere maintained at 50% r.h. at 25°C),
A = area (m²) of the surface of the desiccated enclosure,
D = weight (g) of hygroscopic blocking, cushioning and other material inside the barrier (including cartons, etc.),
M = maximum time of storage (months),
R = water-vapour transmission rate of the barrier ($\text{g m}^{-2}\text{d}^{-1}$) measured at 90% r.h. differential and 38°C and
V = volume (litre) of the air inside the barrier.

Dunnage Factor is $D/5$ for timber with moisture content higher than 14%, $D/8$ for felt, carton board and similar materials, and $D/10$ for plywood and timber with moisture content less than 14%.

Rates of transmission may be affected by creasing, scoring, etc. especially for waxed papers, and of course also strongly depend on thickness. Information can be obtained from suppliers of materials, or measurements can be made according to a method given in BS 3177:1959[6] which includes a table of representative values. General guidance on materials is also given in Sections 7 and 21 of the B.S.I. Packaging Code.

With desiccants with absorptive capacities differing from 27%, the weight calculated from these formulae will need to be proportionately adjusted. Packs of desiccant are obtainable commercially containing quantities stated in terms of basic desiccant.

Dry Storage and Dry Rooms

Storage rooms are similar in principle to packages, but the rate of entry of moisture is less predictable. Replacement of the air and diffusion of water vapour will have a considerable effect on the atmosphere with building materials other than glass and metals, and will vary markedly with weather conditions.

Desiccating agents can be exposed on open trays in store rooms, but in some cases, continuous circulation of the air through the desiccant may be preferable. Finely divided desiccant should be prevented from reaching exposed metal surfaces.

In most cases, however, the air is dried by condensation on a cooled surface, or the relative humidity is lessened without actually removing water vapour by heating the store (Section 2.2). Some practical points need to be considered in these cases:

1. Ventilation is necessary in heated stores, even if the heaters do not themselves produce water vapour, for otherwise the relative humidities will probably rise because water vapour is desorbed from building materials. Ventilation is even more important if gas or kerosine heaters are used.
2. The relative humidity of the air must be measured in relation to the temperature of the metal surfaces to be protected. If incoming air at 83% relative humidity at 13°C is heated to 18°C, its relative humidity will fall to 60%, but if it then comes into contact with surfaces at 10°C or below, condensation will occur until their temperatures rise sufficiently to prevent it. This situation can arise with massive metallic objects during a sudden change in the weather, or if temperature is allowed to fluctuate between day and night. It may thus be necessary to keep a store heated in summer as well as in winter, and to heat sufficiently to keep the average relative humidity as low as 30% if the maximum is not often to exceed 50%. The relative humidity and temperature of the store should be measured and recorded regularly if this method of preventing corrosion is to be operated economically and effectively.
3. Condensation may lead to corrosion when components are placed in relatively impervious wrappings in warm and humid workrooms or stores and then transferred to cold surroundings, and this should be taken into account in choosing the packaging technique.

Elimination of Contaminants

Many common materials are not severely corroded even at high relative humidity so long as the surfaces are clean, and dust particles and gaseous contaminants are eliminated from the air. It is seldom practicable to rely entirely on this method of protection, although copper and silver can be protected from tarnishing by wrappings impregnated with salts of copper, lead or zinc[7], which react with hydrogen sulphide. Elimination of contaminants is nevertheless desirable, since it will minimise damage if other measures (such as desiccation) become ineffective during storage, and also because it will often improve the performance of the object in its ultimate application.

Surface cleaning as a preparation for coatings is discussed in Sections 12.1 and 12.2. It is important to control degreasing baths to prevent accumulation of water and formation of corrosive products which will contaminate the atmosphere as well as the objects being degreased. In the case of trichlorethylene, stabilisers are added to prevent formation of hydrochloric acid[8]. Exclusion of dust is beneficial, and may necessitate filtering the air or use of a temporary protective.

Sweat residues   These contain fatty acids and sodium chloride, and increase the risk of corrosion after handling. Components should be washed in a solution of 5% water in methanol.

Packaging materials   Materials to be used in contact with metals should be as free as possible from corrosive salts or acid. BS 1133, Section 7:1967 gives limits for non-corrosive papers as follows: chloride, 0·05% (as sodium chloride); sulphate, 0·25% (as sodium sulphate) and pH of water extract 5·5–8·0. Where there is doubt, contact corrosion tests may be necessary in conditions simulating those in the package.

Organic materials   Corrosive vapours are sometimes emitted by organic materials used either in packaging or in the manufactured article, and may be troublesome in confined spaces. Some woods, particularly unseasoned oak and sweet chestnut, produce acetic acid (see Section 19.10), and certain polymers used in paints, adhesives and plastics may liberate such corrosive vapours as formic acid and hydrogen sulphide[9]. It may be necessary to carry out exposure trials, particularly where materials capable of liberating formaldehyde or formic acid are involved. Most corrosion problems of this kind can be prevented by using desiccants, and in many cases they are confined to imperfectly cured materials. For an excellent recent review see Reference 10.

Volatile Corrosion Inhibitors

Atmospheric corrosion can be prevented by using volatile inhibitors which need not be applied directly to the surfaces to be protected. Most such inhibitors are amine nitrites, benzoates, chromates, etc. They are mainly used with ferrous metals. There is still some disagreement as to the mechanism of action. Clearly, any moisture that condenses must be converted to an inhibitive solution. It appears that in most cases, the effective inhibitor could equally well be applied as an ester or the sodium salt. On this view, amine salts would be useful in practice for avoiding acid conditions, or because their volatility makes them convenient (see below), rather than for any specific effect of the amine, e.g. in preventing adsorption. Certain free amines have considerable effect as volatile inhibitors. It should be said, however, that a large variety of substances, such as β-naphthol or *m*-dinitrobenzene, have some inhibitive action[11] and some of these may act by hindering wetting of the metal surface. A fairly recent development is the use of compounds containing reducible nitro groups, which are thought to act by stimulating the cathodic process, thus assisting anodic polarisation. An inhibitor of this type, hexamethyleneimine 3:5 dinitrobenzoate, is in use in the USSR, and appears to be effective with a wide range of metals[12].

Commercially available inhibitors differ in respect of volatility, the pH of the aqueous solution, and in attacking some metals while protecting others. The choice of inhibitor may therefore involve a compromise. In order to secure protection in rather aggressive conditions, it may be necessary to choose a relatively volatile inhibitor, which is quickly transferred into the vapour, so that condensed moisture is made innocuous as it forms. This will be particularly necessary with large structures. Such a material, however, will also be quickly lost from the enclosure compared with one less volatile and therefore slower acting. It may be advantageous to use a more alkaline inhibitor where there is contamination by acid fumes, and mixed inhibitors have been employed on this basis. It has been suggested that inhibitors could be designed to control volatility, alkalinity, etc.

Di*cyclo*hexylammonium nitrite[13] (DCHN) has a solubility of 3·9 g in 100 g of aqueous solution at 25°C, giving a solution pH of about 6·8. Its vapour pressure at 25°C appears to be about $1·3 \times 10^{-3} \text{ N/m}^2$, but the value for commercial materials depends markedly on purity. It may attack lead, magnesium, copper and their alloys and may discolour some dyes and plastics. *Cyclo*hexylammonium *cyclo*hexyl carbamate (the reaction product of *cyclo*hexylamine and carbon dioxide, usually described as *cyclo*hexylamine carbonate or CHC)[14] is much more volatile than DCHN (vapour pressure 53 N/m^2 at 25°C), and much more soluble in water (55 g in 100 cm^3 of solution at 25°C, giving a pH of 10·2). It may attack magnesium, copper, and their alloys, discolour plastics, and attack nitrocellulose and cork. It is said to protect cast iron better than DCHN, and to protect rather better in the presence of moderate concentrations of aggressive salts.

Both these materials are available commercially as powders and in impregnated wrapping papers and bags. Various modified inhibitors are also available, containing mixtures of the two, or more alkaline materials such as guanidine carbonate. Other proprietary inhibitors contain volatile amines, e.g. morpholine, combined with solution inhibitors. Certain solution inhibitors have been reported to act to some extent as volatile inhibitors, e.g. sodium nitrite[15]. On the whole, the use of these materials appears to be consistent with the principles stated above, and they provide a very convenient means of protection, particularly for complex not-too-large equipment, where the surfaces are not too heavily contaminated, and conditions of enclosure are reasonably good. Dosages of 35 g/m^3 of free space, or 11 g/m^2 of surface have been recommended for packages. CHC may have some advantage in large, impervious structures, such as boilers, box girders, etc. if openings can be fitted with caps. Volatile inhibitors containing borate (for zinc) and chromate (for copper and its alloys) have been discussed in the literature, but little commercial development appears to have taken place in the UK. A review of inhibitors against atmospheric corrosion is given by Rosenfel'd and Persiantseva[16].

Volatile inhibitors can be applied as loose powder in trays, by insufflation, in sachets, in sprays, or in impregnated wrappings. The application needs to be carefully considered in the light of the design and materials of construction of the equipment and its package, and the cleanliness of the surfaces.

Acknowledgement

Extracts from the British Standards Packaging Code BS 1133, Section 7:1967 and Section 19:1968 quoted in this section are reproduced by permission of the British Standards Institution, 2, Park Street, London, W1A 2BS, from whom copies of the complete standard may be obtained.

G. O. LLOYD

REFERENCES

1. *Silica Gel for Use as a Desiccant for Packages*, BS 2 540:1960, British Standards Institution, London
2. *Activated Alumina for Use as a Desiccant for Packages*, BS 2 541:1961, British Standards Institution, London
3. *Fundamentals of Packaging*, Institution of Packaging and Blackie, London (1967)
4. *Modern Packaging Encyclopaedia*, McGraw Hill, New York, published annually
5. *Packaging Code*, BS 1 133. Section 6:1966, Temporary Protection of Metal Surfaces Against Corrosion (During Transport and Storage); Section 7:1967, Paper and Board Wrappers, Bags and Containers; Section 19:1968, Use of Desiccants in Packaging; Section 21:1964, Transparent Cellulose Films, Plastics Films, Metal Foil and Flexible Laminates. British Standards Institution, London
6. *Permeability to Water Vapour of Flexible Sheet Materials*, BS 3 177:1959, British Standards Institution, London
7. Chemistry Research 1957, Report of the Director of the Chemical Research Laboratory, D.S.I.R., H.M.S.O., London, 17 (1958)
8. *Trichlorethylene*, BS 580:1963, British Standards Institution, London
9. Rance, V. E. and Cole, H. G., *Corrosion of Metals by Vapours from Organic Materials*, H.M.S.O. London (1958)
10. Donovan, P. D. and Stringer, J., *British Corrosion Journal*, **6**, 132 (1971)
11. Rajagopalan, K. S., Subramanyan, N. and Sundaram, N., *Proc. 3rd Int. Cong. Metallic Corrosion, Moscow 1966*, **2**, Mir, Moscow, 179 (1969)
12. Rozenfel'd, I. L., Persiantseva, V. P. and Terentiev, P. B., *Corrosion*, **20**, 222t (1964)
13. Shell VPI, Technical Bulletin ICS/70/9, Shell Chemicals, London (1970)
14. *Machinery*, **85**, London, 630 (1954)
15. Gars, I. and Schwabe, K., *Werkstoffe u. Korr.*, **14**, 842 (1963)
16. Rozenfel'd, I. L. and Persiantseva, V. P., *Zaschita Metallov.*, **2**, 5 (1966); English Translation: *Protection of Metals*, **2**, 3, Scientific Information Consultants, London (1966)

BIBLIOGRAPHY

Paine, F. A. (Ed.), *Packaging for Environmental Protection*, Newnes-Butterworths (1974)

18.2 Corrosion Inhibition: Principles and Practice

Introduction

Corrosion may be defined as 'the undesirable reaction of a metal or alloy with its environment' and it follows that control of the rate of process may be effected by modifying either of the reactants. In 'corrosion inhibition', additions of certain chemicals are made to the environment, although it should be noted that the environment can, in some cases, be made less aggressive by other methods, e.g. removal of dissolved oxygen or adjustment of pH.

Environments are either gases or liquids, and inhibition of the former is discussed in Section 18.1. In some situations it would appear that corrosion is due to the presence of a solid phase, e.g. when a metal is in contact with concrete, coal slurries, etc. but in fact the corrosive agent is the liquid phase that is always present*. Inhibition of liquid systems is largely concerned with water and aqueous solutions, but this is not always so since inhibitors may be added to other liquids to prevent or reduce their corrosive effects— although even in these situations corrosion is often due to the presence of small quantities of an aggressive aqueous phase, e.g. in lubricating oils and hydraulic fluids (see Section 2.11).

The majority of inhibitor applications for aqueous, or partly aqueous, systems are concerned with three main types of environment:

1. Natural waters, supply waters, industrial cooling waters, etc. in the near-neutral (say 5–9) pH range.
2. Aqueous solutions of acids as used in metal cleaning processes such as pickling for the removal of rust or rolling scale during the production and fabrication of metals, or in the post-service cleaning of metal surfaces.
3. Primary and secondary production of oil and subsequent refining processes.

Following a brief discussion of inhibitor classifications and of types of chemicals used as inhibitors, the principles and practice of inhibition are considered in terms of the principal factors affecting inhibitor performance (Principles) and the systems in which inhibitors are used (Practice).

---

*Attack of metal surfaces by the mechanical action of solid materials is properly described as erosion and is not discussed here.

Inhibitor Classifications

A number of methods of classifying inhibitors into types or groups are in use but none of these is entirely satisfactory since they are not mutually exclusive and also because there is not always general agreement on the allocation of an inhibitor to a particular group. Some of the main classifications—used particularly for inhibitors in near-neutral pH aqueous systems—are as follows.

'Safe' or 'dangerous' inhibitors   Each inhibitor must be present above a certain minimum concentration for it to be effective (*see* Principles), and this classification relates to the type of corrosion that will occur when the concentration is below the minimum, or critical, value. Thus, when present at insufficient concentration a 'safe' inhibitor will allow only a uniform type of corrosion to proceed at a rate no greater than that obtaining in an uninhibited system, whereas a 'dangerous' inhibitor will lead to enhanced localised attack, e.g. pitting, and so in many cases make the situation worse than in the absence of an inhibitor.

Anodic or cathodic inhibitors   This classification is based on whether the inhibitor causes increased polarisation of the anodic reaction (metal dissolution) or of the cathodic reaction, i.e. oxygen reduction (near-neutral solutions) or hydrogen discharge (acid solutions).

Oxidising or non-oxidising inhibitors   These are characterised by their ability to passivate the metal. In general, non-oxidising inhibitors require the presence of dissolved oxygen in the liquid phase for the maintenance of the passive oxide film, whereas dissolved oxygen is not necessary with oxidising inhibitors.

Organic or inorganic inhibitors   This distinction is based on the chemical nature of the inhibitor. However, in their inhibitive action many compounds that are organic in nature as, for example, the sodium salts of carboxylic acids, often have more similarities with inorganic inhibitors.

Other classifications   Soviet authors[1] classify inhibitors as Type A to include film-forming types, or Type B which act by de-activating the medium, e.g. by removal of dissolved oxygen. Type A inhibitors are then further sub-divided into A(i) inhibitors that slow down corrosion without suppressing it completely, and A(ii) inhibitors that provide full and lasting protection. From the practical aspect, a useful classification is perhaps one based on the concentration of inhibitor used. It is usually the case that inhibitors are used either at low concentrations, say less than approximately 50 p.p.m., or at rather higher levels of greater than 500 p.p.m. The determining factors in the selection of the concentration used, and hence the type of inhibitor, are the economics, disposal (effluent) problems, and the facilities available for monitoring the inhibitor concentration.

Types of Chemicals Used as Corrosion Inhibitors

Before discussing the nature of chemicals that are used specifically as corrosion inhibitors, reference must be made to two methods of water treatment that are sometimes included in descriptions of inhibitive treatments. These are, respectively, de-aeration techniques and pH control. Since the presence of dissolved oxygen is necessary to sustain the corrosion process in most aqueous systems the removal of this gas by mechanical or chemical methods is an obvious method of corrosion control. The chemicals commonly used are sodium sulphite or hydrazine. There are two distinct mechanisms involved in controlling corrosion by controlling the pH. Firstly, the pH is adjusted to ensure that the metal is exposed to a solution of a pH value at which corrosion is minimal. In the case of ferrous metals, corrosion tends to decrease with pH values higher than approximately 9·0. Hence, simple additions of alkali, such as caustic soda, lime, soda ash, etc. can reduce the corrosion rate of iron and steel. On the other hand such treatment will increase the corrosion rate of other metals, particularly of aluminium and its alloys, and so pH adjustment is not advisable in mixed metal systems. Secondly, the pH is adjusted to give deposition of thin protective carbonate scales from waters of suitable composition. For water saturated with calcium bicarbonate a rise in pH will cause precipitation of calcium carbonate. The pH adjustment to achieve this can be determined from the Langelier (*see* Section 2.3) or Ryzner Stability Indices; these require a knowledge of the pH of the actual system and of the pH of the water when it is saturated with calcium carbonate. It must be emphasised that such calculations measure only the scale forming propensity of the water, and are not direct measurements of the extent of corrosion reduction since other factors can influence the degree of protection afforded by the scale.

A common feature of both these methods is that the quantity of treatment chemical can be calculated from stoichiometric relationships* in the reactions involved. This is not so with conventional inhibitor treatments. With these the concentration of inhibitive chemicals can only be determined on the basis of experimental laboratory studies, service trials and overall practical experience.

The scientific and technical corrosion literature has descriptions and lists of numerous chemical compounds that exhibit inhibiting properties. Of these only a very few are ever actually used in practical systems. This is partly due to the fact that in practice the desirable properties of an inhibitor usually extend beyond those simply relating to metal protection. Thus cost, toxicity, availability, etc. are of considerable importance as well as other more technical aspects (*see* Principles). Also, as in many other fields of scientific development, there is often a considerable time lag between laboratory development and practical application. In the field of inhibition the most notable example of this gap between discovery and application is the case of sodium nitrite. Originally reported in 1899[2] to have inhibitive properties, it remained effectively unnoticed until the 1940s[3]; it is now one of the most widely employed inhibitors.

Some examples from recently published review papers will indicate the wide range of chemicals that show inhibitive properties. Hersch et al.[4] in an extensive laboratory study examined over 70 compounds many of which were good inhibitors. Trabanelli et al.[5] in discussing organic inhibitors list

---

*In practice an excess over the stoichiometric requirement, e.g. of sulphite for de-aeration, is used.

and discuss some 150 compounds. Recent extensive reviews by Indian workers include those on inhibitors for aluminium and its alloys[6] (225 references) and for copper[7] (93 references). Corrosion inhibitors in industry have been reviewed by Rama Char[8] (134 references). More detailed studies of the properties and uses of individual inhibitors also yield much useful data as, for example, that by Walker[9] who gives 92 references in discussing the use of benzotriazole as an inhibitor of copper corrosion. In addition to reviews of this type there are two books entirely devoted to the subject of corrosion inhibition[1, 10].

For near-neutral aqueous solutions the function of inhibitors of the anodic class is generally considered to be that of assisting in the maintenance, repair or reinforcement of the natural oxide film that exists on all metals and alloys. Typical examples of such inhibitors for mild steel include the soluble chromates, dichromates, nitrites, phosphates, borates, benzoates and salts of other carboxylic acids. Some (nitrites and chromates) are oxidising compounds, whereas others show no oxidising capability. The 'safe' or 'dangerous' aspect of these inhibitors varies considerably and depends very much on circumstances. In the presence of aggressive ions, i.e. those that oppose the action of inhibitors (*see* The Composition of the Liquid Environment), the oxidising type tend, when present in insufficient quantity for complete protection, to give localised attack. However, the non-oxidising type, e.g. benzoate[11], can also show this type of behaviour but to a less marked extent. Other compounds used in near-neutral aqueous solutions include polyphosphates, silicates, zinc ions, tannins and soluble oils. These are usually assigned to the cathodic class although some are reported to affect the anodic reaction. Their function is to precipitate thin adherent films on cathodic areas of the corroding metal surface thus preventing access of oxygen to these sites. Zinc ions can react with cathodically produced hydroxyl ions to produce insoluble hydroxides that are partially protective. Similar reactions lead to the formation of films incorporating phosphates and silicates. In general these cathodic inhibitors are considered safe, i.e. not giving rise to localised attack in non-protective conditions.

The extent of inhibition afforded to metals other than mild steel depends on the metal and the inhibitor (*see* The Nature of the Metal, and Dissimilar Metals in Contact). The cathodic type of inhibitor is perhaps less susceptible than the anodic type to the nature of the metal. However, cathodic inhibitors are usually less efficient (although performing quite satisfactorily in many systems) in terms of reduction in corrosion rate, than are anodic inhibitors. The latter, when used in adequate concentrations, can often achieve 100% protection.

In a very few cases there are inhibitors that have been developed for the protection of specific metals, e.g. sodium mercaptobenzothiazole and benzotriazole for preventing the corrosion of copper.

In acid conditions oxide films are not usually present on the metal surface and the cathodic reaction is primarily that of hydrogen discharge rather than oxygen reduction. Thus, inhibitors are required that will adsorb or bond directly onto the bare metal surfaces and/or raise the overpotential for hydrogen ion discharge. Inhibitors are usually organic compounds having N, S or O atoms with free (donor) electron pairs. There are exceptions to this bonding principle: some quaternary ammonium compounds with no donor electrons have inhibitive properties in acid solutions.

In modern practice, inhibitors are rarely used in the form of single compounds—particularly in near-neutral solutions. It is much more usual for formulations made up from two, three or more inhibitors to be employed. Three factors are responsible for this approach. Firstly, because individual inhibitors are effective with only a limited number of metals the protection of multi-metal systems requires the presence of more than one inhibitor. (Toxicity and pollution considerations frequently prevent the use of chromates as 'universal' inhibitors.) Secondly, because of the separate advantages possessed by inhibitors of the anodic and cathodic types it is sometimes of benefit to use a formulation composed of examples from each type. This procedure often results in improved protection above that given by either type alone and makes it possible to use lower inhibitor concentrations. The third factor relates to the use of halide ions to improve the action of organic inhibitors in acid solutions. The halides are not, strictly speaking, acting as inhibitors in this sense, and their function is to assist in the adsorption of the inhibitor on to the metal surface. The second and third of these methods are often referred to as *synergised* treatments.

Principles

The nature of the metal   Since the majority of inhibitors are specific in their action towards particular metals, an inhibitor for one metal may have no effect and even an adverse effect on other metals. Table 18.1 is a general guide to the effectiveness of various inhibitors for metals in the near-neutral pH range. In addition, the compound dodecamolybdophosphate is re-

Table 18.1   General guide to the effectiveness of various inhibitors in the near-neutral pH range

| | *Inhibitor* | | | | | | |
Metal	*Chromates*	*Nitrites*	*Benzoates*	*Borates*	*Phosphates*	*Silicates*	*Tannins*
Mild steel	Effective	Effective	Effective	Effective	Effective	Reasonably effective	Reasonably effective
Cast iron	Effective	Effective	Ineffective	Variable	Effective	Reasonably effective	Reasonably effective
Zinc and zinc alloys	Effective	Ineffective	Ineffective	Effective	—	Reasonably effective	Reasonably effective
Copper and copper alloys	Effective	Partially effective	Partially effective	Effective	Effective	Reasonably effective	Reasonably effective
Aluminium and aluminium alloys	Effective	Partially effective	Partially effective	Variable	Variable	Reasonably effective	Reasonably effective
Lead–tin soldered joints	—	Aggressive	Effective	—	—	Reasonably effective	Reasonably effective

ported[12] as approaching chromates in its ability to prevent the corrosion of a number of metals. However, there is at present only one reported application in practical systems (*see* Inhibitors in Practice: Central Heating Systems).

It must be emphasised that anions usually considered aggressive towards some metals can actually reduce or even prevent corrosion of other metals in certain situations, thus effectively becoming inhibitors. For example, although nitrates[11,13,14] can prevent the inhibitive action of benzoate, chromate, nitrite, etc. towards mild steel they can be incorporated into some inhibited antifreeze formulations to reduce the corrosion of aluminium alloys. Nitrates have also been reported[15] as the only inhibitors capable of preventing the stress-corrosion cracking of type 304 stainless steel. On the other hand inhibitors are necessary to prevent the corrosion of mild steel in ammonium nitrate solutions[16]. Sulphates generally behave as aggressive ions towards mild steel and other metals in waters, but can inhibit the chloride-induced pitting of stainless steels[17] and caustic embrittlement in boilers.

Dissimilar metals in the same system   Because of the specific action of many inhibitors towards particular metals, problems arise in systems containing more than one metal. In the majority of cases these problems can be overcome by the choice of a formulation incorporating inhibitors for the protection of each of the metals involved. With this procedure it is necessary not only to maintain an adequate concentration of each of the inhibitors but also to ensure that they are present in the correct proportion. This is because of two effects: firstly, failure to inhibit the corrosion of one metal may intensify the attack on the other metal; the best example of this is with aluminium and copper in the same system, and failure to inhibit copper corrosion—usually achieved with sodium mercaptobenzothiazole or benzotriazole—can lead to increased corrosion of the aluminium as a result of deposition of copper from copper ions in solution on to the aluminium surface. Secondly, an inhibitor of the corrosion of one metal may actually intensify the corrosion of another metal. Thus, benzoate is usually used to prevent the corrosion of soldered joints by nitrite inhibitor added to protect cast iron in the same system. A benzoate:nitrite ratio of greater than 7:1 is necessary in these cases.

Inhibitors can also lead to the co-called 'polarity-reversal' effects. In corrosive environments the zinc coating on galvanised steel acts sacrificially in preventing the corrosion of any exposed steel. However, in the presence of sodium benzoate[18,19] or sodium nitrite[20] steel exposed at breaks in the zinc coating may corrode quite readily.

Nature of the metal surface   Clean, smooth, metal surfaces usually require a lower concentration of inhibitor for protection than do rough or dirty surfaces. Relative figures for minimum concentrations of benzoate, chromate and nitrite necessary to inhibit the corrosion of mild steel with various types of surface finish have been given in a recent laboratory study[11,13,14]. These results show that benzoate effectiveness is particularly susceptible to surface preparation. It is unwise, therefore, to apply results obtained in laboratory studies with one type of metal surface preparation to other surfaces in practical conditions. The presence of oil, grease or corrosion products on metal surfaces will also affect the concentration of inhibitor required with the added danger of a marked depletion of inhibitor during service as a result of its chemical reaction with these contaminants. It is thus advisable to remove such contaminants before commencing inhibitor treatment. This can be done mechanically, but chemical cleaning may often be necessary. A particular method of preparing rusted surfaces is that involving the phosphate-delayed-chromate (P.D.C.) technique[21,22], in which the system is first treated with an acid phosphate solution to remove rust prior to the introduction of chromate inhibitor. The latter can then be used at a lower concentration than might otherwise have been necessary.

Nature of the environment   This is usually water, an aqueous solution or a two- (or more) component system in which water is one component. Inhibitors are, however, sometimes required for non-aqueous liquid systems. These include pure organic liquids (Al in chlorinated hydrocarbons); various oils and greases: and liquid metals (Mg, Zr and Ti have been added to liquid Bi to prevent mild steel corrosion by the latter[23]). An unusual case of inhibition is the addition of NO to N_2O_4 to prevent the stress-corrosion cracking of Ti–6Al–4V fuel tanks when the N_2O_4 is pressurised[24].

In at least one case water may actually act as an inhibitor, such as in the methanol corrosion of titanium[25].

In all circumstances it is important to ensure that the inhibitor is chemically compatible with the liquid to which it is added. Chromates, for example, cannot be used in glycol antifreeze solutions since oxidation of glycol by chromate will reduce this to the trivalent state which has no inhibitive properties.

Composition of the liquid environment   The ionic composition, arising from dissolved salts and gases, has a considerable influence on the performance of inhibitors. In near-neutral aqueous systems the presence of certain ions tends to oppose the action of inhibitors. Chlorides and sulphates are the most common examples of these aggressive ions, but other ions, e.g. halides, sulphides, nitrates, etc. exert similar effects. The concentration of inhibitor required for protection will depend on the concentrations of these aggressive ions. Laboratory tests[11,13,14,26] have given some quantitative relationships between inhibitor (C_i) and aggressive ion (C_a) concentrations that will provide protection for mild steel. These are of the form

$$\log C_i = K \log C_a + \text{constant}$$

where K is related to the valencies of the respective ions.

Although halide ions are aggressive in near-neutral solutions they can be utilised to improve the action of inhibitors in acid corrosion (*see* Practice: Acid Solutions). Variations exist among the halides, e.g. chloride ions favour the stress-corrosion cracking of Ti in methanol whereas iodide ions have an inhibitive action[27].

Dissolved solid and gaseous impurities can also affect the pH of the system and this may often lead to decreased inhibitor efficiency. In industrial plant, cooling waters can take up SO_2, H_2S or ammonia and pH control of inhibited waters will be necessary. The leakage of exhaust gases into engine coolants is an example in which corrosion can occur despite the presence of inhibitors.

pH of the system   All inhibitors have a pH range in which they are most effective and even in nominally 'neutral' solutions close pH control is often necessary to ensure the continued efficiency of inhibitive treatments. Nitrites lose their effectiveness below a pH of $5\cdot5$–$6\cdot0$; polyphosphates should be used between pH $6\cdot5$ and $7\cdot5$; chromates, although less susceptible to pH changes, are generally used at about pH $8\cdot5$; silicates can be used over a wide pH range but the $Na_2O:SiO_2$ ratio depends on the pH value of the water.

Temperature of the system   When inhibitors are used in the 0–100°C range it is usually found that higher concentrations become necessary at the higher temperatures[11, 13, 14]. Other inhibitors can lose their effectiveness altogether as the temperature is raised. A prime example of this is the polyphosphate type of inhibitor. This is effective in circulating systems at temperatures below about 40°C, but at higher temperatures reversion to orthophosphate can occur and this species is ineffective at the concentrations at which it will then be present. If calcium ions are present, additional loss of inhibitor will occur due to calcium phosphate precipitation.

Inhibitor concentration   To be fully effective all inhibitors require to be present above a certain minimum concentration. In many cases the corrosion that occurs with insufficient inhibitor may be more severe than in the complete absence of inhibitor (see 'Safe' and 'Dangerous' inhibitors). Not only is the initial concentration of importance but also the concentration during service. Inhibitor depletion may occur for a variety of reasons. In the initial stages of use, i.e. after the first application, the inhibitor concentration may fall off rapidly due to its reaction with contaminants in the system and also as a result of protective film formation. Thus, initial concentrations of inhibitor are often recommended to be at higher levels than those subsequently to be maintained. Losses may also occur due to mechanical rather than chemical effects as, for example, with windage losses in cooling towers, blow-down in boilers, and leakages generally.

Maintenance of a correct inhibitor concentration (level) is particularly important where low-level treatments, e.g. less than 100 p.p.m. are used. Such treatments are, however, usually applied (for economic and effluent reasons) in large capacity systems, and plants of this nature will usually have skilled personnel available for control purposes. In smaller closed systems, e.g. automobile engines, higher concentrations of more than approximately $0\cdot1\%$ are commonly used, but in these applications there is usually a good reserve of inhibitor allowed for in the recommended concentration and routine checking is of less importance. Nevertheless, since these inhibitors are often of the 'dangerous' type, gross depletion may lead to enhanced corrosion.

Inhibitor control can be effected by conventional methods of chemical analysis, inspection of test specimens or by instrumentation. The application of instrumental methods is becoming of increasing importance particularly for large systems. The techniques are based on the linear (resistance) polarisation method and the use of electrical resistance probes. They have the advantage that readings from widely separated areas of the plant can be brought together at a central control point. (See Section 20.3.)

Mechanical effects   Corrosion can often be initiated or intensified by the conjoint action of mechanical factors. Typical examples include the presence of inherent or applied stresses, fatigue, fretting or cavitation effects. Inhibitors that are effective in the absence of some or all of these phenomena may not be so in their presence. In fact it may not always be possible to use inhibitors successfully in these situations and other methods of corrosion prevention will be required.

Aeration and movement of the liquid   For the majority of inhibitors in the near-neutral aqueous systems an adequate supply of dissolved oxygen is necessary for them to function properly. The dissolved oxygen present in solutions that are in equilibrium with atmospheric air is adequate for this purpose, but in systems that have become de-aerated the non-oxidising type of inhibitor may not be fully effective. Even in aerated systems the transport of oxygen and inhibitor to the metal surface is assisted by the movement of the solution. In fact, quiescent solutions may require higher concentrations of inhibitor than do circulation systems. Butler[28] has shown, for example, that polyphosphates (normally applied only to flowing solutions) can inhibit under quiescent conditions but at much higher concentrations. However, there are reported instances of excessive aeration having an adverse effect on inhibitor performance*. The action of tannins is partly associated with their effects at the metal surface, i.e. as conventional inhibitors, and partly with their ability to react with and remove dissolved oxygen. In heavily aerated systems these inhibitors may be less effective due to depletion by this latter effect.

Presence of crevices, dead-ends, etc.   Effective protection by inhibitors relies on the continued access of inhibitor to all parts of the metal surface (*see* Aeration and Movement of the Liquid). It frequently happens that this condition is difficult to achieve due to the presence of crevices at joints, dead-ends in pipes, gas pockets, deposits of corrosion products, etc. Corrosion will then occur at these sites even though the rest of the system remains adequately protected.

Effects of micro-organisms   There are three main effects that can arise as a result of the presence of micro-organisms in aqueous solutions: (*a*) direct bacterial participation in metal corrosion usually due to the action of sulphate-reducing bacteria in anaerobic conditions or of the *Thiobacillus* and *Ferrobacillus* genera in aerobic conditions; the action of these organisms can lead to the accumulation of large amounts of corrosion product and pitting of the metal; (*b*) accumulation of flocculent fungal growths that can impede water flow and (*c*) breakdown and hence depletion of inhibitors by bacterial attack.

Many inhibitors will lose their effectiveness in the presence of one or more of these effects. Indeed inhibitors may act as nutrient sources for some microbial organisms. In these circumstances it will be necessary to incorporate suitable bactericides in the inhibitor formulations.

---

*Apart from extreme cases involving cavitation effects.

Scale formation   Controlled scale deposition by the Langelier approach or by the proper use of polyphosphates or silicates is a useful method of corrosion control, but uncontrolled scale deposition is a disadvantage as it will screen the metal surfaces from contact with the inhibitor, lead to loss of inhibitor by its incorporation into the scale and also reduce heat transfer in cooling systems. Apart from scale formation arising from constituents naturally present in waters, scaling can also occur by reaction of inhibitors with these constituents. Notable examples are the deposition of excess amounts of phosphates and silicates by reaction with calcium ions. The problem can be largely overcome by suitable pH control and also by the additional use of scale-controlling chemicals.

Toxicity, disposal and effluent problems   With the increasing awareness of environmental pollution problems, the use of and disposal of all types of treated waters is receiving greater attention than ever before. This often places severe restrictions on the choice of inhibitor, particularly where disposal of large volumes of treated water is involved. The disposal of chromate and phosphate inhibitor formulations is important in this respect and there is an increasing move towards the low-chromate-phosphate types of formulation. In fact for some applications even this approach is not acceptable and inhibitor formulations containing bio-degradable chemicals are being introduced.

Other considerations   In addition to the above general factors affecting inhibitor application and performance, there will be other special effects relating to particular types of systems, e.g. in oil-production technology. Some of these are referred to in appropriate cases in the following section.

Inhibitors in Practice

A difficulty arises in describing the precise chemical nature of many inhibitor formulations that are actually used in practice. With the advancing technology of inhibitor applications there are an increasing number of formulations that are marketed under trade names. The compositions of these are, for various reasons, frequently not disclosed. A similar problem arises in describing the composition of many inhibitor formulations used in the USSR. Here the practice is to use an abbreviated classification system and it is often difficult to trace the actual composition, although in many cases a judicious literature search will provide the required information.

The following discussion is thus restricted to inhibitor formulations that can be described in chemical terms.

Aqueous Solutions and Steam

Potable Waters

In these waters there is a severe limitation on inhibitor choice because of the potability and toxicity factors. As pointed out by Hatch[29], the possibilities are limited to calcium carbonate scale deposition, silicates, polyphosphates and zinc salts. Silicates do not prevent corrosion completely and their inhibitive effect is more marked in soft waters. The molar ratio $Na_2O:SiO_2$ is important. For example, Stericker[30] has proposed $Na_2O:3\cdot3SiO_2$ at 8 p.p.m. for most waters, but $Na_2O:2\cdot1SiO_2$ is preferred if the pH is below 6·0. Concentrations of 4–10 p.p.m. are recommended, and the method of application is often by by-passing part of the flow through a silicate (water-glass) reservoir, the slow dissolution giving the required inhibitor concentration in the main flow. With polyphosphates the most efficient inhibition is obtained in the presence of divalent ions such as Ca^{2+} or Zn^{2+}; in fact the Ca^{2+}:polyphosphate ratio is more important than the actual concentration. A minimum value of 1:5 has been given for this ratio with an overall concentration of up to 10 p.p.m. The optimum pH is in the 5–7 range and the inhibitive action is often improved by the addition of zinc salts. Hatch[29] points out that the treatment concentration depends on the nature of the water distribution system. Thus, with small towns a feed of 5 p.p.m. is needed to provide a residual of 0·5–1 p.p.m. whereas for the more compact systems in cities a feed of 1 p.p.m. is often sufficient. The action of the inhibitor is affected by existing deposits in the mains, and higher initial doses of about 10 p.p.m. are often required. Even higher dosages, say 50–100 p.p.m., can be used for cleaning old mains.

Cooling Systems

For the purposes of corrosion inhibition these may be broadly divided into three types: (*a*) 'Once-through', in which the cooling water runs continually to waste as in the condenser systems of power plants; (*b*) 'open', in which cooling towers are used to dissipate heat taken up by the cooling water elsewhere; (*c*) 'closed' in which the cooling water is retained in the system, the heat being given up via a heat-exchanger as in refrigeration plant, vehicle cooling systems, etc. Systems (*a*) and (*b*) are generally much larger in terms of water-capacity and metal area than those of type (*c*).

Once-through systems   Where mild steel is the primary metal of construction, this usually being so for low-chloride waters, simple treatments with lime and soda can be effective in making the water less aggressive. Of the conventional inhibitors, polyphosphates at 2–10 p.p.m. with small amounts of zinc ions will reduce tuberculation but not necessarily the overall corrosion rate[31]. The use of 9 p.p.m. of an organo-activated zinc–phosphate–chromate inhibitor has been described[32] and this can be replaced, although with some loss in effectiveness, by 10 p.p.m. of polyphosphate if effluent problems exist. Effluent and economic problems in fact limit the choice of inhibitors, and the solution to the corrosion problem may lie in selecting a more suitable material of construction. Mild steel is often avoided and non-ferrous alloys such as the cupro-nickels and aluminium brasses are employed. These alloys are normally resistant even in aerated saline waters but corrosion problems can arise. Small amounts of iron, arising from the alloy or from elsewhere in the system, contribute towards the resistance of these alloys. Bostwick[33] showed the advantage of adding $FeSO_4$ to condenser systems, and recently Brooks[34] confirmed that 1 p.p.m. of this chemical added three times daily for about

1 h to power-house intakes has contributed 25–30% to the life of the condenser tubes. More recently high-molecular-weight water-soluble polymers of, for example, the non-ionic polyacrylamide type have been described[35] for inhibiting the corrosion of cupro-nickel condenser tubes.

Open recirculating systems   These are more amenable to inhibition since it is possible to maintain a closer control on water composition. Corrosion inhibition in these systems is closely allied to a number of other problems that have to be considered in the application of water treatment. Most of these arise from the use of cooling towers, ponds, etc. in which the water is subject to constant evaporation and contamination leading to accumulation of dirt, insoluble matter, aggressive ions and bacterial growths, and to variations in pH. A successful water treatment must therefore take all these factors into account and inhibition will often be accompanied by scale prevention and bactericidal treatments.

The controlled deposition of thin adherent films of calcium carbonate is probably the cheapest method of reducing corrosion, but may not always be entirely satisfactory because local variations in pH and temperature will affect the nature and extent of film deposition.

Treatment with conventional inhibitors may be obtained at two levels of concentration. At the higher levels the choice is probably restricted to either chromate or nitrite. The use of chromates has been described by Darrin[36] who emphasises the need for a high initial dosage of 1 000 p.p.m. which can subsequently be lowered to 300–500 p.p.m. The principal drawbacks of this method are the possibility of localised attack if chloride or sulphate contents rise during operation and the problems associated with chromate disposal. Sodium nitrite used at about 500 p.p.m. is also susceptible to chloride and sulphate and the pH control (7·0–9·0) is probably more important than with chromates. Nitrite is susceptible to bacterial decomposition and can give rise, particularly if reduced to ammonia, to stress-corrosion cracking of copper-base alloys. However, nitrite is used with success in cooling tower systems. Bacterial decomposition of nitrite can be controlled with bactericides. In air-cooling systems Conoby and Swain[37] quote the use of a shock treatment of 2,2′methylene bis (4-chloro-phenol) at 100 p.p.m. and a weekly addition of sodium penta-chlor phenate to control algae formation.

On the low-level treatment side, polyphosphates, variously described as glassy phosphates, hexametaphosphate, etc. have been used as corrosion inhibitors. The concentrations recommended are somewhat above those used in 'threshold treatment' to control scale deposition. The most effective protection is obtained in the presence of an adequate quantity of calcium, magnesium or zinc ions. In general a polyphosphate : calcium ratio expressed as P_2O_5 : Ca of not greater than 3·35 : 1 is recommended. The overall concentration will vary with conditions, but for cooling towers this falls in the 15–37 p.p.m. (as P_2O_5) range. When starting treatment, a higher initial dosage is required, this may be as high as 100 p.p.m. Fisher[32] suggests an initial dosage of 20 p.p.m. for a corroded steel cooling system dropping to 10 p.p.m. after one week's operation. The application and properties of polyphosphates have been reviewed by Butler[38] and by Butler and Ison[39]. Polyphosphate inhibitors are subject to some limitations that are mainly concerned with reversion to orthophosphate and subsequent scale deposition if the calcium concentration is high. Some difficulties with their use have been summarised by Beecher et al.[40].

Silicates at about 20–40 p.p.m. are also used in cooling-water treatment although the build-up of protection can be slow. At higher temperatures calcium silicate may be deposited from hard waters.

Modern practice, on grounds of economy and avoidance of pollution, is towards the use of a combination of inhibitors at low concentration levels. Four main types of compound are involved, viz. chromates, polyphosphates, zinc salts and organic materials, and these are used in various combinations[31, 41]. The principle involved is to combine a cathodic with an anodic type of inhibitor, e.g. zinc ions and/or polyphosphate with chromate. These mixed inhibitor systems usually require an operating pH of 6–7[(42)] and thus should only be used where pH control facilities are available. Typical formulations include 10–12 p.p.m. of a 1:4 $Na_2Cr_2O_7$:Zn mixture[43] which provides good inhibition of copper as well as of steel corrosion, and 35 p.p.m. of a zinc-chromate-organic mixture. The latter introduces 12 p.p.m. of CrO_4^{2-} and 3·5 p.p.m. of Zn^{2+} (added as $ZnSO_4$) into the water; the organic compound is described as a powerful surface-active agent[40]. The zinc-dichromate method is further improved by adding phosphate and sometimes organic compounds such as lignosulphonates and synthetic polymers. Comeaux[41] has listed the constituents of nine commercially available inhibitors. Each of these contains chromate and zinc with the CrO_4:Zn ratio varying from 0·92 to 30·0, five contain phosphate with Zn:PO_4 from 0·1 to 3·24 and three contain organic compounds. In some formulations of the zinc-phosphate type the organic compound will be of the mercaptobenzothiazole type to inhibit corrosion of copper[31]. Five to ten p.p.m. of polyphosphate is said[31] to assist the inhibitive action of 20–40 p.p.m. of silicate but it is still important to avoid calcium silicate deposition on heat transfer surfaces. However, in recommending a silicate–complex phosphate inhibitor (25 p.p.m. at pH 6·5–8·0) Ulmer and Wood[44] state that scale formation is not a problem if the silicate is below 100 p.p.m., except if film boiling occurs, when scaling would occur in any case. The use of 100 p.p.m. of orthophosphate plus 40 p.p.m. of chromate plus 10 p.p.m. of polyphosphate has also been recommended[45].

As anti-pollution requirements become more demanding the use of even these low-level chromate-phosphate treatments is not always approved. New inhibitor formulations employ more acceptable bio-degradable organic compounds, often in conjunction with zinc ions. A formulation consisting of an organic heterocyclic compound plus zinc salt plus an 'alkalinity stabilising agent' has been described[46] applied initially at 500 and then at 100 p.p.m. Organic phosphorus-containing compounds have been introduced for scale control but also for corrosion inhibition. These are salts of aminomethylenephosphonic acid (AMP)[31, 47]. In conjunction with zinc salts they can be used in place of other treatments and have the advantage that close pH control is not required. Corrosion inhibitors compounded from zinc salts and derivatives of methanol phosphonic acid are described in a US patent[48].

Closed recirculating systems This type of system is most commonly encountered in the cooling of internal combustion engines. Inhibitors are required for engine coolants in order to prevent corrosion of the constructional metals, to prevent blockage of coolant passages by corrosion products and to maintain heat-transfer efficiency by keeping metal surfaces free from adherent corrosion products. The problem is often associated with inhibition of antifreeze solutions which are almost invariably ethanediol*–water solutions. When uninhibited these can become acid due to oxidation of the ethanediol in operating conditions. However, inhibition is also important with water coolants that are used in the summer months[49]. The best practice is to use inhibited antifreeze throughout the year changing it annually. Three types of systems may be distinguished, although many problems, and inhibitors, are common to all three.

Road vehicles Numerous formulations exist for coolant inhibition in road vehicles. The inhibitors most frequently encountered are nitrite, benzoate, borax, phosphate, and the specific copper inhibitors sodium mercaptobenzothiazole (NaMBT) and benzotriazole. Various combinations of these are in use. In the UK three compositional British Standards[50] are widely used, i.e. BS 3 150, 3 151 and 3 152. The inhibitors are incorporated into the neat, undiluted, ethanediol in amounts such that when the ethanediol is diluted 1:3 with water the required inhibitor concentration will be obtained. BS 3 150 contains triethanolammonium orthophosphate (T.E.P.), which is prepared by neutralising $0.9–1.0\%$ H_3PO_4 with triethanolamine so that the pH of a 50% aqueous solution is $6.9–7.3$, and NaMBT ($0.2–0.3\%$). This formulation is based on the original work of Squires[51]. The T.E.P. protects ferrous metals and aluminium alloys and the NaMBT protects copper and copper alloys. In the absence of NaMBT corrosion of copper can occur leading to marked attack on aluminium alloys. The NaMBT concentration becomes depleted with time, but experience indicates that with normal usage in road vehicles an annual replacement of the whole coolant will give satisfactory results. BS 3 151 contains $5.0–7.5\%$ sodium benzoate plus $0.45–0.55\%$ sodium nitrite in the undiluted ethanediol and is based on the original work of Vernon *et al.*[52]. The nitrite is for protection of cast iron with the benzoate to protect other metals, but more importantly to protect soldered joints against the adverse action of nitrite. The nitrite concentration depletes in service, but again a one year period of satisfactory inhibition is provided. BS 3 152 contains borax ($2.4–3.0\%$ $Na_2B_4O_7 \cdot 10H_2O$). Some controversy exists as to the efficiency of borax used alone, particularly with aluminium alloys. Nevertheless, borax has been much used and service experience has shown satisfactory inhibition. More recent formulations include other inhibitors, e.g. 3% borax plus 0.1% mercaptobenzothiazole plus 0.1% sodium metasilicate ($Na_2SiO_3 \cdot 5H_2O$) plus 0.03% lime (CaO)—the percentages being $\%$ by weight of the ethanediol which is then used at 33 vol. $\%$ dilution[53].

Alkali-metal phosphates have been incorporated in antifreeze solutions but there are indications of unfavourable behaviour with aluminium alloys under heat-transfer conditions. Soluble oils have also been used as inhibitors but they can cause deterioration of rubber hose connections.

Since new formulations are continually being proposed and introduced there is now a tendency towards performance rather than compositional

---

*Ethylene glycol.

specifications although acceptable performance criteria will need to be agreed.

Locomotive diesels   As larger volumes of coolant are required in railway locomotives than in road vehicles, the cost of inhibition is proportionally greater. An additional factor is the possibility of cavitation attack of cylinder liners. These considerations place a restriction on the choice of inhibitors. In the past, chromates have been used at concentrations of up to 0·4%, but their use presents handling and disposal problems. Chromates cannot be used with ethanediol antifreeze solutions. A 15:1 borate–metasilicate at a concentration of 1% has been used in the UK. Nitrate is added to this to improve inhibition of aluminium alloy corrosion. Tannins and soluble oils are also used, but probably to a lesser extent than in the past. The benzoate–nitrite formulation (BS 3 151) is effective and has been used by continental railways[54].

Marine diesels   Again a wide number of formulations are in use. The inhibitors commonly employed include nitrites, borates and phosphates. Typical formulations include a 1:1 nitrite:borax mixture at 1 250–2 000 p.p.m. and pH 8·5–9·0; and 1 250–2 000 p.p.m. of nitrite with addition of tri-sodium phosphate to give phenolphthalein alkalinity.

The factors affecting railway diesels apply also to marine diesels but with the additional restriction that the inhibitors must not present a toxicity hazard when the cooling system is associated with equipment for producing drinking water. This is because of the possibility of accidental leakage between the two systems.

Central Heating Systems

The principal components in these systems are a cast iron or steel boiler, copper or steel pipework, pressed steel or cast-iron radiators and a copper hot-water tank or calorifier to supply heat to domestic water. If systems are properly designed, installed and maintained, the concentration of dissolved oxygen in the circulating water—which should be subject to little make-up —is low and corrosion is minimal. Nevertheless, corrosion problems occasionally arise in these systems. Often these are associated with ingress of oxygen but this is not always so. The main problems are the perforation of pressed-steel radiators and the necessity for the frequent release of hydrogen gas from radiators. The latter effect is associated with the production of excess amounts of magnetite (Fe_3O_4) as a result of the Schikorr reaction:

$$3Fe(OH)_2 \rightarrow Fe_3O_4 + H_2 + 2H_2O$$

Thin adherent films of magnetite form on the steel surfaces in the initial stages of operation of the system, but in troublesome situations the magnetite becomes non-adherent and in extreme cases can lead to pump blockages.

These difficulties can often be overcome with inhibitive treatments although the procedure is not acceptable where there is any possibility of inadvertent mixing of the heating water with domestic water.

The excess magnetite problem has been associated with the catalytic action of copper ions on the Schikorr reaction[55,56]. Hence, inhibitors that will prevent copper dissolution should reduce magnetite formation. For this purpose 0·01% benzotriazole can be added to the water. For general corrosion inhibition a mixture of 1·0% sodium benzoate with 0·1% sodium nitrite has been successfully used in a number of installations[57-59]. Sodium metasilicate has been used with success, but usually in softened waters. Other workers suggest that it is not reliable due to the possibility of localised attack (results with $Na_2Si_2O_5$)[58] and because of possible pipe and pump blockage by gel-formation or precipitation of hardness salts. The use of a silicate–tannic acid treatment has also been described. A further development is the introduction, based on test rig results, of a four-component formulation containing sodium benzoate, sodium nitrite, sodium dodecamolybdophosphate and benzotriazole[60].

Some corrosion inhibitors can encourage the production of fungal growths in the relatively static cold water in the header tanks of these systems. Biocides will then often need to be included in the inhibitor formulation.

Steam-condensate Lines

The causes and inhibition of corrosion in steam-condensate lines have been reviewed by Obrecht[61]. The major causes of corrosion are carbon dioxide and oxygen and the problems are associated not only with damage to the pipes, which may be of steel but often of copper-base alloys, but also with iron and copper pick-up which will be deposited elsewhere in the circuit. Neutralisation treatments can be employed to keep the pH in the 8·5–8·8 region[62]. Typical compounds used for this purpose are ammonia, cyclohexylamine, morpholine and benzylamine. An important requirement is that these agents should condense at the same rate as the steam. This is not necessarily so with ammonia and pockets of unneutralised condensate may occur. Furthermore, ammonia can cause attack of copper-base alloys. The amines, except at high concentrations, are less aggressive in this respect, they have better distribution characteristics and condense at the same rate as the steam. An important disadvantage with these materials is their cost, since about 3 p.p.m. are needed per p.p.m. of CO_2[62] and so they tend to be used only in high recovery systems.

Inhibitive (as opposed to purely neutralisation) techniques now employ long-chain aliphatic amines with alkyl groups containing 8–22 carbon atoms[61-63]. The most effective are the straight-chain aliphatic primary amines with C_{10-18}, the best known example being octadecylamine and its acetate salt. They are used at a total concentration of only 1–3 p.p.m. and are effective against carbon-dioxide and oxygen-induced corrosion. They function by producing a non-wettable film on the metal surfaces. The acetate salt is used to facilitate dispersion and solubilisation[57]. The most effective distribution is achieved by injection into the boiler or the main steam header. The protective film ceases to form at about 200°C[63] and in a condensing turbine system inhibition will be provided through the feed system up to the point where the feed reaches this temperature. These inhibitors have been successfully applied to prevent exfoliation of 70 Cu–30 Ni tubes[61].

Contrary to the method of application of inhibitors to water systems, the initial addition of filming amine should be at a lower concentration than that subsequently used. This is because the surface-active nature of the amine will loosen and remove existing corrosion products and these will accumulate elsewhere in the system. A cleaning-up phase of up to a month may therefore be necessary to avoid these effects.

High-chloride Systems

(sea-water, desalination, refrigerating brines, road de-icing salts, etc.)

Complete inhibition of corrosion in waters containing high concentrations of chloride is difficult, if not impossible to achieve economically. Despite this, many such systems make use of inhibitors to give marked reductions in corrosion rates.

In refrigerating brines, chromates at a pH of about 8–8·5 have been widely used. Concentrations recommended are between 2 000 and 3 300 p.p.m. corresponding to the 125 and 200 lb (56·7 and 90·7 kg) of sodium dichromate per 1 000 ft^3 (28·32 m^3) for calcium or sodium chloride brines, respectively, recommended by the American Society of Refrigerating Engineers.

In diluted sea-water high concentrations of sodium nitrite can bring about a reduction in the corrosion of steel. For example, corrosion in 50% sea-water can be inhibited with 10% sodium nitrite[64], and 3–7% of this inhibitor has been recommended for preventing the corrosion of turbine journals due to sea-water ingress[65]. The beneficial effects of mixtures of chromates and phosphates have been reported[66]. Combinations of these inhibitors have been examined with respect to preventing corrosion in desalination plant. Oakes et al.[67] have reported good results with 5 p.p.m. of chromate plus 30–45 p.p.m. of Na_2HPO_4. Legault et al.[68] conclude that three mixtures are effective for mild steel in oxygen-saturated sea-water at 121°C, i.e. dichromate plus phosphate at 50 p.p.m., chromate plus phosphate plus zinc plus iodide at 100 p.p.m. and chromate plus phosphate at 50 p.p.m. The chromate: phosphate ratio is usually 1:1 with Na_3PO_4 as the phosphate.

Various opinions exist as to the value of inhibitors[69] in road de-icing salts. Chromates have been advocated but to a limited extent because of the toxicity effects. In general, the most widely used inhibitors are the polyphosphates, either alone or in conjunction with other inhibitors. The use of polyphosphates on roads in Scandinavia has been reported[70], although difficulties arise with loss of inhibitor by absorption in the sand that is mixed with the salt. Extensive laboratory tests conducted in the UK showed[71] that polyphosphates were more effective in preventing further rusting of damaged painted panels than in preventing the corrosion of bare steel. A further development is to compound the polyphosphate with an organic-type inhibitor[72, 73].

Acid Solutions

Probably the major use of inhibitors in acid solutions is in pickling processes. The chief requirements of the inhibitor are that it should not decompose during the life of the pickle, not increase hydrogen absorption by the metal and not lead to the formation of surface films with electrically insulating properties that might interfere with subsequent electroplating or other surface treatments.

A wide variety of compounds are used in acid inhibition. These are now mainly organic compounds usually containing N, S or O atoms, although inorganic arsenic and antimony compounds have been used in the past. In general, for the pickling of steel, as pointed out for example by Machu[74], sulphur-containing compounds are preferred for sulphuric acid solutions and nitrogen-containing compounds for hydrochloric acid solutions. Every and Riggs[75] list 76 individual compounds and 32 mixtures that were subjected to laboratory tests and concluded that often a mixture of N— and S— compounds was better than either type alone. The superiority of S— compounds for inhibition in sulphuric acid is borne out in the list of 112 compounds quoted by Uhlig[76]. Twelve of the fourteen most effective compounds listed contain S atoms in the molecule. Typical of these are phenylthiourea, di-ortho-tolyl-thiourea, mercaptans and sulphides, 90% protection being provided with 0·003–0·01% concentrations.

N— compounds used as acid inhibitors include heterocyclic bases, such as pyridine, quinoline and various amines. Carassiti[77] describes the inhibitive action of decylamine and quinoline, as well as phenylthiourea and dibenzyl-sulphoxides for the protection of stainless steels in hydrochloric acid pickling. Hudson et al.[78, 79] refer to coal tar base fractions for inhibition in sulphuric and hydrochloric acid solutions. Good results are reported with 0·25 vol. % of distilled quinoline bases with addition of 0·05M sodium chloride in 4N sulphuric acid at 93°C. The sodium chloride is acting synergistically, e.g. 0·05M NaCl raises the percentage inhibition given by 0·1% quinoline in 2N H_2SO_4 from 43 to 79%. Similarly, potassium iodide improves the action of phenylthiourea[80].

Acetylenic compounds have been described for inhibition in acid solutions[81–83]. Typical inhibitors include 2-butyne-1,4-diol, 1-hexyne-3-ol and 4-ethyl-1-octyne-3-ol.

An exception to the 'lone pair' or 'donor' electron requirement of organic inhibitors is provided by the quaternary ammonium compounds. Meakins[84,85] reports the effectiveness of tetra-alkyl ammonium bromides with the alkyl group having C $\geqslant$ 10. Comparative laboratory tests of commercial inhibitors of this type have been described[86]. The inhibiting action of tetra-butyl ammonium sulphate for iron in H_2S-saturated sulphuric acid has been described, better results being achieved than with mono-, di- or tri-butylamines[87].

In the USSR much use is made of industrial by-products to prepare acid inhibitors. The PB class is obtained by treating technical butyraldehyde with ammonia and polymerising the resulting aldehyde-ammonia. PB-5, for example, with 0·01–0·15% of an arsenic salt is used in 20–25% HCl. A mixture of urotropine (hexamethyleneimine, hexamine) with potassium iodide, a regulator and a foaming agent is the ChM inhibitor. BA-6 is prepared from the condensation product of hexamine with aniline. A more recent development is the Katapin series which consists of p-alkyl benzyl pyridine chlorides; Katapin A, for example, is the p-dodecyl compound.

The beneficial effect of chloride ions on inhibitor action is brought out in the acid descaling of ships' tanks while at sea[88]. Inhibited 0·75% sulphuric acid prepared with sea-water can be used for this process at ambient temperatures, the chloride present in the sea-water acting synergistically with the inhibitor.

In the practice of acid pickling a foaming agent is often added to the pickling bath in order to facilitate penetration of the rust and scale by the inhibited acid and also to provide a foam blanket to prevent spray coming from the bath. After removal from the bath the metal is rinsed well, finally in hot water and then often dipped in a mild alkali, phosphate or chromate bath, to provide short term protection before the next operation. A suggested[89] variation on this procedure is to follow the acid pickling by a hot water rinse in a bath with a 35 mm layer of stearic acid on the bath surface. As the metal is withdrawn through this a water repellent film is left on its surface.

The Oil Industry

Corrosion problems requiring the application of inhibitors exist in the oil industry at every stage of production from initial extraction to refining and storage prior to use. Comprehensive reviews of these inhibitors have been given by Bregman[10,90]. Four main processes are involved. (a) Primary production, (b) secondary production, (c) refining and (d) storage, and each of these may be further subdivided.

Primary production   Although the technology of the process has many variations, the common factor is that oil flows from the deposit through steel tubing to the surface. Corrosion problems arise due to the presence of water that invariably accompanies the oil. It has been shown[91] that corrosivity is related to the water content which can vary over a wide range. This water can contain a variety of corrosive agents including carbon dioxide, hydrogen sulphide, organic acids, chlorides, sulphates, etc. Wells containing H_2S are referred to as *sour* and those free from H_2S as *sweet*; the former are the more corrosive. In some sweet wells the crude oil itself can provide protection of the metal if the oil: water ratio is suitable, but this effect will not be found in sour wells.

Most of the inhibitors in use are organic nitrogen compounds and these have been classified by Bregman[90] as (a) aliphatic fatty acid derivatives, (b) imidazolines, (c) quaternaries, (d) rosin derivatives (complex amine mixtures based on abietic acid); all of these will tend to have long-chain hydrocarbons, e.g. $C_{18}H_{37}$ as part of the structure, (e) petroleum sulphonic acid salts of long-chain diamines (preferred to the diamines), (f) other salts of diamines and (g) fatty amides of aliphatic diamines. Actual compounds in use in classes (a) to (d) include: oleic and naphthenic acid salts of n-tallow propylenediamine; diamines $RNH(CH_2)_xNH_2$ in which R is a carbon chain of 8-22 atoms and $x = 2$-10; and reaction products of diamines with acids from the partial oxidation of liquid hydrocarbons. Attention has also been drawn to polyethoxylated compounds in which the water solubility can be controlled by the amount of ethylene oxide added to the molecule.

The method of inhibitor *application* varies considerably since so many factors have to be considered. These include the oil: water ratio, the types of oil and the water composition, the fluid velocity, temperature, type of geological formation, emulsion formation, economics, method of well completion, solubility and specific gravity of the inhibitor, etc. It has been stated[92] that there are over 20 methods of introducing an inhibitor into the well to ensure that it enters the producing stream. These include: 'slug' treatment in which regular injections of inhibitors are made with automatic injection equipment; 'batch' treatments in which sufficient inhibitor is added to last for a month or longer; 'weighted' treatments[93] in which organic weighting agents can raise the density of the inhibitor formulation thus assisting its dispersal; 'micro-encapsulation'[94] methods with a liquid inhibitor weighted and coated with a water-soluble sheath to give controlled release at a given temperature; 'squeeze'[95, 96] technique where the inhibitor is displaced under pressure into the geological formulation whence it is absorbed into the rock and then gradually desorbed into the deposit. All these and other methods are subject to their particular advantages and disadvantages which are discussed in the relevant technical literature.

Secondary recovery   In this, water is forced down into the strata to displace further quantities of oil. This water can be that initially obtained from the well or it can be taken from other convenient sources. In either case the probability is that the water will be of an aggressive nature. As the water is now being forced down into the deposit there is the danger of blockage of the geological formation by corrosion products and this is an added reason for inhibition. Apart from the presence of dissolved salts there are the major problems of the oxygen and bacterial contents. Sulphite additions may be made to deal with dissolved oxygen but the method is not so straightforward[97] as, for example, in boiler-water treatment. Thus, care is required in brines containing H_2S as the catalyst* may be precipitated as sulphide. The sulphite may be lost in the deposition of calcium sulphite hemihydrate if the calcium concentration is high.

As in primary production, organic nitrogen compounds are often used since many of these have dispersant properties that will prevent the formation of adherent deposits[90]. It has been suggested that dissolved oxygen can prevent long-chain amines from being fully effective as corrosion inhibitors. Nevertheless some inhibitors of this type appear to be immune to this effect, for example see the results of Jones and Barrett[98]. Oxygen removal has been combined with long-chain amine treatment by using a 40% methanol solution of the oleyldiamine adduct of SO_2—the so-called ODASA method[99]. A concentration of 25 p.p.m. of this inhibitor is quoted for the scavenging of 1 p.p.m. of oxygen and field trials have shown reductions in oxygen from 0·5 to less than 0·1 p.p.m. Inorganic inhibitors can also be used in waterflood treatment to limit oxygen corrosion; a zinc–glassy phosphate type at 12–15 p.p.m. and pH 7·0–7·2 has been described[100]. Silicates at 100 p.p.m. have also been used.

A particular problem in oil recovery arises in the acidising process for stimulating well production in limestone formations[74, 101, 102]. For many years 15% hydrochloric acid for this process has been successfully inhibited with commercially available organic inhibitors to minimise attack on the

---

*Small amounts (less than 1 p.p.m.) of cobalt salts are usually added to the sulphite to catalyse its reaction with oxygen.

steel equipment. Sodium arsenite with a surfactant has been used[74], although problems can occur subsequently at the refining stage due to catalyst poisoning. In a discussion of acetylenic-type inhibitors Tedeschi et al.[103] show that the action of compounds such as hexynol and ethyl octynol for this type of application can be improved by the use of nitrogen synergists such as ethylene diamine, dimethyl formamide, urea or ammonia. However, with the advent of deeper wells this concentration of acid is not so effective and 28–30% concentrations become necessary. These higher concentrations, and the higher temperatures at the well-bottom, together place a limitation on the existing inhibitors. Research is active in this field, e.g. in one case a survey of some 20 000 compounds was made from which it was concluded that acetylenic compounds and some nitrogen compounds offered promise[101]. Russian workers[104] have described the inhibitor ANP-2 for use with 20% HCl in acidising. This is the HCl salt of aliphatic amines with an amine number of 15·75 obtained from the nitration of paraffins. At 0·1–0·15%, ANP-2 reduced the corrosion of steel at 43°C in 20% HCl by 20 times.

Refining   Inhibitors are necessary in the processing of crude oil—particularly where steel is involved—since many of the process fluid constituents are corrosive. Copper-bearing alloys, e.g. admiralty metal, are also used and the problem of controlling steel corrosion is often made more difficult by the need to use methods that will not enhance the corrosion of non-ferrous parts of the system. In general the corrosive agent is the water in the oil stream and its corrosivity is increased by the presence of H_2S, CO_2, O_2, HCl and other acids (naphthenics can be a source of corrosion). As in so many other situations the problem of inhibition cannot be considered in isolation. Problems concerned with fouling and scaling must be taken into account and comprehensive reviews of these problems have been published[90, 105]. Since many of the corrosion problems are due to the presence of acids one remedy is to adjust the pH to 7·0–7·5 by adding sodium hydroxide, sodium carbonate or ammonia. At higher pH values ammonia can lead to corrosion, and possibly stress-corrosion cracking of copper-base alloys. The neutralisation of hydrochloric acid with ammonia will produce ammonium chloride and deposits of solid NH_4Cl can be corrosive even in the absence of water[106]. Other disadvantages of alkali treatment are those of expense and the necessity for pH control to prevent scale formation.

Nitrogen-containing organic inhibitors, often in conjunction with ammonia, are now widely used. These compounds are usually of similar types to those for primary production, because, as Bregman has pointed out, the corrosive agents are often the same in the two cases. This author has reviewed the compounds used and points out that most are imidazoline derivatives. He cites Brooke[107] for specific applications. Thus, 6 p.p.m. of an imidazoline with ammonia to pH 7·5 added to the overhead of a crude topping unit increased the length of a run from 6 to 18 months. In another application, corrosion of overhead condensers and the top tray of a distillation column was prevented by the use of 4 p.p.m. of an amino alkyl aryl phosphate soluble in light hydrocarbons. This had to be changed to 4 p.p.m. of a methylene oxide rosin amine type of inhibitor after the phosphate was found to cause deposits when the produced fuel was used in internal combustion engines.

This last observation is a further example of factors other than those relating only to the metal-environment reaction influencing the selection of an inhibitor. The possible adverse effect of inhibitors on process catalysts in refineries must also be considered.

Storage   Corrosion is again mainly associated with the presence of water which separates at the bottom of storage tanks. Inhibition in this water layer can be achieved with the highly soluble inorganic inhibitors. Nitrites, silicates, polyphosphates, etc. have been used as well as oil-soluble inhibitors. Organic inhibitors include imidazolines alone or with other inhibitors, itaconic salts, oleic acid salts of various amines and polyalkene glycol esters of oleic acid. Again there are other requirements that must be fulfilled apart from prevention of corrosion.

Reinforced Concrete

Inhibitors to prevent or retard the corrosion of steel reinforcing bars in concrete have been discussed on a number of occasions. Treadaway and Russell[108] consider the important considerations to be (a) the extent of inhibition, (b) the rate of inhibitor consumption, (c) the type of attack if inhibition fails and (d) the effect of the inhibitor on concrete strength. Of these (d) is of considerable importance[108,109]. The best practice appears to be the coating of the bars with a strong inhibitive slurry rather than a general incorporation into the concrete mix as a whole. The inhibitors generally considered applicable are sodium benzoate[110] (2–10% in a slurry coating); sodium nitrite[111,112] and sodium benzoate plus sodium nitrite[108]. A mixture of grease with Portland cement, sodium nitrite, casein and water applied as a ⩾1 mm layer coating for reinforcing bars has been described[113]. Sodium mercaptobenzothiazole, stannous chloride and various unidentified proprietary compounds have also been described for inhibition in concrete. Laboratory tests have been reported by Gouda et al.[114].

Miscellaneous

In nitrogenous fertiliser solutions of the $NH_4NO_3-NH_3-H_2O$ type corrosion of steel can be prevented by 500 p.p.m. of sulphur-containing inhibitors, e.g. mercaptobenzothiazole, thiourea and ammonium thiocyanate. However, these inhibitors are not so effective where most of the NH_3 is replaced by urea. For these solutions phosphate inhibitors such as $(NH_4)_2HPO_4$ and polyphosphates were more effective[115].

In the hydraulic transport of solids through steel pipelines, inhibitors of the sodium–zinc–phosphate glass type have been shown[116] to be effective. In the case of coal slurries the polyphosphate type was rejected because the de-oxygenating action of the coal lowered the inhibitor effectiveness. Hexavalent chromium compounds at 20 p.p.m. were more effective[117].

In gas reforming plants, e.g. the hot potassium carbonate process for CO_2 removal, sodium metavanadate is used to prevent mild-steel corrosion[118]. Banks reports[119] that this treatment does not reduce the rather high corrosion rate of Cu–30Ni in these plants.

Conclusions

The principles and practice of corrosion inhibition have been described in terms of the factors affecting inhibitor performance and selection (principles) and the more important practical situations in which inhibitors are used (practice). For the latter a brief account is given of the nature of the system, the reasons for inhibitor application and the types of inhibitor in use.

Tabulated results have been avoided since these are either obtained from carefully controlled laboratory tests or from specific systems and would thus require much qualification before their application to other systems.

The very wide use of inhibitors is obvious, but emphasis must always be placed on the factors affecting their performance and on the specific circumstances and other requirements relating to particular applications.

A. D. MERCER

REFERENCES

1. Putilova, I. N., Balezin, S. A. and Barannik, V. P., *Metallic Corrosion Inhibitors*, Pergamon, Oxford (1960)
2. *Proc. Royal Artillery Assoc. (Woolwich)*, **26** No. 5 (1899); Moody, G. T., *Proc. Chem. Soc.*, **19**, 239 (1903)
3. Wachter, A. and Smith, S. S., *Industr. Engng. Chem.*, **35**, 358 (1943)
4. Hersch, P., Hare, J. B., Robertson, A. and Sutherland, S. M., *J. Appl. Chem.*, **11**, 246 (1961)
5. Trabanelli, G. and Carassiti, V., *Advances in Corrosion Science and Technology*, Vol. 1, edited by Fontana, M. G. and Staehle, R. W., Plenum Press, New York–London (1970)
6. Desai, M. N., Desai, S. M., Gandhi, M. H. and Shah, C. B., *Anti-corrosion Methods and Materials*, **18** No. 4, 8–13 (1971); *ibid.* No. 5, 4–10 (1971)
7. Desai, M. N., Rana, S. S. and Gandhi, M. H., *ibid.*, **18** No. 2, 19–23 (1971)
8. Rama Char, T. L. and Padma, D. K., *Trans. Inst. Chem. Engnrs.*, **47**, T177–182 (1969)
9. Walker, R., *Anti-corrosion Methods and Materials*, **17** No. 9, 9–15 (1970)
10. Bregman, J. J., *Corrosion Inhibitors*, Macmillan, New York–London (1963)
11. Brasher, D. M. and Mercer, A. D., *Br. Corrosion J.*, **3** No. 3, 120–129 (1968)
12. Brasher, D. M. and Rhoades-Brown, J. E., *ibid.*, **4**, 74–79 (1969); **8**, 50 (1973)
13. Mercer, A. D. and Jenkins, I. R., *ibid.*, **3** No. 3, 130–135 (1968)
14. Mercer, A. D., Jenkins, I. R. and Rhoades-Brown, J. E., *ibid.*, 136–144 (1968)
15. Couper, A. S., *Mat. Prot.*, **8** No. 10, 17–22 (1969)
16. Gherhardi, D., Rivola, L., Troyli, M. and Bombara, G., *Corrosion*, **20** No. 3, 73t–79t (1964)
17. Rozenfel'd, I. L. and Maksimchuk, V. P., *Proc. Acad. Sci.* (USSR), Phys. Chem. Section, **119** No. 5, 986 (1958)
18. Wormwell, F. and Mercer, A. D., *J. Appl. Chem.*, **2**, 150 (1952)
19. Gilbert, P. T. and Hadden, S. E., *ibid.*, **3**, 545 (1953)
20. Thomas, J. G. N. and Mercer, A. D., 4th Int. Cong. on Metallic Corrosion, Amsterdam, 1969, N.A.C.E., Houston, 585 (1972)
21. Ride, R. N., Symposium on Corrosion (Melbourne Univ.), 267 (1955–56); *J. Appl. Chem.*, **8**, 175 (1958)
22. Edwards, W. T., Le Surf, J. E. and Hayes, P. A., 2nd European Symp. on Corrosion Inhibitors, 1965; Univ. Ferrara, 679–700 (1966)
23. Spinedi, P. and Signorelli, G., 1st European Symp. on Corrosion Inhibitors, Ferrara, 1960; Univ. Ferrara, 643–652 (1961)
24. Vance, R. W., Proc. 1st Joint Aerospace and Marine Corrosion Techn. Seminar, 1968; NACE, 34–35 (1969)
25. Mansfeld, F., *J. Electrochem. Soc.*, **118** No. 9, 1 412–1 415 (1971)
26. Matsuda, S. and Uhlig, H. H., *J. Electrochem. Soc.*, **111**, 156 (1954)
27. Mazza, F. and Trasatti, S., 3rd European Symp. on Corrosion Inhibs., Ferrara, 1970; Univ. Ferrara, 277–291 (1971)

28. Butler, G. and Owen D., *Corrosion Science*, **9**, 603–614 (1969)
29. Hatch, G. B., *Mat. Prot.*, **8** No. 11, 31 (1969)
30. Stericker, W., *Industr. Engnr. Chem.*, **37**, 716 (1945)
31. Cone, C. S., *Mat. Prot. and Perf.*, **9** No. 7, 32–34 (1970)
32. Fisher, A. O., *Mat. Prot.*, **3** No. 10, 8–13 (1964)
33. Bostwick, T. W., *Corrosion*, **17** No. 8, 12 (1961)
34. Brooks, W. B., *Mat. Prot.*, **7** No. 2, 24–26 (1968)
35. Edwards, B. C., *Corrosion Science*, **9** No. 6, 395–404 (1969)
36. Darrin, M., *Industr. Engng. Chem.*, **38**, 368 (1946)
37. Conoby, J. F. and Swain, T. M., *Mat. Prot.*, **6** No. 4, 55–58 (1967)
38. Butler, G., as Ref. 27, 753–776 (1971)
39. Butler, G. and Ison, H. C. K., *Corrosion and its Prevention in Waters*, Leonard Hill, London (1966)
40. Beecher, J. and Savinelli, E. A., *Mat. Prot.*, **3** No. 2, 15–20 (1964)
41. Comeaux, R. V., *Hydrocarbon Processing*, **46** No. 12, 129 (1967)
42. Sussman, S., *Mat. Prot.*, **3** No. 10, 52 (1964)
43. Hatch, G. B., *Mat. Prot.*, **4** No. 7, 52 (1965)
44. Ulmer, R. C. and Wood, J. W., *Corrosion*, **8** No. 12, 402 (1952)
45. Verma, K. M., Gupta, M. P. and Roy, A. K., *Technology Quat.* (Bull. Fertiliser Corp. India), **5** No. 2, 98–102 (1968)
46. Hwa, C. M., *Mat. Prot. and Perf.*, **9** No. 7, 29–31 (1970)
47. Schweitzer, G. W., Paper at the Int. Water Conference, Pittsburgh (1969)
48. US Patent No. 3 532 639 (4.3.68)
49. Mercer, A. D. and Wormwell, F., *J. Appl. Chem.*, **9**, 577–594 (1959)
50. BS 3150, 3151 and 3152, British Standards Institution, London (1959)
51. Squires, A. P. T. B., *The Protection of Motor Vehicles from Corrosion*, Soc. Chem. Ind. Monograph, No. 4 (1958)
52. Vernon, W. H. J., Wormwell, F., Ison, H. C. K. and Mercer, A. D., Motor Ind. Res. Assocn. Bulletin, 4th quarter, 19–20 (1949)
53. Dulat, J., *Brit. Corrosion J.*, **3** No. 4, 190–196 (1968)
54. Cavitation Corrosion and its Prevention in Diesel Engines, Symposium, 10th Nov. 1965, British Railways Board (1966)
55. Cotton, J. B., *Chem. and Indust.*, 11th Feb., 214 (1967)
56. Cotton, J. B. and Jacob, W. R., *Br. Corrosion J.*, **6** No. 1, 42–44 (1971)
57. Spivey, A. M., *Chem. and Indust.*, 22nd April, 657 (1967)
58. Venczel, J. and Wranglen, G., *Corrosion Science*, **7**, 461 (1967)
59. Drane, C. W., *Br. Corrosion J.*, **6** No. 1, 39–41 (1971)
60. von Fraunhofer, J. A., *ibid.*, No. 1, 28–30 (1971)
61. Obrecht, M. F., 2nd International Congress on Metallic Corrosion, New York, 1963; NACE, 624–645 (1966)
62. Maase, R. B., *Mat. Prot.*, **5** No. 7, 37–39 (1966)
63. Streatfield, E. L., *Corrosion Technology*, **4** No. 7, 239–244 (1957)
64. Hoar, T. P., *J. Soc. Chem. Indust.*, **69**, 356–362 (1950)
65. Bowrey, S. E., *Trans. Inst. Marine Eng.*, **61** No. 3, 1–9 (1949)
66. Palmer, W. G., *J. Iron and Steel Inst.*, No. 12, 421–431 (1949)
67. Oakes, B. D., Wilson, J. S. and Bettin, W. J., Proc. 26th NACE Conf., 1969; NACE, 549–552 (1970)
68. Legault, R. A. and Bettin, W. J., *Mat. Prot. and Perf.*, **9** No. 9, 35–39 (1970)
69. *Motor Vehicle Corrosion and Influence of De-Icing Chemicals*, OECD Report, Oct. (1969)
70. Asanti, P., 'Corrosion and its Prevention in Motor Vehicles', *Proc. Inst. Mech. Engns.*, **182**, Part 3J, 73–79 (1967–68)
71. Bishop, R. R. and Steed, D. E., *ibid.*, 124–129 (1967–68)
72. B.P. Applic. No. 61748/69
73. *Road Research 1970*, Dept. of the Environment, Road Research Laboratory, H.M.S.O., London (1971)
74. Machu, W., as Ref. 27, 107–119 (1971)
75. Every, R. L. and Riggs, O. L., *Mat. Prot.*, **3** No. 9, 46–47 (1964)
76. Uhlig, H. H., *The Corrosion Handbook*, John Wiley and Sons, Inc., 910 (1948)
77. Carassiti, V., Trabanelli, G. and Zucchi, F., as Ref. 22, 417–448 (1966)
78. Hudson, R. M., Looney, Q. L. and Warning, C. J., *Brit. Corrosion J.*, **2** No. 3, 81–86 (1967)
79. Hudson, R. M. and Warning, C. J., *Mat. Prot.*, **6** No. 2, 52–54 (1967)

80. Alfandary, M., *et al.*, as Ref. 22, 363–375 (1966)
81. Froment, M. and Desestret, A., as Ref. 22, 223–236 (1966)
82. Funkhouser, J. G., *Corrosion*, **17**, 283t (1961)
83. Putilova, I. N. and Chislova, E. N., *Zashchita Metallov*, **2** No. 3, 290–294 (1966)
84. Meakins, R. J., *J. Appl. Chem.*, **13**, 339 (1963)
85. Meakins, R. J., *Br. Corrosion J.*, **6** No. 3, 109–113 (1971)
86. Riggs, O. L. and Hurd, R. M., *Corrosion*, **24** No. 2, 45–49 (1968)
87. Rozenfel'd, I. L., Persiantseva, V. P. and Damaskina, T. A., *Zashchita Metallov.*, **9** No. 6, 687–690 (1973)
88. Geld, I. and Acampara, M. A., *Mat. Prot.*, **3** No. 1, 42–46 (1964)
89. Geld, I. and D'Oria, F., *Mat. Prot.*, **6** No. 8, 42–44 (1967)
90. Bregman, J. I., as Ref. 27, 339–382 (1971)
91. Greenwell, H. E., *Corrosion*, **9**, 307–312 (1953)
92. Waldrip, H. E., *Mat. Prot.*, **5** No. 6, 8–13 (1966)
93. Patton, C. C., Deemer, D. A. and Hilliard, H. M., *Mat. Prot.*, **9** No. 2, 37–41 (1970)
94. Haughin, J. E. and Mosier, B., *Mat. Prot.*, **3** No. 5, 42–50 (1964)
95. Poetker, R. H. and Stone, J. D., *Petr. Eng.*, **28** No. 5, B–29–34 (1956)
96. Kerver, J. K. and Morgan, F. A., *Mat. Prot.*, **2** No. 4, 10–20 (1963)
97. Templeton, C. C., Rushing, S. S. and Rodgers, J. C., *Mat. Prot.*, **2** No. 8, 42–47 (1963)
98. Jones, L. W. and Barrett, J. P., *Corrosion*, **11**, 217t (1955)
99. Dunlop, A. K., Howard, R. L. and Raifsnider, P. J., *Mat. Prot.*, **8** No. 3, 27–30 (1969)
100. Hatch, H. B. and Ralston, P. H., *Mat. Prot.*, **3** No. 8, 35–41 (1964)
101. Coulter, A. W. and Smithey, C. M., *Mat. Prot.*, **8** No. 3, 37–38 (1969)
102. McDougall, L. A., *Mat. Prot.*, **8** No. 8, 31–32 (1969)
103. Tedeschi, R. J., Natali, P. W. and McMahon, H. C., Proc. NACE 25th Conf., 1969; NACE, 173–179 (1970)
104. Rybachok, I. N., Mikhailov, M. A. and Tarasova, N. A., *Korroziya i Zashchita v Nettegazovoi Prom.*, No. 7, 7–10 (1971)
105. Nathan, C. C., *Mat. Prot. and Perf.*, **9** No. 11, 15–18 (1970)
106. Carlton, R. H., *Mat. Prot.*, **2** No. 1, 15–20 (1963)
107. Brooke, J. M., *Hydrocarbon Processing*, Jan., 121–127 (1970)
108. Treadaway, K. W. J. and Russell, A. D., *Highways and Public Works*, **36** No. 1704, 19–21; No. 1705, 40–41 (1968)
109. Arber, M. G. and Vivian, H. E., *Australian J. Appl. Science*, **12** No. 4, 435–439 (1961)
110. North Thames Gas Board, British Patent No. 706 319 (31.3.54)
111. Moskin, V. M. and Alexseev, S. N., *Beton i Zhelezbeton*, No. 1, 28 (1957)
112. Alexseev, S. N. and Rozenfel'd, L. M., *ibid.*, No. 2, 388 (1958)
113. Dilaktorski, N. L. and Oit, L. V., *Tr. Nauchn.-Issled. Inst. Betona i Zhelez Betona, Akad. Stroit. i Arkhitekt. SSR*, No. 22, 54–60 (1961)
114. Gouda, V. K., *Br. Corrosion J.*, **5** No. 5, 198–208 (1970)
115. Banks, W. P., *Mat. Prot.*, **7** No. 3, 35–38 (1968)
116. Anon., *Mat. Prot.*, **6** No. 2, 61 (1967)
117. Swan, J. D., Bomberger, D. R. and Barthauer, G. L., *Mat. Prot.*, **2** No. 9, 26–34 (1963)
118. Bienstock, D. and Field, J. M., *Corrosion*, **17**, 571t–574t (1961)
119. Banks, W. P., *Mat. Prot.*, **6** No. 11, 37–41 (1967)

18.3 The Mechanism of Corrosion Prevention by Inhibitors

The mechanisms of corrosion inhibition will be described separately for acid and neutral solutions, since there are considerable differences in mechanisms between these two media. Definitions and classifications of inhibitors are given in Section 18.2 and by Fischer[1].

Inhibitors for Acid Solutions

The corrosion of metals in aqueous acid solutions can be inhibited by a very wide range of substances[2, 3]. These include relatively simple substances, such as chloride, bromide or iodide ions and carbon monoxide, and many organic compounds, particularly those containing elements of Groups V and VI of the Periodic Table, such as nitrogen, phosphorus, arsenic, oxygen, sulphur and selenium. Organic compounds containing multiple bonds, especially triple bonds, are effective inhibitors. Organic compounds of high molecular weight, e.g. proteins and polysaccharides, also have inhibitive properties. The primary step in the action of inhibitors in acid solutions is generally agreed to be adsorption on to the metal surface, which is usually oxide-free in acid solutions. The adsorbed inhibitor then acts to retard the cathodic and/or anodic electrochemical processes of the corrosion. The factors influencing the adsorption and the electrochemical reactions will be considered in turn. (See also Section 9.1.)

Adsorption of Corrosion Inhibitors onto Metals

Measurements of the adsorption of inhibitors on corroding metals are best carried out using the direct methods of radio-tracer detection[4−7] and solution depletion measurements[8−10]. These methods provide unambiguous information on uptake, whereas the corrosion reactions may interfere with the indirect methods of adsorption determination, such as double layer capacity measurements[11], coulometry[11], ellipsometry[12] and reflectivity[12]. Nevertheless, double layer capacity measurements have been widely used for the determination of inhibitor adsorption on corroding metals, with apparently consistent results, though the interpretation[13] may not be straightforward in some cases.

Direct measurements on metals such as iron, nickel and stainless steel have shown that adsorption occurs from acid solutions of inhibitors such as iodide ions[14], carbon monoxide[15] and organic compounds such as amines[8,10,16], thioureas[9,16-18], sulphoxides [19-21], sulphides[21,22] and mercaptans[16]. These studies have shown that the efficiency of inhibition (expressed as the relative reduction in corrosion rate) can be qualitatively related to the amount of adsorbed inhibitor on the metal surface. However, no detailed quantitative correlation has yet been achieved between these parameters. There is some evidence[16,23] that adsorption of inhibitor species at low surface coverage θ (for complete surface coverage $\theta = 1$) may be more effective in producing inhibition than adsorption at high surface coverage. In particular, the adsorption of polyvinyl pyridine on iron in hydrochloric acid at $\theta < 0\cdot1$ monolayer has been found to produce an 80% reduction in corrosion rate[10].

In general, the results of direct adsorption measurements provide a basis for the widely used procedure of inferring the adsorption behaviour of inhibitors from corrosion rate measurements. This involves the assumption that the corrosion reactions are prevented from occurring over the area (or active sites) of the metal surface covered by adsorbed inhibitor species, whereas these corrosion reactions occur normally on the inhibitor-free area. The inhibitive efficiency is then directly proportional to the fraction of the surface covered with adsorbed inhibitor. This assumption has been applied to deduce the effects of concentration on the adsorption of inhibitors, and to compare the adsorption of different inhibitors (usually related in structure) at the same concentration. On the whole, the interpretation in this way of the efficiency of inhibitors in terms of their adsorption behaviour has given consistent results, which have clarified the factors influencing inhibition and adsorption. However, some qualifications are necessary in this approach, since this simple relationship between inhibitive efficiency and adsorption will not always apply. As mentioned above, at low surface coverage ($\theta < 0\cdot1$), the effectiveness of adsorbed inhibitor species in retarding the corrosion reactions may be greater than at high surface coverage[10,16,23]. In other cases, adsorption of inhibitors, e.g. thioureas[24,25] and amines[26,27] from solutions of low concentration may cause stimulation of corrosion. Furthermore, in comparing the inhibitive efficiency and adsorption of different inhibitors, possible differences in the mechanism and effectiveness of retardation of the corrosion reactions must be considered[28].

The information on inhibitor adsorption, derived from direct measurements and from inhibitive efficiency measurements, considered in conjunction with general knowledge of adsorption from solution[29-31], indicates that inhibitor adsorption on metals is influenced by the following main factors.

Surface charge on the metal   Adsorption may be due to electrostatic attractive forces between ionic charges or dipoles on the adsorbed species and the electric charge on the metal at the metal/solution interface. In solution, the charge on a metal can be expressed by its potential with respect to the zero-charge potential (see Section 9.1). This potential relative to the zero-charge potential, often referred to as the ϕ-potential[32], is more important with respect to adsorption than the potential on the hydrogen scale, and indeed the signs of these two potentials may be different. As the ϕ-

potential becomes more positive, the adsorption of anions is favoured and as the ϕ-potential becomes more negative, the adsorption of cations is favoured. The ϕ-potential also controls the electrostatic interaction of the metal with dipoles in adsorbed neutral molecules, and hence the orientation of the dipoles and the adsorbed molecules. For different metals at the same ϕ-potential, electrostatic interactions should be independent of the nature of the metal, and this has been used as a basis to compare adsorption of inhibitors on different metals[32-34]. Thus Antropov[32] has shown that the adsorption of some inhibitors on mercury can be related to their adsorption and inhibitive effect on iron, when considered at the same ϕ-potentials for both metals. The differences in behaviour of an inhibitor on various metals can also in some cases be related to the differences in ϕ-potentials at the respective corroding potentials.

The functional group and structure of the inhibitor   Besides electrostatic interactions, inhibitors can bond to metal surfaces by electron transfer to the metal to form a co-ordinate type of link. This process is favoured by the presence in the metal of vacant electron orbitals of low energy, such as occur in the transition metals. Electron transfer from the adsorbed species is favoured by the presence of relatively loosely bound electrons, such as may be found in anions, and neutral organic molecules containing lone pair electrons or π-electron systems associated with multiple, especially triple, bonds or aromatic rings. In organic compounds, suitable lone pair electrons for co-ordinate bonding occur in functional groups containing elements of Groups V and VI of the Periodic Table. The tendency to stronger co-ordinate bond formation (and hence stronger adsorption) by these elements increases with decreasing electronegativity in the order $O < N < S < Se$[35-37], and depends also on the nature of the functional groups containing these elements. The structure of the rest of the molecule can affect co-ordinate bond formation by its influence on the electron density at the functional group[38-42]. The effects of substituents in related inhibitor molecules, e.g. pyridines[38,39,43,48], anilines[39,42-45], aliphatic amines[42], amino acids[46], benzoic acids[47] and aliphatic sulphides[49], on the inhibitive efficiencies have been correlated with changes in electron densities at functional groups, as derived from nuclear magnetic resonance measurements[39], values of Hammett constants (aromatic molecules)[40-47] or Taft constants (aliphatic molecules)[42,46,49], or from quantum mechanical calculations[38,42,48]. The results of these investigations generally indicate that the electron density at the functional group increases as the inhibitive efficiency increases in a series of related compounds. This is consistent with increasing strength of co-ordinate bonding due to easier electron transfer, and hence greater adsorption. An analogous correlation has been demonstrated by Hackerman[50,51] between inhibitive efficiencies in a series of cyclic imines $(CH_2)_nNH$ and changes in hybrid bonding orbitals of the electrons on the nitrogen atom making electron transfer and co-ordinate bond formation easier.

Interaction of the inhibitor with water molecules   Due to the electrostatic and co-ordinate bond interactions described under the previous two headings, the surfaces of metals in aqueous solutions are covered with adsorbed water molecules. Adsorption of inhibitor molecules is a displacement reaction involving removal of adsorbed water molecules from the surface. During adsorption of a molecule, the change in interaction energy with water molecules in passing from the dissolved to the adsorbed state forms an important part of the free energy change on adsorption. This has been shown to increase with the energy of solvation of the adsorbing species, which in turn increases with increasing size of the hydrocarbon portion of an organic molecule[36]. Thus increasing size leads to decreasing solubility and increasing adsorbability. This is consistent with the increasing inhibitive efficiency observed at constant concentrations with increasing molecular size in a series of related compounds[36,52,53].

Interaction of adsorbed inhibitor species   Lateral interactions between adsorbed inhibitor species may become significant as the surface coverage, and hence the proximity, of the adsorbed species increases. These lateral interactions may be either attractive or repulsive. Attractive interactions occur between molecules containing large hydrocarbon components, e.g. *n*-alkyl chains. As the chain length increases, the increasing van der Waals attractive force between adjacent molecules leads to stronger adsorption at high coverage. Repulsive interactions occur between ions or molecules containing dipoles and lead to weaker adsorption at high coverage. The effects of lateral interactions between adsorbed inhibitors on inhibitive efficiency have been discussed by Hoar and Khera[26].

In the case of ions, the repulsive interaction can be altered to an attractive interaction if an ion of opposite charge is simultaneously adsorbed. In a solution containing inhibitive anions and cations the adsorption of both ions may be enhanced and the inhibitive efficiency greatly increased compared to solutions of the individual ions. Thus, synergistic inhibitive effects occur in such mixtures of anionic and cationic inhibitors[54,55]. These synergistic effects are particularly well defined in solutions containing halide ions, I^-, Br^-, Cl^-, with other inhibitors such as quaternary ammonium cations[56], alkyl benzene pyridinium cations[57], and various types of amines[55,58-61]. It seems likely that co-ordinate-bond interactions also play some part in these synergistic effects, particularly in the interaction of the halide ions with the metal surfaces and with some amines[55].

Reaction of adsorbed inhibitors   In some cases, the adsorbed corrosion inhibitor may react, usually by electro-chemical reduction, to form a product which may also be inhibitive. Inhibition due to the added substance has been termed *primary* inhibition and that due to the reaction product *secondary* inhibition[62-65]. In such cases, the inhibitive efficiency may increase or decrease with time according to whether the secondary inhibition is more or less effective than the primary inhibition. Some examples of inhibitors which react to give secondary inhibition are the following. Sulphoxides can be reduced to sulphides, which are more efficient inhibitors[19,20,21,63,64]. Quaternary phosphonium and arsonium compounds can be reduced to the corresponding phosphine or arsine compounds, with little change in inhibitive efficiency[63,64]. Acetylene compounds can undergo reduction followed by polymerisation to form a multimolecular protective film[66,67]. Thioureas can be reduced to produce HS^- ions, which may act as stimulators of corrosion[24,25,54].

Effects of Inhibitors on Corrosion Processes

In acid solutions the anodic process of corrosion is the passage of metal ions from the oxide-free metal surface into the solution, and the principal cathodic process is the discharge of hydrogen ions to produce hydrogen gas. In air-saturated acid solutions, cathodic reduction of dissolved oxygen also occurs, but for iron the rate does not become significant compared to the rate of hydrogen ion discharge until the pH exceeds about 3. An inhibitor may decrease the rate of the anodic process, the cathodic process or both processes. The change in the corrosion potential on addition of the inhibitor is often a useful indication of which process is retarded[24,67]. Displacement of the corrosion potential in the positive direction indicates mainly retardation of the anodic process (anodic control), whereas displacement in the negative direction indicates mainly retardation of the cathodic process (cathodic control). Little change in the corrosion potential suggests that both anodic and cathodic processes are retarded (see Section 1.4 for appropriate potential versus current diagrams).

The effects of adsorbed inhibitors on the individual electrode reactions of corrosion may be determined from the effects on the anodic and cathodic polarisation curves of the corroding metal[24,28,68,69]. A displacement of the polarisation curve without a change in the Tafel slope in the presence of the inhibitor indicates that the adsorbed inhibitor acts by blocking active sites so that reaction cannot occur, rather than by affecting the mechanism of the reaction. An increase in the Tafel slope of the polarisation curve due to the inhibitor indicates that the inhibitor acts by affecting the mechanism of the reaction. However, the determination of the Tafel slope will often require the metal to be polarised under conditions of current density and potential which are far removed from those of normal corrosion. This may result in differences in the adsorption and mechanistic effects of inhibitors at polarised metals compared to naturally corroding metals[24,56,70]. Thus the interpretation of the effects of inhibitors at the corrosion potential from applied current–potential polarisation curves, as usually measured, may not be conclusive. This difficulty can be overcome in part by the use of rapid polarisation methods[56,71]. A better procedure[24] is the determination of 'true' polarisation curves near the corrosion potential by simultaneous measurements of applied current, corrosion rate (equivalent to the true anodic current) and potential. However, this method is rather laborious and has been little used.

Electrochemical studies have shown that inhibitors in acid solutions may affect the corrosion reactions of metals in the following main ways.

Formation of a diffusion barrier The absorbed inhibitor may form a surface film which acts as a physical barrier to restrict the diffusion of ions or molecules to or from the metal surface and so retard the corrosion reactions. This effect occurs particularly when the inhibitor species are large molecules, e.g. proteins, such as gelatine, agar agar; polysaccharides, such as dextrin; or compounds containing long hydrocarbon chains. Surface films of these types of inhibitors give rise to resistance polarisation and also concentration polarisation affecting both anodic and cathodic reactions[72]. Similar effects also occur when the inhibitor can undergo reaction to form a multimolecular surface film, e.g. acetylenic compounds[67] and sulphoxides[19,73].

Blocking of reaction sites   The interaction of adsorbed inhibitors with surface metal atoms may prevent these metal atoms from participating in either the anodic or cathodic reactions of corrosion. This simple blocking effect decreases the number of surface metal atoms at which these reactions can occur, and hence the rates of these reactions, in proportion to the extent of adsorption. The mechanisms of the reactions are not affected and the Tafel slopes of the polarisation curves remain unchanged. Behaviour of this type has been observed for iron in sulphuric acid solutions containing 2,6-dimethyl quinoline[24], β-naphthoquinoline[69], or aliphatic sulphides[49].

It should be noted that the anodic and cathodic processes may be inhibited to different extents[24,26,74]. The anodic dissolution process of metal ions is considered to occur at steps or emergent dislocations in the metal surface, where metal atoms are less firmly held to their neighbours than in the plane surface. These favoured sites occupy a relatively small proportion of the metal surface. The cathodic process of hydrogen evolution is thought to occur on the plane crystal faces which form most of the metal surface area. Adsorption of inhibitors at low surface coverage tends to occur preferentially at anodic sites, causing retardation of the anodic reaction. At higher surface coverage, adsorption occurs on both anodic and cathodic sites, and both reactions are inhibited.

Participation in the electrode reactions   The electrode reactions of corrosion involve the formation of adsorbed intermediate species with surface metal atoms, e.g. adsorbed hydrogen atoms in the hydrogen evolution reaction; adsorbed (FeOH) in the anodic dissolution of iron[75,76]. The presence of adsorbed inhibitors will interfere with the formation of these adsorbed intermediates, but the electrode processes may then proceed by alternative paths through intermediates containing the inhibitor. In these processes the inhibitor species act in a catalytic manner and remain unchanged. Such participation by the inhibitor is generally characterised by a change in the Tafel slope observed for the process. Studies of the anodic dissolution of iron in the presence of some inhibitors, e.g. halide ions[14,77-79], aniline and its derivatives[28,43], the benzoate ion[70] and the furoate ion[80], have indicated that the adsorbed inhibitor I participates in the reaction, probably in the form of a complex of the type $(Fe \cdot I)_{ads.}$ or $(Fe \cdot OH \cdot I_n)_{ads.}$. The dissolution reaction proceeds less readily via the adsorbed inhibitor complexes than via $(Fe \cdot OH)_{ads.}$, and so anodic dissolution is inhibited and an increase in Tafel slope is observed for the reaction.

Adsorbed species may also accelerate the rate of anodic dissolution of metals, as indicated by a decrease in Tafel slope for the reaction. Thus the presence of hydrogen sulphide in acid solutions stimulates the corrosion of iron, and decreases the Tafel slope[25,54,56]. The reaction path through $(Fe \cdot HS^-)_{ads.}$ has been postulated to lead to easier anodic dissolution than that through $(Fe \cdot OH)_{ads.}$. This effect of hydrogen sulphide is thought to be responsible for the acceleration of corrosion of iron observed with some inhibitive sulphur compounds, e.g. thioureas[25,54,81], at low concentrations, since hydrogen sulphide has been identified as a reduction product. However, the effects of hydrogen sulphide are complex, since in the presence of inhibitors such as amines[56], quaternary ammonium cations[56], thioureas[54,81], synergistic enhancement of inhibition is observed due to interaction of adsorbed HS$^-$ ions with the adsorbed inhibitor.

Inhibitors may also retard the rate of hydrogen evolution on metals by affecting the mechanism of the reaction, as indicated by increases in the Tafel slopes of cathodic polarisation curves. This effect has been observed on iron in the presence of inhibitors such as phenyl-thiourea[58,69], acetylenic hydrocarbons[82,83], aniline derivatives[84], benzaldehyde derivatives[85] and pyrilium salts[85]. According to Antropov[86] and Grigoryev[84,85], the rate determining step (which depends on experimental conditions[75]) for the hydrogen evolution reaction on iron in acid solutions (pH less than 2) is the recombination of adsorbed hydrogen atoms to form hydrogen molecules. Grigoryev[84,85] has shown that addition of anilines, benzaldehydes and pyrilium salts to hydrochloric acid tends to retard the discharge of hydrogen ions to form adsorbed hydrogen atoms on iron, so that this step rather than the recombination step tends to control the rate of the overall hydrogen evolution reaction.

Some inhibitors, e.g. amines[26,27,33] and sulphoxides[87], which can add on hydrogen ions in acid solutions to form protonated species, may accelerate the rate of the cathodic hydrogen evolution reaction on metals, due to participation of the protonated species in the reaction. This occurs when the discharge of the protonated species to produce an adsorbed hydrogen atom at the metal surface occurs more easily than the discharge of the hydrogen ion. This effect becomes more significant as the hydrogen overvoltage of the metal increases, and so may be observed to a greater extent on zinc than on iron[33].

Alteration of the electrical double layer   The adsorption of ions or species which can form ions, e.g. by protonation, on metal surfaces will change the electrical double layer at the metal-solution interface, and this in turn will affect the rates of the electrochemical reactions[33,54]. The adsorption of cations, e.g. quaternary ammonium ions[54] and protonated amines[33], makes the potential more positive in the plane of the closest approach to the metal of ions from the solution. This positive potential displacement retards the discharge of the positively charged hydrogen ion. For the inhibition of iron corrosion by pyridines in acid solutions, Antropov[33] has calculated the theoretical inhibition coefficients of the hydrogen ion discharge reaction, due to the effect of adsorbed pyridine cations on the electrical double layer. The calculated values agreed well with observed values at low inhibitor concentrations, indicating that inhibition could be wholly attributed to electrostatic effects, and that blocking of the surface by adsorbed inhibitor is not important.

Conversely, the adsorption of anions makes the potential more negative on the metal side of the electrical double layer and this will tend to accelerate the rate of discharge of hydrogen ions. This effect has been observed for the sulphosalicylate ion[54] and the benzoate ion[70].

Conclusions

Thus, inhibitors of corrosion in acid solution can interact with metals and affect the corrosion reaction in a number of ways, some of which may occur simultaneously. It is often not possible to assign a single general mechanism of action to an inhibitor, because the mechanism may change with experimental conditions. Thus, the predominant mechanism of action of an inhibitor may vary with factors such as: its concentration, the pH of the acid, the nature of the anion of the acid, the presence of other species in the solution, the extent of reaction to form secondary inhibitors and the nature of the metal. The mechanism of action of inhibitors with the same functional group may additionally vary with factors such as the effect of the molecular structure on the electron density of the functional group and the size of the hydrocarbon portion of the molecule. However, the mechanisms of action of a number of inhibitors have now been identified and are beginning to be understood on the molecular level.

Inhibitors in Near-neutral Solutions

The corrosion of metals in neutral solutions differs from that in acid solutions in two important respects (*see* Section 1.4). In air-saturated solutions, the main cathodic reaction in neutral solutions is the reduction of dissolved oxygen, whereas in acid solution it is hydrogen evolution. Corroding metal surfaces in acid solution are oxide free, whereas in neutral solutions metal surfaces are covered with films of oxides, hydroxides or salts, owing to the reduced solubility of these species. Because of these differences, substances which inhibit corrosion in acid solution by adsorption on oxide-free surfaces, do not generally inhibit corrosion in neutral solution*. Inhibition in neutral solutions is due to compounds which can form or stabilise protective surface films. The inhibitor may form a surface film of an insoluble salt by precipitation or reaction. Inhibitors forming films of this type include: (*a*) salts of metals such as zinc, magnesium, manganese and nickel, which form insoluble hydroxides, especially at cathodic areas, which are more alkaline due to the hydroxyl ions produced by reduction of oxygen; (*b*) soluble calcium salts, which can precipitate as calcium carbonate in waters containing carbon dioxide, again at cathodic areas where the high pH permits a sufficiently high concentration of carbonate ions; (*c*) polyphosphates in the presence of zinc or calcium, which produce a thin amorphous salt film. The mechanism of action of these inhibitors seems fairly straightforward[88]. The salt films, which are often quite thick and may be visible, restrict diffusion, particularly of dissolved oxygen to the metal surface. They are poor electronic conductors and so oxygen reduction does not occur on the film surface; these inhibitors are referred to, therefore, as cathodic inhibitors. The mechanism of inhibition by polyphosphates is more complex, and the various theories of their action have recently been described by Butler[89].

Another class of inhibitors in near-neutral solutions act by stabilising oxide films on metals to form thin protective passivating films. Such inhibitors are the anions of weak acids, some of the most important in practice being chromate, nitrite, benzoate, silicate, phosphate and borate. Passivating

*Exceptions are organic compounds of high molecular weight, e.g. gelatine, agar and dextrin. Adsorption of these large molecules is partly effective in shielding the metal surface from reaction in neutral as well as acid solutions[2].

oxide films on metals offer high resistance to the diffusion of metal ions and the anodic reaction of metal dissolution is inhibited; thus these inhibitive anions are often referred to as *anodic* inhibitors, and they are more generally used than cathodic inhibitors to inhibit the corrosion of iron, zinc, aluminium, copper and their alloys, in near neutral solutions.

The conditions under which oxide films are protective on these metals in relation to inhibition by anions may be characterised in terms of three important properties of the oxide film (see also Sections 1.4 and 1.5):

1. The Flade potential, which is the negative potential limit of stability of the oxide film. At potentials more negative than the Flade potential the oxide film is unstable with respect to its reduction or dissolution, or both, since the rates of these two processes exceed that of film formation.
2. The critical breakdown potential, which is the positive potential limit of stability of the oxide film. At this potential and more positive potentials, the oxide film is unstable with respect to the action of anions, especially halide ions, in causing localised rupture and initiating pitting corrosion.
3. The corrosion current due to diffusion of metal ions through the passivating film, and dissolution of metal ions at the oxide–solution interface. Clearly, the smaller this current, the more protective is the oxide layer.

All of these three properties of the oxide films on metals are influenced by the anion composition and pH of the solution. In addition the potential of the metal will depend on the presence of oxidising agents in the solution. Inhibition of corrosion by anions thus requires an appropriate combination of anions, pH and oxidising agent in the solution so that the oxide film on the metal is stable (the potential then lying between the Flade potential and the breakdown potential), and protective (the corrosion current through the oxide being low).

Most of the information available on the mechanism of action of inhibitive anions relates to iron, which will be discussed in some detail, and followed by brief accounts of zinc, aluminium and copper.

Iron

The corrosion of iron (or steel) can be inhibited by the anions of most weak acids under suitable conditions[90-92]. However, other anions, particularly those of strong acids, tend to prevent the action of inhibitive anions and stimulate breakdown of the protective oxide film. Examples of such aggressive anions are the halides, sulphate, nitrate, etc. Brasher[92] has shown that, in general, most anions exhibit some inhibitive and some aggressive behaviour towards iron. The balance between the inhibitive and aggressive properties of a specific anion depends on the following main factors (which are themselves interdependent).

Concentration Inhibition of iron corrosion in distilled water occurs only when the anion concentration exceeds a critical value[90-93]. At concentrations below the critical value, inhibitive anions may act aggressively and stimulate breakdown of the oxide film[92]. Effective inhibitive anions have low critical concentrations for inhibition. Brasher[92] has classified a number of anions in order of their inhibitive power towards steel, judged from their critical inhibitive concentrations. The order of *decreasing* inhibitive efficiency is: azide, ferricyanide, nitrite, chromate, benzoate, ferrocyanide, phosphate, tellurate, hydroxide, carbonate, chlorate, o-chlorbenzoate, bicarbonate, fluoride, nitrate, formate. Thus, normally aggressive anions such as fluoride and nitrate may inhibit steel corrosion at sufficiently high concentrations.

pH   Inhibitive anions are effective in preventing iron corrosion only at pH values more alkaline than a critical value. This critical pH depends on the anion, e.g. approximate critical pH values for the inhibition of iron or steel in about 0·1 M solution of the anion *increase* in the order: chromate[94], 1·0; azelate[95], 4·5; nitrite[96,97], 5·0-5·5; benzoate[98,99], 6·0; phosphate[100], 7·2; hydroxide[91], ≈12 (not 0·1 M). The critical pH value for inhibition depends on the concentration of the inhibitive anion. In azelate[95] and nitrite[96] solutions, there are indications that the critical pH for inhibition decreases as the anion concentration increases. However, in benzoate solutions[99], increasing benzoate concentration displaces the critical pH to more alkaline values.

Dissolved oxygen concentration and supply   Inhibition of the corrosion of iron by anions requires a critical minimum degree of oxidising power in the solution. This is normally supplied by the dissolved oxygen present in air-saturated solutions. Gilroy and Mayne[101] have shown that the critical oxygen concentration for inhibition of iron in 0·1 M sodium benzoate (pH 7·0) is ≈0·3 p.p.m., considerably less than the air-saturated concentration of ≈8 p.p.m. As the oxygen concentration is reduced below this critical value, the rate of breakdown of the passivating oxide film increases. As the pH of 0·1 M sodium benzoate is reduced below 7·0, the critical oxygen concentration for inhibition increases[102]. The critical oxygen concentration for inhibition depends on the nature of the anion[101]. If the inhibitive anion possesses oxidising properties, e.g. chromate[93,103,104], nitrite[93], pertechnetate[104-106], then the presence of dissolved oxygen may not be necessary for inhibition. The critical oxygen concentrations for good inhibitive non-oxidising anions are low[101]. If the dissolved oxygen concentration is increased above that of the air-saturated solution, the inhibition of iron corrosion is facilitated[102], and inhibition may even be achieved in chloride solution[107]. Similarly, increasing the oxygen supply to the iron surface by rapid stirring or aeration of the solution may favour inhibition, resulting in inhibition at lower critical anion concentrations[98,108], and again inhibition in chloride solutions may be obtained[109]. Addition of an oxidising agent may improve the efficacy of inhibitive anions, e.g. Mayne and Page[110] have recently shown that the presence of hydrogen peroxide lowers the critical concentrations of sodium benzoate and sodium azelate required for inhibition of steel, and also lowers the critical pH values for inhibition.

Aggressive anion concentration   When aggressive anions are present in the solution, the critical concentrations of inhibitive anions required for protection of iron are increased[108,111-116]. Brasher and Mercer[112-115] have shown that the relationship between the maximum concentration of aggressive anion C_{agg}, permitting full protection by a given concentration of inhibitive anion $C_{inh.}$ is of the form

$$\log C_{inh.} = n \log C_{agg.} + K$$

where K is a constant dependent on the nature of the inhibitive and aggressive anions, and n is an exponent which is approximately the ratio of the valency of the inhibitive anion to the valency of the aggressive anion. This relationship indicates a competitive action between inhibitive anions and aggressive anions and its significance will be discussed below. In general, the more aggressive the anion, the smaller the concentration which can be tolerated by an inhibitive anion. The order of tolerance[115] of aggressive anions is, with certain exceptions, consistent with the order of aggressiveness of these anions as determined from their tendency to induce breakdown of the oxide film on iron in aerated solutions.

Nature of the metal surface   The critical concentration of an anion required to inhibit the corrosion of iron may increase with increasing surface roughness. Thus, Brasher and Mercer[112] showed that the minimum concentration of benzoate required to protect a grit-blasted steel surface was about 100 times greater than that required to protect an abraded surface. However, surface preparation had little effect on the critical inhibitive concentrations for chromate[113] or nitrite[114]. The time of exposure of the iron surface to air after preparation and before immersion may also affect the ease of inhibition by anions. There is evidence[91,101,117] that the inhibition by anions occurs more readily as the time of pre-exposure to air increases. Similarly, if an iron specimen is immersed for some time in a protective solution of an inhibitive anion, it may then be transferred without loss of inhibition to a solution of the anion containing much less than the critical inhibitive concentration[91].

Temperature   In general, the critical concentrations of anions, e.g. benzoate[108,112], chromate[113] and nitrite[114], required for the protection of steel increase as the temperature increases.

Passivating Oxide Films

Studies of iron surfaces inhibited in solutions of anions have shown by several independent techniques, e.g. examination of *in situ* and stripped films by electron diffraction[118-124], cathodic reduction[123,125,126] and ellipsometry[127], the presence of a thin film (thickness $\approx 3 \times 10^{-9}$ m to 5×10^{-9} m) of cubic iron oxide (Fe_3O_4 or γ-Fe_2O_3), which is rather similar to the air-formed oxide film[128-30]. Immersion of iron bearing its air-formed oxide film into solutions of inhibitive anions usually results in a thickening of the oxide layer[125,131], except at relatively low pH[132]. Oxide film growth on iron in inhibitive solutions of anions as well as in air, follows a direct logarithmic law, the rate constants being generally slightly greater in solution than in air[94,127,131,133−136]. It is generally agreed that inhibition of the corrosion of iron by anions results from their effects on this oxide layer[137−144]. These effects are of several kinds, and they will now be discussed in relation to the theories of inhibition by anions. It seems probable that there is no single mechanism of inhibition, but that a number of factors are involved, their relative importance depending on the nature of the anion and experimental conditions.

Uptake of Anions by Oxide Films

Early studies on oxide films stripped from iron showed the presence of chromium after inhibition in chromate solution[145] and of crystals of ferric phosphate after inhibition in phosphate solutions[121]. More recently, radio-tracer studies using labelled anions have provided more detailed information on the uptake of anions. These measurements of irreversible uptake have shown that some inhibitive anions, e.g. chromate[94,133−136,146,147] and phosphate[117,148], are taken up to a considerable extent on the oxide film. However, other equally effective inhibitive anions, e.g. benzoate[149,150], pertechnetate[151,152] and azelate[153], are taken up to a comparatively small extent. Anions may be adsorbed on the oxide surface by interactions similar to those described above in connection with adsorption on oxide-free metal surfaces. On the oxide surface there is the additional possibility that the adsorbed anions may undergo a process of ion exchange[115,117,147], whereby they replace oxide ions, which leave the oxide lattice for the solution. Adsorption and ion-exchange represent different aspects of the same process. However, it would be expected that an anion would be more firmly bound after ion exchange because of the greater interaction with neighbouring metal ions. Anions taken up by adsorption/ion exchange, e.g. phosphate[117] and chromate[147], would be expected to be distributed fairly uniformly over the surface, though binding energies would vary with different types of adsorption site. There is considerable evidence that uptake of anions may also be concentrated into particles of separate phases located in the main oxide film, e.g. phosphate[117,121], pertechnetate[105] and azelate[153]. The formation of these particles of separate phase has been observed mainly when conditions are relatively unfavourable for inhibition, e.g. low pH[153], thin oxide film due to short air exposure[117] and the presence of aggressive anions[154].

This evidence of the uptake of inhibitive anions into oxide films forms the basis of the 'chemical' or 'pore plugging' theory of inhibition, associated originally with Evans[155] *et al.* In this theory the rôle of the inhibitive anion is to promote the repair of weak points or pores in the oxide film, where corrosion has started, by reacting with dissolving iron cations to form insoluble products of separate phase, which plug the gaps. These insoluble products may contain the inhibitive anion either as a salt, e.g. phosphate[121], or a basic salt, e.g. azelate[153], or as an insoluble oxide, e.g. Cr_2O_3 from chromate[133,145]. Precipitation of such solid products is favoured if the pH in the region of the pores does not become acid. Thus, on the basis of this theory, inhibition by anions such as phosphate, borate, silicate and carbonate, is enhanced by their buffer properties which serve to prevent a fall in pH in the anodic areas. Since ferric salts are usually more insoluble than ferrous salts, the requirement of oxidising power in the solution for inhibition is explained as necessary for oxidation of ferrous to insoluble ferric compounds, either by dissolved oxygen or oxidising anions, e.g. chromate or nitrite.

There is undoubted evidence that pore plugging as described by this theory does occur, particularly when conditions are relatively unfavourable for inhibition. However, this theory does not provide a complete explanation of the action of inhibiting anions. Some inhibitive anions, e.g. azide[91] and pertechnetate[140], do not form insoluble salts with ferrous or ferric ions. Furthermore, the pertechnetate ion has negligible buffer capacity[140]. Oxidising power is not necessarily a criterion of inhibitive efficiency, e.g. permanganate rapidly oxidises ferrous ions to ferric, but is a poor inhibitor[156]. Also, there is little correlation between the extent to which anions are incorporated into oxide films and their inhibitive efficiency[117]. The inhibitive action of anions on iron is soon lost after transfer of the metal from the anion solution to water. The behaviour[111,115] of iron in solutions containing mixtures of inhibitive and aggressive anions indicates that there is a competitive uptake of the inhibitive and the aggressive anions. These facts strongly suggest that the inhibitive effect of anions is exerted through a relatively labile adsorption on the oxide surface, rather than irreversible incorporation into the oxide film.

Effect of Inhibitive Anions on Formation of Passivating Oxide

Inhibitive anions can also contribute to the repair of weak points, pores or damage to the oxide film on iron by promoting the formation of a passivating film of iron oxide at such areas. This was put forward as a mechanism of action of inhibitive anions by Stern[157], who proposed that the formation of passivating iron oxide was easier in the presence of such anions (due to an increase in the rate of the cathodic process, arising from either reduction of an oxidising anion or acceleration of oxygen reduction) so that a greater equivalent anodic current would be available to more easily exceed the critical current density for passivation. Stern also suggested that inhibitive anions might facilitate the anodic process of oxide formation by reducing the magnitude of the critical current density or by making the Flade potential more negative. Subsequent work has shown that inhibitive anions affect mainly the anodic process. Thus in solutions of the oxidising inhibitive anions chromate[103,158], nitrite[158,159] and pertechnetate[106], reduction of dissolved oxygen is the predominant cathodic process. There is evidence[103,160], that some anions can increase somewhat the rate of oxygen reduction, but the effects do not appear sufficiently large to be significant. However, anodic polarisation studies[158,161-163] have shown that the critical current density for passivation is much smaller in the presence of inhibitive anions than aggressive anions. Comparing a number of inhibitive anions[158], the critical current densities for passivation have been found to increase in generally the same order as the inhibitive efficiencies decrease. In solutions of inhibitive anions[158,164], as the pH becomes more acid the critical current density for passivation generally increases. In benzoate solution[164], the presence of dissolved oxygen has been shown to reduce considerably the critical current density for passivation. However, in carbonate solution[163], dissolved oxygen has little effect. The effects of anions on the passivation reaction are related to their adsorption, since radiotracer measurements[165] during passivation of iron in solutions of sodium phosphate and sodium iodohippurate (a substituted benzoate) have indicated that the greater the adsorption of the anion on the active iron surface, the smaller the critical current density for passivation.

Brasher[166,167] has found that a steel specimen which has begun to corrode in a solution of an aggressive anion can be inhibited by addition of a non-oxidising inhibitive anion only if the potential has not become more negative than a certain value, termed the 'critical potential for inhibition'. This critical potential depends on the nature of the anion[167], and on the relative concentrations of inhibitive and aggressive anions in the solution, becoming more positive as the concentration of aggressive anion increases[168]. The critical potential for inhibition has been related to the effects of potential on the adsorption of anions[165,168], and is probably the potential at which adsorption of the inhibitive anion on the active corroding areas has the minimum value necessary to reduce the metal ion dissolution rate to such an extent that oxide film formation can occur. In pure benzoate or phosphate solutions, the critical potentials for inhibition (benzoate -0.28 V, phosphate -0.43 V) are close to the critical passivation potentials[158,164] for iron. This is a further indication that inhibition under these circumstances occurs due to the formation of passivating iron oxide at the corroding areas.

Effects of Inhibitive Anions on the Dissolution of Passivating Oxide

The passivating oxide layer on iron should remain stable and protective provided its rate of formation exceeds its rate of dissolution. Dissolution of the outer layer of γ-Fe_2O_3 can occur in two ways. At potentials more positive than the Flade potential, dissolution occurs by passage of Fe^{3+} ions from the oxide surface into solution[169,170]. The effect of anions on the rate of this process has not been systematically studied, but there is evidence that the rates are considerably smaller in solutions of chromate[94] than of sulphate[169]. However, the rates in sulphate are slightly less than in solutions of phthalate[171], an inhibitive anion, which may be due to some complex formation. The dissolution rates in solutions of these anions decrease considerably as the pH increases. The thickness of the oxide film on iron also controls the Fe^{3+} dissolution rates, which decrease markedly as the oxide film thickness increases[170]. Thus, under adverse conditions, i.e. relatively low pH, low inhibitive power of anions, low oxide thickness (especially at weak points in the film), on immersion of an iron specimen, an appreciable Fe^{3+} dissolution current could flow, which could depress the potential to the vicinity of the Flade potential. In this region, the rate of oxide dissolution increases[170], due to the onset of reductive dissolution[101,172], leading to passage of Fe^{2+} ions into solution. The dissolving Fe^{2+} ions derive from the reduction of Fe^{3+} ions in the surface layer of γ-Fe_2O_3, by electrons supplied from the oxidation of metallic iron to form cations. Gilroy and Mayne[101] have shown that the rate of reductive dissolution of the oxide film on iron is faster in solutions of aggressive anions than in solutions of inhibitive anions. The rate of dissolution increases as the dissolved oxygen content decreases. Gilroy and Mayne have shown further[173] that the rates of oxidation in solution of Fe^{2+} to Fe^{3+} by dissolved oxygen are greater in the presence of inhibitive anions than of aggressive anions. They propose that a function of the inhibitive anion is to stimulate the oxidation by oxygen of any Fe^{2+} ions produced in the surface of the γ-Fe_2O_3 film, thus retarding its dissolution. These effects of anions on the reductive dissolution of the oxide film should correspond to effects on the Flade potential, i.e. a decrease in the rate of reductive dissolution should displace the Flade potential to more negative values, and vice versa. The effects of a number of anions at varying pH on the Flade potential have been described by Freiman and Kolotyrkin[174].

The reductive dissolution of the outer γ-Fe_2O_3 layer exposes the inner magnetite layer of the oxide film. In acid solutions (pH less than 4) the magnetite layer rapidly dissolves[175], but in near neutral solution it may be stable and protective, depending on the nature of the anion present and its concentration[176,177]. The magnetite layer is stable in inhibitive solutions of anions, e.g. benzoate[177], carbonate[163], hydroxide[163], borate[126] (though not bicarbonate[177]). The stability of the magnetite layer controls the inhibition of corrosion of iron when coupled to electronegative metals such as aluminium, zinc or cadmium[177].

Thus inhibitive anions can retard the dissolution of both the γ-Fe_2O_3 and the magnetite layers of the passivating oxide layer on iron. This has the dual effect of preventing breakdown of an existing oxide film and also of facilitating the formation of a passivating oxide film on an active iron surface, as discussed in the previous section.

Inhibitive Anions and Aggressive Anions

An important function of inhibitive anions is to counteract the effects of aggressive anions which tend to accelerate dissolution and breakdown of the oxide films. The relationships[115] (mentioned above) between the concentrations of inhibitive anions and aggressive anions, when inhibition is just achieved, correspond to competitive uptake of the anions by adsorption or ion exchange at a fixed number of sites at the oxide surface. The effects of the valencies of the competing anions are generally consistent with the total charge due to anion uptake being constant. Iron surfaces protected in solutions of inhibitive anions rapidly begin to corrode on addition of aggressive anions or on transfer to distilled water. All these facts indicate that inhibitive anions overcome the effects of aggressive anions through participation in a reversible competitive adsorption such that the adsorbed inhibitive anions reduce the surface concentration of aggressive anions below a critical value.

The reasons why some anions exhibit strong inhibitive properties while others exhibit strong aggressive properties are not entirely clear. The principal distinction seems to be that inhibitive anions are generally anions of weak acids whereas aggressive anions are anions of strong acids. Due to hydrolysis, solutions of inhibitive anions have rather alkaline pH values and buffer capacities to resist pH displacement to more acid values. As discussed above, both these factors are beneficial to the stability and repair of the oxide film. However, the primary difference between inhibitive and aggressive anions must arise from their effects on dissolution reactions at the oxide surface. For the various XO_4^{n-} ions, Cartledge[140,143] considers that the difference between inhibitive and non-inhibitive anions is due to the contrasting internal polarity of the X^+—O^- bond creating different electrostatic interactions in the electrical double layer, thus affecting transfer of metal ions into solution. Another important factor is that the bonds formed between anions of weak acids and metal ions in the oxide surface are of a more co-ordinate character than those formed by anions of strong acids. The mechanism of dissolution of metal ions from the oxide surface is not well understood, but according to Heusler[170] it proceeds by the passage of $Fe(OH)^{2+}$ ions into solution. It seems likely that dissolution of other anion complexes will occur, and it would appear that dissolution of the more co-ordinately bonded complexes with inhibitive anions occurs less readily than that of the more ionically bonded complexes with aggressive anions. In addition, the electron transfer to the ferric ion in the co-ordinate bonds with inhibitive anions will tend to stabilise the ferric state against reduction to the ferrous state, making the oxide more resistant to reductive dissolution.

The inhibitive efficiency of anions tends to increase with size in a homologous series[91], due probably to the increasing tendency to adsorption, and decreasing solubility of the ferric–anion complex.

Zinc

The effects of inhibitive and aggressive anions on the corrosion of zinc are broadly similar to the effects observed with iron. Thus with increasing concentration, anions tend to promote corrosion but may give inhibition above a critical concentration[141,160,178]. Inhibition of zinc corrosion is somewhat more difficult than that of iron, e.g. nitrite[179,180] and benzoate[98,181,182] are not efficient inhibitors for zinc. However, inhibition of zinc corrosion is observed in the presence of anions such as chromate[91,178,179], borate[179] and nitrocinnamate[91,179], which are also good inhibitors for the corrosion of iron. Anions such as sulphate, chloride and nitrate are aggressive towards zinc and prevent protection by inhibitive anions[141,160]. The presence of dissolved oxygen in the solution is essential for protection by inhibitive anions. As in the case of iron, pressures of oxygen greater than atmospheric or an increase in oxygen supply by rapid stirring can lead to the protection of zinc in distilled water[183]. Inhibition of zinc corrosion occurs most readily[184] in the pH range of 9 to 12, which corresponds approximately to the region of minimum solubility of zinc hydroxide.

The ways in which inhibitive anions affect the corrosion of zinc are mainly similar to those described above for iron. In inhibition by chromate, localised uptake of chromium has been shown to occur at low chromate concentrations[160,185] and in the presence of chloride ions[185]. Thus under conditions unfavourable for inhibition, pore plugging occurs on zinc. Inhibitive anions also promote the passivation of zinc, e.g. passivation is much easier in solutions of the inhibitive anion, borate[186-188], than in solutions of the non-inhibitive anions, carbonate and bicarbonate[186,189]. A critical inhibition potential, analogous to that on iron, has been observed for zinc in borate solutions[144]. Thus inhibitive anions promote repair of the oxide film on zinc by repassivation with zinc oxide. The requirement of dissolved oxygen for inhibition indicates that the passivating oxide is stabilised at potentials more positive than the Flade potential by the reduction in dissolution rate due to the inhibitive anion. The passivating film is ZnO, which dissolves as divalent cations[190], and there is no evidence of reductive dissolution. Thus, on zinc the inhibitive anion presumably stabilises the oxide by formation of an adsorbed complex with the zinc ion, the dissolution rate of which is less than that of analogous zinc complexes with water, hydroxyl ions or aggressive anions.

Aluminium

When aluminium is immersed in water, the air-formed oxide film of amorphous γ-alumina initially thickens (at a faster rate than in air) and then an outer layer of crystalline hydrated alumina forms, which eventually tends to stifle the reaction[191-193]. In near-neutral air-saturated solutions, the corrosion of aluminium is generally inhibited[194,195] by anions which are inhibitive for iron, e.g. chromate, benzoate, phosphate, acetate. Inhibition also occurs in solutions containing sulphate or nitrate ions, which are aggressive towards iron. Aggressive anions for aluminium include the halide ions[192,194-196], F^-, Cl^-, Br^-, I^-, which cause pitting attack, and anions which form soluble complexes with aluminium[194], e.g. citrate and tartrate, which cause general attack. Competitive effects[195,197], similar to those observed on iron, are observed in the action of mixtures of inhibitive anions and chloride ions on aluminium. The inhibition of aluminium corrosion by anions exhibits both an upper and a lower pH limit. The pH range for inhibition depends upon the nature of the anion[194].

In near-neutral and de-aerated solutions[194], the oxide film on aluminium is stable and protective in distilled water and chloride solutions, as well as in solutions of inhibitive anions. Thus the inhibition of aluminium corrosion by anions differs from that of iron or zinc in that the presence of dissolved oxygen in the solution is not necessary to stabilise the oxide film, i.e. the Flade potential is more negative than the hydrogen evolution potential. Lorking and Mayne[194,196] observed that inhibition of aluminium corrosion occurred only when the initial rate of dissolution of aluminium oxide in solutions of anions was less than a critical value. If this dissolution rate was decreased by pre-saturation of the solution with aluminium oxide, the corrosion of aluminium could be inhibited in normally aggressive solutions containing chloride or fluoride ions. The oxide film dissolves as Al^{3+} ions, the degree of hydrolysis and rate of dissolution depending on the pH[198-201]. There is no evidence of reductive dissolution. Thus, as with zinc, the inhibitive anions probably act by adsorption on to Al^{3+} ions in the oxide surface to form a surface complex, which has a low dissolution rate. The formation of surface compounds by anions on aluminium oxide has been discussed by Vedder and Vermilyea[202,203] in connection with the inhibition of hydration of anodic oxide films on aluminium. In corrosion inhibition by chromate ions, their interaction with the oxide film on aluminium has been shown by

Heine and Pryor[204] to result in the formation of an outer layer of the film which is more protective due to its high electronic resistance and low dissolution rate. Chromate ions were also found to prevent the uptake and penetration of chloride ions into the aluminium oxide film[204, 205].

Copper

Little work has been carried out on the mechanism of inhibition of the corrosion of copper in neutral solutions by anions. Inhibition occurs in solutions containing chromate[98], benzoate[98] or nitrite[206] ions. Chloride ions[206, 207] and sulphide[208] ions act aggressively. There is evidence[209] that chloride ions can be taken up into the cuprous oxide film on copper to replace oxide ions and create cuprous ion vacancies which permit easier diffusion of cuprous ions through the film, thus increasing the corrosion rate.

Copper corrosion can also be effectively inhibited in neutral solution by organic compounds of low molecular weight, such as benzotriazole[208, 210-212] and 2-mercaptobenzothiazole[208]. Benzotriazole is particularly effective in preventing the tarnishing and dissolution of copper in chloride solutions. In the presence of benzotriazole, the anodic dissolution reaction, the oxide film growth reaction and the dissolved oxygen reduction reaction, are all inhibited[208, 212], indicating strong adsorption of the inhibitor on the cuprous oxide surface.

Conclusions

The mechanism of action of inhibitive anions on the corrosion of iron, zinc and aluminium in near-neutral solution involves the following important functions:

1. Reduction of the dissolution rate of the passivating oxide film.
2. Repair of the oxide film by promotion of the reformation of oxide.
3. Repair of the oxide film by plugging pores with insoluble compounds.
4. Prevention of the adsorption of aggressive anions.

Of these functions, the most important appears to be the stabilisation of the passivating oxide film by decreasing its dissolution rate (Function 1). Inhibitive anions probably form a surface complex with the metal ion of the oxide, i.e. Fe^{3+}, Zn^{2+}, Al^{3+}, such that the dissolution rate of this complex is less than that of the analogous complexes with water, hydroxyl ions or aggressive anions. For iron only, the special mechanism of reductive dissolution enables the ferric oxide film to dissolve more easily as Fe^{2+} ions. Inhibitive anions may retard this process by catalysing the re-oxidation by dissolved oxygen of any Fe^{2+} formed in the oxide surface. Stabilisation of the oxide films by decrease of dissolution rate is also important with respect to repassivation by oxide formation (Function 2). The plugging of pores by formation of insoluble compounds (Function 3) does not appear to be an essential function, but is valuable in extending the range of conditions under which inhibition can be achieved. The suppression of the adsorption of aggressive anions (Function 4) by participation in a dynamic reversible competitive adsorption equilibrium at the metal surface appears to be related to the general adsorption behaviour of anions rather than a specific property of inhibitive anions.

The relative importance of these functions also depends to a considerable extent on the solution conditions. Under favourable conditions of pH, oxidising power and aggressive anion concentration in the solution, Function 1 is probably effective in preventing film breakdown. Under unfavourable conditions for inhibition, localised breakdown will occur at weak points in the oxide film, and Functions 2 and 3 become important in repairing the oxide film.

J. G. N. THOMAS

REFERENCES

1. Fischer, H., *Werkstoffe u. Korrosion*, **23**, 445, 453 (1972)
2. Putilova, J. N., Balezin, S. A. and Barannik, V. P., *Metallic Corrosion Inhibitors*, Pergamon Press, London (1960)
3. Trabanelli, G. and Carassiti, V., *Advances in Corrosion Science and Technology*, **1**, Plenum Press, New York, London, 147 (1970)
4. Lacombe, P., *2nd European Symposium on Corrosion Inhibitors, Ferrara 1965*, University of Ferrara, 517 (1966)
5. Gileadi, E., ibid., 543 (1966)
6. Wormwell, F. and Thomas, J. G. N., *Surface Phenomena of Metals*, Society of Chemical Industry, London, Monograph No. 28, 365 (1968)
7. Thomas, J. G. N., *Werkstoffe u. Korrosion*, **19**, 957 (1968)
8. Conway, B. E. and Barradas, R. G., *Transactions of the Symposium on Electrode Processes*, John Wiley and Sons, New York, 299 (1961)
9. Cavallaro, L., Felloni, L. and Trabanelli, G., *First European Symposium on Corrosion Inhibitors, Ferrara, 1960*, University of Ferrara, 111 (1961)
10. Annand, R. R., Hurd, R. M. and Hackerman, N., *J. Electrochem. Soc.*, **112**, 138 (1965)
11. Gileadi, E., *J. Electroanal. Chem.*, **11**, 137 (1966)
12. *Optical Studies of Adsorbed Layers at Interfaces*, Symp. Faraday Soc., **4** (1970)
13. Epelboin, I., Keddam, M. and Takenouti, H., *J. Applied Electrochem.*, **2**, 71 (1972)
14. Heusler, K. E. and Cartledge, G. H., *J. Electrochem. Soc.*, **108**, 732 (1961)
15. Trabanelli, G., Zucchi, F. and Zucchini, G. L., *Corrosion, Traitement, Protection, Finition*, **16**, 335 (1968)
16. Zucchini, G. L., Zucchi, F. and Trabanelli, G., *3rd European Symposium on Corrosion Inhibitors, Ferrara, 1970*, University of Ferrara, 577 (1971)
17. Ross, T. K. and Jones, D. H., as Ref. 9, 163 (1961)
18. Fedorov, Y. V., Uzlyuk, M. V. and Zelenin, V. M., *Protection of Metals*, **6**, 287 (1970)
19. Schwabe, K. and Leonhardt, W., *Chem. Ing. Tech.*, **38**, 59 (1966)
20. Baldi, L., Carassiti, V., Trabanelli, G., Zucchi, F. and Zucchini, G. L., *Proc. 3rd International Congress Metallic Corrosion, Moscow, 1966*, MIR, Moscow, **2**, 127 (1969)
21. Trabanelli, G., Zucchini, G. L., Zucchi, F. and Carassiti, V., *British Corrosion J.*, **4**, 267 (1969)
22. Schwabe, K., Reinhard, G., Fischer, M., Schaarschmidt, K., *Werkstoffe u. Korrosion*, **22**, 302 (1971)
23. Fujii, S. and Aramaki, K., *Proc. 3rd International Conference on Metallic Corrosion, Moscow, 1966*, MIR, Moscow, **2**, 70 (1969)
24. Hoar, T. P. and Holliday, R. D., *J. Applied Chem.*, **3**, 502 (1953)
25. Makrides, A. and Hackerman, N., *Ind. Eng. Chem.*, **47**, 1 773 (1955)
26. Hoar, T. P. and Khera, R. P., as Ref. 9, 73 (1961)
27. Felloni, L. and Cozzi, A., as Ref. 4, 253 (1966)
28. Donahue, F. M., Akiyama, A. and Nobe, K., *J. Electrochem. Soc.*, **114**, 1 006 (1967)

29. Frumkin, A. and Damaskin, B. B., *Modern Aspects of Electrochemistry, No. 3*, Ed. J. O'M. Bockris, Butterworths, London, 149 (1964)
30. *Electrosorption*, Ed. E. Gileadi, Plenum Press, New York, (1967)
31. Damaskin, B. B., Petrii, O. A. and Batrakov, V. V., *Adsorption of Organic Compounds on Electrodes*, Plenum Press, New York (1971)
32. Antropov, L. I., *Proc. First International Congress on Metallic Corrosion, London, 1961*, Butterworths, London, 147 (1962)
33. Antropov, L. I., *Corrosion Science*, 7, 607 (1967)
34. Fischer, H. and Seiler, W., *Corrosion Science*, 6, 159 (1966)
35. Makrides, A. C. and Hackerman, N., *Ind. Eng. Chem.*, 46, 523 (1954)
36. Blomgren, E., Bockris, J. O'M. and Jesch, C., *J. Phys. Chem.*, 65, 2 000 (1961)
37. Szklarska-Smialowska, Z. and Dus, B., *Corrosion*, 23, 130 (1967)
38. Ayers, R. C. and Hackerman, N., *J. Electrochem. Soc.*, 110, 507 (1963)
39. Cox, P. F., Every, R. L. and Riggs, O. L., *Corrosion*, 20, 299t (1964)
40. Donahue, F. M. and Nobe, K., *J. Electrochem. Soc.*, 112, 886 (1965)
41. Grigoryev, V. P. and Osipov, O. A., *Proc. 3rd International Congress on Metallic Corrosion, Moscow, 1966*, MIR, Moscow, 2, 48 (1969)
42. Altsybeeva, A. I., Levin, S. Z. and Dorokhov, A. P., as Ref. 16, 501 (1971)
43. Donahue, F. M. and Nobe, K., *J. Electrochem. Soc.*, 114, 1 012 (1967)
44. Grigoryev, V. P. and Ekilik, V. V., *Protection of Metals*, 4, 23 (1968)
45. Grigoryev, V. P. and Gorbachev, V. A:, *Protection of Metals*, 6, 282 (1970)
46. Grigoryev, V. P. and Kuznetsov, V. V., *Protection of Metals*, 5, 356 (1969)
47. Akiyama, A. and Nobe, K., *J. Electrochem. Soc.*, 117, 999 (1970)
48. Vosta, J. and Eliasek, J., *Corrosion Science*, 11, 223 (1971)
49. Brandt, H., Fischer, M. and Schwabe, K., *Corrosion Science*, 10, 631 (1970)
50. Hackerman, N., *Corrosion*, 18, 332t (1962)
51. Aramaki, K. and Hackerman, N., *J. Electrochem. Soc.*, 115, 1 007 (1968)
52. Mann, C. A., Lauer, B. E. and Hultin, C. T., *Industr. Engng. Chem.*, 28, 159, 1 048 (1936)
53. Szlarska-Smialowska, Z. and Wieczorek, G., as Ref. 16, 453 (1971)
54. Iofa, Z. A., as Ref. 4, 93 (1966)
55. Hackerman, N., Snavely, E. S. and Payne, J. S., *J. Electrochem. Soc.*, 113, 677 (1966)
56. Iofa, Z. A., Batrakov, V. V. and Cho-Ngok-Ba, *Electrochimica Acta*, 9, 1 645 (1964)
57. Fud Hasan, S. and Iofa, Z. A., *Protection of Metals*, 6, 218 (1970)
58. Cavallaro, L., Felloni, L., Trabanelli, G. and Pulidori, F., *Electrochimica Acta*, 9, 485 (1964)
59. Murakawa, T., Naguara, S. and Hackerman, N., *Corrosion Science*, 7, 79 (1967)
60. Hudson, R. M. and Warning, C. J., *Materials Protection*, 6 No. 2, 52 (1967)
61. Hudson, R. M. and Warning, C. J., *Corrosion Science*, 10, 121 (1970)
62. Horner, L. and Rottger, F., *Korrosion*, 16, Verlag Chemie, Weinheim/Bergstr., 57 (1963)
63. Ertel, H. and Horner, L., as Ref. 4, 71 (1966)
64. Lorenz, W. J. and Fischer, H., *Proc. 3rd International Congress on Metallic Corrosion, Moscow, 1966*, MIR, Moscow, 2, 99 (1969)
65. Horner, L., *Werkstoffe u. Korrosion*, 23, 466 (1972)
66. Putilova, J. N., as Ref. 4, 139 (1966)
67. Poling, G. W., *J. Electrochem. Soc.*, 114, 1 209 (1967)
68. Kaesche, H. and Hackerman, N., *J. Electrochem. Soc.*, 105, 191 (1958)
69. Kaesche, H., *Die Korrosion der Metalle*, Springer-Verlag, Berlin, 159 (1966)
70. Kelly, E. J., *J. Electrochem. Soc.*, 115, 1 111 (1968)
71. Okamoto, G., Nagayama, M., Kato, J. and Baba, T., *Corrosion Science*, 2, 21 (1962)
72. Machu, W., as Ref. 9, 183 (1961)
73. Thibault, S and Talbot, J., as Ref. 16, 75 (1971)
74. West, J. M., *J. Applied Chem.*, 10, 250 (1960)
75. Kelly, E. J., *J. Electrochem. Soc.*, 112, 124 (1965)
76. Lorenz, W. J., Eichkorn, G., Albert, L. and Fischer, H., *Electrochimica Acta*, 13, 183 (1968)
77. Lorenz, W. J., *Corrosion Science*, 5, 121 (1965)
78. Ammar, I. A., Darwish, S., Khalil, M. W., *Electrochimica Acta*, 12, 657 (1967)
79. McCafferty, E. and Zettlemoyer, A. C., *J. Phys. Chem.*, 71, 2 444 (1967)
80. Vaidyanathan, H. and Hackerman, N., *Corrosion Science*, 11, 737 (1971)
81. Iofa, Z. A., *Protection of Metals*, 6, 445 (1970)
82. Duwell, E. J., *J. Electrochem. Soc.*, 109, 1 013 (1962)
83. Zucchi, F., Zucchini, G. L. and Trabanelli, G., as Ref. 16, 121 (1971)

84. Grigoryev, V. P. and Ekilik, V. V., *Protection of Metals*, **4**, 517 (1968)
85. Grigoryev, V. P. and Osipov, O. A., as Ref. 16, 473 (1971)
86. Antropov, L. I. and Savgira, Y. A., *Protection of Metals*, **3**, 597 (1967)
87. Davolio, G. and Soragni, E., as Ref. 16, 219 (1971)
88. Evans, U. R., *The Corrosion and Oxidation of Metals*, Arnold, London, 134 (1960)
89. Butler, G., as Ref. 16, 753 (1971)
90. Heyn, E. and Bauer, O., *Mitteilungen Koniglichen Materialprufungsamt Berlin-Dahlem*, **26**, 74 (1908)
91. Hersch, P., Hare, J. B., Robertson, A. and Sutherland, S. M., *J. Appl. Chem.*, **11**, 251, 265 (1961)
92. Brasher, D. M., *British Corrosion Journal*, **4**, 122 (1969)
93. Pryor, M. J. and Cohen, M., *J. Electrochem. Soc.*, **100**, 203 (1953)
94. Brasher, D. M., Beynon, J. G., Rajagopalan, K. S. and Thomas, J. G. N., *British Corrosion Journal*, **5**, 264 (1970)
95. Mayne, J. E. O. and Ramshaw, E. H., *J. Appl. Chem.*, **10**, 419 (1960)
96. Wachter, A., *Industr. Eng. Chem.*, **37**, 749 (1945)
97. Legault, R. A. and Walker, M. S., *Corrosion*, **20**, 282t (1964)
98. Wormwell, F. and Mercer, A. D., *J. Appl. Chem.*, **2**, 150 (1952)
99. Davies, D. E. and Slaiman, Q. J. M., *Corrosion Science*, **11**, 671 (1971)
100. Pryor, M. J. and Cohen, M., *J. Electrochem. Soc.*, **98**, 263 (1951)
101. Gilroy, D. and Mayne, J. E. O., *British Corrosion J.*, **1**, 102 (1965)
102. Slaiman, Q. J. M. and Davies, D. E., as Ref. 16, 739 (1971)
103. Cartledge, G. H., *J. Phys. Chem.*, **65**, 1 009 (1961)
104. Cartledge, G. H., *J. Electrochem. Soc.*, **113**, 328 (1966)
105. Sympson, R. F. and Cartledge, G. H., *J. Phys. Chem.*, **60**, 1 037 (1956)
106. Cartledge, G. H., *J. Phys. Chem.*, **64**, 1 882 (1960)
107. Bengough, G. D. and Wormwell, F., *Chem. and Ind.*, 549 (1950)
108. Bogatyreva, E. V. and Balezin, S. A., *J. Applied Chem.*, *U.S.S.R.*, **32**, 1 094 (1959)
109. Wormwell, F. and Ison, H. C. K., as Ref. 107
110. Mayne, J. E. O. and Page, C. L., *British Corrosion J.*, **5**, 93 (1970)
111. Matsuda, S. and Uhlig, H. H., *J. Electrochem. Soc.*, **111**, 156 (1964)
112. Brasher, D. M. and Mercer, A. D., *British Corrosion J.*, **3**, 120 (1968)
113. Mercer, A. D. and Jenkins, I. R., *British Corrosion J.*, **3**, 130 (1968)
114. Mercer, A. D., Jenkins, I. R. and Rhoades-Brown, J. E., *British Corrosion J.*, **3**, 136 (1968)
115. Brasher, D. M., Reichenberg, D. and Mercer, A. D., *British Corrosion J.*, **3**, 144 (1968)
116. Legault, R. A., Mori, S. and Leckie, H. P., *Corrosion*, **26**, 121 (1970)
117. Thomas, J. G. N., *British Corrosion J.*, **5**, 41 (1970)
118. Mayne, J. E. O. and Pryor, M. J., *J. Chem. Soc.*, 1 831 (1949)
119. Mayne, J. E. O., Menter, J. W. and Pryor, M. J., *J. Chem. Soc.*, 3 229 (1950)
120. Mayne, J. E. O. and Menter, J. W., *J. Chem. Soc.*, 99 (1954)
121. Mayne, J. E. O. and Menter, J. W., *J. Chem. Soc.*, 103 (1954)
122. Cohen, M., *J. Phys. Chem.*, **56**, 451 (1952)
123. Draper, P. H. G., *Corrosion Science*, **7**, 91 (1967)
124. Foley, C. L., Kruger, J. and Bechtoldt, C. J., *J. Electrochem. Soc.*, **114**, 994 (1967)
125. Hancock, P. and Mayne, J. E. O., *J. Chem. Soc.*, 4 172 (1958)
126. Nagayama, M. and Cohen, M., *J. Electrochem. Soc.*, **109**, 781 (1962)
127. Kruger, J., *J. Electrochem. Soc.*, **110**, 654 (1963)
128. Hancock, P. and Mayne, J. E. O., *J. Chem. Soc.*, 4 167 (1958)
129. Sewell, P. B., Stockbridge, C. D. and Cohen, M., *J. Electrochem. Soc.*, **108**, 933 (1961)
130. Dye, T. G., Fursey, A. and Lloyd, G. O., *Mem. Scient. Revue Metall.*, **65**, 383 (1968)
131. Brasher, D. M., *British Corrosion J.*, **1**, 183 (1966)
132. Mayne, J. E. O. and Page, C. L., *British Corrosion J.*, **7**, 111 (1972)
133. Brasher, D. M. and Kingsbury, A. H., *Trans. Faraday Soc.*, **54**, 1 214 (1958)
134. Kubaschewski, O. and Brasher, D. M., *Trans. Faraday Soc.*, **55**, 1 200 (1959)
135. Brasher, D. M. and Mercer, A. D., *Trans. Faraday Soc.*, **61**, 803 (1965)
136. Brasher, D. M., De, C. P. and Mercer, A. D., *British Corrosion J.*, **1**, 188 (1966)
137. Hoar, T. P., *Corrosion*, A Symposium, Melbourne 1955, University of Melbourne, 124 (1956)
138. Brasher, D. M., as Ref. 9, 313 (1961)
139. Mayne, J. E. O., as Ref. 9, 273 (1961)
140. Cartledge, G. H., *Corrosion*, **18**, 316t (1962)

141. Brasher, D. M., Beynon, J. G., Mercer, A. D. and Rhoades-Brown, J. E., as Ref. 4, 559 (1966)
142. Gilroy, D. and Mayne, J. E. O., as Ref. 4, 585 (1966)
143. Cartledge, G. H., *British Corrosion J.*, **1**, 293 (1966)
144. Brasher, D. M., *Tribune de Cebedeau*, No. 300, 1 (1968)
145. Hoar, T. P. and Evans, U. R., *J. Chem. Soc.*, 2 476 (1932)
146. Cohen, M. and Beck, A. F., *Z. Elektrochem.*, **62**, 696 (1958)
147. Cartledge, G. H. and Spahrbier, D. H., *J. Electrochem. Soc.*, **110**, 644 (1963)
148. Pryor, M. J., Cohen, M. and Brown, F., *J. Electrochem. Soc.*, **99**, 542 (1952)
149. Brasher, D. M. and Stove, E. R., *Chemistry and Industry*, 171 (1952)
150. Gatos, H. C., as Ref. 9, 257 (1961)
151. Cartledge, G. H., *J. Phys. Chem.*, **59**, 979 (1955)
152. Spitsin, V. I., Rozenfeld, I. L., Persiantseva, V. P., Zamochnikova, N. N. and Kuzina, A. F., *Corrosion*, **21**, 211, (1965)
153. Mayne, J. E. O. and Page, C. L., *British Corrosion J.*, **7**, 115 (1972)
154. Mellors, G. W., Cohen, M. and Beck, A. F., *J. Electrochem. Soc.*, **105**, 332 (1958)
155. Evans, U. R., *J. Chem. Soc.*, 1 020 (1927)
156. Cartledge, G. H., *J. Electrochem. Soc.*, **114**, 39 (1967)
157. Stern, M., *J. Electrochem. Soc.*, **105**, 635 (1958)
158. Thomas, J. G. N. and Nurse, T. J., *British Corrosion J.*, **2**, 13 (1967)
159. Rozenfeld, I. L., *Dokl. Akad. Nauk SSSR.*, **78**, 523 (1951)
160. Gouda, V. K., Khedr, M. G. A. and Shams El Din, A. M., *Corrosion Science*, **7**, 221 (1967)
161. Hancock, P. and Mayne, J. E. O., *J. Applied Chem.*, **9**, 345 (1959)
162. Freiman, L. I. and Kolotyrkin, Y. M., *Protection of Metals*, **1**, 135 (1965)
163. Thomas, J. G. N., Nurse, T. J. and Walker, R., *British Corrosion J.*, **5**, 87 (1970)
164. Slaiman, Q. J. M. and Davies, D. E., *Corrosion Science*, **11**, 683 (1971)
165. Thomas, J. G. N., *British Corrosion J.*, **1**, 156 (1966)
166. Brasher, D. M., *Nature, Lond.*, **185**, 838 (1960)
167. Brasher, D. M., *Proc. 1st International Congress on Metallic Corrosion, 1961*, Butterworths, London, 156 (1962)
168. Antropov, L. I. and Kuleshova, N. F., *Protection of Metals*, **3**, 131 (1967)
169. Vetter, K. J., *Z. Elektrochem.*, **59**, 67 (1955)
170. Heusler, K. E., *Ber. Bunsenges. Phys. Chem.*, **72**, 1 197 (1968)
171. Weil, K. G. and Bonhoeffer, K. F., *Z. Phys. Chem. N.F.*, **4**, 175 (1955)
172. Pryor, M. J. and Evans, U. R., *J. Chem. Soc.*, 1 259, 1 266, 1 274 (1950)
173. Gilroy, D. and Mayne, J. E. O., *British Corrosion J.*, **1**, 107 (1965)
174. Freiman, L. I. and Kolotyrkin, Y. M., *Protection of Metals*, **5**, 113 (1969)
175. Vetter, K. J. and Klein, G., *Z. Phys. Chem. N.F.*, **31**, 405 (1962)
176. Heusler, K., Weil, K. G. and Bonhoeffer, K. F., *Z. Phys. Chem. N.F.*, **15**, 149 (1958)
177. Mercer, A. D. and Thomas, J. G. N., as Ref. 16, 777 (1971)
178. Abu Zahra, R. H. and Shams El Din, A. M., *Corrosion Science*, **5**, 517 (1965) and **6**, 349 (1966)
179. Wormwell, F., *Chemistry and Industry*, 556 (1953)
180. Thomas, J. G. N., Mercer, A. D. and Brasher, D. M., *Proc. 4th International Congress on Metallic Corrosion, 1969*, N.A.C.E., Houston, 585 (1972)
181. Gilbert, P. T. and Hadden, S. E., *J. Applied Chem.*, **3**, 545 (1953)
182. Brasher, D. M. and Mercer, A. D., *Proc. 3rd International Congress on Metallic Corrosion, 1966*, MIR, Moscow, **2**, 21 (1969)
183. Evans, U. R. and Davies, D. E., *J. Chem. Soc.*, 2 607 (1951)
184. Lorking, K. F. and Mayne, J. E. O., *Proc. 1st International Congress on Metallic Corrosion, 1961*, Butterworths, London, 144 (1962)
185. McLaren, K. G., Green, J. H. and Kingsbury, A. H., *Corrosion Science*, **1**, 161, 170 (1961)
186. El Wakkad, S. E. S., Shams El Din, A. M. and Kotb, H., *J. Electrochem. Soc.*, **105**, 47 (1958)
187. Davies, D. E. and Lotlikar, M. M., *British Corrosion J.*, **1**, 149 (1966)
188. Lotlikar, M. M. and Davies, D. E., *Proc. 3rd International Congress on Metallic Corrosion, 1966*, MIR, Moscow, **1**, 167 (1969)
189. Kaesche, H., *Electrochimica Acta*, **9**, 383 (1964)
190. Armstrong, R. D. and Bulman, G. M., *J. Electroanal. Chem.*, **25**, 121 (1970)
191. Hart, R. K., *Trans. Faraday Soc.*, **53**, 1 020 (1957)
192. Pryor, M. J., *Z. Elektrochem.*, **62**, 782 (1958)

193. Godard, H. P. and Torrible, E. G., *Corrosion Science*, **10**, 135 (1970)
194. Lorking, K. F. and Mayne, J. E. O., *J. Appl. Chem.*, **11**, 170 (1961)
195. Bohni, H. and Uhlig, H. H., *J. Electrochem. Soc.*, **116**, 906 (1969)
196. Lorking, K. F. and Mayne, J. E. O., *British Corrosion J.*, **1**, 181 (1966)
197. Anderson, P. J. and Hocking, M. E., *J. Appl. Chem.*, **8**, 352 (1958)
198. Kaesche, H., *Werkstoffe u. Korrosion*, **14**, 557 (1963)
199. Plumb, R. C., *J. Phys. Chem.*, **66**, 866 (1962)
200. Straumanis, M. E. and Poush, K., *J. Electrochem. Soc.*, **112**, 1 185 (1965)
201. Heusler, K. E. and Allgaier, W., *Werkstoffe u. Korrosion*, **22**, 297 (1971)
202. Vedder, W. and Vermilyea, D. A., *Trans. Faraday Soc.*, **65**, 561 (1969)
203. Vermilyea, D. A. and Vedder, W., *Trans. Faraday Soc.*, **66**, 2 644 (1970)
204. Heine, M. A. and Pryor, M. J., *J. Electrochem. Soc.*, **114**, 1 001 (1967)
205. Heine, M. A., Keir, D. S. and Pryor, M. J., *J. Electrochem. Soc.*, **112**, 24 (1965)
206. Hoar, T. P., *J. Soc. Chem. Ind. (Lond.)*, **69**, 356 (1952)
207. Gatty, O. and Spooner, E. C. R., *The Electrode Potential Behaviour of Corroding Metals in Aqueous Solutions*, Clarendon Press, Oxford, 199 (1938)
208. Bonora, P. L., Bolognesi, G. P., Borea, P. A., Zucchini, G. L. and Brunoro, G., as Ref. 16, 685 (1971)
209. North, R. F. and Pryor, M. J., *Corrosion Science*, **10**, 297 (1970)
210. Cotton, J. B. and Scholes, I. R., *British Corrosion J.*, **2**, 1 (1967)
211. Poling, G. W., *Corrosion Science*, **10**, 359 (1970)
212. Mansfeld, F., Smith, T. and Parry, E. P., *Corrosion*, **27**, 289 (1971)

18.4 Boiler Feed-Water Treatment

Introduction

An increasing awareness of the need for the more efficient use of fuel has in the past two decades led to the development of steam plants of greatly increased size, operating at increased steam pressure. This trend continues, and it emphasises the great importance of the correct control of feed-water quality and its influence in ensuring continuous trouble-free operation of boiler plant. As steam pressures approach the critical (approximately $22\,MN/m^2$), the properties of steam and water phases become more nearly alike; since above the critical pressure there is no change of state, the conventional concept of 'boiler water' vanishes. 'Once-through' boilers (consisting virtually of a nest of tubes in parallel, fed at one end with water and delivering superheated steam at the other) have been developed for this duty, and although there is as yet little operational experience in the supercritical pressure range, such boilers have given good service at lower pressures. The once-through boiler demands feed water of extreme purity; experience indicates that such pure water also gives good results as feed for conventional boilers.

The two main objectives of feed-water treatment are the prevention of scale formation and the prevention of corrosion, both in the feed system and in the boiler plant which it supplies. The formation of scale usually (although not always) results from the ingress of calcium or magnesium salts into the feed system and leads to a reduction in the ability of the boiler metal to transfer heat to the water. This in turn leads to an increase in the temperature of the metal (which alone may cause failure by overheating) and may increase its susceptibility to corrosion. Again, corrosion can cause failure by reduction in mechanical strength resulting from the loss of metal, but often it results in the build-up of scabs of corrosion product which can, as in the case of hard-water scales, lead to failure by overheating.

Corrosion can similarly cause loss of metal in the feed system with consequent reduction in mechanical strength of the components of the system. A less obvious process is the carrying of metals dissolved by the feed water in its passage through the system into the boiler where they may be deposited, chemically or electrochemically, as oxides, metallic particles, or even as plated films. Such metallic impurities have been considered[1] to be a cause of corrosion (or at least of its aggravation) in the boiler.

Methods for the prevention of scale formation are well-known[2, 3] and serve to protect the heat-transfer surfaces of the plant from the effects of some adventitious solutes in the raw water, but in modern high-pressure plants there is a need for a further control of water quality. In the electricity supply industry, the use of distilled water or water of even higher purity as feed for high-pressure boilers is almost universal. There are, however, many industrial plants (a few at pressures even in excess of $7 MN/m^2$) operating satisfactorily with water containing appreciable amounts of dissolved solids; these plants are used mainly for supplying process steam, either directly or through back-pressure or pass-out turbines, under conditions often precluding the recovery of condensate.

It is essential to consider the treatment of feed water in relation to the parameters and duty of the boiler plant and of the steam-using equipment. It is, moreover, difficult to separate a study of corrosion processes in the boiler from a study of the associated feed system.

Principles of Feed-water Treatment

In the selection of a scheme for the preparation of make-up and feed water for a particular plant, the principles of removal, toleration and addition must be applied in appropriate combination. It is clearly desirable to remove from a raw water the ions of calcium and magnesium, which are known to play a part in scale formation even at low heat-transfer rates. Other ions such as copper have been thought to play a part in the corrosion of boiler tubes[1], while silica may contribute to scale formation in boilers[4-7] and on turbine blades[8, 9]; these contaminants should likewise be removed. Since, however, many of the available processes leave a residue of the ion concerned, it is necessary to know the concentration which can be tolerated. Thus in a boiler operating at $1 MN/m^2$ it may be acceptable to provide a feed water with a hardness of 40 p.p.m., expressed as $CaCO_3$ (equivalent to 16 p.p.m. of Ca^{2+}), whereas in the make-up water for a boiler operating at $10 MN/m^2$ even 1 p.p.m. of hardness cannot be tolerated (see also Section 2.3). The effects of such residual concentrations are often offset by the addition of other ions to the feed water, and again the principle of toleration applies.

Softening and Purification Treatments

Removal of Hardness

A consideration in detail of methods of water softening would be out of place in the present context. Technical details of many of the available processes have been published[10]. Since the choice of softening process may influence the corrosive properties of the water, mention will be made here of some of the principles involved. It is generally considered that water pretreated by softening alone is acceptable as direct feed only for low-pressure boilers (up to about $1·4 MN/m^2$) or, where it forms a small percentage of the total feed, as make-up for boiler systems operating at pressures up to $3 MN/m^2$, unless the boilers are specially designed for feed water of high salt content.

Precipitation processes The removal of calcium and magnesium salts by precipitation with lime and soda ash is widely practiced in industry. With plant of suitable design correctly operated, it is possible to reduce the hardness to 10 p.p.m. or less. The process gives some reduction in the dissolved solid content of most raw waters, removes much of the dissolved and combined carbon dioxide, and gives a water with an alkaline reaction. All these features are considered advantageous in reducing the corrosive tendency of the water.

If the process is carried out at elevated temperatures—and some modern plants are operated at a slight positive steam pressure—bicarbonates are decomposed thermally (resulting, with many waters, in an appreciable economy of chemicals), and settling takes place more rapidly, allowing a greater through-put for a plant of any given size. Under these conditions, much of the dissolved air present in the softened water can also be removed. This process can produce water acceptable as 100% feed for high-pressure boilers of suitable design when no condensate is available.

Ion-exchange processes In its original form, the 'base-exchange' process provided a simple substitution of sodium ions for calcium and magnesium ions. While the removal of hardness by this process is usually more complete than in the cold precipitation process, there is no reduction in dissolved solids and no removal of carbon dioxide. The corrosive tendency of the raw water is not reduced and (since the tendency of the water to deposit scale is reduced) it may even be increased. Nevertheless, when used in conjunction with one or other of the precipitation processes, base exchange gives a water of low hardness and low carbon dioxide, having reduced dissolved solids content and alkalinity. It is, however, possible in such plants to replace bicarbonate alkalinity by a (lower) caustic alkalinity which may be equally undesirable; some modern plants make provision for the removal of this excess alkalinity.

In more modern ion-exchange processes, such as the sodium–hydrogen blend process, similar or better results are obtained without the uncertainties and effluent difficulties of precipitation plants. These modern ion-exchange plants are, moreover, amenable to automatic control and are less affected by changes in the quality of the raw water than are precipitation plants.

Purification

Distillation It is normally considered necessary to operate high-pressure boilers with water of lower dissolved solids content than can be achieved by softening alone. For many years, distillation proved to be the only practicable means of achieving this reduction. With modern evaporator plants, the concentration of dissolved solids in the distillate can readily be reduced to 0·1 p.p.m., while with the latest vapour purification equipment concentrations of 0·01 p.p.m. are attainable. Moreover, by correct design and operation of the venting system, dissolved gases can be eliminated. From a modern plant, the dissolved oxygen content of the distillate should be no more than 0·03 p.p.m.

It is usual to clarify and filter surface waters as a first step before distillation, but this step is often omitted when the water is drawn from a public supply. With most waters, further treatment is given by one of the processes already described to reduce or prevent the formation of scale in the evaporators.

Purification by ion exchange   In the ion-exchange processes used for soften-ing, the calcium and magnesium ions (and other cations) in the raw water are replaced by sodium ions (or, in the 'blend' and 'starvation' processes, partially by sodium ions and partially by hydrogen ions). Developments from the simple exchange process have resulted in elaboration of new processes for the complete removal of ions from the raw water. In the original form of the process, the water was passed first through a bed of cation-exchange material in which all cations were replaced by hydrogen ions, and then through a similar bed containing an anion exchanger which removed the acids produced in the first bed. Carbon dioxide and silica were not removed, but the former could be scrubbed out with a stream of air.

In a later form, a third bed, of strongly basic anion-exchange material, was added to remove silica and residual carbon dioxide, but this tended to 'bleed' caustic soda into the final water. In the most modern plants, the final unit contains a mixture of a strongly basic anion-exchange material with a strongly acidic cation exchanger. Such plants yield a water of very high purity; 'mixed bed' units may in fact be used as 'polishing' units to improve the quality of evaporator distillate.

Removal of Dissolved Gases

A comprehensive theoretical study of the removal of dissolved gases from boiler feed water has been made by Tietz[11]. It is generally considered neces-sary to remove dissolved oxygen down to about 0·03 p.p.m. or less for water to be fed to boilers operating at $4 \, MN/m^2$; higher boiler pressures are con-sidered to require correspondingly lower levels of residual oxygen concentra-tion. In British power stations having boiler pressures of $10 \, MN/m^2$ or over, the dissolved oxygen concentration in the feed water is normally maintained below 0·007 p.p.m.[12].

Removal of dissolved oxygen by physical means (usually scrubbing with steam under a slight positive pressure) is practicable at these levels. The desirability of de-aerating to even lower levels of dissolved oxygen is the subject of controversy. While physical de-aeration may be possible, chemical scavengers are commonly added to the water in an attempt to remove the last traces of oxygen; these scavengers are discussed later.

Other gases may also play a part in corrosion in the feed system and boiler. Of these, carbon dioxide and ammonia are considered to be the most important. Carbon dioxide can enter in the free or combined state in the raw water, but much of it can be removed during the pretreatment of the water. Ammonia may enter the system through the accidental ingress of polluted cooling water (and in such a case will be associated with other impurities necessitating remedial action), it may arise from the decomposition of hydrazine (discussed later), or it may be added deliberately. In the absence of carbon dioxide and oxygen, ammonia is not now considered to be harmful at low concentrations, but if these gases are present it may lead to the corrosion of copper and its alloys. Emphasis is now placed on control of ammonia concentration rather than complete exclusion.

Hydrogen sulphide may enter the system with polluted cooling water but again the fact of cooling water ingress will itself dictate the need for remedial action. Hydrogen sulphide may also be produced from the breakdown of sodium sulphite, but this is discussed later.

Additions to Feed and Boiler Water

Some addition of chemicals to boiler or feed water (or both) is usually considered necessary to achieve the optimum condition for freedom from both corrosion and scale formation. The nature and amount of the substances added depend on the design and working pressure of the boiler and the operating conditions of the entire system. It is, however, considered to be good practice to add any chemicals required for the conditioning of the boiler water directly to the boiler, so that additions to the feed water can properly be used for ensuring optimum conditions in the feed train. Since, however, any non-volatile solutes added to the feed water must reach the boiler, their effect on the boiler and their ultimate removal from it must also be considered.

Sodium Hexametaphosphate

Sodium hexametaphosphate finds frequent application in feed systems serving boilers operating at moderate pressures. At temperatures below about 50°C it prevents the precipitation of calcium as sludge or scale by forming a stable complex anion. At higher temperatures, hydrolysis to ortho-phosphates occurs and the calcium is precipitated as a sludge of hydroxy-apatite[3].

Sodium hexametaphosphate is, in dilute solution, almost neutral (pH of 0.25% solution about 7.2), but when alkalinity control is required caustic soda may be used at the same time. Without this addition, sodium hexametaphosphate will, by virtue of its hydrolysis to sodium dihydrogen ortho phosphate, reduce the alkalinity of the boiler water.

Orthophosphates

Trisodium orthophosphate is widely used as an additive to both feed and boiler water. As an addition to feed water, it provides a ready means of increasing the alkalinity and avoids the need for adding caustic alkali (which is difficult to handle and to control). It has the drawback, however, that sludge is produced when hard water enters the system, and this may be objectionable. In such circumstances, the use of hexametaphosphate with the appropriate controlled amount of caustic alkali may be preferred, provided that the temperature of the system is low enough to retard the hydrolysis of the metaphosphate. With the advent of more modern methods of alkalinity control (see p. **18**.64), the use of orthophosphate for feed-water treatment appears to be both unnecessary and undesirable.

For scale prevention, orthophosphates are best added directly to the boiler where (provided that temperatures and heat-transfer rates are not too high) a reserve of soluble phosphate can be maintained by regular additions. Provided that the sludge produced is sufficiently mobile, it can be removed by an appropriate schedule of blowing down. If it is desired to reduce the alkalinity of the boiler water when using orthophosphates, disodium hydrogen orthophosphate may replace part or all of the trisodium orthophosphate.

Loss of phosphate (and other salts of similar solubility characteristics) has been reported[13] as a source of difficulty of control even at boiler pressures of $4 \, MN/m^2$.It has been suggested[14] that at $12 \, MN/m^2$ (corresponding to a saturation temperature of about $325°C$), other phosphates of low solubility may be produced with liberation of caustic soda. The available experience indicates that it is prudent to operate such boilers without phosphate wherever possible, paying rigid attention to feed water purity.

Alkalinity Control with Solid Chemicals

A survey of the water conditions in British power stations[15] indicated that the incidences of corrosion were significantly lower in boilers receiving alkali-treated feed water than was the case in boilers receiving untreated feed. Mention has already been made of the use of orthophosphates in controlling the alkalinity of feed and boiler water, and of the use of caustic soda in conjunction with metaphosphates to give alkalinity control. Caustic soda may be used alone to increase the alkalinity of feed water, but the amount required is small and must be carefully controlled to avoid overdosing. Excessive caustic alkalinity in the boiler water is undesirable; it may result in the disruption and dissolution of the protective magnetite film from the steel surfaces[16], and may lead to stress corrosion ('caustic cracking') in riveted seams and similar areas (see also Section 8.2). Sodium carbonate is sometimes used in conditioning feed water for low-pressure boilers, but its use is not now favoured on account of its decomposition in the boiler to caustic soda and carbon dioxide.

In addition to their use in conditioning feed water, the three alkalis mentioned above may be used for the direct conditioning of boiler water. In installations in which feed-line conditioning is practiced, direct additions of alkali to the boiler water are not normally required, but some 'balancing' treatment may be needed when several boilers are fed from a common range. While the intermittent addition of chemicals direct to a boiler drum is of value, especially when the required additions are small (or as an emergency treatment), continuous addition is preferable as it avoids sharp concentration changes.

At pressures below about $1·4 \, MN/m^2$ sodium carbonate is often used, both to control scale in the boiler[2] and to provide alkalinity. Even at $1·4 \, MN/m^2$, however, the rate of hydrolysis to caustic soda and carbon dioxide is such that it is difficult to maintain an adequate reserve of carbonate ion for scale prevention[17]. Moreover, the liberated carbon dioxide passes forward with the steam, giving an acid and corrosive condensate. Severe corrosion may occur also in the stagnant pockets sometimes found on the vapour side of feed-water heaters and in other parts of the feed system, the turbine, and even the boiler.

Sodium hydroxide, whether formed by the hydrolysis of sodium carbonate or added as such, is of value in providing the alkalinity control required to combat corrosion, but must be used in moderation. It can be shown that an excess of sodium hydroxide leads to aggravated corrosion in some circumstances[16, 18]. It is claimed that stress corrosion brought about by the concentration of sodium hydroxide in seams by leakage can be offset by the use of sulphate, nitrate and other substances[19]. The best preventive for stress corrosion appears, however, to be the avoidance of riveted seams and other stressed areas by the use of welding and forging construction methods.

The use of phosphates has already been discussed in some detail. Some mention should be made here, however, of the method of 'co-ordinated phosphate–pH control' for boiler water[20], in which the alkalinity is provided entirely by phosphate and caustic alkali is completely eliminated from the boiler water. This method appears to be of considerable value in eliminating the undesirable effects of excessive hydroxide alkalinity, but requires the maintenance of an adequate reserve of phosphate.

In plants of high evaporative capacity, and especially those operating at high pressures ($6\ MN/m^2$ and upwards), more modern methods of alkalinity control of feed water (and possibly even also of boiler water) which employ volatile chemicals, appear to offer considerable advantages. These methods will be discussed later.

Chemical Oxygen Scavengers

Although proposed much earlier, sodium sulphite did not come into general use as an oxygen scavenger until about 1935. The use of this chemical was investigated in 1938[21, 22], when it was found that side reactions could occur. At high pressures and high heat-transfer rates[21] the primary decomposition products were found to be sodium sulphide and sodium sulphate:

$$4Na_2SO_3 \rightarrow Na_2S + 3Na_2SO_4$$

A trace of thiosulphate was formed by a secondary reaction:

$$2Na_2S + 4Na_2SO_3 + 3H_2O \rightarrow 3Na_2S_2O_3 + 6NaOH$$

Other workers[22] have found evidence of a simple hydrolysis:

$$Na_2SO_3 + H_2O \rightarrow 2NaOH + SO_2$$

leading to contamination of the steam with sulphur dioxide and the formation of an acid condensate.

Steels containing nickel are being increasingly used in the fabrication of power plant for service at high temperatures and pressures. The copper–nickel alloys are also widely used in heat exchangers for service in such plants. The presence of hydrogen sulphide in the steam can cause rapid corrosion of these nickel-bearing alloys. The formation of sulphur dioxide by simple hydrolysis can also result in corrosion in wet areas of the turbine and in the feed system. For these reasons, the use of sodium sulphite is not now favoured in high-pressure plants.

In recent years, hydrazine has found increasing use as an oxygen scavenger by virtue of the reaction

$$N_2H_4 + O_2 \rightarrow 2H_2O + N_2$$

It offers several advantages over sodium sulphite. Being volatile, it does not affect the concentration of dissolved solids in the boilers to which it may be fed. Both products of its reaction with oxygen are harmless; the side reaction whereby excess hydrazine is decomposed:

$$3N_2H_4 \rightarrow 4NH_3 + N_2$$

also yields products now considered harmless under the conditions in which hydrazine is normally used.

It has been found[23] that at temperatures below about 140°C, hydrazine and oxygen can co-exist in solution for an appreciable time. In a boiler feed system, therefore, dissolved oxygen may persist in the presence of hydrazine until the water has reached some point in the high-pressure heating system at which the temperature is high enough for it to react rapidly; the reaction is normally complete before the water enters the boiler. Hydrazine is not usually added until after physical de-aeration, by which time the water has reached a temperature of at least 105°C and the reaction between hydrazine and oxygen is fairly rapid.

It is claimed that in addition to its reaction with oxygen, hydrazine can also act as a corrosion inhibitor. This property is likely to be of most value in feed systems operating at low temperatures.

Volatile Alkaline Additives

Reference has already been made to the benefits derived from alkaline feed water and to the limits imposed by the boiler plant on the addition of solid chemicals to the feed. Hydrazine provides some alkalinity and is volatile, but its main use is as an oxygen scavenger; nevertheless, its decomposition product, ammonia, can play a useful part in providing alkalinity. Ammonia itself has been used extensively as a volatile alkali, notably in Germany and in America, and good results have been claimed for its use[24]. In the presence of oxygen and carbon dioxide, however, ammonia can promote the corrosion of copper and its alloys. It is, therefore, important, when relying on ammonia alone for conditioning, to adjust the pH of the water to about 9 to avoid these effects.

More recently, organic derivatives of ammonia have been used successfully. These compounds offer several advantages in the field of water treatment and they are being used on an ever-increasing scale. Being both water-soluble and volatile, they pass into the steam leaving the boiler and circulate with it. The condensate formed is therefore alkaline and protection is afforded to all parts of the system. Several compounds having widely different constitutions and properties, and affording varying degrees of protection to the condenser, feed heaters, and other parts of the system, have been used.

The two properties of greatest significance are the dissociation constant (determining the effect of the additives on the pH of the water), and the partition coefficient (and its variation with temperature) between water and steam. Although various workers have studied the partition coefficients of various proposed additives, their results are often conflicting[25, 26]. It has, however, been shown that ammonia and cyclohexylamine have partition coefficients which are very sensitive to temperature and favour relatively high concentrations in the vapour phase, while morpholine is less affected by temperature and favours the liquid phase under condensing conditions.

The dissociation constants for ammonia and morpholine at room temperature have been quoted[25] as $1\cdot6 \times 10^{-5}$ and 3×10^{-6} respectively, while their molecular weights are 17·03 and 87·12. It is evident that, weight for weight, morpholine will produce a much smaller pH change than will ammonia. Owing to the difference in partition coefficients, the losses to be expected with morpholine are much less than those to be expected with ammonia, so that despite the greater cost of morpholine, the cost of treatment for a closed system still remains practicable. It is, however, not considered practicable to use morpholine to raise the pH value of the water (as measured at room temperature) above 9·0. The amount of morpholine or other additive required to give this pH depends on the amount of carbon dioxide entering the system. With good conditions and small losses, the concentration required should be less than 5 p.p.m.

The thermal stability of the volatile alkaline additives is of importance, but information concerning this property is scanty. It has been claimed[27] that morpholine is stable in superheated steam at temperatures up to 649°C, but it is evident that more work on the decomposition temperatures of these substances and the nature and properties of their decomposition products is desirable.

Film-forming Amines

In addition to the volatile substances used for oxygen scavenging and alkalinity control, another group of volatile chemicals is playing an increasing part in water treatment. Typified by octadecylamine, these water-insoluble compounds form a film on the metal surfaces, insulating them from attack by water and by water-borne impurities. It has also been claimed that by producing an unwettable surface they help to promote dropwise condensation and assist heat transfer.

Experience with filming amines is limited, but they give promise of providing a simple and effective treatment for the feed water to small low-pressure boilers where elaborate control is impracticable. More knowledge of their chemical and physical properties (including thermal stability) appears desirable before their use on a large scale in high-pressure boilers can be recommended. It is essential, however, that they should be fed continuously either to the steam or to the feed main and precautions are necessary to ensure adequate and rapid mixing with the circulating fluid at the point of injection. It is desirable that the water temperature at the point of injection should be above 66°C to prevent premature precipitation. As they are volatile, however, distillation from the boiler then ensures protection at points of condensation.

It must be noted that these substances are powerful detergents which will loosen and remove existing deposits and corrosion products. Care must, therefore, be exercised in applying them to systems in which scale or corrosion products have already been formed.

Other Solutes in Feed and Boiler Water

Additions of sulphate and nitrate have been proposed for the control of stress corrosion in boilers[19], but there appears to be no sound theoretical reason for their use. Emphasis is laid on the maintenance of a ratio of sulphate to caustic alkali of at least 2·5 to 1, and of nitrate to caustic alkali of 0·4 to 1, in boilers with riveted seams. In boilers of welded and forged construction, it appears preferable to run without the deliberate addition of these solutes, but for operation at low and moderate steam pressures (up to about $4\,MN/m^2$) there appears to be no strong reason against their use. The corrosion survey of British power station boilers[15] suggested, without being conclusive, that the presence of all solutes other than alkalis and phosphates might be undesirable.

The deliberate addition of chloride has been suggested as a means of combating the corrosion which sometimes occurs as a result of the concentration of caustic alkali in dry areas. This addition does not appear to have found much favour; indeed efforts have been made to prevent ingress of chlorides so as to reduce corrosion in boilers.

Metallic Impurities in Feed Water

In a modern steam-raising plant, the feed water is brought into contact with a large area of both ferrous and non-ferrous metals in heat exchangers, pumps and pipework. It is obviously desirable to protect such materials from the corrosive effects of an incorrectly treated water, but present-day operating results show that even with the best known methods of conditioning, some pick-up of metallic ions such as copper and iron does occur. Even when the pick-up of metals is so slight as not seriously to affect the performance or life of the feed system, it is considered that the metallic impurities may initiate or at least influence corrosion processes in the associated boiler plant.

The knowledge that iron and copper can pass into solution in feed water suggests that tin, zinc and nickel, which are often present in alloys, may also occur in feed water. Knowledge concerning the presence and effects of these metals is, however, scanty.

Copper in Feed Water

Concentrations of copper of 0·02 p.p.m. in feed water are not uncommon in stations operating at moderate pressures and with the type of water treatment commonly used at such pressures, but the more modern methods of treatment and control used in stations operating at higher pressures appear to be capable of reducing the dissolved copper concentration to about one-tenth of this value. Nevertheless, even with this lower concentration, a boiler evaporating 1 000 t of water hourly (a not uncommon size for a modern large unit) will receive the equivalent of about 15 kg of metallic copper in the course of a year.

The effects of copper in boilers and their association with corrosion have been well reviewed[3], but the action of copper as a direct cause of corrosion has not been established.

Iron in Feed Water

From a consideration of the solubility product and dissociation constant of ferrous hydroxide, it can be shown that a saturated solution will have a pH of 9·6 at room temperature (see Section 1.4). In the presence of dissolved oxygen, however, the less soluble ferric hydroxide is produced and the pH value falls. It is commonly found that pure water in contact with steel and in the absence of air assumes a pH above 8. The addition of alkaline substances to the feed water to raise its pH above 8·5 may be expected to reduce the solubility of iron and so reduce corrosion in the feed and boiler systems. Transport of dissolved iron into the boiler system is considered to be a possible cause of corrosion in zones of high heat transfer in the boiler[28].

Limits of Impurities in Feed Water and Boiler Water

It is apparent that the tolerance of any boiler to impurities in its feed water depends on the pressure and temperature conditions in the boiler. It also depends largely on design conditions such as rate of heat transfer, ratio of free water surface to evaporation rate, and the design of the steam separators and baffles. The required standard of steam purity is governed by the use to which the steam is put, but this in turn influences the design of separators and the allowable concentration of the boiler water. In many plants (and this practice is becoming increasingly common) the final steam temperature is controlled by the direct injection of feed water; any non-volatile impurities in the feed water pass forward with the steam and may cause trouble in the final superheating stages or external plant. Moreover, at pressures approaching or exceeding the critical, the solution of salts in the steam must be considered as a source of steam (and feed-water) contamination. More information is desirable concerning the solubility of salts in high-density steam and the partition coefficient of salts between the steam and water phases. The development of boilers of the once-through type introduces a new emphasis on feed-water purity. In such boilers, solids in the feed water must either deposit in the 'transition zone' of the boiler itself, or pass forward as solids or in solution in the steam; they cannot be held in solution in the boiler water to be removed by blowdown. Solids passing into the steam are likely to be deposited when the steam expands and may then cause trouble in the turbine.

Table 18.2 Feed-water conditions

Boiler pressure (MN/m²)	pH value	Dissolved oxygen (p.p.m.)	Hardness as CaCO₃ (p.p.m.)	Chloride as NaCl (p.p.m.)	Phosphate as Na₃PO₄* (p.p.m.)	Sulphite as Na₂SO₃ (p.p.m.)	Hydrazine* (p.p.m.)	Metallic impurities (p.p.m.)§	Total dissolved solids (p.p.m.)
1·4	≮8·5	≯0·05	≯2	≯3	≯3	≯0·5			≯8
1·4†	≮8·5	≯0·05	Nil	Nil	≯1	≯0·5			≯2
3†	≮8·5	≯0·03	≯2	≯3	≯3	≯0·3			≯8
3‡	≮8·5	≯0·03	Nil	Nil	≯1	≯0·3			≯1·5
4	≮8·5	≯0·02	Nil	Nil		≯0·2	≯0·04		≯1·0
6	8·5–9·0	≯0·01	Nil	Nil		Not recommended	≯0·02	≯0·02	≯0·2
10	8·5–9·0	≯0·007	Nil	Nil		Not recommended	≯0·015	≯0·02	≯0·05
17	8·5–9·0	≯0·005	Nil	Nil		Not recommended	≯0·010	≯0·02	≯0·02

*If used. †Softened water make-up (5%). ‡Distilled or de-ionised make-up. §Copper, iron, nickel, zinc, etc.

Note. These figures relate primarily to water-tube boilers. For low-pressure fire-tube boilers of high water capacity and low evaporative capacity they may be increased substantially. For once-through boilers at any pressure, the figures for 17 MN/m² should be used.

Table 18.3 Boiler-water conditions

Boiler-water characteristics

Boiler pressure (MN/m²)	pH value	Total alkalinity as CaCO₃* (p.p.m.)	Hydroxide alkalinity as CaCO₃ (p.p.m.)	Phosphate as PO₄³⁻ (p.p.m.)	Sulphate as SO₄²⁻ (p.p.m.)	Carbonate as CO₃²⁻ (p.p.m.)	Nitrate as NO₃⁻ (p.p.m.)	Chloride as Cl⁻ (p.p.m.)	Sulphite as SO₃²⁻ (p.p.m.)	Hydrazine (p.p.m.)	Silica (p.p.m.)	Total dissolved solids (p.p.m.)
1·4	≯11	400	200	100 (if used)	†	§	‖	≯100	30			≯2 000
3	≯11	200	100	60	†	Nil	‖	≯50	20			≯1 000
6	10·5–11·5	120	60	50	‡	Nil	Not recommended	≯20	5			≯250
9	10·5–11·0	70	40	30	≯20	Nil	Not recommended	≯20	Not recommended	Trace	≯1·0	≯150
10	10·5–11·0	50	20	30	≯10	Nil	Not recommended	≯10	Not recommended	Trace	≯0·5	≯100
17	10·5	20	10	10	≯5	Nil	Not recommended	≯5	Not recommended	Trace	≯0·1	≯50

*Often designated as 'M'. Determined by titration to pH4.
†Na₂SO₄:NaOH ratio 2·5:1.
‡Sulphate concentration will depend on rate of sulphite addition and oxidation or hydrolysis, when this treatment is used. The limit will in this case be set by the total dissolved solids. In the absence of sulphite, sulphate should not exceed 30 p.p.m.
§Balance of total alkalinity. Na₂CO₃:Na₂SO₄ ratio about 0·25:1.
‖NaNO₃:NaOH ratio 0·4:1.
Note. These figures relate primarily to water-tube boilers; for low-pressure fire-tube boilers they may be relaxed considerably.

Feed- and Boiler-water Conditions

Tables 18.2 and 18.3 show appropriate feed- and boiler-water conditions. The figures quoted relate to water-tube boilers in which water capacities are low in proportion to steam output; boilers of similar design operated at low load factors may be expected to be more tolerant of impurities in the feed water. In the case of boilers supplying process steam (from which, in many cases, condensate is not recoverable), economic conditions may, at least at the lower steam pressures, dictate the need to adopt a lower standard of feed-water purity; such installations should be designed on the basis of this requirement.

The maintenance of the appropriate conditions will ensure freedom from corrosion in the feed system and will do much to promote freedom from corrosion in the boiler plant. The figures quoted must, however, be accepted only as a guide to good practice. They may need modification in the light of operating experience for any particular plant.

E. W. F. GILLHAM

REFERENCES

1. Corey, R. C., *Combustion (Int. Combust. Engng. Corp.)*, No. 6, 43 (1943)
2. Hall, R. E., Smith, G. W., Jackson, H. A., Robb, J. A. and Hertzell, E. A., *Bull. Carneg. Inst. Tech.*, No. 24 (1927)
3. Clarke, L. M., Hunter, E. and Gerrard, W. F., *J. Chem. Soc.*, **57**, 295 (1938)
4. Imhoff, C. E. and Burkardt, L. A., *Industr. Engng. Chem.*, **35**, 843 (1943)
5. Holmes, J. A. and Jacklin, C., *Proc. 4th Ann. Water Conf. Engrs' Soc., West Pennsylvania, 1943.* [*Combustion (Int. Combust. Engng. Corp.)*, No. 7, 35 (1944)]
6. Holmes, J. A. and Jacklin, C., *Proc. 6th Ann. Water Conf. Engrs' Soc., West Pennsylvania, 1945*, 1–13
7. Holmes, J. A. and Jacklin, C., *Proc. 11th Ann. Water Conf. Engrs' Soc., West Pennsylvania, 1950*, 99–112
8. Straub, F. G. and Grabowski, H. A., *Trans. Amer. Soc. Mech. Engrs.*, **67**, 309 (1945)
9. Straub, F. G., *Univ. Ill. Bull.*, No. 364, June (1946)
10. *Water Treatment Handbook*, Degremont, Paris (1955). (1st English Edition, translated by D. F. Long, Elliott, London (1955)
11. Teitz, H., *Mitt. Ver. Grosskesselb.*, No. 9, 76; No. 10, 103 (1950)
12. Rees, R. Ll. and Taylor, F. J. R., *Trans. Instn. Chem. Engrs., Lond.*, **37**, 65 (1959)
13. Hall, R. E., *Trans. Amer. Soc. Mech. Engrs.*, **66**, 457 (1944)
14. Straub, F. G., *Trans. Amer. Soc. Mech. Engrs.*, **72**, 479 (1950)
15. Gillham, E. W. F. and Rees, R. Ll., *Proc. Amer. Power Conf.*, **14**, 438 (1952)
16. Partridge, E. P. and Hall, R. E., *Trans. Amer. Soc. Mech. Engrs.*, **61**, 597 (1939)
17. Larson, R. F., *Trans. Amer. Soc. Mech. Engrs.*, F.S.P.–54–14, 171 (1932)
18. Kaufman, C. E., Marcy, V. M. and Trautman, W. H., *Proc. 6th Ann. Water Conf. Engrs' Soc., West Pennsylvania, 1945*, 23–42
19. Parr, S. W. and Straub, F. G., *Proc. Amer. Soc. Test. Mater.*, **27** No. 2, 52 (1927)
20. Purcell, T. E. and Whirl, S. F., *Trans. Electrochem. Soc.*, **83**, preprint 25 (1943)
21. Taff, W. O., Johnstone, H. F. and Straub, F. G., *Trans. Amer. Soc. Mech. Engrs.*, **60**, 261 (1938)
22. Hitchens, R. M. and Purssell, J. W., Jr., *Trans. Amer. Soc. Mech. Engrs.*, **60**, 469 (1938)
23. *Hydrazine Conference Report*, Whiffen & Sons Ltd., London (1957)
24. Straub, F. G. and Ongman, H. D., *Corrosion*, **8**, 312 (1951)
25. Homig, H. E. and Richter, H., *Mitt. Ver. Grosskesselb.*, No. 36, 615 (1955)
26. Samuel, T. and Begheim, J., *Cebedeau.*, **35**, 28 (1957)
27. Jacklin, C., *Trans. Amer. Soc. Mech. Engrs.*, **77**, 449 (1955)
28. Straub, F. G., *Combustion (Int. Combust. Engng. Corp.)*, No. 1, 34 (1957)

19 NON-METALLIC MATERIALS

19.1 Carbon   **19**:3

19.2 Glass and Glass-ceramics   **19**:10

19.3 Vitreous Silica   **19**:25

19.4 Glass Linings and Coatings   **19**:28

19.5 Stoneware   **19**:34

19.6 Plastics and Reinforced Plastics   **19**:41

19.7 Rubber and Synthetic Elastomers   **19**:64

19.8 Corrosion of Metals by Plastics   **19**:74

19.9 Wood   **19**:81

19.10 The Corrosion of Metals by Wood   **19**:87

19.1 Carbon

Carbon is an element which does not easily enter into chemical combination at low temperatures. It is therefore, despite some physical disadvantages, used widely as a component construction material for employment under corrosive conditions.

Manufacturing Processes

1. The carbon raw material in the form of coke, coal or natural or synthetic graphite is ground and sieved to suitable particle size.
2. The graded material is mixed with a pitch or tar binder to produce a paste.
3. The paste is formed by pressing or extrusion into a 'green' shape.
4. The shape is kilned at a high temperature in an atmosphere which gives protection from oxidation. This drives off volatile matter, and sinters together carbon particles by the carbonisation of the binding material.
In this form the shape is neither purely amorphous nor purely crystalline, and can be best described as *mesomorphous*.
5. Further heat treatment at a more elevated temperature can then be given to develop graphite crystal growth and form a synthetic graphite. Since this is usually carried out by using the ware as an electrical heating element the product is normally termed *electrographite*. The temperature and the duration of application of heat determine the degree of graphitisation, and the degree of orientation of the graphite crystals. In general, crystal formation will be in planes parallel with the direction of extrusion, or perpendicular to the direction of pressing.

Various extra treatments may be applied. In the production of electrographite, for example, the carbon is often impregnated between kilning and graphitising stages with a material which will completely carbonise on high-temperature kilning. In order to increase electrical and thermal conductivity and strength, the carbon pores may be filled with metal, a treatment which is applied in the production of motor and generator brushes, and for mechanical sealing where the seal must be impervious. Other impregnations used to render carbon products impervious include waxes, oils and synthetic and natural resins. Such impregnations also normally have the effect of improving physical strength (the physical strengths given in Table 19.1 can be increased by as much as 50% by this means) and reducing residual porosities to a very low percentage. The choice of impregnant generally depends on the chemical conditions likely to be encountered.

It will be appreciated that by means of selection of raw material, and variation of pressing force and duration and temperature of the kilning and graphitising processes, it is possible to produce a great variety of carbons, carbon/graphites, natural graphites and electrographites having physical properties suited to a variety of uses. For this reason, only approximate physical properties can be given.

The properties of the so-called 'porous carbons' are not included in Table 19.1. Such components are formed at low pressures, and are therefore characterised by low bulk density, strength, elasticity and thermal conductivity, and by high electrical resistivity and thermal expansion; they are always friable in nature.

Definitions   In the ensuing text the following definitions should be assumed:

Mesomorphous carbon: formed carbon using coke or coal as a raw material, not subjected to an 'electrographitising' process.

Carbon graphite: a mesomorphous carbon using as raw material coke or coal with additions of natural or synthetic graphite.

Natural graphite: formed carbon using natural graphite as a raw material and not subjected to an 'electrographitising' process.

Electrographite: any of the above after subjection to an 'electrographitising' process.

Carbon: general description of a formed component covering any of the above.

Physical Properties

The main physical properties of carbon are given in Table 19.1.

Hardness   It is not possible to use standard indentation methods for evaluating the hardness of carbon, since the nature of the surface leads to formation of an irregular indentation. Hardness measurements are therefore made with a Shore scleroscope, but normally with a non-standard missile. The hardness figure is useful to the manufacturer for material control, but since there is no universally accepted standard apparatus for carbon, comparison of published values should be avoided. It can be generally assumed that the hardest grades are mesomorphous carbons, followed by carbon/graphites, electrographites and natural graphites in order of decreasing hardness.

Heat transfer   Owing to its high thermal conductivity and its non-wetting surface properties, electrographite is widely selected for heat-transfer purposes under arduous corrosive conditions. For such duties the electrographite is normally rendered impervious by resin impregnation.

Table 19.2 indicates the order of overall coefficients of heat transfer obtainable under various conditions. The values are based on performances

Table 19.1 Physical properties of carbon

Property / Form of carbon	Bulk density* (kg/m³)	Porosity		Ultimate physical strength (MN/m²)								Elasticity Young's modulus (GN/m²)	Coefficient of thermal expansion‖ (per °C)	Coefficient of thermal conductivity (W m⁻² K⁻¹)	Specific heat over range 0–100°C	Specific resistivity (Ω cm)
		Degree of porosity† (%)	Pore size‡ (μm)	Compressive Mean range	Max.	Transverse Mean range	Max.	Shear Mean range	Max.	Tensile§ Mean range	Max.					
Mesomorphous carbon	1 440–1 600	10–50	0–10	70–34	182	21–42	70	21–28	42	10·5–17·5	28	2·8–21	1–5 × 10⁻⁶	4·6–4·6	0·17–0·18	0·013–0·003
Carbon/graphite	1 600–1 770	10–50	0–10	70–34	182	21–42	70	21–28	42	10·5–17·5	28	2·8–21	1–5 × 10⁻⁶	17–115	0·17–0·18	0·008–0·002
Electro-graphite	1 600–1 770	10–50	0–10	28–32	84	14–21	49	7–14	21	3·5–8·5	21	2·8–21	1–5 × 10⁻⁶	115–570	0·17–0·18	0·003–0·0008
Natural graphite	—	10–50	0–10	14–21	35	7–14	28	7–10·5	17·5	2–5·5	8·5	2·8–21	1–5 × 10⁻⁶	115–570	0·17–0·18	0·003–0·0008

*Quite large increases in density can be obtained by impregnation with a material which will carbonise in the pores on subsequent heat treatment. Densities in excess of 2 000 kg/m³ can be obtained by this means.
†Most carbons for mechanical and electrical applications are in the range 10–30%. Porosities > 30% are mainly confined to carbons for filtration and similar purposes.
‡Most carbons for mechanical and electrical applications are in the range 0–10 μm. Filter carbons are made with pore sizes ranging from 10–100 μm.
§Elastic limit and yield point can be taken to be coincident with the ultimate tensile stress.
‖Carbon undergoes a reversible linear thermal expansion, except when taken above the max. temperature employed in its manufacture, in which case permanent shrinkage occurs. For most carbons the coefficient of thermal expansion lies in the range $2 \times 10^{-6}/°C$ to $3 \times 10^{-6}/°C$.

in 'tube in shell' type heat exchangers, taking account of the internal tube surface area.

Table 19.2 Overall coefficients of heat transfer

Nature of duty		Tube side flow rate (m/s)	Overall coefficient of heat transfer (W m^{-2}K^{-1})
Shell side	Tube side		
Condensing vapour	Liquid	0·9	1 700–2 270
Liquid	Gas	12	28–85
Liquid	Liquid	0·9	680–1 140

Effect of temperature on physical properties In general the physical strength of carbon is unaffected by temperatures up to 3 000°C provided that protection from oxidation is ensured. If, however, mesomorphous carbons, natural graphites, and carbon–graphites are taken above their kilning temperature, 'graphitisation' will occur and the range of physical strengths will move towards that given for electrographite. Provided the temperature of manufacture is not exceeded, the bulk density, porosity and coefficient of thermal expansion will remain approximately constant with temperature. The coefficient of thermal conductivity of mesomorphous carbons and most carbon–graphites increases with increase in temperature, but natural and electrographites normally suffer a decrease. The temperature coefficient is variable between specific grades.

The electrical resistivity variation with temperature is indicated for a typical electrographite in Table 19.3. The values refer to an extruded rod and are given for directions parallel and perpendicular to the direction of extrusion. There is evidence that at temperatures higher than those given in the table the resistivity decreases, and the curve of resistivity vs temperature becomes parabolic.

Table 19.3 Variation of electrical resistivity with temperature for a typical electrographite

Direction	Resistivity [Ω cm × 10³ at temp. (°C)]														
	20	100	200	300	400	500	600	700	800	900	1 000	1 100	1 200	1 300	1 400
Parallel	2·0	1·82	1·68	1·56	1·50	1·42	1·38	1·34	1·31	1·29	1·28	1·28	1·28	1·28	1·28
Perpendicular	4·2	3·85	3·50	3·22	3·00	2·85	2·75	2·64	2·58	2·52	2·48	2·44	2·41	2·40	2·38

A typical variation of specific heat with temperature for an electrographite is given in Table 19.4.

Table 19.4 Variation of specific heat with temperature

Temperature (°C)	−240	−200	−70	25	230	480	730	890	1 230	1 830
Specific heat	0·005	0·015	0·05	0·17	0·29	0·37	0·43	0·46	0·48	0·50

Corrosion Resistance of Carbon

Carbon is most unreactive at low temperatures, although at high temperatures it readily combines with oxygen. As a general guide it can be assumed that the mesomorphous carbons and carbon–graphites are completely stable in oxygen up to about 350°C, while electrographites and natural graphites are unaffected by oxygen up to 450–500°C.

Carbon is oxidised by highly oxidising solutions at low temperatures. Nitric acid and sulphuric acid in sufficiently high concentrations will, for instance, attack carbon according to the following reactions:

$$4HNO_3 + C \rightarrow CO_2 + 4NO_2 + 2H_2O$$
$$2H_2SO_4 + C \rightarrow CO_2 + 2SO_2 + 2H_2O$$

At high temperatures, carbon, usually in the form of charcoal or coke, is used extensively for the reduction of metallic oxides, nitrates and sulphates. Oxides are normally reduced to the metal, sulphates to the sulphide, and nitrates to the carbonate, with the formation of carbon monoxide and/or carbon dioxide.

Unfortunately, no comprehensive work has been carried out on the corrosion rates of carbon under various chemical conditions. The following data indicate the conditions under which carbon is known to be completely resistant to corrosion and those where caution is advisable. It is always desirable to consult the manufacturer on specific problems, particularly under chemical conditions where caution is advised. The data, in general, refer to all types of carbon; it should be borne in mind that specially treated

Table 19.5
Resistance of carbon to acid conditions

Acid	Boiling point (°C)
Acetic	118
Acetic anhydride	140
Adipic	264
Benzoic	249
Boric	300
Butyric	164
Carbonic	—
Citric	205
Formic	101
Fumaric	290
Glutaric	200
Hydrobromic	—
Hydrochloric	110
Lactic	122
Maleic	135
Oleic	285
Oxalic	150
Propionic	141
Stearic	291
Tartaric	—

carbons which give increased resistance over the basic material, and in some cases total resistance, are sometimes available.

Resistance of carbon to acid conditions   In the acid conditions given in Table 19.5 carbon is completely stable at all concentrations and at temperatures up to the boiling point.

In the acid conditions indicated in Table 19.6 chemical attack on carbon may be expected. The use of carbon under these conditions is thus advised only when some wastage can be tolerated.

Table 19.6   Acid conditions in which chemical attack on carbon may be expected

Acid	Concentration (%)	Temperature (°C)
Nitric	<15	>70
Nitric	15–30	>50
Nitric	>30	all temperatures
Sulphuric	>70	b.p.
Sulphuric	>80	>80
Sulphuric	>90	all temperatures
Chrome-plating solutions	all concentrations	all temperatures
Hydrofluoric	>60	all temperatures
Phosphoric	>85	b.p.
Phosphoric	100	>80

Resistance of carbon to alkaline conditions   Carbon is unattacked by the common alkaline solutions at all concentrations and at temperatures up to boiling point.

Resistance of carbon to inorganic salts, and organic compounds, gases and vapours   Carbon has a good resistance to all inorganic and organic solutions up to their boiling points. Care is recommended in using carbon with solutions of sodium hypochlorite, and hydrogen peroxide in all concentrations, and also sulphur chloride and bromide.

Among gases and vapours, fluorine, bromine, iodine and chlorine dioxides are likely to attack carbon. Other gases and vapours will not normally attack carbon at temperatures below the normal temperature of oxidation.

Carbon at High Temperatures

Combination with oxygen   The oxidation of a carbon surface is assumed to be in the nature of a primary reaction to give carbon monoxide followed by a gas-phase reaction to give dioxide:

$$2C + O_2 \rightarrow 2CO$$
$$2CO + O_2 \rightarrow 2CO_2$$

Further contact of the CO_2 with carbon can, according to temperature conditions, cause reduction of the dioxide to monoxide.

Although the temperature at which oxidation commences varies between grades, as indicated previously, the surface oxidation rate is more or less constant above about 700°C.

Table 19.7 gives oxidation rates of some typical samples of mesomorphous carbons and electrographites at various temperatures.

Table 19.7 Oxidation rates of carbons

Sample	Type	Rate of oxidation $(gm^{-2}h^{-1})$		
		500°C	600°C	700°C
1	Mesomorphous	90	845	1 330
2	Mesomorphous	43	528	—
3	Mesomorphous	31	315	1 490
4	Mesomorphous	8	89	1 178
5	Electrographite	1	87	458
6	Electrographite	1·3	48	580
7	Electrographite	0·9	28	320

Further treatments can be given to carbon to elevate the temperature of oxidation slightly and to bring about a substantial reduction in the rate of oxidation.

Combination with other gases The gases normally used to protect carbon from oxidation at high temperatures are hydrogen, nitrogen or argon. The two first named will, however, start to combine with carbon at temperatures of the order of 1 700°C to form methane and cyanogen, respectively. There is some evidence that argon can combine with carbon under certain circumstances but stability is normally found up to 3 000°C. Oxidising gases such as sulphur dioxide and water vapour will oxidise carbon at approximately the same temperature as will oxygen.

Combination with elements Carbon combines with many elements at high temperatures. Well-known examples include combination with sulphur to produce carbon disulphide, with hydrogen to form methane or acetylene, and with silicon to form silicon carbide.

Iron and many non-ferrous metals form the carbide by direct combination with carbon.

K. F. ANDERSON

BIBLIOGRAPHY

Mantell, C. L., *Carbon and Graphite Handbook*, Interscience, John Wiley, New York (1968)
Carbon and its Uses, Morganite Carbon Ltd., Battersea, London
Carbon (an International Journal), Pergamon Press, Oxford, England
Hove, J. E. and Riley, W. C., *Ceramics for Advanced Technologies*, John Wiley, New York (1965) (deals with graphite and silica)

19.2 Glass and Glass-ceramics

General

One of the most important properties of commercial glasses is their great resistance to corrosion; any chemical laboratory apparatus, any window or windscreen provides an excellent illustration. Windows remain virtually unchanged for centuries, resisting the influences of atmosphere and radiation. A vast range of products may be safely stored in glass for decades at ordinary temperatures, and the fact that glass can be used with alkaline, neutral and acid environments allows the same equipment to be used for a variety of processes.

Glass is one of the engineer's most useful and versatile materials. There are many types of glass to choose from to provide a wide range of physical, mechanical, electrical and optical properties for practically every type of environmental condition. The transparency of glass facilitates inspection of process operations and minimises the risk of failure due to unsuspected corrosion, while the hardness and smoothness contribute to easy cleaning.

In recent years the development of glass-ceramics has further extended the range of glassy engineering materials. Glass-ceramics combine the formability of glasses with many of the advantageous properties of ceramics. They are finding increasing application by virtue of their strength and high chemical durability at elevated temperatures.

The principal difficulty associated with the use of glass equipment is the fact that glass will break rather than deform on severe impact; thermal toughening or ion-exchange techniques may be used to mitigate this disadvantage in some applications. Glass is also more prone than metals to damage by thermal shock, although this difficulty can be largely avoided by the use of low-expansion glass formulations. Finally, the size of glassware which can readily be fabricated is sometimes below the needs of a particular process.

Commercial Glasses

Glass Compositions

The term *glass* defines a family of materials that exhibit as wide a range of differences among themselves as exists among metals and alloys. The great variety of physical and chemical properties available arises from the possibility of including almost all the stable oxides, sulphides, halides, etc. throughout the periodic table in different glass formulations[1]. Many branches of the family, e.g. those borates, silicates and phosphates which are water soluble, are of little interest in the present context. It is, however, worth bearing in mind that glasses can be designed to combine particular physical properties with good chemical resistance.

For most of the commercial glass families, silica sand is the main ingredient. However, greater melting economy and flexibility of properties are achieved with the addition of other oxides and modifiers. Depending upon the choice of these additional constituents, glasses are classified into groups, including fused silica, soda-lime, lead, aluminosilicate, borosilicate, etc. A cross-section of commercial glass compositions is given in Table 19.8, the glasses listed being:

1. Fused silica
2. Window glass
3. Container glass
4. Fluorescent tubing
5. Neutral glass
6. Hard borosilicate
7. Lead tubing
8. TV tube and screen
9. Textile glass fibre
10. Glass wool insulation
11. Superfine glass wool

Table 19.8 Typical glass compositions, %

	1	2	3	4	5	6	7	8	9	10	11
SiO_2	100	72·7	72·8	71·4	71·5	80·3	57·2	67·5	54·2	63·6	57·1
Al_2O_3		1·1	1·7	2·2	5·5	2·8	1·0	4·8	14·3	2·9	4·6
B_2O_3				10·0	12·3				8·3	5·0	11·8
MgO		3·8		3·9					4·5	3·2	
CaO		8·4	10·5	4·6	0·2			0·1	17·7	7·3	0·4
BaO				0·8	3·0			12·0		2·6	
Na_2O		13·1	14 5	15·0	8·0	4·0	4·0	7·2	0·6	14·5	14·2
K_2O		0·5		1·7	1·2	0·4	8·5	6·9	0·1	0·6	0·8
ZrO_2											3·8
TiO_2											7·5
PbO							29·0				
Li_2O								0·5			
Others		0·4	0·5	0·4	0·6	0·2	0·3	0·3	0·3	0·3	0·4

Note. The designations 1–11 represent the various types of glasses listed numerically in the text above this table, and they will be referred to in this way in later tables in this section.

Fused silica is a general classification within which is a range of varieties and types with differences in purity, transmission and grade. This glass may be used up to 900°C in continuous service; it resists attack by a great many chemical reagents, rapid attack occurring only in hydrofluoric acid and concentrated alkali solutions.

The container glass is suitable for the storage of beverages, medicines, cosmetics, household products and a wide range of laboratory reagents.

The tubing glass is suitable for general laboratory use and chemical apparatus construction, though neutral or hard borosilicate are preferred for more severe conditions, these representing the most resistant glasses available in bulk form.

The neutral glasses are generally less resistant than the hard borosilicate type, but are more easily melted and shaped. They are formulated so that the pH of aqueous solutions is unaffected by contact with the glass, making it particularly suitable in pharmaceutical use for the storage of pH-sensitive drugs.

Borosilicates, in terms of different types available, are the most versatile glasses produced. In general, the borosilicates are grouped into six types, viz. low expansion, low electrical loss, sealing, ultraviolet transmitting, laboratory apparatus and optical grade glasses. The example given, i.e. hard borosilicate glasses, is used for ovenware, pipelines, sight glasses and laboratory ware, and combines low expansion and high chemical resistivity with chemical stability. They generally require high founding and fabrication temperatures compared with soft soda glasses.

Glasses for electrical and electronic components are represented by the lead tubing and cathode-ray-tube screen and cone glasses. These glasses do not operate under severe corrosion conditions, but surfaces must not leach excessive alkali under damp conditions or electrical breakdown can occur. The glass compositions are formulated to give the maximum electrical resistivity and moisture resistance compatible with other necessary properties.

Glass fibres present particular problems in corrosive environments due to their very high surface/volume ratios. Glasses for electrical insulation are formulated from alkali-free aluminoborosilicate glasses (generally known as E-glass) and are frequently specified as containing less than 1% alkali (Na_2O and K_2O). This type of glass is also used extensively for the reinforcement of plastics where its high resistivity to moisture attack ensures a durable product. The glass wools are used for less demanding applications and generally contain some alkali. Superfine wool contains zirconia and titania to enhance the chemical resistance while retaining the properties necessary for economic fine-fibre formation.

Physical Properties

Glass has been defined as 'an inorganic product of fusion which has cooled to a rigid condition without crystallising'. The atomic structure of glasses is more closely related to liquids than to crystals. The properties of glasses are manifestations of this structure, being governed in particular by the random liquid-like disposition of the network-forming ions (commonly Si^{4+} and B^{3+}), the presence of mobile, interstitial alkali ions and the 'single-molecule' nature of the lattice. The bonding within the atomic network is partly covalent, partly ionic; the network bonds are highly directional with a range of inter-bond angles, lengths and bond energies; the bonding electrons are restricted to particular energy levels within the bonds. The

Table 19.5 Physical property data for some commercial glasses*

	Units	1	2	3	4	5	6	7	8	9	10	11
Density	g/cm^3	2·20	2·49	2·46	2·49	2·42	2·24	3·03	2·62	2·58	2·57	2·54
Strain point	°C	987	520	490	495	518	515	—	448	616	—	—
Annealing point	°C	1082	545	540	524	565	565	437	470	657	—	—
Littleton softening point	°C	1594	735	720	705	780	820	631	670	843	688	710
Resistivity	$\Omega\,m$	10^{15} (20°C)	$1·1\times10^{13}$ (20°C)	—	$9·6\times10^{14}$ (20°C)	—	10^{15} (20°C)	$3·2\times10^{10}$ (150°C)	$1·6\times10^{6}$ (350°C)	$6·0\times10^{13}$ (155°C)	2×10^{-3} (953°C)	$4·8\times10^{-3}$ (862°C)
Dielectric constant	at 1 kHz	3·8	7·4	—	7·8	—	5·1	7·0	—	6·4 (50 Hz)	—	—
Tan δ	at 1 MHz	very small	0·03 (50 Hz)	—	0·008	—	0·02	0·0011	—	0·0009 (50 Hz)	—	—
Refractive index		1·48	1·52	—	1·51	1·49	1·47	1·56	1·51	1·55	—	—
Thermal conductivity	$W\,m^{-2}\,K^{-1}$	1·38	1·05	1·02	1·04	1·04	1·13	0·84	1·01	0·97	—	—
Thermal expansion	$\times10^{7}\,K^{-1}$	5·4	79·3	87	85·5	50	33	84	85·5	49	83	—
Specific heat	$J\,kg^{-1}$	775	987 (200°C)	821	833	819	794	—	733 (23°C)	796 (23°C)	—	—
Young's modulus	$\times10^{-10}\,Nm^{-2}$	7·3	7·4	—	—	—	6·3	5·75	7·4	7·2	—	—

*Compositions are given in Table 19.8.

network-modifying ions (commonly alkali and alkaline-earth ions) are ionically bound to the network although the field strength and diameter of the alkali ions allow them some mobility.

The structural features are reflected in the characteristic properties of inorganic glasses and bring about a broad overall similarity in behaviour as summarised below. Values for the physical properties of the commercial glasses listed in Table 19.8 are given in Table 19.9.

Viscosity　The random nature of the glass structure imparts a range of bond energies in the network, hence a characteristic feature of glasses is a continuous softening over a range of temperature, a continuous viscosity/temperature curve and the absence of a true melting point. For convenience in comparing the viscosity behaviour of different glasses, arbitrary temperatures at which the glass has specific viscosities are often quoted[2].

Softening temperature　The Littleton Softening Point[3] is most commonly used. At this temperature the glass has a viscosity of $10^{6.6}$ Ns/m^2.

Transformation temperature, T_g　T_g corresponds to a viscosity from 10^{12} to 10^{13} Ns/m^2 depending on the definition and on the method of measurement.

Annealing point and strain point　An important range in practice is that from 10^{11} to $10^{13.5}$ Ns/m^2 known as the *annealing range*. The annealing point and the strain point are the temperatures at which the glass has a viscosity of $10^{12.4}$ Ns/m^2 and $10^{13.6}$ Ns/m^2 respectively. Within this range the glass is effectively a solid, but internal stresses can be relieved within a practical time scale. Rapid cooling of glass articles through the annealing range eventually results in permanent and sometimes catastrophic thermal stresses. However, it is possible to cool the glass relatively quickly from the lower end of the annealing range.

Glass behaves as a Newtonian liquid at temperatures well above the glass transition. It is this behaviour which prevents the necking observed during gross deformation of metals and which allows glass to be formed into such a large number of useful configurations.

Thermal shock resistance　The ability of a glass article to withstand sudden changes of temperature depends primarily on its thermal expansion co-efficient, its thickness and its design. For articles of identical shape a low-expansion glass (such as a commercial borosilicate) will withstand appreciably greater temperature shocks than will glass of a higher expansion. Thermal shock-resistance testing is usually carried out by transferring the articles from a hot environment to a cold vessel containing water at a predetermined temperature[4].

In general, transitions from a hot to a cold environment are more likely to produce failure than those in the opposite direction since they tend to induce tensile stresses at the surface.

Stress birefringence　The presence of stress in glass articles may be monitored readily since stressed glass is birefringent. Standard methods exist for these measurements.

Thermal expansion　Glasses having coefficients of linear thermal expansion of from 0.5×10^{-6}/deg C to over 10×10^{-6}/deg C are available. High expansion glass compositions do not generally have long-term chemical durability however. Glass-ceramics are remarkable for the very wide range of thermal expansion coefficients which can be observed. At one extreme, materials having negative coefficients are available while for other compositions very high positive coefficients can be obtained. Between these two extremes there exist glass-ceramics having thermal expansion coefficients practically equal to zero and others whose expansion coefficients are similar to those of ordinary glasses or ceramics or to those of certain metals and alloys. This range of expansion coefficients is allied with good chemical durability.

Mechanical Properties

Characteristically, glasses are brittle solids which in practice break only under tension. The ionic and directional nature of the bonds and the identification of electrons with particular pairs of atoms preclude bond exchange. This, coupled with the random nature of the atomic lattice, i.e. the absence of close-packed planes, makes gross slip or plastic flow impossible.

Strength If flaws and stress concentrators, which emphasise the brittle nature of glass, can be avoided, then a glass article behaves as a single molecule in which the strength is governed by the very high interatomic bond strength. Glasses are therefore inherently very strong materials, theoretically capable of exhibiting a tensile strength of about $7 \, GNm^{-2}$. In practice however, surface flaws act as stress concentrators under tensile loading and commercial glasses in bulk form show a mean strength in tension of only about $40 \, MNm^{-2}$. The statistical variation of strength about this figure makes it desirable to allow a substantial safety margin and to design using a figure of about $7 \, MNm^{-2}$[5].

The strength of glass can be increased to about $200 \, MNm^{-2}$ by commercial toughening processes. The use of such glasses is not possible, however, at elevated temperatures since detoughening will occur. Commercial glass fibres display a strength of about $2 \, GNm^{-2}$; this high figure is dependent upon surface protection, given usually by organic coatings. Removal of the coating will result in a marked decrease in strength.

Elastic modulus Up to the fracture stress, glass behaves, for most practical purposes, as an elastic solid at ordinary temperatures. Most silicate-based commercial glasses display an elastic modulus of about $70 \, GNm^{-2}$, i.e. about 1/3 the value for steel. If stress is applied at temperatures near the annealing range, then delayed elastic effects will be observed and viscous flow may lead to permanent deformation.

The brittle nature of glasses at normal temperatures makes them inappropriate for use in locations where severe impacts are likely to be encountered. In the design of pipelines or other equipment it is possible to use normal engineering assembly techniques provided that suitable gaskets or cushioning are provided at joints and supports and that care is taken in tightening bolts to avoid unequal or localised stresses.

Chemical Properties

Technical glasses are now used so extensively and in such widely varying circumstances that it is necessary to be as accurate as possible in describing their chemical properties. The deterioration of individual glasses is dependent on composition, founding process and use and, unless the degradation processes are accelerated, may only be observable after very long periods of time. A typical figure for the corrosion rate of an ordinary soda-lime-silica glass would be below 0·008 mm/y. The effect is accelerated when the exposure takes place at higher temperatures, e.g. in boiling water or in an autoclave. Table 19.10 compares the corrosion resistance of some commercial glasses.

Table 19.10   The corrosion resistance of some commercial glasses

Glass	1	2	3	6	7
Water	1	2	2	1–2	2–3
Acid	1	2	2	1–2	2–4
Weathering	1	3	3	1–2	2–3

Key   1. Will virtually never show effects.
        2. May occasionally show effects.
        3. Will probably show effects.

Notes   1. See Table 19.8 for compositions.
        2. Table after Hauck, J. E., *Mats. Engng.*, 85, Aug. (1967).

One of the most commonly used measures of durability, i.e. the loss of sodium from the glass, is important to the pharmaceutical and chemical industries, but other changes such as loss of surface quality, are of equal importance for optical and window glasses. The properties of a wide range of technical glasses are well catalogued[5-7], but the data are often inadequate when considering a particular application and where possible non-standard 'whole article' tests are advisable.

In selecting a glass for chemical durability or weatherability regard must be paid to the temperature and concentration of the corrosive agent, length of exposure, the ratio of reagent volume to surface exposed and to the mechanical operating conditions. Guidelines on the durability of many commercial glasses in some attacking media are available from standard durability tests.

Glass durability tests   There are two types of durability tests for glassware, viz. 'grain' or 'powder' tests and 'whole article' tests.

Grain tests   In these tests, samples of glass, crushed and graded to a specified sieve size, are exposed under standard conditions of time and temperature to the attacking medium. The temperatures commonly used are 98°C (water bath) and 121°C (autoclave) and the attacking media are water, acid and alkali. The amounts of a particular glass constituent (usually soda or total alkali) removed from a standard weight of grains in a given time are determined.

Standard grain tests have been established by various national standards bodies and by some pharmaceutical authorities. The most important of these standard tests are the American[8] and the German[9]. Several other continental standards are essentially based on the German. The present British standard[10] relates only to laboratory glassware. The German and American standards differ in a number of details and to try to establish an international uniformity the I.S.O. have issued recommended procedures[11]. A new British standard in preparation will be based on these procedures.

Careful comparison of results from different laboratories using a particular grain test has shown considerable divergence, and it appears that to obtain consistent results very close adherence to the details of the standard procedures regarding grain preparation, extracting media and analysis is necessary. Nevertheless, grain tests are extensively used, especially in USA and Germany where a further step has been taken of classifying glasses by their 'hydrolytic resistance', a glass being placed in one of four classes according to the titre of the aqueous extract.

The following examples describe the extraction of alkali by water from 2 g samples at $97 \cdot 3°C$[12] for three glasses of different hydrolytic class.

Glass						
Class I	Time (h)	1	2	3	4	6
	Titre (ml of 0·01N HCl)	0·12	0·20	0·29	0·40	0·54
Class III	Time (h)	0·5	1	2	4	
	Titre (ml of 0·01N HCl)	0·65	0·96	1·39	2·0	
Class IV	Time (h)		1	2	4	
	Titre (ml of 0·01N HCl)		3·0	4·4	6·3	

Whole article tests   Grain tests are open to the criticism that they do not necessarily reflect the behaviour of the finished product in service, hence various tests on complete glass articles have been developed. These are normally carried out under accelerated conditions, and on completion various relevant factors are determined, such as loss in weight, alkali or other constituents extracted, the weight of soluble and insoluble materials in the extract and an assessment of surface condition. The advent of the electron microscope as a standard tool has made the latter study much more objective.

Whole article tests are particularly useful in the evaluation of window and optical glasses. Various tests have been proposed for window glass, but no standards exist. The usual procedure is to subject the glass to an accelerated humidity/temperature weathering cycle and to assess the surface conditions after a given period of treatment. The degree of haze formation has been suggested as a method of measuring surface damage, but generally visual comparison with a standard is used. Figure 19.1 illustrates the application of such a test to various optical glasses.

Many optical glasses are much less resistant to attack than are container and window glasses, and less severe tests are necessary. A commonly used method is to immerse specimens in either dilute nitric acid or standard acetate solution of pH 4·6 for specified periods at room temperature, then to examine the surfaces either visually or by interferometry.

Glass fibres present a particular problem. The water resistance of the base glass can of course be measured by a grain test, but this is unlikely to be representative of the performance of the final product. Generally, purely empirical methods are used to test the glass fibres *in situ* in a composite

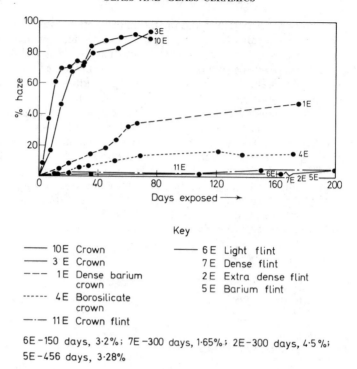

Fig. 19.1   Comparison of optical glasses after exposure to an alkaline solution (after Simpson, H. E., *Glass Tech.*, **37**, 249 (1953)[13])

material, e.g. the fibres are made up into rods or rings with the appropriate partly-polymerised plastic, the composites are then cured under specified conditions and the breaking strength determined after various exposure to water or steam.

A statement of standard tests for glass durability is given in Reference 14.

Mechanisms of Glass Corrosion

General corrosion properties   The glass surface may react with a corrosive agent in one or a combination of the following ways[15]:

   (*a*) By forming new compounds on the surface.
   (*b*) By selectively losing material from a leached porous layer.
   (*c*) By continuous dissolution leaving a freshly exposed surface.

Under certain circumstances components may be leached out of the bulk of the glass to leave a new material.

The nature of the glass surface   It is widely accepted that the composition of the glass surface is different to that of the interior of the glass, but it is difficult to quantify the difference. Alkali loss during forming, grinding, polishing and surface treatments, affects the structure of the surface, but a more basic difference is brought about by the effect of the unbalanced force fields at the surface on the ions within the glass.

Glass is composed of glass-forming cations (e.g. B^{3+}, Si^{4+}, P^{5+}) surrounded by polyhedra of oxygen ions in the form of triangles or tetrahedra. Two types of oxygen ions exist, viz. bridging and non-bridging. The former, bonded to two network-forming ions, link polyhedra, and the latter, bonded to one network-forming ion only, carry an excess negative charge. To compensate for this, charged cations of low positive charge and large size (e.g. Na^+, K^+, Ca^{2+}) are located within the structure. Silicon may be substituted by other cations of large positive charge and small size which are collectively known as *network-formers*.

The difference in size and field strength between ions is reflected in the polarisabilities of each ion, and their final position relative to the glass surface. Since the force field is unbalanced, ions of low polarisability will remain near the surface and ions of higher polarisability will move towards the interior of the glass. A strong feature of chemical reactions associated with the surface is the need to screen adequately (and not merely to neutralise) those cations which have strong electric fields. If the unbalanced force field is removed by the presence of materials, liquid, adsorbed vapour or solid in contact with the glass, there is sufficient mobility within the glass surface zone for it to revert to a more normal structure by the diffusion of ions towards the surface.

Corrosion Mechanisms[16]

There is no serious challenge to the view that the alkali or alkaline earth ions are removed from glass in water by an ion exchange process in which H^+ ions diffuse into the glass to preserve the electrical neutrality of the system. However, only under certain circumstances can the rate-controlling process be directly related to the diffusion of sodium in the glass. Most glass/corrosive agent systems are treated as unique cases, since in addition to the concentration of the attacking agent, temperature, rate of flow, and reaction time contribute to what is observed. General chemical principles of electrophilic and nucleophilic types of general attack can be applied to glasses[17]. The first is considered as an attack on non-bridging oxygen atoms by reagents with an electron deficiency, and the second as an attack on bridging oxygens by reagents with an electron excess.

For the most common series of corrosive agents, water, steam, acids, alkalis and salts, the hydrolytic processes peculiar to each determine the mechanism of attack. Thus, under the right circumstances, hydrolytic attack on the bridging oxygens* can occur in the following way:

$$-Si-O-Si- \rightarrow -Si \quad O-Si- \rightarrow -Si-OH \quad O^--Si- \quad ...(19.1)$$
$$OH^- \quad OH^{\delta-}$$
(free radical)

*It should be noted that the OH^- is partly bonded to the network and has lost part of its charge. The oxygen has acquired a partial negative charge and is only partially bonded to the network. Thus the δ^- indicates a transition state in which the charges on the OH and O lie between 0 and -1.

This is an irreversible reaction resulting in permanent damage to the glass network.

The corrosion process is modified by the physical state of the surface. Grinding and polishing processes, in particular, leave the structure in a more open state and with a degree of roughness and residual stress; all can contribute to accelerated corrosion.

The action of water and acids   During attack the alkali and alkaline earth network-modifying ions are exchanged by H^+ or H_3O^+ from acid solution. In some glasses the exchange process can go to completion causing only a small degree of network damage. In water, the exchange process proceeds at a very much slower rate relative to the acid conditions and some attack of the network is possible due to the presence of alkali ions from the glass moving into solution. This is most pronounced when glass is attacked by steam at high temperature and there is no mechanism for the removal of alkali from the hydrolysed zone. Acid solutions mitigate this form of attack by neutralising the alkali as it is formed.

The attack of most glasses in water and alkali is diffusion controlled and the thickness of the porous layer formed on the glass surface consequently depends on the square root of the time. There is ample evidence that the diffusion of alkali ions and basic oxides is thermally activated, suggesting that diffusion occurs either through small pores or through a compact body. The reacted zone is porous and can be further modified by attack and dissolution, if alkali is still present, or by further polymerisation. Consolidation of the structure generally requires thermal treatment.

Attack by alkali solution, hydrofluoric acid and phosphoric acid   A common feature of these corrosive agents is their ability to disrupt the network. Equation 19.1 shows the nature of the attack in alkaline solution where un-limited numbers of OH^- ions are available. This process is not encumbered by the formation of porous layers and the amount of leached matter is linearly dependent on time. Consequently the extent of attack by strong alkali is usually far greater than either acid or water attack.

Both acids form compounds of silicon as a result of attack on the network, silicon fluoride from hydrofluoric acid and silicyl phosphate from phosphoric acid.

Chemical attack by other agents   If the hydrogen ion concentration is high enough, the glass loses a substantial amount of weight by leaching, but these reactions are very dependent on the nature of the ions in solution. Certain salts, especially those of Zn, Al and Be, if present as trace amounts, can have a beneficial effect by poisoning the process and preventing the occurrence of leaching.

The Cleaning of Glass

There is no universally ideal technique for either cleaning glass or avoiding contamination of the surface. In the most severe circumstances of corrosion the only methods capable of restoring an acceptable surface finish consist of grinding and polishing or removing the contamination and corroded layers by strong etching agents such as hydrofluoric acid. Less severe conditions may respond to treatment by various detergent solutions or organic solvents, these being considerably aided by ultrasonic vibration. Manual washing or ultrasonic cleaning can be used to remove massive dirt accumulations. The vapour degreasing process, which uses isopropyl alcohol, has minimal corrosive action on the glass.

To restore old stocks of corroded glass, treatment in hot 1% sodium hydroxide solution followed by rinsing in 5% hydrochloric acid and a final rinse in pure water at room temperature is recommended[18].

Glass-Ceramics

Definition and Properties

Glass-ceramics are a family of materials that are polycrystalline in nature and are formed from the liquid or glassy state. A glass-ceramic article is made by the heat treatment of a vitreous body in two stages:
   1. Nucleation. The glass is held at a temperature below its softening point for a period of minutes or hours to allow nuclei to develop.
   2. Crystallisation. The temperature of the nucleated glass is raised to just below the softening point when crystals form and grow around the nuclei.

Ideally the product is a fine-grained ceramic containing interlocking crystals with sizes ranging from less than 10 nm in transparent glass-ceramics to several micrometres, with a residual, usually small, glass content. The behaviour of the material is largely determined by the choice of the crystalline phase; by suitable choice a range of useful properties has been obtained[19].

As a class of materials, glass-ceramics have the following general characteristics:
   1. Impervious, with moderate densities similar to those of glasses.
   2. Rather high strengths for oxygen-rich solids, accompanied by some scatter of individual values.
   3. Stiff, elastic, fully Hookean behaviour.
   4. Considerable hardness and resistance to abrasion.
   5. Chemical stability and resistance to corrosion.
   6. Resistance to medium-high temperatures (higher temperatures than most glasses but lower than the refractory oxides) and low thermal conductivity.
   7. Resistance to the passage of electrical current.

Special characteristics can be developed in individual materials depending on the cations present and their arrangement relative to each other and to the oxygen anions. The most important of these characteristics is low, medium or high reversible thermal expansion. The properties of some commercially available glass-ceramics are summarised in Table 19.11.

Chemical Durability of Glass-ceramics

Although the factors which govern the chemical stability of glasses are fairly well-known, there is little information concerning this aspect of glass-ceramics. While the chemical behaviour of a glass-ceramic is strongly

Table 19.11 Property data for some commercially available glass-ceramics

	Units	A	B	C	D	E	F
Specific gravity	g/cm³	2·55	2·5	2·53	2·5	2·5	2·7
Average thermal expansion coefficient	$\times 10^{-7}\,\text{deg}\,\text{C}^{-1}$ (100°C–300°C)	6	12·7	97	4·2	– 3	90
Transition point	°C	820					
Softening point	°C	950					
Maximum service temperature	°C	800	1 000	1 500	1 250	814	
Bending strength	$\times 10^{-7}\,\text{N m}^{-2}$	12	18	8·8	11	13	25
Volume specific resistance	log (Ω cm)	10·6	10·6	6·7	10	11·6	
Compositional type		lithium alumino silicate	lithium alumino silicate	potash magnesia alumino silicate	lithium alumino silicate	lithium alumino silicate	sodium baria silicate

Key: *A* Heatron T (Trade name of Fuji Photo Film Co. Ltd.)
B Neoceram-15 (Trade name of Nippon Electric Glass Co. Ltd.)
C Corning code 9650 (Machinable glass-ceramic)
D Pyroceram 9608 (Trade name of Corning Glass Works)
E Hercuvit (Trade name of Pittsburgh Plate Glass Co.)
F Centura ware (Trade name of Corning Glass Works)

influenced by the chemical composition of the parent glass, several different crystalline compounds together with a residual glass phase are likely to be present. The relative resistances of these phases to attack by water or other reagents will determine the chemical stability. In general, a glass which exhibits poor chemical stability is unlikely to give rise to a glass-ceramic of high stability. To this extent the factors which govern the stability of glass-ceramics can be equated to those which determine the chemical stability of glasses.

In most cases, glass-ceramics possess good chemical stability and certainly compare favourably in this respect with other ceramic materials. Table 19.12 summarises makers' data for chemical attack on commercially available materials.

Table 19.12 Chemical resistance data for some commercially available glass-ceramics

Chemical resistance	Units	A	B	C	D	
Powder method						
Water solubility	mg as Na_2O per 1 mg sample	0·29				
Acid solubility	loss in wt., %	0·06				
Alkali solubility	loss in wt., %	0·13				
Surface method						
H_2O (90°C × 24 h)	mg/cm^2	0·00				0·12
5% HCl (90° × 24 h)	mg/cm^2	0·13	0·8	128	0·1–0·3	0·20
5% NaOH (90° × 24 h)	mg/cm^2	0·04	3·5	5	0·02–0·1	4·0

Key: A Heatron T
B Neoceram-15
C Corning code 9650
D Pyroceram 9608
E Hercuvit

Certain types of glass-ceramic have good resistance to attack by corrosive chemical reagents. Low-expansion glass-ceramics derived from lithium-aluminosilicate glasses are only slightly inferior to borosilicate chemically-resistant glass with regard to attack by strong acids and are somewhat more resistant to attack by alkaline solutions. Materials derived from magnesium-aluminosilicate glass compositions are slightly less resistant to attack by strong acids and alkalis than are chemically-resistant borosilicate glasses. Even at high temperatures these types of glass-ceramic retain resistance to attack by corrosive gases.

For certain applications it is important that the glass-ceramic should be unaffected by contact with reducing gases at high temperatures. In such cases the composition must not include any oxides such as lead which are easily reduced to the metal.

D. S. OLIVER

REFERENCES

1. Rawson, H., *Inorganic Glass Forming Systems*, Academic Press, London (1967)
2. *Viscosity-temperature Relations in Glass*, International Commission on Glass (1970)
3. *Standard Method of Test for Softening Point of Glass*, ASTM Designation C338–57 (1965)
4. *Thermal Shock Testing*, BS 3517:1962

5. *Glass Pipelines and Fittings*, BS 2598:1966
6. Volf, M. B., *Technical Glasses*, Pitman, London (1961)
7. Wessel, H., *Silikattechnik*, **18**, 205–211 (1967); *ibid.*, **19**, 6–10 (1968)
8. *Standard Methods of Test for Resistance of Glass Containers to Chemical Attack*, ASTM Method C 225–68
9. *Testing of Glass: Determination of Resistance to Acid*, DIN 12116:1955;
   Testing of Glass: Determination of Resistance to Caustic Liquors, DIN 12122:1955;
   Testing of Glass: Determination of Resistance to Water, DIN 52322:1967
10. *Method of Testing Chemical Resistance of Glass Used in the Production of Laboratory Glassware*, BS 3473:1962
11. *Determination of Resistance of Glass to Attack by Boiling Aqueous Solution of Mixed Alkali*, ISO R695:1968;
   Determination of Hydrolytic Resistance of Glass Grains at 98°C, ISO R719:1968
12. Wiegel, E., *Glas. Berichte*, **29**, 137 (1956)
13. Simpson, H. E., *J. Soc. Glass Tech.*, **37**, 249 (1953)
14. Fletcher, W. W., BGIRA Information Circ. No. 96
15. Holland, L., *The Properties of Glass Surfaces*, Chapman & Hall (1964)
16. Das, C. R. and Douglas, R. W., *Phys. & Chem. of Glasses*, **8**, 178–184 (1967)
17. Budd, S., *Phys. & Chem. of Glasses*, **2**, 111–114 (1961)
18. Tichane, R. M., *Amer. Cer. Soc. Bulletin*, **42**, 441 (1963)
19. McMillan, P. W., *Glass-ceramics*, Academic Press, London (1964)

19.3 Vitreous Silica

Vitreous silica consists of pure silica fused into a homogeneous product containing over 99.8% SiO_2. Its excellent thermal properties and its high resistance to chemical attack have made it a useful material over a wide range of temperatures. Two main forms of vitreous silica are produced: the translucent, made from high-grade glass sand, and the transparent, made from rock quartz crystal, which in addition to visible light transmits ultra-violet and infra-red radiation. Both types are manufactured with a glazed surface (the translucent is also available with a sand or satin finish); the glaze is integral and therefore not liable to spalling or crazing. The transparent form has greater mechanical strength than the translucent.

Physical and Thermal Properties

Coefficient of thermal expansion The coefficient of thermal expansion of vitreous silica is very small, being about one-sixth that of porcelain $(0.054 \times 10^{-5}$ over the range 0–$1\,000°C)$. It is thus highly resistant to thermal shock. In addition it is almost free from thermal hysteresis (the value for annealed vitreous silica is -3×10^{-9} to $-4 \times 10^{-9}/°C)$.

Heat resistance The melting point of vitreous silica is in the same region as that of platinum $(1\,700$–$1\,800°C)$; the actual point is not well defined, since softening takes place at about $1\,500°C$, especially in the translucent form. On heating for long periods at temperatures in excess of $1\,200°C$, vitreous silica tends to become brittle, with a change of physical characteristics, owing to reversion to the crystalline state. The rate of strength loss from these causes rises very steeply with temperature increase upwards of about $1\,200°C$; whereas heating for $8\,h$ at $1\,188°C$ in tests produced little serious weakening, $4\,h$ at $1\,350°C$ led to a 40–50% reduction in strength. The material can safely be used under ordinary conditions at temperatures up to $1\,050°C$, and may also be used indefinitely at temperatures up to $1\,350°C$, provided the operating temperature is never allowed to fall below $300°C$. It should, however, be noted that certain compounds, including potassium and lithium salts, sodium tungstate, ammonium fluoride, phosphates and radioactive substances, cause devitrification at temperatures below $1\,050°C$.

Thermal conductivity The thermal conductivity of fused silica is low. The transparent form passes infra-red radiation with little loss up to wavelengths of $3·5\,\mu m$.

Electrical characteristics The insulating properties are excellent. At ordinary temperatures the resistivity of the translucent form is over $200\,000\,000\,M\Omega\,cm$, and it is capable of withstanding high frequency discharges at high voltages.

Resistance to Chemical Attack

Fused silica is highly resistant to many acids which attack even the most inert metals, e.g. platinum can be dissolved in aqua regia at elevated temperatures in vessels of fused silica. Its acidic character precludes its application for concentrated solutions of strong alkalis, and fluorides are corrosive owing to the affinity of silicon for fluorine.

In certain cases contamination of the chemical by silica, rather than significant deterioration of the fused silica vessel, may preclude its use.

Boiling water and steam There is no reaction with water and steam at moderate temperatures and pressures. At pressures of the order of $3·5\,MN/m^2$ and temperatures in the range of 400–500°C, the solubility of the translucent form is $0·14\%$ and of the transparent $0·035\%$.

Acids and halogens Vitreous silica is completely resistant to all halogens and acids, with the exception of hydrofluoric and phosphoric acids, irrespective of temperature and concentration. Below 200°C fused silica is suitable for use with concentrated phosphoric acid. At 217°C the concentrated acid takes about 8 months to remove 1 mm of fused silica, while at 270°C this thickness is removed in about 6 weeks. Some service has been obtained with concentrated acid (s.g. 1·75) to 300°C, and fused silica has been used for concentrating phosphoric acid. At temperatures up to 1 000°C, sulphuric, nitric and hydrochloric acids, and mixtures of these acids, do not cause attack.

Alkalis In general, alkalis react with fused silica, the rate of attack increasing with temperature and concentration (5% caustic soda solution can, however, be contained in fused silica at room temperature, provided that there is no serious objection to contamination by silica). Similar considerations apply to sodium carbonate and sodium cyanide, which are, however, less corrosive than caustic soda. Sodium metaphosphate has an appreciable reaction and sodium polyphosphate reacts vigorously with fused silica.

Neutral salts These are not generally corrosive, and fused silica can be used for solutions of halides, sulphates and nitrates.

Basic oxides At elevated temperatures basic oxides, e.g. PbO, react with silica.

Fused salts Fused alkalis (e.g. caustic soda, sodium carbonate), as would be expected, attack silica vigorously. Similar considerations apply to fused alkali halides (fluorides are the most detrimental) and phosphates.

Metals Fused silica is generally inert but is decomposed by metals which have a high affinity for oxygen, and the metal oxide formed causes devitrification at high temperatures. Sodium vapour appears to be more detrimental than the molten metal, while aluminium readily attacks fused silica at 700–800°C.

Applications of Vitreous Silica

The high thermal and electrical resistance of vitreous silica, and its imperviousness to chemical attack, make it suitable for a wide range of applications. These include chemical and physical laboratory ware, tubes and muffles for gas and electric furnaces (including vacuum furnaces), pyrometers, insulators for high-frequency and high-tension electrical work, mercury-vapour and hydrogen-discharge lamps, high-vacuum apparatus, plants (complete or partial) for chemical and related industries, equipment for the manufacture of pure chemicals, tubes, chimneys, and radiants for the gas- and electric-heating industries, component material in refractory and ceramic mixtures, etc.

It is used for pipes to carry hot gases and acids, acid distillation units, condensing coils, S-bend coolers, hydrochloric acid cooling and absorption systems, nitrating pots, and cascade basin concentrators for sulphuric acid. The inertness of vitreous silica to most acids is also utilised in the manufacture of electric immersion heaters and plate heaters for acidic liquors in chemical processes and electroplating baths. Vitreous silica wool is used for filtration of acidic liquids and for filtering hot gases. The resistance of vitreous silica to water and steam at normal temperatures and pressures makes it applicable in the production of pure water for the manufacture of highly purified chemicals, while it is used on account of its thermal properties for the construction of muffles of oval cross-section used for the bright annealing of metal strip and wires.

Special products include transparent vitreosil springs which are ideal for continuous measurement in corrosive atmospheres, and quick-immersion thermocouple protection sheaths for rapid temperature measurement.

Acknowledgment

Acknowledgment is made to The Thermal Syndicate Limited for the information contained in this section, which was derived originally from *About Vitreosil*, Thermal Syndicate Ltd (1958).

C. MAY

BIBLIOGRAPHY

Mackenzie, J. D., *Modern Aspects of Vitreous State*, Vol. 3, Butterworth Inc., Washington (1964)
McMillan, P. W., *Glass-Ceramics*, Academic Press (1964)
Burke, J. E., *Progress in Ceramic Science*, Vol. 4, Pergamon Press, London (1966)
Gulaev, V. M., 'The Strength and Deformation Properties of Vitreous Silica', *Glass Ceramics*, **30**(6), 279 (1973)

19.4 Glass Linings and Coatings

While glass alone can provide many of the desirable features of an ideal inert material, fabrication difficulties prevent its use for large-diameter chemical process equipment, and mechanical considerations would in any case make it necessary to treat any such equipment with great care.

The high chemical resistance, and the non-toxic, non-flavouring and thermal-resistance properties of glass, can however be combined with the mechanical strength of metals by covering metal surfaces exposed to corrosive media with a layer of suitable glass. It thus becomes feasible to produce large storage or transport tanks of over 10 000 litre capacity, reaction vessels, valves, pipes, silos, smoke stacks, etc. which have this interesting and serviceable combination of properties.

The principal advantages of glass linings are the increased size and mechanical strength that are possible compared with all-glass equipment, and the flexibility of operation with different chemicals compared with all-metal equipment. The increased heat transmission, in comparison with glass equipment, can also be an advantage.

The principal disadvantages are the loss of transparency and the potential vulnerability of the lining to mechanical damage unless sensible precautions are taken in handling, installation and service. It is, however, possible to seal small flaws with special plugs or in extreme cases to re-line the equipment.

A variety of metals can be protected in this way, including copper, gold, stainless steel, titanium and uranium, but by far the most extensive use of the technique is for steel equipment.

Glass Preparation

There are a number of proprietary glass formulations for coating steel; in most cases the chemical constitution has not been disclosed but it is understood that the more successful types are of the borosilicate family, containing more aluminium and alkali oxides than the typical heat-resisting borosilicates discussed in Section 19.2. In many formulations, particularly for ground-coating the steel, the glasses contain a much greater proportion of cobalt oxide than is found in ordinary glasses; this constituent is included to encourage the formation of a chemical bond between the metal and the glass. The coefficient of linear expansion of a typical glass for this application is about $10 \times 10^{-6}/°C$. The disparity between this figure and the higher expansion coefficient of the steel is quite deliberate and results in the development of compressive stresses in the glass layer after processing. The stress pattern which develops gives more resistance to thermal shocks or to external stresses than if the expansion of the glass and steel were accurately matched.

The raw materials required to produce the particular glass composition are intimately mixed and melted at a temperature near 1 300°C. The melt is quenched with water, yielding the glass in a granular form known as *frit*, which is subsequently either ground to give dry glass powder or wet-milled with ball clay to form a creamy slip. It can be applied to the steel in either of these forms, the particular method depending on the design and size of the structure.

Metal Preparation

A low-carbon (less than 0·2% C) steel is usually employed in the manufacture of chemical storage or process vessels, water heaters, and the like; one writer refers more precisely to "a rimmed steel (ASTM A285), grade *A* or *B* flange quality" as being suitable, and states that with this base, defects such as fishscaling*, blistering, crazing and poor adhesion are avoided. Cast iron is frequently employed in such equipment as valve bodies and pipe unions.

Conventional welding techniques are used in fabrication and assembly of the steel body, but special precautions must be taken to ensure that the weld metal is of similar composition to the base metal, and that moisture- (hydrogen-) free electrodes are used. Lap-welded joints and riveted structures are not suitable for the subsequent enamelling operations; butt welding is much preferred. The completed assembly is examined in detail to check for cracks, welding faults or other defects likely to impair serviceability, and is then normalised at 900°C to relieve stresses set up during fabrication. The heat treatment also has the advantage of burning-off any organic contamination. After cooling, the vessel is sand-blasted with a silica-type grit to remove oxide scale developed during normalising and to promote, by providing a roughened surface, mechanical keying of the glass ground-coat layer to be applied.

Points which manufacturers note as being important in the design of these vessels are as follows:

1. The metal thickness should be as uniform as possible throughout, avoiding heavy bosses, lugs or brackets. Marked changes of section at such points will seriously modify the heating and cooling rates and can thus give difficulty during the enamelling processes.
2. The minimum radius recommended for all curved surfaces is 6·4 mm; sharp edges must be entirely avoided. On small radii there is a risk that shaling† or other difficulties might arise. This is, in essence, a consequence of the deliberate mismatch of the expansion coefficients of glass and metal: a closer match would permit effective coating of smaller radii but would decrease the performance of the unit in other respects.

---

*Development of roughly circular fractures in the glass lining.
†Flaking away of fragments of the lining, with a characteristic conchoidal fracture surface.

3. The risk of distortion on firing is greater if the design calls for numerous apertures in the vessel or its cover. As few openings as possible should be employed.

Lining Processes

The processes used for coating the metal surfaces are basically similar to those used in porcelain enamelling.

The ground coat, where used, is sprayed on in the form of a wet slip and after drying is fired at approximately 900°C. With open vessels or small cast-iron units, dry glass powder can be dusted directly on to the hot, fused surface of the ground coat; larger vessels such as storage tanks are allowed to cool, inspected and then sprayed uniformly with the wet slip. The coating is thoroughly dried and then fused by heating the vessel in a furnace at approximately 850°C. The vessel is cooled, and further coats of glass enamel may be applied as required in the same way. Thicker layers may be built up by immediately dusting the coating of wet slip with dry glass powder. It is clearly important that the vessels being treated should not distort under their own weight during furnace treatment, and it may be necessary to design the article with slightly greater thickness than usual to allow for the softening effect of high temperatures. Another way of avoiding the difficulty with closed vessels is to force inert gas into the tanks at controlled pressure. The integrity of the linings is tested electrically.

Properties of Glass Linings on Steel

Mechanical Properties

With correctly formulated glass, the lining has the ability to withstand stresses up to the elastic limit of the steel without breaking. Impacts sufficiently severe to dent the steel will probably cause fractures in the lining. The hardness and abrasion resistance of the lining are similar to those of all-glass equipment and a smooth, easily cleaned surface is produced. The combined effect of mechanical keying of the glass to the metal and the chemical bond developed between the two results in a very high adhesive strength. Test figures indicate that the bond strength is of the order of 35–70 MN/m².

Thermal Properties

The thermal shock resistance of glassed steel, i.e. the safe limit of temperature difference between the glass surface of the vessel and any charge introduced, varies according to the general operating temperature. This is because the desired compressive stress on the glass is reduced as the temperature increases. With a typical coating formulation on a vessel operating at 120°C, the recommended maximum thermal shock would be about 93°C, while at an operating temperature of 205°C the corresponding figure would be 55°C.

Recent improved formulations have made it possible for vessels to withstand thermal shocks of some 30% greater than this.

The heat-transfer coefficient for heating in glass-lined equipment is of the order of 340–455 W m^{-2} °C^{-1}, but can be increased by agitation. For cooling, corresponding figures would be 200–285 W m^{-2} °C^{-1}. The numerical values depend on the thickness of the glass coating, increasing with decreasing glass thickness, and the figures quoted represent the behaviour of an average coating.

Chemical Properties

The general pattern of chemical resistance of glass linings is very similar to that of all-glass equipment. Water absorption is negligible, resistance is very high to all acids except hydrofluoric and (at high temperatures and concentrations) phosphoric acid, attack by water is measurable only with difficulty, most organic liquids produce no measurable effect and strongly alkaline solutions are satisfactorily handled at near-ambient temperatures. but at higher temperatures there may be appreciable reaction.

Priest[1] has given the results of 15-day tests on a particular glassed steel, both for immersion and vapour-phase conditions, in a variety of media. Some of this information is extracted in Table 19.13 below.

Table **19.13** Corrosion of glassed steel in boiling acid/ distilled water systems

Test solution	Corrosion rates (mm/y)	
	Liquid phase	Vapour phase
Distilled H$_2$O	<0·025	0·241
50 p.p.m. HCl	<0·025	0·163
500 p.p.m. HCl	<0·025	0·033
1·0% HCl	<0·025	<0·025
10·0% HCl	0·040	0·036
14·0% HCl	0·030	0·079
18·0% HCl	<0·025	0·013
50 p.p.m. H$_2$SO$_4$	<0 025	0·015
500 p.p.m. H$_2$SO$_4$	<0·025	0·043
1·0% H$_2$SO$_4$	<0·025	<0·025
10·0% H$_2$SO$_4$	0·043	<0·025
20% H$_2$SO$_4$	0·033	<0·025

Data on liquid-phase corrosion in alkaline media are given in the same paper, where it is stated that below 37·8°C "glassed steel is completely resistant to the strongest alkalis, 20% sodium hydroxide, for example". The effect of temperature on alkali corrosion is illustrated by data from this paper in Table 19.14.

There are various glass compositions in service and it is important to ensure that the lining is appropriate for the particular processes to be operated. An example of the behaviour of two different types of glass is given in Table 19.15.

Table 19.14   Corrosion of glassed steel in alkaline media as a function of temperature

Test solution	Approximate corrosion rates (mm/y)				
	27°C*	38°C	54°C	66°C	82°C
1% NaOH	0·0114	0·0254	0·0889	0·2032	0·533
1% Na$_3$PO$_4$	0·0102	0·0203	0·0635	0·1143	0·457
1% Na$_2$CO$_3$	0·0076	0·0152	0·0457	0·0889	0·305

*Extrapolated using Arrhenius relationship.

Table 19.15   Corrosion resistance of glass linings

Test solution	Temp. (°C)	Duration of test (days)	Corrosion rate (mm/y)		Estimated life of 0·91 mm-thick lining (years)	
			AR	AAR	AR	AAR
Buffered NaOH, pH 11·5	100	30	0·305	0·0762	3	12
0·5% NaOH solution, pH 13	75	30	1·168	0·305	0·8	3
20% HCl solution	Boiling	30	0·01626	0·02108	56	43

Clearly the AAR glass is much more preferable to AR for alkaline reactions but is only marginally less resistant to the acid solution; glass AAR can in fact be used up to pH 12 at 100°C.

While the usual classification of corrosion rates applied to metal (i.e. 'A' corrosion rate <0·127 mm/y) is often helpful, the extremely high resistance of glass or glass lining would appear to justify a still higher category than 'A' (perhaps 'A'* <0·0127 mm/y). Thus with 40% nitric acid, at boiling point, a given lining was found to have a rate of attack of 0·000 20 mm/y, while with 80% sulphuric acid the rate was immeasurably small. With 20% sulphuric acid, however, this lining would come within the usual 'A' category, as the rate of attack at boiling point was found to be 0·003 8 mm/y.

D. K. HILL

REFERENCE

1. Priest, D. K., *Bull. Amer. Ceram. Soc.*, **39**, 507 (1960)

BIBLIOGRAPHY

London, H., *Glassed-steel for Chemical Plant*, Enamelled Metal Products Corp. (1933) Ltd., London (Agents for Pfaudler Co.) (Trade literature)
Materials for Construction in the Chemical Industry, 'Applications of Glass Linings in Chemical Process Plant', Society of Chemical Industry, 245 (1950)
Vincent, G. L., 'Apply Glascote 778 to Chemical Reactors', *Ceramic Ind.*, **74** No. 6, 65 (1960)

Pfaudler Co., 'Testing of Glass-coated Steel', *Mater. in Des. Engng.*, **49**, 176 (1959)

Bullock, C. E. and Nelson, F., 'Glass Coatings', *Mater. in Des. Engng.*, **47**, 106 (1958)

Pfaudler Co., 'Thermal Shocking News About the New Pfaudler Glassed-steel 59', *Chem. Engng. News*, **36** No. 27, 15 (1958)

Goulds Pumps, 'Glass-lined Pump—Goulds Pumps and Pfaudler Together Design Centrifugal Pumps with all Inner Surfaces Lined with Glass', *Chem. Engng. News*, **35** No. 26, 84 (1957)

Pfaudler Co., 'Glassed Steel for the Petroleum Industry', *Chem. Engng. News*, **35** No. 17, 61 (1957)

Priest, D. K., 'Observations on the Physical Appearance of Chemically Attacked Glassed Steel Surfaces', *Bull. Amer. Ceram. Soc.*, **36**, 416 (1957)

Hall, J. G., 'Glass-coated Steel', *Glass*, **28**, 82 (1951)

Smith, A. O., Corp., 'Production of Glass-lined Tanks', *Industr. Heat.*, **17**, 1942 (1950)

Carborundum Co., UK Pat. 578 280 (21.6.46)

Carpenter, V. W., US Pat. 2 492 682 (27.12.49)

19.5 Stoneware

Introduction

Owing to their almost complete resistance to corrosion, stoneware and porcelain occupy a unique position as materials in the chemical industry. Stoneware bodies can be compounded from many different raw materials to give the properties required and to permit the manufacture of articles up to a considerable size. Porcelain, on the other hand, is not generally used for very large articles, but owing to its completely vitreous nature it normally shows a much greater resistance to corrosion than does stoneware.

The composition of stoneware can vary very widely. In its simplest form it may consist simply of a clay, or a clay mixed with a pre-fired inert material known as *grog*. This grog may be clay which has been fired to a high temperature and then crushed and sieved to give the correct grading. On the other hand, the body may be compounded from clays, felspar, quartz, etc. all materials which are in normal use in the ceramic industry. This type of body normally has improved strength over the simple type. The grog may be of a special nature, such as alumina, silicon carbide or zircon, according to the service conditions to which the material will be subjected in use. The appropriate type of body mix must be decided on in relation to materials which will be in contact with the ware, any thermal shock to which the ware may be subjected, and the mechanical forces the ware will have to withstand.

Mechanical and Physical Properties

While mechanical properties of ceramic materials are usually quite adequate for the duties which they have to perform, it is essential to realise the limitations of the material, and to design and install any articles made from it in such a way as to minimise any weakness. Table 19.16 gives typical values for the mechanical properties of the different materials which are available.

Tensile Strength

This depends very largely on the composition of the body and the temperature to which it is fired. The addition of a clay grog will normally reduce the tensile strength but the degree of reduction depends on the proportion used, the grain-size distribution of the grog, and the particle shape. The

Table 19.16 Some typical values for the physical and mechanical properties of available stoneware and porcelain bodies

Body	Tensile strength (MN/m²)	Compressive strength (MN/m²)	Modulus of rupture (MN/m²)	Apparent porosity (%)	Thermal conductivity (W m^{-1} K^{-1})	Young's modulus (GN/m²)	Thermal expansion coefficient 0–500°C (mm/mm°C)
Porcelain	34	690	90	0	0·0047	69	7×10^{-6}
Stoneware:							
Porous	7	103	34	0·5–3·0	0·0035	55	5×10^{-6}
Vitreous	21	207	70	0–0·5	0·0047	69	6×10^{-6}
Thermal-shock-resisting stoneware	14	138	55	0–1·0	0·0035	83	6×10^{-6}
High alumina	117	1 380	276	0	0·013	276	7×10^{-6}

degree of vitrification also affects the tensile strength. By using special grogs it is possible to produce a body which has a reasonably high strength.

Since all ceramic materials are weaker in tension than in compression, it is good practice in design to avoid tensile forces wherever possible, and where they are unavoidable to reduce them to a minimum and to distribute them evenly over the article, e.g. a circular tank is better than a rectangular one.

Compressive Strength

This is usually about ten times greater than the tensile strength and once again depends on the properties of the grog and the degree of vitrification of the body.

Modulus of Rupture

All ceramic materials are elastic, and hence show very little bending under load. They do not exhibit any creep under load. The modulus of rupture type of test[1] is the routine test most commonly used in the ceramic industry, and gives the figure generally quoted for the strength of the material. It must be remembered that the value obtained for any particular body depends on the cross-sectional area of the test piece; thus figures quoted from test results may be higher than those obtained on actual articles, which usually have a thicker section than the test piece.

Young's Modulus

This is normally very high and very little can be done by the manufacturer to vary it. An important related property is the critical strain, defined as the ultimate tensile strength divided by the elasticity. Most ceramic materials have similar values for critical strain (approx. 0·001), and this must always be remembered whenever a combination of stoneware and other materials is being considered.

Porosity

This is very important as several other properties are dependent upon it. If the porosity is too high, the article will be weak and will not retain liquid. The pore structure should also be taken into account. When a ceramic material is fired, although the internal surface area decreases as the material approaches zero porosity, the mean radius of the pores increases. Thus, when the internal surface area is $3 \, m^2/g$ the mean pore radius may be of the order of $10^{-7} \, m$, while when the internal surface has dropped to $0·5 \, m^2/g$ the mean pore radius may be about $4·5 \times 10^{-7} \, m$. The mean pore radius may reach a value as high as $9 \times 10^{-7} \, m$ as the ware approaches zero porosity during firing. It is thus obvious that at some point the pores must start to close up. This closing of the pores with the approach of vitrification is borne out by results of permeability measurements.

It is generally true that the lower the porosity the higher the mechanical strength and the acid resistance of the body, but unfortunately low porosity also usually results in reduced thermal shock resistance.

Thermal Shock Resistance

The thermal shock resistance of any stoneware material depends particularly upon thermal expansion, strength, Young's modulus and thermal conductivity.

Many of these properties depend upon others which may themselves be governed by yet other factors. Thus, as mentioned above, increased porosity usually gives better thermal shock resistance, but it may be necessary for reasons of watertightness to employ a body with a very low porosity. The size of an article is also closely related to the degree of thermal shock which it will withstand. For this reason it is very difficult to give accurate figures for the thermal shock resistance of stoneware bodies. In practice, if precautions are taken to heat up any stoneware articles slowly and evenly no trouble will be experienced. This is a matter on which the ceramic manufacturer should be consulted.

Dimensional Accuracy

All ceramic articles (except those for highly specialised uses) exhibit high shrinkage, usually varying between 7 and 15%, during manufacture. During the firing process the rate of shrinkage of different parts of an article may vary, and the body may slump under its own weight at the maximum temperature attained. Thus it can be seen that the manufacture of an article to a close dimensional specification is a very difficult task. Manufacturers will normally contract to supply to dimensions with a tolerance of $\pm2\%$. If greater accuracy than this is required, it is necessary to grind the article, which will increase the price.

Abrasion Resistance

This property is very hard to define, as articles may be subjected to very varied forms of abrasion, and in general a given ceramic body will react quite differently to different types of abrasion. This is a question on which the manufacturer should be able to give considerable guidance. Many types of standard abrasion test have been proposed, but none has proved satisfactory and experience must continue to be the main guiding factor.

Thermal Expansion

The thermal expansion coefficient of ceramic material can vary from 2×10^{-6} to 7×10^{-6} mm/mm°C. The values obtained for bodies normally supplied are about 5×10^{-6} to 6×10^{-6} mm/mm°C; values outside this range will be obtained only from special bodies.

Impact Strength

This is another property for which it is very difficult to obtain a reliable figure. In general, ceramic materials are not very resistant to impact and should be guarded to prevent breakage by accidental blows.

Chemical Resistance

Stoneware and porcelain are attacked only by hydrofluoric acid and hot concentrated caustic alkalis and are almost completely unaffected by all concentrations of mineral and organic acids, acid salts and weak alkalis[2]. The degree of susceptibility depends upon composition, firing temperature and porosity.

As stoneware and porcelain can be given a glazed finish on both interior and exterior surfaces, articles made from these materials can be very easily cleaned, even after years of use. The glaze also has an effect on the strength of the ware, increasing it by up to 20%, but certain types of body are difficult to glaze satisfactorily.

Jointing

In the majority of stoneware constructions some form of joint has to be provided, and in runs of piping there will be very many joints, which frequently fail owing to failures in the jointing material. It is essential that all joints should be able to withstand corrosion to the same extent as the stoneware, and that they should be leakproof and capable of withstanding any conditions of mechanical or thermal shock which may be present. For these reasons, very careful thought must be given to the type of joint employed and to the jointing media. It should also be remembered that any joints should be made by skilled operatives, as in most cases one badly made joint could cause failure of the whole installation.

Various methods of jointing are discussed below.

Hydraulic Cement

Three types of hydraulic cement are in use, viz. Portland, supersulphated and high-alumina. Portland cement is satisfactory in solutions with a pH of 7 and upwards, high-alumina will withstand solutions of pH 5·5 and upwards but will be attacked by alkaline solutions greater than pH 9, while supersulphated cement is resistant to solutions of pH 3 and upwards and also to alkaline solutions. All these cements are resistant to solvent solutions. Another advantage in the use of high-alumina cements is that they will attain their maximum strength in about 24 h. If Portland cement is used for the foundations of acid plants, care should be taken to insulate it from the surrounding earth.

Acid-resisting Cement

This class of cement, which will withstand acids of all concentrations, with the exception of hydrofluoric acid, consists of an inert aggregate bonded with either potassium or sodium silicate solution, with the addition of a setting agent, e.g. sodium silicofluoride or ethyl acetate. The aggregate may consist of graded quartz, acid-resisting brick, etc. It is important to remember that this type of cement will be attacked by water and dilute alkalis, and therefore an acid solution must be kept in contact with the cement during use. Sodium silicate is the normal bond used, mainly on account of its cheapness. In the presence of sulphuric acid, however, potassium silicate is preferable, since if sodium silicate is used, there is a danger that the decahydrate of sodium sulphate will be formed, with a consequent large increase in volume. Many trials of the use of sodium silicate in contact with sulphuric acid have however shown very little evidence that this disintegration will actually take place.

Rubber Latex and Resin Cements

Rubber latex cement Rubber latex cement consists of mixtures of sand and other fillers which are gauged with rubber latex solution. These cements are suitable for dilute acid conditions and are particularly useful in conditions where dilute acid alternates with water or dilute alkalis. They remain very slightly resilient and adhere very well to stoneware. They are not of course resistant to organic solvents which would normally attack rubber, nor to strongly oxidising acid solutions which cause ageing of the rubber.

Resin cements Many different resin cements are now available, and types which give satisfactory service under most conditions can be obtained[3].

Many formulae have been suggested for cement using sulphur, e.g. sulphur and sand, sulphur bitumen and sand, etc. These materials are heated in a vat until molten and then poured into the joints where they set to a hard, strong solid. They are usually very brittle and contract considerably on cooling, which may give rise to cracks. They are, however, resistant to acid, water and dilute alkalis.

Lutes

This class of material does not set and must be held in position by a rigid cement or some mechanical means. A material commonly used is asbestos fibre mixed with various other substances such as china clay, sand, etc. and plasticised with some form of oil or tar.

Fixed-flange Piping

In this method of jointing stoneware piping, the joint is formed by a gasket between ground ends of the pipe, the joint being put under compression by a metal clamp. The choice of gasket material will of course depend upon service conditions.

Rubber Sleeves

This is a fairly modern development in which the joint is formed from a rubber sleeve which rolls along the pipe as it is pushed into the spigot of the next pipe, to form a very strong joint, which is highly resistant to vibration*.

Chief Applications

Stoneware is used mainly for storage, transportation and processing, where its great resistance to corrosion is important. As it can be glazed and is completely non-reactive, it is also used in situations where cleanliness is important, as in the food industry.

Storage vessels can be made in a large variety of shapes and sizes. Absorption and distillation towers are available in several different forms and many sizes. Reaction vessels such as coppers, stills and Cellarius receivers can be supplied in the special materials to withstand thermal shock. Valves and pumps of many types are available to suit all conditions. Piping can be

*Flexadrain Joint, developed by the National Salt Glazed Pipe Man. Assn. and British Ceramic Research Assn.

supplied in a great number of bore sizes, complete with the necessary bends, and with any desired form of joint.

Porous articles for filtration, aeration, diffusion, etc. can also be made from materials having the same chemical inertness as those used for the low-porosity articles such as storage vessels. The pore size of these articles can be controlled to suit the operating conditions encountered.

Costs

Costs of stoneware articles are usually lower than those of competitive materials. The cost per pound is always much lower, but the degree of complexity of the required article and its availability or otherwise in standard stock sizes will, of course, influence the total cost. Owing to the manufacturing methods employed for stoneware, special designs may easily be produced, and hence mould charges may be very low.

G. S. SHIPLEY

REFERENCES

1. *Methods of Test for Chemical Stoneware*, British Standard 784, British Standards Institution, London (1974)
2. Corrosion Chart published as annual supplement to December issue of *Chem. Processing*
3. Roff, W. J., Scott, J. R. and Pacitti, J., *Fibres, Films, Plastics and Rubbers*, Butterworths, London (1971)

19.6 Plastics and Reinforced Plastics

For very many years organic materials have been used for coating metals to protect them against corrosion. During the present century two complementary trends may clearly be discerned. The first is the progressive replacement of natural materials with synthetic materials of high molecular weight belonging to a chemical class referred to as *polymers*. The second is the increased use of these materials, not simply as a protective coating for some material of construction, but as a material of construction itself. Thus such materials have extended from two-dimensional applications as surface coatings to three-dimensional forms, which are referred to as *plastics*.

Hence whilst in the past corrosion control has involved, in the main, the use of metal alloys, protective coatings, inhibitors, etc. corrosion problems may now often be circumvented by the use of self-supporting organic polymers in the form of either rubbers or plastics. It must however be immediately stressed that such materials are not invariably inert to chemicals and they display their own particular response to such materials. A consideration of such behaviour will be a prior object of this section.

Definition of 'Plastics'

It is in fact exceedingly difficult to provide a satisfactory definition of the term *plastics*, since attempts at reasonably concise definitions tend to include certain materials such as rubbers, adhesives, fibres, glasses and surface coatings that are not usually considered as plastics, and to exclude a number of somewhat non-typical materials such as bituminous plastics, shellac and polytetrafluorethylene which usually are. In reality then, the term becomes bounded by common usage rather than by a physical-chemical description. However, in general, it may be said that plastics are usually high polymers that at some stage in their existence are capable of flow, but may also may be brought into a non-fluid form in which they have sufficient toughness and strength to be useful in self-supporting applications. Although they may be self-supporting this does not exclude the possibility of reinforcing the plastics with fibres or of laminating them with other materials. Sometimes metals are coated with plastics but usually at a greater thickness than is common with traditional surface coatings such as paint films. (See Section 17.2.)

The rapid rise of the plastics industry since World War II may be attributed to a number of factors. Foremost has been the fact that whilst many materials of construction have been subjected to continual increases in their price, the development of the petrochemicals industry and economies of scale have, for most of the time, led to reductions in the prices of plastics materials. With the passage of time more and more products constructed from traditional materials have become cheaper to produce from plastics. Whilst economies of scale have probably almost reached their limits, and whilst the low profitability of many plastics-producing plants may cause companies to retard increases in plant and production, the trend of increased plastics usage seems bound to continue.

Such an increase also requires that plastics possess properties suitable for the end-use envisaged. Whilst it is today possible, by chemical modification or the selective use of additives, to make products varying widely in their properties, certain properties are common to the vast bulk of plastics materials. These include:

1. Tenacity. Whilst some plastics are rigid and others flexible, all commercial materials show a degree of strength and toughness superior to simple crystals and common glass when rapidly stressed.
2. Low thermal conductivity.
3. Low electrical conductivity.
4. Low heat resistance compared with common metals. The vast bulk of plastics produced will not withstand 100°C and only a very few highly specialised products will withstand 400°C.

The Chemical Nature of Plastics

Most plastics are based on polymers, which may be prepared by a variety of techniques that are briefly as follows.

Double-bond polymerisation   In this case the double bond in a small molecule opens up and allows it to join to another similar molecule, e.g. with ethylene:

$$n\mathrm{CH_2}{=}\mathrm{CH_2} \rightarrow [-\mathrm{CH_2}{-}\mathrm{CH_2}{-}]_n$$

The molecules join together to form a long chain-like molecule which may contain many thousands of ethylene units. Such a molecule is referred to as a *polymer*, in this case polyethylene, whilst in this context ethylene is referred to as a *monomer*. Styrene, propylene, vinyl chloride, vinyl acetate and methyl methacrylate are other examples of monomers which can polymerise in this way. Sometimes two monomers may be reacted together so that residues of both are to be found in the same chain. Such materials are known as *copolymers* and are exemplified by ethylene–vinyl acetate copolymers and styrene–acrylonitrile copolymers.

Condensation polymerisation   In this case reaction between two groups occurs which leads to the production of a polymer and also a simple molecule, e.g. reaction between adipic acid and hexamethylene diamine yields nylon 66 and water:

$$nHOOC(CH_2)_4COOH + nH_2N(CH_2)_6NH_2$$
$$\rightarrow [-OC(CH_2)_4CONH(CH_2)_6NH-]_n + 2nH_2O$$

Most nylons, polyesters, phenolics and a number of other plastics are produced by this route.

Rearrangement polymerisation Here the mechanism resembles condensation polymerisation but no small molecule is split out. In the first example, 1:4-butane diol reacts with hexamethylene di-isocyanate to give 6,4-polyurethane:

$$nHO(CH_2)_4OH + nOCN(CH_2)_6NCO$$
$$\rightarrow [-O(CH_2)_4OOCNH(CH_2)_6NHCO-]_n$$

In the second, ε-caprolactam, a ring compound, opens up to give nylon 6:

$$n\begin{bmatrix} CH_2 \begin{matrix} \diagup CH_2-CH_2 \diagdown \\ CO \\ | \\ NH \\ \diagdown CH_2-CH_2 \diagup \end{matrix} \end{bmatrix} \rightarrow [-(CH_2)_5CONH-]_n$$

The polymerisation of ε-caprolactam is sometimes known as a *ring-opening* polymerisation, a technique also used with ethylene oxide, tetrahydrofuran and a number of other monomers.

Polymers may also be obtained from another source—nature. Cellulose, the principal constituent of cotton and a major constituent of wood, is a polymer. So also are lignin, natural rubber, gutta percha and proteins. Sand (silica) may be considered as an inorganic polymer. Whilst many of these are of value unmodified, some, like cellulose, cannot be considered as plastics in their natural state, but if chemically modified by man useful plastics materials such as cellulose acetate, celluloid and ethyl cellulose may be obtained.

The Physical Nature of Plastics

Many polymers such as polystyrene consist of long chain-like molecules of very high molecular weight. A typical molecular weight for a polystyrene molecule is about 200 000, and since the molecular weight of the monomer is 104 there may well be 2 000 monomer units joined together in this way. Since the backbone carbon–carbon bonds can rotate freely, such molecules are most unlikely to be stretched out in straight lines but are more likely to be coiled up into a random configuration.

In the case of polystyrene the molecules at room temperature do not have enough energy to twist and move around and so in the mass the polymer is rigid. On heating above a certain temperature range sufficient energy for movement is obtained and on application of a shearing stress the polymer molecules partly uncoil, slip past each other and, in the mass, flow occurs. On cessation of stress, slippage ceases, the chains again coil up and on cooling the mass again hardens. If desired, the whole process of heating, shearing and cooling may be repeated and materials which behave in this way are known as *thermoplastics*. Two points should, however, be noted. Firstly, if cooling precedes chain recoiling then a frozen-in molecular orientation will result which can grossly affect the polymer properties, in some cases adversely. Secondly, repeated heating and shearing may be accompanied by changes such as oxidation and polymer degradation which will limit in practice the number of times heating and cooling can be undertaken on a particular polymer sample.

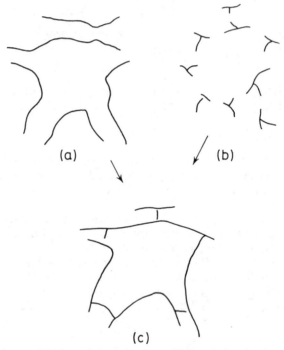

Fig. 19.2   Joining up of (*a*) long-chain molecules or (*b*) branched molecules to produce (*c*) a cross-linked polymer

In terms of tonnage the bulk of plastics produced are thermoplastics, a group which includes polyethylene, polyvinyl chloride (p.v.c.), the nylons, polycarbonates and cellulose acetate. There is however a second class of materials, the thermosetting plastics. They are supplied by the manufacturer either as long-chain molecules, similar to a typical thermoplastic molecule or as rather small branched molecules. They are shaped and then subjected to either heat or chemical reaction, or both, in such a way that the molecules link one with another to form a *cross-linked network* (Fig. 19.2). As the molecules are now interconnected they can no longer slide extensively one past the other and the material has *set*, *cured* or *cross linked*. Plastics materials behaving in this way are spoken of as *thermosetting* plastics, a term which is now used to include those materials which can in fact cross link with suitable catalysts at room temperature.

Important thermosetting plastics include the phenolics, melamine-formaldehyde, epoxides and polyester resins used in glass-reinforced plastics. (See also Sections 15.5 and 15.9.)

Thermoplastics

Thermoplastics may themselves be considered in four sub-classes: (*a*) amorphous thermoplastics, (*b*) rubber-modified amorphous thermoplastics, (*c*) plasticised amorphous thermoplastics and (*d*) crystalline thermoplastics.

Amorphous thermoplastics These are made from polymers which have a sufficiently irregular molecular structure to prevent them from crystallising in any way. Examples of such materials are polystyrene, polymethyl methacrylate and polyvinyl chloride.

At very low temperatures these materials are glass-like and rigid. On heating, a temperature is eventually reached when the material softens. If the polymer is of sufficiently high molecular weight it does not melt but becomes rubbery. The temperature at which this occurs is known as the glass transition (T_G), and is in effect the point at which the molecules have sufficient energy to be able to coil and uncoil; however, chain entanglements prevent flow. At higher temperatures there are two possibilities. Polymers of moderate molecular weight may achieve such energy that they can flow, whilst high molecular weight materials may decompose before the flow point is reached. This is shown schematically in Fig. 19.3 which indicates the phases in which this type of polymer can occur. It should be stressed that the boundary lines will change position with change of polymer species.

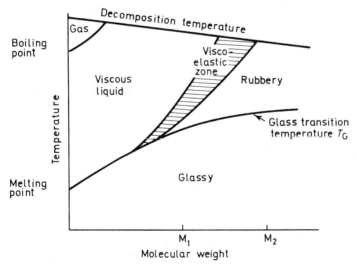

Fig. 19.3   Temperature–molecular weight phase diagram for amorphous polymers

It will be seen that a polymer of molecular weight M_1 may be processed either as a liquid (by injection moulding, extruding, etc.) or as a rubber (by vacuum forming, sheet blowing, warm forging). In the case of the polymer of molecular weight M_2 it will be seen that it can only be shaped in the rubbery state, and if it is intended to process it by say injection moulding then the molecular weight will first have to be reduced. An important illustration of this is with the well known acrylic materials such as Perspex and Oroglas which have too high a molecular weight to be injection moulded. However, lower molecular weight polymers (e.g. of molecular weight M_1 in Fig. 19.3) of similar chemical structure are available (e.g. Diakon) which are suitable for this purpose.

It is possible to make some generalisations about the properties of amorphous thermoplastics:

1. The T_G will determine the maximum temperature of use of the material as a rigid thermoplastic. For amorphous rubbers the T_G will determine the minimum temperature.
2. Below T_G most amorphous polymers show a more or less linear stress-strain curve with no yield point (Fig. 19.4).

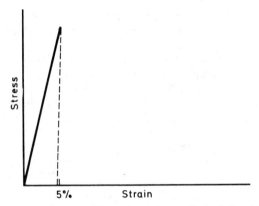

Fig. 19.4 Typical stress–strain curve of amorphous thermoplastics below their glass transition temperature. Area under curve is small compared with many crystalline plastics and hence the impact strength is usually low

The tensile strengths are about 55 MN/m², the elongations at break usually less than 10% and the modulus of elasticity about 2·7 GN/m². Since the area under the curve provides a measure of the energy required to break the bonds, and since this area is small such polymers will have a low impact strength (which is closely related to energy to break) and will break with a brittle fracture.

3. They are generally more permeable to gases than crystalline polymers, more so above T_G than below (see later).
4. They usually have a much wider range of solvents than crystalline polymers (see later).

Rubber-modified amorphous polymers   The brittleness of amorphous polymers has been a hindrance in their commercial development. Fortunately, for reasons still not fully understood, the addition of rubbery polymers as dispersed droplets, or sometimes in a network form, into the glassy polymer can often lead to substantial increases in impact strength, albeit usually at some cost to tensile strength and, in many cases, clarity. Such materials still, however, have a phase diagram similar to Fig. 19.3 except that the terms 'glassy' and 'glass-transition' temperature are no longer strictly accurate.

The most important polymers of this important class are:

1. High-impact polystyrene (polystyrene modified with styrene–butadiene rubber (SBR) or polybutadiene rubber).
2. ABS—a complex polymer based on acrylonitrile, butadiene and styrene.
3. Methacrylate–butadiene–styrene polymers (MBS) and related materials chemically similar to ABS but often available in transparent form.
4. High-impact p.v.c.—in this case the impact modifiers are not always rubbers but the mechanism of their action is probably similar.

Plasticised amorphous thermoplastics   Certain plastics may be mixed with high-boiling low-volatility liquids to give products of lower T_G. The most important example occurs with p.v.c. which is often mixed with liquids such as di-iso-octyl phthalate, tritolyl phosphate or other diesters to bring the T_G below room temperature. The resultant plasticised p.v.c. is flexible and to some degree quite rubbery. Other commonly plasticised materials are cellulose acetate and cellulose nitrate.

It is essential to appreciate that such plasticisers will considerably modify the chemical properties of the plastics material since the plasticiser may be readily extracted by certain chemicals and chemically attacked by others whilst the base polymer may be unaffected.

Crystalline thermoplastics   Whilst polymers such as polyethylene and nylon 66 do not show any regular external form which is characteristic of crystals, closer examination reveals that they have many properties common to crystalline materials. Although the exact nature of this crystallinity is still a matter of dispute it would appear that segments of polymer molecules at least pass through zones in which molecular arrangement is highly ordered, i.e. crystalline. In some ways these zones act like knots or cross-links holding the materials together.

The effect of heating crystalline polymers from low temperatures is more complex than with amorphous polymers. Initially the material is rigid and hard. As the temperature goes through the T_G, lightly crystalline material softens slightly and becomes leatherlike, but highly crystalline materials show little change in properties. Further heating results in the crystals melting over a temperature range and the polymer becomes rubbery. Whether or not it melts or decomposes first on further increases in temperature will depend on molecular weight as with amorphous polymers.

A typical phase diagram for such polymers is given in Fig. 19.5. With such crystalline polymers the melting point T_M replaces the T_G as the factor usually determining the maximum service temperature of thermoplastics and minimum service temperature of rubbers. However, being more complicated than amorphous polymers it is more difficult to make generalisations about properties. The following remarks may, however, be pertinent for crystalline polymers:

1. Below T_G, tensile strengths are usually at least as strong for crystalline polymers as for amorphous polymers. Between T_M and T_G the strength and rigidity will be very dependent on the degree of crystallinity and

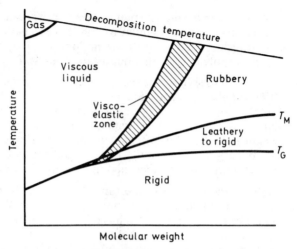

Fig. 19.5   Temperature–molecular weight phase diagram for crystalline polymers

to some extent on molecular weight, e.g. tensile strengths can range from 7 to 84 MN/m^2.

2. In most cases crystal densities differ from the densities of amorphous polymers. This leads to differences in refractive index, which in turn cause scatter of light at boundaries between amorphous and crystalline zones. Such materials are opaque except in certain instances where the crystal structure can be carefully oriented to prevent such scatter of light.

3. The close molecular packing makes diffusion more difficult than with amorphous polymers compared in similar circumstances, i.e. both below T_G or both above T_G (but below T_M of the crystalline polymer).

4. Thermodynamic considerations lead to considerable restriction in the range of solvents available for such polymers.

Thermosetting Plastics

If glycerol is reacted with phthalic anhydride three ester links can be made from each glycerol unit. Continued reaction will eventually cause the molecules to link up in a three-dimensional network in which, theoretically at least, the whole polymer mass becomes one giant molecule.

For reasons of production feasibility such cross-linked plastics are normally prepared in two stages. In the first stage either a low molecular weight branched polymer or a higher molecular weight linear (unbranched) polymer is produced. Such materials are thermoplastic and in most cases soluble in appropriate solvents. By activation of heat and/or additional chemical reactants, the branched molecules join together, or the linear polymers cross link, to produce an infusible, insoluble material. Since, in the early days of the plastics industry, the cross-linking processes required heat these materials became known as *thermosetting* plastics. Today this term is commonly extended to materials which can cross link at room temperature.

Because thermosetting plastics have an irregular form they are amorphous and because of the network structure are invariably rigid. They do not dissolve without decomposition but may swell in appropriate solvents, the amount of swelling decreasing with increased cross-link density.

Well-known thermosetting plastics include the phenolics, urea-formaldehyde and melamine-formaldehyde plastics, polyesters and epoxides.

Reinforced Plastics

The mechanical properties of plastics materials may often be considerably enhanced by embedding fibrous materials in the polymer matrix. Whilst such techniques have been applied to thermoplastics the greatest developments have taken place with the thermosetting plastics. The most common reinforcing materials are glass and cotton fibres but many other materials ranging from paper to carbon fibre are used. The fibres normally have moduli of elasticity substantially greater than shown by the resin so that under tensile stress much of the load is borne by the fibre. The modulus of the composite is intermediate to that of the fibre and that of the resin.

In addition to the nature of resin and fibre, the laminate properties also depend on the degree of bonding between the two main components and the presence of other additives including air bubbles. Because of this some parts, fabricated by simple hand building techniques, may exhibit strengths no better or even worse than unreinforced materials. This problem is often worst with glass fibres which are therefore normally treated with special finishes to improve the resin–glass bond.

The highest mechanical strengths are usually obtained when the fibre is used in fine fabric form but for many purposes the fibres may be used in mat form, particularly glass fibre. The chemical properties of the laminates are largely determined by the nature of the polymer but capillary attraction along the fibre–resin interface can occur when some of these interfaces are exposed at a laminate surface. In such circumstances the resistance of both reinforcement and matrix must be considered when assessing the suitability of a laminate for use in chemical plant. Asbestos and glass fibres are most commonly used for chemical plant, the former usually in conjunction with phenolic resins and the latter with furane, epoxide and, sometimes, polyester resins.

Polymer Orientation

It is not very difficult to appreciate that if polymer molecules are aligned as in Fig. 19.6a then a much higher tensile strength will be obtained if a test is carried out in the X–X direction as opposed to the Y–Y direction. It is also not difficult to understand why such a material has a lower impact strength than a randomly coiled mass of molecules (Fig. 19.6b) because of the ease of cleavage of the material parallel to the X–X direction.

Similar remarks may also be made where crystal structures rather than the individual molecules are aligned.

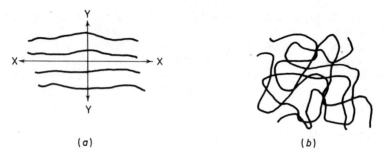

Fig. 19.6　Polymer alignment. (*a*) Parallel and (*b*) random

For fibres and filaments such orientation is desirable, but for solid objects where impact strength is often more important than tensile strength such orientation is usually unwelcome. It can also have further unwanted effects. This arises from the fact that oriented molecules are basically unstable and will at the first opportunity try to coil up. Thus on heating samples up to temperatures near T_G severe distortion can occur leading to warped mouldings.

Another serious effect occurs with liquids which are not in themselves solvents but which may wet the polymer surfaces. These facilitate relief of frozen-in stresses by surface cracking which can be a severe problem in using many injection and blow mouldings with specific chemicals. Examples of this are white spirit with polystyrene, carbon tetrachloride with polycarbonates and soaps and silicone oils with low molecular weight polyethylenes.

In addition to orientation in one direction (mono-axial orientation), biaxial orientation is possible. This is achieved when sheet is stretched in two directions resulting in layering of the molecules. This can increase the impact strength, tensile strength and solvent cracking resistance of polymers and with crystalline plastics the polymer clarity may also be improved.

To summarise this section let it be said that it is necessary for the technologist to control orientation, not the least because of the effect that this might have on stress-cracking that may occur in aggressive chemical environments.

The Chemical Properties of Polymers

It is common practice to talk about the chemical resistance of polymers without it always being appreciated that this can mean different things. To avoid this it is probably wiser to differentiate between:
   1. Resistance of a polymer to chemical attack resulting in breakdown of some covalent bonds and formation of new ones. This could involve breakdown initiated by heat and radiation including u.v. light.
   2. Resistance to dissolution by liquids.
   3. Resistance to cracking in aggressive environments.
   4. Permeability to gases and liquids.

It is also important to bear in mind that, for end use, polymers may be mixed with a number of additives such as plasticisers, stabilisers, antioxidants, fillers, fire-retardants, pigments and so on, and that this may well have an important influence on chemical properties. The following discussion will however confine itself to pure polymers.

Resistance to Chemical Attack

Expressed simply the resistance of a polymer to chemical attack may be said to be determined by the following factors:
(a) Nature of the chemical bonds present.
(b) Interaction between chemical groups which occur repeatedly along the molecular chain.
(c) Presence of occasional 'weak-links'.

In commercial polymers there are a rather limited number of chemical bonds to be found and it is possible to make a number of general observations about the chemical reactivity in the following tabulated list of examples[1]:

1. Polyolefins such as polyethylene and polypropylene contain only C—C and C—H bonds and may be considered as high molecular weight paraffins. Like the simpler paraffins they are somewhat inert and their major chemical reaction is substitution, e.g. halogenation. In addition, the branched polyethylenes and all the higher polyolefins contain tertiary carbon atoms which are reactive sites for oxidation. Because of this it is necessary to add antioxidants to stabilise the polymers against oxidation. Some polyolefins may be cross linked by peroxides.
2. Polytetrafluoroethylene contains only C—C and C—F bonds. These are both very stable and the polymer is exceptionally inert. A number of other fluorine-containing polymers are available which may contain in addition C—H and C—Cl bonds. These are somewhat more reactive, and those containing C—H bonds may be cross linked by peroxides and certain diamines and di-isocyanates.
3. Many polymers, such as the diene rubbers, contain double bonds. These will react with many agents such as oxygen, ozone, hydrogen halides and halogens. Ozone, and in some instances oxygen, will lead to scission of the main chain at the site of the double bond and this will have a catastrophic effect on the molecular weight. The rupture of one such bond per chain will halve the average molecular weight.
4. Ester, amide and carbonate groups are susceptible to hydrolysis. When such groups are found in the main chain, their hydrolysis will also result in a reduction of molecular weight. Where hydrolysis occurs in a side chain the effect on molecular weight is usually insignificant. The presence of benzene rings adjacent to these groups may offer some protection against hydrolysis except where organophilic hydrolysing agents are employed.
5. Hydroxyl groups are extremely reactive. These occur attached to the backbone of the cellulose molecule and polyvinyl alcohol.
6. Benzene rings in both the skeleton structure and on the side groups can be subjected to substitution reactions. Such reactions do not normally cause great changes in the fundamental nature of the polymer, e.g. they seldom lead to chain scission or cross linking. (N.B. The phenolic resins provide an important exception here.)

There are numerous examples of chemical reactions consequent upon chemical groups which occur repeatedly along a chain. In some cases the reaction occurs randomly between adjacent pairs of groups such as in the reaction between aldehydes and polyvinyl alcohol and of zinc dust with polyvinyl chloride:

$$----CH-CH_2-CH-CH_2-CH-CH_2-CH-CH_2-CH----$$

with pendant OH groups and $R-CHO$ reacting to form

and

$$----CH_2-CH-CH_2-CH-CH_2----$$
$$ClCl$$

$$\xrightarrow[\text{dust}]{Zn} \ ----CH_2-CH\overset{\overset{\textstyle CH_2}{\diagdown\diagup}}{}CH-CH_2---- \ +ZnCl_2$$

In other instances the reactions appear to occur in sequence down the chain, for example in the depolymerisation reaction of polyformaldehyde (polyacetal) and polymethyl methacrylate which are referred to as *zippering* or sometimes *unzippering* reactions. In other cases cyclisation reactions can occur such as on heating polyacrylonitrile:

$$----CH_2-CH-CH_2-CH----$$
$$CNCN$$

$$\longrightarrow$$

ring structure with CH_2, CH, C, N groups.

It is commonly found that polymers are less stable particularly to molecular breakdown at elevated temperatures than low molecular weight materials containing similar groupings. In part this may be due to the constant repetition of groups along a chain as discussed above, but more frequently it is due to the presence of weak links along the chain. These may be at the end of the chain (terminal) arising from specific mechanisms of chain initiation and/or termination, or non-terminal and due to such factors as impurities which become built into the chain, a momentary aberration in the *modus operandi* of the polymerisation process, or perhaps, to branch points.

The combination of weak links and unzipping can be catastrophic and has been a particular problem in the commercial development of some polymers, in particular polyacetals.

Polymer Solubility

The solution properties of polymers have been subjected to intensive study, in particular to highly complex mathematical treatment[2,3]. This section will, however, confine discussion to a qualitative and practical level[1,6].

One chemical will be a solvent for another if the molecules are able to co-exist on a molecular scale, i.e. the molecules show no tendency to separate. In these circumstances we say that the two species are *compatible*. The definition concerns equilibrium properties and gives no indication of the rate of solution which will depend on other factors such as temperature, the molecular size of the solvent and the size of voids in the solute.

Molecules of two different species will be able to co-exist if the force of attraction between different molecules is not less than the forces of attraction between two like molecules of either species. This is shown more clearly by reference to Fig. 19.7 which shows two types of molecules, A and B. The average forces between the like molecules are F_{AA} and F_{BB}, and the average forces between dissimilar molecules are F_{AB}. If F_{AA} was the largest of these three forces then the A molecules would tend to congregate or cohere, pushing away the B molecules. A similar phase separation would occur if F_{BB} was the greatest.

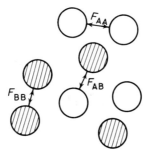

Fig. 19.7   Two different molecular species will be compatible if $F_{AB} \geqslant F_{AA}$ and $F_{AB} \geqslant F_{BB}$. In other circumstances the molecules will tend to separate if they have sufficient energy for molecular movement

It is therefore seen that only when

$$F_{AB} \geqslant F_{AA} \qquad \qquad \ldots(19.2)$$

and

$$F_{AB} \geqslant F_{BB} \qquad \qquad \ldots(19.3)$$

will co-existence or compatibility be possible. Obviously if it is possible to obtain some measure of these forces it should be possible to make predictions about polymer solubility. What then is a suitable measure of the forces holding like molecules together? One would expect the latent heat of vaporisation L to exceed that cohesion energy by an amount corresponding to the mechanical work done by evaporation, an amount approximating to RT where R is the gas constant and T the absolute temperature. Such a figure of $(L - RT)$ might be a sufficient measure if all of the molecules were of about the same size.

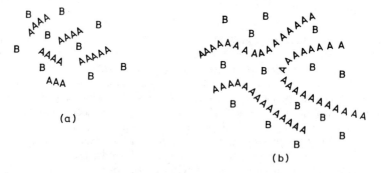

Fig. 19.8   Polymer molecules. (a) Short and (b) long

However, it is reasonable to suppose that compatibility should not be greatly affected by molecular size and that the shorter polymer molecules in Fig. 19.8a should be just as compatible as the longer molecules in Fig. 19.8b, although their latent heats of vaporisation will be greatly different. In such circumstances a reduced figure of $(L - RT)/M$ will give a measure of intermolecular energy per unit weight. Similarly, a measure of the intermolecular or cohesion energy per unit volume will be given by the expression:

$$\frac{L - RT}{M/D}$$

This expression is known as the *cohesive energy density* and in S.I. is expressed in units of joules per cubic centimetre. The square root of this expression is more commonly encountered in quantitative studies and is known as the *solubility parameter* and given the symbol δ, i.e.:

$$\delta = \sqrt{\frac{L - RT}{M/D}} \qquad (\text{J cm}^{-3})^{\frac{1}{2}}$$

The solubility parameter is thus an experimentally determinable property, at least for low molecular weight materials, and a number of methods have been reviewed by Burrell[4]. In the case of polymers which cannot normally be vaporised without decomposition, a method of calculating from a knowledge of structural formula has been devised by Small[5]. An excellent review of the whole subject is given by Gardon[6].

It is now possible to provide an estimate of F_{AA} and F_{BB}, but the magnitude of F_{AB} will have to be considered separately for different systems.

Amorphous non-polar polymers and amorphous non-polar solvents It is assumed in these circumstances, by analogy with gravitational and electrostatic attraction, that F_{AB} will be equal to the geometric mean of F_{AA} and F_{BB}. If by definition we take $F_{AA} > F_{BB}$ then:

$$F_{AA} > F_{AB} > F_{BB} \qquad \ldots(19.4)$$

If we now consider the inequalities given in equations 19.2 to 19.4 it will be seen that compatibility can only occur between amorphous non-polar polymers and solvents when:

$$F_{AA} = F_{AB} = F_{BB}$$

i.e., when polymer and solvent have similar solubility parameters (in practice within about $2(\text{J cm}^{-3})^{\frac{1}{2}}$.

Reference to the values of δ in Tables 19.17 and 19.18 provides a number of examples of this. For example, natural rubber (unvulcanised) ($\delta = 16\cdot5$) is dissolved by toluene ($\delta = 18\cdot2$) and carbon tetrachloride ($\delta = 17\cdot5$), but not ethanol ($\delta = 25\cdot9$). Cellulose diacetate ($\delta = 23\cdot2$) is soluble in acetone ($\delta = 20\cdot4$), but not methanol ($\delta = 29\cdot6$) or toluene ($\delta = 18\cdot2$).

It should be noted that apart from the problem of achieving a molecular level of dispersion it is not necessary for the solvent to be liquid—it may be an amorphous solid.

Such tables are of greatest use with non-polar materials with values of δ not exceeding $19\cdot4$ and when the polymers are amorphous. It is now necessary to consider other systems.

Table 19.17 Solubility parameters of polymers

Polymer	$\delta \, (\text{J cm}^{-3})^{\frac{1}{2}}$	Polymer	$\delta \, (\text{J cm}^{-3})^{\frac{1}{2}}$
Polytetrafluorethylene	12·6	Polyethyl acrylate	18·8
Polychlorotrifluorethylene	14·7	Polysulphide rubber	18·4–19·2
Polydimethyl siloxane	14·9	Polystyrene	18·8
Ethylene–propylene rubber	16·1	Polychloroprene rubber	18·8–19·2
Polyisobutylene	16·1	Polymethyl methacrylate	18·8
Polyethylene	16·3	Polyvinyl acetate	19·2
Polypropylene	16·3	Polyvinyl chloride	19·4
Polyisoprene (natural rubber)	16·5	Bisphenol A polycarbonate	19·4
Polybutadiene	17·1	Polyvinylidene chloride	20·0–24·9
Styrene–butadiene rubber	17·1	Ethylcellulose	17·3–21·0
Poly-t-butyl methacrylate	16·9	Cellulose dinitrate	21·5
Poly-n-hexyl methacrylate	17·5	Polyethylene terephthalate	21·8
Poly-n-butyl methacrylate	17·7	Acetal resins	22·6
Polybutyl acrylate	18·0	Cellulose diacetate	23·1
Polyethyl methacrylate	18·4	Nylon 66	27·7
Polymethylphenyl siloxane	18·4	Polymethyl α-cyanoacrylate	28·8
		Polyacrylonitrile	28·8

Note: because of difficulties in their measurement, published figures for a given polymer can range up to 3% on either side of the average figure quoted. More comprehensive data are given in Reference 6.

Table 19.18   Solubility parameters and partial polarities (P) of some common solvents

	δ (J cm^{-3})$^{\frac{1}{2}}$	P		δ (J cm^{-3})$^{\frac{1}{2}}$	P
Neo-pentane	12·9	0	Tetrachlorethane	19·2	0·01
Isobutylene	13·7	0	Tetralin	19·4	—
n-hexane	14·9	0	Carbitol	19·6	—
Diethyl ether	15·1	0·03	Methyl chloride	19·8	—
n-octane	15·5	0	Methylene chloride	19·8	—
Methyl cyclohexane	15·9	0	Ethylene dichloride	20·0	0
Ethyl isobutyrate	16·1	—	Cyclohexanone	20·2	—
Di-isopropyl ketone	16·3	0·3	Cellosolve	20·2	—
Methyl amyl acetate	16·3	—	Dioxane	20·2	0·01
Turpentine	16·5	0	Carbon disulphide	20·4	0
Cyclohexane	16·7	0	Acetone	20·4	0·69
2,2-dichloropropane	16·7	—	n-octanol	21·0	0·04
Sec. amyl acetate	16·9	—	Butyronitrile	21·4	0·72
Dipentene	17·3	0	n-hexanol	21·8	0·06
Amyl acetate	17·3	0·07	Sec. butanol	22·0	0·11
Methyl n-butyl ketone	17·5	0·35	Pyridine	22·2	0·17
Pine oil	17·5	—	Nitroethane	22·6	0·71
Carbon tetrachloride	17·5	0	n-butanol	23·3	0·10
Methyl n-propyl ketone	17·7	0·4	Cyclohexanol	23·3	0·08
Piperidine	17·7	—	Isopropanol	23·4	—
Xylene	18·0	0	n-propanol	24·3	0·15
Dimethyl ether	18·0	—	Dimethyl formamide	24·7	0·77
Toluene	18·2	0	Hydrogen cyanide	24·7	—
Butyl cellosolve	18·2	—	Acetic acid	25·7	0·30
1,2-dichloropropane	18·4	—	Ethanol	25·9	0·27
Mesityl oxide	18·4	—	Cresol	27·1	—
Isophorone	18·6	—	Formic acid	27·5	—
Ethyl acetate	18·6	0·17	Methanol	29·6	0·39
Benzene	18·8	0	Phenol	29·6	0·06
Diacetone alcohol	18·8	—	Glycerol	33·7	0·47
Chloroform	19·0	0·02	Water	47·7	0·82
Trichloroethylene	19.0	0			

Note: a comprehensive list of solubility parameters is given in Reference 6.

Crystalline non-polar polymers and amorphous solvents   Most polymers of regular structure will crystallise if cooled below a certain temperature, i.e. the melting point T_{M}. This is in accordance with the thermodynamic law that a process will only occur if there is a decrease in Gibbs free energy $(-\Delta F)$ in going from one state to another. Such a decrease occurs on crystallisation as the molecules pack regularly.

Since a process only occurs when accompanied by a decrease in free energy there is no reason why a crystalline non-polar polymer should dissolve in a solvent at temperatures well below the melting point. However, as the melting point is approached the $T\Delta S$ term in the equation:

$$\Delta F = \Delta H - T\Delta S$$

increases (where T is the absolute temperature, ΔS the entropy change and ΔH the enthalpy change) so ΔF can become negative and solution may occur.

Hence at room temperature there are no solvents for polyethylene, polypropylene, poly-4-methyl-pentene-1, polyacetal or polytetrafluoroethylene, but at temperatures of about 30°C below their melting point solvents of similar solubility parameter are effective. It should also be noted that at room temperature swelling may occur in the amorphous zones of the polymer in the presence of solvents of similar solubility parameter.

Amorphous non-polar polymers and crystalline solvents This situation is identical to the previous one and occurs, for example, when paraffin wax is mixed into rubber above its melting point. On cooling the paraffin wax tends to crystallise, some of it on the surface of the rubber. Such a bloom is one way of protecting a diene rubber from ozone attack.

Amorphous polar polymers and solvents Molecules are held together by one or more of four types of forces, *viz.* dispersion, dipole, induction and hydrogen bonding. In the case of aliphatic hydrocarbons the so-called *dispersion* forces predominate. However, with many chemicals certain of the covalent bonds present are unevenly balanced, with one end being positively charged and the other negatively charged. Such *dipoles*, as they are called, can interact with dipoles on other molecules and lead to enhancement of the total inter-molecular attraction. Molecules which possess such dipoles and interact in this way by dipole forces are said to be *polar*. Many well-known solvents and polymers are polar, and it is generally considered that for interaction both the solubility parameter and their degrees of polarity should match. This is usually expressed in terms of partial polarity[6] which expresses the fraction of total forces due to dipole bonds. Some figures for partial polarities of solvents are given in Table 19.18 but there is a serious lack of quantitative data on the partial polarities of polymers. At the present time a comparison of polarities has to be made by a common sense rather than a quantitative approach. For example, hydrocarbon polymers would be expected to have a negligible polarity and would then be more likely to dissolve in toluene rather than diethyl ketone, although both have similar solubility parameters.

Crystalline polar polymers and solvents It has already been pointed out that at temperatures well below their melting point crystalline non-polar polymers will not interact with solvents, and similar considerations can apply to a large number of polar crystalline polymers. It has, however, been possible to find solvents for some polar, crystalline polymers such as the nylons, polyvinyl chloride and the polycarbonates. This is because of specific interactions between polymer and solvent that may often occur, e.g. by hydrogen bonding.

For example, nylon 66 will dissolve in formic acid, glacial acetic acid, phenol and cresol—four solvents which not only have similar solubility parameters but also are capable of acting as proton donors whilst the carbonyl groups on the nylon act as proton acceptors:

A more interesting example is given with p.v.c. and the polycarbonate of bis-phenol *A*, both slightly crystalline polymers. It is noticed here that whilst methylene chloride is a good solvent and tetrahydrofuran a poor solvent for the polycarbonate, the reverse is true for p.v.c. yet all four materials have similar solubility parameters. It would seem that the explanation is that a form of hydrogen bonding occurs between the polycarbonate and methylene chloride and between p.v.c. and tetrahydrofuran. In other words there is a specific interaction between each solvent pair.

$$
\begin{array}{ccccc}
 & CH_2 & CH_2{-}CH_2 & & R & \\
 & | & \diagup & & | & \\
 & | & & O & H \\
Cl{-}C{-}H\cdots O & & | & | & | \\
 & | & \diagdown & C{=}O\cdots H{-}C{-}Cl \\
 & CH_2 & CH_2{-}CH_2 & O & Cl \\
 & | & & | &
\end{array}
$$

Many studies have been made to try to assess the propensity to hydrogen bonding of chemical structures. As a result the following broad generalisations may be made:

1. *Proton donors* include highly halogenated compounds such as chloroform and pentachlorethane. Less halogenated materials are weaker donors.
2. *Polar acceptors* include, in roughly descending order of strength, amines, ethers, ketones, aldehydes and esters, with aromatic materials usually being more powerful than aliphatic.
3. Some materials such as water, alcohols, carboxylic acids and primary and secondary amines may be able to act simultaneously as proton donors and acceptors. Cellulose and polyvinyl alcohol are two polymers which also function in this way.
4. A number of solvents such as the hydrocarbons, carbon disulphides and carbon tetrachlorides are quite incapable of forming hydrogen bonds.

Rubbers and thermosetting plastics   The conventionally covalently cross-linked rubbers and plastics cannot dissolve without chemical change. They will, however, swell in solvents of similar solubility parameter, the degree of swelling decreasing with increase in cross-link density. The solution properties of the thermoelastomers which are two-phase materials are much more complex and dependent on whether or not the rubber phase and the resin domains are dissolved by the solvent.

Resistance to Cracking in Aggressive Environments

It has been found with many rigid plastics materials that under stresses well below the normal yield stress, cracking occurs in environments where when free of stress the polymer will be unaffected. The mechanism for this stress-cracking phenomenon is not well understood and indeed there may well be many different mechanisms in different circumstances. There do, however, appear to be four main types:

1. 'Solvent' cracking of amorphous polymers.
2. 'Solvent' cracking of crystalline polymers.
3. Environmental stress cracking.
4. Thermal cracking.

Of the instances of so-called 'solvent' cracking of amorphous polymers known to the author, the liquid involved is not usually a true solvent of the polymer but instead has a solubility parameter on the borderline of the solubility range. Examples are polystyrene and white spirit, polycarbonate and methanol and ethyl acetate with polysulphone. The propensity to solvent stress cracking is however far from predictable and intending users of a polymer would have to check on this before use.

In the case of crystalline polymers it may be that solvents can cause cracking by activity in the amorphous zone. Examples of this are benzene and toluene with polyethylene. In polyethylene, however, the greater problem is that known as 'environmental stress cracking', which occurs with materials such as soap, alcohols, surfactants and silicone oils. Many of these are highly polar materials which cause no swelling but are simply absorbed either into or on to the polymer. This appears to weaken the surface and allows cracks to propagate from minute flaws.

Thermal cracking appears to be similar, but in this case air is the aggressive environment, becoming active at 70–80°C with some polyethylenes.

The whole of this important subject is reviewed in detail by Howard[7].

Diffusion

The efficacy of polymers when used to protect metals from corrosive environments is influenced by their efficiency as barrier materials. When applied to metals by some techniques, such as fluidised bed coating, there is always the danger of 'macro-diffusion' through 'pinholes' which are gross imperfections in the surface and which do not have to be visible to be very much greater than the dimension of penetrating molecules.

Assuming, however, that the film is continuous then the concern is with the permeability of the polymer to the corrosive fluids. This involves both the dissolution of fluid into the polymer, which will be determined by the conditions discussed previously, and the rate of diffusion of the fluid through the polymer. This has been discussed elsewhere[1,8] in detail, but may be summarised as follows:

1. The lowest diffusion rates occur with crystalline polymers below the T_G, since there is very little space through which diffusing molecules may pass.
2. Amorphous polymer below the T_G has a somewhat higher permeability, but diffusion is still difficult.
3. For amorphous polymers above the T_G, i.e. in the flexible and rubbery states there is more space available through which diffusing molecules may pass, and so these materials show comparatively high diffusion rates with diffusing fluids.
4. For crystalline polymers between T_G and T_M the diffusion rate is very dependent on the degree of crystallisation.

Review of Commercial Plastics[1]

Amorphous Thermoplastics

Polyvinyl chloride (p.v.c.) P.V.C. is one of the two most important plastics in terms of tonnage and shows many properties typical of rigid amorphous thermoplastics. More individually, it softens at about 70°C, burns only with difficulty and is thermally unstable. To reduce this instability, stabilisers are invariably compounded into the polymer.

Being slightly crystalline, there are few good solvents, the best known of which are nitrobenzene, cyclohexanone and tetrahydrofuran. When mixed with certain non-volatile solvents such as some phthalates, adipates and phosphates, flexible materials are obtained and which are referred to as *plasticised p.v.c.*

In order to improve toughness many rubbers and other soft polymers may be used as additives to modify the compound. Some copolymers based on vinyl chloride are available of which the most important are the vinyl chloride–vinyl acetate materials used in gramophone records, flooring compositions and surface coatings.

Polystyrene (PS) The volume of expanded polystyrene produced probably exceeds the volume production of all other plastics (excluding the polyurethanes) put together. At least half the weight of polystyrene produced is in the form of high impact polystyrene (HIPS)—a complex blend containing styrene-butadiene rubber or polybutadiene.

Styrene-acrylonitrile has improved heat resistance, oil resistance and slightly better impact strength.

Acrylonitrile-styrene-butadiene polymers (ABS) These are complexes of blends and copolymers of excellent toughness. Some recent modifications show a degree of transparency.

Poly(methyl methacrylate) Used mainly where high light transmission and excellent weathering properties are of greatest importance. This polymer is the most well known of a very wide range of acrylic polymers which find uses as rubbers, fibres, plastics, adhesives and surface coatings. The sheet forms (Perspex, Plexiglas, Oroglas) are often of high molecular weight and dissolve only with some difficulty.

Methacrylates are components of AMBS and MBS plastics which are similar to ABS; they have improved clarity.

Polyvinyl acetate and derivatives Polyvinyl acetate is used largely for coating applications, but the derivative polyvinyl alcohol, will, providing there are some residual acetate groups, dissolve in water. Reaction products of polyvinyl alcohol with aldehydes such as polyvinyl formal and polyvinyl butyral are highly specialised materials.

Cellulose plastics These old established materials have limited chemical resistance. Ethyl cellulose is, however, often used in conjunction with mineral oil for hot melt strippable coatings for protecting metal parts against corrosion and marring during shipment and storage.

Crystalline Plastics

Polyethylene With p.v.c., polyethylene vies as the most important plastics material in terms of tonnage. It is attacked by only a few chemicals, it is not swollen by water or solvents, but it is susceptible to environmental stress-cracking in the presence of certain detergents, esters, alcohols and silicones. Commercial materials vary in the regularity of their structure, the more regular grades having a higher density, rigidity and a lower gas permeability.

Polypropylene It has similar chemical properties to polyethylene but is less susceptible to environmental stress cracking. It may also be used at somewhat higher temperatures.

Other polyolefins A variety of other crystalline polyolefins are available such as polybutene-1 (improved creep resistance over polyethylene), poly-4-methyl pentene-1 (excellent temperature deformation resistance) and ethylene-vinyl acetate (greater flexibility).

Polytetrafluorethylene (p.t.f.e.) This polymer does not absorb water, has no solvents and is almost completely inert to chemical attack; molten alkali metals and sodium in liquid ammonia are the rare exceptions. Furthermore it does not soften below 320°C, is electrically inert and has a very low co-efficient of friction. It is more expensive than general purpose plastics, requires special fabrication techniques, is degraded by high energy radiation, and has a low creep resistance.

Other fluorine-containing plastics These materials, in general, attempt to compromise between the exceptional end-use properties of p.t.f.e. and the processability of ordinary thermoplastics. Examples include polychlortrifluorethylene, tetrafluorethylene–hexafluorpropylene copolymers (FEP resins) and polyvinylidene fluoride. Polyvinyl fluoride is available in film form (Tedlar) with excellent weathering resistance.

Polyamides (nylons) The main types of nylon are oil and petrol resistant, but on the other hand susceptible to high water absorption and to hydrolysis. There are a few solvents such as phenol, cresol and formic acid. Special grades include a water-soluble nylon, amorphous copolymers and low molecular weight grades used in conjunction with epoxide resins. An aromatic amorphous polyamide Trogamid T is also now available.

Polyformaldehydes (polyoxymethylenes, polyacetals) These are physically similar to general purpose nylons but with greater stiffness and lower water absorption. There are no solvents, but swelling occurs in liquids of similar solubility parameter. Poor resistance to u.v. light and limited thermal stability are two disadvantages of these materials.

Other polyethers The chlorine-containing polyether 'Penton' has excellent resistance to mineral acids, strong alkalis and most common solvents. It is not recommended for use with oxidising acids such as fuming nitric acid.

Poly-2,6-dimethyl phenylene oxide (PPO) and certain related materials are similar to the nylons but have superior heat resistance. These polymers are somewhat liable to stress-cracking problems.

Several polyethers are used as intermediates in the preparation of polyurethane foams whilst others such as polyethylene glycol are water soluble.

Linear polyesters   Polyesters may be obtained in a wide variety of forms including rubbers, fibres, films, laminating resins, surface coatings and thermoplastic moulding powders. The last named are somewhat similar to the nylons but are more rigid. Chemical applications would appear to be limited because of their sensitivity to alkaline solutions and hot water.

Polycarbonates and polysulphones   These are tough materials with heat resistance better than most thermoplastics. They are resistant to attack by acids and alcohols but the polycarbonates are sensitive to alkalis.

Thermosetting Resins

These materials often have better heat resistance than thermoplastics. Thermosetting resins are used in a variety of guises including surface coatings, but as plastics they are most frequently used in moulding compositions and in laminates. The tensile strength of unfilled rigid thermosetting resins is of the same order as for amorphous thermoplastics, i.e. about $55\,MN/m^2$, but this figure may be greatly affected by the choice of fillers. For example, polyester-glass chopped-mat laminates often have tensile strengths in excess of $100\,MN/m^2$ whilst figures several-fold higher may be achieved by using carbon or boron fibres with epoxide resins.

Phenol-formaldehyde (phenolic) plastics   The chemical resistance is affected by the phenol used, cresols giving the best acid resistance whilst xylenols are often used to obtain the best alkali resistance. For chemical-resistant applications the fillers used in moulding powder and reinforcing material in laminates should be inorganic, e.g. asbestos or glass. The resins are usually dark in colour.

Aminoplastics   In this group, melamine-formaldehyde resins with their good heat resistance, scratch resistance and stain resistance, are usually preferred to urea-formaldehyde resins where chemical resistance is important. Unlike the phenolics these materials are not restricted to dark colours.

Polyesters   It is possible to prepare polyester-based laminates without application of external heat and pressure thus facilitating the manufacture of large objects using simple equipment. The laminates have somewhat limited chemical resistance, being attacked by many acids, alkalis and organic solvents. Glass fibre is the common reinforcing agent and in some products there is a tendency for capillary action to occur up the bunches of fibre strands. Distilled water in general is more active in this respect than aqueous salt solutions including sea-water.

Epoxide and furan resins   These materials, somewhat more expensive than the polyesters, may also be fabricated without the use of pressure, and at ambient temperatures if so desired. They have markedly superior chemical resistance to the polyesters, particularly to alkalis.

Silicones   This term is given to a wide range of polymers including fluids, rubbers and thermosetting resins. Although rather expensive relative to most other plastics they are particularly noted for their thermal stability and their water repellency.

Polyurethanes   These are another class of polymers that are available in a wide range of forms, including rigid and flexible rubbers, surface coatings and adhesives. The solid polymers including the rubbers have particularly good abrasion resistance. The polyurethanes as a class are somewhat lacking in resistance to acids and alkalis, and the prolonged action of water and steam.

J. A. BRYDSON

REFERENCES

1. Brydson, J. A., *Plastics Materials*, 3rd Edn, Butterworths, London (1974)
2. Tompa, H., *Polymer Solutions*, Butterworths, London (1956)
3. 'Polymer Solutions', *Discussions of the Faraday Society*, No. 49, London (1971)
4. Burrell, H., *Interchemical Review*, **14**, 3 (1955)
5. Small, P. A. J., *J. Appl. Chem.*, **3**, 71 (1953)
6. Gardon, J. L., 'Cohesive Energy Density', article in *Encyclopaedia of Polymer Science and Technology*, **3**, Interscience, New York, 833 (1969)
7. Howard, J. B., *Engineering Design for Plastics*, Ed. by E. Baer, Reinhold, New York (1964)
8. Crank, J. and Park, J. S., *Diffusion in Polymers*, Academic Press, London and New York (1968)

19.7 Rubber and Synthetic Elastomers

The distinctive property of rubbers or elastomers is a special type of resilience which is difficult to define but easy to recognise in such typical applications as the catapult and the squash ball. In addition, rubbers tend to be chemically inert, and it is this property which makes them useful for the protection of metals against corrosion.

Natural rubber is derived from a white milky fluid (latex) which oozes out of a number of trees, shrubs and plants when their stems are cut. The most familiar in Great Britain is the dandelion, which, however, is of no commercial value for this purpose. The chief commercial source is a tree, *Hevea brasiliensis*, native to the Amazon valley, now cultivated chiefly in Malaya and Indonesia, but also in west Africa and south America. The bark is cut (tapped) and the latex collected; on the addition of acid the raw rubber coagulates in a form similar to the crêpe rubber used for shoe soles.

In 1839 Charles Goodyear in America discovered that raw rubber when mixed with finely divided sulphur and heated for some time to temperatures between 100 and 200°C turned into the tough resilient material which has since become familiar. The process is called *vulcanisation*. It can be speeded up by the addition of accelerators (e.g. cyclohexyl benzthiazyl sulphonamide, etc.). The final product can be made harder or stiffer and its characteristics altered by the addition of reinforcing agents or fillers, such as carbon black, or finely divided silica. The tendency of the rubber to deteriorate with age or exposure to sunlight (i.e. to perish) can be counteracted by the addition of antioxidants (e.g. phenyl-β-naphthylamine). Improvements in processing behaviour can be obtained by the addition of plasticisers such as pine tar, etc. Other chemicals which may be added include activators working in conjunction with the vulcanising agent and the accelerator, antiozonants, antiodorants, pigments, etc. The whole technique is called *compounding*, and the properties of the final compound can be varied over a wide range by altering the nature and proportions of the compounding ingredients.

Raw natural rubber is a *cis*-polyisoprene which structurally consists of long chains of the repeating unit:

$$-CH_2-\underset{\underset{CH_3}{|}}{C}=\underset{\underset{H}{|}}{C}-CH_2-$$

There are anything from 4 000 to 5 000 of these units per molecule, and the whole chain is very much kinked and twisted.

During vulcanisation, sulphur atoms combine with carbons at the double bonds, forming occasional crosslinks between adjacent chains. Thus the whole structure, from being a tangle of kinked but sliding threads, becomes a three-dimensional elastic network. Normally about 2 or 3% of sulphur is used. If, however, 20% or more is used, so that practically all the possible crosslinks are formed, the material becomes hard and black, and is known as *ebonite* (formerly vulcanite) or hard rubber.

Synthetic rubbers are also used for protection against corrosion, the chief ones being:

1. *Cis*-polyisoprene or 'synthetic natural' rubber, which has the same chemical structure as natural rubber.

2. Styrene-butadiene rubber (SBR). Molecules are made up of units derived from butadiene

$$-CH_2-CH=CH-CH_2-$$

and from styrene

$$-CH-CH=CH-CH_2-$$
$$|$$
$$C_6H_5$$

roughly in the proportion 7:1.

3. Polychloroprene or neoprene, in which the repeating unit is

$$-CH_2-CCl=CH-CH_2-$$

4. Nitrile rubber. A copolymer of butadiene and acrylonitrile in which the repeating units are

$$-CH_2-CH=CH-CH_2-$$

and

$$-CH-CH=CH-CH_2-$$
$$|$$
$$CN$$

5. Butyl rubber. A copolymer of isobutylene units

$$CH_3$$
$$|$$
$$-C-CH_2-$$
$$|$$
$$CH_3$$

with 1–2% of isoprene.

6. Chlorosulphonated polyethylene ('Hypalon'). A polymer formed by the reaction of polyethylene

$$-CH_2-CH_2-CH_2-CH_2-$$

with chlorine and sulphur dioxide, so that about one hydrogen atom in thirteen is replaced by a chlorine atom and about one in two hundred by sulphonyl chloride groups (SOCl.).

Bonding Rubber to Metal

The idea of bonding a layer of rubber to a metal surface as a protection against corrosion dates back to 1855, when J. H. Johnson took out a patent for making spinning components out of metal 'covered with hard india-rubber, thereby combining the advantages of the strength and durability of the metal with the non-liability to oxidation of the india-rubber'. Hard rubber or ebonite when vulcanised in contact with a clean steel surface forms a strong chemical bond with the metal. It was this discovery which lay behind Johnson's patent and led to the introduction, before the end of the century, of ebonite-lined tanks and other equipment for handling corrosive chemicals. Soft rubber linings had to wait until after World War I when special cements (based at first on rubber derivatives such as cyclised rubber, chlorinated rubber, rubber hydrohalide, and more recently on isocyanates) were developed for the direct bonding of soft rubber as well as ebonite to steel.

The bond, which is a true chemical bond, welds rubber and metal into a continuous unit so that any attempt to pull the two apart normally results in the rubber itself tearing before the bond yields. The upshot is a unique and highly versatile material which combines the corrosion and abrasion resistance of rubber with the rigidity and structural strength of steel. It is used for storage tanks of all kinds, pickling and plating vats, reaction vessels, road and rail tankers, pumps, valves and pipelines, hoods, ducts and fume stacks, fans, stirrers, filter parts, centrifuge baskets, and many other items of chemical equipment.

Properties of Natural and Synthetic Elastomers

Natural rubber is resistant to a remarkably wide range of chemicals which include all the alkalis and most of the acids, salts, dyes and bleaches commonly used in industry. It has, however, two main limitations: it cannot be used satisfactorily in contact with mineral oils and solvents, or with strong oxidising agents, including nitric acid, chromic acid and concentrated sulphuric acid. Synthetic rubbers have proved particularly valuable in overcoming these limitations and in extending the range of applications for rubber linings generally.

Chloroprene rubber, for instance, which was first developed before World War II as an elastomer which would not lose its essential properties in contact with mineral oils, is normally used for chemical linings in conditions where there may be traces of oil present with other chemicals, although nitrile rubber is also available for this purpose. Chloroprene rubber is more resistant to ozone, sunlight and weathering generally than natural rubber, and is sometimes employed on sites where equipment is likely to be exposed for long periods in the open air.

Butyl rubber also has good weather resistance, and because of its very low gas permeability it is frequently used for gas-liquor phases where its high temperature resistance is an added advantage. Besides these, it is resistant to dilute nitric acid and other oxidising agents and to a number of animal and vegetable oils.

Chlorosulphonated polyethylene (Hypalon) is used for some strongly oxidising conditions, e.g. it withstands sulphuric acid up to 85% and nitric acid up to 20% at room temperature, 50% chromic acid up to 50°C, and 85% phosphoric acid up to 90°C.

Generally speaking, the hard rubbers or ebonites have somewhat greater chemical resistance than the corresponding soft rubbers. Natural ebonites, and ebonites made from styrene-butadiene rubber (SBR), give satisfactory service with demineralised water, and with acetic, formic, lactic and sulphurous acids, where soft rubbers would not be recommended. Nitrile ebonite can be used with nitrobenzene where a soft nitrile rubber would swell and weaken. If necessary, ebonite can be polished to a smooth finish, which is invaluable in the food industry or when dealing with sticky or oily fluids. In pumps, valves and cocks it has the advantage of being easily machined to close tolerances after vulcanisation. It is also less likely to discolour water-white liquors, and it has a low water absorption.

Impact Strength

Where ebonite linings are likely to be exposed to impact hazards such as in transport vessels, or to rapid changes of temperature such as in open-air storage tanks, the likelihood of cracking can be avoided by sandwiching a flexible layer of soft rubber between the metal and the ebonite. Alternatively, one of the new flexible ebonites can be used. These are made by mixing a proportion of chloroprene or butyl rubber with natural rubber or SBR. Neither of these synthetics will combine with large quantities of sulphur during curing to make a hard ebonite; instead they remain largely uncombined and act as a plasticiser in the final product. Flexible ebonites of this sort will give an impact strength of 0·5 J (Izod) or even higher.

Abrasion Resistance

For high abrasion resistance, as required for handling such materials as corrosive slurries or suspended crystals, soft rubber linings can be extremely effective. The compounds used for this purpose, e.g. the anti-abrasive linings of sand- and shot-blast hose, grain hose and dredger hose, are usually very soft. Such linings are also widely used as protection against abrasion when little or no question of corrosion arises. Typical examples are the rubber-lined chutes used for handling coal and crushed stone in quarries. Hoppers may be similarly lined, and the rubber lining of ball mills is common practice for ordinary wet grinding as well as for grinding in acid conditions.

Chemical Resistance

In Tables 19.19 and 19.20 an attempt has been made to set out systematically the resistance of various forms of linings to different chemicals. Such tables, however, can never be either final or exhaustive, and the present ones are intended as a guide only. The indications given refer to chemical resistance

Table 19.19   A selection of chemicals to which all types of rubber or elastomer linings are resistant

Acids	Copper cyanide
Arsenic	(in solution with alkali cyanides)
Benzoic	Copper sulphate
Boric	Magnesium sulphate
Carbonic	Magnesium chloride, 50%
Citric	Potassium (or sodium) acid sulphate
Fluoboric	Potassium cuprocyanide
Fluosilicic	Potassium (or sodium) antimonate
Gallic	Potassium (or sodium) chloride
Hydrocyanic	Potassium (or sodium) cyanide
Maleic, 10%	Potassium (or sodium) sulphide
Salicylic, 15%	Potassium (or sodium) thiosulphate
Tannic	Tin chloride (either stannous or
Tartaric	stannic)
	Zinc chloride
	Zinc sulphate
Plating solutions	
Brass	*Organic materials*
Cadmium	Amyl alcohol
Copper	Beet sugar liquors
Lead	Butyl alcohol
	Casein
Inorganic salts and alkalis	Detergents
Aluminium chloride	Ethyl alcohol
Aluminium sulphate	Ethylene glycol
Alums	Glucose
Ammonia, 25%	Methyl alcohol
Ammonium chloride	Propyl alcohol
Ammonium persulphate	Soap solution
Ammonium sulphate	Sugar
Barium sulphide	Triethanolamine

at room temperatures. They do not necessarily apply to higher temperatures, especially for acids. Moreover, rapid alternations of temperature, mixtures or alternations of chemicals, contaminants and other chemicals which may be present in trace quantities, and the likelihood of abrasion or mechanical damage all affect the choice of lining material and the design of the particular compound required. To be sure of the serviceability of any given lining, the fullest possible details of operating conditions should be supplied.

Chemical resistance is rarely a matter of absolutes, and difficult conditions can sometimes be met by accepting a slight sacrifice in physical properties or a somewhat shorter life than would normally be expected.

Much also depends on actual service conditions. A lining which would withstand both an acid and an alkali if exposed continuously to either in isolation might break down under cyclic exposure to the two. The presence of ingredients in trace quantities in the liquor may have a cumulative effect, and small amounts of contaminants may have an importance out of all proportion to the minute percentage of them present. Alternatively, the problem may be to ensure that pure liquors are not stained or otherwise contaminated by the leaching out from the linings of small traces of compounding ingredients.

Table 19.20 Resistance of various rubber and elastomer linings to chemicals and natural products

Chemical / Lining material	Natural or styrene–butadiene rubber		Nitrile		Butyl	Chloroprene rubber	Chlorosulphonated polyethylene
	Soft	Ebonite	Soft	Ebonite			
Acids							
Acetic	X	S	L	S	L	X	S
Acetic anhydride	X	S	L	L	L	L	S
Chlorine water	L	S	L	S	X	X	S
Chromic, 50%	X	X	X	X	X	X	S
Formic	X	S	S	S	S	L	S
Hydrobromic	S	S	L	S	S	L	S
Hydrochloric	S	S	L	S	S	L	S
Hydrofluoric	S	S	L	L	S	L	S
Lactic	L	S	S	S	S	S	S
Malic	S	S	S	S	S	S	—
Nitric, 8%	X	X	X	X	S	X	S
Nitric, 20%	X	X	X	X	S	X	S
Oxalic, 25%	L	S	S	S	S	L	S
Oxalic, 50%	L	S	S	S	S	L	S
Phosphoric, 80%	L	S	L	S	S	L	S
Picric	S	S	L	—	L	L	S
Sulphuric, 20%	S	S	L	S	S	S	S
Sulphuric, 25%	S	S	L	S	S	S	S
Sulphuric, 50%	S	S	L	S	S	L	S
Sulphurous	X	S	S	S	S	S	S
Inorganic salts and alkalis							
Ammonium hydroxide	S	S	S	S	S	L	S
Calcium bisulphite	L	S	S	S	S	S	S
Calcium chloride	S	S	S	S	S	S	—
Calcium hypochlorite	S	L	L	L	S	L	S
Copper chloride	L	S	S	S	S	S	S
Ferric chloride	L	S	S	S	S	L	S
Ferrous sulphate ('copperas')	S	S	S	S	S	S	—
Nickel acetate	L	S	S	S	S	L	S
Potassium (or sodium) bisulphite	L	S	S	S	S	L	S
Potassium (or sodium) hydroxide	S	S	L	S	S	S	S
Potassium (or sodium) hypochlorite	L	S	L	S	L	L	S
Potassium (or sodium) sulphite	S	S	S	S	S	L	S
Silver nitrate	L	S	L	S	S	S	—

Table 19.20 (*continued*)

Chemical \ Lining material	Natural or styrene–butadiene rubber		Nitrile		Butyl	Chloroprene rubber	Chlorosulphonated polyethylene
	Soft	Ebonite	Soft	Ebonite			
Organic materials							
Acetone	S	S	X	S	S	X	S
Alcohols	S	S	S	S	S	L	S
Aniline hydrochloride	X	S	S	S	L	X	X
Aniline oil	X	S	L	L	L	X	X
Beer	S	S	X	X	S	S	S
Bleach liquor	S	S	S	S	S	L	S
Buttermilk	X	S	X	X	X	S	X
Castor oil	L	S	S	S	S	S	S
Catsup	X	S	S	S	S	S	S
Coconut oil	X	S	S	S	S	S	S
Cottonseed oil	X	S	S	S	S	S	S
Ethers	X	X	S	X	L	L	X
Ethyl acetate	X	X	X	X	X	X	X
Fatty acids	X	L	L	L	L	L	L
Formaldehyde	L	S	S	S	L	L	S
Fruit juices	S	S	S	S	L	L	S
Furfural	X	S	X	L	—	L	L
Glues	S	S	S	—	S	S	S
Glycerine	S	S	S	—	S	S	S
Hexane	X	X	S	S	X	S	S
Hydroquinone	L	S	S	S	S	S	S
Linseed oil	L	L	S	S	L	S	S
Methyl chloride	X	X	X	X	X	X	X
Mineral oils	X	L	S	S	X	L	S
Nitrobenzene	X	X	X	S	L	X	X
Olive oil	X	L	S	S	S	L	S
Petroleum oils	X	L	S	S	X	L	L
Phenol	L	S	X	S	L	X	X
Sweet milk	X	S	—	—	—	—	—
Turpentine	X	X	S	S	X	X	X
Vegetable oils	X	L	S	S	S	L	—
Wines	S	S	S	—	L	L	S
Zeolites	S	S	S	S	S	S	—
Plating solutions							
Chromium	X	X	X	X	X	X	S
Gold	S	S	X	S	S	S	S
Nickel (dull)	S	S	L	S	S	S	S
Nickel (bright)	S	S	L	L	S	S	S
Silver	S	S	X	S	S	S	S
Tin	S	S	X	S	S	S	S
Zinc	S	S	X	S	S	S	S

S = Can be expected to give satisfactory service.
X = Not generally recommended.
L = May be suitable in limited range of applications.
No entry means that insufficient information is available at present.

Temperature Resistance

There is also the question of temperature resistance. There is no simple answer to this problem. In early days the limit was apt to be set by the strength of the bond, which was originally thermoplastic. Since the development of thermosetting bonds, however, this is no longer a problem, and the limit is now more frequently set by the temperature resistance of the rubber compound itself. Moreover, recent developments in compounding techniques have considerably extended the temperature range over which both natural and synthetic rubber linings can satisfactorily be used, the upper limit in particular having been raised by up to 10°C. Thus natural rubber ebonites can now be used within the temperature range of 0–95°C, and nitrile rubber ebonites from 20–110°C. The main drawback of the nitrile ebonites, however, is that they are particularly difficult to process, and are apt to be expensive in consequence. Soft linings of natural rubber can be used continuously from temperatures of −30°C up to 90°C. Butyl rubbers are satisfactory at continuous temperatures up to 110°C and down to −20°C. Linings of neoprene can also be used up to 110°C, but are not recommended for use below 0°C. In special cases, soft nitrile and butyl rubber linings can be made resistant up to 120°C.

The nature of the chemicals to be handled will also frequently have to be taken into account in deciding the maximum temperature at which a particular compound can be used; alternatively, the temperature of operation may affect the maximum concentrations of the chemicals that can be handled. All these considerations are in fact interrelated. No two lining jobs are precisely alike, and each lining should be specially designed for the conditions it has to meet.

Methods of Applying Rubber Linings to Steel

The process of applying a rubber lining, though in principle simple, is in practice a highly skilled job. The metal must first be thoroughly cleaned and degreased, usually by sweating in steam and shotblasting. All surfaces to be covered should therefore be left free from grease and unpainted, and above all should not be coated with zinc, e.g. galvanised steel. Although zinc can be removed by shotblasting, there is always a possibility of some traces remaining, with a consequent reduction in bond strength.

Tanks

When the metal surfaces have been properly prepared, the bonding agent is sprayed or painted on. The rubber compound is then applied by hand in sheet form and carefully rolled down with specially shaped tools. Special precautions must be taken at the seams between one sheet and another, and round all angles and corners. The sheets are usually from 3 to 6 mm thick, but thicknesses of up to 12 mm are used if particularly severe abrasive conditions are involved.

It is essential that no air should be trapped under the rubber, as this will weaken the bond, and the lining may blister and eventually peel off. For this reason the metal should everywhere be smooth and free from weld spatter, burrs, porous spots, depressions or pits. Flaws should be filled with weld and ground flat. Sharp bends and corners must be eliminated, and internal projections, struts, tie-bars, cleats, baffles, weirs or strengthening ribs must be reduced to an absolute minimum. Where necessary, fillet welds should be used on corners, and angles should be smoothed to a radius not less than the thickness of the lining.

Mild steel vessels are best, since problems of surface blowholes or porosity frequently arise with castings. Welded construction is preferable to riveting, but if rivets must be used, they should be flat, tight fitting, and countersunk flush with the adjacent surface, with no undercutting or crevices round the rivet heads. Single or double 'V'-butt joints are preferred to lap joints, but in either case all welds should be ground flush with the surrounding surface.

Mild steel tanks should conform to BS 2594:1955 and pressure vessels to BS 1500. Closed tanks should if possible have a detachable flanged end to provide easy access for shotblasting, lining, testing and inspecting the work. Failing this, circular manholes 450 mm in diameter, or oval or rectangular manholes not less than 450 mm by 400 mm should be provided. If only one manhole can be arranged, an outlet branch for a fume-extraction pipe will be necessary; this should be located as far away from the manhole as possible. All branches and outlets should be designed as flanged pipes, to allow rubber covering of the flanged face. Pads should be avoided if possible, but if they are used, the studs should not penetrate through the main shell to the rubber-covered surface.

Pipework

The design of pipework for lining is equally important. Except where the bore size permits dressing of welded seams, seamless tubing is undoubtedly best, and should be manufactured in accordance with BS 806:1967. Where it is impracticable to use seamless material, tubing in accordance with BS 1387:1967 is acceptable, provided that the butt-welded joint seam does not restrict the bore by more than 0·25 mm.

The maximum length of pipe which can be conveniently lined depends on the bore. Generally speaking, for 75 mm bore and less, a maximum length of 2·75 m is advisable; for larger bores, lengths of up to 6 m can usually be accommodated, although shorter lengths are preferable. Flanges should be flat-faced and welded on square, with the weld seal on the front face continuous and radiused to a minimum of 3 mm.

The design of piping systems should, as far as possible, be based on the use of straight pipes with standard bends and tees. Allowances should be made for the thickness of the lining in specifying the bore. Similarly, in calculating the length of pipework, allowance should be made for the thickness of the lining carried over the flange plus the thickness of the flange gasket.

All lining work is examined for faults both before and after vulcanisation by trailing a flexible wire at high-frequency potentials of 20 000 or 25 000 V over the whole surface. The smallest fault or pinhole is then revealed by a clear bright spark of from 25 to 30 mm long. In special circumstances a vacuum test for adhesion may also be applied.

Vulcanisation

Vulcanisation is preferably carried out in a special vulcanising pan or auto-clave. Saturated steam or hot air is used at temperatures between 110 and 153°C, and pressures of 275–400 kN/m^2 are employed. Firms which specialise in plant lining have pans up to 4·6 m in diameter and from 6 to 7·5 m in length, large enough to accommodate most transport or storage tanks. Longer vessels, such as strip pickling tanks, can be built in sections with flanged ends. Otherwise components that are too large to be transported or to go into an autoclave can be vulcanised on site by introducing hot air, hot water or low-pressure steam into the vessel itself. If the component has to be vulcanised other than in an autoclave, then the choice of lining compound may be limited because of the method of vulcanisation employed. Concrete vessels can be lined with raw rubber sheeting attached by adhesives, and cured on site with hot air, hot water or steam; alternatively, a layer of self-vulcanising rubber compound can be applied.

The amount of industrial equipment lined or covered with rubber or ebonite has increased steadily over the last 45 years and is increasing still. Such equipment is found in all branches of the chemical industry, in dyeing and bleaching and the textile trades (particularly rayon spinning), in the metal industries for pickling and plating vessels, in food processing, in the photographic trades, and in the storage and transport of corrosive fluids. The work of lining and covering has become an expert job, which should be entrusted only to firms specialising in it. It is, moreover, wise to consult the specialists at an early stage of design, in order to ensure not only that plans are practicable, but also that they reflect the most effective—and economical —practice now available.

D. D. AITKEN

BIBLIOGRAPHY

Brazier, S. A., *Trans. Instn. Chem. Engrs., Lond.*, **30**, 46 (1952)
Brooke, T. H., *Chem. Age, Lond.*, **80**, 867 (1958)
Buchan, S., *Rubber in Chemical Engineering*, Plant Lining Group of the British Rubber Manu-facturers' Association and the Natural Rubber Producers' Research Association, London (1965)
Heath, W. A., *Corros. Prev. Contr.*, **5**, 49 (1958)
Jaray, F. F. and Lever, A. E., *Corros. Prev. Contr.*, **1**, 334, 354, 477 and 484 (1954)
Krishnamurti, I. V., *Rubb. J.*, **136**, 502 and 504 (1959)
Wilson, B. J., *Corros. Technol.*, **2**, 107 (1955)
A Code of Practice for the Design and Preparation of Rubber Lined Plant, Plant Lining Group of the British Rubber Manufacturers' Association, London (1962)

19.8 Corrosion of Metals by Plastics

One important advantage of plastics over many other materials is their great resistance to deterioration; however, in some applications this inertness of the bulk of the polymer may be deceptive, as practically all plastics contain quantities of non-polymerised residues or by-products which remain after processing. Many also depolymerise or degrade slowly releasing reactive fragments which migrate readily from the plastic to neighbouring items (either to items in contact or confined in an enclosed air space with the plastic); these ageing reactions may often be induced or accelerated, e.g. by increasing temperatures, ultra-violet irradiation, aerial oxidation and the presence of certain impurities, especially acids and alkalis. Contaminants derived from plastics have produced many problems including the spoilage of foods, the build-up of toxic atmospheres and the corrosion of metals. It is only in recent years that the extent of the last of these has been realised, and with the widening use of plastics and the increasing sophistication of modern equipment new problems are developing rapidly within this general context.

Corrosion by plastics may be conveniently sub-divided into two categories, firstly that resulting from contact of the metal with plastic (contact corrosion) and secondly that resulting from enclosure of the metal with a plastic in a confined air space but not in direct contact (vapour corrosion). A phenomenon related to the latter category is corrosion caused by the decomposition products of heated plastics during processing, during disposal by incineration, in electric circuitry (particularly associated with overloading or short circuits) or in fires. Plastics giving the most corrosive products on thermal degradation are those containing organic halogens, typically p.v.c., halogenated synthetic rubbers, fluoro-plastics, certain fire retardant polyesters and chlorinated rubber paints. The special problems posed by vapours from fires were discussed in a recent symposium in Stockholm[1]. A summary[2] of the work of the Australian D.S.I.R. on corrosion by plastics, lists the degradation products which result from heating a variety of plastics and discusses the corrosive and toxic effects of the vapours. Certain plastics may also be degraded by microbiological action to corrosive products—such reactions are usually restricted to polymers derived from naturally occurring materials, typically cellulose and drying oil derivatives or to susceptible additives which may include plasticisers, fatty acids and organic fillers.

As well as their action on metals, plastics may also actively corrode other polymers—in the author's experience for example, plasticisers from p.v.c. have softened and degraded paint coatings, vapours from a two-stage-cured phenolic moulding produced cracking and degradation of an adjacent polycarbonate switch mechanism, cresols from a cold-curing low-exotherm epoxide softened an ABS casing, and several instances of crazing of polymethyl methacrylate have been attributed to the action of vapours from adjacent plastics. These effects are, however, outside the scope of this section and will not be considered further.

Vapour Corrosion

Vapour corrosion may be defined as 'the acceleration of atmospheric corrosion of metals by traces of volatile contaminants'; typically these are volatile organic acids (particularly formic and acetic acids), but corrosion by other volatiles from plastics including hydrogen chloride, oxides of nitrogen, sulphur dioxide, hydrogen sulphide, phenol, amines and ammonia, has also frequently been reported. The subject was surveyed by Rance and Cole[3] in 1958, and more recently by Knotkova-Germakova and Vlckova[4], and by Donovan and Stringer[5]. These three papers deal largely with the effects of vapours from plastics in an undegraded (freshly manufactured) condition; additionally Knotkova-Germakova and Vlckova[4] refer to corrosion arising during manufacture. Kennett[2] has reported on the effects of vapours arising from thermal degradation of plastics, and Cotton and Jacob[6] have referred to corrosion by vapours from polyvinylchloride (p.v.c.) at moderate temperatures (above 70°C) and from fluorinated polymers at higher temperatures (above 350°C). Corrosion can also be caused by acids produced by microbiological decomposition[7]. Clarke and Longhurst[8], investigating the effects of humidity, showed that below 70% r.h. corrosion in air containing acetic acid vapour was very slow but accelerated rapidly as the relative humidity was increased above 80%.

Table 19.21 summarises the reported vapour corrosion effects of a variety of plastics; references to the source of the information is given at appropriate places.

The available range of plastics is very wide with a variety of compositions and related properties within any one type description. Even when physical properties and formulation are specified, minor variations in trace additives, release agents, moulding cycles, etc. may have a considerable influence on corrosivity. Table 19.21 can therefore be considered only as a guide and even in this sense much more experimentation is needed to provide the full picture. In the present state of technology precise information is only likely to result from tests carried out on the material of interest; various suitable test methods have been described in the literature[4,8,12,18,21].

There are, however, certain materials used as ingredients in the manufacture of plastics, which almost invariably give a corrosive product. Included in this category are wood[8], which is frequently used as a filler or as part of a composite; drying oils, used in paints[18], adhesives, jointing compounds and linoleum[22]; and esters of volatile acids frequently retained in certain cold setting formulations, especially some paints[18].

Table 19.21 Summary of vapour corrosion effects of plastics

Material	Severity of corrosion*	Volatiles evolved and remarks
Thermoplastics		
1. Polyvinyl chloride (p.v.c.) (and other chlorinated thermoplastics)	Non-corrosive at ambient temperature (but see column 3); moderately very corrosive at 70°C	Hydrogen chloride (HCl). May become corrosive at ambient temperature if irradiated with u.v. radiation or in the presence of certain contaminants, e.g. zinc ions[9]
2. Fluorinated thermoplastics (e.g. p.t.f.e.)	Non-corrosive at ambient and moderate temperatures; very corrosive above about 350°C[10]	Decompose to release HF and F_2
3. Nitrocellulose	Slightly–very corrosive[11]	Oxides of nitrogen may be evolved progressively with ageing
4. Nylons		
(a) Nylon 6	Corrosive[11]	Acetic acid; formulations frequently contain acetic acid additions as molecular weight regulators
(b) Nylon 66	Non-corrosive[5]	
5. P.V.A. (polyvinyl acetates and alcohols)	Non-corrosive–very corrosive	Acetic acid released; corrosivity dependent on conditions and formulation (degree of hydrolysis and presence of stabilisers and inhibitors)
6. Cellulose acetate	Slightly corrosive[4]	Acetic acid may be released
7. Polyacetals		
(a) Homopolymer	Slightly corrosive[4,5] at ambient temperature, more corrosive above 40°C	Acetic acid and formic acid evolved (acetic acid may be used as an end-stopper)
(b) Copolymer (formaldehyde and 10% ethylene oxide)	Usually non-corrosive at ambient temperatures, corrosive above 45°C[5]	Formic acid evolved (if arduous moulding conditions have been used, the polymer may be corrosive at ambient temperatures)
8. Polyolefines, polyesters, polycarbonates, polystyrene, polysulphone, polyphenylene oxide and polymethylmethacrylate	Non-corrosive[5,11] at ambient temperatures	
Thermosetting resins		
1. Cross-linked polyesters[3,5,12,13]		
(a) Cold cured polyesters	MEKP catalyst and cobalt naphthenate accelerator—very corrosive. Other peroxide catalyst systems slightly–moderately corrosive. Irradiation or non-oxidising catalyst } non-corrosive	Formic and acetic acids evolved. Corrosivity is determined largely by the catalyst used, but is also affected by the formulation, in particular diethylene glycol gives more corrosive resins than does propylene glycol
(b) Hot cured polyesters	Non-corrosive–moderately corrosive	

*Refers directly to Zn, Mg and steel; but see **19.**79.

Table 19.21 (*continued*)

Material	Severity of corrosion	Volatiles evolved and remarks
2. Phenolformaldehyde[3-5]		
(*a*) Two-stage resins (via novolaks)	Moderately–very corrosive	Evolve ammonia and formaldehyde ⎫ very corrosive if undercured or if they are compounded with wood, flour fillers
(*b*) Single-stage cure (via resols)	Non-corrosive (if fully cured and containing only inert fillers)	
3. Amino plastics[3-5]	Non-corrosive–slightly corrosive	
4. Epoxides[4-6]		
(*a*) Cold cured	Non-corrosive–slightly corrosive	Amines[6] may be evolved from amine catalysed resins, and cresol[11] from certain cold-curing low-exotherm formulations
(*b*) Hot cured	Non-corrosive	
5. Polyurethanes	Non-corrosive[5] at ambient temperatures	

Rubbers[4], elastomers and adhesives
1. Natural rubber

Material	Severity of corrosion	Volatiles evolved and remarks
(*a*) Non-vulcanised[5]	Slightly corrosive on prolonged exposure	Formic and acetic acid evolved
(*b*) Vulcanised[5,14]	Slightly–moderately corrosive	Hydrogen sulphide and sulphur dioxide evolved
2. (*a*) Synthetic rubbers[5,6]	Non-corrosive at ambient condition—most are corrosive above 100°C	Many are chlorinated and evolve HCl on heating; Hypalon may also emit sulphur dioxide[4]
(*b*) Polysulphide rubbers (cold curing)[5]	Moderately corrosive–very corrosive	Formic acid; the catalysts used are peroxides
3. Silicone polymers[5,15-17]	Non-corrosive–very corrosive	Acetic and formic acids. Some single-pack silicone sealants cure by hydrolysis of acetoxy groups releasing acetic acid and are very corrosive; some two-pack formulations evolve formic acid and are corrosive, and others are reputed to be among the most inert polymers
4. Phenol and ureaformaldehyde glues[3,5]	Slightly corrosive–very corrosive	Formaldehyde, phenol, ammonia and HCl may be evolved. Various acids and salts that yield acids (e.g. formic acid and hydrochloric acid) are used in cold-set formulations. Volatiles evolved during cure may be absorbed by the materials being bonded

Table 19.21 (*continued*)

Material	Severity of corrosion	Volatiles evolved and remarks
Paints and lacquers[3, 5, 18 – 20]		
1. Oleo-resinous (drying oil) type		Formic acid, other volatile acids and aldehydes — corrosivity may be much reduced by certain neutralising pigments, e.g. zinc oxide, calcium plumbate
(*a*) Air drying (inert pigments)	Moderately corrosive–very corrosive	
(*b*) Stoving (inert pigment)	Slightly corrosive–very corrosive	Formic acid
2. Air-drying synthetic-resin paints		
(*a*) Chlorinated rubber (hydrocarbon solvent)	Normally non-corrosive but may become corrosive after exposure to u.v. light or high temperature	HCl
(*b*) P.V.A. based	Non-corrosive–very corrosive	Acetic acid
(*c*) Cyclised rubber, nitro-cellulose, shellac, acrylics, two-pack epoxides and polyurethanes	Non-corrosive–slightly corrosive (if solvent fully removed—typically 2 weeks drying)	May have volatile ester solvents, or contain traces of volatile acids and these may be retained for some time and be corrosive
(*d*) Synthetic stoving paints (epoxides, formaldehyde condensation polymers)	Non-corrosive (if fully cured)	

Contact Corrosion

Although less widely reported than the effects of vapours, contact corrosion has been a serious problem in packaging and in electronics[15–17]. As miniaturisation and sophistication of electronic devices has increased, the hazard presented by corrosion is often the limiting factor inhibiting the attainment of expected levels of reliability.

Semiconducting devices, switches and miniaturised v.h.f. circuits are all particularly sensitive to the slightest reaction on critical surfaces, and in devices calling for the highest levels of reliability even the most inert of the phenolic, epoxide and silicone resins are not considered to be fully acceptable[17]; corrosion of electronic assemblies may often be enhanced by migration of ions to sensitive areas under applied potentials, and by local heating effects associated with current flows.

Little comprehensive work has been carried out on contact corrosion, but some results on a range of polymers have been reported by Czech workers[4]. In general, plastics that give rise to vapour corrosion (Table 19.21) will also cause contact corrosion. Some qualification is needed to this statement, however, as much depends on the type of contact and the other ingredients in the polymer, e.g. a paint may give good protection to the metal to which it is applied, but the vapour may cause corrosion of adjacent metal items within an enclosed space.

Access of air and water will also affect the corrosion rate. Metal inserts in corrosive plastics are most actively attacked at the plastic/metal/air interfaces; with certain metals, notably aluminium[23], titanium[10] and stainless steel, crevice effects (oxygen shielding and entrapment of water) frequently accelerate attack. Acceleration of corrosion by 'bimetallic' couples between carbon-fibre-reinforced plastics and metals presents a problem[24] in the use of these composites.

The chloride ion is the most frequent cause of contact corrosion, since chlorine is present in the many chlorinated plastics, and is also frequently retained in residual amounts from reactive intermediates used in manufacture. Thus epoxides usually contain chloride derived from the epichlorhydin used as the precursor of the epoxide. In addition to the contaminants referred to in Table 19.21, various metal and ammonium cations, inorganic anions and long-chain fatty acids (present as stabilisers, release agents or derived from plasticisers) may corrode metals on contact.

Susceptibility of Metals to Attack by Contaminants from Plastics

The relative susceptibility of metals to atmospheric corrosion varies widely with the type of contaminant, e.g. zinc and cadmium, two metals that are used for the protection of steel in exposed environments, are both rapidly attacked by organic acids[20]; on the other hand, aluminium alloys resist attack by organic acids but may be rapidly corroded by chlorides, especially at crevices or areas of contact.

Copper alloys are particularly prone to attack by long-chain fatty acids which are often present in sealing compositions, temporary protectives[25] and as trace additives in many plastics; under acid conditions ester plasticisers may saponify in the presence of copper giving rapid corrosion of the copper and accelerating degradation of the polymer.

Copper and silver tarnish readily in sulphide atmospheres, and copper in contact with sulphur-vulcanised rubber will sometimes react with the sulphur, devulcanising it in the process. The growth of conducting sulphide whiskers on silver is noteworthy as these whiskers may give rise to short circuits across silver-plated contacts. Ammonia has little effect on most metals, but traces will tarnish many copper alloys and cause stress-corrosion cracking of certain stressed brasses.

Various authors[5,8,18] have investigated the relative susceptibility of a variety of metals to attack by the lower fatty acids; the results show that magnesium, lead, steel, zinc and cadmium are all rapidly attacked (Donovan and Stringer[5], for instance, showed that zinc corrodes at a rate of 5 μm per surface per week at 30°C and 100% r.h. in air that contains 0·5 parts per million of acetic acid); copper and nickel are attacked less rapidly, and aluminium, tin and silver are resistant to attack.

P. D. DONOVAN

REFERENCES

1. International symposium entitled *Corrosion Risks in Connection with Fire in Plastics*, Stockholm, April 24 (1969)
2. Kennett, A. C., *Australian Plastics and Rubber J.*, **26**, 8 (1971)
3. Rance, V. E. and Cole, H. G., *Corrosion of Metals by Vapours from Organic Materials*, H.M.S.O., London (1958)
4. Knotkova-Germakova, D. and Vlckova, J., *Br. Corros. J.*, **6** No. 1, 17 (1971)
5. Donovan, P. D. and Stringer, J., *Br. Corros. J.*, **6** No. 3, 132 (1971)
6. Cotton, J. B. and Jacob, W. R., *Br. Corros. J.*, **5** No. 4, 144 (1970)
7. *Biodeterioration of Materials: Microbiological and Allied Aspects*, edited by Walters, A. H. and Elphick, J. P., Elsevier Publishing Co. (1968)
8. Clarke, S. G. and Longhurst, E. E., *J. Appl. Chem.*, London, **11**, 435 (1961)
9. Randall, F. G., *Br. Corros. J.*, **5** No. 4, 144 (1970)
10. Cotton, J. B., *Br. Corros. J.*, **7** No. 2, 59 (1972)
11. Donovan, P. D. and Stringer, J., unpublished M.O.D. report (1973)
12. Cawthorne, R., Flavell, W. and Ross, N. C., *J. Appl. Chem.*, London, **16**, 281 (1966)
13. Cawthorne, R., Flavell, W., Ross, N. C. and Pinchin, F. J., *Br. Corros. J.*, **4**, 35 (1969)
14. Chaston, J. C. (and discussion), J.I.E.E., **88**, Pt. 2, 276 (1941)
15. Macleod, R., Proc. of Conference entitled *Protection of Metal in Storage and in Transit*, London, June, 1970, Brintex Publ., 57 (1971)
16. Heinle, P. J., *S.P.E. J.*, **24**, 51 (1968)
17. Hakim, E. B., Ninth International Electronic Circuit Packaging Symp. (1968)
18. Donovan, P. D. and Moynehan, T. M., *Corros. Sci.*, **5**, 803 (1965)
19. Vernon, W. H., *J. Trans. Faraday Soc.*, **19**, 839 (1924)
20. Defence Guide 3A, *The Prevention of Corrosion of Cadmium and Zinc Coating by Vapours from Organic Materials*, H.M.S.O., London (1966)
21. Burns, R. M. and Campbell, W. E., 55th General Meeting of the Am. Electrochem. Soc., 271, May (1929)
22. Fritz, F. Z., *Angew. Chem.*, **28** No. 1, 272 (1915)
23. Scott, D. J. and Skerrey, E. W., *Br. Corros. J.*, **5**, 239 (1970)
24. Brown, A. R. G. and Coomber, D. E., *Br. Coros. J.*, **7** No. 5, 232 (1972)
25. Andrew, J. F. and Donovan, P. D., Proc. of Conference entitled *Protection of Metal in Storage and in Transit*, London, June, 1970, Brintex Publications, 25 (1971)

19.9 Wood

Wood, one of the oldest of man's constructional materials, retains an important place in the modern industrial world. It is particularly valuable for certain conditions which are corrosive to common metals (acids, external exposure) and for contact with foodstuffs or beverages. Wood is not subject to corrosion in the electrochemical sense of the term applied to metals, but it can be attacked by chemicals (particularly alkalis and oxidising agents) and is subject to biological attack, a hazard not encountered with most other structural materials.

Physical Properties of Wood

Timber, being a botanical material, shows a wide variation in physical properties both between different botanical species and even within the same species. Uniformity can be achieved only be grading and selection within a single species. Timber is anisotropic and most of its properties differ for the three growth directions—longitudinal, tangential and radial. Most strength properties are highest in the longitudinal direction[1]. Density, which can vary from 40 kg/m^3 in some balsa to 1 230 kg/m^3 for lignum vitae, is often related to the other properties.

The thermal conductivity of dry timber is low and it will resist for short periods temperatures which are comparatively high for an organic material, e.g. 200°C.

The most important and most troublesome physical property of wood is its absorption or loss of water with changes of humidity conditions. Not only does this result in swelling or shrinkage of the wood (which varies in amount with the species) but the dimensional change is also anisotropic, so that dimensional distortion and even, in extreme cases, splitting and disintegration are produced.

Moisture absorption or loss occurs largely through the end grain so that coating of end-grain surfaces with a moisture barrier greatly reduces movement.

The timber from the outside part of a tree (the sapwood) has certain properties differing from those of timber from the inner part (the heartwood). It is often a different colour, and is usually more permeable to liquids and susceptible to biological attack.

The availability of timbers of different species and qualities varies considerably with economic and political conditions. Little-known timbers may often be preferred to well-known ones which may have become scarce, of poor quality, or expensive.

Resistance of Timber to Damp and Weather

Timber has very good resistance to exterior exposure. Some slow attack of rain and sun, with surface disintegration resulting from wetting and drying cycles, occurs, giving a surface loss of about 6 mm per century, i.e. a corrosion rating 'A'. Even this can be avoided by painting.

The main hazard is biological attack. In Europe this is mainly 'rot' or fungal attack, although several insects can contribute to breakdown. In tropical countries termites may be of overriding importance and for marine work various marine borers become important. There are two ways of avoiding biological attack: use of a timber with a degree of natural durability, and treatment of the timber with preservative.

The sapwood of almost all timbers is susceptible to attack, so that to use a durable timber implies using one that has had the sapwood removed. On the other hand, since the most effective form of treatment with a preservative is impregnation by vacuum and pressure and since sapwood is more readily permeable by liquids, the presence of sapwood makes for more effective treatment and is an advantage if preserved timber is used.

Timbers exhibit a wide range of resistance to biological attack. Table 19.22 lists those found to resist fungal decay on partial burial in the ground

Table 19.22   Recommended timbers for use in corrosive conditions

Use	Recommended timbers
External exposure or damp conditions	*Very durable:* Afrormosia, afzelia, albizzia*, east African camphorwood, ekki, greenheart, iroko, ironbark*, jarrah, kapur, makore, mansonia, okan, opepe, padauk, purpleheart, pyinkado, 'Rhodesian teak', rosewood*, tallowood, teak, wallaba.
	Durable: Agba, 'central American cedar', sweet chestnut, eng, freijo, guarea, idigbo, jacereuba*, jequitiba*, kempas, red louro*, American mahogany, dark red meranti, mora, oak, rauli, robinia*, thitka, utile, cedar*, 'pencil cedar'*, 'Port Orford cedar'*, sequoia, 'Southern cypress'*, 'Western red cedar', 'yellow cedar', yew.
	After preservative treatment: European redwood, other timbers with high sapwood, exterior grade plywood, beech, birch, maple.
Structures exposed to acid fumes	Pitch pine, teak, kauri*, oak, greenheart, Douglas fir, European redwood (sapwood-free).
Containers for acids	Pitch pine, Douglas fir, teak, oak, kauri*, sequoia, 'Southern cypress'*, ayan, white peroba, larch, afzelia, greenheart.
Plant requiring acid resistance	Ekki, greenheart, iroko, kauri*, pitch pine, purpleheart, pyinkado, Sydney blue gum*, 'Southern cypress'*, teak, brush box, jarrah, karri, opepe.
Containers for mildly corrosive liquids	Pitch pine, Douglas fir, oak, Scots pine/redwood, sapele, idigbo, robinia*, afzelia, parana pine, krabak, podo, agba, danta, mansonia, east African camphorwood, muninga, 'yellow pine', plywood (exterior grade).

*These timbers are not at present readily available in the United Kingdom.

in the UK. 'Very durable' indicates a lifetime of 25 years or more in contact with the ground and 'durable' 15–25 years life in contact with the ground[2,3].

Timbers well known for exterior durability are teak, oak and Western red cedar. The two latter, however, fall only into the 'durable' class, whereas several lesser known but readily available woods such as afrormosia, afzelia, iroko, jarrah, kapur, makore, mansonia, okan, opepe and purple-heart are in the 'very durable' class with teak and could be more widely employed for exterior use.

Preservative treatment is the other method of providing timber resistant to biological attack[4]. Treatment in a cylinder under a vacuum and pressure cycle is the most effective method. Permeable timbers are most readily protected and in practice this usually means European redwood (Scots pine) which is the most commonly available timber and usually has a high proportion of sapwood. Other softwoods are frequently treated; if they are difficult to treat, the timber may be incised to improve penetration. This is frequently done with Douglas fir which is creosoted for use as railway sleepers and telegraph poles. Creosote is the traditional preservative which still has widespread applications, but water-borne preservatives have the advantages of not causing staining or affecting paint adhesion, while they change the colour of the timber only slightly. In the absence of pressure preservation facilities, immersion in hot and cold preservative induces good penetration, and even surface treatment may give good protection if the hazard of biological attack is not great.

Timber has good resistance to polluted atmospheres and is particularly useful in the presence of acid fumes which would rapidly corrode steel. Constructional softwoods such as Douglas fir, pitch pine, larch, or better grades of redwood (Scots pine) give the best acid resistance, although certain hardwoods such as teak or jarrah may find particular uses.

Recent developments in glued-laminated structures raise problems of the durability of adhesives as well as of the wood. Resorcinol adhesives, the only reliable exterior structural glues, will resist all conditions that timber can be expected to resist. Urea and casein glues may serve well under dry conditions but not under full weathering. The gluing of preservative-treated wood presents problems, particularly where oily preservatives have been used.

Corrosive Liquids

There is a steady market for timber containers for liquids, particularly acids and salts which would corrode steel or concrete[5]. The rate of attack on wood by acids is very slow below concentrations of about 25% of non-oxidising mineral acids or about 60% of organic acids at room temperature. Higher concentrations, higher temperatures or alkaline solutions, lead to some breakdown, but a useful life of several years may still be possible, and breakdown products from wood are often less harmful than metallic corrosion products.

Chemical vats and tanks are made in such a way that the swelling of the wood in contact with liquids serves to keep joints tight. The timber used has to be very carefully selected as well as being of a suitable species. It must not be permeable to liquids, and thus sapwood must be absent, and end-grain surfaces must not be exposed, i.e. straightness of grain and freedom from knots is important.

The best woods for acid resistance are those having a high cellulose and lignin content and a correspondingly low pentosan content[6]. Softwoods in general fulfil these requirements best, but some hardwoods, such as teak and jarrah, also meet them.

Low permeability to liquids is as important as acid resistance, since the rate of progress of attack depends on both. Timbers used for this purpose are listed in Table 19.22. Some traditional timbers such as kauri are not now available even in the country of origin. Others such as Southern cypress are practically unobtainable in this country. Still others may be difficult to obtain in the right grades. European redwood (Scots pine), for instance, almost always contains too much sapwood for vat work and the only suitable material is in some of the Russian imports, cut from large trees. Douglas fir has become a standard timber for vats because it has been available, sap-free, in good sizes. Recently it is becoming less sap-free, however, and it lacks durability, particularly if undetected incipient rot is present.

Pitch pine is at present the main high-quality vat timber. It varies very greatly in the amount of resin it contains and selection can be made of highly resinous wood for low-temperature work and low-resin wood for high temperatures and other purposes where resin is a nuisance.

Oak remains a valuable and traditional timber for chemical work.

The painting of vats externally is to be discouraged as it permits build-up of moisture and possible premature decay. Other chemical-plant apparatus such as filter-press plates and frames, stirrers, etc. can be fabricated from the timbers given in Table 19.22.

Linings can be provided for vats or tanks. Such materials as lead, copper, steel, glass, asphalt or rubber are the most popular. It is difficult to protect the interior of a timber vessel by means of a surface coating because the swelling and shrinking of the timber when in use tends to crack the coating. Bituminous, chlorinated rubber, phenolic or furfural resin paints may, however, sometimes be used successfully. Surface impregnation of the timber with materials like paraffin wax or creosote may be valuable as an added protection against oxidising agents.

Mildly Corrosive Liquids

The chief advantage of timber for containers is its lack of harmful con-taminants. It is, therefore, used a great deal in the food and beverage industries. Oak has long been the preferred material for the dying craft of tight cooperage. Other suitable woods are mentioned in Table 19.22. Plywood, provided it is made from a durable glue, is usually suitable.

Liquids which may be satisfactorily contained in wooden vessels include solutions of salts of weak bases such as aluminium, ammonium and iron, copper sulphate, potassium salt solutions, alcohol liquids, hydrocarbons, formaldehyde, concrete, and synthetic polymer lattices. Liquids which can be contained with a minimum of surface treatment of the wood include brine and hydrogen peroxide solutions. Podo, white seraya, lime, maple, or sycamore are recommended for the latter use.

Liquids unsuitable for wooden containers are solutions of nitric acid, nitrates, chlorates, sulphates and other oxidising salts, alkalis, phenol, calcium and zinc salts, and salts of weak acids and strong bases such as sodium sulphide or sodium carbonate[7].

Timber used in cooling towers may be attacked by the weakly alkaline water resulting from the use of base-exchange water softeners, or by chlorine used to kill organisms[8]. Timbers suitable for use in cooling towers include European redwood (Scots pine), whitewood (spruce), Western hemlock, or Douglas fir. These must be treated with 'Celcure' or 'Tanalith C' preservatives to particularly high retentions in order to protect the timber against attack by 'soft rot', a form of decay by microfungi which has only recently been found to be quite widespread and to occur even in the sea.

Corrosion Problems Associated with Timber

Timber is attacked by acids, particularly when they have oxidising properties. Thus sulphuric and sulphurous acids deposited from polluted atmospheres cause slow disintegration of surface fibres.

Alkalis attack wood more readily than do acids, and even weak alkalis and such materials as domestic washing solutions attack the lignin in the wood. For this reason the scrubbing of timber floors, furniture, or draining boards with too much alkaline washing agent must be avoided.

Several timbers give coloured extracts with alkalis or organic solvents. Afzelia, ayan, idigbo, opepe and padauk are among those which cannot be used for draining boards for this reason. Purpleheart gives a coloured extract with acid[5].

Staining of Wood

Some woods are stained by iron compounds resulting from corrosion caused by a reaction with certain tannins in the wood. Woods with a tendency to stain with iron include sweet chestnut, oak, rauli, coigue, afrormosia, elm, alder, idigbo, kapur, African mahogany, walnut, 'Rhodesian teak', gedu nohor, sapele, utile, Honduras cedar, ash, cherry and katsura. Staining is often much more noticeable when the wood is acid, so that woods like Western red cedar, Douglas fir and sequoia which give a much lighter stain than oak are often observed to be stained. Copper salts also give a slight stain with most woods.

Metal stains can usually be removed by bleaching agents or complexing materials such as oxalic acid or salts of diaminoethane tetra-acetic acid.

Timber Names

The names of timbers mentioned in this section conform with the British Standard recommendations[9] used in the F.P.R.L. handbooks on hardwoods[2] and softwoods[3] where botanical names and much additional information may be found.

V. R. GRAY

REFERENCES

1. Armstrong, F. H., 'The Strength Properties of Timber'. *For. Prod. Res. Bull.*, **28** (1953) and **34** (1955). H.M.S.O. London
2. *A Handbook of Hardwoods*, H.M.S.O., London (1973)
3. *A Handbook of Softwoods*, H.M.S.O., London (1968)
4. *Timber Preservation*, British Wood Preserving Association and Timber Development Association (1957)
5. Pearson, F. G. O. and Webster, C., *Timbers Used in Cooperage and the Manufacture of Vats and Filter Presses*, D.S.I.R., Forest Products Research Laboratory (1958)
6. Campbell, W. G. and Bamford, K. F., *J. Soc. Chem. Ind., Lond.*, **58**, 180 (1939)
7. Kollmann, F., *Technologie des Holzes u. die Holzwerkstoffe*, Vol. I., Springer-Verlag, Berlin (1951)
8. Ross, F. F. and Wood, M. J., 'The Preservation of Timber in Water Cooling Towers', Annual Convention of the British Wood Preserving Association, 171 (1957)
9. *Nomenclature of Commercial Timbers*, BS 589 and 881:1955, British Standards Institution, London

19.10 The Corrosion of Metals by Wood

Wood can cause corrosion of metals by direct contact and, in confined spaces, also by the emission of corrosive vapour. With rare exceptions, all woods are acid, and the principal corroding agent in both types of attack is volatile acetic acid.

Acetylated polysaccharides form part of the structure of wood, the acetyl radical constituting some 2–5% by weight of the dry wood. Hydrolysis to free acetic acid occurs in the presence of moisture at a rate varying from one species to another; a wood of lower acetyl content can liberate acetic acid much faster under given conditions than another wood of higher content[1,2]. Small quantities of formic, propionic and butyric acids are also formed[3], but their effects can be neglected in comparison with those of acetic acid. There is a broad, but only a broad, correlation between the corrosivity of a wood and its acidity. The chemistry of acetyl linkage in wood and of its hydrolysis has been examined in some detail[4].

Contact corrosion may be reduced by the presence of natural inhibitors, such as tannins, in the wood, and will be promoted by sulphates and chlorides in it*, especially if mineral preserving processes involving these ions have been applied.

Influence of Moisture

The influence of moisture is fundamental, as it is with other forms of corrosion. Long-term contact tests[5] with ponderosa pine, some treated with zinc chloride, in atmospheres at 30, 65 and 95% r.h. showed that at 30 and 65% r.h. plain wire nails were not very severely corroded even in zinc chloride-impregnated wood. At 95% r.h. plain wire nails were severely corroded, though galvanised nails were attacked only by impregnated wood. Brass and aluminium were also attacked to some extent at 95% r.h. Some concurrent outdoor tests at Madison, Wisconsin, showed that the outdoor climate there was somewhat more severe than a 65% r.h. laboratory test.

---

*Woods contain from 0·2 to 4% of mineral ash. This consists largely of calcium, potassium and magnesium as carbonate, phosphate, silicate and sulphate. Aluminium, iron, sodium and chloride are also present. Sulphate contributes 1 to 10% by weight, usually 2 to 4%, and chloride 0·1 to 5%.

These are useful quantitative results, but they will cause little surprise, since the user of wood will expect metals in contact with damp wood to corrode.

A degree of corrosion acceptable on the nails and fastenings on the outside of a packing case is not, however, acceptable on metal components inside, which the box is supposed to be protecting. Vapour corrosion is also governed by relative humidity and can occur whenever the internal humidity exceeds a critical value, as may happen for a few hours in the cold of the night even in quite good storage conditions. The critical humidity for corrosive attack has been reported[6, 7] as 75%.

Less and More Corrosive Woods

Table 19.23 lists woods whose aggressiveness by vapour corrosion has been quoted in a survey[8], together with typical pH values of the aqueous extracts of these woods reported in another investigation[9].

Table 19.23   Relative corrosivity of woods by vapour corrosion

Wood	Classification in Defence Guide-3A[8]	Typical pH values
Oak	Most corrosive	3·35, 3·45, 3·85, 3·9
Sweet chestnut	Most corrosive	3·4, 3·45, 3·65
Steamed European beech	Moderately corrosive	3·85, 4·2
Birch	Moderately corrosive	4·85, 5·05, 5·35
Douglas fir	Moderately corrosive	3·45, 3·55, 4·15, 4·2
Gahoon	Moderately corrosive	4·2, 4·45, 5·05, 5·2
Teak	Moderately corrosive	4·65, 5·45
Western red cedar	Moderately corrosive	3·45
Parana pine	Least corrosive	5·2 to 8·8
Spruce	Least corrosive	4·0, 4·45
Elm	Least corrosive	6·45, 7·15
African mahogany	Least corrosive	5·1, 5·4, 5·55, 6·65
Walnut	Least corrosive	4·4, 4·55, 4·85, 5·2
Iroko	Least corrosive	5·4, 6·2, 7·25
Ramin	Least corrosive	5·25, 5·35
Obeche	Least corrosive	4·75, 6·75

While certain reservations must be kept in view (i.e. there is not necessarily a correlation between pH and corrosivity, and different samples of the same species of wood show a wide scatter of pH values, which might well be even wider if differences in duration of seasoning were taken into account), the results of vapour corrosion tests nevertheless indicate a general correlation between quoted pH values and the corrosiveness of wood vapours. It may reasonably be concluded that a strongly acid wood, pH less than 4·0, is potentially dangerous, and a less acid wood, pH more than 5·0, is likely to be relatively safe.

Heat treatments of wood are dangerous, for although existing acid vapours may be expelled, further vapours are formed by accelerated hydrolysis.

Volatile acid hardeners such as hydrochloric acid and formaldehyde (which oxidises to formic acid) present in glues in plywood contribute to vapour corrosion, as can varnishes and paints[8].

Wood preservatives appear not to affect emission of corrosive vapours from wood, suggesting that the hydrolysis of acetyl polysaccharides is chemical, not biochemical. Some copper-base preservatives can give enough leachable copper ions to cause galvanic corrosion of other metals, notably aluminium and steel.

Metals Affected

The metals most susceptible to corrosion by wood are steel, zinc, cadmium, magnesium alloy and lead. The susceptibility of zinc and cadmium is no argument against the galvanising or cadmium plating of steel, since these coatings much reduce the rate of corrosion of steel by contact with wood or wood vapours, although they will not give the high degree of protection which they provide in open exposure to marine or tropical atmospheres.

Aluminium is relatively resistant[10]. So also are copper, brass, tin and stainless steel, but these metals should not be used as thin coatings on mild steel as they promote rusting at any points of breakdown.

Bimetallic corrosion between two different metals (see also Section 1.7) embedded in damp wood, e.g. in the hull of a boat, can occur in two ways[11]. If the metals are joined by a metallic conductor, then the formation of the cell

metal A/damp wood/metal B

will result in accelerated corrosion of the metal which has the more negative potential in this electrolyte. Savory and Packman[11] point out that even if two metals are not connected by a metallic path, and project from wood into sea-water, then two opposing cells are set up, the first as described above, and the second

metal A/sea-water/metal B.

These cells are unlikely to have the same potential so that a net potential will exist and one metal will corrode preferentially and the other will tend to be protected. However, although this situation may occur in practice it is difficult to see how the explanation given by Savory and Packman is tenable. In addition, accelerated corrosion can occur on an individual metal, by the action of a concentration cell

metal A/damp wood/sea-water/metal A.

Iron salts from rusting steel, e.g. a nail, have a strongly deleterious effect on wood, causing 'charring' and complete loss of strength.

Practical Conclusions

Contact corrosion   Nails and fastenings in many non-durable wooden articles exposed to damp will outlive the useful lives of the articles, and their corrosion is of no great importance. Corrosion is, however, important in tile and batten nails in roofs, fences and other more permanent structures. Unprotected steel should never be used. Galvanised steel is much better, and brass, copper, the more corrosion-resistant alloys of aluminium, and stainless steel, are likely to give even longer service.

Vapour corrosion   The best way to pack articles made of metals susceptible to vapour corrosion is in boxes made of metal or of those plastics which do not themselves emit corrosive vapours[8]. If wood cannot be avoided, then the less corrosive kinds should be chosen. Dryness, good ventilation and the inclusion of water-vapour barriers should be sought. Other obvious measures are the avoidance of susceptible metals and the use of protective treatments and paints.

H. G. COLE

REFERENCES

1. Packman, D. F., *Holzforschung*, **14**, 178 (1960)
2. Arni, P. C., Cochrane, G. C. and Gray, J. D., *J. Appl. Chem.*, **15**, 305 (1965)
3. Ibid., **15**, 463 (1965)
4. Cochrane, G. C., Gray, J. D. and Arni, P. C., *Biochem. J.*, **113**, 243 and 253 (1969)
5. Baechler, R. H., *American Wood Preservers' Association*, 1 (1949)*
6. Schikorr, G., *Werk. u. Korr.*, **12**, 1 (1961)
7. Clarke, S. G. and Longhurst, E. E., *J. Appl. Chem.*, **11**, 435 (1961)
8. Rance, V. E. and Cole, H. G., *Corrosion of Metals by Vapours from Organic Materials*, H.M.S.O., London (1958) and DG–3A, *Defence Guide for the Prevention of Corrosion of Cadmium and Zinc Coatings by Vapours from Organic Materials*, H.M.S.O., London (1966)
9. Gray, V. R., *J. Inst. Wood. Sci.*, **1**, 58 (1958)*
10. Farmer, R. H. and Porter, F. C., *Metallurgia*, **68**, 161 (1963)
11. Savory, J. G. and Packman, D. F., DSIR Forest Products Bulletin No. 31 entitled *Prevention of Decay of Wood in Boats*, H.M.S.O. (1954)

*References 5 and 9 can be seen at the Library of the Timber Research and Development Association, Hughenden Valley, High Wycombe, Bucks.

20 CORROSION TESTING, MONITORING AND INSPECTION

20.1   Corrosion Testing and Determination of
        Corrosion Rates   **20**:3
            Test procedures   **20**:4
            Laboratory corrosion tests   **20**:16
            Electrochemical measurements   **20**:28
            Polarisation Resistance   **20**:35
            Tests for bimetallic corrosion   **20**:40
            Accelerated tests—electrolyte tests   **20**:43
            Accelerated tests—simulated environments   **20**:47
            Intergranular attack of Cr—Ni—Fe alloys   **20**:55
            Crevice corrosion and pitting   **20**:67
            Impingement tests   **20**:71
            Corrosion fatigue   **20**:73
            Cavitation—erosion   **20**:76
            Fretting corrosion   **20**:78
            Corrosion testing on liquid metals and
            fused salts   **20**:79
            Tests in plants   **20**:85
            Atmospheric tests   **20**:87
            Tests in natural waters   **20**:93
            Field tests in soil   **20**:95
            Corrosion testing of organic coatings   **20**:95

20.1A   Appendix—Chemical and Electrochemical
        Methods for the Removal of Corrosion
        Products   **20**:106

20.1B   Appendix—Standards for Testing   **20**:109

20.2 The Potentiostat and its Application
to Corrosion Studies **20**:123

20.3 Corrosion Monitoring and Chemical Plant **20**:144

20.4 Inspection of Paints and Painting Operations **20**:158

20.1 Corrosion Testing and Determination of Corrosion Rates*

Corrosion tests provide the basis for the practical control of corrosion and therefore deserve a more exhaustive discussion than limitations of space will permit. A detailed description of all the procedures and devices that have been employed in corrosion studies in many countries will not be attempted. Instead, attention will be directed principally to underlying principles and to comments on the significance and limitations of the results of the test methods that are considered. Further details may be obtained from the references and from the comprehensive works by Champion[1] and Ailor[2].

Tests may be classified conveniently as follows:

1. Laboratory tests, in which conditions can be precisely defined and controlled but are usually remote from those that prevail in practice. Laboratory tests are usually accelerated tests.
2. Field tests, in which a number of replicates of different types of specimens are subjected to the environment experienced in service under the same conditions of environmental exposure.
3. Service tests, in which the specimens or protective treatments are incorporated in the actual plant or structure under consideration, with a consequent lower degree of control than that obtained in field tests.

Laboratory corrosion tests are useful in:

1. Studying the chemistry and mechanism of corrosion.
2. Indicating the environments in which a particular metal or alloy may be used satisfactorily.
3. Determining the possible effects of metals and alloys on the characteristics of an environment to which they may be exposed; for example, contamination by corrosion products of chemicals and foodstuffs in process, transportation or storage.
4. Serving as a control test in making a homogeneous, corrosion-resisting metal or alloy.

*Abbreviations for specifications are as follows: BS, British Standard; ASTM, American Society for Testing Materials Standard; DEF, UK Ministry of Defence, Defence Specifications; DIN, West German Standard. A list of specifications relevant to corrosion testing is given in Section 20.1B.

5. Determining the value of changes in composition or treatment in developing corrosion-resisting alloys.
6. Determining whether a metal, an alloy, or a protective coating conforms to a specification requiring a certain performance in a specified corrosion test.

Plant, field and service corrosion tests are most suitable for:

1. Selection of the most suitable material to withstand a particular environment, and estimation of its probable durability in that environment.
2. Assessing the effectiveness of methods of preventing corrosion.

Irrespective of the method of test or the purpose for which it may be made, there are certain practical features which require attention. These will be discussed in turn.

Test Procedures

Preparation of Surface

When the test is to be made to predict the performance of a material in a particular service, the ideal procedure would be to have the surfaces of the test-pieces duplicate the surface of the material as it would be used. Here, however, a complication is presented by the fact that materials in service are commonly used in several forms with different conditions of surface. Where the number of materials to be compared is large, it will usually be impractical to test them all in all the conditions of surface treatment of possible interest. The best practical procedure, then, is to choose some condition of surface more or less arbitrarily selected to allow the materials to perform near the upper limits of their ability. If all the materials to be tested are treated in this way, and preferably with uniform surface treatment, the results of the test will indicate the relative abilities of the different materials to resist the test environment when in a satisfactory condition of surface treatment. Then, if it should be considered prudent or desirable to do so, the most promising materials can be subjected to further tests in a variety of surface conditions so that any surface sensitivity can be detected.

These remarks apply as well to the treatment of the surfaces of specimens to be used in tests in corrosion research projects, except that here selection of a particular method of surface preparation is required so as to achieve reproduceability of results from test to test and amongst different investigators. Methods of preparing specimens are described in an N.A.C.E. Standard[3] and in DEF 1053, Method 2.

The final step in surface preparation should ordinarily be a cleaning and degreasing treatment to remove any dirt, oil or grease that might interfere with the inception or distribution of the corrosion to be studied. The simplest test of a satisfactory surface condition in this respect is for the specimens to be free from 'water break' when rinsed with water after cleaning. As a final treatment for specimens to be weighed prior to exposure, a dip in a mixture of water and acetone or of alcohol and ether will facilitate quick drying and avoid water-deposited films. Specimens to be stored prior to weighing should be placed in a desiccator. In fundamental studies, elaborate precautions must be taken to ensure a perfectly clean surface free from any oxide or other films that might disturb normal reactions. Techniques for accomplishing this have been described by Bockris, *et al.*[4].

In addition to the preparation of the principal surfaces of the specimen it is essential to machine or grind any cut or sheared edges, since these could become sites of preferential attack. As a general rule, edge effects should be kept to a minimum by using specimens in which the ratio of surface area to edge area is large. With flat specimens a disc is best from this point of view, but other shapes may be more convenient and acceptable in many practical instances. When weight loss is to be used as a measure of corrosion, precision will be improved by providing a large ratio of exposed area to mass, and thin flat specimens or fine wires have obvious advantages.

For accuracy of weighing, it is usually necessary to restrict the dimensions of specimens to what can be accommodated on the common analytical balances. It must be borne in mind that where attack occurs in the form of a very few pits or in crevices under supports, the extent of this localised attack may be determined by the total area of the test-piece, as it establishes the area of passive metal acting as a cathode to the few anodic areas. Larger specimens, or much larger surfaces involved in the use of the material under working conditions, may give rise to much more severe localised attack under the same conditions of exposure.

In certain tests it is sometimes desirable to eliminate any effects of a mechanically achieved surface condition by chemical treatment or pickling of the surface prior to test. This may be done in a pickling solution; alternatively, the test itself may be interrupted after sufficient corrosion has occurred to remove the original surface, the specimen then being cleaned and reweighed and the test started over again. Wesley[5] found it to be desirable to pickle off about 0·008 mm from the surface of specimens in acid to improve the reproduceability of the tests.

With materials like the stainless steels, which may be either active or passive in a test environment, it is common practice to produce a particular initial level of passivity or activity by some special chemical treatment prior to exposure. With stainless steels this objective may be subsidiary to eliminating surface contamination, such as iron from processing tools, by treatment in a nitric acid solution[6], which might also be expected to achieve substantial passivity incidental to the cleaning action.

In studies of the behaviour of materials that may be either active or passive in the test environment, there would seem to be a real advantage in starting with specimens in an activated state to see if they will become passive, and to ascertain how fast they are corroded if they remain active. If passivity should be achieved after such an activated start, the material can be considered to be more reliable in the test environment than would be the case if by chance it managed to retain an originally induced passivity for all, or most of, the test period. It may also be valuable to know how fast the metal will be corroded by the test medium if activity should persist.

A procedure for testing previously activated specimens applied in studies of titanium was described by Bayer and Kachik[7]. Renshaw and Ferree[8] also employed prior activation in their studies of the passivation characteristics of stainless steels.

Marking Specimens for Identification

The simplest way to identify a specimen is to mark it with letters or numbers applied by stamping with a stencil. There is, of course, always the danger that the identification marks will be obliterated by corrosion. To guard against this, the several specimens in a test should be identified further by a record of their positions relative to each other or to their supporting device. Before specimens are taken from test their identity should be established in this manner unless inspection has already shown that the identification marks have been preserved.

Other means of identification can be used on steel specimens exposed to atmospheric corrosion. For example, where stamped letters cannot be expected to persist, identification may be provided by holes drilled in particular positions, or by notching the edges of specimens in particular places both in accordance with a template. Where severe corrosion is encountered, the identification by drilled holes is more permanent than that achieved by notching edges.

Other means of identification sometimes used satisfactorily involve chemical etching of the surface, or the formation of letters or numbers by means of a vibrating stylus. The former is advantageous in studies of stress-corrosion cracking in which stamped symbols could lead to regions of stress concentration.

Number of Replicate Specimens

Practical considerations usually limit the number of replicate specimens of each kind that can be exposed for each period of test. At least two are recommended for obvious reasons, and if a larger number can be accommodated in the programme more valuable results can be secured—especially when it is desired to establish the reality of small differences in performance. For statistical analysis, five replicates are desirable. An account of statistical planning and analysis is given by F. H. Haynie in Reference 2.

In providing replicates for tests to be subjected to statistical analysis, it is necessary in the original sampling of the materials to be tested to ensure that normal variations in those qualities of the metals that might affect the results are represented in each set of samples.

In order to secure information as to changes in corrosion rates with time, as in atmospheric exposure tests, it is necessary to expose sufficient specimens to allow sets to be taken from test after at least three time intervals.

For *preliminary* tests where the number of test specimens that can be accommodated is limited, yet numerous materials are of possible interest, it is in order to expose single specimens. This may be more advantageous than limiting the compositions that can be investigated by exposing half the number of materials in duplicate. Probably the greatest advantage in exposing two specimens of a material instead of only one is in detecting gross errors, as in weighing, etc. rather than in any considerable improvement in the precision of the observations that may be made as to the relative behaviours of the metals tested.

Test of Welds

In view of the widespread use of welded joints in equipment and structures exposed to corrosion, it is necessary to know whether such welded joints will demonstrate satisfactory resistance to attack.

It is not necessary to include welded specimens of all materials in a preliminary study to discover which of them have satisfactory resistance to a particular environment. Weld tests can be postponed until the preliminary selection has been made, or, alternatively, those materials expected in advance to be most likely to be resistant can be exposed in the welded condition so as to expedite the final answer.

There are two reasons for testing welded specimens. The first is to discover whether the weld itself will resist corrosion satisfactorily. The second, and usually more critical, purpose is to discover whether the heat effects associated with welding operations have been in any way detrimental to the corrosion resistance of the parent metal near the weld—as in the case of the so-called 'weld decay' of stainless steels (*see* Section 10.5). Since weld deposits may themselves be subject to sensitisation it is necessary to include cross welds in the design of welded specimens for such corrosion tests.

A weld bead included in a test-piece is, to some extent, peculiar to itself and may or may not be representative of presumably similar welds to be made by other welders under other circumstances. To this extent, results of tests on welds must be subject to some qualification in interpretation, having in mind that what will be disclosed principally will be the overall ability of the composition of the weld metal to resist the corrosive environment. In some cases, entrapped flux, craters, fissures, folds, etc. may introduce peculiar localised corrosion that may or may not occur with all welds of the type studied (*see* Section 10.5).

The heat effects of welding are to an even greater extent peculiar to the particular test specimens used. They will be influenced by the skill of the welder, by the thickness of the metal welded, by the type of joint made, and by the geometry and mass of the surrounding structure in so far as they affect heating and cooling rates and areas over which these effects apply. Consequently, what happens to a particular welded test-piece has questionable general significance, especially when the result shows no apparent damage to a material known to be susceptible to welding heat effects in corrosive environments.

With materials of the types under discussion here, there are some heat treatments that are known to reproduce the worst effects of the heat of welding. It is recommended, therefore, that in tests made to qualify a material for a particular service environment, in addition to the exposure of welded test specimens in order to observe effects of welding heat, specimens should be included that have been given a controlled sensitising or abusive heat treatment (such as by holding 18/8 stainless steel at 677°C for 1 h) as provided for in ASTM *Recommended Practice for Conducting Acidified Copper Sulfate Test for Intergranular Attack in Austenitic Stainless Steel*[9] (for further details see *Intergranular Attack* in this section). If such sensitised specimens remain as free from accelerated corrosion as the welded specimens do, then it can be concluded that no detrimental effects of the heat of welding need be anticipated in the environment covered by the test. However, if the sensitised specimens are corroded while the welded specimens are not, there will remain the possibility that under some conditions of welding difficulties due to the effects of the welding heat may be encountered, and appropriate action or the substitution of more reliable compositions will be required. Having in mind the effect of time in damage of this sort, it will be necessary to make a careful examination of the corroded specimens to detect the first signs of intergranular attack before it can be concluded that none has occurred. In assessing the significance of attack observed on drastically sensitised specimens it is necessary to keep in mind that no similar sensitisation may result from good welding practice. Likewise, it should not be concluded that intergranular attack in an arbitrarily chosen test environment known to be capable of causing intergranular corrosion will occur to a similar extent or at all in some quite different environment.

The evaluation of heat treatments or the effectiveness of stabilisation by limiting carbon content of these stainless steels can be determined by subjecting specimens to the ASTM standardised acid copper sulphate test[9] or the boiling nitric acid test[10] (*see also* Sections 10.5 and 1.3).

Duration of Exposure

The duration of a particular test is likely to be determined by practical factors such as the need for some information within a particular limit of time, or the nature of the operation or process with which the test is concerned. Tests are rarely run too long; however, this can happen, particularly in laboratory tests where the nature of the corrosive environment may be changed drastically by the exhaustion of some important constituent initially present in small concentration, or by the accumulation of reaction products that may either stifle or accelerate further attack. In either case, the corrosivity of the environment may be altered considerably. Gross errors may result from the assumption that the results apply to the original conditions of the test rather than to some uncertain and continually changing conditions that may exist during the course of too extended a test period.

Rates of corrosion rarely remain constant with time. More often than not, rates of attack tend to diminish as a result of the formation of adherent insoluble corrosion products or other protective films originating in the environment (Fig. 20.1). Therefore, extrapolation of results of tests that are too short are more likely to indicate a lower resistance to attack than will actually be observed over a prolonged period of exposure. To this extent, such extrapolation may be considered as conservative. At the worst, it may lead to the use of a more resistant material or a heavier section than is actually needed, or to the exclusion from consideration of some materials that might be much better than the short-time test results would indicate.

Having in mind that the principal interest is to run a test long enough to permit demonstration of the possible protective value of films, etc. one basis may be suggested for deciding—provided attack has been uniform—whether a laboratory test has been run long enough. This rests on an assumption that if corrosion has not been stifled before a certain thickness of metal has been corroded away, the self-protection that develops with time can hardly be depended upon. It has thus been suggested that a test can be considered to

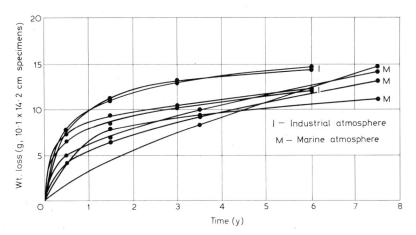

Fig. 20.1   Rate *vs.* time curve showing diminishing rate of attack (after *Proc. A.S.T.M.*, **51**, 500, 1951)

have been run long enough if the observed rate of corrosion in $mg\,dm^{-2}\,d^{-1}$ (mdd), multiplied by the number of hours of test, exceeds 10 000. That is, where the rate is 500 mdd, a 20 h test is adequate; where the rate is 50 mdd, a 200 h test is indicated.

Considerable research on the atmospheric corrosion of steels in different atmospheres has shown that with the standard 102×152 mm (4×6 in) specimens frequently used, about 11 g of metal has to be converted into corrosion products—rust—before a steady rate of corrosion has been established. For the better steels, this will require about 4 years in the most corrosive industrial atmospheres. Therefore such tests should be made for at least this length of time in industrial atmospheres and for even longer periods in marine and rural atmospheres where the full protective effects of rusts take longer to develop.

Tests in waters and soils should ordinarily be allowed to run for extended periods in excess of 3 years, with removals of specimens in groups after different time intervals. A desirable schedule for any extended test in a natural environment is one in which the interval between successive removals is doubled each time. For example, the first removal would be after 1 year, the second after 3 years, and the third after 7 years, and so on.

In any event, the actual duration of a test must be reported along with the results, so that those who may wish to make predictions based on them will have an accurate idea of the extent to which they may undertake any extrapolation or interpolation.

Heat Treatment

Many alloys are subject to drastic changes in their response to the effects of corrosive media when they have undergone certain heat treatments. The principal effect of interest is a loss of corrosion resistance to some degree. This commonly takes the form of concentration of corrosion in particular regions or along certain paths—as in the vicinity of grain boundaries, where phases formed by heat treatment are most likely to be concentrated. In other instances, and particularly in castings, homogenising heat treatments may improve corrosion resistance by eliminating 'coring' or major differences in composition from point to point in the original dendritic cast structure. Heat treatments that eliminate internal stresses are obviously helpful in connection with stress corrosion, but may induce structural changes that can affect corrosion in other forms.

Heat treatments involving heating to a temperature high enough to take harmful phases into solution, followed by cooling (e.g. by quenching) at a rate fast enough to hold such phases in solution, may also be helpful in improving resistance to corrosion by avoiding attack that would otherwise be associated with a precipitated phase or compound.

Obviously, some knowledge of the possible effects of such heat treatments is essential for a complete understanding of the corrosion behaviour of an alloy. Studies along this line should follow upon the initial selection of a material as possibly useful for a particular service by means of tests of it in what should be its best condition to resist corrosion—sometimes the obtaining of this condition may require annealing at a temperature sufficiently high to take any possibly harmful phases or compounds into solution followed by quenching to prevent them precipitating. Following this preliminary selection, it would be prudent to carry out additional corrosion tests on specimens that have been deliberately subjected to any possibly detrimental heat treatments to which the material may be subjected in processing, fabrication, or use.

Heat treatment may also affect the extent and distribution of internal stresses. These may be eliminated by appropriate annealing treatments which can remove susceptibility to stress-corrosion cracking. This must be explored in any studies of the performance of materials in environments where stress-corrosion cracking is a hazard. In particular cases, stress-relief annealing treatments may result in the appearance of new phases which, while eliminating the stress-corrosion effects, will induce another type of path of attack. This possibility must be kept in mind in assessing the overall benefits of heat treatments applied primarily for stress relief.

In other instances, heat treatments involving quenching, tempering, or holding at some temperature to precipitate an age-hardening compound are employed to secure some desired level of hardness or other mechanical properties. It is obviously necessary to explore what effects such heat treatments may have on the corrosion resistance of the material in the condition, or conditions, of heat treatment in which it is to be used.

Stress Effects

Techniques for studying effects of stress on corrosion are covered in some detail elsewhere in this work (*see* Section 8.10). So far as attention to stress effects in a general materials-selection programme is concerned, it is suggested that this should be a supplement to the initial selection of processing materials by exposing specimens in what approaches their best condition to resist corrosion, that is, free from stresses. Materials found to be worthy of further consideration in this way can be subjected to tests for stress effects. Where it is desired to discover whether severe internal stresses can be satisfactorily accepted, it will suffice to expose specimens in such a condition of stress. For example, a crucial test can be made by using a specimen in the form of a heavily cold-drawn tube in the as-drawn condition flattened on one end to introduce some additional multiaxial stresses. If such a severely cold-worked specimen suffers no stress-corrosion cracking in a test, then the danger of this occurring on any structure of that metal in the environment represented by the test is extremely remote.

Appraisal of Damage

There are many ways of determining the extent or progress of corrosion. The choice may be determined either by convenience or on the basis of some special interest in a particular result of corrosion or in a particular stage of a corrosion process.

Probably the most frequently made observation is the change in weight of a test-piece. This may take the form of a weight gain or a weight loss.

Weight-gain determinations are most common in studies of the extent and rate of oxidation or scaling at elevated temperatures. Very precise studies of this sort can be made by continuous observation of weight changes, as in the use of micro-balances, such as are used and described by Gulbransen[11]. Such data have quantitative significance only when the exact composition (metal content) of the scale is known or can be determined and when there has been no loss of loose scale during the test and afterwards. Fundamental studies of the initial stages of corrosion when films of a few monolayers are formed have made use of an ellipsometer to follow the increase of thickness of corrosion products without disturbing the specimen[12].

In most other cases, data on gains in weight due to the accumulation of corrosion products have little quantitative significance since there is usually a question as to how much of the corroded metal is represented in the corrosion products that remain attached to the specimen at a particular time. There are also uncertainties as to the chemical composition of corrosion products, which may consist of mixtures of several compounds with varying amounts of combined or uncombined water, depending on the humidity of the atmosphere at the time.

For these reasons, it is much better to determine the amount of metal removed by corrosion by weighing what is left after removal of all adherent corrosion products by some method that will not cause further attack in the process, or by making a proper correction for losses in the cleaning process. (Removal of corrosion products is dealt with in detail in the Appendix to this section.) Subtracting this final weight from the original weight will give the loss in weight during the test. Since the extent of this loss in weight will be influenced by the area exposed, as well as by the duration of exposure, it is desirable, in order to facilitate comparisons between different tests and different specimens, to report the loss in weight in a unit which includes both area and time. The most commonly used unit of this sort is milligrams weight loss per square decimetre of exposed surface per day (24 h) (mdd), and although the unit $gm^{-2}d^{-1}$ (gmd) has been adopted in the present work the mdd is in current use. It must be recognised that these units embody two assumptions that may not in fact be true. The first is that corrosion has occurred at a constant rate throughout the test period. This is rarely the case, since most rates of attack tend to diminish with time. But if the duration of the test and the actual loss in weight are also reported, the user of the data can take this into account. The second probable error in a weight loss/unit area unit is that it implies that corrosion has proceeded uniformly over the whole surface. These units, therefore, will give the wrong impression as to the probable depth of attack if corrosion has occurred at only a few spots on the surface of the specimen. Obviously, the mdd and gmd units have limited significance when corrosion has taken the form of scattered pits or has been confined to the crevices where the specimen was supported. This should be covered by appended notes describing the nature and location of the corrosion represented and should be supplemented by data on the actual depths of the pitting or crevice attack. Here, again, the report should include data on the actual weight losses and duration of exposure.

Expression of weight loss in terms of a percentage of the original weight of a test-piece is usually meaningless except for comparing specimens of the same size and shape, since it does not take into account the important relationship between surface and mass—on this basis a spherical specimen would appear to resist corrosion better than any other shape.

As indicated, it is necessary to measure and report the depths of any pitting or other localised corrosion, such as in crevices, that may have occurred. It is also useful to provide information on the frequency of occurrence, distribution, and shape of pits, since these features are likely to have practical significance. Champion[1] has produced charts in which the number of pits/unit area, the size of pits, the depth of pitting, cracking and general attack can each be rated by the numbers 1 to 7. Where the number of pits is very large, it is obviously impracticable to measure the depths of all of them. Consequently, the practice has developed of choosing 10 of the deepest pits and reporting their average depth and that of the deepest of them. All surfaces of the specimen should be examined in selecting the 10 deepest pits.

There are several ways of measuring pit depths. If the pits are large enough, their depths may be measured directly with a pointed micrometer or with an indicating needle-point depth gauge. Otherwise, they may be measured optically with a microscope by focusing in turn on the surface of the specimen and on the bottom of the pit using a calibrated wheel on the fine-focus adjustment rack for this focusing operation. In some instances the small dimensions or shapes of pits may require metallographic examination of a cross-section for a precise measurement of depth. Such metallographic examination may also be useful in detecting an association of pitting with a structural feature of the metal.

Since it is often difficult to visualise the extent of attack in terms of depth from such weight-loss units as mdd, it is common practice to convert these mdd figures into others to indicate depth of penetration, i.e. inches per year (ipy), mils (0·001 in) per year (mpy), inches per month (ipm) or $mm\,y^{-1}$ (mpy). Such calculations suffer from the same defects as the mdd figures in that they take into account neither changes in corrosion rates with time nor non-uniform distribution of corrosion· However, since such conversions are often made desirable for the initial reporter of the test results to make the calculations accurately and to report corrosion rates in both mdd and ipy or similar units.

The basic formula for making such calculation is:

$$\text{mdd} \times \frac{(0 \cdot 001\ 437)}{\rho} = \text{ipy}$$

where ρ is the density of the metal in gcm^{-3}; also ipy $\times 25 \cdot 4 = \text{mpy (mmy}^{-1})$. Some values showing the relationship between mdd, and ipy and mpy are given in Table 20.1, and the factors used for converting one form of corrosion rate into another are listed in Table 21.27 in Section 21.1.

Table 20.1 Relationship between corrosion rate in $\text{mg dm}^{-2}\text{d}^{-1}$ (mdd) and penetration in iny^{-1} (ipy) and mmy^{-1} (mpy)

Material	Density (g/cm^3)	Penetration equivalent to a corrosion rate of 1 mdd	
		ipy $\times 10^3$	$\text{mmy}^{-1} \times 10^{-3}$
Aluminium 2S	2·72	0·528	1·346
Ambrac (Cu–6·5Si)	8·86	0·162	0·412
Brass (admiralty)	8·54	0·168	0·427
Brass (red)	8·75	0·164	0·416
Brass (yellow)	8·47	0·170	0·432
Bronze, phosphor (5% Sn)	8·86	0·162	0·412
Bronze (silicon)	8·54	0·168	0·427
Bronze, cast (85–5–5–5)	8·70	0·165	0·419
Cast iron	7·20	0·200	0·508
Copper	8·92	0·161	0·409
Cu–30Ni	8·95	0·161	0·409
Hastelloy *A*	8·80	0·163	0·414
Hastelloy *B*	9·24	0·155	0·394
Hastelloy *C*	8·94	0·161	0·409
Inconel	8·51	0·169	0·429
Iron–silicon alloy	7·0	0·205	0·521
Lead (chemical)	11·35	0·127	0·323
Monel	8·84	0·163	0·414
Nickel	8·89	0·162	0·412
Nickel silver (18% Ni)	8·75	0·164	0·417
Ni-resist	7·48	0·192	0·488
Silver	10·50	0·137	0·348
Stainless steel Type *304*	7·92	0·181	0·462
Stainless steel Type *430*	7·61	0·189	0·480
Steel (mild)	7·86	0·183	0·465
Tin	7·29	0·197	0·500
Zinc	7·15	0·201	0·510

Losses in weight will also not disclose the extent of deterioration that may result from the distribution of a very small amount of attack concentrated along grain boundaries or in transgranular paths (as in some cases of stress-corrosion cracking). In such instances, an apparently trivial or even undetectable loss in weight may be associated with practically complete loss of the strength or ductility of the corroded metal. Where this may be suspected, or in any doubtful cases, the weight-loss determinations must be supplemented by other means of detecting this sort of damage, including simple bend tests followed by visual or metallographic examination to disclose surface cracking, quantitative tension tests, and direct metallographic examination of cross-sections. Changes in electrical resistance have been used as a measure of intergranular attack[13]. Because of the nature of such resistance determinations[14, 15] they have been more useful for comparing specimens of a particular kind and size than as a basis for quantitative expression of rates of attack.

The characteristic mode of corrosion of some alloys may be the formation as a corrosion product of a redeposited layer of one of the alloy constituents, as in the case of the brasses that dezincify, or of a residue of one of the components, as in the case of the graphitic corrosion of cast iron. Particularly in the case of the dezincified brass, the adherent copper is not likely to be removed with the other corrosion products, and therefore the weight-loss determination will not disclose the total amount of brass that has been corroded. This is especially important because the copper layer has very little strength and ductility and the extent of weakening of the alloy will not be indicated by the weight loss. In these cases, also, the weight-loss determinations must be supplemented by, or replaced by, mechanical tests or metallographic examination, or both, to reveal the true extent of damage by corrosion.

Whenever changes in mechanical properties, such as performance in tension tests, fatigue tests, and impact tests, are to be used as a measure of corrosion damage, it is obviously necessary to provide test data on the relevant properties of the uncorroded metal. When tests extend over long periods during which the alloys being tested may be subject to changes in mechanical properties due to ageing effects, entirely aside from corrosion, it will be necessary to provide sets of specimens that may be subjected to similar ageing in a non-corrosive environment so that by direct comparison with corroded specimens of the same age the changes due to corrosion can be separated from those due to ageing. Preferably the control specimens should be stored so that they will be subjected to the same thermal experience as the specimens undergoing corrosion. This is usually very difficult to accomplish while maintaining the control specimens completely protected from corrosion.

In calculating the strength properties of the corroded specimens and comparing them with those of the uncorroded control specimens after appropriate mechanical tests, it will be necessary to take into account the actual area of the cross-section of the corroded metal and report results on this basis instead of, or as well as, on the basis of the original cross-section prior to exposure such as would be represented by the uncorroded control specimens.

In view of possible or probable variations in mechanical properties among different specimens of the same metal cut from different sheets or other pieces, or even from different sections of the same sheet or piece, it is necessary to pay careful attention to the initial sampling of stock to be used for control, as well as exposure, specimens. An interesting case in which several of these considerations were involved was provided by the long-time atmospheric exposure tests of non-ferrous metals carried out by Subcommittee VI of ASTM Committee B-3 on Corrosion of Non-Ferrous Metals and Alloys[16] in which changes in tensile properties were used as one of the means of measuring the extent of corrosion.

Tests carried out for particular purposes may make use of other special means to measure the progress of corrosion. For example, changes in the reflectivity of polished surfaces[17, 18] have been used as a sensitive means of following changes in the very early stages of corrosion in laboratory studies. A similar technique has been applied on a practical scale in connection with the direct evaluation of the relative merits of different alloys as used for mirrors in searchlights exposed to corrosive natural atmospheres.

Kruger[19] at the U.S. National Bureau of Standards has used an ellipsometer to follow the growth of very thin corrosion-product films (oxides) during the initial stages of corrosion. This requires knowledge of the composition of the oxide and its refractive index. An outline of modern physical techniques for studying the nature and kinetics of the growth of oxide films and scales is given in Table 1.6, Section 1.2.

In some cases, the principal interest is in the possibility of undesired contamination or other alteration of an environment rather than in the rate of destruction of the metals being tested. Here, in addition to paying attention to the usual factors that influence rates of corrosion, it is necessary also to consider the ratio of the area of the test specimen to the volume or mass of test solution, and the time of contact. All of these factors may be quite different in a test from what would obtain in a practical case, and any distortions of the test in these ways must be taken into account in planning the test and in interpreting the results.

In cases such as this, the possible contamination of the solution by corrosion products may be estimated from the loss in weight of the test specimen. This, however, does not make any distinction between soluble and insoluble corrosion products, which may have different effects and which can be studied best by chemical analysis of the test solution and the materials filtered from it. Similarly, chemical analysis may be required to detect any other changes in the composition of the test solution that may be of interest.

Particularly in theoretical studies of corrosion processes, it has been useful to measure the progress of corrosion in terms of the rate or extent of consumption of oxygen in the corrosion reactions. This technique has been very useful in following the progress of wet corrosion or of oxidation in its initial stages[20].

Somewhat along the same lines is the measurement of the volume of hydrogen generated as corrosion proceeds[21, 22]. This technique has been used not only in theoretical studies, but as a means of comparing some corrosion-resisting characteristics of different lots of steel which seem to affect their behaviour when used as a base metal for tin cans[23-25].

The polarograph has been found to be a very useful tool for following the progress of corrosion, especially in its early stages, by measuring minute changes in the composition of the solution, as in the consumption of some constituent, such as oxygen, or by the accumulation of metal salts or other reaction products, such as hydrogen peroxide[26].

An electrical resistance method which directly measures loss of metal from a probe installed in the corrosive system under study is described in Section 20.3. It is reported that corrosion equivalent to a thickness loss of as little as $2 \cdot 5 \times 10^{-7}$ cm can be detected[27, 28]. This technique is most useful as a means of monitoring steps taken to reduce corrosion, e.g. by inhibitors, or to detect changes in the corrosivity of process streams. Electrical methods of determining corrosion rates are considered subsequently.

Removal of Corrosion Products

An ideal method for removing corrosion products would be one that would remove them completely without causing any further corrosion or other deterioration of a test specimen in the process. Procedures that achieve this ideal or approach it very closely have been developed for many of the common alloys. Steels, for example, have been cleaned in such a manner that the loss due to cleaning is about 0·01%.

There are numerous satisfactory methods[3] of cleaning corroded specimens, but whatever the method its effect in removing base metal should be determined for each material[29-31]. The various methods may be classified as follows:

  1. Mechanical treatment.
    (a) Scrubbing with bristle brush.
    (b) Scraping.
    (c) Wire brushing.
    (d) Grit, shot and sand blasting.
  2. Chemical treatments.
    (a) Organic solvents.
    (b) Chemical reagents.
  3. Electrolytic treatments as cathode in the following.
    (a) Sulphuric acid.
    (b) Citric acid.
    (c) Potassium cyanide.
    (d) Caustic soda.

Further details of removing corrosion products are given in Appendix 20.1A.

Laboratory Corrosion Tests

Total-immersion Tests

The total-immersion corrosion test is most adaptable to rigorous control of the important factors that influence results. This control may be achieved in different ways and it is unnecessary and undesirable to seek a standardised method or apparatus for universal use. All that is required is a recognition of what is essential, as covered, for example, by the ASTM procedure[31a]. This represents a code of *minimum* requirements without insisting on the use of any particular kind of apparatus or specifying the exact conditions of aeration, temperature or velocity to be used. Since different metals respond differently to effects of aeration, temperature and velocity, the setting up of standard test conditions in terms of these factors would be inappropriate. Depending on the environment, such standardised testing conditions would favour maximum corrosion of some materials and minimum corrosion of others and thus lead to gross errors in indicating any general order of merit applicable under conditions differing from those of a standardised test.

In some instances it may be possible, though it is usually very difficult, to undertake laboratory corrosion tests under conditions that will be the same as those encountered in some practical application, and thus to secure some directly applicable data. More often, the conditions of service are so variable or so difficult to appraise accurately and duplicate in the laboratory that it is impractical and probably unwise to attempt to do so. A better procedure is to examine the individual effects of the several controlling factors by varying them one at a time so as to provide a picture of their influence on the behaviour of the materials of interest in the corrosive medium being investigated. This information will be helpful in deciding whether the conditions of a particular use are favourable or unfavourable to the materials being considered. It will also serve as a guide to account for behaviour in service and to suggest changes in the operating conditions that may be expected to reduce corrosion of a material being used.

Various devices have been used for providing control of the three major factors: *aeration*, *temperature* and *velocity*. Where only moderate test velocities are required, an apparatus in which the specimens are moved in a vertical circular path, all portions of the surface of a specimen moving at the same speed, has been used very successfully.

Statistical analysis of data from tests with this apparatus by Wesley[5] demonstrated satisfactory reproduceability of results not only among specimens in a particular test, but from test to test undertaken at different times.

Temperature control Of the factors mentioned, temperature is probably the easiest to control; this can be accomplished by means of a thermostat or by operating at the boiling point of the testing solution with an appropriate reflux condenser to maintain the solution at a constant concentration. Control to $\pm 1°C$ is not hard to accomplish.

Aeration Control of aeration is more difficult. Aeration here means the amount of oxygen supplied either as such or, more commonly, in air. It requires a surprising amount of air bubbled through a solution to accommodate even a modest rate of corrosion of a small test-piece. Figure 20.2 shows the relationship between the rate of supply of air used for aeration and the rate of corrosion of Monel alloy in 5% sulphuric acid.

To facilitate rapid solution of oxygen from air bubbles it is desirable to make these as small as possible, e.g. by having the air enter through a porous thimble or sintered glass disc. Much less satisfactory results are secured by simply letting air escape into the solution from a tube drawn to a fine tip.

It is also undesirable to permit air bubbles to impinge directly on the test-pieces. This can be avoided by placing the aerator inside a chimney.

When it is desired to study effects of various degrees of aeration, it is better to do this by varying the oxygen content of the saturating gas (e.g. by using controlled mixtures of oxygen and nitrogen) introduced at a constant and adequate rate than by attempting to vary the rate of admission of a gas (such as air) of constant composition. This extends as well to zero aeration, which can best be accomplished by saturating the test solution with de-oxygenated nitrogen or other inert gas. It is unwise to assume that, because no air is

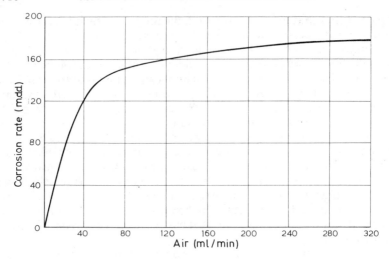

Fig. 20.2   Effect of rate of supply of air used for aeration on corrosion of Monel alloy in 5% sulphuric acid

purposely added, oxygen has been excluded from a test solution in a vessel open to the air. Such a practice provides a low oxygen availability that is not sufficiently under control to ensure reproduceable results.

Velocity   The precise control of velocity and the study of effects of velocity on corrosion are extremely difficult, especially when high velocities are involved. A major problem is to prevent, or to take into account properly, the tendency of a liquid to follow the motion of a specimen moved through it, e.g. by rotation at high velocity. This can be controlled to some extent by proper baffling, but uncertainties as to the true velocity remain—as they do also when the test liquid is made to pass at some calculated velocity over a stationary test-piece or through a test-piece in the form of a tube or pipe[32].

Velocity effects can be achieved either by having the test-piece move through a presumably stationary liquid or by having a moving liquid come into contact with a stationary test-piece. Occasionally tests may involve both types of exposure. Details of test procedures are given in NACE TM-02-70 *Method of Conducting Controlled Velocity Laboratory Corrosion Tests*.

The achievement of zero velocity in a test set-up is about as difficult as the accurate control of some high velocity. It is a common mistake to assume that by not making any attempt to move either the specimen or the testing liquid, the relative velocity between them will be zero. This neglects such effects as convection currents and the agitation due to the effects of corrosion products streaming under the influence of gravity. The most common difficulty arising from this situation is that these uncontrolled effects in tests made under presumably quiet or stagnant conditions make it very difficult to secure reproduceable results from test to test. Therefore, even when there is no practical interest in the effects of any appreciable velocity, it is desirable to provide for some controlled movement of either the specimens or the solution at some velocity such as $7 \cdot 5\,\mathrm{cm\,s^{-1}}$, readily achieved with the vertical circular-path machine already referred to.

Where effects of much higher velocities are to be studied, various devices have been used to move test-pieces through the testing solution at high velocity.

One procedure is to use test specimens in the form of discs which can be rotated at the desired speed while either wholly or partly immersed in the testing solution, and Freeman and Tracy described a device of this sort in a contribution to the ASTM *Symposium on Corrosion Testing Procedures*[29]. With their apparatus the specimen discs were mounted on horizontal shafts and were partially immersed in the testing solution.

A similar method of test is used at the International Nickel Company's Corrosion Laboratory at North Carolina. The specimen discs are mounted on insulated vertical spindles and submerged in sea-water, which is supplied continuously to the tank in which the specimens are immersed. The maximum peripheral speed of the spinning disc is about $760 \, cm \, s^{-1}$, and the characteristic pattern of attack is shown in Fig. 20.3a. Studies of variation of depth of attack with velocity indicate that at low velocities (up to about $450 \, cm \, s^{-1}$) alloys such as admiralty brass, Cu–10Ni and cupro-nickel alloys containing iron maintain their protective film with a consequent small and similar depth of attack for the different alloys. At higher velocities the rate increases due to breakdown of the film.

These tests of this sort indicate a sort of critical velocity for each material that marks the boundary between the maintenance and loss of protective films. These apparently 'critical' velocities must be considered as only relative and applicable only to the conditions of test by which they are measured. Because of the complex effects associated with the differences in velocity from point to point on such rotating specimens, the apparent 'critical velocity' obtained in a given test may be quite different from what might be indicated by another test in which the same velocity might be achieved in some other way—as by moving the liquid past a stationary specimen at a uniform velocity from point to point. The apparent 'critical velocity' indicated by this latter method of test will likely be higher for many materials than that shown by the spinning disc test. Thus the establishment of critical velocities by a particular method of test will afford only qualitative data regarding the relative abilities of a number of materials to resist the destructive effects of high velocity. Furthermore, the critical velocity at which severe attack commences has been found to depend on the diameter of the disc so that no quantitative significance can be attached to it. This restriction extends as well to tests with iron discs, where attack is concentrated at the centre of the disc rather than at the periphery, irrespective of its diameter.

Somewhat similar tests may be made by attaching specimens to discs that can be rotated at some desired velocity in the testing medium. A machine of this sort that is used extensively in studying corrosion of metals by sea-water at high velocity was developed by the staff of the US Naval Engineering Experiment Station at Annapolis, Maryland[33]. A typical assembly of discs and specimens is shown in Fig. 20.3b.

The action of the rotating discs with their attached specimens causes violent agitation of the liquid in the tank. Depending on the height of liquid above the specimens, as determined by the location of the overflow pipe, there may be considerable whipping of air bubbles into the liquid or none at all, as desired. The heat of agitation causes the temperature to rise. This may

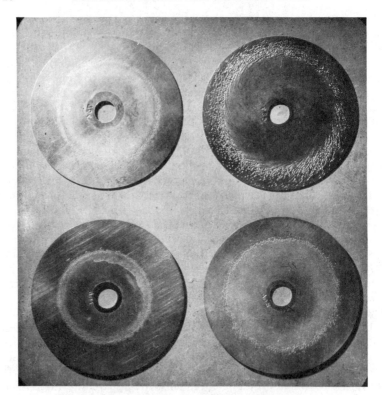

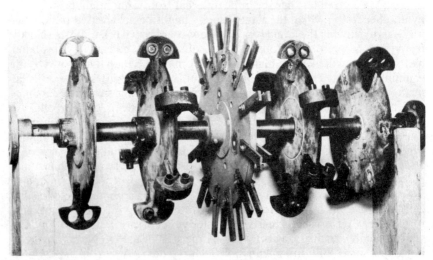

Fig. 20.3   (*a*) Distribution of corrosion on surfaces of rotating disc specimens and (*b*) assembly
of specimens attached to rotating discs

be controlled readily by adjusting the amount of fresh cold liquid, for example
sea-water, allowed to pass into the tank and out through the overflow pipe.
It is not difficult to hold the temperature within 1–2°C of the desired value.

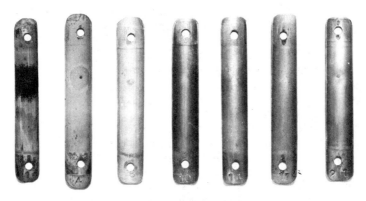

Sample no.	Alloy	Weight loss after 1 000 h (g)	Penetration of impingement pit (% of cross-section)	Appearance
A	Arsenical 70/30 brass	0·262	40	Deep pit opposite jet, surrounding area darkly stained
B	Arsenical Admiralty brass	0·422	44	Deep pit opposite jet
C	Arsenical aluminium brass	0·010	1	Very slightly attacked opposite jet
D	Cu–10Ni	0·026	<1	No measurable attack opposite jet
E	Cu–30Ni–1Fe–1Mn	0·019	<1	No measurable attack opposite jet
F	Cu–30Ni–2Fe–2Mn	0·014	<1	Very slight pitting attack beneath brown tubercles of corrosion products. Also pitting in areas next to sample holder. No measurable attack opposite jet
G	Cu–30Ni	0·061	9	Pitted opposite jet

Materials A–E conform to BS 2871 and ASTM B111.

Fig. 20.4   (top) Pattern of corrosion of jet-impingement test specimens and (bottom) sample test data and results

The use of rotating discs to carry test specimens has been extended to studies of protective coatings in what are considered to be 'accelerated' tests of such coatings for service underwater[34].

Velocity effects involving a high differential in velocity between adjacent areas are achieved simply by exposing a test specimen to the action of a submerged jet. This sort of test has been very popular and very useful in studying impingement attack or erosion of condenser tube alloys. It was introduced originally by Bengough and May[35] and later modifications were described subsequently by May and Stacpoole[36]. The appearances of typical specimens from this test are shown in Fig. 20.4. In this test the dimensions of test specimens should be standardised, since the depth of attack has been found to be influenced by the extent of the immersed area of the specimen that is outside the impingement zone.

Along the same general lines is an apparatus employed by Brownsdon and Bannister[37] in which a stream of air at high velocity is directed against the surface of a submerged test specimen (*see* also p. **20**.71).

A straightforward way to study velocity effects is to force the testing liquid through tubular specimens, which may be arranged to form model piping systems for studying the peculiar corrosion that may result from severe turbulence effects downstream of valves, reducers, branch connections, elbows, and other fittings. In such systems the rates of flow can be measured by suitable orifice meters and regulated by control valves. A somewhat similar technique applied to condenser-tube alloys is to test them as installed in model tube-bundle assemblies[38].

Other methods involve holding specimens in suitable fixtures so that they form the walls of channels through which the test solution can be passed at controlled rates of flow. Such devices are used at the Harbor Island Test Station primarily for studying the electrode potential and polarisation characteristics of metals and alloys, but they are also suitable for observing effects of velocity on corrosion. This is illustrated in Fig. 20.5 in which the specimen and Pt electrode are of the same size and are placed parallel to one another in the holder. When required potentials are measured by inserting a capillary through the hole in the Pt; it is then removed to avoid shielding effect.

Effects of velocity are sometimes aggravated by the presence of abrasive solids in suspension, which increases deterioration by straight mechanical

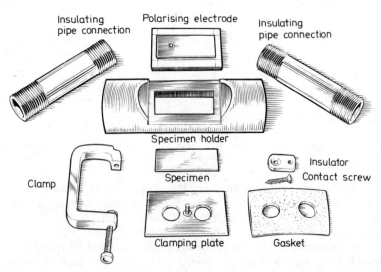

Fig. 20.5   Components of apparatus in which specimens form walls of channel for test solution

abrasion as well as accelerating corrosion by continually exposing fresh surfaces to attack. Such attack is especially serious with pumps, agitators and piping systems. Special apparatus has been designed to measure the performance of materials under such conditions, as described, for example, by Fontana[39].

Special devices have been used to study erosion-corrosion by boiler water moving at high velocity, and an example is the method used by Wagner, Decker and Marsh[40].

Where high rates of flow are desired with a small volume of testing liquid a specimen may be mounted in the form of a tube inside a large glass tube, and a small mass of liquid may be forced to flow through the restricted annular space between the two tubes. Such a method was used successfully in studying corrosion by milk where the volume of milk was small, the required movement being achieved simply by the use of an air lift to return the milk to an overhead reservoir from which it flowed by gravity through the test set-up[41]. Velocities as high as 0.6 m s^{-1} were studied in this way.

Volume of testing solution   If exhaustion of corrosive constituents that may be present in minute concentrations and the accumulation of reaction products which may either accelerate or stifle further attack are to be avoided, the volume or mass of testing solution must be sufficiently large to avoid effects caused by these factors. In laboratory tests, however, practical considerations limit the volume of testing solution that can be provided for. A minimum of 250 ml of testing solution for each 6.3 cm^2 of specimen area is suggested.[3]

Support of specimens   Since crevices set up where specimens are in contact with their supports may become the seats of accelerated corrosion by concentration cell effects, special attention should be given to this detail in setting up tests. The area screened by the supporting members should be kept to a minimum, for example by making contact at a point or along a line rather than over any appreciable area. In some instances it may be desirable to apply some protective coating to the areas that are in contact with the supporting members. In any event, any corrosion that has occurred in the area of the supports should be taken into account in appraising and reporting the results of a test.

Somewhat along the same lines are techniques that have been employed to avoid edge effects by having the specimen come into contact only with a pool of testing solution which does not cover its complete surface—as described, for example, by Brennert[42]. A more elaborate technique in which the pool of testing solution was circulated by thermal currents was described by Smith[43].

Alternating-immersion Tests

One means of ensuring aeration of a testing solution in contact with a specimen is provided by an alternating-immersion corrosion test in which the specimen is alternately immersed in a solution and withdrawn from it in some predetermined cycle. This procedure also has the effect of allowing the test solution that clings to the specimen to become concentrated by evaporation while the specimen is out of the liquid, and in addition it permits corrosion products to remain and reach greater concentrations and undergo more chemical changes in immediate contact with the metal than can occur in continuous-immersion tests. In these ways, an alternating-immersion test may simulate certain circumstances of practical corrosion better than a continuous-immersion test and may, therefore, be preferred.

Since the conditions of this test can be standardised fairly readily, it has also been used as a routine test in comparing different alloys of the same general kind in the course of studies of effects of composition on properties, as, for example, in the researches by Hanawalt, Nelson and Peloubet on the corrosion of magnesium[44].

The ASTM established a recommended procedure for alternating-immersion tests as covered by Tentative Method B 192[45], but this is now obsolete, since it was recognised that different degrees of control had to be placed on the several test conditions in studying various materials and corrosion phenomena.

The alternating immersion may be accomplished either by moving specimens held in a suitable suspension rack into and out of containers holding the test solution, or by leaving the specimens fixed and raising and lowering the solution containers around them so as to immerse them or leave them suspended above the solution.

To favour reproduceability of results, the cycles of immersion and withdrawal must be kept the same from test to test. It is necessary to control the temperature and humidity of the atmosphere surrounding the test set-up as these affect the rate of evaporation of the solution and of drying of the specimens when they are out of the solution. It is also necessary to provide for replenishing losses of water from the test solution resulting from evaporation.

Water-line Tests

Materials may be subject to intense localised attack at the liquid level when they are partially immersed in a solution under conditions where the water line remains at a fixed position for long periods. This attack may be the result of concentration cell effects complicated by differences in the nature and adherence of corrosion-product films as they form in the water-line region as compared with those that form above or below this region.

The testing technique is very simple since it involves no more than providing means of supporting a specimen or specimens in a fixed position of partial immersion, and of maintaining the liquid level constant by the continuous addition of distilled water to make up for evaporation losses. For maximum reproduceability of results, the dimensions of the specimens, and especially the ratio of areas above and below the liquid level, should be held constant, as should be the depth of immersion.

Heat-flux Effects

When heat flows into or out of a fluid through a containing wall, the wall surface reaches a temperature which differs from that of the bulk of the fluid. The wall's corrosion resistance at this temperature may be significantly different from its resistance at the bulk-fluid temperature. Tubes or tank walls heated by steam or direct flame have failed in service in which similar materials, not so heated, performed acceptably.

The name 'hot wall effect' was given to this phenomenon by Benedicks[46] who observed separation of dissolved gas from aerated water in boiler tubes. The metal wall was insulated from the cooler boiler water by the gas, its temperature rose substantially, and the more severe corrosion took the form of pitting. More recently it was recognised that even without such gas formation, a hot-wall effect resulted when heat flowed through a wall into a fluid. The temperature difference which had to exist in order for heat to flow was increased by the insulating effect of the thin film of almost stagnant fluid at the wall surface. This film is thinned by rapid flow of a fluid through a tube, but is not eliminated at any finite velocity. Boiling of a liquid, by either bubble nucleation or coverage of the heating surface by a vapour film, increases the skin temperature further; the second mechanism provides much more severe insulation and greater temperature rise[47].

High rates of heat flow through heat-transfer surfaces in atomic energy installations studied by Groves caused him to develop an appropriate corrosion test method[48]. In this, a small sheet specimen in contact with a hot

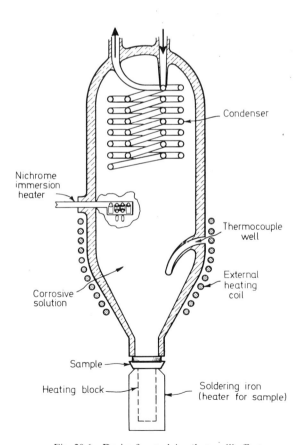

Fig. 20.6   Device for studying 'hot-wall' effects

liquid, usually boiling, is heated externally so that its surface reaches the desired test temperature. The surface or skin temperature is close to the temperature at mid-thickness, which is measured by a thermocouple inserted in a drilled hole. The heating source is an electric soldering iron in which the tip is replaced by a flat-surfaced metal block. The voltage to the heat source is controlled by a variable transformer. Supplementary heating may be provided to the liquid in the test vessel by resistance-wire winding of the liquid container, or by an immersion heater (Fig. 20.6).

A similar unit, modified in details such as locations of condenser, use of an agitator and shape of the vessel, was used by Fisher and Whitney[49]. Further substantial modifications to permit interface location of specimens, cooling of specimens and operation under applied pressure, have been described by Fisher[50]. Earlier laboratory test methods tried by Fisher and Whitney[49] included exposure of specimens heated by their own electrical resistance and of tubular specimens containing a pencil-type resistance-wire heater in a quartz tube.

To investigate corrosion in heated crevices filled with wetted paste a sandwich test assembly was designed by Gleekman and Swandby[51] simulating a slotted cylindrical steam-heated drier. Two plates are bolted together, the lower being heated by an electric hot plate.

It has been concluded from data reported in these studies that the skin temperature is the major controlling factor in corrosion, not the rate of heat flow through the metal[49]. It has also been concluded, however, that corrosion rates at a given mid-specimen temperature do depend on the presence or absence of thermal flux[52]. The difference between temperatures at skin and mid-specimen positions may account for this discrepancy.

Heat-flux corrosion rates can also be determined in plant tests using steam-heated tubular specimens which are weighed or callipered.

In addition to the direct effect of film temperature on corrosion rate, an indirect effect has been observed in the heating of some foods and chemicals, in which insulating solid corrosion films form on different metals. By raising the metal surface temperature, these films may, when pervious, lead to further corrosion.

Apparatus and procedures for testing the corrosion resistance of alloys in brines at temperatures up to 120–150°C are described by Hart[53].

Composition of Testing Solution

There is not much that need be said about control of the composition of the testing solution beyond that it should be what it is supposed to be. Carefully made-up solutions of pure chemicals may not act in the same way as nominally similar solutions encountered in practice, which may, and usually do, contain other compounds or impurities that may have major effects on corrosion. This applies particularly to 'artificial' sea-water, which is usually less corrosive than natural sea-water. Compositions of artificial sea-water used in various countries are given in Table 20.2. Suspected impurities may be added to the pure solutions in appropriate concentrations or, better still, the testing solutions may be taken directly from plant processes whenever this is practical.

Table 20.2 Compositions of artificial sea-waters*

Mark	A	B1‡	B2§	C	D	E‡	F‖	G	H‖	J	K	L
Year	1930	1938	1942	1933	1934	1938	1947	1947	1948	1948	1961	1952
Country	Germany	Germany	Germany	England	France	Germany	USA	England	England	Germany	Germany	England
$NaCl$†	15·6	29·6	29·6	29·3	27·0	27·3	24·5	23·0	27·26	27·4	28	26·5
KCl	—	—	—	—	0·5	—	0·69	—	0·69	—	—	0·73
$MgCl_2$	2·0	3·6	3·6	4·09	3·5	3·3	5·20	4·6	3·51	2·4	2·3	2·4
$CaCl_2$	—	—	—	—	—	—	1·16	1·2	—	1·2	1·2	1·1
$SrCl_2$	—	—	—	—	—	—	0·02	—	—	—	—	—
$NaBr$	—	—	—	—	—	0·2	—	—	—	—	—	0·28
KBr	—	—	—	—	—	—	0·10	—	0·09	—	—	—
NaF	—	—	—	—	—	—	0·003	—	—	—	—	—
Na_2SO_4	0·5	—	—	—	—	—	4·09	3·9	—	—	—	—
K_2SO_4	—	—	—	0·94	—	—	—	—	—	0·2	—	—
$MgSO_4$	1·0	2·4	2·4	3·65	2·5	2·2	—	—	1·84	—	3·5	3·3
$CaSO_4$	0·7	—	1·3	1·72	0·5	—	—	—	1·29	—	—	—
$NaHCO_3$	—	—	—	—	—	—	0·20	—	0·11	0·2	0·2	0·20
$CaCO_3$	0·1	—	—	—	—	—	—	—	—	—	—	—
Na_2HPO_4	—	—	—	—	—	—	—	—	—	—	—	—
H_3BO_3	—	—	—	—	—	—	0·03	—	—	—	—	—
pH	—	—	8·0	—	—	—	7·8–8·2	—	8·1–8·2	—	7·8	—

* Data after Champion[1].
† Concentrations in g/l.
‡ Dissolved in tap water.
§ pH adjusted by additions of NaOH.
¶ Na₂CO₃ added to adjust pH.
‖ Aerated.

It should also be pointed out that in exploring effects of the concentration of a particular acid or other chemical on its corrosivity, it is necessary to cover the full possible variation of concentrations thoroughly, since it frequently happens that particular ranges of concentration are especially corrosive to some metals. This extends to the highest degrees of concentration where sometimes the complete elimination of water may increase corrosion a great deal—as in the case of aluminium in acetic acid. On the other hand, the presence of a trace of water may make other chemicals much more corrosive —as in the case of bromine and other halogens.

It should be noted, also, that exposing a specimen to a solution of some chemical while it is being concentrated by evaporation practically to dryness will not suffice to explore the effects of the complete range of concentration, simply because the period in which any particular concentration range exists is not likely to be long enough to permit any especially corrosive effects to be detected in the overall result.

Changes in corrosivity with time may be observed by exposing fresh specimens to a solution that has already been used for testing. Where such changes are known to occur, or are suspected, it will be necessary to arrange for replacement of the testing solution after appropriate intervals or replenishment of constituents that may be consumed in the corrosion processes.

Electrochemical Measurements

In view of the electrochemical nature of corrosion it is not surprising that measurements of the electrical properties of the interface metal/solution (electric double layer) are used extensively in fundamental studies of the mechanism of corrosion, in corrosion testing and in monitoring and control in service. In the context of this section electrical measurements in the laboratory are used to assess the corrosion behaviour of metals and alloys in service, and to avoid the more tedious and prolonged field testing. Determinations of the corrosion rate, susceptibility of a metal to bimetallic corrosion, pitting, intergranular attack, stress-corrosion cracking, etc. are examples of corrosion phenomena that are studied in the laboratory by means of electrochemical methods in order to anticipate behaviour in service.

Progress in this field has been made possible with increase in knowledge of the detailed mechanism of corrosion and by the developments that have taken place in instrumentation. The widespread use of potentiostatic control (Section 20.2) and the availability of a range of commercial potentiostats have given a tremendous impetous to electrochemical testing, and have perhaps led to the unfortunate belief that corrosion testing in the laboratory and in the field can be replaced completely by electrochemical measurements in the laboratory under conditions of controlled potential. Indeed, La Que[54] in 1969 was prompted to express concern about the proliferation of publications describing electrochemical techniques for corrosion testing, and to advise caution regarding the extrapolation of results obtained in the laboratory with a potentiostat to the performance of metals in service.

Although important contributions in the use of electrical measurements in testing have been made by numerous workers it is appropriate here to refer to the work of Stern and his co-workers[55, 56] who have developed the important concept of *linear polarisation*, which has led to a rapid electro-chemical method for determining corrosion rates, both in the laboratory and in plant. Pourbaix[57] and his co-workers on the basis of a purely thermo-dynamic approach to corrosion have constructed potential–pH diagrams for the majority of M–H_2O systems, and by means of a combined thermo-dynamic and kinetic approach have developed a method of predicting the conditions under which a metal will (*a*) corrode uniformly, (*b*) pit, (*c*) passi-vate or (*d*) remain immune. Laboratory tests for crevice corrosion and pitting, in which electrochemical measurements are used, are discussed later.

Techniques

Electrochemical methods of testing involve the determination of specific properties of the electrical double layer formed when a metal is placed in contact with a solution (*see* Section 9.1), and these can be summarised as follows.

1. The potential difference across the electric double layer $\Delta\psi$. This cannot be determined in absolute terms but must be defined with reference to another charged interface, i.e. a reference electrode (*see* Section 1.9B). In the case of a corroding metal the potential is the corrosion potential which arises from the mutual polarisation of the anodic and cathodic reactions constituting the overall corrosion reaction (*see* Section 1.4).
2. The reaction rate per unit area i. For a corroding metal the partial anodic and cathodic current densities cannot be determined directly by means of an ammeter unless the anodic and cathodic areas can be separated physically, e.g. as in a bimetallic couple. If the metal is polarised a *net* current i_c for cathodic polarisation, and i_a for anodic polarisation, will be obtained and can be measured by means of an ammeter.
3. The capacitance. The electrical double layer may be regarded as a resistance and capacitance in parallel (*see* Section 9.1), and measure-ments of the electrical impedance by the imposition of an alternating potential of known frequency can provide information on the nature of a surface. In testing, impedance measurements have been largely con-fined to determinations of the protective nature of anodic films on aluminium.

The most commonly used measurements are as follows:

1. Determination of the steady-state corrosion potential $E_{corr.}$.
2. Determination of the variation of $E_{corr.}$ with time.
3. Determination of the $E-i$ relationships during polarisation at constant current density (galvanostatic) the potential being the variable.
4. Determination of the $E-i$ relationships during polarisation at constant potential (potentiostatic) the current being the variable.

Instruments

The techniques and instruments used may be classified as follows:

1. Potential measurements—a reference electrode and a potentiometer or electrometer which require only a small current to give a measurement of e.m.f. and thus minimise polarisation of the electrodes.
2. Current measurements—milliammeters or the measurements of the *IR* drop across a conductor of known resistance.
3. Galvanostatic polarisation—constant direct current power units, or banks of accumulators or dry cells used in conjunction with a variable resistance.
4. Potentiostatic polarisation—potentiostats with varying output currents.
5. Determination of impedance—a.c. bridges of various frequencies.

Impedance measurements, although widely used in fundamental studies of anodic oxidation, have had only a limited application in studies of corrosion owing to the sophistication of the electrical equipment required, limitations imposed on the design of the cell and above all on the complexity of the analysis of the results obtained for a corroding electrode. However, Armstrong, *et al.*[58] have used impedance measurements for studying the active–passive transition of chromium, and Epelboin, *et al.*[59] describe its use for determining the instantaneous corrosion rate of a metal. Sathyaharayana[59a] has described a method using faradaic rectification to determine the instantaneous corrosion rate, in which no reference electrode is required; the electrodes consist of the metal under study and a counter electrode of large area of the same metal.

The potentiostatic technique has a number of variations and the potential may be increased or decreased incrementally, changed continuously at a predetermined rate (potential sweep) or applied as pulses of very short duration. The applications of the potentiostatic technique are considered in detail in Sections 1.4, 1.5 and 20.2, and will not be considered here.

Electrochemical Cells

Since the single potential of a metal cannot be measured it is necessary to use a suitable reference electrode such as the $Hg/Hg_2Cl_2/KCl$ electrode or the $Ag/AgCl/KCl$ electrode, and although potentials are frequently expressed with reference to the standard hydrogen electrode (S.H.E.) the use of this electrode in practice is confined to fundamental studies rather than testing. Details of the preparation of reference electrodes, salt bridges, capillaries, etc. are given in the book by Ives and Janz[60].

Measurements of the corrosion potential of a single metal corroding uniformly do not involve an *IR* drop, but similar considerations do not apply when the metal is polarised by an external e.m.f., and under these circumstances the *IR* drop must be minimised by using a Luggin capillary placed close to the surface of the electrode (*see* Fig. 1.22, Section 1.4). Even so, the *IR* drop is not completely eliminated by this method, and a further error is introduced by the capillary shielding the surface from the current flow with a consequent decrease in c.d. At high c.ds. this error due to the *IR* drop can conceal the Tafel region by distorting the measured overpotentials, a difficulty that can be overcome by determining the resistance of the solution at the capillary tip and by making an appropriate correction for each value of

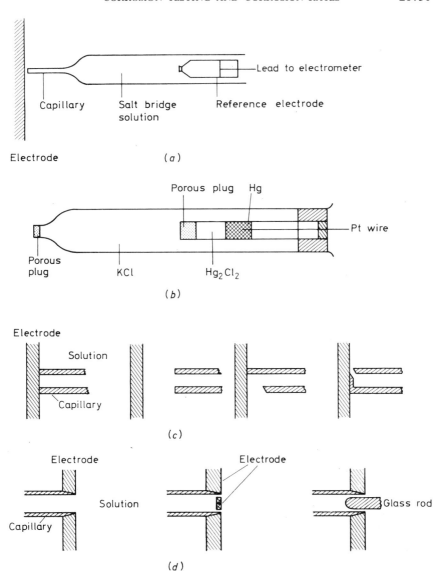

Fig. 20.7   Reference electrodes and capillaries. (*a*) Reference electrode, salt bridge and Luggin capillary; (*b*) calomel electrode; (*c*) frontal types of capillaries and positions; (*d*) rearside capillaries (after von Fraunhofer and Banks[66])

the current density. Alternatively, electronic feedback circuits may be employed for automatic compensation of the *IR* drop, and this method is attractive if rapid variations in overpotential are being studied[61-64]. Other methods are by using a Piontelli capillary[65] or a rearside capillary[66] (Fig. 20.7). However, in testing in electrolyte solutions of low resistivity these errors are normally small, and the conventional Luggin capillary is used in conjunction with a salt bridge and reference electrode.

For polarisation studies the cell must make provision for the metal electrode under study, an auxilary or counter electrode, and a Luggin capillary. Provision must also be made for introducing a gas such as oxygen-free nitrogen or argon, which serves to remove dissolved oxygen and to prevent its introduction during the test (or to introduce it if required at predetermined partial pressures) and to agitate the solution; additional agitation if required can be obtained by means of a stirrer (electric or magnetic).

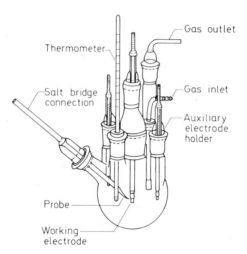

Fig. 20.8   All-glass cell for studies of polarisation of metal electrodes (after ASTM G5:1972)

Figure 20.8 shows the design of an all-glass cell, which has been listed as the standard polarisation cell in the ASTM Recommended Practice G.5 (1970 Book of ASTM Standards, Part 31), which makes provision for the essential requirements listed above; this cell is typical of those used for fundamental studies and for testing, although details of design may vary.

The metal electrode to be studied must be carefully prepared, attached to an electrical lead and mounted so that a known surface area of one face is presented to the solution. Several procedures are used such as mounting in a cold setting resin (Araldite) or inserting into a close-fitting holder of p.t.f.e. In the case of metal–solution systems that have a propensity for pitting care must be taken to avoid a crevice at the interface between metal specimen and the mounting material, and this can be achieved effectively by mounting the specimen in a cold-setting resin with approximately 1 mm thickness on the working face; the latter is then removed by carefully abrading on emery paper to expose the specimen. In view of the widespread use of the cell shown in Fig. 20.8 suitable electrode holders, based on the use of a compression gasket of p.t.f.e. or a similar inert polymer, have been designed for bulk metal specimens[67, 68], wires[69], tubes[70], sheets and foil[71] and for high-temperature high-pressure assemblies[72]. Examples of methods of mounting specimens to give a defined area of surface are shown in Fig. 20.9[66].

Various types of reference electrodes have been considered in Section 9.1B, and the potentials of these electrodes and their variation with the activity of the electrolyte are listed in Table 21.7, Chapter 21. It is appropriate, however,

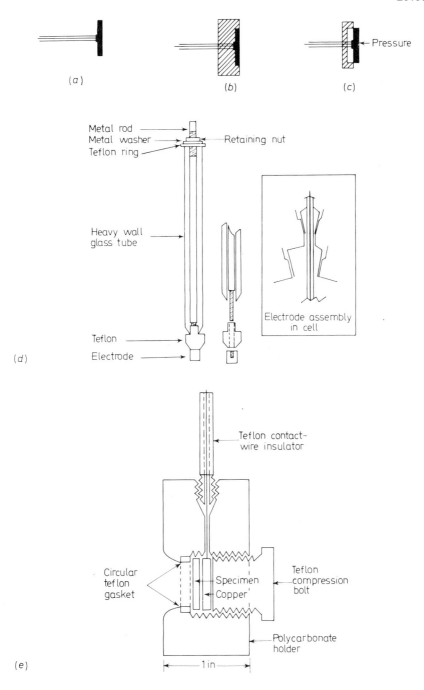

Fig. 20.9   Methods of mounting specimens. (*a*) Wire soldered to metal specimen, wire being enclosed in glass tube; (*b*) specimen completely encapsulated in cold-setting resin and resin ground down to expose one face; (*c*) specimen clipped into machined p.t.f.e. holder; (*d*) Stern-Makrides[67] pressure gasket for cylindrical specimen; (*e*) pressure gasket for sheet or foil[71]

to point out here that the saturated calomel electrode (S.C.E.), the silver–silver chloride electrode and the copper–copper sulphate electrode are the most widely used in corrosion testing and monitoring.

To avoid contamination of the solution under study, and to minimise the liquid-junction potential, it is usual to use a salt bridge, but in many cases this can be dispensed with; thus if corrosion in a chloride-containing solution is being studied a Ag/AgCl electrode immersed directly on the solution could be used; similarly a Pb/PbO$_2$ electrode could be used for studies of corrosion in H$_2$SO$_4$.

Measurements of the Corrosion Potential

The significance of the corrosion potential in relation to the equilibrium potentials and kinetics of anodic and cathodic reactions has been considered in Section 1.4, but it is appropriate here to give some examples of its use in corrosion testing. Pourbaix[57] has provided a survey of potential measurements in relation to the thermodynamics and kinetics of corrosion, and an example of how they can be used to assess the pitting propensity of copper in Brussels water is given in Section 1.6.

The determination of the corrosion potential of the two metals constituting a bimetallic couple will provide information on which one of the two will be predominantly anodic and will suffer enhanced corrosion when they are coupled. Similarly the effect of microscopical heterogeneities in alloys on corrosion has been investigated by measuring the corrosion potentials of selected small areas of the surface of the metal. Smith and Pingel[73] coated the surface of the metal with a lacquer of ethyl cellulose (deposited from a volatile solvent), allowed it to dry and then perforated selected areas with a micro-hardness tester using a steel stylus ground to a truncated cone, and Budd and Booth[74] using a similar technique were able to produce punctures in the film down to 35 µm diameter. Microelectrodes for potential measurements have been described by Cleary[75] who used a Ag/AgCl/Cl$^-$ electrode with capillary tips down to 15 µm; Cleary also describes micro-glass electrodes of approximately 30 µm diameter for the determination of pH.

Potential–time relationships have been widely used for studying film formation and film breakdown, as indicated by an increase or decrease in the corrosion potential, respectively. May[76] studied the corrosion of 70/30 brass and aluminium brass in sea-water and showed how scratching the surface resulted in a sudden fall in potential to a more negative value followed by a rapid rise due to re-formation of the film; conversely, the pitting of stainless steel in chemical plant may be detected by a sudden decrease in potential[77].

Hoar and his co-workers[78] used potential changes to study film breakdown and repair during the stress-corrosion cracking of austenitic stainless steels in boiling saturated MgCl$_2$ solution. More recently, Horst, et al.[79] have used potential measurements as a test to predict the stress-corrosion susceptibility of 2219 aluminium alloy products (alloys containing approximately 6% Cu and tempered to give maximum strength and resistance to stress-corrosion cracking). The test solution used was methanol plus carbon tetrachloride and it was shown that susceptible alloys were 200–500 mV more positive than non-susceptible alloys, and that this difference in potential was revealed in less than 1 h. They claim that the test is more sensitive than that in which sodium chloride plus hydrogen peroxide is used as the test solution, since the latter gives potential differences of only 20 mV.

A detailed and well-referenced account of electrochemical methods of testing has been written by Dean, France and Ketcham in a section of the book by Ailor[2].

Polarisation Resistance

It is evident from previous considerations (*see* Section 1.4) that the corrosion potential $E_{corr.}$ provides no information on the corrosion rate, and it is also evident that in the case of a corroding metal in which the anodic and cathodic sites are inseparable (*c.f.* bimetallic corrosion) it is not possible to determine $i_{corr.}$ by means of an ammeter. The conventional method of determining corrosion rates by weight-loss determinations is tedious and over the years attention has been directed to the possibility of using instantaneous electrochemical methods. Thus based on the Pearson derivation[80], Schwerdtfeger, *et al.*[81,82] have examined the logarithmic polarisation curves for 'potential breaks' that can be used to evaluate the corrosion rate; however, the method has not found general acceptance.

Skold and Larson[83] in studies of the corrosion of steel and cast iron in natural water found that a linear relationship existed between potential and the applied anodic and cathodic current densities, providing the values of the latter were low. However, the recognition of the importance of these observations is due to Stern and his co-workers[55,56] who used the term 'linear polarisation' to describe the linearity of the η–i curve in the region of $E_{corr.}$, the corrosion potential. The slope of this linear curve, ΔE–ΔI or ΔE–Δi, is termed the *polarisation resistance*, R_P, since it has dimensions of ohms, and this term is synonymous with 'linear polarisation' in describing the 'Stern–Geary' technique for evaluating corrosion rates.

Stern and Geary[55,56], on the basis of a detailed analysis of the polarisation curves of the anodic and cathodic reactions involved in the corrosion of a metal, and on the assumption that both reactions were charge-transfer controlled (transport overpotential negligible) and that the iR drop involved in determining the potential was negligible, derived the expression

$$\frac{1}{R_P} = \left(\frac{\Delta i}{\Delta E}\right)_{E_{corr.}} = 2 \cdot 3\left(\frac{b_a + |b_c|}{b_a|b_c|}\right)i_{corr.} \qquad \ldots(20.1)$$

where R_P is the polarisation resistance determined at potentials close to $E_{corr.}$, and b_a, b_c are the Tafel constants; note that in the case of b_c the negative sign is disregarded. This equation shows that the corrosion rate is inversely proportional to R_P (or directly proportional to the reciprocal slope of the ΔE – Δi curve) at potentials close to $E_{corr.}$ ($\ngtr 10$ mV), and that $i_{corr.}$ can be evaluated providing the Tafel constants are known. For a process that is controlled by diffusion of the cathode reactant (transport control) and in which the anodic process is under activation control a similar linear relationship applies:

$$\frac{1}{R_P} = \left(\frac{\Delta i}{\Delta E}\right)_{E_{corr.}} = \frac{2 \cdot 3 i_L}{b_a} = \frac{2 \cdot 3 i_{corr.}}{b_a} \qquad \dots(20.2)$$

where i_L is the limiting current density of the cathodic reaction and it is assumed that $i_L = i_{corr.}$.

Stern and Weisert[84] by taking arbitrary values of the Tafel constants showed that corrosion rates determined by the polarisation resistance techniques are in good agreement with corrosion rates obtained by weight loss methods.

The importance of the method in corrosion testing and research has stimulated other work, and since Stern's papers appeared there have been a number of publications many of which question the validity of the concept of linear polarisation. The derivation of linearity polarisation is based on an approximation involving the difference of two exponential terms, and a number of papers have appeared that have attempted to define the range of validity of polarisation resistance measurements. Barnartt[85] derived an analytical expression for the deviations from linearity and concluded that it varied widely between different systems. Leroy[86], using mathematical and graphical methods, concluded that linearity was sufficient for the technique to be valid in many practical corrosion systems. Most authors emphasise the importance of making polarisation resistance measurements at both positive and negative overpotentials.

Oldham and Mansfeld[87] have approached the problem of linearity in a different way and their derivation avoids the approximation used by Stern and Geary. They conclude that although linearity is frequently achieved this is due to three possible causes: (a) ohmic control due to the IR drop rather than control according to linear polarisation, (b) the similarity of the values of b_a and b_c and (c) a predisposition by the experimenter to assume that the $\Delta E - \Delta i$ curves near $E_{corr.}$ must be linear. In a later paper Oldham and Mansfeld[88] show that linearity of the $\Delta E - \Delta i$ curve is not essential and that $i_{corr.}$ can be evaluated from the slopes of the tangents of the non-linear curve determined at potentials of about 20–30 mV more positive and negative than $E_{corr.}$.

Hickling[89], in attempting to study the corrosion of steels under thin film conditions that simulate atmospheric exposure, has taken into account the time-dependence of polarisation measurements, and has developed a technique using galvanostatic transients.

Tafel Constants

It is evident from equations 20.1 and 20.2 that the evaluation of $i_{corr.}$ from R_p determinations requires a knowledge of the Tafel constants b_a and b_c, which may not be available for the system under study and which may change in value during the progress of the determination. The determination of the Tafel constants from complete $\eta-i$ curves for each system studied is time consuming, and may not be particularly accurate owing to resistance and mass transfer effects. Hoar[90] has criticised the method on these grounds and has pointed out that the complete Tafel equations for the anodic and cathodic reactions, which have to be determined to evaluate the Tafel slopes, can be used to calculate $i_{corr.}$ without resorting to the polarisation resistance technique (see Section 1.4). Mansfeld[91, 91a] suggests that polarisation curves obtained in the R_p region can be fitted to various theoretical curves, preferably by computer analysis, to give the separate value of both b_a and b_c, which since they are determined simultaneously with the R_p values avoids the criticism that they may change substantially during the corrosion test (*see* p. 20.38).

The controversy that arises owing to the uncertainty of the exact values of b_a and b_c and their variation with environmental conditions, partial control of the anodic reaction by transport, etc. may be avoided by substituting an empirical constant for $(b_a + |b_c|/|b_a|b_c|)$ in equation 20.1, which is evaluated by the conventional weight-loss method. This approach has been used by Makrides[92] who monitors the polarisation resistance continuously, and then uses a single weight-loss determination at the end of the test to obtain the constant. Once the constant has been determined it can be used throughout the tests, providing that there is no significant change in the nature of the solution that would lead to markedly different values of the Tafel constants.

Applications

The method in spite of its limitations has a number of significant advantages and provides a method of rapidly monitoring the instantaneous corrosion rates; furthermore, it has the advantage that the small changes in potential required in the determination do not disturb the system significantly. It is capable of measuring both high and low corrosion rates with accuracy, and may be used as a laboratory tool for testing or research, or for monitoring corrosion rates of plant (*see* Section 20.3).

Stern[56] pointed out that the polarisation resistance method could be of value for determining the effect of changes of environment (composition, temperature, velocity) and alloy composition on the corrosion rate and for evaluating inhibitors, and since his original publications the method has been widely used for a variety of studies. Thus Legault and Walker[93] used the method for studying the inhibition of the corrosion of steel in chloride solutions by $NaNO_2$, and France and Walker[94] extended this approach to a study of the *in situ* corrosion of the various metals in automotive-engine cooling systems. Jones and Greene[95] developed the theory of transient linear polarisation to study very low corrosion rates, such as occur with surgical implant materials, and have shown how polarisation resistance data can be used to monitor the onset of pitting or other forms of localised corrosion.

Wilde[96] has applied the Jones d.c.-bridge technique[97] to compensate for errors due to the *IR* drop, and has obtained meaningful corrosion rates from polarisation resistance data in high-temperature high-purity water in nuclear reactors.

Bureau[98] and others[99] have to apply the technique for evaluating the corrosion rate of painted metals, and although the results are controversial the method has also been used successfully in the study of canning materials and lacquered surfaces[100-102].

Rowlands and Bentley[103] have provided an account of the possibilities for continuously monitoring corrosion rates by polarisation resistance measurements, and they also describe the development of a commercial instrument, which uses low-frequency square-wave current to polarise the test specimens.

Derivation of Linear Polarisation Method for Determining Corrosion Rates

It is assumed that
1. The corrosion current $i_{corr.}$ (it is also assumed that the area of the metal is $1\,cm^2$ so that $I_{corr.} = i_{corr.}$) occurs at a value within the Tafel region for the anodic and cathodic reaction, i.e. transport overpotential is negligible.
2. $E_{corr.}$ is remote from the reversible potentials of the anodic and cathodic reactions.
3. The IR drop in measuring the polarised potential is negligible.

Following Oldham and Mansfeld[87], but using the symbols that have been adopted in the present work, it is required to show that*

$$\left(\frac{di}{dE}\right)_{E_{corr.}} = i_{corr.}\left[\frac{1}{b_a} + \frac{1}{b_c}\right] \qquad \text{...(20.3)}$$

where b_a, b_c are the Tafel slopes of the anodic and cathodic reactions constituting the overall corrosion reaction, i.e. $i_{corr.}$ is linearly related to the polarisation resistance (dE/di) at potentials close to $E_{corr.}$.

At any potential E the net current is given by

$$i = \overset{\leftarrow}{i_1} - |\overset{\rightarrow}{i_1}| + \overset{\leftarrow}{i_2} - |\overset{\rightarrow}{i_2}| \qquad \text{...(20.4)}$$

where $\overset{\leftarrow}{i_1}$ is the anodic current for metal dissolution and $\overset{\rightarrow}{i_1}$ is the reverse cathodic current, and $\overset{\rightarrow}{i_2}$ is the cathodic current for reduction of the cathode reactant (dissolved O_2, H_3O^+, H_2O, etc.) and $\overset{\leftarrow}{i_2}$ the reverse current.

The rate of the anodic reaction at a potential E is given by

$$\overset{\leftarrow}{i_1} = i_{0,1}\exp\left\{\frac{E - E_{r,1}}{b_a}\right\} \qquad \text{...(20.5)}$$

where $E_{r,1}$ is the reversible potential for the anodic dissolution reaction, b_a is the Tafel slope and $i_{0,1}$ is the exchange current density. Similarly for the cathodic reaction

$$|\overset{\rightarrow}{i_2}|t_2| = i_{0,1}\exp\left\{\frac{E_{r,2} - E}{b_c}\right\} \qquad \text{...(20.6)}$$

Similar expressions may be written for the partial reverse rates $|\overset{\rightarrow}{i_1}|$ and $\overset{\leftarrow}{i_2}$, but under the conditions assumed here they may be neglected. Hence substituting equations 20.5 and 20.6 in equation 20.4

$$i = i_{0,1}\exp\left\{\frac{E - E_{r,1}}{b_a}\right\} - i_{0,2}\exp\left\{\frac{E_{r,2} - E}{b_c}\right\} \qquad \text{...(20.7)}$$

At the corrosion potential $E_{corr.}$ the *net* current i becomes zero, since $\overset{\leftarrow}{i_1} = |\overset{\rightarrow}{i_2}|$. Thus the two terms on the right-hand side of equation 20.7

---

*Note that the inclusion of the factor 2·3 (*see* equation 20.1) for converting $\ell n\, b$ to $\log b$ is not necessary for this proof.

become equal to one another and equal to $i_{corr.}$ the corrosion density. Thus, replacing E in equation 20.7 by $E_{corr.}$ gives

$$i_{corr.} = i_{0,1} \exp\left\{\frac{E_{corr.} - E_{r,1}}{b_a}\right\} = i_{0,2} \exp\left\{\frac{E_{r,2} - E_{corr.}}{b_c}\right\} \quad ...(20.8)$$

Differentiating equation 20.7 with respect to E gives

$$\frac{di}{dE} = \frac{i_{0,1}}{b_a} \exp\left\{\frac{E - E_{r,1}}{b_a}\right\} + \frac{i_{0,2}}{b_c} \exp\left\{\frac{E_{r,2} - E}{b_c}\right\} \quad ...(20.9)$$

which for $E = E_{corr.}$ becomes

$$\left(\frac{di}{dE}\right)_{E_{corr.}} = \frac{i_{0,1}}{b_a} \exp\left\{\frac{E_{corr.} - E_{r,1}}{b_a}\right\} + \frac{i_{0,2}}{b_c} \exp\left\{\frac{E_{r,2} - E_{corr.}}{b_c}\right\} \quad ...(20.10)$$

Combining equations 20.8 and 20.10 gives equation 20.3 the Stern–Geary equation by simple algebra. However, Oldham and Mansfeld point out that further differentiation of equation 20.9 gives

$$\frac{d^2 i}{dE^2} = \frac{i_{0,1}}{b_a^2} \exp\left\{\frac{E - E_{r,1}}{b_a}\right\} - \frac{i_{0,2}}{b_c^2} \exp\left\{\frac{E_{r,2} - E}{b_c}\right\} \quad ...(20.11)$$

an equation which demonstrates that there is only one point (a point of inflection, corresponding to a minimum slope) at which the di–dE curve has no curvature and is linear. It follows that

$$\left(\frac{d2i}{dE^2}\right)_{E_{corr.}} = \frac{i_{0,1}}{b_a^2} \exp\left\{\frac{E_{corr.} - E_{r,1}}{b_a}\right\} - \frac{i_{0,2}}{b_c^2} \exp\left\{\frac{E_{r,2} - E_{corr.}}{b_c}\right\} \quad ...(20.12)$$

and combining this equation with equation 20.8 gives

$$\left(\frac{\partial^2 i}{\partial E^2}\right)_{E_{corr.}} = i_{corr.}\left[\frac{1}{b_a^2} - \frac{1}{b_c^2}\right] \quad ...(20.13)$$

For the E–i plot to be linear at $E \approx E_{corr.}$

$$\left(\frac{d^2 i}{dE^2}\right)_{E_{corr.}}$$

must be zero, but equation 20.13 shows that this will be true only if $b_a = b_c$.

Simultaneous determination of Tafel Slopes and Corrosion Rates from R_P Determinations

Mansfeld[91a] points out that a major limitation of the polarisation resistance is that the factor $b_a b_c/2 \cdot 3(b_a + b_c)$ must be determined in order to evaluate $I_{corr.}$, and has devised a procedure in which this can be achieved by a graphical method.

The Stern–Geary equation can be written in the form

$$I_{corr.} = \frac{b_a b_c}{2.3(b_a + b_c)} \times \frac{1}{R_P} = \frac{B}{R_P} \quad ...(20.14)$$

where $B = b_a b_c/2 \cdot 3(b_a + b_c)$ and $R_p = (dE/dI)_{E_{corr.}}$. Equation 20.14 is valid only if the relationship between I and E can be expressed as

$$I = I_{corr.} \left\{ \exp\left(\frac{2 \cdot 3(E - E_{corr.})}{b_a}\right) - \exp\left(\frac{-2 \cdot 3(E - E_{corr.})}{b_c}\right) \right\} \quad \ldots(20.15)$$

Combining equations 20.14 and 20.15 and rearranging gives

$$2 \cdot 3 R_p I = \frac{b_a b_c}{b_a + b_c} \left\{ \exp\left(\frac{2 \cdot 3\Delta E}{b_a}\right) - \exp\left(-\frac{2 \cdot 3\Delta E}{b_c}\right) \right\} \quad \ldots(20.16)$$

where $\Delta E = E - E_{corr.}$. Since the right-hand side of equation 20.16 depends only upon the Tafel slopes it should be possible to evaluate b_a and b_c from plots of $R_p I$ vs. ΔE.

Figure 20.10a shows a theoretical plot of the right-hand side of equation 20.16 vs. ΔE in which the cathodic Tafel slope has been assumed to be constant at 120 mV and the anodic Tafel slope to have the arbitrary slopes of 40, 60 and 120 mV. It can be seen that linearity over a range of positive and negative potentials ΔE is achieved only when $b_a = b_c$ and that linearity is confined to $\Delta E \approx 0$ when b_a and b_c differ.

In Fig. 20.10b it has been assumed that the Tafel slopes are equal, i.e. $b_a = b_c = b$ and the modified expression for the right-hand side of equation 20.16 has been plotted against ΔE for different values of b (30, 60 and 120 mV). Comparison of Figs. 20.10a and 20.10b shows how the curvature of the plots differ at cathodic potentials, i.e. $\Delta E < 0$. Thus the kinetic behaviour of a corroding metal, as expressed by different combinations of Tafel slopes, can be recognised by this method of plotting curves. This theoretical approach has been confirmed experimentally by Mansfeld for the system Fe/H_2SO_4.

Mansfeld points out that $I_{corr.}$ can be calculated from the measured polarisation curve by the following four steps which are based on equations 20.14 and 20.15.

1. Determine R_p from

$$\left(\frac{dI}{dE}\right)_{E_{corr}} = R_p^{-1}$$

by drawing a tangent at $\Delta E = 0$ i.e. at $E_{corr.}$.
2. Multiply the current I measured at a certain value of ΔE by $2 \cdot 3 R_p$ and plot $2 \cdot 3 R_p I$ vs. ΔE for various values.
3. Determine from this plot the Tafel slopes b_a and b_c by curve fitting using the theoretical curves calculated for various values of b_a and b_c.
4. Calculate $I_{corr.}$ from equation 20.14 using the R_p value evaluated in Step 1 and the Tafel slopes determined in Step 3.

Tests for Bimetallic Corrosion

The extent of galvanic effects will be influenced by, in addition to the usual factors that affect corrosion of a single metal, the potential relationships of the metals involved, their polarisation characteristics, the relative areas of anode and cathode, and the internal and external resistances in the galvanic circuit (see Section 1.7).

The results of a galvanic corrosion test on a small scale are as a general rule no more than semi-quantitative. A principal reason for this is that the

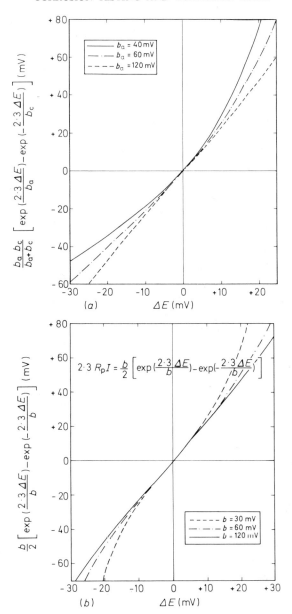

Fig. 20.10 Plots of the right-hand side of equation 20.16 *vs.* $\Delta E = E - E_{corr.}$ for various combinations of Tafel slopes. (*a*) b_c constant at 120 mV, b_a varied, and (*b*) $b_a = b_c = b$ (after Mansfeld[91a])

magnitude of the galvanic effect is a function of galvanic current density which is usually determined by the relative areas of the metals forming the couple. There may also be major differences in circuit resistances in tests as compared with practice—especially if current-measuring shunts of substantial resistance are made part of the circuit in the test. The geometric relationship between the metals in the test will also influence the result through effects on electrolyte resistance and the distribution of the galvanic currents.

The simplest procedure in studying galvanic corrosion is a measurement of the open-circuit potential difference between the metals in a couple in the environment under consideration. This will at least indicate the probable direction of any galvanic effect although no information is provided on the rate. A better procedure is to make similar open-circuit potential measurements between the individual metals and some appropriate reference electrode, which will yield the same information and will also permit observations of any changes in potential of the individual metals with time that will affect the overall potential difference in the couple. For most practical laboratory testing, the saturated calomel half cell is most convenient. The precision of the determinations is adequate and it is easy to maintain a constant concentration of potassium chloride.

The preferred potential-measuring instruments are potentiometers or electrometers, either of which permit measurements to be made without flow of sufficient current to polarise the electrodes during the determinations. It is also possible to use millivoltmeters if the internal resistance of the instrument is high enough to avoid any appreciable flow of current.

Open-circuit potential measurements do not indicate the all-important effects of continued current flow, and much more information is derived from frequent or continuous determinations of the magnitude of the galvanic current. In making these measurements it is necessary to avoid the use of instruments that will introduce sufficient resistance to exert a controlling effect on the magnitude of the galvanic current being measured. Instruments (zero-resistance ammeters) are available that permit current measurements to be made with zero resistance in the measuring circuit[2]. This may also be achieved by connecting the two metals to the input of a potentiostat and setting the control potential to zero volts; the output current of the potentiostat will then be equivalent to the galvanic current flowing between the two metals.

In many cases it will suffice to include in the circuit a shunt of appropriately low resistance over which IR-drop potential measurements can be made for ready calculation of the magnitude of the current flow. This technique permits measurements to be made as required without opening the circuit even momentarily for the introduction of current-measuring devices. It is also possible to arrange instruments in a circuit so that no measuring resistance is introduced in the galvanic current circuit[104].

An obvious method for studying galvanic corrosion either with or without supplementary electrical measurements is to compare the extent of corrosion of coupled and uncoupled specimens exposed under identical conditions. Such measurements may use the same techniques for estimating corrosion damage, such as weight-loss determinations, as have been described in connection with ordinary corrosion tests.

A convenient method of carrying out such a galvanic test in the laboratory has been described by Wesley[105] in which the vertical circular-path machine is used. Each assembly includes two pairs of dissimilar metals—one pair coupled galvanically while the other pair is left uncoupled in order to determine the normal corrosion rates under the same environmental conditions.

The type of motion provided (specimens moving in a vertical circular path) enables electrical connections to be made without mercury cup or commutator and the leads can be connected to a calibrated resistance for current measurements attached to the specimen carrier.

It is often of interest also to measure both the external and internal resistances of the galvanic circuit by the use of appropriate resistance-measurement bridges or by even more elaborate techniques such as have been described by Pearson[80].

It is often desirable to know something about the probable distribution of galvanic effects in a galvanic couple. This will be determined, of course, by the size and shape of the different metals and how they are placed relative to each other—whether more or less parallel in the electrolyte, close together or far apart, or joined along some line of contact. The distribution from such a line of contact may be observed directly if the test couples are exposed in this way and for long enough for sufficient corrosion to occur for it to be observed and measured. Alternatively, the distribution of the galvanic currents in terms of the current density on different portions of both the anode and cathode surfaces may be estimated from data derived from surveys of the potential field in the electrolyte around the couple. Such a potential survey may be made using a fixed and a movable reference electrode so that equipotential lines in several planes may be measured and plotted as was done by Copson[106] using a technique originally proposed by Hoar. By laborious analysis of the data from the potential surveys, it is possible to calculate the current distribution over different areas near to and remote from the contact of the dissimilar metals. This technique has been used by Rowe[107] to study the corrosion behaviour of coated and uncoated couples.

Soil Tests

Soil corrosion does not lend itself readily to direct study in the laboratory. However, indirect methods involving the action of differential aeration cells have yielded valuable information in comparing the probable corrosivities of different soils towards steel. The details of this technique were described by Denison[108], Ewing[109], Schwerdtfeger[110], and by Logan, Ewing and Denison[111].

The Schwerdtfeger[110] 'polarisation break' and the polarisation resistance methods have been studied by Jones and Lowe[103a] in relation to their effectiveness in evaluating corrosion rates of buried metals. A Holler bridge circuit was used to remove IR contributions during the measurement of the polarised potential. Jones and Lowe, on the basis of their studies of buried steel and aluminium specimens, concluded that the polarisation resistance was the most useful, and that the polarisation break had the serious limitation that it was difficult to identify the breaks in the curve.

Accelerated Tests—Electrolytic Tests

In view of the electrochemical nature of corrosion, it has seemed reasonable to many investigators to assume that suitable accelerated corrosion tests could be made by observing the response to electrolytic stimulation of the corrosion processes, or by attaching particular significance to the results of quickly made electrode potential and current measurements.

Acceleration of corrosion by electrolytic stimulation has in fact been found to distort normal corrosion reactions to such an extent that the results bear no consistent relationship to ordinary corrosion and are, therefore, quite inconsistent and unreliable. This was shown, for example, by a series of tests sponsored by ASTM Committee B-3[112, 113].

Some investigators[114] have advocated a type of accelerated test in which the specimens are coupled in turn to a noble metal such as platinum in the corrosive environment and the currents generated in these galvanic couples are used as a measure of the relative corrosion resistance of the metals studied. This method has the defects of other electrolytic means of stimulating anodic corrosion, and, in addition, there is a further distortion of the normal corrosion reactions and processes by reason of the differences between the cathodic polarisation characteristics of the noble metal used as an artificial cathode and those of the cathodic surfaces of the metal in question when it is corroding normally.

Measurements of open-circuit potentials relative to some reference electrode have been assumed on occasion to provide a means of rating metals as to their relative resistance to corrosion on the basis that the more negative the measured potential, the higher will be the rate of corrosion, but this assumption is obviously invalid, since it disregards polarisation of the anodic and cathodic areas.

Some examples have been given of the use of potential measurements in corrosion tests and it is of interest to outline here certain test procedures that are used industrially to supplement or replace the more tedious and prolonged laboratory and field tests. These tests frequently rely on changing the potential as a means of accelerating the test, and although, as emphasised above, this is capable of distorting the mechanism, it is less likely to do so than a change in the nature of the environment, an increase of temperature, etc. The majority of these tests are used for evaluating electrodeposits, anodised coatings and paint films.

Electrolytic Oxalic Acid Etching Test

This test has been developed and used by Streicher[115-117] as a screening test to be used in conjunction with the tedious boiling nitric acid test for assessing the susceptibility of stainless steels to intergranular attack as specified in ASTMA 262:1970(3), and will be considered subsequently in the section concerned with intergranular attack of Cr–Ni–Fe alloys.

The Electrolytic Corrosion (EC) Test

The EC test was developed by Saur and Basco[118, 119] for decorative Cu (optional) + Ni + Cr electrodeposits. After an appropriate area is masked off and cleaned with a slurry of MgO, the specimen is immersed in test solution A or B (Table 20.3). It is held by means of a potentiostat at +0·3 V (vs. S.C.E.) and taken through cycles of 1 min anodically polarised, 2 min unpolarised.

Table 20.3 Electrolytic corrosion test; composition of test solutions (*A*, *B*) and indicators (*C*, *D*)[118, 119]

	Concentration			
	A	*B*	*C*	*D*
NaNO$_3$ (g/l)	10	10		
NaCl (g/l)	1·3	1·0		
HNO$_3$ (conc.; g/l)	5	5		
1,10-Phenanthroline hydrochloride (g/l)		1·0		
KCNS			3	3
Acetic acid (glacial, ml/l)			2	2
Quinoline (ml/l)			8	—
H$_2$O$_2$(30%) ml/l			—	3

The extent of pitting is estimated by a special microscopical technique, or by the attack on the substrate using an appropriate indicator. Thus in the case of steel 1,10-phenanthroline hydrochloride is added to the electrolyte (solution *B*) to detect the formation of Fe^{2+} ions. Alternatively, the specimens can be removed from corrosion test solution and placed in an indicator solution, i.e. solution *C* for zinc-base die castings and solution *D* for steels.

The test is much faster than the CASS test and is probably more reproduceable; more important is the fact that it has been correlated with service exposure.

FACT Test

The Ford Anodised Aluminium Corrosion Test (FACT)[120] was developed as a rapid test for anodised aluminium, and involves cathodic polarisation of the specimen by means of small glass cell that is placed on the surface of the anodised metal (Fig. 20.11); a rubber gasket seals the cell to the specimen to expose a circular area of approximately 3 mm dia. The electrolyte is 5% NaCl containing 0·2 gdm^{-3} CuCl$_2$ adjusted to pH 3·1 with acetic acid. An

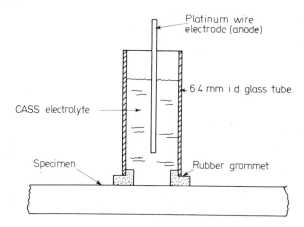

Fig. 20.11   Glass cell for FACT test (after Store and Tuttle[120])

e.m.f. of 34 V is applied using a Pt anode, and the alkali generated by the cathodic reaction at local imperfections in the coating causes dissolution of the latter. This results in a decrease in the resistance of the coating and a consequent increase in the voltage drop across the cell. The FACT number in volt-seconds (Vs) is obtained by integrating the voltage across the cell over a 3 min period. A disadvantage of the method is that the FACT numbers for poor coatings and for those that are acceptable vary by only 15%, i.e. 650 Vs for a poor coating and 750 Vs for an acceptable coating, and this gives an element of uncertainty to the test.

Cathodic Breakdown Test

The cathodic breakdown test[121] has been proposed for evaluating anodised aluminium, and consists of holding the specimen at a constant potential of -1.6 V (vs. S.C.E.) by means of a potentiostat for 3 min in 5% NaCl adjusted to pH 3·5 with HCl. Any weak points in the film are the sites at which alkali is formed owing to the high current density, and these areas become whitened and pitted during the test. Evaluation of the coating is by counting the number of spots of breakdown per unit area and whereas a satisfactory coating will show only 1–5 spots dm^{-2}, a marginal finish will reveal 25–50 spots dm^{-2} and thousands when the anodised film is tested in the unsealed condition.

Impedance (Aztac) Test

The impedance test [122] for anodised aluminium (ASTM B457:1973(7)) uses a similar test cell to that for the FACT test, but employs a 1 V r.m.s. 1 kHz source with the cell in an impedance bridge; the electrolyte solution is 3·5% NaCl. The results are expressed as kilo-ohms, and whereas a bare Al specimen will give a value of about 1 kΩ, a well-sealed anodised coating will give a value of 100 kΩ. The admittance test (BS 1615:1972 and BS 3987:1974) is essentially the same as the impedance test but uses 3·5% K_2SO_4 rather than 3·5% NaCl. An admittance of $< 500/t$ μS (where t is the thickness of the film in μm) denotes good sealing. It should be noted that thickness of the sealed coating should be specified in both tests.

PASS Test

In the PASS test[123] (Paint Adhesion on a Scribed Surface) a small adhesive-backed ring gasket is applied to the painted surface, and a scribe line is made inside the gasket through the paint film so that the underlying metal is exposed. The specimen is then made cathodic in 5% NaCl at a current of 9 mA (the diameter of the exposed area of paint is approximately 9 mm) for 15 min using a Pt wire as a counter-electrode. At the end of the test a piece of pressure-sensitive adhesive tape is applied over the scribe line and then stripped off, a procedure that is repeated until all the loose paint is removed. The width of the basis metal exposed gives the PASS value of the paint film.

The test simulates the generation of alkali, which softens the paint film and results in loss of adhesion, but correlation of the test with salt-spray tests has been sketchy, although it is claimed to be more reproduceable than the latter.

Accelerated Tests—Simulated Environments

Spray Tests

The most common of the spray tests is the salt-spray or salt-fog test which was developed originally by Capp[124] for studying the protective values of metallic coatings on steel under conditions that he hoped would simulate exposure to a sea-coast atmosphere. Since then the test has been used for a number of purposes, for many of which it is not well suited[125-127].

Although there is no standard size or shape of salt-spray box certain other features of the test have been standardised in ASTM B117:1973 and DIN 50907–1952. Various factors affect the rate of attack and Fig. 20.12 shows the effect of angle of exposure of the specimen to the salt droplets, which fall vertically from the spray nozzle, based on early work by May and Alexander[128]. It can be seen that maximum corrosion occurs at angles between about 30 and 80° to the horizontal; vertical exposures was found to give erratic results.

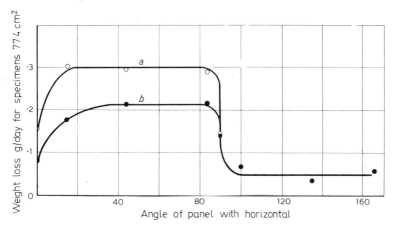

Fig. 20.12 Effect of specimen position on corrosion in salt-spray tests; specimens of cold-rolled steel. (a) 20% NaCl, (b) synthetic sea-water

There may be variations from box to box, depending on differences in fog characteristics as influenced by the design of the spray nozzles, the pressure at which the brine enters the nozzle, and the constancy of this pressure[129].

The results will also be influenced by the concentration of NaCl solution sprayed—some metals are affected more by one concentration than another —for example, zinc is corroded most by a concentrated brine (20%), while iron is corroded most by a dilute brine (3%); synthetic sea-water is less corrosive to these metals than either brine. In view of the many other ways by which the conditions within a salt-spray box differ from those of exposure to a natural sea-coast environment, there seems to be no great advantage in making-up complicated synthetic sea-waters for use in salt-spray testing. However, tablets for this purpose are commercially available (B.D.H., London) and compositions used in various countries are given in Table 20.2.

Some modified brines have been used in salt-spray boxes for particular purposes, such as the acetic acid-modified brine developed by Nixon[130] to reproduce the type of blistering frequently encountered on chromium-plated zinc-base die castings. An acetic acid–salt-spray test has been adopted by the ASTM and BS[131], and it is possible that other modifications of the spray solution will be developed for particular purposes. For example, the original acetic acid–salt-spray test has been modified by including cupric chloride in the brine. This is called the CASS (Copper Accelerated Acetic Acid Salt-Spray Test; ASTM B368:1968[7,21] test. It is used extensively for testing nickel-chromium coatings on steel and zinc. The original acetic acid–salt-spray test, is modified in an important aspect by the addition of 0·25 g/l of $CuCl_2 \cdot 2H_2O$, to the 5% NaCl test solution, which substantially increases the corrosivity of the solution, especially to nickel. The addition of $FeCl_3$ to the acetic acid–salt-spray solution, such as is used in the Corrodkote test, was early noted to be troublesome in that it tended to precipitate. For this reason, ferric iron is not included in the CASS test solution. Essential details[132,133] include control of cabinet temperature at 49°C, control of saturation temperature at 57°C, control of collection rate at 1·5 ± 0·5 m h^{-1} per 80 cm^2 of specimen surface, control of pH by addition of acetic acid to 3·2 ± 0·1, and an operating air pressure of 103·4 ± 6·9 kN/m^2. Higher pressure may be required to achieve specified collection rate in 'walk in' cabinets.

It has been shown that chromium is virtually unattacked by the CASS test solution[134]. Nickel, on the other hand, is corroded at a substantial rate (about 0·072 mm/y), the presence of the copper ions tending to maintain the nickel in an active state[134,135]. Thus, in the CASS test (and in the Corrod-kote test as well) accelerated galvanic corrosion of the nickel occurs at any discontinuities in the chromium layer. Good correlation between the results of the CASS test and the performance of plated parts in service has been reported[134].

Corrodkote Test[132,136]

This is a refinement of an earlier test in which melted street slush together with its contained dirt, salts, etc. was splashed upon plated parts by means of a rotating paddle wheel. Parts soiled in this manner were then exposed to a warm, humid atmosphere. The results were striking and significant in that they closely paralleled service experience. The 'paddle-wheel test' was intended to simulate the conditions to which plated parts on automobiles are subjected when cars are garaged, unwashed, after being driven over salted slush-covered city streets on typical winter days. Platings of inadequate thickness and quality have frequently been observed to show signs of failure after only a few weeks' or even a few days' use under such circumstances. Despite certain limitations, good correlation has been reported between the results of the Corrodkote test and service performance of plated components[136], and is now included in ASTM B380-65 and BS 1224:1970.

Reagents for use in this test are prepared as follows.

1. The cupric nitrate reagent contains 2·5 g $CuNO_3·3H_2O$, dissolved in 500 ml distilled water.
2. The ferric chloride reagent contains 2·5 g $FeCl_3·6H_2O$, dissolved in 500 ml distilled water. This reagent should not be kept longer than two weeks.
3. The ammonium chloride reagent contains 50 g NH_4Cl, dissolved in 500 ml distilled water.
4. The Corrodkote slurry is prepared by mixing 7 ml of the cupric nitrate reagent, 33 ml of the ferric chloride reagent, and 10 ml of the ammonium chloride reagent with 30 g of kaolin to form a homogeneous slurry, which is sufficient slurry to cover about 2·79 m^2 of plated surface. A fresh batch of slurry should be made up each day.
5. The surfaces to be tested should be coated with the slurry by brushing with a circular motion, finishing with brush strokes in one direction. The coating should then be allowed to dry for 1 h, after which the coated specimen should be put into a non-condensing humidity cabinet at 38°C and from 90 to 95% r.h. After 20 h in the cabinet the specimen should be removed for inspection. Zinc die castings should be cleaned with running water and dried before inspection. Steel specimens should be examined before cleaning and the number of rust spots counted. Since most of the rust will come off with the Corrodkote coating, it may be difficult to distinguish after cleaning between surface pits and pin holes reaching the basis metal. Steel parts may be returned to a condensing humidity cabinet for 24 h or to a salt-spray cabinet for 4 h. Either supplementary exposure will bring out rust spots again.

One cycle of the Corrodkote test will reliably reveal coatings that will not endure one winter's normal use in a typical city which uses salt to de-ice its streets. In contrast, several cycles of the Corrodkote test are generally required to 'fail' coatings which will withstand one or more such winter's use.

In this connection, there is some indication that while the Corrodkote test can be depended upon to reveal coatings of unsatisfactory durability, there is presently some question as to its ability to distinguish between, or to predict the relative protective value or length of useful life of, different coating systems in the very good or excellent durability range.

Also of questionable significance is the practice of shortening the Corrodkote cycle, to say 4 h, for the purpose of evaluating the durability of relatively thin coatings intended for use under comparatively mild conditions such as indoors or the interiors of automobiles, since by far the greatest amount of corrosion (of the nickel) appears to occur during the early part of the Corrodkote humidity cycle. Good correlation between the Corrodkote test and service performance has been obtained by Bigge[136].

The A.R.E. Salt Droplet Test[136a] was developed as a simple test for painted steel specimens. The specimens are sprayed manually each day to give complete coverage with non-coalescing drops of artificial sea-water, and are placed in a cabinet containing water so that drying is prevented. BS 1391:1952 describes the salt droplet test and also a test in which specimens are exposed to a humid atmosphere containing SO_2 (*see also* Reference 1).

Sulphur Dioxide Tests

Two tests in which sulphur dioxide is the principal corroding agent have been used i.e. the BNFMRA* Sulphur Dioxide Test[137,138] and the Kesternich Test[139].

These tests were investigated by the American Electroplaters Society Research Project 15 Committee early in its search for an acceptable accelerated corrosion test. They were soon abandoned, however, largely because the types of corrosion failures developed did not resemble those which occurred in actual service. Furthermore, the extreme corrosivity of the test environment to nickel (some 8·38 mm/y) appeared to place an undue premium on the integrity of the overlying chromium deposit which is virtually unattacked in the test. Thus coatings which were substandard in respect of nickel or copper–nickel thicknesses might well pass the test with flying colours provided the chromium top coat was completely continuous and remained so for the duration of the test. Conversely, coatings of proven merit on the basis of service experience, such as 0·039 mm of semi-bright plus bright nickel (duplex) with 0·000 25 mm of conventional chromium, could be expected to fail in these SO_2 tests relatively quickly at any discontinuities in the chromium. In this connection, it is well to keep in mind that even though the chromium may be non-porous initially, it can hardly be expected to remain so in service on an automobile where it is subject to impact from sand, gravel, etc.

The BNFRA test was used in Europe for testing Ni + Cr coatings but was omitted from the 1970 revision of BS 1224. However, although it is no longer used for Ni + Cr coatings it is being used as a sealing test for anodised aluminium (BS 1615:1972). An SO_2 test, in which the SO_2 is generated by reaction of sodium thiosulphate with H_2SO_4, is used for testing tinplate[139a] and for tin-nickel coatings (BS 3597:1963). Another version with a much more aggressive atmosphere is used for Au coatings on Cu, Ni or Ag (ASTM B583-73).

General Considerations of Spray and SO_2 Tests

In Germany much attention has been given to efforts to standardise corrosion testing methods. This has been described by Wiederholt[140].

The salt-spray test has seemed to yield the most consistent results when used to establish the relative merits of different aluminium alloys in resisting attack by marine atmospheres. Here, the best results have been secured when the spray has been interrupted for so many hours each day[141].

Salt-spray boxes are also used for studying the deterioration and protective value of organic coatings, although this test is of doubtful value for such purposes, since it fails to include many factors, e.g. sunlight, which affect the life of such coatings. Methods of testing organic coatings are discussed in a later section.

The variable responses of different metals and coatings to the conditions that can be set up in salt-spray boxes, as well as to the conditions that exist in natural atmospheres, together with the impossibility of defining any

*British Non-Ferrous Metals Research Association.

natural atmosphere in terms of its corrosivity towards even a single metal, render nonsensical any notion that an hour in a salt-spray box is equivalent to so many hours in some natural atmosphere.

For additional information on some of the features of the salt-spray test and its limitations in respect of certain of the purposes for which it may be used, the 1951 ASTM Marburg Lecture entitled 'Corrosion Testing'[125] should be consulted. Reference may also be made to the several papers listed in a bibliography on salt-spray testing[142].

There have been several attempts to develop rather elaborate testing machines in which specimens may be subjected to various sprays or fogs with cycles of condensation, heating and drying. The object has been to reproduce the conditions encountered by metals exposed in polluted industrial atmospheres. Such devices have been experimented with in the UK[143] and the USA[144]. While it is sometimes possible by such tests to rate steels in a rough order of resistance to atmospheric corrosion, it should be appreciated that the nature of rust formed may differ from that obtained during actual. exposure. It is only in rare cases that the resistance of steels to attack by the sprays is analogous to their resistance in the natural atmosphere which the tests seek to simulate. Such parallelism is not common enough to make these tests very reliable[145].

Dennis and Such[146] point out that the BNFRA SO_2 test was really a means of detecting discontinuities in the chromium layer of a Cr–Ni coating system, and it gave therefore unfavourable results when used for testing micro-cracked or micro-porous Cr, since the Cr was rapidly undermined with consequent flaking. Conversely, the test exaggerates the beneficial effect of crack-free Cr. The test also fails to indicate the improved corrosion resistance of duplex Ni as compared with bright Ni. A critical account of laboratory corrosion testing methods for Ni–Cr coatings is given by Dennis and Such[147] who also give details of the ASTM (ASTM B537–70) and BSI (BS3745:1970) methods of rating the extent of corrosion produced by corrosion tests.

Accelerated Tests for Weathering Steels

Recent interest in weathering steels has stimulated work on accelerated laboratory tests which can be used to investigate the effect of alloy composition on performance. It is well established that a wetting and drying cycle should be an integral part of any laboratory test in which the characteristic properties of weathering steels are revealed[148], and Bromley, Kilcullen and Stanners[149] have designed a test rig (Fig. 20.13) which provides results that can be correlated with actual atmospheric exposure data. The rig has been designed to investigate a wide range of alloying elements in a development programme on slow-weathering steels for which it was essential to have a rapid, reliable and reproduceable test that incorporated the specific atmospheric factors responsible for rust formation.

The rig incorporates a time cycle (usually of 1 h duration) in which the specimens are immersed in a solution for a few minutes, removed from the solutions and allowed to dry under controlled conditions. This method has the advantage that solutions of different composition may be used to simulate, for example, natural rain water; furthermore, immersion gives uniform wetting of the specimens. During the drying period conditions can be varied

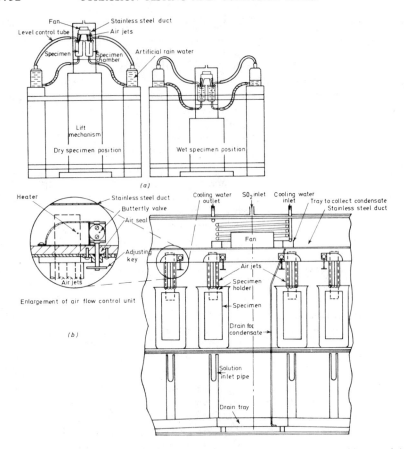

Fig. 20.13   Rig used for a laboratory evaluation of weathering steels. (*a*) General layout of rig showing unimmersed and immersed position and (*b*) detailed view of central portion of cabinet (after Bromley, Kilcullen and Stanners[149])

with respect to temperature, humidity, and degree of air circulation and SO_2 content of the atmosphere. Thus the following six atmospheric factors may be varied and studied: (*a*) frequency of wetting and drying, (*b*) composition of the solution used for wetting, (*c*) temperature of air, (*d*) humidity of air, (*e*) run of wind and (*f*) SO_2 content of air.

The rig contains a central sealed cabinet (Fig. 20.13*b*) fitted with a stainless steel duct through which air is blown by a fan into the individual specimen chambers. The volume and temperature of air supplied to each chamber is regulated by equipment within the duct; SO_2 may be injected into the air stream at the back of the fan, which is surrounded by a cooling coil on which moisture condenses thus lowering the humidity.

Typical test specimens are $126 \times 50 \times 3$ mm and each rig (5 in all) can accommodate 24 such specimens. The cabinet rests on a motor-driven scissor lift so that it can be raised above or lowered below the level of the surrounding stands that hold the water reservoirs. After the wetting period, the cabinet is raised and the beakers are emptied thus exposing the specimens to the air; the concentration of SO_2 in the air is monitored by an electro-conductivity method.

The drying of the specimens can be regulated and varied by suitable operation of the fan, heaters and cooling coil. At the end of each cycle the fan and heaters are switched off and another cycle is commenced with the re-immersion of the specimens. The cycle is controlled automatically by a micro-process computer system, which can also be used for continuous measurement, if required, of the corrosion potentials of the specimens. The reservoirs contain artificial rain water solution, but other natural waters may also be used.

The results obtained show that the corrosion rate in the rig is about four times that encountered in an industrial UK atmosphere. This acceleration, however, is not achieved by accentuating any of the environmental factors, but rather by holding them near to the worst natural conditions for as long as possible. The procedure used ensures that the rust film is completely dried for short periods, thus simulating the conditions that bring out the beneficial effects of protective rust films on the steels under study.

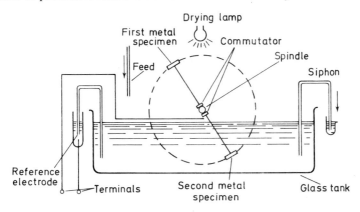

Fig. 20.14   Equipment for studying 'patina' formation on low-alloy steels (after Pourbaix[150])

The use of electrochemical tests for rapid assessment of the performance of these steels has attracted interest, and Pourbaix[150] has devised an apparatus in which potential measurements are used to evaluate the protective nature of corrosion products formed on low-alloy steels, such as the weathering steels, during periodic wetting and drying. The apparatus (Fig. 20.14) consists of a glass tank containing an appropriate electrolyte, such as a natural or artificial water. Two specimens of the metal or alloy under study are attached to a spindle which rotates slowly (about 1 rev/h) so that the specimen is immersed in the solution for approximately half the time and exposed to the atmosphere for the remainder. An electric lamp is placed above the tank so that the specimens remain wet for a time after withdrawal from the solution, but are completely dried during the cycle. Measurement of the potentials of the specimens at the beginning and end of the immersion period is effected by means of the commutators, which are attached to the spindle but electrically insulated from it, and a reference electrode. The e.m.f. taken from the terminals can be fed to a multipoint recorder so that a recording of the E–time relationship may be obtained for each specimen. The solution can be made to circulate slowly by allowing it to drip in from a feed and overflow via a siphon. In a variation of the apparatus a Luggin capillary is attached to the sample so that the potentials can be measured during the period when the specimen has emerged from the solution but is still wet.

Figure 20.15 shows results obtained from the apparatus for different steels some of which (Nos. 1, 2, 3 and 7) form a protective patina of corrosion products, whilst others (Nos. 4, 5 and 6) form patinas that are non-protective; the criterion adopted is that the more positive the potential the more protective is the rust patina[150].

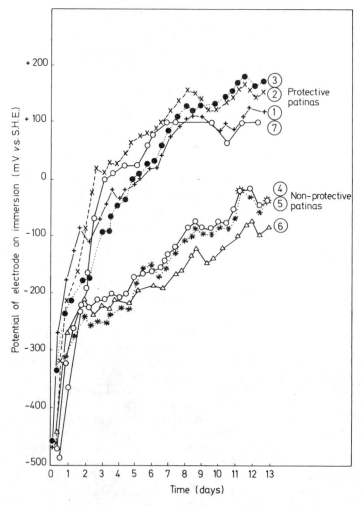

Fig. 20.15   Relationship between E (*vs.* S.H.E.) and time for steels that form a protective patina (Nos. 1, 2, 3 and 7) and those that form a non-protective patina (Nos. 4, 5 and 6). E is determined by the equipment shown in Fig. 20.14 and is determined during initial immersion of the specimen (after Pourbaix[150])

Legault, Mori and Leckie[151] have used open-circuit potential *vs.* time measurements and cathodic reduction of rust patinas for the rapid laboratory evaluation of the performance of low-alloy weathering steels. The steel specimens are first exposed for 48 h to the vapour of an $0.001 \, mol \, dm^{-3}$ sodium bisulphite solution maintained at $54°C$ (humid SO_2-containing atmosphere) to simulate corrosion under atmospheric conditions. They are then subjected to two types of test: (*a*) open-circuit potential–time tests for periods up to $3\,000 \, s$ in either distilled water or $0.1 \, mol \, dm^{-3} \, Na_2SO_4$ and (*b*) cathodic reduction in $0.1 \, mol \, dm^{-3} \, Na_2SO_4$ at $1 \, mA \, cm^{-2}$ c.d. In the cathodic reduction experiments, which provide a means of evaluating the degree of rusting[152], both the potential and time are recorded as a function of time, the onset of hydrogen evolution at constant potential being taken as the end point and giving the *oxide-reduction time.*

In order to evaluate the tests determinations were carried out on the steels that had been exposed to the atmosphere for 1, 2, 3, 4 and 6-month periods. It was established that the initial open-circuit potential and the decrease in potential (more negative) with time varied with the nature of the steel and the time of exposure to the atmosphere, and the maximum negative potential was taken as a measure of corrosion resistance; the more negative the potential the lower the resistance of the alloy. In the case of three alloy steels that differed only in copper content it was found that the open-circuit potential was related to the corrosion rate as assessed by conventional weight loss.

A relationship was also established between the oxide-reduction time and time of exposure, and the results for a mild steel and a 1Cu–3Ni weathering steel were similar to those obtained by weight loss. The authors give various expressions that relate oxide-reduction time (min) with corrosion rate (mm/y), and claim that a short exposure to a laboratory SO_2 atmosphere followed by determining the *E vs.* time and oxide-reduction time provides a rapid method of evaluating weathering steels.

Intergranular Attack of Cr–Ni–Fe Alloys

Early in the history of stainless steels it was recognised that they were highly susceptible to intergranular attack resulting from the precipitation of Cr–Fe carbides with the consequent depletion in the chromium content at grain boundaries when the alloy was heated in a specific range of temperature (*see* Sections 1.3, 3.3 and 10.5). It was necessary, therefore, to develop methods of testing that would detect susceptibility to intergranular attack as influenced by variations in processing and/or composition. As will be seen, most reagents used for these tests are highly aggressive, and it is important to note that an alloy found to be susceptible during testing will not necessarily be attacked intergranularly under the milder environmental conditions that may prevail in service.

Brown[153], in a recent paper, points out that Du Pont use evaluation tests for (*a*) as-received unstabilised alloys containing more than 0.03% C to check the effectiveness of the final heat treatment and (*b*) stabilised or special low-carbon grades after a sensitising treatment (1 h at $677°C$) to determine whether susceptibility might develop during a subsequent welding operation.

Table 20.4   Summary of chemical tests used for the determination of

Test name	Ref.	Usual solution composition	Test procedure
1. Nitric acid test	†	65 wt.% HNO_3	Five 48 h exposures to boiling solution; solution refreshed after period
2. Acid ferric sulphate (Streicher) test	†‡	50 wt.% $H_2SO_4 + 25$ g/l ferric sulphate	120 h exposure to boiling solution
3. Acid copper sulphate test	†§	16 wt.% $H_2SO_4 + 100$ g/l $CuSO_4$ (+ metallic copper)	72 h exposure to boiling solution
4. Oxalic acid etch	†	100 g $H_2C_2O_4 \cdot 2H_2O$ + 900 ml H_2O	Anodically etched at 1 A/cm^2 for 1·5 min
5. Nitric-hydrofluoric acid test	†	10% $HNO_3 + 3\%$ HF	4 h exposure to 70°C solution
6. Hydrochloric acid test	¶	10% HCl	24 h in boiling solution
7. Nitric acid–Cl^{6+} test	‖	5 N $H_2SO_4 + 0.5$ N KCr_2O_7	Boiling with solution renewed every 2–4 h for up to 100 h

*Data after Cowan and Tedmon[154].
†ASTM Tentative Practice A262–68 and Practice G28–71.
‡M. A. Streicher, ASTM Bulletin No. 229, 77–86 (1958).
§ASTM Recommended Practice A393–63 and Practice A262–68.
¶D. Warren, ASTM Bulletin No. 230, 45–56 (1958).
‖J. S. Armijo, *Corrosion*, **24** (1968).

Intergranular corrosion of Fe–Ni–Cr alloys has recently been the subject of a comprehensive review by Cowan and Tedmon[154] who have summarised the various tests used for determining susceptibility (Table 20.4). Of these tests, Nos. 1–5, which are regarded as reliable test procedures by the ASTM, have been incorporated into ASTM A262:1970 'Recommended Practice for Detecting Susceptibility to Intergranular Attack in Stainless Steel' as follows:

Practice *A*—10% oxalic acid, electrolytic etching at ambient temperatures
Practice *B*—Boiling 50% $H_2SO_4 + 25$ g/l $Fe_2(SO_4)_3$
Practice *C*—Boiling HNO_3
Practice *D*—10% $HNO_3 + 3\%$ HF at 70°C
Practice *E*—Boiling 16 wt.% $H_2SO_4 + 5.7\%$ $CuSO_4$ + metallic copper

It should be noted that although ASTM A262:1970 provides details of test procedures no information is given on typical corrosion rates or acceptable limits for various heat-treated alloys, which are regarded as outside the province of a specification that describes test procedures. Table 20.5, taken from a recent paper by Brown[153], shows the maximum acceptable evaluation test rates specified by Du Pont for various alloys tested by the acid ferric sulphate test and by the Huey test. It should be noted that evaluation tests susceptibility to intergranular corrosion of iron–nickel–chromium alloys*

Quantitative measure	Potential range (V vs. S.H.E.)	Species selectively attacked
Average weight loss per unit area of five testing periods	+0·99 to +1·20	1. Chromium-depleted areas 2. σ-phase 3. Chromium carbide
Weight loss per unit area	+0·7 to +0·9	1. Chromium-depleted areas 2. σ-phase in some alloys
1. Appearance of sample upon bending 2. Electrical resistivity change 3. Change in tensile properties	+0·30 to +0·58	Chromium-depleted area
1. Geometry of attack on polished surface at ×250 or ×500	+1·70 to +2·00 or greater	Various carbides
Comparison of ratio of weight loss of laboratory annealed and as-received samples of same material	Corrosion potential of *304* steel = +0·14 to +0·54	1. Chromium-depleted areas 2. Not for σ-phase 3. Used only for Mo-bearing steels
1. Appearance of sample after bending around mandril 2. Weight loss per unit area	(a) Redox potential = +0·32 (b) Corrosion potential = −0·2±0·1	1. Alloy-depleted area 2. Not for σ-phase
1. Weight loss per unit area 2. Electrical resistivity 3. Metallographic examination	(a) Redox potential = +1·37 (b) Corrosion potential of *304* steel = +1·21	Solute segregation to grain boundaries

Table 20.5 Maximum acceptable evaluation test rates specified by Du Pont for services where susceptible material would be intergranularly attacked*

Type	Condition	Max. corrosion rate (in/month)
120 h acid $Fe_2(SO_4)_3$ test (ASTM A-262, Practice *B*)		
304	As received	0·0040
304L	20 min at 677°C	0·0040
316	As received	0·0040
316L	20 min at 677°C	0·0040
317L	20 min at 677°C	0·0040
CF–8	As received	0·0040
CF–8M	As received	0·0040
240 h HNO_3 test (ASTM A-262, Practice *C*)		
304	As received	0·0015
304L	20 min at 677°C	0·0010
304L	1 h at 677°C	0·0020
309S	As received	0·0010
316	As received	0·0015
347	1 h at 677°C	0·0020
CF–8	As received	0·0020
CF–8M	As received	0·0025

*Data after Brown[153].   *Note:* 1 in/month ≡ 305 mm/y.

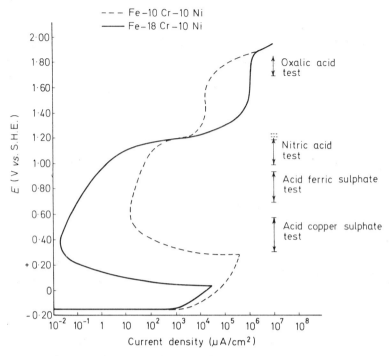

Fig. 20.16 Schematic representation of range of corrosion potentials expected from various chemical tests for sensitisation in relation to the anodic dissolution kinetics of the matrix (Fe–18Cr–10Ni stainless steel) and grain-boundary alloy (assumed to be Fe–10Cr–10Ni) owing to depletion of Cr by precipitation of Cr carbides of a sensitised steel in a hot reducing acid (after Cowan and Tedmon[154])

are specified by Du Pont only when it is known or suspected that the environmental conditions in service are conducive to intergranular attack of susceptible material.

The reagents used in Practice *B* to *E* all have a high redox potential and Cowan and Tedmon[154] have presented schematic $E - \log i$ curves (Fig. 20.16) showing the range of potentials of the various tests and the relative rates of attack on the matrix (Fe–18Cr–10Ni) and the chromium-depleted alloy at the grains boundaries, which has been assumed for this purpose to have a composition Fe–10Cr–10Ni, in a hot reducing acid. Although this diagram cannot show the effect of alloy composition, nature of test solution, conditions of test, etc. on intergranular attack, it serves to illustrate the electrochemical principles involved in the test procedures, all of which are based on reagents that attack the intergranular sensitised areas at a higher rate than the matrix; this may lead to the dislodgement of whole grains with a consequent high weight loss.

Boiling HNO₃ Test

This test, which is frequently referred to as the Huey test, was first described and used by W. R. Huey[155] in 1930, and since that time it has had wide application, particularly in the USA. The test consists of exposing the specimens (20–30 cm^2) in fresh boiling 65% HNO_3 (constant boiling mixture) for five successive periods of 48 h each under a reflux condenser. The specimens are cleaned and weighed after each period, and the corrosion rate (as a rate of penetration) is calculated for each period of test and for the average over the five periods; corrosion rates are expressed as inches per month (ipm or in/month) or as mm/y. The reason for the above procedure is due to the fact that Cr(VI) ions, produced from the oxidation of Cr^{2+} and Cr^{3+} by the HNO_3, if allowed to accumulate in the HNO_3 markedly increase its aggressiveness so that severe intergranular attack with grain dislodgement can occur even with solution-annealed steel free from precipitated carbides. Hence the necessity for the periodic changing of the solution and for a minimum ratio of solution volume to area of specimen (at least 20 ml HNO_3/cm^2 of stainless steel). Brown[153] points out that during a normal test the Cr(VI) content will not reach a level where an acceleration in rate occurs unless the specimen is in the sensitised condition, and under these circumstances the presence of Cr(VI) is an advantage in discriminating between sensitised and unsensitised material.

Maximum corrosion rates used by Du Pont for various alloys are given in Table 20.5, and most users of the test consider average corrosion rates of 0·015–0·002 ipm and 0·0025 ipm to represent the upper limits for satisfactory resistance for wrought austenitic alloys and cast austenitic alloys, respectively. Streicher[116] considers that if the corrosion rate for each period increases over that for the previous period the alloy is susceptible.

The mechanism of the corrosion reaction is not clear, particularly in view of the changes in composition of the HNO_3 that take place during the 48 h period of the test. Streicher[116] reports that the corrosion potential of the steel ranges from 1·00 to 1·20 V (*vs.* S.H.E.) during the test owing to the accumulation of Cr(VI), and it can be seen from Fig. 20.16 that the sensitised areas will have a higher corrosion rate than the matrix throughout this potential range, although they will become similar at the higher potentials. The high corrosion rates obtained in the test are due partly to intergranular attack and partly to the undermining of grains and grains dislodgement.

Stainless steels and Ni-base alloys containing Mo, such as type *316L* (0·03% C max.) and Hastelloy *C*, are found to give very high corrosion rates in the HNO_3 test even when they are immune to intergranular attack when subjected to other tests that reveal sensitisation due to chromium-depleted zones; furthermore, such alloys even after being subjected to a sensitising heat-treatment do not give rise to intergranular attack in most conditions of service. This high corrosion rate is considered to be due to the formation of a submicroscopic σ-phase, and although positive proof is not available its presence is substantiated by the fact that the phase becomes identifiable after longer periods at the sensitising temperatures, although in this form it has little effect on the corrosion rate. It would appear that the σ-phase dissolves rapidly during the HNO_3 test, and since it has a high chromium content the solution becomes enriched in Cr(VI) with a consequent increase in the corrosion rate of the alloy. It follows that the test is unsuitable for evaluating the behaviour of stainless steels that may precipitate σ-phase, unless the alloy is to be used in service for nitric acid plant.

Henthorne[156], in considering the corrosion testing of weldments, points out that the test will also give high rates due to (*a*) end-grain attack, which is particularly prevalent in resulphurised or heavily cold-worked material and (*b*) dissolution of Ti(C,N) such as occurs in Type *321* weldments and leads to knife-line attack. Since most service conditions do not cause attack on the alloy in these conditions the test can be misleading.

Thus under the circumstances already outlined the test can be misleadingly severe, but it is particularly valuable for evaluating alloys for use in HNO_3 or in other strongly oxidising acid solutions to ensure that they have received the correct heat treatment and have an appropriate composition, i.e. a low carbon content or the correct ratio of (Ti or Nb)/C.

Boiling $H_2SO_4 + CuSO_4$ Tests

The use of boiling $H_2SO_4 + CuSO_4$ for detecting intergranular sensitivity was first described by Strauss, *et al.*[157] in 1930, and is frequently referred to as the Strauss test, although the conditions of the test have been modified; whereas the Huey test is most widely used in the USA the Strauss test has been the preferred test in Europe. The test is mild compared to the Huey test and intergranular attack takes place with little grain dislodgement.

The use of metallic copper chips placed in contact with the steel to speed up the test and thus decrease the time of testing, was first described by Rocha[158], and subsequent work by Streicher[116] showed that its presence significantly increased the rate of intergranular attack even when it was not in contact with the steel. Approximate weight losses for a sensitised Type *316* stainless steel during a 240 h testing in boiling $H_2SO_4 + CuSO_4$ are as follows:

No metallic Cu present	$0{\cdot}1\,g\,dm^{-2}$
Metallic Cu present, but not in contact with the steel	$1{\cdot}0\,g\,dm^{-2}$
Metallic Cu in contact with steel	$4{\cdot}0\,g\,dm^{-2}$

As used in Germany the composition of the solution is $110\,g\ CuSO_4{\cdot}5H_2O$, 100 ml H_2SO_4 (s.g. 1·84) and 1 *l* of water, the test being conducted for 168 h in the boiling solution. The ASTM Tentative Procedure A393–63T specified a similar composition containing $100\,g\ CuSO_4{\cdot}5H_2O$, 100 ml H_2SO_4 (s.g. 1·84) with water added to make a total volume of 1 *l*. The test time was 72 h, and with the high carbon contents of the earlier steels this was adequate for detecting susceptibility. However, with the decrease in the carbon contents of stainless steels a more prolonged boiling time was found to be necessary, and Scharfstein and Eisenbrown[159] showed that a Type *304* stainless steel containing 0·068% C would pass the 72 h Strauss test even after a sensitising treatment of up to 4 h at 677°C. For this reason A393–63T was discontinued[156] in 1972 in favour of ASTM A262:1970 Practice *E* in which the specimens are placed in contact with metallic copper chips to increase the rate of intergranular attack[116, 153]. This test is of comparable sensitivity to the other tests, and is far more discriminating than the older tentative standard; furthermore, it is more severe so that the testing time is decreased from 72 to 24 h. This test is now being incorporated in the new international standard ISO/TC–17/SC–7.

Figure 20.16 shows that the corrosion potential of stainless steel in the $H_2SO_4 + CuSO_4$ test lies in the range 0·30–0·58 V, and that whereas the corrosion rate of the unsensitised alloy is approximately $10^{-1}\,\mu A/cm^2$, that of the sensitised material is $10 \mu A/cm^2$; for heavily sensitised material the ratio of rates[154] of sensitised/unsensitised alloy may be as high as $10^5/l$. This large difference in rates leads to rapid attack, which is confined to the depleted zone having a thickness of the order of 1 μm, and under these circumstances there will be little grain dislodgement. Thus the weight change will be so small that it cannot be used as a criterion of susceptibility. For this reason assessment of intergranular attack is normally carried out (ASTM 262–Practice *E*) by bending the specimen around a mandrel through 180° and inspecting the bend surface for cracks. Measurements of changes in electrical resistivity[116] and in ultimate tensile strength[160] are used as quantitative methods of assessment, but according to Ebling and Scheil[161] they are not as discriminating as the qualitative bend test.

The H_2SO_4–$CuSO_4$ test, unlike the Huey test, is specific for susceptibility due to chromium depletion and is unaffected by the presence of submicroscopic σ-phase in stainless steels containing molybdenum or carbide stabilisers. It can be used, therefore, with confidence to test susceptibility in austenitic (*300* series) and ferritic (*400* series) stainless steels and in duplex austeno–ferritic stainless steels such as Types *329* and *326*.

The mechanism of the action of metallic copper was investigated by Streicher[116] who determined the potential of a Type *314* stainless steel, the redox potential of the solution (as indicated by a platinised–Pt electrode) and the potential of the copper. The actual measurements were made with a saturated calomel electrode, but the results reported below are with reference to S.H.E. In the absence of metallic copper the corrosion potential of the stainless steel was 0·58 V, whereas the potential of the Pt electrode was approximately 0·77 V. When metallic copper was introduced into the solution (not in contact with the steel) both the corrosion potential of the steel and that of the Pt electrode attained the same more negative potential of 0·37 V,

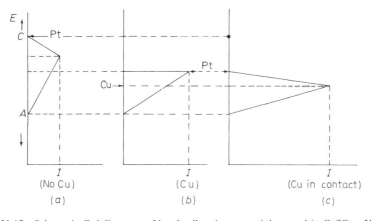

Fig. 20.17   Schematic E–I diagrams of local cell action on stainless steel in $CuSO_4 + H_2SO_4$ solution showing effect of metallic copper on corrosion rate. *C* and *A* are the open-circuit potentials of the local cathodic and anodic areas and *I* is the corrosion current. The electrode potentials of a platinised–platinum electrode and metallic copper immersed in the same solution as the stainless steel are indicated by arrows. (*a*) represents the corrosion of stainless steel in $CuSO_4 + H_2SO_4$, (*b*) the rate when copper is introduced into the acid, but is not in contact with the steel, and (*c*) the rate when copper is in contact with the stainless steel (after Streicher[116])

the copper attaining a steady value of 0·30 V. Finally, when the stainless steel was placed in contact with the copper it took up a more negative potential of 0·30 V, the potential of the copper being unaffected. These potentials have been interpreted by Streicher and have been expressed in $E-I$ diagrams (Fig. 20.17) showing how the corrosion potential and the corrosion rate varies with conditions of the test. Introduction of metallic copper into the solution results in the disproportionation reaction

$$Cu + Cu^{2+} \rightarrow 2Cu^+$$

and the accumulation of the Cu^+ ions in the solution produces a decrease in the polarisation of the local anodes on the stainless steel, which are polarised to the redox potential of the solution (Fig. 20.17b). Contact of copper with the stainless steel results in a further decrease in the corrosion potential of the stainless steel to that of copper, indicating that cathodic polarisation of the steel has occurred since the steel is the cathode of the stainless steel/copper bimetallic couple. There is simultaneously a reduction in the anodic polarisation of the susceptible grain boundaries, and a consequent increase in the corrosion rate (Fig. 20.17c). Thus contact of the steel with the copper results in intergranular attack of the steel at constant potential, the copper acting in the same way as a potentiostat.

The Huey test is widely used in the USA whilst the H_2SO_4–$CuSO_4$ test is preferred in the UK, with an increasing tendency to use the metallic copper variant. The H_2SO_4–$CuSO_4$ test procedure is given in BS 1449:Part 4:1967, and although there is no current British Standard for the H_2SO_4–$CuSO_4$–Cu test, its adoption as an alternative to the normal test is proposed for the next issue.

HNO$_3$–HF Test

This test was first described by Warren[162] in 1958, and consists of two 2 h periods in 10% HNO_3 + 3% HF solution at 70°C using fresh solution for each period. The test is therefore more rapid than the others, and it is specific for chromium depletion by carbide precipitation since it is unaffected by the submicroscopic σ-phase formed in molybdenum-bearing steels; as described in A262:70 its use is confined to Types *316*, *316L*, *317* and *317L* stainless steels[153]. Since the corrosion rates of stainless steels in the acid are high and vary greatly from test to test, it is necessary to run two tests and to compare the corrosion rates of the specimen to be evaluated ('as received' for Types *316* and *317* and in the sensitised condition for Types *316L* and *317L*) and another laboratory-annealed specimen of the same alloy shown to be free from precipitated carbides by the step structure produced after electrolytic etching in oxalic acid. Intergranular attack is assessed by the rate of penetration evaluated from the weight loss, and if the weight loss of the specimen to be evaluated is greater than 1·5 times that of the standard, the former is considered to be susceptible.

The solution has a low redox potential and the corrosion potentials for austenitic stainless steels will be in the range 0·14–0·54 V, according to composition. Thus it can be seen from Fig. 20.16 that all but the highest chromium steels will be in the active region, so that the test relies on vigorous corrosion of the grain boundary zones whilst the matrix remains somewhat passive and corrodes at a slower rate[154]. Although the test gives constant and reliable results it has not been used widely for routine evaluations for the following reasons: (*a*) the need to use a ratio of two test rates, (*b*) inconvenience of handling solution containing HF and (*c*) the availability of the $H_2SO_4 + Fe_3(SO_4)_2$ test.

$H_2SO_4 + Fe_2(SO_4)_3$ Test (Streicher Test)

This was described in 1959 by Streicher[116], and consists of one period of exposure to a boiling solution of 50 wt.% $H_2SO_4 + 25$ g/l $Fe_2(SO_4)_3$ for 120 h, assessment being based on weight loss (*see* Table 20.5). Streicher, however, usually reports a ratio of weight loss of sample to be assessed/weight loss of annealed sample (g/dm^2), and as for the HNO_3–HF test considers that a ratio > 1.5–2.0 indicates susceptibility; for Type *304* Streicher considers a rate > 0.0025 in/month to indicate susceptibility, but Brown considers a higher figure to be acceptable (*see* Table 20.5).

Accumulation of corrosion products does not stimulate attack so that several specimens may be tested in the same solution, but additional $Fe_2(SO_4)_3$ may have to be added (or the solution changed) if there is considerable attack on severely sensitised specimens, as is indicated by a colour change of the solution from brown to dark green.

The redox potential of the solution is that of the Fe^{3+}/Fe^{2+} equilibrium and lies within the range 0.80–0.85 V (*vs.* S.H.E.). The high weight loss of susceptible alloys is due to undermining and grain dislodgement at the sensitised zones, which occurs at about twice the rate of that in the Huey test. Another difference is that whereas in the Huey test corrosion products [Cr(VI)] increase the rate by raising the potential of the alloy into the trans-passive region, the converse applies in the acid ($Fe_2(SO_4)_3$ test, since reduction of Fe^{3+} to Fe^{2+} during the test will result in a decrease in the redox potential and the potential of the steel will enter the active region, and the whole sample will corrode with hydrogen evolution.

According to Cowan and Tedmon[154] the test can selectively attack some types of σ-phase. Those of Types *321* and *247* are readily attacked, whereas the molybdenum-bearing σ-phase of Type *316* is unattacked. The test will also show Hastelloys and Inconels to be susceptible to intergranular attack when there are either chromium- (or molybdenum-) depleted grain boundaries or grain-boundary σ-phase present. Ferritic (*200* series) and austeno-ferritic stainless steels can also be tested for chromium-depletion sensitisation in this reagent, but whether σ-phases formed in these alloys affect the test has not been established.

In conclusion it must be emphasised again that all the tests used are accelerated tests and only provide information on susceptibility to inter-granular attack under the precise test conditions prevailing. They are quality control tests that may be used to demonstrate either that heat treatment has been carried out adequately or that a steel will withstand the test after a certain sensitising heat treatment.

Martensitic steels Martensitic steels used for knife blades can suffer a deterioration in corrosion resistance after heating during hard soldering to the knife handle or grinding during which the metal may 'scorch'. In both cases carbides precipitate at grain boundaries and the metal is liable to pit in service. By careful attention to design and control of heating during silver soldering this problem can be greatly reduced.

A test that has been devised by the Cutlery and Allied Trades Research Association is the darkening that occurs at malheatreated areas when the blade is immersed for 5–10 min in a solution containing 5% HNO_3 and 1% HCl.

Electrolytic Oxalic Acid Etching Test

This test, which was developed by Streicher[116], is used as a preliminary screening test to be used in conjunction with the more tedious testing procedures such as the boiling HNO_3 test. The specimens are polished (3/0 grit paper) and then anodically polarised for 1·5 min at 1 A/cm² at room temperature in a solution prepared by dissolving 100 g of $H_2C_2O_4 \cdot 2H_2O$ in

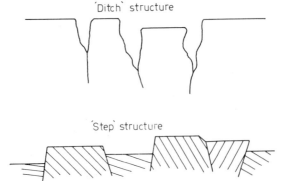

Fig. 20.18 'Ditch' and 'step' structures (after Streicher[116])

900 ml of distilled water. The surface is then examined at about × 500 magnification and the structure is classified as 'step', 'ditch' or 'dual' (both 'step' and 'ditch'). If the surface shows a 'step' structure it is immune to intergranular attack and no further testing is necessary; if the structure is 'ditch', further testing by the Huey test or some other chemical test is necessary; if 'dual' further testing may be necessary. Thus the test, by identifying structures that are immune to intergranular attack, eliminates unnecessary testing, although where a 'ditch' (or possibly a 'dual' structure) is obtained, final confirmation by the Huey test is essential. Figure 20.18 shows diagrammatically the 'ditch' and 'step' structures, and Fig. 20.19 photomicrographs of these structures and a 'dual' structure[163].

The test operates at a potential above 2·00 V (*vs.* S.H.E.), and the 'ditch' structure obtained with sensitised alloys must be due, therefore, to the high

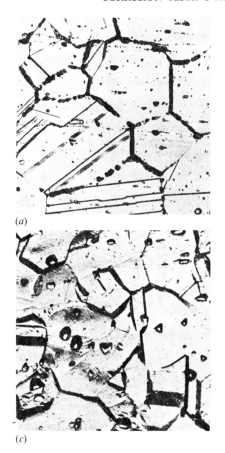
(a)

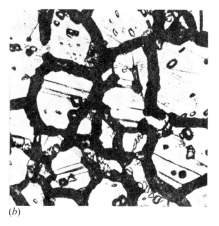
(b)

(c)

Fig. 20.19 Micrographs of (a) a 'step' structure, (b) a 'ditch' structure and (c) a 'dual' structure (after Streicher[116], and Cowan and Tedmon[154])

rate of dissolution of the sensitised areas as compared with the matrix. The 'step' structure is due to the different rates of dissolution of different crystal planes, and the 'dual' structure is obtained when chromium carbides are present at grain boundaries, but not as a continuous network.

Electrochemical Tests

Attempts have been made to use potentiostatic techniques to detect sensitisation, and Clerbois, et al.[164] observed that a sensitised 18–8 stainless steel when anodically polarised potentiostatically in $1 \cdot 0 \, mol \, dm^{-3}$ H_2SO_4 gave rise to a secondary active peak in the range $0 \cdot 14$–$0 \cdot 24$ V (vs. S.H.E.) that was not present in the curve for the annealed alloys. This observation has been criticised by France and Greene[165], who consider that the active peak is due to the dissolution of Ni that had accumulated at the surface during active dissolution at lower potentials. Clerbois, et al.[164] also noted that if a sensitised sample is held at $0 \cdot 14$ V in $1 \cdot 0 \, mol \, dm^{-3}$ H_2SO_4 for 24 h and then bent around a mandrel, it fissures and cracks, and it can be seen from Fig. 20.16 that at this potential the chromium-depleted grain boundary will corrode actively, whereas the matrix will be passive. The potentiostatic test using cracking to detect suceptibility is thus analogous to the acid–copper sulphate test.

France and Greene[165] consider that it should be possible to predict *service* performance by potentiostatic studies of steels in the environments encountered in practice coupled with metallographic examination of the surfaces.

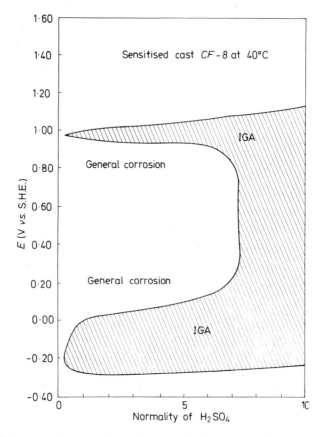

Fig. 20.20   Intergranular corrosion plot for a sensitised cast *CF-8* stainless steel (0·08% C max., 8–11% Ni, 18–21% Cr) in H_2SO_4 at 40°C as a function of potential and concentration of acid (after France and Greene[165])

They argue that many environments do not selectively attack the grain boundaries of sensitised stainless steels so that the use of costly preventative measures is unnecessary. Since the intergranular attack of austenitic stainless steels occurs only in limited potential regions it should be possible to predict service performance providing these regions are precisely characterised.

In their studies, specimens of different sensitised steels were held at various constant potentials in different concentrations of the acid under study at

various temperatures and the surfaces were then examined metallographic-
ally for intergranular attack. Data obtained in this way enabled E–concentration of acid diagrams to be produced showing the zones of general
corrosion, fine intergranular corrosion and coarse intergranular corrosion
for a given sensitised stainless steel in a given acid at various constant
temperatures (Fig. 20.20).

Streicher[166], however, considers this approach to be unsound and points
out that the short duration of the potentiostatic studies carried out by France
and Greene cannot be used to predict long-term behaviour in service. There
has been a prolonged dialogue between these workers[167], which is well summarised in the review article by Cowan and Tedmon[154] who conclude that
these particular potentiostatic tests cannot be regarded as *accelerated* tests
for services environments and that predicting future industrial service for
periods longer than the test is not advisable.

Crevice Corrosion and Pitting

Crevice corrosion and pitting have been dealt with in some detail in Section
1.6, and it is not appropriate here to discuss the nature of the phenomena nor
the methods that have been used to determine the mechanisms of these forms
of localised attack. However, it should be noted that many of the methods
of testing follow directly from the concepts that have been discussed in
Section 1.6, and in particular the potentials E_b (the critical pitting potential)
and E_p (the protection potential) have been investigated by a number of
workers as possible criteria for the resistance of metals and alloys to pitting
and crevice corrosion in service. It should also be noted that since crevice
corrosion and pitting have similar mechanisms, and since the presence of a
crevice is conducive to pitting of alloys that have a propensity to this form of
attack, it is appropriate to consider them under the same heading.

In general, the tests may be classified as follows:

1. Laboratory tests in which the specimen is immersed in a solution conducive to pitting such as an acidified $FeCl_3$ solution (redox potential
   above the critical pitting potential E_b).
2. Laboratory tests in which the specimen is anodically polarised in a
   chloride-containing solution to evaluate E_b and E_p.
3. Field tests in which the specimen (with or without a crevice) is exposed
   to the environment that it will encounter in service.

As far as tests for crevice corrosion are concerned all that is required is a
geometrical configuration that simulates a crevice, which may be achieved
in a variety of ways using either the metal itself or the metal and a non-
metallic material. Figure 1.49 (Section 1.6) shows the testing arrangement
used by Streicher[168] to study the crevice corrosion of Cr–Ni–Fe alloys, in
which two plastic cylinders are held on the two opposite faces of a sheet
metal specimen by two rubber bands, thus providing three different types of
crevice in duplicate. A simple method of testing for crevice corrosion produced by contact with different materials is to use a horizontal strip of the
metal under study and place on its upper surface at intervals small piles of
sand, small piles of sludge, pieces of gasket material, rubber, etc. More

precise crevices can be produced by bolting together two discs of the metal, which are machined on the facing surfaces so that there is a flat central portion followed by a taper to the periphery of the disc, the flat central portion providing a very fine crevice and the tapered portion a coarser one[169].

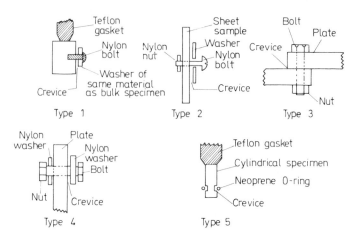

Fig. 20.21   Various types of crevices used for investigating crevice corrosion of Cr–Ni–Fe alloys (after Wilde[170])

Figure 20.21 shows the types of crevices used by Wilde[170] for studying crevice corrosion and pitting of Cr–Ni–Fe alloys in the laboratory and in the field. Types 1 and 5 were used for anodic polarisation studies in nitrogen-saturated 1 mol dm^{-3} NaCl and in aerated 3·5 wt.% NaCl, respectively, and it can be seen that attachment to the conducting lead is by means of a Stern–Makrides pressure gasket; Types 3 and 4 were used for field tests in sea-water for periods up to $4\frac{1}{2}$ years; Type 2 was used for laboratory studies in which the specimens were immersed in acidified FeCl$_3$ (108 g/l FeCl$_3$·6H$_2$O with the pH adjusted to 0·9 with HCl).

The value of electrochemical evaluation of the critical pitting potential as a rapid method of determining pitting propensity is controversial. France and Greene[171] studied the pitting of a ferritic steel (Zype 430) using a con-trolled potential test in 1 mol dm^{-3} NaCl and a conventional immersion test in oxygen-saturated 1 mol dm^{-3} NaCl, but found that at the same potential ($-0·17$ to $0·09$ V vs. S.C.E.) the corrosion rates were 390 and 5·2 mm/y, respectively. Similar studies were carried out on Zr using 0·5 mol dm^{-3} H$_2$SO$_4$ + 1 mol dm^{-3} NaCl for the controlled potential test and 0·5 mol dm^{-3} H$_2$SO$_4$ + FeCl$_3$·6H$_2$O for the immersion test, and again the former gave a much higher corrosion rate than the latter. France and Greene con-clude that these two types of test give rise to significantly different results under identical test conditions. To explain the results obtained with the ferritic stainless steel they pointed out that during the controlled potential test the anodic reaction occurs at the metal's surface whereas the inter-dependent cathodic reaction takes place at the counter-electrode. Under these circumstances the M^{2+} ions produced anodically result in increased migration of Cl^- to maintain electroneutrality, and this in turn results in a higher concentration of Cl^- at the metal/solution interface with consequent increase in the rate of pitting. A similar situation does not arise during the immersion test where the anodic and cathodic sites are in close proximity and charge balance is maintained without migration of Cl^- from the bulk solution (*see* Fig. 20.37, Section 20.2).

Table 20.6   Variation in $E_b(V)$ for stainless steels in $1 \cdot 0$ mol dm^{-3} NaCl at 25°C with nature of dissolved gas (E_b *vs.* S.C.E.)*

Gas	Type 430 stainless steel	Type 304 stainless steel
Hydrogen	−0·185	−0·050
Nitrogen	−0·130	−0·020
Argon	−0·100	+0·050
Oxygen	−0·035	+0·065

*Data after Wilde and Williams[172].

Potentiostatic tests[172-174] have been used and Wilde and Williams[172] in potentiokinetic studies of the critical breakdown potential of stainless steels (Types *430* and *304*) in $1 \cdot 0$ mol dm^{-3} NaCl, and have shown that the nature of the gas used to purge the solution has a pronounced effect on the value of E_b (Table 20.6). In particular, they have established that the presence of dissolved O_2 enhances passivity thus causing E_b to become more positive, and consider that this explains the failure of France and Greene to obtain accord between controlled potential tests in hydrogen-saturated chloride solutions and immersion tests in oxygenated chloride solutions at the same potentials.

Wilde and Williams[172] have used the redox system $Fe(CN)_6^{3-}/Fe(CN)_6^{4-}$ in $0 \cdot 1$ mol dm^{-3} for their immersion tests, which for Type *403* stainless steel gives a corrosion potential of $-0 \cdot 100$ V (*vs.* S.C.E.); selection of this system was based on the premise that being large anions they would be less likely than dissolved O_2 to be involved in the adsorption processes that stabilise the passive state. Pitting occurred within 60 s, and equivalent tests on the same alloy conducted potentiostatically at $-0 \cdot 1$ V (*vs.* S.C.E.) in hydrogen-saturated $1 \cdot 0$ mol dm^{-3} NaCl gave similar results. They conclude that these two tests give comparable results, but that extreme caution must be used in utilising E_b as an index of pitting, since its value is dependent upon environmental variables and in particular the nature of the dissolved gas in the corrodent. Wilde and Williams[175] have also shown that the critical pitting potential can be used to predict the behaviour of alloys exposed for long periods to sea-water or to industrial chemical environments.

In a more recent paper Wilde[170] points out that although E_b is qualitatively related to resistance of a material to breakdown of passivity and pit initiation, it is of questionable value in predicting performance when crevices are present. Wilde found that although the Fe–30Cr–3Mo alloy appeared to indicate total immunity to breakdown when tested anodically in 1 mol dm^{-3} NaCl and in the freely corroding condition in 10% FeCl$_3$, it pitted within the crevice when an artificial crevice was present. Exposure in sea-water for a 16 month period showed that AISI Types *304* and *316* stainless steels and the Fe–30Cr–3Mo alloy all pitted to the same extent when a crevice was present, although the former two alloys are considered to be less resistant to pitting than the Fe–30Cr–3Mo alloy. Pourbaix, *et al.* have defined the protection potential E_p (*see* Section 1.6) as the potential below which no pits can initiate and pre-existing pits cannot propagate, since they are passive at that potential. However, Wilde using cyclic potentiodynamic sweeps at varying sweep rates has established that E_p is not a unique parameter and that it varies in a semi-logarithmic manner with the extent of localised attack produced during the anodic polarisation, i.e. E_p–log(extent of pit propagation) is linear. Thus at a sweep rate of 10 V/h, E_p was found to be -0.290 V (*vs.* S.C.E.) whilst it fell to a more negative value of -0.410 V at the slower sweep rate of 1 V/h. This was explained by Wilde as being due to the chemical changes that occur in the growing pit by hydrolysis of corrosion products and by the increased migration of Cl^- ions. Since E_p is a variable that depends upon experimental procedures it cannot be used on its own as a criterion for protection against the propagation of pre-existing pits or crevices in an engineering structure. Wilde considers that a more useful parameter appears to be the 'difference potential' $(E_b - E_p)$, which is used as a rough measure of the hysteresis loop area produced during the cyclic determination of E_b and E_p. The area of the hysteresis loop obtained in a potentiodynamic sweep using a specimen with an artificial crevice provides a measure of the resistance to crevice corrosion in service, i.e. the greater the area the lower the resistance. Figure 20.22 shows the linear relationship between the 'difference potential' and the weight losses of various stainless alloys containing an artificial crevice that have been exposed to sea-water for $4\frac{1}{4}$ years.

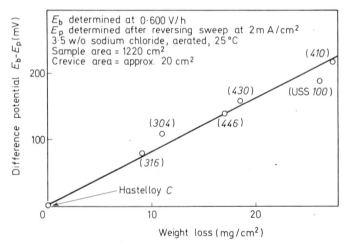

Fig. 20.22   Linear correlation between difference potential and the weight loss obtained for various stainless alloys containing a crevice, and after exposure to sea-water for $4\frac{1}{4}$ years (after Wilde[170])

The above considerations show that although considerable advances have been made in recent years in developing laboratory controlled potential tests for evaluating crevice corrosion and pitting, the results must be interpreted with caution.

Impingement Tests

The method most commonly used for testing condenser materials is the BNFRA May jet impingement test[176] in which small sections of tube, abraded to a standard finish, are immersed in sea-water and subjected to an underwater jet of sea-water containing air bubbles. Resistance to impingement attack is also assessed by the Brownsdon and Bannister test[177] in which a stream of air bubbles is directed onto the surface of the test specimens immersed in sea-water or sodium chloride solution. Special tests for resistance to corrosion under localised heat transfer conditions (hot-spot corrosion) have been described by Breckon and Gilbert[178] and by Bem and Campbell[179], but temperature effects are usually ignored when comparing condenser tube materials.

Campbell[180] points out that in evaluating condenser tube materials a test apparatus is required that will include all the principal hazards likely to be encountered in service and should thus cater for the following conditions: impingement, slow moving water, heat transfer and shielded areas. Furthermore, the internal surfaces should not be abraded, as in the jet impingement test, but should be tested in the 'as-manufactured' condition, particularly in view of the deleterious effect of carbon films produced during manufacture (*see* Sections 1.6 and 4.2).

The general arrangement of the apparatus is shown in Fig. 20.23. It accommodates 10 vertical 200 mm lengths of condenser tube spaced equally around a 125 mm diameter circle. Water enters the bottom of each tube through an inlet nozzle (Part No. 6 in Fig. 20.23) which fits inside the tube and also locates it. The nozzle has a 5 mm diameter blind hole up the centre connecting with a 2·4 mm diameter hole, set at 45° to the vertical, through which the water emerges at a velocity of 10 m/s to impinge on the wall of the tube. The water then rises up through the tube at a mean velocity of 0·1 m/s (in a 22–24 mm dia. condenser tube) and leaves through an outlet nozzle (Part No. 1) fitted into the top end of the tube. Half the length of each outlet nozzle has a 2° taper on the outside to provide a reproduceable annular crevice between it and the inside of the condenser tube. Neoprene 'O'-rings (Part No. 3) provide seals between the tube and the top and bottom nozzles, and the tubes are held in place by a common clamping plate (Part No. 2) at the top. The 10 inlet nozzles are fed with water through a distributor (Parts Nos. 7, 8 and 15) of the design used in the May jet impingement apparatus, which ensures equal distribution of water between them. The distributor and nozzles are all of non-metallic materials. The part of each test piece between 40 and 65 mm from the top is fine-machined externally to fit a semi-circular notch in a 15 mm thick brass heater block (Part No. 4), the tubes being held in contact with the block by a circumferential clip to ensure efficient and equal heat transfer between the block and each tube. The diameter of the inlet and outlet nozzles and that of the semi-circular notches in the heater block are made to suit the size of condenser tube to be tested.

The common heater block shown in Fig. 20.23 can itself be subject to corrosion leading to different heat transfer conditions for different tubes, and in some later versions of the apparatus individual short heating jackets are used for each tube, which are heated with oil from either a steam-heated or electrically heated heat exchanger. This modification not only avoids corrosion

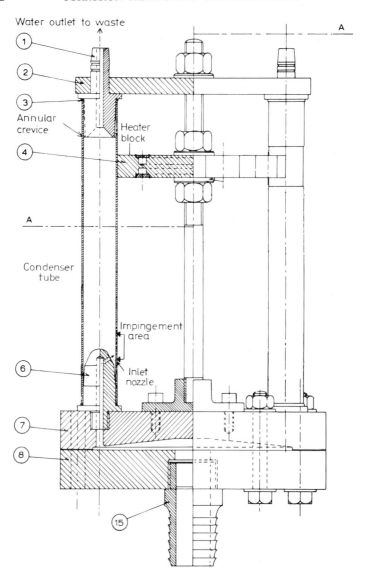

Fig. 20.23   Campbell test apparatus for determining the various forms of attack that condenser-tube materials are subjected to in service (after Campbell[180])

problems but also obviates the necessity to machine a length of the outside of each tube to fit the semi-circular notches in the single heater block. The oil flow is adjusted to give an oil temperature of 95°C at each outlet.

The test usually lasts 8 weeks, after which the tubes are sectioned longitudinally and their interiors inspected for accumulated deposits. Loose deposits are then removed by washing in water and the internal surfaces are examined for impingement attack, pitting and blistering or flaking of the corrosion-product film, using a low-power binocular microscope. After cleaning the section in 10% H_2SO_4, the depth of impingement attack, pitting or other localised corrosion, is determined. Observations and measurements are recorded for each of the following five areas of the section: (*a*) impingement area opposite the inlet nozzle where water velocity and turbulence are greatest, (*b*) the slow-moving cold water area from the impingement area upwards to the heated area, (*c*) the heated area including the heat-transfer area itself and the warm water area above and (*d*) the two annular crevices formed between the tapered portions of the cold-water inlet and warm-water outlet nozzles and the tube wall.

The Campbell apparatus is cheap to construct and easy to use, and can be installed on site to assist the selection of condenser or heat-exchanger tube materials, or to monitor changes in the corrosivity of the cooling water. The information that it provides on the various forms of attack is more comprehensive than that of any other existing apparatus for corrosion testing condenser tubes, and it is therefore particularly suitable also for assessing new materials or the effect of surface conditions arising from changes in manufacture*.

Corrosion Fatigue

The simultaneous action of alternating stresses and corrosion usually has a greater effect than when either are operating separately, and in this respect corrosion fatigue is analogous to stress-corrosion cracking. The important factors in corrosion fatigue (*see* Section 8.6) are as follows:

1. Magnitude of the alternating stress.
2. Magnitude of mean stress.
3. Frequency of reversal of the stress.
4. Characteristics of the metal.

It is extremely difficult in the laboratory to make tests in which both the conditions of corrosion and of stress approximate to those encountered in service. This applies particularly to the corrosive conditions, since it is difficult to find means of applying cyclic stresses that will also permit maintenance around the stressed areas of a corrosive environment in which the several factors that influence corrosion may be controlled. A common technique is to use a conventional Wöhler rotating cantilever beam modified so as to permit the specimen to be brought into contact with the corrodent; this may be achieved by surrounding the specimen with a cell through which the corrosive solution is circulated or by applying it by a pad[181], wick[182] or drip feed[183]. A four-point loading machine may be used in a similar way, and has the advantage over the Wöhler machine that the length of the test-piece between the two points of loading is subjected to an approximately uniform stress.

*Further information is given in Reference 180, which may be obtained from the BNF Metals Technology Centre, Grove Laboratories, Denchworth Road, Wantage, Berks., OX12 9BJ, England.

Fig. 20.24 (a) Rig for a laboratory study of the corrosion fatigue of welded joints in sea-water and (b) view of test-pieces showing welded joint (after Dias, Jarman and Smith[188])

Rawdon[184] used flat specimens that were subjected to repeated flexure while they were being immersed periodically in the corrosive solution. Kenyon[185] used a rotating wire specimen in the form of a loop, the upper part of which was attached to the motor whilst the lower part of the loop passed through the corrodent, and a somewhat similar device was developed by Haigh-Robertson and used in several studies[186, 187]; Gough and Sopwith[188] used this machine in their studies, the corrodent being applied as a spray.

Figure 20.24 shows a slow fatigue machine [189] that has been developed to study the performance of welded butt and fillet joints for steels used in the construction of North Sea oil drilling-rigs; the bending stress and frequency have been selected to simulate the forces produced by the wave motion. The specimens, $1500 \times 100 \times 12.5$ mm with the weld 25 cm from the base, are clamped at the lower end and the stress is applied as a variable bending moment at the upper end by rams. The rams, which are attached to a sliding frame, are activated by a pneumatic cylinder that can be automatically programmed for stroke and frequency and the stress level is monitored by strain gauges. The maximum amplitude of stress is 150 mm, the stress range is up to 300 MN m^{-2} and the frequency can be varied from 1 cycle/2 s to 1 cycle/20 s. The corrodent is artificial sea-water, and provision is made for studying the effect of cathodic protection by means of Zn anodes.

Hoeppner[190] points out that until recently most investigators conducted fatigue tests utilising rotating bending, flat plate bending or torsion-type loading configurations, which have the disadvantage that tests at positive or negative mean-stress values are difficult to achieve. In addition, the rotating bending and flat-plate-bending tests create complex stress states upon crack initiation, e.g. a shifting neutral axis. For these reasons, axial load fatigue machines, as recommended by the ASTM Committee E9, are preferred.

The results obtained from the tests described above are presented in the form of the conventional $S-N$ curve, where S is the stress and N the number of cycles to cause fracture. Curves of this type are obtained for the metal in air and for the metal in the corrodent, and comparison provides information on the effect of the corrosive environment on the fatigue life. Hoeppner points out that even though the $S-N$ curve for either notched or unnotched specimens may be useful for certain applications it cannot always be employed to evaluate the effect of the environment on the fatigue life. This is because in some materials the inherent metallurgical and fabrication discontinuities, which may be undetectable by N.D.T., will be so large that the only factor of engineering significance will be the rate of propagation of a crack from the initial defect, i.e. the fatigue-crack propagation rate may play the dominant rôle in the useful life of the component. For this reason it is important to conduct fatigue crack growth tests on pre-flawed specimens, and the data is then presented in the form of curves showing cyclic crack growth *vs.* stress intensity K_I.

The NACE publication *Corrosion Fatigue*[190] gives a comprehensive account of all aspects of the subject, and in this work a review of the application of fracture mechanics for studying the phenomenon has been presented by McEvily and Wei[191], whilst Kitagawa[192] has given a detailed account of crack propagation in unnotched steel specimens. This work should be consulted for details of testing and interpretation of results.

Cavitation–erosion

The phenomena and mechanisms of cavitation–erosion have been considered in Section 8.8 and here it is only necessary to consider laboratory test methods that have been designed to simulate conditions that prevail in practice and which may be used to evaluate the performance of materials. The methods used have been classified by Lichtman, Kallas and Rufolo[193] as follows:

1. High-velocity flow.
   (a) Venturi tubes.
   (b) Rotating discs.
   (c) Ducts containing specimens in throat sections.
2. High-frequency vibratory devices.
   (a) Magnetostriction devices,
   (b) Piezoelectric devices.
3. Impinging jet.
   (a) Rotating specimens pass through continuous, stationary jets or droplets.
   (b) Stationary specimens exposed to high-speed jet or droplet impact.

All tests are designed to provide high erosion rates on small specimens so that the test can be conducted in a reasonable time, and although vibratory and high-velocity jet methods may not simulate flow conditions they give rise to high-intensity erosion and can be used, therefore, for screening materials.

The essential component of many high-velocity flow rigs is a Venturi-type section in which cavitation occurs in the low-pressure high-velocity region created by the Venturi throat (*see* Figs. 8.52 and 8.53, Section 8.8). Typical of this type is the double-weir arrangement used by Schrofer[194], but since this technique requires very large volumes of water it is not readily adaptable to laboratory use. Hobbs[195] and others have used a uniform-area rectangular-cross-section duct in which a cylinder of small diameter is inserted; cavitation occurs in the wake of the cylinder, which may be used as the test specimen or the specimen may be set in the side wall of the duct near the cylinder. The cavitation intensity will be dependent on the configuration of the test section and the velocity, pressure, temperature, viscosity, surface tension, corrosivity, gas content and density of the liquid.

More recently, devices in which cavitation is achieved by vibrating a test specimen at high frequencies have been popular. The original apparatus was developed by Gaines[196], and was adapted for cavitation–erosion studies by Hunsaker and Peters as described in the paper by Kerr[197]; it has been used also by Beeching[198], Rheingans[199] and Leith, *et al.*[200]. In this method cavitation is produced by attaching the specimen to the vibrating source or by means of a partially immersed probe vibrating axially at a high velocity and low amplitude and placed close to the test specimen. Although originally magnetostriction oscillators were used[196], these have now been largely superseded by piezoelectric oscillators, which are more efficient. The apparatus consists basically of a conventional ultrasonic generator, a piezoelectric transducer and a resonating horn or probe, and the majority of tests are carried out at a frequency of 20 kHz.

Originally the test specimen was fastened to the end of the ultrasonic probe, and this is still specified in ASTM D2966–71 which describes a method of testing aluminium in antifreeze solution. However, this arrangement also subjects the test-piece to high alternating stresses as a result of the high accelerations associated with vibration at ultrasonic frequencies, which may be overcome by using a stationary test-piece and locating it immediately below a dummy tip placed on the end of the ultrasonic probe.

Vibratory test apparatuses are relatively cheap to build and run, and have low power consumption, while flow rigs are bulky, expensive to build and run, and have high power consumptions but have the advantage that they simulate more closely practical conditions of hydrodynamic cavitation. On the other hand, the damage rate is higher in the vibratory tests than in the flow test, although whether this is advantageous depends on the objectives of the test. A further criticism of the vibratory test is that the mechanical component is over-emphasised in relation to the effect produced by corrosion. For this reason Plesset[201] uses a technique in which cavitation is intermittent with short bursts of vibration followed by longer static periods, which significantly increases the erosion rate of materials with poor corrosion resistance but has little effect on materials with good corrosion resistance. Tests of this type have distinguished readily between materials having the same hardness but different resistances to corrosion, and between corrosive and non-corrosive solutions.

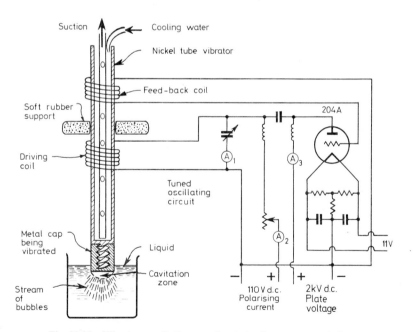

Fig. 20.25   Vibratory cavitation–erosion test using magnetostriction

Figure 20.25 shows an apparatus for studying cavitation–corrosion using the magnetostriction principle for vibration. A nickel tube is made the core of a magnetic field tuned to the natural frequency of the tube assembly, and since nickel changes its length as it is magnetised and demagnetised it will vibrate with the frequency of the magnetising current. The specimen under test vibrates with the nickel tube, and a commonly used frequency is 6 500 Hz with an amplitude of 0·008–0·009 cm. Damage is increased by the amplitude of vibration, and the more resistant the material the greater the amplitude to achieve substantial attack. Increase in temperature decreases damage by increasing the vapour within the cavitation bubbles, thus reducing the force of their collapse, but in opposition to this effect is the increased damage resulting from the lower solubility of gases which cushion the collapse of the cavitation bubbles. Consequently, under many circumstances damage reaches a maximum at a test temperature of about 46–52°C.

Assessment of cavitation-erosion is based on weight loss and the results are expressed as curves showing cumulative weight (or volume) loss *vs.* the time of the test. Eisenberg, *et al.*[202] have expressed the cumulative weight loss plot on the basis of the rate *vs.* time curve as follows:

1. Incubation zone (little or no weight loss).
2. Accumulation zone (increasing rate to a maximum).
3. Attenuation zone (decreasing loss rate to a steady-state value).
4. Steady-state zone (loss rate at a constant value).

It has been proposed that evaluation of the resistance of materials, or the study of experimental variables, should be based on the results obtained for the attenuation zone. Other methods of assessment have been proposed by Hobbs[203], and by Plesset and Devine[204].

Fretting Corrosion

The deterioration of surfaces that occurs when parts supposedly tightly fitted together nevertheless move slightly relative to each other in some sort of cycle under load is called *fretting* corrosion (*see* Section 8.7). With ferrous materials the characteristic corrosion product is a finely divided cocoa-coloured oxide. The general state of knowledge of the subject was reviewed in a symposium on fretting corrosion held by the ASTM in 1952[205] and more recently in the work by Waterhouse[206].

Several techniques for reproducing fretting corrosion have been used. All involve some means for controlling contact pressure, and for achieving and measuring small-amplitude cyclic motion or slip between the contacting surfaces; some control of the environment, particularly moisture which has a considerable effect on the extent of damage, is also desirable. Fink[207] used an Amsler wear machine. Another early series of tests on fretting corrosions arose from a study of the bottom bearings of electricity meters by Shotter[208]. Tomlinson, Thorpe and Gough[209] adapted a Haigh alternating-stress machine by which annular specimens were pressed together under load while being subjected to vibration to achieve the required slip. These investigators also used apparatus in which a specimen having a spherical surface was moved cyclically through a small amplitude while in contact under load with a plane surface. A further modification involved an upper specimen machined to provide an annulus which was oscillated under load in contact with a lower plane specimen. A similar technique was used by Wright[210]. The area of damage was measured optically and the maximum depth of damage was calculated by carefully lapping the lower surface and determining the change in mass. In addition, the amount of oxidised débris was determined chemically.

Uhlig, Tierney and McClellan[211] measured fretting damage by weight loss of recessed 25·4 mm diameter steel cylinders subjected to radial oscillating motion. The specimens were loaded pneumatically, frequency was varied, and slip was adjusted up to 0·020 mm. Weight loss was determined after débris had been removed by pickling the specimens in inhibited acid.

McDowell[212] used a set-up which took advantage of the elastic modulus of one of the test materials to provide a definite deflection subject to control. A rotating-beam fatigue-testing machine was used to produce an alternating compressive and tensile deflection on the surface of the rotating specimen. A sliding specimen slipped back and forth on the rotating specimen as the outer fibres were strained alternately in tension and compression in proportion to the extent of deflection of the rotating specimen.

Horger[205] undertook rotating-beam fatigue tests of press-fitted assemblies using specimens as large as 305 mm diameter shafts.

Warlow-Davies[213] used a technique in which specimens were subjected to fretting corrosion and then tested in fatigue to show the effect of fretting damage in lowering resistance to fatigue.

Herbeck and Strohecker[205] used machines designed particularly for comparing the merits of lubricants in preventing fretting corrosion of anti-friction bearings. One provided for both oscillating conditions and combination radial and thrust loads to simulate service. Another was concerned primarily with thrust bearings and correlated satisfactorily with the radial load tester.

An interesting approach involved microscopic observation of fretting corrosion; a glass slide mounted on the stage of a microscope was used for the bearing surface which pressed against a spherical specimen being vibrated by a solenoid[214].

Other testing machines and techniques have been described by Gray and Jenny[215], de Villemeur[216], Wright[217], Barwell and Wright[217], Field and Waters[218], and Waterhouse[219].

Corrosion Testing in Liquid Metals and Fused Salts*

Liquid metals have high heat capacities and heat transfer coefficients, and these and other properties make them attractive as coolants for high-temperature nuclear reactants and as heat-transfer and working fluids in power-generation systems that operate in conjunction with nuclear reactors.

Corrosion by liquid metals, unlike aqueous corrosion, does not usually involve galvanic effects, and, even where electrochemical phenomena are known to occur, it has not, in general, been demonstrated that they have been responsible for a significant portion of the corrosion observed[220]. In fused salts there is evidence that electrochemical factors are involved[221,222].

See also Section 2.10.

Nevertheless, the corrosion process in relation to liquid metals and fused salts may conveniently be considered under one of the classifications below which do not directly include electrochemical factors.

1. Chemical reaction.
2. Simple solution.
3. Mass transfer.
4. Impurity reactions.

Several of the foregoing may, of course, be involved in a single corrosion reaction but for simplicity they shall be treated separately.

Chemical reaction   This involves the formation of distinct compounds by reaction between the metallic container and the fused metal or salt. If such compounds form an adherent, continuous layer at the interface they tend to inhibit continuation of the reaction; if, however, they are non-adherent or soluble in the molten phase, no protection will be offered, they are non-adherent or soluble in the molten phase, no protection will be offered. In some instances the compounds form in the matrix of the alloy, for example as a grain-boundary intermetallic compound, and result in harmful embrittlement although no corrosion loss can be observed.

Simple solution   The liquid phase may simply dissolve the metal container or the liquid may go into solid solution with the metal to form a new phase. In some instances only a single constituent of an alloy will dissolve in the liquid phase; in this case a network of voids extending into the metal will result, with obvious deleterious effects.

Mass transfer   This phenomenon manifests itself as the physical transport of a metal from one portion of the system to another, and may occur when there is an alloy compositional difference or a temperature gradient between parts of the unit joined by the flowing liquid phase. An exceedingly small solubility of the metal component or corrosion product in the molten metal or salt appears sufficient to permit mass transfer to proceed at a fairly rapid pace.

Impurity reactions   Small amounts of impurities in the liquid phase or on the surface of the container may result in the initiation of attack or in increased severity of attack by one of the mechanisms just outlined.

In general, it is fair to state that one of the major difficulties in interpreting, and consequently in establishing definitive tests of, corrosion phenomena in fused metal or salt environments is the large effects resulting from very small, and therefore not easily measured and controlled, variations in solubility, impurity concentration, temperature gradient, etc.[223]. For example, the solubility of iron in liquid mercury is of the order of 5×10^{-5} at 649°C and static tests show iron and steel to be practically unaltered by exposure to mercury. Nevertheless, in mercury boiler service, severe operating difficulties were encountered owing to the mass transfer of iron from the hot to the cold portions of the unit. Another minute variation was found substantially to alleviate the problem; the presence of 10 p.p.m. of titanium in the mercury reduced the rate of attack to an inappreciable value at 650°C; as little as 1 p.p.m. of titanium was similarly effective at 454°C[224]. In the case of the alkali metals impurities such as oxygen and carbon can have a significant effect on the corrosion of steel and refractory metals.

Testing

As in all corrosion testing, the procedure which most nearly duplicates the conditions anticipated in service will provide the most satisfactory and useful information for those aspects of corrosion under consideration here. In fact, in view of the extraordinary sensitivity of fused metal and salt corrosion phenomena to minute variations in operating conditions and purity of components, as already discussed, failure to reproduce these conditions with considerable accuracy may well make any test results completely unrealistic and worthless. In all of the following, then, it should be understood, if not explicitly stated, that all extraneous matter must be carefully excluded from the system and that only materials closely simulating those to be employed in service (including prior history and surface preparation of the metals) should be used. If, however, screening tests to establish the compatibility of a relatively large number of metals with a given molten metal or salt are to be run, it is often useful to commence with static tests even though the ultimate application involves a dynamic system. This is desirable because static tests are comparatively simple to conduct and interpret, and considerably more economical to operate, and because experience has shown that a metal which fails a static test is not likely to survive the more severe dynamic test[225].

Static Tests

Ideally a static test would consist of immersing a test sample in the liquid medium held in an inert container under isothermal conditions. Tests in mercury, for example, may be contained in glass at temperatures of several hundred degrees[226]. Unfortunately, at the higher temperatures and with the aggressive metals and salts of interest there are few readily available inert container materials and results will often vary according to the nature of the container. The most satisfactory solution to this difficulty is to make the container of the same material as the test sample, or even in some cases to let it be the sample. Klueh used small capsules for determining the effect of oxygen on the compatibility of Nb and Ta with sodium[227] and potassium[228]. For the Nb–K tests the Nb specimen was approximately $2 \cdot 5 \times 1 \cdot 4 \times 0 \cdot 1$ cm and was contained in a Nb capsule surrounded by another capsule of welded Type *304* stainless steel. It was demonstrated that the oxygen concentration, added as K_2O, markedly increased the solubility of the Nb in the molten K. DiStefano[229] studied the interaction of Type *316* stainless steel with Nb (or Nb–1Zr) in Na and NaK by exposing tensile specimens of Nb (or Nb–1Zr) to the liquid metal in a stainless steel container. Carbon and nitrogen from the stainless steel were transferred to the Nb resulting in carbide–nitride at the surface and diffusion of nitrogen into the metal, thus producing an increase in tensile strength and a decrease in ductility. Close control of temperature is also essential if reproduceable results are to be obtained, because of differential solubility as a function of temperature; for example, the corrosion rate for Cu–Bi at $500 \pm 5°C$ is several times its rate at $500 \pm 0.5°C$[230].

Refluxing capsules   In systems where a liquid metal is used as the working fluid, the liquid is converted to vapour in one part of the system whilst the converse takes place in another, and the effect of a boiling–condensing metal on the container materials is most readily studied in a refluxing capsule. DiStefano and De Van[231] used a system in which the lower part of the capsule was surrounded by a heating coil whilst the upper part was water cooled. Specimens were inserted in the upper part of the capsule and thus exposed to the condensing vapour, the rate of condensation being controlled by the water flow rate.

When close control of purity is essential it may be necessary to assemble the test specimens in a dry box under an inert atmosphere and to weld the containers shut under inert gas or vacuum before placing in test. With some environments even the small amount of oxygen and moisture adsorbed on the component surfaces will significantly affect the test results. In one laboratory this problem has been eliminated by maintaining within the dry box a container of molten sodium at $250°C$[225]—a rather cumbersome procedure, but one which emphasises again the importance of purity.

Static test results may be evaluated by measurement of change of weight or section thickness, but metallographic and X-ray examination to determine the nature and extent of attack are of greater value. Changes in the corrodent, ascertained by chemical analysis, are often of considerable value also. In view of the low solubility of many construction materials in liquid metals and salts, changes in weight or section thickness should be evaluated cautiously. A limited volume of liquid metal could become saturated early in the test and the reaction would thus be stifled when only a small corrosion loss had occurred, whereas with a larger volume the reaction would continue to destruction[230, 232].

Dynamic Tests

Various tests have been devised to study the effects of dynamic conditions, and one of the simplest tests is to use a closed capsule that contains a sample at each end and is partially filled with the liquid metal or salt[232a]. A temperature gradient is maintained over the length of the tube, and the capsule is rocked slowly so that the liquid metal passes from one end to the other. After the test the extent of mass transfer is determined from the two specimens placed at each end of the capsule. Tests of this type are useful to establish whether thermal-gradient mass transfer (or concentration-gradient mass transfer if dissimilar metals are incorporated in the system) will occur, but although the method is useful for screening purposes, the dynamic nature of the heating and cooling cycles prevents a rigorous analysis of mass transfer in terms of time and temperature.

High-velocity effects can be studied in spin tests using cylindrical specimens of the solid metal and rotating them at high velocities in an isothermal-metal bath. Although strictly speaking only a single alloy should be tested at a time, it is generally satisfactory to include a variety of alloys since the velocity effects become manifest at considerably shorter times than does mass transfer.

Kassner[233] used a rotating disc, for which the hydrodynamic conditions are well defined, to study the dissolution kinetics of Type *304* stainless steel in liquid Bi–Sn eutectic. He established a temperature and velocity dependence of the dissolution rate that was consistent with liquid diffusion control with a transition to reaction control at 860°C when the speed of the disc was increased.

Loop Tests

Loop-test installations vary widely in size and complexity, but they may be divided into two major categories: (*a*) thermal-convection loops and (*b*) forced-convection loops. In both types the liquid medium flows through a continuous loop or harp mounted vertically, one leg being heated whilst the other is cooled to maintain a constant temperature across the system. In the former type, flow is induced by thermal convection, and the flow rate is dependent on the relative heights of the heated and cooled section, on the temperature gradient and on the physical properties of the liquid. The principle of the thermal convective loop is illustrated in Fig. 20.26. This method has been used by De Van and Sessions[234] to study mass transfer of niobium-based alloys in flowing lithium, and by De Van and Jansen[235] to determine the transport rates of nitrogen and carbon between vanadium alloys and stainless steels in liquid sodium.

The thermal-convection loops are limited to low flow velocities (about 6 cm/s maximum), and where higher velocities are required the liquid must be pumped, either mechanically or electromagnetically; the latter is usually preferred as it avoids the problem of leakage at the pump seal. Basically these forced-convection systems[235-238] consist of (*a*) a hot leg, where the liquid metal is heated to the maximum temperature, (*b*) an economiser or regenerative heat exchanger and (*c*) a cold leg, where the liquid is cooled to its minimum temperature. The economiser consists of concentric tubes with the hotter liquid flowing through the inner tube whilst the cooler liquid flows in a counter-current direction through the annulus between the two tubes, thus minimising power requirements. The material under test may be used for constructing all parts of the loop, and the loop is then destructively examined after a given period of test. However, this is costly and it is now usual practice to use the loop as a permanent testing facility and to use test specimens that are generally placed in the hot leg. Assessment of corrosion is based on changes in weight, dimensions, composition, mechanical properties and microstructure.

The final stage in a testing programme is the design, construction and testing of loops that simulate the type of system for which data are required.

Evaluation of loop-test results Although the thermal loop test approximates the conditions which obtain in a dynamic heat-transfer system, in evaluating the results it is necessary to be aware of those aspects in which the test differs from the full-scale unit, as otherwise unwarranted confidence may be placed

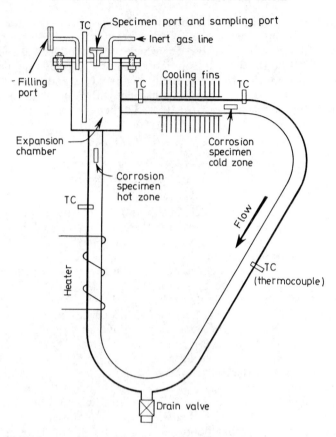

Fig. 20.26   Loop test for studying the corrosion produced by molten metals or salts

in the data. Assuming that adequate attention has been paid to the purity and condition of components, etc. the following factors will influence the observed corrosion behaviour[230]:

1. Temperature.
    (*a*) Maximum temperature.
    (*b*) Temperature gradient.
2. Surface-area/volume ratio of solid metal/liquid metal.
3. Flow velocity.

The relation between corrosion, and maximum temperature and temperature gradient is obvious, since solubility varies as a function of temperature. If the results are to be useful, these factors should match those anticipated in service. Erratic temperature cycling should be avoided as this can also modify the corrosion behaviour. The effect of surface-to-volume ratio will be more pronounced in thermal-convection than in pump loops. It can readily be seen that if a relatively small volume of liquid passes through a given iso-thermal segment of loop per unit time it will become saturated quickly and the corrosion rate will appear lower than would be the case if a substantially larger volume of liquid were passing at the same velocity. In a pumped loop the velocity can be maintained sufficiently high to prevent the attaining of equilibrium between the solid and liquid phases, and the rate of dissolution of the solid will be the controlling step. The flow velocity, or Reynolds number, will affect this step too, in that increased velocity will decrease the stagnant or lamellar layer adjacent to the tube wall and decrease the diffusion path that particles must negotiate to enter the rapidly moving stream[230]. The turbulence of the flow may also be modified by the manner in which test specimens are inserted in the loop and this should also be considered carefully in designing a test unit.

The corrosion rates of the materials of construction are always of importance but it has been found that whereas the uniform removal of metal from the hot leg may not impair the load-carrying ability of the container, the deposition of the metal in the cold leg causes cessation of flow, and the measure of the suitability of an alloy is often the time, under given conditions, which it takes for plugging to occur. Again, the flow velocity and the cross-sectional area are of primary importance in relating test results to operating conditions.

The ultimate test, short of constructing a full-scale unit, is to build a small-scale system in which each item to be incorporated in the final device is represented. Such programmes are too specialised to warrant discussion here, and are fully described in the literature[239-242].

The influence of the liquid environment on the mechanical and physical properties of structural materials is also outside the scope of this discussion but must, of course, be considered in the selection of materials of construction.

A concise account of methods of testing and evaluation of results has been provided by Klueh and De Van[243].

Tests in Plants

Although laboratory tests [244] are obviously of value in selecting materials they cannot simulate conditions that occur in practice, and although an initial sorting may be made on the basis of these tests ultimate selection must be based on tests in the plant. This is particularly important where the process streams may contain small concentrations of unknown corrosive species whose influence cannot be assessed by laboratory trials. Testing is also important for monitoring various phenomena such as embrittlement, hydrogen uptake, corrosion rates, etc. which are considered in Section 20.3.

Corrosion Racks

Exposure of coupons or specimens to the process stream cannot be achieved satisfactorily unless they are rigidly supported in a rack, although in some cases it may be possible to simply hang them in by means of a wire. Methods of exposure coupons were described in ASTM Standard A224-46[245], but this has been replaced by the Recommended Practice Method G4-68, which embodies tests that are a considerable improvement on the methods described in the original standard.

In the birdcage rack, disc specimens are mounted on a central rod, and are insulated from each other and from the rod by insulating spacers and an insulating tube, respectively. Teflon has been found to be suitable for this purpose in aggressive media, particularly at high temperatures. Plates at the end of the rack act as bumpers to prevent the specimens touching the side walls, and the assembly is constructed from a corrosion resistant material such as Monel. Advantages of this method are (1) electrical insulation avoids galvanic effects and (2) the method of holding the specimen at the centre avoids losses due to corrosion around the point of support. The disadvantages are (1) specimens are not subjected to either heating or cooling effects and thus will not disclose 'hot-wall' effects and may also escape corrosive condensates when the specimens are in a vapour stream above the dew point, and (2) the corrosivity of the environment may be affected by the presence of corrosion products of the construction material of the racks or by corrosion products of adjacent specimens. A further disadvantage is that because of its size and shape it must be inserted into the process stream when the plant is out of service. Special devices are required for mounting specimens within pipelines so that they will be subjected to velocity effects.

The *insert rack* is designed for easy installation and removal through an unused nozzle. The supporting rods (one for each specimen) are welded to a single support plate that is of a width that enables it to be introduced through the nozzle. However, this too cannot be inserted unless the equipment is out of service, although its introduction does not require removal of gas.

A slip-in rack is described by Ditton, *et al.*[246] (Fig. 20.27) that is designed to be inserted and removed during the operation of the plant through a full-

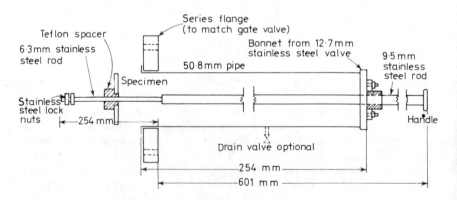

Fig. 20.27   Slip-in corrosion test rack (after Dillon, Krisher and Wissenburg[246])

port gate valve attached to a nozzle of suitable diameter (3·8–5·1 cm). It consists of a short length of pipe flanged at one end to match the gate valve and having a backing-gland arrangement at the other. The coupons are mounted on a rod of small diameter welded to a long heavier rod. The valve is opened and the support rod is pushed through the packing gland so that the specimen is introduced into the process stream. The specimens are removed by withdrawing the rod until they are again within the pipe section, the gate valve is closed and the rack removed from the valve.

Access fittings are available (e.g. Cosasco* access fitting) that enable specimens to be introduced into plant that is operating at high pressures, but can also be used for ambient pressures (*see* Section 20.3). In some instances it is possible to secure valuable information by substituting experimental materials for parts of the operating equipment, a practice that is used most frequently with condenser tubes, evaporators or other heat exchangers or sections of piping systems.

Specimens

A convenient size for a circular coupon is 3·8 cm dia., a thickness of 0·32 cm and a central hole of 1·1 cm. Although inherent in the philosophy of corrosion testing, the use of coupons with surfaces that simulate those in service has been found to be unsatisfactory owing to irreproduceability, and the standard procedure normally adopted is to abrade down to 120-grit. ASTM Method G4–68 gives details of preparation of specimens, evaluation of replicate exposures and the application of statistical methods.

Atmospheric Tests

More or less standardised techniques have been developed for the exposure of specimens to atmospheric weathering. The usual practice in the USA[247] is to mount bare specimens on racks that slope 30° from the horizontal and painted specimens on racks that slope 45° from the horizontal. The usual orientation is to have the specimens face south. In coastal exposures it is not uncommon to have the specimens face the ocean. Steel specimens exposed vertically have been found to corrode about 25% more than similar specimens exposed at the 30° angle[248]. Vertical exposure was used in the large-scale tests of non-ferrous metals undertaken by Subcommittee VI of ASTM Committee B-3[249]. Vertical exposure is also favoured by Hudson[250].

A typical test installation uses a frame to support racks on which the specimens are mounted by means of porcelain insulators. The insulators may be spaced to take specimens varying in size from 10·1 × 13·4 cm to 10·1 × 32 cm and even larger specimens may be used for certain tests. Special types of exposure have been devised to take into account important effects of partial shelter and accumulation of pools of water, as in the case of the specimen and method of support used by Pilling and Wesley[251] to compare steels for roofing.

Copson[252] has described in considerable detail the several factors that require attention in studying atmospheric corrosion, particularly of steels.

Several sizes and shapes of specimens have been used in addition to the common ones already mentioned. In the long-time test of bare and zinc-coated steels undertaken by ASTM Committee A-5 on Corrosion of Iron

*Grant Oil Tool Company.

and Steel, full-size sheets were used[253]. This Committee has also exposed specimens in the form of hardware[254] and wire and fencing[255].

The extent of deterioration may be measured by one or more of the following methods: visual examination, change in weight, change in tensile properties. Visual inspection was depended upon primarily in the A-5 tests of steel sheets[255]. Here, visible perforation more than 6 mm from an edge was the criterion of failure. This leaves much to be desired for close comparisons because of the frequency with which perforations may be obscured by heavy coats of rust[248]. Other shortcomings of the use of time to visible perforation as the criterion of corrosion resistance are as follows.

1. The removal of rust films or other corrosion products to facilitate inspection for perforation prior to termination of the exposure will change the natural performance of the material, and is therefore not tolerable.
2. The recording of a perforation establishes only the time to failure and provides no idea of the progress of corrosion up to the point of failure.
3. The time to perforation may be influenced considerably by the random occurrence of pits that happen to meet after starting from opposite sides of a sheet. This chance meeting of pits may be determined only to a slight extent by the composition of the material and, therefore, will interfere with observations of the effects of composition.

Where changes in appearance are of paramount interest, as in the case of metallic and organic coatings on steel or other metals, visual examination is most desirable. To facilitate ratings on such a basis, photographic standards have been employed, as, for example, in tests on chromium-plated steel undertaken by ASTM Committee B-8 on Electrodeposited Metallic Coatings[256]. These ratings are supplemented by a shorthand description of the nature of the deterioration observed.

Similarly, photographic standards have been established for rating organic coatings with respect to different modes of deterioration[257].

The most precise measurements of corrosion resistance require the use of specimens that can be weighed accurately after careful removal of corrosion products by the techniques described earlier.

A sufficient number of specimens should be exposed initially to permit their withdrawal from test in appropriate groups, for example 3 to 5 duplicates after at least three time intervals. For long-time tests, a suitable schedule would call for removals after 1 year, 3 years, 7 years and 15 years.

It is good practice to determine depths of pitting as well as weight loss.

As is the case with other types of corrosion testing, weight-loss determinations may fail to indicate the actual damage suffered by specimens that are attacked intergranularly or in such a manner as dezincification. In such cases, mechanical tests will be required as discussed already in the section on evaluation techniques.

It would be desirable if the reports of weathering tests could include a precise description of the climatic conditions that prevailed at the test site during the test. Unfortunately, this does not seem to be practical, since the weather factors as they are usually recorded cannot be tied in readily with the results of exposure[258]. The most important missing factors have to do with contaminants of the atmosphere and the time that specimens are actually wetted by condensed moisture other than the recorded rain. In the case of organic coatings, the interlocking effects of sunlight and moisture and their sequence complicate this problem even more[259].

Sereda[260] has described a method of determining time of wetness in which a strip of platinum foil (0.8×7 cm) is mounted on a zinc panel (10.1×13.4 cm) on both the skyward and groundward face. Condensed moisture from dew, or rain or snow, results in a galvanic cell whose potential is monitored on a recorder, thus giving the time of wetness. Guttman and Sereda[261] found that if the SO_2 content remained essentially constant the corrosion rate of zinc was related to time of wetness; furthermore, the dew detector registered the presence of moisture on the panel when the relative humidity ranged between 82 and 89%, thus providing a means of estimating from long-term weather data, such as temperature and relative humidity, the time a specimen is likely to be wet.

Atmospheric Galvanic Tests

Studies of galvanic corrosion in the atmosphere are experimentally simpler than those conducted in solution in the laboratory. The environment is taken as it comes and the relatively high electrical resistance of the rain and moisture films that serve as electrolytes restricts the distance through which the galvanic action can extend, and thus limits the relative area effects that complicate galvanic corrosion in solutions of high conductivity (*see* also Section 1.7).

Because of the limited proportion of the areas of a couple that actually participates in the galvanic action, it is difficult to make quantitative measurements that separate the galvanic action from the total effects of exposure. Thus many of the observations are likely to be qualitative ones, and often no more than what can be determined by visual inspection or measurements of changes in strength, etc. as a result of any localised galvanic action.

An idea of the distribution of galvanic corrosion in the atmosphere is provided by the location of the corrosion of magnesium exposed in intimate contact with steel in the assembly shown in Fig. 20.28 after exposure in the salt atmosphere 24.4 m from the ocean at Kure Beach, North Carolina, for 9 years. Except where ledges or crevices may serve to trap unusual amounts of electrolyte, it may be assumed that, even with the most incompatible metals, simple galvanic effects will not extend more than about 4–5 mm from the line of contact of the metals in the couple.

The extent of galvanic action in atmospheric exposure may also be restricted by the development of corrosion products of high electrical resistance between the contacting surfaces—this is especially likely to occur if one of the metals in the couple is an iron or steel that will rust. In long-time tests such possible interruptions in the galvanic circuit should be checked by resistance measurements from time to time so as to determine the actual periods in which galvanic effects could operate.

The test assembly used originally by Subcommittee VIII of ASTM Committee B-3 in its comprehensive studies of atmospheric galvanic corrosion[262] had the disadvantage that it depended on paint coatings to confine corrosion

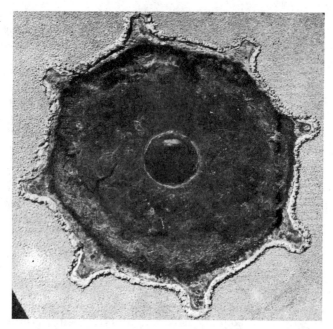

Fig. 20.28   Distribution of galvanic effects around contact of a magnesium casting and a steel core

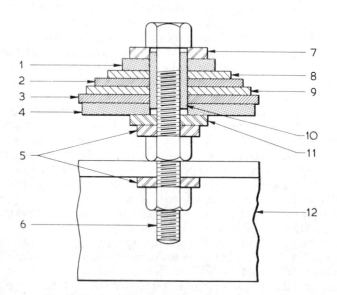

Fig. 20.29   Atmospheric galvanic couple test assembly

KEY
(1) Bakelite washer 19·0 × 3·2 mm. (2) Metal *B* disc 30 × 1·6 mm. (3) Metal *B* disc 36·6 × 1·6 mm. (4) Bakelite washer 35·5 × 3·2 mm. (5) Stainless steel lock washer. (6) Stainless steel bolt 4·8 × 38·1 mm. (7) Stainless steel washer 15·9 mm o.d. (8) Metal *A* disc 25·4 × 1·6 mm. (9) Metal *A* disc 35·5 × 1·6 mm. (10) 11·1 mm Bakelite bushing, 5·2 mm i.d. × 7·9 mm o.d. (11) Stainless steel washer, 15·9 o.d. (12) Galvanised angle support.

to the surfaces in actual contact with each other. In interpreting the results, it was frequently difficult to decide how much corrosion was due to galvanic action and how much to a variable amount of normal corrosion through failure of the paint system.

These difficulties were overcome in a design developed by Subcommittee VIII of ASTM Committee B-3[263] (Fig. 20.29). In this assembly each of the two middle specimens has a specimen of the other metal each side of it and only these middle specimens are considered in appraising the results.

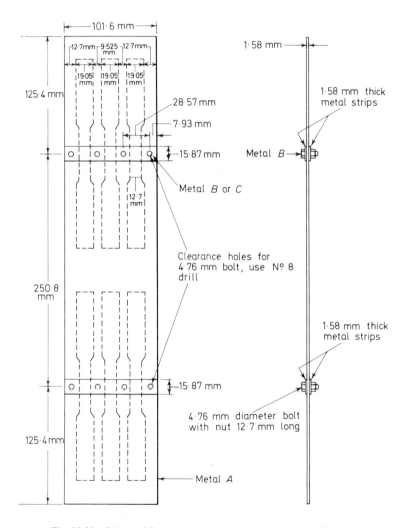

Fig. 20.30   Plate and fastening type galvanic couple test specimen

A fairly direct way of observing galvanic effects, which also permits changes in mechanical properties to be measured, involves the preparation of a composite specimen formed by attaching a strip, or strips, of one metal to a panel of another one. Tensile test specimens that include the areas of galvanic action can be cut from these panels after exposure, as shown in Fig. 20.30.

A modification of the specimen shown in Fig. 20.30 may be made simply by lapping a panel of one material over a panel of another one. The greatest effects may be observed when such panels are exposed with the laps facing up so as to favour retention of corrosive liquids along the line of contact. To permit observations of secondary effects of corrosion products, or exhaustion of corrosive constituents, the relative positions of the dissimilar metals should be changed from top to bottom in duplicate test assemblies.

Where the practical interest is in possible galvanic effects of fastenings, it is simple to make up specimens to include such couple assemblies as illustrated in Fig. 20.31.

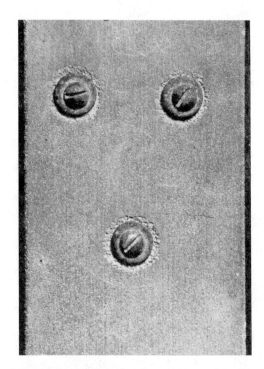

Fig. 20.31   Specimen for studying galvanic corrosion resulting from fasteners

A type of assembly calculated to favour maximum galvanic action was developed by the Bell Telephone Laboratories and is illustrated in Fig. 20.32. Here, the less noble metal is in the form of a wire wound in the grooves of a threaded specimen of the metal believed to be more noble. Good electrical contact is achieved by means of set screws covered with a protective coating. This assembly favours accumulation of corrosive liquids around the wire in the thread grooves. Corrosive damage is also favoured by the high ratio of surface to mass in the wire specimens.

To determine whether a protective metallic coating will retard or accelerate corrosion of a basis metal, and to what distance either effect will extend,

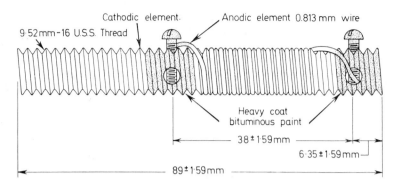

Cathodic element. Anodic element 0.813 mm wire

9·52mm–16 U.S.S. Thread

Heavy coat
bituminous paint

38 ± 1·59mm

6·35 ± 1·59mm

89 ± 1·59mm

Fig. 20.32 Bolt and wire type atmospheric galvanic couple test specimen

specimens in which strips of various widths are left bare or made bare have been used by Subcommittee II of ASTM Committee B-8[264]. The extent of corrosion in and near the bare strips as compared with that on a completely bare or completely coated specimen will provide a measure of the extent of galvanic action and the distance through which the effect is able to extend from the edges of the bare strips.

Tests in Natural Waters

The type of test to be discussed here involves exposure to waters under natural conditions, such as by immersing specimens in the sea or in rivers.

The principal requirements for such tests are as follows:

1. The specimens should be mounted so that they are insulated from their supporting racks and from each other. Such insulation can be achieved by the use of fastening assemblies, such as illustrated in Fig. 20.33. Occasional difficulties have been encountered with this sort of assembly for tests of copper and high-copper alloys because of deposition of copper from corrosion products along the surfaces of the insulating tubes which provided a metallic bridge between the specimens and the rack and introduced un-desired galvanic effects. The required insulation and support can be provided by use of porcelain or plastics knob insulators in much the same manner as used on atmospheric test racks. A modified design has the advantage of offer-ing less resistance to the flow of water and is less likely to serve as a form of screen to catch débris floating on, or suspended in, the water. Additional details of rack design may be found in the section on sea-water tests in *The Corrosion Handbook* (Uhlig, Selective Bibliography).

2. All specimens to be compared should be suspended at the same depth or should pass through the same range of depths. Isolated specimens exposed at different depths will not be corroded in the same way as continuous speci-mens that extend through the total range of depth to be studied. This is especially the case with specimens exposed to sea-water from above high tide to below low tide. Where the behaviour of structures, such as piling, that

pass through these zones is to be investigated, the test specimens must be continuous and large enough to extend through the total range in order to take into account differential aeration and other possible concentration cells that may have such a tremendous effect on the results secured[265]. For example, in sea-water exposure, isolated specimens of steel exposed in the tidal zone have corroded 10 times as fast as portions of continuous specimens of the same steel in the same zone that extended also below low-tide level[248].

3. The specimens should be oriented so that their flat surfaces are parallel to the direction of water flow and so that one specimen will neither shield an adjacent specimen from effects of water velocity nor create any considerable extra turbulence upstream of it.

4. In tests in sea-water where accumulations of marine organisms are likely, specimens exposed parallel to each other should be spaced far enough apart to ensure that the space between specimens will not become completely clogged by fouling organisms. A minimum spacing of 100 mm is suggested.

5. Wooden racks used in sea-water tests are likely to be subject to severe damage by marine borers. The wood used, therefore, must be treated with an effective preservative, for example creosote applied under pressure, if the test is to extend for several years. Organic copper compound preservatives may suffice for shorter tests, for example 2 or 3 years. Since the leaching of such preservatives may have some effects on corrosion, metal racks fitted with porcelain insulators have an advantage over wooden racks.

6. Where constant depth of immersion is desired in spite of tidal action, it is necessary to support the test racks from a float or raft.

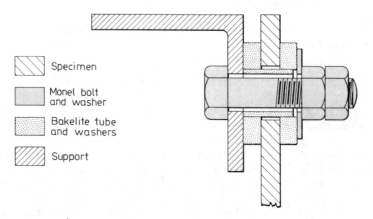

Fig. 20.33   Scheme for insulating specimens from metal test racks

An extensive study of the corrosion of metals in tropical environments has been carried out by Southwell, *et al.*[266]. Tests have included atmospheric exposure, and exposure in sea-water under mean tide and fully immersed conditions for a range of ferrous and non-ferrous metals and alloys.

Field Tests in Soil

Exposure Tests

The procedure for field tests in soils is simply to bury the test specimens, observing certain precautions listed by Logan, Ewing and Denison[267] as follows:

1. A sufficient number of specimens to yield a reliable coverage should be included.
2. The test site should be typical of the type of soil to be investigated.
3. The depth of burial should be that which will be occupied by the structure of interest. Specimens to be compared should be buried at the same depth. Ideally, tests for structures, such as piling, that will extend through several horizons would require the use of test specimens long enough to extend to the same depth.
4. Specimens should be separate so that they will not affect the corrosion of each other. A minimum spacing of two diameters was proposed.
5. Cylindrical specimens should be laid horizontally.
6. Sheet or plate specimens should be placed on edge.
7. The ends of pipe specimens should be closed to prevent internal corrosion.
8. Sufficient specimens should be provided to allow withdrawals after several time intervals so as to permit observations of changes in corrosion rates with time.
9. A portion of the original surface should be protected so as to provide a datum line for the measurement of pit depths.
10. In applying results of tests on small specimens to estimating corrosion, particularly by pitting on large structures, the effect of the increased area in increasing the depth of pitting must be taken into account[268].

Other Tests

Tests to determine the activity of sulphate-reducing bacteria are described in Section 2.6, and the various tests used in cathodic protection such as soil resistivity, soil redox potential and potentials of buried metals, are described in Section 11.7.

Corrosion Testing of Organic Coatings*

Programmes to evaluate the corrosion protection of organic coatings on metals are intended to establish relationships between coating properties and performance. Such knowledge is essential to the most effective use of organic coating systems in corrosion control. Depending on the detail with which such studies are performed, light may be shed on the mechanism of coating deterioration as well.

---

*Only a brief account of standardised procedures is given and for further details Reference 2 should be consulted. Standards for various paint tests are given in Appendix 20.1B.

If valid and useful relationships are to be established, it is essential that the factors affecting performance be recognised and form part of the test record. Since the performance is determined by the interaction among the coating, the substrate and the surrounding environment against which protection is sought, significant factors and their interrelationships will vary with the nature of the service.

Regard for the complex relationships between coating characteristics and performance in different environments warns that care in designing and conducting the test in no way reduces the need for discrimination on the part of the person using the test data in the selection of a coating for a particular purpose. Test environments must reflect the deteriorating influences of the service for which they are applicable. A coating system cannot reliably be selected for service in a chemical plant on the basis of performance determined in a rural atmosphere.

Thus, both the proper conduct of the testing programme and the valid use of the data depend on an understanding of the nature of organic coatings and of the forces through which they are degraded.

Behaviour of Organic Coatings

An organic coating provides corrosion protection through the interposition of a continuous, adherent, high-resistance film between the metal surface and its environment (*see* Section 15.3). In principle, its function is the mechanical exclusion of the environment from the metal surface. It seldom, if ever, succeeds practically in achieving this since all continuous organic films are permeable to some degree to moisture and many coatings either have occasional physical defects or acquire them in service. Surface conversion treatments, such as phosphate and chromate dips, are used to supplement the physical protective properties of coatings, as are chemically inhibiting primers and wash primers. When such treatments are used, they must be included in the record as constituting a part of the coating system.

Paints are considered in detail in Chapter 15, paint failures being discussed in Section 15.4.

Critical parts of the test programme are the preparation of test specimens, the selection of the exposure conditions and the selection of significant coating properties to be evaluated as a measure of deterioration with time.

Preparation

Specimens will normally be flat panels, large enough to avoid any effects caused by nearby edges of the specimen. Edges and backs are usually coated unless the effect of uncoated edges is an intended test variable. Panels may include the structural features of plates, channels, welds, sharp edges, pits or depressions, depending on the service for which the data are to be applicable.

The composition of the basis metal has been found to influence the performance of organic finishes in many cases. Thus, composition is a significant test variable and must be considered in comparing test data.

It is particularly important that surface roughness and cleanness, which greatly affect adhesion, should be carefully controlled and that the procedures used to achieve them be a part of the test record. A high degree of cleanliness is normally sought. If, however, the data are to be applicable to the painting of outdoor structures, a certain amount of outdoor weathering becomes a part of the specimen preparation prior to coating. Specimens again will be of the basis metal appropriate for the related service application.

The thickness of a coating plays an important part in determining its physical characteristics. Uniformity of thickness among specimens therefore is necessary, particularly when coating deterioration is to be assessed by changes in such properties. For the preparation of reproduceable specimens, methods of applying coatings in uniform thicknesses are available, as are methods for accurately measuring film thickness.

Exposure

To account for structural variables, service tests of organic coatings are sometimes conducted on actual structures in service. Service tests, the results of which can be assessed only by visible failure, have the disadvantage that there is little opportunity of controlling the specimen geometry or ambient conditions during the application of the coating. The results obtained are, therefore, specific for the particular test, and the use of test panels incorporating the appropriate structural features in the environment of interest, is preferable. Field exposure of test panels offers the benefit of a high degree of control over surface preparation and application. Moreover, through standardised exposure conditions, broader comparisons between both paint systems and locations are possible. More importantly, since replicates may be removed and laboratory tested periodically, changes in properties can be followed in considerable detail. At least four replicates should be examined for each exposure period to minimise the effects of atypical specimens.

The exposure site is selected according to the service for which the data are to be applicable. For atmospheric service, such factors as marine and industrial contaminants, sunlight, dew and sand abrasion, must be considered. Atmospheric specimens are normally mounted at 45°, facing south. This has been shown to provide about a 2:1 acceleration of failure compared with a vertical exposure. Whether this or other standardised positions are used, the details of the exposure are an important part of the test record.

The degree of deterioration experienced over a given test period varies with climatic conditions. Since these differ significantly from one season to another, a standard specimen, the performance of which is well known, should be included with each exposure to increase the validity of coating comparisons.

For other environments, such as in sea-water or in chemical plants, exposure conditions that most nearly duplicate those of the related service and are at the same time reproduceable, are used. Impingement by water or water carrying entrained solids, thermal effects and physical abuse are among the factors to be considered.

The obvious desirability of early information on the performance of protective paint systems has led to the development and use of accelerated tests.

If intelligently used, with proper respect for their limitations, such tests are of value in determining the probable order of durability of a group of paints, or in assessing the quality of a range of similar compositions where there is already some knowledge of the performance of the general composition. Efforts to correlate accelerated test results and service performance have been characteristically unsuccessful and such accelerated tests cannot be used to predict the type of failure likely to occur from natural exposure. Perhaps the greatest disservice provided by the accelerated test approach is encouragement of the fallacy in the minds of the inexperienced that the rougher the treatment the better the test.

The standardised salt-spray test is used to some extent for accelerated paint evaluation. In more common use are the various types of weatherometers in which the paint-coated panels are subjected to various cycles of wetting and exposure to ultra-violet light to simulate atmospheric conditions of exposure. BS 3900:Part F3:1971 describes a weatherometer consisting of a 1·2 m dia. drum that rotates at 1 rev/20 min, and has facilities for spraying the panels (100 × 150 mm) periodically during a 24 h cycle and exposing them to ultra-violet light by means of an enclosed carbon arc. Spraying with distilled water is effected by means of an atomiser and fan using the following 24 h cycle: 4 h off, 2 h on, 10 h off, 2 h on and 1 h off; the final 1 h is used for checking the arc. The test is continued for 7 days at the end of which the panels are examined visually for change of colour, loss of gloss and blistering, and for checking, cracking and chalking by means of a lens (× 25).

An appraisal of artificial weathering methods is given in a recent report by Hoey and Hipwood[269] who describe the effectiveness of various weatherometer tests such as are described in BS 3900:Part F3; ASTM E42 (1964) Type *E*; Dew cycle (Atlas *XW-R*); Zenotest *WL* apparatus; etc. for various paint systems. They conclude that ASTM E42 produces more severe film deterioration than that produced by BS 3900:Part F3 within a shorter time of exposure, and is the preferred test for paints that are to be exposed to tropical and hot sunny climates. Although these tests simulate atmospheric exposure it is not possible to obtain a direct correlation owing to variation in outdoor exposure conditions from place to place, but they serve a very useful purpose in providing a preliminary sorting of paints that can then be tested in the field.

Coating Evaluation

The performance of organic finishes on test is evaluated by visual observation and by physical tests made upon coated specimens that have been exposed for various periods of time to natural or accelerated weathering conditions. Electrical tests are sometimes used on immersed specimens.

Inspection of test panels at the test sites consists of visual observations of blistering (*see* ASTM D714:1956(1970)(21)) and the appearance of rust (*see* ASTM 610:1968(21)). For these, photographs showing various degrees of degradation which serve as observational standards greatly reduce variations between observers. Results of consecutive observations entered on charts provide visual records of the trend of these features with time.

These more serious evidences of degradation, however, are preceded by invisible physical alterations within the coatings which can be detected readily and quantitatively by suitable physical tests of replicate specimens removed from tests at periodic intervals. The use of such tests to reveal incipient coating changes is, in a sense, a means of accelerating the test programme without distortion of the test environment. This approach is especially dependent on uniformity of properties among replicates, hence on reproduceable application techniques. Moreover, since many coating properties are highly sensitive to changes in temperature and relative humidity, equilibrium of the specimens during testing is necessary. Testing conditions are commonly 25°C and 50% r.h.

Physical tests appropriate for this type of evaluation are not necessarily limited to those properties which the coating may be called upon to display in service. A coating that shows a decrease in distensibility from 20 to 10% is still quite capable of withstanding the expansion and contraction of the substrate in atmospheric temperature cycling, yet such a coating can be expected to fail in service earlier than one which shows no decrease. Thus, the properties of value are those that have been established as reliable indices of deterioration.

Besides providing early comparative data on coating performance, physical tests in dealing with intrinsic coating properties provide much-needed quantitative information on the relationships between the several factors affecting the ageing of organic films. The tests cited below are those which have been shown to indicate reliably significant changes in the condition of coatings on tests.

Distensibility This property is very sensitive to chemical changes within the coating. Its measurement thus shows the beginning of normal ageing or of deterioration through reaction with the environment. Distensibility is generally determined by bending the best panels over a conical mandrel of known radius and calculating the % elongation at first rupture.

Abrasion tests In these tests the end point is normally taken as the amount of abrasion required to penetrate the coating. The results thus reflect the strength of the coating, its cohesion, and in some cases its adhesion to the basis metal as well as resistance to abrasion.

Hardness Coating hardness is related to the method of measurement. Results reflect the resistance to scratching as well as to indentation.

Impact tests Such tests reveal the resistance of coatings to deformation and destruction by concentrated sudden stresses. They thus throw considerable light on the integrity of the metal–coating bond. Changes in adhesion through chemical reaction at the paint/metal interface will be reflected in the impact-test values.

Adhesion This is one of the most important coating properties as far as performance is concerned. Its influence on distensibility tests has been mentioned. Other tests aimed at evaluating adhesion are based on the force necessary to separate the coating from its substrate. These forces may be applied in tension, ultrasonically, with a specially designed ivory knife or by parallel closely spaced cuts through the coating. Technologists continue to search for a method to measure adhesion as a basic property. Among the current approaches being investigated is the ultracentrifugal adhesion tester based on the centrifugal force necessary to dislodge a paint spot from a rapidly rotating rotor.

The above tests for characterising coating properties necessarily continue to involve a certain amount of empiricism. The intelligent use of these tests, however, has shown that wide variations of physical characteristics of coatings as a function of composition may be obtained, and further, that significant changes in these characteristics, that can be measured before the usual evidence of failure appears, occur upon natural and accelerated ageing.

L. L. SHREIR
F. L. LaQUE

Acknowledgements

Acknowledgements are made to Mr. K. A. Chandler, Mr. H. Campbell, Dr. M. Clarke, Dr. V. F. Lucey and Mr. J. E. Truman for their helpful comments.

REFERENCES

1. Champion, F. A., 'Corrosion Testing Procedures', Chapman and Hall, 2nd edn (1964)
2. Ailor, W. H. (ed.), 'Handbook on Corrosion Testing and Evaluation', John Wiley (1971)
3. NACE Standard TM–01–69 *Test Method: Laboratory Corrosion Testing of Metals for the Process Industries*
4. Azzam, A. M., Bockris, J. O'M., Conway, B. E. and Rosenberg, H., *Trans. Faraday Soc.*, **46**, 918 (1950)
5. Wesley, W. A., *Proc. Amer. Soc. Test. Mater.*, **43**, 649 (1943)
6. ASTM Designation A380–57, ASTM Standards, Pt. 1, 1167 (1958)
7. Bayer, R. O. and Kachik, E. A., *Corrosion*, **5**, 308 (1949)
8. Renshaw, W. G. and Ferree, J. A., *Corrosion*, **7**, 353 (1951)
9. ASTM Designation A393:1963(3)
10. ASTM Designation A262:1970(3)
11. Gulbransen, E. A., *Min. and Metall., N.Y.*, **25**, 172 (1944)
12. Kruger, J., *J. Electrochem. Soc.*, **106**, 847 (1959), see also *Ellipsometry in Corrosion Testing* by J. Kruger and P. C. S. Hayfied in Reference 2
13. Rutherford, J. J. B. and Aborn, R. H., *Trans. Amer. Inst. Min. (Metall.) Engrs.*, **100**, 293 (1932)
14. Burns, R. M. and Campbell, W. E., *Trans. Electrochem. Soc.*, **55**, 271 (1929)
15. Hudson, J. C., *Proc. Phys. Soc. Lond.*, **40**, 107 (1928)
16. Finkeldey, W. H., *Proc. Amer. Soc. Test. Mater.*, **32**, 226 (1932)
17. Kenworthy, L. and Waldram, J. M., *J. Inst. Met.*, **55**, 247 (1934)
18. Tronstad, L., *Trans. Faraday Soc.*, **29**, 502 (1933)
19. Kruger, J., *J. Electrochem. Soc.*, **106**, 847 (1959) and **108**, 504 (1961); *see also* Kruger, J., *Recent Developments in Ellipsometry*, Symp. Proc., University of Nebraska, N. M. Bashara, A. B. Buckman and A. C. Hall (Eds.), published in *Surface Science*, **16** (1969)
20. Bengough, G. D., Stuart, J. M. and Lee, A. R., *Proc. Roy. Soc. A*, **116**, 425 (1927); *A*, **121**, 89 (1928)
21. Shipley, J. W., McHaffie, I. R. and Clare, N. D., *Industr. Engng. Chem.*, **17**, 381 (1925)
22. Bloom, M. C. and Krulfeld, M., *J. Electrochem. Soc.*, **104**, 264 (1957)
23. Vaurio, V. W., Clark, B. S. and Lueck, R. H., *Industr. Engng. Chem. (Anal.)*, **10**, 368 (1938)

24. Hudson, R. M. and Stragand, G. L., *Corrosion*, **15**, 135t (1959)
25. Willey, A. R., Krickl, J. L. and Hartwell, R. R., *Corrosion*, **12**, 433t (1956)
26. Burns, R. M., *J. Appl. Phys.*, **8**, 398 (1937)
27. Dravnieks, A. and Cataldi, H. A., *Corrosion*, **10**, 224 (1954)
28. Marsh, G. A. and Schaschl, E., *Corrosion*, **14**, 155t (1958)
29. Amer. Soc. Test. Mater. Spec. Tech. Publ. No. 32 (1937)
30. Knapp, B. B., in *The Corrosion Handbook* (ed. H. H. Uhlig), Wiley, New York; Chapman and Hall, London, 1077 (1948)
31. Teeple, H. O., Amer. Soc. Test. Mater. Spec. Tech. Publ. No. 175, 89 (1956)
31a. *Recommended Practice for Laboratory Immersion Corrosion Testing of Metals*, ASTM G31:1972(31)
32. Romeo, A. J., Skrinde, R. T. and Eliassen, R., *Proc. Amer. Soc., Civ. Engrs.*, **84**, No. SA4 (1958)
33. LaQue, F. L. and Stewart, W. C., *Métaux et Corros.*, **23**, 147 (1948)
34. Vernon, W. H. J., *J. Soc. Chem. Ind., Lond.*, **66**, 137 (1947); *Corrosion*, **4**, 141 (1948)
35. Bengough, G. D. and May, R., *J. Inst. Met.*, **32**, 81 (1924)
36. May, R. and Stacpoole, R. W. de Vere, *J. Inst. Met.*, **77**, 331 (1950)
37. Brownsdon, H. W. and Bannister, L. C., *J. Inst. Met.*, **49**, 123 (1932)
38. Freeman, J. R., Jr. and Tracy, A. W., *Corrosion*, **5**, 245 (1949)
39. Fontana, M. G., *Industr. Engng. Chem.*, **39**, 87A (1947)
40. Wagner, H. A., Decker, J. M. and Marsh, J. C., *Trans. Amer. Soc. Mech. Engrs.*, **69**, 389 (1947)
41. Trebler, H. A., Wesley, W. A. and LaQue, F. L., *Industr. Engng. Chem.*, **24**, 339 (1932)
42. Brennert, S., *J. Iron St. Inst.*, **135**, 101P (1937)
43. Smith, H. A., *Metal Progr.*, **33**, 596 (1938)
44. Hanawalt, J. D., Nelson, C. E. and Peloubet, J. A., *Trans. Amer. Inst. Min. (Metall.) Engrs.*, **147**, 273 (1942)
45. ASTM Designation B192–44T, ASTM Standards, Pt. 3, 243:1958
46. Benedicks, C., *Trans. Amer. Inst. Min. (Metall.) Engrs.*, **71**, 597 (1925)
47. McAdams, D. J., *Heat Transmission*, McGraw-Hill, New York, 3rd edn, 370 (1954)
48. Groves, N. D. and Eisenbrown, C. M., *Metal Progr.*, **75** No. 5, 78 (1959)
49. Fisher, A. O. and Whitney, F. L., Jr., *Corrosion*, **15**, 257t (1959)
50. Fisher, A., O., *Corrosion*, **17**, 215t (1961)
51. Gleekman, L. W. and Swandby, R. K., *Corrosion*, **17**, 144t (1961)
52. Groves, N. D., Eisenbrown, C. M. and Scharfstein, L. R., *Corrosion*, **17**, 173t (1961)
53. Hart, R. J., Reference 2, p. 367
54. La Que, F. L., 'Electrochemistry and Corrosion (Research and Tests)', Acheson Memorial Address, *J. Electrochem. Soc.*, **116**, 73C (1969)
55. Stern, M. and Geary, A. L., *J. Electrochem. Soc.*, **104**, 56 (1957)
56. Stern, M., *Corrosion*, **14**, 440t (1958)
57. Pourbaix, M., 'Lectures on Electrochemical Corrosion', Plenum Press (1973); *see also* references to potential–pH diagrams given in Section 1.4
58. Armstrong, R. D., Henderson, M. and Thirsk, H. R., *J. Electroanal. Chem.*, **35**, 119 (1972)
59. Epelboin, I., Keddam, M. and Takenouri, H., *J. App. Electrochem.*, **2**, 71 (1972)
59a. Sathyaharayana, S., *Electroanal. Chem. and Interfacial Electrochem.*, **50**, 411 (1974)
60. Ives, D. J. G. and Janz, G. J., *Reference Electrodes*, Academic Press (1961); *see also* Compton, K. G., Materials Research and Standards, **10** No. 1, 13 (1970); Covington, A. K., *Electrochemistry*, Vol. 1, The Chemical Society, London, 56 (1970); Meites, L. and Moros, S. A., *Analyt. Chem.*, **31**, 25 (1959)
61. Berzins, T. and Delahay, P., *J. Am. Chem. Soc.*, **77**, 6448 (1955)
62. Pouli, D., Huff, J. R. and Pearson, J. C., *Anal. Chem.*, **38**, 382 (1966)
63. Kooijman, D. J. and Sluyters, J. H., *Electrochim. Acta.*, **11**, 1147 (1966)
64. Bewick, J., *Electrochim. Acta*, **13**, 825 (1968)
65. Piontelli, R. and Bianchi, G., *Proc. 2nd Meeting C.I.T.C.E.*, Milan (1951); Piontelli, R., *Proc. 4th Meeting C.I.T.C.E.*, London–Cambridge (1952)
66. von Fraunhofer, J. A. and Banks, C. A., *Potentiostat and its Applications*, Butterworths, London (1972)
67. Stern, M. and Makrides, A. C., *J. Electrochem. Soc.*, **107**, 782 (1960)
68. Greene, N. D., France, W. D. and Wilde, B. E., *Corrosion*, **21**, 275 (1965)
69. Greene, N. D., Acello, S. J. and Greif, A. J., *J. Electrochem. Soc.*, **109**, 1001 (1962)

70. Cleary, H. J. and Greene, N. D., *Electrochim. Acta.*, **10**, 1107 (1965)
71. France, W. D. (Jr.), *J. Electrochem. Soc.*, **114**, 818 (1967)
72. Wilde, B. E., *Corrosion*, **23**, 331 (1967)
73. Smith, L. W. and Pingel, V. J., *J. Electrochem. Soc.*, **98**, 48 (1951)
74. Budd, M. K. and Booth, F. F., *Metallurgia*, **66**, 245 (1962)
75. Cleary, H. J., *Corrosion*, **24**, 159 (1968)
76. May, R., *J. Inst. Metals*, **40**, 141 (1928)
77. Hines, J., private communication
78. Hoar, T. P. and Hines, J. G., *J. Iron Steel Inst.*, **182** No. 124, 156; Hoar, T. P. and West, J. M., *Proc. Roy. Soc.*, **A268**, 304 (1962)
79. Horst, R. L. (Jr.), Hollingsworth, E. H. and King, W., *Corrosion*, **25**, 199 (1969)
80. Pearson, J. M., *Trans. Electrochem. Soc.*, **81**, 485 (1942)
81. Schwerdtfeger, W. J., *Corrosion*, **19**, 17t (1963)
82. Schwerdtfeger, W. J. and Manuele, R. J., *Corrosion*, **19**, 59t (1963)
83. Skold, R. V. and Larson, T. E., *Corrosion*, **13**, 139t (1957)
84. Stern, M. and Weisert, E. D., *Proc. Am. Soc. Test. Mater.*, **59**, 1280 (1959)
85. Barnartt, S., *Corrosion*, **27**, 467 (1971); *Corros. Sci.*, **9**, 145 (1969)
86. Leroy, R. L., *Corrosion*, **29**, 272 (1973)
87. Oldham, K. B. and Mansfeld, F., *Corrosion*, **27**, 434 (1971)
88. Oldham, K. B. and Mansfeld, F., *Corros. Sci.*, **13**, 811 (1973)
89. Hickling, J., Ph.D. Thesis, University of Cambridge (1974)
90. Hoar, T. P., *Corros. Sci.*, **7**, 455 (1967)
91. Mansfeld, F., *Corrosion*, **29**, 397 (1973); *Corrosion*, **30**, 92, 320 (1974); *J. Electrochem. Soc.*, **118**, 545 (1971)
91a. Mansfeld, F., *J., Electrochem. Soc.*, **120**, 515 (1973)
92. Makrides, A. C., *Corrosion*, **25**, 455 (1969)
93. Legault, R. A. and Walker, M. S., *Corrosion*, **19**, 222 (1963)
94. France, W. D. and Walker, M. S., *Mat. Protect.*, **47** (1969)
95. Jones, D. A. and Greene, N. B., *Corrosion*, **25**, 367 (1969)
96. Wilde, B. E., *Corrosion*, **23**, 379 (1967)
97. Jones, D. A., *Corros. Sci.*, **8**, 19 (1968)
98. Bureau, M., 9th Fatipec Congress, 79 (1968)
99. Mikhailovskii, Y. N., Leonov, V. V. and Tomashov, N. D., 'Korrozya Metallov i Splavov Sbornik 2', Metallurgizdat, Moscow (1965)
100. Butler, T. J. and Carter, P. R., *Electrochem. Tech.*, **1**, 22 (1963)
101. Walpole, J. F., *Bull. Inacol.*, **23**, 22 (1972)
102. Bird, D. W., *Bull. Inacol.*, **22**, 149 (1971)
103. Rowlands, J. C. and Bentley, M. N., *Brit. Corros. J.*, **7**, 42 (1972)
103a. Jones, D. A. and Lowe, T. A., *J. Materials*, **4** (3), 600 (1969)
104. Makar, D. R. and Francis, H. T., *J. Electrochem. Soc.*, **102**, 669 (1955)
105. Wesley, W. A., *Trans. Electrochem. Soc.*, **73**, 539 (1938)
106. Copson, H. R., *Industr. Engng. Chem.*, **37**, 721 (1945)
107. Rowe, L. C., *J. Mater.*, **5**, 323 (1970)
108. Denison, I. A., *J. Res. Nat. Bur. Stand.*, **17**, 363 (1936)
109. Ewing, S. P., *Amer. Gas Ass. Mon.*, **14**, 356 (1932)
110. Schwerdtfeger, W. J., *J. Res. Nat. Bur. Stand.*, **50**, 329 (1953); **52**, 265 (1954); **58**, 145 (1957); **65C**, 271 (1961)
111. Logan, R. H., Ewing, S. P. and Denison, I. A., Amer. Soc. Test. Mater. Spec. Tech. Publ. No. 32, 95 (1937)
112. Fuller, T. S., *Proc. Amer. Soc. Test. Mater.*, **27**, 281 (1927)
113. Lathrop, E. C., *Proc. Amer. Soc. Test. Mater.*, **24**, 281 (1924)
114. Todt, F., *Z. Elektrochem.*, **34**, 586 (1928); *Z. Ver. Dtsch. Zuckerind.*, **79**, 680 (1929)
115. Streicher, M. A., ASTM Bull. No. 188, 35, Feb. (1953); ASTM Bull. No. 195, 63, Jan. (1954)
116. Streicher, M. A., *J. Electrochem. Soc.*, **106**, 161 (1959)
117. Streicher, M. A., *Corrosion*, **19**, 272t (1963)
118. Saur, R. L. and Basco, R. P., *Plating*, **53**, 33, 320, 981 (1966)
119. Saur, R. L., *Plating*, **54**, 393 (1966)
120. Store, J. and Tuttle, H. A., *Plating*, **53**, 877 (1966); ASTM B556, 1971 (7)
121. Michelson, C. E., Dean, S. W. (Jr.) and Stransky, R. D., *Proc. 25th Conf. NACE*, 648, March (1969)
122. Englehart, E. T. and George, D. J., *Mater. Prot.*, **3** No. 11, 25 (1964)

123. Store, J., *J. Paint Technol.*, **41**, 661 (1969)
124. Capp, J. A., *Proc. Amer. Soc. Test. Mater.*, **14**, 474 (1914)
125. LaQue, F. L., *Proc. Amer. Soc. Test. Mater.*, **51**, 495 (1951)
126. LaQue, F. L., *Mater. and Meth.*, **35** No. 2, 77 (1952)
127. Sample, C. H., *Bull. Amer. Soc. Test. Mater.*, No. 123, 19 (1943)
128. May, T. P. and Alexander, A. L., *Proc. Amer. Soc. Test. Mater.*, **50**, 1131 (1950)
129. Darsey, V. M. and Cavanaugh, W. R., *Proc. Amer. Soc. Test. Mater.*, **48**, 153 (1948)
130. Nixon, C. F., *Mon. Rev. Amer. Electropl. Soc.*, **32**, 1105 (1945)
131. ASTM B287:1962 (1968) (7, 21, 31); BS 1224:1970
132. Pinner, W. L., *Plating*, **44**, 763 (1957)
133. Nixon, C. F., Thomas, J. D. and Hardesty, D. W., *46th Ann. Tech. Proc. Amer. Electropl. Soc.*, 159 (1959)
134. Thomas, J. D., Hardesty, D. W. and Nixon, C. F., *47th Ann. Tech. Proc. Amer. Electropl. Soc.*, 90 (1960)
135. LaQue, F. L., *46th Ann. Tech. Proc. Amer. Electropl. Soc.*, 141 (1959)
136. Bigge, D. M., *46th Ann. Tech. Proc. Amer. Electropl. Soc.*, 149 (1959)
136a. *J. Iron Steel Inst.*, **158**, 463 (1948); BS 1391:1952; DEF-29, Amendment PD1942 (1954) PD3656 (1960)
137. Edwards, J., *46th Ann. Tech. Proc. Amer. Electropl. Soc.*, 154 (1959); *Trans. Inst. Metal Finishing*, **35**, 55 (1958)
138. BS 1224:1959, British Standards Institute, London
139. Kesternich, W., *Stahl u. Eisen*, **71**, 587 (1951); German Standards Specification DIN-50018 (1960)
139a. Britton, S. C. and Michael, E. D. G., *Sheet Metal Ind.*, **32**, 576 (1955)
140. Wiederholt, W., *Metalloberfläche*, **12**, 305, 337, 369 (1958)
141. Dix, E. H. (Jr.) and Bowman, J. J., Amer. Soc. Test. Mater. Spec. Tech. Publ. No. 32, 57 (1937)
142. Voigt, L. R., *Corrosion*, **4**, 492 (1948)
143. Swinden, T. and Stevenson, W. W., *J. Iron St. Inst.*, **142**, 165P (1940)
144. Lloyd, T. E., *J. Metals, N.Y.*, **188**, 1092 (1950)
145. Evans, U. R. and Britton, S. C., *J. Iron St. Inst.*, Spec. Rept. No. 1, 139 (1931)
146. Dennis, J. K. and Such, T. E., *Trans. Inst. Metal Finishing*, **40**, 60 (1963); Dennis, J. K., *Proc. Symposium on Decorative Nickel-Chromium Plating*, Inst. of Metal Finishing, London (1966)
147. Dennis, J. K. and Such, T. E., *Nickel and Chromium Plating*, Newnes-Butterworths, London (1972)
148. Chandler, K. A. and Kilcullen, M. B., *Brit. Corros. J.*, **5**, 1 (1970)
149. Bromley, A. F., Kilcullen, M. B. and Stanners, J. F., *5th European Congress of Corrosion*, Paris, Sept. (1973)
150. Pourbaix, M., CEBELCOR RT 160, Aug. (1969); *see also* U.K. Pat. 1 261 544, Jan. (1969)
151. Legault, R. A., Mori, S. and Leckie, H. P., *Corrosion*, **26**, 121 (1970); **29**, 169 (1973)
152. Okada, H., Hosoi, U. and Naito, H., *Corrosion*, **26**, 429 (1970)
153. Brown, M. H., *Corrosion*, **30**, 1 (1974)
154. Cowan, R. L. and Tedmon, C. S. (Jr.), 'Intergranular Corrosion of Iron-Nickel-Chromium Alloys', in *Advances in Corrosion Science and Technology*, Vol. 3 (eds. M. G. Fontana and R. W. Staehle), Plenum Press (1973)
155. Huey, W. R., *Trans. Amer. Soc. Steel Treat.*, **18**, 1126 (1930)
156. Henthorne, M., *Corrosion*, **30**, 39 (1974)
157. Strauss, B., Schottky, H. and Hinnüber, J., *Z. Anorg. Allgem. Chem.*, **188**, 309 (1930)
158. Rocha, H. J. in discussion of paper by Brauns, E. and Pier, G., *Stahl u. Eisen*, **75**, 579 (1955)
159. Scharfstein, L. R. and Eisenbrown, C. M., ASTM STP No. 369, 253 (1963)
160. Tedmon, C. S. (Jr.), Vermilyea, D. A. and Rosolowski, J. H., *J. Electrochem. Soc.*, **118**, 192 (1971)
161. Ebling, H. and Scheil, M. A., ASTM Special Tech. Publ. No. 93, 121 (1949)
162. Warren, D., ASTM Bulletin No. 230, 45, May (1958)
163. Streicher, M. A., ASTM Bulletin No. 188, 35 (1953)
164. Clerbois, L., Clerbois, F. and Massart, J., *Electrochim. Acta.*, **1**, 70 (1959)
165. France, W. D. and Greene, N. D., *Corros. Sci.*, **8**, 9 (1968)
166. Streicher, M. A., *Corros. Sci.*, **9**, 55 (1969)
167. France, W. D. and Greene, N. D., *Corros. Sci.*, **10**, 379 (1970); Streicher, M. A., *Corros. Sci.*, **11**, 275 (1971)

168. Streicher, M. A., *Corrosion*, **30**, 77 (1974)
169. Fontana, M. G. and Greene, N. D., *Corrosion Engineering*, McGraw-Hill (1967)
170. Wilde, B. E., *Corrosion*, **28**, 283 (1972)
171. France, W. D. and Greene, N. D., *Corrosion*, **26**, 1 (1970)
172. Wilde, B. E. and Williams, E., *J. Electrochem. Soc.*, **117**, 775 (1970)
173. Wilde, B. E. and Greene, N. D., *Corrosion*, **25**, 300 (1969)
174. Henry, W. D. and Wilde, B. E., *Corrosion*, **25**, 515 (1969)
175. Wilde, B. E. and Williams, E., *J. Electrochem. Soc.*, **118**, 1058 (1971)
176. May, R. and Stacpoole, R. W. de V., *J. Inst. Met.*, **77**, 331 (1950)
177. Brownsdon, H. W. and Bannister, L. C., *J. Inst. Met.*, **49**, 123 (1932)
178. Breckon, C. and Gilbert, P. T., *1st Int. Congress on Met. Corrosion*, Butterworths, London, 624 (1962)
179. Bem, R. S. and Campbell, H. S., ibid., 630 (1962)
180. Campbell, H. S., MP 577, BNFRA, Feb. (1973)
181. Haigh, B. P., *J. Inst. Metals*, **18**, 55 (1917)
182. Huddle, A. U. and Evans, U. R., *J. Iron and Steel Inst.*, **149**, 109P (1944)
183. Inglis, N. and Lake, G. F., *Trans. Faraday Soc.*, **27**, 803 (1931)
184. Rawdon, H. S., *Proc. Amer. Soc. Test. Mater.*, **29**, 314 (1929)
185. Kenyon, J. N., ibid., **40**, 705 (1940)
186. Gould, A. J. and Evans, U. R., *Iron and Steel Inst.*, Spec. Report No. 24, 325 (1939)
187. Evans, U. R. and Simnad, M. T., *Proc. Roy. Soc.*, **A188**, 372 (1947)
188. Gough, H. J. and Sopwith, D. G., *J. Iron Steel Inst.*, **127**, 301 (1933); *Engineering*, **136**, 75 (1933)
189. Dias, M., Jarman, R. A. and Smith, S. A., to be published.
190. Hoeppner, D. W., *Corrosion Fatigue*, NACE-2, University of Connecticut, 3 (1972)
191. McEvily, A. J. and Wei, R. P., ibid., 381 (1972)
192. Kitagawa, H., ibid., 521 (1972)
193. Lichtman, J. Z., Kallas, D. H. and Rufola, A., in Ref 2, 453 (1971)
194. Schroter, H., *Z. Ver. Dtsch. Ing.*, **78**, 349 (1934)
195. Hobbs, J. M., *Proc. Cavitation Forum*, ASME, 1 (1966)
196. Gaines, N., *Physics*, **3**, 209 (1932)
197. Kerr, S. L., *Trans. Amer. Soc. Mech. Engrs.*, **59**, 373 (1937)
198. Beeching, R., *Trans. Instn. Engrs. Shipb. Scot.*, **90**, 203 (1946)
199. Rheingans, W. J., in *Engineering Approach to Surface Damage*, (eds. C. Lipson and L. V. Colwell), University of Michigan, 249 (1958)
200. Leith, W. C. and Thompson, A. L., *Trans. Amer. Soc. Mech. Engrs.*, *J. Basic Engng.*, **82**, Ser. D, 795 (1960)
201. Plesset, M. S., Trans. ASME Series D, *J. Basic Engng.*, **85**, 360 (1963)
202. Eisenberg, P., Preiser, H. S. and Thiruyengadam, A., *Trans. SNAME*, **73**, 241 (1965)
203. Hobbs, J. M., ASTM STP 408, ASTM, 159 (1967)
204. Plesset, M. S. and Devine, R. D., *J. Basic Eng.*, Trans. ASME, 691, Dec. (1966)
205. ASTM Spec. Tech. Pub. No. 144 (1953)
206. Waterhouse, R. B., *Fretting Corrosion*, Pergamon Press (1972)
207. Fink, M., *Trans. Amer. Soc. Steel. Treat.*, **18**, 1026 (1930)
208. Shotter, G. F., *J. Inst. Elec. Engrs.*, **75**, 755 (1934)
209. Tomlinson, G. A., Thorpe, P. L. and Goufh, H. J., *J. Instn. Mech. Engrs.*, **141**, 233 (1939)
210. Wright, K. H. R., *Proc. Instn. Mech. Engrs.*, **1B**, No. 11, 556 (1952–53)
211. Uhlig, H. H., Tierney, W. D. and McClellan, A., Ref. 205, 71 (1953)
212. McDowell, J. R., Ref. 205, 24 (1953)
213. Warlow-Davies, E. J., *J. Inst. Mech. Engrs.*, **146**, 32 (1941)
214. Godfrey, D., Tech. Note 2039, *Natl. Advisory Comm. Aeronaut.* (1950)
215. Gray, A. C. and Jenny, R. W., *S.A.E.J.*, **52**, 511 (1944)
216. de Villemeur, Y., *Metaux*, Paris, **34**, 413 (1959)
217. Wright, K. H. R., *Proc. Instn. Mech. Engrs.*, **1B**, 556 (1952–53); ibid., **181**, Pt. 30, 256 (1966–67); Barwell, F. T. and Wright, K. H. R., *J. Res. Brit. Cast. Iron Ass.*, **7**, 190 (1958)
218. Field, J. E. and Waters, D. M., N.E.L. Rep. No. 275 (1967)
219. Waterhouse, R. B., *J.I.S.I.*, **197**, 301 (1961)
220. Epstein, L. F., *Proc. Int. Conf. Peaceful Uses of Atomic Energy*, New York, **9**, 311 (1956)
221. Bakish, R. and Kern, F., *Corrosion*, **11**, 533t (1960)
222. Edeleanu, C. and Gibson, J. G., *J. Inst. Met.*, **88**, 321 (1960)
223. Brasunas, A. de S., *Corrosion*, **9**, 78 (1953)

224. Miller, E. C., in *Liquid Metals Handbook* (Ed.-in-Chief, R. N. Lyon), Atomic Energy Comm. and Dept. of the Navy, Washington, DC, June, 144 (1952)
225. Vreeland, D. C., Hoffman, E. E. and Manly, W. D., *Nucleonics*, **11**, 36 (1953)
226. Strachan, J. F. and Harris, N. L., *J. Inst. Met.*, **85**, 17 (1956–57)
227. Klueh, R. L., in *Proc. Int. Conf. Sodium Technol. Large Fast Reactor Design*, Nov. 7–9 (1968); ANL-7520, Pt. 1, p. 171, Argonne National Laboratory
228. Klueh, R. L., *Corrosion*, **25**, 416 (1969)
229. DiStefano, J. R., ORNL-4028, Oak Ridge Laboratory (1966)
230. Manly, W. D., *Corrosion*, **12**, 336t (1956)
231. DiStefano, J. R. and De Van, J. H., *Nuclear Appl. Tech.*, **8**, 29 (1970)
232. Koenig, R. F. and Vandenberg, S. R., *Metal Prog.*, **61** No. 3, 71 (1952)
232a. Hoffman, E. E., *Corrosion of Materials by Lithium at Elevated Temperatures*, ORNL-2924, Oak Ridge National Laboratory (1960)
234. De Van, J. H. and Sessions, C. E., *Nucl. Appl.*, **3**, 102 (1967)
235. De Van, J. H. and Jansen, D. H., *Fuels and Materials Development Program Quart. Progr. Rept.*, Sept. 30 (1968); ORNL-4350, Oak Ridge National Laboratory, p. 91
236. Bonilla, C. F., in *Reactor Handbook*, Vol. IV (ed. S. McLain), Interscience, New York, 107 (1964)
237. Romano, A. J., Fleitman, A. H. and Klamut, C. J., Proc. AEC-NASA Liquid Metals Inform. Meeting, CONF-650411 (1965)
238. Fuller, L. C. and MacPherson, R. E., ORNL-TM-2595, Oak Ridge National Laboratory (1967)
239. Roy, P., Wozaldo, G. P. and Comprelli, F. A., in *Proc. Internat. Conf. Sodium Technolog. Large Fast Reactor Design*, Nov. 7–9 (1968); ANL-7520, Pt. 1, Argonne National Laboratory, 131
240. Roy, P. and Gebhardt, M. F., GEAP-13548, General Electric Company (1969)
241. Hoffman, E. E. and Harrison, R. W. in *Metallurgy and Technology of Refractory Metal Alloys*, Plenum Press, New York, 251 (1969)
242. Harrison, R. W., GESP-258, General Electric Company (1969)
243. Klueh, R. L. and De Van, J. H., Ref. 2, 405 (1971)
244. Thompson, D. H., in Ref. 2; *Laboratory Corrosion Testing of Metals for the Process Industries*, NACE Standard TM–01–69
245. ASTM A224-46, ASTM Standards, Pt. 3, 257 (1958)
246. Dillon, C. P., Krisher, A. S. and Wissenburg, H., Ref. 2, 599 (1971)
247. Rawdon, H. S., Amer. Soc. Test. Mater. Spec. Tech. Pub. No. 32, 36 (1937)
248. LaQue, F. L., *Proc. Amer. Soc. Test. Mater.*, **51**, 495 (1951)
249. Finkeldey, W. H., *Proc. Amer. Soc. Test. Mater.*, **32**, 226 (1932)
250. Hudson, J. C., *J. Iron St. Inst.*, **148**, 161P (1943)
251. Pilling, N. B. and Wesley, W. A., *Proc. Amer. Soc. Test. Mater.*, **40**, 643 (1940)
252. Copson, H. R., *Proc. Amer. Soc. Test. Mater.*, **48**, 591 (1948)
253. Gibboney, J. H., *Proc. Amer. Soc. Test. Mater.*, **19**, 181 (1919)
254. Mendizza, A., *Proc. Amer. Soc. Test. Mater.*, **50**, 114 (1950)
255. Passano, R. F., *Proc. Amer. Soc. Test. Mater.*, **34**, 159 (1934)
256. Pinner, W. L., *Proc. Amer. Soc. Test. Mater.*, **53**, 256 (1953)
257. ASTM Designation D610-43, ASTM Standards, Pt. 8, 875 (1958)
258. Ellis, O. B., *Proc. Amer. Soc. Test. Mater.*, **49**, 152 (1949)
259. Wirshing, R. J. and McMaster, W. D., *Paint Varn. Prod.*, **41** No. 9, 13 (1951)
260. Sereda, P. J., *Bull. Amer. Soc. Test. Mater.*, No. 228, 53 (1958)
261. Guttman, H. and Sereda, P. J., ASTM Special Publication No. 435 (1968)
262. Gorman, L. J., *Proc. Amer. Soc. Test. Mater.*, **39**, 247 (1939)
263. Gorman, L. J., *Proc. Amer. Soc. Test. Mater.*, **48**, 167 (1948)
264. Pray, H. A., *Proc. Amer. Soc. Test. Mater.*, **44**, 280 (1944)
265. Humble, H. A., *Corrosion*, **5**, 292 (1949)
266. Southwell, C. R., NRL Reports, Naval Research Laboratory, Washington, D.C.
267. Logan, R. H., Ewing, S. P. and Denison, I. A., ASTM Tech. Pub. No. 32, 95 (1937)
268. Scott, G. N., *Proc. Amer. Petrol. Inst.*, 95 (1937)
269. Hoey, C. E. and Hipwood, H. A., *J. Oil. Col. Chem. Assoc.*, **57**, 151 (1974)

20.1A Appendix—Chemical and Electrochemical Methods for the Removal of Corrosion Products

This appendix provides information on chemical and electrochemical treatments which have been recommended for the removal of corrosion products. In using these methods the following points need to be borne in mind:

1. The duration of chemical or electrochemical treatment should be kept to the minimum necessary to remove the corrosion product. Loosely adherent material should be removed beforehand by suitable mechanical means, e.g. scrubbing.
2. The combined action of chemical (or electrochemical) treatment and scrubbing is often more effective than either method alone. It is frequently advantageous to alternate short periods of immersion with scrubbing to remove any corrosion product that has become loosened by the action of the chemical reagent.
3. The rate of attack of the chemical reagent on sound metal should be determined on a separate uncorroded sample of the material being cleaned, and if necessary a correction should be applied to the loss in weight of the corroded specimen.
4. The possibility of redeposition of metal from the dissolved corrosion product or, if electrochemical treatment is employed, from the anode material should always be kept in mind. If there is reason to believe this has occurred during removal of the corrosion product, further treatment to remove the redeposited metal will be necessary before the weight loss due to corrosion is measured.

Procedures for Removing Corrosion Products

The removal of corrosion products from metal specimens is described in ASTM G1:1972(31), *Recommended Practice for Preparing, Cleaning and Evaluating Corrosion Test Specimens*, and certain of these procedures are described below.

Electrolytic cathodic cleaning

After scrubbing to remove loosely attached corrosion products, cathodically polarise in hot dilute sulphuric acid under the following conditions:

Electrolyte–sulphuric acid (5 wt.%) plus an inhibitor (0·5 kg/m^3) such as diorthotolyl thiourea, quinoline ethiodide or β-naphthol quinoline. The temperature should be 75°C, the cathode current density 2000 Am^{-2} and the time of cathodic polarisation 3 min. The anode should be carbon or lead. If lead anodes are used, lead may deposit on the specimens and cause an error in the weight loss. If the specimen is resistant to nitric acid the lead may be removed by a flash dip in 1:1 nitric acid. Except for this possible source of error, lead is preferred as an anode, as it gives more efficient corrosion product removal.

After the electrolytic treatment, scrub the specimen with a brush, rinse thoroughly and dry.

Electrolytic treatment may result in the redeposition of a metal, such as copper, from reduceable corrosion products, and thus decrease the apparent weight loss.

Chemical cleaning

Copper and nickel alloys Dip for 1–3 min in 1:1 HCl or 1:10 H_2SO_4 at room temperature. Scrub lightly with bristle brush under running water, using fine scouring powder if needed.

Aluminium alloys Dip for 5–10 min in an aqueous solution containing 2 wt.% chromic acid (CrO_3) plus 5 vol.% orthophosphoric acid (H_3PO_4, 85%) maintained at 80°C. Ultrasonic agitation will facilitate this procedure.

Rinse in water to remove the acid, brush very lightly with a soft bristle brush to remove any loose film, and rinse again. If film remains, immerse for 1 min in concentrated nitric acid and repeat previous steps. Nitric acid may be used alone if there are no deposits.

Tin alloys Dip for 10 min in boiling trisodium phosphate solution (15%). Scrub lightly with bristle brush under running water, and dry.

Lead alloys Preferably use the electrolytic cleaning procedure just described. Alternatively, immerse for 5 min in boiling 1% acetic acid. Rinse in water to remove the acid and brush very gently with a soft bristle brush to remove any loosened matter.

Alternatively, immerse for 5 min in hot 5% ammonium acetate solution, rinse and scrub lightly. This removes PbO and $PbSO_4$.

Zinc Immerse the specimens in warm (60–80°C) 10% NH_4Cl for several minutes. Then rinse in water and scrub with a soft brush. Then immerse the specimens for 15–20 s in a boiling solution containing 5% chromic acid and 1% silver nitrate. Rinse in hot water and dry.

Note: in making up the chromic acid solution it is advisable to dissolve the silver nitrate separately and add it to the boiling chromic acid to prevent excessive crystallisation of the silver chromate. The chromic acid must be free from sulphate to avoid attack on the zinc. Immerse each specimen for 15 s in a 6% solution of hydriodic acid at room temperature to remove the remaining corrosion products. Immediately after immersion in the acid bath, wash the samples first in tap water and then in absolute methanol, and dry in air. This procedure removes a little of the zinc and a correction may be necessary.

Magnesium alloys   Dip for approximately 1 min in boiling 15% chromic acid to which has been added with agitation 1% silver chromate solution.

Iron and steel   Preferably use the electrolytic cleaning procedure, or else immerse in Clark's solution (hydrochloric acid 100 parts, antimonious oxide 2 parts, stannous chloride 5 parts) for up to 25 min. The solution may be cold but it should be vigorously stirred. Remove scales formed under oxidising conditions on steel in 15 vol.% concentrated phosphoric acid containing 0·15 vol.% of an organic inhibitor at room temperature.

Stainless steels   Clean stainless steels in 20% nitric acid at 60°C for 20 min. In place of chemical cleaning, use a brass scraper or brass bristle brush or both, followed by scrubbing with a wet bristle brush and fine scouring powder.

   Other methods of cleaning iron and steel include immersion in molten sodium hydride and cathodic treatment in molten caustic soda. These methods may be hazardous to personnel, and should not be carried out by the uninitiated, or without professional supervision.

General Note

Whatever cleaning method is used, the possibility of removal of solid metal is present. This will result in error in the determination of the corrosion rate. One or more cleaned and weighed specimens should be recleaned by the same method and reweighed. Loss due to this second treatment may be used as a correction to that indicated by the first weighing.

F. L. LaQUE

20.1B Appendix—Standards for Testing

British Standards*

BS 1328:1969, *Methods of sampling water used in industry*
   32 page   Gr 6
Methods of sampling of water, mainly for steam generation, for the purpose of chemical and physical testing, excluding biochemical testing. General instructions for the sampling procedures and detailed recommendations for sampling of raw, treated or softened, condensate, distillate and de-ionised feed-boiler waters.

BS 1344:1965, *Methods of testing vitreous enamel finishes*; Part 1, 1965, *resistance to thermal shock*
   8 page   Gr 2
Determination of the resistance to thermal shock when the method is applied to a component part of a vitreous enamelled appliance.

   Part 2, 1965, *resistance to culinary acids*
   8 page   Gr 2   Amendment PD 6211, July 1967
Determination of the resistance to culinary acids when the method is applied to a vitreous enamelled appliance or specially prepared test specimens, with method for classifying results.

   Part 3, 1967, *resistance to products of combustion containing sulphur compounds*
   8 page   Gr 2
Procedure for testing of parts of what would be subjected to products of combustion containing sulphur compounds. A method of classifying the results is given.

   Part 4, 1968, *resistance to abrasion*
   8 page   Gr 2
Details the procedure for determining and comparing the resistance to abrasion of vitreous enamel finishes applied to test specimens.

   Part 5, 1965, *resistance to detergents*
   8 page   Gr 2
Determination of detergent resistance of vitreous enamel finishes intended to be in contact with detergent solution near boiling point, and a grading system for assessing the results in terms of the time before loss of gloss is detected.

---

*Abbreviations used for BSI and ASTM Standards are given in Section 21.2 on pages **21**.72 and **21**.79 respectively.

Part 6, 1971, *resistance to alkali*
　8 page　Gr 2

Procedure for determining the resistance to alkali of vitreous enamelled finishes on baths, sinks and similar domestic equipment; apparatus, testing solution, procedure and criterion for acceptance.

Part 7, 1967, *resistance to heat*
　8 page　Gr 2

Procedure for determining the resistance to heat of any part that would in service come into direct contact with a source of heat.

BS 1391:1952, *Performance tests for protective schemes used in the protection of*
　　　　　　light-gauge steel and wrought iron against corrosion
　30 page　Gr 5　Amendments PD 1942, Aug. 1954; PD 3656, Jan. 1960

Provides methods of testing the performance of protective schemes under corrosive conditions to determine if a desired standard has been reached. It includes two methods, one being a salt droplet test and the other a test in which the specimens are exposed to an atmosphere of humid sulphur dioxide. The tests are mainly suitable for single coats of stoving paint applied to base, phosphated or metal-coated steel. Both tests are made on special test specimens but the salt droplet test can be modified for application to individual articles. Details are included of the test specimens, the method of conducting the tests and the method of assessing the standard of performance.

BS 1427:1962, *Routine control methods of testing water used in industry*
　52 page　Gr 7　Amendments PD 5807, March 1966; PD 6063, March 1967;
　PD 6344, Jan. 1968 (Gr 0 separately)

Covers methods of test for the following:
1. Alkalis; 2. Aluminium; 3. Ammonia (free and saline); 4. Calcium; 5. Chloride; 6. Residual chlorine (free and total); 7. Colour; 8. Dissolved oxygen; 9. Dissolved solids; 10. Electrical conductivity; 11. Free carbon dioxide; 12. Hardness; 13. Hydrazine; 14. Iron: 15. Magnesium; 16. Nitrite; 17. pH value; 18. Phosphates; 19. Silica; 20. Specific gravity; 21. Sulphate; 22. Sulphite; 23. Fluoride.

BS 1747, *Methods for the measurement of air pollution*; Part 1, 1969, *deposit gauges*　M
　16 page　Gr 4

Construction, installation and use of the deposit gauge for the collection and measurement of atmospheric impurities deposited by their own weight or with the assistance of rain. The gauge also provides an estimate of the rainfall.

Part 2, 1969, *determination of concentration of suspended matter*　M
　16 page　Gr 4

Concentration and use of apparatus for the determination of fine suspended particles (smoke) in the atmosphere, both by comparison with an arbitrary standard and by an absolute (weighing) method.

Part 3, 1969, *determination of sulphur dioxide*　M
　12 page　Gr 3

Construction and use of apparatus for the determination of atmospheric sulphur dioxide.

Part 4, 1969, *the lead dioxide method*　M
　16 page　Gr 4

Construction and use of apparatus for the lead dioxide method of measuring the reactivity of atmospheric sulphur compounds.

Part 5, 1972, *directional dust gauges*　M
　20 page　Gr 4

Gives requirements for construction, installation and use of a directional dust gauge for the collection and measurement of atmospheric impurities, whether precipitated as a liquid or as fine dust.

BS 2690, *Methods of testing water used in industry*; Part 1, 1964, *coppers and iron*   M NZ
    16 page   Gr 4
Details three methods for copper and two methods for iron. These methods supersede those given in the 1956 edition of BS 2690.

Part 2, 1965, *dissolved oxygen, hydrazine and sulphate*   M NZ
    24 page   Gr 5
Gives three methods for dissolved oxygen, one method for sulphite. These methods supersede those given in the 1965 edition of BS 2690.

Part 3, 1966, *silica, phosphate*   M NZ
    16 page   Gr 4
Two methods of silica and two methods for phosphate. These methods supersede those given in the 1956 edition of BS 2690.

Part 4, 1967, *aluminium, calcium, magnesium and fluoride*   M NZ
    28 page   Gr 5   Amendment AMD 227, March 1969
Spectrophotometric methods for aluminium, calcium, magnesium and fluoride, and visible titration methods for two concentration levels of calcium and magnesium. These methods supersede those given in the 1956 edition of BS 2690 for the elements concerned.

Part 5, 1967, *alkalinity, acidity, pH value and carbon dioxide*   M NZ
    24 page   Gr 5
Titrimetric methods for determination of alkalinity and acidity, details of electrometric methods of measuring the pH of high purity water, and methods for two levels of carbon dioxide. These methods supersede those given in the 1956 edition of BS 2690.

Part 6, 1968, *chloride and sulphate*   M NZ
    24 page   Gr 5
Gives two methods for chloride and three methods for sulphate, superseding those given in the 1956 edition.

Part 7, 1968, *nitrite, nitrate and ammonia (free, saline and albuminoid)*   M NZ
    20 page   Gr 4
Gives one method each for nitrite and nitrate, and three methods of increasing sensitivity for ammonia. Includes details of distillation procedure for ammonia determination. These methods are intended to supersede those given in the 1956 edition.

Part 8, 1969, *cyclohexylamine, morpholine and long-chain fatty amines*   M NZ
    16 page   Gr 4
Gives one method for each of the items up to levels of $250\,\mu g$ for cyclohexylamine, $100\,\mu g$ for morpholine and (as an example) $200\,\mu g$ for octadecylamine.

Part 9, 1970, *appearance (colour and turbidity) odour, suspended and dissolved solids and electrical conductivity*   M
    16 page   Gr 4   Amendment AMD 1012, Aug. 1972 (Gr 0)
Gives visual and instrumental methods for colour, one method each for turbidity, odour and dissolved solids, two methods for suspended solids, one method for electrical conductivity of high purity water and one method for electrical conductivity of water of conductivity greater than 1 mS/m.

Part 10, 1970, *sodium, potassium and lithium*   M
    16 page   Gr 4
Flame photometer and flame spectrophotometer methods for sodium suitable for mg/l and μg/l levels respectively. Flame photometer methods for potassium and lithium.

Part 11, 1971, *anionic, cationic and non-ionic detergents and oil*   M
   20 page   Gr 4

Gives colorimetric methods for trace qualities for each of the three types of detergent
and a thin-layer chromatographic method for non-ionic detergents. Extraction methods
for oil supersede that given in the 1956 edition of BS 2690.

Part 12, 1972, *nickel, zinc, chromates and total chromium and manganese*   M
   20 page   Gr 4

Gives spectrophotometric methods for each of these constituents when present as trace
impurities in water and supersedes those given in the 1956 edition of BS 2690.

Part 13, 1972, *dichromate value (chemical oxygen demand), non-volatile
   organic carbon, tannins and chlorine*   M
   20 page   Gr 4

Gives two methods for total chlorine and one for free chlorine and the other determinations named. The ranges are 2·5 mg of oxygen upwards for dichromate value, 0·1 mg
to 15 mg of carbon for non-volatile organic carbon, 10 μg to 250 μg for tannins and up
to 0·4 mg of chlorine.

Part 14, 1972, *arsenic, lead and sulphide*   M
   16 page   Gr 4

Gives spectrophotometric methods for lead and sulphide when present as trace impurities in water and extends the method for sulphide to raw, discoloured or turbid water.
Gives the preparation of samples for arsenic determination by the method in BS 4404
to which reference must be made for the remainder of the procedure.

BS 3745:1970, *The evaluation of results of accelerated corrosion tests on metallic
            coatings*
   20 page   Gr 4

Gives a rating system which provides a means of defining acceptable levels of performance in accelerated corrosion tests where required by other relevant British
standards. Agrees with ISO/R 1462.

BS 3900:1971, *Methods of test for paints*
   A.4 size (Binder Gr 5 separately)

Procedures, apparatus and related information on widely used test methods for paints,
varnishes and similar products. The general introduction setting out the scope of the
series is intended to be read in conjunction with each of the parts, which are issued in
loose-leaf form and can be obtained separately.

   General Introduction: 1969
      8 page   Gr 3

   Lists 32 general test methods for paints published or in course of preparation as
   parts of BS 3900, and also over 50 more specialised methods published or in course
   of preparation in other British standards.

   Group A: tests on liquid paints (excl. chemical tests)

      Part A1:1970, *sampling*   M
         8 page   Gr 3

      Part A2:1970, *examination and preparation of samples for testing*   M
         2 page   Gr 1

      Part A3:1970, *standard panels for paint testing*   M
         4 page   Gr 2

      Part A4:1966, *notes for guidance on paint application*   M
         4 page   Gr 2

      Part A5:1968, *large scale brushing test*   M
         4 page   Gr 2

      Part A6:1971, *determination of flow time*   M
         8 page   Gr 3

Part A7:1968, *determination of the viscosity of paint at a high rate of shear*   M
 4 page   Gr 2   AMD 331, Oct. 1969
Part A8:1968, *danger classification by flashpoint (closed cup method)*   M
 2 page   Gr 1
Group B: tests involving chemical examination of liquid paints and dried paint films
Part B1:1965, *determination of water by the Dean and Stark method*   M
 2 page   Gr 1
Part B2:1970, *determination of volatile matter and non-volatile matter*
 2 page   Gr 1
Part B3:1965, *determination of 'soluble lead'*   M
 2 page   Gr 1
Part B4:1967, *determination of lead in low-lead paints and similar materials*
 4 page   Gr 2   AMD 33, Aug. 1968
Group C: tests associated with paint film formation
Part C1:1965, *wet edge time*
 2 page   Gr 1   Amendment AMD 724, April 1971
Part C2:1971, *surface-drying time (Ballotine method)*   M
 2 page   Gr 1   Agrees with ISO/R 1517
Part C3:1971, *hard-drying time*   M
 2 page   Gr 1
Part C4:1965, *freedom from residual tack*
 2 page   Gr 1
Part C5:1970, *determination of film thickness*   M
 8 page   Gr 3
Part C6:1970, *determination of fineness of grind*   M
 4 page   Gr 2
Group D: optical tests on paint films
Part D1:1970, *colour comparison*   M
 4 page   Gr 2
Part D2:1967, *gloss (specular reflection value)*   M
 4 page   Gr 2
Group E: mechanical tests on paint films
Part E1:1970, *bend test (cylindrical mandrel)*   M
 4 page   Gr 2
Part E2:1970, *scratch test*   M
 4 page   Gr 2
Part E3:1966, *impact (falling weight) resistance*
 6 page   Gr 3
Part E4:1969, *cupping test*
 2 page   Gr 1
Group F: durability tests on paint films
Part F1:1966, *alkali resistance of plaster primer-sealer*
 2 page   Gr 1
Part F2:1966, *resistance to humidity under condensation conditions*
 2 page   Gr 1
Part F3:1971, *resistance to artificial weathering (enclosed carbon arc)*   M
Part F4:1968, *resistance to continuous salt spray*   M
 4 page   Gr 2
Part F5:1972, *determination of light fastness of paints for interior use (exposed to artificial light sources)*   M
 5 page   Gr 2
Part F6:1972, *notes for guidance on the conduct of natural weathering tests*   M
 4 page   Gr 2

*Group G: environmental tests on paint films (including tests for resistance to corrosion
        and chemicals)*
   Part G1:1965, *resistance to organic liquids (withdrawn as methods are covered by
            BS 3900, Part G5)*
   Part G2:1965, *resistance to aqueous liquids (withdrawn as methods are covered by
            BS 3900 Part G5)*
   Part G3:1966, *resistance to spotting by liquids*
      2 page   Gr 1
   Part G4:1967, *resistance to hot fats*   M
      2 page   Gr 1
   IP 154/69 with explanatory notes.
   Part G5:1972, *resistance to liquids*   M
      2 page   Gr 1
   This method is intended to supersede those published as Part G1 and Part G2.
   AU 148, *methods of test for motor vehicle paints*
   Part 1:1969, *visual colour matching*   M
      2 page   Gr 1
   Part 2:1969, *resistance to continuous salt spray*   M
      8 page   Gr 3
   Part 3:1969, *flexibility and adhesion*   M
      4 page   Gr 2
   Part 4:1969, *resistance to abrasion*   M
      4 page   Gr 2
   Part 5:1969, *gloss measurement*   M
      6 page   Gr 3
   Part 6:1969, *hardness*   M
      6 page   Gr 3
   Part 7:1969, *hiding power*   M
      6 page   Gr 3
   Part 8:1969, *measurement of paint film thickness*   M
      4 page   Gr 2
   Part 9:1969, *resistance to dry heat*   M
      2 page   Gr 1
   Part 10:1969, *resistance to heat and corrosion*   M
      2 page   Gr 1
   Part 11:1969, *resistance to blistering*   M
      8 page   Gr 3
   Part 12:1969, *resistance to accelerated weathering*   M
      8 page   Gr 3
   Part 13:1969, *resistance to deterioration by contact with other materials*   M
      6 page   Gr 3
   Part 14:1969, *resistance to outdoor exposure*   M
      2 page   Gr 1
   Part 15:1969, *resistance to chipping*   M
      4 page   Gr 2

HR 100:1971, *inspection and testing procedure for aerospace material. Wrought heat
         resisting alloys*   M+I
   36 page   Gr 8

BS 4232:1967, *surface finish of blast-cleaned steel for painting*
   16 page   Gr 4
First, second and third qualities of surface finish (cleanliness and roughness). Diagrammatic indications of first and second qualities. Recommendations on selection of qualities, methods of control and inspection and blast-cleaning procedures.

BS 4351:1971, *method for detection of copper corrosion from petroleum products by the copper strip tarnish test*
    16 pages   Gr 4
Identical with method ASTM D 130-68.

ASTM Standards

*A 90:1969 (3)
Weight of coating on zinc-coated (galvanised) iron or steel articles. Test for.
A 143:1972 (3)
Safeguarding against embrittlement of hot-dip galvanised structural steel products and procedure for detecting embrittlement. Rec. practice for.
A 239:1941 (1965) (3)
Uniformity of coating by the Preece test (copper-sulphate dip) on zinc-coated (galvanised) iron or steel articles. Test for.
A 262:1970 (3)
Detecting susceptibility to intergranular attack in stainless steels. Rec. practice for.
A 279:1963 (3)
Total immersion corrosion test of stainless steels.
*A 309:1954 (1971) (3)
Weight and composition of coating on long terne sheets by the triple spot test. Test for.
A 393:1963 (3)
Conducting acidified copper sulphate test for intergranular attack in austenitic stainless steel. Rec. practice for.
A 428:1968 (3)
Weight coating on aluminium-coated iron or steel articles. Tests for.
A 630:1968 (3) ASTM. Tin coatings weight for hot-dip and electrolytic tin plate; determination of.
*B 117:1973 (7,21,31)
Salt spray (fog) testing. Test for. (Method is intended for ferrous and non-ferrous metals with or without inorganic or organic coatings)
B 136:1972 (6,7)
Resistance of anodically coated aluminium to staining by dyes. Measurement of. (Sealing test for anodised aluminium.)
B 137:1945 (1972) (6,7) ASTM. Weight of coating on anodically coated aluminium, measurement of.
*B 154:1971 (5)
Standard method for mercurous nitrate test for copper and copper alloys. (Stress-corrosion cracking.)
*B 201:1968 (7)
Chromate coating on zinc and cadmium surfaces. Rec. practice for testing.
B 244:1968 (1972) (6,7)
Thickness of anodic coatings on aluminium with eddy-current instruments. Measurement of.
B 287:1962 (1968) (7,21,31)
Method for acetic acid–salt-spray testing. (This standard is intended for the same purpose as B 117 with differences.)
*B 368:1968 (7,21)
Standard method of test for copper accelerated acetic acid–salt-spray (fog) testing.

*An asterisk preceding the designation indicates that the standard has been approved as an American National Standard by the American National Standards Institute.

(Cass test.) Gives the copper-accelerated acetic-acid salt-spray (fog) test as primarily applicable to rapid testing of decorative copper–nickel–chromium coatings on steel and zinc-base die castings. Similar to B 287.
*B 380:1965 (1972) (7)
Corrosion testing of decorative chromium plating by the Corrodkote procedure.
*B 457:1967 (1972) (6,7)
Impedance of anodic coating on aluminium. Measurement of.
*B 487:1973 (7)
Metal and oxide coating thickness by microscopical examination of a cross-section. Measurement of.
*B 489:1968 (7)
Bend test for ductility of plated metals. Rec. practice for.
*B 490:1968 (7)
Micrometer bend test for ductility of electrode deposits. Rec. practice for.
*B 499:1969 (7)
Measurement of coating thicknesses by the magnetic method. Non-magnetic coatings on magnetic basis metals.
B 504:1970 (7)
Measuring the thickness of metallic coatings by the coulometric method.
B 529:1970 (7)
Coating thicknesses by the eddy-current test method. Non-conductive coatings on non-magnetic basis metals. Measurement of.
B 530:1970 (7)
Measurement of coating thicknesses by the magnetic method. Electrodeposited nickel coatings on magnetic and non-magnetic substrates.
*B 537:1970 (7)
Rating of electroplated panels subjected to atmospheric exposure. Rec. practice for.
*B 538:1970 (7)
F.A.C.T. (Ford anodised aluminium corrosion test). Testing method of.
B 556:1971 (7)
Thin chromium coatings by the spot test. Guidelines for measurement of.
B 567:1972 (7)
Coating thickness by the β-backscatter principle. Measurement of.
B 568:1972 (7)
Coating thickness by X-ray spectrometry. Measurement of.
B 571:1972 (7)
Adhesion of metallic coatings. Test for.
*C 313:1959 (1972) (13)
Adherence of porcelain enamel and ceramic coatings to sheet metal. Tests for.
*C 464:1964 (1970) (14)
Corrosion effect of thermal insulating cements on base metal. Test for.
C 633:1969 (13)
Adhesion or cohesive strength of flame-sprayed coatings. Test for.
D 130:1968 (1973) (17)
Copper corrosion from petroleum products by the copper strip test. Test for.
*D 610:1968 (21)
Evaluating degree of rusting on painted steel surfaces.
D 659:1944 (1970) (21)
Evaluating degree of resistance to chalking of exterior paints.
D 660:1944 (1970) (21)
Evaluating the degree of resistance to checking of exterior paints.
D 661:1944 (1970) (21)
Evaluating degree of resistance to cracking of exterior paints.
D 662:1944 (1970) (21)
Evaluating degree of resistance to erosion of exterior paints.

*D 665:1960 (1973) (17)
Rust-preventing characteristics of steam-turbine oil in the presence of water. Test for.
D 714:1956 (1970) (21)
Evaluating degree of blistering of paints.
D 772:1947 (1970) (21)
Evaluating degree of resistance to flaking (scaling) of exterior paints.
*D 807:1952 (1970) (23)
Corrosivity test of industrial water (USBM Embrittlement Detector method).
*D 849:1947 (1971) (20)
Copper corrosion of industrial aromatic hydrocarbons. Test for.
D 870:1954 (1968) (21)
Water immersion test of organic coatings on steel.
D 930:1967 (1972) (22)
Total immersion test of water-soluble aluminium cleaners.
D 1014:1966 (21)
Conducting exterior exposure tests of paint on steel.
*D 1275:1967 (1971) (18,29)
Corrosive sulphur in electrical insulating oils. Test for.
D 1280:1967 (1972) (22)
Total immersion corrosion test for soak tank cleaners. (Method intended to determine the corrosive effects on the metal to be cleaned, but excluding aluminium and its alloys; removal of corrosion products is outlined.)
D 1374:1957 (1971) (22)
Aerated total immersion corrosion test for metal cleaners. (A method intended to determine the corrosive effects on the metal to be cleaned, but excluding aluminium and its alloys.)
*D 1384:1970 (22)
Corrosion test for engine antifreezes in glassware.
*D 1611:1960 (1970) (15)
Corrosion produced by leather in contact with metal. Test for.
D 1616:1960 (1973) (20)
Copper corrosion by mineral spirits (copper strip test). Test for.
D 1654:1961 (1968) (21)
Evaluation of painted or coated specimens subjected to corrosive environments.
D 1735:1962 (1973) (21)
Water fog testing of organic coatings.
D 1743:1973 (17)
Rust preventive properties of lubricating greases. Tests for.
*D 1748:1970 (17)
Rust protection by metal preservatives in the humidity cabinet. Test for.
*D 1838:1964 (1973) (18,19)
Copper strip corrosion by liquified petroleum (LP) gases. Test for.
D 2247:1968 (1973) (21)
Coated metal specimens at 100% r.h. testing.
D 2251:1967 (1972) (22)
Metal corrosion of halogenated organic solvents and their admixtures. Test for.
D 2688:1970 (23)
Corrosivity of water in the absence of heat transfer (weight loss methods). Test for.
D 2776:1972 (23)
Corrosivity of water in the absence of heat transfer (electrical methods). Tests for.
D 2803:1970 (21)
Filiform corrosion resistance of organic coatings on metal. Test for.
D 2809:1972 (22)
Cavitation-erosion–corrosion characteristics of aluminium automotive water pumps with coolants. Test for.

D 2933:1972T (21)
Coated steel specimens dynamically for resistance to corrosion testing.
D 2939:1973 (11)
Bituminous base emulsions for use as protective coatings. Testing.
D 2939:1971 (22)
Cavitation-erosion characteristics of aluminium in engine antifreeze solutions using ultrasonic energy. Test for.
D 3170:1973 (21)
Chip resistance of coatings. Test for.
E 243:1971 (5,31)
Electromagnetic (eddy-current) testing of seamless copper and copper-alloy heat-exchanger and condenser tubes. Rec. practice for.
E 376:1969 (3,21,31)
Coating thicknesses by magnetic-field or eddy-current (electromagnetic) test methods. Rec. practice for.
*F 64:1969 (28)
Corrosive and adhesive effect of gasket materials on metal surfaces. Test for.
F 361:1972 (7)
Experimental testing for biological compatibility of metals for surgical implants. Rec. practice for.
*G 1:1972 (31)
Preparing, cleaning and evaluating corrosion test specimens. Rec. practice for.
*G 2:1967 (7,31)
Aqueous corrosion testing of samples of zirconium and zirconium alloys.
G 3:1968 (31)
Conventions applicable to electrochemical measurements in corrosion testing.
G 4:1968 (3,31)
Conducting plant corrosion tests. Rec. practice for.
G 5:1972 (31)
Standard reference method for making potentiostatic and potentiodynamic anodic polarisation measurements. Rec. practice for.
G 6:1972 (21,30)
Abrasion resistance of pipeline coatings. Test for.
G 7:1969T (30)
Atmospheric environmental exposure testing of non-metallic materials. Rec. practice for.
G 8:1972 (21,30)
Cathodic disbonding of pipeline coatings. Test for.
G 9:1972 (21,30)
Water penetration into pipeline coating. Test for.
G 10:1972 (21,30)
Bendability of pipeline coating. Test for.
G 11:1972 (21,30)
Effects of outdoor weathering on pipeline coatings. Test for.
G 12:1972 (21,30)
Non-destructive measurement of film thickness of pipeline coatings on steel.
G 13:1972 (21,30)
Impact resistance of pipeline coatings (limestone drop test). Test for.
G 14:1972 (21,30)
Impact resistance of pipeline coatings. Test for.
G 15:1971 (31)
Corrosion and corrosion testing. Definition of terms relating to.
G 16:1971 (31)
Applying statistics to analysis of corrosion data.
Rec. practice for.

G 17:1972 (21,30)
Penetration resistance of pipeline coatings. Test for.
G 18:1972 (21,30)
Joints, fittings and patches in coated pipelines. Test for.
G 19:1972 (21,30)
Disbonding characteristics of pipeline coatings by direct soil burial. Test for.
G 20:1972 (21,30)
Chemical resistance of pipeline coatings. Test for.
G 21:1970 (26,30)
Resistance of synthetic polymeric materials to fungi. Rec. practice for determining.
G 22:1969 (26,30)
Resistance of plastics to bacteria. Rec. practice for determining.
G 23:1969 (24,27,30)
Operating light and water exposure apparatus (carbon-arc type) for exposure of non-metallic materials. Rec. practice for.
G 26:1970 (30)
Operating light and water exposure apparatus (xenon-arc type) for exposure of non-metallic materials. Rec. practice for.
*G 28:1972 (31)
Detecting susceptibility to intergranular attack in wrought nickel-rich chromium-bearing alloys. Test for.
G 30:1972 (31)
Making and using U-bend stress-corrosion test specimens. Rec. practice for.
G 31:1972 (31)
Laboratory immersion corrosion testing of metals. Rec. practice for.
G 32:1972 (31)
Vibratory cavitation erosion test.
G 33:1972 (31)
Recording data from atmospheric corrosion test of metallic-coated steel specimens. Rec. practice for.
G 34:1972 (6,31)
Exfoliation corrosion susceptibility in 7 XXX series copper-containing aluminium alloys. (EXCO test.) Test for.
G 35:1973 (31)
Susceptibility of stainless steels and related Ni–Cr–Fe alloys to stress-corrosion cracking in polythionic acids. Rec. practice for.
G 36:1973 (31)
Performing stress-corrosion cracking tests in a boiling magnesium chloride solution. Rec. practice for.
G 37:1973 (31)
Mattson's solution of pH 7·2 to evaluate the stress-corrosion cracking susceptibility of Cu–Zn alloys. Rec. practice for use of.

DIN (West German Standards)*

50900 11.60
Corrosion of metals; definitions (S).
50010-1961
Testing under climatic conditions. (Emphasis is on atmospheric attack, indoors and outdoors, but reference is also made to soil corrosion; serves as introduction to other DIN specifications (50011–50019).)

*English translations of German standards may be obtained from the Beuth-Vertrieb GmbH, Berlin 30.

50014-1959
Standard (laboratory) atmosphere conditions. (Standard conditions for laboratory atmospheres in the temperate zone are defined as $20 \pm 2°C$, $65 \pm 3\%$ r.h. (20/65), or $23 \pm 3\%$ r.h. (23/50); room temperature is defined as $18-28°C$.)
50905-1959
Corrosion tests. (Specification provides a summary of the factors to be considered and of the method of conducting corrosion tests in general.)
50907-1952
Resistance to marine climate and sea-water. (This specification is concerned with both field and laboratory tests on light metals with tensile tests and visual and metallographic examination as the general methods of assessment of corrosion.)
50908-1957
Stress corrosion of light alloys. (A draft standard which defines stress corrosion simply as cracking while exposed to a corroding agent and to static stress; this standard should be used with caution.)
50951 5.57
Testing of electroplated coatings; determination of coat thickness by jet methods (7).
50901 8.57
Corrosion of metals; corrosion factors, formula, symbols, units.
50909-1954
Testing of metallic materials; corrosion (tests in soil in the absence of electric earth currents).
50910-1955
Testing of metallic materials; influences (affecting corrosion in soil in the presence of electric earth currents, and methods of measuring such influences).
50912-1958
Pressure vessel test. (Where stresses produced by high pressure are to be included as a factor in the corrosion tests it is proposed that the pressure vessel (3 litre capacity) be made of the metal to be tested; applies to ferrous and non-ferrous metals and alloys).
50906 10.58
Testing of metallic materials; corrosion testing in boiling liquids (boiling test).
51779-1954
Diesel fuel. (Standard is concerned with the possibility of corrosion of zinc or galvanised steel during storage, transportation or usage of diesel fuel oil.)
50952 12.63
Testing of metallic coatings; determination of the weight per unit area of zinc coatings on steel by chemical dissolution of the coating, gravimetric method.
50954 2.65
Testing of metallic coatings; determination of the average coating weight of tin coatings on steel by chemical dissolution of the coating.
50953 12.67
Testing of electroplated coatings; determination of thickness of thin chromium coatings by the spot test.
50950 4.68
Testing of electroplated coatings; microscopic measurement of coat thickness.
50914 9.70
Testing of resistance of stainless steels to intercrystalline corrosion; copper sulphate and sulphuric acid method.

National Association of Corrosion Engineers (NACE Standards)

Test Methods

TM–01–69   *Laboratory corrosion testing of metals for the process industries*
TM–01–70   *Visual standard for surfaces of new steel airblast cleaned with sand abrasive*
TM–02–70   *Method of conducting controlled velocity laboratory corrosion tests*
TM–01–71   *Autoclave corrosion testing of metals in high temperature water*
TM–01–72   *Antirust properties of petroleum products pipeline cargoes*
TM–01–73   *Methods for determining water quality for subsurface injection using membrane filters*
TM–01–74   *Laboratory methods for the evaluation of protective coatings used as lining materials in immersion service*
TM–02–74   *Dynamic corrosion testing of metals in high temperature water*
TM–03–74   *Laboratory screening tests to determine the ability of scale inhibitors to prevent the precipitation of calcium sulfate and calcium carbonate from solution*
TM–01–75   *Visual standard for surfaces of new steel centrifugally blast cleaned with steel grit and shot*

International Standards Organisation (I.S.O.)

ISO 1247 1971, *Aluminium pigments*
     16 page   Gr 9
Requirements and the corresponding methods of test for aluminium pigments used in paints, including general purpose, decorative and protective paints and speciality finishing paints.
ISO/R 1460 1970, *Determination of weight per unit area of hot-dip galvanised coatings on ferrous materials by chemical dissolution of the coating; gravimetric method*
     4 page   Gr 7
Recommends the stripping solution and procedure for determining the coating weight of hot-dip galvanised coatings on ferrous materials.
ISO/R 1462 1970, *A method for the evaluation of the results of accelerated corrosion tests on coatings other than those anodic to the basis metal*
     8 page   Gr 7
Rating system that provides a means of defining levels of performance of coatings, other than those anodic to the basis metal, that have been subjected to accelerated corrosion tests.
ISO/R 1463 1970, *Measurements of metal and oxide coating thickness by microscopical examination of cross-sections*
     8 page   Gr 7
Method for the measurement of the thickness of metallic coatings, oxide layers and vitreous or porcelain enamel coatings, by microscopical examination and preparation of specimens. Recommends suitable etchants.
ISO/R 2063 1971, *Metal spraying of zinc and aluminium for the protection of iron and steel against corrosion*
     7 page   Gr 8
Gives definitions, characteristics and methods of test for sprayed coatings. Some parts may be adapted to apply to metals other than zinc and aluminium.

ISO/R 2085 1971, *Surface treatment of metals. Anodisation of aluminium and its alloys. Check of continuity of thin coatings. Copper sulphate test*

1 page Gr 6

Enables a rapid check to be made of the continuity of a thin coating of aluminium oxide, i.e. in case of doubt regarding the presence of a visible fault on the surface of the coating it makes it possible to verify whether such a fault corresponds to a local gap in the coating (base metal). The use of this method is limited to thin oxide coatings (less than 5 μm thickness).

ISO/R 2064 1971, *Metallic and other non-organic coatings. Definition of terms concerning the measurement of thickness*

2 page Gr 7

Does not apply to coatings on machine screw threads, on sheet, strip, or wire in unfabricated forms, or on coil springs.

ISO 2360 1972, *Non-conductive coatings on non-magnetic basis metals; measurement of coating thickness. Eddy current method*

3 page Gr 7

Specifies the method of using eddy current instruments for the non-destructive measurement of the thickness of non-conductive coatings of non-magnetic basis metals.

ISO/R 1879 1970, *Instruments for the measurement of oxide coating thickness by microscopical examination of cross-sections*

ISO/R 2106 1971, *Surface treatment of metals. Anodisation (anodic oxidation) of aluminium and its alloys. Measurement of the mass of oxide coatings, gravimetric method.*

ISO/R 2128 1971, *Surface treatment of metals. Anodisation (anodic oxidation) of aluminium and its alloys. Measurement of thickness of oxide coating. Non-destructive measurement by light section microscope.*

ISO/R 2160 1972, *Petroleum products corrosiveness to copper. Copper strip test.*

20.2 The Potentiostat and its Application to Corrosion Studies

The potentiostatic technique discussed here involves the polarisation of a metal electrode at a series of predetermined constant potentials. Potentiostats have been used in analytical chemistry for some time[1]; Hickling[2] was the first to describe a mechanically controlled instrument and Roberts[3] was the first to describe an electronically controlled instrument. Greene[4] has discussed manual instruments and basic instrument requirements.

The determination of polarisation curves of metals by means of constant potential devices has contributed greatly to the knowledge of corrosion processes and passivity. In addition to the use of the potentiostat in studying a variety of mechanisms involved in corrosion and passivity, it has been applied to alloy development, since it is an important tool in the accelerated testing of corrosion resistance. Dissolution under controlled potentials can also be a precise method for metallographic etching or in studies of the selective corrosion of various phases. The technique can be used for establishing optimum conditions of anodic and cathodic protection. Two of the more recent papers have touched on limitations in its application[5], and differences between potentiostatic tests and exposure to chemical solutions[6].

In this section an attempt is made to give a more detailed introduction to experimental procedures, as well as to some of the ideas where the use of the potentiostat has helped in the understanding of corrosion processes.

Experimental Apparatus

Instruments very suitable for corrosion work are readily available, with several different models produced commercially. Although most, if not all, of the available potentiostats are properly designed, it should be kept in mind that corrosion studies require the instrument to have a low internal resistance and to react quickly to changes of potential of the working electrode.

A basic circuit is shown schematically in Fig. 20.34(a). The specimen C., or working electrode W.E. is the metal under study, the auxiliary electrode A.E. is usually platinum and R.E. is the reference electrode, for instance a saturated calomel electrode. The desired potential difference between the specimen and the reference electrode is set with the backing circuit B. Any

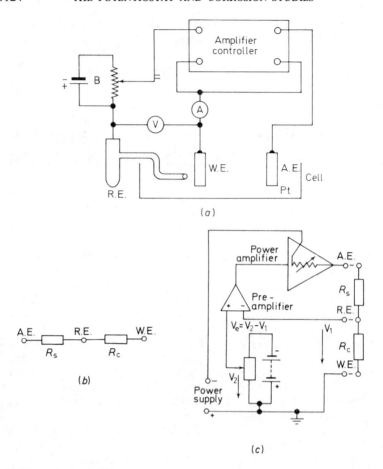

Fig. 20.34   Basic circuit for a potentiostat. (*a*) Basic circuit for a potentiostat and electro-chemical cell; (*b*) equivalent circuit; (*c*) circuit of a basic potentiostat. A.E. is the auxiliary electrode, R.E. the reference electrode and W.E. is the working electrode (Fig. 20.34*b* and *c* have been taken from *Potentiostat and its Applications*, by J. A. von Fraunhofer and C. H. Banks, Butterworths (1972))

unbalance between the electrode potential and the backing potential pro-duces an error signal at the input of the amplifier-controller circuit. The latter rapidly adjusts the cell current between the specimen and the auxiliary electrode until the error signal is reduced to zero.

The electrical characteristics of the cell and electrode will comprise both capacitative and resistive components, but for simplicity the former may be neglected and the system can be represented by resistances in series (Fig. 20.34*b* and *c*). The resistance R_s simulates the effective series resistance of the auxiliary electrode A.E. and cell solution, whilst the potential developed across R_c by the flow of current between the working electrode W.E. and A.E. simulates the controlled potential W.E. with respect to R.E.

Figure 20.34*c* shows a basic circuit of a potentiostat in which the difference between the desired potential (V_2) and the actual potential of the working electrode (V_1) is amplified by a high gain differential pre-amplifier. The output is an error signal $AV_e = A(V_2 - V_1)$, where A is the gain of the amplifier, which is arranged to control the power amplifier in such a way that the potential of A.E. is continuously adjusted to minimise V_e. If the loop gain is high, V_e can be made to approach zero very closely, the limit being determined by the electrical noise in the system. The potential of W.E. with respect to R.E. is thus held constant at the desired potential, V_2.

When potential setting is varied manually during the determination of polarisation, each change can be made after a constant time interval or when the rate of current change reaches a predetermined low level. A number of instruments for programmed potential changes have been introduced, permitting a variety of continuous sweeps or stepwise traverses over a desired range of potential. This, together with a suitable electrometer, recorder, noise filter (when necessary) and logarithmic converter, provide an automated procedure for plotting E-log i curves.

An ASTM recommended practice (*A Standard Reference Method for Making Potentiostatic and Potentiodynamic Anodic Polarisation Measurements*, G5:1972) has been issued. It provides a means of checking experimental technique and instrumentation using a specimen from a single heat of AISI Type *430* stainless steel, which is available from ASTM*.

Scanning Rate

The time factor in stepwise potentiostatic or potentiodynamic polarisation experiments is very important, because large differences can be caused by changes in the scanning rate. Since the steady state depends on the particular system and conditions of exposure, no set rule exists for the magnitude or frequency of potential changes. Chatfield et al.[6a] have studied the Ni/H_2SO_4 system and have shown how $E_{pass.}$ becomes more passive with increase in sweep rate.

In order for the potentiostatic technique to provide an accelerated test, whether for general or localised corrosion, it is obvious that an accelerating factor is needed. Merely duplicating service conditions by substituting the potentiostat for chemical potential control does not necessarily shorten the required testing time. When employing an accelerating factor, such as higher temperature, change in chemistry of the environment, or a greater driving force (potential), care should be taken to ensure that the mechanism of the reaction(s) under examination is not altered.

iR Corrections, Probe Positioning, Specimen Masking and Mounting

There is no difference between galvanostatic and potentiostatic polarisation experiments regarding the iR potential drop between the specimen and the tip of the probe used for measuring the electrochemical potential. In either case corrections should be made for accuracy. These could be quite large if the current density is high and/or the conductivity of the electrolyte is low.

*American Society for Testing and Materials, Headquarters, 1916 Race Street, Philadelphia, Pa., USA.

The position of the probe relative to the test specimen surface can cause differences in potential readings[7].

Adequate specimen masking is one of the major problems in corrosion testing. Crevices with non-uniform current distribution and in which changes in the chemistry of the electrolyte can take place rapidly, are particularly undesirable[8]. In potentiostatic work, pitting or crevice attack is frequently found near the masking interface (or under it when seepage and undermining occur). Unless this factor is specifically being investigated, it should be avoided. The use of a partially immersed specimen can eliminate the need for masking, although it leaves the water line to be dealt with. A method for mounting a cylindrical specimen has been described[9,10], which avoids crevices when the holder is properly tightened (*see* Section 20.7). Specimens can also be rotated during the test, if desired, when this method is used.

Applications

Studies of Passivity

The potentiostat is particularly useful in determining the behaviour of metals that show active–passive transition. Knowledge of the nature of passivity and the probable mechanisms involved has accumulated more rapidly since the introduction of the potentiostatic technique. Perhaps of more importance for the subject at hand are the practical implications of this method. We now have a tool which allows an 'operational' definition of passivity and a means of determining the tendency of metals to become passive and resist corrosion under various conditions.

The use of the potentiostatic method has helped to show that the process of self-passivation is practically identical to that which occurs when the metal is made anodically passive by the application of an external current[11-15]. The polarisation curve usually observed is shown schematically in Fig. 20.35a. Without the use of a potentiostat, the active portion of the curve AB would make a sudden transition to the curve DE, e.g. along curve AFE or AFD, and observation of the part of the curve $BCDE$ during anodic polarisation was not common until the potentiostat was used.

The current–potential relationship $ABCDE$, as obtained potentiostatically, has allowed a study of the passive phenomena in greater detail and the operational definition of the passive state with greater preciseness. Bonhoeffer, Vetter and many others have made extensive potentiostatic studies of iron which indicate that the metal has a thin film, composed of one or more oxides of iron, on its surface when in the passive state[16-19]. Similar studies have been made with stainless steel, nickel, chromium and other metals[20-25]. Some of the more recent papers on this subject were presented at the International Conference on Passivity in Cambridge, England, in 1970, and are now in the press[26] (*see also* Section 1.5).

Since the corrosion potential of a metal in a particular environment is a mixed potential—where the total anodic current is equal to the total cathodic current—the potentiostatic curve obtained by external polarisation will be influenced by the position of the local cathodic current curve. (Edeleanu[27] and Mueller[28] have discussed the details which must be considered in the

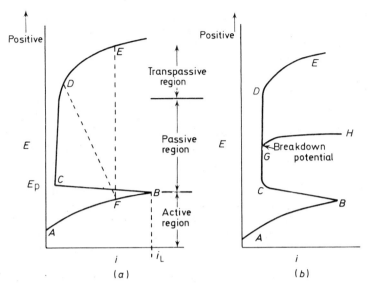

Fig. 20.35   Schematic polarisation curves from anodic potentiostatic polarisation

analysis and interpretation of the curves.) For this reason, residual oxygen in the test solution can cause a departure from the usual curve; in such a case, a 'negative loop' B' corresponding with cathodic reduction of dissolved oxygen occurs after passing the critical point, with the normal passive region and low positive currents only being resumed at E_{C_2}, as shown in Fig. 20.36. Other ions can also interfere with the currents observed. If they are oxidising, then they will have much the same effect as dissolved oxygen. However,

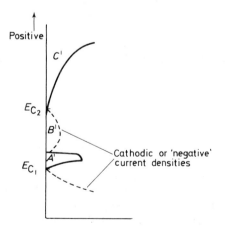

Fig. 20.36   Schematic polarisation from potentiostatic polarisation. B' shows the 'negative loop' and represents cathodic reduction of dissolved oxygen. The dashed curves in the diagram are cathodic currents and are frequently drawn on the left-hand side of the E axis

some of them may increase the observed current to values that suggest corrosion rates much higher than are actually taking place. The polarisation curve for titanium[29] indicated current levels in the passive region that did not agree with the (lower) corrosion rates determined gravimetrically, and this was found to be due to the presence of Ti^{3+} ions in solution. In the case of iron in neutral water[30], the passive anodic current densities were found to be proportional to the concentration of Fe^{2+}. Indig[31] observed similar current increases in stagnant high-temperature water tests due to the formation of Fe_2O_3 from Fe^{2+}; he largely eliminated this by changing to a flowing electrolyte system, and achieved correlation with actual corrosion rates.

The critical current and primary passivation potential E_{pp} will not appear on an anodic polarisation curve when the steady-state potential already is higher than E_{pp}. In such a case the potentiostat is unable to provide direct data for constructing the full polarisation curve. If that portion of the curve below the steady-state potential is desired, then the potential has to be held constant at several points in this range and corrosion currents calculated from corrosion rates as determined from solution analyses and/or weight losses.

The potentiostat is very useful for determining the effects of composition and heat treatment on the corrosion resistance of alloys. Sometimes it is possible to understand what appear to be discrepancies in practice. Edeleanu[27] used the method for determining the resistance of stainless steels to acids. The potentiostatic curves showed that the current in the transpassive region (at a very high potential) increased with the chromium content, while the current in the passive region (at lower potentials) decreased. This explained the behaviour of several steels in service, where steels of higher chromium content showed poor resistance to corrosion in environments of high redox potentials—nitric acid plus chromic acid mixtures—but greater resistance in nitric acid. Edeleanu also discussed potentiostatic curves which showed the beneficial effects of nickel, copper and molybdenum on the corrosion resistance of stainless steel in sulphuric acid. This paper should be consulted for an excellent discussion on the use of these techniques for determining the effects of alloy composition on corrosion resistance.

Cihal, et al.[32] presented early data on the effects of chromium, nickel, molybdenum, titanium, niobium and silicon on the passive behaviour of stainless steel.

Pitting

Some stainless steels and aluminium alloys are examples of metals that show pitting corrosion when exposed to aqueous solutions containing halide ions, although the phenomenon is not confined to these alloys. Various factors influence the onset of pitting, one of which is the interfacial potential; pits are thought to form only at potentials more positive than a certain critical value. This can be demonstrated by electrochemical measurements using potentiostatic techniques[33-42], and Fig. 20.35b which is similar to that shown by Kolotyrkin's[43], represents a typical curve. In the absence of aggressive (pitting) ions, *ABCDE* represents the usual polarisation behaviour, where *DE* is the region of transpassivity. However, when conditions suitable for pitting prevail, the curve $ABCGH$ is typically found, and the breakout (or breakthrough) GH from the passive region is accompanied by pitting. The first point of departure from the passive region has been referred to as the 'critical pitting potential', but it should be noted that its value is time dependent and that it varies with the rate of potential change, i.e. if the potential sweep rate or stepping rate is too rapid, a more positive 'critical' value is obtained. In general, the shorter the time the more positive is the 'critical pitting potential', and in order to be certain that a reliable estimate of the pitting potential has been made, it is necessary to hold the potential at a constant value just below the critical point for a suitably long time (in practice, several days*), in order to demonstrate retention of passivity and the absence of pits in the specimen surface (*see also* Section 1.6).

The correlation between the redox potential of a system and the occurrence of pitting attack was established some 30 years ago[44]. Also, the use of passivity breakdown as a screening method for alloy resistance was described over the years by several workers, for example, Brennert[45], Mahla and Nielsen[46], and Pourbaix[47]. For a metal in a given solution, it may appear that the electrochemical potential, regardless of its origin, will be the only determinant of whether or not pitting will take place. While this is generally expected, France and Greene[6] suggested that a potentiostatically controlled corrosion test could be more severe than a conventional one (chemically controlled potential). Their reason is related to local chemistry changes required to preserve charge neutrality; during anodic polarisation migration of Cl^- ions occurs to balance the excess positive charge produced by the Fe^{2+} ions and this results in an increase of Cl^- ions at the metal/solution interface and a consequent increase in pitting propensity. A similar movement of anions in, for example, a solution of $FeCl_3$, does not occur (*see also* page **21**.68). Figure 20.37 is taken from the work of France and Greene to illustrate the movement of Cl^- ions across a boundary line.

Correspondence between electrochemical tests and field exposures has not always been found. Therefore, it is difficult to interpret potentiostatic data in terms of service performance. Disagreement between results can be caused by factors that are not immediately obvious. For example, gases such as hydrogen, argon or nitrogen are generally used to remove oxygen from solution before proceeding with a potentiostatic test. At first sight, it might appear immaterial which gas is used, but Wilde and Williams[48] found differences in the breakout (critical pitting) potential of stainless steels, depending on the selection of gas.

More details of other factors that affect the critical pitting potential have been discussed by Uhlig and his co-workers[49,50]. They indicated that for stainless steel the critical pitting potential decreased with increasing concentration of chloride ion. At a fixed chloride level, passivating ions in solution, such as sulphate and nitrate, etc., cause the pitting potential to become more positive; at a sufficient concentration these ions totally inhibited pitting, as shown in Fig. 20.38 for SO_4^{--} and ClO_4^-.

Lizlovs and Bond[51] reported a molar ratio of 5:1 (SO_4^{--}:Cl^-) for inhibiting pitting in ferritic stainless steels. A plot of critical potential *vs.*

*Even after several days of no localised attack, the question could legitimately be asked whether the test was long enough to establish a true value, below which pitting would *never* occur.

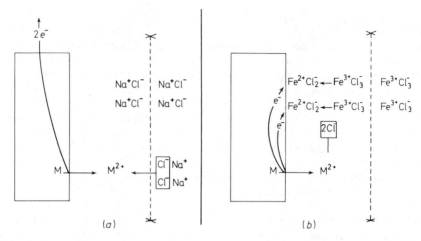

Fig. 20.37 Schematic representation of reactions during (a) controlled potential and (b) conventional corrosion tests in acidic chloride solutions. In (a) charge balance must be maintained by migration of Cl^- ions, since the cathodic reaction occurs elsewhere at the counter-electrode. In (b) the anodic and cathodic sites are in close proximity, and charge balance is maintained without migration of Cl^- ions from the bulk solution (after France and Greene[6])

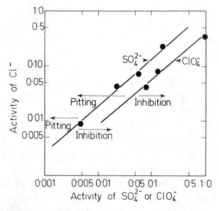

Fig. 20.38   Activity of SO_4^{2-} or ClO_4^- required to inhibit pitting as a function of Cl^- activity; 25°C (after Leckie and Uhlig[49])

ratio of $SO_4^{--}:Br^-$ is reproduced in Fig. 20.39 from the work of Kolo-tyrkin[43]. Lowering the temperature causes the critical potential to become more positive[49]. Changes of pH in the acid range did not affect the pitting potential appreciably, but in the alkaline region it increased markedly with pH, as shown in Fig. 20.40.

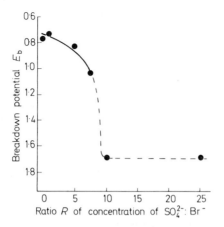

Fig. 20.39 Dependence of breakdown potential of Fe–Cr alloys (containing 13% Cr, in 0.1 mol dm^{-3} HBr + K_2SO_4 solution) on the ratio of sulphate and bromide concentrations in solution (after Kolotyrkin[43])

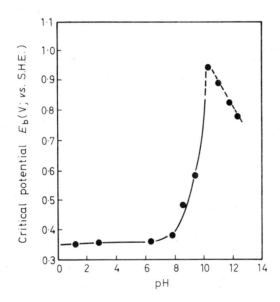

Fig. 20.40 Effect of pH on critical potential for pitting in 0.1 mol dm^{-3} NaCl; 25°C (after Leckie and Uhlig[49])

The fact that scanning speed can affect polarisation behaviour has already been mentioned. In the case of stainless steel a plot of critical potential E_b vs. rate shows how E_b becomes more positive with potential change rate (Fig. 20.41)[52]. When a specimen was held at a fixed passive potential while aggressive ions (Cl⁻) were added to determine the concentration required

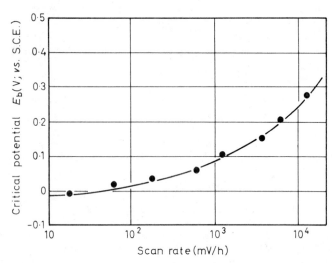

Fig. 20.41   Effect of potential scan rate on the value of E_b for Type *304* stainless steel in 0·1 mol dm^{-3} NaCl (after Leckie[52])

for loss of passivity (pitting), then a similar time effect was observed[53]; a higher apparent resistance, i.e. greater apparent Cl$^-$ ion tolerance, was found when making more rapid additions.

Numerous references regarding alloying additions have been published, and the reader should look up specific effects for relevant alloying systems.

Potential–pH Diagrams

Pourbaix has contributed substantially to the science of corrosion through plotting thermodynamic data of systems as a function of electrode potential and pH. Numerous publications of his have appeared in the literature, as well as an atlas of potential–pH diagrams. A recent paper[54] exemplifies the usefulness of potentiostatic polarisation curves in the experimental plotting of various domains, such as protection, pitting, general corrosion and passivity, in these diagrams. This particular procedure, which has been dealt with in some detail in Section 1.6, is a very powerful tool which is now available for studying corrosion.

De-alloying

The selective net loss of a component such as zinc, aluminium or nickel from copper-base alloys sometimes occurs when these alloys corrode. Early studies of the phenomenon were done by simple immersion. More recently, however, the potential–pH dependence of de-alloying has been examined[55], and it appears that this approach can provide a much more detailed understanding of the mechanism. Future experimental work is expected to include potentiostatic and potentiodynamic techniques to a much greater extent.

Selective Etching

Dissolution kinetics are influenced by pH, potential and the ions present in the test solution, and this forms the basis of selective metallographic etching techniques that have been used for some time[32,56,57]. The potentiostat is often used to hold the potential of a multi-phase alloy constant at a level

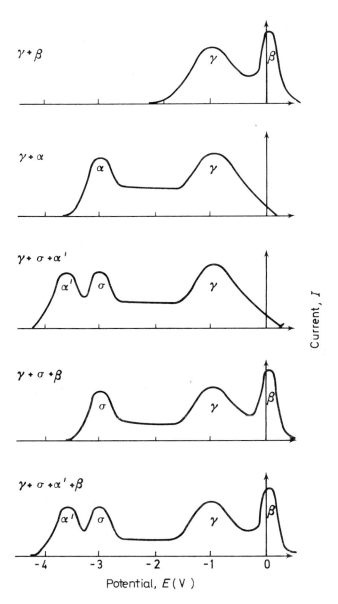

Fig. 20.42   Typical current *vs.* potential curves for alloys of various phase combinations (after Jones and Hume-Rothery[58])

suitable for attack on a specific phase while the rest of the surface remains passive. A scan over a potential range can be done as a preliminary step to ascertain whether the electrochemical properties of various phases differ sufficiently in the chosen electrolyte for selective etching or phase extraction. One such procedure was reported by Jones and Hume-Rothery[58] for austenitic steels alloyed with aluminium, and Fig. 20.42 is reproduced from their work. Another example is the behaviour of Fe–Fe$_3$C as a function of potential, pH and anion[59]. In this latter work, the conditions were defined under which four different modes of attack took place, i.e. general-, matrix-, interface- and carbide attack, and the data were interpreted in terms of thermodynamics, kinetics and the influence of complex formation. These potentiostatic experiments were coupled with detailed electron transmission microscopy and the comprehensive nature of the results demonstrate the effectiveness of such a combined approach.

The extraction of precipitates for further examination is also possible by the same techniques. Conditions would have to be chosen to give matrix or interface attack.

Grain-boundary Corrosion

Intergranular corrosion is encountered in many metal systems, often associated with the presence of precipitates at grain boundaries. In the case of stainless steels, one widely accepted theory states that the precipitation of chromium carbides leads to a chromium-denuded zone which undergoes rapid corrosion.

Potentiostatic methods, being capable of detecting differences in corrosion and passivation behaviour of various parts of a heterogeneous surface, have been applied to the electrochemical determination of grain boundary corrosion[60,63,64,66,68].

Cihal and Prazak[60] determined the resistance of 18/8 stainless steel to this type of corrosion. They claimed that the technique could be used on steels which are difficult to test by other methods, including steels of low carbon content, and steels in which stabilising elements are present. By means of potentiostatic curves and light etching at constant potential they confirmed that the extent of intergranular corrosion depended upon the amount of precipitated chromium carbide.

Corradi and Gasperini[61] claimed that the potentiostatic method was more effective and simpler than the Strauss test for determining intergranular corrosion of stainless steels, and suggested that the method may lend itself for use on finished equipment in service as a 'non-destructive' test.

Cihal, et al.[62] tabulated potentials which may be used for the selective etching of the various phases in several stainless steels. Bergholtz[67] suggested a potential of +0·160 V vs. S.H.E. for grain-boundary etching of stainless steel, while Desestret[64,65] favoured various levels of potential, depending on its chromium content. He concluded that potentiostatic etching was more sensitive for determining susceptibility to intergranular corrosion than chemical tests in boiling nitric acid or acidified copper sulphate.

Budd and Booth[68] found the potentiostatic test best for investigating the intergranular and layer corrosion of aluminium alloys.

Not all test methods are necessarily accelerated by the use of a potentiostat. France and Greene[69a] used a potentiostat to hold sensitised 18/8 stainless steels at various constant potentials in 1 N H_2SO_4 in order to determine the range of potentials at which intergranular attack occurred (*see* Section 20.1, Fig. 20.20). However, this method of testing for sensitivity has been criticised by Streicher[69b], who points out that the duration of the potentiostatic test is too short, and that alloys found to be immune during this test will suffer intergranular attack when the duration of exposure is more prolonged.

Streicher's work[69] indicates how useful the potentiostat has been in studying intergranular corrosion. Ideally, future data would be expanded to provide Pourbaix-type diagrams that also contain kinetic information showing various rates of attack within the general domain of intergranular corrosion. (Similar data for cases other than intergranular attack would be equally valuable.)

Stress-corrosion Cracking (s.c.c.)

S.C.C. has received a share of the potentiostatic approach to corrosion. Barnartt and van Rooyen[70] reported that potentiostatically controlled corrosion in a potential range 50–100 mV above the corrosion potential provided an accelerated test for the s.c.c. of stainless steels. The elevation of the potential by means of a potentiostat eliminated the incubation period, and also increased the density of cracks. Booth and Tucker[71] used potentiostatic methods in the s.c.c. of Al–Mg alloys.

Hoar and his co-workers[72,73] at first used galvanostatic equipment in their investigations into the 'mechano-chemical' dissolution of metal during plastic deformation; subsequently, the potentiostatic rather than galvanostatic control of potential was reported to give better results, and it enabled them to show that high corrosion rates were possible without appreciable elevation of the driving potential. This mechano-chemical theory has recently been refined in work reported for copper-base alloys[74,75]. In the latter case, the potential dependence of the reactions leading to cracking has been analysed very carefully as a function of pH[76].

Staehle, *et al.*[77] have considered several aspects of s.c.c. from the electrochemical standpoint, including a feature of a recently suggested cracking mechanism which relates to the amount of corrosion that takes place each time that a slip step emerges at a surface. They recorded the shape of the current rise and decay curve that accompanied instantaneous straining (impact load), while the potential was controlled with a potentiostat. The number of coulombs of charge indicated the magnitude of metal dissolution. They believe that s.c.c. would be likely in a metal that showed the right amount of corrosion per slip step event.

Another contribution of the potentiostatic technique to s.c.c. studies has been the report[78] that cracking prevails essentially at two potential levels for metals showing an active–passive transition. These potentials are located near the top and bottom of the passive region. Along the same lines, Uhlig and his co-workers have determined critical ranges of potential for s.c.c.[79,80], although their theoretical interpretation differs from that of the other references cited.

High-temperature Water

Pressurised water nuclear reactors require metals that will have a high degree of corrosion resistance to pure water at around 300°C. Laboratory testing of materials for this application have included potentiostatic polarisation experiments designed to clarify the active–passive behaviour of alloys as well as to establish corrosion rates. Since pressure vessels are used for this work, it is necessary to provide sealed insulated leads through the autoclave head[83].

Care should be taken to avoid short circuits; for instance, an insulated specimen, being common with the ground point of some potentiostats, can become electrically reconnected to the autoclave if the latter is not separated from ground by using an isolation transformer.

Not all reference electrodes are suitable for use at high temperatures, and in addition, they may cause contamination if placed directly in pure water. A liquid junction consisting of a pressure-reducing tube with a wet string (or plug) has been employed[81,84]; this enabled locating a reference electrode at room temperature and low pressure outside the autoclave, but compensation for contact and other potentials was difficult. Platinum has been used directly in the high-temperature electrolyte[85,86], and it functioned as a hydrogen electrode as long as hydrogen was present. Most recently, a canned electrode has been described[87], in which a silver–silver chloride reference is used inside a small container suspended in the hot water. Tiny pinholes and a long diffusion distance permitted a continuous electrolyte path while avoiding contamination of the test medium with chloride.

A review article on techniques for electrochemical measurements in pressurised water has been written by Jones and Masterson[88], which describes many of the experimental ramifications involved.

The low conductivity of high-purity water makes it difficult to study electrode processes potentiostatically, since too high an electrical resistance in the circuit can affect the proper functioning of a potentiostat, and it can also introduce large iR errors. The increase in conductivity of water with temperature has been measured[89] and iR-corrected polarisation data have been obtained[86] in hot water that originally had very low conductivity at room temperature. Other results[31,83,85,90,91] in high-temperature water are all for tests where the conductivity was deliberately increased through the addition of electrolytes.

The interpretation of the polarisation curves requires care. Wilde[89] determined corrosion rates by linear polarisation in pure degassed water. Indig and Groot[90], however, found that electrochemical methods were unsuccessful in measuring the corrosion rate of Ni–Cr–Fe alloy 600 in the presence of hydrogen, owing to the kinetic ease of the redox reaction on the metal surface. In another paper, the same authors[85] also indicated that linear polarisation generally gave corrosion rates that were not accurate, as a result of competing half-reactions, under conditions that were different from those used by Wilde. In a test with stainless steel they found that the removal of hydrogen could reduce the problems.

Corrosion products are another source of error in the potentiostatic determination of polarisation curves in high-temperature water. In stagnant tests Fe^{2+} could be converted to Fe_2O_3[31], causing a false anodic current reading in the passive region. This effect was eliminated by using a flowing electrolyte.

S.C.C. has been examined as a function of potential[82] in high-temperature water with chlorides present and an increased susceptibility of stainless alloys to intergranular attack was found as the potential was increased. Additional work[91] reported that no intergranular cracking was observed in tests of short duration.

Hydrogen Permeation

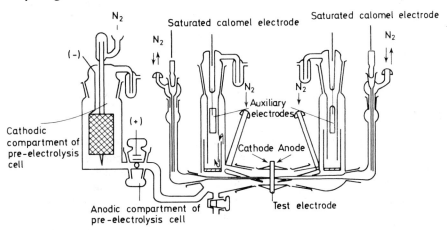

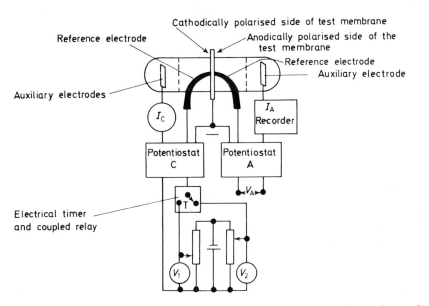

Fig. 20.44   Electrical circuit of cell and pre-electrolysis vessel (after Devanathan and Stachewski[93])

A very sensitive technique using the potentiostat was developed at the University of Pennsylvania[92] for studying the permeation of hydrogen through thin metal foils. Such studies have and will continue to contribute important information in areas where hydrogen embrittlement is a problem. The technique involves the use of a double cell coupled by a thin metal membrane. Hydrogen is generated on the input side of the membrane which is maintained at a cathodic potential. Upon diffusion through the membrane, the hydrogen is electrochemically oxidised at the exit surface by an anodic potential that must be maintained constant by a potentiostat, and the anodic current provides a direct measure of the hydrogen flux. The technique is capable of detecting fluxes of 3×10^{-2} cm^3 of H$_2$ per second and can provide information on diffusivity, permeability, solubility and the interaction of hydrogen with metallic lattices. The apparatus is shown in Figs. 20.43 and 20.44[93] and further details of the underlying theory are given in Section 9.1.

Molten Salts

The potentiostatic technique has been used in the investigation of the behaviour of metals in molten salts. In principle, the experimental method is the same as the one for aqueous media. Results are also capable of interpretation in the same way as those in aqueous solutions, and typical active–passive behaviour as well as anodic and cathodic Tafel lines have been observed. One of the more recent papers[94] also contains several references to earlier work. These authors state that 'the potentiodynamic method, so successful in evaluating corrosion-resistant materials for aqueous systems, appears to be quite suitable also in selecting materials to be employed in molten salts'. In addition to the plotting of individual polarisation curves, it is possible to construct stability diagrams for molten salt systems resembling the well-known Pourbaix diagrams. The main difference is that the oxygen anion potential pO^{2-} replaces the pH function, since the former is more important in molten salts.

It has been established that salts can deposit or form on metals during gas–metal reactions. Molten layers could then develop at high operating temperatures. Consequently, the laboratory testing of corrosion resistance in molten salts could yield valuable results for evaluating resistance to some high-temperature gaseous environments.

Inhibitors

There are many published papers dealing with the electrochemical investigation of the effects of inhibitors and surfactants on corrosion processes, using the potentiostat[95, 96]. Adsorption of organic and inorganic ions on metal surfaces is found to be important, since it is related to their positive or negative charges as well as the potential of the metal surface. Some details regarding the use of polarisation techniques for examining specific effects on anode and cathode kinetics are described in References 95 and 96; the reader will find that numerous other papers are also available.

Rotating Disc-ring Electrodes

Frumkin and Nekrasov[97] introduced a rotating disc-ring electrode suitable for the detection of intermediates in corrosion reactions, and its theory was considered by Ivanov and Levich[98]. In this method, a disc electrode (specimen) can be corroded under controlled conditions, and a metal ring around it is held potentiostatically at a predetermined potential E_r in order to measure the rate at which an ionic species arrives at its surface; this is proportional to the current flowing in the potentiostat circuit. E_r is varied to appropriate values for particular ions of interest. Figures 20.45 and 20.46 are taken from the work of Pickering and Wagner[99], who applied the technique in their study of the de-alloying phenomenon. Modifications used by other workers include the 'split-ring' technique.

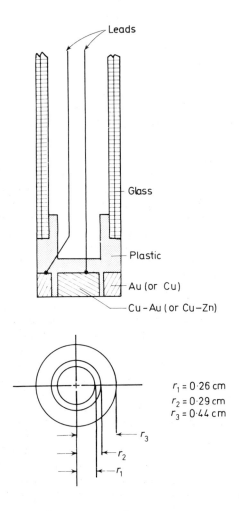

Leads

Glass

Plastic

Au (or Cu)

Cu–Au (or Cu–Zn)

$r_1 = 0.26$ cm
$r_2 = 0.29$ cm
$r_3 = 0.44$ cm

r_3

r_2

r_1

Fig. 20.45   Disc-ring electrode assembly

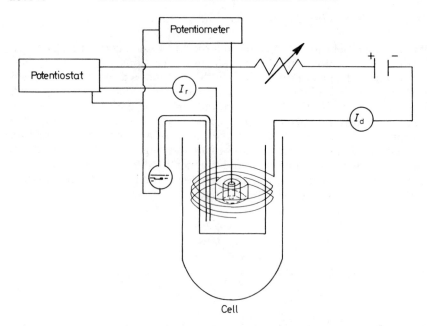

Fig. 20.46   Circuit used for the ionisation and redeposition experiments

Anodic Protection

The potentiostat has supplied an experimental tool for the study of anodic protection. The elucidation of passive behaviour made possible by potentiostatic anode polarisation curves allowed investigators to determine the conditions necessary for maintaining a metal in a stable passive condition by provision of a suitable environment, addition of cathodic alloying elements[100,101], and/or maintenance of the required potential by means of external anodic polarisation[27,29,106].

Edeleanu[102,103] made use of potentiostatic curves to determine the optimum conditions for the protection of stainless steel in sulphuric acid. A pilot plant was then used to determine the practicability of anodic protection at a constant potential. He pointed out several factors necessary for proper control and indicated the spectacular results obtained.

Stern, et al.[29] obtained potentiostatic polarisation curves for titanium alloys in various solutions of sulphuric acid and showed that the mixed potentials of titanium–noble metal alloys are more positive than the critical potential for the passivity of titanium. This explains the basis for the beneficial effects of small amounts of noble metals on the corrosion resistance of titanium in reducing-type acids. Hoar's review of the work on the effect of noble metals on including anodic protection should also be consulted[104].

The use of potentiostatic curves has also facilitated the study of the rôle of oxidising agents and inhibitors in corrosion processes. Stern[105] discussed the rôle of passivating-type inhibitors and used potentiostatic curves to explain their action. Posey[106] used the potentiostatic technique for determining the reduction of cupric ion on stainless steel. Both of these references should be consulted for an extensive bibliography on these subjects. The reader will also find many subsequent papers dealing with this subject, and they generally confirm the principles that were set out in the aforementioned work.

Fast Electrode Reactions

A recent paper by Cahan, Nagy and Genshaw[107] examines design criteria for an electrochemical measuring system to be used for potentiostatic transient investigation of fast electrode reactions. They emphasise the importance of co-design of the experimental cell and electronics. This particular paper deals only with the cell design, and a second paper treating the potentiostat is to follow.

Accurate control of potential, stability, frequency response and uniform current distribution required the following: low resistance of the cell and reference electrode; small stray capacitances; small working electrode area; small solution resistance between specimen and point at which potential is measured; and a symmetrical electrode arrangement. Their design appears to have eliminated the need for the usual Luggin capillary probe.

D. van ROOYEN

REFERENCES

1. Lingane, J. J., *Electroanalytical Chemistry*, Interscience, N.Y., 2nd edn (1958)
2. Hickling, A., *Trans. Faraday Soc.*, **38**, 27 (1942)
3. Roberts, M. H., *J. Appl. Phys.*, **5**, 351 (1954)
4. Greene, N. D., *Corrosion*, **15**, 369t (1959)
5. France, W. D., *Mats. Res. and Stds.*, 21, August (1969)
6. France, W. D. and Greene, N. D., *Corrosion*, **26**, 1 (1970)
6a. Chatfield, C. J. and Shreir, L. L., *Electrochim. Acta.*, **12**, 563 (1972)
7. Mears, D. C. and Rothwell, G. P., *J. Electrochem. Soc.*, **115**, 36 (1968)
8. Greene, N. D., France, W. D. and Wilde, B. E., *Corrosion*, **21**, 275 (1965)
9. ASTM Recommended Procedure, Committee G5 (1969)
10. Stern, M. and Makrides, A. C., *J. Electrochem. Soc.*, **107**, 728 (1960)
11. Tomashov, N. D., *Adv. Chem.*, *Moscow*, **24**, 453 (1955)
12. Heumann, T. and Rosener, W., *Z. Elektrochem.*, **57**, 17 (1953)
13. Batrakov, V. P., *C. P. Acad. Sci. U.R.S.S.*, **99**, 797 (1954)
14. Edeleanu, C., *Metallurgia, Manchr.*, **50**, 113 (1954)
15. Stern, M., *J. Electrochem. Soc.*, **105**, 638 (1958)
16. Bonhoeffer, K. F. and Gerischer, H., *Z. Elektrochem.*, **52**, 149 (1948)
17. Weil, K. G. and Bonhoeffer, K. F., *Z. Physik. Chem.*, **4**, 175 (1955)
18. Franck, U. F. and Weil, K. G., *Z. Elektrochem.*, **56**, 814 (1952)
19. Vetter, K. J., *Z. Elektrochem.*, **55**, 274 (1951) **56**, 106 (1952); **58**, 230 (1954); **59**, 67 (1955)
20. Kolotyrkin, Y. M., *Z. Elektrochem.*, **62**, 664 (1958)
21. Frankenthal, R. P., *J. Electrochem. Soc.*, **114**, 542 (1967); **116**, 680 (1969); Pickering, H. W. and Frankenthal, R. P., *J. Electrochem. Soc.*, **112**, 761 (1965)
22. Prazak, M., Prazak, V. and Cihal, V., *Z. Elektrochem.*, **62**, 739 (1958)
23. Schwabe, K. and Dietz, G., *Z. Elektrochem.*, **62**, 751 (1958)
24. Okamoto, G., Kobayashi, H., Nagayama, M. and Sato, N., *Z. Elektrochem.*, **62**, 775 (1958)
25. Weidinger, H. and Lange, E., *Z. Elektrochem.*, **64**, 468 (1960)
26. To be published in *Electrochim. Acta*.
27. Edeleanu, C., *J. Iron St. Inst.*, **188**, 122 (1958)
28. Mueller, W. A., *Canad. J. Chem.*, **38**, 576 (1960)

29. Stern, M. and Wissenberg, H., *J. Electrochem. Soc.*, **106**, 755 (1959)
30. Nagayama, M. and Cohen, M., *J. Electrochem. Soc.*, **110**, 670 (1963)
31. Indig, M. E., Ph.D. Thesis, RPI, Troy, N.Y. (1971)
32. Cihal, V. and Prazak, M., *JISI*, **193**, 360 (1959)
33. Franck, U. F., *Werkstoffe u. Korrosion*, **9**, 504 (1958); **11**, 401 (1960)
34. Weil. K. G. and Manzel, D., *Z. Elektrochem.*, **63**, 669 (1959)
35. Engell, H. J. and Stolica, N. D., *Z. Phys. Chem.*, **20**, 113 (1959)
36. Stoffils, H. and Schwenk, W., *Werkstoffe u. Korrosion*, **12**, 493 (1961)
37. Kolotyrkin, Ja. M., *Progress in Chemistry (USSR)*, **21**, 322 (1962); *J. Electrochem. Soc.*, **108**, 209 (1961)
38. Kolotyrkin, Ja. M. and Gilman, V. A., *Proc. Acad. Sci. U.S.S.R.*, **137**, 642 (1961)
39. Vanleugenhaghe, C., Klimzack-Mathieiu, I., Meunier, J. and Pourbaix, M., *International Conf. Corr. Reactor Materials*, Salzburg (1962)
40. Popova, T. I. and Kabaniov, B. N., *J. Physic. Chem. (U.S.S.R.)*, **35**, 1296 (1961)
41. Acello, S. J. and Greene, N. D., *Corrosion*, **18**, 286t–290t, Aug. (1962)
42. Trumpler, G. and Keller, R., *Helv. Chim. Acta.*, **44**, 1785 (1961)
43. Kolotyrkin, Ja. M., *Corrosion*, **19**, 261t (1963)
44. Uhlig, H. H., *Trans. Hur. Met. Min. Met. Eng.*, **140**, 422 (1940)
45. Brennert, S., *J. Iron Steel Inst.*, **135**, 101 (1937)
46. Mahla, E. M. and Nielsen, N. A., *Trans. Electrochem. Soc.*, **89**, 167 (1946)
47. Pourbaix, M., Klimzack-Mathieiu, L., Mertens, Ch., Meunier, J., Vanleugenhaghe, Cl., deMunk, L., Laureys, L., Neelemans, L. and Warzee, M., *Corrosion Sci.*, **3**, 239 (1963)
48. Wilde, B. E. and Williams, E., *J. Electrochem. Soc.*, **116**, 1539 (1969)
49. Leckie, H. P. and Uhlig, H. H., *J. Electrochem. Soc.*, **113**, 1262 (1966)
50. Horvath, J. and Uhlig, H. H., *J. Electrochem. Soc.*, **115**, 791 (1968)
51. Lizlovs, E. A. and Bond, A. P., *J. Electrochem. Soc.*, **116** No. 5, 574 (1969)
52. Leckie, H. P., *J. Electrochem. Soc.*, **117**, 1152 (1970)
53. Jackson, R. P. and van Rooyen, D., *Corrosion*, **27**, 203 (1971)
54. Pourbaix, M., *Corrosion*, **26**, 431 (1970)
55. Verink, E. D. and Parrish, P. A., *Corrosion*, **26**, 214 (1970)
56. Edeleanu, C., *JISI*, **185**, 482 (1957)
57. Greene, N. D., Rudaw, P. S. and Lee, L., *Corrosion Science*, **6**, 371 (1966)
58. Jones, J. D. and Hume-Rothery, W., *JISI*, **204**, 1 (1966)
59. Cron, C. J., Payer, J. H. and Staehle, R. W., *Corrosion*, **27**, 1 (1971)
60. Cihal, V. and Prazak, M., *Hutn. List.*, **11**, 225 (1956); *Corrosion*, **16**, 530t (translation) (1960)
61. Corradi, B. and Gasperini, R., *Metallurg. Ital.*, **52**, 249 (1960)
62. Cihal, V. and Prazak, M. J., *J. Iron St. Inst.*, **193**, 360 (1959)
63. Voeltzel, J. and Plateau, J., *Compt. Rend.*, **254**, 1791 (1962)
64. Desestret, A., *Mem. Sci. Rev. Met.*, **59**, 553 (1962)
65. Desestret, A., *Scuola Azione*, **13**, 165 (1964)
66. Schueller, H. J. *et al.*, *Arch. Eisenhüttenw.*, **33**, 853 (1962)
67. Bergholtz, G., *(ZIS) Mitt.*, **4**, 1134 (1962)
68. Budd, M. K. and Booth, F. F., see Ref. 95, New York, 44 (1963)
69a. France, W. D. and Greene, N. D., *Corrosion Science*, **8**, 9 (1968)
69b. Streicher, M. A., *Corrosion Science*, **11**, 275 (1971)
70. Barnartt, S. and van Rooyen, D., *J. Electrochem. Soc.*, **108**, 222 (1961)
71. Booth, F. F. and Tucker, G. E. G., *Corrosion*, **21**, 173 (1965)
72. Hoar, T. P. and West, J. M., *Proc. Roy. Soc.*, **A268**, 304 (1962)
73. Hoar, T. P. and Scully, J. C., *J. Electrochem. Soc.*, **111**, 348 (1964). See also Ref. 95, New York, 184 (1963)
74. Hoar, T. P. *et al.*, *Corr. Science*, **11**, 231 and 241 (1971)
75. Hoar, T. P., *NATO Conference on The Theory of Stress Corrosion in Alloys*, Ericeira, Portugal, to be published
76. Hoar, T. P. and Rothwell, G. P., *Electrochim. Acta.*, **15**, 1037 (1970)
77. Staehle, R. W., Royuela, J. J., Raredon, T. L., Serrate, E., Morin, C. R. and Farrar, R. V., *Corrosion*, **26**, 451 (1970)
78. Staehle, R. W., *NATO Conference on The Theory of Stress Corrosion in Alloys*, Ericeira, Portugal, to be published
79. Uhlig, H. H. and Cook, E. W., *J. Electrochem. Soc.*, **116**, 173 (1969)
80. Lee, H. H. and Uhlig, H. H., *J. Electrochem. Soc.*, **117**, 18 (1970)

81. Bacarella, A. L. and Sutton, A. L., *J. Electrochem. Soc.*, **112**, 546 (1965)
82. Hübner, W., de Pourbaix, M. and Ostberg, G., Papers at *4th Int. Cong. Met. Corr.*, Amsterdam (1969)
83. Taylor, A. H. and Cocks, F. H., *Br. Corrosion J.*, **4**, 287 (1969)
84. Cowan, R. L., Ph.D. Thesis, Ohio State University (1969)
85. Indig, M. E. and Groot, C., *Corrosion*, **25**, 455 (1969)
86. Wilde, B. E., *Corrosion*, **23**, 331 and 379 (1967)
87. Indig, M. E. and Vermilyea, D. A., *Corrosion*, **27**, 312 (1971)
88. Jones, D. de G. and Masterson, H. G., *Advances in Corrosion Science and Technology* Vol. 1 (Ed. M. G. Fontana and R. W. Staehle), Plenum Press (1970)
89. Wilde, B. E., *Electrochim. Acta.*, **12**, 737 (1967)
90. Indig, M. E. and Groot, C., *Corrosion*, **26**, 171 (1970)
91. Hübner, W., Johansson, B. and de Pourbaix, M., Paper presented at Annual NACE Conference, Chicago, March (1971)
92. Devanathan, M. A. V. and Stachewski, Z., *Proc. Roy. Soc.*, **A270**, 90 (1962)
93. Devanathan, M. A. V. and Stachewski, Z., *Journal of Electrochemical Soc.*, **111**, 619 (1964)
94. Baudo, G., Tamba, A. and Bombara, G., *Corrosion*, **26**, 293 (1970)
95. International Congress on Metallic Corrosion (first—London, 1961; second—New York, 1963; third—Moscow, 1966; fourth—Amsterdam, 1969)
96. *Symposium on Corrosion Inhibitors*, Ferrara, Italy (1961 and 1965)
97. Frumkin, A. N. and Nekrasov, L. I., *Dokl. Akad. Nauk. S.S.S.R.*, **126**, 115 (1959); *Proc. Acad. Sci. U.S.S.R.*, Phys. Chem. Sect., **126**, 385 (1959)
98. Ivanov, Yu. B. and Levich, V. G., *Dokl. Akad. Nauk. S.S.S.R.*, **126**, 1029 (1959); *Proc. Acad. Sci. U.S.S.R.*, Phys. Chem. Sect., **126**, 505 (1959); Levich, V. G., *Physiochemical Hydrodynamics*, Prentice-Hall, Inc., Englewood Cliffs, N.J., 329 (1962)
99. Pickering, H. W. and Wagner, C., *J. Electrochem. Soc.*, **114**, 702 (1967)
100. Tomashov, N. D., *Adv. Chem.*, **24**, 453 (1955)
101. Tomashov, N. D., *Corrosion*, **14**, 229t (1958)
102. Edeleanu, C., *Nature, Lond.*, **173**, 739 (1954)
103. Edeleanu, C., *Metallurgia, Manchr.*, **50**, 113 (1954)
104. Hoar, T. P., *Platinum Metals Rev.*, **4**, 59 (1960)
105. Stern, M., *J. Electrochem. Soc.*, **105**, 638 (1958)
106. Posey, F. A., Cartledge, G. H. and Yaffe, R. P., *J. Electrochem. Soc.*, **106**, 582 (1959)
107. Cahan, B. D., Nagy, Z. and Genshaw, M. A., *J. Electrochem. Soc.*, **119**, 64 (1972)

20.3 Corrosion Monitoring and Chemical Plant

Monitoring of corrosion has received increased emphasis in the operation of chemical and refinery plant in the last few years. Plants such as oil refineries are now expected to operate continuously for periods of up to three years without shut-down, and in such operating conditions it is essential to have some idea of the internal corrosion rates of the metals used in the plant construction.

With little planned down-time for visual inspection the plant operator naturally welcomes any inspection data which can be obtained when the plant is on-stream. The cost of unscheduled down-time can run into tens of thousands of pounds per day due to loss of product. Selling prices of products and costs of raw materials are often outside management responsibility and therefore avoidance of unscheduled down-time can help considerably in holding costs. In addition, the smooth operation of the plant without presenting any hazard or injury to operating personnel can be facilitated by the inspection and corrosion-monitoring programme.

Plant designers are well supplied with corrosion data by materials manufacturers, which are based on both experience and laboratory data, but the information is usually based on specific parameters such as concentration of chemical or temperature. Edeleanu[1] has emphasised this problem with regard to stainless steel corrosion in sulphuric acid, and although the designer may select the correct choice of steel, it is certain that many times in the life of the plant changes in concentration and/or temperature (as well as other parameters) will occur with a consequent increase in the corrosion rate above that anticipated.

No one method for corrosion inspection is sufficient in itself and it is extremely dangerous to rely on data provided by one method only. A study is required of all methods available and the most suitable are then chosen— usually two or three methods are necessary. These are then used and additional methods can be called upon to supplement data if excessive corrosion is experienced or requires verification.

The Hoar report on Corrosion and Protection[2] estimated that a potential saving of £15 000 000 could be made by the oil and chemical industry by 'improving effectiveness in selection of materials and protection', and a recent estimate[3] showed that 20% of this figure could be achieved by increased use of corrosion-monitoring techniques.

Selection of Inspection Points

The selection of inspection points is of paramount importance and factors to be considered have been outlined by Abramchuk[4] as follows:

1. Abrupt changes in direction of flow such as elbows, tees, return bends and changes in pipe size which create turbulence or changes in velocity.
2. Presence of 'dead-ends', loops, crevices, obstructions or other conditions which may produce turbulent flow causing erosion or stagnant flow which will allow debris or corrosive media to accumulate and set up corrosion cells.
3. Junctions of dissimilar metals which might promote severe galvanic (bimetallic) action.
4. Stressed areas such as those at welds, rivets, threads or areas which undergo cyclic temperature or pressure changes.

Selection of inspection points should be based on a thorough knowledge of process conditions, materials of construction, geometry of the system, external factors and historical records. Some of these factors may not be present, for example, when new plant is commissioned.

Variables Affecting Corrosion Monitoring

The shortcomings of plant testing are considerable. Variables which affect the rate and type of corrosion are chemical composition, temperature, pressure, trace compounds or contaminants, velocity, presence of insoluble metal compounds, presence of insoluble materials (either as abrasives or deposits), crevices, stress (both magnitude and type are important), interface effects, phase changes (vapourising or condensing), chemical composition of the metal, metallurgical condition of the metal and galvanic effects[5]. No single corrosion test can include all of these variables and the corrosion data obtained should only be considered in this context. Mechanical phenomena, localised corrosion, stress-corrosion cracking and thermal effects are some of the factors that cannot be assessed with accuracy.

Dillon, *et al.*[5] refer to the current recommended practice for plant corrosion testing designated by Method G 4-68, which evolved from ASTM and NACE technical committees.

Methods Available

Visual inspection   This is regarded by Abramchuk[4] as so fundamental that it should be a logical prelude to most other inspection methods. A properly executed visual inspection can accomplish the following:

1. Aids the analysing of causes of failure.
2. Indicates the need for further exploration.
3. Helps to define the search area if further exploration is warranted.
4. Aids in suggesting techniques for further exploration.

5. Aids in determining the measures needed to prevent the recurrence of damage to equipment, or else minimises it.
6. Reduces the possibility of installing faulty fabrications by ensuring that the right materials and procedures are used and that workmanship is of the proper quality.

Obvious signs of possible damage can include rust staining, bulging, cracked or distorted insulation, and hot spots that are indicative of possible corrosion damage[6]. Internal inspection can only be carried out when plant is shut down but can determine the condition of many components.

The equipment utilised can range from simple types, such as callipers, pit gauges, scrapers, mirrors and magnifying glasses, to complex equipment such as miniaturised TV cameras and fibre optics.

'Tell-tale' holes   This method has been often used by the oil industry and comprises drilling small holes usually 6·5 mm into vessels or pipe walls to a depth which equals the design pressure thickness[7]. The depth of wall remaining is equal to the corrosion allowance, and when this thickness has been consumed a small leak develops. If on-stream repairs cannot be carried out, shut-down of plant will be required for repair and further non-destructive testing.

The above method has long been in favour by the oil industry but even a 'controlled' leak of highly inflammable liquid can only be regarded as hazardous.

Coupons   These can be of different geometry but are usually cylindrical rods or strips and mounted in suitable racks. The coupons are electrically insulated from each other and the rack material with Teflon or other suitable insulating material. There are several types of rack described by Dillon, *et al.*[5] One kind of rack (referred to as the *slip-in*) can be fitted or removed from on-stream plant by means of a gate valve and packing-gland fitting (*see* Fig. 20.27).

Coupons can also be inserted and retracted using Cosasco* access fittings. A high-pressure balanced type of retriever is used and it can handle other types of sensing devices such as resistance probes, polarisation probes and hydrogen probes.

Comprehensive guidance to the preparation of coupons is given by Champion[8] and ASTM Method G4-68. Coupons can be heat treated to represent plant material and it is important to avoid cold work which can be caused by identification stamping. A suitable heat treatment is often necessary. Post-exposure techniques for cleaning of coupons using electrolytic and chemical treatments are given in ASTM Method G3.

A disadvantage of coupon techniques is that the response to severe corroding conditions that may occur for short periods of time is not detected but only averaged over the period of exposure.

Coupons can be sectioned for conventional metallographic examination and can be withdrawn at intervals of time if a sufficient number are exposed at the start of exposure.

Electrical resistance probes   Resistance probes such as the Corrosometer†

---

*Cosasco, Division of Grant Oil Tool Co., London.
†Magnachem (UK) Limited, Oxford.

system are a well-proven method for the detection and measurement of corrosion[9-24]. A wire element (made of the metal of interest) is mounted in a suitable casing and exposed to the corrosive medium, which can be either liquid or gas. The wire decreases in cross-sectional area due to corrosion and as most corrosion products have greater electrical resistance than the metals from which they were formed, an accurate measurement of the increase in resistance can be equated to metal loss (Fig. 20.47).

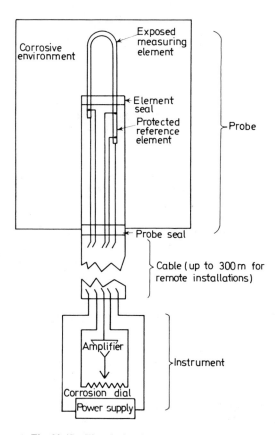

Fig. 20.47   Electrical resistance probe–circuit

A variety of elements in different geometric forms, e.g. wire, tube and strip, which correspond to the commercial metals and alloys used in the process plant, are available. Also a variety of casings or housings are available depending on temperature and pressure requirements. The most popular form of fitting is the 'retractable' probe which can be inserted or retracted into a plant via a standard gate valve (Fig. 20.48). The retractable probe is fitted with a packing gland with a maximum rating of 68 atm. Probes are available which can be screwed into plant (Fig. 20.49) or fitted to flanges. These probes are available up to 204 atm but with the limitation that fitting can take place or elements can be replaced only when the plant is off-stream; therefore retractable probes should be fitted whenever possible.

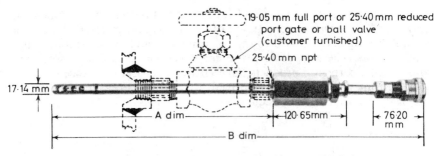

Fig. 20.48   Electrical resistance probe with retractable fitting

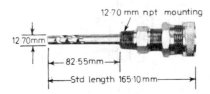

Fig. 20.49   Electrical resistance probe with screw-in fitting

Retraction of probes at pressures above 68 atm can be achieved using the Cosasco access system. The maximum pressure rating using this system is 680 atm.

Consideration should be given to the compatibility with the corrosive environment of the probe casing materials as well as seals used in construction.

Temperature compensation for resistance changes is provided by reference elements mounted in the stem of the probe and protected from the corrosive environment.

A complete range of instrumentation is available including portable units (carried from probe to probe) to rack-mounted units installed in control rooms giving continuous recording of corrosion rate. The latter units, when combined with a multiplexer, can record data from as many as 12 probes.

The data obtained from electrical resistance probes, and that provided by test coupons, are similar in giving an integrated or average rate, but the former has the advantage that the data are obtained whilst the plant is operating. The time periods can be decided by the frequency of measurement, and periods of changing corrosion rate can be detected and measured whilst the plant is on-stream. It is usual to plot the probe reading against time, the slope of the line giving the corrosion rate at any particular time.

Pitting of the wire element increases the slope of the graph as the instrumentation cannot discriminate between general and localised attack. Pitting should be suspected if an increase in slope does occur and no changes have occurred in plant process conditions that would increase the general corrosion rate. Inspection of the element (another advantage in using retractable probes) can usually confirm whether pitting is occurring or not.

A recent probe development utilises a 'flush' strip element (Fig. 20.50) which can be used in pipelines. This element design more closely represents

Fig. 20.50   Electrical resistance probe 'flush' mounting (courtesy Magna Corporation)

the inner wall of a pipe and is useful in pipeline applications as withdrawal of the probe is not necessary when cleaning of the pipeline is carried out. Due to manufacturing limitations, this type of probe is only available with steel elements at the time of writing.

Probes fitted with resistance elements can also be fitted with coupon holders. Another type of housing can be used either for resistance monitoring or for measurement with a conventional coupon.

Exploratory work is in progress with a view to computer storage of corrosion data provided by resistance probes. Large refineries or chemical plants sometimes have spare computer capacity, and this could lead to much improved data handling and possible control of plant variables to mitigate against corrosion.

Successful applications using electrical resistance monitoring have been described in the atomic energy industries[25,26] and in multi-stage flash-desalting-plant environments[27].

For a description of resistance probes presented in the general context of oil-refinery inspection, reference should be made to papers by Wilkinson[28] and Hildebrand[29]. A description of on-line monitoring in plant is given by Bovankovich[30].

Polarisation resistance   Unlike resistance probes, polarisation-resistance probes can only be used in liquids; and the liquids must have certain conductive requirements. An increasing number of publications describing polarisation resistance techniques for corrosion measurement in plant have been published in recent years[5, 14, 22, 30-40] (*see also* pages **20**.33–39).

The advantage of this technique is that measurements are instantaneous and are therefore eminently suitable for providing signals for both alarm and control. In addition to the general corrosion rate, an indication of pitting or localised corrosion rate can be obtained, referred to as the 'pitting index'. This gives a numerical value but cannot be quantified for obvious reasons, and is therefore assessed by comparison with the general corrosion rate, which is evaluated from the *mean* of the two readings obtained with each electrode.

The pitting index is obtained by a *difference* of reading between two identical electrodes. The general corrosion rate, the pitting index, or a combination of both these parameters, can be utilised for alarm and control indications. For example, the addition of a biocide such as chlorine to a recirculating cooling-water system may only slightly increase the general corrosion rate but will indicate a large 'kick' in the pitting index signal which would be more suitable for automatic control of inhibitor than the signal indicating the rate of general attack.

Probes are available in both two- and three-electrode forms (three-electrode probes can extend measurement in higher resistivity solutions) but for many industrial applications the two-electrode is adequate (Fig. 20.51).

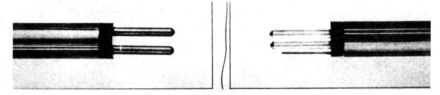

Fig. 20.51   Two- and three-electrode polarisation probes. In the two-electrode probe both electrodes are composed of the metal of interest; in the three-electrode probe the third electrode (the reference electrode) is Type *316* stainless steel

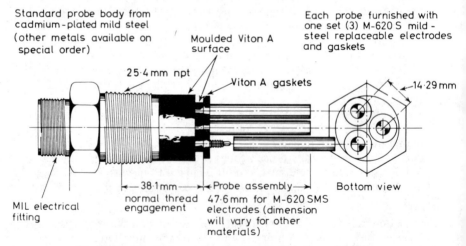

Fig. 20.52   Polarisation probe—three-electrode in 'screw-in' fitting

A variety of probe designs is available on the market which can be retracted or have screw-in fittings (Fig. 20.52). The range of instrumentation* includes portable meters, control equipment (Fig. 20.53) and control-room installations recording many channels. Units are available which combine measurement of corrosion with measurement of pH and 'total dissolved solids'

*Supplied by Magnachem (UK) Ltd., Petrolite., London and Waverley Instruments, Weymouth.

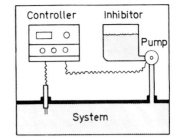

Fig. 20.53   Automatic addition of inhibitor

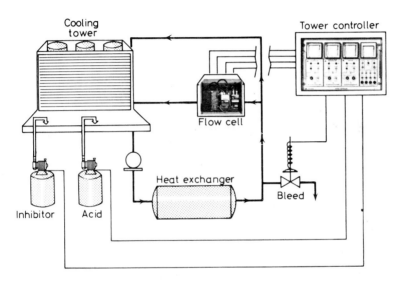

Typical installation

Fig. 20.54   Automatic control for recirculating cooling-water system

content in recirculating cooling-water circuits. Such instruments can automatically control the addition of corrosion inhibitor, acid level or blow down (Fig. 20.54)[41]. In addition to water systems, other industrial applications are increasing, such as the large multi-stage desalting plants, detection of water leakage into anhydrous organic reactions, inhibition control in monoethanolamine systems, petroleum waterfloods and oil refineries[38].

Ultrasonics   Use of ultrasonics involves the transmission of very high frequency sound waves through the metal whose thickness is required[6, 28, 42-45]. An advantage of this technique is that access is only required to one side of the vessel or pipe wall. Two methods are commonly used, referred to as 'pulse-echo' and 'resonance' methods (Figs. 20.55 and 20.56). With 'pulse-echo' a short burst of energy is transmitted (via a transducer probe), into the metal. The time taken for the sound to traverse the thickness of metal and return to the probe is usually displayed on an oscilloscope although modern instruments now provide digital read-out of thickness.

The 'resonant' method varies the frequency of ultra-sound to a value equal to twice the thickness of the test material. A condition of resonance

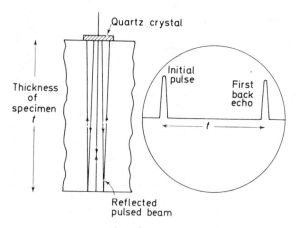

Fig. 20.55   Pulse-echo method (after Wilkinson[28])

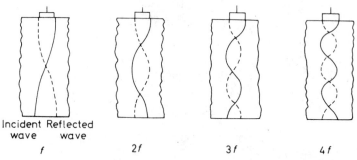

Fig. 20.56   Representation of standing wave patterns due to ultrasonic stress waves; thickness can be determined from $t = v/2f$, where t is the thickness, f is the fundamental frequency and v is the velocity of sound in the medium (after Wilkinson[28])

then occurs resulting in standing waves in the metal causing a resonance at a greater amplitude which will be recorded by the transmitting probe. Metal thickness can be determined from its sound velocity properties after a number of resonant frequencies are determined.

A problem associated with plant inspection is the coupling of the probe to the metal under examination. The surfaces of steel are covered in rust, which causes problems in coupling the probe to the metal and results in spurious and inaccurate results. Often operators move the probe to an area where adequate coupling is obtained, circumventing the real objective of the measurement[46]. It is often impossible to obtain any coupling in severely corroded steel surfaces.

Temperature can destroy the piezoelectric properties of the probe, although techniques for cooling probes and development of temperature-resistant piezoelectric materials are extending the temperature range. One manufacturer claims operating temperatures up to 500°C using a special probe and coupling paste.

Pitting can cause problems as it will be impossible to obtain resonance in some circumstances. Thickness measurement on pitted surfaces will be the average of the pit depth and the wall thickness.

Surface preparation before measurement is important as sound travels about twice as fast in steel as in paint. Many years accumulation of paint can indicate a thicker steel wall than exists in reality[6].

Eddy currents   The examination of non-ferrous tubing using external coils is a well-tried and successful inspection technique, due mainly to the pioneering work of Forster in Germany. The adoption of this method for *in-situ* inspection of condenser tubes by mounting eddy-current coils in probes (or bobbins) which can be inserted in condenser tubes, was a logical development of the technique. Suitable apparatus was developed in the immediate post-war period more or less independently by several oil and chemical companies. The principle of operation has been described by a number of authors and references and bibliography have already been reported[47].

In-situ inspection is obviously concerned with corrosion in its many forms such as pitting or more general attack and thinning. The double coil probe, however, does not provide any output if general uniform thinning should occur in a length of tube exceeding the probe length. This can be overcome by winding a different coil or coil factor and the centre pole-piece may be moved to obtain a magnetic impedance between coils of equal numbers of turns.

Probes are available for a wide range of alloys and tube sizes, ranging from 6·3 mm diameter to around 50 mm diameter, which are suitable for tubes from 11·1 mm o.d. × 1·8 mm wall to 57 mm o.d. × 1·6 mm wall. The probe is propelled through the tube by either winching or compressed air. Large and small holes, pits, larger areas of localised attack and cracks, are easily detected. In addition, areas of selective attack such as intergranular corrosion or dezincification can be identified.

Results of inspection are normally shown on a pen recorder. The pen is balanced in the centre of the paper tape and any change in tube dimension, such as already outlined, will cause a deflection of the pen. An example of a

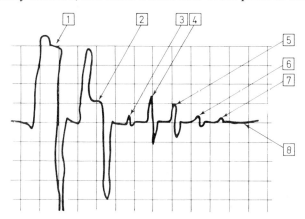

Fig. 20.57   Artificial defects in stainless steel tube

KEY
(1) 50% wall reduction. (2) 10% wall reduction. (3) Circumferential slot 0·254 mm wide × 12·7 mm 1 × 50% gauge depth. (4) Longitudinal slot 0·254 mm wide × 12·7 mm 1 × 50% gauge depth. (5) 1·59 mm diameter through hole. (6) 1·59 mm diameter hole to 50% gauge depth. (7) 0·78 mm diameter through hole. (8) 0·78 mm diameter hole to 50% gauge depth

tape recording artificial defects in a standard 25·4 mm o.d. × 2·6 mm wall, Type *316L* stainless steel tube is shown in Fig. 20.57. A different method of presentation utilises Lissajou patterns which are displayed on an oscilloscope. This display is claimed to lead to easier interpretation.

Many materials can be tested, and they include copper, cupro-nickel alloys, brasses, stainless steels, zirconium, zircaloy, tungsten, molybdenum, lead, beryllium and titanium. A strip-chart record of each tube surveyed is obtained (Fig. 20.57) from which the defects in question can be determined. It is usual to check the calibration of the instrument with a tube having calibrated defects. The baffles supporting the condenser tubes may mask any corrosion occurring close to the baffles.

Infra-red   Commercial instrumentation for recording infra-red radiation has been available for some years[48] and has been explored by the CEGB*[49] for assessing corrosion in boiler tubes at power-station shut-down. An external heat source is played onto the outside of boiler tubes at the same time as cold water is circulated inside the tubes. Hot spots due to poor heat conductivity caused by excessive corrosion product indicated areas of high corrosion.

Radiography   With its background of success in detecting weld defects and cracks, radiography can successfully reveal corrosion in plant[6,28,50-52]. Various techniques are available such as conventional radiography, back-scattered radiography and transmitted radiation.

Advantages are that lagging need not be removed, and permanent records obtained permit comparison with subsequent exposures and several boiler tubes, for example, can be examined using one radiation source; also, spot checks can be made to units on-stream.

Disadvantages are the radiation hazards, the time required to complete exposure, and the fact that access is required to both sides of the item to be examined.

Hydrogen probe   Hydrogen probes for detecting hydrogen entry into steels

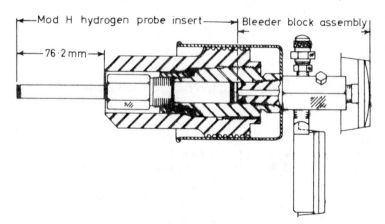

Fig. 20.58   Cosasco hydrogen probe assembly

*Central Electricity Generating Board.

are commercially available (Fig. 20.58) and the Lawrence Hydrogen Detection Gauge* (ASTM STP543, 1974) consists of a steel tube through which hydrogen diffuses into an evacuated valve; the current gives a measure of the rate of hydrogen diffusion[5-7]. This technique is essentially a laboratory tool that is used to evaluate hydrogen entry during cadmium plating; the higher the current the more porous the deposit. Subsequently the cadmium plate probe is baked at 200°C and the decay in current provides a measure of the rate of removal of hydrogen; the greater the rate of decay the more porous the deposit and the more readily hydrogen is removed during the baking operation.

Retractable hydrogen probe The Cosasco retractable hydrogen probe is a method of determining relative hydrogen atom release by mechanical means as opposed to the electronic method used in the Lawrence gauge. It consists of a slender steel tube, with an intentionally fabricated internal (annular) lamination. The hydrogen atoms, which are liberated at the outer surface of the tube, migrate through the shell and reach the annular space, where they unite to form gaseous hydrogen molecules ($H + H \rightarrow H_2$). Continued accumulation of hydrogen molecules results in increased pressure, which registers on the pressure gauge. The Cosasco hydrogen probe, being sensitive to conditions favouring production of nascent hydrogen, is of value when used to monitor the effectiveness of any procedure that is employed to minimise the formation of atomic hydrogen with the possibility of hydrogen blistering and cracking of steels.

Other methods Analysis of process streams can indicate corrosion trends, e.g. analysis of iron in refinery streams, but it can only be related to the total surface area of steel present, whereas the corrosion could well be localised[7].

Test heat-exchangers can be fitted to side-stream units in plants for corrosion measurement and can be fitted or removed whilst plant is operating.

Special holders can be used for stressing coupons in a 'U'-bend or in a controlled three-point loading system[5].

A number of special instruments have been developed for internal corrosion inspection of pipelines[53-57]. One example is self-contained and travels in a pipeline similar to a conventional 'cleaning pig'. The tool contains its own power supply, excitor, electronics and recording devices for self-contained operation of 20 h duration. A fluctuating magnetic field is produced and variations in a relatively remote portion of the field are detected. Careful consideration must be given to line pressure, and skilled interpretation of results is necessary. The instrument will not detect defective longitudinal seams and cannot detect the type of preferential seam corrosion which can occur at the mill weld. It is claimed that pitting attack is easily detected.

A technique for measuring the presence or absence of an inhibitor film (amine type) in a refinery application has recently been reported[58]. Steel electrodes immersed in most produced brines exhibit an apparent resistance of $10\,\Omega$ or less without inhibitor. When a good inhibitive film is present the apparent resistance of the same steel electrodes is over $1\,k\Omega$.

C. F. BRITTON

*Lawrence Electronics Co., Seattle, USA.

Bibliographical Note

The reader's attention is drawn to a meeting entitled *Corrosion Monitoring of Process Equipment*, organised by C.A.P.A.* and the Institute of Corrosion Technology and held at Teeside, UK in April, 1972. Only abstracts of papers were published.

REFERENCES

1. Edeleanu, C., 'Corrosion Monitoring for Chemical Plant', *Corrosion Technology*, **2**, 7, July (1955)
2. Hoar, T. P., *Report of the Committee on Corrosion and Protection*, London, HMSO (1971)
3. Elliott, P., *Corrosion Survey Report*, Suppl. to the *Chemical Engineer*, Sept. (1973)
4. Abramchuk, J., 'Basic Inspection Methods', *Materials Protection*, **1** No. 3, 60, March (1962)
5. Dillon, C. P., Krisher, A. S. and Wissenberg, H., 'Plant Corrosion Tests', *Handbook on Corrosion Testing and Evaluation*, Wiley (1971)
6. Lautzenheiser, C. E., 'Nondestructive Test Methods for Corrosion Detection', *Materials Protection*, **2** No. 8, 72, Aug. (1963)
7. Couper, A. S., 'Methods in Oil Refining Industries', *Handbook on Corrosion Testing and Evaluation*, Wiley (1971)
8. Champion, F. A., *Corrosion Testing Procedures*, Wiley, New York (1952)
9. Trascinski, E. S., Couper, A. S. and Dravnieks, A., 'Continuous Corrosion Monitoring', *Mechanical Engineering*, **47**, July (1961)
10. Wingate, R. H., 'Monitoring Corrosion', *Materials Protection*, **1** No. 7, 64, July (1962)
11. Krisher, A. S., 'Measuring Corrosion', *Materials Protection*, **4** No. 10, 8, Oct. (1965)
12. Schmitt, R. J. and Christian, R. R., 'Monitoring Corrosion', *Industrial and Engineering Chemistry*, **52**, 11, 57A, Nov. (1960)
13. Winegartner, E. C., 'Recording Electrical Resistance Corrosion Meters', *Corrosion*, **16**, 265t, June (1960)
14. Alvis, R. L., 'Comparison of Water Flood Corrosion Detection and Monitoring Devices', *Materials Protection*, **8**, 39, Feb. (1969)
15. Finley, H. F., 'Statistical Correlation of Corrosion Rates Obtained with Electrical Resistance Corrosion Probes', *Corrosion*, **18**, 70t (1962)
16. Fochtman, E. G., Langdon, W. M. and Howard, D. R., 'Continuous Corrosion Measurements', *Chemical Engineering*, 140, January 21 (1963)
17. Freedman, A. J. and Canapary, R. C., 'Corrosion Monitoring by the Electrical Resistance Method', *Oil in Canada*, August 24 (1959)
18. Lanaman, J. H. and Azar, L., 'A New Instrument for the Measurement of Corrosion by Electrical Resistance Techniques', paper presented at N.A.C.E., *Corrosion*, **73**, Anaheim, California, March (1973)
19. Krisher, A. S., 'Measuring Corrosion', *Materials Protection*, **4**, Oct. 8 (1965)
20. Maller, G. E. and Patrick, J. P., 'Electrical Resistance Measuring Device Plus Statistical Analysis Yield Unique Approach to Corrosion Problems', *Corrosion*, **16**, 155t, March (1960)
21. Morbe, K., 'Corrosion Tests by Electrical Resistance Measurements', *Chemische Technik*, **21**, 155, March (1969)
22. *Modern Electrical Methods for Determining Corrosion Rates*, NACE Task Group T-3D-1 on Instruments for Corrosion Rate Measurement (1972)
23. Wingate, R. H., 'Corrosion Inhibitor Tests', *Petro./Chem. Engineer*, Oct. (1964)
24. Dravnieks, A., 'Probes Prove Talented Corrosion Sleuths', *Oil Gas J.*, **58** No. 20, 135 (1960)
25. Richman, R. B. and Pollock, C. W., Report BNWL-726, Battelle Memorial Institute, Richland, Washington (1968)
26. Richman, R. B., 'Continuous On-reactor Measurement of Instantaneous Corrosion Rates', *Nuclear Applications*, **1**, 267, June (1965)
27. Saline Water Conversion Report P146, Office of Saline Water, U.S. Dept. of Interior (1966)

*Corrosion and Protection Association.

28. Wilkinson, L., Paper presented at The Institution of Chemical Engineers (London and South Eastern Branch), *Symposium on Non-destructive Testing*, Dec. 1 (1971)
29. Hildebrand, E. L., 'Materials Selection for Petroleum Refineries and Petrochemical Plants', *Materials Protection*, **11** No. 7, 19, July (1972)
30. Bovankovich, J. C., 'On-line Corrosion Monitoring', *Materials Protection*, **12** No. 6, 20, June (1973)
31. Evans, U. R., *The Corrosion and Oxidation of Metals*, 1st Sup. Volume, Arnold, 323 (1968)
32. Neufeld, P., 'Application of the Polarisation Resistance Technique to Corrosion Monitoring', *Corrosion Science*, **4**, 245 (1964)
33. Rowlands, J. C. and Bently, M. N., *Br. Corros. J.*, **2**, 92, May (1967)
34. Meany, J. J., 'Prediction of Corrosive Rates in Sea-water-cooled Condensers using Segmented Condenser Tube and Polarisation Measurements', Paper No. 62, NACE 1969 Conf., Houston (1969)
35. Paul, R. and Shirley, W. L., 'Determination of Refinery Corrosion Rates using the PAIR Technique', *Materials Protection*, **8** No. 1, 25, Jan. (1969)
36. Baker, W. C. and Bergen, C. R., 'Linear Polarisation Instruments Provide Key to Corrosion Problems in Reactor', *Materials Protection*, **8** No. 4, 39, April (1969)
37. McCallion, J., 'Measuring Instantaneous Corrosion Rates', *Chemical Processing*, Nov. (1966)
38. Britton, C. F., Cebelcor *Raport Technique* V.122, RT2095 (1973)
39. Henthorne, M., 'Chemical Engineer Refresher Course No. 4', *Chemical Engineering*, 89, Aug. (1971)
40. Hines, J. G., 'Monitoring Corrosion', *Chemical Processing*, Dec. (1969)
41. Feiteer, H. and Townsend, C. R., 'New Automatic System Optimises Cooling Tower Corrosion Control', *Materials Protection*, **8** No. 3, 19, March (1969)
42. Wilkinson, L., 'On-stream Ultrasonic Examination in the Oil Industry', *British J. of Nondestructive Testing*, **10** No. 3, 59, Sept. (1968)
43. Ostrofsky, B. and Parrish, C. B., 'Ultrasonic Inspection of On-stream Units', *Petro./Chem. Engineer*, **34**, 226, Sept. (1962)
44. Evans, D. J., 'Ultrasonic Inspection in the Oil-refinery and Related Industries', *Nondestructive Testing*, **15**, 156, May–June (1957)
45. Erdman, D., 'Ultrasonic Pulse-echo Techniques for Evaluating Thickness Bonding and Corrosion', *Nondestructive Testing*, **18**, 408, Nov.–Dec. (1960)
46. Minton, W. C., 'Wall Thickness Measurement', *Materials Protection*, **1** No. 10, 27, Jan. (1971)
47. Britton, C. F., *In-situ Eddy Current Inspection of Condenser Tubes*, Nondestructive Testing Centre, UK Atomic Energy Authority, AERE Harwell, Dec. (1968)
48. Britton, C. F., *Non-destructive Testing Using Infrared Radiation*, ibid., NDT-40, Nov. (1969)
49. McIntosh, D. R., private communication
50. 'Gamma Ray Devices for On-stream Refinery Inspections', *Materials Protection*, **1** No. 9, 18, Sept. (1962)
51. Berk, S., 'Measuring Metallic Corrosion by Radiation Backscattering and Radiation Induced X-rays', *Materials Protection*, **4** No. 11, 39, Nov. (1965)
52. McGregor, A., 'On Stream Radiography is Worth the Money', *Materials Protection*, **4** No. 10, 26, Oct. (1965)
53. Love, F. H., 'Camera Spots Pipe Corrosion from Inside', *Pipeline Engineer*, **34** No. 12, 252, Dec. (1962)
54. 'Electronic Pig Inspects Pipe Wall', *Pipe Line Industry*, **18**, 53, March (1963)
55. Sparkes, H. S., 'Instrumented Pig Finds, Measures Pipeline Pitting', *The Oil and Gas Journal*, **64**, 130, April (1966)
56. Beaver, R. C. and Mason, C. P., 'In-place Detection of Pipeline Corrosion', *Mechanical Engineering*, **90** No. 6, 25, June (1968)
57. Schmidi, T. R., 'Measuring Wall Thickness of Pipelines in Place', *Materials Protection*, **2** No. 1, 9, Jan. (1963)
58. Fincher, D. F. and Nestle, A. C., 'New Developments in Monitoring Corrosion Control', *Materials Protection*, **12** No. 7, 17, July (1973)

20.4 Inspection of Paints and Painting Operations

Improvements in process and quality control made significant contributions to the transition from iron to steel as the major ferrous construction material nearly a century and a half ago. For most of that time red lead was relied upon, and not without a remarkable degree of success, as the rust-inhibitive pigment in anti-corrosive paints. It is only in recent years that there has been a similar dramatic change from such simple paints as red lead to synthetic polymer coatings which have as complex a technology as steel manufacture itself.

Improved processes and quality control have helped to establish these new coating materials but the care necessary for successful use has only just begun to be fully appreciated. Sections 12.1 and 12.2 have shown how necessary it is to remove millscale before coating and how even scale-free surfaces may still retain seeds of further corrosion even when apparently well cleaned. Although under these conditions conventional primers may have an advantage over their modern equivalents, the latter, unlike the old paints, are capable of providing such a basis for long-life abrasion-, corrosion- and weather-resistant systems as to leave the designer little option but to use them. Unfortunately, the percentage of premature failures with these sophisticated systems was high to start with, even on apparently well-prepared surfaces. However, it gradually became established that the traditional arbitrary and rudimentary cleaning and application operations were inadequate for modern coating materials. Furthermore, only when adequate quality control is carried out is their full potential realised and the higher initial cost justified.

With the older protective coatings for steel, such as those obtained by galvanising, vitreous enamelling and blueing, the necessity for quality control of surface preparation was quickly appreciated, since deficiencies in this respect frequently became apparent even before the coated steel was put into service. This care for surface preparation was continued to other coating processes as they were developed, together with better control of application, which ultimately became automated. From motor vehicles and household equipment to rolling stock and the vast array of electrical control panels, processes for repeat painting have been continually improved.

Concomittant with advances in surface treatment and application has been the improved quality control by the producers of materials (not only of finished paints but of their raw materials) and also by the manufacturers of the end products, especially in situations where one coating error could mean enormous financial losses because of the lost production of finished goods.

Painting of Structural Steel

All the advances already mentioned have been slow to be adopted into protective coating operations for structural steelwork. Capital has not been available for special coating shops, automatic equipment, air-conditioning, etc. and often the ultimate economy of providing better facilities has not been appreciated.

Although some specialist contractors are now taking the new sophisticated coating technology seriously, it is not easy to apply the same concepts of quality control that have now become routine on coating production lines to single steel structures. For instance, in ordinary steelwork fabrication shops the service conditions for which a significant proportion of the throughput is required, simply do not call for maximum performance systems to be applied. This generally means at least two levels of quality in one works. The problem is further complicated by the multiplicity of techniques involved with a seemingly never-ending variety of coatings and by the near-uniqueness of every steel design, all of which makes smooth organisation difficult even for the specialist coating contractor. Nevertheless, many conditions of environment, use, maintenance and safety, exist where it is essential to produce long-life protective coating for structural steel. Under existing circumstances it has been suggested that only by continuous and specialist inspection of initial surface preparation and coating application, with prior laboratory tests and regular checks of materials in process, can such high performance from modern systems be predictably achieved.

Case Histories and Cost

To illustrate the above, it is relevant to quote two case histories[1] where no specialist inspection was provided for long-life protection-coating operations.

Figure 20.59 gives some indication of the disruption by corrosion of surfaces which had been specified to be blast cleaned to a high standard and coated with a zinc-rich epoxy primer at a works with multicoat application of a two-pack paint at site, to $175\,\mu$m. After exposure for 18 months to its marine environment with sea-spray, flaking millscale from beneath the paint was observed, and a survey showed the thickness varied between only 50 and $140\,\mu$m and that some parts were without primer. Although the remedial re-blast cleaning of all surfaces had to be carried out *in-situ*, and a suitable system applied, the resulting protection (Fig. 20.60) after exposure for 4 years is of the high order to be expected of such first-class treatment. The only difference between the two contracts was that full-time specialist inspection was given at every stage of the site remedial work.

Fig. 20.59   Rusting surface after marine exposure for 18 months; no inspection during the coating process

Fig. 20.60   Same section shown in Fig. 20.59 photographed 4 years after remedial work; full inspection during recoating on site

Figures 20.61 and 20.62 were taken at the time of investigation into the failure of some 1000 t of steel for what had been intended as a nearly maintenance-free system, for reasons of difficult access. Zinc metal spray and four coats of paint had been ordered but, for reasons of economy, no special measures for inspection were taken. Within months of erection areas were flaking off where millscale had not been removed (Fig. 20.61) and other areas were blistering to reveal a rusted substrate. Figure 20.62 shows one of the many areas investigated where there were no visible signs of failure until the whole system was removed; the dark surface in the cut-out area represents the rusted metal found underneath the metal spray. It transpired

Fig. 20.61   Result of millscale not being removed from steelwork before zinc spraying

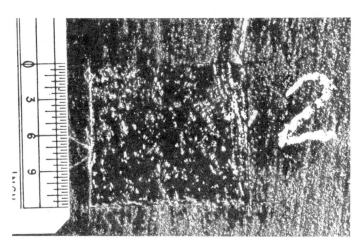

Fig. 20.62   Random check of the surface of the steelwork shown in Fig. 20.61 showing the surface completely oxidised under the coating

that here was an example of the steel members being too great in length to be housed completely in the workshops concerned, and of the blast-cleaned surface being left exposed outside overnight for metal spraying during the day; under these conditions a rust film quickly formed on the steel exposed to the atmosphere. Not only had there been no full-time inspection, but no one had visited the works before the order had been placed to see if the contractor would be able to cope with this very long steelwork. The benefit of the protective system specified had been lost for good for want of initial expenditure on inspection—a sum that rarely amounts to more than 5% of the total contract cost for protective coating. The compensation finally paid, of many times that sum in this particular case, could never properly recompense the owners as the circumstances prevented the system from ever being reinstated as originally specified.

In short, the type of protective system ordered, the environment that it has to withstand, the anticipated life, and the full implications of any necessary maintenance (in terms of delays to production, interruption of the service provided, contamination of product and/or limitations due to difficulty of access), should always be taken into account before deciding that full-time inspection is essential, that it can be safely reduced or that it can be disregarded.

Prior Inspection of Works and Site Facilities

Before a works painting or metal spraying contract for steelwork is awarded, the workshops concerned should be inspected by qualified personnel. It is essential that some check be made to ensure that the necessary facilities and equipment for carrying out all the terms of the order or specification are available. Cleanliness of the painting areas and order in the paint store form a good barometer of the seriousness with which works personnel take protective-coating operations.

Similar prior inspection should be made on site to ensure scaffolding and platforms have been erected to give the operators easy access to the work, i.e. at the correct level to, and distance from, the surface to be treated; temporary covers and equipment, including compressors, should be included in the survey.

The safety of operators, the rigging of life-lines for scaffolding, and the proper removal of dust and solvent fume should be discussed with the management since such measures against the operational hazards are not infrequently left for the protective-coating inspector to check and report upon.

Type of Inspection

The British Standard Code of Practice for protecting iron and steelwork from corrosion, CP 2008[2], is in the process of revision and will contain a new section on inspection which may be helpful. A detailed, tabulated guide of typical defects to be found during protective coating operations may be published later but would only confuse the reader if included here in summarised form. The Zinc Development Association[3] and the Association of Metal Sprayers[4] have their own very full inspection guides with similar tables where, as indicated above for all long-life systems, continuous inspection is recommended. This type of inspection is entirely unnecessary for the simpler, traditional systems. For example, for two coats of red lead and two weather-coats to be applied over wire-brushed steel where the millscale has already rusted off, occasional unheralded visits by an inspector could be sufficient to ensure acceptable results once initial standards of surface preparation and paint application have been established.

For the many systems which lie between these two extremes, some will require the maximum level of surface preparation and demand continuous inspection of this stage, while for others this type of inspection will be reserved for the onerous tasks involved in measuring coating thickness or in holiday detection. A thoughtful approach is required as to what type or types of inspection are most appropriate, taking each stage of each job on its own merits.

Quality of Inspection

This will depend partly on the experience and personal integrity of the inspector and partly on the organisation to which he belongs. Whatever the type of inspection ordered, the inspector himself should always be experienced in the processes concerned. Although he must have tact he should be authorised to interrupt the schedule of work if necessary, but this should only occur in situations of non-co-operation or incompetence on the part of the operators or through misunderstandings, which are largely avoidable if preliminary discussions are held (*see* below).

The organisation behind the inspector must be sufficiently competent in coating technology to give him the backing he requires, to answer his queries by telephone if necessary to prevent delays (although anticipation is the hallmark of the competent organisation) and to provide qualified technologists to investigate and report where necessary. Some of the larger owner organisations, e.g. government departments, public utilities and major contractors do have such staff but, where a long-life protective-coating contract is awarded, it is becoming more and more frequent for them to employ independent inspection firms who specialise in this field and can even provide teams of inspectors where necessary.

Independent Inspection Organisations

On its appointment to provide inspection for a coating contract, the organisation concerned might be expected to include in their service the following:

1. Constructive criticism of the coating specification, with any necessary warning as to where it may be difficult to implement it fully.
2. Initiation of the preliminary discussions for examining the contractor's proposed methods and suggesting any improvements in equipment layout, handling or work flow.
3. Close liaison with the coating manufacturer regarding his materials and their use and where necessary ensuring that contractor's staff have been fully informed of special application requirements.
4. An introductory report or schedule showing precisely how and at what stages inspection will be carried out, followed by regular progress reports and records of the performance of the contract; they should be as detailed and as comprehensive as was agreed beforehand.

Inspection schedule   It is not unusual for those who employ such firms to ask for an inspection schedule as suggested in (4) above, prior to making an appointment. Some large organisations lay down in general terms the numbers of inspectors and how they require the work to be inspected. However, the inspection schedule is not only the key to what may be expected from individual firms, it is also useful in summarising his duties to the inspector and in informing the contractor as to how his work will be checked.

Preliminary Discussions

Preliminary discussions of all those involved in each protective-coating contract are essential if certain difficulties that can arise in these operations are to be prevented. Although the ultimate client may not wish to be present, the engineer concerned and/or his own or appointed inspectors, the main contractor, steel fabricator and coating contractor, should meet to discuss and agree action on such subjects as the following:

1. Ambiguities, inaccuracies or omissions in the coating specification.
2. Details of design which may impede or prevent the implementation of the specification, possible changes in design or alternative treatment for special areas.
3. The methods of inspection to be used; the instruments that are to be used and pass figures (9) for certain instruments have to be agreed by trial beforehand.
4. The programme for all operations, bearing in mind pot-life capability, recoating intervals, approximate ageing before transport, etc.
5. Special measures to reduce coating damage during handling, transport or erection and measures for maintenance of incomplete systems stored in the open.
6. Arrangements and facilities for testing materials.
7. Availability of manufacturers' data sheets.
8. Any special skills required for operating equipment.
9. The procedure and responsibility for 'stopping the job'.
10. The chain of authority with client and contractor, to be followed in cases of difficulty.

The minutes of such a meeting should be detailed enough to be agreed by all parties and serve as a rider to the contract.

Inspectors' Duties

Apart from his liaison, progressing and reporting duties, the inspector of protective-coating operations has three quite distinct areas of operation:

1. The materials to be used (which includes the substrate).
2. The surface preparation of the substrate.
3. The application of the coating or coatings.

Given below are the points to watch, the measurements to be taken and the standards for comparison, but it must be emphasised that they are for long-life systems only.

Materials

The supplies of the specified paint, metal wire or powder, plastic coating powder, blasting abrasive or other materials involved in the coating process should be checked to make sure that they are adequately stored and available in sufficient quantity to permit operations to begin. Sometimes it is part of the inspection contract to sample and check to a predetermined schedule in the laboratory that supplies are up to specification. It is always the inspector's duty to ensure to the best of his ability that what he has checked in the stores is being used on his job (*see* Sampling and Testing, later), and that the information on the container labels agrees with the specification requirement. The basic substrate offered for treatment in the case of steel will normally be subjected to steelwork inspection, but the painting inspector has the duty to point out to the contractor for rectification any omissions that he may have noticed, e.g. that it should be free from pits, that it should be machined or that edges should be radiused, that certain areas should be masked from any treatment, etc.

Additionally the inspector should ensure that the specified measures for removing oil and grease from the surface with water-rinsable solvent or emulsion cleaners are implemented, and that all weld-spatter and other asperities have been removed as a preliminary to the surface preparation proper.

Surface Preparation

Certain British Standards, e.g. for galvanising[5] or metal spraying[6], contain detailed sections on surface preparation. Others are concerned with acid pickling but the chief method of preparing iron and steel for a long-life protective system is by blast cleaning (*see* Section 13.4).

The inspector's working standard for the surface finish of blast-cleaned steel usually relates to the British one[7], but great use is made both in specification and in practice of the colour pictorial standard of the Swedish IVA Corrosion Committee[8]. These two specifications are to some extent complementary but neither defines the chemical cleanliness (necessary for some systems), which does not relate to the visual appearance. In the special circumstances where absolute freedom from soluble contaminants such as ferrous sulphate is necessary, the specification should include reference to a test for removal of such residual salts, e.g. by checking with filter papers soaked in 5% potassium fericyanide solution, and drying. The test area should be wetted with a fine spray of distilled water and the indicator paper held against it with a rubber pad. The development of blue spots on the paper indicates the presence of iron salts on the surface.

In most circumstances, a lower standard of blast cleaning is adequate and the Surclean* instrument, developed under the auspices of C.I.R.I.A.[9], may be used. As the instrument does not provide an absolute measure, there must be prior agreement with the contractor. At the commencement of the contract, trial readings are taken with the instrument of the level of surface reflectance of typical pieces of the contract steelwork cleaned to the specified standard. Pass readings may then be agreed at the preliminary discussions (*see* earlier) which the inspector will then use as his criterion of cleanliness. This method has also been most successful in showing where dust has

*Elcometer Instruments Ltd., Edge Lane, Droylesden, Manchester, M35 68U.

subsequently settled into a blast-cleaned surface before it could be coated, or for determining the onset of post-blasting rusting.

Methods are available for checking on the profile of blast cleaning which is of particular importance where a prefabrication or holding primer is used or where the total protective system has a slender margin of thickness over the profile height. Peaks far in excess of the normal profile (rogue peaks) may be produced by insufficiently controlled sweep of the blasting nozzle, by using the wrong approach angle altogether or by using an abrasive with excessive particle size, and the inspector has to be aware of this. One instrument, the Roughtector*, has been developed by the B.S.C. Corporate Engineering Laboratory at Battersea to help the inspector measure the nominal profile, and, with further experience, may prove useful in locating and measuring rogue peaks. Measurements with the dial-type of depth gauge, however often repeated, can prove somewhat misleading. Where profile is important the method of measurement should be stated in the specification and agreed at the preliminary discussions when the ways and means of surface preparation are examined. A retained sample of the abrasive used in the process during the early trials for setting acceptance levels in the works can be used as a standard against which to make *ad hoc* checks later, or when prompted by high-profile measurements.

It frequently happens that after surface preparation 'shelling' or other imperfections in the steel surface are revealed. It is preferable to deal with these by grinding or filling as necessary before any other coating is applied to the bare steel.

Application of Coating

Galvanising and metal spraying are well covered by inspection guides (References 3 and 4), but paint demands some special attention, particularly in its modern application.

Paints have their own individual data sheets, prepared by the manufacturer as the result of extensive testing including laboratory tests, field trials and experiences in use. These instructions should be followed closely in respect of type of application equipment, operating air pressure, nozzle size, pot life, curing time at various temperatures, recoating interval, etc. The inspector should have the data sheets available at all times and however familiar he or the contractor may be with the material, he should constantly refer to them, particularly when any abnormal results are obtained. Complete inability to obtain the required result in film thickness, non-sag, gloss, aeration or spray pattern, when following data sheets will require the attention of the paint manufacturer and, as already indicated, the inspector should already be aware of the right person to contact in such an event.

Holiday detection or 'spark testing' may be specified in the contract with details of the appropriate instrument and testing procedure. In certain cases, where the conditions of service include exposure to chemicals or other hostile environments, the test is essential to secure the desired performance, although normally unnecessary for weather exposure.

*Elcometer Instruments Ltd., Edge Lane, Droylesden, Manchester, M35 68U.

Thickness Measurements

In the case of a paint coating the film thickness may be determined when the film is dry or when it is still wet; both methods are described and discussed in detail in CP 2008[2], but some discussion is appropriate here.

However the paint film thickness is specified in the contract, it is an obvious advantage to all parties for the painter himself to have a means of checking whether or not he is applying sufficient paint, and this can be achieved for each coat by means of a wet-film-thickness gauge. The relationship between wet- and dry-film thickness can be provided by the paint manufacturer, and the wise contractor will train his applicators to use the simple wet-thickness 'comb gauge'. Besides reducing the risk that extra coatings may have to be applied in the end to bring the thickness up to specification, the gauge also provides the painter with job satisfaction. Wet gauges must be used immediately the paint is applied if they are to give reliable results, and the method is of course a valuable means of inspection that permits correction without additional operations. Many non-destructive instruments are available for measurement of the dry-film thickness and in general these are to be preferred to the destructive methods, which may however be necessary for an official investigation or for agreed certifying tests. The main group of non-destructive instruments employ a magnetic principle, and although these are satisfactory certain types employ a strong magnetic field and give low readings when used for softer films owing to indentation of the film by the probes. All non-destructive instruments should be calibrated frequently and regularly by the inspector, preferably using a procedure agreed upon by the parties concerned. Film thickness is so often a source of argument that all relevant details should be decided upon during the preliminary discussions, e.g. criteria of film thickness, instruments to be used and how they are to be calibrated. Time spent on this will be recouped many times over in delays avoided during the running of the contract.

Wet-thickness measurement, then, is basically the method in process for the contractor, but assurance of adequate dry thickness should be the final job of the inspector on any piece of work. He should make his measurements at numerous points, evenly distributed not more than 300 mm apart, and the results should be so recorded and arranged that they may be examined statistically if necessary.

Working Conditions

Good conditions of work, e.g. adequate ventilation, control of temperature, shelter from the elements, etc. which should be provided for the operators in any case, are, in the majority of coating operations, also vital to the success of long-life systems. Appropriate conditions are now specified for most contracts and it is for the inspector to see that they are not exceeded. He should be provided with means to measure steel temperature, wet-and-dry bulb hydrometer and any air-sampling instruments that may be necessary for work in special areas or confined spaces. In particular, he will be watching for moisture, dust or debris (falling or wind-borne) which may contaminate painting which is in process, or before it is sufficiently dry.

Sampling and Testing

There are two ways of checking materials for a protective coating contract. Firstly, by sampling at the point of manufacture and/or after delivery at the coating works, and secondly by taking samples of the materials in use.

Testing of Supplies

For many government orders and for most major projects, or where the performance of the coating is vital to some process or storage facility, supplies of paint are often sampled at the manufacturer's works and checked in a laboratory to ensure that they conform to the standard specified. This may be a written standard for composition or performance or a proprietary paint which has been approved by long experience or special trials. Delegated inspection by one of the manufacturer's qualified staff is often used for batch testing of government orders, but specialist inspectors are sometimes engaged to sample each batch of paint before delivery and to send them to an independent laboratory for testing.

How comprehensive such tests will be is decided by the authority or engineer concerned and, in some instances, complete analysis is called for. However, much information may readily be gained in the laboratory at far less cost by such tests as viscosity, specific gravity, drying time, fineness of grind, adhesion to steel and percentages of the volatile constituent and non-resin solids. If these tests have been carried out on a sample of the paint as originally approved, then the results can comprise a suitable standard for any batch testing. Approved batch numbers or passing certificates may then be forwarded to the inspector for checking against deliveries.

Testing of Materials in Use

As already stated, it must be one of the inspector's duties to see that the materials delivered conform to the specification, or to any subsequent modification which may have been agreed. It is usually also his job to do whatever testing of the materials used in the process is thought to be necessary. The adulteration of paint has often been responsible for early failure and one good but not infallible guide is the weight per litre of samples taken from the painter's kettle or spray pot. If the inspector can be supplied with a 100 ml specific gravity cup and weighing facilities, he can readily make what regular or spasmodic checks he thinks necessary to reassure himself on this point.

Modern paints are frequently non-Newtonian liquids so that viscosity checks have to be made with appropriate instruments, requiring more know-how, capital outlay and bench space than the simple flow cup. Only where the application viscosity has to be held within narrow limits for correct performance is this test as valuable as the one for specific gravity. Where the viscosity appears too thin or too thick to the inspector, his first check should indeed be the weight per litre.

If the inspector has been present as he should be at the mixing of two-component paints, then he may be less worried about viscosity and specific gravity, but wherever failure to cure may have a disastrous effect on programme, a painted steel panel, heated to accelerate cure, may be introduced as a regular test to help him cover this point.

The abrasive in use should not be ignored by the inspector. He should have a standard reference sample and will soon recognise when the grade is wrong or the grit is dirty. Where surface profile is important, he should be supplied with the sieves appropriate to the grade of approved abrasive.

Whatever samples are taken from materials in use, whether tested or not, they should be preserved for a reasonable time, at least until the end of any maintenance period under the contract.

Conclusion

It can be seen from what has been said that the technology of surface preparation and coating application has become rather complex. Attention to details that could safely be ignored in the past is now part of the normal procedure for long-life maintenance-free coatings. With a few more years' experience of such systems, with better training and some qualification of those involved, with improved automation and instrumentation, there can be some confidence that the time will come when intensive independent specialist inspection will no longer be the prime requirement that it appears to be at present.

K. J. DAY
F. G. DUNKLEY

REFERENCES

1. Day, K. J. and Deacon, D. H., *Proc. CIRIA Symposium*, Paper 3, I.C.E., London, Nov. 26 (1970)
2. *Code of Practice for the Protection of Iron and Steel Structures from Corrosion*, CP 2008:1966, British Standards Institution, London (1966)
3. *Galvanising Guide* and *Inspection of Zinc Sprayed Coatings*, Zinc Development Association, London (1971)
4. *Inspection of Sprayed Aluminium Coatings*, The Association of Metal Sprayers, Birmingham (undated)
5. *Hot Dip Galvanised Coatings on Iron and Steel Articles*, BS 729:1971, British Standards Institution, London (1971)
6. *Sprayed Metal Coatings*, BS 2569, Parts 1 and 2:1964/65, British Standards Institution, London (1964 and 1965)
7. *Surface Finish of Blast-cleaned Steel for Paintings*, BS 4232:1967, British Standards Institution, London (1967)
8. *Pictorial Surface Preparation Standards for Painting Steel Surfaces*, Swedish Standard SIS 05-59-00-1967
9. Report No. 38, Construction Industry Research and Information Association, London, May (1972)

21 USEFUL INFORMATION

21.1   Tables   **21**:3

21.2   British and American Standards   **21**:71

21.3   Calculations Illustrating the Economics of
       Corrosion Protection   **21**:88

21.1 Tables

Table 21.1   Periodic table

21.2   Properties of some metals and alloys

21.3   Values of important constants in SI units

21.4 $(RT/F)\log x$ at various temperatures

21.5   Standard chemical potentials

21.6   Standard electrode potentials (reduction potentials) for inorganic systems at 25°C

21.7   Reference electrodes

Fig. 21.1   Practical galvanic series of metals and alloys indicating approximate potentials

Fig. 21.2   Galvanic series in hot domestic water at 71°C

Table 21.8   Galvanic series of some commercial metals and alloys in sea-water

21.9   Stoichiometric mean molal activity coefficients $(\gamma_\pm)$ for aqueous inorganic electrolytes at 25°C

21.10   Differential diffusion coefficients for dilute aqueous solutions at 25°C

21.11   Ionisation constants for water and weak electrolytes and variation with temperature

21.12   Tafel constants for hydrogen evolution from aqueous solution

21.13   Exchange current densities i_0 for the hydrogen-evolution reaction

21.14   Exchange current densities and transfer coefficients for evolution of gases at 20–25°C at different anodes

21.15   Exchange current densities i_0 for some electrode reactions

21.16   Exchange current densities for several noble metals and a platinum–rhodium alloy in the reduction of oxygen from perchloric acid solution

21.17   Exchange current densities for M^{z+}/M equilibria in different solutions

21:4

21.18 Structure, thermal data and molecular volumes of metal oxides and hydroxides, and of some double oxides

21.19 Solubility of gases in water

21.20A Solubility of air in water

21.20B Solubility of oxygen in certain electrolyte solutions

21.21 Oxygen dissolved in sea-water in equilibrium with a normal atmosphere (760 mm) of air saturated with water vapour

21.22 Coefficients of saturation of atmospheric gases in sea-water

21.23 Properties of sea-water of different salinities

Fig. 21.3 Distribution of temperature, salinity and oxygen in the Pacific Ocean

Fig. 21.4 Distribution of temperature, salinity and oxygen in the western Atlantic Ocean

Table 21.24 Resistivity of waters

21.25 Soil resistivities in UK and corrosiveness

Fig. 21.5 Distribution of dissolved constituents in UK fresh waters

Table 21.26 Compositions of natural waters arranged in increasing chloride concentrations

Fig. 21.6 Variation of hardness of water in England, Scotland and Wales

Table 21.27 Corrosion rate conversion factors

21.28 Corrosion rates of metals

21.29 Corrosion rates of iron at various current densities

21.30 Corrosion rates of copper at various current densities

21.31 Relationship between current density and $1\,g\,m^{-2}\,d^{-1}$ or $1\,mm\,y^{-1}$ for various metals

21.32 Weight and thickness of certain metals deposited by 1 A h

21.33 Composition of industrial metals and alloys

Table 21.1 Periodic Table

At. no.	Element	Symbol	At. wt.	Mass number of common isotopes	Periodic group	Valency − +	Density*† at 20°C (g cm⁻³)	Boiling* point (°C)
1	Hydrogen	H	1·0080	1, 2	1	1 1	0·07	−252·8
2	Helium	He	4·003	3, 4	0	Inert	0·147	−268·9
3	Lithium	Li	6·940	6, 7	1	1	0·534	1330
4	Beryllium	Be	9·013	9	2	2	1·848	2400, 2770
5	Boron	B	10·82	10, 11	3	3	2·34	2550
6	Carbon	C	12·011	12, 13	4	4 2, 4	3·52(D)	4830
7	Nitrogen	N	14·008	14, 15	5	3 1–5	0·808	−195·8
8	Oxygen	O	16·0000	16, 17, 18	6	2 6	1·14	−182·970
9	Fluorine	F	19·00	19	7	1 7	1·108	−188·2
10	Neon	Ne	20·183	20, 21, 22	0	Inert	1·204	−245·9
11	Sodium	Na	22·991	23	1	1	0·9712	883, 892
12	Magnesium	Mg	24·32	24, 25, 26	2	2	1·741	1107
13	Aluminium	Al	26·98	27	3	3	2·70	2400, 2450
14	Silicon	Si	28·09	28, 29, 30	4	4 4	2·33	2480, 3240
15	Phosphorus	P	30·975	31	5	3 3, 5	1·83(w)	280
16	Sulphur	S	32·066	32–34, 36	6	2 4, 6	2·07	444·600
17	Chlorine	Cl	35·457	35, 37	7	1 1, 5, 7	1·56	−34·7
18	Argon	A	39·944	36, 38, 40	0	Inert	1·40	−185·8
19	Potassium	K	39·100	39, 40n, 41	1	1	0·86	760
20	Calcium	Ca	40·08	40, 42–44, 46, 48	2	2	1·55	1440
21	Scandium	Sc	44·96	45	3	3	2·50, 3·0*	2500, 2730
22	Titanium	Ti	47·90	46–50	4	3, 4	4·507	3260
23	Vanadium	V	50·95	51	5	2, 4, 5	5·96, 6·1*	3400
24	Chromium	Cr	52·01	50, 52–54	6	2, 6	7·19	2665
25	Manganese	Mn	54·94	55	7	2, 3, 4, 6, 7	7·42	2150
26	Iron	Fe	55·85	54, 56–58	8	2, 3	7·88	2900, 3000
27	Cobalt	Co	58·94	59	8	2, 3	8·85	2900
28	Nickel	Ni	58·71	58, 60–62, 64	8	2, 3	8·85	2820, 2730
29	Copper	Cu	63·54	63, 65	1	1, 2	8·95	2595
30	Zinc	Zn	65·38	64, 66–68, 70	2	2	7·133	906
31	Gallium	Ga	69·72	69, 71	3	3	5·91, 5·99*	2237
32	Germanium	Ge	72·60	70, 72–74, 76	4	4 4	5·36	2880, 2830
33	Arsenic	As	74·92	75	5	3 3, 5	5·73	613
34	Selenium	Se	78·96	74, 76–78, 80, 82	6	2 4, 6	4·81	685
35	Bromine	Br	79·916	79, 81	7	1 1, 5, 7	3·12	58·78
36	Krypton	Kr	83·80	78, 80, 82–84, 86	0	Inert	2·16	−152
37	Rubidium	Rb	85·48	85, 87	1	1	1·53	710, 679
38	Strontium	Sr	87·63	84, 86–88	2	2	2·60	1460, 1380
39	Yttrium	Y	88·91	89	3	3	5·51, 4·6*	3030
40	Zirconium	Zr	91·22	90–92, 94, 96	4	4	6·489	4400, 3580
41	Niobium	Nb	92·91	93	5	3, 5	8·57	5100, 4927
42	Molybdenum	Mo	95·95	92, 94–98, 100	6	3–5, 6	10·2	4600, 5560
43	Technetium	Te	[99]	99a	7		11·46	
44	Ruthenium	Ru	101·1	96, 98–102, 104	8	3, 4, 6, 8	12·45	3900
45	Rhodium	Rh	102·91	103	8	3, 4	12·41	3900
46	Palladium	Pd	106·4	102, 104–106, 108, 110	8	2, 4	12·0	3200, 3980
47	Silver	Ag	107·880	107, 109	1	1, 2	10·5	2180, 2210
48	Cadmium	Cd	112·41	106, 108, 110–114, 116	2	2	8·65	765
49	Indium	In	114·82	113, 115	3	3	7·30	2000
50	Tin	Sn	118·70	112, 114–120, 122, 124	4	4 2, 4	7·31	2750, 2270
51	Antimony	Sb	121·76	121, 123	5	3 3, 5	6·62	1675, 1380

Notes. *Isotopes:* n—naturally radioactive isotopes, a—isotopes capable of being rendered artificially radioactive. *Valency:* principal valency is shown in bold type. *Magnetic properties* (at ordinary temperatures): d—diamagnetic, p—paramagnetic, f—ferromagnetic. *Crystal-structure classification:* 1—c.p. hexagonal, 2—f.c. cubic, 3—b.c. cubic, 4—diamond cubic, 5—rhombohedral, 6—orthorhombic, 7—cubic, 8—hexagonal, 9—monoclinic, 10—tetragonal.

At. no.	Melting* point (°C)	Work function (eV)	Ionisation potential (eV) I	II	Mag. props.	Crystal struct.	1s	2s	2p	3s	3p	3d	4s	4p	4d	4f	5s	5p
							$n=1$ $l=0$	2 0	2 1	3 0	3 1	3 2	4 0	4 1	4 2	4 3	5 0	5 1
1	−259·19		13·59		d	8	1											
2	−269·7		24·56	54·1	d	1	2											
3	180·54	2·4	5·4	75·7	p	3	———	1										
4	1277	3·9	9·32	18·2	d	1		2										
5	2030, 2300	4·6	8·28	25·1	d	6	He	2	1									
6	3727	4·4	11·27	24·8	d	4, 8	(2)	2	2									
7	−209·97		14·55	29·6	d	7, 8		2	3									
8	−218·8		13·62	35·2	p	5, 6, 7		2	4									
9	−219·6		17·43	34·9				2	5									
10	−248·67		21·56	40·9	d	2		2	6									
11	97·82	2·2	5·14	47·3	p	3	———			1								
12	650	3·6	7·64	15·0	p	1				2								
13	660·1, 659·7	4·0	5·97	18·8	p	2	Ne			2	1							
14	14·80	4·0	8·15	16·4	d	4	structure			2	2							
15	44·25		10·9	19·7	d	6, 7	(10)			2	3							
16	119·0	5·4	10·36	23·4	d	9, 6				2	4							
17	−101·0		12·90	23·7	d	10				2	5							
18	−189·4		15·76	27·5	d	2				2	6							
19	63·0	2·2	4·34	31·7	p	3	———————						1					
20	850, 838	3·2	6·11	11·9	p	2							2					
21	1450, 1539	3·3	6·7	12·8		1						1	2					
22	1668	4·1	6·84	13·6	p	1, 3						2	2					
23	1900	4·1	6·71	14·1	p	3						3	2					
24	1875	4·6	6·74	16·7	p	1, 3						5	1					
25	12·4	4·0	7·43	15·6	p	3, 7						5	2					
26	1539, 1536	4·4	7·83	16·5	f	2, 3	A					6	2					
27	1492, 1495	4·2	7·84	17·4	f	1, 2	structure					7	2					
28	1453	5·1	7·63	18·2	f	1, 2	(18)					8	2					
29	1083	4·4	7·72	20·2	d	2						10	1					
30	419·5	4·2	9·39	18·0	d	1						10	2					
31	29·8	4·1	5·97	20·5	d	6						10	2	1				
32	958, 937	4·8	8·13	16·0	d	4						10	2	2				
33	814	4·8	10·5	20·1	d	5						10	2	3				
34	217	4·7	9·73	21·5	d	8, 9						10	2	4				
35	−7·2		11·76	19·2	d	6						10	2	5				
36	−157·3		14·20	24·5	d	2						10	2	6				
37	38·9	2·2	4·17	27·3	p	3	———————										1	
38	768	2·4	5·69	11·0	p	2											2	
39	1509	3·3	6·5	12·3	p	1, 2									1		2	
40	1852	4·0	6·95	14·0	p	1, 3									2		2	
41	2468	4·0	6·77		p	3									4		1	
42	2610	4·3	7·06		p	3									5		1	
43	2130	4·4	7·1				K								6		1	
44	2400, 2310	4·5	7·5	16·0	p	1	structure								7		1	
45	1960	4·6	7·7	18·0	p	2	(36)								8		1	
46	1552	4·9	8·1	19·8	p	2									10			
47	960·8, 960·880	4·6	7·58	21·4	d	2									10		1	
48	320·9	4·1	8·99	16·9	d	1									10		2	
49	156·2	4·0	5·79	18·9	d	10									10		2	1
50	231·9	4·3	7·30	14·6	d	4, 10									10		2	2
51	630·5	4·6	8·64	18·0	d	5									10		2	3

*Values of densities, and of boiling and melting points, obtained from various sources do not always correspond.
†To convert density in g cm^{-3} to kg m^{-3} multiply by 10^3.

Table 21.1 (continued)

At. no.	Element	Symbol	At. wt.	Mass number of common isotopes	Per-iodic group	Valency − +	Density* at 20°C (g cm^{-3})	Boiling point (°C)
52	Tellurium	Te	127·61	120, 122–6, 128, 130	6	2 **4, 6**	6·24	990
53	Iodine	I	126·91	127	7	**1** 5, 7	4·94	183
54	Xenon	Xe	131·30	124, 126, 128–32, 134, 136	0	Inert	3·52	112
55	Caesium	Cs	132·91	133	1	1	1·90	713, 690
56	Barium	Ba	137·36	130, 132, 134–8	2	2	3·5	1770, 1640
57	Lanthanum	La	138·92	138, 139	3	3	6·19	4200, 3470
58	Cerium	Ce	140·13	136, 138, 140, 142	3	**3, 4**	6·9	2900
59	Praseodymium	Pr	140·92	141	3	**3, 4**, 5	6·77	3026
60	Neodymium	Nd	144·27	142–6, 148, 150	3	3, 4	6·96	3180
61	Prometheum	Pm	[147][145]	145a, 147a	3	3		
62	Samarium	Sm	150·35	144, 147, 148n, 149, 150, 152	3	**2, 3**	7·49	1600
63	Europium	Eu	152·0	151, 153	3	**2, 3**	5·24	
64	Gadolinium	Gd	157·26	152, 154–8, 160	3	3	7·95	2700
65	Terbium	Tb	158·93	159	3	3, 4	8·33	2500
66	Dysprosium	Dy	162·51	158, 160–4	3	3	8·56	2300
67	Holmium	Ho	164·94	165	3	3	10·12, 8·8	2300
68	Erbium	Er	167·27	162, 164, 166–8, 170	3	3	9·16	2600
69	Thulium	Tm	168·94	169	3	3	9·35	2100
70	Ytterbium	Yb	173·04	168, 170–4, 176	3	2, **3**	7·01	1500
71	Lutecium	Lu	174·99	175, 176n	3	3	9·85	1930
72	Hafnium	Hf	178·50	174, 176–80	4	4	13·09	5100, 5400
73	Tantalum	Ta	180·95	181	5	5	16·6	6000, 5425
74	Tungsten	W	183·86	180, 182–4, 186	6	4, 6	19·3	5400, 5930
75	Rhenium	Re	186·22	185, 187	7	1 4, 7	21·04	5900
76	Osmium	Os	190·2	184, 186–90, 192	8	**4, 6**, 8	22·57	4600, 5500
77	Iridium	Ir	192·2	191, 193	8	3, **4**, 6	22·65	4500
78	Platinum	Pt	195·09	192, 194–6, 198	8	2, 4	21·45	3800, 4530
79	Gold	Au	197·0	197	1	**1**, 3	19·32	2660, 2970
80	Mercury	Hg	200·61	196, 198–202, 204	2	1, **2**	13·55	356·58
81	Thallium	Tl	204·39	203, 205, 206–8n, 210n	3	**1**, 3	11·85	1457
82	Lead	Pb	207·21	204, 206–8, 210–12n, 214	4	2, 4	11·36	1750, 1725
83	Bismuth	Bi	208·99	209, 210–12n, 214n	5	3, 5	9·80	1530, 1680
84	Polonium	Po	210	210–216n, 218n	6	2, 4	9·24	900, 965
85	Astatine	At	[210]	210n, 215n, 216n, 218n	7			
86	Radon	Rn	222	219n, 220n, 222n	0	Inert	4·40	−61·8
87	Francium	Fr	[223]	223n	1	1		
88	Radium	Ra	226·05	223–4n, 226n, 228n	2	2	5·0	1140
89	Actinium	Ac	227·05	227n, 228n	3	3		
90	Thorium	Th	232	227–8n, 230–2n, 234n	3	3, 4	11·66	3800 ± 400
91	Protoactinium	Pa	231	231n, 234n	3	3, 4, 5	15·4*	
92	Uranium	U	238·07	234n, 235n, 238n	3	3, 4, 5, 6	19·07	3818
93	Neptunium	Np	[237]	231a, 234a, 239a	3	3, 4, 5, 6		
94	Plutonium	Pu	[242]	238n, 242n	3	3, 4, 5, 6	1900 to 19·81	3235
95	Americium	Am	[243]	238–43a	3	2, 3, 4, 5	11·7	
96	Curium	Cm	[247]	238a, 240–3a	3	3		
97	Berkelium	Bk	[249]	244a, 245a	3	3, 4		
98	Californium	Cf	[251]	244a, 246a	3			
99	Einsteinium	Es	[254]		3			
100	Fermium	Fm	[253]		3			
101	Mendelevium	Md	[256]		3			
102	Nobelium	No	[254]		3			

*To convert density in g cm^{-3} to kg m^{-3} multiply by 10^3.

At. no.	Melting point (°C)	Work function (eV)	Ionisation potential (eV) I	Ionisation potential (eV) II	Mag. props.	Crystal struct.	4s	4p	4d	4f	5s	5p	5d	5f	6s	6p	6d	7
52	450	4·8	8·96	19·0	d	8	K		10		2	4						
53	113·7		10·44	19·0	d	6	structure		10		2	5						
54	−108		12·13	21·2	d	2	(36)		10		2	6						
55	28·7	1·9	3·89	23·4	d	3									1			
56	714	2·3	5·21	10·0	p	3									2			
57	920	3·3	5·61	11·4	p	8, 2, 3							1		2			
58	804, 785	2·8	6·54	12·3	p	8, 2, 3				1			1		2			
59	940	2·7	5·76		p	8, 3				2			1		2			
60	1024, 840	3·3	6·31		p	8, 3				3			1		2			
61										4			1		2			
62	1052, 1350	3·2	5·6	11·4		5, 3				5			1		2			
63	1400		5·64	11·2	p	3				6			1		2			
64	1320		6·70		f	1			Xe	7			1		2			
65	1450		6·74			1			structure	8			1		2			
66	1500		6·82		f	1			(54)	9			1		2			
67	1500					1				10			1		2			
68	1525					1				11			1		2			
69	1600					1				12			1		2			
70	824		6·2	12·1		1, 2				13			1		2			
71	1700, 1652		5·0			1				14			1		2			
72	2000, 2222	3·6	5·5	14·8		1				14			2		2			
73	3000 ± 50	4·2	6·0		p	3				14			3		2			
74	3380, 3410	4·5	7·94		p	3				14			4		2			
75	3170	4·9	7·87	13·0		1				14			5		2			
76	2700	4·6	8·7	15·0	p	1				14			6		2			
77	2443	4·6	9·2	16·0	p	2				14			7		2			
78	1769	5·3	8·96	18·5	p	2				14			8		2			
79	1063·0	4·7	9·23	20·0	d	2				14			10		1			
80	−38·87, −38·36	4·5	10·44	18·8	d	5				14			10		2			
81	300	3·8	6·12	20·3	d	1, 3				14			10		2	1		
82	327·3, 327·425	4·0	7·42	15·0	d	2				14			10		2	2		
83	271·3	4·4	8·8	16·6	d	5				14			10		2	3		
84	254	4·6	8·2	19·0		9, 2				14			10		2	4		
85	approx. 302		9·6	18·0						14			10		2	5		
86	−71		10·75	20·0						14			10		2	6		
87	approx. 27	1·5		22														1
88	700		5·27	10·1														2
89	1050																1	2
90	1700, 1800	3·4			p	2											2	2
91	approx. 1230	3·3			p								Rn	2			1	2
92	1133	3·5	4·0			3, 10, 6							structure	3			1	2
93	637												(86)	4			1	2
94	640					9								5			1	2
95														6			1	2
96														7			1	2
97														8			1	2
98														9			1	2
99														10			1	2
100														11			1	2
101														12			1	2
102														13			1	2

Table 21.2 Properties of some metals and alloys

	Thermal conductivity (W m⁻¹ K⁻¹; 273–373 K)	Thermal expansion per kelvin (K⁻¹)	Elect. resistivity (μΩ m)*	Temp. coef. of res. per kelvin (in range 273–373 K) (K⁻¹)	Mean specific heat in range 273–373 K (J kg⁻¹ K⁻¹)	Young's modulus (MN m⁻²† (or MPa))
Al	221·752	$23·5 \times 10^{-6}$	0·0269	$4·2 \times 10^{-3}$	916·296	$0·06894 \times 10^{6}$
Be	221·752	$13·0 \times 10^{-6}$	0·04–0·06	$6·0 \times 10^{-3}$	2012·504	$0·29647 \times 10^{6}$
Co	69·036	$12·3 \times 10^{-6}$	0·0624	$6·0 \times 10^{-3}$	435·136	$0·20684 \times 10^{6}$
Cu	393·296	$16·5 \times 10^{-6}$	0·01673	$4·3 \times 10^{-3}$	384·928	$0·11031 \times 10^{6}$
Pb	34·309	$29·0 \times 10^{-6}$	0·206	$3·36 \times 10^{-3}$	129·704	$0·01378 \times 10^{6}$
Mg	153·553	$27·0 \times 10^{-6}$	0·0440	$4·2 \times 10^{-3}$	1037·632	$0·04481 \times 10^{6}$
Mn	—	$22·0 \times 10^{-6}$	1·60(α)	—	489·528	$0·15857 \times 10^{6}$
Mo	142·256	$5·1 \times 10^{-6}$	0·0570	$4·6 \times 10^{-3}$	259·408	$0·29647 \times 10^{6}$
Ni	92·048	$12·8 \times 10^{-6}$	0·0684	$6·8 \times 10^{-3}$	451·872	$0·20684 \times 10^{6}$
Nb	52·300	$7·2 \times 10^{-6}$	0·169	$3·95 \times 10^{-3}$	255·224	$0·10342 \times 10^{6}$
Pd	70·291	$11·7 \times 10^{-6}$	0·108	$3·8 \times 10^{-3}$	246·856	$0·11721 \times 10^{6}$
Pt	69·036	$9·0 \times 10^{-6}$	0·106	$3·92 \times 10^{-3}$	133·888	$0·15168 \times 10^{6}$
Ag	418·400	$18·9 \times 10^{-6}$	0·016	$4·1 \times 10^{-3}$	225·936	$0·07584 \times 10^{6}$
Ta	54·392	$6·5 \times 10^{-6}$	0·145	$3·8 \times 10^{-3}$	142·256	$0·18615 \times 10^{6}$
Sn (white)	62·760	$23·5 \times 10^{-6}$	0·128	$4·2 \times 10^{-3}$	225·936	$0·04688 \times 10^{6}$
Ti	17·029	$8·4 \times 10^{-6}$	0·550	$3·5 \times 10^{-3}$	527·184	$0·11721 \times 10^{6}$
W	166·105	$4·6 \times 10^{-6}$	0·055	$4·6 \times 10^{-3}$	138·072	$0·35852 \times 10^{6}$
U	25·104–27·196	‡	0·29(α)	$3·4 \times 10^{-3}$	117·152	$0·18960 \times 10^{6}$
V	30·962	$8·3 \times 10^{-6}$	0·195	$2·8 \times 10^{-3}$	497·896	$0·13789 \times 10^{6}$
Zn	110·876	$31·0 \times 10^{-6}$	0·059	$4·2 \times 10^{-3}$	393·296	$0·09652 \times 10^{6}$
Zr	16·736	$5·9 \times 10^{-6}$	0·446	$4·4 \times 10^{-3}$	288·696	$0·09307 \times 10^{6}$
Al alloys	167·360(±)	$21·6 \times 10^{-6}$	0·035(+)	—	962·320	$0·06895 \times 10^{6}$
Brass (70/30)	125·520	$19·6 \times 10^{-6}$	0·062	$1·5 \times 10^{-3}$	380·744	$0·11032 \times 10^{6}$
Bronze (95/5)	86·680	$18·0 \times 10^{-6}$	0·096	$1·9 \times 10^{-3}$	376·560	$0·11032 \times 10^{6}$
Monel (70/30)	25·104	$14·4 \times 10^{-6}$	0·482	$1·1 \times 10^{-3}$	543·920	$0·17926 \times 10^{6}$
Fe–18Cr–8Ni stainless steel	14·644	$9·5 \times 10^{-6}$	0·70	—	502·08	$0·19305 \times 10^{6}$

*To convert into μΩ cm multiply by 100.
†To convert into kN cm⁻² multiply by 10^{-1}.
‡α-uranium 23×10^{-6} parallel to a-axis; $-3·5 \times 10^{-6}$ parallel to b-axis; 17×10^{-6} parallel to c-axis. β-uranium 46×10^{-6} parallel to c-axis; 23×10^{-6} perpendicular to c-axis.

Table 21.3 Values of important constants in SI units

Quantity	Symbol	SI value
Molar gas constant	R	$8 \cdot 314\,3 \times$ J mol^{-1} K^{-1}
Boltzmann constant	k	$1 \cdot 380\,54 \times 10^{-23}$ J K^{-1}
Avogadro's constant	N_A	$6 \cdot 022\,52 \times 10^{23}$ mol^{-1}
Planck constant	h	$6 \cdot 625\,6 \times 10^{-34}$ J s
Faraday constant	$F = N_A e$	$9 \cdot 648\,70 \times 10^4$ C mol^{-1}
Mass of hydrogen atom	m_H	$1 \cdot 673\,43 \times 10^{-27}$ kg
Proton mass	m_p	$1 \cdot 672\,52 \times 10^{-27}$ kg
Neutron mass	m_n	$1 \cdot 674\,82 \times 10^{-27}$ kg
Electron mass	m_e	$9 \cdot 109\,1 \times 10^{-31}$ kg
	m_p/m_e	$1 \cdot 836\,10 \times 10^3$
Charge of positron	e	$1 \cdot 602\,10 \times 10^{-19}$ C
Electron volt	$1\,\text{eV}$	$1 \cdot 602\,10 \times 10^{-19}$ J
Charge to mass ratio	e/m	$1 \cdot 758\,796 \times 10^{11}$ C kg^{-1}
Molar volume of ideal gas at s.t.p. (101 325 Nm^{-2} and 273·15 K)	V_m	$2 \cdot 241\,36 \times 10^{-2}$ m^3 mol^{-1}
Speed of light	c	$2 \cdot 997\,925 \times 10^8$ m s^{-1}
1st radiation constant	$c_1 = 2\pi h c^2$	$3 \cdot 741\,5 \times 10^{-16}$ W m^2
2nd radiation constant	$c_2 = hc/k$	$1 \cdot 438\,79 \times 10^{-2}$ m K
Wien's radiation law $T_{max} = c_2/4 \cdot 955\,11423$		$2 \cdot 897\,8 \times 10^{-3}$ m K
Stefan–Boltzmann constant	σ	$5 \cdot 669\,7 \times 10^{-8}$ W m^{-2} K^{-4}
Standard gravity acceleration	g	$9 \cdot 806\,65$ m s^{-1}
Absolute temperature of ice point	$(0°C)$	$273 \cdot 16$ K
$RT_{298 \cdot 16} \ln x$		$5709 \cdot 4 \log x$ J mol^{-1}
$(RT_{298 \cdot 16}/F) \ln x$		$0 \cdot 059\,173 \log x$ V

Data selected from 'Changing to the Metric System', *Metals and Materials*, Inst. of Metallurgists, 370, Dec. (1968).

Table 21.4 Values of $(RT/F) \log x$ at various temperatures

$$(RT/F) \ln x = \frac{8 \cdot 314\,3\,T \times 2 \cdot 303}{96\,487} \log x$$

T (°C)	$(RT/F) \log x$ (V)
0	0·054 195
10	0·056 179
20	0·058 163
25	0·059 173
30	0·060 147
40	0·062 131
50	0·064 115
60	0·066 099
70	0·068 083
80	0·070 067
90	0·072 051
100	0·074 035

Table 21.5 Standard chemical potentials*

Notes. In the case of a single oxide (or hydroxide) existing in different allotropic states, indicated by the letters *a, b, c, d,* etc. the oxides are arranged in descending order of stability, i.e. in ascending order of standard chemical potentials (expressed for an identical chemical formula).

With reference to the degree of precision which can be attached to the numerical values given in this table, it may be taken that for each of these values all figures but the last are reasonably exact, while the concluding figure must always be treated with reserve.

Element	Symbol, denomination, crystal structure	$\mu^{\ominus}$ (kJ)	Element	Symbol, denomination, crystal structure	$\mu^{\ominus}$ (kJ)
Actinium	Ac	0	Antimony	Sb	
	Ac^{3+}	$-753\cdot12$		trig.	0
Aluminium	Al	0		Sb_2O_3	
	$Al_2O_3.3H_2O$			senarmontite,	
	hydrargillite *a*	$-2320\cdot45$		cub. *a*	$-623\cdot42$
	$Al_2O_3.3H_2O$			Sb_2O_3	
	bayerite *b*	$-2311\cdot53$		valentinite,	
	$Al_2O_3.H_2O$			orthorhomb. *b*	$-615\cdot05$
	bohmite *c*	$-1820\cdot04$		Sb_2O_4	
	Al_2O_3			orthorhomb.	$-694\cdot13$
	corundum α *d*	$-1576\cdot41$		Sb_2O_5	
	$Al(OH)_3$			cub.	$-838\cdot89$
	amorph. *e*	$-1137\cdot63$		$SbCl_3$	$-324\cdot76$
	$AlCl_3$	$-636\cdot80$		Sb_2S_3	
	Al_2S_3	$-492\cdot46$		amorph.	$-133\cdot89$
	$Al_2(SO_4)_3$	$-3091\cdot93$		SbO^+	$-175\cdot73$
	AlN	$-209\cdot62$		$HSbO_2$	$-407\cdot94$
	Al_4C_3	$-121\cdot34$		SbO_2^-	$-345\cdot18$
	Al_2SiO_5			or $Sb(OH)_4^-$	$-819\cdot56$
	sillimanite	$-2573\cdot16$		SbO_2^+	$-274\cdot05$
	Al_2SiO_5			SbO_3^+	$-514\cdot34$
	andalusite	$-2543\cdot04$		or $Sb(OH)_6^-$	$-1225\cdot91$
	Al_2SiO_5			SbS_2^-	$-54\cdot39$
	disthene (kyanite)	$-2539\cdot69$		SbH_3	
	Al^{3+}	$-481\cdot16$		stibine	$147\cdot70$
	AlO_2^-	$-839\cdot77$		$SbCl_3$	$-302\cdot50$
	or $H_2AlO_3^-$	$-1076\cdot96$	Arsenic	As	
Americium	Am	0		trig.	0
	$Am(OH)_3$			As_2O_3	
	a	$-1255\cdot20$		cub.	$-576\cdot05$
	Am_2O_3			As_2O_5	
	b	$-1681\cdot97$		amorph.	$-772\cdot37$
	$Am(OH)_4$			As_2S_2	$-134\cdot52$
	a	$-1451\cdot85$		As_2S_3	$-135\cdot81$
	AmO_2			AsO^+	$-163\cdot59$
	cub. *b*	$-966\cdot50$		$HAsO_2$	$-402\cdot71$
	AmO_2OH			or H_3AsO_3	$-639\cdot90$
	or $\frac{1}{2} Am_2O_5.H_2O$	$-1066\cdot90$		AsO_2^-	$-350\cdot20$
	$AmO_2(OH)_2$	$-1117\cdot96$		or $H_2AsO_3^-$	$-587\cdot39$
	or $AmO_3.H_2O$	$-1117\cdot96$		$HAsO_3$	$-531\cdot83$
	$Am^{2+}(?)$	$-584\cdot70$		or H_3AsO_4	$-769\cdot02$
	Am^{3+}	$-671\cdot53$		AsO_3^-	$-511\cdot33$
	Am^{4+}	$-461\cdot08$		or $H_2AsO_4^-$	$-748\cdot52$
	AmO_2^+	$-813\cdot79$		$HAsO_4^{2-}$	$-707\cdot10$
	AmO_2^{2+}	$-655\cdot63$		AsO_4^{3-}	$-635\cdot97$
				AsH_3	$175\cdot73$

*Data after M. Pourbaix, Enthalpies Libres de Formations Standards, à 25°C, Cebelcor Rapport Technique, No. 87 (1960). Abbreviations: aq., aqueous; cub., f.c.c.; cub.b, CaF$_2$ cubic type; cub.c, b.c.c.; g., gaseous; hex., hexahedral; liq., liquid; orthorh., orthorhombic; monocl., monoclinic; trig., trigonal; tetr., tetragonal.

Table 21.5 (continued)

Element	Symbol, denomination, crystal structure	$\mu^{\ominus}$ (kJ)	Element	Symbol, denomination, crystal structure	$\mu^{\ominus}$ (kJ)
Barium	BaH_2 orthorh.	−132·21	Boron	$B_{10}H_{14}$	271·96
	Ba cub.	0		B	0
	$Ba(OH)_2 \cdot 8H_2O$ monocl. a	−2789·89		B_2O_3 a	−1198·42
	BaO cub. b	−528·44		B_2O_3 vitreous b	−1187·80
	$BaO_2 \cdot H_2O$ tetr. a	−815·88		HBO_2	−713·37
	BaO_2 b	−568·19		or H_3BO_3	−963·11
	BaF_2	−1140·14		$H_3BO_3^-$	−962·99
	$BaCl_2$	−810·86		$H_2BO_3^-$	−910·44
	$BaSO_4$	−1353·11		HBO_3^{2-}	−838·01
	$Ba(NO_3)_2$	−794·96		BO_3^{3-}	−759·31
	$Ba_3(PO_4)_2$	−3951·37		BO_2^-	−709·61
	$BaCO_3$	−1138·88		$H_2B_4O_7$	−2651·48
	$Ba(CN)_2$	−195·39		$HB_4O_7^-$	−2628·68
	Ba^{2+}	−560·66		$B_4O_7^{2-}$	−2577·34
Berkelium	Bk	0		B_2H_6	82·76
				B_5H_9	168·40
Beryllium	Be hex.	0		$B_{10}H_{14}$	−297·06
	$BeO \cdot Be(OH)_2$ 'precipitated' a	−1414·15		BH	471·12
	$Be(OH)_2$ 'β', orthorh. b	−820·90		BO	−81·67
	BeO hex.	−581·58	Bromine	Br_2(liq.)	0
	$Be(OH)_2$ 'α' d	−817·97		Br^-	−102·82
	$BeCl_2$	−467·77		Br_3^-	−105·73
	$BeSO_4$	−1088·68		Br_5^-	−101·84
	Be^{2+}	−356·48		Br_2(aq.)	4·09
	Be_2O^{2+}	−912·11		HBrO	−83·26
	$Be_2O_3^{2-}$	−1246·83		BrO^-	−33·47
	BeO_2^{2-}	−649·78		$HBrO_3$	5·65
	BeH	298·32		BrO_3^-	9·62
Bismuth	Bi hex.	0		HBr	−53·22
	BiO	−182·00		Br_2(g.)	3·14
	Bi_2O_3 monocl. tetr.	−496·64	Cadmium	Cd cadmium α, hex.	0
	$Bi(OH)_3$ amorph. b	−573·21		Cd cadmium β	0·59
	Bi_4O_7	−973·83		$Cd(OH)_2$ 'inactive' trig. a	−473·34
	Bi_2O_4	−456·06		$Cd(OH)_2$ 'active' b	−470·03
	Bi_2O_5	−383·11		CdO cub. c	−225·06
	Bi^{3+}	62·05		$CdCl_2$	−342·59
	$BiOH^{2+}$	−163·72		$CdBr_2$	−293·47
	BiO^+	−144·52		CdS	−140·58
	BiH_3 bismuthine	231·54		$CdSO_4$	−820·02
				$CdCO_3$	−670·28
				$Cd(CN)_2$	207·94
				Cd^{2+}	−77·74
				$HCdO_2^-$	−361·92
				$CdCl^+$	−216·73
				$CdCl_2$(aq.)	−352·71
				$Cd(NH_3)_4^{2+}$	−224·81
				$Cd(CN)_4^{2-}$	464·42
				CdH	233·17

Table 21.5 (continued)

Element	*Symbol, denomination, crystal structure*	$\mu^{\ominus}$ (kJ)
Calcium	CaH_2	
	orthorh.	−149·79
	Ca	
	cub.	*0*
	$Ca(OH)_2$	
	rhomb. *a*	−896·76
	CaO	
	cub. *b*	−604·17
	CaO_2	−598·31
	CaF_2	−1161·90
	$CaCl_2$	−750·19
	$CaBr_2$	−656·05
	CaI_2	−529·69
	CaS	−477·39
	$CaCO_3$	
	calcite rhomb.	−1128·76
	$CaCO_3$	
	aragonite, orthor.	−1127·71
	$CaC_2O_4.H_2O$	
	precipitated	−1508·75
	$CaSO_4$	
	anhydrite	−1320·30
	$CaSO_4$	
	soluble α	−1311·77
	$CaSO_4$	
	soluble β	−1307·33
	Ca^{2+}	−553·04
Californium	Cf	*0*
Carbon	C	
	graphite, hex.	*0*
	C	
	diamond, cub.	2·87
	HCN	121·34
	CS_2	63·60
	CCl_4	−68·74
	CH_3OH	−174·47
	HCHO	−129·70
	HCO_2H	−356·06
	HCO_2^-	−334·72
	$H_2C_2O_4$	−697·89
	$HC_2O_4^-$	−690·86
	$C_2O_4^{2-}$	−666·93
	H_2CO_3	−623·42
	HCO_3^-	−587·06
	CO_3^{2-}	−528·10
	HCN	112·13
	CN^-	165·69
	HCNO	−120·92
	CNO^-	−98·74
	CH_4	−50·79
	CO	−137·27
	CO_2	−394·38
	C_2N_2	296·27
	HCN	120·08
	CS_2	65·06
	CNCl	137·65
	CCl_4	−64·22
	$COCl_2$	−210·50
Cerium	Ce	*0*
	$Ce(OH)_3$	−1303·86
	CeO_2	
	cub.	−916·30
	Ce^{3+}	−718·60
	$Ce(OH)^{3+}$	−790·36
	CeO^{2+}	−788·73
	or $Ce(OH)_2^{2+}$	−1025·92
Caesium	CsH	
	cub.	−30·54
	Cs	
	cub. *c*	*0*
	CsOH	
	a	−355·22
	Cs_2O	
	trig. *b*	−274·47
	Cs_2O_2	−327·19
	Cs_2O_3	−360·24
	Cs_2O_4	
	tetr.	−387·02
	Cs^+	−282·04
Chlorine	Cl_2(g.)	*0*
	HCl	−131·17
	Cl^-	−131·17
	Cl_2(aq.)	6·90
	HClO	−79·96
	ClO^-	−37·24
	$HClO_2$	0·29
	ClO_2^-	11·46
	$HClO_3$	−2·59
	ClO_3^-	−2·59
	$HClO_4$	−10·33
	ClO_4^-	−10·33
	HCl	−95·27
	Cl	105·40
	Cl_2O	93·72
	ClO_2	123·43
Chromium	Cr	*0*
	$Cr(OH)_2$	−587·85
	$Cr(OH)_3$	
	a	−900·82
	Cr_2O_3	
	b	−1046·84
	$Cr(OH)_3.nH_2O$	
	c	−859·81
	$Cr(OH)_4$	−1014·12
	CrO_3	−502·08

Table 21.5 (continued)

Element	Symbol, denomination, crystal structure	$\mu^{\ominus}$ (kJ)	Element	Symbol, denomination, crystal structure	$\mu^{\ominus}$ (kJ)
	$CrCl_3$	$-493\cdot71$		$Dy(OH)_3$	$-1277\cdot79$
	Cr^{2+}	$-176\cdot15$		$DyCl_3$	$-920\cdot06$
	Cr^{3+}	$-215\cdot48$		DyI_3	$-599\cdot15$
	$CrOH^{2+}$	$-430\cdot95$		Dy^{3+}	$-681\cdot16$
	$Cr(OH)_2^+$	$-632\cdot66$	Erbium	Er	
	CrO_2^-	$-535\cdot93$		hex.	0
	CrO_3^{3-}	$-603\cdot42$		$Er(OH)_3$	
	H_2CrO_4	$-777\cdot89$		cub.	$-1266\cdot92$
	$HCrO_4^-$	$-773\cdot62$		$ErCl_3$	$-895\cdot38$
	CrO_4^{2-}	$-736\cdot80$		ErI_3	$-580\cdot74$
	$Cr_2O_7^{2-}$	$-1319\cdot63$		Er^{3+}	$-664\cdot67$
Cobalt	Co		Europium	Eu	0
	hex.	0		$Eu(OH)_3$	$-1291\cdot18$
	$Co(OH)_2$			Eu^{2+}	$-655\cdot21$
	trig. a	$-456\cdot06$		Eu^{3+}	$-696\cdot64$
	CoO				
	cub. b	$-205\cdot02$	Fluorine	F_2	0
	Co_3O_4			HF(aq.)	$-294\cdot60$
	cub.	$-702\cdot22$		HF_2^-	$-574\cdot88$
	$Co(OH)_3$	$-596\cdot64$		F^-	$-276\cdot48$
	CoO_2	$-216\cdot90$		HF(g.)	$-270\cdot70$
	$CoCl_2$	$-274\cdot05$			
	CoS			F	$117\cdot95$
	α precipitated	$-82\cdot84$		F_2O	$40\cdot58$
	$CoSO_4$	$-753\cdot54$	Gadolinium	Gd	
	$Co(NO_3)_2$	$-230\cdot54$		hex.	0
	$CoCO_3$	$-650\cdot90$		$Gd(OH)_3$	$-1288\cdot67$
	Co^{2+}	$-53\cdot56$		Gd^{3+}	$-693\cdot71$
	$HCoO_2^-$	$-347\cdot15$		Gd	$322\cdot17$
	Co^{3+}	$120\cdot92$			
Copper	Cu	0	Gallium	Ga	
	Cu_2O			tern.	0
	cuprite, cub.	$-146\cdot36$		Ga_2O	$-314\cdot64$
	CuO			Ga_2O_3	
	mon. or cub. a	$-127\cdot19$		anh. monocl. or	
	$Cu(OH)_2$			orthorh. a	$-992\cdot44$
	b	$-356\cdot90$		$Ga(OH)_3$	
	CuCl	$-117\cdot99$		or $\frac{1}{2}Ga_2O_3.3H_2O$	
	Cu_2S	$-86\cdot19$		hydroxide b	$-832\cdot62$
	Cu_2SO_4	$-652\cdot70$		Ga^{2+}	$-87\cdot86$
	$CuCl_2$	$-175\cdot73$		Ga^{3+}	$-153\cdot13$
	CuS	$-48\cdot95$		$GaOH^{2+}$	$-375\cdot72$
	$CuSO_4$	$-661\cdot91$		GaO^+	$-357\cdot31$
	$CuCO_3$	$-517\cdot98$		or $Ga(OH)_2^+$	$-594\cdot50$
	Cu^+	$50\cdot21$		GaO_2^-	$-507\cdot52$
	Cu^{2+}	$64\cdot98$		or $H_2GaO_3^-$	$-744\cdot75$
	$HCuO_2^-$	$-256\cdot98$		$HGaO_3^{2-}$	$-686\cdot18$
	CuO_2^{2-}	$-182\cdot00$		GaO_3^{3-}	$-619\cdot23$
	$CuCl_2^-$	$-242\cdot25$	Germanium	Ge	0
	CuH(?)	$267\cdot78$		GeO hydr.	
Curium	Cm	0		a	$-292\cdot46$
				GeO hydr.	
Dysprosium	Dy			b	$-262\cdot34$
	hex.	0			

Table 21.5 (continued)

Element	Symbol, denomination, crystal structure	$\mu^{\ominus}$ (kJ)	Element	Symbol, denomination, crystal structure	$\mu^{\ominus}$ (kJ)
Germanium (continued)	GeO_2 quad. a	$-569 \cdot 44$		In^{2+}	$-51 \cdot 88$
	GeO_2 'precipitated' hex. b	$-552 \cdot 29$		In^{3+}	$-99 \cdot 16$
	$GeCl_4$	$-497 \cdot 90$		$InOH^{2+}$	$-314 \cdot 22$
	Ge^{2+}	0		InO_2^-	$-435 \cdot 14$
	$HGeO_2^-$	$-385 \cdot 35$		InH	$-188 \cdot 28$
	H_2GeO_3	$-781 \cdot 57$	Iodine	I_2 orthorh.	0
	$HGeO_3^-$	$-733 \cdot 04$		ICl	$-13 \cdot 56$
	GeO_3^{2-}	$-660 \cdot 65$		ICl_3	$-22 \cdot 59$
	GeH_4	$334 \cdot 72$		I^-	$-51 \cdot 67$
Gold	Au cub.	0		I_3^-	$-51 \cdot 51$
	Au_2O_3 a	$163 \cdot 18$		I_5^-	$-28 \cdot 87$
	$Au(OH)_3$ b	$289 \cdot 95$		I_2 in solution	$16 \cdot 43$
	AuO_2	$200 \cdot 83$		I^+	$130 \cdot 92$
	Au^+	$163 \cdot 18$		or H_2IO^+	$-106 \cdot 27$
	Au^{3+}	$433 \cdot 46$		HIO	$-98 \cdot 32$
	H_3AuO_3	$-258 \cdot 57$		IO^-	$-35 \cdot 56$
	$H_2AuO_3^-$	$-191 \cdot 63$		HIO_3	$-139 \cdot 49$
	$HAuO_3^{2-}$	$-115 \cdot 48$		IO_3^-	$-134 \cdot 93$
	AuO_3^{3-}	$-24 \cdot 27$		HIO_4	$-62 \cdot 84$
Hafnium	Hf hex.	0		or H_5IO_6	$-537 \cdot 23$
	$HfO(OH)_2$ or $HfO_2 \cdot H_2O$	$-1361 \cdot 89$		IO_4^-	$-53 \cdot 14$
	HfO_2 monocl. tetr.	$-1055 \cdot 20$		$H_4IO_6^-$	$-518 \cdot 31$
	Hf^{4+}	$-656 \cdot 05$		HIO_5^{2-}	$-243 \cdot 13$
	HfO^{2+}	$-902 \cdot 49$		or $H_3IO_6^{2-}$	$-480 \cdot 32$
Holmium	Ho hex.	0		IO_5^{3-}	$-180 \cdot 37$
	$Ho(OH)_3$ or $\frac{1}{2}Ho_2O_3 \cdot 3H_2O$	$-1272 \cdot 35$		or $H_2IO_6^{3-}$	$-417 \cdot 56$
	Ho^{3+}	$-671 \cdot 11$		ICl	$-16 \cdot 74$
Hydrogen	$H_2(g.)$	0		HI	$1 \cdot 30$
	H_2O	$-237 \cdot 19$		I_2	$19 \cdot 37$
	H^-	$217 \cdot 15$	Iridium	Ir cub. f.c.	0
	$H_2(aq.)$	$17 \cdot 70$		$Ir_2O_3 \cdot xH_2O$ hydr. sesquiox.	
	H^+	0		$\mu^{\ominus}$ Ir_2O_3	$-175 \cdot 73$
	OH^-	$-157 \cdot 30$		$Ir(OH)_4$	$-591 \cdot 53$
	H_1	$203 \cdot 24$		Ir^{3+}	$334 \cdot 72$
Indium	In tetr. f.c.	0		IrO_4^{2-}	$-196 \cdot 65$
	In_2O_3 a	$-821 \cdot 74$	Iron	Fe	0
	$In(OH)_3$ b	$-761 \cdot 49$		$Fe(OH)_2$ a	$-483 \cdot 54$
	In^+	$-13 \cdot 39$		FeO wüstite b	$-244 \cdot 35$
				Fe_3O_4 magnetite, cub.	$-1014 \cdot 20$
				Fe_2O_3 haematite, trig. or cub. a	$-740 \cdot 99$
				$Fe(OH)_3$ b	$-694 \cdot 54$
				$FeCl_2$	$-302 \cdot 08$
				FeS	$-97 \cdot 57$
				$FeSO_4$	$-829 \cdot 69$

Table 21.5 (continued)

Element	Symbol, denomination, crystal structure	$\mu^{\ominus}$ (kJ)	Element	Symbol, denomination, crystal structure	$\mu^{\ominus}$ (kJ)
	$FePO_4$	$-1138\cdot05$		$LiNO_3$	$-389\cdot53$
	$FeCO_3$	$-673\cdot88$		Li_2CO_3	$-1132\cdot44$
	$FeCl_3$	$-336\cdot39$		Li^+	$-293\cdot80$
	Fe_2S_3	$-246\cdot86$		$LiOH$	$-451\cdot12$
	FeS_2	$-166\cdot69$		$LiCl$	$-424\cdot97$
	Fe^{2+}	$-84\cdot94$		$Li_2SO_4(aq.)$	$-1329\cdot59$
	FeO_2H^-	$-379\cdot18$		$LiNO_3(aq.)$	$-404\cdot30$
	Fe^{3+}	$-10\cdot59$		$Li_2CO_3(aq.)$	$-1115\cdot71$
	$FeOH^{2+}$	$-233\cdot93$		LiH	$105\cdot44$
	$Fe(OH)_2^+$	$-444\cdot34$	Lutetium	Lu	
	$FeO_4^{2-}(?)$	$-467\cdot29$		hex.	0
	$FeCl^{2+}$	$-150\cdot21$		$Lu(OH)_3$	$-1259\cdot38$
Lanthanum	La	0		Lu^{3+}	$-652\cdot70$
	$La(OH)_3$		Magnesium	Mg	
	amorph. a	$-1310\cdot43$		hex.	0
	La_2O_3			$Mg(OH)_2$	
	trig. b	$-1786\cdot15$		trig. a	$-833\cdot75$
	La^{3+}	$-730\cdot11$		MgO	
Lead	Pb			b	$-569\cdot57$
	cub.	0		MgO	
	PbO			finely divided c	$-566\cdot14$
	tetr. a	$-189\cdot33$		$MgCl_2$	$-592\cdot33$
	PbO			$MgOHCl$	$-732\cdot20$
	orthorh. b	$-188\cdot49$		$MgBr_2$	$-499\cdot15$
	$Pb(OH)_2$			MgS	$-349\cdot78$
	c	$-420\cdot91$		$MgSO_4$	$-1173\cdot61$
	Pb_3O_4	$-617\cdot56$		$Mg(NO_3)_2$	$-588\cdot40$
	Pb_2O_3	$-411\cdot78$		$Mg_3(PO_4)_2$	$-3782\cdot34$
	PbO_2	$-218\cdot99$		$MgCO_3$	$-1029\cdot26$
	$PbCl_2$	$-313\cdot97$		$MgNH_4PO_4$	$-1631\cdot76$
	$PbBr_2$	$-260\cdot41$		Mg^{2+}	$-456\cdot01$
	PbI_2	$-173\cdot76$		MgH	$142\cdot26$
	PbS	$-92\cdot68$	Manganese	Mn	
	$PbSO_4$	$-811\cdot24$		α cub.	0
	$Pb(NO_3)_2$	$-252\cdot30$		Mn	
	$PbCO_3$	$-626\cdot34$		γ tetr. f.c.	$1\cdot38$
	$PbCrO_4$	$-851\cdot86$		$Mn(OH)_2$	
	Pb^{2+}	$-24\cdot31$		a	$-614\cdot63$
	$HPbO_2^-$	$-338\cdot90$		MnO	
	Pb^{4+}	$302\cdot50$		b	$-363\cdot17$
	PbO_3^{2-}	$-277\cdot57$		Mn_3O_4	
	PbO_4^{4-}	$-282\cdot09$		tetr.	$-1280\cdot30$
	PbH_2	$290\cdot79$		Mn_2O_3	
Lithium	Li	0		cub. a	$-888\cdot26$
	LiH			$Mn(OH)_3$	
	cub.	$-69\cdot96$		monocl. b	$-757\cdot30$
	$LiOH$			MnO_2	
	tetr. a	$-443\cdot09$		β, pyrolusite	$-464\cdot84$
	Li_2O			$MnCl_2$	$-441\cdot41$
	cub. b	$-560\cdot24$		MnS	$-208\cdot78$
	Li_2O_2	$-564\cdot84$		MnS	
	$LiCl$	$-383\cdot67$		precipitated	$-223\cdot01$
	Li_2SO_4	$-1324\cdot65$		$MnSO_4$	$-955\cdot96$

Table 21.5 (continued)

Element	Symbol, denomination, crystal structure	$\mu^\ominus$ (kJ)	Element	Symbol, denomination, crystal structure	$\mu^\ominus$ (kJ)
Manganese (continued)	$MnCO_3$	$-817\cdot55$	Neptunium	Np	
	Mn^{2+}	$-227\cdot61$		orthorh.	0
	$HMnO_2^-$	$-505\cdot85$		$Np(OH)_3$	$-1248\cdot92$
	Mn^{3+}	$-82\cdot01$		NpO_2	
	MnO_4^{2-}	$-503\cdot75$		cub. f.c. a	$-979\cdot06$
	MnO_4^-	$-449\cdot36$		$Np(OH)_4$	
Mercury	HgO			b	$-1450\cdot17$
	red, orthorh. a	$-58\cdot53$		Np_2O_5	
	HgO			a	$-2008\cdot32$
	yellow, orthorh. b	$-58\cdot40$		$NpO_2\cdot OH$	
	Hg_2Cl_2	$-210\cdot54$		amorph. or	
	Hg_2Br_2	$-178\cdot72$		crystalline b	$-1095\cdot37$
	Hg_2I_2	$-111\cdot29$		$NpO_2(OH)_2$	$-1206\cdot25$
	Hg_2S	$-6\cdot69$		Np^{3+}	$-537\cdot23$
	Hg_2SO_4	$-623\cdot92$		Np^{4+}	$-522\cdot58$
	Hg_2CO_3	$-442\cdot67$		NpO_2^+	$-924\cdot66$
	$HgCl_2$	$-185\cdot77$		NpO_2^{2+}	$-813\cdot79$
	$HgBr_2$	$-147\cdot36$	Nickel	Ni	0
	HgI_2			$Ni(OH)_2$	
	a	$-100\cdot71$		trig. a	$-453\cdot13$
	HgI_2			NiO	
	b	$-96\cdot65$		cub. b	$-214\cdot64$
	HgS			$Ni_3O_4\cdot2H_2O$	$-1186\cdot29$
	cinnabar a	$-48\cdot83$		$Ni(OH)_3$	
	HgS			a	$-541\cdot83$
	b	$-46\cdot23$		$Ni_2O_3\cdot H_2O$	
	$HgSO_4$	$-589\cdot94$		b	$-706\cdot93$
	Hg	0		$NiO_2\cdot2H_2O$	$-689\cdot52$
	Hg_2^{2+}	$152\cdot09$		NiS α	
	Hg^{2+}	$164\cdot77$		a	$-114\cdot22$
	$Hg(OH)_2$	$-274\cdot89$		NiS γ	
	$HHgO_2^-$	$-190\cdot04$		b	$-74\cdot06$
	HgH	$220\cdot08$		$NiSO_4$	$-773\cdot62$
	Hg	$31\cdot76$		Ni^{2+}	$-47\cdot49$
	HgCl	$58\cdot58$		$HNiO_2^-$	$-349\cdot22$
Molybdenum	Mo		Niobium	Nb	
	cub. c	0		cub.	0
	MoO_2			NbO	$-378\cdot65$
	tetr.	$-502\cdot08$		NbO_2	$-736\cdot38$
	H_2MoO_4			Nb_2O_5	$-1766\cdot49$
	or $MoO_3\cdot H_2O$		Nitrogen	N_2O_5	
	hex. a	$-1186\cdot96$		hex.	$133\cdot89$
	MoO_3			$HNO_3(liq.)$	$-79\cdot91$
	orthorh. b	$-677\cdot60$		NH_4^+	$-79\cdot50$
	Mo^{3+}	$-57\cdot74$		NH_4OH	$-263\cdot80$
	$HMoO_4^-$	$-893\cdot70$		$N_2H_4H_2^{2+}$	$94\cdot14$
	MoO_4^{2-}	$-859\cdot48$		N_2H_4	$127\cdot86$
Neodymium	Nd	0		$NH_2OH_2^+$	$-56\cdot65$
	$Nd(OH)_3$			NH_2OH	$-23\cdot43$
	a	$-1294\cdot11$		$HN_3(aq.)$	$298\cdot32$
	Nd_2O_3			N_3^-	$325\cdot10$
	tetr. or cub. b	$-1759\cdot79$		$N_2(aq.)$	$12\cdot53$
	Nd^{3+}	$-703\cdot75$		$H_2N_2O_2$	$35\cdot98$

Table 21.5 (continued)

Element	Symbol, denomination, crystal structure	$\mu^{\ominus}$ (kJ)	Element	Symbol, denomination, crystal structure	$\mu^{\ominus}$ (kJ)
	$NH_2O_2^-$	76·15	Phosphorus	P_4H_2	66·94
	$N_2O_2^{2-}$	138·91		P	
	HNO_2	−53·64		red, cub. a	−13·81
	NO_2^-	−34·52		P	
	HNO_3(aq.)	−110·58		white, cub. b	0
	NO_3^-	−110·58		P_2H_4	37·66
	NH_3	−16·64		PCl_3	−287·02
	HN_3(g.)	328·44		H_3PO_2	−523·42
	N	340·90		$H_2PO_2^-$	−512·12
	N_2(g.)	0		H_3PO_3	−856·88
	N_2O	103·60		$H_2PO_3^-$	−846·63
	NO	86·69		HPO_3^{2-}	−811·70
	NO_2	51·84		$H_4P_2O_6$	−1640·13
	N_2O_4	98·29		$H_3P_2O_6^-$	−1627·58
	NOCl	66·36		$H_2P_2O_6^{2-}$	−1611·68
	NOBr	82·42		$HP_2O_6^{3-}$	−1570·26
Osmium	Os			$P_2O_6^{4-}$	−1513·35
	hex.	0		HPO_3	−902·91
	$Os(OH)_4$	−683·58		H_3PO_4	−1147·25
	OsO_4			$H_2PO_4^-$	−1135·12
	monocl. a	−295·81		HPO_4^{2-}	−1094·12
	OsO_4			PO_4^{3-}	−1025·50
	b	−294·97		PH_3	18·24
	OsO_4^{2-}	−373·42		P_2	102·93
	H_2OsO_5	−529·86		P_4	24·35
	$HOsO_5^-$	−472·79		PCl_3	−286·27
	OsO_5^{2-}	−390·16		PCl_5	−324·55
	OsO_4	−284·09	Platinum	Pt	0
Oxygen	H_2O(liq.)	−237·19		$Pt(OH)_2$	−285·35
	H_2O_2(liq.)	−113·97		PtO_2 hydrated	
	OH^-	−157·30		$\mu^{\ominus}$ PtO_2	−83·68
	H_2O_2(aq.)	−131·67		PtO_3 hydrated	
	OH(aq.)	35·69		$\mu^{\ominus}$ PtO_3	−66·94
	HO_2^-	−65·31		Pt^{2+}	229·28
	HO_2(aq.)	12·55	Plutonium	Pu	0
	O_2^-	54·39		$Pu(OH)_3$	−1172·36
	O_2(aq.)	16·53		PuO_2	−979·06
	H_2O(g.)	−228·59		$Pu(OH)_4$	−1422·56
	OH(g.)	37·36		$PuO_2.OH$	
	O_1(g.)	230·09		or $\frac{1}{2}Pu_2O_5.H_2O$	−1032·19
	O_2(g.)	0		$PuO_2(OH)_2$	−1166·92
	O_3(g.)	163·43		or $PuO_3.H_2O$	
Palladium	Pd_2H	−4·59		Pu^{3+}	−587·85
	Pd			Pu^{4+}	−494·55
	cub. f.c.c.	0		PuO_2^+	−857·30
	$Pd(OH)_2$			PuO_2^{2+}	−767·76
	a	−301·25	Polonium	Po	0
	PdO			PoO_2	> −194·97
	tetr. b	−60·25		PoO_3	−138·07
	$Pd(OH)_4$	−528·02		Po^{2+}	125·52
	PdO_3	100·83		PoO_3^{2-}	−422·58
	Pd^{2+}	−190·37		PoH_2	192·46

Table 21.5 (continued)

Element	Symbol, denomination, crystal structure	$\mu^\ominus$ (kJ)	Element	Symbol, denomination, crystal structure	$\mu^\ominus$ (kJ)
Potassium	KH			Re^{3+}	86·84
	cub.	−37·24		ReO_4^{2-}	−631·62
	K			ReO_4^-	−699·15
	cub. c	0	Rhodium	Rh	0
	KOH			Rh_2O	−83·68
	orthorh. a	−374·47		RhO	−75·31
	K_2O			Rh_2O_3	
	cub. b	−193·30		trig.	−219·66
	K_2O_2	−418·82		RhO_2	−62·76
	K_2O_3	−418·40		Rh^+	−58·58
	K_2O_4			Rh^{2+}	117·15
	tetr.	−416·73		Rh^{3+}	230·12
	KF	−533·13		RhO_4^{2-}	−62·76
	KCl	−408·32	Rubidium	RbH	
	$KClO_3$	−289·91		cub.	−30·54
	$KClO_4$	−304·18		Rb	
	KBr	−379·20		cub. c	0
	KI	−322·29		RbOH	
	K_2S	−404·17		trig. a	−364·43
	K_2SO_4	−1316·37		Rb_2O	
	KNO_3	−393·13		cub. b	−290·79
	KCN	−83·68		Rb_2O_2	
	K^+	−282·25		cub.	−349·78
	KOH	−439·58		Rb_2O_3	−386·60
Praseodymium	Pr			Rb_2O_4	−395·81
	hex. or cub. c	0		Rb^+	−282·21
	$Pr(OH)_3$	−1295·78		RbOH	−439·53
	Pr_2O_3		Ruthenium	Ru	0
	trig. or cub.	−1770·25		$Ru(OH)_3$	−497·90
	PrO_2			$Ru(OH)_4$	−644·67
	cub.	−920·48		RuO_4	−109·20
	Pr^{3+}	−712·54		Ru^{2+}?	87·86
	Pr^{4+}	−436·60		RuO_4^{2-}	−257·73
Promethium	Pm	0		RuO_4^-	−200·83
	$Pm(OH)_3$	−1292·86		H_2RuO_5	−341·41
	Pm^{3+}	−701·24		$HRuO_5^-$	−277·40
Protoactinium	Pa	0	Samarium	Sm	
	tetr.			trig. or hex.	0
	PaO_2^+	958·14		$Sm(OH)_3$	−1291·60
Radium	Ra	0		Sm^{2+}	−602·24
	RaO	−491·62		Sm^{3+}	−698·73
	Ra^{2+}	−562·75	Scandium	Sc	
Rhenium	Re			hex.	0
	hex.	0		$Sc(OH)_3$	
	Re_2O_3 (hydr.)			amorph. a	−1228·00
	$\mu^\ominus$ Re_2O_3	−579·94		Sc_2O_3	
	ReO_2	−372·38		b	−1631·76
	ReO_3			Sc^{3+}	−601·24
	cub.	−532·62		$ScOH^{2+}$	−801·44
	Re_2O_7		Selenium	Se	
	monocl.	−1057·30		γ	0
	Re^-	38·49			

Table 21.5 (continued)

Element	Symbol, denomination, crystal structure	$\mu^{\ominus}$ (kJ)	Element	Symbol, denomination, crystal structure	$\mu^{\ominus}$ (kJ)
	SeO_2	$-173\cdot64$		Ag_2CO_3	$-437\cdot14$
	H_2Se	$76\cdot99$		Ag_2CrO_4	$-647\cdot26$
	HSe^-	$98\cdot32$		$AgCN$	$164\cdot01$
	Se^{2-}	$178\cdot24$		$AgCNS$	$97\cdot49$
	H_2SeO_3	$-425\cdot93$		Ag^+	$77\cdot11$
	$HSeO_3^-$	$-411\cdot29$		AgO^-	$-22\cdot97$
	SeO_3^{2-}	$-373\cdot76$		Ag^{2+}	$268\cdot19$
	H_2SeO_4	$-441\cdot08$		AgO^+	$225\cdot52$
	$HSeO_4^-$	$-452\cdot71$		$Ag(S_2O_3)_2^{3-}$	$-1035\cdot96$
	SeO_4^{2-}	$-441\cdot08$		$Ag(SO_3)_2^{3-}$	$-943\cdot07$
	H_2Se	$71\cdot13$		$Ag(NH_3)_2^+$	$-17\cdot41$
	Se_2	$88\cdot49$		$Ag(CN)_2^-$	$301\cdot46$
Silicon	Si cub.	0	Sodium	NaH cub.	$-37\cdot66$
	SiO_2 quartz, hex. a	$-805\cdot00$		Na cub. c	0
	SiO_2 cristobalite b cub.	$-803\cdot75$		$NaOH.H_2O$ a	$-623\cdot42$
	SiO_2 tridymite c cub.	$-802\cdot91$		$NaOH$ cub. b	$-376\cdot98$
	SiO_2 vitreous d	$-798\cdot73$		Na_2O cub. c	$-376\cdot56$
	H_2SiO_3 amorph. e	$-1022\cdot99$		Na_2O_2 quad.	$-430\cdot12$
	$SiCl_4$	$-572\cdot79$		NaO_2	$-194\cdot56$
	H_2SiO_3	$-1012\cdot53$		NaF	$-540\cdot99$
	$HSiO_3^-$	$-955\cdot46$		$NaCl$	$-384\cdot03$
	SiO_3^{2-}	$-887\cdot01$		$NaBr$	$-347\cdot69$
	SiF_6^{2-}	$-2138\cdot02$		NaI	$-237\cdot23$
	SiH_4	$-39\cdot33$		Na_2S	$-362\cdot33$
	SiO	$-137\cdot11$		Na_2SO_4	$-1266\cdot83$
	SiF_4	$-1506\cdot24$		$NaNO_3$	$-365\cdot89$
	$SiCl_4$	$-569\cdot86$		Na^+	$-261\cdot87$
Silver	Ag			$NaOH$	$-419\cdot17$
	cub.	0	Strontium	SrH_2 orthorh.	$-138\cdot49$
	Ag_2O cub. a	$-10\cdot82$		Sr cub.	0
	$AgOH$ b	$-91\cdot97$		$Sr(OH)_2$	$-869\cdot44$
	AgO cub.	$21\cdot76$		SrO cub.	$-559\cdot82$
	Ag_2O_3 cub.	$87\cdot23$		SrO_2	$-581\cdot58$
	$AgCl$	$-109\cdot72$		$SrCl_2$	$-781\cdot15$
	AgI	$-66\cdot32$		$SrSO_4$	$-1334\cdot28$
	Ag_2S rhomb. α	$-40\cdot25$		Sr^{2+}	$-557\cdot31$
	Ag_2S rhomb. β	$-39\cdot16$	Sulphur	S orthorh.	0
	Ag_2SO_4	$-615\cdot76$		S monocl.	$0\cdot10$
	$AgNO_3$	$-32\cdot17$		H_2S	$-27\cdot36$
				HS^-	$12\cdot59$
				S^{2-}	$91\cdot87$
				S_2^{2-}	$82\cdot63$

Table 21.5 (continued)

Element	Symbol, denomination, crystal structure	$\mu^{\ominus}$ (kJ)	Element	Symbol, denomination, crystal structure	$\mu^{\ominus}$ (kJ)
Sulphur (continued)	S_3^{2-}	75·18		Te^{2-}	220·50
	S_4^{2-}	69·52		Te_2^{2-}	162·13
	S_5^{2-}	65·64		Te^{4+}	219·16
	$H_2S_2O_3$	−543·50		$HTeO_2^+$	−261·54
	$HS_2O_3^-$	−541·83		$HTeO_3^-$	−436·56
	$S_2O_3^{2-}$	−532·20		TeO_3^{2-}	−392·42
	$S_5O_6^{2-}$	−956·04		H_2TeO_4	−550·86
	$S_4O_6^{2-}$	−1022·15		or H_6TeO_6	−1025·24
	$H_2S_2O_4$	−585·76		$HTeO_4^-$	−515·75
	$HS_2O_4^-$	−591·65		or $H_5TeO_6^-$	−990·14
	$S_2O_4^{2-}$	−577·39		TeO_4^{2-}	−456·42
	$S_3O_6^{2-}$	−958·14		or $H_4TeO_6^{2-}$	−930·81
	H_2SO_3	−538·44		H_2Te	138·49
	HSO_3^-	−527·18		Te	159·41
	SO_3^{2-}	−485·76		Re_2	121·34
	$S_2O_5^{2-}$	−790·78			
	$S_2O_6^{2-}$	−966·50	Terbium	Tb	
	H_2SO_4	−741·99		hex.	0
	HSO_4^-	−752·87		$Tb(OH)_3$	
	SO_4^{2-}	−741·99		cub.	−1287·84
	$S_2O_8^{2-}$	−1096·21		Tb^{3+}	−692·03
	H_2S	−33·02			
	SO	53·47	Thallium	Tl	
	SO_2	−300·37		hex.	0
	SO_3	−370·37		TlOH	
	SF_6	−991·61		a	−190·37
				Tl_2O	
Tantalum	Ta			b	−138·49
	cub. c	0		$Tl(OH)_3$	
	Ta_2O_5			a	−514·63
	orthorh.	−1910·00		Tl_2O_3	
				b	−263·59
Technetium	Tc			TlCl	−184·89
	hex.	0		TlBr	−166·10
	TcO_2	−369·41		TlI	−124·26
	TcO_3	−460·53		$TlIO_3$	−199·16
	$Tc_2O_7 . H_2O$			Tl_2SO_4	−823·41
	or $2HTcO_4$			Tl^+	−32·45
	a	−1182·15		Tl^{3+}	209·20
	Tc_2O_7			TlH	179·91
	b	−931·11	Thorium	Th	
	Tc^{2+}	−77·19		cub.	0
	$HTcO_4$	−629·50		ThO_2	
	TcO_4^-	−630·17		cub. a	−1164·83
Tellurium	Te	0		$Th(OH)_4$	
	TeO_2			b	−1585·74
	tetr. a	−273·30		Th^{4+}	−733·04
	H_2TeO_3				
	or $TeO_2 . H_2O$		Thulium	Tm	
	b	−478·48		hex.	0
	H_6TeO_6			$Tm(OH)_3$	−1265·24
	or $TeO_3 . 3H_2O$			Tm^{3+}	−659·40
	cub. or monocl.	−1025·24	Tin	Sn	
	H_2Te	142·67		tetr.	0
	HTe^-	157·74		Sn	4·60

Table 21.5 (continued)

Element	Symbol, denomination, crystal structure	$\mu^{\ominus}$ (kJ)	Element	Symbol, denomination, crystal structure	$\mu^{\ominus}$ (kJ)
	SnO			U_3O_8	$-3363\cdot94$
	a	$-257\cdot32$		$UO_3.H_2O$	
	$Sn(OH)_2$			a	$-1435\cdot11$
	b	$-492\cdot04$		$UO_3.2H_2O$	
	SnO_2			b	$-1668\cdot75$
	a	$-515\cdot47$		UO_3	
	$Sn(OH)_4$			c	$-1142\cdot23$
	b	$-951\cdot86$		U^{3+}	$-520\cdot49$
	$SnCl_2$	$-302\cdot08$		U^{4+}	$-579\cdot07$
	SnS	$-82\cdot42$		UOH^{3+}	$-809\cdot60$
	$Sn(SO_4)_2$	$-1451\cdot01$		UO_2^+	$-994\cdot12$
	$SnCl_4$	$-474\cdot05$		UO_2^{2+}	$-989\cdot10$
	Sn^{2+}	$-26\cdot25$	Vanadium	V	0
	$SnOH^+$	$-253\cdot55$		V_2O_2	$-790\cdot78$
	SnO_2H^-	$-410\cdot03$		V_2O_3	$-1133\cdot86$
	$Sn_2O_3^{2-}$	$-590\cdot28$		V_2O_4	$-1330\cdot51$
	Sn^{4+}	$2\cdot72$		V_2O_5	
	SnO_3^{2-}	$-574\cdot97$		'evolved' a	$-1439\cdot30$
	SnH_4	$414\cdot22$		V_2O_5	
Titanium	Ti			'non-evolved' b	$-1431\cdot08$
	hex.	0		V^{2+}	$-226\cdot77$
	TiO	$-489\cdot19$		V^{3+}	$-251\cdot37$
	Ti_2O_3			$V(OH)^{2+}$	$-471\cdot91$
	trig. a	$-1018\cdot01$		VO^+	$-451\cdot83$
	$Ti(OH)_3$			VO^{2+}	$-456\cdot06$
	b	$-1049\cdot79$		HVO_2^+	$-662\cdot66$
	Ti_3O_5	$-2314\cdot25$		$HV_2O_5^-$	$-1508\cdot96$
	TiO_2			VO_2^+	$-596\cdot43$
	a	$-888\cdot39$		$H_3V_2O_7^-$	$-1886\cdot64$
	$TiO_2.H_2O$			$H_2VO_4^-$	$-1040\cdot87$
	b	$-1058\cdot51$		HVO_4^{2-}	$-986\cdot59$
	Ti^{2+}	$-314\cdot22$		VO_4^{3-}	$-921\cdot00$
	Ti^{3+}	$-349\cdot78$		VO_4^-	$-853\cdot12$
	TiO^{2+}	$-577\cdot39$	Ytterbium	Yb	
	$HTiO_3^-$	$-955\cdot88$		cub. c	0
	TiO_2^{2+}	$-467\cdot23$		$Yb(OH)_3$	
Tungsten	W			cub.	$-1262\cdot31$
	cub. c and cub.	0		Yb^{2+}	$-539\cdot74$
	WO_2			Yb^{3+}	$-656\cdot05$
	tetr.	$-520\cdot49$	Yttrium	Y	
	W_2O_5	$-1284\cdot07$		hex.	0
	WO_3			$Y(OH)_3$	
	monocl.	$-763\cdot45$		a	$-1284\cdot91$
	WO_4^{2-}	$-920\cdot48$		Y_2O_3	
Uranium	UH_3	$-127\cdot19$		cub. b	$-1681\cdot97$
	U	0		Y^{3+}	$-686\cdot59$
	UO	$-514\cdot63$	Zinc	Zn	0
	$U(OH)_3$	$-1101\cdot23$		$Zn(OH)_2$	
	UO_2			orthorh. ε, a	$-559\cdot09$
	a	$-1031\cdot77$		ZnO	
	$U(OH)_4$			'inactive'	
	b	$-1471\cdot09$		orthorh. b	$-321\cdot65$

Table 21.5 (continued)

Element	Symbol, denomination, crystal structure	$\mu^{\ominus}$ (kJ)	Element	Symbol, denomination, crystal structure	$\mu^{\ominus}$ (kJ)
Zinc	$Zn(OH)_2$			$Zn(OH)^+$	$-329\cdot28$
(continued)	γ, white c	$-557\cdot78$		$HZnO_2^-$	$-464\cdot01$
	$Zn(OH)_2$			ZnO_2^{2-}	$-389\cdot24$
	β, orthorh. d	$-557\cdot04$	Zirconium	Zr	
	ZnO			cub.	0
	'active' e	$-316\cdot67$		$Zr(OH)_4$	
	$Zn(OH)_2$			a	$-1548\cdot08$
	α, f	$-552\cdot02$		$ZrO(OH)_2$	
	$Zn(OH)_2$			or $ZrO_2 . H_2O$	
	amorph. g	$-551\cdot68$		b	$-1303\cdot32$
	$ZnCl_2$	$-369\cdot26$		ZrO_2	
	$ZnBr_2$	$-310\cdot21$		zircon monocl. c	$-1036\cdot38$
	ZnS			$ZrCl_4$	$-874\cdot46$
	sphalerite	$-198\cdot32$		Zr^{4+}	$-594\cdot13$
	$ZnSO_4$	$-871\cdot57$		ZrO^{2+}	$-843\cdot08$
	$ZnCO_3$	$-731\cdot36$		$HZrO_3^-$	$-1203\cdot74$
	Zn^{2+}	$-147\cdot21$			

Table 21.6 Standard electrode potentials* (reduction potentials) for inorganic systems at 25°C†

A Aqueous Acid Solutions

Electrode reaction	$E^{\ominus}$ (V)	Electrode reaction	$E^{\ominus}$ (V)
$\frac{3}{2}N_2 + e^- \rightarrow N_3^-$	$-3\cdot09$	$Sc^{3+} + 3e^- \rightarrow Sc$	$-2\cdot08$
$Li^+ + e^- \rightarrow Li$	$-3\cdot045$	$Pu^{3+} + 3e^- \rightarrow Pu$	$-2\cdot07$
$K^+ + e^- \rightarrow K$	$-2\cdot925$	$AlF_6^{3-} + 3e^- \rightarrow Al + 6F^-$	$-2\cdot07$
$Rb^+ + e^- \rightarrow Rb$	$-2\cdot925$	$Th^{4+} + 4e^- \rightarrow Th$	$-1\cdot90$
$As^+ + e^- \rightarrow As$	$-2\cdot923$	$Np^{3+} + 3e^- \rightarrow Np$	$-1\cdot86$
$Ra^{2+} + 2e^- \rightarrow Ra$	$-2\cdot92$	$Be^{2+} + 2e^- \rightarrow Be$	$-1\cdot85$
$Ba^{2+} + 2e^- \rightarrow Ba$	$-2\cdot90$	$U^{3+} + 3e^- \rightarrow U$	$-1\cdot80$
$Sr^{2+} + 2e^- \rightarrow Sr$	$-2\cdot89$	$Hf^{4+} + 4e^- \rightarrow Hf$	$-1\cdot70$
$Ca^{2+} + 2e^- \rightarrow Sr$	$-2\cdot87$	$Al^{3+} + 3e^- \rightarrow Al$	$-1\cdot66$
$Na^+ + e^- \rightarrow Na$	$-2\cdot714$	$Ti^{2+} + 2e^- \rightarrow Ti$	$-1\cdot63$
$La^{3+} + 3e^- \rightarrow La$	$-2\cdot52$	$Zr^{4+} + 4e^- \rightarrow Zr$	$-1\cdot53$
$Ce^{3+} + 3e^- \rightarrow Ce$	$-2\cdot48$	$SiF_6^{2-} + 4e^- \rightarrow Si + 6F^-$	$-1\cdot2$
$Nd^{3+} + 3e^- \rightarrow Nd$	$-2\cdot44$	$TiF_6^{2-} + 4e^- \rightarrow Ti + 6F^-$	$-1\cdot19$
$Sm^{3+} + 3e^- \rightarrow Sm$	$-2\cdot41$	$Mn^{2+} + 2e^- \rightarrow Mn$	$-1\cdot18$
$Gd^{3+} + 3e^- \rightarrow Gd$	$-2\cdot40$	$V^{2+} + 2e^- \rightarrow V$	$(-1\cdot18)$
$Mg^{2+} + 2e^- \rightarrow Mg$	$-2\cdot37$	$Nb^{3+} + 3e^- \rightarrow Nb$	$(-1\cdot1)$
$Y^{3+} + 3e^- \rightarrow Y$	$-2\cdot37$	$TiO^{2+} + 2H^+ + 4e^- \rightarrow Ti + H_2O$	$-0\cdot89$
$Am^{3+} + 3e^- \rightarrow Am$	$-2\cdot32$	$H_3BO_3 + 3H^+ + 3e^- \rightarrow B + 3H_2O$	$-0\cdot87$
$Lu^{3+} + 3e^- \rightarrow Lu$	$-2\cdot25$	$SiO_2 + 4H^+ + 4e^- \rightarrow Si + 2H_2O$	$-0\cdot86$
$\frac{1}{2}H_2 + e^- \rightarrow H^-$	$-2\cdot25$	$Ta_2O_5 + 10H^+ + 10e^- \rightarrow$	
$H^+ + e^- \rightarrow H(g.)$	$-2\cdot10$	$2Ta + 5H_2O$	$-0\cdot81$

*This table of standard electrode potentials (or redox potentials) includes equilibria of the type $M^{z+} + ze \rightleftharpoons M$, i.e. the e.m.f. series of metals. Brackets indicate that the value of $E^{\ominus}$ is unreliable.

†Data after Parsons, *Handbook of Electrochemical Constants*, Butterworths, London (1959).

Table 21.6 (continued)

Electrode reaction	$E^\ominus$ (V)	Electrode reaction	$E^\ominus$ (V)
$Zn^{2+}+2e^-\rightarrow Zn$	$-0{\cdot}763$	$HCOOH(aq.)+2H^++2e^-\rightarrow$	
$TlI+e^-\rightarrow Tl+I^-$	$-0{\cdot}753$	$\qquad HCHO(aq.)+H_2O$	$0{\cdot}056$
$Cr^{3+}+3e^-\rightarrow Cr$	$-0{\cdot}74$	$P+3H^++3e^-\rightarrow PH_3(g.)$	$0{\cdot}06$
$Te+2H^++2e^-\rightarrow H_2Te$	$-0{\cdot}72$	$AgBr+e^-\rightarrow Ag+Br^-$	$0{\cdot}095$
$TlBr+e^-\rightarrow Tl+Br^-$	$-0{\cdot}658$	$TiO^{2+}+2H^++e^-\rightarrow Ti^{3+}+H_2O$	$0{\cdot}1$
$Nb_2O_5+10H^++10e^-\rightarrow$		$Si+4H^++4e^-\rightarrow SiH_4$	$0{\cdot}102$
$\qquad 2Nb+5H_2O$	$-0{\cdot}65$	$C+4H^++4e^-\rightarrow CH_4$	$0{\cdot}13$
$U^{4+}+e^-\rightarrow U^{3+}$	$-0{\cdot}61$	$CuCl+e^-\rightarrow Cu+Cl^-$	$0{\cdot}137$
$As+3H^++3e^-\rightarrow AsH_3$	$-0{\cdot}60$	$S+2H^++2e^-\rightarrow H_2S$	$0{\cdot}141$
$TlCl+e^-\rightarrow Tl+Cl^-$	$-0{\cdot}557$	$Np^{4+}+e^-\rightarrow Np^{3+}$	$0{\cdot}147$
$Ga^{3+}+3e^-\rightarrow Ga$	$-0{\cdot}53$	$Sn^{4+}+2e^-\rightarrow Sn^{2+}$	$0{\cdot}15$
$Sb+3H^++3e^-\rightarrow SbH_3(g.)$	$-0{\cdot}51$	$Sb_2O_3+6H^++6e^-\rightarrow 2Sb+3H_2O$	$0{\cdot}152$
$H_3PO_2+H^++e^-\rightarrow P+2H_2O$	$-0{\cdot}51$	$Cu^{2+}+e^-\rightarrow Cu^+$	$0{\cdot}153$
$H_3PO_3+2H^++2e^-\rightarrow$		$BiOCl+2H^++3e^-\rightarrow$	
$\qquad H_2PO_2+H_2O$	$-0{\cdot}50$	$\qquad Bi+H_2O+Cl^-$	$0{\cdot}16$
$Fe^{2+}+2e^-\rightarrow Fe$	$-0{\cdot}440$	$SO_4^{2-}+4H^++2e^-\rightarrow H_2SO_3+H_2O$	$0{\cdot}17$
$Eu^{3+}+e^-\rightarrow Eu^{2+}$	$-0{\cdot}43$	$HCHO(aq.)+2H^+_+2e^-\rightarrow$	
$Cr^{3+}+e^-\rightarrow Cr^{2+}$	$-0{\cdot}41$	$\qquad CH_3OH(aq.)$	$0{\cdot}19$
$Cd^{2+}+2e^-\rightarrow Cd$	$-0{\cdot}403$	$HgBr_4^{2-}+2e^-\rightarrow Hg+4Br^-$	$0{\cdot}21$
$Se+2H^++2e^-\rightarrow H_2Se$	$-0{\cdot}40$	$AgCl+e^-\rightarrow Ag+Cl^-$	$0{\cdot}222$
$Ti^{3+}+e^-\rightarrow Ti^{2+}$	$(-0{\cdot}37)$	$HAsO_2(aq.)+3H^++3e^-\rightarrow$	
$PbI_2+2e^-\rightarrow Pb+2I^-$	$-0{\cdot}365$	$\qquad As+2H_2O$	$0{\cdot}247$
$PbSO_4+2e^-\rightarrow Pb+SO_4^{2-}$	$-0{\cdot}356$	$ReO_2+4H^++4e^-\rightarrow Re+2H_2O$	$0{\cdot}252$
$In^{3+}+3e^-\rightarrow In$	$-0{\cdot}342$	$BiO^++2H^++3e^-\rightarrow Bi+H_2O$	$0{\cdot}32$
$Tl^++e^-\rightarrow Tl$	$-0{\cdot}336$	$HCNO+H^++e^-\rightarrow \frac{1}{2}C_2N_2+H_2O$	$0{\cdot}33$
$PtS+2H^++2e^-\rightarrow Pt+H_2S$	$-0{\cdot}30$	$UO_2^{2+}+4H^++2e^-\rightarrow U^{4+}+2H_2O$	$0{\cdot}334$
$PbBr_2+2e^-\rightarrow Pb+2Br^-$	$-0{\cdot}280$	$Cu^{2+}+2e^-\rightarrow Cu$	$0{\cdot}337$
$Co^{2+}+2e^-\rightarrow Co$	$-0{\cdot}277$	$AgIO_3+e^-\rightarrow Ag+IO_3^-$	$0{\cdot}35$
$H_3PO_4+2H^++2e^-\rightarrow$		$Fe(CN)_6^{3-}+e^-\rightarrow Fe(CN)_6^{4-}$	$0{\cdot}36$
$\qquad H_3PO_3+H_2O$	$-0{\cdot}276$	$VO^{2+}+2H^++e^-\rightarrow V^{3+}+H_2O$	$0{\cdot}361$
$PbCl_2+2e^-\rightarrow Pb+2Cl^-$	$-0{\cdot}268$	$ReO_4^-+8H^++7e^-\rightarrow Re+4H_2O$	$0{\cdot}363$
$V^{3+}+e^-\rightarrow V^{2+}$	$-0{\cdot}255$	$\frac{1}{2}C_2N_2+H^++e^-\rightarrow HCN(aq.)$	$0{\cdot}37$
$V(OH)_4^++4H^++5e^-\rightarrow V+4H_2O$	$-0{\cdot}253$	$2H_2SO_3+2H^++4e^-\rightarrow$	
$SnF_6^{2-}+4e^-\rightarrow Sn+6F^-$	$-0{\cdot}25$	$\qquad S_2O_3^{2-}+3H_2O$	$0{\cdot}40$
$Ni^{2+}+2e^-\rightarrow Ni$	$-0{\cdot}250$	$RhCl_6^{3-}+3e^-\rightarrow Rh+6Cl^-$	$0{\cdot}44$
$N_2+5H^++4e^-\rightarrow N_2H_5^+$	$-0{\cdot}23$	$Ag_2CrO_4+2e^-\rightarrow 2Ag+CrO_4^{2-}$	$0{\cdot}446$
$2SO_4^{2-}+4H^++2e^-\rightarrow$		$H_2SO_3+4H^++4e^-\rightarrow S+3H_2O$	$0{\cdot}45$
$\qquad S_2O_6^{2-}+2H_2O$	$-0{\cdot}22$	$Sb_2O_5+2H^++2e^-\rightarrow Sb_2O_4+H_2O$	$0{\cdot}48$
$Mo^{3+}+3e^-\rightarrow Mo$	$(-0{\cdot}2)$	$Ag_2MoO_4+2e^-\rightarrow 2Ag+MoO_4^{2-}$	$0{\cdot}49$
$Co_2+2H^++2e^-\rightarrow HCOOH(aq.)$	$-0{\cdot}196$	$H_2N_2O_2+6H^++4e^-\rightarrow 2NH_3OH^+$	$0{\cdot}496$
$CuI+e^-\rightarrow Cu+I^-$	$-0{\cdot}185$	$ReO_4^-+4H^++3e^-\rightarrow$	
$AgI+e^-\rightarrow Ag+I^-$	$-0{\cdot}151$	$\qquad ReO_2+2H_2O$	$0{\cdot}51$
$Sn^{2+}+2e^-\rightarrow Sn$	$-0{\cdot}136$	$4H_2SO_4+4H^++6e^-\rightarrow$	
$O_2+H^++e^-\rightarrow HO_2$	$-0{\cdot}13$	$\qquad S_4O_6^{2-}+6H_2O$	$0{\cdot}51$
$Pb^{2+}+2e^-\rightarrow Pb$	$-0{\cdot}126$	$C_2H_4+2H^++2e^-\rightarrow C_2H_6$	$0{\cdot}52$
$GeO_2+4H^++4e^-\rightarrow Ge+2H_2O$	$-0{\cdot}15$	$Cu^++e^-\rightarrow Cu$	$0{\cdot}521$
$WO_3(colourless)+6H^++6e^-\rightarrow$		$TeO_2(colourless)+4H^++4e^-\rightarrow$	
$\qquad W+3H_2O$	$-0{\cdot}09$	$\qquad Te+2H_2O$	$0{\cdot}529$
$2H_2SO_3+H^++2e^-\rightarrow$		$I_2+2e^-\rightarrow 2I^-$	$0{\cdot}536$
$\qquad HS_2O_4^-+2H_2O$	$-0{\cdot}08$	$I_3^-+2e^-\rightarrow 3I^-$	$0{\cdot}536$
$HgI_4^{2-}+2e^-\rightarrow Hg+4I^-$	$-0{\cdot}04$	$Cu^{2+}+Cl^-+e^-\rightarrow CuCl$	$0{\cdot}538$
$2H^++2e^-\rightarrow H_2$	$0{\cdot}000$	$AgBrO_3+e^-\rightarrow Ag+BrO_3^-$	$0{\cdot}55$
$Ag(S_2O_3)_2^{3-}+e^-\rightarrow Ag+2S_2O_3^{2-}$	$0{\cdot}01$	$TeOOH^++3H^++4e^-\rightarrow Te+2H_2O$	$0{\cdot}559$
$CuBr+e^-\rightarrow Cu+Br^-$	$0{\cdot}033$	$H_3AsO_4+2H^++2e^-\rightarrow$	
$UO_2^{2+}+e^-\rightarrow UO_2^+$	$0{\cdot}05$	$\qquad HAsO_2+2H_2O$	$0{\cdot}559$

Table 21.6 (continued)

Electrode reaction	$E^{\ominus}$ (V)	Electrode reaction	$E^{\ominus}$ (V)
$AgNO_2 + e^- \rightarrow Ag + NO_2^-$	0·564	$ICl_2^- + e^- \rightarrow \frac{1}{2}I_2 + 2Cl^-$	1·06
$MnO_4^- + e^- \rightarrow MnO_4^{2-}$	0·564	$Br_2(l.) + 2e^- \rightarrow 2Br^-$	1·065
$PtBr_4^{2-} + 2e^- \rightarrow Pt + 4Br^-$	0·58	$N_2O_4 + 2H^+ + 2e^- \rightarrow 2HNO_2$	1·07
$Sb_2O_5 + 6H^+ + 4e^- \rightarrow$ $2SbO^+ + 3H_2O$	0·581	$Cu^{2+} + 2CN^- + e^- \rightarrow Cu(CN)_2^-$	1·12
$CH_3OH(aq.) + 2H^+ + 2e^- \rightarrow$ $CH_4 + H_2O$	0·586	$PuO_2^+ + 4H^+ + e^- \rightarrow Pu^{4+} + 2H_2O$	1·15
$PdBr_4^{2-} + 2e^- \rightarrow Pd + 4Br^-$	0·6	$SeO_4^{2-} + 4H^+ + 2e^- \rightarrow$ $H_2SeO_3 + H_2O$	1·15
$RuCl_5^{2-} + 3e^- \rightarrow Ru + 5Cl^-$	0·60	$NpO_2^{2+} + e^- \rightarrow NpO_2^+$	1·15
$UO_2^{2+} + 4H^+ + 2e^- \rightarrow U^{4+} + 2H_2O$	0·62	$CCl_4 + 4H^+ + 4e^- \rightarrow C + 4Cl^- + 4H^+$	1·18
$PdCl_4^{2-} + 2e^- \rightarrow Pd + 4Cl^-$	0·62	$ClO_4^- + 2H^+ + 2e^- \rightarrow ClO_3^- + H_2O$	1·19
$Cu^{2+} + Br^- + e^- \rightarrow CuBr$	0·640	$IO_3^- + 6H^+ + 5e^- \rightarrow \frac{1}{2}I_2 + 3H_2O$	1·195
$AgC_2H_3O_2 + e^- \rightarrow Ag + C_2H_3O_2^-$	0·643	$ClO_3^- + 3H^+ + 2e^- \rightarrow HClO_2 + H_2O$	1·21
$Ag_2SO_4 + 2e^- \rightarrow 2Ag + SO_4^{2-}$	0·653	$O_2 + 4H^+ + 4e^- \rightarrow 2H_2O$	1·229
$Au(CNS)_4^- + 3e^- \rightarrow Au + 4CNS^-$	0·66	$S_2Cl_2 + 2e^- \rightarrow 2S + 2Cl^-$	1·23
$PtCl_6^{2-} + 2e^- \rightarrow PtCl_4^{2-} + 2Cl^-$	0·68	$MnO_2 + 4H^+ + 2e^- \rightarrow Mn^{2+} + 2H_2O$	1·23
$O_2 + 2H^+ + 2e^- \rightarrow H_2O_2$	0·682	$Tl^{3+} + 2e^- \rightarrow Tl^+$	1·25
$HN_3 + 11H^+ + 8e^- \rightarrow 3NH_4^+$	0·69	$AmO_2^+ + 4H^+ + e^- \rightarrow Am^{4+} + 2H_2O$	1·26
$Te + 2H^+ + 2e^- \rightarrow H_2Te$	0·70	$N_2H_5^+ + 3H^+ + 2e^- \rightarrow 2NH_4^+$	1·275
$2NO + 2H^+ + 2e^- \rightarrow H_2N_2O_2$	0·71	$ClO_2 + H^+ + e^- \rightarrow HClO_2$	1·275
$H_2O_2 + H^+ + e^- \rightarrow OH + H_2O$	0·72	$PdCl_6^{2-} + 2e^- \rightarrow PdCl_4^{2-} + 2Cl^-$	1·288
$PtCl_4^{2-} + 2e^- \rightarrow Pt + 4Cl^-$	0·73	$2HNO_2 + 4H^+ + 4e^- \rightarrow N_2O + 3H_2O$	1·29
$C_2H_2 + 2H^+ + 2e^- \rightarrow C_2H_4$	0·73	$Cr_2O_7^{2-} + 14H^+ + 6e^- \rightarrow$ $2Cr^{3+} + 7H_2O$	1·33
$H_2SeO_3 + 4H^+ + 4e^- \rightarrow Se + 3H_2O$	0·74	$NH_3OH^+ + 2H^+ + 2e^- \rightarrow$ $NH_4^+ + H_2O$	1·35
$NpO_2^+ + 4H^+ + e^- \rightarrow Np^{4+} + 2H_2O$	0·75	$Cl_2 + 2e^- \rightarrow 2Cl^-$	1·360
$(CNS)_2 + 2e^- \rightarrow 2CNS^-$	0·77	$2NH_3OH^+ + H^+ + 2e^- \rightarrow$ $N_2H_5^+ + 2H_2O$	1·42
$IrCl_6^{3-} + 3e^- \rightarrow Ir + 6Cl^-$	0·77	$Au(OH)_3 + 3H^+ + 3e^- \rightarrow Au + 3H_2O$	1·45
$Fe^{3+} + e^- \rightarrow Fe^{2+}$	0·771	$HIO + H^+ + e^- \rightarrow \frac{1}{2}I_2 + H_2O$	1·45
$Hg_2^{2+} + 2e^- \rightarrow 2Hg$	0·789	$PbO_2 + 4H^+ + 2e^- \rightarrow Pb^{2+} + 2H_2O$	1·455
$Ag^+ + e^- \rightarrow Ag$	0·799	$Au^{3+} + 3e^- \rightarrow Au$	1·50
$2NO_3^- + 4H^+ + 2e^- \rightarrow N_2O_4 + 2H_2O$	0·80	$HO_2 + H^+ + e^- \rightarrow H_2O_2$	1·5
$Rh^{3+} + 3e^- \rightarrow Rh$	(0·8)	$Mn^{3+} + e^- \rightarrow Mn^{2+}$	1·51
$OsO_4(colourless) + 8H^+ + 8e^- \rightarrow$ $Os + 4H_2O$	0·85	$MnO_4^- + 8H^+ + 5e^- \rightarrow Mn^{2+} + 4H_2O$	1·51
$2HNO_2 + 4H^+ + 4e^- \rightarrow$ $H_2N_2O_2 + 2H_2O$	0·86	$BrO_3^- + 6H^+ + 5e^- \rightarrow \frac{1}{2}Br_2 + 3H_2O$	1·52
$Cu^{2+} + I^- + e^- \rightarrow CuI$	0·86	$HBrO + H^+ + e^- \rightarrow \frac{1}{2}Br_2 + H_2O$	1·59
$AuBr_4^- + 3e^- \rightarrow Au + 4Br^-$	0·87	$Bi_2O_4 + 4H^+ + 2e^- \rightarrow 2BiO^+ + 2H_2O$	1·59
$2Hg^{2+} + 2e^- \rightarrow Hg_2^{2+}$	0·920	$H_5IO_6 + H^+ + 2e^- \rightarrow IO_3^- + 3H_2O$	1·6
$NO_3^- + 3H^+ + 2e^- \rightarrow HNO_2 + H_2O$	0·94	$Bk^{4+} + e^- \rightarrow Bk^{3+}$	1·6
$PuO_2^{2+} + e^- \rightarrow PuO_2^+$	0·93	$Ce^{4+} + e^- \rightarrow Ce^{3+}$	1·61
$NO_3^- + 4H^+ + 4e^- \rightarrow NO + 2H_2O$	0·96	$HClO + H^+ + e^- \rightarrow \frac{1}{2}Cl_2 + H_2O$	1·63
$AuBr_2^- + e^- \rightarrow Au + 2Br^-$	0·96	$AmO_2^{2+} + e^- \rightarrow AmO_2^+$	1·64
$Pu^{4+} + e^- \rightarrow Pu^{3+}$	0·97	$HClO_2 + 2H^+ + 2e^- \rightarrow HClO + H_2O$	1·64
$Pt(OH)_2 + 2H^+ + 2e^- \rightarrow Pt + 2H_2O$	0·98	$NiO_2 + 4H^+ + 2e^- \rightarrow Ni^{2+} + 2H_2O$	1·68
$Pd^{2+} + 2e^- \rightarrow Pd$	0·987	$PbO_2 + SO_4^{2-} + 4H^+ + 2e^- \rightarrow$ $PbSO_4 + 2H_2O$	1·685
$IrBr_6^{3-} + e^- \rightarrow IrBr_6^{4-}$	0·99	$AmO_2^{2+} + 4H^+ + 3e^- \rightarrow$ $Am^{3+} + 2H_2O$	1·69
$HNO_2 + H^+ + e^- \rightarrow NO + H_2O$	1·00	$MnO_4^- + 4H^+ + 3e^- \rightarrow MnO_2 + 2H_2O$	1·695
$AuCl_4^- + 3e^- \rightarrow Au + 4Cl^-$	1·00	$Au^+ + e^- \rightarrow Au$	(1·7)
$V(OH)_4^+ + 2H^+ + e^- \rightarrow VO^{2+} + 3H_2O$	1·00	$AmO_2^+ + 4H^+ + 2e^- \rightarrow Am^{3+} + 2H_2O$	1·725
$IrCl_6^{2-} + e^- \rightarrow IrCl_6^{3-}$	1·017	$H_2O_2 + 2H^+ + 2e^- \rightarrow 2H_2O$	1·77
$H_6TeO_6 + 2H^+ + 2e^- \rightarrow$ $TeO_2 + 4H_2O$	1·02	$Co^{3+} + e^- \rightarrow Co^{2+}$	1·82
$N_2O_4 + 4H^+ + 4e^- \rightarrow NO + 2H_2O$	1·03	$FeO_4^{2-} + 8H^+ + 3e^- \rightarrow Fe^{3+} + 4H_2O$	1·9
$PuO_2^{2+} + 4H^+ + 2e^- \rightarrow Pu^{4+} + 2H_2O$	1·04		

Table 21.6 (continued)

Electrode reaction	$E^{\ominus}$ (V)	Electrode reaction	$E^{\ominus}$ (V)
$HN_3 + 3H^+ + 2e^- \rightarrow NH_4^+ + N_2$	1·96	$O(g.) + 2H^+ + 2e^- \rightarrow H_2O$	2·42
$Ag^{2+} + e^- \rightarrow Ag^+$	1·98	$F_2 + 2e^- \rightarrow 2F^-$	2·65
$S_2O_8^{2-} + 2e^- \rightarrow 2SO_4^{2-}$	2·01	$OH + H^+ + e^- \rightarrow H_2O$	2·8
$O_3 + 2H^+ + 2e^- \rightarrow O_2 + H_2O$	2·07	$H_2N_2O_2 + 2H^+ + 2e^- \rightarrow N_2 + 2H_2O$	2·85
$F_2O + 2H^+ + 4e^- \rightarrow 2F^- + H_2O$	2·1	$F_2 + 2H^+ + 2e^- \rightarrow 2HF(aq.)$	3·06
$Am^{4+} + e^- \rightarrow Am^{3+}$	2·18		

B   Aqueous Basic Solutions

Electrode reaction	$E^{\ominus}$ (V)	Electrode reaction	$E^{\ominus}$ (V)
$Ca(OH)_2 + 2e^- \rightarrow Ca + 2OH^-$	−3·03	$Te + 2e^- \rightarrow Te^{2-}$	−1·14
$Sr(OH)_2 8H_2O + 2e^- \rightarrow Sr + 2OH^- + 8H_2O$	−2·99	$PO_4^{3-} + 2H_2O + 2e^- \rightarrow HPO_3^{2-} + 3OH^-$	−1·12
$Ba(OH)_2 8H_2O + 2e^- \rightarrow Ba + 2OH^- + 8H_2O$	−2·97	$2SO_3^{2-} + 2H_2O + 2e^- \rightarrow S_2O_4^{2-} + 4OH^-$	−1·12
$H_2O + e^- \rightarrow H(g.) + OH^-$	−2·93	$ZnCO_3 + 2e^- \rightarrow Zn + CO_3^{2-}$	−1·06
$La(OH)_3 + 3e^- \rightarrow La + 3OH^-$	−2·90	$WO_4^{2-} + 4H_2O + 6e^- \rightarrow W + 8OH^-$	−1·05
$Lu(OH)_3 + 3e^- \rightarrow Lu + 3OH^-$	−2·72	$MoO_4^{2-} + 4H_2O + 6e^- \rightarrow Mo + 8OH^-$	−1·05
$Mg(OH)_2 + 2e^- \rightarrow Mg + 2OH^-$	−2·69	$Cd(CN)_4^{2-} + 2e^- \rightarrow Cd + 4CN^-$	−1·03
$Be_2O_3^{2-} + 3H_2O + 4e^- \rightarrow 2Be + 6OH^-$	−2·62	$Zn(NH_3)_4^{2+} + 2e^- \rightarrow Zn + 4NH_3$	−1·03
$Sc(OH)_3 + 3e^- \rightarrow Sc + 3OH^-$	(−2·6)	$FeS(\alpha) + 2e^- \rightarrow Fe + S^{2-}$	−1·01
$HfO(OH)_2 + H_2O + 4e^- \rightarrow Hf + 4OH^-$	−2·50	$In(OH)_3 + 3e^- \rightarrow In + 3OH^-$	−1·0
$Th(OH)_4 + 4e^- \rightarrow Th + 4OH^-$	−2·48	$PbS + 2e^- \rightarrow Pb + S^{2-}$	0·98
$Pu(OH)_3 + 3e^- \rightarrow Pu + 3OH^-$	−2·42	$CNO^- + H_2O + 2e^- \rightarrow CN^- + 2OH^-$	−0·97
$UO_2 + 2H_2O + 4e^- \rightarrow U + 4OH^-$	−2·39	$Tl_2S + 2e^- \rightarrow Tl + S^{2-}$	−0·96
$H_2AlO_3^- + H_2O + 3e^- \rightarrow Al + 4OH^-$	−2·35	$Pu(OH)_4 + e^- \rightarrow Pu(OH)_3 + OH^-$	−0·95
$H_2ZrO_3 + H_2O + 4e^- \rightarrow Zr + 4OH^-$	−2·36	$SnS + 2e^- \rightarrow Sn + S^{2-}$	−0·94
$U(OH)_4 + e^- \rightarrow U(OH)_3 + OH^-$	−2·2	$SO_4^{2-} + H_2O + 2e^- \rightarrow SO_3^{2-} + 2OH^-$	−0·93
$U(OH)_3 + 3e^- \rightarrow U + 3OH^-$	−2·17	$Se + 2e^- \rightarrow Se^{2-}$	−0·92
$H_2PO_2^- + e^- \rightarrow P + 2OH^-$	−2·05	$HSnO_2^- + H_2O + 2e^- \rightarrow Sn + 3OH^-$	−0·91
$H_2BO_3^- + 3e^- \rightarrow B + 4OH^-$	−1·79	$HGeO_3^- + 2H_2O + 4e^- \rightarrow Ge + 5OH^-$	−0·9
$SiO_3^{2-} + 3H_2O + 4e^- \rightarrow Si + 6OH^-$	−1·70	$Sn(OH)_6^{2-} + 2e^- \rightarrow HSnO_2^- + H_2O + 3OH^-$	−0·90
$Na_2UO_4 \mid 4H_2O \mid 2e^- \rightarrow U(OH)_4 + 2Na^+ + 4OH^-$	−1·61	$P + 3H_2O + 3e^- \rightarrow PH_3 + 3OH^-$	−0·89
$HPO_3^{2-} + 2H_2O + 2e^- \rightarrow H_2PO_2^- + 3OH^-$	−1·57	$Fe(OH)_2 + 2e^- \rightarrow Fe + 2OH^-$	−0·877
$Mn(OH)_2 + 2e^- \rightarrow Mn + 2OH^-$	−1·55	$NiS(\alpha) + 2e^- \rightarrow Ni + S^{2-}$	−0·83
$MnCO_3 + 2e^- \rightarrow Mn + CO_3^{2-}$	−1·48	$2H_2O + 2e^- \rightarrow H_2 + 2OH^-$	−0·828
$ZnS + 2e^- \rightarrow Zn + S^{2-}$	−1·44	$Cd(OH)_2 + 2e^- \rightarrow Cd + 2OH^-$	−0·809
$Cr(OH)_3 + 3e^- \rightarrow Cr + 3OH^-$	−1·3	$FeCO_3 + 2e^- \rightarrow Fe + CO_3^{2-}$	−0·756
$Zn(CN)_4^{2-} + 2e^- \rightarrow Zn + 4CN^-$	−1·26	$CdCO_3 + 2e^- \rightarrow Cd + CO_3^{2-}$	−0·74
$Zn(OH)_2 + 2e^- \rightarrow Zn + 2OH^-$	−1·245	$Co(OH)_2 + 2e^- \rightarrow Co + 2OH^-$	−0·73
$H_2GaO_3^- + H_2O + 3e^- \rightarrow Ga + 4OH^-$	−1·22	$HgS + 2e^- \rightarrow Hg + S^{2-}$	−0·72
$ZnO_2^{2-} + 2H_2O + 2e^- \rightarrow Zn + 4OH^-$	−1·216	$Ni(OH)_2 + 2e^- \rightarrow Ni + 2OH^-$	−0·72
$CrO_2^- + 2H_2O + 3e^- \rightarrow Cr + 4OH^-$	−1·2	$Ag_2S + 2e^- \rightarrow 2Ag + S^{2-}$	−0·69
$CdS + 2e^- \rightarrow Cd + S^{2-}$	−1·21	$AsO_2^- + 2H_2O + 3e^- \rightarrow As + 4OH^-$	−0·68
$HV_6O_{17} + 16H_2O + 30e^- \rightarrow 6V + 33OH^-$	−1·15	$AsO_4^{3-} + 2H_2O + 2e^- \rightarrow AsO_2^- + 4OH^-$	−0·67
		$Fe_2S_3 + 2e^- \rightarrow 2FeS + S^{2-}$	−0·67
		$SbO_2^- + 2H_2O + 3e^- \rightarrow Sb + 4OH^-$	−0·66
		$CoCO_3 + 2e^- \rightarrow Co + CO_3^{2-}$	−0·64

Table 21.6 (continued)

Electrode reaction	$E^\ominus$ (V)	Electrode reaction	$E^\ominus$ (V)
$Cd(NH_3)_4^{2+} + 2e^- \rightarrow Cd + 4NH_3$	-0.597	$N_2H_4 + 4H_2O + 2e^- \rightarrow$	
$ReO_4^- + 2H_2O + 3e^- \rightarrow$		$\qquad\qquad 2NH_4OH + 2OH^-$	0.1
$\qquad\qquad ReO_2 + 4OH^-$	-0.594	$Ir_2O_3 + 3H_2O + 6e^- \rightarrow Ir + 6OH^-$	0.1
$ReO_4^- + 4H_2O + 7e^- \rightarrow Re + 8OH^-$	-0.584	$Co(NH_3)_6^{3+} + e^- \rightarrow Co(NH_3)_6^{2+}$	0.1
$2SO_3^{2-} + 3H_2O + 4e^- \rightarrow$		$Mn(OH)_3 + e^- \rightarrow Mn(OH)_2$	0.1
$\qquad\qquad S_2O_3^{2-} + 6OH^-$	-0.58	$Pt(OH)_2 + e^- \rightarrow Pt + 2OH^-$	0.15
$ReO_2 + H_2O + 4e^- \rightarrow Re + 4OH^-$	-0.576	$Co(OH)_3 + e^- \rightarrow Co(OH)_2 + OH^-$	0.17
$TeO_3^{2-} + 3H_2O + 4e^- \rightarrow Te + 6OH^-$	-0.57	$PbO_2 + H_2O + 2e^- \rightarrow$	
$Fe(OH)_3 + e^- \rightarrow Fe(OH)_2 + OH^-$	-0.56	$\qquad\qquad PbO(red) + 2OH^-$	0.248
$O_2 + e^- \rightarrow O_2^-$	-0.56	$IO_3^- + 3H_2O + 6e^- \rightarrow I^- + 6OH^-$	0.26
$Cu_2S + 2e^- \rightarrow 2Cu + S^{2-}$	-0.54	$PuO_2(OH)_2 + e^- \rightarrow PuO_2OH + OH^-$	0.26
$HPbO_2^- + H_2O + 2e^- \rightarrow Pb + 3OH^-$	-0.54	$Ag(SO_3)_2^{2-} + e^- \rightarrow Ag + 2SO_3^{2-}$	0.30
$PbCO_3 + 2e^- \rightarrow Pb + CO_3^{2-}$	-0.506	$ClO_3^- + H_2O + 2e^- \rightarrow ClO_2^- + 2OH^-$	0.33
$S + 2e^- \rightarrow S^{2-}$	-0.48	$Ag_2O + H_2O + 2e^- \rightarrow 2Ag + 2OH^-$	0.344
$Ni(NH_3)_6^{2+} + 2e^- \rightarrow Ni + 6NH_3(aq.)$	-0.47	$ClO_4^- + H_2O + 2e^- \rightarrow ClO_3^- + 2OH^-$	0.36
$NiCO_3 + 2e^- \rightarrow Ni + CO_3^{2-}$	-0.45	$Ag(NH_3)_2^+ + e^- \rightarrow Ag + 2NH_3$	0.373
$Bi_2O_3 + 3H_2O + 6e^- \rightarrow 2Bi + 6OH^-$	-0.44	$TeO_4^{2-} + H_2O + 2e^- \rightarrow$	
$Cu(CN)_2^- + e^- \rightarrow Cu + 2CN^-$	-0.43	$\qquad\qquad TeO_3^{2-} + 2OH^-$	0.4
$Hg(CN)_4^{2-} + 2e^- \rightarrow Hg + 4CN^-$	-0.37	$O_2^- + H_2O + e^- \rightarrow OH^- + HO_2^-$	0.4
$SeO_3^{2-} + 3H_2O + 4e^- \rightarrow Se + 6OH^-$	-0.366	$O_2 + 2H_2O + 4e^- \rightarrow 4OH^-$	0.401
$Cu_2O + H_2O + 2e^- \rightarrow 2Cu + 2OH^-$	-0.358	$Ag_2CO_3 + 2e^- \rightarrow 2Ag + CO_3^{2-}$	0.47
$Tl(OH) + e^- \rightarrow Tl + OH^-$	-0.345	$NiO_2 + 2H_2O + 2e^- \rightarrow$	
$Ag(CN)_2^- + e^- \rightarrow Ag + 2CN^-$	-0.31	$\qquad\qquad Ni(OH)_2 + 2OH^-$	0.49
$CuCNS + e^- \rightarrow Cu + CNS^-$	-0.27	$IO^- + H_2O + 2e^- \rightarrow I^- + 2OH^-$	0.49
$HO_2^- + H_2O + e^- \rightarrow OH + 2OH^-$	-0.24	$2AgO + H_2O + 2e^- \rightarrow Ag_2O + 2OH^-$	0.57
$CrO_4^{2-} + 4H_2O + 3e^- \rightarrow$		$MnO_4^- + 2H_2O + 2e^- \rightarrow$	
$\qquad\qquad Cr(OH)_3 + 5OH^-$	-0.13	$\qquad\qquad MnO_2 + 4OH^-$	0.60
$Cu(NH_3)_2^+ + e^- \rightarrow Cu + 2NH_3$	-0.12	$RuO_4^- + e^- \rightarrow RuO_4^{2-}$	0.60
$2Cu(OH)_2 + 2e^- \rightarrow Cu_2O + 2OH^-$	-0.080	$BrO_3^- + 3H_2O + 6e^- \rightarrow Br^- + 6OH^-$	0.61
$O_2 + H_2O + 2e^- \rightarrow HO_2^- + OH^-$	-0.076	$ClO_2^- + H_2O + 2e^- \rightarrow ClO^- + 2OH^-$	0.66
$Tl(OH)_3 + 2e^- \rightarrow TlOH + 2OH^-$	-0.05	$H_3IO_6^{2-} + 2e^- \rightarrow IO_3^- + 3OH^-$	0.7
$AgCN + e^- \rightarrow Ag + CN^-$	-0.017	$2NH_2OH + 2e^- \rightarrow N_2H_4 + 2OH^-$	0.73
$MnO_2 + H_2O + 2e^- \rightarrow$		$Ag_2O_3 + H_2O + 2e^- \rightarrow 2AgO + 2OH^-$	0.74
$\qquad\qquad Mn(OH)_2 + 2OH^-$	-0.05	$BrO^- + H_2O + 2e^- \rightarrow Br^- + 2OH^-$	0.76
$NO_3^- + H_2O + 2e^- \rightarrow NO_2^- + 2OH^-$	0.01	$HO_2^- + H_2O + 2e^- \rightarrow 3OH^-$	0.88
$HOsO_5^- + 4H_2O + 8e^- \rightarrow Os + 9OH^-$	0.02	$ClO^- + H_2O + 2e^- \rightarrow Cl^- + 2OH^-$	0.89
$Rh_2O_3 + 3H_2O + 6e^- \rightarrow 2Rh + 6OH^-$	0.04	$FeO_4^{2-} + 2H_2O + 3e^- \rightarrow$	
$SeO_4^{2-} + H_2O + 2e^- \rightarrow$		$\qquad\qquad FeO_2^- + 4OH^-$	0.9
$\qquad\qquad SeO_3^{2-} + 2OH^-$	0.05	$ClO_2 + e^- \rightarrow ClO_2^-$	1.16
$Pd(OH)_2 + 2e^- \rightarrow Pd + 2OH^-$	0.07	$O_3 + H_2O + 2e^- \rightarrow O_2 + 2OH^-$	1.24
$S_4O_6^{2-} + 2e^- \rightarrow 2S_2O_3^{2-}$	0.08	$OH + e^- \rightarrow OH^-$	2.0
$HgO(red) + H_2O + 2e^- \rightarrow Hg + 2OH^-$	0.098		

Table 21.7 Reference electrodes

Electrode	Electrode equilibrium	Potential at 25°C (vs. S.H.E.; V)
Calomel (Hg/Hg$_2$Cl$_2$, Cl$^-$)	Hg$_2$Cl$_2$ + 2e ⇌ 2Hg + 2Cl$^-$	$E = 0.2677 - 0.0591 \log a_{Cl^-}$

Solution	$E_{calomel}$	$E_{calomel + liquid\ junction}$	Temp. coeff.
0.1 mol dm^{-3} KCl	0.3337	0.336	−0.06 mV/°C
1.0 mol dm^{-3} KCl	0.280	0.283	−0.24 mV/°C
Sat. KCl	0.241	0.244	−0.65 mV/°C

Electrode	Electrode equilibrium	Potential at 25°C (vs. S.H.E.; V)
Mercury/mercurous sulphate (Hg/HgSO$_4$, SO$_4^{2-}$)	HgSO$_4$ + 2e ⇌ Hg + SO$_4^{2-}$	$E = 0.6151 - 0.0295 \log a_{SO_4^{2-}}$ Average temp. coeff. ≈ −0.6 mV/°C*
Silver/silver chloride (Ag/AgCl, Cl$^-$)	AgCl + e ⇌ Ag + Cl$^-$	$E = 0.2224 - 0.0591 \log a_{Cl^-}$ 0.1 mol dm^{-3} KCl, $E = 0.2881$ V 1.0 mol dm^{-3} KCl, $E = 0.2224$ V Sea-water $E \approx 0.250$ V
Copper/copper sulphate (Cu/CuSO$_4$, Cu^{2+})	Cu^{2+} + 2e ⇌ Cu	$E = 0.340 + 0.0295 \log a_{Cu^{2+}}$; for sat. CuSO$_4$, $E = +0.318$ V; for practical electrodes $E \approx 0.30$ V
Quinhydrone	Quinone + H$_2$ ⇌ hydroquinone	$E = E_x^{\ominus} - 0.0591$ pH, and $E_x^{\ominus} = 0.6990$ at 25°C $E_x^{\ominus}$ contains a term due to diffusion potentials and is not a thermodynamic constant
Antimony/antimony oxide (Sb/Sb$_2$O$_3$, H$^+$)	Sb$_2$O$_3$ + 6H$^+$ + 6e ⇌ 2Sb + 3H$_2$O	$E = 0.1445 - 0.0591$ pH
Mercury/mercuric oxide (Hg/HgO, OH$^-$)	HgO + H$_2$O + 2e ⇌ Hg + 2OH$^-$	$E = 0.0974 - 0.0591$ pH (for pH determinations in alkaline solution)
Lead dioxide/lead sulphate (Pb/PbO$_2$/PbSO$_4$, SO$_4^{2-}$)	PbO$_2$ + 4H$^+$ + SO$_4^{2-}$ + 2e ⇌ PbSO$_4$ + 2H$_2$O	$E = 1.685 + 0.0295 \log 4m^3\gamma_{\pm}^3/\alpha_w^2$, where $\gamma_{\pm}$ and α_w are the stoichiometric mean activity coefficient of sulphuric acid and the activity of water, respectively, at molality m of H$_2$SO$_4$

Solution	E
0.1 mol dm^{-3} H$_2$SO$_4$	+1.565
1.1 mol dm^{-3} H$_2$SO$_4$	+1.632
6.1 mol dm^{-3} H$_2$SO$_4$	+1.735

Electrode	Electrode equilibrium	Potential at 25°C (vs. S.H.E.; V)
Zn/ZnSO$_4$	Zn^{2+} + 2e ⇌ Zn	$E = -0.763 + 0.0295 \log a_{Zn^{2+}}$
Zn/sea-water	Mixed potentials approximating to $E^{\ominus}_{M^{z+}/M}$	$E \approx -0.80$ V
Zn/artificial sea-water	Mixed potentials approximating to $E^{\ominus}_{M^{z+}/M}$	$E \approx 0.81$ V
Cd/sea-water	Mixed potentials approximating to $E^{\ominus}_{M^{z+}/M}$	$E \approx -0.52$ V
Cd/artificial sea-water	Mixed potentials approximating to $E^{\ominus}_{M^{z+}/M}$	$E \approx 0.54$ V

*Variation of $E^{\ominus}_{Ag/AgCl}$ with temperature:

Temp., °C	35	45	55	70	95	125	150	200
$E^{\ominus}$, V	0.21570	0.20828	0.20042	0.18782	0.1651	0.1330	0.1032	0.0348

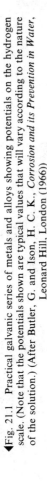

NOBLE OR CATHODIC END

Passive

Stainless steels

Active

Gold
Graphite
Titanium
Silver
Molybdenum
Nickel
Monel
70/30 cupro-nickel
Copper
67/33 nickel-copper
Nickel (active)
Aluminium bronze
70/30 brass
Gunmetal
60/40 brass
Chromium
Ni-resist

Tin
2/1 tin-lead solder
Lead

Aluminium and aluminium alloys

Steel
Grey cast iron
Cadmium

Galvanised iron

Zinc

Magnesium

BASE OR ANODIC END

Potential (vs. S.H.E.; mV)

+200 · 0 · -200 · -400 · -600 · -800 · -1000 · -1200 · -1400

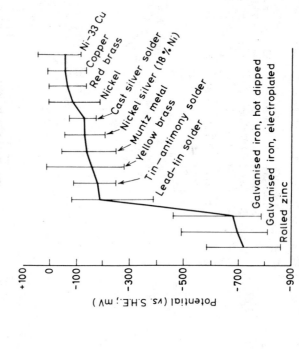

Ni-33 Cu
Copper
Red brass
Nickel
Cast silver solder
Nickel silver (18% Ni)
Muntz metal
Yellow brass
Tin-antimony solder
Lead-tin solder

Galvanised iron, hot dipped
Galvanised iron, electroplated
Rolled zinc

Potential (vs. S.H.E.; mV)

+100 · 0 · -100 · -300 · -500 · -700 · -900

◄Fig. 21.1 Practical galvanic series of metals and alloys showing potentials on the hydrogen scale. (Note that the potentials shown are typical values that will vary according to the nature of the solution.) (After Butler, G. and Ison, H. C. K., *Corrosion and its Prevention in Water*, Leonard Hill, London (1966))

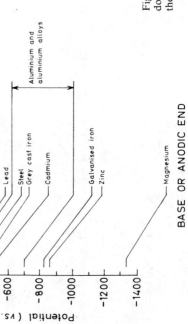

Fig. 21.2 Galvanic series showing ranges of potentials of metals and alloys in flowing hot domestic water at 71°C (Long Island, N.Y.). Potentials measured weekly for three months and then monthly for a period of ten months. (After Butler, G. and Ison, H. C. K., *Corrosion and its Prevention in Water*, Leonard Hill, London (1966))

Table 21.8 Galvanic series of some commercial metals and alloys in sea-water*

↑	Noble or cathodic	Platinum
		Gold
		Graphite
		Titanium
		Silver
		⎰ Chlorimet 3 (62Ni–18Cr–18Ni)
		⎱ Hastelloy C (62Ni–17Cr–15Mo)
		⎰ 18/8 Mo stainless steel (passive)
		⎱ 18/8 stainless steel (passive)
		Chromium stainless steel 11–30% Cr (passive)
		⎰ Inconel (passive) (Ni–13Cr–7Fe)
		⎱ Nickel (passive)
		Silver solder
		⎰ Monel (Ni–30Cu)
		Cupro-nickels (Cu–10 to 40Ni)
		Bronzes (Cu–Sn)
		Copper
		⎱ Brasses (Cu–Zn)
		⎰ Chlorimet 2 (66Ni–32Mo–1Fe)
		⎱ Hastelloy B (60Ni–30Mo–6Fe–1Mn)
		⎰ Inconel (active)
		⎱ Nickel (active)
		Tin
		Lead
		Lead–tin solders
		⎰ 18/8 Mo stainless steel (active)
		⎱ 18/8 stainless steel (active)
		Ni-Resist (high nickel cast iron)
		Chromium stainless steel, 13% Cr (active)
		⎰ Cast iron
		⎱ Steel or iron
		2024 aluminium (Al–4·5Cu–1·5Mg–0·6Mn)
	Active or anodic	Cadmium
		Commercially pure aluminium (1100)
↓		Zinc
		Magnesium and magnesium alloys

*Data after Fontana, M. G., and Greene, N. D., *Corrosion Engineering*, McGraw Hill (1967).

Table 21.9 Stoichiometric mean molal activity coefficients ($\gamma_\pm$) for aqueous inorganic electrolytes at 25°C*

Electrolyte	Molality									
	0·001	0·005	0·01	0·05	0·1	0·2	0·5	1·0	3	7
$AgNO_3$		0·925	0·897	0·793	0·734	0·657	0·536	0·429	0·252	0·142
$AlCl_3$				0·447	0·337	0·305	0·331	0·539		
$Al(NO_3)_2$					0·204	0·157		0·190	0·102	
$Al_2(SO_4)_3$					0·035	0·023	0·014	0·018		
$Ba(OH)_2$		0·773	0·712	0·526	0·443	0·370				
$CaCl_2$	0·889	0·789	0·731	0·583	0·518	0·472	0·448	0·500	1·483	18·28
$Ca(NO_3)_2$	0·88	0·77	0·71	0·545	0·485	0·426	0·363	0·336	0·380	0·690

*Data after Parsons, *Handbook of Electrochemical Constants*, Butterworths, London (1959).

Table 21.9 (continued)

Electrolyte	Molality									
	0·001	0·005	0·01	0·05	0·1	0·2	0·5	1·0	3	7
$Cd(NO_3)_2$					0·513	0·464	0·425	0·433		
$CdCl_2$	0·819	0·623	0·524	0·304	0·228	0·164	0·101	0·0669	0·0352	
$CdSO_4$	0·726	0·505	0·399	0·206	0·150	0·102	0·061	0·041	0·033	
$CoCl_2$					0·522	0·479	0·462	0·531		
$Co(NO_3)_2$					0·518	0·471	0·445	0·490		
$CrCl_3$					0·331	0·298	0·314	0·481		
$Cr(NO_3)_3$					0·319	0·285	0·291	0·401		
$Cr_2(SO_4)_3$					0·0458	0·0300	0·0190	0·0208		
$CuCl_2$	0·888	0·783	0·723	0·577	0·508	0·455	0·411	0·417	0·520	
$Cu(NO_3)_2$					0·511	0·460	0·426	0·455	0·903	4·21
$CuSO_4$	0·74	0·573	0·438	0·217	0·154	0·104	0·062	0·043		
$FeCl_2$					0·518	0·473	0·450	0·506		
HCl	0·965	0·928	0·904	0·830	0·796	0·767	0·757	0·809	1·316	4·37
HNO_3	0·965	0·927	0·902	0·823	0·791	0·754	0·720	0·724	0·909	
H_2SO_4	0·830	0·639	0·544	0·340	0·265	0·209	0·156	0·132	0·142	0·317
$HClO_4$					0·803	0·778	0·769	0·823	1·448	7·44
KCl	0·965	0·927	0·902	0·816	0·770	0·718	0·649	0·604	0·569	
KNO_3	0·965	0·926	0·898	0·799	0·739	0·663	0·545	0·443	0·269	
K_2SO_4	0·885	0·777	0·711	0·525	0·441	0·360	0·264			
KOH				0·824	0·798	0·760	0·732	0·756	1·081	2·88
$MgCl_2$					0·529	0·489	0·481	0·570	2·32	
$Mg(NO_3)_2$	0·882	0·771	0·712	0·554	0·523	0·481	0·470	0·537	1·452	
$MnCl_2$					0·516	0·469	0·440	0·479	0·934	2·25
$MnSO_4$					0·150	0·106	0·064	0·044	0·038	
NH_4Cl		0·924	0·896	0·808	0·770	0·718	0·649	0·603	0·561	0·566
NH_4NO_3		0·925	0·897	0·799	0·740	0·677	0·582	0·504	0·368	0·261
$NaCl$	0·965	0·928	0·903	0·822	0·778	0·735	0·681	0·657	0·714	
$NaNO_3$	0·966	0·929	0·905	0·821	0·762	0·703	0·617	0·548	0·437	
Na_2SO_4	0·877	0·778	0·714	0·536	0·445	0·365	0·266	0·201	0·137	
$NaCNS$					0·787	0·750	0·715	0·712	0·814	
NaH_2PO_4					0·744	0·675	0·563	0·468	0·320	
$NaOH$			0·905	0·818	0·766	0·727	0·690	0·678	0·784	1·603
$NiCl_2$					0·522	0·479	0·464	0·536	1·692	
$NiSO_4$					0·150	0·105	0·063	0·042	0·035	
$PbCl_2$	0·859	0·704	0·612							
UO_2Cl_2					0·544	0·510	0·517	0·620	1·551	
$UO_2(NO_3)_2$					0·551	0·520	0·542	0·689	2·03	
UO_2SO_4					0·150	0·102	0·0611	0·0439	0·0383	
$ZnCl_2$	0·88	0·77	0·71	0·56	0·515	0·462	0·394	0·339	0·287	0·499
$Zn(NO_3)_2$					0·531	0·489	0·473	0·535	1·363	
$ZnSO_4$	0·700	0·477	0·387	0·202	0·150	0·104	0·063	0·043	0·041	

Note. Where experimental data are unavailable, mean ionic activity coefficients (up to an ionic strength I of 0·1) in water at 25°C can be calculated from the formula

$$-\log \gamma_{\pm} = 0.509 z_+ z_- \left(\frac{\sqrt{I}}{1+\sqrt{I}} - 0.020 I \right)$$

where z_+, z_- are valencies of the ions, and I (the ionic strength) = $\frac{1}{2}\Sigma m_i z_i^2$.

Table 21.10 Differential diffusion coefficients for dilute aqueous solutions at 25°C*

Solution	Range of concentration (mol dm^{-3})	Range of $D \times 10^5$ $(\text{cm}^2\,\text{s}^{-1})$†
$AgNO_3$	$0-6\cdot28 \times 10^{-3}$	$1\cdot768-1\cdot701$
$BaCl_2$	$0-5\cdot42 \times 10^{-3}$	$1\cdot387-1\cdot261$
$CaCl_2$	$0-5\cdot01 \times 10^{-3}$	$1\cdot336-1\cdot179$
$CaCl_2$	$0-3\cdot5$	$1\cdot336-1\cdot195$
$CsCl$	$0-12\cdot87 \times 10^{-3}$	$2\cdot406-1\cdot946$
Cs_2SO_4	$0-4\cdot72 \times 10^{-3}$	$1\cdot569-1\cdot424$
HBr	$0-1\cdot0$	$3\cdot403-3\cdot869$
HCl	$0-4\cdot0$	$3\cdot339-5\cdot17$
KBr	$0-4\cdot0$	$2\cdot018-2\cdot434$
KCl (4°C)	$16\cdot6-558 \times 10^{-3}$	$1\cdot080-1\cdot042$
(20°C)	$0-11\cdot21 \times 10^{-3}$	$1\cdot765-1\cdot689$
(25°C)	$0-527\cdot6 \times 10^{-3}$	$1\cdot996-1\cdot852$
(30°C)	$0-12\cdot36 \times 10^{-3}$	$2\cdot233-2\cdot139$
KCl	$0-3\cdot5$	$1\cdot995-2\cdot152$
KI	$0-3\cdot5$	$2\cdot001-2\cdot533$
KNO_3	$0-9\cdot19 \times 10^{-3}$	$1\cdot931-1\cdot855$
$K_4Fe(CN)_6$	$0-5\cdot56 \times 10^{-3}$	$1\cdot473-1\cdot178$
$LaCl_3$	$0-26 \times 10^{-3}$	$1\cdot294-1\cdot021$
$LiBr$	$0-3\cdot5$	$1\cdot379-1\cdot693$
$LiCl$	$0-11\cdot00 \times 10^{-3}$	$1\cdot368-1\cdot313$
$LiCl$	$0-3\cdot5$	$1\cdot368-1\cdot464$
$LiNO_3$	$0-4\cdot0$	$1\cdot337-1\cdot292$
Li_2SO_4	$0-5\cdot73 \times 10^{-3}$	$1\cdot041-0\cdot946$
$MgCl_2$	$0-4\cdot00 \times 10^{-3}$	$1\cdot251-1\cdot164$
$MgSO_4$	$0-6\cdot36 \times 10^{-3}$	$0\cdot849-0\cdot702$
$NaBr$	$0-2\cdot5$	$1\cdot627-1\cdot702$
$NaCl$	$0-14\cdot73 \times 10^{-3}$	$1\cdot612-1\cdot542$
$NaCl$	$0-4\cdot5$	$1\cdot612-1\cdot607$
NaI	$0-3\cdot0$	$1\cdot616-1\cdot992$
Na_2SO_4	$0-4\cdot79 \times 10^{-3}$	$1\cdot230-1\cdot124$
NH_4Cl	$0-4\cdot5$	$1\cdot996-2\cdot257$
$RbCl$	$0-11\cdot10 \times 10^{-3}$	$2\cdot057-1\cdot969$
$SrCl_2$	$0-7\cdot74 \times 10^{-3}$	$1\cdot336-1\cdot208$

*Data after Parsons, *Handbook of Electrochemical Constants*, Butterworths, London (1959).
†$D \times 10^9\,\text{m}^2\,\text{s}^{-1}$.

Table 21.11 Ionisation constants of water and weak electrolytes and variation with temperature

A. Ionisation constants of water ($pK_w = -\log K_w$)

Temperature (°C)	$-\log K_w$	Temperature (°C)	$-\log K_w$
0	14·9435	35	13·6801
5	14·7338	40	13·5348
10	14·5346	45	13·3960
15	14·3463	50	13·2617
20	14·1669	55	13·1369
25	13·9965	60	13·0171
30	13·8330		

B. Ionisation constants of weak electrolytes and their temperature variation
$$pK_\alpha = -\log K_\alpha = A_1/T - A_2 + A_3 T$$

Aqueous solution	pK_α at 25°C	A_1	A_2	A_3
Acetic acid	4·756	1170·48	3·1649	0·013 399
Ammonium ion	9·245	2835·76	0·6322	0·001 225
Benzoic acid	4·201	1590·2	6·394	0·017 65
Boric acid	9·234	2237·94	3·305	0·016 883
n-Butyric acid	4·820	1033·39	2·6215	0·013 334
Carbonic acid K_1	6·352	3404·71	14·8435	0·032 786
Carbonic acid K_2	10·329	2902·39	6·4980	0·023 79
Chloroacetic acid	2·861	1049·05	5·0273	0·014 654
Citric acid K_1	3·128	1255·6	4·5635	0·011 673
Citric acid K_2	4·761	1585·2	5·4460	0·016 399
Citric acid K_3	6·396	1814·9	6·3664	0·022 389
Formic acid	3·752	1342·85	5·2743	0·015 168
Glycine K_1	2·350	1332·17	5·8870	0·012 643
Glycine K_2	9·780	2686·95	0·5103	0·004 286
Lactic acid	3·860	1286·49	4·8607	0·014 776
Malonic acid K_1	2·855	—	—	—
Malonic acid K_2	5·696	1703·31	6·5810	0·022 014
Oxalic acid K_1	1·271	—	—	—
Oxalic acid K_2	4·266	1423·8	6·5007	0·020 095
Phosphoric acid K_1	2·148	799·31	4·5535	0·013 486
Phosphoric acid K_2	7·198	1979·5	5·3541	0·019 840
o-Phthalic acid K_1	2·950	561·57	1·2843	0·007 883
o-Phthalic acid K_2	5·408	2175·83	9·5508	0·025 694
Succinic acid K_1	4·207	1206·25	3·3266	0·011 697
Succinic acid K_2	5·638	1679·13	5·7043	0·019 153
Sulphamic acid	0·988	3792·8	24·122	0·041 544
Sulphanilic acid	3·227	1143·71	1·2979	0·002 314
Tartaric acid K_1	3·033	1525·59	6·6558	0·015 336
Tartaric acid K_2	4·366	1765·35	7·3015	0·019 276
Trimethylammonium ion	9·800	541·4	$-12·611$	$-0·015 525$

Note. All values are with reference to the molarity scale. Data for bases are expressed as acidic ionisation constants, e.g. for ammonia we quote pK at 25° = 9·245 for the ammonium ion

$$NH_4^+ + H_2O \rightarrow NH_3 + H_3O^+$$

The basic ionisation constant of the reaction

$$NH_3 + H_2O \rightarrow NH_4^+ + OH^-$$

is obtained from the relation

$$pK_a (\text{acidic}) + pK_b (\text{basic}) = pK_w (\text{water})$$

pK_w (water) being 13·9965 at 25°C.

Table 21.12   Tafel constants for hydrogen evolution from aqueous solution*

The Tafel equation for a cathodic reaction is $\eta_c = a - b\log_{10}i = b\log_{10}(i_0/i)$, where η_c is the overpotential (mV), i is the c.d. (A cm^{-2}) and i_0 is the exchange c.d. (A cm^{-2}). Note that η will always be negative

Metal	Electrolyte		Temp. (°C)	$-\log_{10}i$ range	$-a$ (mV)	b (mV)	$-\log_{10}i_0$ (i_0 in A cm^{-2})
Ag	0·001 N	HCl	20	6·0 to 2·0	810	125	6·5
	0·01 N	HCl	20	6·0 to 2·0	820	130	6·3
	0·1 N	HCl	20	6·0 to 3·3	570	90	6·3
	0·1 N	HCl	20	3·3 to 1·0	670	120	5·6
	1·0 N	HCl	20	6·0 to 2·3	320	60	5·4
	1·0 N	HCl	20	2·3 to 1·0	480	130	3·7
	5·0 N	HCl	20	6·0 to 2·8	470	70	6·7
	5·0 N	HCl	20	2·8 to −2	630	120	5·3
	7·0 N	HCl	20	6·0 to 3·4	640	90	7·1
	7·0 N	HCl	20	3·4 to 1·0	740	110	6·7
Al	2 N	H$_2$SO$_4$	25	3·0 to 0·7	1000	100	10·0
Au	0·001 N	HCl	20	7·0 to 2·0	524	72	7·32
	0·01 N	HCl	20	6·0 to 2·0	558	84	6·63
	0·1 N	HCl	20	6·0 to 3·0	468	71	6·59
	0·1 N	HCl	20	3·0 to 2·0	548	97	5·64
	0·001 N	NaOH	20	6·0 to 4·5	832	118	7·05
	0·01 N	NaOH	20	6·5 to 3·7	836	119	7·04
	0·1 N	NaOH	20	4·8 to 3·0	856	123	6·95
Be	1·0 N	HCl	20	3·0 to 1·3	1080	120	9·0
Bi	1·0 N	HCl	20	3·0 to 1·0	840	120	7·0
Cd	1·7 N	H$_2$SO$_4$	20	4·0 to 3·0	1450	120	12·1
	10 N	H$_2$SO$_4$	20	4·0 to 2·0	1400	120	11·7
Cu	0·001 N	HCl	20	5·0 to 3·3	802	122	6·61
	0·01 N	HCl	20	4·5 to 2·3	786	118	6·71
	0·1 N	HCl	20	5·0 to 2·5	790	117	6·76
	0·005 N	NaOH	16		890	139	6·40
	0·02 N	NaOH	16	6·0 to 3·7	710	114	6·29
	0·15 N	NaOH	16		690	117	5·99
Fe	0·001 N	HCl	20	4·0 to 3·8	787	127	6·19
	0·01 N	HCl	20	4·1 to 3·2	741	118	6·29
	1 N	HCl	16	3·0 to 0·0	770	130	5·9
	0·01 N	NaOH	20	4·5 to 3·8	776	117	6·62
	0·1 N	NaOH	20	4·1 to 3·2	726	120	6·06
	4·8 N	KOH	20	4·0 to 3·0	350	70	5·0
	10·5 N	KOH	20	4·0 to 3·0	340	70	4·9
Ga	0·2 N	H$_2$SO$_4$	87		800	120	6·7
Hg	0·001–0·1 N	HCl	20	7·0 to 1·0	1410	116	12·2
	1 N	HCl	20	6·0 to 2·5	1390	119	11·7
	3 N	HCl	20	6·0 to 2·5	1420	141	10·1
	5 N	HCl	20	6·0 to 2·5	1320	127	10·4
	7 N	HCl	20	6·0 to 2·5	1130	108	10·5
	10 N	HCl	20	6·0 to 2·5	1020	95	10·7
	0·1 N	H$_2$SO$_4$	20	6·0 to 2·5	1440	114	12·7
	0·25 N	H$_2$SO$_4$	20	6·5 to 3·0	1403	116	12·1
	5 N	H$_2$SO$_4$	20	6·5 to 3·0	1400	116	12·05
	0·1 N	LiOH	20	6·0 to 4·0	1598	102	15·7
Hg	0·2 N	LiOH	20	6·0 to 4·0	1545	100	15·5
	0·1 N	NaOH	20	6·0 to 4·0	1457	100	14·6
	0·2 N	NaOH	20	6·0 to 4·0	1405	97	14·5
	0·002 N	KOH	20	6·0 to 4·0	1682	98	17·1
	0·02 N	KOH	20	6·0 to 4·0	1545	90	17·3
	0·1 N	KOH	20	6·0 to 4·0	1430	93	15·4

*Data after Parsons, *Handbook of Electrochemical Constants*, Butterworths, London (1959).

Table 21.12 (continued)

Metal	Electrolyte		Temp. (°C)	$-\log_{10}i$ range	$-a$ (mV)	b (mV)	$-\log_{10}i_0$ (i_0 in A cm^{-2})
Hg (cont.)	0·01 N	Ba(OH)$_2$	20	6·0 to 4·0	1170	45	26·0
	0·02 N	Ba(OH)$_2$	20	6·0 to 4·0	1220	65	18·8
	0·1 N DCl in D$_2$O		20	5·0 to 2·4	1485	119	12·19
Mo	0·001 N	HCl	20	5·6 to 4·2	557	81	7·12
	0·01 N	HCl	20	5·2 to 3·7	543	76	7·19
	0·1 N	HCl	20	6·0 to 3·5	586	80	7·30
	0·1 N	HCl	20	3·5 to 2·0	671	104	6·45
	0·001 N	NaOH	20	5·9 to 4·4	667	92	7·27
	0·01 N	NaOH	20	4·9 to 3·6	664	103	6·42
	0·1 N	NaOH	20	4·7 to 3·7	641	87	7·35
	0·1 N	NaOH	20	3·6 to 2·1	739	116	6·37
Nb	1·0 N	HCl	20	3·0 to 1·0	900	80	11·0
Ni	0·000 04 N	HCl	20	6·0 to 5·0	650	100	6·5
	0·001 N	HCl	20	5·8 to 3·3	617	93	6·6
	0·01 N	HCl	20	5·5 to 3·3	611	91	6·7
	0·1 N	HCl	20	5·0 to 2·0	626	104	6·0
	1·0 N	HCl	20	4·3 to 2·0	594	109	5·4
	0·001 N	NaOH	20	6·8 to 4·8	720	103	7·0
	0·006 N	NaOH	20	6·3 to 3·8	660	101	6·6
	0·1 N	NaOH	20	6·0 to 3·0	650	101	6·4
Pb	0·1 N	HCl	20	5·8 to 2·5	1524	116	13·2
	1 N	HCl	20	5·8 to 2·5	1531	119	12·9
	3 N	HCl	20	5·1 to 2·5	1573	142	11·1
	5 N	HCl	20	4·9 to 2·5	1495	140	10·7
	7 N	HCl	20	4·7 to 2·5	1417	138	9·76
	10 N	HCl	20	4·6 to 2·0	1195	135	8·84
	0·1 N	H$_2$SO$_4$	20	7·0 to 2·5	1533	118	13·0
	1 N	H$_2$SO$_4$	20	6·5 to 2·0	1536	119	12·9
	8 N	H$_2$SO$_4$	20	5·9 to 2·0	1530	120	12·8
	15 N	H$_2$SO$_4$	20	5·3 to 2·0	1469	121	12·1
	20 N	H$_2$SO$_4$	20	5·0 to 2·0	1411	119	11·9
	1 N	HBr	20	5·3 to 2·3	1484	116	12·7
	3 N	HBr	20	5·1 to 2·3	1467	123	11·9
	6 N	HBr	20	4·7 to 2·3	1377	130	10·6
	8·5 N	HBr	20	4·3 to 2·3	1285	140	9·17
	1 N	HClO$_4$	20	4·8 to 1·6	1537	118	13·0
	3 N	HClO$_4$	20	4·8 to 1·6	1517	118	12·8
	7 N	HClO$_4$	20	4·8 to 1·6	1504	121	12·4
	9 N	HClO$_4$	20	4·8 to 1·6	1453	122	11·9
	11·6 N	HClO$_4$	20	4·8 to 1·6	1446	132	11·0
Pd	1·0 N	H$_2$SO$_4$	20	3·0 to 2·0	240	80	3·0
	0·01 N	HCl	20	3·9 to 3·1	447	107	4·18
	0·1 N	HCl	20	2·9 to 1·4	321	99	3·25
	0·001 N	NaOH	20	5·0 to 3·9	589	100	5·88
	0·01 N	NaOH	20	5·4 to 4·0	610	110	5·56
	0·1 N	NaOH	20	4·1 to 3·1	637	125	5·01
Pt	0·5 N	HCl	25	2·0 to 0·7	73	28	2·6
Rh	0·01 N	HCl	20	3·4 to 3·1	209	55	3·80
	0·01 N	NaOH	20	4·2 to 3·5	551	119	4·64
Sb	2 N	H$_2$SO$_4$	20	3·0 to 0	900	100	9·0
Sn	1 N	HCl	20	3·0 to 0	1100	140	8·0
Ta	1 N	HCl	20	3·0 to 1·0	550	120	4·6
W	5 N	HCl	20	2·0 to −2·0	550	110	5·0

Table 21.13 Exchange current densities i_0 for the hydrogen evolution reaction

Metal	$-\log i_0$ [A cm^{-2}] in approx. 1 mol dm^{-3} H$_2$SO$_4$
Palladium	3·0
Platinum	3·1
Rhodium	3·6
Iridium	3·7
Nickel	5·2
Gold	5·4
Tungsten	5·9
Niobium	6·8
Titanium	8·2
Cadmium	10·8
Manganese	10·9
Thallium	11·0
Lead	12·0
Mercury	12·3

After Bockris, J. O'M. and Reddy, A. K. N., *Modern Electrochemistry*, Macdonald (1970).

Table 21.14 Exchange current densities and transfer coefficients α for evolution of gases at 20–25°C at different anodes

Gas	Metal	Solution	α	i_0 (A cm^{-2})
O$_2$	Au	0·1 mol dm^{-3} NaOH	0·74–1·2	5×10^{-13}
	Pt	0·5 mol dm^{-3} H$_2$SO$_4$	0·45	10^{-8}–10^{-11}
	Pt	HNO$_3$ + NaOH pH 0·5–14	0·51	$0·6$–1×10^{-10}
	Pt	Phosphate buffer pH 6·8	0·29	
	Pt	0·1 mol dm^{-3} NaOH	0·81	4×10^{-13}
	PbO$_2$	0·5 mol dm^{-3} H$_2$SO$_4$	0·50	
		1·0 mol dm^{-3} KOH	0·50	
Cl$_2$	Pt	1·0 mol dm^{-3} HCl	0·48	5×10^{-3}
	Pt	Various solutions	0·5–0·7	10^{-3}
	Ir	1·0 mol dm^{-3} HCl	0·73	4×10^{-5}
	PbO$_2$	0·5–2 mol dm^{-3} NaCl	0·17–0·27	10^{-7}
Br$_2$	Ir	1 mol dm^{-3} HBr	0·6	3×10^{-3}
	Pt	1 N KBr	0·5–0·7	3×10^{-3}
N$_2$	Pt	1·0 mol dm^{-3} NaN$_3$	0·98	10^{-76}
	Ir	1·0 mol dm^{-3} NaN$_3$	1·0	10^{-75}
	Pd	1·0 mol dm^{-3} NaN$_3$	1·1	10^{-81}

Data after Parsons, *Handbook of Electrochemical Constants*, Butterworths, London (1959).

Table 21.15 Exchange current densities i_0 at 25°C for some electrode reactions*

Metal	System	Medium	$-\log i_0$ (A cm^{-2})
Mercury	Cr^{3+}/Cr^{2+}	KCl	6·0
Platinum	Ce^{4+}/Ce^{3+}	H_2SO_4	4·4
Platinum	Fe^{3+}/Fe^{2+}	H_2SO_4	2·6
Rhodium	Fe^{3+}/Fe^{2+}	H_2SO_4	2·76
Iridium	Fe^{3+}/Fe^{2+}	H_2SO_4	2·8
Palladium	Fe^{3+}/Fe^{2+}	H_2SO_4	2·2
Gold	H^+/H_2	H_2SO_4	3·6
Platinum	H^+/H_2	H_2SO_4	3·1
Mercury	H^+/H_2	H_2SO_4	12·1
Nickel	H^+/H_2	H_2SO_4	5·2
Tungsten	H^+/H_2	H_2SO_4	5·9
Lead	H^+/H_2	H_2SO_4	11·3

*Data after Bockris, J. O'M. and Reddy, A. K. N., *Modern Electrochemistry*, Macdonald (1970).

Table 21.16 Exchange current densities for several noble metals and a platinum–rhodium alloy in the reduction of oxygen from perchloric acid solution*

Metal or alloy	Exchange current density i_0 (A cm^{-2})
Platinum	10^{-9}
Platinum and 40 atomic % rhodium	10^{-9}
Rhodium	6×10^{-9}
Iridium	10^{-11}

*Data after Bockris, J. O'M. and Reddy, A. K. N., *Modern Electrochemistry*, Macdonald (1970).

Table 21.17 Exchange current densities for M^{z+}/M equilibria in different solutions*

Metal	Solution	i_0 (A cm^{-2})
Zn	Perchlorate	3×10^{-8}
Pb	Perchlorate	8×10^{-4}
Tl	Perchlorate	10^{-3}
Ag	Perchlorate	1·0
Bi (amalgam)	Perchlorate	10^{-5}
Ni	Sulphate	2×10^{-9}
Fe	Sulphate	$10^{-8}, 2 \times 10^{-9}$
Zn	Sulphate	3×10^{-5}
Cu	Sulphate	$4 \times 10^{-5}, 3 \times 10^{-2}$
Tl	Sulphate	2×10^{-3}
Sb	Chloride	2×10^{-5}
Zn	Chloride	$3 \times 10^{-4}, 7 \times 10^{-1}$
Sn	Chloride	2×10^{-3}
Bi	Chloride	3×10^{-2}
Hg	$Hg_2(NO_3)_2 + HClO_4$	2×10^{-1}

*Data after West, J. M., *Electrodeposition and Corrosion Processes*, 2nd edn, Van Nostrand Reinhold (1970).

Table 21.18 Structures, thermal data, and molecular volumes of metal oxides and hydroxides, and of some double oxides*

A. Metal oxides and hydroxides

Compound		Remarks	Structure	m.p. etc. (°C)	b.p. etc. (°C)	$\Delta H_{dec.}$ (kJ)	Volume (cm³)	φ†
Li₂O			CaF₂	sbl.	s.p. 1300	1194·1	15·0	0·58
Li₂O₂	α		hexag.	t.p. 225		77·5	21·45	0·83
	β	Li₂O₂₋₁·₈		d. 160				
LiOH			SnO	m.p. 462	d. 770	137·3	16·4	1·26
Na₂O			CaF₂		d. vol.	843·3	25·9	0·55
Na₂O₂			tetrag.		d. 585	186·7	31·6	0·67
NaO₃			NaCl (Cl = O₂)			8·4	24·9	1·05
NaOH	α		tetrag.: B33	t.p. 300		149·0	(18·8)	0·83
	β			m.p. 320	d. 1000	140·3		
K₂O			CaF₂	d. vol.	d.	723·5	40·4	0·45
K₂O₂			tetrag.	m.p. 490		173·3	33·0	0·73
KO₂	α		NaO₂	t.p. 80			34·0	0·75
	β		monocl.	m.p. 380	d.	117·2		
KOH	α			t.p. 249		205·2	26·5	0·58
	β		NaCl	m.p. 400		192·6		
Rb₂O			CaF₂		b.p. 1330	660·7	29·0	0·64
Rb₂O₂			Th₃P₄	m.p. 570	d.	192·6	46·4	0·42
Rb₂O₃			CaC₂	m.p. 489	d. 920	129·8	55·6	0·50
RbO₂				m.p. 412	d. 770	83·7	62·1	0·56
RbOH	α		(NaCl)	t.p. 245		211·9	38·3	0·69
	β			m.p. 300	d. 730	197·6	(32·0)	
Cs₇O				m.p. 4				
Cs₄O				d. 10				
Cs₃O			hexag.	m.p. 165			151·2	0·74
Cs₂O			CdCl₂	m.p. 490	d. vol.	636·4	60·0	0·44
Cs₂O₂				m.p. 594	d. 960	318·2		
Cs₂O₃			Th₃P₄	m.p. 502	d. 870	175·9	73·8	0·54

*Data after Kubaschewski and Hopkins, *The Oxidation of Metals and Alloys*, 2nd edn, Butterworths, London (1962).
†φ is the volume ratio.
Abbreviations are given on page **21**.52.

Table 21.18 (continued)

Compound	Remarks	Structure	m.p. etc. (°C)	b.p. etc. (°C)	$\Delta H_{dec.}$ (kJ)	Volume (cm³)	ϕ
CsO₂ α, β		CaC₂	m.p. 432; t.p. 223	d.	142·4		1·68
CsOH			m.p. 272		253·3	(40·8)	
BeO		wurtzite	m.p. 2530	b.p. 3850	1198·3	8·25	4·6
Be(OH)₂ α	unst.	ε-Zn(OH)₂	d. ⟨250⟩		(55·3)	22·4	4·6
Be(OH)₂ β		NaCl			(58·6)	22·35	
MgO			sbl.	s.p. 2770	1203·3	11·25	0·81
MgO₂	unst.		d. ⟨50⟩		84·6		
Mg(OH)₂		Cd(OH)₂	d. 283			24·4	1·74
CaO		NaCl	m.p. 2600	b.p. (3500)	1268·7	16·7	0·64
CaO₂	unst?	CaC₂	d. ⟨400⟩		50·2	22·4	
Ca(OH)₂		Cd(OH)₂	d. 550		118·9	33·2	1·28
SrO		NaCl	m.p. 2450		1180·7	20·4	0·61
SrO₂		CaC₂	d. 170		100·5		
Sr(OH)₂			d. 700		126·0	25·1/26·9	0·80
BaO	BaO₀.₉₉₇₋₁.₀₀	NaCl	m.p. 1925	b.p. (2750)	1113·7	25·3	0·67
BaO₂		CaC₂	d. 720		163·3	29·9/31·2	
Ba(OH)₂			m.p. 408	d. 890	147·0		0·82
Sc₂O₃		cubic: D5₃			(1214·2)	35·4	1·19
ScOOH		γ-FeOOH				25·5	1·82
Sc(OH)₃		cubic: ca. ReO₃				36·5	2·4
Y₂O₃		Sc₂O₃	m.p. 2420	b.p. (4300)	1271·2	44·9	1·39
Y(OH)₃		La(OH)₃	d. ⟨190⟩			36·6	2·27
La₂O₃ A		hexag.: D5₂	m.p. 2320		1244·8	49·7	1·10
Sc₂O₃ C	st.: h.t.?	Sc₂O₃					
La(OH)₃		hexag.: ca. Do19	d. 260		100·5	36·8/42·0	1·16
Ce₂O₃ A	unst.	A-La₂O₃			(1277·0)	47·9	
Sc₂O₃ B		Sc₂O₃					
CeO₂		CaF₂			(523·4)	25·3	1·22
ThO·	unst.?	NaCl	d. 2700			20·7–22·2	

Table 21.18 (continued)

Compound	Remarks	Structure	m.p. etc. (°C)	b.p. etc. (°C)	$\Delta H_{dec.}$ (kJ)	Volume (cm³)	φ
Th_2O_3	unst.						
ThO_2		CaF_2	m.p. (3000)		1227·6	26·7	1·35
$Th(OH)_4$		amorph.			1084·4	24·7	1·98
UO_2	>1120°C complete sol. solns $UO_{2·6-2·67}$	CaF_2	m.p. 2820				
U_4O_9		cubic			351·7	97·0	1·94
U_3O_8					305·6	100·4	2·77
UO_3		various			233·6	(34·3)	
TiO α	$TiO_{0·95-1·26}$	NaCl, dist.	t.p. 990		1035·4	11·0/13·0	1·20
TiO β	$TiO_{0·3-1·33}$	NaCl	m.p. [1750]				
Ti_2O_3 α		Cr_2O_3	t.p. 200		1030·8	31·5	1·46
Ti_2O_3 β			m.p. 1800		963·8		
Ti_3O_5 α	st?	orhomb.	t.p. 177		971·4	(52)	
Ti_3O_5 β		tricl.					
Ti_5O_9					783·0	92·0	1·70
rutile	$TiO_{1·91-2·0}$ 915° ↑ rutile	tetrag.: C4	m.p. 1920		737·7	18·8	1·73
anatase		tetrag.				19·3	1·78
brookite	unst.	orhomb.				19·8/20·5	1·89
ZrO_2 α	$ZrO_{1·83-2·0}$	monocl.: C43	t.p. 1170		1086·5	21·9	1·56
ZrO_2 β		tetrag.	m.p. 2700			20·2	1·45
VO	$VO_{0·9-1·1}$	NaCl	m.p. 1900		854·1	10·5/11·4	1·51
V_2O_3		Cr_2O_3			753·7	29·2/30·8	1·82
V_3O_5							
VO_2 α		rutile	t.p. 72		401·9	18·0	2·12
VO_2 β			m.p. [1360]		385·2		
$V_{12}O_{26}$	$VO_{2·495-2·50}$	monocl.	m.p. 670				
V_2O_5	unst.	orhomb.: $D8_7$	m.p. 674	d.*	251·2	54·0	2·60
Nb_4O		tetrag. (interst.)	d. ⟨350⟩				3·19
NbO	$NbO_{0·93-1·0}$	NaCl	m.p. 1945		(816·5)	15·0	1·37

*Decomposition begins below the m.p., proceeds gradually, and should be almost complete at 1700–1800°C when VO_2 is formed.

Table 21.18 (continued)

Compound		Remarks	Structure	m.p. etc. (°C)	b.p. etc. (°C)	$\Delta H_{dec.}$ (kJ)	Volume (cm³)	ϕ
NbO_2	α	unst.	ca. rutile	m.p. 1915		783·0	20·5	1·87
Nb_2O_5	β	unst. $NBO_{2\cdot43-2\cdot50}$	orhomb.	m.p. 1495	d.†	614·6	58·3	2·68
Ta_4O		unst.	monocl.				45·7	1·05
Ta_xO			tetrag. (interst.)					
Ta_2O_3	α	unst.	rutile	t.p. 1350		819·0	54·0	2·50
	β	$TaO_{2\cdot35-2\cdot50}$	orhomb.	m.p. 1872		753·7	52·8	2·43
Cr_2O_3			tricl.	m.p. 2400			29·0	2·07
CrO_2		$CrO_{1\cdot5-1\cdot54}$	rhombohed.: D5$_1$			75·4		
CrO_3		unst.?	rutile?	m.p. (187)		−6·3	35·6	5·1
MoO_2		unst.	orhomb.	dpr. 1780		588·7	19·7	2·10
Mo_4O_{11}		$MoO_{1\cdot97-2\cdot08}$	monocl.			311·5	134·0	3·57
Mo_9O_{26}		$MoO_{2\cdot65-2\cdot75}$	orhomb.	t.p. 650				3·5
Mo_8O_{23}			monocl.					
MoO_3		$MoO_{2\cdot95-3\cdot00}$	monocl.	m.p. 795	b.p. 1100	324·9	31·3	3·3
WO_2			MoO_2	m.p. 1580	dpr.	589·9	19·8	2·08
W_4O_{11}		$WO_{2\cdot65-2\cdot72}$	monocl.	dpr.			118·0	3·03
$W_{20}O_{58}$		$WO_{2\cdot88-2\cdot92}$	monocl.	dpr.				3·12
WO_3	α		tricl.	t.p. 735		561·5	31·5	3·35
	β		orhomb.	m.p. 1473				
$WO_3.H$			ReO_3				(52·5)	5·5
MnO			NaCl	m.p. 1875	d.	770·4	13·15	1·79
Mn_3O_4	α	$MnO_{1\cdot0-1\cdot12}$	spinel, tetrag. def.	t.p. 1170		463·9	47·3	2·15
	β	$MnO_{1\cdot33-1\cdot45}$	cubic	m.p. 1560		422·0		
Mn_2O_3	α	$MnO_{1\cdot5-1\cdot6}$	Sc_2O_3	t.p. 600		212·7	(35·0)	
	γ		γ-Fe_2O_3	d. 900				
MnO_2	α	$MnO_{1\cdot96-2\cdot0}$	orhomb.	t.p. 250		160·8		
	β	unst.						
	γ		rutile	d. 480			16·6/17·2	2·27

† Decomposition to NbO_2 begins just above the m.p. but would not be complete below 2000 C.

Table 21.18 (continued)

Compound		Remarks	Structure	m.p. etc. (°C)	b.p. etc. (°C)	$\Delta H_{dec.}$ (kJ)	Volume (cm³)	φ
Mn₂O₇			(FeS₂)			−207·3	20·2	2·75
MnOOH	α		monocl.: Eo6				27·3	3·71
	γ	unst.						
Mn(OH)₂		manganite	Cd(OH)₂	d.		(38·1)		
ReO₂		pyrochroite	MoO₂	m.p. 160		425·0	31·5	3·38
ReO₃			cubic: Do9	m.p. 296		372·6	(79·2)	
Re₂O₇			cubic: ca. c.p.	m.p. [1424]	b.p. 362	39·8		
FeO		FeO₁·₀₅₅₋₁·₁₉	NaCl	m.p. 1597		529·7	11·9/12·5	
Fe₃O₄			spinel	t.p. 675		605·0	44·7	2·10
Fe₂O₃	α	haematite	Cr₂O₃					
	β	unst.	β-Al₂O₃			461·4	30·4	2·14
	γ	unst.	cubic: D5₇	m.p. 1457		457·6		
	δ		hexag.					
FeOOH	α	goethite	orhomb: Eo2	d. (230)*		(75·4)*	21·3	3·0
	β	unst.	orhomb.				(27·5)	
	γ	lepidocrite	orhomb: Eo4	d. (400)†		(102·6)	22·4	
Fe(OH)₂			Cd(OH)₂			54·4	26·4	3·7
Fe(OH)₃		unst.	amorph.					
CoO		CoO₁·₀₀₁₇	NaCl	m.p. 1805		478·2	11·6/12·3	1·86
Co₃O₄			spinel	d. 910			39·8	2·01
Co₂O₃		unst.	hexag?				31·1/32·6	2·46
CoOOH		α, β, γ					20·6–22·2	
Co(OH)₂			Cd(OH)₂	d.			(25·0)	
NiO	n	NiO₁·₀₀₅	NaCl	m.p. 1960	d.	67·8	10·9	1·65
	u	unst.				481·5		
NiOOH	α		hexag?				26·3/28·6	4·33
	β	unst.	Cd(OH)₂				19·9/22·1	
	γ		CdCl₂				23·9	
Ni(OH)₂	α		hexag				24·3	
	β		Cd(OH)₂			56·9	25·7	

*d↑α-Fe₂O₃. †d↑γ-Fe₂O₃.

Table 21.18 (continued)

Compound	Remarks	Structure	m.p. etc. (°C)	b.p. etc. (°C)	$\Delta H_{dec.}$ (kJ)	Volume (cm³)	ϕ
RuO_2		rutile			439·6	(26·8)	
RuO_4	unst.		m.p. 27		0·0	45·5	
RhO			d. 1020		181·7		
Rh_2O_2		Cr_2O_3	d. 990		208·5	31·3	1·88
PdO		tetrag.: B17	d. 790		182·1	14·7	1·65
$Pd(OH)_2$			d.⟨100⟩ dpr.				
OsO_2	yellow	rutile	m.p. 56	b.p. 130	257·5	28·2	3·24
IrO_2		monocl.	d. 1124		195·9	51·3/49·5	
IrO_3	st.: gas	rutile			221·9	19·2	2·23
PtO	unst.	PdO					
Pt_3O_4	unst.?	b.c. cubic					
PtO_2	st.: gas	hexag.		a.			
Cu_2O		cubic: C3	m.p. 1230		335·0	23·3	1·64
CuO		monocl.: B26	d. 1100		286·4	12·2	1·72
$Cu(OH)_2$			d. 35		44·0	29·0	4·0
Ag_2O		Cu_2O	d. 185		61·1	32·1	1·56
AgO	unst.	monocl.					
ZnO	unst.	wurtzite / Zn blende	d. vol.		698·3	14·2	1·55
$Zn(OH)_2$	unst.	Cd(OH)₂				14·5	1·59
α		orthomb.: C31	d. 80		61·1		
β		orthomb.: C31	d. 85		51·5		
γ		orthomb.	d. 90		51·9		
ε					52·8		
CdO	$CdO_{0·999-1·000}$	NaCl	d. vol.		511·6	15·65	1·21
$Cd(OH)_2$	2 modifications	hexag.: C6	d. 190		62·8	29·9	2·30
HgO	red	orthomb.: SnO, def.	d. 430		180·9	19·3	1·30
Al_2O	st.: gas						
AlO	st.: h.t.?		m.p. (2050)				

Table 21.18 (continued)

Compound		Remarks	Structure	m.p. etc. (°C)	b.p. etc. (°C)	$\Delta H_{dec.}$ (kJ)	Volume (cm³)	ϕ
Al_2O_3	α	corundum	Cr_2O_3	m.p. 2030		1117·1	25·6	1·28
	β	unst.	hexag.: $H2_8$				30·7	1·54
	γ	unst.	defect-spinel			1059·7	29·8	1·49
	δ	st.: electrolyt. layers	tetrag.					
	γ'		cubic					
$AlOOH$	χ	unst.	similar to γ				17·8	1·78
	α	diaspore	α-FeOOH	d. ⟨300⟩‡		21·8‡	19·5	1·95
	γ	böhmite	γ-FeOOH			51·5‡	30·9	3·09
$Al(OH)_3$	α	bayerite	amorph.	d. ⟨150⟩‡		54·0‡	32·0	3·20
	γ	gibbsite	monocl.: Do7			686·7	32·6	1·38
Ga_2O							29·1	1·23
Ga_2O_3	α	unst.	Cr_2O_3	m.p. 1725		734·8	31·6	1·35
	β		monocl.				37·7	
	γ	unst.						
$GaOOH$			α-FeOOH					
In_2O							35·1	1·12
In_2O_3			Sc_2O_3			620·9	39·3	1·26
$In(OH)_3$			$Sc(OH)_3$				37·5	2·4
Tl_2O				m.p. (300)		355·9	(44)	1·27
Tl_2O_3		unst.	Sc_2O_3	m.p. (715)	d.	175·8	45·2	1·31
TlO_2				d. ⟨490⟩				
$TlOH$				d. 140		57·8		
SiO		st.: gas	amorph.			(833·2)	(20·7?)	
$Quartz$	α	SiO_2	hexag.: C8	t.p. 575		880·1	22·6	1·88
	β		hexag.	m.p. 1610		879·2	22·5	
$Cristobalite$	α	unst.	tetrag.: C30	m.p. 1713	d. vol.	876·3	25·8	
	β		cubic: C9				27·0	2·15

‡d↑γ-Al_2O_3.

Table 21.18 (continued)

Compound	Remarks	Structure	m.p. etc. (°C)	b.p. etc. (°C)	ΔH$_{dec.}$ (kJ)	Volume (cm³)	φ
Tridymite α	unst.	orhomb.					
Tridymite β		hexag.: C10	t.p. 1470		875·5	26·5	
GeO		amorph.	s.p. 810			20·0	
GeO₂ α		α-quartz	t.p. 1033			16·6	1·23
GeO₂ β		rutile	m.p. 1116		540·1	(20·6)	1·32
SnO	unst.	tetrag.: B10	d. 1100		572·8		
Sn₃O₄			t.p. 410				
SnO₂		rutile	t.p. 540	d.	580·7	21·5	
Sn(OH)₂					44·4		
Sn(OH)₄					32·7		
Pb₂O	unst.	Cu₂O				47·0	1·29
PbO α	red	SnO	t.p. 489		438·8	23·7	1·31
PbO β	yellow	orhomb.	m.p. 885	b.p. 1470	441·7	23·0	1·26
Pb₃O₄	PbO$_{1·33-1·57}$	tetrag.			154·1	(76)	1·4
Pb₂O₃		monocl.	d. 550		95·0	45·6/46·6	
PbO₂ α	unst.?	orhomb.	d. 315		100·5	49·6/50·4	1·37
PbO₂ β	PbO$_{1·87-2·0}$	rutile			53·6	25·0	1·57
Pb(OH)₂		hexag. c.p.				(31·8)	
Sb₂O₃ α	senarmontite	cubic: D6₁	t.p. 573		465·6	52·5/53·0	1·44
Sb₂O₃ β	valentinite	orhomb.: D5₁₁	m.p. 656	b.p. 1425	460·6	50·6	
SbO₂ α		cubic: D6₂			209·3	(20·5)	
SbO₂ β		orhomb.	d. (1080)				
Sb₆O₁₃		cubic	d. 700			77·9	2·12
Sb₂O₅		cubic	d. (400)		385·2	49·7	1·17
Bi₂O₃ α		monocl.: D5₅	t.p. 710			50·9	1·20
Bi₂O₃ β	unst.	tetrag.: D5₁₂	m.p. 817	b.p. 1890		51·8	1·22
Bi₂O₃ γ	unst.	cubic b.c.				50·3	1·18
		simple cubic					

Table 21.18 (continued)

B. Some double oxides

Oxide		Remarks	Structure	m.p. etc. (°C)	Heat of decomp. (kJ)	Volume (cm³)
Si, Ti, Zr						
Be_2SiO_4		phenakite	rhombohed.: SI_3	m.p. [1560]	(50·2)	37·1
$MgSiO_3$	α	enstatite	orhomb.		36·4	(29·5)
	β	unst.?: clinoenstat.				
	γ	st. 1150°C				
Mg_2SiO_4		forsterite	orhomb.: SI_2	m.p. [1560]	63·2	43·7
$CaSiO_3$	α	woilastonite	monocl.: $S3_3$	m.p. 1890	90·0	39·5
$Ca_3Si_2O_7$				m.p. 1540		
Ca_2SiO_4	α	also denoted γ	orhomb.: HI_2	m.p. [1460]	126·4	52·1
	β	also denoted α′	orhomb.	t.p. 800	122·3	54·2
	γ	also denoted α	monocl. or hexag.	t.p. 1430	118·9	
Ca_3SiO_5		st. 1250°C		m.p. 2130		
				d. 1900		
$MgTi_2O_5$			Ti_3O_5	m.p. 1660		31·8
$MgTiO_3$			Cr_2O_3	m.p. [1640]	sl. exo.	44·7
Mg_2TiO_4			spinel	m.p. 1750		33·6
$CaTiO_3$	α	perowskite	$E2_1$, orhomb. def.	t.p. 1260		38·5
	β		$CaTiO_3$	m.p. 1970		39·0
$Ca_3Ti_2O_7$				m.p. 1750		
$CaZrO_3$				m.p. 2400		
$ZrSiO_4$		zircon	tetrag.: HO_3	m.p. 2430	24·7	46·7
$MnSiO_3$		rhodonite	tricl.	m.p. [1270]		33·3
Mn_2SiO_4		tephroite	Mg_2SiO_4	m.p. [1340]	49·4	49·1
$MnTiO_3$		pyrophanite	Cr_2O_3	m.p. [1360]		
Mn_2TiO_4			spinel	m.p. 1450		
Fe_2SiO_4		fayalite	Mg_2SiO_4	m.p. 1205	35·2	47·6
Co_2SiO_4			Mg_2SiO_4	m.p. 1420		45·3

Table 21.18 (continued)

Oxide	Remarks	Structure	m.p. etc. (°C)	Heat of decomp. (kJ)	Volume (cm³)
$FeTi_2O_5$		spinel			55·3
$FeTi_2O_4$					45·6
$FeTiO_3$	ilmenite	Cr_2O_3			31·7
Fe_2TiO_5	pseudobrookite	Ti_3O_5	m.p. 1370	(16·7)	54·8
$CoTiO_3$		Cr_2O_3			44·9
Co_2TiO_4		spinel			52·9
Zn_2SiO_4	willemite	Be_2SiO_4	m.p. 1510	8·4*	45·9
Zn_2TiO_4		spinel	m.p. 1550	29·3	
Al_2SiO_5	kyanite	tricl.	m.p. [1810]	sl. exo.	(45·0)
	andalusite	orhomb.		166·2	(50·5)
	sillimanite	orhomb.		164·5	50·0
$Al_6Si_2O_{18}$	mullite	orhomb.	m.p. 1900	192·6	118·3
Al_2TiO_5 α		Ti_3O_5	t.p. 1820		49·4
β			m.p. 1890		
$PbSiO_3$		monocl.	m.p. 765	10·5	45·2
Pb_2SiO_4			m.p. 743	29·3	
Pb_4SiO_6 α			t.p. 137		
β			t.p. 720		
γ			m.p. 727		
$PbTiO_3$ α		$CaTiO_3$, def.	t.p. 490	sl. exo.	38·5
β					
$PbZrO_3$		tetrag. ca. $E2_1$	m.p. 1170		42·4
Al					
$BeAl_6O_{10}$			m.p. 1910		
$BeAl_2O_4$	chrysoberyl	orhomb.	m.p. 1870		34·0
$MgAl_2O_4$	spinel	cubic: $H1_1$	m.p. 2135		39·7

*Free-energy values at 1000°C approximately equal to the heats of decomposition, assuming that $\Delta S \approx 0$.

Table 21.18 (continued)

Oxide	Remarks	Structure	m.p. etc. (°C)	Heat of decomp. (kJ)	Volume (cm³)
$Ca_5Al_6O_{14}$		cubic: $K7_4$	m.p. 1455		218·0
$CaAl_2O_4$		cubic	m.p. 1600	15·5	
$Ca_3Al_2O_6$		$CaTiO_3$	m.p. [1535]	6·7	32·5
$LaAlO_3$					
$NbAlO_4$		spinel	m.p. 1510		42·5
$MnAl_2O_4$		spinel	m.p. [1560]		39·6
$FeAl_2O_4$	hercynite	spinel	d. 1440		
$Fe_2Al_2O_6$			d. 1230		
$CoAl_2O_4$		spinel	m.p. 1960	20·5*	39·4
$NiAl_2O_4$		spinel	m.p. 2020	21·8*	39·1
$CuAlO_2$				8·4*	
$CuAl_2O_4$					39·5
$ZnAl_2O_4$		spinel	m.p. (1950)		39·7
V, Nb, Ta					
Mg_2VO_4		spinel			44·7
MgV_2O_4		spinel			44·85
$BeNb_8O_{21}$			m.p. 1365		
$BeNb_6O_{16}$			m.p. 1365		
$BeNb_4O_{11}$			m.p. 1445		
$BeNb_2O_6$			m.p. 1445		
$MgNb_2O_6$ α		columbite	m.p. 1480		58·8
β		rutile			62·1
$Mg_4Nb_2O_9$		haematite			
$Ca_3Nb_2O_8$		Na_xWO_3 $(x < 1)$			
$Ca_4Nb_2O_9$		tetrag.: perowskite			
$TiNb_2O_7$			m.p. 1490		113·6

* Free-energy values at 1000°C approximately equal to the heats of decomposition, assuming that $\Delta S \approx 0$.

Table 21.18 (continued)

Oxide		Remarks	Structure	m.p. etc. (°C)	Heat of decomp. (kJ)	Volume (cm³)
$CrNbO_4$			rutile			39·35
FeV_2O_4			spinel			45·6
$FeNbO_4$	α		columbite			76·1
	β		rutile			83·2
$FeTa_2O_6$	α		tetrag.			62·8
$Co_2Nb_4O_{14}$	β		columbite			101·7
$NiNb_2O_6$	α		rutile			103·3
	β		columbite			55·2
Zn_3VO_4			rutile			61·0
ZnV_2O_4			spinel			
$ZnNb_2O_6$			spinel			45·0
$ZnNbO_3$			columbite			
UTa_2O_8			U_3O_8	m.p. 1150		
UTa_2O_7			cubic f.c.			
Cr, Mo, W						
$MgCr_2O_4$			spinel	m.p. 2200	20·9	43·3
$CaMoO_4$		powellite	$CaWO_4$			46·7
$MgWO_4$			monocl.		(14·2)	48·0
$CaWO_4$		scheelite	tetrag.: HO_4		(166·6)	47·5
$CeCrO_3$			$CaTiO_3$	m.p. 2430		35·2
$LaCrO_3$			$CaTiO_3$			46·0
$MnCr_2O_4$		chromite	spinel			44·0
$FeCr_2O_4$			spinel	m.p. 2180	34·3*	
$CoCr_2O_4$			spinel			43·3
$NiCr_2O_4$			spinel		8·4*	43·0

*Free-energy values at 1000°C approximately equal to the heats of decomposition, assuming that $\Delta S \approx 0$.

Table 21.18 (continued)

Oxide	Remarks	Structure	m.p. etc. (°C)	Heat of decomp. (kJ)	Volume (cm³)
$FeWO_4$	ferberite	$MgWO_4$		(40.2)	(42.7)
$NiWO_4$		$MgWO_4$			43.8
$CuCr_2O_4$		$CuFe_2O_4$ spinel			43.3
$ZnCr_2O_4$		spinel			52.8
$PbMoO_4$	wulfenite	$CaWO_4$	m.p. 1065	(138.6)	(56.0)
Pb_2MoO_5			m.p. 950		
$PbWO_4$ α	stolzite	$CaWO_4$	t.p. 870	(110.9)	
$PbWO_4$ β	raspite	monocl.	m.p. 1120		
			m.p. 900		
Pb_2WO_5		$CaTiO_3$			35.2
$LaMnO_3$					31.9
$FeMnO_3$		spinel			46.0
Fe_2MnO_4					47.4
$CuMn_2O_4$		tetrag.: Mn_3O_4			45.9
$ZnMn_2O_4$					
Fe, Co, Ni					
$MgFe_2O_4$		spinel	m.p. [1750]	18.8*	44.3
$MgCo_2O_4$		spinel			40.1
$CaFe_2O_4$		V_2CaO_4	m.p. [1240]	31.0	
$Ca_2Fe_2O_5$			m.p. [1480]		
$LaFeO_3$		$CaTiO_3$	m.p. 1890		35.5
Fe_2CoO_4	trf. 500°C	spinel			44.6
Fe_2NiO_4	trf. 582°C	spinel			44.8
Co_2NiO_4		spinel			40.2
$CuFe_2O_4$		spinel, tetrag. def.	m.p. [900]		45.3
$CuFeO_2$		$NaHF_2$			27.5
$CuCo_2O_4$		spinel			39.1

*Free-energy values at 1000°C approximately equal to the heats of decomposition, assuming that $\Delta S \approx 0$.

Table 21.18 (continued)

Oxide	Remarks	Structure	m.p. etc. (°C)	Heat of decomp (kJ)	Volume (cm³)
$ZnFe_2O_4$		*spinel*	m.p. 1720	*exo.*	45·2
$GeNi_2O_4$		*spinel*			41·5
$SnCo_2O_4$		*spinel*			47·9
$PbFe_4O_7$		*hexag.*			82·8
Others					
Mg_2SnO_4		*spinel*			47·5
$CaSnO_3$		$CaTiO_3$			36·8
$MgLa_2O_4$			m.p. 2030		
$CaCeO_3$		$CaTiO_3$			34·4
$CaUO_3$		Sc_2O_3	m.p. [1850]		
$CaUO_4$	$Ca(UO_2)O_2$	*rhombohed.*			46·0
Ca_2UO_4		*tetrag.*	m.p. [1800]		
Zn_2SnO_4		*spinel*			48·0
$PbSnO_3$		$CaTiO_3$, *def.*			

Abbreviations: amorph., amorphous; b.p., boiling point; b.c., body centred; c.p., close packed; d., dissociation temperature calculated; d.⟨ ⟩, dissociation temperature observed; *dpr*., disproportionation; def., distorted, deformed; d.vol., dissociation, forming gaseous dissociation products; endo., endothermic; exo., exothermic; f., form; f.c., face centred; hexag., hexagonal; h.t., at high temperature; h.t.f., high temperature form; l.t., at low temperature; l.t.f., low temperature form; m.p., melting point; m.[ ], melting associated with decomposition; m.β, melting point under pressure; max., maximum solubility; monocl., monoclinic; orhomb., orthorhombic; ppt, precipitated; qu., quartz; r.t., room temperature; rhombohed., rhombohedral; sl.exo., slightly exothermic; s.*sbl*., sublimation point; sol., solid; st., stable; sur.lay., surface layer; tetrag., tetragonal; tricl., triclinic; unst., unstable; v.small (solubility); ↑, converted into; ( ) inaccurate or unreliable data.

Table 21.19 Solubility of gases in water

		\multicolumn{8}{c}{Temperature of water ($\theta/°C$)}
		0	10	15	20	30	40	50	60
Ammonia	S	1130	870	770	680	530	400	290	200
Argon	A	0·054	0·041	0·035	0·032	0·028	0·025	0·024	0·023
Carbon dioxide	A	1·676	1·163	0·988	0·848	0·652	0·518	0·424	0·360
Carbon monoxide	A	0·035	0·028	0·025	0·023	0·020	0·018	0·016	0·015
Chlorine	S	4·61	3·09	2·63	2·26	1·77	1·41	1·20	1·01
Helium	A	0·0098	0·0091	0·0089	0·0086	0·0084	0·0084	0·0086	0·0090
Hydrogen	A	0·0214	0·0195	0·0188	0·0182	0·0170	0·0164	0·0161	0·0160
Hydrogen sulphide	A	4·53	3·28	2·86	2·51	1·97	1·62	1·37	1·18
Hydrochloric acid	S	512	475	458	442	412	385	362	339
Nitrogen	A	0·0230	0·0183	0·0165	0·0152	0·0133	0·0119	0·0108	0·0100
Nitrous oxide	A	—	0·88	0·74	0·63	—	—	—	—
Nitric oxide	A	0·071	0·055	0·049	0·046	0·039	0·034	0·031	0·029
Oxygen	A	0·047	0·037	0·033	0·030	0·026	0·022	0·020	0·019
Sulphur oxide	S	79·8	56·6	47·3	39·4	27·2	18·8	—	—

Values of A for 20°C for other rare gases are: Ne, 0·0101; Kr, 0·0594; Xe, 0·126. S indicates the number of m³ of gas measured at 0°C and 101·325 kN m⁻² which dissolve in 1 m³ of water at the temperature stated, and when the pressure of the gas plus that of the water vapour is 101·325 kN m⁻². A indicates the same quantity except that the gas itself is at the uniform pressure of 101·325 kN m⁻² when in equilibrium with water.

Table 21.20A Solubility of air in water*

A kilogram of water saturated with air at a pressure of 101·325 kN m⁻² contains the following volumes of dissolved oxygen, etc., in cm³ at 0°C and 101·325 kN m⁻²

Gas	\multicolumn{7}{c}{Temperature of water ($\theta/°C$)}
	0	5	10	15	20	25	30
Oxygen	10·19	8·9	7·9	7·0	6·4	5·8	5·3
Nitrogen, argon, etc.	19·0	16·8	15·0	13·5	12·3	11·3	10·4
Sum of above	29·2	25·7	22·9	20·5	18·7	17·1	15·7
% of oxygen in dissolved air (by vol.)	34·9	34·7	34·5	34·2	34·0	33·8	33·6

*After Kaye, G. W. C. and Laby, T. H., *Tables of Physical and Chemical Constants*, 14th ed., Longmans (1973).

Table 21.20B Solubility of oxygen in certain electrolyte solutions*

Electrolyte	Conc. of electrolyte (mol/ℓ) 0·5	1·0	2·0	
	\multicolumn{3}{c}{Solubility (mℓ/ℓ)†}	
HNO₃	27·67	27·03	26·03	
HCl	27·13	26·30	24·47	
H₂SO₄	25·20	23·00	19·15	25°C
NaCl	24·01	20·44	14·48	
KOH	23·09	18·88	—	
NaOH	22·91	18·69	12·19	
HNO₃	33·00	31·86	29·87	
HCl	32·62	31·01	28·35	
H₂SO₄	32·05	31·76	22·09	15°C
NaCl	29·20	24·65	17·26	
KOH	27·59	22·19	—	
NaOH	27·31	21·90	14·41	

*Data after Uhlig, H. H. (Ed.), *Corrosion Handbook*, Wiley (1953).
†Solubility given in cubic centimetres of gas (°C, 1 atm) dissolved in 1 litre of solution when partial pressure of the gas equals one atmosphere.

Table 21.21 Oxygen dissolved in sea-water in equilibrium with a normal atmosphere (101 325 N m^{-2}) of air saturated with water vapour*

Parts per million

Chlorinity $^o/_{oo}$	0	5	10	15	20
Salinity $^o/_{oo}$	0	9·06	18·08	27·11	36·11

Temperature (°C)					
0	14·62†	13·70	12·78	11·89	11·00
5	12·79	12·02	11·24	10·49	9·74
10	11·32	10·66	10·01	9·37	8·72
15	10·16	9·67	9·02	8·46	7·92
20	9·19	8·70	8·21	7·77	7·23
25	8·39	7·93	7·48	7·04	6·57
30	7·67	7·25	6·80	6·41	5·37

Note: the table gives the quantity of oxygen dissolved in sea-water at different temperatures and chlorinities when in equilibrium with a normal atmosphere saturated with water vapour. It thus represents the condition approached by the surface water when biological activity is not excessive.
*Data after C. J. J. Fox, Conseil Permanent International pour l'Exploration de la Mer, Copenhagen, *Publication de Ciconstance*, 41 (1907).
†The values of solubility in water of zero chlorinity differ slightly from those for fresh water.

Table 21.22 Saturated solubilities of atmospheric gases in sea-water at various temperatures*
Concentrations of oxygen, nitrogen and carbon dioxide in equilibrium with 1 atm (101 325 N m^{-2}) of designated gas

Gas	Chlorinity ($^o/_{oo}$)	Temperature (°C)	Concentration	
			mℓ/ℓ	Parts per million
Oxygen	0	0	49·2†	70·4
		12	36·8	52·5
		24	29·4	42·1
	16	0	40·1	56·0
		12	30·6	42·9
		24	24·8	34·8
	20	0	38·0	52·8
		12	29·1	40·4
		24	23·6	32·9
Nitrogen	0	0	23·0†	28·8
		12	17·8	22·7
		24	14·6	18·3
	16	0	15·0	18·4
		12	11·6	14·2
		24	9·36	11·5
	20	0	14·2	17·3
		12	11·0	13·4
		24	8·96	10·9

Table 21.22   (continued)

Gas	Chlorinity ($^0/_{00}$)	Temperature (°C)	Concentration	
			mℓ/ℓ	Parts per million
Carbon dioxide‡	0	0	1715†	3370
		12	1118	2198
		24	782	1541
	16	0	1489	2860
		12	980	1888
		24	695	1342
	20	0	1438	2746
		12	947	1814
		24	677	1299

*Calculated from data in Sverdrup, H. U., Johnson, M. W., and Fleming, R. H., *The Oceans*, Prentice-Hall, Inc., New York (1942).
†These values differ slightly from those for fresh water.
‡Includes CO_2 present as H_2CO_3 but not as HCO_3^- or CO_3^{2-}.
Note: atmospheric gases are present in sea-water in approximately the following quantities:

	mℓ/ℓ	Parts per million
Oxygen	0–9	0–12
Nitrogen	8–15	10–18
Carbon dioxide*	33–56	64–107
Argon	0·2–0·4	0·4–0·7
Helium and neon	$1·7 \times 10^{-4}$	$0·3 \times 10^{-4}$†

*Includes CO_2 present as H_2CO_3, HCO_3^- and CO_3^{2-}
†Estimated as helium.

Table 21.23   Properties of sea-water of different salinities*

Salinity ($^0/_{00}$)	Freezing point (°C)	Temperature of maximum density (°C)	Osmotic pressure (atm)	Specific heat (J kg^{-1})
0	0·00	3·95	0	$4·184 \times 10^3$
5	−0·27	2·93	3·23	$4·109 \times 10^3$
10	−0·53	1·86	6·44	$4·050 \times 10^3$
15	−0·80	0·77	9·69	$4·008 \times 10^3$
20	−1·07	−0·31	12·98	$3·979 \times 10^3$
25	−1·35	−1·40	16·32	$3·954 \times 10^3$
30	−1·63	−2·47	19·67	$3·929 \times 10^3$
35	−1·91	−3·52	23·12	$3·899 \times 10^3$
40	−2·20	−4·54	26·59	$3·874 \times 10^3$

*Data after Subow, N. N., *Oceanographical Tables*, p. 208, Moscow (1931). Thompson, T. G., 'The Physical Properties of Sea Water', *Bull.* 85, 'Physics of the Earth. V. Oceanography', p. 63, National Research Council of the National Academy of Sciences, Washington (1932).

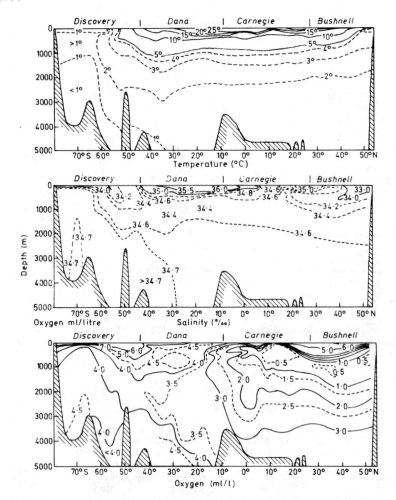

Fig. 21.3   Vertical sections showing distribution of temperature, salinity, and oxygen in the Pacific Ocean, approximately along the meridian of 170°W. (After Sverdrup, H. U., *Ocean-ography for Meteorologists*, Allen and Unwin (1945))

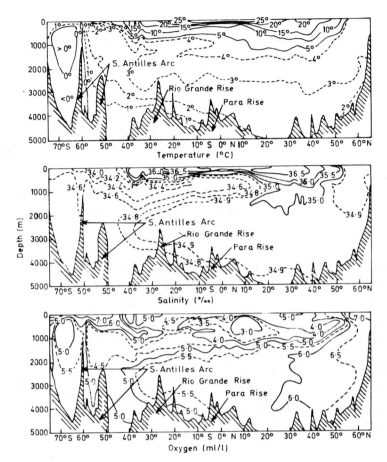

Fig. 21.4 Vertical sections showing distribution of temperature, salinity, and oxygen in the Western Atlantic Ocean (after Wüst). (After Sverdrup, H. U., *Oceanography for Meteorologists*, Allen and Unwin (1945))

Table 21.24 Resistivity of waters (approximate
values $\Omega\,cm$)*

Pure water	20 000 000
Distilled water	500 000
Rain water	20 000
Tap water	1–5000
River water (brackish)	200
Sea-water (coastal)	30
Open sea	20–25

*Data after Morgan, J. H., *Cathodic Protection*, Leonard Hill,
London (1959).
Note: the resistivity of sea-water drops as the chlorinity and
temperature rise and in open sea-water (chlorinity 19‰) it
varies from about 16 Ω cm in the tropics to 35 Ω cm in the Arctic.

Table 21.25 Soil resistivities and corrosiveness

Range of resistivity (Ω cm)*	Location	Soil type and classification
10 000–100 000 and above	Perth, Scotland	Comparatively non-corrosive. Red sand-stone
8000–10 000	West Durham	Mildly corrosive. Sandstone and shale
Varying 1000–20 000	Staffordshire	Many built-up areas. Possibly very corrosive
1000–1500	Eastbourne, Sussex	Marshy ground. Very corrosive
15 000–20 000	Sussex Downs	Chalk. Non-corrosive
750–1500	Port Clarence, S.E. Durham	Salt marsh. Very corrosive
600–1500	S. Essex	Essex clay. Very corrosive
1400–3200	Newport, Gwent	Grey, yellow and blue clays. Corrosive
12 000–15 000	North Devon	Millstone grit. Comparatively non-corrosive
1000–2500	Gloucester	Generally clay. Corrosive
25 000–250 000	West Hampshire	Sandy gravel. Not generally corrosive

*To obtain Ω m divide by 100.

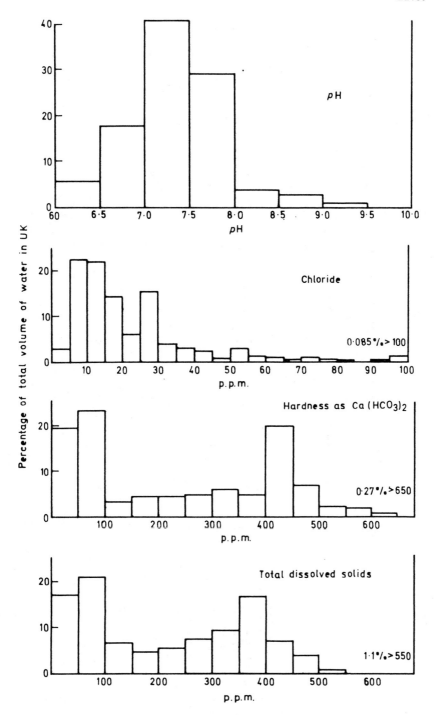

Fig. 21.5 Distribution of dissolved constituents in UK fresh waters. (After Butler, G. and Ison, H. C. K., *Corrosion and its Prevention in Water*, Leonard Hill, London (1966))

Table 21.26 Compositions of natural waters arranged in increasing chloride concentration (concentration in p.p.m.)

Type	Source	Total dissolved solids	Cl⁻	SO_4^{2-}	Ca^{2+}	Mg^{2+}	Alkalinity ($CaCO_3$)	pH	Hardness ($CaCO_3$) Total	Hardness ($CaCO_3$) Temporary	Free CO_2
Reservoir	Vaich Res., Scot.	37	Trace	17	3	<1	—	6·2	—	—	—
Bay	Shawinigan Falls, Quebec	34	2	0	—	—	—	6·8	15	—	—
Lake	Lake Vyrnwy, Wales	40	7	5	—	—	8	7·0	16	—	3
Mains	Bristol	283	14	31	105	10	—	7·6	210	154	5
Mains	Teddington	360	51	32	118	20	200	7·5–8·0	320	240	5
Mains	Kuwait	336	108	48	37	10	—	9·3	—	—	—
Deep well	Reading	511	117	39	42	29	—	—	233	216	(+ HCO₃) 152
Borehole	New Windsor	870	215	—	—	—	265	7·7	195	195	7
Artesian well	New Windsor	1250	320	—	—	—	271	—	700	150	—
Mains	Bledington, Oxon.	2968	540	1080	35	12	210	8·1	165	165	—
Mains	Benghazi	—	540	112	82	64	160	7·8	470	120	—
Mains	Aden	1860	582	423	103	113	—	7·8	715	160	4
Mains	Malta	1525	681	65	142	37	—	—	393	—	—
Well	Damman, Saudi Arabia	2656	1041	489	241	89	—	—	—	—	—
Borehole	Harwich	2577	1132	136	—	—	160	7·8	—	305	—
Well	Awali, Bahrain	15 000	7310	561	688	309	—	7·1	602	—	—
Sea	Average figure	34 800	19 500	2380	380	1250	—	8·2	2990	—	—

Data after Butler, G. and Ison, H. C. K., *Corrosion and its Prevention in Water*, Leonard Hill, London (1966).

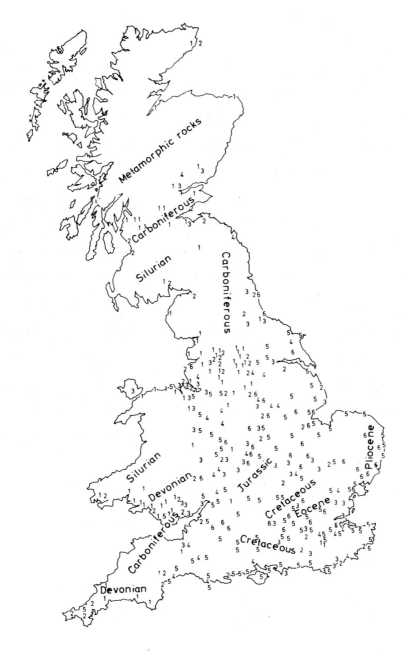

Fig. 21.6 Variation of hardness of water in England, Scotland and Wales in relation to geological formations. (After Butler, G. and Ison, H. C. K., *Corrosion and its Prevention in Water*, Leonard Hill, London (1966))

1. Up to 50 p.p.m.—very soft
2. 50–100 p.p.m.—moderately soft
3. 100–150 p.p.m.—slightly hard
4. 150–200 p.p.m.—moderately hard
5. 200–300 p.p.m.—hard
6. Over 300 p.p.m.—very hard

Table 21.27 Corrosion-rate conversion factors

Multiply	by	To obtain
Milligrams* per square decimetre per day ($mg\,dm^{-2}\,d^{-1}$ or mdd)	10	grams per square metre per day ($g\,m^{-2}\,d^{-1}$ or gmd)
Inches per year ($in\,y^{-1}$ or 25·4 ipy)	25·4	millimetres per year ($mm\,y^{-1}$ or mpy)
Milligrams per square decimetre per day (mdd)	$0·001\,44/\rho$ (ρ in $g\,cm^{-3}$)	inches per year (ipy)
Milligrams per square decimetre per day (mdd)	$0·036\,52/\rho$ (ρ in $g\,cm^{-3}$)	millimetres per year (mpy)
Grams per square metre per day (gmd)	$0·365\,25/\rho$ (ρ in $g\,cm^{-3}$)	millimetres per year (mpy)
Grams per square inch per hour	372 000	milligrams per square decimetre per day ($mg\,dm^{-2}\,d^{-1}$)
Grams per square metre per year	0·0274	milligrams per square decimetre per day ($mg\,dm^{-2}\,d^{-1}$)
Milligrams per square decimetre	0·000 327 7	ounces per square foot ($oz\,ft^{-2}$)
Milligrams per square decimetre per day	0·000 002 69	grams per square inch per hour ($g\,in^{-2}\,h^{-1}$)
Milligrams per square decimetre per day	36·5	grams per square metre per year ($g\,m^{-2}\,y^{-1}$)
Milligrams per square decimetre per day	0·007 48	pounds per square foot per year ($lb\,ft^{-2}\,y^{-1}$)
Ounces per square foot	3052	milligrams per square decimetre ($mg\,dm^{2}$)
Pounds per square foot per year	133·8	milligrams per square decimetre per day ($mg\,dm^{-2}\,d^{-1}$)
Grams per square metre per day	$0·365\,25/\rho$ (ρ in $g\,cm^{-3}$)	millimetres per year ($mm\,y^{-1}$)
Grams per square metre per day	$365·25/\rho$ (ρ in $kg\,m^{-3}$)	millimetres per year ($mm\,y^{-1}$)

*Latest spelling, recommended at the time of going to press.

Table 21.28 Corrosion rates of metals

The corrosion rate of a metal in terms of weight loss per unit area $(g\,m^{-2}\,d^{-1})$ or rate of penetration $(mm\,y^{-1})$ can be calculated from Faraday's law if the current density is known. Conversely, the corrosion current density can be evaluated from the weight loss per unit area or from the rate of penetration. The following symbols and units have been adopted in deriving these relationships in which it is assumed that corrosion is uniform and the rate is linear:

m = Mass of metal corroded (g)
M = Molecular mass $(g\,mol^{-1})$
z = Number of electrons involved in one act of the corrosion reaction
F = Faraday's constant, $96\,487\,C$ $(1\,A \equiv 1\,C\,s^{-1})$
I = Current (A)
i = Current density $(A\,cm^{-2})$ and $i = I/S$
t = Time (s)
ρ = Density of metal $(g\,cm^{-3})$
S = Area of metal involved (cm^2)
d = Thickness of metal removed (cm).

From Faraday's law the weight loss per unit area is

$$\frac{m}{S} = \frac{Mit}{zF} \qquad (g\,cm^{-2}) \qquad\qquad ...(21.1)$$

Since $m = \rho Sd$, then from equation 21.1

$$\rho Sd = \frac{MiSt}{zF} \text{ or } \rho d = \frac{Mit}{zF} \qquad\qquad ...(21.2)$$

and from equation 21.2 the rate of penetration d/t when i is in $A\,cm^{-2}$ is given by

$$\frac{d}{t} = \frac{Mi}{\rho zF} \qquad (cm\,s^{-1}) \qquad\qquad ...(21.3)$$

If the c.d. i' is in $mA\,cm^{-2}$, d in mm and t in years, then

$$\frac{d}{t} \times \frac{10^{-1}}{365 \times 24 \times 60 \times 60} = \frac{Mi'}{\rho zF} \times 10^{-3} \qquad (mm\,y^{-1})$$

$$\therefore \frac{d}{t} = 3\cdot2706\frac{Mi'}{z\rho} \qquad (mm\,y^{-1}) \qquad\qquad ...(21.4)$$

To convert rate of penetration into weight loss per unit area per unit time:

$$(g\,m^{-2}\,d^{-1}) \times \frac{0\cdot36525}{\rho} = (mm\,y^{-1})$$

To convert c.d. $(mA\,cm^{-2})$ into gmd:

$$(g\,m^{-2}\,d^{-1}) = 3\cdot2706\frac{Mi'}{z\rho} \times \frac{\rho}{0\cdot36525} = 8\cdot954\frac{Mi'}{z} \qquad\qquad ...(21.5)$$

If i'' is the current density in $\mu A\,cm^{-2}$

$$(g\,m^{-2}\,d^{-1}) = 8\cdot954\frac{Mi''}{z} \times 10^{-3} \qquad\qquad ...(21.6)$$

Table 21.29 Corrosion rates of iron at various current densities

Current density	mm y^{-1}	g m^{-2} d^{-1}
1 μA cm^{-2}	0·0116	0·250
10	0·1159	2·501
100	1·1591	25·006
200	2·3181	50·011
300	3·4772	75·017
400	4·6362	100·022
500	5·7953	125·028
600	6·9543	150·034
700	8·1134	175·039
800	9·2724	200·045
900	10·4315	225·050
1 mA cm^{-2}	11·5905	250·056
2	23·1810	500·113
3	34·7715	750·169
4	46·3620	1000·226
5	57·9525	1250·282
6	69·5430	1500·339
7	81·1335	1750·395
8	92·7240	2000·452
9	104·3145	2250·508
10	115·9050	2500·565

Data: molecular weight of Fe = 55·85 g mol^{-1}; density = 7·88 g cm^{-3}.
Fe → Fe^{2+} + ze, ∴ z = 2.
Relationships: (mm y^{-1}) = 11·5905 i' mm y^{-1} (i' in mA cm^{-2}).
or (mm y^{-1}) = 0·011 590 5 i'' mm y^{-1} (i'' in μA cm^{-2}).
(g m^{-2} d^{-1}) = 250·056 i'.

Table 21.30 Corrosion rates of copper at various current densities

Current density	(mm y^{-1})	(g m^{-2} d^{-1})
1 μA cm^{-2}	0·0116	0·284
10	0·1161	2·845
100	1·1610	28·449
200	2·3220	56·897
300	3·4830	85·346
400	4·6440	113·795
500	5·8050	142·244
600	6·9659	170·692
700	8·1269	199·141
800	9·2879	227·590
900	10·4489	256·038
1 mA cm^{-2}	11·6099	284·487
2	23·2198	568·974
3	34·8298	853·460
4	46·4397	1137·947
5	58·0496	1422·434
6	69·6595	1706·921
7	81·2695	1991·408
8	92·8794	2275·895
9	104·4893	2560·381
10	116·0992	2844·868

Data: molecular weight of Cu = 63·54 g mol^{-1}; density = 8·95 g cm^{-3}.
Cu → Cu^{2+} + ze, ∴ z = 2.
Relationships: (mm y^{-1}) = 11·6099 i'. (i' in mA cm^{-2}).
(mm y^{-1}) = 0·0116 i''. (i'' in μA cm^{-2}).
(g m^{-2} d^{-1}) = 284·4868 i'.

Table 21.31 Relationship between current density and 1 g m^{-2} d^{-1} or 1 mm y^{-1} for various metals

	Reaction	M (g mol^{-1})	A cm^{-2} for 1 g m^{-2} d^{-1}	ρ (g cm^{-3})	A cm^{-2} for 1 mm y^{-1}
Al	Al^{3+} + 3e	26·98	12·42 × 10^{-6}	2·70	91·811 × 10^{-6}
Mg	Mg^{2+} + 2e	24·32	9·18 × 10^{-6}	1·741	43·76 × 10^{-6}
Fe	Fe^{2+} + 2e	55·85	4·00 × 10^{-6}	7·88	86·30 × 10^{-6}
Fe	Fe^{3+} + 3e	55·85	6·00 × 10^{-6}	7·88	129·45 × 10^{-6}
Zn	Zn^{2+} + 2e	65·38	3·42 × 10^{-6}	7·133	66·79 × 10^{-6}
Cd	Cd^{2+} + 2e	112·41	1·99 × 10^{-6}	8·65	47·13 × 10^{-6}
Ni	Ni^{2+} + 2e	58·71	3·80 × 10^{-6}	8·85	92·07 × 10^{-6}
Sn	Sn^{2+} + 2e	118·70	1·88 × 10^{-6}	7·31	37·63 × 10^{-6}
Pb	Pb^{2+} + 2e	207·21	1·08 × 10^{-6}	11·36	33·59 × 10^{-6}
Cu	Cu^{2+} + 2e	63·54	3·51 × 10^{-6}	8·95	86·01 × 10^{-6}
Ti	Ti^{3+} + 3e	47·90	6·99 × 10^{-6}	4·507	86·25 × 10^{-6}
Ti	Ti^{4+} + 4e	47·90	9·33 × 10^{-6}	4·507	115·13 × 10^{-6}
Ag	Ag^{+} + e	107·873	1·04 × 10^{-6}	10·5	29·90 × 10^{-6}

Calculation:
From equation 21.4 and replacing i' by i (A cm^{-2})

$$i = \frac{3·0575 z \rho}{M} \times 10^{-4} \quad \text{(mm y}^{-1}\text{)}$$

From equation 21.5

$$i = \frac{1·1167 z}{M} \times 10^{-4} \quad \text{(g m}^{-2}\text{ d}^{-1}\text{)}$$

Table 21.32 Weight and thickness of certain metals deposited by 1 A h

Metal	Density* (g cm⁻³)	Molecular weight (g mol⁻¹)	Theoretical weight for 100% efficiency (g)	Theoretical thickness at 100% current efficiency (mm)	Typical current efficiency (C.E.) obtained in practice (%)	Actual thickness for typical current efficiency (mm)
Cadmium, Cd^{2+}	8·64	112·40	2·097	2·427	90 (cyanide)	2·184
Chromium, Cr^{6+}	7·19	51·996	0·323	0·450	14	0·063
Copper(cyan.), Cu^+	8·95	63·54	2·372	2·649	45	1·192
Copper (acid), Cu^{2+}	8·95	63·54	1·186	1·324	97	1·285
Gold, Au^+	19·3	196·967	7·356	3·808	77	2·932
Iron, Fe^{2+}	7·88	55·874	1·042	1·322	93	1·230
Lead, Pb^{2+}	11·36	207·19	3·865	3·402	95	3·232
Nickel, Ni^{2+}	8·85	58·71	1·095	1·238	96	1·188
Platinum, Pt^{4+}	21·4	195·09	1·821	0·850	35 (approx.)	0·298
Silver, Ag^+	10·5	107·870	4·025	3·833	100	3·833
Tin, Sn^{2+}	7·31	118·69	2·214	3·029	93 (acid) / 78 (stannate)	2·817 / 2·363
Zinc, Zn^{2+}	7·2	65·37	1·219	1·694	98 (acid) / 87 (cyanide)	1·660 / 1·474

*To obtain kg m⁻³ multiply by 1000.

Calculation:

Since 1 Ah = 3600 C, the theoretical weight deposited $m = \dfrac{3600M}{96\,500z}$ g(Ah)⁻¹

The theoretical thickness $= \dfrac{3600M}{96\,500z\rho}$ cm (Ah)⁻¹ or $\dfrac{36\,000M}{96\,500z\rho}$ mm (Ah)⁻¹

The actual thickness $= \dfrac{0.3731\,M \times \text{C.E.}}{z\rho}$ mm (Ah)⁻¹

Table 21.33 Composition of industrial metals and alloys

Material	Description and composition (%)
Admiralty brass	Cu–29Zn–1Sn
AISI type *4130*	Cr–Mo steel: 0·27–0·33% C, 0·75–1·20% Cr, 0·15–0·25% Mo
AISI type *4140*	Cr–Mo steel: 0·37–0·44% C, 0·75–1·20% Cr, 0·15–0·25% Mo
Al–Bz9	Cast Al bronze with 8·0–10·5% Al
Alclad *24S–T3*	Al–4·5Cu–1·5Mg–0·6Mn, plated with 99·7% Al
Alcoa *3S*	Al–1·2Mn; German Werkstoff 3.0515
Alcoa *13* or *13S* (Alcan)	Al–12·0Si; German Werkstoff 3.2572
Alcoa *52S*	Al–2·5Mg–0·25Cr; German Werkstoff 3.3535
Alcoa *54S*	Al–3·5Mg–0·25Cr
Alcoa *61S*	Al–1Mg; German Werkstoff 3.2315
Alcoa *356*	Al–7Si–0·3Mg; German Werkstoff 3.2341
Aloyco *20*	Austenitic steel with 19–21% Cr, 28–30% Ni, 4·0–4·5% Cu, 2·5–3·0% Mo
Aloyco *31*	Austenitic steel with 25% Cr, 10% Ni, 3% Mo, 1% Si
Ambrac	Cu–Si alloy with 6·5% Si
Ambrac *850*	Cu–20Ni–5Zn
Ampco *8, 12* or *16*	Cu–8 to 10Al
Anticorodal *1*	Al–1·0Mg+Si
Antinit	see Böhler Antinit and Böhler Superantinit
AT-Nickel	min. 99% Ni; welding quality
BA-28	Al–5Mg–0·5Mn (British alloy)
Bergit *B*	Ni–Mo alloy with 62% Ni, 32% Mo
Bergit *C*	Ni–Mo alloy with 53% Ni, 19% Mo, 13% Cr, 5% W
Böhler Antinit *AS5W*	High-alloyed austenitic steel with 17% Cr, 13% Ni+Mo
Böhler Antinit *SAS 8*	High-alloyed austenitic steel with 18% Cr, 18% Ni, 2% Mo, 2% Cu+Nb; especially for sulphuric acid
Böhler Antinit *SAS 10*	High-alloyed austenitic steel with 18% Cr, 22% Ni+Mo+Cu+Nb
Böhler Superantinit *B* (Antinit *HB*)	64% Ni, 28% Mo+Fe
Böhler Superantinit *C* (Antinit *HC*)	60% Ni, 17% Mo, 16% Cr+Fe
BS-Seewasser	Al–3 to 7Mg or Al–3Mg+Si
Cadag	Ag–4Cd
Carpenter *20*	Fe–29Ni–20Cr–2Mo–3Cu
Carpenter *20Cb*	Fe–29Ni–20Cr–2Mo–3Cu+Nb
Carpenter *20Cb3*	High-alloyed steel with 20·1% Cr, 33·9% Ni, 2·3% Mo, 3·3% Cu+Nb
Chlorimet *2*	62% Ni, 32% Mo, max. 3% Fe
Chlorimet *3*	60% Ni, 18% Mo, 18% Cr, max. 3% Fe
Copper, O.F.H.C. or P.D.C.P.	min. 99·9% Cu (oxygen-free high conductivity*)
Corrix metal	Cu–8 to 10Al–3 to 4Fe
Corronel *220*	Ni–Mo alloy with 28% Mo, max. 3% Fe, 2% V
Corronel *B* (Corronel *210*)	Ni–Mo alloy with 66% Ni, 28% Mo, 6% Fe
Croloy *5 Si*	Low-C steel with 4–5% Cr, 1–2% Si, 0·45–0·65% Mo
Cron. *2525 Ti*	High-alloyed austenitic steel with 24–26% Cr, 24–26% Ni, 2·0–2·5% Mo, max. 2% Si, max. 0·06% C (similar to German Werkstoff 4577)
Cunifer	Cu–10 to 30Ni–2Fe
DIN AlMn	Al–1·25Mn
DIN X5 CrNiMo 1812	Steel with 16·5–18·5% Cr, 12–14% Ni, 2·5–3·0% Mo (similar to German Werkstoff 4436)

*See *Vergleich US-Amerikanischer, Britischer und Deutscher Normen auf dem Gebiet der Kupfersorten, Deutsches Kupferinstitut,* Berlin (1962).

Table 21.33   (continued)

Material	Description and composition (%)
DIN X5 CrNiMoCuNb 1818	High-alloyed austenitic steel with 17–18% Cr, 19–21% Ni, 2·0–2·5% Mo, 1·8–2·2% Cu, max. 0·07% C, Nb > 8 × %C (similar to German Werkstoff 4505)
Dowmetal *H*	Mg–6Al–3Zn
Duracid	Cast Si–Fe (min. 14·5% Si) with or without Mo
Duranalium	Al–1 to 5Mg–0·3 to 0·6Mn
Duranickel	Ni–4·5Al–0·5Ti
Durichlor	Cast Si–Fe with 14·5% Si and Mo
Durimet *20*	Cast steel with 29% Ni, 20% Cr, 2%Mo, 3% Cu, 1% Si
Duriron	Cast Fe, 14·5% Si, 0·8% C
Elcomet *K*	High-alloyed steel with 22% Ni, 23% Cr, 4% Cu, 2% Mo, 1·3% Si, max. 0·12% C
Elektron	Mg alloy with Zn or Zn + Al and other elements
Elektron *AZM*	Mg–6Al–1Zn
Euzonit *50*	Cast Ni–Mo alloy with 55% Ni, 20% Mo, 20% Fe
Euzonit *50S*	Wrought Ni–Mo alloy with 55% Ni, 20% Mo, 20% Fe
Euzonit *60*	Cast Ni–Mo alloy with 60% Ni, 20% Mo, 17% Cr
Euzonit *60S*	Wrought Ni–Mo alloy with 60% Ni, 20% Mo, 17% Cr
Euzonit *70*	Cast Ni–Mo alloy with 67% Ni, 30% Mo
Euzonit *70S*	Wrought Ni–Mo alloy with 68% Ni, 30% Mo
Euzonit *85*	Cast Ni–Si alloy with 85% Ni, 9% Si
Everbrite	Cu–10 to 30Ni–8Fe
Everbrite *90*	Cu–35Ni
Everdur *1000*	Cu–4Si–1Mn
Everdur *1010*	Cu–3·1Si–1·1Mn
Ferry-Metal	Cu–20Ni or Cu–45Ni
FK-Silver	Ag, little Ni
German silver	Cu–20Ni–5Sn–4·5Pb–6·5Zn
German Werkstoff No. 4000	Steel with 12–14% Cr, 0·08% C
German Werkstoff No. 4122	Stainless steel with 15·5–17·5% Cr, 0·33–0·43% C
German Werkstoff No. 4300	Steel with 18·0% Cr, 9·0% Ni, max. 0·12% C
German Werkstoff No. 4301	Steel with 18·0% Cr, 10·0% Ni, max. 0·07% C
German Werkstoff No. 4401	Steel with 17·5% Cr, 11·0% Ni, 2·2% Mo, max. 0·07% C
German Werkstoff No. 4410	Steel casting with 17–19% Cr, 9–11% Ni, 2·0–2·5% Mo, max. 0·15% C
German Werkstoff No. 4436	Steel with 16·5–18·5% Cr, 12–14% Ni, 2·5–3% Mo, max. 2% Si, max. 0·07% C
German Werkstoff No. 4449	Steel with 17% Cr, 13·5% Ni, 4·5% Mo, max. 0·07% C
German Werkstoff No. 4472	Cast alloy with 60% Ni, 17% Mo, 16·5% Cr, max. 0·10% C
German Werkstoff No. 4505	Steel with 17·5% Cr, 20% Ni, 2·2% Mo, 2% Cu, stabilised with Nb
German Werkstoff No. 4506	Steel with 17·5% Cr, 20% Ni, 2·2% Mo, 2% Cu, stabilised with Ti
German Werkstoff No. 4510	Stainless steel with 17% Cr, stabilised with Ti
German Werkstoff No. 4511	Steel with 16·5–18% Cr, stabilised with Nb
German Werkstoff No. 4523	Stainless steel with 16–18% Cr, 1·5–2·0% Mo, stabilised with Nb
German Werkstoff No. 4541	Steel with 18·0% Cr, 10·5% Ni, max. 0·10% C, stabilised with Ti
German Werkstoff No. 4550	Steel with 18·0% Cr, 10·5% Ni, max. 0·10% C, stabilised with Nb
German Werkstoff No. 4571	Steel with 17·5% Cr, 11·5% Ni, 2·2% Mo, max, 0·10% C, stabilised with Ti

Table 21.33 (continued)

Material	Description and composition (%)
German Werkstoff No. 4577	High-alloyed austenitic steel with 24–26% Cr, 24–26% Ni, 2·0–2·5% Mo, max. 0·06% C, stabilised with Ti
German Werkstoff No. 4578	High-alloyed austenitic steel with 24–26% Cr, 24–26% Ni, 2·0–2·5% Mo, max. 0·06% C, stabilised with Nb
German Werkstoff No. 4580	Steel with 17·5% Cr, 11·5% Ni, 2·2% Mo, max. 0·10% C, stabilised with Nb
German Werkstoff No. 4590	Steel with 18% Cr, 20% Ni, 2% Mo, 2% Cu, stabilised with Nb
Gun-metal	Cu–8Sn–4Zn
Guronit GS2	Alloyed cast iron with 25–30% Cr
Hastelloy B	Ni–2·5Co–28Mo–6Fe–1Cr–0·05C
Hastelloy C	Ni–16Mo–15Cr–6Fe–4W–2·5Co–0·08C
Hastelloy D	Ni–10Si–3Cu
Hastelloy F	44–47% Ni, 21–23% Cr, 5·5–7·5% Mo, 1·75–2·5% Ta + Nb, max. 2·5% Co, Fe, other elements; C (wrought) 0·05%, C (cast) 0·12%
H-Monel	see Monel H
Hybnickel D	Ni–Cr steel with 20–30% Cr, 5–10% Ni, 0·25–0·50% C
Hybnickel S	Ni–Cr cast steel with 25% Ni, 20% Cr
Hydronalium	Al alloy with Mg
Illium G	Ni–22Cr–6Fe–6Mo–6Cu
Inalium	Al–2Cd–0·8Mg–0·4Si
Incoloy 825 (formerly Ni-o-nel)	Ni–31Fe–21Cr–3Mo–1·8Cu–0·05C
Inconel	Ni–15Cr–8Fe
Inconel 600	Ni–14 to 17Cr–6 to 10Fe–max. 0·2C
Inconel X	Ni–15Cr–7Fe–3Ti–1Co
Irrubigo 25	High-alloyed austenitic steel with 18% Cr, 25% Ni, 4% Mo, 0·08% C, some Cu
K-Monel	see Monel K
KS-Seewasser	Al–1·5Mg–1·5Mn–max. 1Sb
Lang alloy 5R	Ni–15 to 18Cr–17Mo–5W–5Fe
LC-Nickel	Low-C nickel, min. 99% Ni, max. 0·02% C
Mangal	Al–1·5Mn
Märker 1818 (SN18)	High-alloyed austenitic cast steel with 18% Cr, 18% Ni, 2%Mo, 2%Cu, 0·08%C
Märker SN25	High-alloyed austenitic cast steel with 18% Cr, 25% Ni, 4% Mo, 2% Cu, 0·1% C
Märker SN42	Cast alloy with 42% Ni, 18% Cr, 5% Mo, 2% Cu, balance Fe
Meehanite CC (formerly KC)	Flake graphite–pearlitic cast iron for general use for solutions with pH less than 2
Meehanite HE	Flake graphite–pearlitic cast iron; all-round material for general use with good thermal shock resistance
Midvale 2024	Fe–26Cr–4Mo
Monel 400	Ni–30Cu–1Mn–max. 0·5Fe
Monel H	Ni–30Cu–2·5 to 3·0Si (cast alloy)
Monel K	Ni–30Cu–3Al–0·5Ti
Monel S	Ni–30Cu–2Fe–4Si
Muntz metal	Cu–40Zn
Nichrotherm NCT 1A	Steel with 20% Cr, 12% Ni, 2% Si
Nichrotherm NCT 3	Steel with 25% Cr, 20% Ni, 2% Si
Nimonic 75	Ni–20Cr–8Fe–0·1C–1Si–0·4Ti (stabilised)
Ni-o-nel	see Incoloy 825
Ni-Resist 1	Fe–15Ni–6Cu–2Cr–2·8C

Table 21.33 (continued)

Material	Description and composition (%)
Ni-Resist 2	Alloyed cast iron with 20% Ni, 2·3% Cr, max. 2·0% C
Ni-Resist 3	Alloyed cast iron with 30% Ni, 3% Cr, max. 2·6% C
Packfong	Ni–Cu–Zn-alloy with 40–44% Cu, 31–40% Ni, 15–25% Zn
Pallacid	Au–Ag–Pd alloy with 30% Au, 40% Ag, 30% Pd
Pantal	Al–1·4Mg–0·70Si–0·90Mn–0 to 0·2Ti
Remanit 1880 SSW	Austenitic Cr–Ni steel with 17·5% Cr, 11·5% Ni, 2·3% Mo, max. 0·06% C (similar to U.S. type 316)
Remanit 1880 SW	Austenitic Cr–Ni steel with 18·5% Cr, 11% Ni, max. 0·06% C (similar to U.S. type 304)
Remanit 1990 SS	Stainless steel with 18% Cr, 9·5% Ni, 2% Mo, 2·4% Si
Remanit 2525 SST	High-alloyed austenitic steel with 25% Cr, 25% Ni, 2·2% Mo, Ti-stabilised
Remanit HB	Ni–Mo alloy, Ni–32Mo–1Si–0·15C + Fe
Remanit HC	Ni–Cr–Mo alloy, Ni–15Cr–17Mo–5W–0·15C + Fe
*SAE 1020	Steel with 0·20% C, 0·45% Mn
*SAE 4130	Steel with 0·31% C, 1·0% Cr, 0·20% Mo
*SAE 8630	Steel with 0·30% C, 0·55% Ni, 0·50% Cr, 0·20% Mo
*SEL 4500	Cast high-alloyed austenitic steel with 19–21% Cr, 24–26% Ni, 2·5–3·5% Mo, 1·5–2·5% Cu, Nb-stabilised
*SEL 4505	High-alloyed austenitic steel with 16·5–18·5% Cr, 19–21% Ni, 2·0–2·5% Mo, 1·8–2·2% Cu, Nb-stabilised
*SEL 4541	Austenitic Cr–Ni steel with 17–19% Cr, 9–11% Ni, max. 0·1% C, Ti-stabilised
*SEL 4571	Austenitic Cr–Ni steel with 16·5–18% Cr, 10·5–12·5% Ni, 2·0–2·5% Mo, max. 0·1% C, Ti-stabilised
*SEL 4580	Cast austenitic Cr–Ni steel with 16·5–18·5% Cr, 10·5–12·5% Ni, 2·0–2·5% Mo, max. 0·1% C, Nb-stabilised
*SEL 4585	Cast high-alloyed austenitic steel with 16·5–18·5% Cr, 19–21% Ni, 2·0–2·5% Mo, 1·8–2·4% Cu, 0·08% C, Nb-stabilised
Sicromal 5S	Low-C steel with 4·6% Cr, 1–2% Si, 0·45–0·64% Mo
Sicromal M 9	Steel with 0·12% C, 1·0–1·3% Si, 12–14% Cr, 0·8–1·1% Al
Sicromal M 10	Steel with 0·12% C, 0·8–1·1% Si, 17–19% Cr, 0·8–1·1% Al
Sicromal M 11	Steel with 0·15–0·25% C, 0·8–1·3% Si, 24–26% Cr, 3·5–4·5% Ni
Sicromal M 12	Steel with 0·12% C, 1·3–1·6% Si, 23–25% Cr, 1·3–1·6% Al
Sicromal M 20/10	Steel with 0·20% C, 1·8–2·3% Si, 19–21% Cr, 11–13% Ni
Sicromal M 23/20	Steel with 0·20% C, 1·8–2·3% Si, 24–26% Cr, 19–21% Ni
Silumin	Cast Al with 12–13% Si
S-Monel	see Monel S
Steel	see also under AISI, DIN, German Werkstoff, SAE, SEL and U.S. type
Steel APS 10 M4	Steel with 0·12% C, 2% Cr, 0·35% Al, 0·35% Mo
Stellite 1	Co–33Cr–13W–2·5C
Stellite 6	Co–26Cr–6W–1·0C
Stellite 6B	58% Co, 30% Cr, 4–5% W, 1·0% C, plus some Mo, Ni, Fe
Stellite 12	Co–29Cr–9W–1·8C
Sterling silver	Ag–7·5 to 10Cu
Super Anoxin	Ni–Mo alloy, similar to Hastelloy B
Superantinit	see Böhler Superantinit
Thermisilid E	Cast Si–Fe with 18% Si
Tombac	Cu–Zn alloy with 72–90% Cu

*SAE = Society of Automotive Engineers.
SEL = Stahl und Eisen-Liste.

Table 21.33 (continued)

Material	Description and composition (%)
Toncan enamelling iron	Pure iron with max. 0·03% C
AISI type 201	Austenitic steel with 16–18% Cr, 3·5–5·5% Ni, 5–7·5% Mn, max. 0·15% C
AISI type 202	Austenitic steel with 17–19% Cr, 4–6% Ni, 7·5–10% Mn, max. 0·15% C
AISI type 302	Austenitic steel with 17–19% Cr, 8–10% Ni, 0·08–0·15% C
AISI type 302 B	Austenitic steel with 17–19% Cr, 8–10% Ni, 2–3% Si, max. 0·15% C
AISI type 304	Austenitic steel with 18–20% Cr, 8–11% Ni, 0·08% C
AISI type 304 L	Austenitic steel with 18–20% Cr, 8–11% Ni, max. 0·03% C
AISI type 308	Austenitic steel with 19–21% Cr, 10–12% Ni, 0·08% C
AISI type 309	Austenitic steel with 22–24% Cr, 12–15% Ni, 0·2% C
AISI type 309 Nb	Austenitic steel with 22–24% Cr, 12–15% Ni, 0·2% C, Nb-stabilised
AISI type 310	Austenitic steel with 24–26% Cr, 19–22% Ni, 0·25% C
AISI type 316	Austenitic steel with 16–18% Cr, 10–14% Ni, 2% Mo, 0·1% C
AISI type 316 Nb	Austenitic steel with 16–18% Cr, 10–14% Ni, 2% Mo, 0·1% C, Nb-stabilised
AISI type 317	Austenitic steel with 17–19% Cr, 9–12% Ni, max. 0·1% C
AISI type 330	Austenitic steel with 13–17% Cr, 33–37% Ni, 0·2–0·5% C
AISI type 347	Austenitic steel with 17–19% Cr, 9–12% Ni, max. 0·1% C
AISI type 347 Nb	Austenitic steel with 17–19% Cr, 9–12% Ni, max. 0·1% C, Nb-stabilised
AISI type 430	Ferritic steel with 14–18% Cr, max. 0·12% C
AISI type 430 F	Ferritic steel with 14–18% Cr, max. 0·12% C, some Zr, Mo
AISI type 446 L	Alloyed cast iron with 23–27% Cr, max. 0·35% C, max. 0·25% N
AISI type 410	Martensitic steel with 11–13·5% Cr, max. 0·15% C
AISI type 416	Martensitic steel with 12–14% Cr, max. 0·15% C, Se, Mo or Zr (free machining)
AISI type 420	Martensitic steel with 12–14% Cr, 0·35–0·45% C
AISI type 431	Martensitic steel with 15–17% Cr, 1·25–2·5% Ni, max. 0·2% C
AISI type 44 A	Martensitic steel with 16–18% Cr, 0·60–0·75% C
V2A-E	Austenitic steel with 17–19% Cr, 9–11% Ni, Ti-stabilised
V4A-E	Austenitic steel with 16·5–18·5% Cr, 10·5–12·5% Ni, 2·0–2·5% Mo, Ti-stabilised
V4AX-E	Austenitic steel with 18% Cr, 11·5% Ni, 2–2·5% Mo, Nb-stabilised
V16A-E	High-alloyed austenitic steel with 16·5–18·5% Cr, 19–21% Ni, 2·0–2·5% Mo, 1·8–2·2% Cu, Ti-stabilised
V24A-E	High-alloyed austenitic steel with 24–26% Cr, 24–26% Ni, 2·0–2·5% Mo, Ti-stabilised
Wegucit	Cast iron with 30% Cr
Worthite	Cast steel with 20% Cr, 24% Ni, 3% Mo, 3·3% Si, 1·8% Cu
Zircalloy 3	Zr alloy with 0·2–0·3% Sn, 0·2–0·3% Fe, max. 0·05% C
Zircalloy 4	Zr alloy with 1·5% Sn, 0·2% Fe, 0·1% Cr, 0·005% Ni

Data after Rabald, E., *Corrosion Guide*, 2nd ed., Elsevier (1968).

P. WOOD
J. SADOWSKA-MAZUR

21.2 British and American Standards*

British Standards†

Materials

BS 534:1966, *Steel pipes, fittings and specials for water, gas and sewage*
    32 page   Gr 6   AMD 476, Mar. 1970
Seamless and welded carbon steel pipes $2\frac{3}{8}$ in (60·3 mm)–84 in (2 134 mm) outside diameter; associated butt-welded, sleeve welding, slip on, fixed flange and screwed and coupled joints; even curvature and gusseted bends, tees and collars; external protection by bitumen sheathing, reinforced bitumen sheathing, bitumen wrapping and coal tar enamel wrapping; internal protection by bitumen and concrete lining; guidance on information to be given by the purchaser when ordering.

BS 715:1970, *Sheet metal flue pipes and accessories for gas-fired appliances*   M
    32 page   Gr 6
Single- and twin-wall flue pipes, fittings and accessories with welded or folded seams, finished as necessary to resist heat and minimise corrosion. Tabulated dimensions with appendices on testing certain coatings.

BS 1554:1949, *Rust, acid and heat resisting steel wire*   NZ
    12 page   Gr 3   Amendment PD 1199, June 1951
Rust-, acid- and heat-resisting steels commonly supplied in the form of wire for general engineering purposes. Section 1 includes details of manufacture, methods of test and tolerances. Section 2 gives the chemical composition and mechanical properties of each steel.

BS 2540:1960, *Silica gel for use as desiccant*   NZ
    12 page   Gr 3   Amendment PD 6174, May 1967
Specifies a grade of silica gel suitable for use in packaging, as indicated in BS 1133, Section 19. Covers particle size, absorptive capacity, pH reaction, loss on ignition, and includes limits for ammonia and ammonium compounds, water soluble chlorides and sulphates, particle breakdown and dust. Test methods are described.

---

*Acknowledgements are made to Dr P. Boden for supplying details of specifications, which have been selected from the comprehensive list he is preparing for the Institute of Corrosion Science and Technology.
†British Standards Institution, 2 Park Street, London W1A 2BS.
Abbreviations: BS, standard; CP, Code of Practice; AMD, Amendment; PD, Published Document; ISO, International Organisation for Standards; M, standard in which the requirements are specified only in metric units; M + I, standard in which both metric and imperial units are specified; N, standards which are independent of units. 'AS', 'NZ' and 'CA' signify that the standards have been endorsed for use as Australian, New Zealand and Canadian standards. Gr refers to the price that is rated from 0 to 16, e.g. 0 = 16p, 1 = 30p, 2 = 45p, 3 = 60p, 4 = 75p, 5 = 90p, 6 = £1.05, 8 = £20, 16 = £28.50 (1973 prices).

BS 3100:1967, *Steel castings for general engineering purposes*
   68 page   Gr 7   AMD 731, May 1971
   1398   *Low alloy steel castings for use at elevated temperatures*
Manufacture, fettling and dressing, heat treatment, sampling and testing, non-destructive testing, rectification and inspection. Chemical composition and mechanical properties of 20 specifications covering 59 grades. Appendices include welding procedures and a tabular summary of compositions and mechanical properties.

BS 3761:1970, *Non-flammable solvent-based paint remover*
   16 page   Gr 4
Description; solvent extractable matter, consistency, volatility, water-insolubility, effect on wood, flammability, acidity or alkalinity, corrosive effects, toxicity, storage properties, marking of containers, packing requirements.

BS 3972:1966, *Yacht ropes*   M + I
   28 page   Gr 5
Part I. *Steel wire ropes (galvanised and stainless) and stainless steel strands*
In all cases material, construction, lay, size, lubrication, protective treatment, breaking load, testing, inspection, marking, packing.

Coatings

BS 282–389:1963, *Lead chromes, zinc chromes for paints*
   36 page   Gr 6
Description, composition, residue on sieve, matter volatile at 100°C, oil absorption value, colour staining power, colour on reduction, tar zinc chromes, a requirement and method of test for freedom from impurities is included.

BS 388:1972, *Aluminium pigments*
   24 page   Gr 5
Covers four types. Type 1 aluminium powder, leafing. Type 2 aluminium paste, leafing. Type 3 aluminium powder, non-leafing. Type 4 aluminium paste, non-leafing. Types 1 and 2 are further classified according to water-covering capacity.

BS 443:1969, *Galvanised coatings on wire*   M   NZ
   16 page   Gr 4
Weight of coating from 0·23 to 10 mm nominal diameter of coated wire in two ranges; tensile strength below 54 hbar, and 54 hbar and above. Uniformity and adhesion of coating. The (referee) gravimetric (control) volumetric methods for determining weight of coating and the copper sulphate (Preece) dip method for determining uniformity of coating are appended.

BS 729:1971, *Hot-dip galvanised coatings on iron and steel articles*
   16 page   Gr 4
Composition of zinc in galvanising bath, appearance and uniformity of coating and coating weight distortion cracking—embrittlement of basis metals, repair of damaged area.

BS 1070:1973, *Black paint (tar based)*
   8 page   A4 size   Gr 5
Gives requirements and methods of test for two types of material: Type A (normal drying) and Type B (quick drying).

BS 1133:1966, *Packaging Code*, Section 6, *Temporary protection of metal surfaces*
               against corrosion (during transport and storage)
   120 page   Gr 8
Guidance on cleaning and drying of metal surfaces prior to the application of temporary

(i.e. easily removable) protectives against corrosion and on selection and application. Performance specifications for eight types of protective together with methods of test.

BS 1224:1970, *Electroplated coatings of nickel and chromium*
    24 page   Gr 5   AMD 963, May 1972 (Gr 0)
Nickel (with a copper undercoat if required) with and without a chromium topcoat electrodeposited on to steel (or iron) zinc alloy, copper or copper alloys and aluminium and alloys. Specifies appearance, type and thickness of nickel and thickness of chromium, adhesion and corrosion resistance.

BS 1615:1972, *Anodic oxidation coatings on aluminium*
    32 page   A4 size   Gr 8
Specifies seven grades of anodic oxidation coatings on aluminium. Gives numerous methods of testing including those for coating, thickness, efficiency of sealing, corrosion resistance, abrasion resistance, fastness to light and resistance to heat and weathering of coloured coatings.

BS 1706:1960, *Electroplated coatings of cadmium and zinc on iron and steel*
    20 page   Gr 4   Amendment PD 4403, Dec. 1961
Specifies purity, appearance, thickness and adhesion of electroplated coatings of cadmium and zinc on iron and steel, and specifies methods of test, local and average thickness tests, test for adhesion by burnishing, humidity test for passivated coatings and tests for presence and adherence of a chromate film. Recommendations are also appended for heat treatment for removal of brittleness for plated articles of steel of nominal tensile strength of five tons and over.

BS 1822:1952, *Nickel-clad steel plate*
    12 page   Gr 3   Amendment PD 4111, Apr. 1961
Requirements of steel plate to which a layer of nickel is integrally and continuously bonded on one or both sides. The base metal is required to comply with BS 1501–151 or BS 1501–161. Two qualities of nickel are specified. Other clauses deal with thickness of cladding metal, rolling tolerances, tensile and bend testing and a special test for liability to surface embrittlement.

BS 2015:1965, *Glossary of paint terms*
    72 page   Gr 7
Definitions of over 500 terms used in the paint industry including terms describing types of finishes, application and film defects. In general definitions of materials of processes used in the manufacture of paint have not been included.

BS 2521 and 2523:1966, *Lead based priming paints*
    12 page   Gr 3   Amendment PD 6074, Mar. 1967
Scope, composition, sample, agreed sample, consistency, drying time, finish, water content and keeping properties. Methods for determination of consistency and of drying time are given in the appendices.

BS 2524:1966, *Red oxide-linseed oil priming paint*
    8 page   Gr 2
Scope, composition, sample, agreed sample, consistency drying time, finish, water content and keeping properties. Methods for determination of consistency and of drying time are given in the appendices.

BS 2525–27:1969, *Undercoating and finishing paints for protective purposes (white lead-based)*
BS 2525, *Undercoating paints for protective purposes*
BS 2526, *White finishing paints for protective purposes*
BS 2527, *Tinted white finishing paints for protective purposes*
Compositions for white lead-based undercoating and finishing paints with a table giving colours selected from BS 2660 for 53 finishing paints and appropriate undercoating

paints. Requirements also for consistency, drying time, finishing colour, water content capacity and storage properties.

BS 2569:1964 Part 1, *Protection of iron and steel by aluminium and zinc against atmospheric corrosion*
   12 page   Gr 3   AMD 55, Aug. 1968
Composition of coating metal, methods of surface preparation and application of coating, requirements for thickness and adhesion. Methods of test in appendices.

BS 2569:1965 Part 2, *Protection of iron and steel against corrosion and oxidation at elevated temperatures*
   16 page   Gr 4
Composition of coating metal, methods of surface preparation and application of coating, requirements for thickness and adhesion, method of test of subsequent treatment in appendices.

BS2816:1973, *Electroplated coatings of silver for engineering purposes*
   12 page   Gr 6
Requirements for electroplated silver coatings on metallic and non-metallic basis materials: appearance, thickness (thickness of undercoats), silver content, adhesion. Methods of test for thickness, adhesion and hardness given in appendices.

BS 2989:1967, *Hot-dip galvanised plain steel sheet and coil*
   12 page   Gr 3
Five classes of sheet and coil produced on a wide strip mill, and coated by the hot-dip galvanising process. Thicknesses up to and including 0·118 in (under 3 mm) with a maximum width of 60 in (1 524 mm).

BS 3289:1959, *Phosphate treatment of iron and steel for protection against corrosion*
   16 pages   Gr 4   Amendment PD 6405, May 1968
Five classes of protective coatings as used for the protection of iron and steel against corrosion. The weight and composition of phosphate coating for each class is specified, general treatment requirements as well as requirements for creviced components and composite articles. Special requirements for high-tensile steels and springs; details of recommended heat-treatment. Inspection and testing procedure.

BS 3382, *Electroplated coatings on threaded components;* Part 1 *cadmium on steel components;* Part 2 *zinc on steel components*
   32 page   Gr 6   Amendments PD 6279, Nov. 1967; AMD 926, Mar. 1972 (Gr 0)
Cadmium and zinc thicknesses, sampling and other inspection procedures, methods of thickness determination for bolts, screws and nuts specified in BS 57, 450, 1083, 1768, 1981 and 3155, having threads conforming to the requirements of BS 84, 93, 811, 1095, 1580 and 2779. Includes an explanatory introduction and definition of terms, and specifies requirements for resistance to corrosion, purity, adhesion, porosity, reduction of hydrogen embrittlement, finish and appearance, plating thickness and passivation for cadmium and zinc electroplated coatings.

      Part 3, *Nickel or nickel plus chromium on steel components*
   32 page   Gr 6   Amendment PD 6280, Nov. 1967
Explanatory introduction, definitions, requirements for resistance to corrosion, adhesion, reduction of hydrogen embrittlement, finish and appearance, plating thickness, sampling and other inspection procedures, methods of thickness determination. Applicable to bolts, screws and nuts specified in BS 57, 450, 1083, 1768, 1981 and 3155 having threads conforming to the requirements of BS 84, 93, 811, 1095, 1580 and 2779.

      Part 7, 1966, *Thicker platings for threaded components*
   24 page   Gr 5   Amendment PD 6231, Aug. 1967
Plating thicknesses which provide a greater degree of protection than is possible on standard threads without interference. Sampling, inspection procedures, thickness

determination for bolts, screws and nuts. Does not apply to components for aircraft, self-tapping screws or wood screws.

BS 3416:1961, *Black bitumen coating solutions for cold application*
    20 page   Gr 4   Amendment PD 4905, May 1963
Two types of black bitumen coating solutions: Type 1 and Type 2. Type 1: for general purposes for the protection of iron and steel; Type 2: for the coating of drinking water tanks. Requirements of the coatings, methods of test for application properties, drying time, finish, protection against corrosion, flexibility, volatile matter, flashpoint and effect on water.

BS 3597:1963, *Electroplated coatings of 65/35 tin–nickel alloy*
    16 page   Gr 4   Amendment PD 5576, July 1965
Essential qualities of electroplated coatings of tin–nickel alloys on articles of steel, copper and copper alloys and zinc alloys other than articles having threads of basic major diameter up to and including 0·75 in (19 mm). Specifies requirements for three classes of coating on each basis metal type, according to the conditions that the coating is required to withstand. Test methods for the determination of local thickness and average thickness, for adhesion, and for resistance to corrosion in a sulphur dioxide atmosphere.

BS 3740:1964, *Steel plate clad with corrosion-resisting steel*
    16 page   Gr 4
For corrosion-resisting chromium and austenitic chromium nickel steel clad plate with cladding layer integrally and continuously bonded to one side of a carbon or low alloy steel plate. Eight qualities of cladding steel, tolerances, tests on bond, on the base steel plate and on the corrosion-resisting layer.

BS 3830:1964, *Vitreous enamelled steel building components*
    8 page   Gr 2   Amendment PD 5514, Apr. 1965
Requirements for manufacture, including base metal, design and fabrication, vitreous enamel finish, colour and texture, processing.

BS 3831:1964, *Vitreous enamel finishes for domestic and catering appliances*
    8 page   Gr 2
Requirements for the ground coat and cover coat of enamel, enamel thickness, and imperfections that may or may not be acceptable. Definitions of imperfections appended. Applicable only to sheet steel or cast iron with the exception of baths.

BS 3981:1966, *Iron oxide pigments for paints*   M   NZ
    28 page   Gr 5
Classification requirements and test methods for 12 classes of natural and synthetic iron oxide pigments for paints. Incorporates BS 272, 305, 312, 313, 337, 339 and 851.

BS 3982:1966, *Zinc dust pigment*   M   NZ
    24 page   Gr 5
Requirements for sampling, composition, residue on sieve and bulk density. Methods of test for total zinc, metallic zinc, lead, arsenic, matter insoluble in acid, residue on sieve and bulk density.

BS 4164:1967, *Coal tar based hot applied coating material for protecting iron and steel,*
              including suitable primers where required
    44 page   Gr 7   AMD 119, Oct. 1968 (Gr 0)
Coatings and associated primers for prevention of corrosion in building and utility services such as pipelines and civil engineering structures. Marking of the containers containing the coating material is obligatory.

BS 4292:1968, *Electroplated coatings of gold and gold alloy*   M
    32 page   Gr 6
Requirements for basis metal, finish and appearance, sampling and testing, undercoats

and for thickness, corrosion resistance, purity, adhesion, hardness and marking for engineering applications, general decorative applications and specialised applications in the jewellery trade. Comprehensive appendices give methods of test for coating thickness, resistance to corrosion, purity, adhesion and hardness.

BS 4479:1969, *Recommendations for the design of metal articles that are to be coated*
    52 page   Gr 7
Design recommendations in respect of features of various coating processes for protective and decorative purposes. Covers electroplating, anodic oxidation, chemical conversion, hot dipping and metal spraying.

BS 4495:1969, *Recommendations for the flame-spraying of ceramic and cement*
                        coatings   M
    12 page   Gr 3
Coatings applied to metals and other suitable materials for wear resistance, corrosion resistance, heat insulation and electrical insulation. Recommendations for coating material, surface preparation, application of coating, sealing, surface finishing.

BS 4601:1970, *Electroplated coatings of nickel plus chromium on plastics materials*
    20 page   Gr 4
Appearance, type and thickness of nickel, type and thickness of chromium, thermal cycling and corrosion resistance tests for various service conditions.

BS 4641:1970, *Electroplated coatings of chromium for engineering purposes*
    24 page   Gr 5
Coatings of chromium on ferrous and non-ferrous metals for engineering applications, specifies appearance and methods of testing, coating thickness, surface finish, hardness, adhesion and porosity; heat treatment for steels before and after plating.

BS 4761:1971, *Sprayed unfused metal coatings for engineering purposes*
    16 page   Gr 4
Requirements for sprayed unfused coatings of metals and metallic compounds applied by combustion gas flame, plasma arc, detonation and similar processes; preparation of components; spraying technique, sealing and inspection.

BS 4950:1973, *Sprayed fused metal coatings for engineering purposes*
    12 page   A4 size   Gr 6
Requirements for sprayed and fused coatings of metal and metallic components applied by combustion gas flame and arc processes; preparation of components, conditions for spraying and subsequent fusion, finishing and inspection.

CP 231:1966, *Painting of buildings*
    152 page   Gr 9
Design, organisation, supervision, paint systems, substrates, materials, surface preparation, and maintenance for a wide range of materials and surfaces to be painted; the latter including wood, plywood, building boards, steel, non-ferrous metals, brick, stone, concrete and plastered and rendered surfaces.

CP 2008:1966, *Protection of iron and steel structures from corrosion*
    208 page   Gr 9
Prevention, cathodic protection, new structures, structures and corrosive conditions requiring special attention, maintenance, organisation, inspection and control, paint film thickness.

CP 3003:1970, *Lining of vessels and equipment for chemical processes*, Part 1, 1967,
                        Rubber
    40 page   AMD 1216, July 1973 (Gr 0)
Gives guidance to manufacturers and users of rubber-lined vessels and equipment on the various types of lining available and on the selection, design, application, main-

tenance, inspection and testing together with recommendations on the design of the items to be lined. Brush-on and spray-on rubber coatings are excluded. Deals with natural rubber (soft) ebonite, styrene and butadiene rubber, polychloroprene (neoprene), butyl rubber, nitrile rubber, chlorsulphonated polyethylene and copolymers of hexafluoropropylene and vinylidene fluoride.

Part 2, *Glass enamel*
Selection, design, application, maintenance, inspection, testing, with recommendations on the design of items to be lined. Not intended for less exacting domestic and architectural applications.

Part 3, 1965, *Lead*
28 page   Gr 5
Guidance to manufacturers and users of lead-lined vessels and equipment on types of lining available, selection, design, application, maintenance, inspection and testing of linings. Recommendations on the design of items to be lined. Excludes electrolytically deposited and sprayed lead loadings.

Part 4, 1965, *Plasticised p.v.c. sheet*
24 page   Gr 5
Guidance to manufacturers and users of p.v.c. lined vessels and equipment, selection, design, applications, maintenance, inspection and testing of linings; recommendation on the design of the items to be lined.

Part 5, *Epoxide resins*
20 page   Gr 4
Selection, design, application, maintenance, inspection and testing; recommendations on the design of the items to be lined.

Part 6, *Phenolic resins*
16 page   Gr 4
Guidance to manufacturers and users of phenolic resin-lined vessels and equipment selection, design, application, maintenance, inspection and testing of linings; recommendations on design of items to be lined.

Part 7, 1970, *Corrosion and heat-resistant metals*
36 page   Gr 7
Guidance to manufacturers and users of vessels and equipment lined with corrosion-resisting steels, nickel and nickel alloys; selection, design, application, maintenance, inspection and testing of linings; recommendations on the design of items to be lined.

Part 8, *Precious metals*
28 page   AMD 669, Dec. 1970 (Gr 5)
Guidance to manufacturers and users of vessels and equipment lined with silver, gold, platinum, platinum alloys and platinum group metals; selection, design, application, maintenance and inspection and testing of linings; recommendations on the design of the items to be lined.

Part 9, 1970, *Titanium*   M
40 page   Gr 7
Guidance to manufacturers and users of titanium-lined vessel and equipment; selection, design, applications, maintenance, inspection and testing of linings; recommendation on the design of the items to be lined.

Part 10, *Brick and tile*
28 page   Gr 5
Guidance to manufacturers and users of vessels and equipment lined with brick and tile; selection, design, application, maintenance, inspection and testing of linings; recommendations on design of items to be lined.

CP 3012:1972, *Cleaning and preparation of metal surfaces*
    40 page   A4 size   Gr 8
Section 1: recommended practice for cleaning a variety of metals and alloys when no subsequent coating is required. Section 2: cleaning and preparation prior to the application of a comprehensive range of surface coatings.

Inhibitors

BS 1170:1968, *Methods for treatment of water for marine boilers*
    92 page   Gr 8
Sources of water and limiting impurities for various groups of boilers, pre-commission cleaning, objects of water treatment, treatment of water before the boiler, treatment of water in the boiler and preservation of idle boilers. Appendices give short explanations of terms, chemicals used in marine boiler water treatment, methods of testing, and list of reagents.

BS 3150:1959, *Corrosion-inhibited ethanediol anti-freeze for water-cooled engines.*
        Type A   Triethanolammonium orthophosphate and sodium
        mercaptobenzothiazole inhibited   M
    16 page   Gr 4   Amendment PD 6124, Apr. 1967
Specifies an anti-freeze for water-cooled internal combustion engines and gives requirements for description, colour, specific gravity, freezing point, pH value, sodium, phosphoric acid content, triethanolamine content, and mercaptobenzothiazole content, as well as requirements for the marking of containers and for samples. Test methods are given and notes on the purpose and use of anti-freeze solutions.

BS 3151:1959, *Corrosion-inhibited ethanediol anti-freeze for water-cooled engines.*
        Type B   Sodium benzoate and sodium nitrite inhibited   M
    16 page   Gr 4
Specifies an anti-freeze for water-cooled internal combustion engines and gives requirements for description, specific gravity, freezing-point, pH value, benzoate content and nitrite content, as well as requirements for the marking of containers for the samples. Test methods are given and notes on the purpose and use of anti-freeze solutions.

BS 3152:1959, *Corrosion-inhibited ethanediol anti-freeze for water-cooled engines.*
        Type C   Sodium tetraborate inhibited   M
    12 page   Gr 3
Specifies an anti-freeze for water-cooled internal combustion engines and gives requirements for description, specific gravity, freezing point, pH value and borate content, as well as requirements for the marking of containers and for samples. Test methods are given and notes on the purpose and use of anti-freeze solutions.

BS 4959:1974, *Recommendations for corrosion and scale prevention in engine cooling*
        systems
    16 page   Gr 7
A guide to protective maintenance of engine cooling water systems in use, when idle and when drained. Describes the additives used to prevent corrosion and scale formation, with chemical tests for control of their concentration. Also describes methods for descaling cooling systems.

Cathodic Protection

CP 1021:1973, *Cathodic protection*
    103 page   A4 size   Gr 12
Definitions, related Codes of Practice and Standards; principles; applications to buried

structures, ships, immersed structures, internal protection of plant. Measures to safe-guard neighbouring structures. Electrical measurements. Safety aspects. Operation and maintenance. Tables of potentials, output of galvanic anodes.

Annual Book of ASTM Standards*

Each standard and tentative has a serial designation which is a positive identification, this designation is made up of a letter† indicating the general classification, and a serial number, together with a further number indicating the year of adoption or latest revision. The letter 'a' after the year in the designation denotes the second revision in that year, 'b' the third revision, etc. Tentatives are identified by the letter T, thus A.48–64 fully identifies the standard on grey iron castings (A.48) adopted or revised in 1964; D.3037–71T covers a method for surface area of carbon black (D.3037) which was issued or revised in 1971 and is tentative (T).

A number in parentheses after the designation of a standard indicates the year of last reapproval by the committee concerned. In reference to any given standard, the complete designation should be given. Best practice is to state the title and the designation.

An asterisk preceding the designation indicates that the standard has been approved as American National Standard by the American National Standards Institute. The numbers after the titles refer to the parts of the annual book of A.S.T.M. standards in which the standards appear.

Parts of the Annual Book of the ASTM, 1974

Part 1   Steel piping, tubing and fittings (*April*)
Part 2   Ferrous castings; ferroalloys (*April*)
Part 3   Steel plate, sheet strip, and wire; metallic coated products (*April*)
Part 4   Structural steel; concrete reinforcing steel; pressure vessel plate; steel rails, wheels, and tyres (*April*)
Part 5   Steel bars, chain, and springs; bearing steel; steel forgings (*April*)
Part 6   Copper and copper alloys (including electrical conductors) (*July*)
Part 7   Die-cast metals; light metals and alloys (including electrical conductors) (*July*)
Part 8   Nonferrous metals—nickel, lead, and tin alloys, precious metals, primary metals; reactive metals (*July*)
Part 9   Electrodeposited metallic coatings; metal powders, sintered P/M structural parts (*July*)
Part 10   Metals—mechanical, fracture, and corrosion testing; fatigue; erosion; effect of temperature (*November*)
Part 11   Metallography; nondestructive tests (*November*)
Part 12   Chemical analysis of metals; sampling and analysis of metal bearing ores (*April*)
Part 13   Cement; lime; ceilings and walls (including Manual of Cement Testing) (*November*)
Part 14   Concrete and mineral aggregates (including Manual of Concrete Testing) (*November*)
Part 15   Bituminous materials for highway construction, waterproofing and roofing, and pipe; skid resistance (*April*)

---

*American Society for Testing and Materials, 1916 Race Street, Philadelphia, Pa. 19103, USA.
†General classification: A, ferrous metals; B, non-ferrous metals; C, cementitious, ceramic, concrete and masonry materials; D, miscellaneous materials; E, miscellaneous subjects; F, end-use materials; G, corrosion, deterioration and degradation of materials.

Part 16   Chemical-resistant nonmetallic materials; clay and concrete pipe and tile; masonry mortars and units; asbestos-cement products (*April*)

Part 17   Refractories, glass, and other ceramic materials; manufactured carbon and graphite products (*April*)

Part 18   Thermal and cryogenic insulating materials; building joint sealants; fire tests; building constructions; environmental acoustics (*November*)

Part 19   Natural building stones; soil and rock; peats, mosses, and humus (*April*)

Part 20   Paper; packaging; business copy products (*April*)

Part 21   Cellulose; casein; leather; flexible barrier materials (*April*)

Part 22   Wood; adhesives (*November*)

Part 23   Petroleum products and lubricants (I) (*November*)

Part 24   Petroleum products and lubricants (II) (*November*)

Part 25   Petroleum products and lubricants (III); aerospace materials (*November*)

Part 26   Gaseous fuels; coal and coke; atmospheric analysis (*November*)

Part 27   Paint—tests for formulated products and applied coatings (*April*)

Part 28   Paint—Pigments, resins and polymers (*April*)

Part 29   Paint—fatty oils and acids, solvents, miscellaneous; aromatic hydrocarbons; naval stores (*April*)

Part 30   Soap; engine coolants; polishes; halogenated organic solvents; activated carbon; industrial chemicals (*November*)

Part 31   Water (*November*)

Part 32   Textile materials—yarns, fabrics, and general methods (*November*)

Part 33   Textile materials—fibres, zippers; high modulus fibres (*November*)

Part 34   Plastic pipe (*November*)

Part 35   Plastics—general test methods, nomenclature (*April*)

Part 36   Plastics—materials, films, reinforced and cellular plastics; fibre composites (*April*)

Part 37   Rubber—test methods (*July*)

Part 38   Rubber—specifications; carbon black; gaskets; tyres (*July*)

Part 39   Electrical insulating materials—test methods (*July*)

Part 40   Electrical insulating materials—specifications; electrical insulating liquids and gases (*July*)

Part 41   General test methods (nonmetal); statistical methods; space simulation; particle size measurement; deterioration of nonmetallic materials (*July*)

Part 42   Emission, molecular, and mass spectroscopy; chromatography; resinography; microscopy (*July*)

Part 43   Electronics (*November*)

Part 44   Magnetic properties; metallic materials for thermostats and for electrical resistance, heating, and contacts; temperature measurement (*November*)

Part 45   Nuclear standards (*November*)

Part 46   End use products—aerosols and closures, surgical implants, resilient floor coverings; appearance of materials; sensory evaluation; forensic sciences (*November*)

Part 47   Index (*October*)

Materials

A 296:1972 (2) ASTM
Corrosion-resistant, iron–chromium, iron–chromium–nickel, and nickel-base alloy castings for general application. Spec. for.

A 297:1967 (2) ASTM
Heat-resistant iron–chromium and iron–chromium–nickel alloy castings for general application. Spec. for.

*A 351:1972 (1,2) ASTM
Ferritic and austenitic steel castings for high temperature service. Spec. for.

*A 409:1973 (1) ASTM
Welded large outside diameter light wall austenitic chromium–nickel alloy steel pipe for corrosive or high temperature service. Spec. for.

A 430:1973 (1) ASTM
Austenitic steel forged and bored pipe for high temperature service. Spec. for.

*B 418:1967 (7) ASTM
Cast and wrought galvanic zinc anodes for use in saline electrolytes. Spec. for.

E 46:1966 (1972) (32) ASTM
Lead and tin-base solder metal. Chemical analysis of.

E 87:1958 (1972) (32) ASTM
Lead, tin, antimony and their alloys, photometric method for chemical analysis.

F 55:1971 (7) ASTM
Stainless steel bars and wire for surgical implants. Spec. for.

F 56:1971 (7) ASTM
Stainless steel sheet and strip for surgical implants. Spec. for.

F 75:1967 (7) ASTM
Cast cobalt–chromium–molybdenum alloy for surgical implants. Spec. for.

F 90:1968 (7) ASTM
Wrought cobalt–chromium alloy for surgical implants. Spec. for.

F 136:1970 (7) ASTM
Ti–6Al–4V–ELi alloy for use in clinical evaluations as a surgical implant material. Spec. for.

F 138:1971 (7) ASTM
Stainless steel bars and wire for surgical implants (special quality). Spec. for.

F 139:1971 (7) ASTM
Stainless steel sheet and strip for surgical implants (special quality). Spec. for.

Cleaning and Surface Preparation

*A 380:1972 (3) ASTM
Cleaning and descaling stainless steel parts, equipment and systems. Rec. practice for.

*B 183:1972 (7) ASTM
Low carbon steel for electroplating. Rec. practice for preparation of.

B 242:1971 (7) ASTM
Preparation of high-carbon steel for electroplating. Rec. practice for.

*B 252:1969 (7) ASTM
Preparation of zinc alloy die castings for electroplating. Rec. practice for.

*B 281:1958 (1972) (5,7) ASTM
Preparation of copper and copper-base alloys for electroplating. Rec. practice for.

*B 319:1960 (1971) (7) ASTM
Preparation of lead and lead alloys for electroplating. Rec. practice for.

*B 320:1960 (1971) (2,7) ASTM
Iron casting for electroplating. Rec. practice for preparation of.

B 322:1968 (7) ASTM
Cleaning metals prior to electroplating. Rec. practice for.

*B 480:1968 (7) ASTM
Preparation of magnesium and magnesium alloys for electroplating. Rec. practice for.

B 481:1968 (1973) (7) ASTM
Preparation of titanium and titanium alloys for electroplating. Rec. practice for.

B 482:1968 (1973) (7) ASTM
Preparation of tungsten and tungsten alloys for electroplating. Rec. practice for.

D 609:1961 (1968) (21) ASTM
Preparation of steel panels for testing paint, varnish, lacquer and related products.

*D 1669:1962 (1973) (11) ASTM
Preparation of test panels for accelerated outdoor weathering of bituminous coatings.

D 1730:1967 (1973) (21) ASTM
Preparation of aluminium and aluminium alloy surfaces for painting. Rec. practice for.

D 1731:1967 (1973) (21) ASTM
Preparation of hot-dip aluminium surfaces for painting. Rec. practice for.

D 1732:1967 (1973) (21) ASTM
Preparation of magnesium alloy surfaces for painting. Rec. practices for.

D 1733:1963 (1969) (21) ASTM
Preparation of aluminium alloy panels for testing paint, varnish, lacquer and related products.

D 2092:1968 (21) ASTM
Preparation of zinc-coated steel surfaces for painting. Rec. practices for.

*D 2200:1967 (1972) (21) ASTM
Pictorial surface preparation standards for painting steel surfaces.

D 2201:1965 (1970) (21) ASTM
Preparation of hot-dipped non-passivated galvanised steel panels for testing paint, varnish, lacquer and related products.

Coatings

*A 123:1973 (3) ASTM
Zinc (hot galvanised) coatings on products fabricated from rolled, pressed, and forged steel shapes, plates, bars and strip. Spec. for.

*A 153:1973 (3) ASTM
Zinc coating (hot dip) on iron and steel hardware. Spec. for.

*A 164:1971 (3,7) ASTM
Electrodeposited coatings of zinc on steel. Spec. for.

*A 165:1971 (3,7) ASTM
Electrodeposited coatings of cadmium on steel. Spec. for.

*A 263:1972 (3,4) ASTM
Corrosion-resisting chromium steel clad plate, sheet and strip. Spec. for.

A 384:1972 (3) ASTM
Safeguarding against warpage and distortion during hot-dip galvanising of steel assemblies. Rec. practice for.

*A 386:1973 (3) ASTM
Zinc-coating (hot-dip) on assembled steel products. Spec. for.

*A 446:1972 (3) ASTM
Steel sheet zinc-coated (galvanised) by the hot-dip process. Physical (structural) quality. Spec. for.

A 591:1968 (3) ASTM
Electrolytic zinc-coated steel sheets. Spec. for.

A 603:1970 (3) ASTM
Zinc-coated steel structural wire rope. Spec. for.

*B 177:1968 (7) ASTM
Recommended practice for chromium plating on steel for engineering use.

*B 200:1970 (7) ASTM
Electrodeposited coatings of lead on steel. Spec. for.

*B 242:1954 (1971) (7) ASTM
Preparation of high-carbon steel for electroplating. Rec. practice for.

*B 253:1968 (6,7) ASTM
Preparation of and electroplating on aluminium alloys by the zincate process. Rec. practice for.

*B 254:1970 (7) ASTM
Preparation of and electroplating on stainless steel. Rec. practice for.

*B 374:1967 (1972) (7) ASTM
Standard definitions of terms relating to electroplating.

B 432:1971 (5) ASTM
Copper and copper alloy clad steel plate. Spec. for.

B 449:1967 (1972) (6,7) ASTM
Chromate treatments on aluminium. Rec. practice for.

B 454:1970 (7) ASTM
Standard specification for mechanically deposited coatings of cadmium and zinc on ferrous metals.

*B 456:1971 (7) ASTM
Standard specification for electrodeposited coatings of nickel plus chromium.

*B 488:1971 (7) ASTM
Electrodeposited coatings of gold for engineering uses. Spec. for.

*B 507:1970 (7) ASTM
Design of articles to be plated on racks. Rec. practice for.

B 579:1973 (7) ASTM
Electrodeposited coatings of tin–lead alloy (solder plate). Spec. for.

B 580:1973 (7) ASTM
Anodic oxide coatings on aluminium. Spec. for.

*C 286:1973 (13) ASTM
Porcelain enamel and ceramic-metal systems. Definition of terms relating to.

D 16:1973 (20,21) ASTM
Paint, varnish, lacquer and related products: definition of terms relating to (including tentative revision).

DIN (West German) Standards

55928 6.59
Protective coatings for steel structures; directions (5).

50961 3.63
Protection against corrosion; electroplated zinc coatings on steel.

50962 3.63
Protection against corrosion; electroplated cadmium coatings on steel.

50965 10.62
Protection against corrosion; electroplated tin and copper, tin coatings on steel.

509 6.63
Protection against corrosion; electroplated coatings; symbols, coat thicknesses, general directions.

50903 1.67
Metallic coatings; pores, inclusions, blisters and cracks; definitions.

51213 12.70
Testing of metallic coatings on wires; coatings of tin or zinc.

Inhibitors

50940 8.52
Protection against corrosion; testing of derusting agents and inhibitors for iron and steel laboratory tests (5).

International Standards Organisation (ISO)

Materials

ISO/R 429:1965, *Classification of copper–nickel alloys*
    8 page   Gr 7   (Part of this recommendation is included in BS 374, 378, 1464, 2579, 2870–1 and 2875)
Chemical composition, forms of semi-manufactured products, classification principles appended.

ISO/R 2092:1971, *Light metals and their alloys. Code of designation*
    2 page   Gr 7
Applies to all light metals and their alloys specified in ISO recommendations.

Coatings

ISO/R 2079:1971, *Surface treatment and metallic coatings. General classification of terms*

ISO/R 2080:1971, *Electroplating and related processes. Vocabulary* (a British Standard is in course of preparation)
    36 page   Gr 12
Defines a large number of terms widely used in the science and industry of electroplating and allied processes.

ISO/R 2081:1971, *Electroplated coatings of zinc on iron and steel* (BS 1706 will be
          revised to implement the requirements of this recommendation)
     6 page   Gr 8
Specifies thickness of zinc coatings in relation to service conditions, their appearance
and adhesion to iron and steel basis metals. Passivation by suitable chromate conversion coatings should be applied unless there is agreement to the contrary.

ISO/R 2082:1971, *Electroplated coatings of cadmium on iron and steel* (BS 1706 will be
          revised to implement the requirements of this recommendation)
     6 page   Gr 8
Specifies thickness of cadmium coatings in relation to service conditions, their appearance and adhesion to iron and steel basis metals. Passivation by suitable chromate
conversion coatings should be applied unless there is agreement to the contrary.

ISO/R 2093:1971, *Electroplated coatings of tin* (BS 1872 will be revised to implement
          the requirements of this recommendation)
     9 page   Gr 8
Specifies thickness of tin coatings in relation to service conditions, their appearance and
adhesion to the basis metal. Some coatings are subject to porosity test; a solderability
test may be applied when requested by the purchaser.

ISO/R 1457:1970, *Electroplated coating of copper plus nickel plus chromium on steel
          (or iron)*
     20 page   Gr 9
Refers to composite coatings of copper plus nickel plus chromium electroplated on to
steel (or iron). Recommends appearance of coating, thickness of copper, type and thickness of nickel, type and thickness of chromium, adhesion and corrosion resistance.

ISO/R 1456:1970, *Electroplated coatings of nickel plus chromium*
     24 page   Gr 10
Refers to nickel plus chromium coatings electroplated on to steel (or iron) zinc alloys
and copper or copper alloys. Recommends appearance of coating, type of thickness of
nickel and thickness of chromium, adhesion and corrosion resistance.

ISO/R 1458:1970, *Electroplated coatings of nickel*
     12 page   Gr 8
Refers to nickel coatings electroplated on to steel (or iron) zinc alloys and copper and
copper alloys. Recommends appearance of coating, type and thickness.

ISO/R 1459:1970, *Guiding principles for protection against corrosion by hot dip
          galvanising*

ISO/R 1461:1970, *Requirements for hot dip galvanised coatings on fabricated ferrous
          products*

British Government Standards

Reference standards (DEF) may be obtained free of charge from the Director
of Standardisation, Ministry of Defence, First Avenue House, High Holborn,
London WC1.
    Defence Guides (DG) and DTD specifications may be bought from H.M.
Stationery Office.

Surface Preparation

DEF STAN 03–2, *Cleaning and preparation of metal surfaces*

DTD 935, *Surface sealing of magnesium rich alloys*

DEF STAN 03–4, *The pretreatment and protection of steel parts of tensile strength exceeding* 1400 N/mm²

Coatings

DEF 60, *Selection and treatment of aluminium-base materials for use with concentrated hydrogen peroxide (HTP)*

DEF 61, *Selection and treatment of corrosion-resisting steels for use with concentrated hydrogen peroxide (HTP)*

DEF 130, *Chromate passivation of cadmium and zinc surfaces*

DEF 151, *Anodising of aluminium and aluminium alloys*

DEF 160, *Chromium plating for engineering purposes*

DEF–STAN 03–3, *Protection of aluminium alloys by sprayed metal coatings*

DEF–STAN 03–5, *Electroless nickel coating of metals*

DEF–STAN 03–6 (1972), *Guide to flame spraying processes; includes metal and ceramic coatings*

DEF STAN 03–7 (1973), *Painting of metal and wood*

DEF STAN 03–8, *Electrodeposition of tin*

DEF STAN 03–11, *Phosphate treatment of iron and steel*

DG 8, *Treatment for the protection of metal parts of service stores and equipment against corrosion*

DG 3–A (1966), *Prevention of corrosion of cadmium and zinc coatings by vapours from organic materials such as paints, varnishes, insulating materials, adhesives and wood*

DG 10 (1967), *Metallic springs—protection against corrosion*

DG 13 (1968), *Design aspects of chromium plating for engineering purposes*

DTD 903 D, *Zinc plating*

DTD 904 C, *Cadmium plating*

DTD 905 A, *Nickel plating (heavy)*

DTD 911 C, *Protection of magnesium-rich alloys against corrosion*

DTD 919 B, *Electroplating of aluminium, steel and copper with silver and nickel*

DTD 927 A, *Tin–zinc alloy plating .*

DTD 931, *Rhodium plating*

DTD 938, *Gold plating*

DTD 939, *Silver plating of heat-resisting threaded parts for anti-seizure purposes*

DTD 940, *The cadmium coating of very strong steel parts by vacuum evaporation*

DTD 942, *Anodising of titanium and titanium alloys*

DTD 943, *Electrodeposited cobalt/chromium carbide composite coatings*

DEF 1053, *Standard methods of testing paint, varnish laquer and related products*

National Association of Corrosion Engineers (NACE) Standards

Recommended Practices

RP–01–69, *Control of external corrosion on underground or submerged metallic piping systems*

RP–01–70, *Protection of austenitic stainless steel in refineries against stress-corrosion cracking by use of neutralising solutions during shut down*

RP–01–72, *Surface preparation of steel and other hard materials by water blasting prior to coating or recoating*

RP–02–72, *Direct calculation of economic appraisals of corrosion control measures*

RP–03–72, *Method for lining lease production tanks with coal tar epoxy coating*

RP–04–72, *Methods and controls to prevent in-service cracking of carbon steel (P-1) welds in corrosive petroleum refining environments*

RP–05–72, *Design, installation, operation and maintenance of impressed-current deep groundbeds*

RP–01–73, *Collection and identification of corrosion products*

RP–02–73, *Handling and proper usage of inhibited oilfield acids*

RP–01–74, *Corrosion control of electric underground residential distribution systems*

RP–02–74, *High-voltage electrical inspection of pipeline coatings prior to installation*

RP–01–75, *Control of internal corrosion in steel pipelines and piping systems*

RP–02–75, *Application of organic coatings to the external surface of steel pipe for underground service*

Test Methods

TM–01–69, *Laboratory corrosion testing of metals for the process industries*

TM–01–70, *Visual standard for surfaces of new steel airblast cleaned with sand abrasive*

TM–02–70, *Method of conducting controlled velocity laboratory corrosion tests*

TM–01–71, *Autoclave corrosion testing of metals in high-temperature water*

TM–01–72, *Anti-rust properties of petroleum products pipeline cargoes*

TM–01–73, *Methods for determining water quality for subsurface injection using membrane filters*

TM–01–74, *Laboratory methods for the evaluation of protective coatings used as lining materials in immersion service*

TM–02–74, *Dynamic corrosion testing of metals in high-temperature water*

TM–03–74, *Laboratory screening tests to determine the ability of scale inhibitors to prevent the precipitation of calcium sulfate and calcium carbonate from solution*

TM–01–75, *Visual standard for surfaces of new steel centrifugally blast cleaned with steel grit and shot*

Material Requirements

MR–01–74, *Recommendations for selecting inhibitors for use as sucker rod thread lubricants*

MR–02–74, *Material requirements in prefabricated plastic films for pipeline coatings*

21.3 Calculations Illustrating the Economics of Corrosion Protection

Each protective system has its own pattern of capital and running costs. Choosing the optimum alternative requires care because some of the factors can be measured precisely while others cannot, and the costs arise at different points in time.

A quick comparison of systems is to compare present first cost + maintenance costs (including access and disruption) in each case. This can be elaborated as follows:

Step 1 Set out the alternatives, ensuring that there is the same basis of comparison. Eliminate capital and revenue costs common to all protection systems. Write in the extra capital arising from the need to protect the structure. This capital cost may be a direct quotation from a processor or be calculated (*see* Note 3 below). Make sure quotations or calculations refer to the same data.

Step 2 Add extra transport costs (e.g. when work goes to specialist processing plant instead of direct to site) and extra mark-up (the contractor will add a percentage on subcontracted work) where appropriate and where not included in Step 1.

Step 3 Deduct tax allowances. Typically for limited liability companies in the UK these comprise initial grants for certain projects in development areas plus first year allowances, adjusted for corporation tax.

Step 4 Estimate the time required for each system to be complete on site. For the longer time systems enter the cost in interest of this delay assuming the capital sums involved could have been otherwise employed during the extra time required (compared with the quickest system).

Step 5 Estimate maintenance costs (after making allowance for tax relief) over life of structure including access costs such as scaffolding. In general the future maintenance costs must be estimated making allowance for inflation and then discounted to present values (*see* maintenance terms in equation below).

Step 6 Estimate disruption during maintenance (or erection). As with maintenance costs these can be entered at present costs of such disruption whenever interest rates equal inflation rate.

Table 21 A1   Model cost calculation

	Competing systems (conforming to systems under same environment and same time to first maintenance)	
	first system	*second system etc.*
Extra capital cost (Step 1)		
Add extra transport costs (Step 2)		
Add extra fabricator's mark-up (Step 2)		
Total gross extra capital cost		
Deduct total taxed allowances (Step 3)		
Total gross extra capital cost net of tax allowances etc.		
Delay factor (Step 4)		
Maintenance costs (Step 5)		
Disruption factor (Step 6)		
Total cost of system, over and above costs common to all systems, and expressed in present value terms		

The maintenance costs and the initial cost are frequently assessed as the Net Present Value (N) which represents the sum of money that must be set aside to now cover both initial and maintenance costs over the total life required.

$$N = F + \frac{M_1}{(1+r)^{P_1}} + \frac{M_2}{(1+r)^{P_2}}$$

where F = first cost of the protective system
M_1 = the cost of maintenance in the year P_1
M_2 = the cost of maintenance in the year P_2 etc.
r = the interest (or 'discount') rate (NB: 12% rate is written 0·12, etc.)
P_1 = the number of years to first maintenance
P_2 = the number of years to second maintenance, etc.

Maintenance costs are steadily rising and we must adjust maintenance costs at present-day values (M) to allow for inflation at an annual rate of r_1 (Note: r_1 is expressed like r).

i.e.
$$N = F + \frac{M(1+r_1)^{P_1}}{(1+r)^{P_1}} + \frac{M(1+r_1)^{P_2}}{(1+r)^{P_2}}$$

In the simplest case where the rate of inflation of maintenance costs is the same as the interest (or 'discount') rate available on money i.e. where there is no time value of money $r_1 = r$ and the formula simplifies to

$N = F +$ the sum of all maintenance costs at present-day rates
    = the aggregate of all initial and maintenance costs at present-day rates

Similarly other costs can be treated as for the maintenance costs or for the first cost, as appropriate.

Note 1   Present day economics usually favour a protective scheme adequate for the full design life of the structure.

Note 2   For many protective schemes initial cost in the factory is roughly proportional to total thickness but life to first maintenance is not necessarily proportional to thickness e.g. relatively thin metal coatings in certain environments often have the longest lives to first maintenance.

Note 3   Calculation of protection costs from fundamentals has been dealt with in industry orientated handbooks e.g. *Galvanising for Profit*, Galvanisers Association, London, UK. In principle first costs include:
Materials
Labour
Overheads (including inspection; access equipment and tools for painting)
These are usually of the same order of magnitude, with materials being most costly for metal coatings and labour most costly for paint coatings.

Note 4   Surface preparation for painting is often analysed separately as it may typically form one-third of the total cost. Poor surface preparation can more than halve the life expectancy of a coating (and add over 100% to the total life cost) but save less than 20% of the total capital cost. Correct choice of blasting equipment (e.g. nozzle size, grade of abrasive) can save up to half the surface preparation cost.

Note 5   In calculating paint requirements up to 50% more than the theoretical dry film weight requirement should be allowed to cover thick coatings, wastage, repairs, losses (typically 30% for air spray; 15% airless spray; 10% electrostatic spray; 5% roller or brush) etc. Manufacturers should be asked to quote percentage volume solids in their paint to facilitate calculations.

Note 6   Labour costs for the application of paint increase in the order

	airless spray	normal air spray	roller	brush
ratio	1	2	3	4

Brush application is very desirable for inhibitive primers applied directly to steel and for toxic materials. Roller application is not recommended for structural steel surfaces.

Note 7   Comparative costs (unlike actual costs) change little with time. When labour costs are rising more rapidly than material costs, painting will become slightly more expensive relative to metal coatings and vice versa.

Note 8   Metal spray, paint powder and tape coatings are usually costed by the area covered which should be known reasonably accurately. Galvanising is costed by weight of steel coated i.e. by thickness of steel. Costs also vary by product.

Note 9   The most economical system for one job is not necessarily the most economical for another, e.g. galvanised steel may be best for a bolted job but not so where much site welding is necessary.

Note 10 First costs include access costs e.g. scaffolding. Surfaces treated before erection do not require scaffolding.

F. C. PORTER

22 TERMS AND ABBREVIATIONS

22.1   Glossary of Terms   **22**:3
22.2   Symbols and Abbreviations   **22**:17

22.1 Glossary of Terms*

Activation Overpotential: that part of an overpotential (polarisation) that exists across the electrical double layer at an electrode/solution interface and thus directly influences the rate of the electrode process by altering its activation energy.

Active: freely corroding; the antithesis of passive.

Active–passive Metal (alloy): a metal that exhibits a transition from the active (freely corroding) state to the passive state when its potential is raised above the passivation potential (or a transition from the passive to the active state when its potential falls below the Flade potential).

Activity (thermodynamic): the thermodynamic activity of an entity, i, is $a_i = \exp(\Delta G/RT)$ where ΔG is the free energy increase of 1 mol of i when it is converted from a standard state (defined as when $a_i = 1$) to any other activity $a_i = x$.

Addition Agent: a substance added to an electroplating solution to produce a desired change in the physical properties of the electrodeposit.

Adhesion: the attractive force that exists between an electrodeposit and the substrate.

Alclad: a composite in which a thin layer of aluminium, or an aluminium alloy of good corrosion resistance, is bonded metallurgically to a high-strength aluminium alloy (of lower corrosion resistance) to provide a combination of these two properties.

Anion: a negatively charged ion; it migrates to the anode of a galvanic or voltaic cell.

Anode: the electrode of a galvanic or voltaic cell where the positive electrical current flows from the electrode to the solution (for example, transfer of cations from electrode to solution or anions from solution to electrode).

Anode Corrosion Efficiency: the ratio of the actual corrosion rate of the anode to the theoretical rate according to Faraday's law, expressed as a percentage.

Anode Polarisation: difference between the potential of an anode passing current and the equilibrium (or steady-state) potential of the electrode having the same electrode reaction.

Anodic Protection: reduction of the corrosion rate by making the potential

---

*Terms used in chemical thermodynamics have not been included and reference should be made to Appendix 1.9B. For paint terms, see Section 15.10.

of the metal sufficiently more electropositive by an external source of e.m.f., so that the metal becomes passive.

Anaerobic: air or uncombined oxygen being absent.

Anaerobic Bacteria (anaerobes): group of bacteria that are unable to multiply in any but a minute trace of oxygen.

Angstrom Unit (Å): $1\text{ Å} = 10^{-8}\text{ cm} = 10^{-4}\,\mu\text{m} = 10^{-1}\text{ nm}$ (preferred unit).

Anodising: the formation of oxide films on metals by anodic oxidation of the metal in an electrolyte solution. The term can be used for thin dielectric films but is more particularly applied to thick porous films formed on aluminium.

Anolyte: the electrolyte solution adjacent to the anode.

Anti-pitting Agent: an addition agent (q.v.) which is used to prevent the formation of pits or large pores in an electrodeposit.

Austenite: the γ-modification of iron, having an f.c.c. lattice, which is stable above about 700°C; the term is also applicable to solid solutions of carbon, chromium, nickel, etc. in γ-iron.

Bainite: a structure produced in carbon and alloy steels by rapid cooling to a temperature above that at which martensite is formed, followed by slower cooling.

Base Potential: *see under* **Potential**.

Bimetallic Corrosion: corrosion of two metals in electrical contact, in which one metal stimulates attack on the other and may itself corrode more slowly than when it is not in such contact; *galvanic* is often used in place of *bimetallic*.

Breakaway Corrosion: a sudden increase in corrosion rate, especially in high-temperature 'dry' oxidation, etc.

Breakdown Potential: *see* **Critical Pitting Potential**.

Bright Plating: electroplating under conditions whereby the electrodeposit has a high degree of specular reflectivity.

Brightener: an addition agent used specifically to produce an electrodeposit of high specular reflectivity.

Brush Plating: a method of electrodeposition in which the plating solution is applied to the article to be plated by means of an absorbent pad or brush which contains the anode.

Buffer: a substance, or mixture of substances, which when present in an electrolyte solution tends to diminish fluctuations in pH.

Burnt Deposit: a rough, poorly coherent, electrodeposit that results from the application of an excessively high current density.

Calcareous Scale: a scale consisting largely of calcium carbonate and magnesium hydroxide which may be precipitated from a hard water.

Cathode: the electrode of a galvanic or voltaic cell where positive current flows from the solution to the electrode (for example, by transfer of cations from solution to electrode).

Cathodic Protection*: reduction of the corrosion rate by making the potential of the metal to be protected more negative, so that the current at any anode is reduced while that at any cathode is increased.

Cathodic Reactant: species which is reduced at a cathode.

Catholyte: the electrolyte solution adjacent to the cathode.

*See also list of underground-corrosion and cathodic-protection terms, p. **22**.13.

Cation: a positively charged ion; it migrates to the cathode in a galvanic or voltaic cell.

Caustic Embrittlement: stress-corrosion cracking of carbon steels caused by the presence of caustic alkali.

Cavitation Damage: erosion of a solid surface caused by the collapse of vacuum bubbles formed in a fluid.

Cementite: the iron–carbon compound of formula very close to Fe_3C.

Chemical Conversion Coating: a protective or decorative coating which is produced deliberately on a metal surface by a chemical environment.

Chemical Polishing: improvement in the brightness and levelness of a surface finish of a metal by a chemical dissolution reaction.

Composite Plate: an electrodeposit consisting of two or more layers of metals deposited separately.

Concentration Cell: a galvanic cell in which the e.m.f. is due to differences in the concentration of one or more electrochemically reactive constituents of the electrolyte solution.

Concentration (diffusion or transport) Overpotential: change of potential of an electrode caused by concentration changes near the electrode/solution interface produced by an electrode reaction.

Corrosion: the transformation of a metal from the elementary to the combined state by reaction with a non-metal (or liquid metal).

For the purpose of the present work—the transformation of a metal, *used as a material of construction*, from the elementary to the combined state. (*See also* **Bimetallic Corrosion, Breakaway Corrosion, Filiform Corrosion, Fretting Corrosion, Galvanic Corrosion, Graphitic Corrosion, Intergranular Corrosion, Localised Attack, Pitting, Stress-corrosion Cracking, Uniform Corrosion.**)

Corrosion Control: control of the corrosion rate and form of attack of a metal of a given metal/environment system at an acceptable level and at an economic cost.

Corrosion–Erosion (or erosion–corrosion): conjoint action of corrosion and erosion resulting from the abrasive action of a fluid (solution or gas) moving at a high velocity, resulting in continuous removal of the protective film from the metal surface.

Corrosion Fatigue: failure by cracking caused by reversing alternating stress in the presence of a corrosive environment.

Corrosion Potential (mixed potential, compromise potential): potential resulting from the mutual polarisation of the interfacial potentials of the partial anodic and cathodic reactions that constitute the overall corrosion reaction.

Corrosion Product: metal reaction product resulting from a corrosion reaction; although the term is normally applied to solid compounds it is equally applicable to gases and ions resulting from a corrosion reaction.

Corrosion Rate: the rate at which a corrosion reaction proceeds; it may be expressed as a rate of penetration $mm\,y^{-1}$ (in y^{-1} or i.p.y. are still widely used) or as a weight loss per unit area per unit time, i.e. $g\,m^{-2}\,d^{-1}$ $mg\,dm^{-2}\,d^{-1}$ or m.d.d. are still widely used).

Couple (bimetallic, galvanic): two dissimilar metals in electrical contact.

Covering Power: the ability of a plating solution to produce an electrodeposit (irrespective of thickness) at low current densities on all significant areas.

Crevice Corrosion: localised corrosion resulting from a crevice formed between two surfaces—one at least of which is a metal.

Critical Anode Current Density: anodic current density that must be exceeded in order to produce an active to passive transition (for a given metal it varies with the nature of the solution, temperature, velocity, etc.).

Critical Humidity: the relative humidity (r.h.) at and above which the atmospheric corrosion rate of a metal increases markedly.

Critical Pitting Potential: the most negative potential required to initiate pits in the surface of a metal held within the passive region of potentials (it varies with the nature of solution, temperature, time, etc.).

Current (I): the rate of transfer of electric charge; unit current is the ampere (A) which is the transfer of 1 coulomb/second (1 C/s).

Current Density (i): the current per unit area (geometric) of surface of an electrode (units: $A\,m^{-2}$, $mA\,m^{-2}$, $A\,cm^{-2}$, etc.).

Current Efficiency: the ratio of the rate of the actual electrochemical change at the electrode (anode or cathode) to the theoretical rate according to Faraday's law, expressed as a percentage. (*The particular electrode and electrode reaction must be specified.*)

Deactivation: prior removal of the constituent (of a liquid) that is active in causing corrosion. The term is usually applied to the removal of oxygen by physical and/or chemical methods.

Dealloying: selective removal by corrosion of a constituent of an alloy.

Delayed Failure: *see under* **Sustained Load Failure.**

Depolarisation: reduction or elimination (by physical or chemical methods) of the electrode polarisation needed to produce a specified current.

Deposit Attack: localised corrosion (a form of crevice corrosion) under and resulting from a deposit on a metal surface.

Dezincification: preferential corrosion of zinc from brass resulting in mechanically weak copper-rich areas in the form of plugs or layers; sometimes both zinc and copper corrode, but copper is redeposited.

Differential Aeration: the stimulation of corrosion at a localised area by differences in oxygen concentration in the electrolyte solution in contact with the metal surface; the area in contact with the solution of lower oxygen concentration is the anode.

Diffusion Coating: a coating produced by diffusion at elevated temperatures.

Diffusion Layer: the thin layer of solution adjacent to an electrode through which transport of species to or from the electrode surface occurs by diffusion rather than by convection.

Electrochemical Cell: a cell in which chemical energy is transferred into electrical energy.

Electrolytic Cell: a cell in which electrical energy is used to bring about electrode reactions and is thus converted into chemical energy. (Note: the term 'electrochemical cell' is frequently used to describe both types of cells.)

Electrochemical Equivalent: theoretical number of moles of electrochemical change resulting from the passage of 1 Faraday (1 F) of electrical charge; if M is the molar mass (kg) and z the number of electrons required for one act of the reaction then the electrochemical equivalent is M/z kg.

Electrode: an electron conductor by means of which electrons are provided for, or removed from, an electrode reaction.

Electrode Potential (E): the difference in electrical potential between an electrode and the electrolyte with which it is in contact. It is best given with reference to the standard hydrogen electrode (S.H.E.), when it is equal in magnitude to the e.m.f. of a cell consisting of the electrode and the S.H.E. (with any liquid-junction potential eliminated). When in such a cell the electrode is the cathode, its electrode potential is positive; when the electrode is the anode, its electrode potential is negative. When the species undergoing the reaction are in their standard states, $E = E^\ominus$, the *standard electrode potential*.

Electrogalvanising: galvanising by electroplating.

Electroless Plating: formation of a metallic coating by chemical reduction catalysed by the metal deposited.

Electrolysis: decomposition by the passage of electric current.

Electrolyte: a substance which in solution gives rise to ions; the term is also used for a solution of an electrolyte or for a molten ionic salt.

Electrolyte Solution: a solution in which the conduction of electric current occurs by the passage of dissolved ions.

Electrolytic Cleaning: cleaning obtained at cathode or anode of a cell containing a suitable solution.

Electron Acceptor: a species in solution that accepts one or more electrons from a cathode for each act of the cathodic reaction and is thus reduced to a lower valence state.

Electrophoretic Plating: production of a layer or deposit by discharge of colloidal particles in a solution on an electrode.

Electroplating: electrodeposition of a thin adherent layer of a metal or alloy on a substrate of desirable chemical, physical and mechanical properties.

Electropolishing: surface finishing of a metal by making it the anode in an appropriate solution, whereby a bright and level surface showing specular reflectivity is obtained.

e.m.f. Series: a table of the standard equilibrium electrode potentials of systems of the type $M^{z+}(\text{aq.}) + ze = M$, relative to the standard hydrogen electrode, and arranged in order of sign and magnitude.

Epitaxy: the phenomenon whereby a deposit or coating takes up the lattice habit and orientation of the substrate.

Equilibrium Potential: the electrode potential of an unpolarised reversible electrode.

Evans Diagram: diagram in which the E vs. I relationships for the cathodic and anodic reactions of a corrosion reaction are drawn as straight lines intersecting at the corrosion potential, thus indicating the corrosion current associated with the reaction.

Exchange Current Density (i_0): the rate of exchange of electrons (expressed as an electric current per unit area) between the two species concerned in a reversible electrode process at the equilibrium potential.

Exfoliation: loss of material in the form of layers or leaves from a solid metal or alloy.

Faraday: quantity of electric charge required to bring about a change of one electrochemical equivalent ($1\,\text{F} \approx 96\,500\,\text{C}$).

Fatigue: failure of metal under conditions of repeated alternating stress.

Ferrite: the body-centred cubic form of iron (α-iron) and the solid solutions of one or more elements in b.c.c. iron.

Filiform Corrosion: corrosion in the form of hairs or filaments progressing across a metal surface.

Film: a thin coating of material, not necessarily thick enough to be visible.

Flade Potential: the potential at which a metal which is passive becomes active (*see* **Passivation Potential**).

Fogging: reduction of the lustre of a metal by a film or particulate layer of corrosion product, e.g. the dulling of bright nickel surfaces.

Fouling: deposition of flora and fauna on metals exposed to natural waters, e.g. sea-water.

Fretting Corrosion: deterioration resulting from repetitive rubbing at the interface between two surfaces (fretting) in a corrosive environment.

Galvanic Cell: an electrochemical cell having two electronic conductors (commonly dissimilar metals) as electrodes.

Galvanic Corrosion: corrosion associated with a galvanic cell (often used to refer specifically to **Bimetallic Corrosion**).

Galvanostatic Polarisation (intentiostatic): polarisation of an electrode during which the current density is maintained at a predetermined constant value.

Galvanic Series: a list of metals and alloys based on their relative potentials in a given specified environment, usually sea water.

Galvanising (hot dip): coating of iron and steel with zinc using a bath of molten zinc.

Graphitic Corrosion: corrosion of grey cast iron in which the metallic constituents are removed as corrosion products, leaving the graphite.

Green Rot: carburisation and oxidation of certain nickel alloys at around 1000°C resulting in a green corrosion product.

Half-cell: one of the electrodes, and its immediate environment, in an electro-chemical or electrolytic cell; particularly, an electrode and environment arranged for the passage of current to another electrode or (especially) for the measurement of its electrode potential.

Hematite: an oxide of iron corresponding closely to Fe_2O_3 produced during the oxidation of iron: it is an n-type semiconductor in which diffusion of O^{2-} ions occurs via anion vacancies.

Hydrogen Electrode: an electrode at which the equilibrium $H^+(aq.) + e \rightarrow \frac{1}{2}H_2$ is established. By definition, at unit activity of hydrogen ions and unit fugacity of hydrogen gas the potential of the standard hydrogen electrode $E^{\ominus}_{H^+/\frac{1}{2}H_2} = 0.00\,V$.

Hydrogen Embrittlement: embrittlement caused by the entry of hydrogen into a metal.

Hydrogen Overpotential (Overvoltage): the displacement of the equilibrium (or steady-state) electrode potential of a cathode required for the discharge of hydrogen ions at a given rate per unit area of electrode.

Immunity (thermodynamic): a condition whereby the equilibrium activity of metal ions (simple or complexed) in solution is prevented from exceeding some arbitrary value by making the potential of the metal sufficiently negative to that of the solution. Pourbaix has defined this arbitrary activity at $10^{-6}\,g\,ion/l$.

Impingement Attack: localised corrosion resulting from the action of corrosion and/or erosion (separately or conjoint) when liquids impinge on a surface.

Inhibitor: a substance added to an environment in small concentrations to reduce the corrosion rate.

Inseparable Anodes and Cathodes: anodes and cathodes which cannot be distinguished experimentally although their existence is postulated by theory.

Interference Films: thin transparent films which exhibit interference colours.

Intergranular Corrosion: preferential corrosion at grain boundaries.

Internal Oxidation (subsurface corrosion, subscale formation): formation of isolated particles of corrosion product (or products) beneath the metal surface, which may be coated with a uniform film of corrosion product.

Ion: an electrically charged atom or complex of atoms.

Isocorrosion Chart: a chart having temperature and concentration of the corrodent as co-ordinates and curves (isocorrosion curves) of various specified constant corrosion rates of the metal.

Knife-line Attack: severe highly localised attack (resembling a sharp cut into the metal) extending only a few grains away from the fusion line of a weld in a stabilised austenitic stainless steel, which occurs when the metal comes into contact with hot nitric acid and is due to the precipitation of chromium carbides.

Leveller: a substance which is added to a plating bath to produce a levelling or smoothing action.

Levelling Action: the ability of a plating bath to produce a surface which is smoother than that of the substrate.

Limiting Current Density: the current density at which change of polarisation produces little or no change of current density.

Linear Polarisation: the linear relationship between overpotential and current density that is considered to prevail at potentials very close to the corrosion potential.

Local Anodes and Cathodes: *see* **Inseparable Anodes and Cathodes.**

Localised Attack: corrosion in which one area (or areas) of the metal surface is predominantly anodic and another area (or areas) is predominantly cathodic, i.e. anodes and cathodes are *physically separable*.

Matt Surface: a surface of low specular reflectivity.

Metal Distribution Ratio: the ratio of the thicknesses of metal produced during electroplating on two specified parts of the cathode.

Metal Spraying: application of a metal coating to a surface (metallic or non-metallic) by means of a spray of metal particles. The metal particles may be produced by 'atomising' a metal wire in a flame-gun or by introducing metal powder into a similar gun.

Migration of Ions: movement of ions towards the anode and cathode under the influence of an electrical potential gradient.

Mil: one thousandth of an inch (1 mil = 0·001 in = 25·4 μm).

Millscale: thick oxide layer on metals produced during fabrication by hot-rolling; most of the oxidation occurs as the metal passes from one rolling-mill to the next.

Negative Potential: *see under* **Potential.**

Nernst Equation: the thermodynamic relationship between the reversible potential of a reaction (or half-reaction) and the activities of the species involved in the reaction.

Noble Potential: *see under* **Potential.**

Normal Electrode Potential: *see under* **Electrode Potential.**

Occluded Cell: a corrosion cell of a geometry that prevents intermingling of the anodic reaction products (anolyte) with the bulk solution, resulting in a decrease in pH of the anolyte; shielded areas or pits, crevices or cracks in the surface of the metal are examples.

Open-circuit Potential: the potential of an electrode from which no significant current is flowing, as on open circuit.

Overpotential (overvoltage, polarisation): the displacement of the equilibrium (or steady-state) electrode potential required to cause a reaction to proceed at a given rate.

Oxidation: loss of electrons by a species during a chemical or electrochemical reaction; addition of oxygen or removal of hydrogen from a substance.

Oxidising Agent: a substance that causes oxidation of another species.

Oxygen Concentration Cell: *see under* **Differential Aeration.**

Parabolic Oxidation: an oxidation reaction that conforms kinetically to a parabolic law so that the rate of oxidation decreases with thickness of oxide.

Partial Reactions: anodic reaction (reactions) and cathodic reaction (reactions) constituting a single exchange process or a corrosion reaction.

Partial Current (current densities): the currents (current densities) corresponding with each of the partial reactions.

Parting: selective dissolution of one metal (usually the most electro-reactive) from an alloy leaving a residue of the less reactive constituents.

Passivation Potential: the potential at which a metal in the active state becomes passive.

Passivator: a substance which in solution causes passivity.

Passivity: (difficult to define precisely) a metal or alloy which is thermodynamically unstable in a given electrolyte solution is said to be passive when it remains visibly unchanged for a prolonged period. The following should be noted:

(i) *During passivation* the appearance *may* change if the passivating film is sufficiently thick (e.g. interference films on Ti).

(ii) The electrode potential of a passive metal is always appreciably more noble than its potential in the active state.

(iii) Passivity is an *anodic* phenomenon, and control of corrosion by decreasing cathodic reactivity (e.g. amalgamated zinc in sulphuric acid) or by cathodic protection is not passivity.

An alternative definition: a metal in a given solution is said to become passive when there is a significant decrease in the corrosion rate on raising the potential of the metal (by an external e.m.f. or by adding an oxidant to the solution) above a certain critical value.

Patina: a green coating of corrosion products of copper (basic sulphate, carbonate and chloride) which forms on copper or copper alloys after prolonged atmospheric exposure.

pH: a measure of the hydrogen ion activity defined by $pH = -\log a_{H^+}$ where a_{H^+} is the activity of the hydrogen ion.

Pickle, Pickling: a solution (usually acidic) used to remove or loosen corrosion products from the surface of a metal.

Anodic and *cathodic* pickling are forms of *electrolytic* pickling in which the metal is anodically or cathodically polarised in the pickle.

Pitting–Bedworth Ratio: ratio of volume of oxide/volume of metal converted into oxide during a high-temperature oxidation reaction.

Pits (in electroplating): macroscopic channels in an electrodeposit which may or may not extend to the substrate.

Pitting (corrosion): localised corrosion in which appreciable penetration into the metal occurs resulting in the formation of cavities.

Pitting Potential: potential at which a metal pits (*see also* **Critical Pitting Potential**).

Polarisation (overpotential, overvoltage): difference of the potential of an electrode from its equilibrium or steady-state potential.

Potential–pH Equilibrium Diagram (Pourbaix diagram): diagram with equilibrium potential and pH as co-ordinates that show the phases that are at equilibrium when a metal reacts with water or an aqueous solution of a specified electrolyte.

Pores, Porosity (in coatings): microscopic channels in coatings (metallic or non-metallic) which extend to the substrate.

Positive Potential: *see under* **Potential.**

Potential: potential difference at an electrode defined with reference to another specified electrode.

    base potential: a potential towards the negative end of a scale of electrode potentials. The potential of an electrode which is made cathodic is said to become 'more base' or 'more negative' (preferred term).

    negative potential: a potential more negative than the potential of the standard hydrogen electrode, e.g. $E^{\ominus}_{Zn^{2+}/Zn} = -0.76$ V.

    noble potential: a potential towards the positive end of a scale of electrode potentials. The potential of an electrode which is made anodic is said to become 'more noble' or 'more positive' (preferred term).

    positive potential: a potential more positive than the potential of the standard hydrogen electrode, e.g. $E^{\ominus}_{Cu^{2+}/Cu} = +0.34$ V.

(*See also* **Corrosion Potential, Electrode Potential, Equilibrium Potential, Flade Potential, Open-circuit Potential, Passivation Potential, Protection Potential, Redox Potential.**)

Polarisation Resistance: slope of the 'linear' overpotential *vs.* current density curve at potentials close to E_{corr} or the tangent of the curve if it is non-linear; $R_p = \Delta E/\Delta i \, \Omega$.

Potentiostatic Polarisation: polarisation of an electrode during which the potential is maintained at a predetermined constant value by means of a *potentiostat.*

Protection Potential for Pitting: potential below which new pits cannot be initiated nor pre-existing pits continue to propagate.

Rack: a frame for suspending and carrying current to articles during plating and related operations.

Rate-determining Step (r.d.s.): the step in the sequence of steps of a reaction that has the highest activation energy and is thus rate-controlling.

Redox Potential: the electrode potential of a reversible oxidation–reduction system, e.g. Cu^{2+}/Cu, Fe^{3+}/Fe^{2+}, MnO_4^-/Mn^{2+}, Cl_2/Cl^-.

Redox System: a reversible oxidation–reduction system.

Reducing Agent: a substance that causes reduction of another species.

Reduction: a chemical or electrochemical reaction in which a species gains electrons; the removal of oxygen or the addition of hydrogen.

Reference Electrode: a half-cell of reproduceable potential by means of which an unknown electrode potential can be determined on some arbitrary scale (e.g. hydrogen scale, $Cu/CuSO_4$ scale, $Ag/AgCl$ scale, etc.).

Relative Humidity: the ratio of the amount of water vapour present in the atmosphere at a given temperature to the amount required for saturation at the same temperature, expressed as a percentage.

Reversible Electrode: an electrode in which a small increase or decrease in potential can reverse the direction of the electrode reaction.

Reversible Potential: *see under* **Equilibrium Potential.**

Rusting (rust): corrosion of iron or ferrous alloys resulting in a corrosion product which consists largely of hydrous ferric oxide.

Saturation Index: an index which shows if a water of given composition and pH is at equilibrium, supersaturated or unsaturated with respect to calcium carbonate (or to magnesium hydroxide).

Scale: a thick visible oxide film formed during the high-temperature oxidation of a metal (the distinction between a film and a scale cannot be defined precisely).

Season Cracking: cracking resulting from the combined effect of corrosion and stress, which is usually confined to the stress-corrosion cracking of brass in ammoniacal environments.

Selective Leaching: *see* **Parting.**

Sensitisation: susceptibility to intergranular attack in a corrosive environment resulting from heating a stainless steel at a temperature and time that results in precipitation of chromium carbides at grain boundaries.

Sherardising: the coating of iron and steel with zinc by heating in contact with zinc powder at a temperature below the melting point of zinc.

Shield: a non-conducting medium which is used for altering the current distribution on a cathode or anode.

Spalling: the break-up of a surface through the operation of internal stresses, often caused by differential heating or cooling.

Standard Electrode Potential: *see under* **Electrode Potential.**

Steady-state Potential: the potential (not varying with time) of an electrode that is operating under steady-state conditions of zero or constant current density.

Stoichiometric Number: number of times the rate-determining step must occur for one act of the overall reaction.

Stray-current Corrosion: corrosion caused by stray currents flowing from another source of e.m.f. (usually d.c.).

Stress Intensity Factor $K_{Is.c.c.}$: a fracture toughness parameter used for evaluating susceptibility to stress-corrosion cracking (the subscript I signifies a tensile mode of stressing).

Stress-corrosion Cracking: cracking produced by the combined action of corrosion and static tensile stress (internal or applied).

Strike (n.): an electrolyte solution used to deposit a thin initial film of a metal.

Strike (vb): to electroplate for a short time at a higher initial current density than is normally used.

Stripping: removal of a metal coating by means of an electrolyte solution (or an electrolyte solution and external e.m.f.).

Subscale Formation: *see under* **Internal Oxidation.**

Substrate: the basis metal on which a coating is present.

Subsurface Corrosion: *see under* **Internal Oxidation.**

Sulphate-reducing Bacteria (S.R.B.): a species of anaerobic bacteria (*devibrio desulphuricans*) that is capable of causing rapid corrosion of iron and steel in near-neutral solutions in the absence of dissolved oxygen.

Sustained Load Failure: delayed failure due to the presence of hydrogen in stressed high-tensile steels.

Symmetry Factor: ratio of distance across the double layer to summit of energy barrier/distance across the whole double layer.

Tafel Equation: the linear relationship between the overpotential and the logarithm of the current density for an electrode reaction in which charge transfer is rate determining.

Tarnish: dulling, staining or discoloration of metals due to the formation of thin films of corrosion products. (The term can also be applied to thin transparent film which may give rise to interference colours.)

Thief: an auxiliary cathode which is placed in a position relative to the article to be electroplated so that the current density on certain areas is reduced.

Throwing Power: a measure of the ability of an electroplating solution to give a uniform plate on an irregularly shaped cathode.

Transfer Coefficient α: the transfer coefficient of a cathodic process is the fraction of the electrical energy difference $zF\Delta\phi$ that assists the transfer of an ion through the double layer towards the electrode and inhibits its transfer in the reverse direction; frequently experimental values of α are approximately 0·5.

Transpassivity: active behaviour of a metal at potentials more positive than those leading to passivity.

Transport (transference): *see under* **Migration of Ions.** The term 'transport' is frequently used to include diffusion as well as migration of an ion to an electrode.

Transport Number: the proportion of the current carried by a particular ion (transfer number).

Tuberculation: localised attack in which the corrosion products form wart-like mounds over the corroded areas.

Uniform Corrosion (general corrosion): corrosion in which no distinguishable area of the metal surface is solely anodic or cathodic, i.e. anodes and cathodes are *inseparable*, cf. localised corrosion.

Voltaic Cell: a term sometimes used for an electrochemical cell; it is sometimes used to refer to a cell in which chemical changes are caused by the application of an external e.m.f.

Weld Decay: localised attack of austenitic stainless steels at zones near a weld, which results from precipitation of chromium carbides.

Terms used in Cathodic Protection and Underground Corrosion*

Anode Lead: the electrical connection between the anode and the power unit

*These terms are derived from various sources, principally from the *List of Definitions of Essential Telecommunications Terms* (*Part I, Corrosion*), International Telecommunications Union, Geneva, 1957, 1st Supplement, 1960, and the B.S.I. Code of Practice for Cathodic Protection, CP1021:1973; terms taken from the Code of Practice are marked with an asterisk.

in impressed-current schemes; normally a copper-cored plastics-sheathed standard electric cable.

Anode Shield: protective covering of insulating material placed on a painted structure in the immediate vicinity of the anode to reduce the cathodic current density in that area, thus preventing the development of excessive alkalinity and stripping of the paint (*see* **Saponification**).

Attenuation*: the decrease in potential and current density along a long buried or immersed structure from the drainage point.

Attenuation Curves: a graph of interfacial electrical potentials between the pipe and surrounding soil in cathodic protection schemes *vs.* the length of the pipe.

Backfill: the soil replaced over the pipe in the trench (general connotation). In cathodic protection, special backfills are packed around the anodes. These backfills are selected to lower circuit resistance of the anode; for sacrificial anodes a gypsum/bentonite mixture is used, and for impressed-current anodes, coke breeze.

Bond*: a conductor, usually of copper, that connects two points on the same structure or on different adjacent structures, and thus prevents significant differences in potential between two points.

Cantilever Anode: anode supported as a cantilever.

Cathode Lead: the electrical connection from the negative terminal of a d.c. power unit.

Cathodic Protection Rectifier: transformer–rectifier arrangement for supplying the direct current which flows between a groundbed and a buried structure which is receiving cathodic protection.

Continuity Bond: electrical connection made to connect together adjacent sections of a buried structure in order to ensure its electrical continuity.

Corrosion Interaction (or interaction): increase (or decrease) in the rate of corrosion of a buried or immersed structure caused by interception of part of the cathodic protection current applied to another buried or immersed structure.

Current Drainage Survey: a survey to determine current requirements for cathodic-protection schemes. A temporary groundbed is installed at a suitable location and an impressed current is applied between it and the pipe or other structure. A survey of potentials along the structure will produce attenuation curves, and permits fairly accurate estimation of initial current requirements.

Direct Electric Drainage: a means of electric drainage comprising the employment of permanent metallic connections.

Drainage Bond*: bond to effect drainage (*see* **Drainage**, definition (*b*)).

Drainage (electric drainage)*: (*a*) flow of positive current through the soil or the electrolyte solution from the cathodically protected structure to the groundbed of the impressed current system, or (*b*) protection of an immersed structure from electrochemical corrosion by making an electrical connection between the structure and the negative return circuit (rail, feeder, busbar) of a d.c. electric traction system.

Drainage Point: the location in a cathodic-protection installation where the cathode lead is attached to the pipe; normally the point of maximum negative potential of the pipe, though this is not always the case when an area of low-resistance soil is adjacent to the drainage point.

Earth*: (a) the conducting mass of earth or of any other conductor in direct electrical connection with earth, (b) a connection, intentional or unintentional, between a conductor and earth.

Electric Drainage: means of electric protection of an underground system against the corrosive action of stray currents arising from a d.c. electric traction system employing one or more connections (drainage bonds) made between the system to be protected and the return circuit of the traction system (rail, return current feeder, negative busbar of the sub-stations).

Electro-osmosis: passage of a liquid through a porous medium (such as a soil) under the influence of a potential difference.

Flood Coat: application of coal-tar or bituminous coatings to pipelines by flooding the heated coating materials over the pipe surface; wrappings and slings are used to pull the coat underneath the pipe.

Forced Drainage*: form of drainage in which the connection between the protected structure and a traction system includes a source of direct current.

Groundbed: in cathodic protection of underground structures, a buried mass of inert material (e.g. carbon), or scrap metal connected to the positive terminal of a source of e.m.f. to a structure.

Holiday: a flaw, often in the form of a pinhole, in a protective organic coating.

Holiday Detector: a high-voltage low-current output generator with a built-in spark gap used for detecting pinholes (holidays) in pipeline coatings.

Impressed Current: cathodic current supplied by a d.c. source to a structure in order to lower the potential to the protective potential for cathodic protection.

Insulating Flange*: flanged joint between adjacent lengths of pipe in which the nuts and bolts are electrically insulated from one or both of the flanges and the jointing gasket is non-conducting, so that there is an electrical discontinuity in the pipeline at that point.

Interaction: *see* **Corrosion Interaction.**

Interaction Testing: routine investigation carried out when installing cathodic protection schemes on pipelines. The accepted criterion in the UK is that when the secondary structure potential has been moved more than 0·02 V in a positive direction, steps must be taken to eliminate the interaction.

Line Wrapping: the technique of wrapping a pipeline over the pipeline trench; applied only to welded steel pipelines. It is carried out by a line-wrapping machine which travels along the pipe, cleaning it, priming it, pulling in the reinforcement, and, if necessary, applying the outer wrap.

Locating and Bonding: when cathodic protection is applied to an existing jointed pipeline, all joints must be located without digging up the pipe. On location, each joint is exposed and an electrically conducting bond (usually galvanised steel strip or copper cable) is welded into position.

Polarised Electric Drainage: a means of electric drainage comprising the employment of metallic connections between the underground cable system and the traction return circuit, with the insertion of a unidirectional system (rectifier or contactor and relays) in the connections.

Primary Structure*: a buried or immersed structure cathodically protected by a system that may constitute a source of corrosion interaction with another (secondary) structure.

Protection Current: current made to flow into a metallic structure in order to effect cathodic protection.

Protective Potential: an optimum negative potential at which pipe protection is accepted as being sufficient in cathodic-protection schemes. It is a negative potential of the pipe/environment interface, and is applicable to steel when the potential is more negative than $-0.85\,V$ *vs.* a $Cu/CuSO_4$ electrode located as close to the pipe as practicable. When sulphate-reducing bacteria are present in soil the value should be $-0.95\,V$. An alternative criterion for steel is to change the pipe potential $0.3\,V$ in a negative direction. Similar criteria are applicable to steel in natural waters.

Reactive (active, sacrificial) Anode: a mass of metal (Mg, Zn, Al) which, buried or immersed and connected to a metallic structure which is to be protected, forms a cell with that structure and has the effect of making it more negative with respect to the surrounding environment.

Remedial Bond*: a bond established between a primary and secondary structure in order to eliminate or reduce corrosion interaction.

Remote Potentials: in dealing with a pipeline cathodic-protection system, it is often advantageous to refer all measurements to a half-cell located at a distance from the pipe. Such measurements are referred to as *remote potentials*.

Resistance Bond*: a bond either incorporating resistors or of adequate resistance in itself to limit the flow of current.

Sacrificial Anode: *see under* **Reactive Anode.**

Safety Bond*: a bond connecting the metallic carcase of a piece of electrical apparatus with earth, in order to limit the rise in potential of the carcase above earth caused by the passage of any fault current or excessive leakage current, and so reduce the risk of electric shock to anyone touching the carcase.

Saponification: deterioration by softening of paint film caused by action of aqueous alkali, resulting from cathodic protection at excessively high current densities, on the fatty-acid constituents of the film.

Sensing Electrode*: a permanently installed reference electrode used to measure the structure/electrolyte solution potential and to control the protection current.

Stray Currents: currents flowing in the soil and arising from electric installations parts of which are not insulated from the soil.

Structure/Soil Potential: potential measured between a buried structure and a suitable non-polarisable electrode, placed in the soil or water as near to the structure as possible; when the value is being indicated the type of electrode used should be clearly stated.

L. L. SHREIR
T. P. HOAR

22.2 Symbols and Abbreviations

Basic SI Units

Quantity	Name	Symbol
length	metre	m
mass	kilogram	kg
time	second	s
force	newton	N
work	joule	J
power	watt	W
current	ampere	A
charge	coulomb	C
e.m.f. and p.d.	volt	V
resistance	ohm	Ω
conductance	siemens	S
inductance	henry	H
capacitance	farad	F
amount of substance	mole	mol

Å	Angstrom unit
a_A	relative activity (dimensionless) of component A
ABS	acrylonitrile butadiene styrene
a.c.	alternating current
Ah	ampere hour
aq.	aqueous, hydrated
α	transfer coefficient
b.p.	boiling point
β	symmetry factor
$C_{p,m}$	molar heat capacity at constant pressure $(JK^{-1}\,mol^{-1})$
c	concentration $(mol\,m^{-3},\ mol\,dm^{-3})$*
CAB	cellulose acetate butyrate
c.d.	current density

*The abbreviations N (normal) and M (molar), although not S.I. units, have been retained in the book when these units have been used by authors of published work.

CHC	cyclohexylamine carbonate
cm	centimetre
°C	degree Celsius (Centigrade)
D	diffusion coefficient ($m^2 s^{-1}$, $cm^2 s^{-1}$)
d	differential
∂	partial differential; thickness of diffusion layer (m, cm)
Δ	difference, finite change
d	day
dia.	diameter
d.c.	direct current
DCHN	dicyclohexylamine nitrate
dia.	diameter
dm	decimetre
DPN	diamond pyramid number
e.m.f.	electromotive force
E	e.m.f., potential (V)
$E^{\ominus}$	standard e.m.f., standard electrode potential
$E_{eq.}$, E_r	equilibrium or reversible e.m.f. or potential
E_{cell}	e.m.f. of cell
E_a	anode potential
E_c	cathode potential
$E_{corr.}$	corrosion potential
E_{pp}	passivation potential
E_b	breakdown or pitting potential (E_c is also used)
E_p	protection potential
$E^{\ddagger}$	activation energy ($J\,mol^{-1}$)
e	elementary charge (proton or electron) (C)
eV	electron volt
exp	exponential function
η_a	anode overpotential (V)
η_c	cathode overpotential
η_A	activation overpotential
η_T	transport, concentration or diffusion overpotential
η_R	resistance overpotential
F	Faraday constant ($C\,mol^{-1}$)
f, p*	fugacity ($N\,M^{-2}$ or Pa)
G	Gibbs free energy (free enthalpy) (J)
$G^{\ominus}$	standard Gibbs free energy (J)
Γ	Gibbs surface excess (free energy)
g	gram
g.m.d.	grams per square metre per day ($gm^{-2}\,d^{-1}$)

H	enthalpy
HB	Brinell hardness number
HV	Vickers hardness number
h	hour
I	current (A)
i, j	current density (Am^{-2}, Acm^{-2})
$i_{corr.}$	corrosion c.d.
i_a	actual anodic c.d. measured
i_c	actual cathodic c.d. measured
i_0	equilibrium exchange current density
$\overrightarrow{i}_1$	cathodic current density for exchange I
$\overleftarrow{i}_1$	anodic current density for exchange I
i_L	limiting (or maximum c.d.)
i.d.	internal diameter
i.p.y.	inches per year (in y^{-1})
I	ionic strength ($mol\,kg^{-1}$)
J	electrochemical equivalent
$K_{p/p}$	equilibrium constant for gaseous system (dimensionless)
$K_{m\gamma/m}$	equilibrium constant for solution (dimensionless)
k	rate constant ($mol\,s^{-1}$)
$\overrightarrow{k}$	rate constant of a cathodic reaction
$\overleftarrow{k}$	rate constant of an anodic reaction
κ	electric conductivity ($S\,m^{-1}$)
K	degree Kelvin
l	length (m, cm)
l, ℓ	litre
ln, ℓn	natural logarithm
log	logarithm to the base 10
m	molecular mass (kg)
M	molar mass ($kg\,mol^{-1}$)
m.d.d.	$mg\,dm^{-2}\,d^{-1}$
m.p.y.	$mm\,y^{-1}$
ml	millilitre
min	minute
m.p.	melting point
mol	moles
N_0, N_A	Avogadro constant
n	number of moles
n_+	transference number of cation

n_-	transference number of anion
v	kinematic viscosity
o.d.	outside diameter
ω	angular speed of rotation
Pa	pascal ($N\,m^{-2}$)
P(J)	permeation rate
p	pressure
pH	$-\log a_{H^+}$
p.p.m.	parts per million
$\psi(\phi)$	potential of a charged interface
p.t.f.e.	polytetrafluoroethylene
p.v.c.	polyvinylchloride
R	electric resistance; gas constant
ρ	density; oxide/metal volume ratio
r.d.s.	rate-determining step of reaction
$S(A)$	area
S	entropy
s	second
S.C.E.	saturated calomel electrode
S.H.E.; N.H.E.	standard (normal) hydrogen electrode
s.g.	specific gravity
σ	surface charge; stress
T	absolute temperature
θ	surface coverage
t	time
t_f	time to fracture
μ	chemical potential (J)
$\mu^{\ominus}$	standard chemical potential ($J\,mol^{-1}$)
U	internal energy
U.T.S.	ultimate tensile strength
v	velocity
V	volume
$\bar{V}$	partial molar volume
VCI	volatile corrosion inhibitor
VPI	Vapour Phase Inhibitor (trade name)
VPN	Vickers pyramid number
v/v	volume for volume
v/wt.	volume for weight
wt %	weight percent
wt./v	weight for volume

| x_A | mole fraction of component A |
| $z(n)$ | number of electrons involved in one act of the reaction (or in the r.d.s.); charge on ion |

INDEX

Abbreviations, **22**:17
Abrasion, nickel coatings, **14**:83
Abrasion tests, coatings, **20**:99
Abrasive solids, **20**:22
Absorption coefficient, **21**:53
Accelerators, phosphate coatings, **16**:22
Access fittings for corrosion racks, **20**:87
Acetic acid, **3**:55; **4**:81
Acetic acid-salt-spray test, **20**:48
Acid copper sulphate test, **20**:8
Acid pickling, hydrogen absorption, **8**:64
Acid solutions,
    inhibitors, **18**:25
    thermodynamics of corrosion
        reactions, **1**:58
Acids, **3**:55, 61
    aluminium and aluminium alloys in,
        4:26, 27
    beryllium in, **5**:6
    carbon in, **19**:7
    cast iron in, **3**:94
    copper and copper alloys in, **4**:53
    glass in, **19**:20
    high-chromium cast irons in, **3**:114
    high-nickel cast iron in, **3**:106
    lead in, **4**:79
    magnesium in, **4**:88
    molybdenum in, **5**:15
    nickel and nickel alloys in, **4**:130
    nickel-iron alloys in, **3**:81
    niobium in, **5**:27
    silicon-iron alloys in, **3**:121
    stainless steels in, **3**:51
    tantalum and tantalum alloys in,
        5:63, 69, 70, 71
    tin in, **4**:145
    titanium in, **5**:37, 38
    wood in, **19**:84, 85
    zinc in, **4**:158
    zirconium in, **5**:53, 54
Acrylic resins, **15**:53, 96
Acrylonitrile-styrene-butadiene polymer
        (ABS), **19**:60

Activation, **9**:29
Activation energy, **2**:19, 90–91, **9**:29
    self-diffusion, **1**:233
Active-passive transitions, **1**:103; **2**:16,
        23
Activity, **9**:64
    relative, **9**:65
Activity coefficient, **7**:127, 128, 129;
        9:64, 65; **21**:31
Addition agents **13**:8, 15, 92; **14**:71, 72
Additives,
    greases, **2**:124
    paint, **15**:22
Adhesion, coatings, **20**:100
Adion, **13**:9
Admiralty brass, **4**:48
Adsorbed electrolyte layers, **2**:33
Adsorption, **9**:13
    electrostatic, **9**:13
    Gibbs equation, **9**:14
    in breakdown of passivity, **9**:27
    in corrosion reactions, **9**:26
    isotherms, **9**:23
    Langmuirean, **9**:49
    of inhibitors, **9**:27; **18**:34
    specific or contact, **9**:13, 22
    van der Waals, **9**:13
Adsorption coefficient, **9**:23
Adsorption law, Freundlich, **2**:5, 6, 18,
        88, 95
Aeration of testing solution, **20**:23
Aeration control in total-immersion
        tests, **20**:17
Affinity, **1**:65
Age hardening. *See* Precipitation
        hardening
Ageing, **9**:122; **20**:14
Air, *see also* Atmosphere
    high-temperature corrosion in, **19**:8
    rusting in, **3**:5
Air-stream corrosion test, **20**:22
Airborne particles, **2**:30
Alcoholic beverages, **2**:99

1

Al-Fin process, **14**:20
Alkaline cleaners, **12**:8–10
Alkalinity of water, **2**:44
Alkalis, **3**:58
  aluminium and aluminium alloys in, **4**:27
  carbon in, **19**:8
  cast iron in, **3**:95
  copper and copper alloys in, **4**:53
  glass in, **19**:20
  high-chromium cast irons in, **3**:116
  high-nickel cast irons in, **3**:108
  lead in, **4**:81
  molybdenum in, **5**:17
  nickel and nickel alloys in, **4**:133
  nickel-iron alloys in, **3**:81
  niobium in, **5**:27
  silicon-iron alloys in, **3**:125
  tantalum in, **5**:66
  tin in, **4**:145
  zinc in, **4**:158
  zirconium in, **5**:55
Alkyd resins, **15**:53, 95
  air-drying, **15**:16
  modified, **15**:18
  paint coatings, **15**:16
  plasticising, **15**:17
  stoving, **15**:17
Alloying elements, **1**:44
  in dilute solutions, **7**:127
  complex systems, **9**:123
  components and phases, **9**:102
  effects of composition and heat treatment on corrosion resistance of, **20**:128
  phase diagrams, **9**:105
  properties of, **1**:3
  single-phase, **1**:44
  solid solutions, **9**:103
  structure of, **9**:102
  two-phase, **1**:45
Alumina, **18**:4
Aluminium, **2**:57; **4**:3; **7**:21, 87, 93
  anodes, **11**:19, 21, 50
  anodising, **1**:22
  aqueous corrosion, **4**:29
  atmospheric corrosion, **4**:19
  bimetallic corrosion, **4**:18
  chromate treatment, **16**:35
    acid, **16**:35
    acid pickles, **16**:36
    alkaline, **16**:35, 36
    sealing of anodic films, **16**:36
  corrosion behaviour, **4**:14
  corrosion inhibition, **18**:50
  dissolution, **4**:15
  electroplating, **13**:22
  filiform corrosion, **4**:17
  high-temperature corrosion, **4**:29
  in acid electrolytes, **1**:30
  in acids, **4**:26, 27

Aluminium *continued*
  in alkalis, **4**:27
  in buildings and structures, **10**:43
  in chemical environments, **4**:23
  in chemical plants, **4**:28
  in contact with concrete, **10**:44
  in contact with other materials, **4**:30, 31
  in fused salts, **4**:30
  in gases, **4**:19
  in liquid metals, **4**:30
  in organic compounds, **4**:28
  in salt solutions, **4**:28
  in steam, **4**:30
  in water, **4**:22
  intercrystalline corrosion, **4**:16
  mechanical properties, **4**:10
  nuclear engineering, **4**:29
  oxide film, **4**:14
  oxides, **1**:235; **7**:21, 98
  physical properties, **4**:10
  pipes, **10**:49
  pitting corrosion, **1**:162; **4**:16
  purity grades, **4**:12, 18
  soil corrosion, **4**:23
Aluminium alloys, **4**:3; **8**:23; **20**:33
  aqueous corrosion, **4**:29
  atmospheric corrosion, **4**:19
  bimetallic corrosion, **4**:18
  British Standard form and heat treatment notation, **4**:9
  cast, **4**:7
  chemical cleaning, **20**:107
  cladding, **4**:3, 23
  composition, **4**:3
  composition effects, **4**:17
  corrosion behaviour, **4**:14
  descaling, **12**:13
  design factors, **4**:19
  dissolution, **4**:15
  electroplating, **13**:22
  filiform corrosion, **4**:17
  high-temperature corrosion, **4**:29
  heat treatable, **4**:5, 6, 9
  in acids, **4**:26, 27
  in alkalis, **4**:27
  in chemical environments, **4**:23
  in chemical plants, **4**:28
  in contact with other materials, **4**:30, 31
  in fused salts, **4**:30
  in liquid metals, **4**:30
  in organic compounds, **4**:28
  in salt solutions, **4**:28
  in steam, **4**:30
  in water, **4**:22
  intercrystalline corrosion, **4**:16
  intergranular corrosion, **1**:42
  layer corrosion, **4**:17
  mechanical properties, **4**:3
  non-heat-treatable, **4**:9

Aluminium alloys *continued*
  nuclear engineering, **4**:29
  oxide film, **4**:14
  pipes, **10**:49
  precipitation hardening, **9**:120
  selection, **4**:12
  soil corrosion, **4**:23
  special purpose, **4**:8
  standard designations, **4**:3, 9
  stress-corrosion cracking, **1**:48; **4**:17;
    8:89, 90, 92, 93
  stress effects, **4**:19
  welding, **10**:79
  wrought, **4**:4
Aluminium brass, **4**:48, 50; **20**:33
  role of iron, **1**:172
Aluminium-bronze, **4**:35, 48
  cavitation damage resistance, **8**:131
Aluminium coatings, **13**:22, 49, 50;
    14:11, 17
  anodic oxidation, **16**:3
  applications, **14**:25
  casting, **14**:20
  chemical deposition, **14**:19
  cladding, **14**:20
  corrosion resistance, **14**:22
  diffusion, **13**:59
    superalloys, **13**:71
  electrophoretic, **14**:19
  electroplating, **14**:19
  hot-dip, **13**:55; **14**:18, 20, 24
    alloy formation, **13**:56
    continuous strip, **13**:55
    heat-resistant, **13**:56
    structure, **13**:57
  methods available, **14**:17
  spray-aluminised, **14**:18
  sprayed, **13**:82, 90; **14**:17, 20, 23, 24,
    25
  vacuum deposited, **14**:18
Aluminium-copper alloys, **4**:13
  phase diagram, **9**:111
  stress-corrosion cracking, **8**:92
Aluminium-copper-magnesium alloys
  corrosion fatigue, **8**:112, 113
  stress-corrosion cracking, **8**:92
Aluminium-iron couple, polarity
    reversal, **1**:206
Aluminium-magnesium alloys, **4**:12
  stress-corrosion cracking, **8**:91
Aluminium-magnesium-silicon alloys,
    4:13
  stress-corrosion cracking, **8**:91
Aluminium-magnesium-zinc alloys,
    corrosion fatigue, **8**:112
Aluminium-manganese alloys, **4**:12, 23
Aluminium-tin alloys, **4**:3
Aluminium-zinc couple, polarity
    reversal, **1**:206
Aluminium-zinc-magnesium alloys,
    4:3, 13

Aluminium-zinc-magnesium alloys
    continued
  stress-corrosion cracking, **8**:92
Aluminium-zinc-magnesium-copper
    alloys, stress-corrosion cracking,
    8:92
Aminoplastics, **19**:62
Amino resins, paint coatings, **15**:18
Ammeters, **11**:104, 105
Ammonia, nickel and nickel alloys in,
    7:104
Ammonium compounds, **2**:30
Ammonium sulphate, **2**:30
Anaerobic corrosion, **2**:79
Anion concentration, effect on corrosion
    rate, **2**:3
Anions, **9**:3
  inhibitive, **18**:46, 47, 48
  uptake by oxide films, **18**:45
Annealing, **9**:100, 107
Annuli, corrosion rate in, **2**:11
Anode potential, **11**:28
Anode reactions,
  active-passive transition, **2**:16
  under film-free conditions, **2**:14
  with film formation, **2**:15
Anodes, **1**:75; **5**:31; **9**:5, 6; **13**:9
  see also Cathodic protection
  aluminium, **11**:19, 21, 50
  carbon, **11**:47
  cast iron, **11**:42
  DSA-titanium, **11**:41
  electroplating, **13**:16
  graphite, **11**:47, 66
  high-silicon iron, **11**:43
  high-silicon/chromium iron, **11**:44
  high-silicon/molybdenum iron, **11**:44
  iron, **11**:42
  lead, **4**:84; **11**:44, 67
  lead alloy, **4**:84
  magnesium, **10**:51; **11**:19, 21, 22
  nickel, **14**:70
  niobium, **11**:40, 67
  platinised, **11**:36, 67, 76
  platinum, **11**:36
  silicon iron, **11**:66
  silver, **11**:40
  stainless steel, **11**:42
  steel, **11**:41, 66
  tantalum **11**:40, 67
  tin, **4**:146
  titanium, **5**:46; **11**:36, 67, 76
  zinc, **4**:160, 161, 162; **11**:3, 19, 20, 51
  zinc alloy, **4**:160
Anodic behaviour,
  nickel, **4**:108, 121
    solution effect, **4**:110
    temperature effect, **4**:109
  nickel alloys, **4**:111
Anodic dissolution under film-free
    conditions, **2**:4

4 INDEX

Anodic *E–i* curves, **1**:102, 103, 105, 137
  active region, **1**:104
  passive region, **1**:107
  transpassive region, **1**:108
Anodic oxidation coatings. *See* Coatings
Anodic passivation, **15**:28
  titanium, **5**:46
Anodic passivity, **1**:115
Anodic polarisation, **1**:203
Anodic processes, **1**:115; **9**:35, 38
Anodic protection, **1**:105, 123; **8**:111;
    11:112
  applications, **11**:119, 122, 123
  economics, **11**:122
  electrolyte agitation, **11**:117
  limitations, **11**:119
  passivity of metals, **11**:113
  potentiostat study of, **20**:140
  practical aspects, **11**:118
  principles, **11**:112
  temperature effects, **11**:117
Anodic reactions, **1**:75, 93, 94, 133; **3**:3;
    9:34, 36
  inhibition, **15**:27–30
  partial, **1**:78
Anodisation comparator, **16**:15
Anodising, **16**:3
  colour treatments, **16**:4, 5
  effect on mechanical properties, **16**:12
  processes, **16**:3
Antifreeze solutions, **4**:28
Aqueous corrosion, **5**:5; **11**:6
  aluminium and aluminium alloys,
    4:14, 29
  ·high-nickel cast irons, **3**:100
  low-alloy steels, **3**:21
  nickel, **4**:106
  tin, **4**:144
  titanium, **5**:37
  zirconium, **5**:53
  surface reaction products, **1**:26
Aqueous solutions, **1**:52; **8**:80; **13**:10
Architecture, **3**:60
Arrhenium equation, **2**:15, 95; **9**:30;
    13:61
Arsenic, **4**:35, 47
Ash. *See* Fuel ashes
ASTM Standards, **21**:78, 79
  cleaning, **21**:81
  coatings, **21**:82
  corrosion testing, **20**:115
  materials, **21**:80
  surface preparation, **21**:81
Atmosphere, **2**:26–37
  composition, **2**:28
  contaminants, **2**:29
  controlled, **7**:100
  impurity concentrations, **2**:27
  moisture precipitation in, **2**:31
Atmospheric corrosion, **3**:8; **5**:14
  aluminium and aluminium alloys, **4**:19

Atmospheric corrosion *continued*
  bimetallic, **1**:204
  buildings, **10**:33
  cast iron, **3**:88
  classification, **2**:26
  copper, **4**:39
  copper alloys, **4**:40
  damp, **2**:27
  electrochemistry of, **2**:34
  high-chromium cast irons, **3**:113
  high-nickel cast iron, **3**:104
  in contaminants from plastics, **19**:79
  in long-chain fatty acids, **19**:79
  inhibitors, **18**:6
  lead, **4**:76
  low-alloy steels, **3**:23; **7**:24
  magnesium alloys, **4**:90
  marine, **10**:55
  nickel-iron alloys, **3**:77
  niobium, **5**:26
  rates of metals, **14**:5
  stainless steels, **3**:48
  steels, **20**:9, 88
  steelwork, costs, **10**:7
  structures, **10**:33
  tantalum, **5**:63
  tin, **4**:143
  tinplate, **14**:49
  uranium, **5**:74
  wet, **2**:27
  zinc, **4**:154
Atomic power production, **3**:63
Atoms, **8**:32; **9**:88
ATR alloy, **5**:52, 53
Austenite, **9**:113, 117
  grain boundaries, **8**:31
Autocatalytic process, **1**:146
Avtur Test, **2**:81
Azelaic acid, **15**:29

Babbitt alloys, **4**:148
Bacteria, sulphate-reducing, **2**:79–80
Bainite, **9**:114, 117
Barrel plating, **13**:30
Barrier layer, **1**:257; **16**:6
Bearing metals, tin, **4**:148
Bend tests, embrittlement, **8**:159
Bending-beam theory, **1**:261
Bending stresses,
  in bent specimens, **8**:166
  specimen gripping to avoid, **8**:162
Bentonite, **2**:64
Beryllium, **1**:239; **5**:3, 5; **7**:106, *see also*
    Copper-beryllium
  alloying element, **5**:3
  break-away reaction, **5**:7
  brittleness, **5**:3, 4
  chemical composition of various
    forms, **5**:5
  corrosion behaviour, **5**:5

Beryllium *continued*
  extraction, **5**:4
  fabrication, **5**:4
  galvanic effects, **5**:7
  in acids, **5**:6
  in aqueous environments, **5**:5
  in gaseous environments, **5**:7
  metallurgy, **5**:4
  protective measures, **5**:8
  stress-corrosion cracking, **5**:7
  surface condition, **5**:6
  temperature effects, **5**:7
Bimetallic corrosion, **1**:91, 192; **13**:41; **15**:8; **20**:40
  aluminium and aluminium alloys, **4**:18
  beneficial effects, **1**:210
  catchment area principle, **1**:200
  compatibility data, **14**:6
  copper and copper alloys, **4**:59
  corrosive atmosphere, **1**:204
  degree of corrosion at bimetallic contacts (table), **1**:215
  dissolved oxygen, **1**:197
  electrolyte composition, **1**:203
  general theory, **1**:192
  hydrogen evolution, **1**:202
  in atmosphere, **20**:89
  in buildings and structures, **10**:41
  in sea-water, **1**:219; **3**:51
  in wood, **19**:89
  magnesium and magnesium alloys, **4**:90, 92, 93
  marine engineering, **10**:62
  molybdenum, **5**:20
  nickel and nickel alloys, **4**:125
  nickel-iron alloys, **3**:82
  protective measures, **1**:207, 209
  tantalum, **5**:68
  testing, **20**:40
  titanium, **5**:43
  variable polarity, **1**:205
  zinc, **4**:153, 154
Bimetallic couples, **20**:32
Birdcage rack, **20**:86
Blast cleaning, **20**:165
Blister formation, **8**:64
Bockris, Devanathan and Muller isotherm, **9**:25, 26
Boiler feed water treatment, **2**:41, 45; **18**:57; **20**:23
  treatment, additions, **18**:61
  alkalinity control with solid chemicals, **18**:62
  chemical oxygen scavengers, **18**:63
  conditions, **18**:70
  distillation, **18**:59
  film-forming amines, **18**:65
  hardness removal, **18**:58
  ion exchange processes, **18**:59, 60
  metallic impurities, **18**:66, 67
  precipitation processes, **18**:59

Boiler feed water treatment *continued*
  principles of, **18**:58
  purification, **18**:59
  removal of dissolved gases, **18**:60
  softening, **18**:58
  solutes, **18**:66
  volatile alkaline additives, **18**:64
Boiler tubes, oxygen pitting of, **10**:18
Boilers, caustic cracking, **8**:44, 45
Boiling heat transfer, **2**:20
Boiling nitric acid test, **20**:8
Boltzmann's equation, **1**:232
Boric oxide, **7**:112
Boudouard reaction, **1**:256; **7**:102
Brasses, **4**:35
  aluminium, **20**:33
  dezincification **1**:45, 166, 167; **4**:36, 41, 46, 51; **20**:14
  season cracking, **4**:36
  70/30, **20**:33
  stress-corrosion cracking, **1**:47; **4**:56; **8**:157
Brazing, **4**:19; **10**:71
Brazing alloys, **6**:16–17
Breakaway corrosion, **5**:7; **7**:44
  mechanism, **1**:254
  zirconium, **1**:225; **5**:55
Breakdown potential, **1**:154
  austenitic irons, **3**:102
  nickel-chromium alloys, **4**:122
  titanium, **11**:37
Bright dipping, **12**:25
Brighteners, **14**:70
British Standards,
  cathodic protection, **21**:78
  coatings, **21**:72
  corrosion testing, **20**:109
  inhibitors, **21**:78
  materials, **21**:67
  stainless steels, **3**:32
Brittle fracture, **8**:137
  Griffith's theory, **8**:133
Brush plating, **13**:100
Buildings,
  atmospheric corrosion, **10**:33
  contact corrosion, **10**:34
  corrosive environment, **10**:33
  design, **10**:33
  ferrous metals in air, **10**:34
  materials of construction, **10**:37
  non-ferrous metals, **10**:42
Burgers vector, **9**:96
Buried metal locating instruments, **11**:111
Bursting discs, **6**:15
Bushings, **6**:20

Cadmium, atmospheric corrosion, contaminants from plastics, **19**:79
Cadmium coatings, **14**:10, 31
  chromate passivation, **16**:36

Cadmium coatings *continued*
  comparison with zinc, **14**:33
  electroplating. *See* Cadmium plating,
    14:34
  nature and degree of protection, **14**:31
  protection of, **14**:33
  vacuum deposition, **14**:34
Cadmium plating, **1**:208; **8**:63, 160;
    10:57; **14**:33
  specifications, **14**:34
Calcium bicarbonate, **2**:42
Calcium carbonate, **2**:41, 44
Calcium plumbate, **15**:21, 30, 39
Calorising, **14**:18, 22, 24
  aluminising, **13**:59
Campbell test apparatus, **20**:71
Canning, **2**:98
Capillaries, **20**:29
Capillary condensation, **2**:32
Carbide formation, **3**:47
Carbide inclusions, **5**:6
Carbide precipitation, **3**:45; **8**:47
Carbides,
  free-energy diagram, **7**:119
  segregation, **7**:127
  from stainless steel containing small
    amounts of carbon, **7**:129
Carbon, **7**:19; **19**:3
  anodes, **11**:47
  combination with elements, **19**:9
  combination with gases, **19**:9
  corrosion resistance, **19**:7
  definitions, **19**:4
  electrical resistivity variation with
    temperature, **19**:6
  forms, **19**:3
  high-temperature behaviour, **19**:8
  high-temperature corrosion, **7**:7
  in acids, **19**:7
  in alkalis, **19**:8
  in liquid-metal corrosion, **2**:107
  manufacturing processes, **19**:3
  nitrogen solubility in, **2**:107
  oxidation, **19**:9
  physical properties, **19**:4
    temperature effects, **19**:6
  solution in iron, **7**:125
  specific heat, variation with
    temperature, **19**:4
Carbon dioxide, **2**:28, 41; **5**:63, 75;
    7:7, 100
  in water, **2**:40–44; **3**:15
  oxidation effect, **7**:41
  saturation coefficient in sea-water,
    21:55
Carbon monoxide, **5**:63, 75; **7**:7, 102
Carbonaceous gases, **7**:98
Carburisation, nickel and nickel alloys,
    7:98
Carnot cycle, **9**:60, 61
Cast iron, **2**:49; **3**:3, 85; **20**:33

Cast iron *continued*
  anodes, **11**:42
  atmospheric corrosion, **3**:88
  austenitic, **3**:98–110
    scaling, **7**:55
  composition, **3**:85
  composition effects, **3**:87
  corrosion fatigue, **3**:96
  flake graphite, **3**:98
    scaling, **7**:54
  graphisation, **7**:48, 49
  graphite in, **1**:45
  graphitic corrosion, **3**:102
  graphitic residue, **3**:86
  high-chromium, **3**:111
    atmospheric corrosion, **3**:113
    composition, **3**:111
    corrosion resistance, **3**:113
    high-chromium cast irons in waters,
      3:113
    high-temperature, **3**:116
    in acids, **3**:114
    in alkalis, **3**:116
    in salt solutions, **3**:116
    mechanical properties, **3**:112
    production, **3**:112
    structure, **3**:111
  high-nickel, **3**:81, 98
    applications, **3**:109
    aqueous corrosion, **3**:100
    atmospheric corrosion, **3**:104
    composition, **3**:98
    in acids, **3**:106
    in alkalis, **3**:108
    in salt solutions, **3**:109
    in waters, **3**:105
    mechanical properties, **3**:98
  high-temperature corrosion, **7**:47
    growth, **7**:47
    scaling, **7**:50
  in acids, **3**:94
  in alkalis, **3**:95
  in industrial environments, **3**:94
  in salt solutions, **3**:95
  in sea-water, **3**:90
  in waters, **3**:90
    velocity effect, **3**:91
  low-alloy, scaling, **7**:54
  nodular graphite, scaling, **7**:54
  oxidation, **7**:47, 54, 55
  oxide penetration, **7**:48
  scaling, **7**:50
  soil corrosion, **3**:91
  spheroidal graphite, **3**:98
  stress corrosion cracking, **3**:94
  structure, **3**:85
  structure effects, **3**:86
  tinning, **13**:51
  types, **3**:85
  vitreous enamel coating, **17**:4
Castings, surface roughness, **12**:38

Catering applications, **3**:60
Cathodes, **1**:75; **9**:5, 6; **13**:9
Cathode potential, **11**:28
Cathode sputtering, **13**:99
Cathodic depolarisation of ferrous
    metals, **2**:77–78
Cathodic overvoltage, **11**:7
Cathodic processes, **1**:115; **9**:35, 40
Cathodic protection, **1**:44, 56; **3**:71, 72;
    4:79, 160; **8**:53, 58, 65, 94, 111,
    127, 128; **15**:27
  application, **11**:57
  attenuation, **11**:72
    field measurements, **11**:72
    measurement between drainage
      points, **11**:73
  attenuation constant, **11**:14
  attenuation curves, **11**:13, 15, 73, 74
  bimetallic corrosion, **1**:208
  British Standards, **21**:78
  buried structures, **11**:67
  cable types, **11**:75
  coating resistance, **11**:74
  comparison of properties of
    materials, **11**:51
  comparison of systems, **11**:60
  constant current, **11**:10
  controlled potential, **11**:10
  criteria of, **11**:8
  current density requirements, **11**:78
  current drain test, **11**:64
  current requirements, **11**:9, 64
  deep well groundbeds, **11**:71
  design of system, **11**:63
  economics, **11**:65, 82
  electrical continuity, **11**:63
  electrical interference, **11**:17
  electrochemical potential, **11**:5
  electrochemical principles, **11**:6
  electrolyte resistivity, **11**:65, 108
  external power supply, **11**:3
  factors affecting corrosion rate, **11**:9
  forced drainage, **11**:62
  groundbed resistance, **11**.70
  groundbeds remote from structure,
    11:80
  impressed-current, **11**:34, 60, 81
    advantages and disadvantages,
      11:62
    anode backfill, **11**:70
    anode materials, **11**:34, 66
    applications, **11**:57
    automatically controlled modular
      system, **11**:78
    automatically-thyristor-controlled
      systems, **11**:78
    carbonaceous backfills, **11**:49
    circulating water systems, **11**:76
    design, **11**:69
    manually controlled system, **11**:77
  in soil, **2**:69–71

Cathodic protection *continued*
  instruments, **11**:97
    ancillary, **11**:110
    basic requirements, **11**:97
    buried metal locating, **11**:111
    conductivity-measuring, **11**:109
    current-measuring, **11**:144
    d.c. indicating, **11**:97
    high-voltage coating-testing
      equipment, **11**:110
    multicombination, **11**:110
    potential-measuring, **11**:98
    recording, **11**:110
    resistance-measuring, **11**:108
    resistivity-measuring, **11**:106
    types, **11**:98
  interaction, **11**:91
    estimation methods, **11**:92
    methods of preventing or reducing,
      11:95
    internal surface pipelines, **11**:16
    marine structures, **11**:78
  measurements, **11**:97
  mechanism, **11**:6
  monitoring, **11**:5
  pH effects, **11**:8
  pipelines, **11**:13, 61, 63, 72, 74
    internal surface, **11**:82
  potential distribution, **11**:12
  power-impressed anodes, **11**:34
  power sources, **11**:75
  principles, **11**:3
  principles of application, **11**:3
  rectifier voltage, **11**:72
  resistivity measurements, **11**:67, 106
  rod anodes, **11**:76
  sacrificial anode, **11**:4, 11, 18, 60
    advantages and disadvantages,
      11:61
    anode capacity, **11**:27
    anode efficiency, **11**:28
    anode potential, **11**:28
    application, **11**:32
    area of steel requiring protection,
      11:28
    backfills, **11**:22
    cathode potential, **11**:28
    circulating water systems, **11**:76
    current densities for protecting steel,
      11:31
    design parameters, **11**:30
    effect of electrolyte solution, **11**:29
    fixing **11**:25
    general anode requirement, **11**:18
    inserts, **11**:24
    manufacturing method, **11**:22
    material composition, **11**:19
    method of choosing anode weight
      and number, **11**:30
    output, **11**:25
    physical shape, **11**:22

Cathodic protection *continued*
  sacrificial anode *continued*
    position in space of anode relative
      to steelwork, **11**:29
    resistivity in waters, **11**:25
    system life, **11**:29
  sheet steel piling, **11**:80
  ships, **11**:80
  stray-current, **11**:62, 91
  structure/electrolyte potentials, **11**:101
  structures applicable, **11**:57
  structures in sea-water, **11**:17
  surface area, **11**:63
  surface coating, **11**:6, 9
  thermo-electric generators, **11**:75
  tubular anodes, **11**:76
  water storage tanks, **11**:81
Cathodic reactions, **1**:75, 93, 94, 133;
    2:13; **3**:4; **9**:30, 36
  inhibition, **15**:25–27
  partial, **1**:78
  simultaneous, **1**:100
Cations, **9**:3
Caustic cracking, **2**:45; **8**:25, 40, 44, 45
Caustic soda solutions, **12**:14
Caustic solutions, austenitic steels in,
    8:48
Cavitation corrosion, **8**:127
Cavitation-damage, **2**:60; **8**:124
  accelerated tests, **8**:128
  factors influencing resistance of
    materials to, **8**:126
  hydraulic tests, **8**:129
  impingement attack, **8**:127
  influence of fluid environment, **8**:125
  inhibitors, **8**:126
  marine engineering, **10**:66
  materials resistant to, **8**:131
  remedial measures, **8**:132
  role of corrosion, **8**:127
  ships' underwater fittings, **8**:124
  temperature effects, **8**:126
  testing procedures, **8**:128
  vibratory tests, **8**:130
Cavitation-erosion, **8**:128
  assessment of, **20**:78
  tests, **20**:76
Cavitation number, **8**:125
Cells,
  concentration, **1**:75
  corrosion, **4**:77
  electrochemical, **1**:74, 80; **20**:28
  electrolytic, **1**:74
  e.m.f. of, **9**:71
  geological, **4**:77
  occluded, **1**:142
  permeation, **9**:52
  reversible, **9**:66, 75, 76
Cellulose acetate butyrate coatings,
    17:18
Cellulose plastics, **19**:60

Cellulose products, **3**:62
Cementite, **9**:112
Cements,
  acid-resisting, **19**:38
  hydraulic, **19**:38
  resin, **19**:38
  rubber latex, **19**:38
Central heating systems, inhibitors,
    18:23
Ceramics, *see also* Glass-ceramics
  high-temperature corrosion, **7**:11
Cerium oxides, **1**:256
Charge transfer, **1**:76, 84
Chemical condensation, **2**:33
Chemical corrosion, pitting corrosion,
    10:18
Chemical deposition,
  aluminium, **14**:19
  copper, **14**:63
Chemical desorption, **9**:42, 43, 44, 45
Chemical interaction, **1**:20
Chemical passivity, **1**:122
Chemical plant, **3**:61; **10**:10; **19**:28;
    20:144
  aluminium and aluminium alloys in,
    4:28
  construction stage checks, **10**:24
  corrosion data, **10**:23
  corrosion monitoring, **10**:26
  corrosion testing, **10**:23; **20**:145
  crack detection, **10**:26
  design, **20**:144
    influence of process variables, **10**:13
  design philosophy, **10**:12
  equipment design, **10**:13, 23
  equipment requirements, **10**:11
  erosion corrosion, **10**:22
  failures, **10**:13, 24
  general corrosion, **10**:14
  localised attack, **10**:16
  maintenance, **10**:25, 26
  materials schedule, **10**:11
  materials selection, **10**:24
  new materials, **10**:23
  process design, **10**:12
  remedial measures, **10**:31, 32
  stress-corrosion cracking, **10**:16, 17, 18
  tantalum in, **5**:71
  thickness measurement, **10**:26
  titanium in, **4**:44
  zirconium in, **5**:56
Chemical potentials, **1**:56, 57; **9**:63, 68,
    75, 78
Chemical rate constant, **9**:29, 33
Chemical rate equation, **9**:31, 32
Chemical reactions,
  first-order, **9**:31
  homogeneous, **9**:28
Chemical reduction, **13**:94
Chemicals, **2**:83–95
  aggressive, **2**:92

Chemicals *continued*
  copper and copper alloys in, **4**:52
  Corrosion Guide, **2**:84
  corrosion rate in. *See* Corrosion rate
  corrosiveness data, **2**:86
  inhibitive, **2**:92
  lead in, **4**:79
  molybdenum in, **5**:22
  pH of, **2**:87
  rationalisation of data, **2**:88
  sources of information, **2**:84
  synergistic action, **2**:83
  tantalum in, **5**:62
  titanium in, **5**:38
  zinc in, **4**:158
Chemisorption, **9**:13
Chloride concentration of natural
    waters, **21**:60
Chloride ions, **2**:95
Chloride melts, aggressiveness, **2**:117–118
Chloride salt baths, **2**:118
Chloride solutions, **4**:95; **8**:54, 75
  austenitic steels in, **8**:48
Chlorides, **2**:30, 45; **7**:73, 74; **10**:17
  inhibitors, **18**:25
  solid, **8**:78
Chlorinated hydrocarbons, **12**:4
Chlorinated rubber, paint coatings,
    15:20
Chromate treatments, **14**:44; **16**:33
  aluminium, **16**:35
  cleaning etch for copper and copper
    alloys, **16**:38
    Cronak process, **16**:37
    etch primers, **16**:41, 42
    iron, **16**:38
    magnesium alloys, **4**:98; **16**:39
    passivation of cadmium and zinc
      coatings, **16**:36
    principles of, **16**:33
    silver, **16**:41
    tin, **16**:41
    types, **16**:33
Chromic acid, **4**:81
Chromising, **13**:59, 65
  carbon steels, **13**:70
  cast irons, **13**:70
Chromium, **3**:23, 25, 27; **7**:22, 58; **20**:48
  carbides, **1**:41; **7**:129, 130, 131
  oxides, **1**:240; **7**:66
Chromium chromate, **16**:33
Chromium coatings, **14**:12, 87
  black deposits, **14**:90
  chromium-nickel-chromium deposits
    14:95
  corrosion resistance, **14**:90, 93
  crack-free, **14**:91
  decorative, **14**:87
  diffusion, **13**:59
  discontinuities, **14**:89

Chromium coatings *continued*
  duplex, **14**:92
  electrodeposition, **14**:87
  electrolyte, **14**:87
  hard, **14**:87, 90
  methods of applying, **14**:87
  microcracked, **14**:93
  microcracked topcoats, **14**:74
  microporous, **14**:93, 95
  porosity, **14**:89, 91, 94
  porous, **14**:95
  properties, **14**:89
  self-regulating solutions, **14**:88
  structure, **14**:89
  tetrachromate electrolytes, **14**:88
  trivalent baths, **14**:89
Chromium-iron alloys, **1**:240
Chromium plating, **13**:16, 17, 30
  lead in, **4**:83
Circulating water systems, cathodic
    protection, **11**:76
Cladding, **14**:6, 20, 97
Clay fraction of soils, **2**:64
Cleaning. *See* Pretreatment
Coatings, **2**:38
  see also Diffusion coatings;
      Electroplating; Hot dipping;
      Sprayed coatings; Vitreous
      enamel coatings
  accelerated testing, **20**:21
  anodic oxidation, **16**:3
    corrosion resistance, **16**:13
    film thickness measurement, **16**:13
    maintenance, **16**:17
    porous structure, **16**:5
    properties of, **16**:8
    sealing treatment, **16**:36
    tests, **16**:14, 15, 16
  ASTM Standards, **21**:82
  bituminous **15**:81
  British Standards, **21**:72
  chemical conversion, **16**:1
  copper alloys, **14**:66
  copper-zinc alloys, **14**:66
  DIN Standards, **20**:121
  electroplated. *See* Electroplating
  epoxide resin, **4**:97
  galvanic, **8**:93
  glass, **19**:28
  holiday detector, **11**:110
  in buildings and structures, **10**:40
  ISO Standards, **21**:83
  marine engineering, **10**:58
  metallic, **1**:208; **5**:77; **13**:1; **14**:1
    see also Sprayed coatings
    anodic, **14**:3
    brush plating, **13**:100
    cathode sputtering, **13**:99
    cathodic, **14**:3
    chemical reduction, **13**:94
    decorative, **14**:4

Coatings *continued*
  metallic *continued*
    discontinuities, **14**:3
    economics, **14**:9
    electrical conductivity, **14**:9
    factors affecting choice of, **14**:5
    gas plating, **13**:98
    immersion plating, **13**:92
    mechanical properties, **14**:8
    miscellaneous methods, **13**:92
    peen plating, **13**:96
    physical properties, **14**:8
    plasma spray-coating process, **13**:100
    practicability of application, **14**:6
    practical use, **14**:10
    protective action of, **14**:3
    resistance to corrosive environment, **14**:5
    temperature resistance, **14**:9
    vacuum evaporation, **13**:97
  organic, **5**:77; **8**:113
    behaviour of, **20**:96
    exposure, **20**:97
    performance evaluation, **20**:98
    physical tests, **20**:99
    service tests of, **20**:97
    specimen preparation, **20**:96
    tests, **20**:95
    thickness of, **20**:97
    weathering methods, **20**:98
  paint. *See* Paint coatings
  phosphate. *See* Phosphate coatings
  pipelines, **2**:69–70
  pretreatment for. *See* Pretreatment
  rubber, cavitation damage resistance, **8**:132
  steel sheet, **10**:39
  temporary, **17**:21–29
    application, **17**:21–26
    failure, **17**:25
    hard-film, **17**:22, 26
    oil-type, **17**:22, 27
    selection, **17**:26
    slushing, **17**:27
    soft-film, **17**:21–22, 26
    special modifications, **17**:23
    strippable, **17**:22–23, 27
    types, **17**:21, 23
    uses, **17**:23–25
  thermoplastic. *See* Thermoplastic coatings
  tin alloys, **14**:56
  tin-cadmium alloys, **14**:57
  tin-copper alloys, **14**:58
  tin-lead alloys, **14**:56
  tin-nickel alloys, **14**:58
  tin-zinc alloys, **14**:57
  underground, **15**:81
    coal-tar epoxy, **15**:90
    cold-applied tapes, **15**:88

Coatings *continued*
  underground *continued*
    foam polyurethane, **15**:91
    laminated tapes, **15**:90
    petrolatum-type tapes, **15**:89
    petroleum asphalt or coal-tar pitch, **15**:86
    plasticised coal tar and petroleum asphalt enamels, **15**:85
    polyethylene sheet, **15**:91
    preparation of metal surface, **15**:83
    pressure-sensitive tapes, **15**:89
    properties required, **15**:82
    straight and filled enamels, **15**:86
    techniques, **15**:84
    types of materials, **15**:85
    weight coatings, **15**:91
    wrapping materials, **15**:87
  visual examination, **20**:88
Coke breeze anode extender, **11**:49
Cold cure process, **15**:58
Cold working, **1**:36, 47, 49
Colmonoy, **13**:90
Combustion of fuels, **7**:8
Concentration and velocity interaction, **2**:14
Concentration cells, **1**:75, 135, 144
Concentration effect on corrosion, **2**:3, 89
Concentration gradient, **9**:6
Concentration overpotential, **2**:8; **9**:37
  contact corrosion, **10**:44
  prestressed reinforcement, **10**:47
  reinforced, inhibitors, **18**:30
  steel reinforcements, **10**:46
  water pipes in contact with, **10**:50
Condensation, **2**:29
  capillary, **2**:32
  chemical, **2**:33
Condenser tubes, **4**:45, 48; **20**:21, 22, 71
Condenser water-boxes, **4**:50
Condensers, **10**:61
  inhibitors, **18**:24
Confectionery, **2**:99
Constants in SI units, **21**:11
Construction materials,
  corrosion rates, **20**:85
  for foodstuffs, **2**:96
Contact corrosion,
  buildings, **10**:34
  concrete, **10**:44
  pipes, **10**:50
  plastics, **19**:78, 79
  structures, **10**:34
  wood, **10**:44; **19**:89
Contaminants, elimination of, **18**:5
Conventional combinations, **2**:46
Cooling systems, inhibition, **18**:19
Co-ordination number, **9**:89, 101, 102; **13**:12
Copper, **2**:49, 57; **3**:21, 23, 27, 87, 88;

Copper *continued*
    4:33
    action on stainless steel, **20**:61
    atmospheric corrosion, **4**:39
        contaminants from plastics, **19**:79
    bimetallic corrosion, **4**:59
    chromate cleaning etch, **16**:38
    corrosion behaviour, **4**:36
    corrosion fatigue, **4**:43
    corrosion inhibition, **18**:50
    corrosion products, **4**:39
    corrosion rates, **21**:64
    electrode potential relationships, **4**:37
    electrodes, **4**:37
    green patina, **4**:39, 40
    green staining, **4**:51
    impingement attack, **4**:45, 51
    in acids, **4**:53
    in alkalis, **4**:53
    in buildings and structures, **10**:42
    in chemical environments, **4**:52
    in contact with concrete, **10**:45
    in feed water, **18**:66
    in graphite, **4**:59
    in sea-water, **4**:45
    in waters, **4**:44, 51
    oxidation, **4**:54
    oxide film, **4**:55
    oxides, **1**:230
    pipes, **10**:49
    pitting corrosion, **1**:163, 165; **4**:47, 52
    properties, **4**:33
    protective measures, **4**:58
    purity grades, **4**:33
    scaling, **4**:54
    soil corrosion, **4**:43
    tanks, **10**:51
    tin coating, **14**:54
    tinning, **13**:52
Copper alloys, **2**:58; **4**:33
    see also Brasses
    applications, **4**:48, 52, 56
    atmospheric corrosion, **4**:40
        contaminants from plastics, **19**:79
    bimetallic corrosion, **4**:59
    cast, **4**:33
    chemical cleaning, **20**:107
    chromate cleaning etch, **16**:38
    coatings, **14**:66
    composition, **4**:33
    condenser tubes, **4**:45, 48
    corrosion behaviour, **4**:36
    corrosion fatigue, **4**:43
    deposit attack, **4**:47
    descaling, **12**:13
    distillation plant, **4**:50
    impingement attack, **2**:60; **4**:45, 51
    in acids, **4**:53
    in alkalis, **4**:53
    in chemical environments, **4**:52
    in sea-water, **4**:45

Copper alloys *continued*
    in waters, **4**:44, 51
    marine applications, **4**:50
    mechanical properties, **4**:33
    oxidation, **4**:55
    pitting corrosion, **4**:47, 52
    protective measures, **4**:58
    season cracking, **4**:56
    selective attack, **4**:47
    soil corrosion, **4**:43
    stress-corrosion cracking, **1**:47; **4**:56, 58
    tin coating, **14**:54
    welding, **10**:80
    wrought, **4**:33
Copper-beryllium, **5**:3
Copper coatings, **13**:16; **14**:12
    applications, **14**:61
    chemical deposition, **14**:63
    copper cyanide bath, **14**:64
    corrosion resistance, **14**:64
    cyanide baths, **14**:62
    mechanical properties, **14**:65
    porosity, **14**:64
    properties of, **14**:63
    pyrophosphate bath, **14**:63
    solutions, **14**:62
    sprayed, **13**:87
    sulphate bath, **14**:62
Copper-nickel alloys, **1**:247; **4**:35, 48
    oxidation, **1**:237
    phase diagram, **9**:106
Copper plating. *See* Copper coatings
Copper steels, **3**:21, 28
Copper-zinc alloys,
    coatings, **14**:66
    phase diagram, **9**:111
Coring, **9**:107
Corrodkote test, **20**:48
Corrosion,
    anaerobic, **2**:79
    atmospheric. *See* Atmospheric
        corrosion
    basic concepts of, **1**:3
    bimetallic. *See* Bimetallic corrosion
    classification of processes, **1**:14, 16
    concept of, **1**:55
    costs of, **10**:3, 6
    crevice. *See* Crevice corrosion
    definition, **1**:4, 6; **18**:9
    deterioration, definition, **1**:5
    dry, **1**:16–18; **7**:3
    electrochemical. *See* Electrochemical
        corrosion
    hot. *See* High-temperature corrosion
    in aqueous solutions. *See* Aqueous
        solutions
    inhibition. *See* Inhibition
    liquid-metal. *See* Liquid-metal
        corrosion
    localised. *See* Localised corrosion

Corrosion *continued*
  'long-line', **2**:63
  mechanism of, **1**:7; **15**:24
  metallic, **1**:4
  metallurgy relevant to, **9**:88
  microbial. *See* Microbial corrosion
  phenomena of, **1**:6
  principles of, **1**:13
  soil. *See* Soil corrosion; Soils
  stray-current, **2**:63, 70
  terminology, **1**:14
  thermogalvanic. *See* Thermogalvanic
    corrosion
  transformation definition, **1**:4
  types of, **1**:11, 131
  wet, **1**:16, 18; **7**:3; **13**:41
Corrosion allowance, **10**:13
Corrosion control, **1**:6
Corrosion-erosion, control of, **1**:173
Corrosion failure, **1**:13
Corrosion fatigue, **8**:96, 112
  cast iron, **3**:96
  constant-stress contours, **8**:104
  constant total damage surface, **8**:103
  copper and copper alloys, **4**:43
  corrosive environment, **8**:106
  curves, **8**:102, 108
  data, **8**:101
  effect of hydrogen in steel, **8**:70
  electrochemical measurements, **8**:105
  endurance limits, **8**:113
  failure detection, **8**:107
  marine engineering, **10**:65
  mechanism, **8**:97
  nickel coatings, **14**:78
  potential measurements, **8**:105
  prevention, **8**:110
  rate of damage, **8**:103
  stainless steels, **3**:59
  test duration, **8**:101
  testing, **8**:109, 110; **20**:73
  titanium, **5**:43
  welded structures, **10**:80
Corrosion Guide, **2**:84
Corrosion interaction, **11**:91
Corrosion interference, **11**:91
Corrosion monitoring, **10**:11; **20**:15,
    144
  chemical plant, **10**:26
  crack detection, **10**:26
  eddy-current methods, **20**:153
  electrical resistance monitors, **10**:28
  electrical resistance probes, **20**:146
  instrumentation, **20**:148
  polarisation-resistance probes, **20**:149
  retractability, **10**:30
  thickness measurement, **10**:26
  ultrasonic methods, **20**:151
  variables affecting, **20**:145
  weight-loss coupons, **10**:27
Corrosion potential, **1**:85, 136, 192, 195,

Corrosion potential *continued*
    196, 203; **8**:40, 41, 43, 54, 156;
    9:81, 83; **20**:37, 126
  electroplating, **13**:23
  measurement, **10**:29; **20**:29, 32
Corrosion products, **1**:8, 22, 27; **2**:87
  accumulation of, **20**:11
  contamination of, **20**:15
  copper, **4**:39
  effect of temperature, **2**:49
  effect on corrosion rate, **2**:35
  formation of, **20**:14
  high-temperature water, **20**:136
  insoluble, **2**:27
  removal of **20**:16, 106
  soluble, **2**:27
  thin films, **20**:15
Corrosion racks, **20**:86, 93, 94
Corrosion rate, **1**:12, 53, 76; **3**:16
  and Tafel slopes, simultaneous
    determinations of, **20**:38
  calculation, **2**:89
  changes with time, **20**:6, 8
  construction materials, **20**:85
  conversion factors, **21**:62
  conversion formula, **20**:13
  copper, **21**:64
  decreasing with increase in
    temperature, **2**:89, 93–95
  decreasing with increasing
    concentration, **2**:89
  determination of, **2**:116–118
  effect of corrosion products, **2**:27, 35
  effect of surface roughness, **2**:21
  effect of temperature, **2**:13, 30
  effect of velocity, **2**:7
  effect on anion concentration, **2**:3
  electrochemical determination of,
    20:27
  evaluation, **1**:80
  factors controlling, **1**:73
  Freundlich adsorption isotherm, **2**:6
  Freundlich equation, **2**:5
  galvanic, **1**:196
  graphical method, **1**:89
  heat-flux, **20**:26
  in pipes and annuli, **2**:11
  in sea water, **2**:55
  increasing with increasing concentra-
    tion and temperature, **2**:89
  iron, **21**:64
  linear polarisation method for, **20**:36
  maximum at certain concentration,
    2:91
  metals, **21**:63
  monitoring, **20**:36
  non-uniform, **2**:95
  prediction of **2**:88, 94
  Stern-Geary technique, **20**:33
  uniform corrosion, **2**:88
  zero, **11**:8

Corrosion reaction, 1:6, 8
  chemical, 1:19, 20
  classification of, 1:20
  detection of intermediates in, 20:139
  electrochemical, 1:19, 20
  kinetics of, 1:53
  thermodynamics of, 1:53, 56
Corrosion resistance, 20:44
  dynamic, 8:106
  effects of composition and heat
    treatment on, 20:128
  static, 8:106
Corrosion testing, 20:1
  accelerated, 20:21, 43
    CASS, 20:48
    cathodic breakdown test, 20:46
    Corrodkote, 20:48
    electrochemical, 20:53
    electrolytic, 20:43, 44
    electrolytic oxalic acid etching,
      20:44
    FACT, 20:44
    impedance (Aztac), 20:46
    PASS, 20:46
    salt droplet test, 20:49
    spray, 20:47, 50
    sulphur dioxide, 20:50
    weathering steels, 20:51
  active or passive materials, 20:5
  air-stream, 20:22
  alternating-immersion, 20:23
  anodic oxidation coatings, 14:14, 15,
    16
  ASTM Standards, 20:115
  atmospheric, 20:14, 87, 89
  bimetallic, 20:40
  British Standards, 20:109
  cavitation-erosion, 20:76
  chemical plant, 10:23; 20:145
  chemical treatment prior to, 20:5
  classification, 20:3
  coatings, 20:88
  corrosion fatigue, 20:73
  corrosion potential, 20:32
  corrosion racks, 20:86
  crevice corrosion, 20:67
  damage appraisal of, 20:11
  DIN Standards, 20:119
  duration of exposure, 20:8
  electrochemical, 20:26
    accelerated, weathering steels, 20:53
    instruments, 20:28
    measurements used, 20:28
    reference electrodes, 20:28, 32
    specimen mounting, 20:32
    stainless steels, 20:65
  electrochemical cells, 20:28
  electrochemical techniques, 20:27
  electrolytic, accelerated, 20:43
  exposure methods, 20:86
  fretting corrosion, 20:78

Corrosion testing *continued*
  fused salts, 20:79
  galvanic, 20:49, 89
  general aspects, 20:3
  heat-flux effects in, 20:24
  heat treatment, 20:9
  impingement, 20:71
  intergranular attack of Cr-Ni-Fe
    alloys, 20:55
    boiling HNO_3 test (Huey test),
      20:58, 63, 64
    boiling $H_2SO_4 + CuSo_4$ tests, 20:60
    electrochemical tests, 20:65
    electrolytic oxalic acid etching test,
      20:64
    HNO_3-HF test, 20:62
    $H_2SO_4 + Fe_2(SO_4)_3$ test (Streicher
      test), 20:63
    I.S.O. Standards, 20:121
  jet-impingement, 20:21, 71
  laboratory methods, 20:16
  liquid metals, 20:79
  loop-test installations, 20:83
  natural waters, 20:93
  martensitic steels, 20:64
  number of replicate specimens, 20:6
  organic coatings, 20:95
  pitting corrosion, 20:67
  plant tests, 20:85
  polarisation resistance, 20:33
    applications, 30:35
    Tafel constants, 20:35
  preliminary, 20:6
  procedures, 20:4
  sea-water, 20:94
  soil, 20:9, 43, 95
  special processes, 20:15
  specimen marking for identification,
    20:6
  specimen requirements, 20:5
  specimens, plant tests, 20:87
  strength properties, 20:14
  stress-corrosion cracking. *See* Stress-
    corrosion cracking
  stress effects, 20:10
  surface preparation, 20:4
  total-immersion, 20:16
    aeration control, 20:17
    support of specimens, 20:23
    temperature control, 20:17
    velocity control, 20:18
    volume of testing solution, 20:23
  water, 20:9
  water-line, 20:24
  weathering, 20:89, 98
  weighing of specimens, 20:5, 11
  weight-gain determinations, 20:11
  weight-loss determinations, 20:11
  welds, 20:7
Corrosion zone. *See* Potential-pH
  diagrams

Costs, corrosion control, **0**:2
  of corrosion, **10**:3, 6
  of protection of steelwork, **10**:7
Coupons, **20**:146
Crack detection, chemical plant, **10**:26
Crack propagation, **3**:72
Cracking,
  caustic. *See* Caustic cracking
  stress-corrosion. *See* Stress-corrosion
    cracking
  transgranular, **3**:74
Crevice corrosion, **1**:134, 143, 151;
    2:58; **8**:112; **13**:43; **20**:12, 23, 26,
    126
  control methods, **1**:148
  mechanism of, **1**:145
  mild steel, **1**:145
  resistance, **1**:148
  stainless steels, **1**:144
  tests, **20**:67
  titanium, **1**:147, 148; **5**:43
  welds, **10**:21
  with and without heat transfer, **10**:19
Critical pitting potential, **20**:129
Cronak process, **16**:37
Crystal defects, **1**:50; **9**:91 *et seq.*
  active dissolution, energy
    considerations, **1**:34
  effects on corrosion, **1**:33
Crystal planes, **9**:89
Crystal structure: **1**:34; **9**:88 *et seq.*
Current density, **1**:79, 87, 94, 116, 136;
    11:31, 78
  and 1 $gm^{-2}d^{-1}$ or 1 mmy^{-1}
    relationship, **21**:64
  critical, **1**:104, 105
  exchange, **9**:81; **11**:7
  maximum or limiting, **9**:40
Cyanide solutions, **1**:60

Damage appraisal of corrosion testing,
    20:11
Daniell cell, **1**:80, 81, 82
Deaeration, bimetallic corrosion, **1**:208
De-alloying, potential-pH dependence
    of, **20**:132
Debye-Hückel limiting law, **9**:65
Degreasing. *See* Pretreatment
Delayed failure, **8**:161
Derusting. *See* Rust removal
Descaling. *See* Scale removal
Desiccating agents, **18**:3, 5
Design,
  buildings, **10**:33
  chemical plant. *See* Chemical plant
  electroplated coatings, **12**:28–34
  joining, **10**:68
  marine engineering. *See* Marine
    engineering
  overdesign, **10**:5

Design *continued*
  paint coatings, **12**:35–43
  structures, **10**:33
  welding, **10**:68
Desulphovibrio, **2**:78, 81
Dezincification, **1**:45, 166, 167; **4**:36, 41,
    46, 51; **20**:14
  mechanism, **1**:167
  of brasses, **4**:46, 51
Dew, **2**:31, 32
Dew point, **2**:29, 31
  depression, **2**:29
Difference potential, **20**:70
Differential aeration, **1**:135
  examples of, **1**:138
  pH in, **1**:137
Diffusion, **1**:99, 243; **4**:54; **5**:75; **7**:14,
    80; **9**:38, 91
  anionic, **1**:244, 246, 248, 258
  cationic, **1**:244, 248
  hydrogen in steel, **8**:67
  path length for, **2**:9
  polymers, **19**:59
  solid-solid, **13**:61
  theory of **13**:60
  voids in, **1**:250
Diffusion barriers, **1**:248; **18**:38
Diffusion coatings, **13**:58
  aluminising, superalloys, **13**:71
  characteristics of, **13**:68
  chromising, **13**:65
    alloy steels, **13**:71
    carbon steels, **13**:70
    cast irons, **13**:70
    coating thickness, **13**:68
    cobalt and cobalt alloys, **13**:71
    nickel and nickel alloys, **13**:71
  deposition methods, **13**:61
  ferrous materials, **13**:68
  gas-phase deposition, **13**:62
  interchange reaction, **13**:62
  iron-aluminium alloys, **14**:25
  metals applied by, **13**:58
  molten bath deposition, **13**:65
  principles of, **13**:58
  processes, **13**:59
  properties of, **13**:72
  rediffusion, **13**:73
  reduction reaction, **13**:64
  solid-solid, **13**:61
  theory, **13**:60
Diffusion coefficient, **1**:227, 232; **8**:67;
    9:50; **13**:60, 61; **21**:33
Diffusion-controlled process, **2**:18
Dilute solutions, **7**:122
  effects of large amounts of alloying
    elements, **7**:127
  in solid iron, **7**:124
Dimensionally Stabilised Anode, **11**:41
DIN Standards,
  coatings, **20**:121

DIN Standards *continued*
  corrosion testing, **20**:119
  inhibitors, **20**:121
Di-phase cleaners, **12**:8
Dislocations, **8**:33, 56, 71, 72; **9**:94, 95,
    97, 98, 104, 121, 122
Dispersion forces, **19**:57
Dispersion hardening. *See* Precipitation
    hardening
Dissimilar metals, **1**:192; **15**:8
Dissolution, **1**:120, 121
  in porous oxide formation, **1**:258
  process, **2**:5
Distensibility, coatings, **20**:99
Distillation process, **4**:50
Ditch structures, **20**:64
Double oxides, structures, thermal data
    and molecular volumes, **21**:47
Dry corrosion, **1**:16–18; **7**:3
Drying oils for paint coatings, **15**:12
Dunnage factor, **18**:4

Ebonite, **19**:65, 67
Economics, **10**:3
  discounted cash flow, **10**:7, 12
  loss of production, **10**:4
  losses resulting from corrosion, **10**:6
  maintenance of standby plant and
    equipment, **10**:5
  net present value, **10**:8
  overdesign, **10**:5
  paint coatings, **15**:69, 79
  paint finishes, **15**:58
  product contamination, **10**:5
  reduction of efficiency, **10**:5
Edible oils, **2**:99
Eddy-current methods of corrosion
    monitoring, **20**:153
Elastomers. *See* Rubber
Electrical double layer, **9**:6, 10, 22;
    18:40; **20**:27
  Gouy-Chapman model, **9**:16
  Helmholz model, **9**:6, 15
  Helmholz model structure, **9**:19
  Stern model, **9**:17
  structure, **9**:19
Electrical potential, **5**:46
Electrical resistance probes, **20**:146
Electrical triple layer, **9**:22
Electrocapillary curves, **9**:9, 11
Electrochemical cells, **1**:74, 80; **20**:28
Electrochemical corrosion, **1**:52, 73;
    2:115
  lead, **4**:77
  mechanism, **2**:52, 73
  zinc, **5**:153
Electrochemical desorption, **9**:42, 43, 44,
    45
Electrochemical measurements in
    corrosion testing, **20**:26

Electrochemical potential, **11**:5; **20**:128
Electrochemical rate constant, **9**:33
Electrochemical reaction rate, **9**:30
Electrochemical reactions, **1**:20
Electrochemistry, **9**:3
Electrocure process, **15**:58
Electrode area, **1**:79
Electrode potential, **1**:82; **9**:75, 79;
    20:22
  see also Standard electrode potential
  copper, **4**:37
  magnesium, **4**:90
Electrode reactions, **21**:38
  fast, potentiostatic transient
    investigation, **20**:141
  inhibitors in, **18**:39
Electrodes, **1**:74
  copper, **4**:37
  first kind, **9**:78
  hydrogen, **9**:77, 78
  kinetics, **9**:28
  oxide, **9**:83
  polarisable and non-polarisable, **9**:76
  polarisation studies, **20**:30
  reference. *See* Reference electrodes
  reversible and irreversible, **9**:75
  rotating disc-ring, **20**:139
  second kind, **9**:78
Electroforming, nickel coatings, **14**:85
Electrolytes, **9**:3; **13**:7, 10; **14**:99
Electrolytic cell, **1**:74
Electrolytic cleaners, **12**:9–10
Electrolytic composition in bimetallic
    corrosion, **1**:203
Electrolytic conduction, **9**:5
Electron acceptor, **1**:93
Electron conductivity, **1**:120
Electronic conduction, **9**:6
Electrophoretic coatings, aluminium,
    14:19
Electroplating, **8**:111; **13**:3
  addition agents, **13**:15
  additional ingredients, **13**:15
  aluminium and aluminium alloys,
    13:22; **14**:19
  anode behaviour, **13**:18
  anodes, **13**:16
  application, **12**:28
  barrel plating, **13**:30
  British Standards, **12**:28, 34
  brush plating, **13**:31
  'burners' or 'robbers', **12**:30
  burnt coatings, **13**:29
  cadmium. *See* Cadmium plating
  cathode corrosion, **13**:19
  chromium. *See* Chromium plating
  conducting salts, **13**:15
  contamination of plating solutions,
    12:33
  copper, **13**:16
  corrosion potentials, **13**:23

Electroplating *continued*
  corrosion resistance, **12**:33
  criterion of failure, **13**:32
  current density, **12**:29–30
  current path geometry, **13**:35
  current supply, **13**:29
  deposit distribution, **12**:28
  deposit uniformity, **12**:32
  design for corrosion protection by,
      12:28–34
  design considerations, **12**:32–34
  electrodeposit properties, **13**:32
    composition, **13**:36
    internal stress, **13**:37
    mechanical properties, **13**:38
    porosity, **13**:41
    structure-dependent, **13**:36
  electrolyte effects, **13**:29
  electrolytes, **13**:10
  epitaxy, **13**:23
  flash, **13**:21
  flow melting, **13**:32
  gold, **13**:17
  hydrogen absorption, **8**:63
  industrial techniques, **13**:30
  mechanical pretreatments, **13**:7
  mechanism, **13**:7
  nickel. *See* Nickel coatings
  noble metals. *See* Noble metal
      coatings
  non-aqueous, **13**:12
  non-epitaxial, **13**:25
  origin, **13**:3
  passive alloys, **13**:21
  periodic reverse current, **13**:30
  plastics, **15**:57
  platinum. *See* Platinum coatings
  porosity, **12**:33
  post-plating treatments, **13**:32, 44
    pretreatment, **13**:4
    metallic substrates, **13**:4
    non-conductors, **13**:6
  processes, **13**:7
  properties of coating, **12**:32
  pseudomorphism, **13**:23, 26
  rinsing, **13**:31
  service corrosion effects, **13**:22
  shields, **12**:31
  silver, **13**:20
  simple and complex ions, **13**:12
  strike baths, **13**:20
  structural factors, **13**:23
  substrates, **13**:4, 23
    interdiffusion with, **13**:40
    metallic, **13**:4
    non-conductors, **13**:6
  thickness of deposit, **12**:28, 32
  throwing index, **13**:33, 35
  throwing power, **12**:32, 33
  tin. *See* Tinplate
  vat plating, **13**:30

Electroplating *continued*
  zinc diecastings, **13**:19
Electropolishing, **12**:25
Embrittlement, **7**:90
  pressure theory, **8**:137
Embrittlement detector, **8**:157
E.m.f. of cell, **9**:71
E.m.f. series of metals, **9**:79
Emulsion cleaners, **12**:7
Enamel coatings, vitreous. *See* Vitreous
      enamel coatings
Energy balance approach to fracture, **8**:3
Enthalpy, **9**:59
  free, **9**:62
Entropy, **7**:123; **9**:60
Environment, **1**:8, 11, 22; **2**:1–60
  micro, **2**:28
Environmental conditions, **1**:4, 6, 8, 22
Epitaxy, **13**:23
Epoxide resins, **19**:62
  magnesium alloys, **4**:97
  paint coatings, **15**:18
Epoxy resins, **15**:53
Equilibrium constant, **9**:68
Equilibrium e.m.f.s., sign convention,
      9:69
Equilibrium phase diagrams. *See* Phase
      diagrams
Equilibrium potentials, **1**:91; **9**:69
Equilibrium state, predictions of, **1**:53
Erosion-corrosion, **1**:169
  by boiler water, **20**:23
  chemical plant, **10**:22
  titanium, **5**:42
Etch primers, **15**:47–48, 55; **16**:41, 42
Etching,
  of grain boundaries, **1**:36
  of metallographic surfaces, **1**:34, 36
  selective, **20**:133
Ethylene glycol, **4**:28
Eutectic reaction, **9**:107, 112
Eutectoid reaction, **9**:111, 112
Evans' diagrams, **1**:90, 91, 92
Excess charge of interface, **9**:8
Exchange current densities, **21**:37, 38
  hydrogen evolution reaction, **21**:37
Exfoliation, **1**:43
Exhaust valves, **7**:108

Fabrication stresses, **8**:44, 45
Failure, **2**:103
False form, **13**:23
Faradaic mass transfer, **2**:116
Faraday's law, **1**:76, 131, 193; **2**:10;
      11:27
Fast electrode reactions, potentiostatic
      transient investigation, **20**:141
Fasteners, **10**:68
Fatigue, corrosion. *See* Corrosion fatigue
Fatigue strength, nickel coatings, **14**:78

Fatty acids, **3**:62; **19**:79
Ferrite, **3**:35, 45; **9**:117, 118
Ferrobacillus ferro-oxidans, **2**:74, 76
Fertilisers, **3**:62
Fick's law, **2**:8; **9**:40
Filiform corrosion, **1**:148; **15**:47
    aluminium alloys, **4**:17
    mechanism of, **1**:150
    relative humidity in, **1**:149
Filigran corrosion, **15**:47
Film. *See also* Oxide films
Film breakdown
    and repair, **20**:33
    by undermining, **1**:126
    chemical, **1**:125
    electrochemical, **1**:124
    mechanical, **1**:126
Film dissolution,
    chemical, **1**:121
    electrical, **1**:120
    mechanical, **1**:121
    oxidative, **1**:125
Film formation, **1**:5; **2**:6, 15, 33, 35, 58,
        87, 94; **8**:10; **15**:28
    effect of flow, **2**:12
    second barrier, **2**:88
Film nucleation and growth, **1**:117
Film rupture, **8**:17
Film thickness, effective, **2**:9
Fine abrasive blasting, **12**:12
Finger-print cracking, **8**:78
Fish products, **2**:100
Flade potential, **1**:105; **11**:113, 119;
        18:42
Flame descaling, **12**:10
Flexure method for oxide stresses, **1**:260
Flow,
    effect on film formation, **2**:12
    laminar, **2**:9
    turbulent, **2**:9, 12
Flow rate,
    effect on corrosion rate, **2**:7
    of water, **2**:49
Fluidised bed process, **15**:5
Fluorides, **7**:111
Fogging, nickel, **4**:126
Food and drink production and
        distribution, **3**:61
Foodstuffs, **2**:96–101
    alcoholic beverages, **2**:99
    confectionery, **2**:99
    construction materials for, **2**:96
    corrosive effects, **2**:96
    edible oils, **2**:99
    fish products, **2**:100
    fruit juices, **2**:97
    liquid, **2**:97
    meat products, **2**:100
    milk, **2**:98
    sugar products, **2**:99
    vegetable processing, **2**:100

Ford Anodised Aluminium Corrosion
        Test (FACT) **16**:15
Fracture,
    delayed failure, **8**:161
    energy balance approach to, **8**:3
    unstable, **8**:134
Fracture mechanics, **8**:45, 56, 69, 82, 133
    hydrogen embrittlement, **8**:137
    stress-corrosion cracking, **8**:136, 150
Fracture toughness, **8**:134
Free energy, **1**:53–56, 82; **7**:115, 117;
        9:77
Free-energy diagrams, **7**:117, 118, 119,
        124, 127
Fretting corrosion, **8**:114
    amplitude of slip, **8**:116
    atmosphere effects, **8**:115
    characteristics of, **8**:115
    definition and terminology, **8**:114
    frequency of oscillation, **8**:116
    hardness effects, **8**:116
    incidence, **8**:114
    load effect, **8**:116
    lubricant effects, **8**:117
    mechanism, **8**:117
    nickel coating, **14**:79
    number of cycles, **8**:116
    prevention measures, **8**:119
    temperature effects, **8**:115
    temperature effects, **8**:119
    tests, **20**:78
Fretting fatigue, **8**:120
    prevention of, **8**:123
Freundlich adsorption law, **2**:5, 6, 18,
        88, 95
Frit, **17**:3, 5, 8
Fruit juices, **2**:97
Frumkin isotherm, **9**:27
Fuel-ash corrosion, **2**:110; **7**:9, 10, 32,
        72, 75, 76
    nickel and nickel alloys in, **7**:105
Fuel combustion, **7**:8
Fungi, isolation and enumeration media,
        2:81
Furan resins, **19**:62
Fused salts, **2**:110–115
    aggressiveness, **2**:115
    aluminium and aluminium alloys in,
        4:30
    behaviour of metals in, **20**:138
    container metal for, **2**:112
    corrosion conditions, **2**:113
    corrosion data, **2**:117
    corrosion mechanism, **2**:111
    corrosion rates, **2**:116–118
    corrosion tests, **20**:79
    displacement reaction, **2**:111
    dynamic tests, **20**:82
    electrochemical corrosion, **2**:115
    high-alloy steels in, **7**:75
    high-temperature corrosion, **7**:11

Fused salts *continued*
    loop tests, **20**:83
    measures to reduce corrosion in, **2**:118
    metal/melt systems, **2**:113
    molybdenum in, **5**:19
    nickel and nickel alloys in, **7**:109
    regenerators, **2**:118
    selective attack, **2**:116
    solubility variations, **20**:81
    special features of corrosion, **2**:115
    stability diagrams, **20**:138
    stainless steels in, **3**:58
    static tests, **20**:81
    thermal mass transfer, **2**:115–116
Fused silica, **19**:11, 25, 26

Galling, nickel coatings, **14**:79
Galvanic cells, **4**:78
Galvanic coatings, **8**:93
Galvanic corrosion. *See* Bimetallic
    corrosion
Galvanic coupling, beneficial effects,
    1:210
Galvanic effects, **20**:92
    beryllium, **5**:7
    in galvanic couple, **20**:41
    of fastening, **20**:92
Galvanic series, **21**:30, 31
Galvanising,
    continuous strip, **13**:54
    sheet steel and fabricated articles,
        13:52
Galvanometer, **11**:105
Galvanostatic methods, **1**:103, 104
Galvanostatic transients, **20**:34
Gas(es),
    aluminium in, **4**:19
    beryllium in, **5**:7
    carbon oxide, **7**:100
    carbonaceous, **7**:98
    combustion, in oxidation, **7**:60
    dissolved in sea water, **2**:54
    dissolved in water, **2**:40
    evolution of, **21**:37
    flue, **7**:93
    flue gas, **7**:31, 66
    in sea-water, **21**:54
    in soil, **2**:65
    in water, **3**:15
    industrial, **7**:72
    lead in, **4**:79
    molybdenum in, **5**:19
    nickel and nickel alloys in, **4**:134
    niobium in, **5**:28
    oxidising, **7**:72
    solubility in metals, **7**:127
    solubility in water, **21**:53
    sulphur-containing, **7**:7
    tantalum in, **5**:63
    uranium in, **5**:75

Gas(es) *continued*
    waste, **7**:8
Gas-flow-rate effects, **7**:6
Gas-metal systems, thermodynamics,
    7:115
Gas plating, **13**:98
Geological cells, **4**:77
Gerber relationship, **8**:123
Gibbs absorption equation, **9**:14
Gibbs-Duhem relation, **7**:132; **9**:65, 66
Gibbs free energy, **9**:62, 75, 76
Gibbs-Helmholz equation, **9**:63
Gibbs' phase rule, **9**:103
Gibbs surface excess, **9**:14
Glass, **19**:10
    chemical attack, **19**:20
    chemical properties, **19**:16
    cleaning, **19**:20
    coatings, **19**:28
    commercial, **19**:10
    compositions, **19**:10
    corrosion mechanism, **19**:, 18, 19, 20
    corrosion properties, **19**:18
    durability tests, **19**:16
    grain tests, **19**:16
    high-temperature corrosion, **7**:11
    in acids, **19**:20
    in alkalis, **19**:20
    in water, **19**:20
    lining processes, **19**:30
    linings, **19**:28
        chemical properties, **19**:31
        mechanical properties, **19**:30
        thermal properties, **19**:30
        thermal shock resistance, **19**:30
    mechanical properties, **19**:15
    network-formers, **19**:19
    physical properties, **19**:12
    preparation, **19**:28
    surface, **19**:18
    whole article tests, **19**:17
Glass-ceramics, **19**:21
    chemical durability, **19**:21
    chemical resistance data, **19**:23
    definition, **19**:21
    general characteristics, **19**:21
    makers' data, **19**:23
    properties, **19**:21
Glass fibres, **19**:21, 17
Glossary of terms, **22**:3
Gold,
    anodic behaviour, **6**:18
    chemical properties, **6**:9–10
    high-temperature properties, **6**:13
    physical and mechanical properties,
        6:5
    standard electrode potentials of gold
        systems, **6**:9
    working, **6**:21
Gold coatings, **14**:15, 97, 100
Gold plating, **13**:17

Gold-platinum alloys, spinnerets, 6:16
Grain boundaries, 9:99, 104, 123
  etching of, 1:36
Grain-boundary corrosion, 8:8, 31
  potentiostatic methods, 20:134
Grain size and stress-corrosion cracking,
  8:12
Grain structure, 9:99
  and stress-corrosion behaviour, 1:43
  effect on corrosion, 1:43
Grains, 9:98
Graphite,
  anodes, 11:47, 66
  copper in, 4:59
  in cast iron, 1:45
Graphite corrosion, 3:17, 90, 102
Graphitic residue, 3:86
Graphitisation, cast iron, 7:48, 49
Greases,
  additives in, 2:124
  removal of 12:3–10
Green-rot, 7:9, 101, 102
Griffith's theory of brittle fracture, 8:133
Grignard reaction, 1:19
Grit blasting, 12:10
Gunmetals, 4:35

Halogens, 3:58
  high-temperature corrosion, 7:11
Hard facing, 13:90
Hardness,
  coatings, 20:99
  effect of hydrogen in steel, 8:71
  water, 21:61
Haring-Blum cathode, 13:33
Haring-Blum cell, 13:34–35
Hastelloy alloys, 7:87
Hastelloy B, 4:111, 119
Hastelloy C, 4:120
Hastelloy F, 4:114
Hastelloy G, 4:114
Hastelloy N, 4:119; 7:111
Hastelloy W, 7:87
Heat content, 9:59
Heat effects of welding, 20:7
Heat exchanger tubes 10:19, 22, 26
Heat exchangers, 2:21; 10:61
Heat flow through heat-transfer surfaces,
  20:25
Heat-flux corrosion rates, 20:26
Heat-flux effects in corrosion testing,
  20:24
Heat transfer, 2:19
  boiling, 2:20
Heat-transfer surfaces, heat flow
  through, 20:25
Heat treatment,
  corrosion testing, 20:9
  effect on stress-corrosion cracking of
    ferritic steel, 8:35

Heat treatment *continued*
  effect on stress-corrosion susceptibility,
    1:48
  evaluation of, 20:8
Henry's Law, 7:122, 124
Hetergeneous system, 9:57
High-strength steels, hydrogen embrittlement, 8:158
High-temperature corrosion, 7:3
  see also Oxidation, high-temperature
  air, 7:4
  aluminium and aluminium alloys, 4:29
  ash effect, 7:9
  ash formation, 7:10
  atmospheres, 7:24
  carbon and oxides of carbon, 7:7; 19:8
  cast iron. *See* Cast iron, high-
    temperature corrosion
  ceramics, 7:11
  combustion products, 7:8
  glasses, 7:11
  green rot, 7:9, 101, 102
  halogens, 7:11
  high-alloy steels. *See* Steels, high-alloy
  high-chromium cast irons, 3:116
  hydrogen, 7:7
  impurity effects, 7:5
  low-alloy steels, 7:22
  maraging steel, 3:75
  molten salts, 7:11
  nickel and nickel alloys, 4:136
  nitrogen, 7:8
  oxygen, 7:4
  pressure effects, 7:6
  refractories, 7:11
  silicon-iron alloys, 3:126
  steam, 7:6
    nickel and nickel alloys in, 4:135
  stress effects, 7:3
  sulphur-containing gases, 7:7
  titanium, 5:44
  uranium, 5:75
Holiday detector, 11:110
Hollow ware, paint coatings, 12:42
Homogeneous system, 9:57
Hot corrosion. *See* High-temperature
  corrosion
Hot dipping, 13:48
  aluminising, 13:55
    alloy formation, 13:56
    continuous strip, 13:55
    heat-resistant coatings, 13:56
    structure, 13:57
  aluminium, 14:18, 20, 24
  galvanising,
    continuous strip, 13:54
    sheet steel and fabricated articles,
      13:52
  limitations, 13:48
  paint coatings, 15:61
  principles of, 13:49

Hot dipping *continued*
  process stages, **13**:48
  terne coatings, **13**:52
  tin, **14**:6
  tinning,
    cast iron, **13**:51
    copper, **13**:52
    fabricated steel articles, **13**:50
    steel strip, **13**:51
  zinc, **14**:6, 39, 40
Hot-wall effect, **20**:25
Huey test, **20**:58, 63, 64
Hull cell cathode, **13**:34
Humidity effects in rusting, **3**:5, 6
Humidity tests, phosphate coatings,
    16:29
Hydraulic cement, **19**:38
Hydraulic tests, cavitation damage,
    8:129
Hydrocarbons, chlorinated, **12**:4
Hydrochloric acid, **3**:53, 121; **4**:80;
    5:53, 54, 65, 69
  for pickling, **12**:19
Hydrogen, **5**:63; **7**:72
  adsorption, **9**:49
  high-temperature corrosion, **7**:7
  liquid-metal corrosion, **2**:108
  in steel,
    diffusion, **8**:67
    effect on mechanical properties,
      8:68
    fatigue properties, **8**:70
    hardness, **8**:71
    internal friction, **8**:71
    sources of, **8**:63
    state and location of, **8**:66
Hydrogen absorption, **8**:63, 65, 67, 79;
    9:49
  acid pickling, **8**:64
  electroplating, **8**:63
Hydrogen adsorption, **8**:42
Hydrogen chloride, **5**:8
Hydrogen cracking, **8**:47
Hydrogen diffusivity, **8**:23
Hydrogen electrode, **9**:77, 78
Hydrogen embrittlement, **1**:50; **3**:71;
    8:7, 14, 15, 23, 62, 64, 65, 68;
    12:9, 17; **14**:34, 66, 79
  bend tests, **8**:70
  dynamic tests, **8**:159
  electrochemical aspects, **9**:52
  failure process, **8**:69
  fracture mechanics, **8**:137
  high-strength steels. **8**:75, 158
  models of, **8**:72
  static tests, **8**:160
  tantalum, **5**:62
  theories, **8**:71
Hydrogen evolution, **1**:202; **2**:13; **8**:64,
    65
  aqueous solution, **21**:35

Hydrogen evolution *continued*
  iron, **9**:49
Hydrogen evolution reaction, **1**:93;
    9:41, 47
  exchange current densities for, **21**:37
Hydrogen permeation,
  potentiostat studies, **20**:137
  through iron, **9**:49, 55
Hydrogen peroxide, **4**:28
Hydrogen probes, **20**:154
Hydrogen sulphide, **2**:30, 40, 78, 79
Hydrogen trapping, **8**:66, 67
Hydrogenase determination, **2**:80
Hydroxyl ion adsorption, **2**:4

Ihrigising, **13**:60
Immersion plating, **13**:92
Immunity, concept of, **1**:55, 56
Immunity zone. *See* Potential-pH
    diagrams
Impact tests, coatings, **20**:99
Impingement attack, **1**:171; **2**:60
  aluminium brass, role of iron, **1**:172
  copper and copper alloys, **2**:60; **4**:45,
    51
  tests, **20**:21
Impurity effects, **1**:44; **9**:124
Impurity reactions, **20**:80
Inclusions, **9**:123
Inconel, **7**:100
Inconel-600, **4**:114, 123, 136
Infra-red radiation, **20**:154
Inhibition, **15**:25; **18**:9
  aluminium, **18**:50
  anion uptake by oxide films, **18**:45
  anodic reaction, **15**:27-30
  cathodic reaction, **15**:25-27
  copper, **18**:50
  effects of inhibitive anions on
    dissolution of passivating oxide,
    18:47
  effect of inhibitive anions on formation
    of passivating oxide, **18**:46
  inhibitive anions and aggressive
    anions, **18**:48
  iron corrosion. *See* Iron, corrosion
    inhibition
  resistance reaction, **15**:30-36
  zinc, **18**:49
Inhibitors, **2**:41, 48; **4**:58; **8**:53, 126,
    132; **14**:50; **16**:22; **18**:6
  acid solutions, **18**:25, 34
  adsorption, **9**:27; **18**:34
  aeration and movement of liquid,
    18:17
  anodic, **18**:42
  aqueous solutions and steam, **18**:18
  bimetallic corrosion, **1**:209
  blocking of reaction sites, **18**:39
  British Standards, **21**:78

Inhibitors *continued*
  central heating systems, **18**:23
  chemicals used, **18**:10
  chlorides, **18**:25
  classifications, **18**:10
  composition of liquid environment, **18**:15
  concentration, **18**:16
  condensers, **18**:24
  cooling systems,
    closed recirculating, **18**:21
    once-through, **18**:19
    open re-circulating, **18**:20
  crevices, dead-ends, **18**:17
  DIN Standard, **20**:121
  dissimilar metals in contact, **18**:14
  effect on corrosion processes, **18**:38
  effectiveness for one-component
    systems in near-neutral pH range, **18**:13
  efficiency of, **12**:21
  electrical double layer, **18**:40
  electrode reactions, **18**:39
  formulations, **18**:18
  functional group and structure of, **18**:36
  inorganic, **12**:23
  interaction of adsorbed species, **18**:37
  interaction with water molecules, **18**:36
  mechanical effects, **18**:17
  mechanisms of **18**:34
  micro-organisms, **18**:17
  miscellaneous uses, **18**:30
  nature of environment, **18**:15
  nature of metal surface, **18**:14
  neutral solutions, **18**:41
  oil industry, **18**:27
  organic, **12**:20
  passivating-type, **20**:140
  pH of system, **18**:16
  polarity-reversal effects, **18**:14
  potable waters, **18**:18
  potentiostat investigations, **20**:138
  principles, **18**:13
  reaction of adsorbed, **18**:37
  reinforced concrete, **18**:30
  scale deposition, **18**:18
  storage, **18**:30
  temperature of system, **18**:16
  toxicity, disposal and effluent
    problems, **18**:18
  volatile, **17**:21, 23, 28
Inner Helmholz plane, **9**:6, 21, 22
Inor 8, **7**:111
Insert rack, **20**:86
Inspection,
  see also Corrosion monitoring
  in-situ, **20**:153
  paints and painting. *See* Paint coatings
  visual, **20**:144, 145

Inspection points, selection of, **20**:145
Instability constant, **13**:12
Integral free energy, **7**:123
Interchange reaction, **13**:62
Intercrystalline corrosion, **3**:47; **7**:90
  aluminium and aluminium alloys, **4**:16
  stress-accelerated, **3**:48
Inter-diffusion, **14**:104
Intergranular corrosion, **1**:210; **2**:102; **8**:8–10, 23, 31, 47; **10**:78; **20**:8
  alloy systems, **1**:42
  aluminium alloys, **1**:42
  austenitic stainless steels, **1**:37
  chromium-nickel-iron alloys, **20**:55
  magnesium alloys, **4**:95
  mechanism, **1**:38
  nickel and nickel alloys, **4**:104, 123, 133
  potentiostatic methods, **20**:134
  sensitisation, **1**:38
Intergranular cracking, **3**:48; **8**:28, 29, 35, 43
Intergranular penetration of oxide, **7**:88
Intergranular stress corrosion, **8**:40
Intermediate phases, **9**:105
Intermetallic compounds, **9**:105
Internal friction, effect of hydrogen in steel, **8**:71
Internal stresses, **8**:27; **20**:10
  electrodeposits, **13**:37
International Standards Organisation,
  coatings, **21**:83
  corrosion testing standards, **20**:121
  materials, **21**:83
Interstitial atoms, **8**:32
Interstitial solutions, **8**:66
Interstitials, **9**:92
Ion,
  complex, **13**:12
  simple, **13**:12
Ion-exchange processes, **18**:59, 60
Ionisation constants, **21**:34
IR drop, **1**:88, 142; **20**:30, 36, 125
Iridium,
  anodic behaviour, **6**:19
  chemical properties, **6**:11
  high-temperature properties, **6**:13
Iridium coatings, **14**:103
Iridium-platinum alloys, **6**:6
  linings, **6**:15
Iron, **2**:38
  see also Cast iron,
  anodes, **11**:42
  carbon solution in, **7**:124
  chemical cleaning, **20**:108
  chromate treatment, **16**:38
  corrosion inhibition, **18**:42
    aggressive anion concentration, **18**:43
    dissolved oxygen concentration and supply, **18**:43

Iron *continued*
  corrosion inhibition *continued*
    nature of metal surface, **18**:44
    temperature effect in, **18**:44
  corrosion mechanism, **15**:24
  corrosion rates, **21**:64
  factors affecting corrosion, **3**:3
  high-silicon anodes, **11**:43
  high-silicon/chromium anodes, **11**:44
  high-silicon/molybdenum anodes,
    11:44
  hydrogen evolution, **9**:49
  hydrogen permeation through, **9**:49,
    55
  in feed water, **18**:67
  impingement attack of aluminium
    brass, **1**:72
  oxidation, **1**:250; **7**:18
    effects of other elements, **7**:19
  rusting. *See* Rusting
  solutions in, **7**:124
  stainless, welding, **10**:76
Iron-aluminium alloys, diffusion
  coatings, **14**:25
Iron carbides, **9**:112
Iron-chromium alloys, phase diagram,
  9:112
Iron-iron carbide phase diagram, **9**:112
Iron-nickel alloys, phase diagram, **9**:112
Iron oxides, **1**:240; **7**:18
Iron phosphate, **16**:23, 24
Irradiation effects, uranium, **5**:77
Iso-corrosion lines, **2**:90
Iso-corrosion rates, **2**:85–86
Isothermal mass transfer, **2**:106

Jet-impingement tests, **20**:21
Joinery, **10**:68
Joining, glossary of terms, **10**:82, 83
Jointing compounds, **15**:7–8
Jones d.c.-bridge technique, **20**:36

Kaolinite, **2**:64
Kesternich Test, **20**:50
Kinetics of corrosion reactions, **1**:53
Kirkendall effect, **2**:116
Knife-line attack, **1**:40; **3**:47; **10**:78

Lacquers, **15**:48
Langelier diagram, **2**:43, 44
Langelier equation, **2**:42
Langmuir constant, **9**:23, 24
Langmuir isotherm, **9**:23, 24
LaQue, F. L., **20**:108
Lattice defects in metal oxides. *See under*
  Oxide films
Lattice distortion, **9**:120
Layer corrosion, **1**:43; **4**:17
Leaching, **2**:102

Lead, **2**:49; **4**:68
  alloying elements, **4**:84
  anodes, **4**:84; **11**:44, 67
  anodic behaviour, **4**:72
  atmospheric corrosion, **4**:76
  cathodic behaviour, **4**:72
  coatings for, **4**:77
  composition, **4**:68
  compounds, **4**:75
  corrosion behaviour, **4**:69
  corrosion products, **4**:75
  electrochemical corrosion, **4**:77
  in acids, **4**:79
  in alkalis, **4**:81
  in buildings and structures, **10**:42
  in chemicals, **4**:79
  in chromium plating, **4**:83
  in contact with concrete, **10**:45
  in gases, **4**:79
  in organic acids, **4**:81
  in salt solutions, **4**:82
  in water, **4**:76
  mechanical properties, **4**:68
  pH/potential diagrams, **4**:72
  pipes, **10**:48
  protective film, **4**:77
  soil corrosion, **4**:77
  standard electrode potential, **4**:69
  stray-current corrosion, **4**:77
  thermodynamics of Pb–H_2O,
    Pb–H_2O–X systems, **4**:72
Lead alloys, **4**:68, 69
  anodes, **4**:84
  chemical cleaning, **20**:107
  composition, **4**:69
  corrosion behaviour, **4**:69
  mechanical properties, **4**:69
Lead-antimony alloys, phase diagram,
  9:107
Lead coatings, **13**:52; **14**:12
  sprayed, **13**:86
Lead compounds, **7**:108
Lead dioxide, **11**:44
Lead hydroxide, **15**:30
Lead linoleate, **15**:29
Lead oxides, **4**:72, 75, 76
Lead peroxide, **11**:44, 46
Lead/platinum bi-electrodes, **11**:46
Levelling agents, **14**:71
Levich equation, **2**:10
Lightning, **2**:70
Line defects, **9**:94
Linear polarisation, **20**:27, 33, 34, 36
Linings, protective, **6**:14
Linoleic acid, **15**:29
Linolenic acid, **15**:29
Linseed oil, **15**:12
Lippman electrometer, **9**:9
Lippmann equation, **9**:9
Liquid-metal corrosion, **2**:102–109
  activity gradient, **2**:106

Liquid-metal corrosion *continued*
  dynamic tests, **20**:82
  loop tests, **20**:83
  metal transfer, **2**:106
  non-metal transfer, **2**:107
  solubility of metals, **2**:102
  solubility variations, **20**:81
  static tests, **20**:81
  temperature effect, **2**:104
  tests, **20**:79
Liquid metals,
  aluminium and aluminium alloys in,
    4:30
  high-alloy steels in, **7**:77
  high-temperature corrosion, **7**:12
  molybdenum in, **5**:19
  nickel and nickel alloys in, **7**:109
  niobium in, **5**:29
  titanium in, **5**:38
Liquid salts. *See* Fused salts
Litharge, **15**:30
Lithium, **1**:236; **7**:110
Localised corrosion, **1**:130; **3**:17
  chemical plant, **10**:16
  heterogeneities and geometrical
    factors, **1**:133
  pH changes, **1**:139
  principles, **1**:133
Long-line currents, **2**:70
Loop tests, **20**:83
Lubricants, **2**:125–127
  additive depletion, **2**:124
  additives and additive interaction,
    2:122, 123
  contamination, **2**:121
  deterioration of, **2**:121
  extreme pressure additives, **2**:122
  function of, **2**:120
  gear, **2**:126
  manufacture, **2**:120
  metal-working, **2**:126
  miscellaneous applications, **2**:126
  nature of **2**:120
  performance characteristics, **2**:121
  steam-turbine, **2**:125
  sulphur additives, **2**:124
Lutes, **19**:39

Magnesium, **4**:86
  accelerated tests, **4**:94
  anodes, **10**:51; **11**:19, 21, 22
  anodic oxidation, **16**:3
  corrosion products, **4**:91
  electrode potentials, **4**:90
  galvanic corrosion, **4**:90
  in acids, **4**:8
  metallurgy, **4**:86
  standard electrode potential, **4**:90
Magnesium alloys, **4**:86
  accelerated, **4**:94

Magnesium alloys *continued*
  applications, **4**:86
  atmospheric corrosion, **4**:90
  chemical cleaning, **20**:108
  chromate sealing, **16**:40
  chromate treatments, **4**:98; **16**:39
  cleaning, **4**:98
  compositions, **4**:86
  corrosion behaviour, **4**:86
  designations, **4**:86
  epoxide resin coatings, **4**:97
  film formation, **4**:88
  galvanic corrosion, **4**:92, 93
  high-purity, **4**:92
  in industrial atmospheres, **4**:91
  in marine atmospheres, **4**:91
  in sea-water, **4**:92
  intergranular corrosion, **4**:95
  nuclear engineering, **4**:86
  painting, **4**:97
  pickling, **4**:98
  protective measures, **4**:96
  stress-corrosion cracking, **8**:86, 87, 89
  surface attack, **4**:94
  surface treatments, **4**:99, 100
  welding, **10**:80
Magnesium chloride, **2**:47
Magnesium-tin alloys, phase diagram,
    9:111
Magnetostriction oscillator test
    apparatus, **8**:130
Maintenance,
  chemical plant, **10**:25, 26
  paint coatings, **15**:70
  standby plant and equipment, **10**:5
Manganese, **3**:24, 36; **7**:21, 93
  pigment properties, **15**:28
Manganese phosphate, **14**:24
Marine engineering, **3**:60; **10**:54
  see also Sea-water
  bimetallic corrosion, **10**:62
  cavitation damage, **10**:66
  coatings, **10**:58
  corrosion fatigue, **10**:65
  design of components and fittings,
    10:61
  design principles, **10**:55, 56
  entrapment of corrosive agents, **10**:56
  paint coatings, **15**:72, 74, 77, 78
  stress-corrosion cracking, **10**:65
Marine fouling, **2**:56
Martensite,
  formation, **9**:115
  tempering, **9**:118
Mass transfer, **20**:80, 83
Materials,
  ASTM Standards, **21**:80
  British Standards, **21**:71
  ISO Standards, **21**:83
Maxwell's Distribution Law, **9**:29
Meat products, **2**:100

Mechanical factors, **8**:1
Mechanical properties, changes in, **20**:14
Melamine formaldehyde resins, **15**:52, 94
Mercury boiler service, **20**:81
Metal deposition, weight and thickness by 1 Ah, **21**:65
Metal/environment interface, **1**:7
Metal finishing. *See* Electro-plating; Paint coatings; Polishing; Pretreatment, etc.
   pretreatment for. *See* Pretreatment
Metal/gas reactions, **1**:17
Metal oxides, **1**:229
  and hydroxides, structures, thermal data, and molecular volumes, **21**:39
  double oxides, structures, thermal data and molecular volumes, **21**:47
Metal/solution interface, **9**:6
  non-polarisable and polarisable, **9**:75
  potential difference, **9**:31, 71
Metal spraying. *See* Sprayed coatings
Metal/vapour reactions, **1**:17
Metallurgy relevant to corrosion, **9**:88
Metals,
  see also Alloys; Crystal
  behaviour in molten salts, **20**:138
  cathodic depolarisation of ferrous, **2**:77–78
  corrosion in tropical environments, **20**:94
  dissimilar, **15**:8
  e.m.f. series of, **9**:79
  heterogeneities in, **1**:9
  immersed in water, **2**:38
  liquid. *See* Liquid-metal corrosion; Liquid metals
  macroscopic defects, **9**:100
  polystalline, **9**:98
  potentials in sea water, **2**:54
  properties of, **1**:3
  solubility in liquid metals, **2**:102
  structure of, **9**:88
Methanolic environments, stress-corrosion cracking in, **8**:84
Microbial corrosion, **2**:73–82
  cathodic depolarisation of ferrous metals, **2**:77–78
  diagnostic tests, **2**:80–81
  Ferrobacillus, **2**:74, 76
  isolation and enumeration, **2**:79
  prevention, **2**:74, 75
Microscopy, microbial corrosion detection by, **2**:80
Milk, **2**:98; **20**:23
Millscale, **3**:4, 5, 11, 17
Mineral constituents of water, **2**:44
Mineral salts in water, **2**:45
Moisture,
  in air, **7**:25
  in marine environment, **10**:56

Moisture *continued*
  in rusting, **3**:4, 5
  in soils, **2**:62, 65
  in stress-corrosion cracking, **8**:79
  in zinc corrosion, **4**:156
  precipitation in atmosphere, **2**:31
Molten salts. *See* Fused salts
Molybdenum, **3**:24, 27, 36; **5**:10; **7**:21, 87, 104
  alloying element, **5**:14
  applications, **5**:21
  corrosion, **5**:14
  corrosion behaviour, **5**:13
  fabrication methods, **5**:12
  galvanic corrosion, **5**:20
  in acids, **5**:15
  in alkalis, **5**:17
  in chemicals, **5**:22
  in fused salts, **5**:19
  in liquid metals, **5**:19
  in salt solutions, **5**:19
  in sea water, **5**:15
  mechanical properties, **5**:10, 12
  refractories compatibility, **5**:19
Molybdenum alloys, **5**:21
Monel alloy, **20**:17
Monel 400, **4**:119, 128
Monel K500, **4**:119
Monitoring, corrosion. *See* Corrosion monitoring
Montmorillonite, **2**:64
Multistep reactions, **9**:36
Mutually protective effect, **1**:58

Naval brass, **4**:49
Nernst equation, **2**:22; **9**:38, 53, 69, 70, 79
Neutral solutions, inhibitors, **18**:41
Nichrome, **7**:87
Nichrome V, **7**:100
Nickel, **3**:24, 27, 88; **4**:102; **7**:21; **20**:48
  anodes, **14**:70
  anodic behaviour, **4**:108, 121
    solution effect, **4**:110
    temperature effect, **4**:109
  applications, **4**:136
  aqueous corrosion, **4**:106
  atmospheric corrosion, **4**:125
  bimetallic corrosion, **4**:125
  carburisation, **7**:98
  fabrication methods, **4**:104
  fogging, **2**:27; **4**:126
  forms, **4**:139
  fuel-ash corrosion, **7**:105
  high-temperature corrosion, **4**:136; **7**:79
  in acids, **4**:130
  in alkalis, **4**:133
  in ammonia, **7**:104
  in chemical environments, **4**:130

Nickel *continued*
  in gases, **4**:134
  in liquid metals and salts, **7**:109
  in nitrogen, **7**:104
  in organic compounds, **4**:134
  in salt solutions, **4**:134
  in sea-water, **4**:128
  in steam, **4**:135
  in water, **4**:128
  intergranular corrosion, **4**:123, 133
  mechanical properties, **4**:102, 104
  oxidation, **4**:136; **7**:79
  oxide layers, **7**:79, 82
  oxides, **1**:236, 247
  passivity, **4**:110
  physical properties, **4**:104
  pitting corrosion, **4**:120
    temperature effects, **4**:122
  soil corrosion, **4**:130
  stress-corrosion cracking, **4**:135
  structure, **4**:104
  sulphidation, **7**:90
Nickel alloys, **4**:102
  anodic behaviour, **4**:111
  applications, **4**:136
  atmospheric corrosion, **4**:125
  bimetallic corrosion, **4**:125
  carburisation, **7**:98
  chemical cleaning, **20**:107
  compositions, **4**:102
  cyclic heating tests, **7**:104
  fabrication methods, **4**:104
  forms, **4**:139
  fuel-ash corrosion, **7**:105
  high-temperature, **4**:136
  high-temperature corrosion, **7**:79
  in acids, **4**:130
  in alkalis, **4**:133
  in ammonia, **7**:104
  in chemical environments, **4**:130
  in gases, **4**:134
  in liquid metals and salts, **7**:109
  in nitrogen, **7**:104
  in organic compounds, **4**:134
  in sea-water, **4**:128
  in steam, **4**:135
  in water, **4**:128
  intergranular corrosion, **4**:104, 123. 133
  intergranular precipitates, **4**:104
  magnetic transformation, **4**:104
  mechanical properties, **4**:102, 104
  oxidation, **4**:136; **7**:79
  physical properties, **4**:104
  pitting corrosion, **4**:120
  potential/anodic current density curves, **4**:111
  scale, **7**:80
  soil corrosion, **4**:130
  stress-corrosion cracking, **4**:135
  sulphidation, **7**:90

Nickel alloys *continued*
  welding, **4**:106; **10**:78
Nickel-aluminium alloys, **4**:114
Nickel-chromium alloys, **4**:111, 114, 122
  applications, **7**:84
  oxidation, **7**:82
  sprayed coatings, **13**:87
Nickel-chromium-iron alloys, **4**:119
  oxidation, **7**:88
Nickel-chromium-iron-molybdenum alloys, **4**:119
Nickel-chromium-iron-molybdenum-copper alloys, **4**:119
Nickel coatings, **8**:111; **13**:16; **14**:11, 69
  applications, **14**:69, 84
  chloride solution, **14**:78
  corrosion fatigue, **14**:78
  decorative, **14**:69, 70, 75, 84
    with improved resistance to corrosion, **14**:72
  double-layer, **14**:72
  electroforming, **14**:85
  electroless, **14**:80
    engineering, **14**:81, 84
    mechanical properties, **14**:82
    plating on plastics, **14**:81
    preparation of basis metals, **14**:81
    resistance to abrasion, **14**:83
    resistance to corrosion, **14**:82
  engineering, **14**:75, 80, 81, 84
  fatigue strength, **14**:78
  fretting corrosion, **14**:79
  galling, **14**:79
  hard, **14**:76
  heat treatment after plating, **14**:79
  mechanical properties, **14**:75
  microporosity in chromium topcoats, **14**:74
  resistance to corrosion, **14**:78
  sulphamate solutions, **14**:77
  triple-layer, **14**:74
  Watts solution, **14**:69, 75
Nickel-copper alloys, **4**:111, 117, 123
  oxidation, **7**:90
  phase diagram, **9**:106
Nickel-germanium alloys, **4**:114
Nickel-iron alloys, **3**:77; **4**:111, 114
  applications, **3**:77
  atmospheric corrosion, **3**:77
  bimetallic corrosion, **3**:82
  electrochemical characteristics, **3**:77
  in acids, **3**:81
  in alkalis, **3**:81
  in salt solutions, **3**:82
  in sea-water, **3**:79
  in water, **3**:80
  oxidation, **7**:88
  pitting corrosion, **3**:78
  stress-corrosion cracking, **3**:82
Nickel-molybdenum alloys, **4**:111, 117, 119, 123, 125

Nickel plating. *See* Nickel coatings
Nickel-silicon alloys, **4**:111, 115
Nickel silvers, **4**:36
Nickel-tin alloys, **4**:104, 114
Nickel-titanium alloys, **4**:104, 114
Nimonic 75, **7**:100
Nimonic 80A, **7**:97, 111
Niobium, **5**:24; **7**:87, 104
  anodes, **11**:40, 67
  applications, **5**:31
  atmospheric corrosion, **5**:26
  corrosion resistance, **5**:26
  fabrication methods, **5**:24
  in acids, **5**:27
  in alkalis, **5**:27
  in gases, **5**:28
  in liquid metals, **5**:29
  in salt solutions, **5**:27
  in water, **5**:26
  mechanical properties, **5**:24
  oxidation, **1**:258
  physical properties, **5**:24
Niobium alloys, **5**:26, 27, 30, 31
Niobium carbide, **1**:41
Niobium-molybdenum alloys, **5**:31
Niobium-tantalum alloys, **5**:30
Niobium-titanium alloys, **5**:30
Niobium-vanadium alloys, **5**:31
Niobium-zirconium alloys, **5**:30
Niobium-zirconium-titanium alloys, **5**:31
Ni-Resist, **3**:98, 104–107, 109
Nitrate environments, **8**:31, 41
Nitrate solutions, **8**:27
Nitrates, **2**:45
Nitric acid, **3**:53, 123; **4**:80; **5**:53, 71
  for pickling, **12**:20
Nitric-sulphuric acid mixtures, **3**:124
Nitrides, free-energy diagram, **7**:119
Nitrocellulose, paint coatings, **15**:20
Nitrogen, **2**:40; **5**:8, 63; **7**:72
  high-temperature corrosion, **7**:8
  in liquid-metal corrosion, **2**:107
  nickel and nickel alloys in, **7**:104
  saturation coefficient in sea-water, **21**:54
Nitrogen compounds, **2**:30
Noble metal coatings, **14**:97
  see also under individual noble metals
  brush plating, **14**:103
  electrodeposited, **14**:98
  electroless plating processes, **14**:98
  hardness, **14**:100
  high-temperature protection, **14**:104
  thickness, **14**:98
Noble metals, **6**:3–23
  see also under individual noble metals
  anodic behaviour, **6**:17–19
  alloying additions to facilitate passivity, **6**:19
  applications, **6**:14–17
  characteristics, **6**:3

Noble metals *continued*
  chemical properties, **6**:6–13
  corrosion ratings in acid, **6**:7
  corrosion ratings in halogens, **6**:7
  economies in use, **6**:22
  exchange current densities, **21**:38
  high-temperature properties, **6**:13
  mechanical properties, **6**:3–6
  physical properties, **6**:3–6
  solders, **6**:16–17
  thermodynamics of corrosion reactions, **1**:60
  uses, **6**:3
  working, **6**:21–22
Non-ferrous metals,
  descaling, **12**:13
  pickling, **12**:23
Notched 'C'-ring specimens, **8**:162
Nuclear engineering, **2**:110
  aluminium and aluminium alloys, **4**:29
  magnesium alloys, **4**:86
  nickel and nickel alloys, **4**:136
  titanium, **5**:45
  uranium, **5**:73
  zirconium, **5**:52, 55
Nuclear reactors, **2**:105–106; **20**:136
Nucleation, **9**:123
Nylon coatings, **17**:18
Nylons, **19**:61

Occluded cells, **1**:142
Ohm's law, **11**:72
Oil in water, **2**:47
Oil industry, inhibitors, **18**:27
Oil length in paint coatings, **15**:17
Oil refining and by-products, **3**:62
Oil varnishes, **15**:16
Order-disorder transitions, **9**:104
Organic compounds, **3**:58
  aluminium and aluminium alloys in, **4**:28
  nickel and nickel alloys in, **4**:134
  zinc in, **4**:160
Organic growths in water, **2**:48
Organic materials, **18**:6
Organic matter,
  in soils, **2**:61
  in water, **2**:47; **3**:15
Organic solvents, **1**:18
Organosols, **15**:61
Orthophosphates, **18**:61
Orthophosphoric acid, **3**:123
Osmium, **6**:13
  anodic behaviour, **6**:19
  chemical properties, **6**:13
  coatings, **14**:103
Osmotic blistering, **15**:59
Outer Helmholz plane, **9**:7, 21, 22
Overpotential, **1**:65, 83–85; **9**:34
  activation, **1**:85

Overpotential *continued*
  resistance, **1**:88
  transport (diffusion or concentration),
      1:87
Oxidation, **1**:16, 24, 53, 60, 227, 231,
      232, 236; **2**:112, 122; **7**:4, 55, 120
  see also Oxide films
  anodic. *See* Coatings
  carbon, **19**:9
  cast iron, **7**:47, 54, 55
  copper, **4**:54
  copper alloys, **4**:55
  dry, **2**:26
  factors governing, **7**:13
  growth laws of, **1**:252
  high-alloy steels, **7**:58, 65
    fuel-ash effect, **7**:72
    in air, **7**:69
    in industrial gases, **7**:72
    in steam, **7**:70
    in flue gases, **7**:66
  in combustion gases, **7**:60
  in presence of subscales, **1**:257
  internal, **7**:5, 49
  iron, **1**:250; **7**:18
    effects of other elements, **7**:19
  kinetics, influence of voids, **1**:248
  low-alloy steels, **7**:13, 19
    alloying elements, **7**:31
    in air, **7**:25
    in carbon dioxide, **7**:41
    in flue gas, **7**:31
    in steam, **7**:38
  nickel and nickel alloys, **4**:136; **7**:79
  nickel-chromium alloys, **7**:82
  nickel-chromium-iron alloys, **7**:88
  nickel-copper alloys, **7**:90
  nickel-iron alloys, **7**:88
  niobium, **1**:258
  non-planar surfaces, **1**:244
  paralinear, **1**:241, 256
  rate constant, **7**:14, 15
  reactions, **1**:25
  sulphur, **2**:77
  tantalum, **1**:258
  tin, **4**:143
  void formation, **1**:248
  volume change on, **1**:242
  zirconium, **1**:258, 262
Oxidation products, **1**:30
Oxidation rate, **1**:235, 241, 252, 257;
      7:14
Oxide electrodes, **9**:83
Oxide films, **1**:22, 24, 26, 114, 199, 200;
      2:87–88; **7**:13, 14; **9**:83
  see also Film; Oxidation
  anodic, **1**:28, 88
  continuous, **1**:227, 229
    alloying effects, **1**:236
    cation vacancies, **1**:232
    growth laws, **1**:235, 236

Oxide films *continued*
  continuous *continued*
    linear rate constant, **1**:229
    parabolic rate constant, **1**:229
    positive holes, **1**:232
    surface reactions, **1**:229
    thin, **1**:234
    very thin, **1**:234
    Wagner's theory of parabolic law,
        1:231, 234
  dielectric, **1**:88
  discontinuous, **1**:241, 242
    breakaway mechanism, **1**:254
    cracking and void cavities, **1**:252
    flat surfaces, **1**:243
    growth rate, **1**:241
    growth laws, **1**:252
    mass transport, **1**:243
    non-planar surfaces, **1**:244
    oxidation in presence of subscales,
        1:257
    oxide drift, **1**:243, 247
    parabolic rate constant, **1**:246
    paralinear oxidation, **1**:256
    Pilling-Bedworth theory, **1**:254
    pore formation, **1**:247
    porous, metal dissolution or
        volatilisation, **1**:258
    rate laws, **1**:241
    stress relief by metal creep, **1**:263
    stresses, **1**:260
    voids, **1**:247
      diffusion barriers, **1**:248
      in 'short-circuit' diffusion, **1**:250
      influence on oxidation kinetics,
          1:248
    volume change, **1**:242
  erosion-corrosion, **1**:170
  examination techniques, **1**:30
  formation, **20**:26
  forms and formations of, **7**:15
  influence of structure, **1**:46
  lattice defects, **1**:222
    anion vacancies, **1**:227
    cation vacancies, **1**:226
    degree of non-stoichiometry, **1**:222
    entropy change, **1**:223
    in metal oxides, **1**:222
    motion of, **1**:226
    n-type oxides, **1**:225
    p-type oxides, **1**:223
    positive holes, **1**:223
  passive, **1**:27
  porous, **1**:258
  protective, **20**:19
  thin, **1**:23; **20**:15
  uptake of anions by **18**:45
  voids, **8**:66, 67, 72
Oxide layers,
  mass transport, **1**:243
  nickel, **7**:79, 82

Oxide layers *continued*
  stresses in, **1**:260
Oxide-reduction time, **20**:55
Oxides, **1**:23, 26, 30, 107
Oxygen, **5**:63
  see also Oxidation
  diffusion through paint films, **15**:26
  dissolved, in bimetallic corrosion,
    1:197
  dissolved in sea-water, **21**:54
  exposure to, **1**:23
  in atmospheric corrosion, **2**:29
  in high-temperature corrosion, **7**:4
  in liquid-metal corrosion, **2**:107
  in rusting, **3**:4
  in sea-water, **2**:54
  in water, **2**:40–41; **3**:15
  saturation coefficient in sea-water,
    21:54
  solubility in electrolyte solutions,
    21:53
  solubility in sodium, **2**:106, 107
  transport of, **1**:97
Oxygen pitting of boiler tubes, **10**:18
Oxygen reduction reaction, **1**:93, 96, 99;
  2:13

Packaging materials, **18**:6
Paint coatings, **0**:5, 6; **8**:65; **15**:1–103;
  16:36
  see also Coatings; Paints
  abrasion resistance, **15**:50
  acrylic resins, **15**:53
  additives, **15**:22
  adhesion, **15**:46–48
  airless spray process, **15**:6
  alkyd resins, **15**:16, 53
  amino resins, **15**:18
  antifouling, **15**:44
  application, **20**:166
    effect on failure, **15**:45
    industrial, **15**:56
    methods, **15**:3–9, 69, 79
  bimetallic corrosion, **1**:209
  binding media, **15**:12–21
  buried metals, **15**:48
  cargo and ballast tanks, **15**:75
  chlorinated rubber, **15**:20
  constituents of, **15**:10
  curing, **15**:60
  curtain coating, **15**:4, 60
  cycling processes, **15**:3–5
  data sheets, **20**:166
  design for particular methods of
    finishing, **12**:39–43
  design principles, **12**:35–43
  dip tank, **15**:3
  dipping, **12**:40; **15**:56
  drying oils, **15**:12
  economics, **15**:58, 69, 79

Paint coatings *continued*
  electrodeposition, **15**:4
  electrostatic spraying, **15**:6, 56
  enamel finishes, **15**:51
  epoxide resins, **15**:18, 53
  etch primer, **15**:47–48, 55; **16**:41, 42
  failure, **15**:38–49, 59, 79
    acid environments, **15**:42
    adhesion, **15**:46–48
    alkaline environments, **15**:43
    anti-oxidising environments, **15**:42
    buried metals, **15**:48
    causes of **15**:38
    climatic effects, **15**:40
    effects of application methods, **15**:45
    forms of, **15**:38
    industrial applications, **15**:42–44
    industrial atmosphere effects, **15**:41
    low-temperature effects, **15**:40
    marine atmosphere, **15**:43
    moisture effects, **15**:41
    oxidising environments, **15**:42
    pretreatment as cause of, **15**:39
    repainting, **15**:46
    salt solutions, **15**:43
    storage conditions, **15**:44
    stoving effects, **15**:44
    sunlight effects, **15**:41
    wet weather effects, **15**:40
  flow coating, **15**:4
  fluidised bed process, **15**:5
  formulation, **15**:10–23
    basic principles, **15**:10–12
    procedure, **15**:23
  gas-checking, **15**:44
  glossary of terms, **15**:100
  holiday detection, **20**:164
  hollow ware, **12**:42
  hot dipping, **15**:61
  industrial applications, **15**:50–61
  inspection, **20**:158
    case histories and cost, **20**:159
    of works and site facilities, **20**:162
  inspection organisations, **20**:163
  inspection schedule, **20**:163
  inspection quality, **20**:163
  inspection type, **20**:162
  inspectors' duties, **20**:164
  lacquers, **15**:48
  machinery and equipment, **15**:22
  maintenance, **15**:70
  marine engineering, **15**:72, 74, 77, 78
  material supplies, **20**:164
  material-supplies test, **20**:168
  materials-in-use tests, **20**:168
  melamine formaldehyde resins, **15**:52
  miscellaneous binders, **15**:20
  moisture resistance, **15**:50
  new developments, **15**:60
  nitrocellulose, **15**:20
  non-cycling processes, **15**:5

Paint coatings *continued*
 non-ferrous metals, **15**:47
 number of coats, **15**:11, 46
 oil lengths, **15**:17
 oil varnishes, **15**:16
 organosols, **15**:61
 oxygen diffusion, **15**:26
 phenol formaldehyde resins, **15**:52
 pigments, **15**:21, 28, 29
  corrosion-promoting, **15**:44
  soluble, **15**:30
 plastics, **15**:51, 57
 polyurethane, **15**:19
 preliminary discussions, **20**:164
 pretreatment, **12**:35; **15**:54, 62, 72;
   20:165
  chemical processes, **15**:63
  mechanical processes, **15**:63
  merits and limitations, **15**:64
  metal coating, **15**:63
  quality control, **20**:158
  weathering, **15**:64
 primers, **15**:6–9, 11, 39, 46–48
  pigments, **15**:21
 process limitations, **15**:8
 properties of liquid films, **12**:36
 protective action, **15**:24–37
  anodic reaction, **15**:27–30
  cathodic reaction, **15**:25–27
  resistance reaction, **15**:30–36
 repainting after failure, **15**:46
 Roto-dip process, **15**:4
 sampling and testing, **20**:168
 ships' bottoms, **15**:76
 slow dipping, **15**:4
 small objects, **12**:42
 solvents, **15**:22
 spark testing, **20**:166
 specific applications, **15**:6–9
 spraying, **12**:39; **13**:89, 90; **15**:56
  air-less, **15**:56
  processes, **15**:5–6
 storage effects, **15**:44
 stoving effects, **15**:44
 structural steel, **15**:64, 67; **20**:159
  for atmospheric exposure, **15**:62
 surface condition effect, **12**:38
 synthetic resin, **15**:52, 93
  acrylic resins, **15**:96
  alkyd resins, **15**:95
  curing of epoxy resin, **15**:98, 99
  melamine formaldehyde resins,
   15:94
  phenol formaldehyde resins, **15**:93
  production of epoxy resins, **15**:97
  urea formaldehyde resins, **15**:94
 thick coatings, **15**:67
 thickness measurements, **20**:167
 thinners, **15**:45
 underground, **15**:81
 urea formaldehyde resins, **15**:52

Paint coatings *continued*
 vinyl resin, **15**:19
 viscosity checks, **20**:168
 water diffusion, **15**:26
 working conditions, **20**:167
Paints,
 see also Paint coatings
 air-drying, **15**:65
 antifouling, **15**:77
 chemically cured, **15**:66
 chlorinated rubber, **15**:66, 74
 epoxy resin, **15**:75
 high-performance, **15**:74
 oil-based, **15**:65
 primers, **15**:64, 67, 73, 74
 types of, **15**:64
 zinc chromate primers, **15**:74
 zinc-rich, **14**:40, 44
Palladium,
 chemical properties, **6**:11–12
 oxide films, **14**:104
Palladium coatings, **14**:97, 102
Parallel plate model, **9**:6
 inconsistencies of, **9**:15
Partial anodic reactions, **1**:78
Partial cathodic reactions, **1**:78
Partial derivatives, **9**:58
Partial molar free energy, **9**:63
Passivation, **1**:105, 114–117, 123; **13**:21;
   16:33
 see also Passivity
Passivation potential, **1**:104, 105
Passivity, **1**:5, 14; **2**:113, 118; **11**:113;
   20:5
 see also Passivation
 active-passive transitions, **1**:103
 adsorption theory, **1**:160; **9**:26
 anodic, **1**:115
 breakdown of, **1**:123, 160; **9**:27
 chemical **1**:122
 concept of, **1**:55, 56
 definition, **1**:114
 effect of noble-metal additions, **6**:19
 maintenance of, **1**:123
 nickel, **4**:110
 solid film theory, **1**:160
Passivity studies, potentiostat, **20**:126
Passivity zone. *See* Potential-pH
   diagrams
Pearlite, **9**:113, 117
Peaty acids, **2**:48
Peen plating, **13**:96
Penton coatings, **17**:16–18
Perforation,
 recording of, **20**:88
 time to, **20**:88
Periodic reverse current, **13**:30
Periodic table, **21**:6
Peritectic reaction, **9**:111, 112
Peritectoid reaction, **9**:111
Permeation cell, **9**:52

Permeation current, **9**:50
Petrochemical plant. *See* Chemical plant
Pfeil-type porous oxide, **1**:252
pH,
  see also Potential-pH equilibrium
    diagrams
  at crack tip, **8**:75
  inhibitors, **18**:11, 16
  of chemicals, **2**:87
  of sea water, **2**:54
  of soil, **2**-67
  of water, **2**:42, 44, 48
pH changes during localised attack,
    1:139
pH effects, **1**:60, 61, 96, 97; **4**:153; **5**:7
  cathodic protection, **11**:8
  changes in **2**:3, 4, 15, 18
  differential aeration, **1**:137
  in iron corrosion inhibition, **18**:43
  in pitting corrosion, **1**:157
  in stress-corrosion cracking, **8**:41
  in stress-corrosion testing, **8**:154
  on wood, **19**:88
  on zinc, **4**:157, 158
pH-potential diagrams, **6**:8, 10
Phase diagrams, **9**:105
  binary isomorphous, **9**:106
  complex binary, **9**:109
  eutectic, **9**:107
Phenol formaldehyde resins, **15**:52, 93;
    19:62
Phosphate coatings, **8**:65; **12**:3; **16**:19
  absorption, **14**:26
  accelerators, **16**:22
  acid ratio, **16**:20
  anodic treatment, **16**:21
  applications, **14**:26; **16**:19
  cathodic treatment, **16**:21
  chemical nature of, **14**:24
  chromate rinsing, **16**:27, 31, 38
  corrosion protection, **16**:27
  corrosion resistance, **16**:27
    assessment, **16**:29
  crystal size, **16**:30
  crystal structure, **14**:24
  formation mechanism, **16**:20
  heating, **14**:26
  humidity tests, **16**:29
  methods, **16**:19
  nature of, **16**:23
  phosphate solution effects, **16**:23
  processes, **16**:29
  reactions, **16**:20
  rinsing, **14**:26
  salt-spray tests, **16**:29
  sealing, **16**:26, 27, 29
  specifications, **16**:30, 31
  surface effects, **16**:23
  tests on, **16**:28–31
  texture, **16**:30
  thin, **16**:24

Phosphate coatings *continued*
  water of hydration loss, **14**:24
  weight determinations, **16**:30, 31
Phosphor-bronzes, **4**:35
Phosphoric acid, **3**:54, 61; **4**:81; **5**:54
  for pickling, **12**:19
Phosphorus, **3**:23; **7**:22, 48
Pickling, **12**:16–23; **13**:5, 52; **16**:36;
    18:25
  acids used, **12**:18–20
  alloy steels, **12**:20
  alternating current, **12**:19
  anodic, **12**:18
  cathodic, **12**:19..
  electrolytic, **12**:18
  magnesium alloys, **4**:98
  mechanism, **12**:16
  non-ferrous metals, **12**:23
Pilling-Bedworth theory, **1**:254, 262;
    5:36
Pipeline travelling machines, **15**:85
Pipelines,
  cathodic protection, **11**:13, 61, 63, 72,
    74
    internal surface, **11**:16, 82
  coatings, **2**:69–70
  corrosion rate in, **2**:11
  ditch environment, **2**:68
  internal coatings, **15**:92
  internal inspection, **20**:155
  long-line currents, **2**:70
  protective coatings, **15**:81
Pit depth measurements, **20**:12
Pitting corrosion, **1**:12, 46, 127, 130, 131,
    150; **3**:17, 44; **13**:16; **20**:12, 35,
    126, 152
  aluminium, **1**:162; **4**:16, 23
  aluminium alloys, **4**:23
  breakdown potential, **1**:154
  carbon steels, **1**:160
  chemical corrosion, **10**:18
  control of, **1**:160
  copper, **1**:163, 165; **4**:47, 52
  copper alloys, **4**:47, 52
  induction period, **1**:157
  mechanism, **1**:160
  mutually protective effect, **1**:158
  nature of solution and temperature,
    1:156
  nickel and nickel alloys, **4**:120, 122
  nickel-iron alloys, **3**:78
  nodular, **1**:163
  pH effect, **1**:157
  pit exterior, **1**:162
  pit interior, **1**:161
  pit mouth, **1**:161
  potentiostatic techniques, **20**:128
  protection potential, **1**:158
  stainless steels, **10**:18
  tests, **20**:67
  wire elements, **20**:148

Pitting index, **20**:150
Pitting potentials, **1**:155, 156
   critical, **20**:129
Plasma-spray-coating process, **13**:81, 100
Plaster, water pipes in contact with,
   10:50
Plastic deformation, **8**:3, 7, 8, 12, 23, 32,
   72, 118, 153
Plastics, **19**:41
   see also Thermoplastic(s); Thermo-
      setting
   amorphous thermoplastics, **19**:60
   cellulose, **19**:60
   chemical nature of, **19**:42
   commercial, review of, **19**:60
   contact corrosion, **19**:78, 79
   corrosion of metals by, **19**:74
   crystalline, **19**:61
   definition, **19**:41
   electroplating, **15**:57
   fluorine-containing, **19**:61
   industrial finishes, **15**:57
   paint coating, **15**:57
   paint finishes, **15**:51
   physical nature of, **19**:43
   pipes, **10**:50, 60
   properties, **19**:42
   reinforced, **19**:49
   susceptibility of metals to attack by
      contaminants from, **19**:79
   thermoplastics, **19**:44–48
   thermosetting, **19**:44, 48, 58, 62
   vacuum metalising, **15**:57
   vapour corrosion effects, **19**:75
Plastisols, **12**:39
Platinised anodes, **11**:36, 67, 76
Platinum, **5**:47
   anodes, **11**:36
   anodic behaviour, **6**:18–19
   applications, **6**:21–22
   grain-growth, **6**:14
   high-temperature properties, **6**:13
   linings, **6**:15
   mechanical properties, **6**:5
   physical properties, **6**:5
   volatilisation, **6**:14
Platinum alloys, wettability of, **6**:20–21
Platinum coatings, **13**:18; **14**:15, 102,
   104
Platinum-gold alloys, spinnerets, **6**:16
Platinum-group metals,
   chemical properties, **6**:10
   high-temperature applications, **6**:20
   high-temperature properties, **6**:13
   oxide-film formation, **6**:13
   oxides, **6**:19
   physical and mechanical properties,
      6:5
   working, **6**:21–22
Platinum-iridium alloys,
   linings, **6**:15

Platinum-iridium alloys *continued*
   physical and mechanical properties,
      6:6
Platinum plating, **13**:18
Platinum-rhodium alloys, **6**:20–21
   exchange current densities, **21**:38
   high-temperature properties, **6**:14
   linings, **6**:15
   physical and mechanical properties,
      6:5–6
Platinum-ruthenium alloys, physical and
      mechanical properties, **6**:6
Point defects, **9**:91
Polarisation, **1**:88, 117; **11**:8; **13**:35
   forward, **1**:104
   linear, **20**:27, 29, 33, 34, 36; **10**:29
   austenitic steel, **8**:52
   of metal in soil, **2**:72
   reverse, **1**:104
Polarisation characteristic, **20**:22
Polarisation curves, **8**:39; **11**:7, 9;
      20:123, 136
   titanium, **20**:128
   titanium alloys, **20**:140
Polarisation resistance, **20**:33
   probes, **20**:149
Polarisation studies, **20**:30
Polarity reversal, **1**:205
Polarograph, **20**:15
Polished surfaces, reflectivity of, **20**:15
Polishing, **13**:32
   chemical, **12**:25, 26
   electrolytic, **12**:24, 26
Pollution, **3**:6, 50; **4**:144
Polyacetal, **19**:61
Polyamides, **19**:61
Polycarbonates, **19**:62
Polycrystals, etching, **1**:34
Polyesters, **19**:62
Polyethers, **19**:61
Polyethylene, **19**:61
Polyformaldehyde, **19**:61
Polymerisation, **19**:42, 43
Polymers, **19**:41, 42
   amorphous, solvent cracking, **19**:59
      solvents, **19**:55, 56, 57
      non-polar, **19**:55, 57
      polar, **19**:57
   chemical properties, **19**:50
   cohesive energy density, **19**:54
   compatibility, **19**:53
   cracking in aggressive environments,
      19:58
   crystalline,
      non-polar, **19**:56
      polar, **19**:57
      solvents, **19**:57
   diffusion, **19**:59
   environmental stress cracking, **19**:59
   molecules, **19**:53
   orientation, **19**:49

Polymers *continued*
   resistance to chemical attack, **19**:51
   solubility, **19**:53
   solubility parameters, **19**:54, 55
   solvent,
      partial polarities, **19**:56
      solubility parameters, **19**:55
Poly(methyl methacrylate), **19**:60
Polyolefins, **19**:61
Polyoxymethylene, **19**:61
Polypropylene, **19**:61
Polystyrene, **19**:43, 60
Polysulphones, **19**:62
Polytetrafluorethylene (p.t.f.e.), **17**:19;
   19:61
Polythene coatings, **17**:19
Polythionic acid, stress corrosion
   cracking in, **8**:58
Polyurethane, **19**:63
   foams, self-skinning, **15**:57
   paint coatings, **15**:19
Polyvinyl acetate and derivatives, **19**:60
Polyvinyl chloride, **17**:15; **19**:60
Porosity, **1**:247; **3**:22
   electrodeposits, **13**:41
Portland cement, **19**:38
Potassium chloride, effect on paint films,
   15:32
Potassium halides, **9**:10
Potential, **1**:103
   electrode, **1**:82
   equilibrium, **1**:91
   of zero charge, **9**:11
   passivation, **1**:104, 105
Potential/anodic current density curves,
   nickel alloys, **4**:111
Potential difference, **9**:6
   metal/solution interface, **9**:71
   surface or adsorption, **9**:7
Potential gradients, **11**:12
Potential measurements, **20**:40
Potential-pH dependence of de-alloying,
   20:132
Potential-pH diagrams, **1**:59, 61, 137,
   158; **2**:3–4, 18; **6**:17–18; **20**:132
   advantages and limitations, **1**:65
   Al–H$_2$O system, **4**:14
   construction, **1**:62
   corrosion zone, **1**:64, 67
   Cr–H$_2$O system, **1**:110
   Cu–H$_2$O, **4**:39
   Cu–H$_2$O–NH$_3$, **1**:70
   Fe–H$_2$O system, **1**:62, 67
   immunity zone, **1**:64, 68
   importance in corrosion science, **1**:71
   Ni–H$_2$O system, **1**:70; **4**:107
   passivation zone, **1**:64, 68
   Pb–H$_2$O system, **4**:73
   Pb–SO$_4$ system, **4**:74
   tin, **4**:142
Potential-time relationships, **20**:33

Potential-time tests, **20**:55
Potentiometers, **11**:101
Potentiostats, **1**:103, 104; **20**:27, 65, 125,
   126
   applications, **20**:126
   basic circuit, **20**:124
   corrosion studies, **20**:123
   experimental apparatus, **20**:123
   iR corrections, **20**:125
   probe positioning, **20**:125
   scanning rate, **20**:125
   specimen masking and mounting,
      20:125
Pourbaix diagrams. *See* Potential-pH
   diagrams
Precious metals, coatings, **14**:15
Precipitation hardening, **9**:120
Precursors, **13**:42
Pressure vessels, **3**:62
Pretreatment, **12**:3–15
   see also Pickling
   alkaline cleaners, **12**:8–10
   and paint failure, **15**:39
   ASTM Standards, **21**:81
   cleaning, **13**:5
      ASTM Standards, **21**:81
      chemical, **20**:107
      electrolytic cathodic, **20**:107
      glass, **19**:20
      magnesium alloys, **4**:98
      methods, **20**:16
   Code of Practice, **12**:3
   cold immersion solvent degreasing,
      12:5
   degreasing, **13**:5
   di-phase cleaners, **12**:8
   electrolytic cleaners, **12**:9–10
   emulsion cleaners, **12**:7
   grease removal, **12**:3–10
   liquor/vapour degreasing, **12**:6
   paint coatings. *See* Paint coatings
   requirements, **12**:3
   rinsing after, **12**:14
   solvent cleaners, **12**:4, 7
   sprayed coatings, **13**:82
   ultrasonic cleaning, **12**:10
   vapour and hot-immersion cleaning,
      12:5
   vapour degreasing, **12**:5
Properties of metals and alloys, **1**:3;
   21:10
Protection potential, **1**:158
Pseudomorphism, **13**:23, 26
P.T.F.E., **17**:19; **19**:61
P.T.F.C.E. coatings, **17**:19
Pumps, marine engineering, **10**:62
Pyrophoricity,
   titanium, **5**:47
   zirconium, **5**:53
Pyrophosphoric acid, **3**:123

Quality control of surface preparation, **20**:158
Quicklime, **18**:4

Radiography, **20**:154
Rainfall, **2**:32
  and soil corrosion, **2**:62, 67
Rate-determining step, **9**:43, 47
Recovery, **9**:100
Recrystallisation, **9**:100
Red lead, **15**:21, 30, 39, 46
Red spot test, **2**:81
Redox potentials, **1**:57, 58, 60, 93, 105; **2**:78, 79, 112; **21**:24
  and pitting corrosion, **20**:129
Redox probe, **2**:71
Redox reaction, **1**:8
Reduction potentials, **21**:24
Reduction reaction, **13**:64
Reference cells, **11**:77
Reference electrodes, **9**:10, 69, 77; **10**:29; **11**:5, 28, 98; **20**:29, 32, 136; **21**:29
Reflectivity of polished surfaces, **20**:15
Refluxing capsules, **20**:82
Refractories,
  high-temperature corrosion, **7**:11
  molybdenum compatibility, **5**:19
Refrigerating systems, **4**:28
Regenerators, **2**:118
Regulus Metal, **4**:69
Relative humidity, **2**:28–29; **18**:3
  critical, **2**:31
  filiform corrosion, **1**:149
Resin cements, **19**:38
Resistance inhibition reaction, **15**:30–36
  electrolytes underneath film, **15**:31
  ionogenic groups in film substance, **15**:31
  water and electrolytes outside film, **15**:32–36
Retractable hydrogen probe, **20**:155
Reversible cells, **9**:66, 75, 76
Reversible potentials, **9**:81
Reversible system, **9**:58
Reynolds number, **2**:11
Rhodium,
  anodic behaviour, **6**:19
  chemical properties, **6**:11
  high-temperature properties, **6**:13
  oxide films, **14**:104
Rhodium coatings, **14**:100
Rhodium-platinum alloys, **6**-5–6, 20
  high-temperature properties, **6**:14
  linings, **6**:15
Rinsing,
  after pretreatment, **12**:14
  phosphate coatings, **14**:26
Rotating disc-ring electrodes, **20**:139
Rotating disc tests, **20**:19, 83

(RT/F) log x at various temperatures, **21**:11
Rubber, **19**:58, 64
  abrasion resistance, **19**:67
  bonding to metal, **19**:66
  chemical resistance, **19**:67
  hard, **19**:65, 67
  impact strength, **19**:67
  linings, **19**:67
    applying to steel, **19**:71
    pipework, **19**:72
    tanks, **19**:71, 72
  natural, **19**:64, 66
  synthetic, **19**:65, 66
  temperature resistance, **19**:71
  vulcanisation, **19**:65, 73
Rubber coatings, cavitation damage resistance, **8**:132
Rubber latex, **19**:38
Rupture potential, **1**:154
Rust and rusting, **2**:35–36, 80; **3**:3; **15**:38
  dotted type, **15**:44
  effect of exposure conditions, **3**:10
  effect of exposure duration, **3**:11
  effect of mass, **3**:10
  effect of surface condition, **3**:11
  electrochemical nature of, **3**:4
  formation, **15**:24–25
  humidity effects, **3**:5, 6
  in air, **3**:5
  in natural waters, **3**:16
  in sea-water, **3**:13, 25
  in soil, **3**:17
    effect of duration of burial, **3**:19
    effect of metal composition, **3**:17
    effect of soil type, **3**:18
  in water, **3**:12
    effect of metal composition, **3**:13
    effect of operating conditions, **3**:15
    effect of surface condition, **3**:14
    effect of water composition, **3**:14
  indoors and in enclosed spaces, **3**:12
  low-alloy steels, **3**:22
  mechanism, **3**:3
  patchy, **15**:47
  patina, **20**:54, 55
  pinpoint, **15**:47
  rates in different climates, **3**:9
  removal, **12**:10–12; **20**:88
  temperature effects on, **3**:8
  under-, **15**:47
  white, **2**:32
Ruthenium, **6**:13
  anodic behaviour, **6**:19
  chemical properties, **6**:13
Ruthenium coatings, **14**:103
Ruthenium-platinum alloys, **6**:6

Saline particles, **2**:30

Salt, **2**:30
Salt droplet test, **20**:49
Salt-fog test, **20**:47
Salt solutions, **8**:59
  aluminium and aluminium alloys in, **4**:28
  cast iron in, **3**:95
  high-chromium cast irons in, **3**:116
  high-nickel cast irons in, **3**:109
  lead in, **4**:82
  molybdenum in, **5**:19
  nickel and nickel alloys in, **4**:134
  nickel-iron alloys in, **3**:82
  niobium in, **5**:27
  silicon-iron alloys in, **3**:126
  tantalum in, **5**:67
  titanium in, **5**:37
  zinc in, **4**:159
Salt-spray tests, **16**:29; **20**:47, 50, 98
Saturation index, water, **2**:42
Saturated vapour pressure of water, **2**:32
Scale deposition, inhibitors, **18**:18
Scale removal,
  see also Pickling
  aluminium alloys, **12**:13
  chemical, **12**:12
  copper alloys, **12**:13
  mechanical **12**:10–12
  non-ferrous metals, **12**:13
Scaling, **1**:22, 24
  nickel alloys, **7**:80
  cast iron, **7**:50
  copper, **4**:54
  high-alloy steels, **7**:65
Schori process, **13**:78
Sea-water, **2**:51–60
  see also Marine engineering
  aluminium alloys in, **4**:22
  artificial, **20**:41
  bimetallic corrosion, **1**:219; **3**:51
  calcareous deposit, **2**:59
  cast iron in, **3**:90
  chlorinity, **2**:51
  composition, **2**:51, 52
  copper and copper alloys in, **4**:45
  corrosion rates, **2**:55
  corrosion tests, **20**:94
  density, **2**:52
  design and layout of systems, **10**:58
  dissolved gases, **2**:54
  dissolved oxygen, **2**:54
  distribution of temperature, salinity, and oxygen, **21**:56, 57
  effect of depth, **2**:58
  effect of movement, **2**:59
  electrical conductivity, **2**:53
  ferrous metal corrosion rates, **2**:55
  freezing point, **2**:54
  gases in, **21**:54
  impingement attack, **2**:60
  magnesium alloys, **4**:92

Sea-water *continued*
  maraging steel in, **3**:67
  molybdenum in, **5**:15
  nickel and nickel alloys in, **4**:128
  nickel-iron alloys in, **3**:79
  non-ferrous metal corrosion rates, **2**:57
  oxygen dissolved in, **21**:54
  pH, **2**:54
  physical properties, **2**:52
  pipes and piping materials, **10**:59, 60
  potentials of metals in, **2**:54
  properties of, **21**:55
  rusting in, **3**:13, 25
  salinity, **2**:51, 52
  speed of movement, **2**:58
  stainless steels in, **20**:69, 70
  temperature, **2**:53
  zinc in, **4**:157
  zinc coatings in, **14**:43
Season cracking, **4**:36, 56
Segregation, **1**:36; **7**:127, 129; **8**:32, 33, 47; **9**:104
Selenium, **3**:36
Self-diffusion, activation energy, **1**:233
Self-diffusion coefficient, **1**:227, 233
Shear lips, **8**:135
Sheaths, protective, **6**:14
Shepard Cane resistivity meter, **11**:107
Sherardising, **13**:59, 61; **14**:39, 40
Shot blasting, **12**:10
Shot peening, **8**:93
Sievert's law, **7**:122
Silica, **2**:47
  fused, **19**:11, 25, 26
  vitreous. *See* Vitreous silica
Silica gel, **18**:4
Silicates, **2**:47
Silicon, **3**:24, 87, 88; **7**:19, 93, 104, 106
Silicon-bronzes, **4**:36
Silicon coatings, diffusion, **13**:60
Silicon-iron alloys, **3**:118
  composition, **3**:118
  corrosion behaviour, **3**:120
  high-temperature corrosion, **3**:126
  in acids, **3**:121
  in alkalis, **3**:125
  in industrial environments, **3**:121
  in natural environments, **3**:121
  in salt solutions, **3**:126
  mechanical properties, **3**:118
  structure, **3**:118
Silicon-iron anodes, **11**:66
Silicones, **19**:63
Silver,
  anodes, **11**:40
  anodic behaviour, **6**:18
  atmospheric corrosion, contaminants from plastics, **19**:79
  brazing alloys, **6**:17
  chemical properties, **6**:8

Silver *continued*
  chromate treatments, **16**:41
  grades, **6**:21
  high-temperature properties, **6**:13
  linings, **6**:14–15
  physical and mechanical properties,
    6:3–4
  standard electrode potentials of silver
    systems, **6**:9
  standard forms, **6**:21
  working, **6**:21
Silver coatings, **13**:20; **14**:97
Silver-copper alloys, phase diagram,
  9:111
Single crystals, etching, **1**:34
Single-phase alloys, **1**:44
Skin temperature, **20**:26
Slip, **9**:96
Slip-in rack, **20**:86
Sodium, **7**:110
  heat-transfer medium, **2**:106
  nitrogen solubility in, **2**:107
  oxygen solubility in, **2**:106, 107
Sodium bicarbonate, **2**:46
Sodium chloride, **7**:112
  diffusion through paint films, **15**:32
Sodium hexametaphosphate, **18**:61
Sodium hydride descaling, **12**:12
Sodium hydroxide, molten, **2**:113
Sodium sulphate, **7**:107, 112
Soil corrosion, **3**:17
  aluminium and aluminium alloys,
    4:23
  cast iron, **3**:91
  copper and copper alloys, **4**:43
  lead, **4**:77
  low-alloy steels, **3**:27
  nickel and nickel alloys, **4**:130
  pipes, **10**:50
  stainless steels, **3**:51
  tests, **20**:43
  zinc and zinc alloys, **4**:158
  zinc coatings, **14**:43
Soil resistivity, **11**:106, 107; **21**:58
Soils, **2**:37–72
  acidity, **2**:62, 67
  aeration, **2**:65
  biological activity, **2**:67
  cathodic protection in, **1**:69–71
  chemical properties, **2**:67
  classification, **2**:63
  clay fraction of, **2**:64
  conductivity, **2**:63
  corrosion process in, **2**:62
  corrosion testing, **20**:9
  corrosivity evaluation, **2**:71–72
  field tests, **20**:95
  gases in, **2**:65
  genesis, **2**:61
  moisture in, **2**:62, 65
  organic matter in, **2**:61

Soils *continued*
  oxidation-reduction potential and
    redox probe, **2**:71
  oxygen-concentration-cell formation,
    2:66
  pH of, **2**:67
  pipe-line ditch environment, **2**:68
  polarisation of metal in, **2**:72
  profile, **2**:63–64
  properties, **2**:63–68
  resistivity, **2**:69, 71; **11**:67, 68, 69
  rusting in, **3**:17
    effect of duration of burial, **3**:19
    effect of metal composition, **3**:17
    effect of soil type, **3**:18
  sampling, **2**:79
  soluble salts, **2**:67
  texture and structure, **2**:63
  water relations, **2**:65
Soldered joints, **4**:19, 148; **10**:70
Solders,
  noble-metal, **6**:16–17
  tin, **4**:147
  tin-silver, **6**:15
Solid solubility, **9**:104, 111
Solid-solution hardening, **9**:120
Solid solutions, **9**:103
Solutions, aqueous. *See* Aqueous
  solutions
Solvent cleaners, **12**:4, 7
Solvents,
  in paint coatings, **15**:22
  organic, **1**:18; **12**:4
Sparking-plug electrodes, **7**:108
Spheroidised structures in steels, **9**:120
Spin tests, **20**:83
Spinel, **7**:82, 84, 88
Spinel structure, **1**:25
Spinnerets, **6**:16
Split-ring technique, **20**:139
Spontaneous reaction, **9**:66, 77
Spray tests, **20**:47
Spray-welding, **13**:90
Sprayed coatings, **8**:93; **13**:78
  aluminium, **13**:82, 90; **14**:17, 20,
    23–25
  copper, **13**:87
  detonation process, **13**:81
  hard facing, **13**:90
  lead, **13**:86
  mechanical properties, **13**:87
  methods of spraying, **13**:78
  molten-metal process, **13**:78
  nickel-chromium alloys, **13**:87
  painting, **13**:89, 90
  physical properties, **13**:88
  plasma process, **13**:81
  powder process, **13**:78
  reclamation, **13**:82
  Schori process, **13**:78
  shrinkage, **13**:88

Sprayed coatings *continued*
  spray-welding, **13**:90
  stainless steels, **13**:87
  surface preparation, **13**:82
  thermoplastic, **17**:14
  thicknesses, **14**:6
  tin, **13**:86
  wire-pistol gas feeds, **13**:80
  wire process, **13**:79
  zinc, **13**:82; **14**:39, 40
Stabilisation, effectiveness of, **20**:8
Stability diagrams for molten salt
    systems, **20**:138
Stacking fault energy, **8**:11, 56
Stacking faults, **9**:92, 99
Stainless steels. *See* Steels
Standard chemical potential, **21**:12
Standard electrode potential, **1**:56, 57;
    9:67, 69; **21**:24
  see also Electrode potential
  lead, **4**:69
  magnesium, **4**:90
Standard hydrogen electrode, **9**:69;
    20:29
Standard polarisation cell, **20**:30
Statistical analysis, **20**:6
Steam, **7**:6
  aluminium and aluminium alloys in,
    4:30
  nickel and nickel alloys in, **4**:135
  oxidation effects, **7**:38
  uranium in, **5**:76
Steels, **2**:38; **20**:33
  age-hardening, **3**:64
  alloy, pickling, **12**:20
  anodes, **11**:41, 66
  atmospheric corrosion, **20**:9, 88
  austenitic, **3**:31, 36, 50, 59; **7**:60, 62,
    77; **10**:76; **20**:33
    See also Steels, stainless
    applications, **3**:60
    behaviour in chloride and caustic
      solutions, **8**:48
    corrosion fatigue, **8**:107
    fabrication, **3**:41
    heat treatment, **3**:47
    intergranular corrosion, **1**:37
    mechanical properties, **3**:40
    pH, chloride ion concentration and
      potential for artificial pits, **1**:141
    pitting corrosion, **1**:47
    stress-corrosion cracking, **8**:47
    stress-corrosion, practical solutions,
      8:56
  carbon
    pitting corrosion, **1**:160
    stress corrosion, **8**:44, 45
    welding, **10**:76
  chemical cleaning, **20**:108
  corrosion in sea water, **2**:55–57
  creep resisting, **7**:65

Steels *continued*
  factors affecting corrosion, **3**:3, 4
  ferritic, **3**:35, 47; **7**:60, 62, 77; **10**:76
    applications, **3**:60
    corrosion fatigue, **8**:107
    ductile/brittle transition, **3**:40
    mechanical properties, **3**:40
    stress-corrosion cracking, **8**:27
  galvanised, **10**:48
    pipes, **10**:50
    tanks, **10**:51
  heat-resisting, **7**:58, 60, 77
  high-alloy
    applications, **7**:77
    in liquid metals, **7**:77
    in molten salts, **7**:76
    mechanical properties, **7**:62
    oxidation, **7**:58, 65
      fuel-ash effect, **7**:72
      in air, **7**:69
      in flue gases, **7**:66
      in industrial gases, **7**:72
      in steam, **7**:70
    scale structure, **7**:65
    scaling indices, **7**:67
  high-strength, **3**:64; **8**:12, 14, 23
    crack propagation rates, **3**:73
    stress-corrosion cracking, **8**:158
  high-tensile
    hydrogen embrittlement, **8**:75
    stress-corrosion cracking, **8**:62, 73
  hypo-eutectoid, **9**:117
  in contact with corrosion, **10**:45
  low-alloy, **3**:3, 21
    alloying additions, **3**:25
    alloying elements, **3**:21, 23, 25
    applications, **3**:28
    atmospheric corrosion, **3**:23; **7**:24
    copper, **3**:21, 28
    corrosion behaviour, **3**:22
    heat-resistant properties and
      applications, **7**:22
    in aqueous environments, **3**:21
    in buildings and structures, **10**:38
    in natural environments, **3**:23
    mechanical properties, **3**:21
    oxidation, **7**:19
      alloying elements, **7**:31
      in air, **7**:25
      in carbon dioxide, **7**:41
      in flue gas, **7**:31
      in steam, **7**:38
    oxidation resistance, **7**:13
    soil corrosion, **3**:27
    welding, **10**:76
  low-carbon, **19**:29
    stress-corrosion cracking, **8**:29
    transgranular cracking, **8**:42
  maraging, **3**:64
    applications, **3**:75
    composition, **3**:64

Steels *continued*
  maraging *continued*
    corrosion behaviour, **3**:67
    cracking resistance in smooth
      materials, **3**:71
    development, **3**:64
    fabrication methods, **3**:66
    heat treatment, **3**:65, 66
    high-temperature corrosion, **3**:75
    in chemicals, **3**:69
    in natural environments, **3**:67
    in sea-water, **3**:67
    mechanical properties, **3**:65
    physical properties, **3**:65
    polarisation tests, **3**:69
    stress-corrosion cracking, **3**:69, 70,
      72, 74
    structural features, **3**:65
  martensitic, **3**:31, 32, 35, 50, 59; **7**:60,
    62, 77; **10**:76
    see also Steels; stainless
    applications, **3**:59
    corrosion testing, **20**:64
    fabrication, **3**:41
    mechanical properties, **3**:37
    stress-corrosion cracking, **8**:47, 58
    tempering temperature, **3**:37
  mild
    crevice corrosion, **1**:145
    heat-resistant properties and
      applications, **7**:22
    in buildings and structures, **10**:37
    stress-corrosion cracking, **8**:28
  rusting. *See* Rusting
  spheroidised structures, **9**:120
  stainless, **2**:97–100, 116; **3**:3, 31;
    7:60, 62
    see also Steels, austenitic and
      martensitic
    anodes, **11**:42
    applications, **3**:59
    atmospheric corrosion, **3**:48
    breakdown potentials, **3**:44
    British Standards, **3**.32
    carbide precipitation, **3**:45
    carbide segregation, **7**:129
    cavitation damage resistance, **8**:131
    chemical cleaning, **20**:108
    classification, **3**:32
    copper action on, **20**:61
    corrosion behaviour, **3**:42
    corrosion in sulphuric acid, **1**:106
    corrosion resistance, **3**:42, 44
    crevice corrosion, **1**:144
    definition, **3**:31
    electrochemical tests, **20**:65
    electropolishing, **12**:25
    fabrication methods, **3**:41
    fatigue, **3**:59
    free-cutting grades, **3**:36
    in acids, **3**:51

Steels *continued*
  stainless *continued*
    in buildings and structures, **10**:38
    in chemical environments, **3**:51
    in fused salts, **3**:58
    in natural environments, **3**:48
    in natural waters, **3**:50
    intergranular corrosion, **1**:37;
      20:55, 134
    mechanical properties, **3**:37
    metallurgical considerations, **3**:44
    pH, chloride ion concentration and
      potential for artificial pits, **1**:141
    physical properties, **3**:36
    pickling, **12**:20
    pitting corrosion, **1**:47; **10**:18
    precipitation hardening, **3**:36, 39
    sea-water corrosion, **20**:69, 70
    semi-austenitic, **3**:36, 39
    soil corrosion, **3**:51
    sprayed coatings, **13**:87
    stabilisation, **1**:41
    stress-corrosion cracking, **1**:50;
      8:47
    tubes, **10**:50
    welding, **10**:76
    stress-corrosion cracking, **1**:50; **3**:59;
      8:47, 54
    structural, **3**:28, 29
    paint finishes, **15**:62
    tempering temperature, **3**:45
    valve, **7**:58, 60
    vitreous enamel coating, **17**:4
    weathering, **3**:3
    accelerated tests for, **20**:51
Steelwork, atmospheric corrosion, costs,
  10:7
Stellite, **13**:90
Step structures, **20**:64
Stern-Geary equation, **10**:29; **20**:38
Stirling's approximation, **1**:224
Stoichiometric number, **9**:45
Stoneware, **19**:34
  abrasion resistance, **19**:37
  applications, **19**:39
  chemical resistance, **19**:37
  compressive strength, **19**:35
  costs, **19**:40
  dimensional accuracy, **19**:36
  impact strength, **19**:37
  jointing, **19**:37, 39
  mechanical properties, **19**:34
  modulus of rupture, **19**:35
  physical properties, **19**:34
  porosity, **19**:36
  tensile strength, **19**:34
  thermal expansion, **19**:37
  thermal shock resistance, **19**:36
  Young's modulus, **19**:35
Storage, inhibitors, **18**:30
Storage rooms, **18**:5

Strain, plane, **8**:134
Strain energy, **8**:3
Straining electrode experiments, **8**:39
Stray-current corrosion, **2**:63, 70; **11**:85
  a.c., **11**:88
  control methods, **11**:88
  d.c., **11**:86
  lead, **4**:77
Stray-current electrolysis, **11**:85
Streicher test, **20**:63
Strength properties, **20**:14
Stress-corrosion cracking, **1**:47, 127, 128,
    131; **3**:48; **8**:69; **10**:78, 79;
    20:10, 33
  active path mechanisms, **8**:6
  aluminium alloys, **1**:48; **4**:17; **8**:89,
    90, 92, 93
  as function of potential in high-
    temperature water, **20**:137
  austenitic steels, applied stress, **8**:49
    cold work, **8**:49
    crack path, **8**:48
    effect of composition, **8**:50
    effect of environment, **8**:53
    effect of microstructure, **8**:51
    effect of polarisation, **8**:52
    practical solution, **8**:56
  beryllium, **5**:7
  brasses, **4**:56
  cast iron, **3**:96
  chemical plant, **10**:16–18
  copper alloys, **1**:47; **4**:56, 58
  crack initiation, **8**:54
  crack propagation mechanisms, **8**:55
  crack tip adsorption model, **8**:14
  crack tip pH, **8**:75
  crack tip properties, **8**:12
  crack tip-solution interface, **8**:18
  crack velocity, **8**:18
  dissolution processes, **8**:40
  effect of additions to cracking
    environments, **8**:40
  effect of environmental composition,
    8:37
  effect of pH, **8**:41
  environmental aspects, **8**:15
  environments promoting
    transgranular cracking, **8**:42
  ferritic steels, **8**:27
    effect of alloying additions, **8**:33, 44
    effect of carbon content, **8**:28
    effect of composition and structure,
      a.c., **8**:27
    effect of grain size, **8**:35
    effect of heat treatments, **8**:35
  fracture mechanics, **8**:45, 136, 150
  grain-size effect, **8**:12
  grain structure, **1**:43
  high-strength steels, **8**:23, 158
  high-tensile steels, **8**:62, 73
  heat-treatment effect, **1**:48

Stress-corrosion cracking *continued*
  in polythionic acid, **8**:58
  induction period, **8**:54
  load-extension curve, **8**:148
  magnesium alloys, **8**:86, 87
  maraging steels, **3**:69
    critical stress intensity factor, **3**:72
    effect of metallurgical variables,
      3:74
    tests for, **3**:70
  marine engineering, **10**:65
  martensitic steels, **8**:58
  mechanisms, **8**:3
  moisture in, **8**:79
  nickel and nickel alloys, **4**:135
  nickel-iron alloys, **3**:82
  potentiostatic approach, **20**:135
  pre-existing active paths, **8**:8
  prevention methods, **8**:43, 80, 89
  propagation models, **8**:4
  propagation rate, **8**:5, 10
  stainless steels, **1**:50; **3**:59; **8**:47, 54
  steels, **1**:50
  strain-generated active paths, **8**:10
  strain rate, **8**:20
  stress factor control, **8**:44
  stress function, **8**:19
  susceptibility, **8**:4
  systems resulting in, **8**:25
  test cells, **8**:157
  test methods, **8**:139
    constant-deflection-rate tests, **8**:144
    constant-load tests, **8**:143
    constant total-deflection tests, **8**:141
    effects of surface finish, **8**:149
    electrochemical aspects, **8**:156
    environmental aspects, **8**:154
    initiation, **8**:157
    pH variations, **8**:154
    pre-cracked specimens, **8**:150, 152
    solution preparation, **8**:154
    stressing systems, **8**:139
    surface finish in, **8**:149
  titanium, **5**:43
  titanium alloys, **5**:48; **8**:78
    hot salt cracking, **8**:78
    mechanism, **8**:82
    room-temperature cracking, **8**:80
  velocity, **8**:145
    carbon steel, **8**:21
    measurements, **8**:153
  visible manifestations, **8**:3
Stress effects, **1**:47
  corrosion testing, **20**:10
  hydrogen solubility and permeation,
    9:53
  high-temperature corrosion, **7**:3
Stress intensity, **8**:134
Stress intensity factors, **3**:70, 72
Stress reducers, **14**:71
Stress relief, **1**:263; **8**:44, 50, 57; **10**:17

Stresses,
  in bent specimens, **8**:166
  in oxide layers, **1**:260
  plane, **8**:135
Structural effects, **1**:50
Structural materials, **20**:85
Structural steel painting, **20**:159
Structures, **20**:94
  atmospheric corrosion, **10**:33
  contact corrosion, **10**:34
  corrosive environment, **10**:33
  design, **10**:33
  ferrous metals, in air, **10**:34
  materials of construction, **10**:37
  non-ferrous metals, **10**:42
Substrates, **13**:4, 23
  interdiffusion with, **13**:40
  metallic, **13**:4
  non-conductors, **13**:6
Sugar products, **2**:99
Sulphate-reducing bacteria, **2**:79–80
Sulphates, **2**:45
Sulphidation, nickel and nickel alloys,
  7:90
Sulphide environments, **8**:58, 59
Sulphides, free-energy diagram, **7**:119
  inclusions, **1**:161
Sulphur, **3**:36, 90; **7**:22
  in lubricants, **2**:121
  oxidation mechanism, **2**:77
Sulphur compounds, **1**:19; **3**:4, 7, 8;
  7:9, 68, 73, 74
  action of, **6**:7, 9
  in lubricants, **2**:124
Sulphur-containing gases, **7**:7
Sulphur dioxide, **2**:29–30, 32, 34, 36, 40;
  3:5, 7, 8; **7**:35, 37
Sulphur dioxide tests, **20**:50
Sulphur oxides, **2**:29–30
Sulphuric acid, **2**:74; **3**:52, 122; **4**:79, 84,
  88; **5**:54, 65, 70; **12**:18
Sulphurous acid, **3**:62, 122
Superlattices, **9**:104
Surface-active agents, **12**:21
Surface condition, effect on rusting, **3**:14
Surface defects, **9**:100
Surface energy, **8**:3, 7, 8, 137
Surface films, **9**:100
  see also Oxide films
Surface preparation. *See* Pretreatment
Surface reaction products, **1**:26, 30
Surface reactions, **1**:23, 229
Surface roughness effect on corrosion
  rate, **2**:21
Surface structure, **9**:100
Surgical instruments, **3**:61
Symbols, **22**:17
Symmetry factor, **9**:33, 45

Tafel constants, **1**:85; **4**:143; **10**:29;
  11:7, 9; **20**:34, 35; **21**:35

Tafel equation, **1**:85, 94, 104; **21**:35
Tafel relationship, **1**:102, 103, 203
Tafel slope, **9**:36, 46–48, 52; **20**:38
Tantalum, **5**:59
  anodes, **11**:40, 67
  applications, **5**:70
  atmospheric corrosion, **5**:63
  bimetallic corrosion, **5**:68
  corrosion behaviour, **5**:59
  corrosion resistance, **5**:62
  fabrication methods, **5**:61
  galvanic effects, **5**:68
  hydrogen embrittlement, **5**:62
  in acids, **5**:63
  in alkalis, **5**:66
  in chemical environments, **5**:62
  in chemical plants, **5**:71
  in gases, **5**:63
  in hydrochloric acid, **5**:65
  in salt solutions, **5**:67
  in sulphuric acid, **5**:65
  in various environments, **5**:67
  mechanical properties, **5**:61
  oxidation, **1**:258
  physical properties, **5**:61
  welding, **5**:62
Tantalum alloys, **5**:69
  in acids, **5**:69–71
Tantalum-molybdenum alloys, **5**:69
Tantalum-niobium alloys, **5**:69
Tantalum pentoxide, **5**:62
Tantalum-titanium alloys, **5**:70
Tantalum-tungsten alloys, **5**:69
'Tell-tale' holes, **20**:146
Temkin isotherm, **9**:24, 25
Temperature–concentration diagrams,
  2:86
Temperature control in total-
  immersion tests, **20**:17
Temperature effects, **1**:99
  anodic polarisation of copper, **2**:22
  corrosion, **2**:3
  corrosion products, **2**:49
  corrosion rate, **2**:13–18, 30, 89, 93–95
  potential-pH diagrams, **2**:19
  rusting, **3**:8
  solubility of metals, in molten metals,
  2:104
Tempering, **9**:118
Temporary protectives. *See* Coatings,
  temporary
Terminology in corrosion, **1**:14
Terms, glossary of, **22**:3
Ternary solutions, **7**:132
Terne coatings, **13**:52
Terne-plate, **14**:56
Testing, corrosion. *See* Corrosion testing
Textile industry, **3**:61
Thermal cycling, **1**:25; **7**:4, 47, 67
Thermal-flux effect, **20**:26
Thermal mass transfer, **2**:115, 116

Thermocouple protection tubes, 7:100
Thermocouple sheaths, 7:112
Thermodynamic efficiency, 9:60
Thermodynamic functions, 7:115, 123
Thermodynamics, 9:57
  1st law of, 9:58
  2nd law of, 9:59
  corrosion reactions, 1:53, 56
  definitions, 9:57
  gas-metal systems, 7:115
  reversible cell, 9:76
  scope, 9:57
Thermo-electric generators, 11:75
Thermogalvanic cell,
  e.m.f. of, 2:22
  roll of, 2:22
Thermogalvanic corrosion, 2:21
Thermogalvanic coupling, 2:23
Thermogalvanic potentials, 2:22
Thermoplastic coatings, 17:13–20
  application, 17:13
    dipping, 17:13
    electrostatic spraying, 17:14
    flame spraying, 17:14
    spraying, 17:14
    vacuum coating, 17:14
  cellulose acetate butyrate, 17:18
  general data on, 17:17
  nylon, 17:18
  Penton, 17:16–18
  polythene, 17:19
  p.t.f.c.e., 17:19
  p.t.f.e., 17:19
  p.v.c., 17:15
Thermoplastics, 19:44–48
  amorphous, 19:60
Thermosetting plastics, 19:44, 48, 58
Thermosetting resins, 19:62
Thickness deposited by 1 Ah, 21:65
Thickness measurement,
  chemical plant, 10:26
  paint coatings, 20:167
  ultrasonic methods, 20:152
Thiobacillus ferroxidans, 2:76
Thiobacillus thio-oxidans, 2:74, 75, 76
Thiourea derivatives, 12:23
Thomson equation, 2:32–33
Threshold stress, 8:152
Throwing power of plating bath, 13:33
Tillmans formula, 2:44
Timber. See Wood
Time-temperature-transformation
    (T-T-T) diagrams, 9:117
Tin, 4:141; 13:50
  anodes, 4:146
  applications, 4:148
  aqueous corrosion, 4:144
  atmospheric corrosion, 4:143
  bearing metals, 4:148
  chromate treatments, 16:41
  corrosion behaviour, 4:142

Tin continued
  in acids, 4:145
  in alkalis, 4:145
  in liquid media, 4:146
  oxidation, 4:143
  oxide, 4:143
  oxide film, 4:144
  physical properties, 4:141
  potential/pH diagram, 4:142
  solders, 4:147
  structure, 4:141
Tin alloys, 4:141
  chemical cleaning, 20:107
  coatings, 14:56
Tin-bronzes, 4:35, 48
Tin-cadmium alloys, coatings, 14:57
Tin coatings, 14:11, 47
  applications, 14:55
  copper and copper alloys, 14:54
  corrosion resistance, 14:48
  galvanic action, 14:54
  hot-dipped, 14:6
    cast iron, 13:51
    copper, 13:52
    fabricated steel articles, 13:50
    steel, 13:51
    steel strip, 13:51
  immersion in aqueous media open to
      air, 14:50
  methods of application, 14:47
  properties, 14:48, 55
  sprayed, 13:86
  thickness, 14:48, 55
Tin-copper alloys, coatings, 14:58
Tin-lead alloys, coatings, 14:56
Tin-nickel alloys, coatings, 14:58
Tin-terne, 14:56
Tin-zinc alloys, coatings, 14:57
Tinplate, 13:16, 32; 14:49, 55
  atmospheric corrosion, 14:49
  containers, 14:51
  lacquered, 14:52, 53
  special quality, 14:53
  yellow-purple staining, 14:51
Titanium, 5:34; 7:87, 104
  anodes, 5:56; 11:36, 67, 76
  anodic passivation, 5:46
  applications, 5:46
  aqueous corrosion, 5:37
  bimetallic corrosion, 5:43
  breakdown potentials, 11:37
  cavitation damage resistance, 8:131
  corrosion behaviour, 5:36
  corrosion fatigue, 5:43
  corrosion resistance, 1:211
  crevice corrosion, 1:147, 148; 5:43
  erosion resistance, 5:42
  fatigue, 5:43
  high-temperature corrosion, 5:44
  in acids, 5:37, 38
  in chemical-environments, 5:37

Titanium *continued*
in chemical plant, **5**:44
in chemicals, **5**:38
in liquid metals, **5**:38
in nuclear energy, **5**:45
in salt solutions, **5**:37
mechanical properties, **5**:36
oxidation, **5**:47, 48
oxides, **5**:36, 46
physical properties, **5**:34
platinised, **5**:47
polarisation curve for, **20**:128
pyrophoricity, **5**:47
stress-corrosion cracking, **5**:43
Titanium alloys, **8**:23
mechanical properties, **5**:48
polarisation curves, **20**:140
stress-corrosion cracking, **5**:48; **8**:78,
80, 82
welding, **10**:80
Titanium carbide, **1**:41
Trade names, **21**:66
Transfer coefficient, **9**:45; **21**:37
Transformation-time-temperature
(TTT) diagrams, **1**:72
Transgranular cracking, **3**:74; **8**:11, 35,
42, 43, 49, 53, 81, 87, 88
Transgranular stress corrosion, **8**:11
Transport, **9**:5, 37
Transport overpotential, **9**:38
Tropical environments, **20**:94
Twins, **9**:92
Two-phase alloys, **1**:45

Udimet 500, **7**:84
Ultrasonic cleaning, **12**:10
Ultrasonic methods of corrosion
monitoring, **20**:151
Underfilm corrosion, **1**:148
Uranium, **5**:73; **7**:111
atmospheric corrosion, **5**:74
coatings, **5**:77
high-temperature corrosion, **5**:75
in air, **5**:76
in gases, **5**:75
in nuclear engineering, **5**:73
in steam, **5**:76
in water, **5**:73
effect of dissolved gases, **5**:73
irradiation effects, **5**:77
oxidation, **5**:76
oxides, **5**:76
Urea formaldehyde resins, **15**:52, 94

Vacancies, **9**:91, 92
Vacuum deposition, aluminium, **14**:18
Vacuum evaporation, **13**:97
Vacuum metalising of plastics, **15**:57
Valve steels, **7**:58, 60

Valves,
exhaust, **7**:108
marine engineering, **10**:62
Van der Waals adsorption, **9**:13
Van der Waals, forces, **2**:33
Vanadium, **7**:105
Vanadium compounds, **7**:75, 76
Vanadium pentoxide, **7**:105
Vant Hoff isotherm, **9**:68
Vapour blasting, **12**:12
Vapour corrosion, **19**:90
definition, **19**:75
plastics, **19**:75
woods, **19**:88
Vapour deposition cadium, **14**:34
Vapour pressure, saturated, of water,
2:32
Varnishes, **15**:16
Vegetable processing, **2**:100
Velocity and concentration interaction,
2:14
Velocity control in total-immersion tests,
20:18
Velocity effect
on corrosion, **2**:3
on dissolution rate, **2**:7
Vibratory tests, cavitation damage,
8:130
Vinyl resin paint coatings, **15**:19
Vitreous enamel coatings, **17**:3–12
abrasion resistance, **17**:6
acid resistance, **17**:8–10
adhesion, **17**:5, 6
alkali resistance, **17**:10
bonding, **17**:5
cast iron, **17**:4
chemical resistance, **17**:8
detergent resistance, **17**:10
dry process, **17**:3, 5
frit composition, **17**:3, 5, 8
glass-bottle industry, **17**:9
hardness, **17**:6
mechanical properties, **17**:5
metal preparation, **17**:3–4
nature of, **17**:3
properties, **17**:5–11
purpose of, **17**:3
resistance to water and atmosphere,
17:10
softening point, **17**:7
steel, **17**:4
thermal expansion coefficient, **17**:5
thermal properties, **17**:7
thermal shock resistance, **17**:7
wet process, **17**:3, 5
Vitreous silica, **19**:25
applications, **19**:27
electrical characteristics, **19**:26
heat resistance, **19**:25
melting point, **19**:25
physical properties, **19**:25

Vitreous silica *continued*
  resistance to chemical attack, **19**:26
  thermal conductivity, **19**:25
  thermal expansion, **19**:25
  thermal properties, **19**:25
Voids. *See* under Oxide films,
    discontinuous
Volatilisation, in porous oxide
    formation, **1**:258
Voltmeters, **11**:101

Water-break test, **13**:5
Water-supply systems, **10**:47
  aluminium pipes, **10**:49
  contact corrosion, **10**:50
  copper pipes, **10**:49
  fittings, **10**:52
  galvanised steel pipes, **10**:48
  lead pipes, **10**:48
  non-ferrous metals, **10**:47
  pipe materials, **10**:47–51
  soil corrosion, **10**:50
  tank materials, **10**:51, 52
Water vapour, **2**:28
Water(s)
  see also Sea-water
  alkalinity of, **2**:44
  aluminium and aluminium alloys in,
    4:22
  analysis, **2**:39
  cast iron in, **3**:90, 91
  chloride concentration, **21**:60
  constituents, **2**:39
  copper and copper alloys in, **4**:44, 51
  corrosion testing, **20**:9, 93
  diffusion through paint films, **15**:26
  dissolved gases in, **2**:40; **3**:15
  dissolved solids in, **3**:14
  distribution of dissolved constituents,
    21:59
  drinking, **2**:39
  flow rate, **2**:49
  glass in, **19**:20
  hardness, **2**:39, 44; **21**:61
  hardness salts, **2**:44
  high-chromium cast irons in, **3**:113
  high-nickel cast irons in, **3**:105
  high-purity, **20**:136
  high-temperature, **20**:136
  impurities, **2**:39
  in soil, **2**:65–66
  lead in, **4**:76
  metals immersed in, **2**:38
  mineral constituents, **2**:44, 46
  mineral salts, **2**:45
  natural, **2**:38–50, 93
  nickel and nickel alloys in, **3**:80;
    4:128
  niobium in, **5**:26
  oil in, **2**:47

Water(s) *continued*
  organic growths in, **2**:48
  organic matter in, **2**:47; **3**:15
  pH of, **2**:42, 44, 48
  pressurised, **20**:136
  resistivity, **11**:67; **21**:58
  rusting in, **3**:12
    effect of metal composition, **3**:13
    effect of operating conditions, **3**:15
    effect of surface condition, **3**:14
    effect of water composition, **3**:14
  saturated vapour pressure of, **2**:32
  saturation index, **2**:42
  sea. *See* Sea-water
  stainless steel in, **3**:50
  temperature effects, **2**:49
  uranium in, **5**:73
  zinc in, **4**:156, 157
  zinc coatings in, **14**:43
Watts solution, **14**:69, 75
Weathering,
  artificial methods, **20**:98
  steel, **15**:64
Weathering steels, **3**:3
  accelerated tests for, **20**:51
Weighing of specimens, **20**:5, 11
Weight deposited by 1 Ah, **21**:65
Weight-gain determinations, **20**:11
Weight-loss determinations, **20**:11
Weight-loss units, conversion of, **20**:12
Weld decay, **1**:40; **3**:47; **8**:47; **10**:77
Weld defects, **10**:74
Welded joints, protection of, **10**:80
Welding, **10**:68, 73
  aluminium alloys, **10**:79
  carbon steels, **10**:76
  copper alloys, **10**:80
  glossary of terms, **10**:82, 83
  heat effects of, **20**:7
  low-alloy steels, **10**:76
  magnesium alloys, **10**:80
  nickel alloys, **4**:106; **10**:78
  primers, **15**:8
  spot, **12**:41
  stainless irons and steels, **10**:76
  strap joints, **12**:33
  tantalum, **5**:62
  titanium alloys, **10**:80
Welds,
  corrosion testing, **20**:7
  crevice corrosion, **10**:21
  nickel-clad steel, **3**:83
  rusting in water, **3**:14
Wenner technique, **11**:67, 106
Wet corrosion, **7**:3; **13**:41
Wetness,
  of metal surface, **2**:31
  time of, **2**:31
Wetting agents, **14**:72
Whiskers, on tin coatings, **14**:48
White rust, **2**:32; **4**:154, 156; **13**:89

Widmanstätten structure, **9**:118
Williams Corfield test, **2**:72
Wood, **19**:81
 acid resistance, **19**:84
 bimetallic corrosion in, **19**:89
 chemical vats and tanks, **19**:83
 contact corrosion, **19**:89
 corrosion of metals by, **19**:87
 corrosion problems associated with, **19**:85
 corrosive-liquid containers, **19**:83, 84
 exterior exposure, **19**:82, 83
 glued-laminated structures, **19**:83
 in acids, **19**:85
 metals susceptible to corrosion, **19**:89
 pH values, **19**:88
 physical properties, **19**:81
 preservative treatment, **19**:83, 89
 recommended for use in corrosive conditions, **19**:82
 resistance to damp weather, **19**:82
 staining by metals, **19**:85
 timber names, **19**:85
 vapour corrosion, **19**:88
 zinc in contact with, **10**:44
Work hardening, **8**:127

Zeta-potentials, **1**:173
Zinc, **4**:150
 anodes, **4**:160, 161, 162; **11**:3, 19, 20, 51
 atmospheric corrosion, **4**:154
 contaminants from plastics, **19**:79
 bimetallic corrosion, **4**:153, 154
 chemical cleaning, **20**:107
 corrosion behaviour, **4**:152
 corrosion inhibition, **18**:49
 electrochemical corrosion, **4**:153
 grades, **4**:150
 high-purity, **4**:150
 in acids, **1**:82; **4**:158
 in alkalis, **4**:158
 in buildings and structures, **10**:43
 in chemical environments, **4**:158
 in contact with concrete, **10**:45
 in organic compounds, **4**:160
 in salt solutions, **4**:159
 in sea-water, **4**:157
 in water, **4**:156, 157
 mechanical properties, **4**:150
 oxide film, **4**:153
 pH effects, **4**:157, 158
 physical properties, **4**:150
 protective film, **4**:153
 soil corrosion, **4**:158
 white rust, **4**:154, 156; **13**:89
Zinc alloys, **4**:150
 anodes, **4**:160

Zinc alloys *continued*
 composition, **4**:151
 corrosion behaviour, **4**:152
 diecastings, **10**:44
 mechanical properties, **4**:150
 physical properties, **4**:150
 soil corrosion, **4**:158
Zinc chromate, **15**:21, 39, 48
Zinc chromate phosphate, **15**:21
Zinc coatings, **2**:49; **4**:157; **13**:49, 50; **14**:10, 36
 applications, **14**:45
 building components, **10**:44
 characteristics, **14**:36
 chromate passivation, **16**:36
 comparison of process, **14**:36
 comparison with cadmium, **14**:33
 corrosion resistance, **14**:40
 diffusion, **13**:59, 62
 economics, **14**:46
 galvanising of continuous strip, **13**:54
 galvanising of sheet steel and fabricated articles, **13**:52
 hot-dip, **14**:6, 39, 40
 in sea-water, **14**:43
 in water, **14**:43
 life in atmosphere, **14**:41
 methods of application, **14**:36
 painted, **13**:89; **14**:44
 protective systems applied subsequently, **14**:44
 soil corrosion, **14**:43
 sprayed, **13**:82; **14**:39, 40
 steel sheet, **10**:39
 relative advantages of methods, **14**:39
Zinc diecastings, electroplating, **13**:19
Zinc dust, **15**:21, 28
Zinc-iron couple, polarity reversal, **1**:205
Zinc linoleate, **15**:29
Zinc oxides, **1**:225, 230, 237
Zinc phosphate, **14**:24; **15**:21, 30, 39; **16**:30
Zinc plating, **8**:111; **14**:39
Zircaloy-2, **1**:261; **5**:52, 53, 56
Zirconium, **4**:92; **5**:34, 48, 52
 applications, **5**:55
 aqueous corrosion, **5**:53
 breakaway corrosion, **1**:255; **5**:55
 corrosion behaviour, **5**:53
 in acids, **5**:53, 54
 in alkalis, **5**:55
 in chemical plant, **5**:56
 in nuclear engineering, **5**:52, 55
 mechanical properties, **5**:52
 oxidation, **1**:258, 262; **5**:56
 oxide, **5**:53
 physical properties, **5**:52
 pyrophoricity, **5**:53
Zirconium alloys, **5**:53
Zirconium-titanium alloys, **5**:56